IL	interleukin
IMP	inosine-5'-monophosphate
IP_3	inositol-1,4,5-triphosphate
K_m	Michaelis constant
kb	kilobases
kD	kilodalton
LDL	low-density lipoprotein
LHC	light harvesting complex
Man	mannose
NAA	nonessential amino acids
NAD^+	nicotinamide adenine dinucleotide (oxidized form)
NADH	nicotinamide adenine dinucleotide (reduced form)
$NADP^+$	nicotinamide adenine dinucleotide phosphate (oxidized form)
NADPH	nicotinamide adenine dinucleotide phosphate (reduced form)
NDP	nucleoside-5'-diphosphate
NMR	nuclear magnetic resonance
NO	nitric oxide
NTP	nucleoside-5'-triphosphate
P_i	orthophosphate (inorganic phosphate)
PAPS	3'-phosphoadenosine-5'-phosphosulfate
PC	plastocyanin
PDGF	platelet-derived growth factor
PEP	phosphoenolpyruvate
PFK	phosphofructokinase
PIP_2	phosphatidylinositol-4,5-bisphosphate
PP_i	pyrophosphate
PRPP	phosphoribosylpyrophosphate
PS	photosystem
PQ(Q)	plastoquinone (oxidized)
PQH_2 (QH_2)	plastoquinone (reduced)
RER	rough endoplasmic reticulum
RF	releasing factor
RFLP	restriction-frament length polymorphism
RNA	ribonucleic acid
dsRNA	double-stranded RNA
hnRNA	heterogenous nuclear RNA
mRNA	messenger RNA
rRNA	ribosomal RNA
snRNA	small nuclear RNA
ssRNA	single-stranded RNA
tRNA	transfer RNA
snRNP	small ribonucleoprotein particles
RNase	ribonuclease
S	Svedberg unit
SAH	S-adenosylhomocysteine
SAM	S-adenosylmethionine
SDS	sodium dodecyl sulfate
SER	smooth endoplasmic reticulum
SRP	signal recognition particle
T	thymine
THF	tetrahydrofolate
TPP	thiamine pyrophosphate
U	uracil
UDP	uridine-5'-diphosphate
UMP	uridine-5'-monophosphate
UTP	uridine-5'-triphosphate
UQ	ubiquinone (coenzyme Q)(oxidized form)
UQH_2	ubiquinone (reduced form)
VLDL	very low density lipoprotein
XMP	xanthosine-5' monophosphate

BIOCHEMISTRY
THE MOLECULAR BASIS OF LIFE
Fourth Edition

Trudy McKee
James R. McKee

New York   Oxford

OXFORD
UNIVERSITY PRESS
2009

Oxford University Press, Inc., publishes works that further Oxford University's
objective of excellence in research, scholarship, and education.

Oxford   New York
Auckland   Cape Town   Dar es Salaam   Hong Kong   Karachi
Kuala Lumpur   Madrid   Melbourne   Mexico City   Nairobi
New Delhi   Shanghai   Taipei   Toronto

With offices in
Argentina   Austria   Brazil   Chile   Czech Republic   France   Greece
Guatemala   Hungary   Italy   Japan   Poland   Portugal   Singapore
South Korea   Switzerland   Thailand   Turkey   Ukraine   Vietnam

Published by Oxford University Press, Inc.
198 Madison Avenue, New York, New York 10016
http://www.oup.com

ISBN: 978-0-19-530575-3

Printing number: 9 8 7 6 5 4 3 2 1

Printed in the United States of America on acid-free paper

This book is affectionately dedicated to
the memory of our mentor and friend

NICHOLAS ROSA, PhD

*"Do not fail to show hospitality to strangers, for thereby
some have entertained angels unawares"*
ST. PAUL

Nick's kindness and generosity of spirit
made him an angel for all who knew him.

TOPIC	QUESTION FOR STUDENTS	KEY POINTS	PAGE NO.
Organelles and human disease	What is the role of biochemistry in modern medicine?	Biochemical analysis of organelles has resulted in significant progress in our understanding of the causes of many human diseases.	61
Water, abiotic stress, and compatible solutes	If water is so important for sustaining life, how can certain organisms survive catastrophic dehydration conditions?	Organisms that can adapt to severe water loss utilize specialized protective molecules, such as compatible solutes, that replace water by forming hydrogen bonds with proteins and other macromolecules and membranes.	98
Nonequilibrium thermodynamics	How does thermodynamic theory relate to energy flow in living organisms?	Living organisms are far-from-equilibrium dissipative structures. They create internal organization via a continuous flow of energy.	114
The extremophiles: organisms that make a living in hostile environments	How can living organisms sustain life in very hostile environments?	Extremophilic organisms utilize an array of oxidation-reduction reactions to generate the energy required to sustain life in hostile environments.	119
Molecular machines	How do living organisms utilize chemical bond energy to perform the thousands of tasks required to sustain life?	Molecular machine function is made possible by conformational changes triggered by the hydrolysis of nucleotides bound to protein subunits called motor proteins.	158
Protein folding and human disease	What are the effects of misfolded proteins on human health?	The accumulation of misfolded proteins impedes cell function. Eventually protein aggregates cause cell death.	164
Enzymes and clinical medicine	How are enzymes used to promote human health?	Specific enzymes are used in the diagnosis or treatment of a variety of human diseases.	220
Sweet medicine	How does carbohydrate biochemistry impact human health?	Carbohydrate biochemistry provides insight into several human diseases. Carbohydrate biotechnology is being used to develop carbohydrate-based drugs and vaccines.	258
Glycolysis and jet engines	Can systems biology improve our understanding of biochemical pathways such as glycolysis?	Certain catabolic pathways are optimized by the use of highly exergonic reactions in an early phase. The ATP product of the pathway is used to drive the pathway forward.	281
The evolutionary history of the citric acid cycle	How and why did the citric acid cycle originate?	The citric acid cycle probably developed in primordial cells as two separate pathways: the reductive branch, which provided a means of reoxidizing NADH, and an oxidative branch, which produced the biosynthetic precursor molecules citrate and α-ketoglutarate.	333

TOPIC	QUESTION FOR STUDENTS	KEY POINTS	PAGE NO.
Ischemia and reperfusion	How are heart and brain cells damaged by the inadequate nutrient and oxygen flow caused by blood clots, and why does the reintroduction of O_2 cause further damage?	Damage to heart and brain cells due to oxygen deprivation originates with inefficient energy production, followed by osmotic pressure increases, lysosomal breakage, and ER stress. In the absence of preventative measures, the reperfusion of damaged cells with O_2 leads to ROS formation, causing further damage.	367
The aquaporins	How do water molecules flow so rapidly across hydrophobic cell membranes?	The aquaporins are a class of membrane channel proteins that are responsible for water flow into and out of cells in response to osmotic pressure changes. The water pore within AQP-1, the best characterized aquaporin, has a structure that allows the movement of water molecule across the plasma membranes, but not those of other small species.	406
Atherosclerosis	What is the biochemical basis of arterial damage in the disease process called atherosclerosis?	Atherosclerosis, which may lead to myocardial infarction, is initiated by damage to the endothelial cells that line arteries. The formation of atherosclerotic lesions begins with the accumulation of LDL and progresses to an inflammatory process that degrades arterial structure and function.	443
Biotransformation	How are potentially toxic hydrophobic molecules metabolized by the body?	Biotransformation, the enzyme-catalyzed process in which toxic, hydrophobic molecules are converted to less toxic, water-soluble molecules, consists of two types of biochemical reaction class: phase I and phase II. Phase I reactions introduce or unmask polar functional groups in hydrophobic molecules. In phase II the water solubility of substrate molecules is substantially improved by the conjugation of functional groups with substances such as glucuronic acid.	461
Photosynthesis in the deep	How can light-dependent photosynthesis occur at the bottom of the ocean, where there is no sunlight?	Dim light, sufficient for a slow-growing type of photosynthetic green sulfur bacterium, is used to fix carbon near hydrothermal vents. This light is emitted by hydro-thermal vents, which are hot enough to cause "blackbody" radiation.	497

TOPIC	QUESTION FOR STUDENTS	KEY POINTS	PAGE NO.
Gasotransmitters	How do gas molecules, previously thought to be toxic at any concentration, act as signal molecules?	At very low concentrations, NO, CO, and H_2S are signal molecules that diffuse easily through cell membranes and whose synthesis is rigorously controlled.	536
Disorders of amino acid catabolism	What are the effects on human health of deficiency of a single enzyme in amino acid metabolism?	The deficiency in humans of a single enzyme in amino acid metabolism has widespread effects that typically include brain damage.	577
Diabetes mellitus	Why does diabetes mellitus, a metabolic disease, damage the entire body?	Diabetes is an example of how a single defect (the inability to synthesize or respond to insulin) in a complex biological system can cause devastating damage.	606
Obesity and the metabolic syndrome	Why are so many humans predisposed to obesity in the modern world?	Natural selection in response to the rigors of chronic food scarcity has left many humans with the propensity to gain weight when food is plentiful. The body's inability to cope with lipotoxicity, caused by excessive body weight, can result in metabolic syndrome.	621
Forensic investigations	How is DNA analysis used in the investigation of violent crime?	Forensic scientists use a technique called PCR to amplify crime scene DNA in order to generate the unique genetic profile that distinguishes one individual from all others.	658
Viral "lifestyles"	How do viruses infect and then disrupt host cells?	Viral infection disrupts cell function. By suppressing some cellular genes and activating others, the viral genome directs the host cell to produce new virus in a process that often results in cell death.	665
Carcinogenesis	What is cancer and what are the biochemical processes that facilitate the transformation of normal cells into those with cancerous properties?	Carcinogenesis is the process whereby cells with a growth advantage over their neighbors are transformed by mutations in the genes that control cell division into cells that no longer respond to regulatory signals.	729
EF-Tu: A motor protein	How do the structural features of the prokaryotic elongation factor EF-Tu facilitate its role as a motor protein?	EF-Tu is an NTPase that binds and hydrolyzes GTP. The binding of GTP to domain 1 of EF-Tu causes a change in conformation of the entire protein that facilitates the binding of an aa-tRNA. Once the	753

Don't forget your favorite Biochemistry in Perspective essays from past editions, now available on the companion website at www.oup.com/us/mckee

Cell volume regulation and metabolism
Protein poisons
Scurvy and ascorbic acid
Fermentation: an ancient heritage
Hans Krebs and the citric acid cycle
Glucose-6-phosphate dehydrogenase deficiency
Membrane fusion and botulism
Starch and sucrose metabolism
Parkinson's disease and dopamine
Lead poisoning
Hyperammonemia
Gout
Hormone methods
Epigenetics and the epigenome: genetic inheritance beyond DNA base sequences

Brief Contents

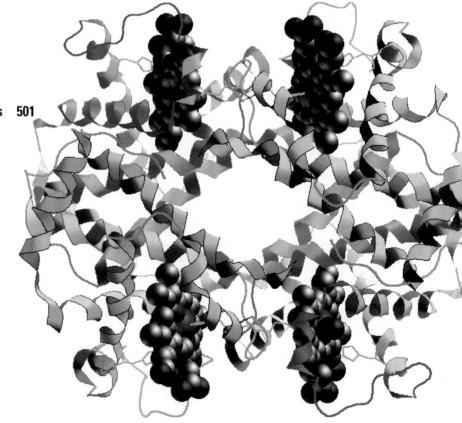

Contents

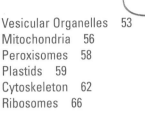

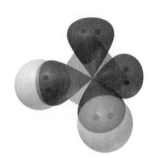

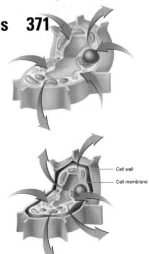
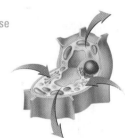

Cell wall
Cell membrane

16 Integration of Metabolism   591

17 Nucleic Acids   625

18 Genetic Information   675

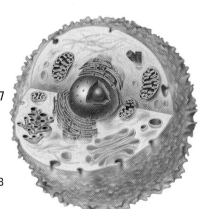

Preface

Welcome to the fourth edition of *Biochemistry: The Molecular Basis of Life*. In the few years since the third edition published, the rapid pace of biochemical and life science research has continued unabated. In fields as diverse as medicine, development, and plant biology, the technologies developed by biochemists and molecular biologists have been utilized to reveal a more detailed understanding of the molecules and mechanisms that sustain life. Although the textbook has been revised and updated to reflect this work, our original mission remains unchanged. We continue to believe that the cornerstone of an education in the life sciences is a coherent understanding of the basic principles of biochemistry. Once biochemical concepts have been mastered, students are prepared to tackle the complexities of their chosen fields of science.

ORGANIZATION AND APPROACH

CHEMICAL AND BIOLOGICAL PRINCIPLES IN BALANCE. As with previous editions, the fourth edition is designed for life science students and chemistry majors. Few assumptions have been made about the chemistry and biology backgrounds of students. To ensure that all students are sufficiently prepared to acquire a meaningful understanding of biochemistry, the first four chapters review the principles of organic functional groups, noncovalent bonding, thermodynamics, and cell structure. Several topics are introduced in these early chapters and further explored throughout the book. Examples include the self-assembly of biopolymers (such as proteins) into supermolecular structures, and the nature and function of molecular machines. Other important concepts that are emphasized include the relationship between biomolecular structure and function, and the dynamic, unceasing, and self-regulating nature of living processes. Students are also provided with overviews of the major physical and chemical techniques that biochemists have used to explore life at the molecular level.

REAL-WORLD RELEVANCE. Because students who take the survey course in biochemistry come from a range of backgrounds and have diverse career goals, the fourth edition consistently demonstrates the fascinating connections between biochemical principles and the worlds of medicine, nutrition, agriculture, bioengineering, and forensics. Features such as the "Biochemistry in Perspective" essays, as well as many other examples integrated into the body of the text, help students see the relevance of biochemistry to their chosen fields of study.

SUPERIOR PROBLEM-SOLVING PROGRAM. Analytical thinking is at the core of the scientific enterprise, and mastery of biochemical principles requires consistent and sustained engagement with a wide range of problems. The fourth edition continues to present students with a complete problem-solving package that includes in-chapter "Worked Problems," which illustrate how quantitative problems are solved, and other in-chapter "Questions," which give students the opportunity to put their knowledge into action as soon as new concepts are introduced.

SIMPLE, CLEAR ILLUSTRATIONS. Biochemical concepts often require a high degree of visualization, and we have crafted over 700 full-color figures that bring complex processes to life.

CURRENCY. The fourth edition has been updated to reflect recent developments in the field, while remaining focused on the "big-picture" principles that are the focus of the one-semester biochemistry course. A detailed list of updated material follows.

WHAT IS NEW IN THIS EDITION

As a result of our commitment to provide students with the highest-quality learning system available, we have revised the fourth edition in the following ways.

MORE RELEVANCE. Thirteen brand-new "Biochemistry in Perspective" essays (formerly titled "Special Interest Boxes") have been added to capture student interest. Topics such as "Sweet Medicine" (carbohydrate-based therapies), "Glycolysis and Jet Engines" (a systems biology perspective on an ancient energy-generating biochemical pathway), and "The Evolutionary History of the Citric Acid Cycle" introduce students to a range of high-interest topics in biochemistry. All the "Biochemistry in Perspective" essays, including several from the third edition, are also now available on the text's companion website. (See the "Guide to Essays" at the front of this book for a complete listing.)

EXPANDED PROBLEM-SOLVING PROGRAM. The fourth edition includes *twice* as many end-of-chapter questions as the previous edition, now approximately 50 per chapter. The expanded problem sets span a wider range of difficulty, from basic practice problems to more challenging integrative exercises.

BRAND-NEW ILLUSTRATIONS. With over 60 new figures, the fourth edition incorporates a superior and expanded art program designed to help students develop a strong visual grasp of biochemical processes and their impact on the living state.

NEW THEMES. As a result of significant research efforts, we have introduced two new themes: macromolecular crowding and systems biology. Macromolecular crowding, the dense packing of vast numbers of proteins and other molecules within cells, has a profound effect on a wide variety of living processes. The concept of macromolecular crowding provides students with a more realistic view of cell structure and function. The relatively new field of systems biology is an approach to biochemical processes that is based on engineering principles. Developed in response to the overwhelming amount of new information now available to life scientists, systems biology is the computer-assisted investigation of the complex interactions among biomolecules. Our accessible introduction to the principles of systems biology provides students with new insight into why biomolecular processes are so complex. In addition, the text includes new content in the areas of proteomics, epigenetics, and protein-folding diseases.

INCREASED ATTENTION TO REACTION MECHANISMS. Catalytic mechanisms provide students with an enhanced understanding of the means by which biochemical reactions occur. We describe the roles of amino acid side chains in the catalytic mechanisms of enzymes and the mechanisms of the nucleic acid polymerases and ribosome-catalyzed peptide bond formation.

CURRENT TOPICS. What follows is an abbreviated sampling of some, but not all, of the updated content that has been introduced in the fourth edition.

Chapter 1 now includes a brief overview of systems biology that provides students with insight into why the regulatory mechanisms for living processes are so complicated, and how life scientists can eventually evaluate the deluge of data derived from the Human Genome Project.

The concept of macromolecular crowding is introduced in Chapter 2 to provide students with a more realistic view of cell structure and function. There is also a new emphasis on the endomembrane system with special reference to ER stress and ER overload response, both of which contribute to numerous human disease processes.

In Chapter 3, discussion of structured water and sol-gel transitions provides students with new insight into the role of water in living organisms, especially in reference to the structural properties of cytoplasmic proteins.

A new discussion of multifunctional (or "moonlighting") proteins in Chapter 5 provides insight into the evolutionary process of protein recruitment.

A new section in Chapter 6 describes the means by which the side chains of amino acid residues contribute to catalytic mechanisms.

In Chapter 8, an expanded discussion of glycolysis regulation includes the role of AMPK, now known to be a metabolic master switch that has a key role in cellular energy homeostasis.

In Chapters 9 and 10 (aerobic metabolism), a new introduction describes why most primordial organisms that existed when large amounts of oxygen first appeared in the atmosphere were already adapted to the toxic effects of this gas. In a revised introduction to Chapter 10, the discussion of oxidative stress focuses on the redox environment in which living processes occur.

An expanded discussion of fatty acids in Chapter 11 includes the omega-3 and omega-6 fatty acids, and an expanded discussion of membrane structure covers the structural and functional properties of membrane microdomains (referred to as lipid rafts). Chapter 12 includes a newly revised discussion of fatty acid metabolism regulation that explores the regulatory effects of AMPK and the transcription factors SREBPs and PPARs.

There is an extensive revision to the discussion of the alternatives to C3 metabolism in Chapter 13.

In Chapters 14 and 15 (nitrogen metabolism) the discussion of ribonucleotide reductase activity has been expanded to include the radical-mediated mechanism, and the material on protein turnover has been updated to cover the mechanism of proteasomal protein degradation.

In Chapter 16, a new discussion of feeding behavior outlines the biochemical processes that control how much food an animal consumes, a topic of clear relevance to young adult students.

The discussion of histones in Chapter 17 is updated to include greater structural detail, as well as material on epigenetic modifications and the assembly of nucleosomes. An updated discussion of eukaryotic genomes provides an overview of types of genes and intergenic sequences. An expanded description of transposons covers LINEs and SINEs. Finally, the section devoted to the structure and functional roles of noncoding RNA is expanded.

In Chapter 18, an expanded section devoted to prokaryotic gene expression now includes riboswitches, and an expanded discussion of gene expression in eukaryotes covers RNA editing.

LEARNING PACKAGE

We have created a set of additional resources designed to help students master the subject matter and to assist instructors in meeting this objective. Here is a list of the available supplements (please see the "Guided Tour" that follows for more details).

For Students:

- Student Study Guide and Solutions Manual, by Patricia DePra, Carlow College
- Companion website at www.oup.com/us/mckee
- Multiple-choice quizzes, by Dan Sullivan, University of Nebraska at Omaha
- Interactive 3-D Molecular Models, by Todd Carlson, Grand Valley State University

For Instructors:

- All text images in electronic format
- Test bank containing over 700 questions
- All "Biochemistry in Perspective" essays, including several from the third edition
- Instructor's Resource CD-ROM containing all of the above
- Companion website (all items from Instructor's Resource CD-ROM, with the exception of the test bank)

ACKNOWLEDGMENTS

We wish to express our appreciation for the efforts of the dedicated individuals who provided detailed content and accuracy reviews of the text and the supplemental materials of the fourth edition:

Kevin Ahern   Oregon State University
Thurston E. Banks   Tennessee Technological University
Ronald Bartzatt   University of Nebraska, Omaha
Werner G. Bergen   Auburn University
Steven M. Berry   University of Minnesota, Duluth
John Brewer   University of Georgia
Martin Brock   Eastern Kentucky University
Sean T. Coleman   University of the Ozarks
Elizabeth S. Critser   Columbia College
Michael A. Cusanovich   University of Arizona
Bansidhar Datta   Kent State University
William Deutschman   State University of New York, Plattsburgh
Gregory W. Grove   Pennsylvania State University
Pui Shing Ho   Oregon State University
Holly A. Huffman   Arizona State University
John R. Jefferson   Luther College

Gail Jones   Texas Christian University
Peter J. Kennelly   Virginia Tech University
Barry Kitto   University of Texas, Austin
James A. Knopp   North Carolina State University
Gary E. Means   Ohio State University
Rakesh Mogul   California Polytechnic State University
Joyce Mohberg   Governors State University
Ann V. Paterson   Williams Baptist College
Jennifer Powers   Kennesaw State University
Gordon S. Rule   Carnegie Mellon University
Andrew K. Shiemke   West Virginia University
Aaron Sholders   Colorado State University
Salvatore A. Sparace   Clemson University
Narasimha Sreerama   Colorado State University
Dan M. Sullivan   University of Nebraska, Omaha
Anthony P. Toste   Missouri State University
Toni Trumbo Bell   Bloomsburg University of Pennsylvania
Harry van Keulan   Cleveland State University

We would also like to thank the individuals who reviewed the first, second, and third editions of this text:

Gul Afshan   Milwaukee School of Engineering
Mark Annstron   Blackburn College
Donald R. Babin   Creighton University
Bruce Banks   University of North Carolina
Allan Bieber   Arizona State University
Brenda Braaten   Framingham State College
Oscar P. Chilson   Washington University
Danny J. Davis   University of Arkansas
Patricia DePra   Carlow University
Robert P. Dixon   Southern Illinois University–Edwardsville
Patricia Draves   University of Central Arkansas
Lawrence K. Duffy   University of Alaska, Fairbanks
Charles Englund   Bethany College
Nick Flynn   Angelo State University
Clarence Fouche   Virginia Intermont College
Terry Helser   State University of New York, Oneonta
Pui Shing Ho   Oregon State University
Charles Hosler   University of Wisconsin
Larry L. Jackson   Montana State University
John R. Jefferson   Luther College
Craig R. Johnson   Carlow College
Ivan Kaiser   University of Wyoming
Michael Kalafatis   Cleveland State University
Paul Kline   Middle Tennessee State University
Hugh Lawford   University of Toronto
Carol Leslie   Union University
Duane LeTourneau   University of Idaho
Robley J. Light   Florida State University
Maria O. Longas   Purdue University, Calumet
Cran Lucas   Louisiana State University–Shreveport
Jerome Maas   Oakton Community College
Arnulfo Mar   University of Texas–Brownsville
Larry D. Martin   Morningside College

Martha McBride Norwich University
Joyce Miller University of Wisconsin–Platteville
Robin Miskimins University of South Dakota
Bruce Morimoto Purdue University
Alan Myers Iowa State University
Harvey Nikkei Grand Valley State University
Treva Palmer Jersey City State College
Scott Pattison Ball State University
Allen T. Phillips Pennsylvania State University
Tom Rutledge Urinus College
Richard Saylor Shelton State Community College
Edward Senkbeil Salisbury State University
Ralph Shaw Southeastern Louisiana University
Ram P. Singhal Wichita State University
David Speckhard Loras College
Ralph Stephani St. John's University
Dan M. Sullivan University of Nebraska, Omaha
William Sweeney Hunter College
Christine Tachibana Pennsylvania State University
John M. Tomich Kansas State University
Anthony Toste Southwest Missouri State University
Craig Tuerk Morehead State University
Shashi Unnithan Front Range Community College
William Voige James Madison University
Alexandre G. Volkov Oakwood College
Justine Walhout Rockford College
Linette M. Watkins Southwest Texas State
  University
Lisa Wen Western Illinois University
Alfred Winer University of Kentucky
Beulah Woodfin University of New Mexico

Kenneth Wunch Tulane University
Les Wynston California State University, Long Beach

We wish to express our appreciation to John Challice, vice president and publisher; Jason Noe, senior editor; and Jane Clayton, development manager. The excellent efforts of the Oxford University Press production team are gratefully acknowledged. We are especially appreciative of the efforts of Barbara Mathieu, production editor; Steven Cestaro, production manager; Running River Design, designer of the book's interior and cover; Paula Schlosser, art director; Evelyn O'Shea, senior designer, and Cassandra Palmer, editorial assistant. In addition, we extend our gratitude to the Oxford marketing team, namely Adam Glazer, director of marketing; Preeti Parasharami, product manager; Jill Crosson, advertising and promotions manager; and Trent Haywood, senior copywriter. We give a very special thank you to Joseph Rabinowitz (Professor Emeritus, University of Pennsylvania) and Ann Randolph, whose consistent diligence on this project has ensured the accuracy of the text.

We also wish to extend our deep appreciation to those individuals who have encouraged us and made this project possible: Ira and Jean Cantor and Joseph and Josephine Rabinowitz.

Finally we thank our son James Adrian McKee, not only for his patience and encouragement, but also for his efforts to ensure the accuracy of the text.

Trudy McKee
James R. McKee

A great new partnership for the one-semester biochemistry course!

Higher Education Group

Dear Professor,

Oxford University Press is proud to announce the fourth edition of *Biochemistry: The Molecular Basis of Life*, by Trudy McKee and James R. McKee. As the new publisher for this trusted and respected textbook, we would like to extend our warmest regards to our friends in the Biochemistry community.

The fourth edition of McKee & McKee provides the ideal balance between chemical and biological principles and shows the relevance of biochemistry to students' own fields of study (including Health, Agriculture, Engineering, and Forensics). It features a renewed emphasis on problem solving and critical thinking and has been thoroughly updated to reflect the most recent developments in the field. It achieves all of this while remaining focused on the "big-picture" that is so essential for the one-semester survey course.

We invite you to take a guided tour of the new edition and peruse the chapters that cover your favorite topics. We believe you will find that our authors have prepared a new edition that is of exceptional quality and value to your students.

Sincerely,

Jason Noe

Jason Noe
Senior Editor
Oxford University Press

Guided Tour of McKee's Robust Biochemistry Learning System

Biochemistry in Perspective essays show you the real-world relevance of the biochemical processes you are studying. The fourth edition includes over a dozen new essays, each of which now begins with a thought-provoking question and concludes with a concept summary statement.

BIOCHEMISTRY IN PERSPECTIVE

Protein Folding and Human Disease

What are the effects of misfolded proteins on human health? The accumulation of insoluble, misfolded proteins is an important feature of several human neurodegenerative diseases. Alzheimer's and Huntington's diseases are prominent examples. Despite differences in their initiating events, specific brain areas affected, and symptoms, both diseases have in common cell-destroying dysfunctional processes set in motion by toxic proteins. A key feature of disease initiation is the conversion of normal protein structure, most commonly α-helices and random coils, into abnormal β-pleated sheet conformations. A brief overview of the molecular basis of each disease is provided.

Alzheimer's Disease

Alzheimer's Disease (AD) is a progressive and ultimately fatal condition that is characterized by seriously impaired intellectual function. AD first manifests itself with short-term memory loss. Eventually, severe memory loss, disorientation, and agitation accompany a total loss of the patient's personality. Caused by

There are inherited and sporadic versions of AD. Most cases of inherited AD have an early onset (i.e. in middle age). Sporadic AD typically is diagnosed after age 65. The genes associated with familial (inherited) forms of AD code for mutant versions of APP, presenilin 1 (PS1), presenilin 2 (PS2), and apolipoprotein E4 (p. xxx). APP is a transmembrane protein of unknown function with a large extracellular region that undergoes several proteolytic processing reactions. PS1 and PS2 are components of secretase γ, one of several proteases involved in APP processing. Mutations in the genes for APP, PS1, and PS2 have been associated with the release of the toxic Aβ40 and Aβ42 fragments. The mechanisms by which AD patients develop the sporadic form of the disease in the absence of known risk factors remain unresolved.

Huntington's Disease

Huntington's disease (HD) is one of a group of inherited neu-

SUMMARY: The accumulation of misfolded proteins impedes cell function. Eventually protein aggregates cause cell death.

Biochemistry in the Lab boxes introduce you to both classic and modern research techniques, reinforcing the fact that the best way to learn science is by actually doing science!

BIOCHEMISTRY IN THE LAB

Protein Technology

Living organisms produce a stunning variety of proteins. Consequently, it is not surprising that considerable time, effort, and funding have been devoted to investigating their properties. Since the amino acid sequence of bovine insulin was determined by Frederick Sanger in 1953, the structures of several thousand proteins have been elucidated.

In contrast to the 10 years required for insulin, current technologies allow protein sequence determination within a few days. In addition to the Edman degradation method and mass spectrometry, the amino acid sequence of a protein can be generated from its DNA or mRNA sequence if this information is available. After a brief review of protein purification methods, the Edman degradation method and mass spectrometry are described. Note that all the techniques for isolating, purifying, and characterizing proteins exploit differences in charge, molecular weight, and binding affinities. Many of these technologies apply to the investigation of other biomolecules.

Purification

Protein analysis begins with isolation and purification. Extraction of a protein requires cell disruption and homogenization (see Biochemistry in the Lab, Cell Technology, Chapter 2). This process is often followed by differential centrifugation and, if the protein is a component of an organelle, by density gradient centrifugation. After the protein-containing fraction has been obtained, several relatively crude methods may be used to enhance purification. **Salting out** is a technique in which high concentrations of salts such as ammonium sulfate [(NH₄)₂SO₄]

In all chromatographic methods the protein mixture is dissolved in a liquid known as the **mobile phase**. As the protein molecules pass across the **stationary phase** (a solid matrix), they separate from each other because they are differently distributed between the two phases. The relative movement of each molecule results from its capacity to remain associated with the stationary phase while the mobile phase continues to flow.

Three chromatographic methods commonly used in protein purification are gel-filtration chromatography ion-exchange chromatography, and affinity chromatography. **Gel-filtration chromatography** (Figure 5D) is a form of size-exclusion chromatography in which particles in an aqueous solution flow through a column (a hollow tube) filled with gel and are separated according to size. Molecules that are larger than the gel pores are excluded and therefore move through the column quickly. Molecules that are smaller than the gel pores diffuse in and out of the pores, so their movement through the column is retarded. Differences in the rates of particle movement separate the protein mixture into bands, which are then collected separately.

Ion-exchange chromatography separates proteins on the basis of their charge. Anion-exchange resins, which consist of positively charged materials, bind reversibly with a protein's negatively charged groups. Similarly, cation-exchange resins bind positively charged groups. After proteins that do not bind to the resin have been removed, the protein of interest is recovered by an appropriate change in the solvent pH and/or salt concentration. (A change in pH alters the protein's net charge.)

Affinity chromatography takes advantage of the unique bio-

Unparalleled problem-solving system

provides you with multiple opportunities and methods for developing strong analytical skills.

Worked Problems throughout the text walk you through the solutions to each problem in a step-by-step manner.

WORKED PROBLEM 5.1

Consider the following amino acid and its pK_a values:

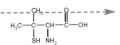

$pK_{a1} = 2.19$ $pK_{a2} = 9.67$ $pK_{aR} = 4.25$

a. Draw the structure of the amino acid as the pH of the solution changes from highly acidic to strongly basic.

Solution

The ionizable hydrogens are lost in order of acidity, the most acidic ionizing first.

b. Which form of the amino acid is present at the isoelectric point?

Solution

Hundreds of in-chapter **Questions** embedded in the text use real-world applications to pique your interest and get you thinking more deeply about biochemical principles.

FIGURE 5.14
Structure of Penicillamine

KEY CONCEPT

- Polypeptides are polymers composed of amino acids linked by peptide bonds. The order of the amino acids in a polypeptide is called the amino acid sequence.
- Disulfide bridges, formed by the oxidation

QUESTION 5.4

In extracellular fluids such as blood (pH 7.2–7.4) and urine (pH 6.5), the sulfhydryl groups of cysteine (pK_a 8.1) are protonated and subject to oxidation to form cystine. In peptides and proteins the nucleophilic character of free protonated thiol groups are used to advantage in stabilizing protein structure and in thiol transfer reactions, but the free amino acid in tissue fluids can be problematic because of the low solubility of cystine. In a genetic disorder known as *cystinuria*, defective membrane transport of cystine results in excessive excretion of cystine into the urine. Crystallization of the amino acid results in formation of calculi (stones) in the kidney, ureter, or urinary bladder. The stones may cause pain, infection, and blood in the urine. Cystine concentration in the kidney is reduced by massively increasing fluid intake and administering D-penicillamine. It is believed that penicillamine (Figure 5.14) is effective because penicillamine–cysteine disulfide, which is substantially more soluble than cystine, is formed. What is the structure of the penicillamine–cysteine disulfide?

Review Questions at the end of each chapter allow you to practice basic skills, while expanded **Thought Questions** challenge you to think about concepts across chapters, testing your ability to synthesize the material you have just studied.

ReviewQuestions

These questions are designed to test your knowledge of the key concepts discussed in this chapter, before moving on to the next chapter. You may like to compare your answers to the solutions provided in the back of the book, and in the accompanying Study Guide.

1. Distinguish between proteins, peptides, and polypeptides.
2. Indicate whether each of the following amino acids is polar, nonpolar, acidic, or basic:
   a. glycine
   b. tyrosine
   c. glutamic acid
   d. histidine
   e. proline
   f. lysine
   g. cysteine
   h. asparagine
   i. valine
   j. leucine
3. Arginine has the following pK_a values:
   $pK_1 = 2.17$, $pK_2 = 9.04$, $pK_R = 12.48$
   Give the structure and net charge of arginine at the following pH values: 1, 4, 7, 10, 12

ThoughtQuestions

These questions are designed to reinforce your understanding of all of the key concepts discussed in the book so far, including this chapter and all of the chapters before it. They may not have one right answer! The authors have provided possible solutions to these questions in the back of the book and in the accompanying Study Guide, for your reference.

28. Residues such as valine, leucine, isoleucine, methionine, and phenylalanine are often found in the interior of proteins, whereas arginine, lysine, aspartic acid, and glutamic acid are often found on the surface of proteins. Suggest a reason for this observation. Where would you expect to find glutamine, glycine, and alanine?
29. Proteins that are synthesized by living organisms adopt a biologically active conformation. Yet when such molecules are prepared in the laboratory, they usually fail to spontaneously adopt their active conformations. Can you suggest why?
30. The active site of an enzyme contains sequences that are conserved because they participate in the protein's catalytic activity. The bulk of an enzyme, however, is not part of the active site. Because a substantial amount of energy is required to assemble enzymes, why are they usually so large?

a. What is the approximate isoelectric point?
b. In which direction will the tripeptide move if placed in an electric field at pH 1, 5,10, and 12?
36. Chymotrypsin is an enzyme that cleaves other enzymes during sequencing. Why don't chymotrypsin molecules attack each other?
37. Most amino acids appear bluish purple when treated with ninhydrin reagent. Proline and hydroxyproline appear yellow. Suggest a reason for the difference.
38. When the multifunction protein glyceraldehyde-3-phosphate dehydrogenase (GAPD) catalyzes a key reaction in glycolysis (a metabolic pathway in cytoplasm), it does so as a homotetramer (four identical subunits). The GAPD monomer is a nuclear DNA repair enzyme. Describe in general terms what structural properties of multifunction pro-

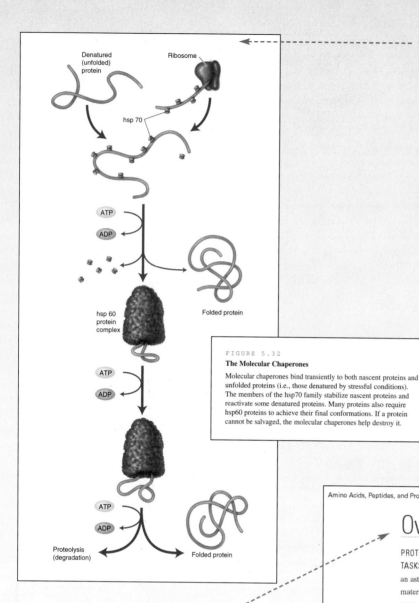

FIGURE 5.32

The Molecular Chaperones

Molecular chaperones bind transiently to both nascent proteins and unfolded proteins (i.e., those denatured by stressful conditions). The members of the hsp70 family stabilize nascent proteins and reactivate some denatured proteins. Many proteins also require hsp60 proteins to achieve their final conformations. If a protein cannot be salvaged, the molecular chaperones help destroy it.

Vivid and accurate figures illuminate the text discussion, illustrating important concepts and processes that you will need to know for your exams.

Concept icons have been included throughout the text to point out coverage of important biochemical applications.

 Medical

 Metabolic Regulation Mechanism

The chapter-opening **Overviews** help you focus on the "big picture," while end-of-chapter **summaries, suggested readings,** and **vocabulary lists** provide you with pedagogical support, reinforcing what you have learned and guiding you toward further inquiry.

Amino Acids, Peptides, and Proteins

Overview

PROTEINS ARE ESSENTIAL CONSTITUENTS OF ALL ORGANISMS. MOST TASKS PERFORMED BY LIVING CELLS REQUIRE PROTEINS, WHICH PERFORM an astonishing variety of functions. In addition to serving as structural materials in all living organisms (e.g., structural components in the muscle and connective tissue of animals or cell wall components of prokaryotes), proteins are involved in such diverse functions as metabolic regulation, transport, defense, and catalysis. The functional diversity exhibited by this class of biomolecules is directly related to the combinatorial possibilities of the monomeric units, the 20 amino acids.

KeyWords

affinity chromatography, 000
aldimine, 000
aldol condensation, 000
aliphatic hydrocarbon, 000
allosteric transition, 000
allostery, 000
Alzheimer's disease, 000
amino acid residue, 000
amphipathic molecule, 000
amphoteric molecule, 000
antigen, 000
apoprotein, 000
aromatic hydrocarbon, 000
asymmetric carbon, 000
chaperonins, 000
chiral carbon, 000
conjugated protein, 000
cooperative binding, 000
denaturation, 000
disulfide bridge, 000
effector, 000

electrophoresis, 000
enantiomer, 000
fibrous protein, 000
fold, 000
gel-filtration chromatography, 000
globular protein, 000
glycoprotein, 000
heat shock protein, 000
hemoprotein, 000
holoprotein, 000
homologous polypeptide, 000
hormone, 000
hsp60, 000
hsp70, 000
Huntington's disease, 000
intrinsically unstructured protein, 000
ion-exchange chromatography, 000
isoelectric point, 000

ligand, 000
lipoprotein, 000
metalloprotein, 000
mobile phase, 000
modular protein, 000
modulator, 000
molecular chaperone, 000
molecular disease, 000
molten globule, 000
motif, 000
motor protein, 000
multifunction protein, 000
natively unfolded protein, 000
neurotransmitter, 000
oligomer, 000
optical isomer, 000
peptide, 000
peptide bond, 000
phosphoprotein, 000
polypeptide, 000
primary structure, 000

prosthetic group, 000
protein, 000
protein family, 000
protein folding, 000
protein superfamily, 000
protomer, 000
quaternary structure, 000
response element, 000
salt bridge, 000
salting out, 000
Schiff base, 000
SDS-polyacrylamide gel electrophoresis, 000
secondary structure, 000
site-directed mutagenesis, 000
stationary phase, 000
stereoisomer, 000
subunit, 000
supersecondary structure, 000
tertiary structure, 000
zwitterion, 000

A powerful learning and teaching package

The fourth edition of *Biochemistry: the Molecular Basis of Life* is accompanied by a variety of ancillary materials for both instructors and students.

Please visit the companion website—**www.oup.com/us/mckee**—to gain access to lecture and assessment resources, as well as interactive student study aids.

FOR INSTRUCTORS

All images from the text available in electronic format

Instructors who adopt the fourth edition of McKee & McKee gain access to every illustration and photo from the text in high-resolution format. Images are available on the Instructor's Resource CD-ROM or via the McKee & McKee website.

Test Bank

Written by the authors, the Test Bank includes over 700 questions that are provided as editable Word files and can be easily customized. Available on the Instructor's Resource CD-ROM.

FOR STUDENTS

Student Study Guide and Solutions Manual

Coauthored by Patricia DePra (Carlow College) and the authors, this manual provides the solutions to all of the exercises not included in the back of the book. Each solution has been independently checked for accuracy by a panel of expert reviewers. (ISBN 9780195342925)

Multiple-Choice Quizzes

Students seeking an online resource to test their knowledge of biochemistry can visit **www.oup.com/us/mckee** to access over 600 questions written by Dan Sullivan (University of Nebraska, Omaha). New to the fourth edition, students now receive feedback on each graded quiz.

Interactive 3-D Molecules

Todd Carlson (Grand Valley State University) has created over 300 interactive 3-D molecules in JMOL format. Students can manipulate and study individual molecules and their structures, take self-guided concept tutorials, and test their molecule-recognition abilities by working through interactive tutorials and quizzes.

AFFORDABLE PACKAGE OPTIONS

Instructors: Save your students 50% on the *Study Guide and Solutions Manual*, or add a unique Oxford University Press science dictionary at a 50% discount, by using one of the following value-package ISBNs when placing your order with your bookstore!

- Textbook and *Study Guide and Solutions Manual* value package: ISBN 9780195372311
- Textbook and *Oxford Dictionary of Chemistry, 6th edition*, value package: ISBN 9780195372304
- Textbook and *Oxford Dictionary of Biology, 6th edition*, value package: ISBN 9780195372298

Biochemistry: An Introduction

OUTLINE

The Living Cell Living organisms consist of one or more cells. The capacity of living cells to perform functions of energy generation, growth, and reproduction is made possible by their complex structures.

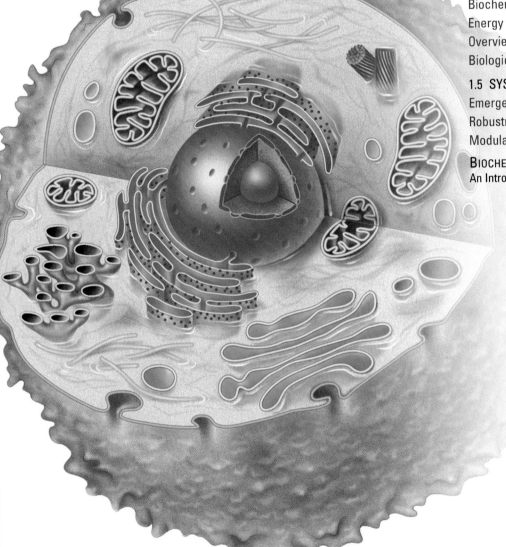

Overview

HOW FAR THE LIFE SCIENCES HAVE COME! IN LITTLE OVER A CENTURY
HUMAN KNOWLEDGE OF THE INNER WORKINGS OF LIVING ORGANISMS HAS
been dramatically transformed. From modest beginnings in the late nineteenth
century, the science of biochemistry has provided increasingly more sophisticated intellectual and laboratory tools for the investigation of living processes.
Today, in the early years of the twenty-first century, we find ourselves in the
midst of an unimagined biotechnological revolution. Immense amounts of information are being generated in life sciences as diverse as medicine, agriculture,
and forensics. The capacity to understand and appreciate the significance of
this phenomenon begins with a thorough knowledge of biochemical principles.
This chapter provides an overview of these principles. The chapters that follow
focus on the structure and functions of the most important biomolecules and
the major biochemical processes that sustain the living state.

n 1977 two geologists were exploring the Galapagos Rift, which is a fissure
in the underwater, earthquake-prone mountain range known as the Galapagos
Ridge. Their objective, to investigate the seafloor, was made possible by *Alvin*,
a deep-sea submarine fitted with external lights and sensors, cameras, and a
sample collection mechanism. As the scientists maneuvered their vessel downward through the pitch-black water to a ridge at a depth of about 2500 meters,
they made an astonishing discovery: the water temperature was unexpectedly
high (7°C instead of the normal 2°C). Proceeding up the ridge slope, they fully
expected to find little more than barren rock and sediment. Instead, they found
hot springs surrounded by dense populations of unique life-forms such as giant
tubeworms, sea anemones, pink fish, shrimp, and crabs (Figure 1.1). The hot
springs, also called *hydrothermal vents*, spew forth superheated (350–400°C)
mineral-laden water that is prevented from boiling by the enormous pressure
at this depth. The heat source is geothermal. As underwater plates in the earth's
crust move apart, the seafloor cracks, and seawater seeps in and becomes superheated. This scalding water dissolves or reacts with minerals (metals such as
iron, manganese, and nickel) and gases (hydrogen sulfide and hydrogen gases)
in the magma, or molten rock. As the minerals are ejected from the vent and
mix with cold seawater, they precipitate, forming chimneys.

The hydrothermal vent habitat was a complete surprise because of its location.
Sunlight does not penetrate to the ocean floor. According to the prevailing view
in the late 1970s, with several seemingly minor exceptions (e.g., microbial inhab-
itants of oil and sulfur deposits and terrestrial geysers), all life on earth derives
energy directly or indirectly from sunlight. Subsequent investigations of the vent
community, however, challenged this traditional notion and revealed the presence
of huge quantities of specially adapted bacteria, capable of converting toxic chemicals into food and energy. Instead of **photosynthesis**, the biochemical mechanism for capturing light energy, vent microbes utilize a process called
chemosynthesis in which chemical energy is extracted from certain minerals.
Other vent organisms either consume the microbes directly or live in a symbiotic relationship with them.

The hydrothermal vent discoveries are instructive because they provide insight
into the scientific process, the powerful, rational methodology that humans use
to investigate the universe. Over the past century and a half, life scientists have

FIGURE 1.1

Life-Forms Near a Hydrothermal Vent

Tubeworms, some as long as 1.8 m (6 ft), and giant white clams are viewed by scientists within the *Alvin*.

built a coherent and detailed understanding of living organisms by rigorous observation and experimentation. One of the principles that resulted from decades of this work is the unique role of the sun as *the* energy source that supports life on our planet. The hydrothermal vent communities, consequently, were certainly disconcerting. Numerous scientists then proceeded to investigate this remote habitat, despite long-held views concerning energy flow in the *biosphere*, the portions of the earth that sustain living organisms. As a result of this work, scientists in fields as wide-ranging as geology, oceanography, chemistry, and biology were compelled to rethink some conventional assumptions about the diversity and capacities of living organisms. For example, research into environments that had been previously dismissed as lifeless, such as deep aquifers and terrestrial sedimentary rock, revealed a hidden subterranean microbial biosphere that has now been shown to extend at least 6 km beneath the earth's surface. Amazingly, it is now estimated that the *biomass* of life underground, representing the total weight of living organisms there, exceeds the biomass on the surface.

As scientists investigated the newly discovered species, substantial amounts of biochemical and molecular data accumulated. It soon became apparent that despite unique adaptations to extreme environmental conditions, these organisms have a great deal in common with species already known. In other words, this work reaffirmed an important concept in the life sciences, the unity of all living organisms, and allowed life scientists to develop a deeper insight into living processes and a more detailed understanding of the history of living organisms.

This opening chapter provides an overview of the major components of living organisms and the processes that sustain the living state. After a brief description of the nature of the living state and a review of the diversity of life on earth, an introduction to the structures and functions of the major biomolecules is provided. This material is followed by a discussion of the most important biochemical processes. The chapter concludes with a brief discussion of the concepts of modern experimental biochemistry and an introduction to *systems biology*, an investigative strategy that is being developed to understand living organisms as integrated systems rather than collections of isolated components and chemical reactions. Throughout this chapter and the ones that follow, it will become apparent that all the topics, whether they concern biomolecular structure, biochemical reactions, or genetic inheritance, are inextricably linked. Understanding of any one of these subjects is required to understand the others.

1.1 WHAT IS LIFE?

What is life? The answer to this deceptively simple question has been elusive despite the work of life scientists over several centuries. Much of the difficulty in delineating the precise nature of living organisms lies in the overwhelming diversity of the living world and the apparent overlap in several properties of living and nonliving matter. Consequently, life has been viewed as an intangible property that defies simple explanation and is usually described in operational terms, such as movement, reproduction, adaptation, and responsiveness to external stimuli. The work of life scientists, made possible by the experimental approaches of biochemistry, has revealed that all organisms obey the same chemical and physical laws that rule the universe. Among the most important insights gained from the work of biochemists are the following.

1. **Life is complex and dynamic.** All organisms are composed of the same set of chemical elements, primarily carbon, nitrogen, oxygen, hydrogen, sulfur, and phosphorus. **Biomolecules**, the molecules synthesized by living organisms, are organic (carbon based). Living processes, such as growth and development, involve thousands of chemical reactions in which vast quantities and varieties of vibrating and rotating molecules interact, collide, and rearrange into new molecules.

2. **Life is organized and self-sustaining.** Living organisms are hierarchically organized systems; that is, they consist of patterns of organization from smallest (atom) to largest (organism) (Figure 1.2). In biological systems, the functional capacities of each level of organization are derived from the structural and chemical properties of the level below it. Biomolecules are composed of atoms, which in turn are formed from subatomic particles. Certain biomolecules become linked to form polymers called **macromolecules**. Examples include nucleic acids, proteins, and polysaccharides, which are formed from nucleotides, amino acids, and sugars, respectively. Cells are composed of a diversity of biomolecules and macromolecules that form into more complex supermolecular structures. At the chemical level, sets

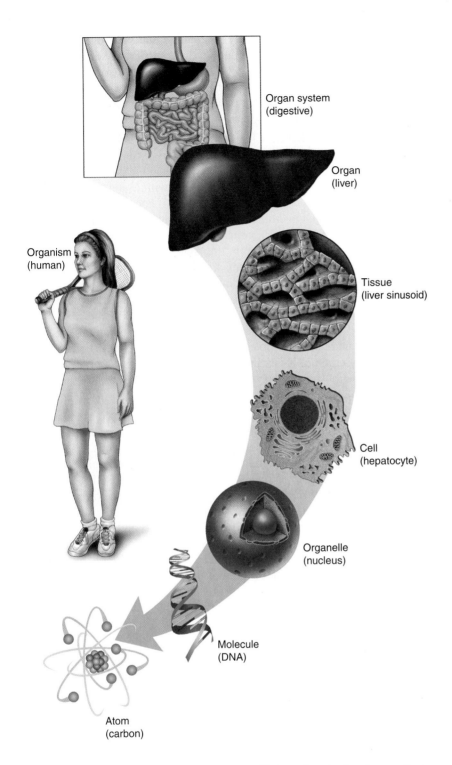

Organ system
(digestive)

Organ
(liver)

Organism
(human)

Tissue
(liver sinusoid)

Cell
(hepatocyte)

Organelle
(nucleus)

Molecule
(DNA)

Atom
(carbon)

FIGURE 1.2

**Hierarchical Organization
of a Multicellular Organism:
The Human Being**

Multicellular organisms have several
levels of organization: organ systems,
organs, tissues, cells, organelles,
molecules, and atoms. The digestive
system and one of its component
organs (the liver) are shown. The liver
is a multifunctional organ that has
several digestive functions. For example,
it produces bile, which facilitates fat
digestion, and it processes and distributes the food molecules absorbed in
the small intestine to other parts of the
body. DNA, one type of molecule
found in cells, contains the genetic
information that controls cell function.

of interdependent molecules create efficient chemical pathways that convert
an entering molecule(s) to a terminal product(s). (A *pathway* is a specific
series of sequential chemical reactions.) In multicellular organisms other
levels of organization include tissues, organs, and organ systems. The organization and ordered functioning of living organisms require the continuous
acquisition of both energy and matter, and the removal of waste molecules.
These tasks are accomplished by hundreds of biochemical reactions that
are catalyzed by biomolecular catalysts called **enzymes**. The sum total of
all the reactions in a living organism is referred to as **metabolism**. The capacity of living organisms to regulate metabolic processes despite variability
in their internal and external environments is called **homeostasis**.

3. **Life is cellular.** Cells differ widely in structure and function, but each is surrounded by a membrane that controls the transport of some chemical substances into and out of the cell. The membrane also mediates the response of the cell to components of the extracellular environment. If a cell is divided into its component parts, it will cease to function in a life-sustaining way. Cells arise only from the division of existing cells.

4. **Life is information-based.** Organization requires information. Living organisms can be considered to be information-processing systems because maintenance of their structural integrity and metabolic processes involves interactions among a vast array of molecules within and between cells. Biological information is expressed in the form of coded messages that are inherent in the unique three-dimensional structure of biomolecules. Genetic information, which is stored in **genes**, the linear sequences of nucleotides in deoxyribonucleic acid (DNA), in turn specifies the linear sequence of amino acids in proteins, and how and when those proteins are synthesized. Proteins perform their function by interacting with other molecules. The unique three-dimensional structure of each type of protein allows it to bind to, and interact with, a specific type of molecule that has a precise complementary shape. Information is transferred during the binding process. For example, the binding of the protein insulin to insulin receptor molecules on the surface of certain cells is a signal that initiates the uptake of the nutrient molecule glucose. The transport of amino acids is insulin-sensitive as well.

5. **Life adapts and evolves.** All life on earth has a common origin, with new forms arising from older forms. When an individual organism in a population reproduces itself, stress-induced DNA modifications and errors that occur when DNA molecules are copied can result in **mutations** or sequence changes. Most mutations are silent; that is, they either are repaired by the cell or have no effect on the functioning of the organism. Some, however, are harmful, serving to limit the reproductive success of the offspring. On rare occasions mutations may contribute to an increased ability of the organism to survive, to adapt to new circumstances, and to reproduce. A principal driving force in this process is the capacity to exploit energy sources. Individuals possessing traits that allow them to better exploit a specific energy source within their habitat may have a competitive advantage when resources are limited. Over many generations, the interplay of environmental change and genetic variation can lead to the accumulation of favorable traits, and eventually to increasingly different forms of life.

KEY CONCEPTS

- All living organisms obey the chemical and physical laws.
- Life is complex, dynamic, organized, and self-sustaining.
- Life is cellular and information-based.
- Life adapts and evolves.

1.2 THE LIVING WORLD

Estimates of the number of living species currently range from several million to tens of millions. All are composed of either **prokaryotic** or **eukaryotic cells**. Most organisms are prokaryotes; that is, their cells lack a nucleus (*pro* = "before," *karyon* = "nucleus" or "kernel"). The eukaryotes (*eu* = "true") are composed of relatively large cells that possess a nucleus, a membrane-bound compartment that contains the genetic material.

The prokaryotes are the oldest forms of life on earth; indeed, from about 3.8 billion years ago until about 1.8 billion years ago, they were the only forms of life. Until the 1980s, the prokaryotes were believed to consist only of bacteria. Analysis of the nucleotide sequences of ribonucleic acid (RNA), a type of nucleic acid involved in the synthesis of proteins, has revealed that there are two quite distinct groups of prokaryotes: the Bacteria and the Archaea. Their external appearances are similar, but the differences in their molecular properties are greater than their differences from the eukaryotes. The single-celled prokaryotes are the smallest living organisms. Nevertheless, their combined biomass exceeds that of the larger eukaryotic organisms (animals, plants, fungi, and

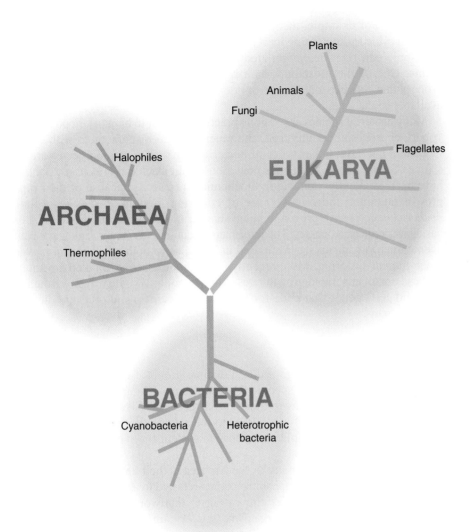

FIGURE 1.3

The Domains of Life on Earth

Molecular evidence indicates that all
life forms investigated so far can be
classified into three domains.

single-celled protists) by at least tenfold. Prokaryotes occupy virtually every
niche on earth. In addition to air, soil, and water, various prokaryotic species live
on the skin and in the digestive tracts of animals, within hot springs, and, as
previously mentioned, to a depth of several kilometers below the earth's surface.

The molecular evidence concerning the evolutionary relationships of living
species is sufficiently compelling that many life scientists now classify all living organisms into three domains: the **Bacteria**, the **Archaea**, and the **Eukarya**
(Figure 1.3). Each domain is briefly discussed below.

Bacteria

Bacteria are so diverse in their habitats and nutritional capacities that only general statements can be made about them. As a group, bacteria are especially
known for their biochemical diversity. Certain species of bacteria can exploit
virtually every conceivable energy source, nutrient, and habitat. For example,
some bacterial species can use light energy to convert carbon dioxide (CO_2) into
organic molecules. Others use energy extracted from inorganic or organic
molecules.

Some bacterial species cause disease (e.g., cholera, tuberculosis, syphilis,
and tetanus). The vast majority, however, play vital roles in sustaining life on
earth. The activity of many types of bacteria is required in *biogeochemical*

cycles, the global cycles of nutrients such as carbon, nitrogen, phosphorus, and sulfur. For example, the bacterium *Rhizobium* plays a critical role in the nitrogen cycle by converting inert molecular nitrogen (N_2) into ammonia (NH_3). Ammonia can then be assimilated by other organisms such as leguminous plants. One of the most important roles of bacteria is in decomposition, a process that releases nutrients from dead organisms so that they can become available to the living.

Many bacterial species are of great practical interest to humans. Foods such as yogurt, cheese, sourdough bread, and sauerkraut are manufactured with the aid of certain bacteria. Other types of bacteria such as the actinomycetes are the source of many of the antibiotics now used to cure bacterial infections. Bacteria have been especially valuable in biochemical research. Because of their rapid growth rates and the relative ease of culturing, certain species (especially *Escherichia coli*) have proven to be invaluable in the investigation into the most basic biochemical processes. The information acquired in research studies of pathogenic microorganisms has been used in medicine to both alleviate and prevent much human suffering. More recently, biotechnologists have taken advantage of these organisms' rapid growth rates and metabolic flexibility by inserting into bacterial cells genes that code for hormones, vaccines, and other products of use to humans.

Archaea

The Archaea were not recognized as a distinct group of organisms until 1977, when Carl Woese analyzed specific RNA molecules. Comparison of the molecular properties of archaeans to those of bacteria and eukaryotes has revealed that archaeans are in many ways closer to the eukaryotes than to the outwardly similar bacteria. For example, the archaean system for synthesizing protein is more like that of eukaryotes.

A prominent feature of many of the Archaea is their capacity to occupy and even thrive in very challenging habitats. Often referred to as **extremophiles**, some archaean species can live under circumstances that would easily kill most lifeforms. Although other types of organisms (e.g., certain bacteria, algae, and fungi) can live in extreme conditions, the archaeans include the most extremophilic species. For example, the Halobacteria are a diverse group of Archaea that live in hypersaline bodies of water (e.g., the Great Salt Lake in Utah, the Dead Sea). These organisms have been observed to thrive in concentrations of between 4 to 5 M NaCl. Extremophiles can be classified according to the types of exotic conditions in which they live: very high or low temperatures, high salt concentrations, or high pressure. In addition to providing considerable insight into the history of life on earth, investigations of the extremophilic archaeans have allowed unique insights into adaptations of biomolecular structure to extreme conditions. The research efforts of biochemists and biotechnologists have been concentrated on the **extremozymes**, enzymes that work under noxious conditions. Examples of industrial applications of this work include enzymes used in food processing and laundry detergents. Along with many bacterial species, archaeans have proven to be useful in **bioremediation**, a process in which microorganisms are used to degrade or remove pollutants from toxic waste sites and oil spills.

Eukarya

The third domain of living organisms, the Eukarya, is composed of all the remaining species on earth. Although the presence or absence of a nucleus is the most notable difference between prokaryotes and eukaryotes, there are other significant distinctions:

1. **Size**. Eukaryotic cells are substantially larger than prokaryotic cells. The diameter of animal cells, for example, varies between 10 and 30 μm. Such

values are approximately 10 times higher than those for prokaryotes. Size disparity between the two cell types is more obvious, however, when volume is considered. For example, the volume of a typical eukaryotic cell such as a liver cell (hepatocyte) is between 6000 and 10,000 μm^3. The volume of *E. coli* cells is several hundred times smaller.

2. **Complexity.** Although the structural complexity of prokaryotes is significant, that of the eukaryotes is greater by several orders of magnitude, largely because of the presence of subcellular compartments called **organelles**. Each organelle is specialized to perform specific tasks. The compartmentalization afforded by organelles permits the concentration of reactant and product molecules at sites where they can be efficiently used. This availability is among the factors that make intricate regulatory mechanisms possible. Consequently, the cells of multicellular eukaryotes are able to respond quickly and effectively to the intercellular communications that are required for growth and development.

3. **Multicellularity.** True multicellularity is found only in the Eukarya. The highly complex single-celled protists make up the largest biomass of the Eukarya. All the remaining categories are multicellular. Some bacteria exhibit the habit of colonial living, especially on solid media. However, the cooperativity and specialization of multicellularity is rarely achieved. Multicellular organisms are not just collections of cells: they are highly ordered living systems that together form a coherent entity. The structural complexity of eukaryotic cells provides the capacity for the intricate mechanisms of regulation and cellular communication required in these organisms.

KEY CONCEPT

Living organisms have been classified into three domains: Bacteria, Archaea, and Eukarya.

1.3 BIOMOLECULES

Living organisms are composed of thousands of different kinds of inorganic and organic molecules. Water, an inorganic molecule, may constitute 50 to 95% of a cell's content by weight, and ions such as sodium (Na^+), potassium (K^+), magnesium (Mg^{2+}), and calcium (Ca^{2+}) may account for another 1%. Almost all the other kinds of molecules in living organisms are organic. Organic molecules are principally composed of six elements: carbon, hydrogen, oxygen, nitrogen, phosphorus, and sulfur, and they contain trace amounts of certain metallic and other nonmetallic elements. It is noteworthy that the atoms of each of the most common elements found in living organisms can readily form stable covalent bonds, the kind that allow the formation of such important molecules as proteins.

The remarkable structural complexity and diversity of organic molecules are made possible by the capacity of carbon atoms to form four strong, single covalent bonds either to other carbon atoms or to atoms of other elements. Organic molecules with many carbon atoms can form complicated shapes such as long, straight structures or branched chains and rings.

Functional Groups of Organic Biomolecules

Most biomolecules can be considered to be derived from the simplest type of organic molecules, called the **hydrocarbons**. Hydrocarbons (Figure 1.4) are

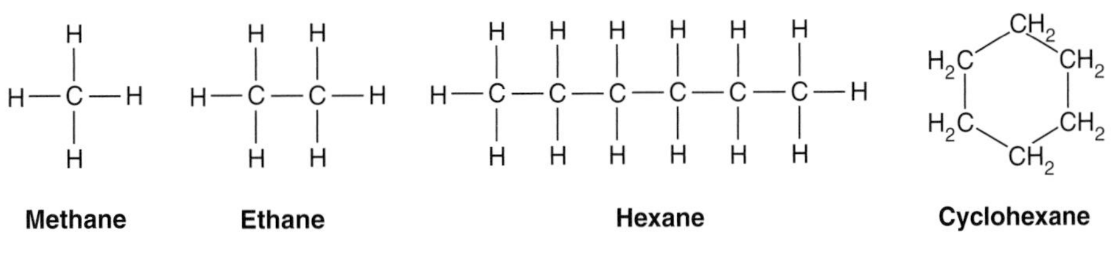

Methane   **Ethane**   **Hexane**   **Cyclohexane**

FIGURE 1.4

Structural Formulas of Several Hydrocarbons

TABLE 1.1 Important Functional Groups in Biomolecules

Family Name	Group Structure	Group Name	Significance
Alcohol	R—OH	Hydroxyl	Polar (and therefore water-soluble), forms hydrogen bonds
Aldehyde	$\overset{\displaystyle O}{\overset{\|}{R—C—H}}$	Carbonyl	Polar, found in some sugars
Ketone	$\overset{\displaystyle O}{\overset{\|}{R—C—R'}}$	Carbonyl	Polar, found in some sugars
Acids	$\overset{\displaystyle O}{\overset{\|}{R—C—OH}}$	Carboxyl	Weakly acidic, bears a negative charge when it donates a proton
Amine	R—NH$_2$	Amino	Weakly basic, bears a positive charge when it accepts a proton
Amide	$\overset{\displaystyle O}{\overset{\|}{R—C—NH_2}}$	Amido	Polar but does not bear a charge
Thiol	R—SH	Thiol	Easily oxidized; can form —S—S— (disulfide) bonds readily
Ester	$\overset{\displaystyle O}{\overset{\|}{R—C—O—R'}}$	Ester	Found in certain lipid molecules
Alkene	RCH=CHR'	Double bond	Important structural component of many biomolecules (e.g., found in lipid molecules)

carbon- and hydrogen-containing molecules that are **hydrophobic**, or insoluble in water. All other organic molecules are formed by attaching other atoms or groups of atoms to the carbon backbone of the hydrocarbon. The chemical properties of these derivative molecules are determined by the specific arrangement of atoms called **functional groups** (Table 1.1). For example, alcohols result when hydrogen atoms are replaced by hydroxyl groups (—OH). Thus methane (CH$_4$), a component of natural gas, can be converted into methanol (CH$_3$OH), a toxic liquid that is used as a solvent in many industrial processes.

Most biomolecules contain more than one functional group. For example, many simple sugar molecules have several hydroxyl groups and an aldehyde group. Amino acids, the building block molecules of proteins, have both an amino group and a carboxyl group. The distinct chemical properties of each functional group contribute to the behavior of any molecule that contains it.

Major Classes of Small Biomolecules

Many of the organic compounds found in cells are relatively small, with molecular weights of less than 1000 daltons (D). (One dalton, 1 atomic mass unit, is equal to $1/12$ of the mass of one atom of ^{12}C.) Cells contain four families of small molecules: amino acids, sugars, fatty acids, and nucleotides (Table 1.2). Members of each group serve several functions. First, they are used in the synthesis of larger molecules, many of which are polymers. For example, proteins, certain carbohydrates, and nucleic acids are polymers composed of amino acids, sugars, and nucleotides, respectively. Fatty acids are components of lipid (water-insoluble) molecules of several types.

Second, some molecules have special biological functions. For example, the nucleotide adenosine triphosphate (ATP) serves as a cellular reservoir of chemical energy. Finally, many small organic molecules are involved in complex reaction pathways. Examples of each class of molecule are described next.

TABLE 1.2   Major Classes of Biomolecules

Small Molecule	Polymer	General Functions
Amino acids	Proteins	Catalysts and structural elements
Sugars	Carbohydrates	Energy sources and structural elements
Fatty acids	N.A.	Energy sources and structural elements of complex lipid molecules
Nucleotides	DNA	Genetic information
	RNA	Protein synthesis

AMINO ACIDS AND PROTEINS There are hundreds of naturally occurring amino acids, each of which contains an amino group and a carboxyl group. Amino acids are classified α, β, or γ according to the location of the amino group in reference to the carboxyl group. In α-amino acids, the most common type, the amino group is attached to the carbon atom (the α-carbon) immediately adjacent to the carboxyl group (Figure 1.5). In β- and γ-amino acids, the amino group is attached to the second and third carbon, respectively, from the carboxyl group. Also attached to the α-carbon is another group, referred to as the side chain or R group. The chemical properties of each amino acid, once incorporated into protein, are determined largely by the properties of its side chain. For example, some side chains are hydrophobic (i.e., low solubility in water), whereas others are **hydrophilic** (i.e., dissolve easily in water). The general formula for α-amino acids is

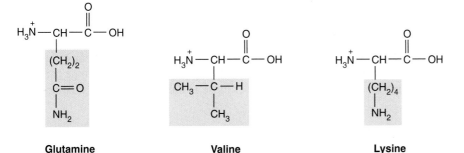

FIGURE 1.5

Structural Formulas for Several α-Amino Acids

An R group (highlighted) in an amino acid structure can be a hydrogen atom (e.g., in glycine), a hydrocarbon group (e.g., the isopropyl group in valine), or a hydrocarbon derivative (e.g., the hydroxymethyl group in serine).

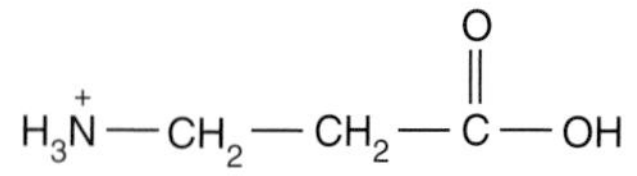

β-Alanine

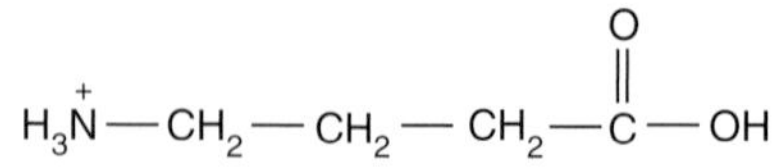

GABA

FIGURE 1.6

Selected Examples of Naturally Occurring Amino Acids That Are Not α-Amino Acids: β-Alanine and γ-Aminobutyric Acid (GABA)

There are 20 standard α-amino acids that occur in proteins. Some standard amino acids have unique functions in living organisms. For example, glycine and glutamic acid function in animals as **neurotransmitters**, signal molecules released by nerve cells. Proteins also contain nonstandard amino acids that are modified versions of the standard amino acids. The structure and function of protein molecules are often altered by conversion of certain amino acid residues to derivatives via phosphorylation, hydroxylation, and other chemical modifications. (The term "residue" refers to a small biomolecule that is incorporated in a macromolecule, e.g., amino acid residues in a protein.) For example, many of the residues of proline are hydroxylated in collagen, the connective tissue protein. Many naturally occurring amino acids are not α-amino acids. Prominent examples include β-alanine, a precursor of the vitamin pantothenic acid, and γ-aminobutyric acid (GABA), a neurotransmitter found in the brain (Figure 1.6).

Amino acid molecules are used primarily in the synthesis of long, complex polymers known as **polypeptides**. Up to a length of about 50 amino acids, these molecules are called either **peptides** or **oligopeptides**. Longer polypeptides are often referred to as **proteins**. Polypeptides play a variety of roles in living organisms. Examples of molecules composed of polypeptides include transport proteins, structural proteins, and the enzymes (catalytic proteins).

The individual amino acids are connected in peptides (Figure 1.7) and polypeptides by the peptide bond. **Peptide bonds** are amide linkages that form in a type of nucleophilic substitution reaction (p. 18) in which the amino group nitrogen of one amino acid attacks the carbonyl carbon in the carboxyl group of another. The final three-dimensional structure, and therefore biological function, of polypeptides results largely from interactions among the R groups (Figure 1.8).

SUGARS AND CARBOHYDRATES Sugars contain alcohol and carbonyl functional groups. They are described in terms of both carbon number and the type of carbonyl group they contain. Sugars that possess an aldehyde group are called *aldoses* and those that possess a ketone group are called *ketoses*. For example, the six-carbon sugar glucose (an important energy source in most living organisms) is an aldohexose; fructose (fruit sugar) is a ketohexose (Figure 1.9).

Sugars are the basic units of carbohydrates, the most abundant organic molecules found in nature. Carbohydrates range from the simple sugars, or **monosaccharides**, such as glucose and fructose, to the **polysaccharides**, polymers that contain thousands of sugar units. Examples of the latter include starch and cellulose in plants and glycogen in animals. Carbohydrates serve a variety of functions in living organisms. Certain sugars are important energy sources. Glucose is the principal carbohydrate energy source in animals and plants. Sucrose is used by many plants as an efficient means of transporting energy throughout their tissues. Some carbohydrates serve as structural materials. Cellulose is the major structural component

FIGURE 1.7

Structure of Met-Enkephalin, a Pentapeptide

Met-enkephalin is one of a class of molecules that have opiate-like activity. Found in the brain, met-enkephalin inhibits pain perception. (The peptide bonds are colored blue. The R groups are highlighted.)

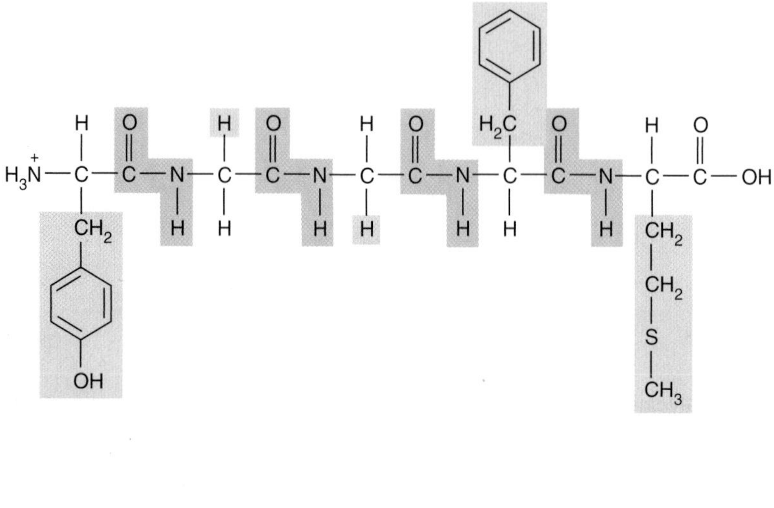

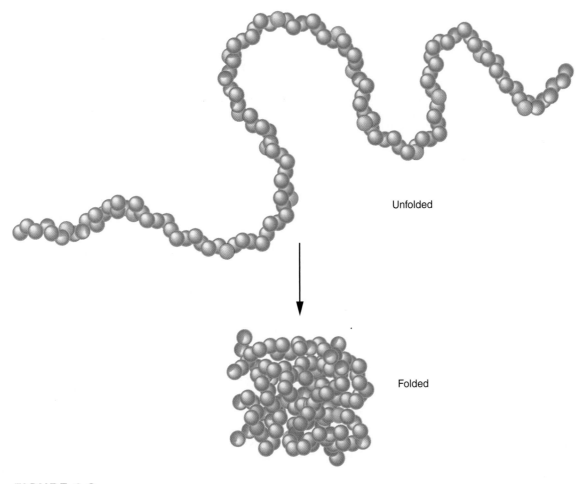

FIGURE 1.8

Polypeptide Structure

As a polypeptide folds into its unique three-dimensional form, at least 50% of hydrophobic R groups (yellow spheres) become buried in the interior away from water. Hydrophilic groups usually occur on the surface.

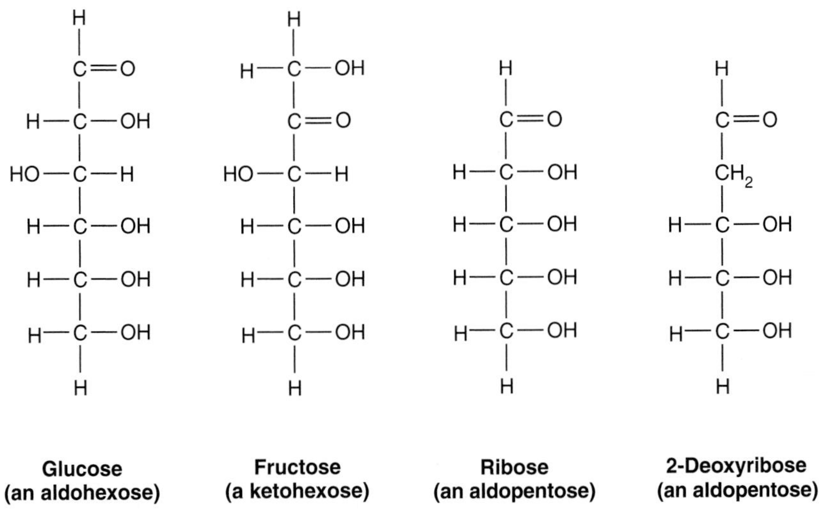

FIGURE 1.9

Some Biologically Important Monosaccharides

Glucose and fructose are important sources of energy in plants and animals. Ribose and deoxyribose are components of nucleic acids. These monosaccharides occur as ring structures in nature.

Glucose
(an aldohexose)

Fructose
(a ketohexose)

Ribose
(an aldopentose)

2-Deoxyribose
(an aldopentose)

of wood and certain plant fibers. Chitin, another type of polysaccharide, is found in the protective outer coverings of insects and crustaceans.

Some biomolecules contain carbohydrate components. Nucleotides, the building block molecules of the nucleic acids, contain either ribose or deoxyribose, which are sugars. Certain proteins and lipids also contain carbohydrate. Glycoproteins and glycolipids occur on the external surface of cell membranes in multicellular organisms, where they play critical roles in the interactions between cells.

FATTY ACIDS Fatty acids are monocarboxylic acids that usually contain an even number of carbon atoms. In some organisms they serve as energy sources. Fatty acids are represented by the chemical formula R—COOH, in which R is an alkyl group that contains carbon and hydrogen atoms. There are two types of fatty acids: **saturated** fatty acids, which contain no carbon-carbon double bonds, and **unsaturated** fatty acids, which have one or more double bonds (Figure 1.10). Under physiological conditions the carboxyl group of fatty acids exists in the ionized state, R—COO$^-$. For example, the 16-carbon saturated fatty acid called palmitic acid usually exists as palmitate, $CH_3(CH_2)_{14}COO^-$. Although the charged carboxyl group has an affinity for water, the long nonpolar hydrocarbon chains render most fatty acids insoluble in water.

Fatty acids occur as independent (free) molecules in only trace amounts in living organisms. Most often they are components of several types of **lipid** molecules (Figure 1.11). Lipids are a diverse group of substances that are soluble in organic solvents such as chloroform or acetone, but are not soluble in water. For example, triacylglycerols (fats and oils) are esters containing glycerol (a three-carbon alcohol with three hydroxyl groups) and three fatty acids. Certain lipid molecules that resemble triacylglycerols, called phosphoglycerides, contain two fatty acids. In these molecules the third hydroxyl group of glycerol is coupled with phosphate, which is in turn attached to small polar compounds such as choline. Phosphoglycerides are an important structural component of cell membranes.

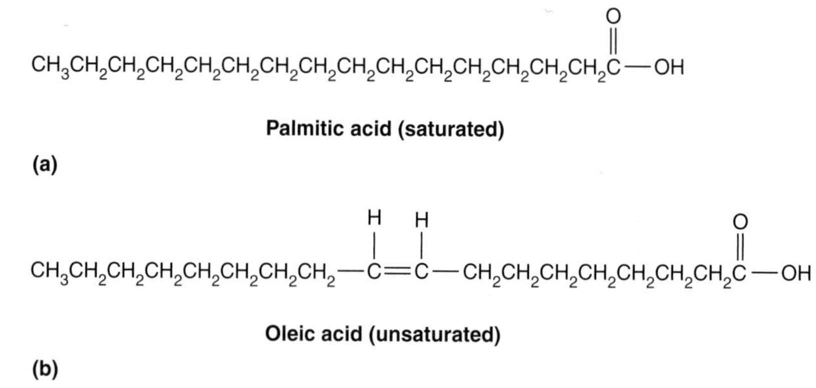

FIGURE 1.10

Fatty Acid Structure

(a) A saturated fatty acid. (b) An unsaturated fatty acid.

Palmitic acid (saturated)

(a)

Oleic acid (unsaturated)

(b)

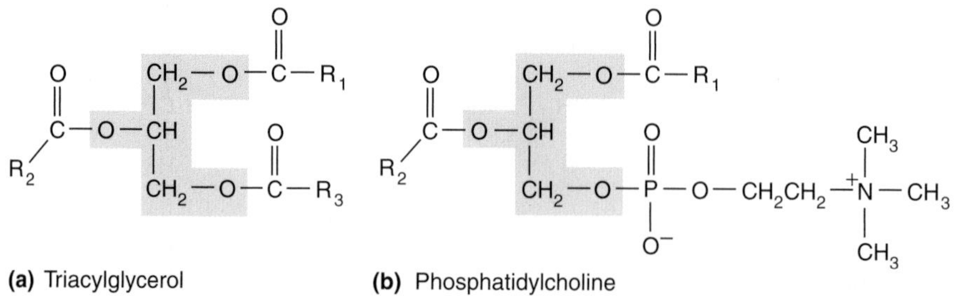

FIGURE 1.11

Lipid Molecules That Contain Fatty Acids

(a) Triacylglycerol. (b) Phosphatidylcholine, a type of phosphoglyceride.

(a) Triacylglycerol   **(b)** Phosphatidylcholine

NUCLEOTIDES AND NUCLEIC ACIDS Each nucleotide contains three components: a five-carbon sugar (either ribose or deoxyribose), a nitrogenous base, and one or more phosphate groups (Figure 1.12). The bases in nucleotides are heterocyclic aromatic rings with a variety of substituents. There are two classes of base: the bicyclic purines and the monocyclic pyrimidines (Figure 1.13).

Nucleotides participate in a wide variety of biosynthetic and energy-generating reactions. For example, a substantial proportion of the energy obtained from food molecules is used to form the high-energy phosphate bonds of adenosine triphosphate (ATP). Nucleotides also have an important role as the building block

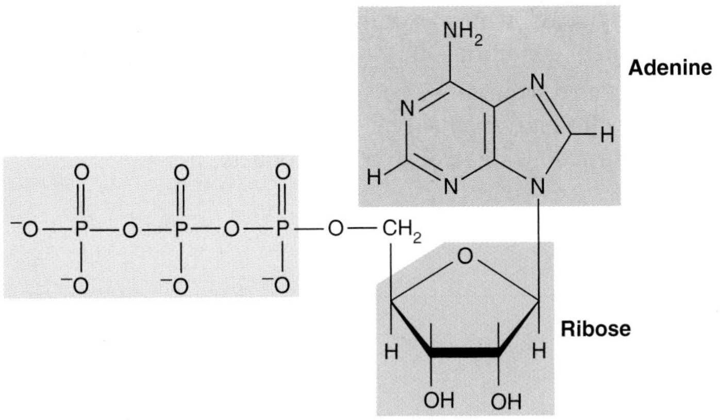

FIGURE 1.12

Nucleotide Structure

Each nucleotide contains a nitrogenous base (in this case, adenine), a pentose sugar (ribose), and one or more phosphates. This nucleotide is adenosine triphosphate.

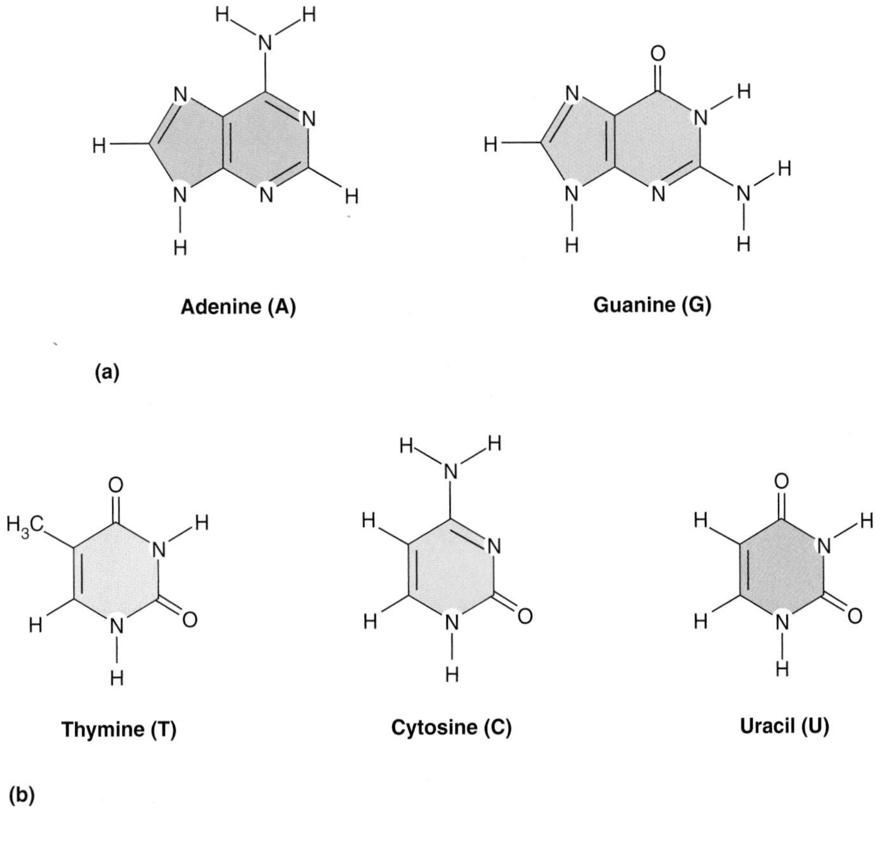

FIGURE 1.13

The Nitrogenous Bases

(a) The purines. (b) The pyrimidines.

molecules of the nucleic acids. In a nucleic acid molecule, from hundreds to millions of nucleotides are linked by phosphodiester linkages to form long polynucleotide chains or strands. There are two types of nucleic acid: DNA and RNA.

DNA. DNA is the repository of genetic information. Its structure consists of two **antiparallel** polynucleotide strands wound around each other to form a right-handed double helix (Figure 1.14). In addition to the pentose sugar deoxyribose and phosphate, DNA contains bases of four types: the **purines** adenine and guanine and the **pyrimidines** thymine and cytosine; adenine pairs with thymine and guanine pairs with cytosine. The double helix forms because of complementary pairing between the bases made possible by the formation of hydrogen bonds. A hydrogen bond is a force of attraction between a polarized hydrogen of one molecular group and the electronegative oxygen or nitrogen atoms of nearby aligned molecular groups.

Each gene is composed of a specific and unique linear sequence of bases. An organism's entire set of DNA base sequences is called a **genome**. Although genes specify the amino acid sequence of polypeptides, DNA is not directly involved in protein synthesis. Instead, another type of nucleic acid, RNA, is used to convert DNA's coded instructions into polypeptide products.

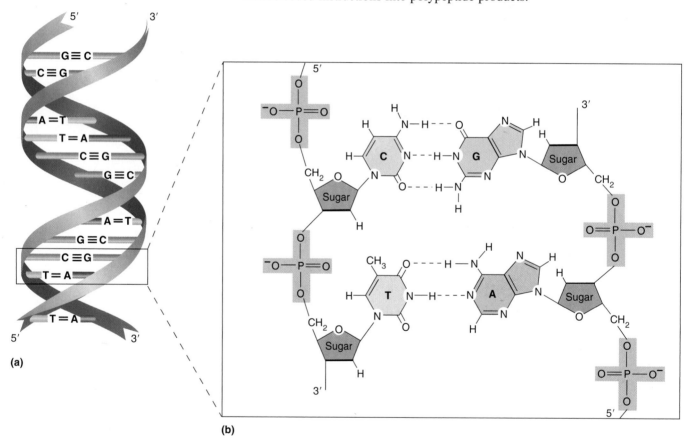

(a)

(b)

FIGURE 1.14

DNA

(a) A diagrammatic view of DNA. The sugar-phosphate backbones of the double helix are represented by colored ribbons. The bases attached to the sugar deoxyribose are on the inside of the helix. (b) An enlarged view of two base pairs. Note that the two DNA strands run in opposite directions defined by the 5′ and 3′ groups of deoxyribose. The bases on opposite strands form pairs because of hydrogen bonds. Cytosine always pairs with guanine; thymine always pairs with adenine.

RNA. Ribonucleic acid is a polynucleotide that differs from DNA in that it contains the sugar ribose instead of deoxyribose, and the base uracil instead of thymine. In RNA, as in DNA, the nucleotides are linked by phosphodiester linkages. In contrast to the double helix of DNA, RNA is single-stranded. RNA molecules fold into complex three-dimensional structures created by local regions of complementary base pairing. During the complex process called **transcription**, the DNA double helix partially unwinds and RNA molecules are synthesized. As the DNA double helix unwinds, one strand serves as a template. Complementary base pairing specifies the nucleotide base sequence of the RNA molecule. There are three major types of RNA: messenger RNA (mRNA), ribosomal RNA (rRNA), and transfer RNA (tRNA). Each unique sequence or molecule of mRNA possesses the information that codes directly for the amino acid sequence in a specific polypeptide. Ribosomes, the large, complex, supramolecular structures composed of rRNA and protein molecules, convert the mRNA base sequence into the amino acid sequence of a polypeptide. Transfer RNA molecules function as adapters during protein synthesis.

In recent years large numbers of RNA molecules have been discovered that are not directly involved in protein synthesis. These molecules, called *noncoding RNAs* (ncRNA), have roles in a great variety of cellular processes. Examples include short interfering RNAs (siRNA), micro RNAs (miRNA), small nuclear RNAs (snRNA), and small nucleolar RNAs (snoRNA). Small interfering RNAs are important components in *RNA interference*, an antiviral defense mechanism. Micro RNAs regulate the timing of mRNA synthesis, and small nuclear RNAs facilitate the process by which mRNA precursor molecules are transformed into functional mRNA. Small nucleolar RNAs assist in the maturation of ribosomal RNA during ribosome formation.

KEY CONCEPTS

- Most molecules in living organisms are organic. The chemical properties of organic molecules are determined by specific arrangements of atoms called functional groups.
- Cells contain four families of small molecules: amino acids, sugars, fatty acids, and nucleotides.
- Proteins, polysaccharides, and the nucleic acids are biopolymers composed of amino acids, sugars, and nucleotides, respectively.

1.4 IS THE LIVING CELL A CHEMICAL FACTORY?

The properties of even the simplest cells are so remarkable that cells have often been characterized as chemical factories. Certainly, like factories, living organisms acquire raw materials, energy, and information from their environment. Components are manufactured, and waste products and heat are discharged back into the environment. However, for this analogy to hold true, human-made factories would not only manufacture and repair all their structural and functional components, they would make all the machines that make these components, and then clone themselves, that is, manufacture new factories. The term **autopoiesis** has been created to describe the remarkable properties of living organisms. In this view, each living organism is considered to be an autopoietic system, an autonomous, self-organizing, and self-maintaining entity. Life emerges from a self-regulating network of thousands of biochemical reactions.

The constant flow of energy and nutrients through organisms and the functional properties of thousands of catalytic biomolecules called enzymes make possible the process of metabolism. The primary functions of metabolism are (1) acquisition and utilization of energy, (2) synthesis of molecules needed for cell structure and functioning (i.e., proteins, carbohydrates, lipids, and nucleic acids), (3) growth and development, and (4) removal of waste products. Metabolic processes require significant amounts of useful energy. This section begins with a review of the primary chemical reaction types and the essential features of energy-generating strategies observed in living organisms. A brief outline of metabolic processes and the means by which living organisms maintain ordered systems follows.

Biochemical Reactions

At first glance the thousands of reactions that occur in cells appear overwhelmingly complex. However, several characteristics of metabolism allow us to simplify this picture:

1. Although the number of reactions is very large, the number of reaction types is relatively small.

2. Biochemical reactions have simple organic reaction mechanisms.

3. Reactions of central importance in biochemistry (i.e., those used in energy production and the synthesis and degradation of major cell components) are relatively few.

Among the most common reaction types encountered in biochemical processes are the following: nucleophilic substitution, elimination, addition, isomerization, and oxidation-reduction. Each is briefly described.

NUCLEOPHILIC SUBSTITUTION REACTIONS In **nucleophilic substitution** reactions, as the name suggests, one atom or group is substituted for another:

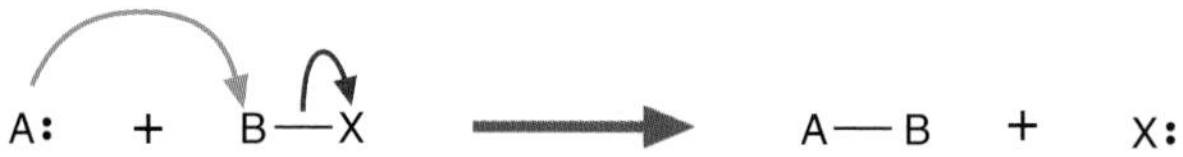

$$A: \ + \ B—X \ \longrightarrow \ A—B \ + \ X:$$

In the general reaction shown, the attacking species (A) is called a **nucleophile** ("nucleus lover"). Nucleophiles are anions (negatively charged atoms or groups) or neutral species possessing nonbonding electron pairs. **Electrophiles** ("electron lovers") are deficient in electron density and are therefore easily attacked by a nucleophile. As the new bond forms between A and B, the old one between B and X breaks. The outgoing nucleophile (in this case, X), called a **leaving group**, leaves with its electron pair.

The reaction of glucose with ATP provides an important example of nucleophilic substitution (Figure 1.15). In this reaction, which is the first step in the

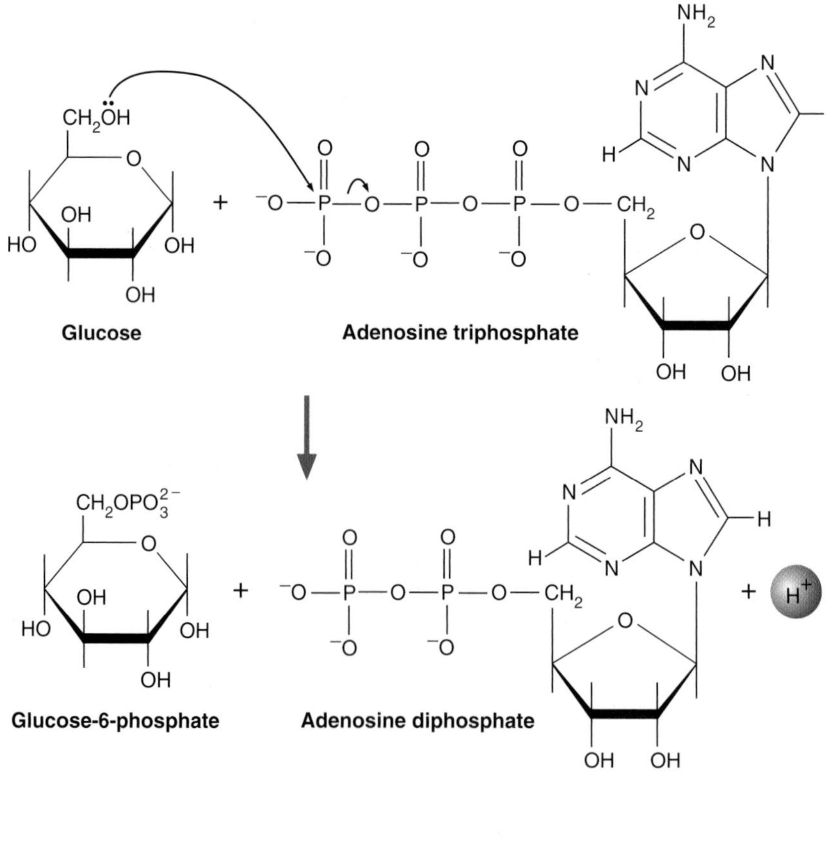

FIGURE 1.15

Example of Nucleophilic Substitution

In the reaction of glucose with ATP, the hydroxyl oxygen of glucose is the nucleophile. The phosphorus atom (the electrophile) is polarized by the oxygens bonded to it so that it bears a partial positive charge. As the reaction occurs the unshared pair of electrons on the CH_2OH of the sugar attacks the phosphorus, resulting in the expulsion of ADP, the leaving group.

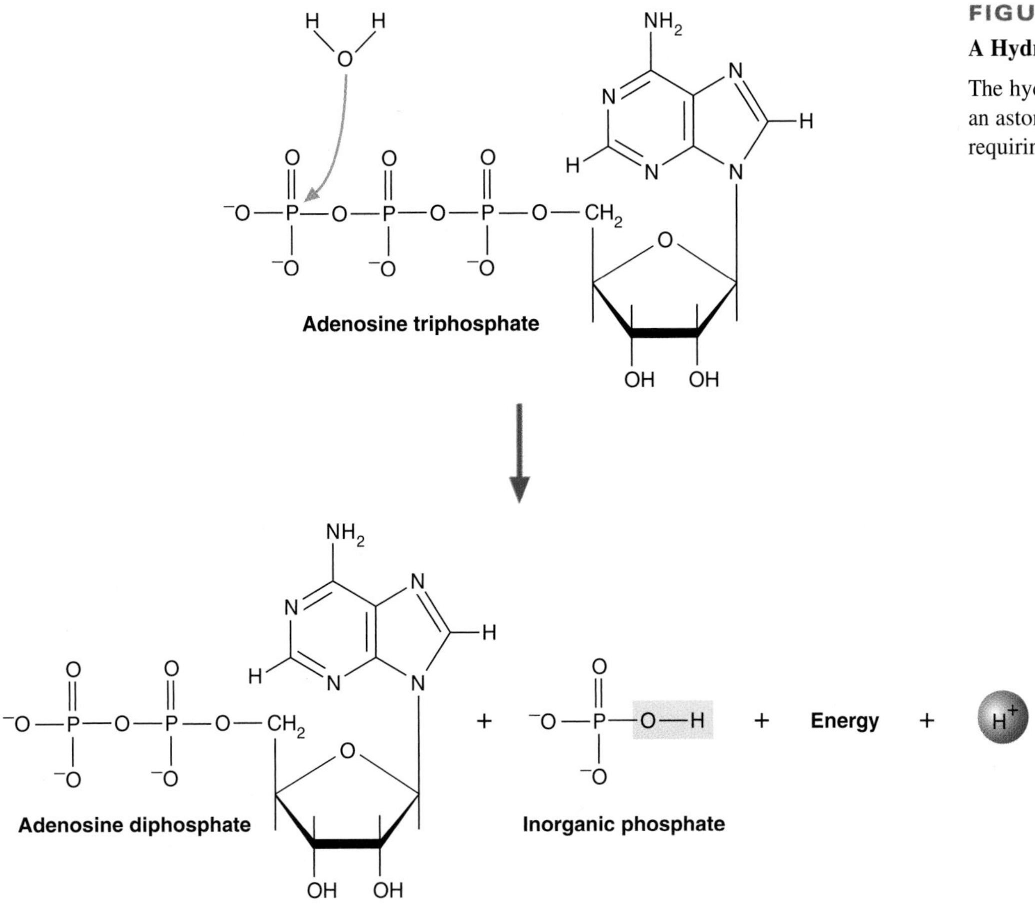

FIGURE 1.16

A Hydrolysis Reaction

The hydrolysis of ATP is used to drive an astonishing diversity of energy-requiring biochemical reactions.

utilization of glucose as an energy source, the hydroxyl oxygen on carbon 6 of the sugar molecule is the nucleophile and phosphorus is the electrophile. Adenosine diphosphate is the leaving group.

Hydrolysis reactions are nucleophilic substitution reactions in which the oxygen of a water molecule serves as the nucleophile. The electrophile is usually the carbonyl carbon of an ester, amide, or anhydride. (An **anhydride** is a molecule containing two carbonyl groups linked through an oxygen atom.)

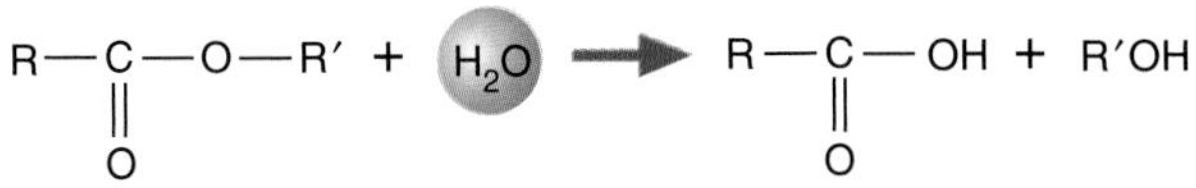

The digestion of many food molecules involves hydrolysis. For example, proteins are degraded in the stomach in an acid-catalyzed hydrolytic reaction. Another important example is breaking the phosphate bonds of ATP (Figure 1.16). The energy obtained during this reaction is used to drive many cellular processes.

ELIMINATION REACTIONS In **elimination reactions** a double bond is formed when atoms in a molecule are removed.

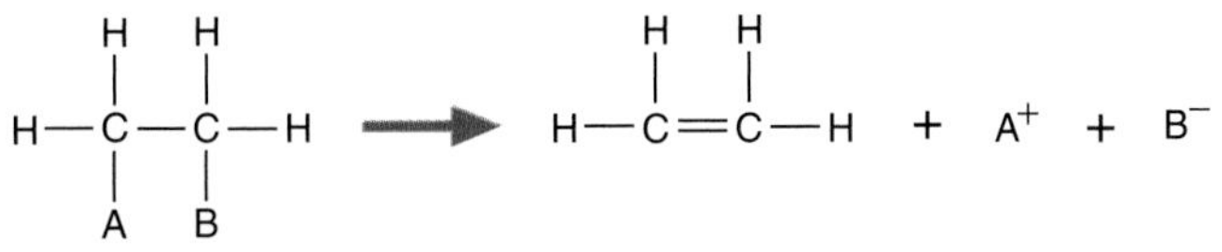

The removal of H_2O from biomolecules containing alcohol functional groups is a commonly encountered reaction. A prominent example is the dehydration of 2-phosphoglycerate, an important step in carbohydrate metabolism (Figure 1.17). Other products of elimination reactions include ammonia (NH_3), amines (RNH_2), and alcohols (ROH).

ADDITION REACTIONS In **addition reactions** two molecules combine to form a single product.

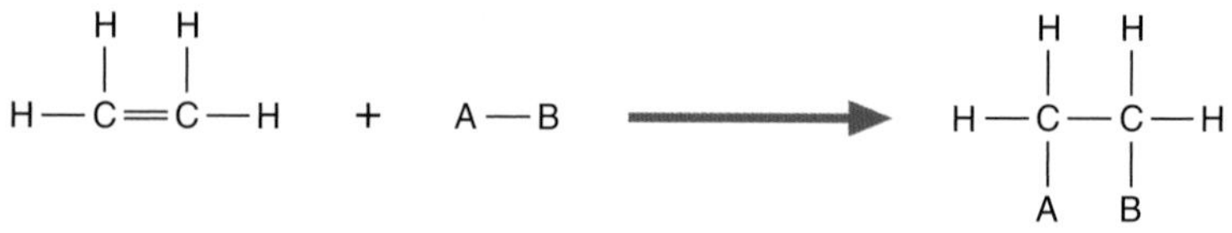

Hydration is one of the most common addition reactions. When water is added to an alkene, an alcohol results. The hydration of the metabolic intermediate fumarate to form malate is a typical example (Figure 1.18).

ISOMERIZATION REACTIONS In **isomerization** reactions, atoms or groups undergo intramolecular shifts. One of the most common biochemical isomerizations is the interconversion between aldose and ketose sugars (Figure 1.19).

FIGURE 1.17

An Elimination Reaction

When 2-phosphoglycerate is dehydrated, a double bond is formed.

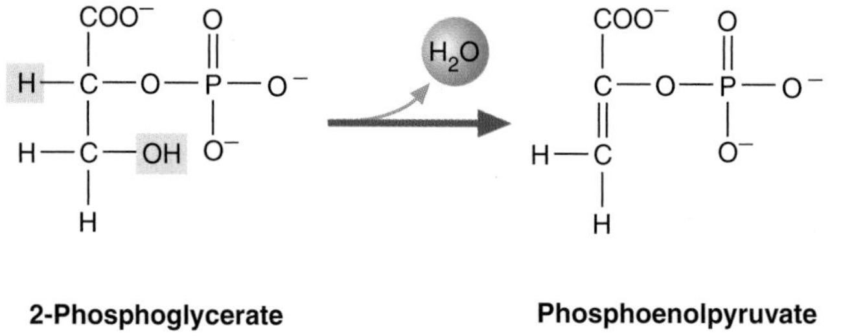

FIGURE 1.18

An Addition Reaction

When water is added to a molecule that contains a double bond, such as fumarate, an alcohol results.

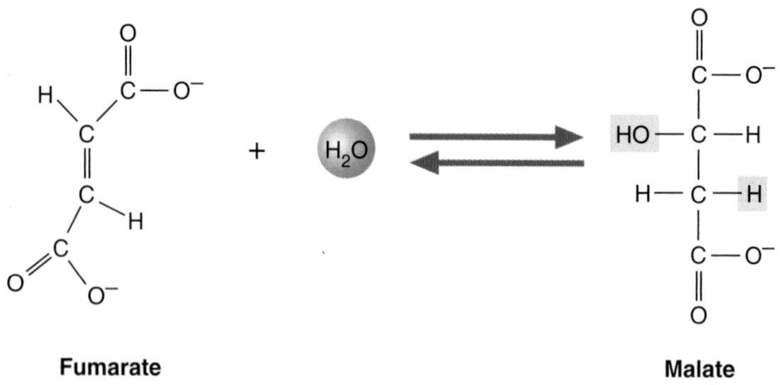

FIGURE 1.19

An Isomerization Reaction

The reversible interconversion of aldose and ketose isomers is a commonly observed biochemical reaction type.

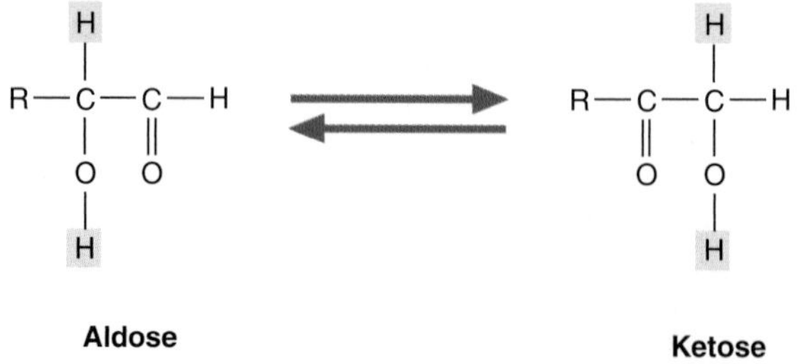

OXIDATION-REDUCTION REACTIONS **Oxidation-reduction (redox) reactions** occur when there is a transfer of electrons from a donor (called the **reducing agent**) to an electron acceptor (called the **oxidizing agent**). When reducing agents donate their electrons, they become **oxidized**. As oxidizing agents accept electrons, they become **reduced**. The two processes always occur simultaneously.

It is not always easy to determine whether biomolecules have gained or lost electrons. However, two simple rules may be used to ascertain whether a carbon atom in a molecule has been oxidized or reduced:

1. Oxidation has occurred if a carbon atom gains oxygen or loses hydrogen:

Ethyl alcohol   **Acetic acid**

2. Reduction has occurred if a carbon atom loses oxygen or gains hydrogen:

Acetic acid   **Ethyl alcohol**

In biological redox reactions, electrons are transferred to electron acceptors such as the nucleotide NAD^+/NADH (nicotinamide adenine dinucleotide in its oxidized/reduced form).

Energy

Energy is defined as the capacity to do work, that is, to move matter. In contrast to human-made machines, which generate and use energy under harsh conditions such as high temperature, high pressure, and electrical currents, the relatively fragile molecular machines within living organisms must use more subtle mechanisms. Cells generate most of their energy by using redox reactions in which electrons are transferred from an oxidizable molecule to an electron-deficient molecule. In these reactions electrons are often removed or added as hydrogen atoms ($H\bullet$) or hydride ions ($H:^-$). The more reduced a molecule is—that is, the more hydrogen atoms it possesses—the more energy it contains. For example, fatty acids contain proportionately more hydrogen atoms than sugars do and therefore yield more energy upon oxidation. When fatty acids and sugars are oxidized, their hydrogen atoms are removed by the redox coenzymes FAD (flavin adenine dinucleotide) or NAD^+, respectively. (Coenzymes are small molecules that function in association with enzymes by serving as carriers of small molecular groups, or in this case, electrons.) The reduced products of this process ($FADH_2$ or NADH, respectively) can then transfer the electrons to another electron acceptor.

Whenever an electron is transferred, energy is lost. Cells have complex mechanisms for exploiting this phenomenon in a way that permits some of the released energy to be captured for cellular work. The most prominent feature of energy generation in most cells is the electron transport pathway, a series of linked membrane-embedded electron carrier molecules. During a regulated process, energy is released as electrons are transferred from one electron carrier molecule to another. During several of these redox reactions, the energy released is sufficient to drive the synthesis of ATP, the energy carrier molecule that directly supplies the energy used to maintain highly organized cellular structures and functions.

Despite their many similarities, groups of living organisms differ in the precise strategies they use to acquire energy from their environment. **Autotrophs** are organisms that transform the energy of the sun or various chemicals into chemical bond

KEY CONCEPT

The most common reaction types encountered in biochemical processes are nucleophilic substitution, elimination, addition, isomerization, and oxidation-reduction.

energy; they are called, respectively, **photoautotrophs** and **chemoautotrophs**. The **heterotrophs** obtain energy by degrading preformed food molecules obtained by consuming other organisms. **Chemoheterotrophs** use preformed food molecules as their sole source of energy. Some prokaryotes and a small number of plants (e.g., the pitcher plant, which digests captured insects) are **photoheterotrophs**; that is, they use both light and organic biomolecules as energy sources.

As described previously, the ultimate source of the energy used by most life-forms on earth is the sun. Photosynthetic organisms such as plants, certain prokaryotes, and algae capture light energy and use it to transform carbon dioxide (CO_2) into sugar and other biomolecules. Chemotrophic species derive the energy required to incorporate CO_2 into organic biomolecules by oxidizing inorganic substances such as hydrogen sulfide (H_2S), nitrite (NO_2^-), or hydrogen gas (H_2). The biomass produced in both types of process is, in turn, consumed by heterotrophic organisms that use it as sources of energy and structural materials. At each step, as molecular bonds are rearranged, some energy is captured and used to maintain the organism's complex structures and activities. Eventually energy becomes disorganized and is released in the form of heat. The metabolic pathways by which energy is generated and used by living organisms are briefly outlined next, under "Overview of Metabolism." Descriptions of the basic mechanisms by which cellular order is maintained make up the subsection entitled "Biological Order".

KEY CONCEPT

In living organisms, energy, the capacity to move matter, is usually generated by redox reactions.

Overview of Metabolism

Metabolism is the sum of all the enzyme-catalyzed reactions in a living organism. Many of these reactions are organized into pathways (Figure 1.20) in which an initial reactant molecule is modified in a step-by-step sequence into a product that can be used by the cell for a specific purpose. For example, glycolysis, the energy-generating pathway that degrades the six-carbon sugar glucose, is composed of ten reactions. All of an individual organism's metabolic processes consist of a vast weblike pattern of interconnected biochemical reactions. There are three classes of biochemical pathways: metabolic, energy transfer, and signal transduction.

METABOLIC PATHWAYS There are two types of metabolic pathway: anabolic and catabolic. In biosynthetic, or **anabolic pathways**, large complex molecules are synthesized from smaller precursors. Building-block molecules (e.g., amino acids, sugars, and fatty acids), either produced or acquired from the diet, are incorporated into larger, more complex molecules. Because biosynthesis increases order and complexity, anabolic pathways require an input of energy. Anabolic processes include the synthesis of polysaccharides and proteins from sugars and amino acids, respectively. During **catabolic pathways** large complex molecules are degraded into smaller, simpler products. Some catabolic pathways release energy. A fraction of this energy is captured and used to drive anabolic reactions.

The relationship between anabolic and catabolic processes is illustrated in Figure 1.21. As nutrient molecules are degraded, energy and reducing power (high-energy electrons) are conserved in ATP and NADH molecules, respectively. Biosynthetic processes use metabolites of catabolism, synthesized ATP and NADPH (reduced nicotinamide adenine dinucleotide phosphate, a source of reducing power), to create complex structure and function.

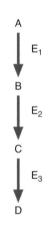

FIGURE 1.20

A Biochemical Pathway

In this three-step biochemical pathway biomolecule A is converted into biomolecule D in three sequential reactions. Each reaction is catalyzed by a specific enzyme (E).

ENERGY TRANSFER PATHWAYS Energy transfer pathways capture energy and transform it into forms that organisms can use to drive biomolecular processes. The absorption of light energy by chlorophyll molecules and the energy-releasing redox reactions required for its conversion to chemical bond energy in a sugar molecule is a prominent example.

SIGNAL TRANSDUCTION Signal transduction pathways allow cells to receive and respond to signals from their surroundings. The signal transduction

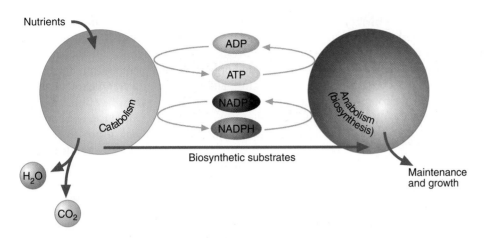

FIGURE 1.21

Anabolism and Catabolism

In organisms that use oxygen to generate energy, catabolic pathways convert nutrients to small-molecule starting materials. The energy (ATP) and reducing power (NADPH) that drive biosynthetic reactions are generated during catabolic processes as certain nutrient molecules are converted to waste products such as carbon dioxide and water.

mechanism consists of three phases: reception, transduction, and response. In the initial or reception phase, a signal molecule such as a hormone or a nutrient molecule binds to a receptor protein. This binding event initiates the transduction phase, a cascade of intracellular reactions that triggers the cell's response to the original signal. For example, glucose binds to its receptor on pancreatic insulin-secreting cells, whereupon insulin is released into the blood. Most commonly, such responses are an increase or decrease in the activity of already existing enzymes or the synthesis of new enzyme molecules.

Biological Order

The coherent unity that is observed in all living organisms involves the functional integration of millions of molecules. In other words, life is highly organized complexity. Despite the rich diversity of living processes that contribute to generating and maintaining biological order, most can be classified into the following categories: (1) synthesis and degradation of biomolecules, (2) transport of ions and molecules across cell membranes, (3) production of force and movement, and (4) removal of metabolic waste products and other toxic substances. Each is discussed briefly.

SYNTHESIS OF BIOMOLECULES Cellular components are synthesized in a vast array of chemical reactions, many of which require energy, which is supplied directly or indirectly by ATP molecules. The molecules formed in biosynthetic reactions perform several functions. They can be assembled into supramolecular structures (e.g., the proteins and lipids that constitute membranes) or serve as informational molecules (e.g., DNA and RNA) or catalyze chemical reactions (i.e., the enzymes).

TRANSPORT ACROSS MEMBRANES Cell membranes regulate the passage of ions and molecules from one compartment to another. For example, the plasma membrane (the animal cell's outer membrane) is a selective barrier. It is responsible for the transport of certain substances such as nutrients from a relatively disorganized environment into the more orderly cellular interior. Similarly, ions and molecules are transported into and out of organelles during biochemical processes. For example, fatty acids are transported into organelles known as mitochondria so that they may be broken down to generate energy.

KEY CONCEPTS

- Metabolism is the sum of all the enzyme-catalyzed reactions in a living organism.
- There are three classes of biochemical pathway: a metabolic (anabolic and catabolic), energy transfer, and signal transduction.

CELL MOVEMENT Organized movement is one of the most obvious characteristics of living organisms. The intricate and coordinated activities required to sustain life require the movement of cell components. Examples in eukaryotic cells include cell division and organelle movement, two processes that depend to a large extent on the structure and function of a complex network of protein filaments known as the *cytoskeleton*. The forms of cellular motion profoundly influence the ability of all organisms to grow, reproduce, and compete for limited resources. As examples, consider the movement of protists as they search for food in a pond, or the migration of human white blood cells as they pursue infectious foreign cells. More subtle examples include the movement of specific enzymes along a DNA molecule during the chromosome replication that precedes cell division and the secretion of insulin by certain pancreatic cells.

WASTE REMOVAL All living cells produce waste products. For example, animal cells ultimately convert food molecules, such as sugars and amino acids, into CO_2, H_2O, and NH_3. These molecules, if not disposed of properly, can be toxic. Some substances are readily removed. In animals, for example, CO_2 diffuses out of cells and (after a brief and reversible conversion to bicarbonate by red blood cells) is quickly exhaled through the respiratory system. Excess H_2O is excreted through the kidneys. Other molecules, however, are so toxic that specific mechanisms have evolved to provide for their disposal. The urea cycle (described in Chapter 15), provides a mechanism for converting free ammonia and excess amino nitrogen into urea, a less toxic molecule. The urea molecule is then removed from the body through the kidney as a major component of the urine.

Living cells also contain a wide variety of complex organic molecules that must be disposed of. Plant cells solve this problem by transporting such molecules into a vacuole, where they are either broken down or stored. Animals, however, must use disposal mechanisms that depend on water solubility (e.g., the formation of urine by the kidney). Hydrophobic substances such as steroid hormones, which cannot be broken down into simpler molecules, are converted during a series of reactions into water-soluble derivatives. This mechanism is also used to solubilize some organic molecules such as drugs and environmental contaminants.

KEY CONCEPT

In living organisms, processes of highly ordered complexity are sustained by a constant input of energy.

1.5 SYSTEMS BIOLOGY

The discovery of the information provided in the overview of biochemical processes that you have just read about was made possible by a method of inquiry based on *reductionism*, a powerful, mechanistic strategy in which a complex, living "whole" is studied by "reducing" it to its component parts. Each individual part is then further broken down so that the chemical and physical properties of its molecules and the connections between them can be determined. Most of the accomplishments of the modern life sciences would have been impossible without the reductionist philosophy. However, reductionism has its limitations. The most prominent is the assumption that detailed knowledge of all the properties of the parts will ultimately provide a complete understanding of the functioning of the whole. Despite intense efforts, a coherent understanding of dynamic living processes continues to elude investigators.

In recent decades a new approach called systems biology has been utilized to achieve a deeper understanding of living organisms. Based on the engineering principles originally developed to build jet aircraft, **systems biology** regards living organisms as integrated systems. Each system allows certain functions to be performed. One such system in animals is the digestive system, which comprises a group of organs that is tasked to break down food into molecules that can be absorbed by the body's cells.

Although human-engineered systems and living systems are remarkably similar in some respects, they are significantly different in others. The most important difference is the design issue. When engineers plan a complex mechanical or electrical system, each component is designed to fulfill a precise function, and there are no unnecessary or unforeseen interactions between network components. For example, the individual electrical wires in the cables that control aircraft are insulated to prevent damage due to short circuits. In contrast, biological systems have evolved by trial and error over several billion years. Evolution, the adaptation of populations of living organisms in response to selection pressure, is made possible by the capacity to generate genetic diversity through various forms of mutation, gene duplications, or the acquisition of new genes from other organisms. The components of living organisms, unlike engineer-designed parts, have no fixed functions, and overlapping functions are permissible. Living systems have become increasingly more complex, in part because of the unavoidability of interactions among established system components and potentially useful new parts (e.g., derived from gene duplications followed by mutations).

The systems approach is especially useful because the human mind cannot analyze the hundreds of biochemical reactions that are taking place at once in a living organism. To tackle this problem, systems biologists have invented mathematical and computer models to derive from biochemical reaction pathways an understanding of how these processes operate over time and under varying conditions. The success of these models is reliant on huge data sets containing accurate information about cellular concentrations of biomolecules and the rates of biochemical reactions as they occur in living, functioning cells. Though these data sets are incomplete, this analytic method has produced some notable successes. The technology required to identify and quantify biomolecules of all types continues to be refined. System biologists have identified three core principles that underpin the complex and diverse biochemical pathways described in this textbook: emergence, robustness, and modularity.

Emergence

As we have discovered, the behavior of complex systems cannot always be understood by knowing the properties of constituent parts. At each level of organization of the system, new and unanticipated properties emerge from interactions among parts. For example, hemoglobin (the protein that carries oxygen in the blood) requires ferrous iron (Fe^{2+}) to be functional. Whereas iron easily oxidizes in the inanimate world, the iron in the hemoglobin does not usually oxidize even though it is linked directly to oxygen during the transport process. The amino acid residues that line the iron-binding site protect Fe^{2+} from oxidation. The protection of ferrous iron in hemoglobin is an **emergent property**, that is, a property conferred by the complexity and dynamics of the system.

Robustness

Systems that remain stable despite diverse perturbations are described as *robust*. Autopilot systems in aircraft, for example, maintain a designated flight path despite expected fluctuations in conditions such as wind speed or the plane's mechanical functions. All robust systems are necessarily complex because failure prevention requires an integrated set of automatic fail-safe mechanisms. The robust (fail-safe) properties of human-made mechanical systems are created by *redundancy*, the use of duplicate parts (e.g., backup electric generators in an airplane). Although the design of living organisms does include some redundant parts, the robust properties of living systems are largely the result of **degeneracy**, the capacity of structurally different parts to perform the same or similar functions. The genetic code is a simple, well-recognized example. Of the 64 possible three-base sequences (called codons) on an mRNA molecule, 61 base triplets code for 20 amino acids during protein synthesis. Since most amino acids have more than one codon, degeneracy of the code provides a measure of protection against base substitution mutations.

FIGURE 1.22

Feedback Mechanisms

(a) **Negative Feedback**. As a product molecule accumulates, it binds to and inhibits the activity of an enzyme in the pathway. The result is the decreased production of the product. (b) **Positive Feedback**. As product molecules accumulate, they stimulate an enzyme in the pathway, thereby causing an increased rate of product synthesis.

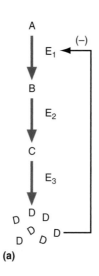

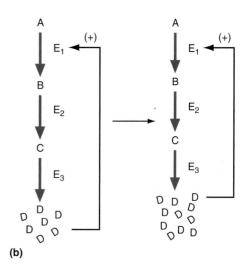

(a) (b)

Robust systems have elaborate control mechanisms. In living organisms the most common type is **feedback control** (Figure 1.22), a self-regulating mechanism in which the product of a process acts to modify the process, either negatively or positively. In negative feedback, the most common form, an accumulating product slows its own production. Many biochemical pathways are regulated by negative feedback. Typically a pathway product inhibits an enzyme near the beginning of the pathway. In positive feedback control, a product increases its own production. Positive feedback control is found less often in living organisms because the mechanism is potentially destabilizing. If not carefully controlled, the amplifying effect of a positive feedback loop can result in the collapse of the system. In blood clotting, for example, the platelet plug that seals a damaged blood vessel does not expand continuously because inhibitors are released by nearby undamaged blood vessel cells.

Fail-safe control mechanisms, whether in human-made systems or living organisms, are expensive. Constraints such as energy availability require priorities in resource allocation. As a result, systems are generally protected from commonly encountered environmental changes, but they are vulnerable to unusual or rare events that cause damage. This vulnerability, referred to as *fragility*, is an inescapable feature of robust systems. Cancer, a group of diseases in which cell cycle control is disrupted, is an example of the "robust, yet fragile" nature of robust systems. Despite the elaborate controls on cell division in animal bodies, mutations in just a few of the genes that code for regulatory proteins can result in the uncontrolled proliferation of the affected cell.

Modularity

Complex systems are composed of *modules*, components or subsystems that perform specific functions. Living organisms utilize modules because they are easily assembled, rearranged, and repaired, and eliminated when necessary. Although modules (e.g., enzymes extracted from cells in the lab) can often be isolated with some or even most of their functional properties, their function is meaningful only within the context of the larger system. In living organisms, modularity occurs at all system levels. Examples within a cell include amino acids, proteins, and biochemical pathways. Modularity is especially important because it provides the capacity to limit damage to components that can be easily removed and replaced. For example, cells possess mechanisms that detect the presence of a damaged protein and then destroy it and synthesize a new one. Functional relationships between modules in a system are managed by *protocols*, or sets of rules that specify how and whether modules will interact. The regulatory mechanism that controls the synthesis of a particular protein is an example of a protocol.

KEY CONCEPTS

• Systems biology is an attempt to reveal the functional properties of living organisms by developing mathematical models of interactions from available data sets.

• The systems approach has provided insights into the emergence, robustness, and modularity of living organisms.

GCGACATCACTCCAGCTTGAAGCAGTTCTTCTCGTCTTCTGTTTTGTCTAACTT TCTTCC
500   510   530   540   550

BIOCHEMISTRY IN THE LAB
An Introduction

Biochemical technologies exploit the chemical and physical properties of biomolecules: chemical reactivity, size, solubility, net electrical charge, movement in an electric field, and absorption of electromagnetic radiation. As life science research has become increasingly more sophisticated, scientists have provided a progressively more coherent view of the living state. The Human Genome Project was a landmark event in this process. The goal of this international research effort, begun in the late 1980s, was to determine the nucleotide base sequence of human DNA. The subsequent development of automated DNA sequencing technology revolutionized life science research because it provided scientists with a "high-throughput" (i.e., rapid, high-volume, relatively inexpensive) means of investigating the information content of genomes, a field now referred to as **genomics**.

Genomics has been especially useful in medical research. A large number of human diseases have been linked to errors in one or more gene sequences or to faulty regulation of gene expression. Among the early benefits of this work are rapid, accurate tests for predisposition to pathological conditions such as cystic fibrosis, breast cancer, and some liver diseases. Several recently developed technologies have created additional opportunities to investigate the molecular basis of disease. For example, DNA microchips (thousands of DNA molecules arrayed on a solid surface) are now routinely used to monitor gene expression of cells. Proteins can also be rapidly analyzed with a combination of gel electrophoresis and mass spectrometry. Among the new fields created by high-throughput methods are **functional genomics** (the investigation of gene expression patterns) and **proteomics** (the investigation of protein synthesis patterns and protein-protein interactions). The science of **bioinformatics** is the computer-based field that facilitates the analysis of the massive amounts of protein and nucleic acid sequence data that are being generated.

In the past, biochemists and other scientists have often benefited from each other's work. For example, technologies created by physicists such as X-ray diffraction, electron microscopy, and radioisotope labeling made biomolecular structure investigations possible. In recent years, the life sciences have also benefited from the services provided by computer scientists, mathematicians, and engineers. As the biological knowledge base has continued to expand, it has become increasingly obvious that future advances in life science and medical research will require the efforts of multidisciplinary teams of scientists. ∎

Chapter**Summary**

1. Biochemistry may be defined as the study of the molecular basis of life. Biochemists have contributed to the following insights into life: (1) life is complex and dynamic, (2) life is organized and self-sustaining, (3) life is cellular, (4) life is information-based, and (5) life adapts and evolves.

2. The molecular evidence concerning the evolutionary relationships of living species is sufficiently compelling that many life scientists now classify all living organisms into three domains: the Bacteria, the Archaea, and the Eukarya.

3. All living things are composed of either prokaryotic cells or eukaryotic cells. Prokaryotes, which include bacteria and the Archaea, lack a membrane-bound cellular organelle called a nucleus. The eukaryotes consist of all the remaining species. These cells contain a nucleus and complex structures that are not observed in prokaryotes.

4. Many eukaryotes are multicelluar. Multicellular organisms have several advantages over unicellular ones. These include the provision of a relatively stable environment for most of the organism's cells, the capacity for greater complexity in an organism's form and function, and the ability to exploit environmental resources more effectively than is possible for individual single-celled organisms.

5. Animal and plant cells contain thousands of different types of molecules. Water constitutes 50 to 90% of a cell's content by weight, and ions such as Na^+, K^+, and Ca^{2+} may account for another 1%. Almost all the other kinds of biomolecules are organic.

6. Many of the biomolecules found in cells are relatively small, with molecular weights of less than 1000 daltons. Cells contain four families of small molecules: amino acids, sugars, fatty acids, and nucleotides.

7. All life processes consist of chemical reactions catalyzed by enzymes. Among the most common reaction types encountered in biochemical processes are nucleophilic substitution, elimination, addition, isomerization, and oxidation-reduction.

8. Living organisms require a constant flow of energy to prevent disorganization. The principal means by which cells obtain energy is oxidation of biomolecules or certain minerals.

9. Metabolism is the sum of all the reactions in a living organism. There are two types of metabolic pathway: anabolic and catabolic. Energy transfer pathways capture energy and transform it into forms that organisms can use to drive biomolecular processes. Signal transduction pathways, which allow cells to receive and respond to signals from their environment, consist of three phases: reception, transduction, and response.

10. The complex structure of cells requires a high degree of internal order. This is accomplished by four primary means: synthesis of biomolecules, transport of ions and molecules across cell membranes, production of movement, and removal of metabolic waste products and other toxic substances.

11. Systems biology is a new field that attempts to provide understanding of the functional properties of living organisms by applying mathematical modeling strategies to amassed biological data. Among the early benefits of the systems approach are the insights associated with emergence, robustness, and modularity.

COMPANION
WEBSITE
Take your learning further by visiting the **companion website** for Biochemistry at **www.oup.com/us/mckee** where you can complete a multiple-choice quiz on this introductory chapter to help you prepare for exams.

Suggested**Readings**

Campbell, N. A., and Reece, J. B., *Biology*, 7th ed., Benjamin Cummings, San Francisco, 2005.

Gibbs, W. W., Cybernetic Cells, *Sci. Am.* 265(2):42–47, 2001.

Goodsell, D. S., *Bionanotechnology: Lessons from Nature*, Wiley-Liss, Hoboken, New Jersey, 2004.

Harold, F. M., *The Way of the Cell*, Oxford University Press, Oxford, 2001.

Lutz, R. A., Shank, T. M., and Evans, R., Life After Death in the Deep Sea, *Am. Sci.* 89:422–431, 2001.

Newman, D. K., and Banfield, J. F., Geomicrobiology: How Molecular-Scale Interactions Underpin Biogeochemical Systems, *Science* 296:1071–1077, 2002.

Rothman, S., *Lessons from the Living Cell: The Limits of Reductionism*, McGraw-Hill, New York, 2002.

Tudge, C., *The Variety of Life: A Survey and a Celebration of All the Creatures That Have Ever Lived*, Oxford University Press, New York, 2000.

Van Regenmortel, M. H. V., Reductionism and Complexity in Molecular Biology, *EMBO Rep.* 5(11):1016–1020, 2004.

Key**Words**

active transport, *23*

addition reaction, *20*

amino acid, *11*

anabolic pathway, *22*

anhydride, *19*

archaea, *7*

autopoiesis, *17*

autotroph, *21*

bacteria, *7*

biogeochemical cycle, *7*

biomolecule, *4*

bioremediation, *8*

catabolic pathway, *22*

chemoautotroph, *22*

chemoheterotroph, *22*

chemosynthesis, *2*

degeneracy, *25*

electrophile, *18*

elimination reactions, *19*

emergent property, *25*

energy, *21*

enzyme, *5*

eukarya, *7*

eukaryotic cell, *6*

extremophile, *8*

extremozyme, *8*

fatty acid, *14*

feedback control, *26*

functional group, *10*

gene, *6*

gene expression, *27*

genome, *16*

heterotroph, *22*

homeostasis, *5*

hydration, *20*

hydrocarbon, *9*

hydrolysis, *19*

hydrophilic, *11*

hydrophobic, *9*

isomerization, *20*

leaving group, *18*

lipid, *14*

macromolecule, *4*

metabolism, *5*

modules, *26*

monosaccharide, *12*

mutation, *6*

negative feedback, *26*

neurotransmitter, *12*

noncoding RNA, *17*

nucleic acid, *15*

nucleophile, *18*

nucleophilic substitution, *18*

nucleotide, *16*

oligopeptide, *12*

organelle, *9*

oxidation-reduction (redox reaction), *21*

oxidize, *21*

oxidizing agent, *21*

peptide, *12*

peptide bond, *12*

photoautotroph, *22*

photoheterotroph, *22*

photosynthesis, *2*

polypeptide, *12*

polysaccharide, *12*

positive feedback, *26*

prokaryotic cell, *6*

protein, *12*

purine, *16*

pyrimidine, *16*

reduce, *21*

reducing agent, *21*

reductionism, *24*

robust, *25*

saturated, *14*

signal transduction, *22*

sugar, *12*

system, *17*

systems biology, *24*

transcription, *17*

unsaturated, *24*

Review Questions

These questions are designed to test your knowledge of the key concepts discussed in this chapter, before moving on to the next chapter. You may like to compare your answers to the solutions provided in the back of the book and in the accompanying Study Guide.

1. Describe the impact of hydrothermal vent discoveries on the life sciences.

2. There are three domains of living organisms. What are they? What unique features do the organisms in each domain possess?

3. Describe the major differences between prokaryotic and eukaryotic cells.

4. Identify the functional groups in the following molecules.

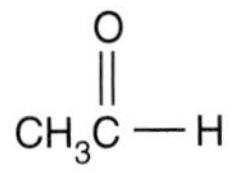

$$CH_3\overset{\overset{\displaystyle O}{\|}}{C}-H$$

(a)

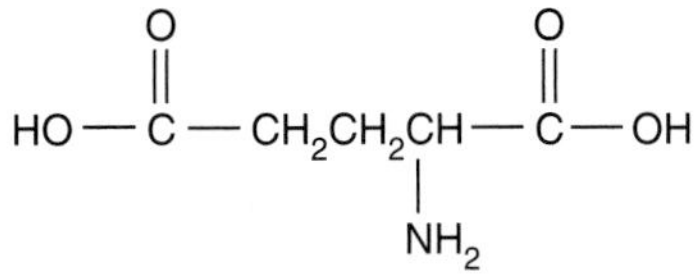

$$HO-\overset{\overset{\displaystyle O}{\|}}{C}-CH_2CH_2\underset{\underset{\displaystyle NH_2}{|}}{CH}-\overset{\overset{\displaystyle O}{\|}}{C}-OH$$

(b)

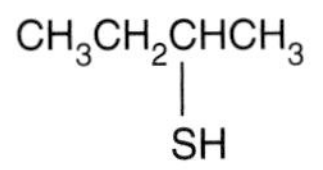

$$CH_3CH_2\underset{\underset{\displaystyle SH}{|}}{CH}CH_3$$

(c)

$$CH_3\overset{\overset{\displaystyle O}{\|}}{C}-O-CH_3$$

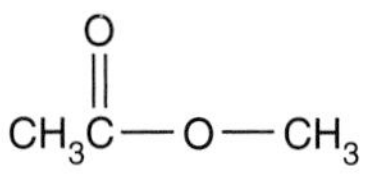

(d)

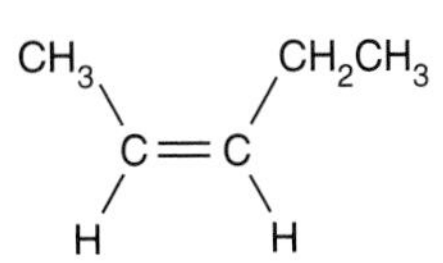

$$\underset{\underset{\displaystyle H}{|}}{\overset{\overset{\displaystyle CH_3}{\diagup}}{C}}=\underset{\underset{\displaystyle H}{|}}{\overset{\overset{\displaystyle CH_2CH_3}{\diagup}}{C}}$$

(e)

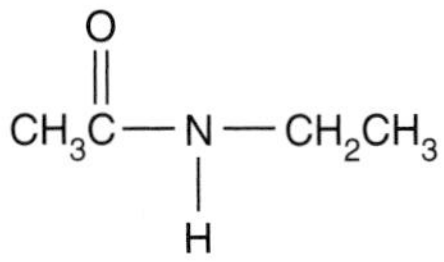

$$CH_3\overset{\overset{\displaystyle O}{\|}}{C}-\underset{\underset{\displaystyle H}{|}}{N}-CH_2CH_3$$

(f)

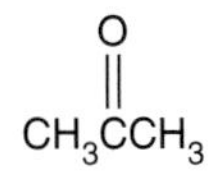

$$CH_3\overset{\overset{\displaystyle O}{\|}}{C}CH_3$$

(g)

$$\begin{array}{c} CH_2OH \\ | \\ CH-OH \\ | \\ CH_2OH \end{array}$$

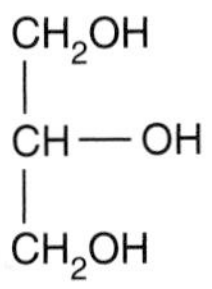

(h)

5. Name four classes of small biomolecules. In what larger biomolecules are they found?

6. Define the following terms:
   a. biochemistry
   b. oxidation
   c. reduction
   d. active transport
   e. leaving group

7. List two functions for each of the following biomolecules:
   a. fatty acids
   b. sugars
   c. nucleotides

   d. amino acids

8. What are the roles of DNA and RNA?

9. How do cells obtain energy from chemical bonds?

10. Define the following terms:
    a. oxidizing agent
    b. elimination
    c. reducing agent
    d. isomerization
    e. nucleophilic substitution

11. How do plants dispose of waste products?

12. What is the difference between an unsaturated and a saturated hydrocarbon?

13. Define the following terms:
    a. module
    b. noncoding RNA
    c. chemoautotrophs
    d. biogeochemical cycles
    e. signal transduction

14. What advantages do multicellular organisms have over unicellular organisms?

15. Assign each of the following compounds to one of the major classes of biomolecule:

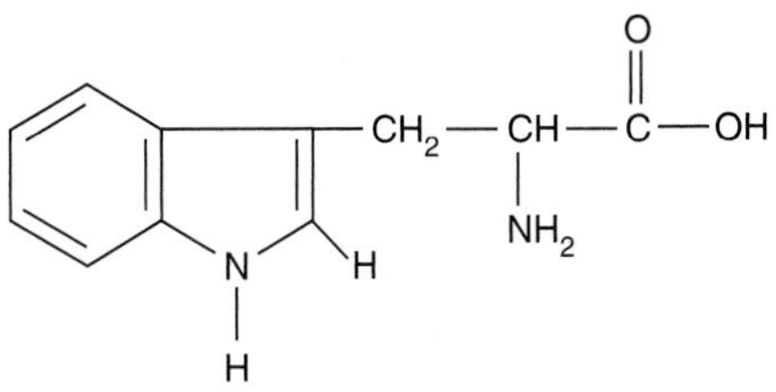

(a)

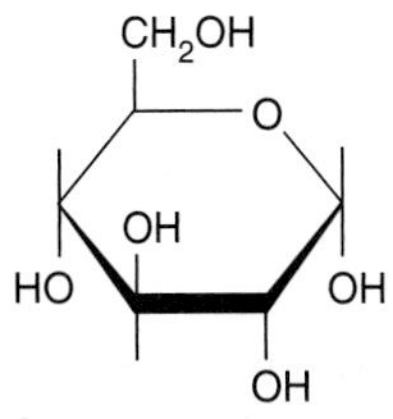

(b)

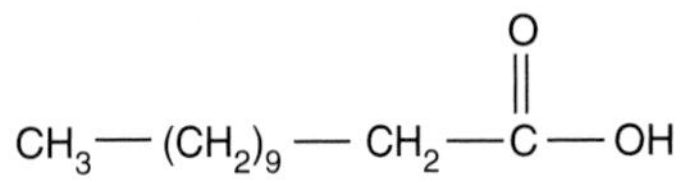

$$CH_3-(CH_2)_9-CH_2-\overset{\overset{\displaystyle O}{\|}}{C}-OH$$

(c)

(d)

16. Define the following terms:
    a. metabolism
    b. nucleophile
    c. reductionism
    d. electrophile
    e. energy
17. What are organelles? In general, what advantages do they provide to eukaryotes?
18. What are the primary functions of metabolism?
19. Define the following terms:
    a. system
    b. emergent properties
    c. robustness
    d. feedback
    e. degeneracy
20. Give an example of each of the following reaction processes:
    a. nucleophilic substitution
    b. elimination
    c. oxidation-reduction
    d. addition
21. List several important ions that are found in living organisms.
22. Compare and contrast the features of an airplane autopilot system with a biological system.
23. Carbohydrates are widely recognized as sources of metabolic energy. What are two other critical roles that carbohydrates play in living organisms?
24. Describe several functions of polypeptides.
25. Define the following terms:
    a. mutation
    b. extremozymes
    c. genome
    d. autopoiesis
26. What are the largest biomolecules? What functions do they serve in living organisms?
27. Nucleotides have roles in addition to being components of DNA and RNA. Give an example.
28. How is order maintained within living cells?
29. Name several waste products that animal cells produce.
30. Provide several examples of emergent properties.
31. Compare the functions of mRNA, rRNA, and tRNA in protein synthesis.
32. Describe the significance of the phrase "robust yet fragile."
33. Compare and contrast the general features of human-designed complex systems and living systems.
34. Compare an autopoietic system with a factory that manufactures airplanes.

ThoughtQuestions

These questions are designed to reinforce your understanding of all of the key concepts discussed in the book so far. They may not have one right answer! The authors have provided possible solutions to these questions in the back of the book and in the accompanying Study Guide, for your reference.

35. Biochemical reactions have been viewed as exotic versions of organic reactions. How do biochemical reactions differ from those used in organic synthesis?
36. It is often assumed that biochemical processes in prokaryotes and eukaryotes are basically similar. Is this a safe assumption?
37. Much of what is known about biochemical processes is a direct result of research on prokaryotic organisms. Most organisms, however, are eukaryotic. Why do you think so many research efforts have used prokaryotes? Why not use eukaryotes directly?
38. Why are fatty acids the principal long-term energy reserve of the body?
39. When a substance such as sodium chloride is dissolved in water, the ions that form become completely surrounded by water molecules, which form structures called hydration spheres. When the sodium salt of a fatty acid is mixed with water, the carboxylate group of the molecule becomes hydrated but the hydrophobic hydrocarbon portion of the molecule is poorly hydrated, if at all. Using a circle to represent the carboxylate group and an attached squiggly line to represent the hydrocarbon chain of a fatty acid, draw a picture of how fatty acids interact in water.

40. The bases of two complementary DNA chains pair with each other because of hydrogen bonding; that is,

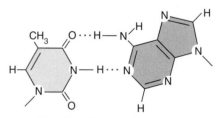

Thymine - Adenine base pair

A new nucleotide has been isolated containing the following purine:

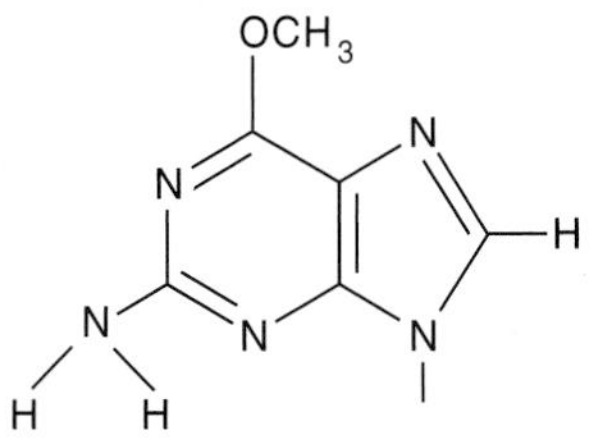

2-Amino-6-methoxypurine

Which of the normal purines and pyrimidines (adenine, guanine, cytosine, or thymine) would you expect to pair with it?

41. Humans synthesize most of the cholesterol required for cell membranes and for the synthesis of vitamin D and steroid hormones. What would you expect to happen if a person's diet is high in cholesterol? Provide a reason for your response.

42. Tay-Sachs disease is a devastating genetic neurological disorder caused by the lack of the enzyme that degrades a specific lipid molecule. When this molecule accumulates in brain cells, an otherwise healthy child undergoes motor and mental deterioration within months after birth and dies by the age of 3 years. In general terms, how would a systems biologist evaluate this phenomenon?

43. The cancerous cells in a tumor proliferate uncontrollably, and treatment often involves the use of toxic drugs in attempts to kill them. Often, however, after initial success (i.e., shrinkage of the tumor), the cancer returns because resistance to the drugs has developed. Biochemists have identified one of the major causes of this phenomenon, called multidrug resistance. One or more cells in the tumor have expressed the gene for P-glycoprotein, a membrane transport protein that pumps the drugs out of the cells. In the absence of the toxic drug molecules, these cells grow uncontrollably and eventually become the dominant cells in the tumor. What features of living organisms does this process illustrate?

44. Hundreds of thousands of proteins have been discovered in living organisms. Yet, as astonishing as this diversity is, these molecules constitute only a small fraction of those that are possible. Calculate the total number of possible decapeptides (molecules with 10 amino acid residues linked by peptide bonds) that could be synthesized from the 20 standard amino acids. [*Hint*: use the function X^n where X is the number of different types of building block molecules and n is the number of building block molecules in the polymer.] If you were to spend 5 minutes writing out the molecular structure of each possible decapeptide, how long would the task take?

Living Cells

OUTLINE

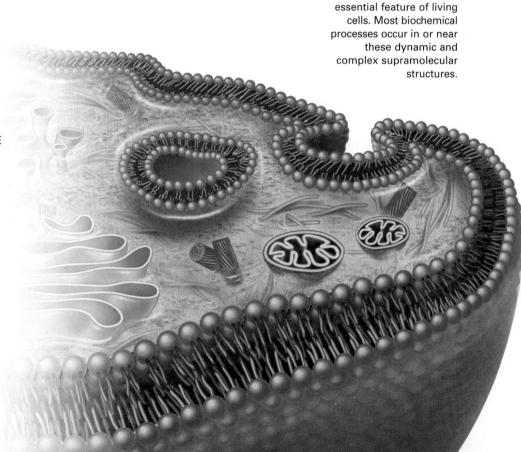

Membranes in Living Cells Membranes are an essential feature of living cells. Most biochemical processes occur in or near these dynamic and complex supramolecular structures.

Overview

CELLS ARE THE STRUCTURAL UNITS OF ALL LIVING ORGANISMS. ONE REMARKABLE FEATURE OF CELLS IS THEIR DIVERSITY: THE HUMAN BODY contains about 200 types of cells. This great variation reflects the variety of functions that cells can perform. However, no matter what their shape, size, or species, cells are also amazingly similar. They are all surrounded by a membrane that separates them from their environment. They are all composed of the same types of molecules.

The structural hierarchy of life on earth extends from the biosphere to biomolecules. Each level is inextricably linked to the levels above and below it. Cells, however, are considered to be the basic unit of life, since they are the smallest entities that are actually alive. Cells are complex, intricate molecular machines that can sense and respond to their environment, transform matter and energy, and reproduce themselves.

There are two types of living cell: prokaryotic and eukaryotic. Most noted for their small sizes, prokaryotes also have remarkably rapid reproduction rates and biochemical diversity across species. Thus prokaryotic species occupy virtually every ecological niche in the biosphere. In contrast, the most conspicuous feature of the eukaryotes is their extraordinarily complex internal structure. Eukaryotes carry out their various metabolic functions via a variety of membrane-bound organelles. The diverse metabolic regulatory mechanisms made possible by this complexity promote two important lifestyle features required by multicellular organisms: cell specialization and intercellular cooperation. Consequently, it is not surprising that the majority of eukaryotes are multicellular organisms composed of specialized cells of numerous types.

Despite their immense diversity of sizes, shapes, and capacities, living cells are also remarkably similar. In fact, all modern cells are believed to have evolved from primordial cells over 3 billion years ago. The common features of prokaryotic and eukaryotic cells include their similar chemical composition and the universal use of DNA as genetic material. This chapter provides an overview of cell structure. This review is a valuable exercise because biochemical reactions do not occur in isolation. Our understanding of living processes is incomplete without knowledge of their cellular context. After a brief discussion of some basic themes in cellular structure and function, the essential structural features of prokaryotic and eukaryotic cells are described in relation to their biochemical roles.

2.1 BASIC THEMES

Each living cell contains millions of densely packed biomolecules that perform at a frenetic pace the thousands of tasks that together constitute life. The application of biochemical techniques to investigations of living processes has provided significant insights into the unique chemical and structural properties of biomolecules that make their functional properties possible. Cellular structure and function can be understood by examining the following key concepts:

1. Water
2. Biological membranes
3. Self-assembly

4. Molecular machines

5. Macromolecular crowding

6. Signal transduction

Water

Water dominates living processes. Its chemical and physical properties (described in Chapter 3) that result from its unique polar structure and its high concentration make it an indispensable component of living organisms. Among water's most important properties is its capacity to interact with a wide range of substances. In fact, the behavior of all other molecules in living organisms is defined by the nature of their interactions with water. **Hydrophilic** molecules, that is, those that possess positive or negative charges or contain relatively large numbers of electronegative oxygen or nitrogen atoms, interact easily with water. Examples of simple hydrophilic molecules include salts such as sodium chloride and sugars such as glucose. In contrast, **hydrophobic** molecules, those that possess few electronegative atoms, do not interact with water. Instead, water excludes them, and they end up confined to nonaqueous regions similar to the droplets of oil that form when oil and water are mixed. When mixed with water, hydrophobic substances spontaneously form clusters, minimizing contact between the hydrocarbon chains and water molecules (Figure 2.1). In between the two extremes is an enormous group of both large and small biomolecules, each of which possesses its own unique pattern of hydrophilic and hydrophobic functional groups. Living organisms exploit the distinctive molecular structure of each of these biomolecules. Phospholipids and proteins are excellent examples. In an aqueous environment phospholipid molecules, which have a relatively large hydrophobic component, form spontaneously into a membranelike bilayer. Similarly the proportion and precise placement of the hydrophilic and hydrophobic side chains in each polypeptide largely determine its structural and functional properties. The charged surface of most proteins attracts and orients water molecules. It is the association of proteins and layers of structured water that gives cytoplasm, the semifluid substance within cells, excluding organelles, its gel-like character.

KEY CONCEPTS

- The chemical and physical properties of water make it an indispensable component of living organisms.

- Hydrophilic molecules interact with water. Hydrophobic molecules do not interact with water.

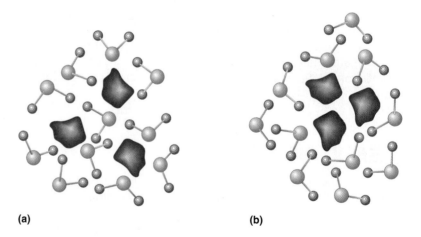

(a)   (b)

FIGURE 2.1

Hydrophobic Interactions Between Water and a Nonpolar Substance

As soon as nonpolar substances (e.g., hydrocarbons) are mixed with water (a), they coalesce into droplets (b). Hydrophobic interactions between nonpolar molecules take effect only when the cohesiveness of water and other polar molecules forces nonpolar molecules or regions of molecules close together.

Biological Membranes

Biological membranes are thin, flexible, and relatively stable sheetlike structures that enclose all living cells and organelles. These membranes can be thought of as noncovalent two-dimensional supramolecular complexes that provide chemically reactive surfaces and exhibit unique transport functions between the extracellular and intracellular compartments. They are also versatile and dynamic cellular components that are intricately integrated into all living processes. Among the numerous crucial functions that have been assigned to membranes, the most basic is to serve as selective physical barriers. Membranes prevent the indiscriminate leakage of molecules and ions out of cells or organelles into their surroundings and allow the timely intake of nutrients and export of waste products. In addition, membranes have significant roles in information processing and energy generation.

Most biological membranes have the same basic structure: a lipid bilayer composed of phospholipids and other lipid molecules, into which various proteins are embedded or attached indirectly (Figure 2.2). Phospholipids have two features that make them ideally suited to their structural role: a hydrophilic charged or uncharged polar group (referred to as a "head group") and a hydrophobic group composed of two fatty acid chains (often called hydrocarbon "tails").

There are two classes of membrane proteins: integral and peripheral. **Integral proteins** are embedded within the membrane because the amino acid residues in the membrane-spanning portions of these proteins are hydrophobic. **Peripheral proteins** are not embedded within the membrane. Rather, they are attached to it either by a covalent bond to a lipid molecule or by noncovalent interaction with a membrane protein or lipid. Membrane proteins perform a variety of functions. **Channel** and **carrier proteins** transport specific ions and molecules, respectively. **Receptors** are proteins with binding sites for extracellular ligands (signal molecules). Ligand-receptor binding triggers a cellular response. **Anchor** proteins attach the membrane to macromolecules on either side of the membrane.

KEY CONCEPTS

- Each biological membrane is composed of a lipid bilayer into which proteins are inserted or attached indirectly.
- Biological membranes are inextricably integrated into all living processes.

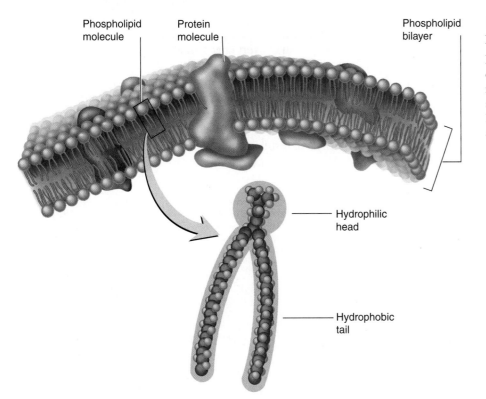

Phospholipid molecule

Protein molecule

Phospholipid bilayer

Hydrophilic head

Hydrophobic tail

FIGURE 2.2

Membrane Structure

Biological membranes are bilayers of phospholipid molecules in which numerous proteins are suspended. In this space-filling model of a typical phospholipid, some proteins extend completely across the membrane.

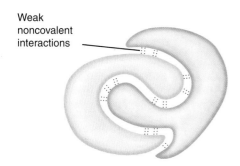

Weak noncovalent interactions

FIGURE 2.3

Self-Assembly

The information that permits the self-assembly of biomolecules consists of the complementary shapes and distributions of charges and hydrophobic groups in the interacting molecules. Large numbers of weak interactions are required for supramolecular structures to form. In this diagrammatic illustration, several weak noncovalent interactions stabilize the binding of two molecules that possess complementary shapes.

KEY CONCEPTS

- In living organisms the molecules in supramolecular structures assemble spontaneously.

- Biomolecules are able to self-assemble because of the steric information they contain.

Self-Assembly

Many of the working parts of living organisms are supramolecular structures composed of individual polymers such as nucleic acids and proteins. Prominent examples include ribosomes (the protein-synthesizing units that are formed from several different types of protein and RNA) and large protein complexes such as the sarcomeres in muscle cells and proteosomes (large protein complexes that degrade certain proteins). According to the principle of self-assembly, most molecules that interact to form stable and functional supramolecular complexes are able to do so spontaneously because they inherently possess the steric information required. They have intricately shaped surfaces with complementary structures, charge distributions, and/or hydrophobic regions that allow numerous relatively weak noncovalent interactions (Figure 2.3). Self-assembly of such molecules involves a balance between the tendency of hydrophilic groups to interact with water and for water to exclude hydrophobic groups from the aqueous regions of the cell. In some cases self-assembly processes need assistance. For example, the folding of some proteins requires the aid of molecular chaperones, protein molecules that prevent inappropriate interactions during the folding process. The assembly of certain supramolecular structures (e.g., chromosomes and membranes) requires preexisting information; that is, a new structure must be created on a template of an existing structure.

Molecular Machines

Researchers now recognize that many of the multisubunit complexes involved in cellular processes function as molecular machines: physical entities with moving parts that perform work, the product of force and distance. Like the mechanical devices used by humans, molecular machines ensure that precisely the correct amount of applied force results in the appropriate amount and direction of movement required for a specific task to be completed. Machines permit the accomplishment of tasks that often would be impossible without them.

Although biological machines are composed of relatively fragile proteins that cannot withstand the physical conditions (e.g., heat and friction) associated with human-made machines, the two device classes do share important features. In addition to being composed of moving parts, both require energy-transducing mechanisms; that is, they both convert energy into directed motion. Despite the wide diversity of types of work performed by biological machines, they all share one key feature: energy-driven changes in the three-dimensional shapes of proteins. One or more components of biological machines bind nucleotide molecules such as ATP or GTP. The binding of nucleotide molecules to these protein subunits, referred to as **motor proteins**, and the release of energy that occurs when the nucleotide is hydrolyzed, result in a precisely targeted change in the subunit's shape (Figure 2.4). This wave of change is transmitted to nearby subunits in a

FIGURE 2.4

Biological Machines

Proteins perform work when motor protein subunits bind and hydrolyze nucleotides such as ATP. The energy-induced change in the shape of a motor protein subunit causes an orderly change in the shapes of adjacent subunits. In this diagrammatic illustration, a motor protein complex moves attached cargo (e.g., a vesicle) as it "walks" along a cytoskeletal filament.

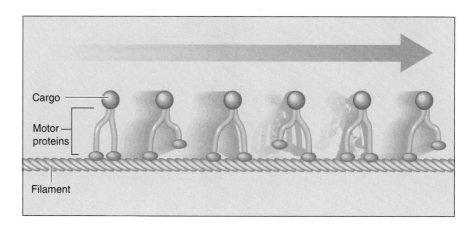

Cargo

Motor proteins

Filament

process that resembles a series of dominoes falling. Biological machines are relatively efficient because the hydrolysis of nucleotides is essentially irreversible; therefore, the functional changes that occur in each machine occur in one direction only.

Macromolecular Crowding

The interior space within cells is dense and crowded. The concentrations of proteins, the dominant type of cellular macromolecules, may be as high as 200 to 400 mg/mL. The term "crowded" rather than "concentrated" is used because macromolecules of each type usually are present in low numbers. Estimates of the volume occupied by macromolecules, called the *excluded volume*, in individual cell types vary between 20 and 40%. As illustrated in Figure 2.5, non-specific steric repulsion prevents the introduction of additional macromolecules under macromolecular crowding conditions. In contrast, the remaining 70% of the space is available to small molecules. The consequences of macromolecular crowding on living systems are significant. It is now believed to be an important factor in biochemical reaction rates, protein folding, protein-protein binding, chromosome structure, gene expression, and signal transduction.

Signal Transduction

If energy is the force that drives biochemical processes, then information is the power to specify what is done. Self-organizing living organisms are so complicated that they must have not only precise structural specifications for each type of biomolecule, but also specifications for how, when, and where each type is to be synthesized, utilized, and degraded. In other words, living organisms require both energy and information to create order. Survival requires that organisms process information from their environment. For example, bacterial cells track down food molecules, plants adapt to changing light levels, and animals seek to avoid predators. Information, or *signals*, comes in the form of molecules (e.g., nutrients) or physical stimuli (e.g., light). Although organisms are bombarded with signals, they can adapt to changing environmental conditions only if they can recognize, interpret, and respond to each type of message. The process that organisms use to receive and interpret information is referred to as **signal transduction**. Although both prokaryotes and eukaryotes process environmental information, most research efforts have been concerned with eukaryotic signal transduction. Consequently, the following discussion focuses on information processing in eukaryotes. Examples of eukaryotic signal molecules include **neurotransmitters**

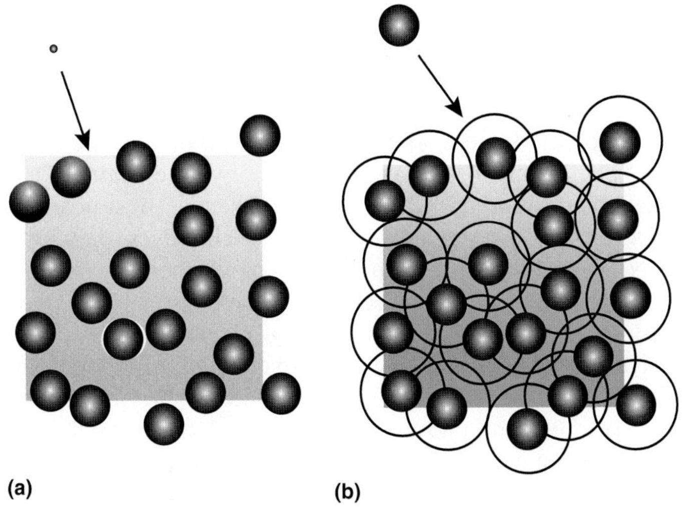

KEY CONCEPT

Many molecular complexes in living organisms function as molecular machines; that is, they are mechanical devices with moving parts that perform work.

KEY CONCEPT

Cells are densely crowded with macromolecules of diverse types. Macromolecular crowding is a significant factor in a wide variety of cellular processes.

FIGURE 2.5

Volume Exclusion

Macromolecules and small molecules are depicted with large balls and small balls, respectively. Within each square, macromolecules occupy 30% of available space. (a) An introduced small molecule can penetrate into virtually all of the remaining 70% of the space. (b) Steric repulsion between macromolecules (open circles) limits these molecules' ability to approach each other. Even though the macromolecules occupy only 30% of the volume, the introduced macromolecule is excluded.

(a)   **(b)**

(products of neurons), **hormones** (products of glandular cells), and *cytokines* (products of white blood cells). All information-processing mechanisms can be divided into three phases:

1. **Reception.** An external signal molecule, called a **ligand**, binds to and activates a receptor.

2. **Transduction.** Ligand binding triggers a change in the three-dimension structure of the receptor that results in the conversion of an extracellular message into an intracellular message.

3. **Response.** Once initiated, the internal signal causes a cascade of events that involve covalent modifications (e.g., phosphorylation) of intracellular proteins. Results of this process include changes in enzyme activities and/or gene expression, cytoskeletal rearrangements, cell movement, or cell cycle progression (e.g., cell growth or division).

The protein hormone insulin is a signal molecule. When released from the pancreas in response to high blood glucose levels, insulin binds to its receptor on a target cell. The insulin receptor is a member of a class of receptors called the *tyrosine kinase receptors*. They are so named because upon activation, these receptors initiate an intracellular response by catalyzing the transfer of phosphate groups to tyrosine (an amino acid residue that contains an OH group) in specific target proteins. Cellular responses triggered by insulin binding to its receptor include uptake of glucose into the cell, and increased fat and glycogen synthesis.

2.2 STRUCTURE OF PROKARYOTIC CELLS

The prokaryotes are an immense and heterogeneous group that includes the Bacteria and the Archaea. Most prokaryotes are similar in external appearance: cylindrical or rodlike (bacillus), spheroidal (cocci), or helically coiled (spirilla). Prokaryotes are also characterized by their relatively small size (a typical rod-shaped bacterial cell has a diameter of 1 μm and a length of 2 μm), their capacity to move (i.e., whether they have flagella, whiplike appendages that propel them), and their retention of specific dyes. Most are identified on the basis of more subtle characteristics. Among the most useful of these are nutritional requirements, energy sources, chemical composition, and biochemical capacities. Despite their diversity, most prokaryotes possess the following common features: cell walls, a plasma membrane, circular DNA molecules, and no internal membrane-enclosed organelles. The anatomical features of a typical bacterial cell are illustrated in Figures 2.6 and 2.7.

Cell Wall

The prokaryotic cell wall (Figure 2.8) is a complex semirigid structure that serves as the primary source of support. It maintains the shape of the organism and protects it from mechanical injury. The cell wall's strength is largely due to the presence of a polymeric network made up of *peptidoglycan*, a covalent complex of short peptide chains linking long carbohydrate chains. The thickness and chemical composition of the cell wall and its adjacent structures determine how avidly a cell wall takes up and/or retains specific dyes.

Most cells can be differentiated on the basis of whether they retain crystal violet stain during the Gram stain procedure. Those that retain the dye are called *Gram-positive*; those that do not are called *Gram-negative*. The wall of a Gram-positive cell consists of a single, relatively thick peptidoglycan layer that lies outside the plasma membrane. Also embedded in this layer are the teichoic acids, polymers of glycerol phosphate and/or ribitol phosphate that give the cell surface a net negative charge. Bacterial viruses often attach to the bacterial cell via teichoic acid polymers prior to infection.

KEY CONCEPTS

• Living organisms receive, interpret, and respond to environmental information by means of the process of signal transduction.

• Signal transduction can be divided into three phases: reception, transduction, and response.

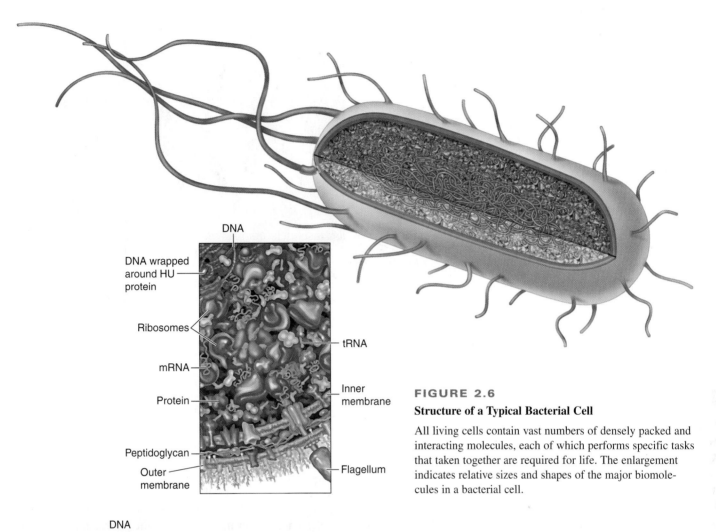

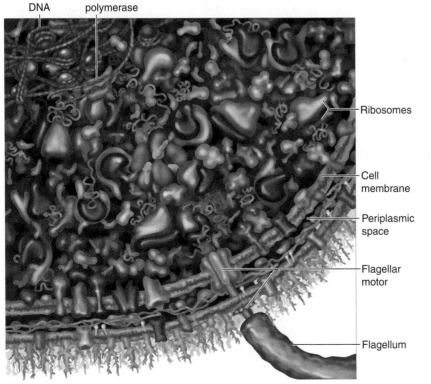

FIGURE 2.6

Structure of a Typical Bacterial Cell

All living cells contain vast numbers of densely packed and interacting molecules, each of which performs specific tasks that taken together are required for life. The enlargement indicates relative sizes and shapes of the major biomolecules in a bacterial cell.

FIGURE 2.7

Bacterial Cell

Bacterial cells are not the bags of protoplasm they were once envisioned to be. Considering that they have no membrane compartments, their internal structure is surprisingly well organized. Note, for example, the spatial separation of the supercoiled DNA molecule (upper left) from other biomolecules. (Also refer to Figure 2.6.)

FIGURE 2.8

Prokaryotic Cell Wall

The cell wall from the Gram-negative organism *Escherichia coli* is complex. Like many Gram-negative bacteria, the cell wall of *E. coli* possess a periplasmic space, a gelatinous layer between the inner and outer membrane. It often contains some peptidoglycan molecules.

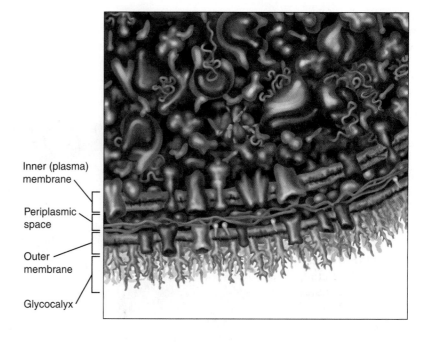

Inner (plasma) membrane

Periplasmic space

Outer membrane

Glycocalyx

The cell walls of Gram-negative bacteria (Figure 2.8) are substantially more complex than those of Gram-positive cells. A thin peptidoglycan layer lies between the outer membrane and the plasma membrane and within the periplasmic space. The lipid component of the *outer membrane* is lipopolysaccharide (LPS), instead of phospholipids. LPS, which is composed of a membrane-bound lipid (lipid A) attached to a polysaccharide, acts as an endotoxin. So called because they are released when the cell disintegrates, *endotoxins* are responsible for symptoms such as fever and shock in animals infected by Gram-negative bacteria. The outer membrane is relatively permeable, and small molecules move across it through *porins*, transmembrane protein complexes that contain channels. The *periplasmic space*, the region between the outer membrane and the plasma membrane, is filled with a gelatinous fluid that contains, in addition to peptidoglycan, a variety of proteins. Many of these proteins participate in nutrient digestion, transport, or chemotaxis.

Some bacteria secrete substances such as polysaccharides and proteins, collectively known as the *glycocalyx*. Depending on the structure and composition of this material, which accumulates on the outside of the cell, the glycocalyx may also be referred to as a capsule or a slime layer. This layer is highly organized and firmly attached to the cell wall. Some bacterial species are especially pathogenic (disease-causing) because the capsule allows them to avoid detection or damage by host immune systems, to attach to host cells to facilitate colonization, and to slough endotoxins that damage host cell. Slime layers are disorganized accumulations of polysaccharides. Slime layers, also referred to as *biofilms*, form when microorganisms adhere to surfaces and grow. In time, as more cells and secreted material accumulate, biofilms become thicker. Biofilms provide microorganisms with a protective barrier and are a significant feature in a variety of medical conditions. The most familiar biofilm is the bacterial plaque that forms on teeth. The bacteria become attached to the tooth surface and secrete a gluelike substance that protects the growing bacterial community from removal by tooth brushing and antibacterial rinses. Only flossing and professional cleaning can successfully remove all of the plaque. Biofilms can also form in catheterized patients and on joint implants as well as on lung tissue in diseases such as cystic fibrosis and tuberculosis. The bacteria in biofilms are very resistant to immune system attack and antibiotic therapy.

The cell walls of the Archaea are quite variable in their composition, and some archaeans lack cell walls. Depending on thickness and chemical composition, the cell walls of the Archaea stain either Gram positive or Gram negative. Some Gram-positive Archaea have cell walls containing pseudopeptidoglycan, a molecule that is similar to peptidoglycan in its composition. Other Gram-positive Archaea have cell walls formed from a thick layer of complex polysaccharides. The chemical compositions of Gram-negative cell walls vary considerably. Most species have one or two layers composed of proteins or glycoproteins. One consequence of the differing cell wall compositions among the Archaea is that none are susceptible to penicillin, an antibiotic that inhibits the cell wall synthesis of Gram-positive bacteria.

Plasma Membrane

Directly inside the cell wall of bacteria is the **plasma membrane** (Figure 2.9), also called the cytoplasmic membrane, a phospholipid bilayer that is reinforced with hopanoids, a group of relatively stiff molecules that resemble sterols, the hydrophobic molecules composed of several attached rings that stiffen membranes in eukaryotes. A diverse group of proteins are embedded in the lipid bilayer.

In addition to acting as a selective permeability barrier, the bacterial plasma membrane possesses receptor proteins that detect nutrients and toxins in their environment. Numerous types of transport proteins involved in nutrient uptake and waste product disposal also occur here. Depending on the species of organism, there may also be proteins involved in energy transduction processes such as **photosynthesis** (the conversion of light energy into chemical energy) and **respiration** (the oxidation of fuel molecules to generate energy).

The composition of membranes in the Archaea is remarkably different from both bacterial species and the eukaryotes. Instead of the straight-chain fatty acids linked to glycerol through ester bonds that are typically found in lipid membranes, the hydrocarbon chains in the membranes of the Archaea are linked by ether linkages. In addition, archaean membrane lipids contain some branched-chain hydrocarbons.

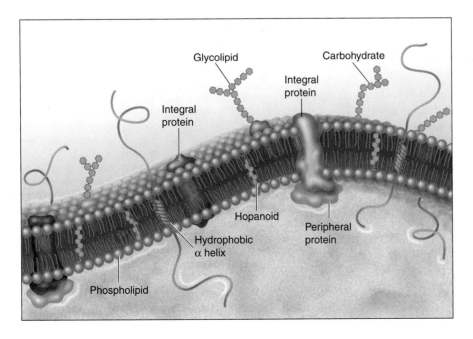

FIGURE 2.9

The Bacterial Plasma Membrane

Simplified view of the plasma membrane illustrating several classes of protein and lipid. Many of these proteins and certain lipids are covalently bound to carbohydrate molecules. (Glycolipids contain carbohydrate groups.) Hopanoids are complex lipid molecules that stabilize bacterial membranes.

Cytoplasm

Despite the absence of internal membranes, prokaryotic cells do appear to have functional compartments (Figure 2.10a). The most obvious of these is the **nucleoid** (Figure 2.10b), a spacious, irregularly shaped region that contains a long, circular DNA molecule called a **chromosome**. The bacterial chromosome is attached to the plasma membrane and typically comprises numerous regions of highly coiled and uncoiled structures. Protein complexes involved in DNA synthesis and regulation of gene expression are also found within the nucleoid. Many bacteria also contain additional small circular DNA molecules called **plasmids** that exist outside the nucleoid. Although they are not required for growth or cell division, plasmids usually provide the cell with a biochemical advantage over cells that lack plasmids. For example, DNA segments that code for antibiotic resistance are often found on plasmids. In the presence of the antibiotic, resistant cells synthesize a protein that inactivates the antibiotic before it can damage the cell. Such cells continue to grow and reproduce, whereas susceptible cells die.

Under low magnification the cytoplasm of prokaryotes has a uniform, grainy appearance except for inclusion bodies, large granules that contain organic or inorganic substances. Some species use glycogen or poly-β-hydroxybutyric acid as carbon storage polymers. Polyphosphate inclusions are a source of phosphate for nucleic acid and phospholipid synthesis. Prokaryotes that derive energy by

FIGURE 2.10

Bacterial Cytoplasm

(a) Cytoplasm is a complex mixture of proteins, nucleic acids, and an enormous variety of ions and small molecules. For clarity, the small molecules appear only in the upper right corner. (b) Close-up view of the nucleoid. Note that DNA is coiled and folded around protein molecules (brown).

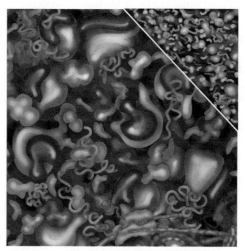

(a)

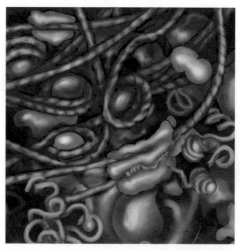

(b)

oxidizing reduced sulfur compounds form sulfur granules. The iron mineral magnetite (Fe_3O_4) forms inclusions, called magnetosomes, which allow some species of aquatic anaerobic prokaryotes to orient themselves with the earth's magnetic field. The remaining space in the cytoplasm is filled with ribosomes and a diverse number of macromolecules and smaller metabolites.

It was long believed that prokaryotes lack the structural protein filaments that are so prominent in eukaryotic organisms. Recent research has revealed, however, that prokaryotes possess proteins that bear a structural and functional resemblance to several eukaryotic cytoskeletal proteins. FtsZ is a filament-forming, tubulin-like protein (p. 62) that forms a constricting ring during cell division. Helical filaments of MreB, an actinlike molecule (p. 63), form spiral structures just underneath the cytoplasmic membrane of all nonspheroidal bacterial cells and aid in maintaining cell shape. Bacterial cells such as *E. coli* in which the gene for MreB has been deleted are spheroidal.

Pili and Flagella

Many bacterial cells have external appendages. *Pili* (singular: pilus) are fine, hair-like structures that may allow cells to attach to food sources and host tissues. Sex pili are used by some bacteria to transfer genetic information from donor cells to recipients, a process called *conjugation* (Figure 2.11). In bacteria, the *flagellum* (plural: flagella) is a flexible corkscrew-shaped protein filament that is used for locomotion (Figure 2.12). Cells are pushed forward when flagella rotate in a counterclockwise direction, whereas clockwise rotation results in a stop-and-tumble motion, allowing the cell to reorient for a forward run. The filament of the flagellum is anchored into the cell by a protein complex, illustrated schematically in the inset to Figure 2.12. Motor proteins in this complex convert chemical energy into rotational motion.

KEY CONCEPTS

- Prokaryotic cells are small and structurally simple. They are bounded by a cell wall and a plasma membrane. They lack a nucleus and other organelles.
- Their DNA molecules, which are circular, are located in an irregularly shaped region called the nucleoid.
- At low magnification ribosomes and inclusion bodies of several types appear to be present in an otherwise featureless cytoplasm.

QUESTION 2.1

A typical, roughly spheroidal, hepatocyte (liver cell) is a widely studied eukaryotic cell that has a diameter of about 20 μm. Calculate the volume of both a prokaryotic and a eukaryotic cell. To appreciate the magnitude of the size difference between the two cell types, estimate how many bacterial cells would fit inside the liver cell. [*Hint:* Use the expression $V = \pi r^2 h$ for the volume of a cylinder and $V = 4\pi r^3/3$ for the volume of a sphere.]

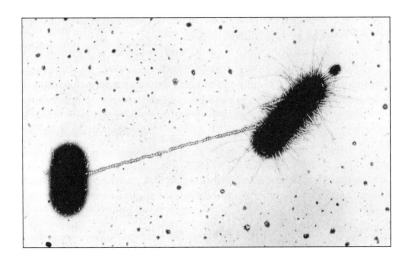

FIGURE 2.11

Bacterial Pili

In this electron micrograph, a sex pilus connects two conjugating *E. coli* cells. Note the numerous smaller pili covering the surface of one of the cells.

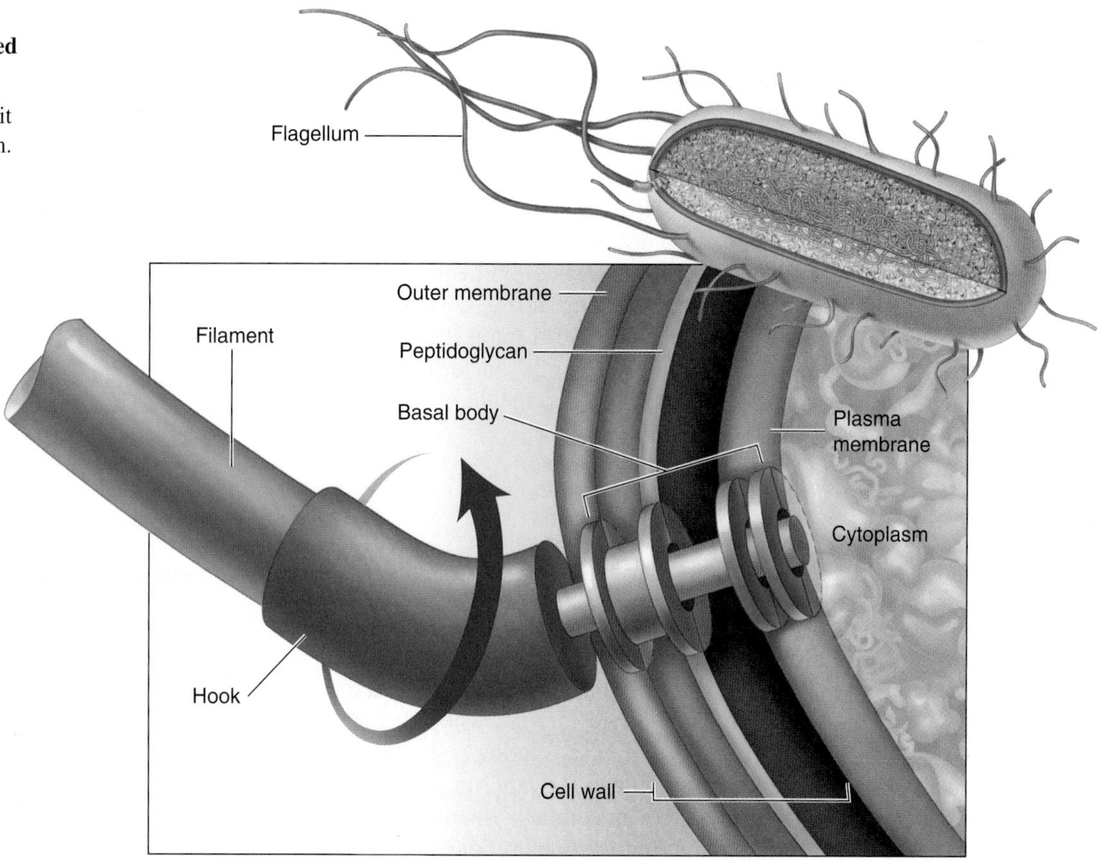

FIGURE 2.12

A Prokaryote Equipped with Flagella

Inset shows how the unit attaches to the organism.

2.3 STRUCTURE OF EUKARYOTIC CELLS

The structural complexity of eukaryotic cells allows more sophisticated regulation of living processes than is possible in the prokaryotes. The most obvious features of eukaryotic cells are their large sizes (diameters of 10–100 μm) in comparison to prokaryotes. More importantly, the membrane surface area is greatly expanded by the presence of membrane-bound organelles. Since most metabolic processes occur within or near membranes, the chemical activity and diversity exhibited by eukaryotic cells is substantially greater than that of prokaryotes. Each organelle within the cell contains a characteristic set of biomolecules and is specialized to perform specific functions. The biochemical processes within an organelle proceed efficiently because of locally high enzyme concentrations and because they can be individually regulated.

Most organelles are components of the **endomembrane system**, an extensive set of interconnecting internal membranes that divide the cell into functional compartments. Other membrane-bound organelles are the mitochondria and peroxisomes. Either through direct physical contact between compartments or by transport vesicles, the endomembrane system transports a vast array of molecules through cells, as well as to and from cell exteriors. **Vesicles** are membranous sacs that bud off from a donor membrane and then fuse with the membrane of another compartment or with the plasma membrane.

The components of the endomembrane system are the plasma membrane, endoplasmic reticulum, Golgi apparatus, nucleus, and lysosomes. Plant cells also contain chloroplasts. In addition to membranous organelles, eukaryotic cells possess several components that are devoid of membranes. Included in this group are ribosomes and the cytoskeleton. The cytoskeleton is a complex, dynamic, and force-generating network of filaments that give eukaryotic cells shape, structural support, and the capacity for the directed movement of molecules and organelles.

Although most eukaryotic cells possess similar structural features, there is no "typical" eukaryotic cell. Each cell type has its own characteristic structural and functional properties. They are sufficiently similar, however, that a discussion of the basic components is useful. The generalized structures of cells from animals and plants, the major forms of multicellular eukaryotic organisms, are illustrated in Figures 2.13 and 2.14. The structure and functional properties of each cellular component are briefly described in the sections that follow.

Plasma Membrane

The plasma membrane isolates the cell from the outside environment. It is composed of a lipid bilayer and an enormous number and variety of integral and peripheral proteins (Figure 2.15). Channels and carriers within the plasma membrane regulate the passage of various ions and molecules in and out of the cell. Immense numbers of receptors play key roles in signal transduction. The extracellular face of a eukaryotic cell is heavily "decorated" with carbohydrate; that is, much of the membrane protein and lipid contains covalently attached

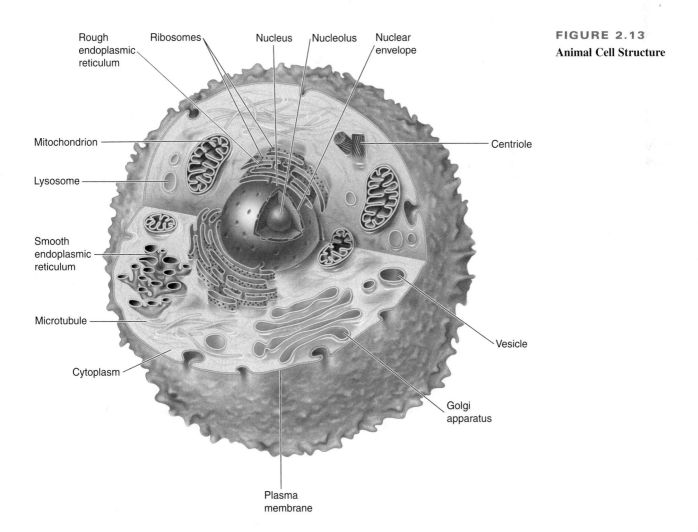

FIGURE 2.13
Animal Cell Structure

FIGURE 2.14

Plant Cell Structure

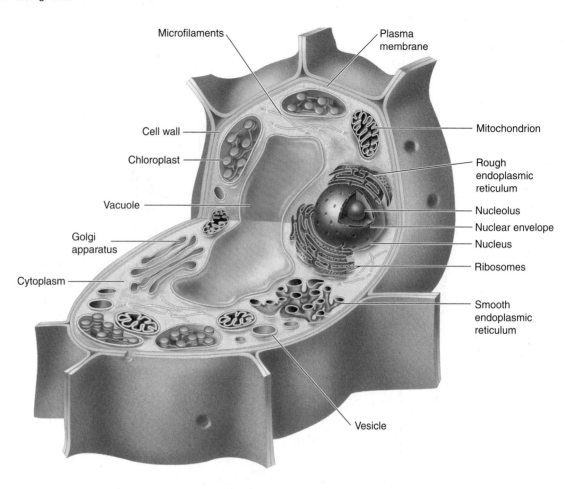

Microfilaments

Plasma membrane

Cell wall

Mitochondrion

Chloroplast

Rough endoplasmic reticulum

Vacuole

Nucleolus

Nuclear envelope

Golgi apparatus

Nucleus

Cytoplasm

Ribosomes

Smooth endoplasmic reticulum

Vesicle

carbohydrate. This carbohydrate "coat" is referred to as the **glycocalyx** (Figure 2.16). This carbohydrate plays important roles in cell-cell recognition and adhesion, receptor specificity and self-identity (an immune system requirement). The basic blood group antigens are an example of this self-identity function.

In most eukaryotes the plasma membrane is protected with extracellular and intracellular structures (Figure 2.15). Within animal tissues, the specialized cells called fibroblasts synthesize and secrete structural proteins and complex carbohydrates that form the **extracellular matrix (ECM)**, a gelatinous material that binds cells together. In addition to its support and protective functions, the extracellular matrix plays roles in cell behavior regulation through the binding of some of its components to specific membrane receptors in chemical and mechanical signaling processes of various types. The inner surface of the eukaryotic plasma membrane is reinforced by a three-dimensional meshwork of proteins called the **cell cortex** that is attached to the membrane by extensive noncovalent bonding to peripheral proteins. In animal cells this protein network provides mechanical strength to the plasma membrane and determines cell shape. The cell cortex consists mainly of cytoskeleton components: actin and several types of actin-binding proteins.

Endoplasmic Reticulum

The **endoplasmic reticulum (ER)** is a system of interconnected membranous tubules, vesicles, and large flattened sacs. A hint of its importance in cell function

KEY CONCEPTS

- Besides providing mechanical strength and shape to the cell, the plasma membrane is actively involved in selecting the molecules that can enter or exit the cell.

- Receptors on the plasma membrane's surface allow the cell to respond to external stimuli.

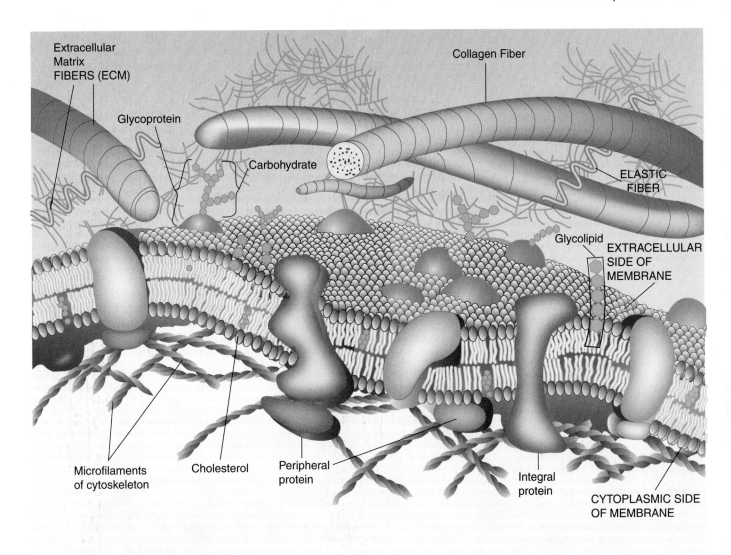

FIGURE 2.15

The Plasma Membrane of an Animal Cell

The plasma membrane (PM) is composed of a lipid bilayer in which a wide variety of integral proteins are embedded. Note that numerous integral proteins and lipid molecules are covalently attached to carbohydrate. Peripheral proteins are attached by noncovalent bonds to the cytoplasmic surface of the PM. Specialized cells of the connective tissue of higher animals called fibroblasts synthesize and secrete glycoproteins of the extracellular matrix (ECM). The inner surface of the PM is reinforced by the cell cortex, which is composed of a meshwork of microfilaments and other proteins.

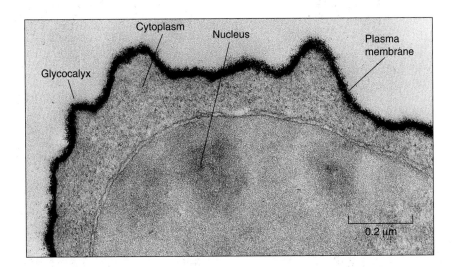

FIGURE 2.16

The Glycocalyx

Electron micrograph of the surface of a lymphocyte stained to reveal the glycocalyx (cell coat).

is that it often constitutes more than half a cell's total membrane. The repeatedly folded, continuous sheets of ER membrane enclose an internal space called the ER *lumen*. This compartment, which is often referred to as the *cisternal space*, is entirely separated from the cytoplasm by ER membrane.

As shown in Figure 2.17, endoplasmic reticulum comes in two forms: **rough ER (RER)** and **smooth ER (SER)**. The precise functional properties and relative sizes of both ER types vary with cell type and physiological conditions. RER is so named because of the numerous ribosomes that stud its cytoplasmic surface. Several protein classes are processed by the RER: membrane proteins, as well as water-soluble proteins destined for retention within the ER, transport to other organelles, or export out of the cell. Polypeptides enter the RER during ongoing protein synthesis as they are threaded, or translocated, through the membrane.

Transmembrane polypeptides (i.e., those that contain one or more hydrophobic sequence segments) remain embedded in the membrane because the translocation process is halted when hydrophobic segments enter the membrane. As water-soluble polypeptides emerge into the ER lumen, the folding process, facilitated by processing enzymes and molecular chaperones, begins. Glycosylation reactions, the attachment of carbohydrate groups to specific amino acid residues, are a prominent example of ER-processing reactions. The binding of molecular chaperones to short hydrophobic segments in partially folded polypeptides facilitates efficient folding and prevents aggregation. Failure of a polypeptide to fold or assemble into a protein complex triggers a mechanism in which the polypeptide is exported back into the cytoplasm, where it is destroyed. In certain stressful circumstances, misfolded proteins begin to accumulate within the ER. *ER stress*, caused by the accumulation of misfolded proteins, is a threat to the entire

FIGURE 2.17

The Endoplasmic Reticulum

There are two forms of endoplasmic reticulum: RER, the rough endoplasmic reticulum, and SER, the smooth endoplasmic reticulum.

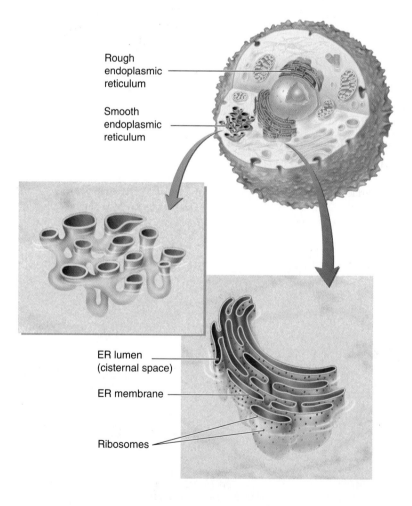

Rough endoplasmic reticulum

Smooth endoplasmic reticulum

ER lumen (cisternal space)

ER membrane

Ribosomes

cell and can potentially disrupt overall cell function. If stress is severe, the ER initiates the *unfolded protein response* (UPR). Signals sent to the nucleus result in the inhibition of new protein synthesis with the exception of additional molecular chaperones. If the damage cannot be repaired, the *ER overload response* (EOR) is triggered. EOR initiates *apoptosis*, a cellular process that culminates in cell death.

Smooth ER lacks attached ribosomes, and its membranes are continuous with those of RER. The size and functional properties of SER vary considerably in different cell types from sparse to abundant. In most cells SER is involved in the synthesis of lipid molecules. SER also stores calcium ions (Ca^{2+}), a commonly used cell signal. The SER is especially noteworthy in hepatocytes and striated muscle cells. Hepatocyte SER performs a wide variety of functions, which include biotransformation and synthesis of the lipid components of very-low-density lipoproteins (VLDL: water-soluble lipid transport complexes that deliver lipids to tissue cells). Biotransformation reactions convert an enormous variety of water-insoluble metabolites and xenobiotics (foreign and potentially toxic molecules) into more soluble products that can then be excreted. The SER in striated muscle is so highly specialized in both structure and function that it has a different name, the *sarcoplasmic reticulum* (SR). The SR membrane extends throughout the muscle cell and is in close proximity to all myofibrils, the organized arrays of contractile proteins. SR is a reservoir for calcium, the signal that triggers muscle contraction.

Golgi Apparatus

The **Golgi apparatus** (also known as the **Golgi complex**) is formed from relatively large, flattened, saclike membranous vesicles that resemble a stack of plates. The Golgi apparatus (called **dictyosomes** in plants) is involved in the packaging and distribution of cell products to internal and external compartments (Figure 2.18).

KEY CONCEPTS

- The RER is primarily involved in protein synthesis. The external surface of RER membrane is studded with ribosomes.
- SER lacks attached ribosomes and is involved in lipid synthesis, biotransformation, and Ca^{2+} storage.

FIGURE 2.18
The Golgi Apparatus

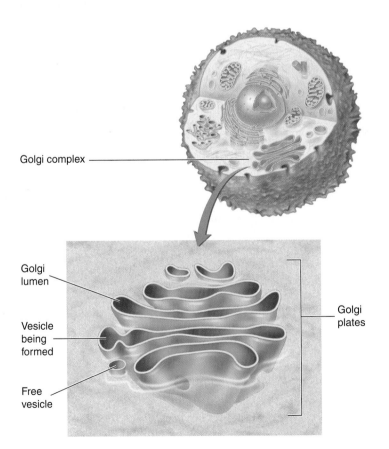

Golgi complex

Golgi lumen

Vesicle being formed

Free vesicle

Golgi plates

The Golgi apparatus has two faces. The plate (or *cisterna*) positioned closest to the ER is on the forming (*cis*) face, whereas the one on the maturing (*trans*) face is typically close to the portion of the cell's plasma membrane that is engaged in secretion. Small membranous vesicles containing newly synthesized protein and lipid bud off from the ER and fuse with the *cis* Golgi membrane. Until recently it was believed that Golgi cisternae are relatively stationary and that protein and lipid vesicles were the mechanism of cargo transport from one Golgi sac to the next. In this view, as cargo molecules proceed through the Golgi system, Golgi enzymes process them further. Recent research has revealed the probability that Golgi cisternae move physically from the cis face position to the trans face as they simultaneously transport and modify their cargo. According to the *cisternal maturation model*, transport vesicles recycle Golgi membrane and enzymes back to the newly formed cis Golgi cisterna. When product molecules reach the *trans*

FIGURE 2.19

Exocytosis

Proteins produced in the ER and processed by the Golgi apparatus are packaged into vesicles that migrate to the plasma membrane and merge with it.

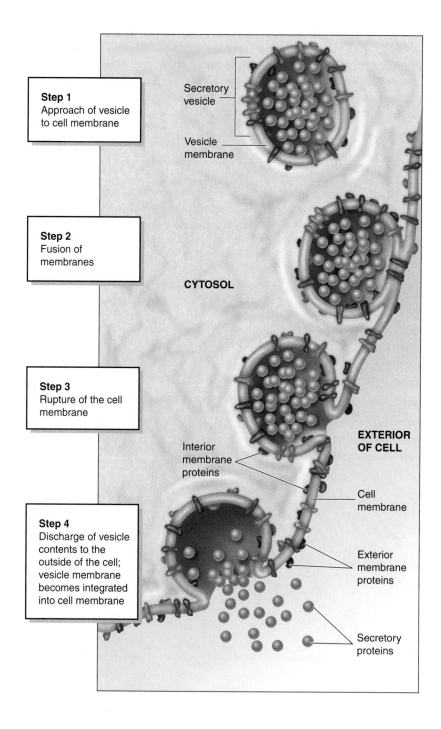

Step 1
Approach of vesicle to cell membrane

Step 2
Fusion of membranes

Step 3
Rupture of the cell membrane

Step 4
Discharge of vesicle contents to the outside of the cell; vesicle membrane becomes integrated into cell membrane

Secretory vesicle

Vesicle membrane

CYTOSOL

EXTERIOR OF CELL

Interior membrane proteins

Cell membrane

Exterior membrane proteins

Secretory proteins

face, they are then targeted to other parts of the cell. Secretory products, such as digestive enzymes or hormones, are concentrated within *secretory vesicles* (also known as *secretory granules*) that bud off from the *trans* face. Secretory granules remain in storage in the cytoplasm until the cell is stimulated to secrete them. The secretory process, referred to as **exocytosis**, consists of the fusion of the membrane-bound granules with the plasma membrane (Figure 2.19). The contents of the granules are then released into the extracellular space. In plants, the functions of the Golgi apparatus include transport of substances into the cell wall and expansion of the plasma membrane during cell growth.

Nucleus

The nucleus (Figure 2.20), the most prominent organelle in eukaryotic cells, contains the cell's genome. Low-resolution micrographs reveal that nuclear structure consists of a seemingly amorphous nucleoplasm surrounded by membrane, the nuclear envelope. **Nucleoplasm** contains a network of **chromatin fibers**, which during the mitotic phase of the cell cycle, are condensed to form the chromosomes that will be distributed to daughter cells. Chromatin is highly structured, consisting of DNA and DNA packaging proteins known as the *histones*. Individual chromosomes occupy separate chromosome territories, with gene-dense chromosomes usually located near the center of the nucleus and gene-poor chromosomes typically having more peripheral locations. The **nuclear matrix** (or nucleoskeleton) is believed to serve as a scaffold within which loops of chromatin are organized. Macromolecular crowding may also be an important organizing factor. Nuclear architecture, the spatial pattern of chromatin and other nuclear macromolecular complexes, is important because of the functional relationship

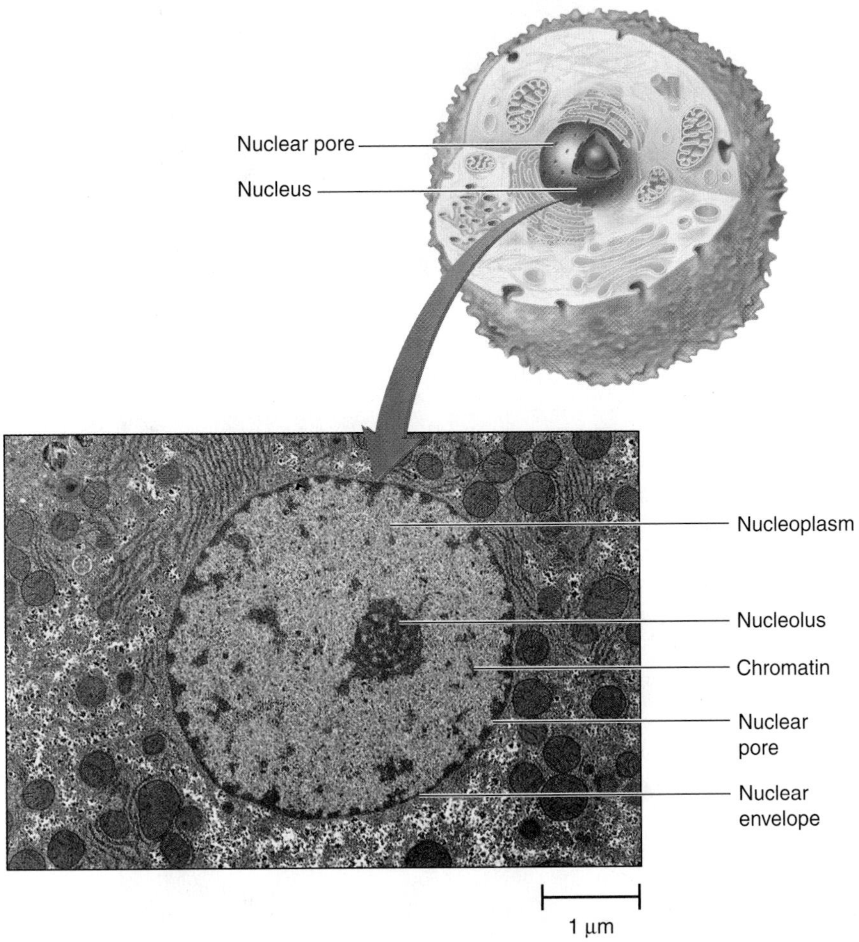

KEY CONCEPT

Formed from relatively large, flattened, saclike membranous vesicles, the Golgi apparatus functions in the packaging and secretion of cell products.

FIGURE 2.20

The Eukaryotic Nucleus

The nucleus is an organelle surrounded by a double membrane, the nuclear envelope.

Nuclear pore

Nucleus

Nucleoplasm

Nucleolus

Chromatin

Nuclear pore

Nuclear envelope

1 μm

between gene expression and chromatin conformation. Interspersed among and between the chromosome territories are numerous functional sites. Examples include nucleoli, speckles, and Cajal bodies. The **nucleolus** (a relatively large, dark-staining spherical structure) is the site of rRNA synthesis, ribosomal assembly, and processing reactions for several types of RNA. *Speckles*, as many as 50 per cell, are storage sites for certain types of transcription components. Cajal bodies (named after Ramón y Cajal, the Spanish histologist, 1852–1934) are sites for mRNA processing reactions.

The **nuclear envelope** separates DNA replication and transcription processes from the cytoplasm. By limiting access of cytoplasmic molecules, the nuclear envelope allows for more sophisticated regulation of gene expression than would be possible otherwise. It is composed of two membranes that fuse at structures called **nuclear pores**. The space between the membranes, the **perinuclear space**, is continuous with the RER lumen. The outer nuclear membrane is continuous with the rough endoplasmic reticulum. Ribosomes are attached to its cytoplasmic surface. Unlike the outer membrane, the **inner membrane** contains integral proteins that are unique to the nucleus. Many of these molecules serve as attachments to the nuclear lamina, a protein meshwork that strengthens the nuclear envelope. The nuclear lamins are structural proteins that resemble the intermediate filament components of the cytoplasmic cytoskeleton (see p. 62). The nuclear pores

FIGURE 2.21

The Nuclear Pore Complex

The nuclear envelope is studded with nuclear pore complex structures, or plugs, one of which is indicated by the arrow in the photo at top.

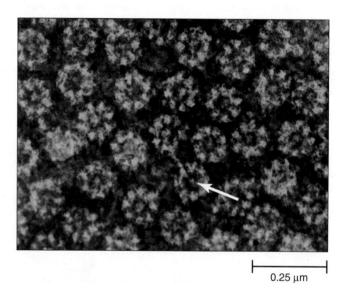

0.25 μm

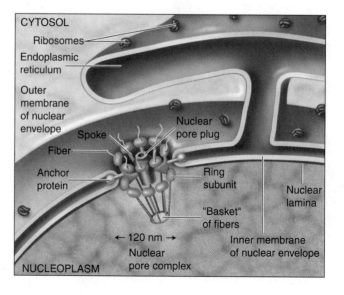

(Figure 2.20) are relatively large and intricate structures through which pass most of the molecules that enter and leave the nucleus. Many of these substances such as ions, small proteins, and other molecules diffuse through the nuclear pore complex. The movement of larger molecules such as RNA and large proteins into and out of the nucleus is believed to be regulated by protein components of a centrally located transport complex, sometimes referred to as the *nuclear pore plug* (see Figure 2.21).

Vesicular Organelles

The eukaryotic cell contains a collection of vesicles. These small spheroidal membranous sacs contain either materials originating in the ER and Golgi and/or material brought into the cell by endocytosis. **Endocytosis** is the process whereby cells internalize exogenous substances. Localized regions of the plasma membrane surround the exogenous material and pinch off to form vesicles. The endocytic process can be unregulated, in bulk **phagocytosis** (e.g., the engulfment of foreign or damaged cells by certain white blood cells), or regulated via receptor-mediated endocytosis in coated pits (Figure 2.22). Some vesicles can participate in exocytosis in which material is secreted from the cell. These vesicles may be intermediate transport vesicles or specialized for a particular cellular function.

Lysosomes (Figure 2.23) are vesicles that contain granules or aggregates of digestive enzymes. These enzymes are called *acid hydrolases* because they catalyze the attack of water molecules on ester and amide linkages under acidic conditions. For example, they cleave lipids into free fatty acids, polysaccharides into free sugars, and proteins into free amino acids. The high proton concentration (low pH) in these organelles is generated by a proton pump in the lysosomal membrane. The function of lysosomes is to degrade debris in the cell by a process called *autophagy* and to play a role in the receptor-mediated endocytic pathway. Plant vacuoles are multifunctional vesicular organelles that contain numerous enzymes, some of which are similar to lysosomal acid hydrolases. Vacuoles store biomolecules required for plant nutrition, growth, and development. They degrade material no longer needed by the cell, accumulate debris, and contribute to cell turgor (rigidity).

Lysosomes are also involved in a variety of secretory processes. The lysosomes of the bone-remodeling cells called osteoclasts can fuse with the plasma membrane and release enzymes onto the bone surface to aid in bone resorption. More recently discovered examples include the release of α-granules from platelets and melanin (the dark pigment that contributes to skin color) from melanocytes. In the early stages of platelet aggregation (initiated during blood vessel injury), platelets release α-granules, which contain a variety of adhesive protein ligands. In melanocytes, which are cells in the basal layer of the skin, melanin accumulates in vesicles called melanosomes. Melanosomes migrate in response to UV light, hormones, and/or neural signals into the epithelial layers of the skin where they are ingested by keratinocytes (skin cells).

KEY CONCEPTS

- The nucleus contains the cell's genetic information and the machinery for converting that information into protein molecules.
- The nucleolus plays an important role in the synthesis of ribosomal RNA.

KEY CONCEPT

The cell contains a number of vesicular organelles that may be specialized for a particular function. Examples include the acid hydrolase–rich lysosomes and plant vacuoles, the α-granules of platelets, and the melanosomes of skin melanocytes.

QUESTION 2.2

In many genetic disorders, a lysosomal enzyme required to degrade a specific molecule is missing or defective. These maladies, often referred to as *lysosomal storage diseases*, include Tay-Sachs disease. Afflicted individuals inherit from each parent a defective gene that codes for an enzyme that degrades a complex lipid molecule. Symptoms include severe mental retardation and death before the age of 5 years. What is the nature of the process that is destroying the patient's cells? [*Hint:* Synthesis of the lipid molecule continues at a normal rate.]

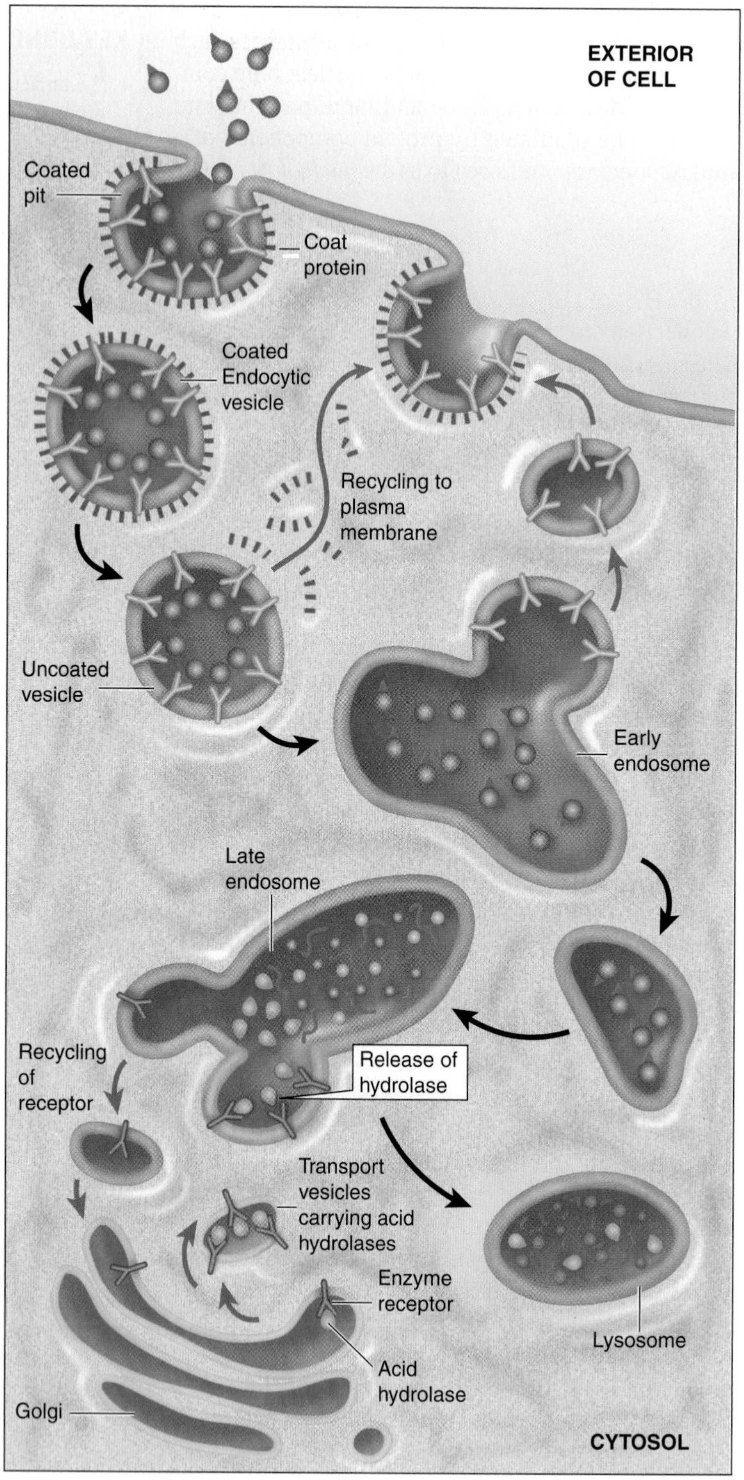

FIGURE 2.22

Receptor-Mediated Endocytosis

(a) Extracellular substances may enter the cell during endocytosis, a process in which receptor molecules in the plasma membrane bind to the specific molecules or molecular complexes called ligands. Specialized regions of plasma membrane called coated pits progressively invaginate to form closed vesicles. After the coat proteins are removed, the vesicle fuses with an early endosome, the precursor of lysosomes. The coat proteins are then recycled to the plasma membrane. During endosomal maturation, the proton concentration rises and the ligands are released from their receptors, which are subsequently also recycled back to the plasma membrane. As endosomal maturation continues, lysosomal hydrolases are delivered from the Golgi apparatus. Lysosomal formation is complete when all the hydrolases have been transferred to the late endosome and the Golgi membrane has been recycled back to the Golgi apparatus. (b) Electron micrographs illustrating the initial events in endocytosis. (See facing page.)

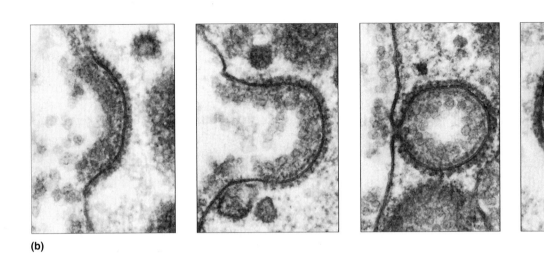

(b)

FIGURE 2.22 (CONT'D)

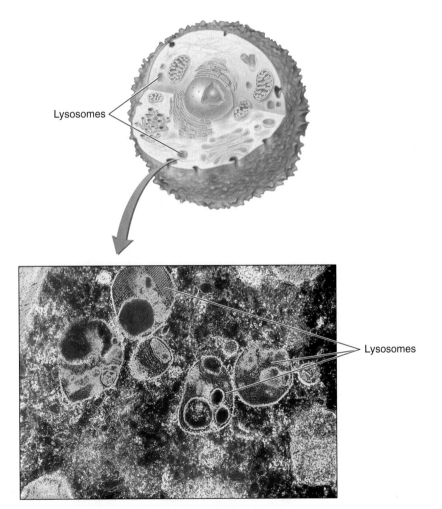

Lysosomes

Lysosomes

FIGURE 2.23

Lysosomes

Lysosomes are membranous sacs that contain hydrolytic enzymes.

QUESTION 2.3

Nearly half the world's chameleon species live on the island of Madagascar. These lizards are known for their ability to change color, exhibiting a variety of colors including brown, black, white, red, green, yellow, and blue. Chameleons use color as a means of communication rather than as a camouflage mechanism, like some fish and amphibians. Suggest a mechanism by which chameleon species change color.

Mitochondria

Mitochondria (singular: mitochondrion) are organelles that have long been recognized as the site of **aerobic metabolism**, the mechanism by which the chemical bond energy of food molecules is captured and used to drive the oxygen-dependent synthesis of adenosine triphosphate (ATP), the cell's energy storage molecule. Mitochondria are also central integrators of other metabolic processes. Prominent examples include the metabolism of amino acids, lipids, and iron as well as calcium homeostasis. In recent years mitochondria have also been recognized as key regulators of **apoptosis**, the genetically programmed series of events that lead to cell death (Figure 2.24). They have traditionally been described as sausage-shaped structures with lengths ranging from 1 to 10 μm. This view has changed considerably because researchers have now learned that mitochondria have no fixed sizes. Rather, they are dynamic organelles that are continuously dividing (*fission*), branching, and merging (*fusion*) to form extended reticular networks.

Each **mitochondrion** is bounded by two membranes (Figure 2.25a). The relatively porous smooth **outer membrane** is permeable to most molecules with masses less than 10,000 D. The **inner membrane**, which is impermeable to ions and a variety of organic molecules, projects inward into folds that are called *cristae* (singular: crista). Embedded in this membrane are structures composed of molecular complexes called respiratory assemblies that are responsible for the synthesis of ATP. Also present are a series of proteins that are responsible for the transport of specific molecules and ions.

Together, both membranes create two separate compartments: (1) the *intermembrane space* and (2) the *matrix*. The intermembrane space contains several enzymes involved in nucleotide metabolism, whereas the gel-like matrix consists of high concentrations of enzymes and ions and a myriad of small organic molecules. The matrix also contains several circular DNA molecules.

Mitochondrial DNA resembles bacterial DNA in that both molecules are "naked" (i.e., not packaged with histones) and are located within a nucleoid. The mitochondrial genome codes for rRNAs, tRNAs, and several respiratory assembly protein components. About 95% of the genes that code for mitochondrial molecules are located on nuclear chromosomes. The number of mitochondria a cell possesses is a function of its energy demands and physiological status. There is considerable variety across cell types. For example, oocytes and hepatocytes may have as many as 100,000 and 1000 mitochondria, respectively. Most cell types have several hundred mitochondria. Erythrocytes (red blood cells) have none. Notably, the configuration of mitochondria changes with the physiological status of the cell. For example, the internal appearance of liver mitochondria has been observed to change dramatically during active respiration

KEY CONCEPTS

- Aerobic respiration, the process that generates most of the energy required in eukaryotes, takes place in mitochondria.
- Embedded in the inner membrane of a mitochondrion are respiratory assemblies, where ATP is synthesized.

FIGURE 2.24

Apoptosis

White blood cells before (left) and during (right) apoptosis. The apoptotic cell is forming blebs that will eventually fragment into apoptotic bodies. Ultimately, the apoptotic bodies will be ingested by phagocytes (immune system cells that digest cell debris).

2 µm

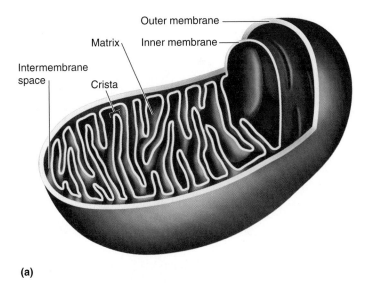

(a)

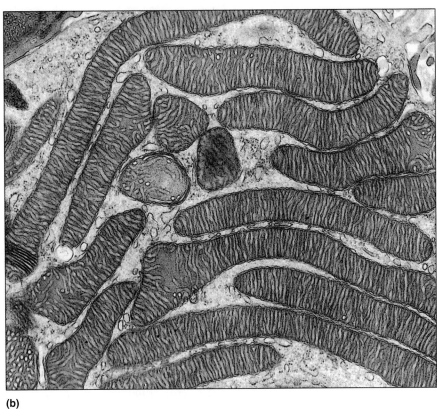

(b)

FIGURE 2.25

The Mitochondrion

(a) Membrane and crista. The internal structure depicted in this diagram is referred to as the baffle model because of the bellow-like shape of the crista. Electron tomography studies (a microscopic technique in which electron beams are used to create three-dimensional reconstructions of specimens) have revealed a more complex anatomy. Complex arrays of fusing and dividing inner membrane tubules have been observed in the mitochondria of some tissues. The functional significance of these structural features is unknown. (b) Mitochondria from adrenal cortex.

FIGURE 2.26

Rat Liver Mitochondria

(a) Low-energy (orthodox) and (b) high-energy (condensed) conformations.

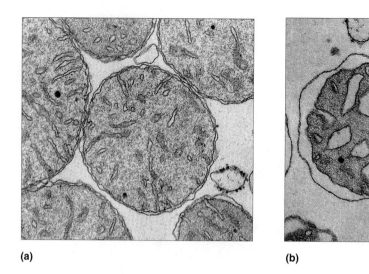

(a)   (b)

(Figure 2.26). Additionally, the fragmentation or inordinate swelling of mitochondria is a very sensitive indicator of cell injury.

QUESTION 2.4

It has been estimated that mitochondria occupy 20% of the volume in the human body. For an average adult, the average number of mitochondria has been estimated to be 1×10^{16} (ten thousand trillion). Assuming that the average person weighs 70 kg, provide a rough estimate of the average mass of a mitochondrion.

Peroxisomes

Peroxisomes are small spherical membranous organelles that contain oxidative enzymes (proteins that catalyze the transfer of electrons). Peroxisomal enzymes are involved in a variety of anabolic and catabolic processes, including degradation of fatty acids, synthesis of certain membrane lipids, and degradation of purine bases. As the name suggests, peroxisomes are most noted for their involvement in the generation and breakdown of toxic molecules known as *peroxides*. For example, hydrogen peroxide (H_2O_2) is generated when molecular oxygen (O_2) is used to remove hydrogen atoms from specific organic molecules. Once formed, H_2O_2 must be immediately destroyed before it damages the cell. Carrying out this destruction is especially critical in liver and kidney cells, which have an important detoxifying role in animal bodies. For example, peroxisomes are involved in the oxidation of ingested ethanol.

Two types of peroxisome have been identified in plants. One, found in leaves (Figure 2.27) is responsible for an oxygen-consuming process known as *photorespiration* that produces CO_2. Peroxisomes of the other type (often called **glyoxysomes**), are found in germinating seeds. In these structures lipid molecules are converted into carbohydrate, which provides energy for growth and development.

The biogenesis (formation) of peroxisomes requires both protein and membrane synthesis. Peroxisomal enzymes and membrane proteins are coded for by nuclear genes, synthesized on cytoplasmic ribosomes, and then imported into preperoxisomes. For many years peroxisomes were thought to be autonomous organelles that proliferate by division of preexisting ones. There

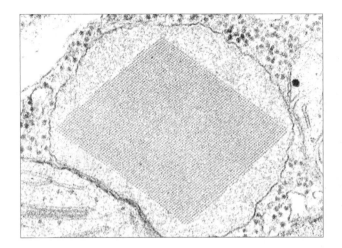

FIGURE 2.27
Peroxisome in a Tobacco Leaf Cell
The granular substance surrounding the crystal-like core is called the matrix.

is now substantial evidence that the ER is the source of peroxisomal membrane. The assembly of peroxisomes, the coordinated acquisition of both membrane and protein components, requires a group of proteins called the peroxins.

Plastids

Plastids, structures that are found only in plants, algae, and some protists, are bounded by a double membrane. Although the inner membrane is not folded as in mitochondria, another separate, intricately arranged internal membrane is often present. In plants, all plastids develop from *proplastids*, which are small, nearly colorless structures found in the meristem (a special region in plants made up of undifferentiated cells from which new tissues arise). Proplastids develop according to the requirements of each differentiated cell. Mature plastids are of two types: (1) *leucoplasts*, which store substances such as starch or proteins in storage organs (e.g., roots or tubers), and (2) **chromoplasts**, which accumulate the pigments that are responsible for the colors of leaves, flower petals, and fruits.

Chloroplasts are a type of chromoplast that are specialized for the conversion of light energy into chemical energy. In this process, called **photosynthesis**, which will be described in Chapter 13, light energy is used to drive the synthesis of carbohydrate from CO_2. The structure of chloroplasts (Figure 2.28) is similar in several respects to that of mitochondria. For example, the outer membrane is highly permeable, whereas the relatively impermeable internal membrane contains special carrier proteins that control molecular traffic into and out of the organelle.

An intricately folded internal membrane system, called the **thylakoid membrane**, is responsible for the metabolic function of chloroplasts. For example, chlorophyll molecules, which capture light energy during photosynthesis, are bound to thylakoid membrane proteins. Certain portions of thylakoid membrane form tightly stacked structures called **grana** (singular: granum), whereas the entire membrane encloses a compartment known as the *thylakoid lumen* (or channel). Surrounding the thylakoid membrane is the **stroma**, a dense enzyme-filled substance, analogous to the mitochondrial matrix. In addition to enzymes, the stroma contains DNA, RNA, and ribosomes. Membrane segments that connect adjacent grana are referred to as *stroma lamellae* (singular: lamella).

Oak leaf

Leaf cross section

Plant cell

Stroma

Inner membrane

Granum (stacks of thylakoids)

Thylakoid

Outer membrane

Chloroplast

FIGURE 2.28

The Chloroplast

Chloroplasts are one type of organelle found in multicellular plants.

BIOCHEMISTRY IN PERSPECTIVE

Organelles and Human Disease

What is the role of biochemistry in modern medicine? The scientific investigation of human disease is only two hundred years old. During Europe's *Age of Enlightenment* (seventeenth and eighteenth centuries), as a result of various political and social factors combined with the discoveries of Galileo, Isaac Newton, Francis Bacon, René Descartes, and other scientists, belief systems began to change. Health concepts originating with Hippocrates (fifth century BCE) and Galen (second century CE) had been unchallenged for over a thousand years. Humoral medicine, in which health was understood in terms of a balance of the "humors" of blood, phlegm, yellow bile, and black bile, was universally accepted, and later supplemented by medieval superstition (sickness caused by divine intervention). Gradually, however, the capacity of human reason to understand the human body gained acceptance. By the end of the nineteenth century, previously unimaginable progress toward disease diagnosis and treatment had been made because of discoveries in fields ranging from anatomy, cellular pathology, and bacteriology to statistics. Today, human disease is investigated at the cellular and molecular levels because of breakthrough work performed in the 1940s and 1950s. Among the most important was the discovery of DNA as the genetic material and its subsequent structure determination (p. 636). The adaptation of the electron microscope (p. 68) by Keith Porter for use with biological specimens, and the centrifugation techniques (p. 67) developed by George Palade, Albert Claude, and Christian DeDuve made the identification of distinct organelles possible. More recent work utilizing DNA technology has profoundly increased our understanding of the molecular basis of disease and vastly improved diagnostic and treatment options.

Organelles can contribute to a disease state in several ways. First, the organelle itself may be dysfunctional either because it contains one or more defective biomolecules that impair function, or because it has been damaged by exposure to harmful substances such as chemicals, heavy metals, or oxygen radicals. Second, an organelle can, through its normal function, exacerbate damage occurring elsewhere in the cell. For example, as we have seen, misfolded proteins in the ER can trigger apoptosis, even in circumstances in which it is counterproductive. The subsections that follow describe diseases associated with the endomembrane system: the ER, Golgi apparatus, vesicular organelles, the nuclear envelope, and the plasma membrane.

THE ENDOPLASMIC RETICULUM. The ER plays such a central role in the synthesis of proteins and lipids that any disturbance in its function can have serious consequences. Misfolded proteins coded for by mutated genes and ER stress cause a vast number of diseases. Cystic fibrosis (CF) is a prominent example of a disease caused by misfolded proteins. CF is an ultimately fatal inherited disorder in which the lack of a specific type of plasma membrane chloride channel, the cystic fibrosis transmembrane regulator (CFTR), causes the accumulation of a thick mucus that compromises several organs, most notably the lungs and pancreas. The misfolded CFTR protein becomes trapped within the ER and is subsequently degraded. The structural and functional properties of CFTR are described in Chapter 11.

ER stress, induced by a variety of conditions such as protein aggregation, Ca^{2+} depletion, glucose deprivation, or fatty acid overload, can result in severe cell dysfunction or death. It is an important feature of such neurodegenerative conditions as Alzheimer's, Huntington's, and Parkinson's diseases, as well as heart disease and diabetes.

GOLGI APPARATUS. The most commonly recognized Golgi-linked diseases are a group of 15 congenital disorders of glycosylation (CDG). (The term *glycosylation* is used to describe the covalent linkage of carbohydrate groups to polypeptide or lipid molecules.) Caused by mutations in genes that encode glycosylation enzymes or glycosylation-linked transport proteins, a CDG is usually lethal by the age of 2. Symptoms include mental retardation, seizures, and liver disease.

NUCLEAR ENVELOPE. Many of the diseases attributed to defects in the nuclear envelope occur in the genes that code for lamin, a cytoskeletal component of the nuclear lamina, and emerin, an inner membrane protein. Examples include a variety of diseases of skeletal and cardiac muscle, neurons, and tendons. Progeria, a fatal childhood disease characterized by premature aging of the musculoskeletal and cardiovascular systems, has been linked to a specific mutation in the lamin A gene. One form of a rare hereditary muscular disease called Emery-Dreifuss muscular dystrophy is caused by the absence or mutation of the gene that codes for emerin. The cellular consequences of nuclear envelope deficits include a fragile nuclear membrane, altered regulation of DNA replication and transcription, and low tolerance to mechanical stress.

VESICULAR ORGANELLES. Diseases associated with vesicular organelles have been linked to lysosomes and peroxisomes. The lysosomal storage diseases (LSD) are a group of

BIOCHEMISTRY IN PERSPECTIVE cont

genetic disorders caused by the absence of one or more lysosomal enzymes. The resulting accumulation of undigested molecules causes irreversible cell damage. The lipid storage diseases Tay-Sachs and Gaucher's, as well as Pompe's disease (glycogen storage disease type II), are caused by the absence of a single enzyme. Death occurs in early childhood. In I-cell disease, the import of all lysosomal enzymes into lysosomes in certain organs is defective. In affected cells, the enzymes are instead secreted into the extracellular matrix. Symptoms include mental deterioration, heart disease, and respiratory failure.

PLASMA MEMBRANE. The plasma membrane occupies a pivotal position in the endomembrane system, as it is both the end point of the secretory pathway and the beginning of the endocytic pathway. Consequently, the PM plays important roles in a wide diversity of diseases. Diseases such as CF, diabetes, and familial hypercholesterolemia (inherited high blood cho-

lesterol levels) are directly caused by defective or missing membrane proteins. In a large number of infectious diseases, microorganisms invade body cells in endocytic processes initiated by binding to certain plasma membrane receptors. Examples of such organisms include bacteria such as *Listeria monocytogenes, Salmonella*, and *Shigella*, and some viruses (e.g., HIV). For viruses like HIV, which are covered in an "envelope" derived from host cell membrane, entry is gained when the virus binds to one or more PM receptors. Following fusion of the host cell membrane, and the viral envelope, the viral genome enters the host cell. Other diseases are caused when certain bacteria release toxins that injure cells. Once the toxin has become bound to a specific PM receptor on a target cell, either a pore is formed through which the toxic protein is transferred or endocytosis is triggered. Examples include cholera, pertussis (whooping cough), and diphtheria toxins.

SUMMARY: Biochemical analysis of organelles has resulted in significant progress in our understanding of the causes of many human diseases.

Cytoskeleton

Cytoplasm was once believed to be a structureless solution in which the nucleus was suspended. Research efforts over many years gradually revealed an intricate supportive network of fibers, filaments, and associated proteins called the **cytoskeleton** (Figure 2.29). Recent findings indicate that the cytoskeleton is far more important in cell function than had been realized. Components of the cytoskeleton include microtubules, microfilaments, and intermediate fibers.

Microtubules (diameter 25 nm), composed of the protein tubulin, are the largest constituent of the cytoskeleton. Tubulin is a dimer that consists of two polypeptides: α-tubulin and β-tubulin, a GTP-binding molecule. **Microtubules** are filaments formed by the reversible polymerization of tubulin dimers that assemble into straight, girderlike hollow tubes. Microtubules are polar; that is, their ends are different. At the plus ($+$) end, polymerization can occur rapidly. The minus ($-$) end grows more slowly. As the microtubule grows at the plus end, it extends toward the cell's periphery. Microtubule dynamics are regulated by microtubule-associated proteins (MAPs), a series of molecules that control microtubule stability. The ATP-dependent motor proteins kinesin and dynein move along microtubules. In general, kinesin moves cargo such as vesicles or organelles toward the plus end and dynein moves toward the minus end. Although found in many cellular regions, microtubules are most prominent in long, thin structures that require support (e.g., the extended axons and dendrites of nerve cells). They are also found in the *mitotic spindle* (the structure formed in dividing cells that is responsible for the equal dispersal of chromosomes into daughter cells) and the slender, hairlike organelles of locomotion known as cilia and flagella (Figure 2.30).

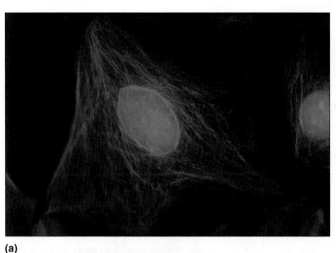

(a)

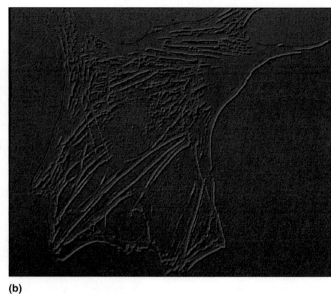

(b)

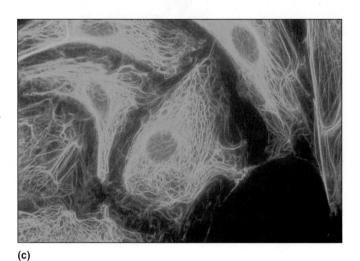

(c)

FIGURE 2.29

The Cytoskeleton

The major components of the cytoskeleton are (a) microtubules, (b) microfilaments, and (c) intermediate filaments (c). The intracellular distribution of each type of cytoskeletal component is visualized by staining with fluorescent dyes.

Microfilaments are small fibers (5–7 nm diameter) composed of polymers of globular actin (G-actin). The filamentous or polymeric form (F-actin) exists as a coil of two actin polymers with a plus end and a minus end. Polymerization, driven by ATP hydrolysis, occurs more rapidly at the plus end. The individual filaments, being highly flexible, are usually cross-linked into bundles of different sizes. A large variety of actin-binding proteins regulate the structure and functional properties of microfilaments. They cross-link, stabilize, sever (cut into fragments), or cap (block polymerization) microfilaments. Microfilaments can exert force simply by polymerizing or depolymerizing. Together with the myosins, a large family of ATP-dependent motor proteins, microfilaments generate contractile forces that create tension. Important roles of microfilaments include involvement in cytoplasmic streaming (a process that is most easily observed in plant cells in which cytoplasmic currents rapidly displace organelles such as chloroplasts), ameboid movement (a type of locomotion created by the formation of temporary cytoplasmic protrusions), and muscle contraction.

Intermediate filaments (8–12 nm in diameter) are a large group of flexible, strong, and relatively stable polymers. They provide cells with significant

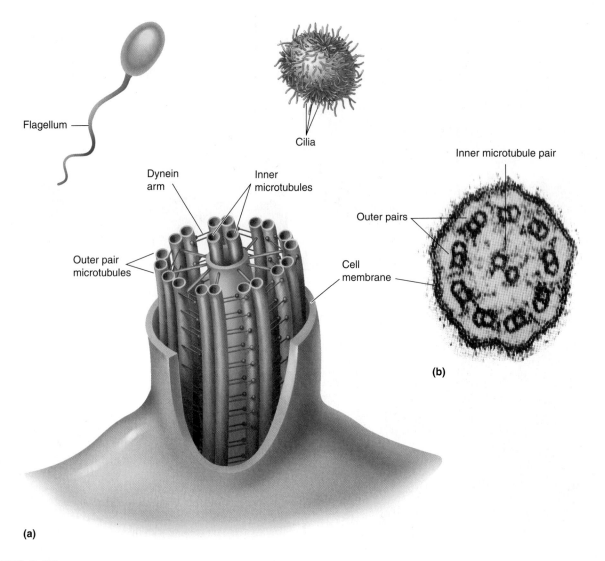

(a)

(b)

FIGURE 2.30

Cilia and Flagella

(a) The microtubules of eukaryotic cells are arranged in the classic 9+2 pattern. Two central microtubules are surrounded by an outer ring of nine pairs of microtubules. The undulating (snakelike) motion of flagella and the beating of cilia are created by the controlled ATP-driven attachment and the detachment of dynein of one microtubule outer pair to an adjacent pair. This "walking" motion is converted to bending because all the outer pairs are attached to the inner microtubule pair by the radial spokes. (b) Transmission electron micrograph of a cross section of a flagellum.

mechanical support. A network of intermediate filaments (IF) extends from a ringlike meshwork around the nucleus to attachment points on the plasma membrane. There are six classes of IF proteins, which differ in their amino acid sequences. Well-known examples are the keratins found in skin and hair cells, and the lamins that reinforce the nuclear envelope. Despite this diversity, each IF type consists of a rodlike domain flanked by globular head and tail domains. IF polypeptides assemble into dimers (two polypeptides), tetramers (four polypeptides), and higher-order structures. Involved primarily in the maintenance of cell shape, IFs are especially prominent in cells that are subjected to mechanical stress.

The cytoskeleton, a dynamic mechanical system, is an integral feature of most cell activities. The unique functional properties of the cytoskeleton are made possible by a balance of mechanical forces between compression-resistant microtubules and tension generated by contractile microfilaments. IFs connect microtubules and microfilaments to each other and to the nucleus and plasma membrane. As a result of this functional "cytoarchitecture," opposing forces are continuously equilibrated throughout all cytoskeletal elements (Figure 2.31). Living cells are, therefore, in a constant state of dynamic instability. It is noteworthy that cytoskeletal reorganization, triggered by a vast array of chemical and physical signals, is a principal feature of most cellular processes.

Among the most important functions made possible by the properties of the cytoskeleton are the following.

1. **Cell shape**. Eukaryotic cells come in a vast variety of shapes including the bloblike amoeba, columnar epithelial cells, and neurons with complex branching architecture. Changes in cell shape result from responses to external signals. Amoebas, for example, rapidly change shape as they move closer to a source of nutrient molecules. During embryonic and fetal development, animal cells acquire their differentiated shapes in response to a progressive and complicated series of signal molecules secreted by nearby cells.

2. **Large- and small-scale cell movement**. Large-scale cellular movements such as cytoplasmic streaming and ameboid movement are made possible

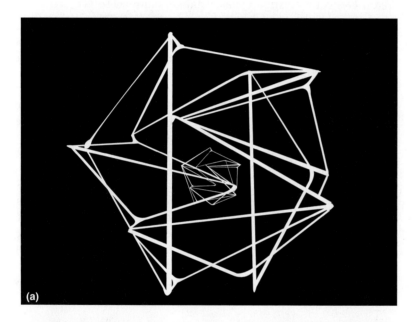
(a)

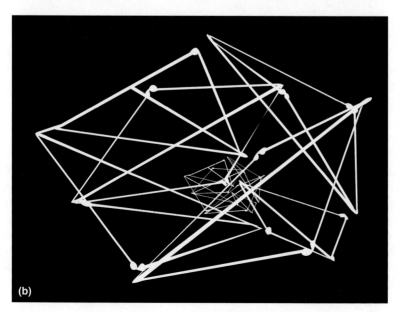
(b)

FIGURE 2.31

Cytoskeletal Reorganization Model

Both these stable structures are held together by balanced mechanical stresses, namely, tensional strings and rigid struts. The "cell" here is composed of aluminum struts and thin elastic cord; the "nucleus," a geodesic sphere, is constructed of wooden sticks and white elastic thread. When an external force is applied to structure (a), it rearranges to form structure (b).

by a dynamic cytoskeleton that can rapidly assemble and disassemble its structural elements according to the cell's immediate needs. Organelles are moved around within cells by being attached to cytoskeletal structures. For example, after cell division, the extension of the endoplasmic reticulum membrane from the newly formed nuclear membrane out to the cell's periphery and the re-formation of the Golgi complex is accomplished by attachment to microtubules. Movement occurs as specific motor proteins linked to microtubules and to the membrane cargo undergo ATP hydrolysis–dependent conformational changes.

3. **Solid state biochemistry**. It is now generally recognized that many of the biochemical reactions previously believed to occur within the liquid phase of the cytoplasm proceed in large measure on a cytoskeletal platform. Biochemical pathways are both more efficient and more easily controlled when enzymes assemble into complexes on a solid surface. Prominent examples are the reactions of glycolysis, an ATP-generating pathway in carbohydrate metabolism. The binding of glycolytic enzymes to cytoskeletal filaments has been observed to vastly increase reaction rates. Drugs that disrupt cytoskeletal structure cause glycolytic enzyme detachment and a rapid decrease in cytoplasmic ATP production.

4. **Signal transduction**. Cells are information-processing systems, and the cytoskeleton provides the structural continuity for signal transduction mechanisms. Signal cascade proteins, from cell surface receptors to target molecules throughout the cytoplasm and within the nucleus, can transmit information because they are either immobilized or transiently bound to cytoskeletal filaments. Several types of accessory protein enhance the sophistication, speed, and precision of information processing. In response to specific signals, *adaptor* and *anchor* proteins facilitate the recruitment and the assembly of specific sets of signal cascade proteins into complexes bound to the cytoskeleton. The information- processing system of cells resembles the *integrated circuits* (microchips) in computers: information-processing devices composed of transistors and capacitors, connected by wires, and driven by electricity. In living cells a vast number of components (signal complexes, biochemical pathways, and gene expression devices) are connected by cytoskeletal filaments. Information flow within cells occurs as the result of sequential protein structure changes triggered by protein-protein interactions.

KEY CONCEPT

The cytoskeleton, a highly structured network of proteinaceous filaments, is responsible for maintenance of cell shape, large- and small-scale cell movement, solid state biochemistry, and signal transduction.

QUESTION 2.5

Cancer is a group of diseases characterized by unregulated cell division. Taxol, a drug used to treat ovarian cancer, attaches to and stabilizes microtubules. Briefly, what is the basis of taxol's anticancer action?

Ribosomes

The cytoplasmic **ribosomes** of eukaryotes are RNA/protein complexes (20 nm in diameter), whose function is the biosynthesis of proteins. Composed of a variety of proteins and ribosomal RNA, these complex structures contain two irregularly shaped subunits of unequal size (Figure 2.32). The subunits come together to form whole ribosomes when protein synthesis is initiated; when not in use, the ribosomal subunits separate. In any cell, the number and distribution of ribosomes depend on the relative metabolic activity and the proteins being synthesized. Although eukaryotic ribosomes are larger and more complex than those of prokaryotes, they are similar in overall shape and function.

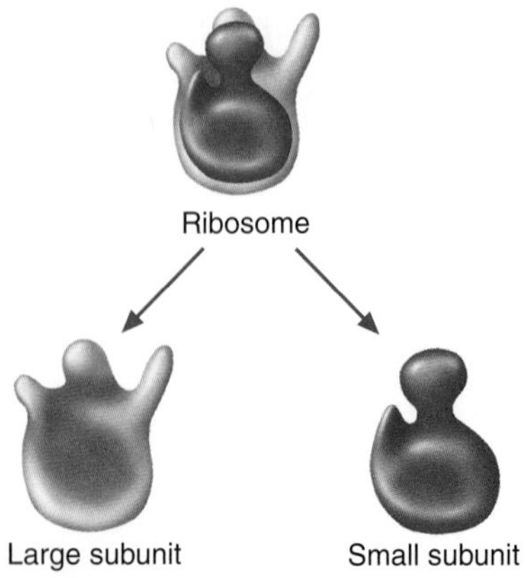

Ribosome

Large subunit   Small subunit

FIGURE 2.32

The Eukaryotic Ribosome

BIOCHEMISTRY **IN THE LAB**

Cell Technology

During the past 50 years, our understanding of the functioning of living organisms has undergone a revolution. Much of our current knowledge of biochemical processes is due directly to technological innovations. Three of the most important cellular techniques used in biochemical research are briefly described: cell fractionation, electron microscopy, and autoradiography.

Cell Fractionation

Cell fractionation techniques (Figure 2A) allow the study of cell organelles in a relatively intact form outside of cells. For example, functioning mitochondria can be used to study cellular energy generation. In these techniques, cells are gently disrupted and separated into several organelle-containing fractions. Cells may be disrupted by several methods, but homogenization is the most commonly used. In this process a cell suspension is placed either in a glass tube fitted with a specially designed glass pestle or into an electric blender. The resulting homogenate is separated into several fractions by means of **differential centrifugation**. In this procedure, a refrigerated instrument, the *ultracentrifuge*, generates enormous centrifugal forces that separate cell components on the basis of size, surface area, and relative density. (Forces as large as 500,000 times the force of gravity, or 500,000 $\times$ g, can be generated in unbreakable test tubes placed in the rotor of an ultracentrifuge.) Initially, the homogenate is spun in the ultracentrifuge at low speed (700–1000 g) for 10 to 20 minutes. The heavier particles, such as the nuclei, form a sediment, or pellet. Lighter particles, such as mitochondria and lysosomes, remain suspended in the *supernatant*, the liquid above the pellet. The supernatant is then transferred to another centrifuge tube and spun at a higher speed (15,000–20,000 g) for 10 to 20 minutes. The resulting pellet contains mitochondria, lysosomes, and peroxisomes. The supernatant, which contains **microsomes** (small closed vesicles formed from ER during homogenization), is transferred to another tube and spun at 100,000 g for 1 to 2 hours. Microsomes are deposited in the pellet, and the supernatant contains ribosomes, various cellular membranes, and granules such as glycogen, a carbohydrate polymer. After this latest supernatant has been recentrifuged at 200,000 g for 2 to 3 hours, ribosomes and large macromolecules are recovered from the pellet.

Often, the organelle fractions obtained with this technique are not sufficiently pure for research purposes. One method that is often employed to further purify cell fractions is **density-gradient centrifugation** (Figure 2B). In this procedure the

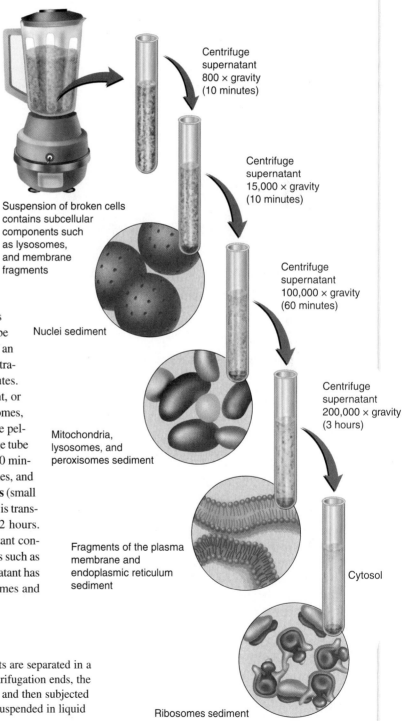

Suspension of broken cells contains subcellular components such as lysosomes, and membrane fragments

Nuclei sediment

Centrifuge supernatant 800 × gravity (10 minutes)

Centrifuge supernatant 15,000 × gravity (10 minutes)

Centrifuge supernatant 100,000 × gravity (60 minutes)

Centrifuge supernatant 200,000 × gravity (3 hours)

Mitochondria, lysosomes, and peroxisomes sediment

Fragments of the plasma membrane and endoplasmic reticulum sediment

Cytosol

Ribosomes sediment

FIGURE 2A

Cell Fractionation

After homogenization of cells in a blender, cell components are separated in a series of centrifugations at increasing speeds. As each centrifugation ends, the supernatant is removed, placed into a new centrifuge tube, and then subjected to greater centrifugal force. The collected pellet can be resuspended in liquid and examined by microscopy or biochemical tests.

▶▶

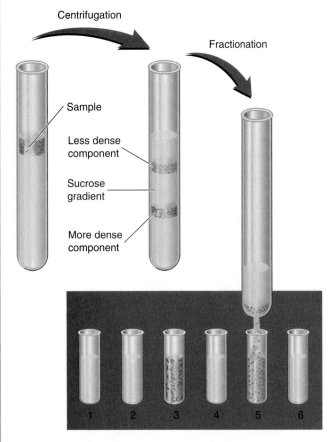

FIGURE 2B

Density-Gradient Centrifugation

The sample is gently layered onto the top of a preformed gradient of an inert substance such as sucrose. As centrifugal force is applied, particles in the sample migrate through the gradient bands according to their densities. After centrifugation, the bottom of the tube is punctured and the individual bands are collected in separate tubes.

fraction of interest is layered on top of a solution that consists of a dense substance such as sucrose. (In the centrifuge tube containing the solution, the sucrose concentration increases from the top to the bottom.) During centrifugation at high speed for several hours, particles move downward in the gradient until they reach a level that has a density equal to their own. Then the plastic centrifuge tube is punctured, and the cell components are collected in drops from the bottom. The purity of the individual fractions can be assessed by visual inspection (electron microscopy). However, assays for **marker enzymes** (enzymes that are known to be present in especially high concentration in specific organelles) are more commonly used. For example, glucose-6-phosphatase, the enzyme responsible for converting glucose-6-phosphate to glucose in the liver, is a marker for liver microsomes. Likewise, DNA polymerase, which is involved in DNA synthesis, is a marker for nuclei.

Electron Microscopy

The electron microscope (EM) permits a view of cell ultrastructure not possible with the more common light microscope. Direct magnifications as high as $1,000,000\times$ have been obtained with the EM. Electron micrographs may be enlarged photographically to $10,000,000\times$. The light microscope, in contrast, magnifies an image to about $1,000\times$. This difference is due to the greater resolving power of the EM. With the light microscope, the limit of resolution, defined as the minimum distance between two points that allows for their discrimination as two separate points, is $0.2~\mu m$. The limit of resolution for the EM is approximately 0.5 nm. The lower resolving power of the light microscope is related to the wavelength of visible light. In general, shorter wavelengths allow greater resolution. The EM uses a stream of electrons instead of light to illuminate specimens. Because the electron stream has a much shorter wavelength than visible light, more detailed images can be obtained.

There are two types of EM: the transmission electron microscope (TEM) and the scanning electron microscope (SEM). Like the light microscope, the TEM is used for viewing thin specimens. Because the image in the TEM depends on variations in the absorption of electrons by the specimen (rather than on variations in light absorption), heavy metals such as osmium or uranium are used to increase contrast among cell components. The SEM is used to obtain three-dimensional views of cellular structure. Unlike the TEM, which uses electrons that have passed through a specimen to form an image, the SEM uses electrons that are emitted from the specimen's surface. The specimen is coated with a thin layer of heavy metal and then scanned with a narrow stream of electrons. The electrons emitted from the specimen's surface, sometimes referred to as *secondary electrons*, form an image on a television screen. Although only surface features can be examined with the SEM, this form of microscopy provides very useful information about cell structure and function.

Autoradiography

Autoradiography is used to study the intracellular location and behavior of cellular components. It has been an invaluable tool in biochemistry. For example, it was used to determine the precise sites of DNA, RNA, and protein synthesis within eukaryotic cells. In this procedure, living cells are briefly exposed to radioactively labeled precursor molecules. The radioisotope tritium (^{3}H) is most commonly used. The tritiated nucleotide thymidine, for example, is used to study DNA synthesis because thymidine is incorporated only into DNA molecules. After exposure to the radioactive precursor, the cells are processed for light or electron microscopy. The resulting slides are then dipped in photographic emulsion. After storage in the dark, the emulsion is developed by standard photographic techniques. The location of radioactively labeled molecules is indicated by the developed pattern of silver grains. ∎

Chapter**Summary**

1. Cells are the structural units of all living organisms. Within each living cell are hundreds of millions of densely packed biomolecules. Insights into their structure and functional behavior have been gained through biochemical research. Among the most significant of these are the following. The unique chemical and physical properties of water are a crucial determining factor in the behavior of all other bio-molecules. Biological membranes are thin, flexible, and relatively stable sheetlike structures that enclose cells and organelles. They are formed from biomolecules such as phospholipids and proteins that together form a selective physical barrier. Membranes also serve as a chemically reactive surface that is integrated into all living processes within the organism. Self-assembly of supramolecular struc-tures occurs within living cells because of the steric infor-mation encoded into the intricate shapes of biomolecules that allows numerous weak, noncovalent interactions between complementary surfaces. Many of the multisubunit complexes involved in cellular processes are now known to function as molecular machines; that is, they are mechanical devices composed of moving parts that convert energy into directed motion. Macromolecular crowding, created by the density of proteins within the cell, is an important factor in the wide variety of cellular phenomena. Signal transduc-tion mechanisms allow cells to process internal and external information.

2. All currently existing organisms contain either prokaryotic or eukaryotic cells. Prokaryotes are simpler in structure than eukaryotes. They also have a vast biochemical diversity across species lines, because almost any organic molecule can be used as a food source by some species of prokaryote. Unlike the prokaryotes, the eukaryotes carry out their meta-bolic functions in membrane-bound compartments called organelles.

3. Although the prokaryotic cell lacks a nucleus, a circular DNA molecule called a chromosome is located in an irregu-larly shaped region called the nucleoid. Many bacteria con-tain additional small circular DNA molecules called plasmids. Plasmids may carry genes for special function proteins that provide protection, metabolic specialization, or reproductive advantages to the organism.

4. The plasma membrane of both prokaryotes and eukaryotes performs several vital functions. The most important of these is controlled molecular transport, which is facilitated by car-rier and channel proteins.

5. The endoplasmic reticulum (ER) is a system of intercon-nected membranous tubules, vesicles, and large flattened sacs found in eukaryotic cells. There are two forms of ER. The rough ER, which is primarily involved in protein syn-thesis, is so named because of the numerous ribosomes that stud its cytoplasmic surface. The second form lacks attached ribosomes and is called smooth ER. Functions of the smooth ER include lipid synthesis and biotransformation.

6. Formed from relatively large, flattened, saclike membranous vesicles that resemble a stack of plates, the Golgi apparatus is involved in the modification, packaging, and release of cell products into the vesicular compartment for delivery to target locations in the cell.

7. The nucleus of any eukaryote contains DNA, the cell's genetic information. Ribosomal RNA is synthesized in the nucleolus, found within the nucleus. Separating DNA repli-cation and transcription processes from the cytoplasm is the nuclear envelope; it is composed of two membranes that fuse at structures called the nuclear pores.

8. The cell contains a system of vesicular organelles involved in processing both endogenous and exogenous materials around, into, and out of the cell and performing specialized biochemical functions. Lysosomes, secretory vesicles, plant vacuoles, and glyoxysomes are vesicular organelles.

9. Aerobic respiration, a process by which cells use oxygen to generate energy, takes place in mitochondria. Each mito-chondrion is bounded by two membranes. The smooth outer membrane is permeable to most molecules with masses less than 10,000 D. The inner membrane, which is impermeable to ions and a variety of organic molecules, projects inward into folds that are called cristae. Embedded in this mem-brane are respiratory assemblies, structures that are responsi-ble for the synthesis of ATP.

10. Peroxisomes are small spherical membranous organelles that contain a variety of oxidative enzymes. These organelles are most noted for their involvement in the generation and breakdown of peroxides. Plant peroxisomes mobilize stored lipid to supply energy during seed germination.

11. Plastids, structures that are found only in plants, algae, and some protists, are bounded by a double membrane. Another separate intricately arranged internal membrane is also often present. Chromoplasts accumulate the pigments that are responsible for the color of leaves, flower petals, and fruits. Chloroplasts are a type of chromoplast that are specialized to convert light energy into chemical energy.

12. The cytoskeleton, a supportive network of fibers and fila-ments, is involved in the maintenance of cell shape, large- and small-scale cellular movement, solid state biochemistry, and signal transduction.

13. The ribosomes are large two-subunit rRNA/protein com-plexes that are responsible for protein synthesis. The assem-bly of the subunits around an mRNA molecule initiates protein synthesis either free in the cytoplasm or bound to the ER, according to the destination of the finished protein.

Suggested Readings

Alberts, B., Bray, D., Hopkin, K., Johnson, A., Lewis, J., Raff, M., Roberts, K., and Walter, P., *Essential Cell Biology*, 2nd ed., Garland Science, New York, 2004.

Becker, W. M., Kleinsmith, L. J., and Hardin, J., *The World of the Cell*, 6th ed., Benjamin Cummings, San Francisco, 2006.

Campbell, N. A., and Reece, J. B., *Biology*, 7th ed., Benjamin Cummings, San Francisco, 2005.

Carballido-Lopez, R., The Bacterial Actin-like Cytoskeleton, *Microbiol. Mol. Biol. Rev.* 70(4):888–909, 2006.

Davis, P., *The Fifth Miracle: The Search for the Origin and Meaning of Life*, Simon & Schuster, New York, 1999.

deDuve, C., The Birth of Complex Cells, *Sci. Am.* 274(4):50–57, 1996.

Goodsell, D. S., *Bionanotechnology: Lessons from Nature*, Wiley-Liss, Hoboken, New Jersey, 2004.

Goodsell, D. S. *The Machinery of Life*, Springer-Verlag, New York, 1998.

Harrison, J. J., Turner, R. J., Marques, L. L. R., and Ceri, H., Biofilms, *Am. Sci.* 93(6):508–515, 2005.

Ingber, D. E., The Architecture of Life, *Sci. Am.* 278(1):48–57, 1998.

Lane, N., *Power, Sex, Suicide: Mitochondria and the Meaning of Life*, Oxford University Press, New York, 2005.

Margulis, L., *What Is Life?* University of California Press, Berkeley, 2000.

Porter, R., *The Greatest Benefit to Mankind: A Medical History of Humanity*, W. W. Norton, New York, 1997.

Key Words

aerobic metabolism, *56*

anchor protein, *35*

apoptosis, *56*

autophagy, *22*

biotransformation, *49*

carrier protein, *35*

cell cortex, *46*

cell fractionation, *67*

channel protein, *35*

chloroplast, *59*

chromatin fiber, *51*

chromoplast, *59*

chromosome, *42*

cytoskeleton, *62*

density-gradient centrifugation, *68*

dictyosome, *49*

differential centrifugation, *67*

endocytosis, *53*

endomembrane system, *44*

endoplasmic reticulum (ER), *46*

exocytosis, *51*

extracellular matrix, *46*

glycocalyx, *46*

glyoxysome, *58*

Golgi apparatus (Golgi complex), *49*

granum, *59*

hormone, *38*

hydrophilic, *34*

hydrophobic, *34*

inner membrane, *52*

integral protein, *35*

intermediate fiber, *62*

ligand, *38*

limit of resolution, *68*

lysosome, *55*

marker enzyme, *68*

membrane potential, *49*

microfilament, *63*

microsome, *67*

microtubule, *62*

mitochondrion, *56*

motor protein, *36*

neurotransmitter, *37*

nuclear envelope, *52*

nuclear matrix, *51*

nuclear pore, *52*

nucleoid, *42*

nucleolus, *52*

nucleoplasm, *51*

nucleus, *51*

outer membrane, *56*

perinuclear space, *52*

peripheral protein, *35*

peroxisome, *58*

phagocytosis, *63*

photosynthesis, *41*

plasma membrane, *41*

plasmid, *42*

plastid, *59*

receptor, *35*

respiration, *41*

ribosome, *66*

rough ER (RER), *48*

signaling cascade, *66*

signal transduction, *37*

smooth ER (SER), *48*

stroma, *59*

thylakoid membrane, *59*

transmembrane protein, *40*

vesicles, *44*

vesicular organelles, *53*

Review Questions

These questions are designed to test your knowledge of the key concepts discussed in this chapter, before moving on to the next chapter. You may like to compare your answers to the solutions provided in the back of the book and in the accompanying Study Guide.

1. Define the term *cell*.

2. Draw a diagram of a segment of biological membrane, indicating the positions of integral and peripheral proteins.

3. Draw a diagram of a bacterial cell. Label and explain the function of each of the following components:
   a. nucleoid
   b. plasmid
   c. cell wall
   d. pili
   e. flagella

4. What are the functions of channel proteins, carrier proteins, receptors, and anchor proteins?

5. The outer boundary of most eukaryotic cells is a cell membrane, whereas the outer boundary of a prokaryotic cell is a cell wall. How do these structures differ in function?

6. Indicate whether the following structures are present in prokaryotic or eukaryotic cells:
   a. nucleus
   b. plasma membrane
   c. endoplasmic reticulum
   d. mitochondria

e. nucleolus

f. cytoskeleton

7. Describe how noncovalent interactions promote the self-assembly of supramolecular structures in living organisms.

8. Briefly define the following terms:
   a. exocytosis
   b. biotransformation
   c. grana
   d. supramolecular
   e. self-assembly

9. Describe the three phases of signal transduction in living organisms.

10. What are the components of the endomembrane system? How are these components functionally connected?

11. Outline the role of cytoskeleton in intracellular signal transduction.

12. How do lysosomes participate in the life of a cell?

13. Plastids, structures found only in _____, are of two types. These are _____, which are used to store starch and protein, and _____, which accumulate pigments.

14. List and describe six diseases linked to organelles.

15. What functions does the cytoskeleton perform in living cells?

16. Define the following terms:
    a. biofilm
    b. vesicle
    c. extracellular matrix
    d. sarcoplasmic reticulum
    e. thylakoid

17. Briefly outline how cell technology has contributed to modern medicine. Provide three specific examples.

18. Some eukaryotes possess a cell wall while others do not. In each condition, specific advantages are gained. Explain.

19. What are the two essential functions of the nucleus?

20. What roles do plasma membrane proteins play in cells?

21. Name the two forms of endoplasmic reticulum. What functions do they serve in the cell?

22. Describe the functions of the Golgi apparatus.

23. What evidence is there that all cells have a common ancestor?

24. Define the following terms:
    a. nucleoplasm
    b. nuclear matrix
    c. autophagy
    d. apoptosis
    e. differential centrifugation

25. Eukaryotic cells possess a system of vesicular organelles. Name and describe three specific examples.

26. Peroxisomes are not included in the population of vesicular organelles because of their specialized function and unique biogenesis. Describe the function of this organelle and explain how it is formed.

27. Describe the function of motor proteins in molecular machine protein complexes.

28. Define the following terms:
    a. membrane potential
    b. transmembrane protein
    c. peripheral protein
    d. receptor protein
    e. anchor protein

29. Define the following terms:
    a. proplastids
    b. motor protein
    c. hydrophobic
    d. hydrophilic

30. Describe the intracellular and extracellular structures that protect the eukaryotic plasma membrane.

31. Define the following terms:
    a. signal transduction
    b. neurotransmitter
    c. hormone
    d. ligand
    e. endotoxin

32. Describe the functions of the smooth endoplasmic reticulum in hepatocytes and muscle cells.

33. Describe the structure and functional properties of the nuclear envelope.

34. Define the following terms:
    a. endomembrane system
    b. extracellular matrix
    c. excluded volume
    d. solid state biochemistry
    e. cell cortex

35. Describe the functions of secretory lysosomes.

Thought Questions

These questions are designed to reinforce your understanding of all of the key concepts discussed in the book so far, including this chapter and the chapter before it. They may not have one right answer! The authors have provided possible solutions to these questions in the back of the book and in the accompanying Study Guide, for your reference.

36. Several pathogenic bacteria (e.g., *Bacillus anthracis*, the cause of anthrax) produce an outermost mucoid layer called a capsule. Capsules may be composed of polysaccharide or protein. What effect do you think this "coat" would have on a bacterium's interactions with a host animal's immune system?

37. Eukaryotic cells are more highly specialized than prokaryotic cells. Can you suggest some advantages and disadvantages of specialization?

38. In addition to providing support, the cytoskeleton immobilizes enzymes and organelles in the cytoplasm. What advantage does this immobilization have over allowing the cell contents to freely diffuse in the cytoplasm?

39. Familial hypercholesterolemia (FH) is an inherited disease characterized by high blood levels of cholesterol, xanthomas (lipid-laden nodules that develop under the skin near tendons), and early-onset atherosclerosis (the formation of yellowish plaques within arteries). In the milder form of this

disease, patients have half the plasma membrane low-density lipoprotein (LDL) receptors needed for cells to bind to and internalize LDL (a plasma lipoprotein particle that transports cholesterol and other lipids to tissues). These patients have their first heart attacks in young adulthood. In the severe form of FH, in which patients have no functional LDL receptors, heart attacks begin at about age 8, with death occurring a few years later. Based on what you have learned in this chapter, briefly describe the cellular processes that are defective in FH.

40. Diphtheria is a potentially fatal infectious disease. It was once one of the most dreaded childhood diseases: epidemics in colonial America had fatality rates as high as 40%. Diphtheria is now relatively rare, however, because of childhood immunization. *Corynebacterium diphtheriae* infection begins with a mild sore throat and then progresses to a stage in which an impenetrable membrane, formed by dead cell tissue, clotted blood, and bacterial cells, obstructs the airway. Patients die from slow suffocation. A diphtheria exotoxin is synthesized from a gene integrated into the bacterial chromosome following infection by a specific bacterial virus. The toxin is released by the bacterium into the respiratory tissue of the patient. The patient's cells possess a receptor for the B-subunit of the A/B toxin dimer. The A subunit, when free of the B subunit, interrupts protein synthesis in the patient's cells. Track the progress of the exotoxin from the cell surface receptor to the ribosomes of the patient's cell. What overall effect would the toxin have on cell function?

41. A particular organelle found in a eukaryote is thought to have arisen from a free-living organism. The finding of what type of molecule in the organelle would strongly support this hypothesis?

42. Mycoplasmas are unusual bacteria that lack cell walls. With a diameter of 0.3 μm, they are believed to be the smallest known free-living organisms. Some species are pathogenic to humans. For example, *Mycoplasma pneumoniae* causes a very serious form of pneumonia. Assuming that mycoplasmas are spherical, calculate the volume of an individual cell. Compare the volume of a mycoplasma with that of *E. coli*.

43. The dimensions of prokaryotic ribosomes are approximately 14 nm by 20 nm. If ribosomes occupy 20% of the volume of a bacterial cell, calculate how many ribosomes are in a typical cell such as *E. coli*. Assume that the shape of a ribosome is approximately that of a cylinder.

44. The *E. coli* cell is 2 μm long and 1 μm in diameter, while a typical eukaryotic cell is 20 μm in diameter. Assuming that the *E. coli* cell is a perfect cylinder and the eukaryotic cell is a perfect sphere, calculate the surface-to-volume ratio for each cell type [cylinder volume, $V = \pi r^2 h$; cylinder area $A = 2\pi r^2 + 2\pi rh$; sphere volume, $V = 4/3(\pi r^3)$; sphere area, $A = 4\pi r^2$]. What do these numbers tell you about the evolutionary changes that would have to occur to generate an efficient eukaryotic cell, considering that most biochemical processes depend on membrane-bound transport processes?

45. The narcotic morphine is a prominent alkaloid, a member of a vast array of nitrogen-containing organic molecules produced primarily by plants. Although the purpose for alkaloid synthesis remains unknown, it is suspected that many of these compounds, especially those with physiological effects on animals, may have a role in defense against predators. Can you suggest a reason why morphine, derived from the oriental poppy, can have such a profound effect on the nervous systems of animals? [*Hint*: Review the concepts of signal transduction.]

Water: The Matrix of Life

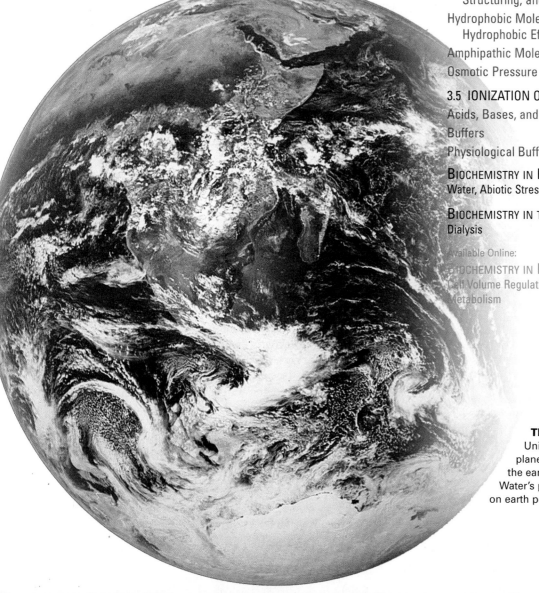

The Water Planet
Unique among the
planets in the solar system,
the earth is an oceanic world.
Water's properties make life
on earth possible.

Overview

EARTH IS UNIQUE AMONG THE PLANETS IN OUR SOLAR SYSTEM PRIMARILY BECAUSE OF ITS VAST OCEANS OF WATER. OVER BILLIONS OF YEARS, WATER was produced during high-temperature interactions between atmospheric hydrocarbons and the silicate and iron oxides in the earth's mantle. Moisture reached the planet's surface as steam emitted during volcanic eruptions. Oceans formed as the steam condensed and fell back to Earth as rain. This first rain may have lasted more than 60,000 years.

Over millions of years, water has profoundly affected our planet. Whether falling as rain or flowing in rivers, water has eroded the hardest rocks and transformed the mountains and continents. Many scientists today believe that life arose in a *primordial pudding* of clay and water. Shallow clay pools can promote the synthesis of macromolecules and accumulate the building blocks of life. It is not an accident that life arose in association with water, for this substance has several unusual properties that suit it to be the matrix of life. Among these are its thermal properties and unusual solvent characteristics. Water's properties are directly related to its molecular structure.

Why is water so vital for life? Water's chemical stability, its remarkable solvent properties, and its role as a biochemical reactant have long been recognized. What has not been widely appreciated is the critical role that *hydration* (the noncovalent interaction of water molecules with solutes) plays in the architecture, stability, and functional dynamics of macromolecules such as proteins and nucleic acids. Water is now known to be an indispensable component of biological processes as diverse as protein folding and biomolecular recognition in signal transduction mechanisms, the self-assembly of supramolecular structures such as ribosomes, and gene expression. Understanding how essential water is in living processes requires a review of its molecular structure and the physical and chemical properties that are the consequences of that structure.

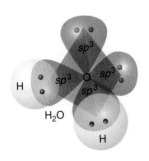

FIGURE 3.1

Tetrahedral Structure of Water

In water, two of the four sp^3 orbitals of oxygen are occupied by two lone pairs of electrons. Each of the other two half-filled sp^3 orbitals is filled by the addition of an electron from hydrogen.

3.1 MOLECULAR STRUCTURE OF WATER

The water molecule (H_2O) is composed of two atoms of hydrogen and one of oxygen. Water has a tetrahedral geometry because its oxygen atom is sp^3 hybridized and at the center of the tetrahedron is the oxygen atom. Two of the corners are occupied by hydrogen atoms, each of which is linked to the oxygen atom by a single covalent bond (Figure 3.1). The other two corners are occupied by the unshared electron pairs of the oxygen. Oxygen is more electronegative than hydrogen (i.e., oxygen has a greater capacity to attract electrons when bonded to hydrogen). Consequently, the larger oxygen atom bears a partial negative charge (δ^-) and each of the two hydrogen atoms bears a partial positive charge (δ^+) (Figure 3.2). The electron distribution in oxygen–hydrogen bonds is displaced toward the oxygen and, therefore, the bond is **polar**. If water molecules were linear, like those of carbon dioxide (O=C=O), then the bond polarities would balance each other and water would be nonpolar. However, water molecules are bent; the bond angle is 104.5°, slightly less than the symmetrical tetrahedral angle of 109°. This is because

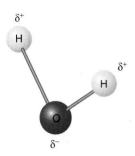

FIGURE 3.2

Charges on a Water Molecule

The two hydrogen atoms in each molecule carry partial positive charges. The oxygen atom carries a partial negative charge.

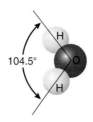

FIGURE 3.3

Space-Filling Model of a Water Molecule

Because the water molecule has a bent geometry, the distribution of charge within the molecule is asymmetric. Water is therefore polar.

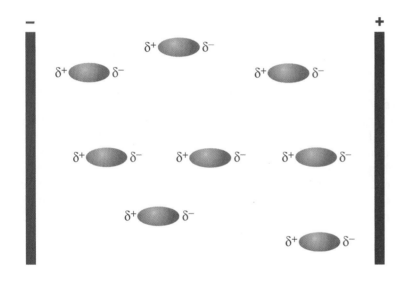

FIGURE 3.4

Molecular Dipoles in an Electric Field

When polar molecules are placed between charged plates, they line up in opposition to the field.

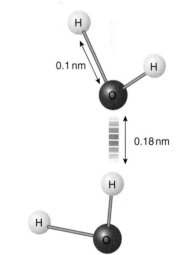

FIGURE 3.5

The Hydrogen Bond

A hydrogen bond results when the electronegative oxygen atoms of two water molecules compete for the same electron-deficient hydrogen atom. The hydrogen bond is represented by short parallel lines designating the weak covalent character and directionality of the bond.

the lone pair electrons occupy more space than the bonding electron pairs of the O—H bonds (Figure 3.3).

Molecules such as water, in which charge is separated, are called **dipoles**. When molecular dipoles are subjected to an electric field, they orient themselves in the direction opposite to that of the field (Figure 3.4).

The electron-deficient hydrogens of one water molecule are attracted to the unshared pairs of electrons of another water molecule because of the large difference in electronegativity of hydrogen and oxygen. (Hydrogens attached to nitrogen and fluorine behave the same way.) In this interaction, called a **hydrogen bond** (Figure 3.5), the hydrogen is unequally shared by the two electronegative centers: oxygen nuclei in the case of a pair of water molecules. The bond has both electrostatic (ionic) and covalent character. **Electrostatic interactions** occur between any two opposite partial charges (polar molecules) or full charges (ions or charged molecules). **Covalent bonds** involve electron sharing with orbital overlap or mixing. Covalent character confers directionality to the bond or interaction, unlike the uniformly spherical force field around an ion.

3.2 NONCOVALENT BONDING

Noncovalent interactions are usually electrostatic; that is, they occur between the positive nucleus of one atom and the negative electron clouds of another nearby atom. Unlike the stronger covalent bonds, individual noncovalent interactions are relatively weak and are therefore easily disrupted (Table 3.1). Nevertheless, they play a vital role in determining the physical and chemical properties of water and the structure and function of biomolecules because the cumulative effect of many weak interactions can be considerable. Large numbers of noncovalent interactions stabilize macromolecules and supramolecular structures, whereas the capacity of these bonds to be rapidly formed and broken endows biomolecules with the flexibility required for the rapid flow of information that occurs in dynamic living processes. In living organisms, the most important noncovalent interactions include ionic interactions, van der Waals interactions, and hydrogen bonds.

Ionic Interactions

The ionic interactions that occur between charged atoms or groups are nondirected (i.e., they are felt uniformly in space around the center of charge). Oppositely charged ions such as sodium (Na^+) and chloride (Cl^-) are attracted to each other. In contrast, ions with like charges, such as Na^+ and K^+ (potassium), repel each other. In proteins, certain amino acid side chains contain ionizable groups. For example, the side chain of the amino acid glutamic acid ionizes at physiological pH as $-CH_2CH_2COO^-$. The side chain group of the amino acid lysine ($-CH_2CH_2CH_2CH_2-NH_2$) ionizes as $-CH_2CH_2CH_2CH_2NH_3^+$ at physiological pH. The attraction of positively and negatively charged amino acid side chains forms **salt bridges** ($-COO^{-+}H_3N-$), and the repulsive forces created when similarly charged species come into close proximity are an important feature in many biological processes, such as protein folding, enzyme catalysis, and molecular recognition. It should be noted that stable salt bridges rarely form between biomolecules in the presence of water; this is because the hydration of ions is preferred, and the attraction between the biomolecules decreases significantly. Most salt bridges in biomolecules occur in relatively water-free depressions or at biomolecular interfaces where water is excluded.

Hydrogen Bonds

Covalent bonds between hydrogen and oxygen or nitrogen are sufficiently polar that the hydrogen nucleus is weakly attracted to the lone pair of electrons of an oxygen or nitrogen, on a neighboring molecule. In the water molecule, each of oxygen's unshared electron pairs can form a hydrogen bond with nearby water molecules (Figure 3.6). Because this bond is partially covalent, the force of attraction has directionality. Maximum attraction occurs when the two O—H bonds of the participating water molecules are colinear. The resulting intermolecular "bonds" act as a bridge between water molecules. Neither hydrogen bond is especially strong (about 20 kJ/mol) in comparison to covalent bonds (e.g., 393 kJ/mol for N—H bonds and 460 kJ/mol

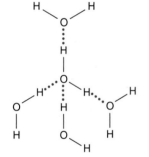

FIGURE 3.6

Tetrahedral Aggregate of Water Molecules

In water, each molecule can form hydrogen bonds with four other water molecules.

TABLE 3.1 Bond Strengths of Bonds Typically Found in Living Organisms*

Bond Type	Bond Strength	
	kcal/mol	kJ/mol*
Covalent	>50	>210
Noncovalent		
Ionic interactions[†]	1–20	4–80
Van der Waals forces	<1–2.7	<4–11.3
Mixed: hydrogen bonds	3–7	12–29

* The actual strength varies considerably with the identity of the interacting species.

[†] 1 cal = 4.184 J.

for O—H bonds). However, when large numbers of intermolecular hydrogen bonds can be formed (e.g., in the liquid and solid states of water), the molecules effectively become large, dynamic, three-dimensional aggregates. In water, the substantial amounts of energy that are required to break up this aggregate explain the high values for its boiling and melting points, heat of vaporization, and heat capacity. Other properties of water, such as surface tension and viscosity, are also due largely to its capacity to form large numbers of hydrogen bonds.

Van der Waals Forces

Van der Waals forces are relatively weak electrostatic interactions. They occur between neutral permanent and/or induced dipoles. They may be attractive or repulsive, depending on the distance between the atoms or groups. The attraction between molecules is greatest at a distance called the *van der Waals radius*. If molecules approach each other more closely, a repulsive force develops. The magnitude of van der Waals forces depends on how easily an atom is polarized. Electronegative atoms with unshared pairs of electrons are easily polarized. The larger the electronegativity difference between two atoms, the larger the polarity of the bond.

There are three types of van der Waals force:

1. **Dipole-dipole interactions**. These forces, which occur between molecules containing electronegative atoms, cause molecules to orient themselves so that the positive end of one molecule is directed toward the negative end of another (Figure 3.7a). Hydrogen bonds are an especially strong type of dipole–dipole interaction.

2. **Dipole–induced dipole interactions**. A permanent dipole induces a transient dipole in a nearby molecule by distorting its electron distribution (Figure 3.7b). For example, a carbonyl-containing molecule is weakly attracted to an aromatic ring because of the ability of the permanent dipole of the carbonyl group to delocalize (shift) the electrons of the π-electron cloud of the aromatic ring. Dipole–induced dipole interactions are weaker than dipole–dipole interactions.

3. **Induced dipole–induced dipole interactions**. The motion of electrons in nearby nonpolar molecules results in transient charge imbalance in adjacent molecules (Figure 3.7c). A transient dipole in one molecule polarizes the electrons in a neighboring molecule. This attractive interaction, often called **London dispersion forces**, is extremely weak. The stacking of the base rings in a DNA molecule, a classic example of this type of interaction, is made possible because of the ability of the loosely held π electrons to distribute unequally above and below closely spaced parallel rings. Although individually weak, these interactions extending over the length of the DNA molecule provide significant stability.

(a) Dipole-dipole interactions

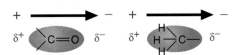

(b) Dipole–induced dipole interactions

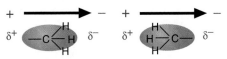

(c) Induced dipole–induced dipole interactions

KEY CONCEPTS

- Noncovalent bonds (i.e., ionic interactions and van der Waals forces) are important in determining the physical and chemical properties of living systems.

- Hydrogen bonds, with both dipole/dipole and covalent character, play a critical role in the properties of water and its place in the structure and function of cells.

FIGURE 3.7

Dipolar Interactions

The three types of electrostatic interaction involving dipoles are (a) dipole-dipole interactions, (b) dipole–induced dipole interactions, and (c) induced dipole–induced dipole interactions. The relative ease with which electrons respond to an electric field determines the magnitude of van der Waals forces. Dipole-dipole interactions are the strongest; induced dipole–induced dipole interactions are the weakest.

3.3 THERMAL PROPERTIES OF WATER

Perhaps the oddest property of water is that it is a liquid at room temperature. If water is compared with related molecules of similar molecular weight, it becomes apparent that water's melting and boiling points are exceptionally high (Table 3.2). If water followed the pattern of compounds such as hydrogen sulfide, it would melt at −100°C and boil at −91°C. Under these conditions, most of Earth's water would be steam, making life unlikely. However, water actually melts at 0°C and boils at +100°C. Consequently, it is a liquid over most of the wide range of temperatures typically found on Earth's surface. Hydrogen bonding is responsible for this anomalous behavior.

Each water molecule can form hydrogen bonds with four other water molecules. Each of the latter molecules can form hydrogen bonds with other water molecules. The maximum number of hydrogen bonds form when water has frozen into ice (Figure 3.8). Energy is required to break these bonds. When ice is warmed to its melting point, approximately 15% of the hydrogen bonds break. The energy required to melt ice (*heat of fusion*) is substantially higher than expected (Table 3.3). Liquid water consists of icelike clusters of molecules whose hydrogen bonds are continuously breaking and forming. As the temperature rises, the movement and vibrations of the water molecules accelerate, and additional hydrogen bonds are broken. When the boiling point is reached, the water molecules break free from one another and vaporize. In addition to the energy absorbed in increasing molecular agitation, a significant amount of energy is dissipated by the rapid vibration of shared hydrogens back and forth between oxygen atoms.

TABLE 3.2 Melting and Boiling Points of Water and Three Other Group VI Hydrogen-Containing Compounds

Name	Formula	Molecular Weight (D)*	Melting Point (°C)	Boiling Point (°C)
Water	H_2O	18	0	100
Hydrogen sulfide	H_2S	34	−85.5	−60.7
Hydrogen selenide	H_2Se	81	−50.4	−41.5
Hydrogen telluride	H_2Te	129.6	−49	−2

* I dalton (D) = 1 atomic mass unit (amu).

FIGURE 3.8

Hydrogen Bonding Between Water Molecules in Ice

Hydrogen bonding in ice produces a very open structure. Ice is less dense than water in its liquid state.

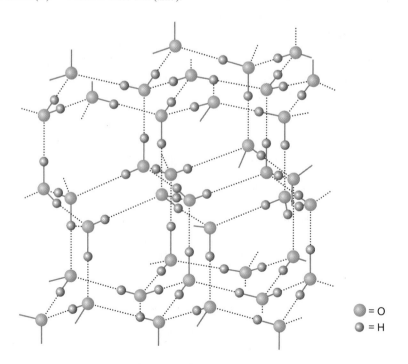

= O
= H

TABLE 3.3 Heat of Fusion of Water and Two Other Group VI Hydrogen-Containing Compounds

Name	Formula	Molecular Weight (D)	Heat of Fusion* cal/g	J/g
Water	H_2O	18	80	335
Hydrogen sulfide	H_2S	34	16.7	69.9
Hydrogen selenide	H_2Se	81	7.4	31

* The heat of fusion is the amount of heat required to change 1 g of a solid into a liquid at its melting point; 1 cal = 4.184 J.

One consequence of water's high *heat of vaporization* (the energy required to vaporize one mole of a liquid at a pressure of one atmosphere) and high *heat capacity* (the energy that must be added or removed to change the temperature by one degree Celsius) is that water acts as an effective modulator of climatic temperature. Water also plays an important role in the thermal regulation of living organisms. Water's high heat capacity, coupled with the high water content found in most organisms (between 50 and 95%, depending on species), helps maintain an organism's internal temperature. The evaporation of water serves as a cooling mechanism because it permits large losses of heat. For example, an adult human may eliminate as much as 1200 g of water daily in expired air, sweat, and urine. The associated heat loss may amount to approximately 20% of the total heat generated by metabolic processes.

KEY CONCEPTS

- Hydrogen bonding is responsible for water's unusually high freezing and boiling points.
- Because water has a high heat capacity, it can absorb and release heat slowly. Water plays an important role in regulating heat in living organisms.

QUESTION 3.1

Water (H_2O), ammonia (NH_3), and methane (CH_4) have approximately the same molecular weight: 18, 17, and 16 g/mol, respectively. Although these molecules are all structurally in the tetrahedral family, they differ significantly in physical properties. For example, the heat of fusion decreases somewhat from water (6.01 kJ/mol) to ammonia (5.66 kJ/mol) and significantly from water to methane (0.94 kJ/mol). Draw the structure of these molecules and explain the difference in properties based on what you know about hydrogen bonding in the solid state. If it were realistic to generate NH_3 ice (melting point $-97.8°C$), would you expect it to be more or less dense than liquid ammonia?

3.4 SOLVENT PROPERTIES OF WATER

Water is the ideal biological solvent. It easily dissolves a wide variety of the constituents of living organisms. Examples include ions (e.g., Na^+, K^+, and Cl^-), sugars, and many of the amino acids. Supramolecular structures (e.g., membranes) and numerous biochemical processes (e.g., protein folding) are possible because water cannot dissolve other substances, such as lipids and certain amino acids. This section describes the behavior of hydrophilic and hydrophobic substances in water. The discussion is followed by a brief review of osmotic pressure, one of the colligative properties of water. Colligative properties are physical properties that are affected not by the specific structure of dissolved solutes, but rather by their numbers.

Hydrophilic Molecules, Cell Water Structuring, and Sol-Gel Transitions

A dipolar structure and the capacity to form hydrogen bonds with electronegative atoms enable water to dissolve both ionic and polar substances. Salts such as sodium chloride (NaCl) are held together by ionic forces. An important aspect of all ionic interactions in aqueous solution is the hydration of ions. Because water molecules are polar, they are attracted to charged ions such as Na^+ and Cl^-. Shells of water molecules, referred to as **solvation spheres**, cluster around both positive and negative ions (Figure 3.9). The size of the solvation sphere depends upon the charge density of the ion (i.e., size of charge per unit volume). As ions become hydrated, the attractive force between them is reduced, and the charged species dissolves in the water. Organic molecules with ionizable groups and many neutral organic molecules with polar functional groups also dissolve in water, primarily because of the solvent's hydrogen bonding capacity. Such associations form between water and the carbonyl groups of aldehydes and ketones and the hydroxyl groups of alcohols. The capacity of a solvent to reduce the electrostatic attraction between charges is indicated by its *dielectric constant*. Water, sometimes referred to as the *universal solvent* because of the large variety of ionic and polar substances it can dissolve, has a very large dielectric constant.

STRUCTURED WATER Sophisticated spectroscopic technologies, which measure electromagnetic absorption by molecules, have revealed that the arrangement of water molecules in living organisms is distinctive. Even though organismal water is in liquid form, most water molecules are not in the "bulk water" state (i.e., they do not flow freely). At any given time, most of a cell's billions of water molecules are noncovalently associated with macromolecules and membrane surfaces throughout its densely packed interior. The surfaces of proteins, for example, are studded with positive and negative charges and polar functional groups. Dipolar water molecules readily form hydrogen bonds with such species (Figure 3.10). Recall that water molecules are tetrahedral and that each one can form hydrogen bonds with four other water molecules. For this reason, a single layer of water molecules attracts additional water molecules and an extended three-dimensional network of water molecules forms. In crowded

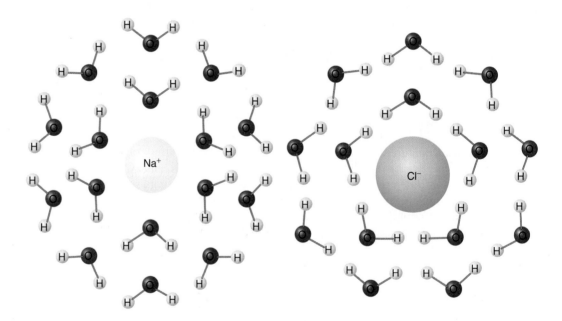

FIGURE 3.9

Solvation Spheres of Water Molecules Around Na^+ and Cl^- Ions

When an ionic compound such as NaCl is dissolved in water, its ions separate because the polar water molecules attract the ions more than the ions attract each other. In reality, the solvation sphere of Na^+ has four times the volume of that of Cl^- because of the higher charge density of the sodium ion (the same unit charge distributed over a smaller volume).

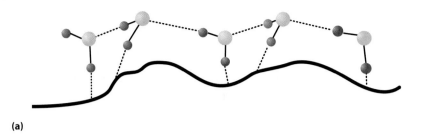

(a)

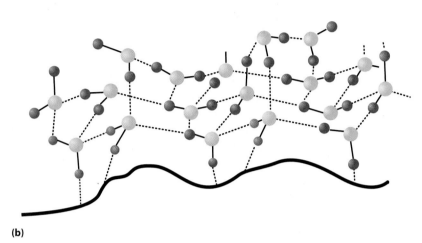

(b)

FIGURE 3.10
Diagrammatic View of Structured Water

Polar surfaces of macromolecules attract
water molecules: (a) a short segment of a
polar macromolecular surface with a single
water layer, which attracts additional water
molecules that form an extended network (b).

cells, numerous water layers bridge the space between adjacent macromolecules. Moreover, the water molecules in these layers, referred to as *structured water*, are in perpetual motion and constantly rearranging. They exchange with bulk water molecules, farther away from the protein's surface, on time scales that range from femtoseconds (10^{-15} s) to picoseconds (10^{-12} s). The pace of the exchange for an individual water molecule depends on how restricted its motion is. In other words, the closer a water molecule is to a polar surface, the slower its motion. The dynamics of structured water contribute to the structural stability of macromolecules such as proteins, while simultaneously facilitating the flexibility required for function.

SOL-GEL TRANSITIONS Cytoplasm, like any water-based material that contains polymers, has the properties of a gel. A *gel* is a colloidal mixture—in the case of cells, consisting of biopolymers with polar surfaces in association with adsorbed water. Gelatin desserts are well-known examples of gels with fibers of the protein collagen suspended and hydrated in a large quantity of water. The highly structured solvation layers on a matrix of protein give the viscoelastic properties we associate with Jell-O. The stability of a gel is very dependent on the length and cross-linking of the polymer and the continuity of the adsorbed water. The freedom of solutes to move within this meshwork or gel matrix varies with the trabecular arrangement of the protein polymers. If you punch wells in a petri dish filled with solidified gelatin, and pour a solution of inorganic ions into the wells, the ions will migrate out into the gel at rates related to their size and degree of hydration. You can observe the results of this sieving effect in just a few minutes. In addition, the water solvating the surface of the gelatin (collagen polymers) is for all practical purposes fixed in position; that is, diffusion is limited.

Changes in temperature (and therefore molecular motion), matrix architecture, and inclusion of solutes can lead to a transition from the gel to a "sol" or liquid state. Cells behave in a similar way because of the highly structured solvation surfaces of the polymeric proteins. Transitions from gel to sol (from more solid to less solid) contribute to many aspects of cell function, most notably cell movement. These transitions are caused by the reversible polymerization of G-actin to form F-actin, and the subsequent cross-linking of actin filaments. These transitions are carefully regulated by signal transduction mechanisms that affect the concentrations and functions of a group of proteins called *actin-binding proteins*. Various actin-binding proteins can inhibit polymerization, or they can cross-link or sever actin filaments.

Amoeboid motion provides an example of the highly regulated nature of cellular sol-gel transitions and the forces that such transformations create. The principal feature of amoeboid motion is the protrusion of a cellular extension called a *pseudopodium* (Figure 3.11). The pseudopodium moves forward because polymerizing actin filaments in the cell cortex in this isolated part of the cell undergo further cross-linking (a sol-to-gel transition). The swelling *ectoplasm* (the gelatinous outer layer of the cytoplasm adjacent to the plasma membrane) combined with a net elongation of actin filaments pushes the leading edge forward. Soon the newly elongated pseudopodium attaches to a surface (e.g., an extracellular matrix or a lab dish). Once this has occurred, actin filaments in the cell's interior (the *endoplasm*) depolymerize, effecting a gel-to-sol transition. Simultaneously, a contractile force created by the binding of actin filaments to myosin (a motor protein) in the trailing end of the cell squeezes the freely flowing endoplasm, causing it to stream forward into the pseudopodium.

FIGURE 3.11

Amoeboid Motion and Sol-Gel Transitions

The cell moves forward because of the coordination of sol-gel transitions in the cell cortex (ectoplasm) and cytoplasm in the cell's interior (endoplasm). A contractile force in the rear of the cell squeezes the fluid endoplasm forward.

Cortex (outer cytoplasm): gel with actin network

Inner cytoplasm: sol with actin subunits

Extending pseudopodium

QUESTION 3.2

Proteins are amino acid polymers. Noncovalent bonding plays an important role in determining the three-dimensional structures of proteins. The noncovalent interactions indicated here by shaded areas are typical of the bonding that occurs between amino acid side chains.

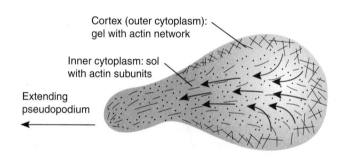

Which noncovalent bond is primarily responsible for the interactions indicated in the figure?

QUESTION 3.3

Collagen, a large fiberlike protein, combined with other molecules, forms a gel-like material found in shock-absorbing body components (e.g., tendons and ligaments). Explain the role of structured water in the function of these tissues. [*Hint*: Water is an incompressible substance.]

Hydrophobic Molecules and the Hydrophobic Effect

Small amounts of nonpolar substances mixed with water are excluded from the solvation network of the water; that is, they coalesce into droplets. This process is called the *hydrophobic effect*. Hydrophobic ("water-hating") molecules, such as the hydrocarbons, are virtually insoluble in water. Their association into droplets (or, in larger amounts, into a separate layer) results from the solvent properties of water, not from the relatively weak attraction between the associating nonpolar molecules. When nonpolar molecules enter an aqueous environment, the water molecules organize into a cagelike structure that drives the hydrophobic region in on itself (a partitioning or exclusion process). The excluded hydrophobic phase is ultimately stabilized by van der Waals interactions between closely spaced nonpolar regions (Figure 3.12). The water-caged structure, or *clathrate*, is stabilized when exposure of water to the hydrophobic material is minimized. The hydrophobic effect is responsible for the generation of stable lipid membranes and contributes to the fidelity of protein folding.

Amphipathic Molecules

A large number of biomolecules, referred to as **amphipathic**, contain both polar and nonpolar groups. This property significantly affects their behavior in water. For example, ionized fatty acids are amphipathic molecules because they contain hydrophilic carboxylate groups and hydrophobic hydrocarbon groups. When they are mixed with water, amphipathic molecules form structures called *micelles* (Figure 3.13) In **micelles**, the charged species (the carboxylate groups), called *polar heads*, orient themselves so that they are in contact with water. The nonpolar hydrocarbon "tails" become sequestered in the hydrophobic interior.

KEY CONCEPTS

- Water's dipolar structure and its capacity to form hydrogen bonds enable water to dissolve many ionic and polar substances.
- Nonpolar molecules cannot form hydrogen bonds with water and are excluded via clathrate formation.
- Amphipathic molecules, such as fatty acid salts, spontaneously rearrange themselves in water to form micelles.

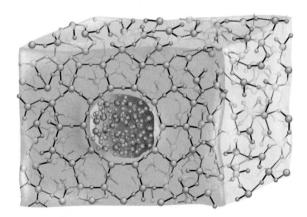

FIGURE 3.12

The Hydrophobic Effect

When nonpolar molecules and water are mixed a cage of organized hydrogen-bonded water molecules forms to minimize exposure to the hydrophobic substance. Nonpolar molecules, when in close proximity, are attracted to each other by van der Waals forces. However, the driving force in the formation of the cage and exclusion of the hydrophobic substance is the strong tendency of water molecules to form hydrogen bonds among themselves. Nonpolar molecules are excluded because they cannot form hydrogen bonds.

FIGURE 3.13

Formation of Micelles

The polar heads of amphipathic molecules orient themselves so that they are hydrogen-bonded to water. The nonpolar tails aggregate in the center, away from water.

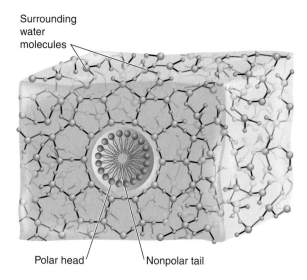

The tendency of amphipathic biomolecules to spontaneously rearrange themselves in water is an important feature of numerous cell components. For example, a group of bilayer-forming phospholipid molecules is the basic structural feature of biological membranes (see Chapter 11).

Osmotic Pressure

Osmosis is the spontaneous passage of solvent molecules through a semipermeable membrane that separates a solution of lower solute concentration from a solution of higher solute concentration. Pores in the membrane are wide enough to allow solvent molecules to pass through in both directions but too narrow for the larger solute molecules or ions to pass. Figure 3.14 illustrates the movement of solvent across a membrane. As the process begins, there are fewer water molecules on the high solute concentration side of the membrane. Over time, more water moves from side A (lower solute concentration) to side B (higher solute concentration). The higher the concentration of water in a solution (i.e., the lower the solute concentration), the greater the amount of water that flows through the membrane.

Osmotic pressure is the pressure required to stop the net flow of water across the membrane. The force generated by osmosis can be considerable. The principal cause of water flow across cellular membranes, osmotic pressure, is a driving force in numerous living processes. For example, osmotic pressure appears to be a significant factor in the formation of sap in trees. Cell membranes are not, strictly speaking, osmotic membranes because they permit molecules other than solvent (water) to move across the membrane. The term *dialyzing membrane* would be more accurate.

Osmotic pressure depends on solute concentration. A device called an *osmometer* (Figure 3.15) measures osmotic pressure. Osmotic pressure can also be calculated using the following equation, keeping in mind that the final osmotic pressure reflects the contribution of all solutes present.

$\pi = iMRT$
where π = osmotic pressure (atm)
$\quad$ i = van't Hoff factor (reflects the extent of ionization of solutes)
$\quad$ M = molarity (mol/L)
$\quad$ R = gas constant (0.082 L·atm/K·mol)
$\quad$ T = temperature (K)

The concentration of a solution can be expressed in terms of *osmolarity*. The unit of osmolarity is osmoles (osmol) per liter. In the equation $\pi = iMRT$, the osmolarity is equal to iM, where i (the van't Hoff factor) represents the degree of ionization of the solute species, which varies with temperature. The

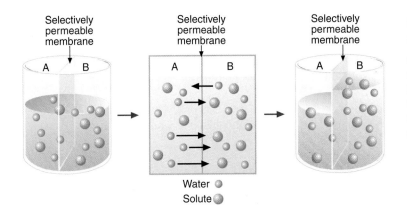

Selectively permeable membrane | Selectively permeable membrane | Selectively permeable membrane

A B | A B | A B

Water ○
Solute ●

FIGURE 3.14

Osmotic Pressure

Over time water diffuses from side A (more dilute) to side B (more concentrated). Equilibrium between the solutions on both sides of a semipermeable membrane is attained when there is no net movement of water molecules from side A to side B. Osmotic pressure stops the net flow of water across the membrane.

FIGURE 3.15

The Measurement of Osmosis Using an Osmometer

Volume 1 contains pure water. Volume 2 contains a solution of sucrose. The membrane is permeable to water but not to the sucrose. Therefore there will be a net movement of water into the osmometer. The osmotic pressure is proportional to the height H of the solution in the tube.

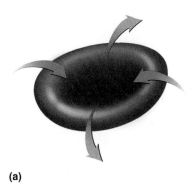

(a)

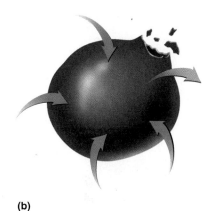

(b)

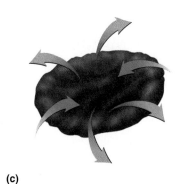

(c)

degree of ionization of a 1 M NaCl solution is 90%, with 10% of the NaCl existing as ion pairs. Thus if we write

$$i = [Na^+] + [Cl^-] + [NaCl]_{un\text{-}ionized} = 0.9 + 0.9 + 0.1 = 1.9$$

the value of i for this solution is 1.9. The value of i approaches 2 for NaCl solutions as they become increasingly more dilute. The value of i for a 1 M solution of a weak acid that undergoes a 10% ionization is 1.1. The value of i for a nonionizable solute is always 1.0. Problems 3.1 and 3.2, which follow shortly, use the concept of osmotic pressure.

Osmotic pressure creates some critical problems for living organisms. Cells typically contain fairly high concentrations of solutes, that is, small organic molecules and ionic salts, as well as lower concentrations of macromolecules. Consequently, cells may gain or lose water because of the concentration of solute in their environment. If cells are placed in an **isotonic solution** (i.e., the concentration of solute and water is the same on both sides of the selectively permeable plasma membrane) there is no net movement of water in either direction across the membrane (Figure 3.16). For example, red blood cells are isotonic to a 0.9% NaCl solution. When cells are placed in a solution with a lower solute concentration (i.e., a **hypotonic solution**), water moves into the cells. When red blood cells are immersed in pure water, for example, they swell and rupture in a process called *hemolysis*. In solutions with higher solute concentrations (i.e., **hypertonic solutions**), cells shrivel because there is a net movement of water out of the cell. The shrinkage of red blood cells in a hypertonic solution (e.g., a 3% NaCl solution) is referred to as *crenation*.

FIGURE 3.16

The Effect of Hypertonic and Hypotonic Solutions on Animal Cells

(a) Isotonic solutions do not change cell volume because water is entering and leaving the cell at the same rate.
(b) Hypotonic solutions cause cell rupture.
(c) Hypertonic solutions cause cell shrinkage (crenation).

WORKED PROBLEM 3.1

When 0.1 g of urea (MW 60) is diluted to 100 mL with water, what is the osmotic pressure of the solution? [Assume room temperature, i.e., 25°C (298 K).]

Solution

Calculate the osmolarity of the urea solution. Urea is a nonelectrolyte, so the van't Hoff factor (i) is 1.

$$\text{Molarity} = \frac{0.1 \, \text{g urea} \times 1 \, \text{mol}}{60 \, \text{g}} \times \frac{1}{0.1 \, \text{L}} = 1.7 \times 10^{-2} \, \text{mol/L}$$

The osmotic pressure at room temperature is given by

$$\pi = iMRT$$

$$\pi = (1) \frac{1.7 \times 10^{-2} \, \text{mol}}{\text{L}} \frac{0.0821 \, \text{L} \cdot \text{atm}}{\text{K} \cdot \text{mol}} (298 \, \text{K})$$

$$\pi = 0.42 \, \text{atm}$$ ∎

WORKED PROBLEM 3.2

Estimate the osmotic pressure of a solution of 0.1 M NaCl at 25°C. Assume 100% ionization of solute.

Solution

A solution of 0.1 M NaCl produces 0.2 mol of particles per liter (0.1 mol of Na^+ and 0.1 mol of Cl^-). The osmotic pressure at room temperature is

$$\pi = \frac{2 \times 0.1 \, \text{mol}}{\text{L}} \frac{0.0821 \, \text{L} \cdot \text{atm}}{\text{K} \cdot \text{mol}} \times 298 \, \text{K}$$

$$\pi = 4.9 \, \text{atm}$$ ∎

Macromolecules have little direct effect on cellular osmolarity because their cellular molar concentrations are relatively low. However, macromolecules such as the proteins contain a large number of ionizable groups. The ions of opposite charge that are attracted to these groups have a substantial effect on intracellular osmolarity. The nature of this effect is determined by the interaction of the structured water associated with proteins with hydrated ions. The size of an ion's solvation sphere is inversely related to its unhydrated size. For example, sodium and potassium ions have nonhydrated diameters of 1.96 and 2.66 Å, respectively. The hydrated diameters of sodium and potassium ions are 9.0 and 6.0 Å, respectively. Consequently, the hydrated volume of Na^+ is 3.4 times that of K^+. In addition, since the solvation sphere of K^+ is much smaller than that of Na^+, the potassium ion is easier to remove to form ion pairs with the surface anions of proteins. It costs significantly more energy to remove the solvation sphere of Na^+, a step that must occur for the sodium ion to move from one aqueous compartment to another. As a result, the ion distribution across the cell membrane is unequal, with the tendency to accumulate inside the cell being much greater for K^+ (159 mM) than for Na^+ (10 mM). This inequality would occur even if specific ion pumps were not present.

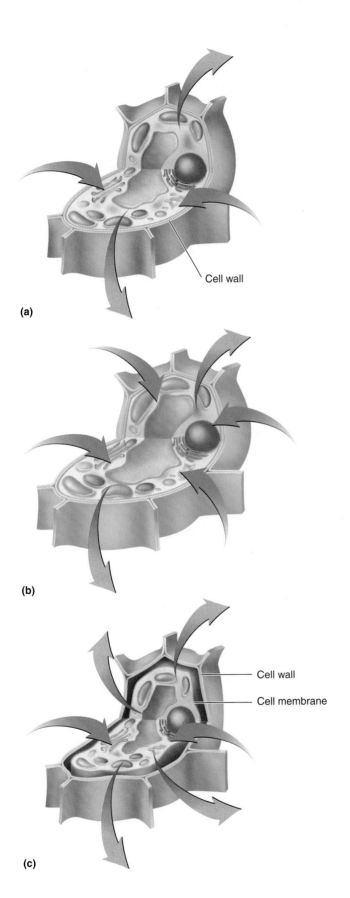

Cell wall

(a)

(b)

Cell wall

Cell membrane

(c)

FIGURE 3.17
Osmotic Pressure and Plant Cells

(a) Isotonic solutions do not change cell volume. (b) Plant cells typically exist in a hypotonic environment. When water enters these cells they become swollen. Rigid cell walls prevent the cells from bursting. (c) In a hypertonic environment the cell membrane pulls away from the cell wall because of water loss and the plant wilts.

COMPANION

WEBSITE **Visit the companion website at www.oup.com/us/mckee to read the Biochemistry in Perspective box on cell volume regulation and metabolism.**

KEY CONCEPTS

- Osmosis is the movement of water across a semipermeable membrane from a dilute solution to a more concentrated solution.
- Osmotic pressure is the pressure exerted by water on a semipermeable membrane as a result of a difference in the concentration of solutes on either side of the membrane.

Unlike most inorganic ions, the ionizable groups of cellular proteins are fixed within the cell, conferring a significant net negative charge to the intracellular environment. As a consequence, there exists an electronegative gradient across the cell membrane: that is, the ions and negative charge are distributed unequally. The cytoplasmic side of the membrane is strongly negative, an effect that is partly offset by potassium ions. The outside of the membrane is positive because of the relatively large number of extracellular sodium ions. The existence of this asymmetry on the surfaces of cell membrane results in the establishment of an electrical gradient, called a **membrane potential**, which provides the means for electrical conduction, active transport, and even passive transport.

Although hydrated sodium ions tend to be excluded from structured water inside cells, leakage of these ions back across the plasma membrane does occur. Small intracellular increases in $[Na^+]$ cause the cytoplasm to be slightly less negative. As a result, small amounts of K^+ move down their concentration gradient out of the cell. Animals and bacteria control cell volume by opposing this process with ATP-driven Na^+-K^+ pumps.

Ion pumping in these cells requires substantial amounts of energy. Several species, such as some protozoa and algae, control cell volume by periodically expelling water from special contractile vacuoles. Because plant cells have rigid cell walls, plants use osmotic pressure to create an internal hydrostatic pressure, called *turgor pressure* (Figure 3.17). This process drives cellular growth and expansion and makes many plant structures rigid.

3.5 IONIZATION OF WATER

Liquid water molecules have a limited capacity to ionize to form a proton, or hydrogen ion (H^+), and a hydroxide ion (OH^-). Protons do not actually exist in aqueous solution. In water a proton combines with a water molecule to form H_3O^+, commonly referred to as *hydronium ion*. For convenience, H^+ will be used in representing the ionization reactions of water.

The disassociation of water

$$H_2O(l) \rightleftharpoons H^+ + OH^-$$

may be expressed as

$$K_{eq} = \frac{[H^+][OH^-]}{[H_2O]}$$

where K_{eq} is the equilibrium constant for the reaction.

The concentration of water is essentially unchanged and can be thought of as a constant. The equilibrium expression can be rewritten combining the two constants as

$$K_{eq}[H_2O] = [H^+][OH^-]$$

The term $K_{eq}[H_2O]$ is referred to as the ion product of water or K_w. After substitution of the term K_w, the preceding equation may be rewritten as

$$K_w = [H^+][OH^-]$$

The K_w for H_2O at 25°C and 1 atm pressure is 1.0×10^{-14} and is a fixed characteristic of water at this temperature and pressure. In pure water, where there are no other contributors of H^+ or OH^-, the concentrations of these ions are equal:

$$[H^+] = [OH^-] = (K_w)^{1/2} = (1.0 \times 10^{-14})^{1/2} = 1.0 \times 10^{-7} \text{ M}$$

A solution that contains equal amounts of H^+ and OH^- is said to be *neutral*. When an ionic or polar substance is dissolved in water, it may change the relative numbers of H^+ and OH^-. Solutions with an excess of H^+ are *acidic*, whereas those with a greater number of OH^- are *basic*. Hydrogen ion concentration varies over a very wide range: commonly between 10^0 and 10^{-14} M, which provides the basis of the pH scale (pH $= -\log [H^+]$).

Acids, Bases, and pH

The concentration of the hydrogen ion, one of the most important ions in biological systems, affects most cellular and organismal processes. For example, the structure and function of proteins and the rates of most biochemical reactions are strongly affected by hydrogen ion concentration. Additionally, hydrogen ions play a major role in processes such as energy generation (see Chapter 10) and endocytosis.

Many biomolecules have acidic and/or basic properties. Large polymers and macromolecular complexes usually have amphoteric surfaces; that is, they possess both acidic and basic groups. A side group of a molecule is said to be an acid if it is a proton donor and a base if it is a proton acceptor.

Strong acids (e.g., HCl) and bases (e.g., NaOH) ionize almost completely in water:

$$HCl \rightarrow H^+ + Cl^-$$

$$NaOH \rightarrow Na^+ + OH^-$$

Many acids and bases, however, do not dissociate completely. Organic acids (compounds with carboxyl groups) do not completely dissociate in water. They are referred to as **weak acids**. Organic bases have a small but measurable capacity to combine with hydrogen ions. Many common **weak bases** contain amino groups.

The dissociation of an organic acid is described by the following reaction:

HA $\rightleftharpoons$ $H^+ + A^-$
Weak acid   Conjugate base of HA

Note that the deprotonated product of the dissociation reaction is referred to as a **conjugate base**. For example, acetic acid (CH_3COOH) dissociates to form the conjugate base acetate (CH_3COO^-).

The strength of a weak acid (i.e., its capacity to release hydrogen ions) may be determined by using the following expression:

$$K_a = \frac{[H^+][A^-]}{[HA]}$$

where K_a is the acid dissociation constant. The larger the value of K_a, the stronger the acid is. Because K_a values vary over a wide range, they are expressed by using a logarithmic scale:

$$pK_a = -\log K_a$$

The lower the pK_a, the stronger the acid. Dissociation constants and pK_a values for several common weak acids are given in Table 3.4.

The **pH scale** (Figure 3.18) can be used to determine hydrogen ion concentration $[H^+]$:

$$pH = -\log[H^+]$$
$$[H^+] = antilog(-pH)$$

On the pH scale, neutrality is defined as pH 7; that is, $[H^+]$ is equal to 1×10^{-7} M. Acidic solutions have pH values less than 7; that is, $[H^+]$ is greater than 1×10^{-7} M. A pH value greater than 7 indicates a solution that is basic, or alkaline.

It is important to note that although pK_a and pH appear to be similar mathematical expressions, they are in fact different. At constant temperature, the pK_a value of a substance is a constant. In contrast, the pH values of a system may vary.

TABLE 3.4 Dissociation Constants and pK_a Values for Common Weak Acids*

Acid	HA	A⁻	K_a	pK_a
Acetic acid	CH_3COOH	CH_3COO^-	1.76×10^{-5}	4.76
Carbonic acid	H_2CO_3	HCO_3^-	4.5×10^{-7}	6.35
Bicarbonate	HCO_3^-	CO_3^{2-}	5.61×10^{-11}	10.33
Lactic acid	$CH_3CHCOOH$ 　　$|$ 　　OH	CH_3CHCOO^- 　　$|$ 　　OH	1.38×10^{-4}	3.86
Phosphoric acid	H_3PO_4	$H_2PO_4^-$	7.25×10^{-3}	2.14
Dihydrogen phosphate	$H_2PO_4^-$	HPO_4^{2-}	6.31×10^{-8}	7.20

* Equilibrium constants should be expressed in terms of activities rather than concentrations (activity is the effective concentration of a substance in a solution). However, in dilute solutions, concentrations may be substituted for activities with reasonable accuracy.

FIGURE 3.18

The pH Scale and the pH Values of Common Fluids

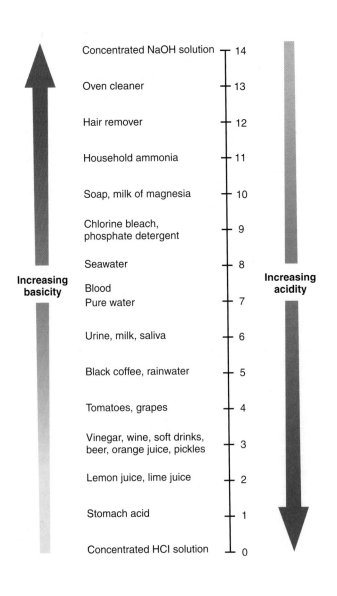

Buffers

The regulation of pH is a universal and essential activity of living organisms. Hydrogen ion concentration must typically be kept within very narrow limits. For example, normal human blood has a pH of 7.4. It may vary between 7.35 and 7.45, depending on the concentrations of acidic and basic waste products and metabolites. Certain disease processes cause pH changes that, if not corrected, can be disastrous. **Acidosis**, a condition that occurs when human blood pH falls below 7.35, results from an excessive production of acid in the tissues, loss of base from body fluids, or the failure of the kidneys to excrete acidic metabolites. Acidosis occurs in certain diseases (e.g., diabetes mellitus) and during starvation. If blood pH drops below 7, the central nervous system becomes depressed. This results in coma and eventually death. When blood pH rises above 7.45, **alkalosis** results. This condition, brought on by prolonged vomiting or by ingestion of excessive amounts of alkaline drugs, overexcites the central nervous system. Muscles then go into a state of spasm. If this situation is uncorrected, convulsions and respiratory arrest develop.

Buffers help maintain a relatively constant hydrogen ion concentration. The most common buffers consist of weak acids and their conjugate bases. A buffered solution can resist pH changes because an equilibrium between the buffer's components is established. Therefore buffers obey **Le Chatelier's principle**, which states that if a stress is applied to a reaction at equilibrium, the equilibrium will be displaced in the direction that relieves the stress. Consider a solution containing acetate buffer, which consists of acetic acid and sodium acetate (Figure 3.19). The buffer is created by mixing a solution of sodium acetate with a solution of acetic acid to create an equilibrium mixture of the correct pH and ionic strength.

$$CH_3-\overset{\overset{\displaystyle O}{\|}}{C}-OH \rightleftharpoons CH_3-\overset{\overset{\displaystyle O}{\|}}{C}-O^- \; + \; H^+$$

If hydrogen ions are added, the equilibrium shifts toward the formation of acetic acid with the $[H^+]$ changing little:

$$H^+ + CH_3COO^- \rightarrow CH_3COOH$$

If hydroxide ions are added, they react with the free hydrogen ions to form water, the equilibrium shifts to the acetate ion, and the pH changes little.

$$CH_3-\overset{\overset{\displaystyle O}{\|}}{C}-OH \rightleftharpoons CH_3-\overset{\overset{\displaystyle O}{\|}}{C}-O^- \; + \; H^+ \qquad OH^-$$
$$H_2O$$

BUFFERING CAPACITY The capacity of a buffer to maintain a specific pH depends on two factors: (1) the molar concentration of the acid–conjugate base pair and (2) the ratio of their concentrations. Buffering capacity is directly proportional to the concentration of the buffer components. In other words, the more molecules of buffer present, the more H^+ and OH^- ions can be absorbed without changing the pH. The concentration of the buffer is defined as the sum of the concentration of the weak acid and its conjugate base. For example, a 0.2 M acetate buffer may contain 0.1 mol of acetic acid and 0.1 mol of sodium acetate in 1 L of H_2O. Such a buffer may also consist of 0.05 mol of acetic acid and 0.15 mol of sodium acetate in 1 L of H_2O. The most effective buffers are those that contain equal concentrations of both components. Biological systems generate acids during metabolism, and buffer capacity for acid neutralization must be maximized. Consequently, biological buffers often contain a higher concentration of the conjugate base. Bicarbonate buffer (p. 96) is an example of such a buffering system.

FIGURE 3.19

Titration of Acetic Acid with NaOH

The shaded band indicates the pH range over which acetate buffer functions effectively. A buffer is most effective at or near its pK_a value.

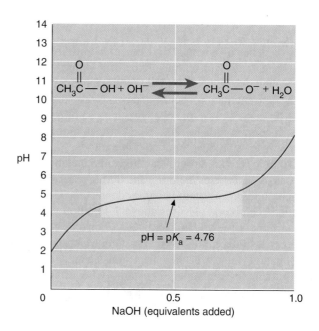

HENDERSON-HASSELBALCH EQUATION In choosing or making a buffer, the pH and pK_a concepts are useful. The relationship between these two quantities is expressed in the Henderson-Hasselbalch equation, which is derived from the following equilibrium expression:

$$K_a = \frac{[H^+][A^-]}{[HA]}$$

Solving for $[H^+]$ results in

$$[H^+] = K_a \frac{[HA]}{[A^-]}$$

Taking the negative logarithm of each side, we obtain

$$-\log [H^+] = -\log K_a - \log \frac{[HA]}{[A^-]}$$

Defining $-\log [H^+]$ as pH and $-\log K_a$ as pK_a gives

$$pH = pK_a - \log \frac{[HA]}{[A^-]}$$

If the log term is inverted, thereby changing its sign, the *Henderson-Hasselbalch equation* is obtained:

$$pH = pK_a + \log \frac{[A^-]}{[HA]}$$

Notice that when $[A^-] = [HA]$, the equation becomes

$$pH = pK_a + \log 1$$
$$= pK_a + 0$$

Under this circumstance, pH is equal to pK_a. Figure 3.19 illustrates that buffers are most effective when they are composed of equal amounts of weak acid and conjugate base. The most effective buffering occurs in the portion of the titration curve that has a minimum slope, that is, 1 pH unit above and below the value of pK_a. In the graph, the abscissa displays the equivalents added. Here, an equivalent is the mass of base that can accept 1 mol of H^+ ions; an acid equivalent gives the mass of acid that can accept a mole of protons.

Problems 3.3 through 3.7 are typical buffer problems.

KEY CONCEPTS

- Liquid water molecules have a limited capacity to ionize to form H^+ and OH^- ions.

- The concentration of hydrogen ions is a crucial feature of biological systems primarily because of their effects on biochemical reaction rates and protein structure.

- Buffers, which consist of weak acids and their conjugate bases, prevent changes in pH (a measure of $[H^+]$).

WORKED PROBLEM 3.3

Calculate the pH of a mixture of 0.25 M acetic acid and 0.1 M sodium acetate. The pK_a of acetic acid is 4.76.

Solution

$$pH = pK_a + \log \frac{[\text{acetate}]}{[\text{acetic acid}]}$$

$$pH = 4.76 + \log \frac{0.1}{0.25} = 4.76 - 0.398 = 4.36$$ ∎

WORKED PROBLEM 3.4

What is the pH in the preceding problem if the mixture consists of 0.1 M acetic acid and 0.25 M sodium acetate?

Solution

$$pH = 4.76 + \log \frac{0.25}{0.1} = 4.76 + 0.398 = 5.16$$ ∎

WORKED PROBLEM 3.5

Calculate the ratio of lactic acid and lactate required in a buffer system of pH 5.00. The pK_a of lactic acid is 3.86.

Solution
The equation

$$pH = pK_a + \log \frac{[\text{lactate}]}{[\text{lactic acid}]}$$

can be rearranged to

$$\log \frac{[\text{lactate}]}{[\text{lactic acid}]} = pH - pK_a$$

$$= 5.00 - 3.86 = 1.14$$

Therefore the required ratio is

$$\frac{[\text{lactate}]}{[\text{lactic acid}]} = \text{antilog } 1.14$$

$$= 13.8$$

For a lactate buffer to have a pH of 5, the lactate and lactic acid components must be present in a ratio of 13.8:1. A good buffer is a mixture of a weak acid and its conjugate base present in near equal concentrations, and the buffered pH should be within 1 pH unit of the pK_a. Thus lactate buffer is a poor choice in this situation. A better choice would be the acetate buffer. ∎

WORKED PROBLEM 3.6

During the fermentation of wine, a buffer system consisting of tartaric acid and potassium hydrogen tartrate is produced by a biochemical reaction. Assuming that at some time the concentration of potassium hydrogen tartrate is twice that of tartaric acid, calculate the pH of the wine. The pK_a of tartaric acid is 2.96.

Solution

$$pH = pK_a + \log \frac{[\text{hydrogen tartrate}]}{[\text{tartaric acid}]}$$
$$= 2.96 + \log 2$$
$$= 2.96 + 0.30 = 3.26$$

WORKED PROBLEM 3.7

What is the pH of a solution prepared by mixing 150 mL of 0.1 M HCl with 300 mL of 0.1 M sodium acetate (NaOAc) and diluting the mixture to 1 L? The pK_a of acetic acid is 4.76.

Solution

The amount of acid present in the solution is found by multiplying the volume of the solution, in milliliters, by M, the molarity of the solution; it is expressed in millimoles (mmol):

$$150 \text{ mL} \times 0.1 \text{ M} = 15 \text{ mmol acid}$$

The amount of sodium acetate is found using the same equation:

$$300 \text{ mL} \times 0.1 \text{ M} = 30 \text{ mmol acid}$$

Each mole of HCl will consume 1 mol of sodium acetate and produce 1 mol of acetic acid. This will give 15 mmol of acetic acid with 15 mmol remaining of sodium acetate (i.e., 30 mmol – 15 mmol). Substituting these values into the Henderson-Hasselbalch equation gives

$$pH = 4.76 + \log \frac{15}{15}$$
$$= 4.76 + \log 1$$
$$= 4.76$$

Because the log term is a ratio of two concentrations, the volume factor can be eliminated and the molar amounts can be used directly.

WORKED PROBLEM 3.8

What would be the effect of adding an additional 50 mL of 0.1 M HCl to the solution in Problem 3.7 before dilution to 1 L?

Solution

Using the same equation as in Problem 3.7, the amount of HCl would be

$$200 \text{ mL} \times 0.1 \text{ M} = 20 \text{ mmol}$$

which is also equal to the concentration of acetic acid.

The amount of sodium acetate would be

$$30 \text{ mmol} - 20 \text{ mmol} = 10 \text{ mmol}$$

Substituting into the Henderson-Hasselbalch equation gives

$$pH = 4.76 + \log \frac{10}{20}$$
$$= 4.76 + \log 0.5$$
$$= 4.76 - 0.3$$
$$= 4.46$$ ∎

WEAK ACIDS WITH MORE THAN ONE IONIZABLE GROUP Some molecules contain more than one ionizable group. Phosphoric acid (H_3PO_4) is a weak polyprotic acid; that is, it can donate more than one hydrogen ion (in this case, three hydrogen ions). During titration with NaOH (Figure 3.20) these ionizations occur in a stepwise fashion with 1 proton being released at a time:

$$H_3PO_4 \;\underset{pK_1 = 2.1}{\rightleftharpoons}\; H^+ + H_2PO_4^- \;\underset{pK_2 = 7.2}{\rightleftharpoons}\; H^+ + HPO_4^{2-} \;\underset{pK_3 = 12.3}{\rightleftharpoons}\; H^+ + PO_4^{3-}$$

The pK_a for the most acidic group is referred to as pK_1. The pK_a for the next most acidic group is pK_2. The third most acidic pK_a value is pK_3.

At low pH most molecules are fully protonated. As NaOH is added, protons are released in the order of decreasing acidity, with the least acidic proton (with the largest pK_a value) ionizing last. When the pH is equal to pK_1, equal amounts of H_3PO_4 and $H_2PO_4^-$ exist in the solution.

Amino acids are biomolecules that contain several ionizable groups. Like all amino acids, alanine contains both a carboxyl group and an amino group. At low pH, both of these groups are protonated. As the pH rises during a titration with NaOH, the acidic carboxyl group (COOH) loses its proton to form a

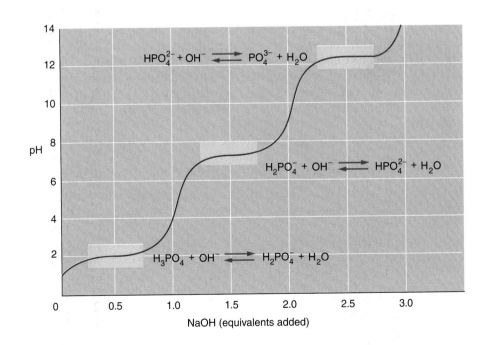

FIGURE 3.20

Titration of Phosphoric Acid with NaOH

Phosphoric acid (H_3PO_4) is a polyprotic acid that releases 3 protons sequentially upon titration with NaOH.

carboxylate group (COO⁻). The addition of more NaOH eventually causes the ionized amino group to release its proton:

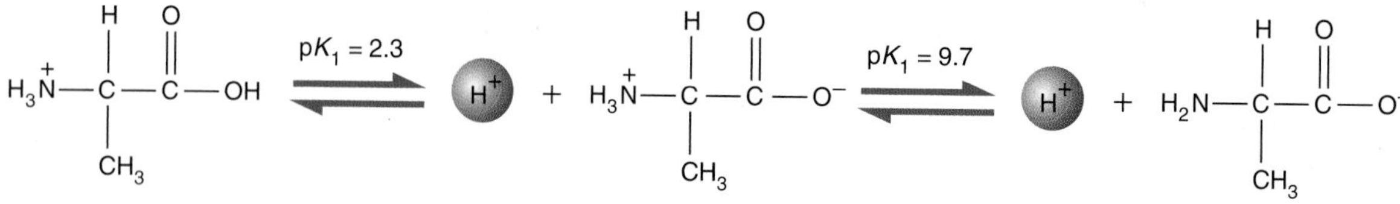

Certain amino acids also possess side chains with ionizable groups. For example, the side chain of lysine possesses an ionizable amino group. Because of their structures, alanine, lysine, and the other amino acids can act as effective buffers at or near their respective pK_a values [e.g., lysine: $pK_1 = 2.0$, $pK_a = 9.0$, pK_3 (R group) = 10.7]. See Chapter 5 for further descriptions of the titration and buffering capacity of amino acids.

Physiological Buffers

The three most important buffers in the body are the bicarbonate buffer, the phosphate buffer, and the protein buffer. Each is adapted to solve specific physiological problems in the body.

BICARBONATE BUFFER Bicarbonate buffer, one of the more important buffers in blood, has three components. The first of these, carbon dioxide, reacts with water to form carbonic acid:

$$CO_2 + H_2O \rightleftharpoons H_2CO_3$$
$$\text{Carbonic acid}$$

Carbonic acid then rapidly dissociates to form H^+ and HCO_3^- ions:

$$H_2CO_3 \rightleftharpoons H^+ + HCO_3^-$$
$$\text{Bicarbonate}$$

Because the concentration of H_2CO_3 is very low in blood, the preceding equations may be simplified to

$$CO_2 + H_2O \rightleftharpoons H^+ + HCO_3^-$$

Recall that buffering capacity is greatest at or near the pK_a of the acid–conjugate base pair. Carbonic acid is a diprotic acid (it can donate two hydrogen ions) with a pK_1 of 6.3. In blood there is a critical need to maintain the pH at the high end of the buffering range of this acid and to maximize buffering capacity for acid. Therefore, it is optimal for the concentration of the conjugate base, bicarbonate, to be high compared to H_2CO_3 (or CO_2), typically 11 to 1.

The uncatalyzed conversion of CO_2 to HCO_3^- and H^+ is a slow process:

$$CO_2 + H_2O \rightleftharpoons H_2CO_3 \rightleftharpoons HCO_3^- + H^+$$

In blood, this reaction is catalyzed by the enzyme carbonic anhydrase. With rates as high as 10^6 molecules of CO_2 converted to bicarbonate per second, carbonic anhydrase is one of the most efficient enzymes known. The CO_2 level is kept low and is regulated through changes in the respiratory rate. The bicarbonate level stays high because the kidneys excrete H^+. When excessive amounts of HCO_3^- are produced, the kidney excretes the bicarbonate. As acid, a metabolic waste product, is added to the body's bicarbonate system, the concentration of HCO_3^- decreases and CO_2 is formed. Because the excess CO_2 is exhaled, the ratio of HCO_3^- to CO_2 remains essentially unchanged.

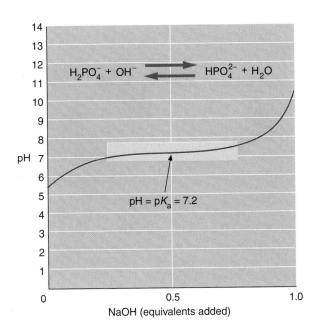

FIGURE 3.21

Titration of $H_2PO_4^-$ by Strong Base

The shaded band indicates the pH range over which the weak acid–conjugate base pair $H_2PO_4^-$/HPO_4^{2-} functions effectively as a buffer.

PHOSPHATE BUFFER Phosphate buffer consists of the weak acid–conjugate base pair $H_2PO_4^-$/HPO_4^{2-} (Figure 3.21):

$$H_2PO_4^- \rightleftharpoons H^+ + HPO_4^{2-}$$
Dihydrogen phosphate   Hydrogen phosphate

With pK_a 7.2, it would appear that phosphate buffer is an excellent choice for buffering the blood. Although the blood pH of 7.4 is well within this buffer system's capability, the concentrations of $H_2PO_4^-$ and HPO_4^{2-} in blood are too low to have a major effect. Instead, the phosphate system is an important buffer in intracellular fluids where its concentration is approximately 75 milliequivalents (mEq) per liter. Phosphate concentration in extracellular fluids such as blood is about 4 mEq/L. Because the normal pH of cell fluids is approximately 7.2 (the range is from 6.9 to 7.4), an equimolar mixture of $H_2PO_4^-$ and HPO_4^{2-} is typically present. Although cells contain other weak acids, these substances are unimportant as buffers. Their concentrations are quite low, and their pK_a values are significantly lower than intracellular pH. For example, lactic acid has a pK_a of 3.86.

PROTEIN BUFFER Proteins are a significant source of buffering capacity. Composed of amino acids linked together by peptide bonds, proteins contain several types of ionizable groups in side chains that can donate or accept protons. Because protein molecules are present in significant concentration in living organisms, they are powerful buffers. For example, the oxygen-carrying protein hemoglobin is the most abundant biomolecule in red blood cells. Hemoglobin plays a major role in maintaining blood pH because of its structure and high cellular concentration. Also present in high concentrations and buffering the blood are the serum albumins and other proteins.

KEY CONCEPT

The most important buffers in the body are the bicarbonate buffer (blood), the phosphate buffer (intracellular fluids), and the protein buffer.

QUESTION 3.4

Severe diarrhea is one of the most common causes of death in young children. One of the principal effects of diarrhea is the excretion of large quantities of sodium bicarbonate. In which direction does the bicarbonate buffer system shift under this circumstance? What is the resulting condition called?

BIOCHEMISTRY IN PERSPECTIVE

Water, Abiotic Stress, and Compatible Solutes

If water is so important for sustaining life, how can certain organisms survive catastrophic dehydration conditions? Abiotic stresses are environmental conditions that can potentially threaten the survival of living organisms. Examples include drought, temperature extremes, and excessive salinity. The adverse effects of these conditions are caused by alterations in the amounts and/or physical properties of water in affected organisms. Desiccation due to drought and high temperature results when environmental vapor pressure is low. Cell dehydration occurs in low temperatures as water moves down its concentration gradient and ice crystals form outside cells. (Since ice is a poor solvent, extracellular fluid becomes hyperosmotic.) The formation of intracellular ice crystals causes membrane rupture and disruption of osmotic gradients. Dehydration also occurs when cells are exposed to high salt concentrations.

Organisms vary widely in their capacity to adapt to environmental stresses. For those that can adapt, protection is afforded by stress-triggered signal transduction mechanisms that result in the accumulation of specialized molecules. Many organisms are protected by *compatible solutes* and/or the synthesis of protective proteins. Compatible solutes are a diverse group of water-soluble organic molecules that are generally considered to be nontoxic even at high concentrations. Commonly observed examples include sugars (e.g., trehalose and sucrose), alcohols (e.g., sorbitol, see p. 235), amino acids such as proline, or amino acid derivatives such as taurine (see p. 455). The protective effects of compatible solutes include interactions with structured water that prevent protein and membrane destabilization, as well as freezing-point depression, and osmoprotection. Some organisms also produce protective proteins. Antifreeze proteins (p. 252) are used by some cold-water fish and freeze-resistant plants and insects. Plants such as cotton and rice use LEA (late embryogenesis abundant) proteins to protect seeds from dehydration damage during the desiccation phase of seed development.

The stress adaptations of some organisms are astonishing. Anhydrobiotes, which can tolerate an almost complete loss of water, are such a group. Nematodes, brine shrimp, yeast, and other such organisms survive desiccation by producing certain compatible solutes that can function as water substitutes. The most common of these is trehalose, which is believed to stabilize cells by transforming cytoplasm into a glassy matrix. (All glasses are viscous liquids that have the mechanical properties of a solid.) As desiccation proceeds, newly synthesized trehalose molecules replace water throughout the cell. In the absence of water, the hydroxyl groups of trehalose form hydrogen bonds with the ionic and polar groups of cellular macromolecules. All metabolic and potentially damaging processes are considerably slowed because all the cell's molecules become immobilized within a sugar glass.

SUMMARY: Organisms that can adapt to severe water loss utilize specialized protective molecules, such as compatible solutes, that replace water by forming hydrogen bonds with proteins and other macromolecules and membranes.

GCGACATCACTCCAGCTTGAAGCAGTTCTTCTCGTCTTCTGTTTTGTCTAACTT TCTTCC
500   510   530   540   550

BIOCHEMISTRY **IN THE LAB**

Dialysis

Artificial semipermeable membranes are routinely used in biochemical laboratories to separate small solutes from larger solutes. For example, this technique (referred to as **dialysis**) is used as an important early step in protein purification. A specimen containing an impure protein is placed in a cellophane dialysis bag (Figure 3A), which is suspended in flowing distilled water or in a buffered solution. After a certain time, all the small solutes will have left the bag. The protein solution, which may contain many high-molecular-weight impurities, will then be ready for further purification.

Hemodialysis is a clinical application of dialysis that removes toxic waste from the blood of patients suffering from temporary or permanent renal failure. All constituents of blood except blood cells and the plasma proteins move freely between blood and the dialyzing fluid. Because dialysis tubing allows passage of nutrients (glucose, amino acids) and essential electrolyes (Na^+, K^+), these and other vital substances are added to the dialyzing fluid. Their inclusion prevents a net loss of these materials from the blood. Because no waste products such as urea and uric acid are in the dialyzing fluid, these substances are lost from the blood in large quantities.

Although hemodialysis is very effective in removing toxic waste from the body, it does not solve all the problems brought on by renal failure. For example, until recently, patients suffering from renal failure often became anemic because they lacked a protein hormone called *erythropoietin*, which is normally secreted by the kidney. (Erythropoietin stimulates red blood cell synthesis.) Because of DNA technology (Chapter 18), erythropoietin can now be readily administered to dialysis patients.

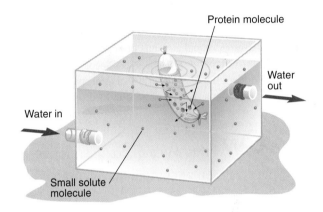

FIGURE 3A

Dialysis

Proteins are routinely separated from low-molecular-weight impurities by dialysis. When a dialysis bag containing a cell extract is suspended in water or a buffered solution, small molecules pass out through the membrane's pores. If the solvent outside the bag is continually renewed, all low-molecular-weight impurities are removed from the inside.

Chapter**Summary**

1. Water molecules (H_2O) are composed of two atoms of hydrogen and one of oxygen. Each hydrogen atom is linked to the oxygen atom by a single covalent bond. The oxygen-hydrogen bonds are polar, and water molecules are dipoles. One consequence of water's polarity is that water molecules are attracted to each other by the electrostatic force between the oxygen of one molecule and the hydrogen of another. This attraction is called a hydrogen bond.

2. Noncovalent bonds are relatively weak and, therefore, easily disrupted. They play a vital role in determining the physical and chemical properties of water and biomolecules. Ionic interactions occur between charged atoms or groups. Although each hydrogen bond is not especially strong in comparison to covalent bonds, large numbers of them have a significant effect on the molecules involved. Van der Waals forces can either be attractive or repulsive, occur between permanent and/or induced dipoles.

3. Water has an exceptionally high heat capacity. Its boiling and melting points are significantly higher than those of compounds of comparable structure and molecular weight. Hydrogen bonding is responsible for this anomalous behavior.

4. Water is also a remarkable solvent. Water's dipolar structure and its capacity to form hydrogen bonds enable it to dissolve many ionic and polar substances.

5. Most water molecules in living organisms are structured; that is, they are noncovalently associated with macromolecules and membrane surfaces. The networks of water molecules that form act as bridges between macromolecules in the densely crowded cytoplasm. Cytoplasm has the properties of a gel, a semisolid viscoelastic substance that resists flow and stores mechanical energy. Gels can undergo reversible transitions to a liquid or sol state.

6. Hydrophobic molecules are virtually insoluble in water. When nonpolar molecules enter an aqueous environment, they form droplets surrounded by water molecules that rearrange into their most energetically favorable configuration.

7. Amphipathic molecules contain both polar and nonpolar groups. Fatty acids are amphipathic molecules that form structures called micelles when they are placed in water.

8. Several physical properties of liquid water change when solute molecules are dissolved. The most important of these for living organisms is osmotic pressure, the pressure that prevents the flow of water across cellular membranes. Macromolecules have little direct effect on cellular osmolarity. The large number of ionizable groups on these molecules attracts ions of opposite charge. The structured water network that surrounds macromolecules such as proteins tends to exclude Na^+ because of its relatively large hydrated volume. The charge asymmetry across the cell membrane (negative on the inside and positive on the outside) creates an electrical gradient called a membrane potential.

9. Liquid water molecules have a limited capacity to ionize to form a hydrogen ion (H^+) and a hydroxide ion (OH^-). When a solution contains equal amounts of H^+ and OH^- ions, it is said to be neutral. Solutions with an excess of H^+ are acidic, whereas those with a greater number of OH^- are basic.

Because organic acids do not completely dissociate in water, they are referred to as weak acids. The acid dissociation constant K_a is a measure of the strength of a weak acid. Because K_a values vary over a wide range, pK_a values ($-\log K_a$) are used instead.

10. The hydrogen ion is one of the most important ions in biological systems. The pH scale conveniently expresses hydrogen ion concentration. pH is defined as the negative logarithm of the hydrogen ion concentration.

11. Because hydrogen ion concentration affects living processes so profoundly, it is not surprising that regulating pH is a universal and essential activity of living organisms. Hydrogen ion concentration is typically kept within narrow limits. Because buffers combine with H^+ ions, they help maintain a relatively constant hydrogen ion concentration. The ability of a solution to resist pH changes is called buffering capacity. Most buffers consist of a weak acid and its conjugate base.

 Take your learning further by visiting the **companion website** for Biochemistry at **www.oup.com/us/mckee** where you can complete a multiple-choice quiz on water to help you prepare for exams.

Suggested**Readings**

Gerstein, M., and Levitt, M., Simulating Water and the Molecules of Life, *Sci. Am.* 279(5):100–105, 1998.

Hochachka, P. W., and Somero, G. N., *Biochemical Adaptation: Mechanism and Process in Physiological Evolution*, Oxford University Press, New York, 2002.

Lang, F., and Waldegger, S., Regulating Cell Volume, *Am. Sci.* 85:456–463, 1997.

Leterrier, J.-F., Water and the Cytoskeleton, *Cell Mol. Biol. (Noisy-le-Grand)* 47(5):901–923, 2001.

Pollack, G. H., *Cells, Gels and the Engine of Life*, Ebner and Sons, Seattle, Washington, 2001.

Segel, I. H., *Biochemical Calculations*, 2nd ed., John Wiley & Sons, New York, 1976.

Key**Words**

acid, *89*
acidosis, *91*
alkalosis, *91*
amphipathic molecule, *83*
base, *89*
buffer, *91*
conjugate base, *89*
covalent bond, *75*

dialysis, *99*
dipole, *75*
electrostatic interaction, *75*
hydrogen bond, *75*
hydrophobic interaction, *83*
hypertonic solution, *85*
hypotonic solution, *85*
isotonic solution, *85*

Le Chatelier's principle, *91*
London dispersion force, *77*
membrane potential, *88*
micelle, *83*
osmosis, *84*
osmotic pressure, *84*
pH scale, *89*

polar, *74*
salt bridge, *76*
solution, *85*
solvation sphere, *80*
van der Waals force, *77*
weak acid, *89*
weak base, *89*

Review**Questions**

These questions are designed to test your knowledge of the key concepts discussed in this chapter, before moving on to the next chapter. You may like to compare your answers to the solutions provided in the back of the book and in the accompanying Study Guide.

1. Which of the following are acid–conjugate base pairs?
   a. H_2CO_3, CO_3^{2-}
   b. $H_2PO_4^-$, PO_4^{3-}
   c. HCO_3^-, CO_3^{2-}
   d. H_2O, OH^-

2. What is the hydrogen ion concentration in a solution at pH 8.3?

3. Consider the following titration curve. Estimate the effective buffer range.

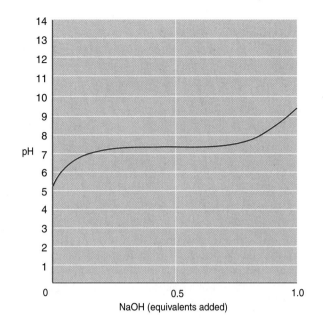

4. Describe how you would prepare a 0.1 M phosphate buffer with a pH of 7.2. What ratio of conjugate base to acid would you use?

5. Which of the following compounds can form hydrogen bonds with like molecules or with water?

$$CH_3—CH_2—\underset{\underset{O}{\|}}{C}—O—CH_3$$

(a)

$$CH_3—\underset{\underset{O}{\|}}{C}—N\underset{CH_3}{\overset{H}{<}}$$

(b)

$$CH_3—\underset{\overset{CH_3}{\underset{CH_3}{|}}}{N}$$

(c)

$$CH_3—O—CH_2—O—H$$

(d)

6. What is the osmolarity of a 1.3 M solution of sodium phosphate (Na_3PO_4)? Assume 85% ionization for this solution.

7. A dialysis bag containing a 3 M solution of the sugar fructose is placed in the following solutions. In each case, give the direction in which water flows.
   a. 1 M sodium lactate
   b. 3 M sodium lactate

c. 4.5 M sodium lactate

$$CH_3—\underset{\underset{OH}{|}}{C}\overset{H}{}—\underset{\overset{\|}{O}}{C}—O^-\quad Na^+$$

8. What interactions occur between the following molecules and ions?
   a. water and ammonia
   b. lactate and ammonium ion
   c. benzene and octane
   d. carbon tetrachloride and chloroform
   e. chloroform and diethylether

9. A solution containing 56 mg of a protein in 30 mL of distilled water exerts an osmotic pressure of 0.01 atm at $T = 25°C$. Determine the molecular weight of the unknown protein.

10. Tyrosine is an amino acid.

$$H—O—\overset{}{\bigcirc}—CH_2—\underset{\underset{+NH_3}{|}}{CH}—\underset{\overset{\|}{O}}{C}—OH$$

Which atoms in this molecule can form hydrogen bonds?

11. Briefly define the following terms:
    a. hydrogen bond
    b. pH
    c. buffer
    d. osmotic pressure
    e. osmolytes

12. Which of the following molecules would you expect to have a dipole moment?
    a. CCl_4
    b. $CHCl_3$
    c. H_2O
    d. CH_3OCH_3
    e. CH_3CH_3
    f. H_2

13. Which of the following molecules would you expect to form micelles?
    a. NaCl
    b. CH_3COOH
    c. $CH_3COO^-NH_4^+$
    d. $CH_3(CH_2)_{10}COO^-Na^+$
    e. $CH_3(CH_2)_{10}CH_3$

14. Briefly define the following terms:
    a. isotonic
    b. amphipathic
    c. hydrophobic interactions
    d. dipole
    e. induced dipole

15. Bicarbonate is one of the main buffers of the blood, and phosphate is the main buffer of the cells. Why might this be?

16. Describe how you can increase the buffering capacity of a 0.1 M acetate buffer.

17. Which of the following molecules or ions are weak acids? Explain.
    a. HCl
    b. $H_2PO_4^-$
    c. CH_3COOH
    d. HNO_3
    e. HSO_4^-

18. Briefly define the following terms:
    a. electrostatic interactions
    b. equilibrium dialysis
    c. salt bridge
    d. micelle
    e. weak acid

19. Which of the following species can form buffer systems?
    a. $NH_4^+ Cl^-$
    b. CH_3COOH, HCl
    c. CH_3COOH, $CH_3COO^- Na^+$
    d. H_3PO_4, PO_4^{3-}

20. What effect does hyperventilation have on blood pH?

21. What is the relationship between osmolarity and molarity?

22. Briefly define the following terms:
    a. solvation sphere
    b. hypertonic
    c. alkalosis
    d. dipole-dipole interaction
    e. structured water

23. Is it possible to prepare a buffer consisting of only carbonic acid and sodium carbonate?

24. Calculate the ratio of dihydrogen phosphate to hydrogen phosphate in blood at pH 7.4. The K_a is 6.3×10^{-8}.

25. Calculate the pH of a solution prepared by mixing 300 mL of 0.25 M sodium hydrogen ascorbate and 150 mL of 0.2 M HCl. The pK_{a1} of ascorbic acid is 4.04.

26. What is the pH of a solution that is 1×10^{-8}M in HCl?

27. Compounds such as the sugar trehalose are used as compatible solutes (i.e., water replacements) in desiccated organisms. What are the requirements for a substance that is to be used in this manner?

28. Calculate the pH for a mixture of one mole of benzoic acid and one mole of sodium benzoate. The pK_a of benzoic acid is 4.2.

29. Detergents that are good micelle formers often have strong antibacterial properties. Considering the structure of the cell membrane, suggest a reason for this antibacterial action.

30. Determine the pH of a solution composed of 1 M acetic acid and 1 M sodium acetate.

31. What would be the pH of the solution in Question 30 if 1 mL of 1M HCl is added?

32. What would be the pH of 1 L of water if 1 mL HCl is added?

33. Many molecules are polar, yet they do not form significant hydrogen bonds. What is so unusual about water that hydrogen bonding becomes possible?

Thought Questions

These questions are designed to reinforce your understanding of all of the key concepts discussed in the book so far, including this chapter and all of the chapters before it. They may not have one right answer! The authors have provided possible solutions to these questions in the back of the book and in the accompanying Study Guide, for your reference.

34. Many fruits can be preserved by candying. The fruit is immersed in a highly concentrated sugar solution, then the sugar is allowed to crystallize. How does the sugar preserve the fruit?

35. Explain why ice is less dense than water. If ice were not less dense than water, how would the oceans be affected? How would the development of life on earth be affected?

36. Why can't seawater be used to water plants?

37. Explain how the acids produced in metabolism are transported to the liver without greatly affecting the pH of the blood.

38. The pH scale is valid only for water. Why is this so?

39. Gelatin is a mixture of protein and water that is mostly water. Explain how the water-protein mixture becomes a solid.

40. Water has been described as the universal solvent. If this statement were strictly true, could life have arisen in a water medium? Explain.

41. Alcohols (ROH) are structurally similar to water. Why are alcohols not as powerful a solvent as water for ionic compounds? [*Hint*: Methanol is a better solvent for ionic compounds than is propanol.]

42. During stressful situations, some cells in the body convert glycogen to glucose. What effect does this conversion have on cellular osmotic balance? Explain how cells handle this situation.

43. Would you expect a carboxylic acid group within the water-free interior of a protein to have a higher or lower K_a than it would have if it occurred on the protein's surface, where it is hydrated?

44. The strength of ionic interactions is weaker in water than in an anhydrous medium. Explain how water weakens these interactions.

45. Water has a significant effect on the shape of proteins. Consider the following segment of a polypeptide. How would the side chains of the amino acid residues interact in (a) an anhydrous environment and (b) a hydrated environment.

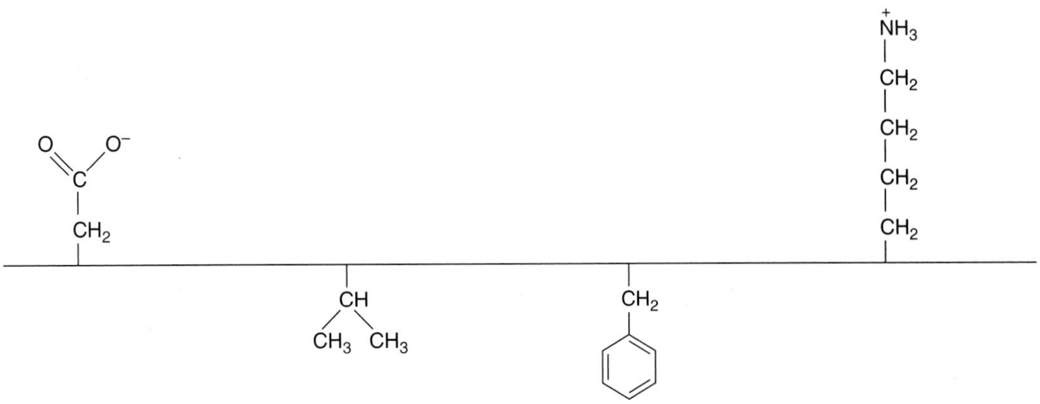

$$\begin{array}{ccc} O{=}C{-}O^- & & \overset{+}{N}H_3 \\ | & & | \\ CH_2 & & CH_2 \\ & & | \\ & & CH_2 \\ & & | \\ & & CH_2 \\ & & | \\ & & CH_2 \\ \end{array}$$

$$\underset{CH_3 \quad CH_3}{CH} \qquad \underset{\text{(phenyl)}}{CH_2}$$

46. In many cells that can survive severe dehydration, certain sugars replace water. These sugars interact with and protect membrane surfaces and prevent protein aggregation. What structural feature of the sugar molecules is responsible for this phenomenon?

47. Consider the following ion series:

$$Mg^{2+} > Ca^{2+} > Na^+ > K^+ > Cl^- > NO_3^-$$

Ions to the left are more strongly hydrated than ions to the right. Indicate whether ions such as Mg^{2+} and Cl^- would move easily into the structured water that is associated with cellular macromolecules.

48. Suggest a structure for the micelle formed by dissolving the following molecule in water.

$$^+Na \; ^-OOC{-}(CH_2)_{16}{-}COO^- \; Na^+$$

49. Pure sugars are often crystalline solids. Frequently, the process used to concentrate aqueous solutions of sugars produces syrups rather than crystals. Explain.

50. Sketch the titration curve of the amino acid tyrosine starting with the following structure.

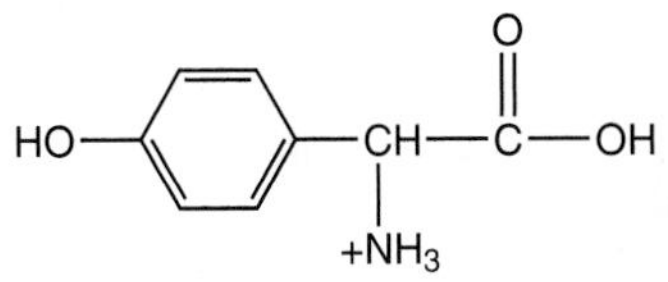

$$HO{-}\langle\text{ring}\rangle{-}\underset{\underset{+NH_3}{|}}{CH}{-}\overset{\overset{O}{\|}}{C}{-}OH$$

The pK_a values are as follows: amino group $= 9.11$, carboxyl group $= 2.2$, and side chain hydroxyl group $= 10.07$.

51. The heat absorbed or liberated by a substance (q) can be calculated using the following equation: $q = mc\Delta T$, where m is the mass in grams, c is the heat capacity per unit mass, and ΔT is the change in temperature. Use the following values to calculate the energy required to convert one gram of ice at 0°C to one gram of steam at 100°C. The heat capacity of water is 4.25 J/g · °C. The heat of fusion (the amount of heat required to change a solid into a liquid at its melting point) of ice is 335 J/g. The heat of vaporization of water is 2258 J/g.

52. Calculate the energy required to convert solid hydrogen sulfide (H_2S) to a gas. The heat capacity of H_2S is 1.03 J/g · °C. Its heats of fusion and vaporization are 69.9 and 549 J/g, respectively. Compare your answer with the value for water.

53. Potassium chloride (KCl) is slightly soluble in methyl alcohol. Draw the hydration sphere of the potassium ion.

54. Consider the following compound:

$$Cl{-}CH_2{-}(CH_2)_{10}{-}CH_2{-}COOH$$

Would this molecule form into a lipid bilayer and, if so, what would it look like?

55. Water forms stronger hydrogen bonds than ammonia. Suggest a reason for this.

56. When acetic acid ionizes under normal conditions, both the departing proton and the acetate anion are solvated by water molecules. In the absence of water, would you expect the pK_a of acetic acid to be larger or smaller? Explain your answer.

Energy

OUTLINE

Energy Transformation.
The cheetah, the fastest land
animal on earth, transforms
the chemical bond energy in
food to the energy required
to track and capture prey.

Overview

ALL LIVING ORGANISMS HAVE AN UNRELENTING REQUIREMENT FOR
ENERGY. THE FLOW OF ENERGY THAT SUSTAINS MOST FORMS OF LIFE ON
earth originates in the sun, where thermonuclear reactions generate radiant
energy. A small amount of the solar energy that reaches the earth is captured
by plants and certain microorganisms. During photosynthesis, organisms called
phototrophs convert solar energy into the chemical bond energy of sugar molecules. This bond energy is used to produce the vast array of organic molecules
found in living organisms and to drive processes such as active transport, cell
division, endocytosis, and muscle contraction. Ultimately, chemical bond
energy is converted into heat, which is then dissipated into the environment.

Every event in the universe, from the collisions among individual atoms in
laboratory test tubes to the explosions of stars in deep space, involves
energy. The energy in the universe comes in many interconvertible forms:
gravitational, nuclear, radiant, heat, mechanical, electrical, and chemical.
Despite its obvious importance, however, a precise definition of energy remains
elusive. According to modern scientific theory, energy is *the* basic constituent
of the universe. The relationship between matter and its energy equivalent is
defined by Einstein's famous equation $E = mc^2$. In other words, energy and matter are interconvertible: matter is condensed energy. The total energy (E) in
joules ($kg \cdot m^2/s^2$) in a particle is equal to the mass (m) in kilograms of the particle multiplied by the speed of light ($c = 3.0 \times 10^8$ m/s) squared. Energy is
more commonly defined, however, as the capacity to do work. Work is organized molecular motion that involves the movement of an object caused by
the use of force, and the various forms of energy differ in the amount of work
they can accomplish. Electromagnetic radiation, electrical energy, and chemical energy are high-quality energy sources. Heat, in contrast, is low-quality
energy. Compare, for example, the work potential of electricity as it enters a
building carried by electrical wire and the heat energy radiating from a light-
bulb. In living organisms, work is performed by thousands of molecular
machines. Each machine repetitively performs a single task powered directly or
indirectly by the energy provided by adenosine triphosphate (ATP). Typical
examples of work in living cells include the maintenance of concentration gradients across membranes and the synthesis of biomolecules. Among the most
common types of molecular machines are contractile proteins, transporter complexes, and enzymes.

The investigation of energy transformations that accompany physical and
chemical changes in matter is called **thermodynamics**. The principles of thermodynamics are used to evaluate the flow and interchanges of matter and energy.
Bioenergetics, a branch of thermodynamics, is the study of energy transformations in living organisms. It is especially useful in determining the direction and
extent to which specific biochemical reactions occur. These reactions are affected
by three factors. Two of these, **enthalpy** (total heat content) and **entropy** (disorder), are related to the first and second laws of thermodynamics, respectively.
The third factor, called **free energy** (energy available to do chemical work and
a measure of the spontaneity of chemical reactions), is explained by a mathematical relationship between enthalpy and entropy.

The chapter begins with a brief description of thermodynamic concepts and
their relationship to biochemical reactions. This is followed by a discussion of

free energy, a useful measure of the spontaneity of chemical reactions. The chapter ends with a description of the structure and function of ATP and other high-energy compounds. The discussion of oxidation-reduction (redox) reactions, the cell's primary mechanism for generating energy, is deferred to Chapter 9 where their roles in metabolism is described.

4.1 THERMODYNAMICS

The modern concept of energy is an invention of the Industrial Revolution. In the nineteenth century, investigations of the relationship between mechanical work and heat by engineers, physicists, mathematicians, physiologists, and physicians led to the discovery of a set of rules, called the *laws of thermodynamics*, that describe energy transformations. The three laws of thermodynamics are as follows:

1. **The first law of thermodynamics:** The total amount of energy in the universe is constant. Energy can neither be created nor destroyed, but it can be transformed from one form into another.

2. **The second law of thermodynamics:** The disorder of the universe always increases. In other words, all chemical and physical processes occur spontaneously only when the disorder of the universe increases.

3. **The third law of thermodynamics:** As the temperature of a perfect crystalline solid approaches absolute zero (0 K), disorder approaches zero.

The first two laws are powerful tools that biochemists use to investigate the energy transformations in living systems.

Thermodynamics is concerned with heat and energy transformations. Such transformations are considered to take place in a "universe" composed of a system and its surroundings (Figure 4.1). A system is defined according to the interests of the investigator: an entire organism, for example, or a single cell or a reaction occurring in a flask. In an *open system* matter and energy are exchanged between the system and its surroundings. If only energy can be exchanged with the surroundings, then the system is said to be *closed*. Living organisms, which consume nutrients from their surroundings and release waste products into it, are open systems.

Knowledge of thermodynamic functions such as enthalpy, entropy, and free energy enables biochemists to predict whether a process is spontaneous (thermodynamically favorable). Conditions of spontaneity alone do not indicate that a reaction will occur (is kinetically favorable), only that it can occur under the right set of conditions. Reactions are kinetically favorable only if there is sufficient energy available to the system potentially undergoing change.

Several thermodynamic properties are state functions. The values of state functions depend only on their initial and final states and are independent of the pathway taken to get from the initial state to the final state. For example, the energy content of a glucose molecule is the same regardless of whether it was synthesized via photosynthesis or the breakdown of lactose (milk sugar). How the energy of a reaction is distributed, however, is not fixed but is governed by the system or pathway undergoing change. For example, living cells use some of the energy in glucose molecules to perform cellular work such as muscle contraction. The remainder is released as disordered heat energy. **Work** (the displacement or movement of an object by a force) and heat are not state functions; that is, their values vary with the pathway. If glucose molecules are instead ignited in a laboratory dish, the overall reaction is the same; however, all of the chemical bond energy in the glucose is transformed directly into heat, and little or no measurable work is performed. The energy content of the glucose molecules is the same in each process. The work accomplished by each process is different.

The exchange of energy between a system and its surroundings can occur in only two ways: heat (q), random molecular motion, may be transferred to or from

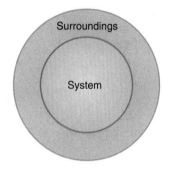

FIGURE 4.1

A Thermodynamic Universe

A universe consists of a system and its surroundings.

the system, or the system may do work (w) on its surroundings or have work done on it by its surroundings. Energy is transferred as heat when the system and its surroundings are at different temperatures. Energy is transferred as work when an object is moved by force.

First Law of Thermodynamics

The first law of thermodynamics expresses the relationship between the internal energy (E) of a closed system and the heat (q, or disorganized motion) and work (w, or organized motion) transferred between the system and its surroundings. It is an alternative statement of the law of conservation of energy, according to which the total energy of an isolated system (e.g., our universe) is constant. With respect to a closed system, the first law can be stated as

$$\Delta E = q + w \tag{1}$$

where ΔE = the change in energy of the system
 q = the heat absorbed by the system
 w = the work done by the system

Chemists have defined the term enthalpy (H), which is related to internal energy by the equation

$$H = E + PV \tag{2}$$

where PV = pressure-volume work, that is, the work done on or by a system that involves changes in pressure and volume

In biochemical systems in which pressure is nearly constant and volume changes are negligible, changes in enthalpy are essentially equal to changes in internal energy:

$$\Delta H = \Delta E \tag{3}$$

If ΔH is negative ($\Delta H < 0$), the reaction or process gives off heat and is referred to as **exothermic**. If ΔH is positive ($\Delta H > 0$), heat is absorbed from the surroundings, and the process by which it is emitted is called **endothermic**. In **isothermic** processes ($\Delta H = 0$), heat is not exchanged with the surroundings.

Equation (3) indicates that the total energy change of a biological system is equivalent to the heat evolved or absorbed by the system. Because the enthalpy of a reactant or product is a state function (independent of pathway), then the enthalpy change for any reaction forming that substance can be used to calculate the ΔH of a reaction involving that substance. If the sum of the ΔH values ($\Sigma \Delta H$) for both the reactants and the products is known, then the enthalpy change for the reaction can be calculated by using the following equation:

$$\Delta H_{\text{reaction}} = \Sigma \Delta H_{\text{products}} - \Sigma \Delta H_{\text{reactants}} \tag{4}$$

The standard enthalpy of formation per mole (25°C, 1 atm), symbolized by ΔH_f°, is commonly used in enthalpy calculations; H_f° is the energy evolved or absorbed when 1 mol of a substance is formed from its most stable elements. Note that Equation (4) cannot predict the direction of any chemical reaction. It determines only the heat flow. Problem 4.1 gives a standard enthalpy calculation.

Second Law of Thermodynamics

The first law accounts for the energy changes that can occur during a process, but it cannot be used to predict whether and to what extent a specific process will occur. In some circumstances, whether certain processes occur appears to be obvious: for example, the behavior of ice at room temperature or gasoline in an internal combustion engine. Experience tells us that ice melts at temperatures above

KEY CONCEPTS

- At constant pressure, a system's enthalpy change ΔH is equal to the flow of heat energy.
- If ΔH is negative, the reaction or process is exothermic. If ΔH is positive, the reaction or process is endothermic. In isothermic processes no heat is exchanged with the surroundings.

WORKED PROBLEM 4.1

Given the following ΔH_f° values, where ΔH_f° is the energy change required to produce a compound from its elements, calculate ΔH_f° for the reaction

$$6CO_2 + 6H_2O \rightarrow C_6H_{12}O_6 + 6O_2$$

	ΔH_f°	
	kcal/mol	kJ/mol
$C_6H_{12}O_6$	−304.7	−1274.9
CO_2	−94.0	−393.3
H_2O	−68.4	−286.2
O_2	0	0

The units in the table have the following definitions: 1 kcal is the energy required to raise the temperature of 1000 g of water 1°C; the joule (J) is a unit of energy that is gradually replacing the calorie (cal) in scientific usage (1 cal = 4.184 J).

Solution

The total enthalpy for a reaction is equal to the sum of enthalpy values of the products minus those of the reactants.

$$
\begin{array}{ccc}
6CO_2 + & 6H_2O & \rightarrow \quad C_6H_{12}O_6 + 6O_2 \\
6(-393.3) + & 6(-286.2) & -1274.9 + 6(0) \\
-2359.8 + & (-1717.2) & -1274.9
\end{array}
$$

$$\Delta H = -1274.9 - (-4077.0) = 2802.1 \text{ kJ/mol}$$

The positive ΔH indicates that the reaction is endothermic. ∎

0°C and that gasoline molecules can be converted to energy in the presence of oxygen to CO_2 and H_2O. Physical or chemical changes that occur with the release of energy are said to be **spontaneous**. Nonspontaneous processes are those that occur when a constant input of energy is required to support a change. Experience convinces us that certain processes will not occur: ice will not form at temperatures above 0°C, and gasoline molecules are not formed from an engine's exhaust fumes. In other words, we intuitively understand that there is directionality to these processes and that predictions about their outcome are easily made. When experience cannot be relied on to allow us to make predictions concerning spontaneity and direction, the second law can be used. According to the second law, all spontaneous processes occur in the direction that increases the total disorder of the universe (a system and its surroundings) (Figure 4.2). As a result of spontaneous processes, matter and energy become more disorganized. Gasoline molecules, for example, are hydrocarbons in which carbon atoms are linked in an orderly arrangement. When gasoline burns, the carbon atoms in the gaseous products are randomly dispersed (Figure 4.3). Similarly, the energy that is released as gasoline burns becomes more disordered; it becomes less concentrated and less useful. In a car engine, increased gas pressure in the cylinders drives the pistons and causes the car to move. When we compare the chemical energy in the gasoline molecules and the kinetic energy that moves the car, it becomes apparent that a significant amount of energy does no useful work. Rather, it is dissipated (dispersed) into the surroundings, as evidenced by the hot engine and exhaust fumes.

The degree of disorder of a system is measured by the state function called **entropy** (S). The more disordered a system is, the greater is its entropy value.

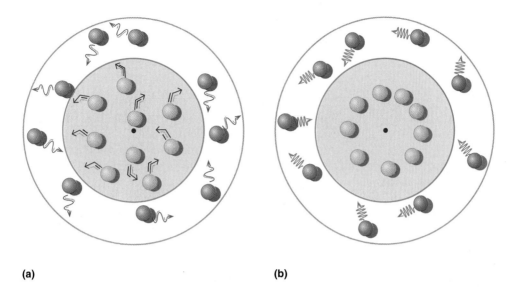

(a)   **(b)**

FIGURE 4.2

A Living Cell as a Thermodynamic System

(a) The molecules of the cell and its surroundings are in a relatively disordered state.
(b) Heat is released from the cell as a consequence of reactions that create order among the molecules inside the cell. This energy increases the random motion, and therefore the disorder, of the the molecules outside the cell (indicated by tighter springs on the outer molecules). This process causes a net positive entropy change. The cell's decrease in entropy is more than offset by an increase in the entropy of the surroundings.

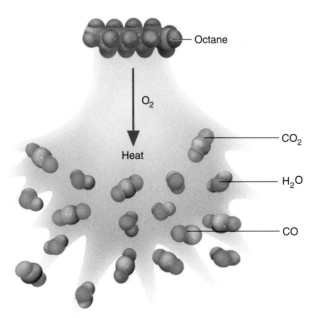

FIGURE 4.3

Gasoline Combustion

When hydrocarbons such as octane are burned, the release of energy is accompanied by the conversion of highly ordered reactant molecules into relatively disorganized gaseous products such as CO_2 and H_2O. However, gasoline combustion is inefficient; that is, other substances such as the environmental pollutant carbon monoxide (CO) are also released.

KEY CONCEPTS

- The second law of thermodynamics states that the universe tends to become more disorganized.
- Entropy increases may take place anywhere in the system's universe.
- For processes in living organisms, the increase in entropy takes place in the surroundings.

$$\Delta G = \Delta H - T\Delta S$$

ΔH negative
Energy released during the reaction

ΔS positive
Randomness or disorder in the system increases. If $T\Delta S$ is sufficiently large, then ΔG will be negative and ΔS_{univ} will increase (favorable reaction).

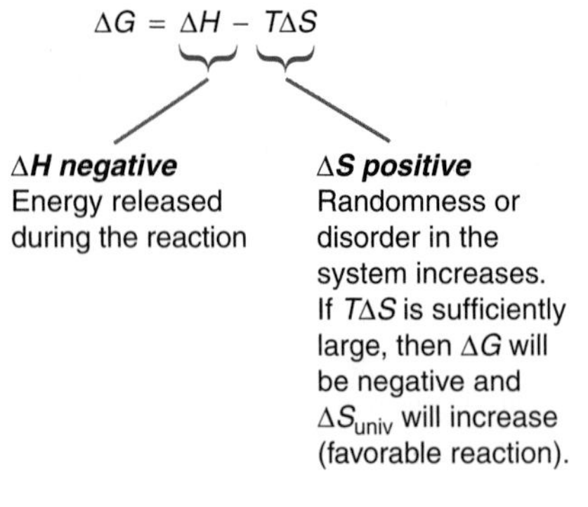

FIGURE 4.4

The Gibbs Free Energy Equation

At constant pressure, enthalpy (H) is essentially equal to the total energy content of the system. A process is spontaneous if it decreases free energy. At constant temperature and pressure, free energy changes (ΔG) are negative if enthalpy decreases or if the entropy term $T\Delta S$ is sufficiently large.

According to the second law, the entropy change of the universe is positive for every spontaneous process. The increase may take place in any part of the universe (ΔS_{sys} or ΔS_{surr}):

$$\Delta S_{univ} = \Delta S_{sys} + \Delta S_{surr}$$

Living cells do not increase their internal disorder when they consume and metabolize nutrients. The organism's surroundings increase in entropy instead. For example, the food molecules that humans consume to provide the energy and structural material needed to maintain their complex bodies are converted into vast amounts of disordered waste products (e.g., CO_2, H_2O, and heat) that are discharged into their surroundings.

Although entropy may be considered to be unusable energy, the formation of entropy is not a useless activity. In most spontaneous processes entropy increases. Some reactions are said to be entropy-driven because the increase in entropy in the system overrides a gain in enthalpy to result in a spontaneous reaction. (By definition, a spontaneous process will occur. The rate at which it occurs, however, may be very rapid or very slow.) In irreversible processes, processes that proceed in only one direction, entropy and enthalpy are driving forces. Entropy directs a system toward equilibrium with its surroundings. Once a process has reached equilibrium (i.e., there is no net change in either direction), there is no longer any driving force to propel it. To predict whether a process is spontaneous, the sign of ΔS_{univ} must be known. For example, if the value of ΔS_{univ} for a process is positive (i.e., the entropy of the universe increases), then the process is spontaneous. If ΔS_{univ} is negative, the process does not occur, but the reverse process takes place spontaneously. If ΔS_{univ} is zero, neither process tends to occur. Organisms that are at equilibrium with their surroundings are dead.

4.2 FREE ENERGY

Although the entropy of the universe always increases in a spontaneous process, measuring it is often impractical because both the ΔS_{sys} and ΔS_{surr} must be known. A more convenient thermodynamic function for predicting the spontaneity of a process is free energy, which can be derived from the expression for ΔS_{univ}:

$$\Delta S_{univ} = \Delta S_{surr} + \Delta S_{sys}$$

The ΔS_{surr} is defined as the quantity of heat exchanged per kelvin (K) of temperature in the course of a specific chemical or physical change. For an exothermic reaction, heat is released and the value of ΔH is a negative number. Therefore,

$$\Delta S_{surr} = -\Delta H/T$$

By substitution

$$\Delta S_{univ} = -\Delta H/T + \Delta S_{sys}$$

Multiply both sides by $-T$:

$$-T\Delta S_{univ} = \Delta H - T\Delta S_{sys}$$

Josiah Gibbs defined $-T\Delta S_{univ}$ as the state function called the *Gibbs free energy change* or ΔG:

$$\Delta G = \Delta H - T\Delta S_{sys}$$

At constant temperature and pressure, the change in free energy is negative when ΔS_{univ} is positive, which reflects a spontaneous reaction said to be **exergonic** (Figure 4.4). If the ΔG is positive, the process is said to be **endergonic** (nonspontaneous). When the ΔG is zero, the process is at equilibrium. As with other thermodynamic functions, ΔG provides no information about reaction rates.

Reaction rates depend on the precise mechanism by which a process occurs and are dealt with under the study of kinetics (Chapter 6).

Standard Free Energy Changes

A convention known as the *standard state* provides a uniform basis for free energy calculations. The standard free energy, $\Delta G°$, is defined for reactions at 25°C (298 K) and 1.0 atm pressure with all solutes at a concentration of 1.0 M.

The standard free energy change is related to the reaction's equilibrium constant, K_{eq}, the value of the reaction quotient at equilibrium when the forward and reverse reaction rates are equal.

The change in free energy for a reaction

$$aA + bB \rightleftharpoons cC + dD$$

is related to the reaction's equilibrium constant:

$$K_{eq} = \frac{[C]^c[D]^d}{[A]^a[B]^b}$$

Based on the observations that the free energy of an ideal gas depends on its pressure (concentration) and that the state function G can be manipulated in the same way as the state function H, the following equation was derived:

$$\Delta G = \Delta G° + RT \ln \frac{[C]^c[D]^d}{[A]^a[B]^b}$$

If the reaction is allowed to go to equilibrium, the ΔG is 0 and the expression reduces to

$$\Delta G° = -RT \ln K_{eq}$$

This equation allows the calculation of $\Delta G°$ if the K_{eq} is known. Because most biochemical reactions take place at or near pH 7 ($[H^+] = 1.0 \times 10^{-7}$ M), this exception is made in the 1.0 M solute rule in bioenergetics and the free energy change is expressed as $\Delta G°'$. Problem 4.2 is a free energy problem.

WORKED PROBLEM 4.2

For the reaction $HC_2H_3O_2 \rightleftharpoons C_2H_3O_2^- + H^+$, calculate $\Delta G°$ and $\Delta G°'$. Assume that $T = 25°C$. The ionization constant for acetic acid is 1.8×10^{-5}. Is this reaction spontaneous? Recall that

$$K_{eq} = \frac{[C_2H_3O_2^-][H^+]}{[HC_2H_3O_2]}$$

Solution

1. Calculate $\Delta G°$.

$$\begin{aligned}
\Delta G° &= -RT \ln K_{eq} \\
&= -(8.315 \text{ J/mol·K})(298 \text{ K}) \ln(1.8 \times 10^{-5}) \\
&= 27,071 = 27.1 \text{ kJ/mol}
\end{aligned}$$

The $\Delta G°$ indicates that under these conditions the reaction is not spontaneous.

2. Calculate $\Delta G°'$. Use the relation between free energy change and standard free energy change.

For this example, the expression becomes

$$\Delta G^{\circ\prime} = \Delta G^{\circ} + RT \ln[H^+]$$

Substituting values, we have

$$\begin{aligned}
\Delta G^{\circ\prime} &= 27071 \text{ J/mol} + (8.315 \text{ J/mol·K})(298 \text{ K})(\ln 10^{-7}) \\
&= 27071 - 39{,}939 \\
&= -12867.54 = -12.9 \text{ kJ/mol}
\end{aligned}$$

Under the conditions specified for ΔG° (i.e., 1 M concentrations for all reactants including H^+) the ionization of acetic acid is not spontaneous, as indicated by the positive ΔG°. When the pH value is 7, however, the reaction becomes spontaneous. A low $[H^+]$ makes the ionization of a weak acid such as acetic acid a more likely process, as indicated by the negative $\Delta G^{\circ\prime}$. ■

Coupled Reactions

Many chemical reactions within living organisms have positive $\Delta G^{\circ\prime}$ values. Fortunately, free energy values are additive in any reaction sequence.

$$A + B \rightleftharpoons C + D \qquad\qquad \Delta G^{\circ\prime}_{\text{reaction 1}} \qquad\qquad (1)$$

$$C + E \rightleftharpoons F + G \qquad\qquad \Delta G^{\circ\prime}_{\text{reaction 2}} \qquad\qquad (2)$$

$$A + B + E \rightleftharpoons D + F + G \quad \Delta G^{\circ\prime}_{\text{overall}} = \Delta G^{\circ\prime}_{\text{reaction 1}} + \Delta G^{\circ\prime}_{\text{reaction 2}} \quad (3)$$

Note that reactions (1) and (2) have a common intermediate, C. If the net $\Delta G^{\circ\prime}$ value ($\Delta G^{\circ\prime}_{\text{overall}}$) is sufficiently negative, forming the products F and G is an exergonic process.

The conversion of glucose-6-phosphate to fructose-1,6-bisphosphate illustrates the principle of coupled reactions (Figure 4.5). The common intermediate in this reaction sequence is fructose-6-phosphate. Because the formation of fructose-6-phosphate from glucose-6-phosphate is endergonic ($\Delta G^{\circ\prime}$ is +1.7 kJ/mol), the reaction is not expected to proceed as written (at least under standard conditions). The conversion of fructose-6-phosphate to fructose-1,6-bisphosphate is strongly exergonic because it is coupled to the cleavage of the phosphoanhydride bond of ATP. (The cleavage of ATP's phosphoanhydride bond to form ADP

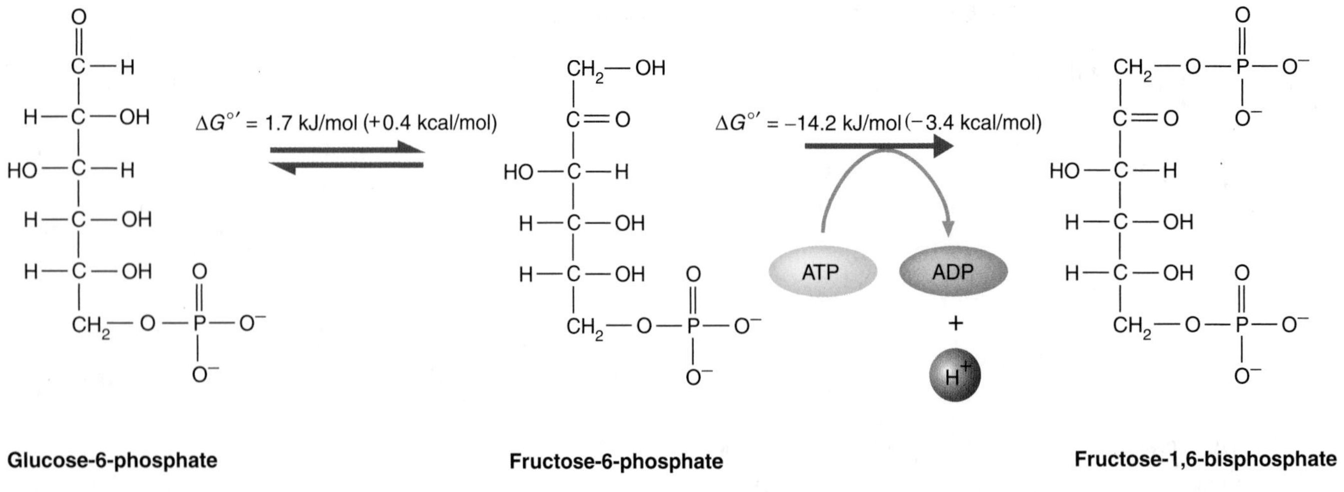

Glucose-6-phosphate   Fructose-6-phosphate   Fructose-1,6-bisphosphate

FIGURE 4.5

A Coupled Reaction

The net $\Delta G^{\circ\prime}$ value for the two reactions is −12.5 kJ/mol (−3.0 kcal/mol).

yields approximately -30.5 kJ/mol. ATP in living organisms is discussed in Section 4.3). Because $\Delta G^{\circ\prime}_{overall}$ for the coupled reactions is negative, the reactions do proceed in the direction written at standard conditions.

WORKED PROBLEM 4.3

Glycogen is synthesized from glucose-1-phosphate. To be incorporated into glycogen, glucose-1-phosphate is converted to a derivative of the nucleotide uridine diphosphate (UDP). The UDP serves as an excellent leaving group in the condensation reaction to form the glycogen polymer. The reaction is

Glucose-1-phosphate + UTP + H_2O → UDP-glucose + PP_i

where PP_i is the inorganic compound pyrophosphate.
If the $\Delta G^{\circ\prime}$ value for this reaction is approximately zero, is this reaction favorable? If PP_i is hydrolyzed, then

PP_i + H_2O → $2P_i$

where P_i is the inorganic compound orthophosphate.
The loss in free energy ($\Delta G^{\circ\prime}$) is -33.5 kJ. How does this second reaction affect the first one? What is the overall reaction? Determine the $\Delta G^{\circ\prime}_{overall}$ value.

Solution
The overall reaction is

Glucose-phosphate + UTP → UDP-glucose + $2P_i$

$$\Delta G^{\circ\prime}_{overall} = \Delta G^{\circ\prime}_{reaction\ 1} + \Delta G^{\circ\prime}_{reaction\ 2}$$
$$= 0 + (-33.5\ kJ)$$
$$= -33.5\ kJ$$

The hydrolysis of PP_i drives the formation of UDP-glucose to the right. ■

QUESTION 4.1

In living cells, the concentrations of ATP and the products of its hydrolysis (ADP and P_i) are significantly lower than the standard 1 M concentrations. Therefore the actual free energy of hydrolysis of ATP ($\Delta G'$) differs from the standard free energy ($\Delta G^{\circ\prime}$). Unfortunately, it is difficult to obtain an accurate measure of the concentrations of cellular components. For this reason, only estimates can be made. The following equation includes a correction for nonstandard concentrations:

$$\Delta G' = \Delta G^{\circ\prime} + RT \ln \frac{[ADP][P_i]}{[ATP]}$$

The temperature is 37°C. Assume that the pH is 7. In a liver cell, the concentrations (mM) are as follows:

ATP = 4.0,   ADP = 1.35, P_i = 4.65
$\Delta G^{\circ\prime}$ = -30.5 kJ/mol

What is the actual $\Delta G'$ for the hydrolysis of ATP under these conditions?

The Hydrophobic Effect Revisited

Understanding the spontaneous aggregation of nonpolar substances in water is enhanced by consideration of thermodynamic principles. When nonpolar

BIOCHEMISTRY IN PERSPECTIVE

Nonequilibrium Thermodynamics

How does thermodynamic theory relate to energy flow in living organisms? The thermodynamic concepts described in this chapter, referred to as classical thermodynamics, were discovered during investigations of internal combustion engines in the nineteenth century. Classical thermodynamics explains energy flow in ideal systems in or near equilibrium. Living organisms, however, are open systems that are never at equilibrium until they die. In contrast to stable systems that are in thermodynamic equilibrium, systems that are far from equilibrium are inherently unstable. Thus, a critical question arises: How can an organized living system (a living organism) not in equilibrium remain structurally stable for an extended period of time?

The properties of a Benard cell provide a clue to the phenomenon of order within a universe that favors disorder. A Benard cell is a fluid-filled insulated container that is fitted with a cold reservoir on top and a heat source at the bottom. The liquid placed in the container at the beginning of the experiment has a uniform temperature. As the temperature of the liquid at the bottom of the container gradually increases, a temperature gradient is created. Warm (less dense) liquid begins to rise and cooler (more dense) liquid moves downward. As a specific temperature threshold is reached, convection currents that are organized, rotating, and dynamically stable form spontaneously. The term used to describe the capacity of far-from-equilibrium systems such as the Benard cell to form ordered structures under the influence of an energy gradient is *dissipative*. Living organisms are **dissipative systems** that facilitate the reduction in the enormous energy gradient between the sun and the earth. Energy dissipation begins when photosynthetic organisms capture a fraction of the total solar radiation (about 10^{18} kJ/day). Phototrophs dissipate a portion of the captured energy by producing their own ordered structure. The dissipation process continues as animals and other heterotrophs consume phototrophs. Eventually all of the energy captured from the sun is released as heat (disorganized energy). Living organisms can be compared to the phenomenon of the Benard cell only in the limited sense that the energy-driven creation of ordered structures is reminiscent of the organized convection currents in the Benard cell.

The critical property of energy flow in living organisms is that equilibrium is never reached because the key feature of the living state is the ability to dissipate energy. The coherent structure of living systems, made possible by the uninterrupted flow of high-quality energy, is maintained by the capacity to release heat and more disordered waste products into the surroundings. The maintenance of dissipative systems requires that continuous work be done on the system because otherwise all natural processes will proceed toward equilibrium. In living organisms this far-from-equilibrium state is maintained by transport, chemical, and mechanical work.

The evolution of living systems is driven by the size of the energy gradient to be dissipated and the dictates of the second law of thermodynamics, namely, the required increase in entropy in the universe. But although the second law determines the direction of living processes, it is insufficient to explain the precise molecular mechanisms that sustain life. The mechanisms by which energy flow is coupled to the performance of work to build and maintain living organisms evolved over several billion years, and the precise details of the mechanism of energy dissipation by living organisms have yet to be resolved. Through trial and error, living organisms, taking advantage of the physical and chemical properties of elements such as carbon, nitrogen, and oxygen, have developed gigantic and complex energy-dissipating biochemical and information-processing networks. The vast species diversification observed on Earth can, therefore, be viewed as a means to provide the greatest number of pathways for energy dissipation. The region of the planet that exhibits the greatest number of species of living organisms is at the equator, where the energy gradient between the sun and Earth is at its largest. Unsurprisingly, nonequilibrium thermodynamics is an active area of research.

SUMMARY: Living organisms are far-from-equilibrium dissipative structures. They create internal organization via a continuous flow of energy.

molecules are mixed with water, they disrupt water's energetically favorable hydrogen-bonded interactions. The hydrogen bonds that stabilize the highly ordered cagelike structures around clusters of nonpolar molecules restrict the motion of the water molecules, thus resulting in a decrease in entropy. Consequently, the free energy of dissolving nonpolar molecules is unfavorable (i.e., ΔG is positive because ΔH is positive and $-T\Delta S$ is strongly positive). The decrease in entropy, however, is proportional to the surface area of contact between nonpolar molecules and water. The aggregation of nonpolar molecules significantly decreases the surface area of their contact with water, and thus the water becomes less ordered (i.e., the entropy change, ΔS, is now positive). Because $-T\Delta S$ becomes negative, the free energy of the process is negative, and therefore, it proceeds spontaneously. The spontaneous exclusion of water by hydrophobic groups and molecules is a major factor in biological processes such as protein folding and the assembly of supramolecular structures such as membranes.

4.3 THE ROLE OF ATP

Adenosine triphosphate is a nucleotide that plays an extraordinarily important role in living cells. The hydrolysis of ATP (Figure 4.6) immediately and directly provides the free energy to drive an immense variety of endergonic biochemical reactions. Produced from ADP and P_i with energy released by the breakdown of food molecules and the light reactions of photosynthesis, ATP drives processes of several types (Figure 4.7). These include (1) biosynthesis of biomolecules, (2) active transport of substances across cell membranes, and (3) mechanical work such as muscle contraction.

ATP is ideally suited to its role as universal energy currency because of its structure (Figure 4.8). ATP is a nucleotide composed of adenine, ribose, and a triphosphate unit. Its two terminal phosphoryl groups ($-PO_3^{2-}$) are linked by phosphoanhydride bonds. Although anhydrides are easily hydrolyzed, the phosphoanhydride bonds of ATP are sufficiently stable under mild intracellular conditions. Specific enzymes facilitate ATP hydrolysis.

The tendency of ATP to undergo hydrolysis, also referred to as its **phosphoryl group transfer potential**, is not unique. A variety of biomolecules can transfer phosphate groups to other compounds. Table 4.1 lists several important examples.

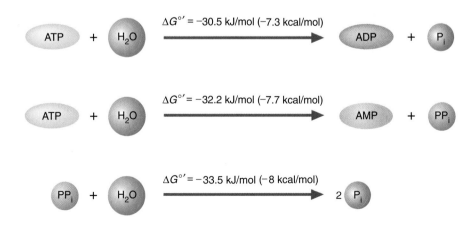

FIGURE 4.6

Hydrolysis of ATP

ATP may be hydrolyzed to form ADP and P_i (orthophosphate) or AMP (adenosine monophosphate) and PP_i (pyrophosphate). Pyrophosphate may be subsequently hydrolyzed to orthophosphate, releasing additional free energy. The hydrolysis of ATP to from AMP and pyrophosphate is often used to drive reations with high positive $\Delta G^{\circ\prime}$ values or to ensure that a reaction goes to completion.

FIGURE 4.7
The Role of ATP

ATP is an intermediate in the flow of energy from food molecules to the biosynthetic reations of metabolism.

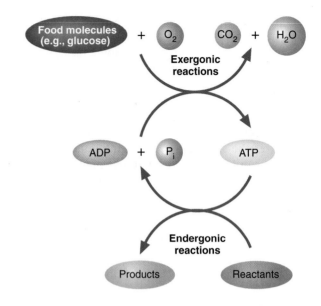

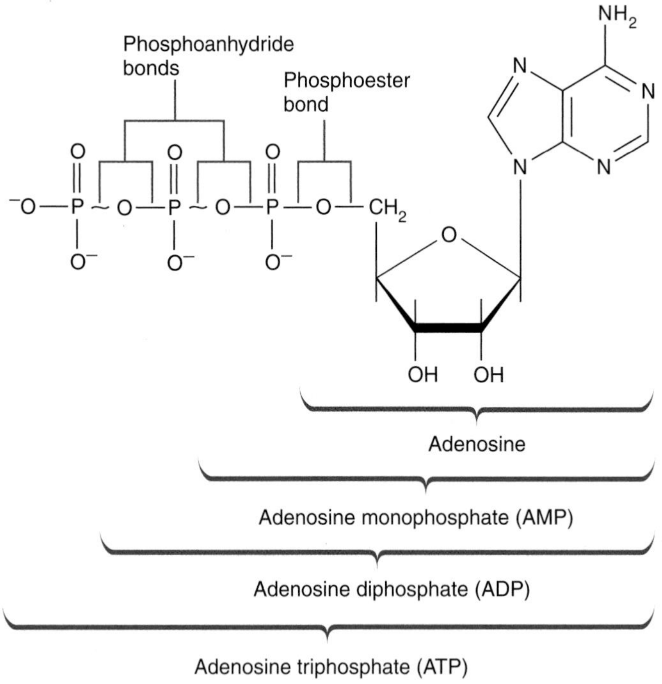

FIGURE 4.8
The Structure of ATP

The squiggles (~) in ATP indicate that the bonds so connected are easily hydrolyzed.

Phosphorylated compounds with high negative $\Delta G^{\circ\prime}$ values of hydrolysis have larger phosphate group transfer potentials than those compounds with smaller, negative values. Because ATP has an intermediate phosphoryl group transfer potential, it can be an intermediate carrier of phosphoryl groups from higher-energy compounds such as phosphoenolpyruvate to low-energy compounds (Figure 4.9). ATP is therefore the "energy currency" for living systems, because cells usually transfer phosphate by coupling reactions to ATP hydrolysis. The two phosphoanhydride bonds of ATP are often referred to as "high energy." The term *high-energy bond* is now considered inappropriate, however, because it denotes instability of the bond and, therefore, its ability to participate in reactions rather

TABLE 4.1 Standard Free Energy of Hydrolysis of Selected Phosphorylated Biomolecules

Molecule	$\Delta G^{\circ\prime}$ kcal/mol	kJ/mol
Glucose-6-phosphate	-3.3	-13.8
Fructose-6-phosphate	-3.8	-15.9
Glucose-1-phosphate	-5	-20.9
ATP $\rightarrow$ ADP + P_i	-7.3	-30.5
ATP $\rightarrow$ AMP + PP_i	-7.7	-32.2
$PP_i \rightarrow 2P_i$	-8.0	-33.5
Phosphocreatine	-10.3	-43.1
Glycerate-1,3-bisphosphate	-11.8	-49.4
Carbamoyl phosphate	-12.3	-51.5
Phosphoenolpyruvate	-14.8	-61.9

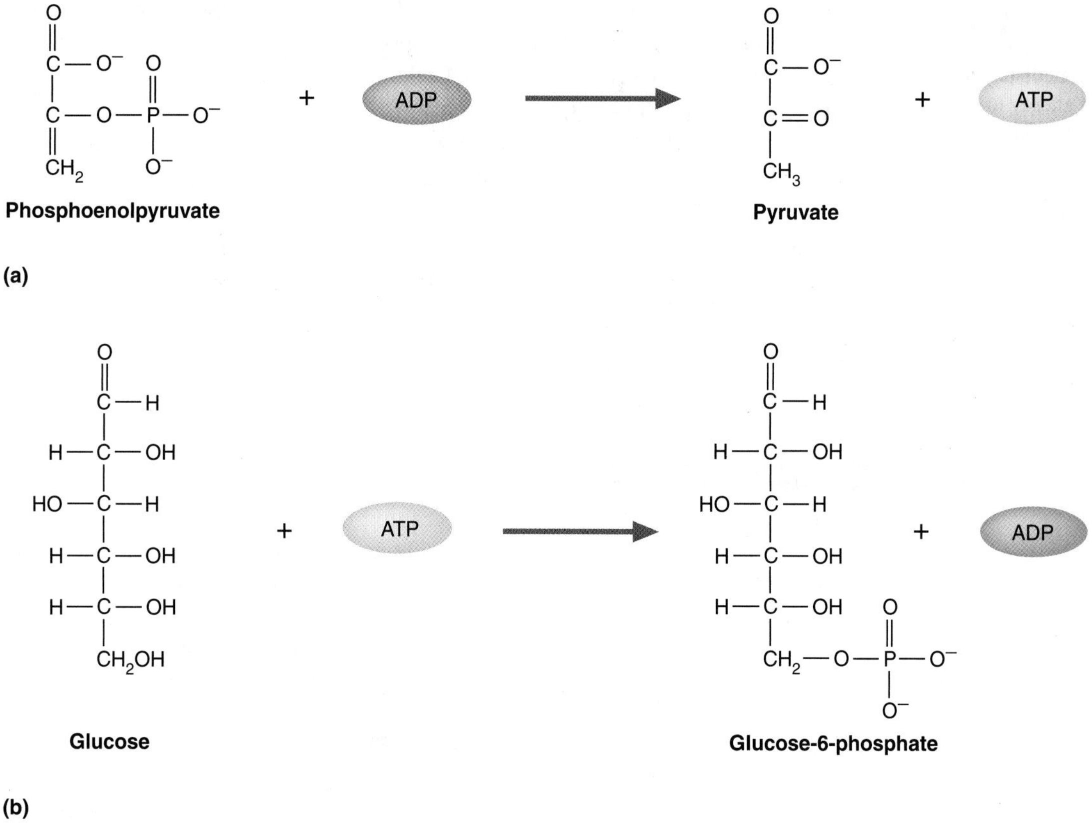

(a)

(b)

FIGURE 4.9

Transfer of Phosphoryl Groups

(a) Transfer of a phosphoryl group from phosphoenolpyruvate to ADP. As discussed in Chapter 8, this reaction is one of two steps that form ATP during glycolysis, a reaction pathway that breaks down glucose. (b) Transfer of a phosphoryl group from ATP to glucose. The product of this reaction, glucose-6-phosphate, is the first intermediate formed during glycolysis.

than the quantitative value of the bond energy. To understand why ATP hydrolysis is so exergonic, several factors must be considered.

1. At typical intracellular pH values, the triphosphate unit of ATP carries three or four negative charges that repel each other. Hydrolysis of ATP reduces electrostatic repulsion.

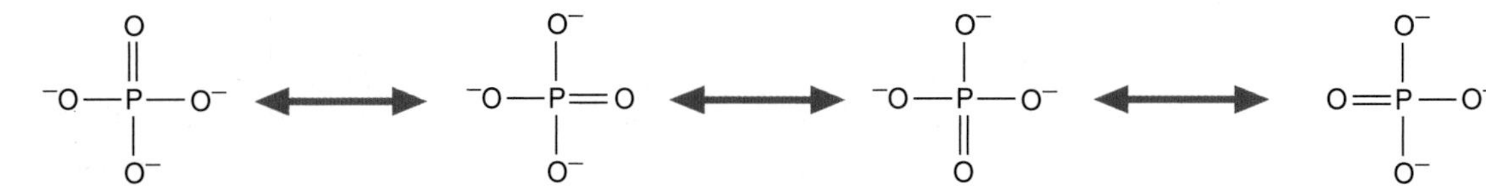

FIGURE 4.10

Contributing Structure of the Resonance Hybrid of Phosphate

At physiological pH, orthophosphate is HPO_4^{2-}. In this illustration, H^+ is not assigned permanently to any of the four oxygen atoms.

KEY CONCEPTS

- The hydrolysis of ATP immediately and directly provides the free energy to drive an immense variety of endergonic biochemical reactions.

- Because ATP has an intermediate phosphoryl group transfer potential, it can carry phosphoryl groups from high-energy compounds to low-energy compounds.

- ATP is the energy currency for living systems.

2. Because of *resonance hybridization*, the products of ATP hydrolysis are more stable than ATP. When a molecule has two or more alternative structures that differ only in the position of electrons, the result is called a **resonance hybrid**. The electrons in a resonance hybrid with several contributing structures possess much less energy than those with fewer contributing structures. The contributing structures of the phosphate resonance hybrid are illustrated in Figure 4.10.

3. The hydrolyzed products of ATP, either ADP and P_i or AMP and PP_i, are more easily solvated than ATP. Recall that the water molecules that form the solvation spheres around ions shield them from one another. The resulting decrease in the repulsive force between phosphoryl groups drives the hydrolytic reaction.

4. There is an increase in disorder caused by an increase in the number of molecules. ATP is converted into two molecules (ADP and P_i), both of which are now moving randomly.

QUESTION 4.2

Walking consumes approximately 100 kcal/mi. In the hydrolysis of ATP (ATP → ADP + P_i), the reaction that drives muscle contraction, $\Delta G^{\circ\prime}$ is −7.3 kcal/mol (−30.5 kJ/mol). Calculate how many grams of ATP must be produced to walk a mile. ATP synthesis is coupled to the oxidation of glucose ($\Delta G^{\circ\prime} = -686$ kcal/mol). How many grams of glucose are actually metabolized to produce this amount of ATP? (Assume that only glucose oxidation is used to generate ATP and that 40% of the energy generated from this process is used to phosphorylate ADP. The gram molecular weight of glucose is 180 g, and that of ATP is 507 g.)

BIOCHEMISTRY IN PERSPECTIVE

The Extremophiles: Organisms That Make a Living in Hostile Environments

How can living organisms sustain life in very hostile environments? The *extremophiles*, the unique organisms that live in harsh or extreme environments, are classified according to the specific conditions to which they are adapted. Examples include thermophiles and hyperthermophiles (high temperature), acidophiles (low pH), piezophiles (high pressure), and halophiles (high salt concentration). Such organisms, many of which are archaeans, often thrive in habitats with several extreme conditions. For example, the hot water near hydrothermal vents is under very high pressure. As with all living organisms, the extremophiles require energy. Energy generation involves oxidation-reduction (redox) reactions (described in Chapter 8) in which free energy is released as electrons are transferred from an **electron donor** to an **electron acceptor**.

Briefly, there are three energy-generating mechanisms: photosynthesis, chemoorganotrophy, and chemolithotrophy. Photosynthesis (Chapter 13) is a process that converts light energy into chemical energy (ATP). Most photosynthetic organisms produce O_2 as a by-product. In certain microorganisms O_2 evolution does not occur. *Chemoorganotrophs* and *chemolithotrophs* generate ATP by oxidizing organic and inorganic compounds, respectively.

Chemotrophs have two possible mechanisms of synthesizing ATP: fermentation and respiration. **Fermentation** is a biochemical process in which energy capture occurs via the oxidation of organic molecules. ATP is synthesized from ADP by substrate-level phosphorylation, the direct transfer of a phosphate group from a phosphorylated organic molecule. Fermentation is inefficient because the product molecules released as waste are only partially oxidized. **Respiration** is a more sophisticated process in which a series of electron carrier molecules reversibly accept and donate electrons. The energy captured is used to create an electrochemical gradient that drives ATP synthesis. In **aerobic respiration**, the low-energy electrons

that result from this process are donated to O_2. In **anaerobic respiration**, terminal electron acceptors other than O_2 are used. Examples of the latter include ferric iron (Fe^{3+}), nitrate (NO_3^-), or certain organic molecules. The following examples of energy-generating mechanisms used by selected extremophiles provide a hint of the diversity among these remarkable organisms.

Methanococcus janaschii is a hyperthermophile. It lives near hydrothermal vents at temperatures of 80°C or greater. It is also a piezophile. As its name suggests, this archaean is a methane (CH_4) producer. Under strictly anaerobic conditions, *M. jannaschii* obtains energy by converting CO_2 to CH_4 in a series of complex electron transfer reactions. The electron donor is hydrogen gas (H_2), a product of geochemical processes. The net reaction is

$$CO_2 + 4H_2 \rightarrow CH_4 + 2H_2O$$

Bacillus infernus (called the "bacillus from hell") is a thermophile that grows at 60°C. This halotolerant (0.6 M Na^+) bacterium was discovered 2700 m below the earth's surface. *B. infernus*, isolated in nutrient-deficient environments, has a very slow rate of reproduction. It derives energy from the fermentation of the sugar glucose or the anaerobic respiration of electron donors such as formate ($HCOO^-$) and lactate ($CH_3CHOHCOO^-$). Electron acceptors include manganese dioxide (MnO_2), ferric iron (Fe^{3+}), and nitrate (NO_3^{2-}).

Thiobacillus ferroxidans is an aerobic, acidophilic (pH 2–4) chemolithotroph that is commonly found in sulfate-containing acid drainage from coal mines. It derives energy from ferrous iron (Fe^{2+}) and reduced forms of sulfur such as H_2S and iron sulfide (FeS). The products of this process are Fe^{3+} and sulfuric acid. Acid drainage pollutes lakes and rivers and kills aquatic life.

SUMMARY: Extremophilic organisms utilize an array of oxidation-reduction reactions to generate the energy required to sustain life in hostile environments.

Chapter**Summary**

1. All living organisms unrelentingly require energy. Bioenergetics, the study of energy transformations, can be used to determine the direction and extent to which biochemical reactions proceed. Enthalpy (a measure of heat content) and entropy (a measure of disorder) are related to the first and second laws of thermodynamics, respectively. Free energy (the portion of total energy that is available to do work) is related to a mathematical relationship between enthalpy and entropy.

2. Energy and heat transformations take place in a "universe" composed of a system and its surroundings. In an open system, matter and energy are exchanged between the system and its surroundings. If energy but not matter can be exchanged with the surroundings, then the system is said to be closed. Living organisms are open systems.

3. Several thermodynamic quantities are state functions; that is, their value does not depend on the pathway used to make or degrade a specific substance. Examples of state functions are total energy, free energy, enthalpy, and entropy. Quantities such as work and heat depend on the pathway and thus are not state functions.

4. Free energy, a state function that relates the first and second laws of thermodynamics, represents the maximum useful work obtainable from a process. Exergonic processes, that is, processes in which free energy decreases ($\Delta G < 0$), are spontaneous. If the free energy change is positive ($\Delta G > 0$), the process is called endergonic. A system is at equilibrium when the free energy change is zero. The standard free energy ($\Delta G°$) is defined for reactions at 25°C, 1 atm pressure, and 1 M solute concentrations. The standard pH in bioenergetics is 7. The standard free energy change $\Delta G°'$ at pH 7 is used in this textbook.

5. ATP hydrolysis provides most of the free energy required for living processes. ATP is ideally suited to its role as universal energy currency because its phosphoanhydride structure is easily hydrolyzed.

 Take your learning further by visiting the **companion website** for Biochemistry at **www.oup.com/us/mckee** where you can complete a multiple-choice quiz on energy to help you prepare for exams.

Suggested**Readings**

Bergethon, P. R., *The Physical Basis of Biochemistry: The Foundations of Molecular Biophysics*, Springer, New York, 1998.

Bustamante, C., Liphardt, J., and Ritort, F., The Nonequilibrium Thermodynamics of Small Systems, *Phys. Today* 58(7):43–48, 2005 [http://www.physicstoday.org].

Hanson, R. W., The Role of ATP in Metabolism, *Biochem. Educ.* 17:86–92, 1989.

Ho, M. W., *The Rainbow and the Worm: The Physics of Organisms*, 2nd ed. World Scientific Publishing, Singapore, 1999.

Pross, A., The Driving Force for Life's Emergence: Kinetic and Thermodynamic Considerations, *J. Theor. Biol.* 220:393–406, 2003.

Rothschild, L. J., and Mancinelli, R. L., Life in Extreme Environments, *Nature* 409:1092–1101, 2001.

Schneider, E. D., and Kay, J. J., Life as a Manifestation of the Second Law of Thermodynamics, *Math, Compu. Modeling* 19(6–8):25–48, 1994.

Schneider, E. D., and Kay, J. J., Order from Disorder: The Thermodynamics of Complexity in Biology, in Murphy, M. P., and Luke, A. J. (eds.), *What Is Life? The Next Fifty Years. Reflections on the Future of Biology*, pp. 161–172, Cambridge University Press, Cambridge, 1995.

Schneider, E. D., and Sagan, D., *Into the Cool: Energy Flow, Thermodynamics and Life*, University of Chicago Press, Chicago, 2005.

Schrödinger, E., *What Is Life?* Cambridge University Press, Cambridge, 1944.

Key**Words**

aerobic respiration, *119*
anaerobic respiration, *119*
bioenergetics, *105*
chemolithotrophs, *119*
dissipative structure, *114*
electron acceptor, *119*

electron donor, *119*
endergonic process, *110*
endothermic reaction, *107*
enthalpy, *105*
entropy, *105*

exergonic process, *110*
exothermic reaction, *107*
fermentation, *119*
free energy, *105*
isothermic reaction, *107*
lithotrophs, *119*

phosphoryl group transfer potential, *115*
resonance hybrid, *118*
respiration, *119*
spontaneous changes, *108*
thermodynamics, *105*
work, *106*

Review**Questions**

These questions are designed to test your knowledge of the key concepts discussed in this chapter, before moving on to the next chapter. You may like to compare your answers to the solutions provided in the back of the book and in the accompanying Study Guide.

1. Define each of the following terms:
   a. thermodynamics
   b. endergonic

   c. enthalpy
   d. free energy
   e. high-energy bond

2. Which of the following thermodynamic quantities are state functions? Explain.
   a. work
   b. entropy
   c. enthalpy
   d. free energy

3. Which of the following reactions could be driven by coupling to the hydrolysis of ATP? (The $\Delta G^{\circ\prime}$ value for each reaction is indicated in parentheses in units of kilojoules per mole.)

   ATP + H_2O → ADP + P_i (−30.5)

   a. Pyruvate + P_i → phosphoenolpyruvate (+31.7)
   b. Glucose + P_i → glucose-6-phosphate (+13.8)
   c. Acetic acid → acetic anhydride (+99.6)
   d. Maltose + H_2O → 2 glucose (–15.5)
   e. Glycylglycine + water → 2 glycine (−9.2)

4. The K_a for the ionization of formic acid is 1.8×10^{-4}. Calculate ΔG° for this reaction.

5. Define each of the following terms:
   a. dissipative structure
   b. respiration
   c. electron donor
   d. high-quality energy
   e. resonance hybrid

6. The following reaction is catalyzed by the enzyme glutamine synthase:

   ATP + glutamate + NH_3 → ADP + P_i + glutamine

   Use the following equations (with $\Delta G^{\circ\prime}$ values given in kJ/mol) to calculate $\Delta G^{\circ\prime}$ for the overall reaction.

   ATP + H_2O → ADP + P_i (−30.5)
   Glutamine + H_2O → glutamate + NH_3 (–14.2)

7. $\Delta G^{\circ\prime}$ values (kJ/mol) for the following reactions are indicated in parentheses.

   Ethyl acetate + water →
                   ethyl alcohol + acetic acid (−19.7) (i)
   Glucose-6-phosphate + water → glucose + P_i (−13.8) (ii)

   Indicate whether each of the following statements is true, false, or undetermined. Explain your answers.
   a. The rate of reaction (i) is greater than the rate of reaction (ii).
   b. The rate of reaction (ii) is greater than the rate of reaction (i).
   c. Neither reaction is spontaneous.
   d. Reaction rates cannot be determined from energy values.

8. Define the thermodynamic term work. Provide two physiological examples of work.

9. Under standard conditions, which statements are true?
   a. $\Delta G = \Delta G^{\circ}$
   b. $\Delta H = \Delta G$
   c. $\Delta G = \Delta G^{\circ} + RT \ln K_{eq}$
   d. $\Delta G^{\circ} = \Delta H - T\Delta S$
   e. $P = 1$ atm
   f. $T = 273$ K
   g. [reactants] = [products] = 1 M

10. Define each of the following terms:
    a. fermentation
    b. anaerobic respiration
    c. redox reaction
    d. chemolithotroph
    e. phosphoryl group transfer potential

11. Which of the following compounds would you expect to liberate the least free energy when hydrolyzed? Explain.
    a. ATP
    b. ADP
    c. AMP
    d. phosphoenolpyruvate
    e. phosphocreatine

12. Which statements are true and which are false? Modify each false statement so that it reads correctly.
    a. In a closed system, neither energy nor matter is exchanged with the surroundings.
    b. State functions are independent of the pathway.
    c. A process is isothermic if $\Delta H = 0$.
    d. The sign and magnitude of ΔG give important information about the direction and rate of a reaction.
    e. At equilibrium, $\Delta G = \Delta G^{\circ}$.
    f. For two reactions to be coupled, they must have a common intermediate.

13. Which statements concerning free energy change are true?
    a. Free energy change is a measure of the rate of a reaction.
    b. Free energy change is a measure of the maximum amount of work available from a reaction.
    c. Free energy change is a constant for a reaction under any conditions.
    d. Free energy change is related to the equilibrium constant for a specific reaction.
    e. Free energy change is equal to zero at equilibrium.

14. Consider the following reaction:
    Glucose-1-phosphate → glucose-6-phosphate
    $\Delta G^{\circ} = -7.1$ kJ/mol

    What is the equilibrium constant for this reaction at 25°C?

15. Describe why ATP, the molecule that serves as the energy currency for the body, has an intermediate phosphoryl group transfer potential.

16. *Methanococcus janaschii* obtains energy by converting carbon dioxide to methane:

    $CO_2 + 4H_2$ → $CH_4 + 2H_2O$

    Considering that CO_2 has a lower energy content than CH_4, how does the organism accomplish this conversion?

17. The $\Delta G^{\circ\prime}$ value for glucose-1-phosphate is −20.9 kJ/mol. If glucose and phosphate are both at 4.8 mM, what is the concentration of glucose-1-phosphate?

18. Given the following equation
    Glycerol-3-phosphate → glycerol + P_i $\Delta G^{\circ\prime} = -9.7$ kJ/mol

    At equilibrium the concentrations of both glycerol and inorganic phosphate are 1 mM. Under these conditions, calculate the final concentration of glycerol-3-phosphate.

19. Magnesium ion (Mg^{2+}) forms complexes with the negative charges of the phosphate in ATP. In the absence of Mg^{2+}, would ATP have more, less, or the same stability as when the ion is present?

20. The free energy of hydrolysis ($\Delta G^{\circ\prime}$) of pyrophosphate (PP_i) at pH 7 is 19.2 kJ/mol. Does the value of $\Delta G^{\circ\prime}$ change with pH? If so, why? If not, why not?

21. The $\Delta G^{\circ\prime}$ value for the hydrolysis of glucose-6-phosphate is −13.8 kJ/mol. Assuming a concentration of 4 mM for this molecule, what would the phosphate concentration be at equilibrium ($\Delta G = 0$)?

22. If glucose, phosphate, and glucose-6-phosphate are combined in concentrations of 4.8, 4.8, and 0.25 mM, respectively, what is the equilibrium constant for the hydrolysis of glucose-6-phosphate?

23. Using the data for the reaction in Question 21, calculate the ΔG value ($\Delta G^{\circ\prime} = -13.8$ kJ/mol).

24. Using the data in Questions 21 and 22 determine the temperature at which $\Delta G = 0$.

25. Consider the following reaction:

    $$ATP \rightarrow AMP + 2P_i$$

    Calculate the equilibrium constant (K_{eq}) given the following $\Delta G^{\circ\prime}$ values:

    $$ATP \rightarrow AMP + PP_i \ (-32.2 \text{ kJ/mol})$$
    $$PP_i \rightarrow 2P_i \ (-33.5 \text{ kJ/mol})$$

ThoughtQuestions

These questions are designed to reinforce your understanding of all of the key concepts discussed in the book so far, including this chapter and all of the chapters before it. They may not have one right answer! The authors have provided possible solutions to these questions in the back of the book and in the accompanying Study Guide, for your reference.

26. Pyruvate oxidizes to form carbon dioxide and water and liberates energy at the rate of 1142.2 kJ/mol. If electron transport also occurs, approximately 12.5 ATP molecules are produced. The free energy of hydrolysis for ATP is -30.5 kJ/mol. What is the apparent efficiency of ATP production?

27. In the reaction

    $$ATP + glucose \rightarrow ADP + glucose\text{-}6\text{-}phosphate$$

    ΔG° is -16.7 kJ/mol. Assume that both ATP and ADP have a concentration of 1 M and $T = 25°C$. What ratio of glucose-6-phosphate to glucose would allow the reverse reaction to begin?

28. Thermodynamics is based on the behavior of large numbers of molecules. Yet within a cell there may only be a few molecules of a particular type at a time. Do the laws of thermodynamics apply under these circumstances?

29. Frequently, when salts dissolve in water, the solution becomes warm. Such a process is exothermic. When other salts, such as ammonium chloride, dissolve in water, the solution becomes cold, indicating an endothermic process. Because endothermic processes are usually not spontaneous, why does ammonium chloride dissolve in water?

30. Of the three thermodynamic quantities ΔH, ΔG, and ΔS, which provides the most useful criterion of spontaneity in a reaction? Explain.

31. What factors make ATP suitable as an "energy currency" for the cell?

32. To determine the $\Delta G^{\circ\prime}$ of a reaction within a cell, what information would you need?

33. Given the following data, calculate K_{eq} for the denaturation reaction of β-lactoglobin at 25°C:

    $$\Delta H^\circ = -88 \text{ kJ/mol}$$
    $$\Delta S^\circ = 0.3 \text{ kJ/mol}$$

34. Which of the following compounds would have the higher phosphoryl group transfer potential? Explain your answer.

35. The free energy of hydrolysis of ATP in systems free of Mg^{2+} is -35.7 kJ/mol. When the concentration of this ion is 5 mM, $\Delta G^\circ_{observed}$ is approximately -31 kJ/mol at pH 7 and 38°C. Suggest a possible reason for this effect.

36. Balance the following reaction and calculate its ΔH value:

    $$C_{17}H_{35}COOH + O_2 \rightarrow CO_2 + H_2O$$

    where the ΔH values (kcal/mol) are as follows:

    $C_{17}H_{35}COOH$ (–211.4)
    O_2 (0)
    CO_2 (–94)
    H_2O (–68.4)

37. The free energy of hydrolysis for acetic anhydride is -21.8 kJ/mol. The conversion of ATP to ADP also involves the cleavage of an anhydride bond. Its free energy of hydrolysis is -30 kJ/mol. Explain the difference in these values.

38. Calculate the temperature for the reaction in Question 37 at which ΔG° is zero.

39. Consider the following reactions and their $\Delta G^{\circ\prime}$ values:

    Ethyl acetate + $H_2O \rightarrow$ ethanol + acetate (–19.6 kJ/mol)
    Acetyl-S–CoA $\rightarrow$ Acetate + CoASH (–31 kJ/mol)

    Explain why thioester hydrolysis reactions have larger $\Delta G^{\circ\prime}$ values than those of esters.

40. Glucose-1-phosphate has a $\Delta G^{\circ\prime}$ value of -20.9 kJ/mol, whereas that for glucose-6-phosphate is 12.5 kJ/mol. After reviewing the molecular structures of these compounds, explain why there is such a difference in these values.

41. Phosphoenolpyruvate has a high phosphoryl group transfer potential, whereas two similar compounds, isopropyl phosphate and allyl phosphate, do not. Explain.

Phosphoenol pyruvate   Isopropyl phosphate   Allyl phosphate

Amino Acids, Peptides, and Proteins

Hemoglobin Within a Red Blood Cell
Human red blood cells are filled almost to bursting with the oxygen-carrying protein hemoglobin. The large pink structures are hemoglobin molecules. Sugar and amino acids are shown in green. Positive ions are blue. Negative are red. The large blue molecule is an enzyme.

Overview

▼

PROTEINS ARE ESSENTIAL CONSTITUENTS OF ALL ORGANISMS. MOST TASKS PERFORMED BY LIVING CELLS REQUIRE PROTEINS, WHICH PERFORM an astonishing variety of functions. In addition to serving as structural materials in all living organisms (e.g., structural components in the muscle and connective tissue of animals or cell wall components of prokaryotes), proteins are involved in such diverse functions as metabolic regulation, transport, defense, and catalysis. The functional diversity exhibited by this class of biomolecules is directly related to the combinatorial possibilities of the monomeric units, the 20 amino acids.

Proteins are molecular tools. They are a diverse and complex group of macromolecules that perform the thousands of tasks that sustain life. One measure of their importance is their abundance: at least 50% of the dry weight of cells is protein. Another is the vast numbers of unique protein molecules that living organisms produce. The genomes of most organisms code for thousands or tens of thousands of proteins. How can proteins be so diverse? The answer lies in their structural composition. Proteins are linear polymers composed of 20 different amino acids, linked by covalent bonds. Amino acids can theoretically be joined to form protein molecules in any imaginable size or sequence. Consider, for example, a protein composed of 100 amino acids. The total possible number of combinations for such a molecule is an astronomical 20^{100}. Not all protein sequences, however, code for useful proteins. Of the trillions of possible protein sequences, only a small fraction (recent estimates range from 650,000 to about 2 million) are actually produced by living organisms. An important reason for this remarkable discrepancy is demonstrated by the complex set of structural and functional properties of naturally occurring proteins that have evolved over billions of years in response to selection pressure. Among these are (1) structural features that make protein folding a relatively rapid and successful process, (2) the presence of binding sites that are specific for one or a small group of molecules, (3) an appropriate balance of structural flexibility and rigidity so that function is maintained, (4) surface structure that is appropriate for a protein's immediate environment (i.e., hydrophobic in membranes and hydrophilic in cytoplasm), and (5) vulnerability of proteins to degradation reactions when they become damaged or no longer useful.

Considering the vital importance of proteins in living organisms, the investigation of the structural and functional properties of proteins has always been a priority with biochemists. Proteins can be distinguished based on their number of amino acids (called **amino acid residues**), their overall amino acyl composition, and their amino acyl sequence. Selected examples of the diversity of proteins are illustrated in Figure 5.1.

Amino acid polymers are also differentiated according to their molecular weights or the number of amino acid residues they contain. Molecules with molecular weights ranging from several thousand to several million daltons are called **polypeptides**. Those with low molecular weights, typically consisting of fewer than 50 amino acids, are called **peptides**. The term **protein** describes molecules with more than 50 amino acids. Each protein consists of one or more polypeptide chains.

Throughout this textbook the terms *peptide* and *protein* will be used just as defined. In the literature, however, the distinction between proteins and peptides is often imprecise. For example, some biochemists define oligopeptides as polymers consisting of two to ten amino acids and polypeptides as having more than

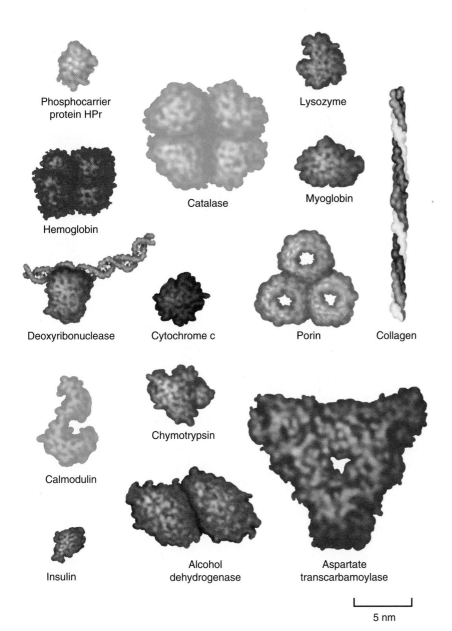

FIGURE 5.1

Protein Diversity

Proteins occur in an enormous
diversity of sizes and shapes.

ten residues. Proteins, in this view, have molecular weights greater than 10,000 D.
In addition, the terms *protein* and *polypeptide* are often used interchangeably.
Here, the term *polypeptide* will be used whenever the topic under discussion
applies to both peptides and proteins.

This chapter begins with a review of the structures and chemical properties of
the amino acids. This is followed by descriptions of the structural and functional
features of peptides and proteins and the protein folding process. The emphasis
throughout is on the intimate relationship between the structure and function of
polypeptides. In Chapter 6 the functioning of the enzymes, an especially important group of proteins, is discussed. Protein synthesis is covered in Chapter 19.

5.1 AMINO ACIDS

The hydrolysis of each polypeptide yields a set of amino acids, referred to as the
molecule's *amino acid composition*. The structures of the 20 amino acids that
are commonly found in naturally occurring polypeptides are shown in Figure 5.2.

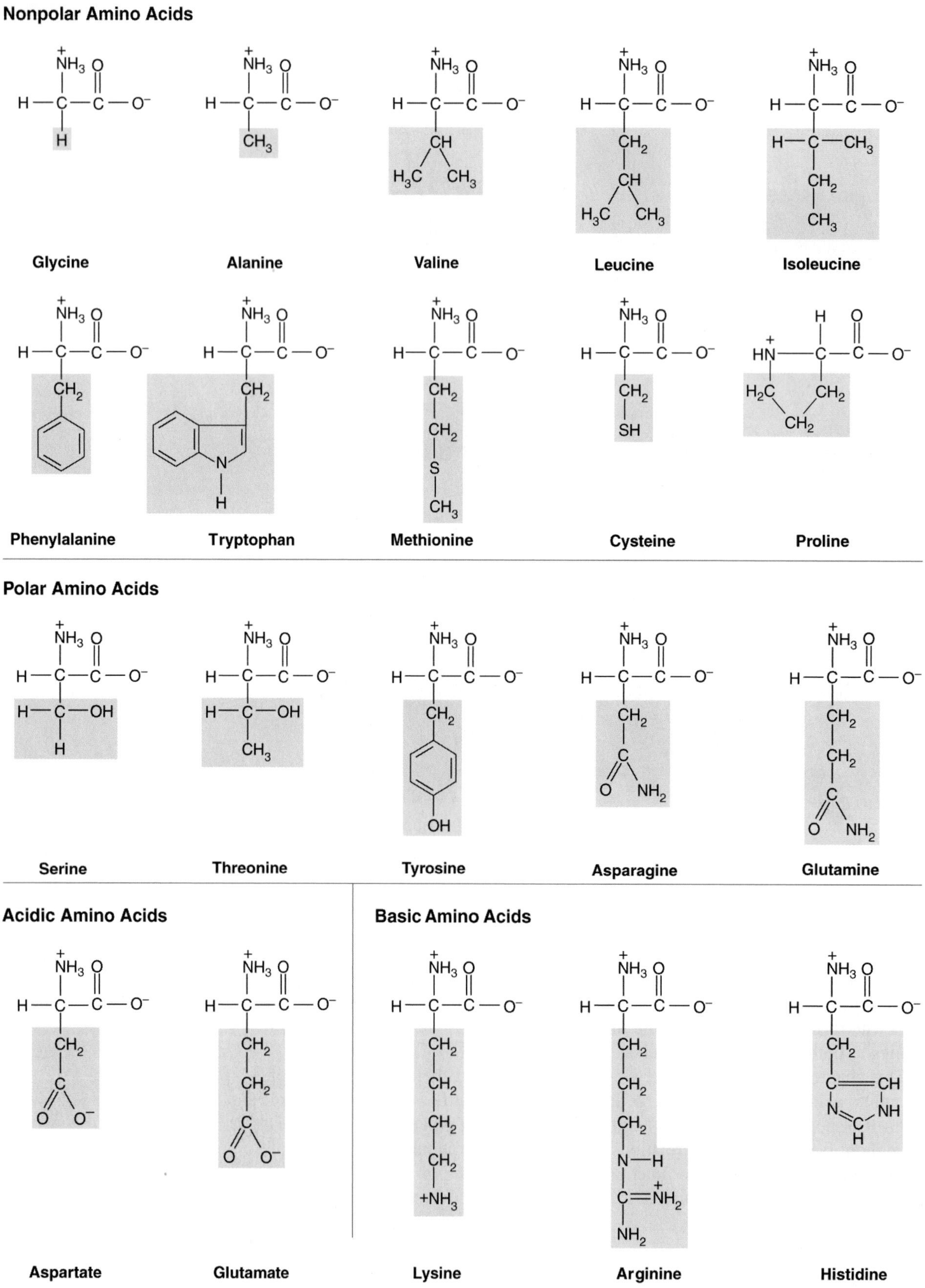

Nonpolar Amino Acids

Polar Amino Acids

Acidic Amino Acids

Basic Amino Acids

FIGURE 5.2

The Standard Amino Acids

The ionization state of the amino acid molecules in this illustration represents the dominant species that occur at a pH of 7. The side chains are indicated by shaded boxes.

TABLE 5.1   Names and Abbreviations of the Standard Amino Acids

Amino Acid	Three-Letter Abbreviation	One-Letter Abbreviation
Alanine	Ala	A
Arginine	Arg	R
Asparagine	Asn	N
Aspartic acid	Asp	D
Cysteine	Cys	C
Glutamic acid	Glu	E
Glutamine	Gln	Q
Glycine	Gly	G
Histidine	His	H
Isoleucine	Ile	I
Leucine	Leu	L
Lysine	Lys	K
Methionine	Met	M
Phenylalanine	Phe	F
Proline	Pro	P
Serine	Ser	S
Threonine	Thr	T
Tryptophan	Trp	W
Tyrosine	Tyr	Y
Valine	Val	V

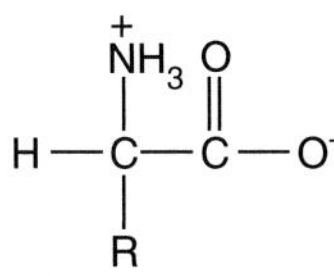

FIGURE 5.3

General Structure of the α-Amino Acids

These amino acids are referred to as *standard* amino acids. Common abbreviations for the standard amino acids are listed in Table 5.1. Note that 19 of the standard amino acids have the same general structure (Figure 5.3). These molecules contain a central carbon atom (the α-carbon) to which an amino group, a carboxylate group, a hydrogen atom, and an R (side chain) group are attached. The exception, proline, differs from the other standard amino acids in that its amino group is secondary, formed by ring closure between the R group and the amino nitrogen. Proline confers rigidity to the peptide chain because rotation about the α-carbon is not possible. This structural feature has significant implications in the structure and, therefore, the function of proteins with a high proline content.

Nonstandard amino acids consist of amino acid residues that have been chemically modified after incorporation into a polypeptide or amino acids that occur in living organisms but are not found in proteins.

At a pH of 7, the carboxyl group of an amino acid is in its conjugate base form ($-COO^-$), and the amino group is in its conjugate acid form ($-NH_3^+$). Thus each amino acid can behave as either an acid or a base. The term **amphoteric** is used to describe this property. Molecules that bear both positive and negative charges on different atoms are called **zwitterions**. The R group, however, gives each amino acid its unique properties.

Amino Acid Classes

Because the sequence of amino acids determines the final three-dimensional configuration of each protein, their structures are examined carefully in the next four subsections. Amino acids are classified according to their capacity to interact with water. By using this criterion, four classes may be distinguished: (1) nonpolar, (2) polar, (3) acidic, and (4) basic.

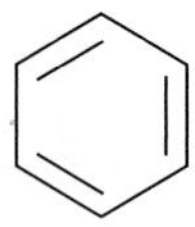

FIGURE 5.4
Benzene

NONPOLAR AMINO ACIDS The nonpolar amino acids contain mostly hydrocarbon R groups that do not bear positive or negative charges. Nonpolar (i.e., hydrophobic) amino acids play an important role in maintaining the three-dimensional structures of proteins, because they interact poorly with water. Two types of hydrocarbon side chains are found in this group: aromatic and aliphatic. **Aromatic** hydrocarbons contain cyclic structures that constitute a class of unsaturated hydrocarbons with unique properties. Benzene is one of the simplest aromatic hydrocarbons (Figure 5.4). The term **aliphatic** refers to nonaromatic hydrocarbons such as methane and cyclohexane. Phenylalanine and tryptophan contain aromatic ring structures. Glycine, alanine, valine, leucine, isoleucine, and proline have aliphatic R groups. A sulfur atom appears in the aliphatic side chains of methionine and cysteine. Methionine contains a thioether group ($-S-CH_3$) in its side chain. Its derivative *S*-adenosyl methionine (SAM) is an important metabolite that serves as a methyl donor in numerous biochemical reactions. The sulfhydryl group ($-SH$) of cysteine is highly reactive and is an important component of many enzymes. It also binds metals (e.g., iron and copper ions) in proteins. Additionally, the sulfhydryl groups of two cysteine molecules oxidize easily in the extracellular compartment to form a disulfide compound called cystine. (See p. 136 for a discussion of this reaction.)

POLAR AMINO ACIDS Because polar amino acids have functional groups capable of hydrogen bonding, they easily interact with water. (Polar amino acids are described as hydrophilic, or "water-loving.") Serine, threonine, tyrosine, asparagine, and glutamine belong to this category. Serine, threonine, and tyrosine contain a polar hydroxyl group, which enables them to participate in hydrogen bonding, an important factor in protein structure. The hydroxyl groups serve other functions in proteins. For example, the formation of the phosphate ester of tyrosine is a common regulatory mechanism. Additionally, the $-OH$ groups of serine and threonine are points for attaching carbohydrates. Asparagine and glutamine are amide derivatives of the acidic amino acids aspartic acid and glutamic acid, respectively. Because the amide functional group is highly polar, the hydrogen-bonding capability of asparagine and glutamine has a significant effect on protein stability.

ACIDIC AMINO ACIDS Two standard amino acids have side chains with carboxylate groups. Because the side chains of aspartic acid and glutamic acid are negatively charged at physiological pH, they are often referred to as aspartate and glutamate.

QUESTION 5.1

Classify these standard amino acids according to whether their structures are nonpolar, polar, acidic, or basic.

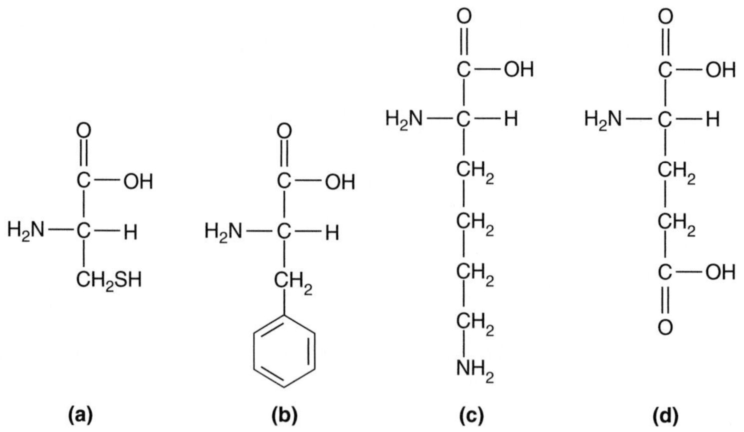

(a)   (b)   (c)   (d)

BASIC AMINO ACIDS Basic amino acids bear a positive charge at physiological pH. They can therefore form ionic bonds with acidic amino acids. Lysine, which has a side chain amino group, accepts a proton from water to form the conjugate acid ($-NH_3^+$). When lysine side chains in collagen fibrils, a vital structural component of ligaments and tendons, are oxidized and subsequently condensed, strong intramolecular and intermolecular cross-linkages are formed. Because the guanidino group of arginine has a pK_a range of 11.5 to 12.5 in proteins, it is permanently protonated at physiological pH and, therefore, does not function in acid-base reactions. Histidine, on the other hand, is a weak base because it is only partially ionized at pH 7. Consequently, histidine residues act as a buffer. They also play an important role in the catalytic activity of numerous enzymes.

KEY CONCEPT

Amino acids are classified according to their capacity to interact with water. This criterion may be used to distinguish four classes: nonpolar, polar, acidic, and basic.

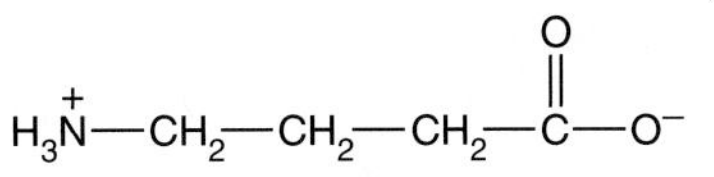

GABA

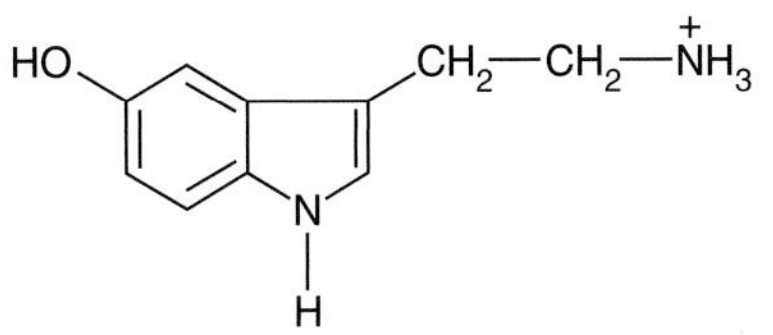

Serotonin

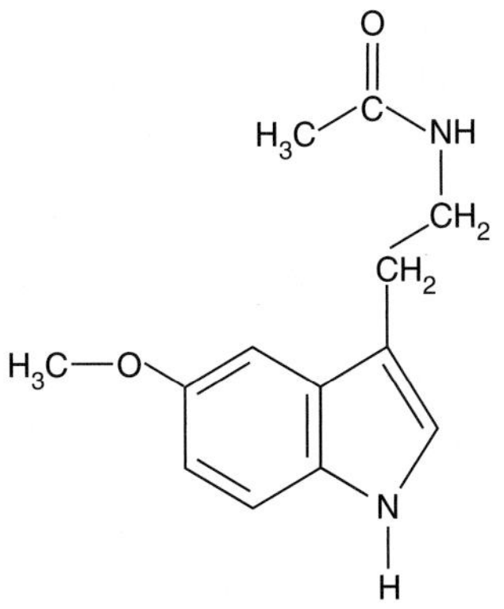

Melatonin

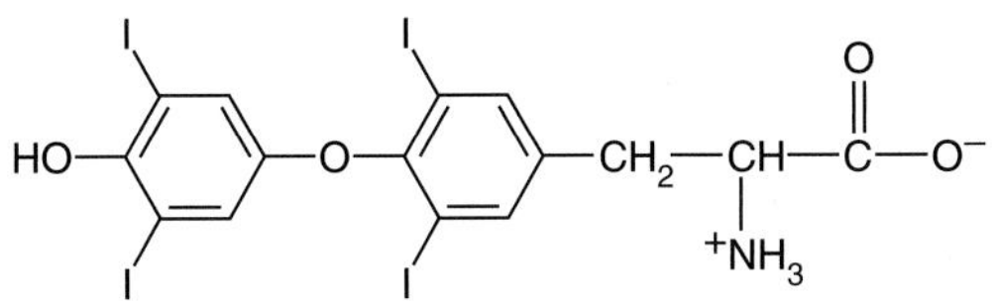

Thyroxine

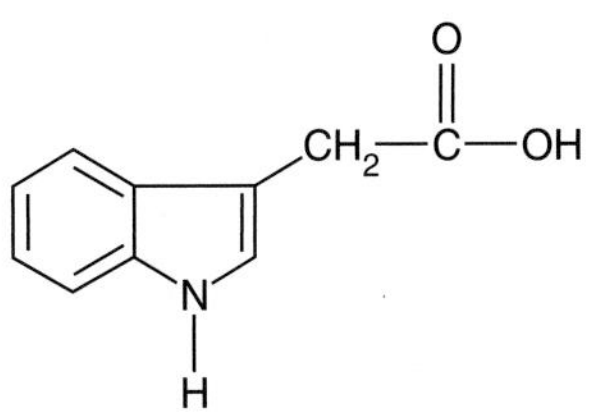

Indole acetic acid

FIGURE 5.5

Some Derivatives of Amino Acids

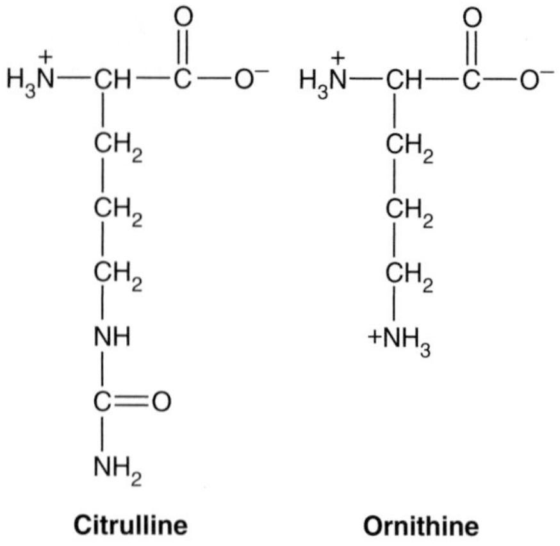

FIGURE 5.6
Citrulline and Ornithine

Biologically Active Amino Acids

In addition to their primary function as components of protein, amino acids have several other biological roles.

1. Several α-amino acids or their derivatives act as chemical messengers (Figure 5.5). For example, glycine, glutamate, γ-amino butyric acid (GABA, a derivative of glutamate), and serotonin and melatonin (derivatives of tryptophan) are **neurotransmitters**, substances released from one nerve cell that influence the function of a second nerve cell or a muscle cell. Thyroxine (a tyrosine derivative produced in the thyroid gland of animals) and indole acetic acid (a tryptophan derivative found in plants) are **hormones**—chemical signal molecules produced in one cell that regulate the function of other cells.

2. Amino acids are precursors of a variety of complex nitrogen-containing molecules. Examples include the nitrogenous base components of nucleotides and the nucleic acids, heme (the iron-containing organic group required for the biological activity of several important proteins), and chlorophyll (a pigment of critical importance in photosynthesis).

3. Several standard and nonstandard amino acids act as metabolic intermediates. For example, arginine (Figure 5.2), citrulline, and ornithine (Figure 5.6) are components of the urea cycle (Chapter 15). The synthesis of urea, a molecule formed in vertebrate livers, is the principal mechanism for the disposal of nitrogenous waste.

Modified Amino Acids in Proteins

Several proteins contain amino acid derivatives that are formed after a polypeptide chain has been synthesized. Among these modified amino acids is γ-carboxyglutamic acid (Figure 5.7), a calcium-binding amino acid residue found in the blood-clotting protein prothrombin. Both 4-hydroxyproline and 5-hydroxylysine are important structural components of collagen, the most abundant protein in connective tissue. Phosphorylation of the hydroxyl-containing amino acids serine, threonine, and tyrosine is often used to regulate the activity of proteins. For example, the synthesis of glycogen is significantly curtailed when the enzyme glycogen synthase is phosphorylated. Two other modified amino acids, selenocysteine and pyrolysine, are discussed in Chapter 19.

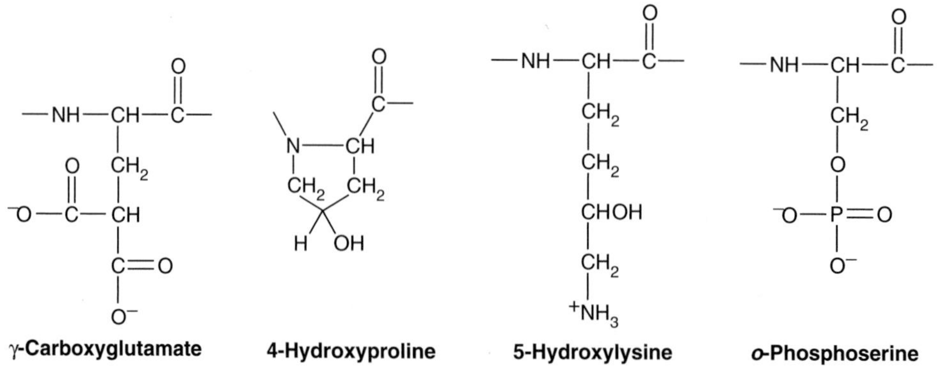

FIGURE 5.7
Some Modified Amino Acid Residues Found in Polypeptides

FIGURE 5.8

Two Enantiomers

L-Alanine and D-alanine are mirror images of each other. (Nitrogen = large red ball; Hydrogen = small red ball; Carbon = black ball; Oxygen = blue balls)

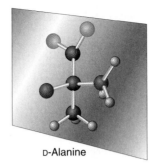

L-Alanine   D-Alanine

Amino Acid Stereoisomers

Because the α-carbons of 19 of the 20 standard amino acids are attached to four different groups (i.e., a hydrogen, a carboxyl group, an amino group, and an R group), they are referred to as **asymmetric**, or **chiral**, **carbons**. Glycine is a symmetrical molecule because its α-carbon is attached to two hydrogens. Molecules with chiral carbons can exist as **stereoisomers**, molecules that differ only in the spatial arrangement of their atoms. Three-dimensional representations of amino acid stereoisomers are illustrated in Figure 5.8. Notice in the figure that the atoms of the two isomers are bonded together in the same pattern except for the position of the ammonium group and the hydrogen atom. These two isomers are mirror images of each other. Such molecules, called **enantiomers**, cannot be superimposed on each other. Enantiomers have identical physical properties except that they rotate plane-polarized light in opposite directions. Plane-polarized light is produced by passing unpolarized light through a special filter; the light waves vibrate in only one plane. Molecules that possess this property are called **optical isomers**.

Glyceraldehyde is the reference compound for optical isomers (Figure 5.9). One glyceraldehyde isomer rotates the light beam in a clockwise direction and is said to be dextrorotatory (designated by +). The other glyceraldehyde isomer, referred to as levorotatory (designated by −), rotates the beam in the opposite direction to an equal degree. Optical isomers are often designated as D or L (e.g., D-glucose, L-alanine) to indicate the similarity of the arrangement of atoms around a molecule's asymmetric carbon to the asymmetric carbon in either of the glyceraldehyde isomers.

Most biomolecules have more than one chiral carbon. As a result, the letters D and L refer only to a molecule's structural relationship to either of the glyceraldehyde isomers, not to the direction in which it rotates plane-polarized light. Most asymmetric molecules found in living organisms occur in only one stereoisomeric form, either D or L. For example, with few exceptions, only L-amino acids are found in proteins.

Chirality has had a profound effect on the structural and functional properties of biomolecules. For example, the right-handed helices observed in proteins result from the exclusive presence of L-amino acids. Polypeptides synthesized in the laboratory from a mixture of both D- and L-amino acids do not form helices. In addition, because the enzymes are chiral molecules, most bind substrate (reactant) molecules in only one enantiomeric form. Proteases, enzymes that degrade proteins by hydrolyzing peptide bonds, cannot degrade artificial polypeptides composed of D-amino acids.

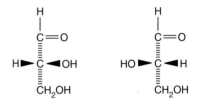

D-Glyceraldehyde   **L-Glyceraldehyde**

FIGURE 5.9

D- and L-Glyceraldehyde

These molecules are mirror images of each other.

KEY CONCEPTS

- Molecules with an asymmetric or chiral carbon atom differ only in the spatial arrangement of the atoms attached to the carbon.
- The mirror-image forms of a molecule are called enantiomers.
- Most asymmetric molecules in living organisms occur in only one stereoisomeric form.

QUESTION 5.2

Certain bacterial species have outer layers composed of polymers made of D-amino acids. Immune system cells, whose task is to attack and destroy foreign cells, cannot destroy these bacteria. Suggest a reason for this phenomenon.

TABLE 5.2 pK_a Values for the Ionizing Groups of the Amino Acids

Amino Acid	pK_1 (—COOH)	pK_2 (—NH$_3^+$)	pK_R
Glycine	2.34	9.6	
Alanine	2.34	9.69	
Valine	2.32	9.62	
Leucine	2.36	9.6	
Isoleucine	2.36	9.6	
Serine	2.21	9.15	
Threonine	2.63	10.43	
Methionine	2.28	9.21	
Phenylalanine	1.83	9.13	
Tryptophan	2.83	9.39	
Asparagine	2.02	8.8	
Glutamine	2.17	9.13	
Proline	1.99	10.6	
Cysteine	1.71	10.78	8.33
Histidine	1.82	9.17	6.0
Aspartic acid	2.09	9.82	3.86
Glutamic acid	2.19	9.67	4.25
Tyrosine	2.2	9.11	10.07
Lysine	2.18	8.95	10.79
Arginine	2.17	9.04	12.48

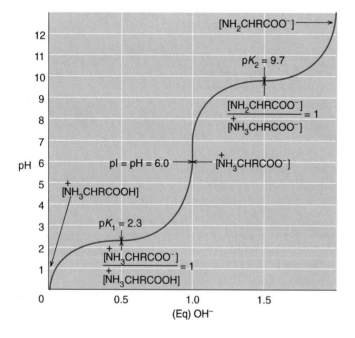

(a)

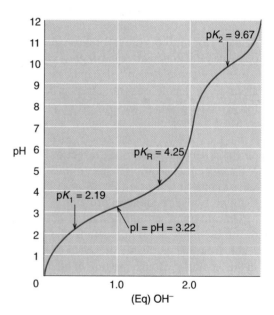

(b)

FIGURE 5.10

Titration of Two Amino Acids

(a) Alanine and (b) Glutamic Acids.

Titration of Amino Acids

Because amino acids contain ionizable groups (Table 5.2), the predominant ionic form of these molecules in solution depends on the pH. Titration of an amino acid illustrates the effect of pH on amino acid structure (Figure 5.10a). Titration is also a useful tool in determining the reactivity of amino acid side chains. Consider alanine, a simple amino acid, which has two titratable groups. During titration with a strong base such as NaOH, alanine loses two protons in stepwise fashion. In a strongly acidic solution (e.g., at pH 0), alanine is present mainly in the form in which the carboxyl group is uncharged. Under this circumstance the molecule's net charge is +1 because the ammonium group is protonated. If the H^+ concentration is lowered, the carboxyl group loses its proton to become a negatively charged carboxylate group. (In a polyprotic acid, the protons are first lost from the group with the lowest pK_a.) Once the carboxyl group has lost its proton, alanine has no net charge and is electrically neutral. The pH at which this occurs is called the **isoelectric point** (pI). The isoelectric point for alanine may be calculated as follows:

$$pI = \frac{pK_1 + pK_2}{2}$$

The pK_1 and pK_2 values for alanine are 2.34 and 9.7 respectively (see Table 5.2). The pI value for alanine is therefore

$$pI = \frac{2.34 + 9.67}{2} = 6.02$$

As the titration continues, the ammonium group loses its proton, leaving an uncharged amino group. The molecule then has a net negative charge because of the carboxylate group.

Amino acids with ionizable side chains have more complex titration curves. Glutamic acid, for example, has a carboxyl side chain group (Figure 5.10b). At low pH, glutamic acid has net charge +1. As base is added, the α-carboxyl group loses a proton to become a carboxylate group. Glutamate now has no net charge.

Titration of Glutamic acid

As more base is added, the second carboxyl group loses a proton, and the molecule has a −1 charge. Adding additional base results in the ammonium ion losing its proton. At this point, glutamate has a net charge of −2. The pI value for glutamate is the pH halfway between the pK_a values for the two carboxyl groups (i.e., the pK_a values that bracket the zwitterions):

$$pI = \frac{2.19 + 4.25}{2} = 3.22$$

The isoelectric point for histidine is the pH value halfway between the pK values for the two nitrogen-containing groups. Problem 5.1 is a sample titration problem.

WORKED PROBLEM 5.1

Consider the following amino acid and its pK_a values:

$$HO-\underset{\underset{O}{\|}}{C}-CH_2-CH_2-\underset{\underset{+NH_3}{|}}{CH}-\underset{\underset{O}{\|}}{C}-OH$$

$pK_{a1} = 2.19$ $pK_{a2} = 9.67$, $pK_{aR} = 4.25$

a. Draw the structure of the amino acid as the pH of the solution changes from highly acidic to strongly basic.

Solution (a)

$$HO-\underset{\underset{O}{\|}}{C}-CH_2-CH_2-\underset{\underset{+NH_3}{|}}{CH}-\underset{\underset{O}{\|}}{C}-OH \xrightarrow{OH^-} HO-\underset{\underset{O}{\|}}{C}-CH_2-CH_2-\underset{\underset{+NH_3}{|}}{CH}-\underset{\underset{O}{\|}}{C}-O^-$$

$$\xrightarrow{OH^-} {}^-O-\underset{\underset{O}{\|}}{C}-CH_2-CH_2-\underset{\underset{+NH_3}{|}}{CH}-\underset{\underset{O}{\|}}{C}-O^- \xrightarrow{OH^-} {}^-O-\underset{\underset{O}{\|}}{C}-CH_2-CH_2-\underset{\underset{NH_2}{|}}{CH}-\underset{\underset{O}{\|}}{C}-O^-$$

The ionizable hydrogens are lost in order of acidity, the most acidic ionizing first.

b. Which form of the amino acid is present at the isoelectric point?

Solution (b)

The form present at the isoelectric point is electrically neutral:

$$HO-\underset{\underset{O}{\|}}{C}-CH_2-CH_2-\underset{\underset{+NH_3}{|}}{CH}-\underset{\underset{O}{\|}}{C}-O^-$$

c. Calculate the isoelectric point.

Solution (c)

The isoelectric point is the average of the two pK_a's bracketing the isoelectric structure:

$$pI = \frac{pK_{a1} + pK_{aR}}{2} = \frac{2.19 + 4.25}{2} = 3.22$$

d. Sketch the titration curve for the amino acid.

Solution (d)

Plateaus appear at the pK_a and are centered about 0.5 equivalent (Eq), 1.5 Eq, and 2.5 Eq of base. There is a sharp rise at 1 Eq, 2 Eq, and 3 Eq. The isoelectric point is midway on the sharp rise between pK_{a1} and pK_{aR}.

e. In what direction does the amino acid move when placed in an electric field at the following pH values: 1, 3, 5, 7, 9, 12?

Solution (e)

At pH values below the pI, the amino acid is positively charged and moves to the cathode (negative electrode). At pH values above the pI, the amino acid is negatively charged and moves toward the anode (positive electrode). At the isoelectric point, the amino acid has no net charge and therefore does not move in the electric field. ■

KEY CONCEPTS

- Titration is useful in determining the relative ionization potential of acidic and basic groups in an amino acid or peptide.
- The pH at which an amino acid has no net charge is called its isoelectric point.

When amino acids are incorporated in polypeptides, the α-amino and α-carboxyl groups lose their charges. Consequently, except for the N- and C-terminal residues (amino acid residues at the beginning and end, respectively, of a polypeptide chain) all the ionizable groups of proteins are the side chain groups of seven amino acids: histidine, lysine, arginine, aspartate, glutamate, cysteine, and tyrosine. It should be noted that the pK_a values of these groups differ from those of free amino acids. The pK_a values of individual R groups are affected by their positions within protein microenvironments. For example, when the side chain groups of two aspartate residues are in close proximity, the pK_a of one of the carboxylate groups is raised. The significance of this phenomenon will become apparent in the discussion of enzyme catalytic mechanisms (Section 6.4).

Amino Acid Reactions

The functional groups of organic molecules determine which reactions they may undergo. Amino acids with their carboxyl groups, amino groups, and various R groups can undergo numerous chemical reactions. Peptide bond and disulfide bridge formation, however, are of special interest because of their effect on protein structure. Schiff base formation is another important reaction.

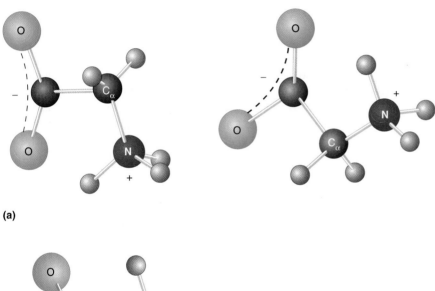

(a)

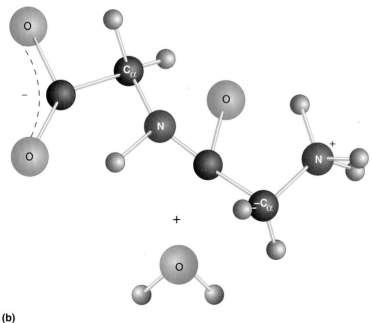

(b)

FIGURE 5.11

Formation of a Dipeptide

(a) A peptide bond forms when the α-carboxyl group of one amino acid reacts with the amino group of another. (b) A water molecule is formed in the reaction.

PEPTIDE BOND FORMATION Polypeptides are linear polymers composed of amino acids linked together by peptide bonds. **Peptide bonds** (Figure 5.11) are amide linkages formed when the unshared electron pair of the α-amino nitrogen atom of one amino acid attacks the α-carboxyl carbon of another in a nucleophilic acyl substitution reaction. A generalized acyl substitution reaction is shown:

$$R-\overset{\overset{\displaystyle O}{\|}}{C}-X + Y^- \longrightarrow R-\overset{\overset{\displaystyle O}{\|}}{C}-Y + X^-$$

Because peptide bond formation is a dehydration (i.e., a water molecule is removed), the linked amino acids are referred to as *amino acid residues*. When two amino acid molecules are linked, the product is called a dipeptide. For example, glycine and serine can form the dipeptides glycylserine or serylglycine. As amino acids are added and the chain lengthens, the prefix reflects the number of residues: a tripeptide contains three amino acid residues, a tetrapeptide four, and so on. By convention, the amino acid residue with the free amino group is called the *N-terminal* residue and is written to the left. The free carboxyl group on the *C-terminal* residue appears on the right. Peptides are named by using their amino acid sequences, beginning from their N-terminal residue. For example,

$$H_2N—Tyr—Ala—Cys—Gly—COOH$$

is a tetrapeptide named tyrosylalanylcysteinylglycine.

Large polypeptides have well-defined, three-dimensional structures. This structure, referred to as the molecule's native conformation, is a direct consequence of its *amino acid sequence* (the order in which the amino acids are linked together). Because all the linkages connecting the amino acid residues consist of single bonds, each polypeptide might be expected to undergo constant conformational changes caused by rotation around the single bonds. However, most polypeptides spontaneously fold into a single biologically active form. In the early 1950s, Linus Pauling (1901–1994, 1954 Nobel Prize in Chemistry) and his colleagues proposed an explanation. Using X-ray diffraction studies, they characterized the peptide bond as rigid and planar (flat) (Figure 5.12). Having discovered that the C—N bonds joining each two amino acids are shorter than other types of C—N bonds, Pauling deduced that peptide bonds have a partial double-bond character. (This indicates that peptide bonds are resonance hybrids.) The rigidity of the peptide bond has several consequences. Because fully one-third of the bonds in a polypeptide backbone chain cannot rotate freely, there are limits on the number of conformational possibilities.

QUESTION 5.3

Considering only the 20 standard amino acids, calculate the total number of possible tetrapeptides.

CYSTEINE OXIDATION The sulfhydryl group of cysteine is highly reactive. The most common reaction of this group is a reversible oxidation that forms a disulfide. Oxidation of two molecules of cysteine forms cystine, a molecule that contains a disulfide bond (Figure 5.13). When two cysteine residues form such a bond, it is referred to as a **disulfide bridge**. This bond can occur in a single chain to form a ring or between two separate chains to form an intermolecular bridge. Disulfide bridges help stabilize many polypeptides and proteins.

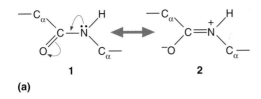

(a)

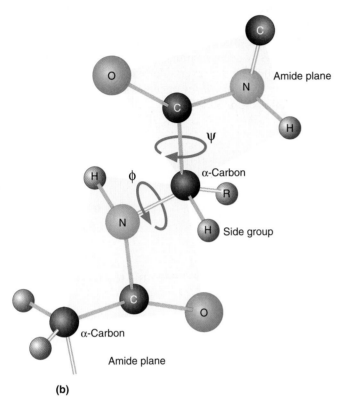

(b)

FIGURE 5.12

The Peptide Bond

(a) Resonance forms of the peptide bond. (b) Dimensions of a dipeptide. Because peptide bonds are rigid, the conformational degrees of freedom of a polypeptide chain are limited to rotations around the $C\alpha$—C and $C\alpha$—N bonds. The corresponding rotations are represented by ψ and ϕ, respectively.

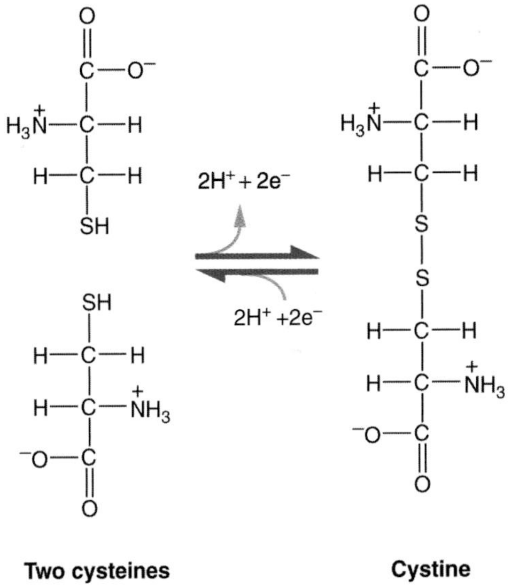

Two cysteines   **Cystine**

FIGURE 5.13

Oxidation of Two Cysteine Molecules to Form Cystine

The disulfide bond in a polypeptide is called a disulfide bridge.

QUESTION 5.4

In extracellular fluids such as blood (pH 7.2–7.4) and urine (pH 6.5), the sulfhydryl groups of cysteine (pK_a 8.1) are protonated and subject to oxidation to form cystine. In peptides and proteins the nucleophilic character of free protonated thiol groups are used to advantage in stabilizing protein structure and in thiol transfer reactions, but the free amino acid in tissue fluids can be problematic because of the low solubility of cystine. In a genetic disorder known as *cystinuria*, defective membrane transport of cystine results in excessive excretion of cystine into the urine. Crystallization of the amino acid results in formation of calculi (stones) in the kidney, ureter, or urinary bladder. The stones may cause pain, infection, and blood in the urine. Cystine concentration in the kidney is reduced by massively increasing fluid intake and administering D-penicillamine. It is believed that penicillamine (Figure 5.14) is effective because penicillamine–cysteine disulfide, which is substantially more soluble than cystine, is formed. What is the structure of the penicillamine–cysteine disulfide?

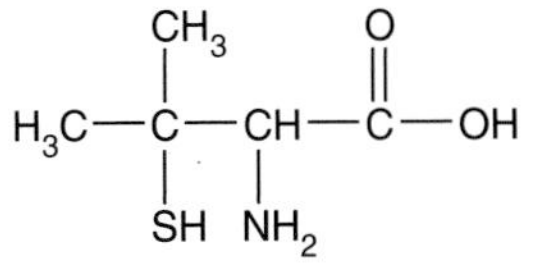

FIGURE 5.14

Structure of Penicillamine

KEY CONCEPTS

- Polypeptides are polymers composed of amino acids linked by peptide bonds. The order of the amino acids in a polypeptide is called the amino acid sequence.
- Disulfide bridges, formed by the oxidation of cysteine residues, are an important structural element in polypeptides and proteins.
- Schiff bases are imines that form when amine groups react reversibly with carbonyl groups.

SCHIFF BASE FORMATION Molecules such as amino acids that possess primary amine groups can reversibly react with carbonyl groups. The imine products of this reaction are often referred to as **Schiff bases**. In a nucleophilic addition reaction, an amine nitrogen attacks the electrophilic carbon of a carbonyl group to form an alkoxide product. The transfer of a proton from the amine group to the oxygen to form a carbinolamine, followed by the transfer of another proton from an acid catalyst, converts the oxygen into a good leaving group (OH_2^+). The subsequent elimination of a water molecule followed by loss of a proton from the nitrogen yields the imine product. The most important examples of Schiff base formation in biochemistry occur in amino acid metabolism. Schiff bases, referred to as **aldimines**, formed by the reversible reaction of an amino group with an aldehyde group, are *intermediates* (species formed during a reaction) in transamination reactions (pp. 508–510).

TABLE 5.3   Selected Biologically Important Peptides

Name	Amino Acid Sequence
Glutathione	
Oxytocin	Cys — Tyr — Ile — Gln — Asn — Cys — Pro — Leu — Gly — NH_2 (S—S bridge between the two Cys)
Vasopressin	Cys — Tyr — Phe — Gln — Asn — Cys — Pro — Arg — Gly — NH_2 (S—S bridge between the two Cys)
Atrial natriuretic factor	Ser[1]—Leu—Arg—Arg—Ser—Ser—Cys—Phe—Gly—Gly[10]—Arg—Met—Asp—Arg—Ile—Gly—Ala—Gln—Ser—Gly—Leu—Gly—Cys—Asn—Ser—Phe—Arg—Tyr[28]

5.2 PEPTIDES

Although less structurally complex than the larger protein molecules, peptides have significant biological activities. The structure and function of several interesting examples, presented in Table 5.3, are now discussed.

The tripeptide glutathione (γ-glutamyl-L-cysteinylglycine) contains an unusual γ-amide bond. (Note that the γ-carboxyl group of the glutamic acid residue, not the α-carboxyl group, contributes to the peptide bond.) Found in almost all organisms, glutathione is involved in protein and DNA synthesis, drug and environmental toxin metabolism, amino acid transport, and other important biological processes. One group of glutathione's functions exploits its effectiveness as a reducing agent. Glutathione protects cells from the destructive effects of oxidation by reacting with substances such as peroxides (R–O–O–R), by-products of O_2 metabolism. For example, in red blood cells, hydrogen peroxide (H_2O_2) oxidizes the iron of hemoglobin to its ferric form (Fe^{3+}). Methemoglobin, the product of this reaction, is incapable of binding O_2. Glutathione protects against the formation of methemoglobin by reducing H_2O_2 in a reaction catalyzed by the enzyme glutathione peroxidase. In the oxidized product GSSG, two tripeptides are linked by a disulfide bond:

$$2\text{ GSH} + H_2O_2 \rightarrow \text{GSSG} + 2H_2O$$

Because of the high GSH:GSSG ratio normally present in cells, glutathione is an important intracellular antioxidant. The abbreviation GSH is used because the reducing component of the molecule is the —SH group of the cysteine residue.

Peptides are one class of signal molecules that multicellular organisms use to regulate their complex activities. The dynamic interplay between opposing processes maintains a stable internal environment, a condition called *homeostasis*. Peptide molecules with opposing functions are now known to affect numerous processes (e.g., blood pressure regulation). The roles of selected peptides in each of these processes are briefly described.

Blood pressure, the force exerted by blood against the walls of blood vessels, is influenced by several factors such as blood volume and viscosity. Two peptides known to affect blood volume are vasopressin and atrial natriuretic factor. Vasopressin, also called antidiuretic hormone, contains nine amino acid residues. It is synthesized in the hypothalamus, a small structure in the brain that regulates a wide variety of functions including water balance, appetite, body temperature, and sleep. In response to low blood pressure or a high blood Na^+ concentration, osmoreceptors in the hypothalamus trigger vasopressin secretion. Vasopressin stimulates water reabsorption in the kidneys by initiating a signal transduction mechanism that inserts aquaporins (water channels) into kidney tubule membrane. Water then flows down its concentration gradient through the tubule cells and back into the blood. The structure of vasopressin is remarkably similar to that of another peptide produced in the hypothalamus called oxytocin, the signal molecule that stimulates the ejection of milk by mammary glands during lactation. Oxytocin produced in the uterus stimulates the contraction of uterine muscle during childbirth. Because vasopressin and oxytocin have similar structures, it is not surprising that the functions of the two molecules overlap. Oxytocin has mild antidiuretic activity and vasopressin has some oxytocin-like activity. Atrial natriuretic factor (ANF), a peptide produced by specialized cells in the heart in response to stretching and in the nervous system, stimulates the production of a dilute urine, an effect opposite to that of vasopressin. ANF exerts its effect, in part, by increasing the excretion of Na^+, a process that causes increased excretion of water, and by inhibiting the secretion of renin by the kidney. (Renin is an enzyme that catalyzes the formation of angiotensin, a hormone that constricts blood vessels.)

> ### QUESTION 5.5
>
> Write out the complete structure of oxytocin. What would be the net charge on this molecule at the average physiological pH of 7.3? At pH 4? At pH 9? Indicate which atoms in oxytocin can potentially form hydrogen bonds with water molecules.

> ### QUESTION 5.6
>
> The structural features of vasopressin that allow binding to vasopression receptors are the rigid hexapeptide ring and the amino acid residues at positions 3 (Phe) and 8 (Arg). The aromatic phenylalanine side chain, which fits into a hydrophobic pocket in the receptor, and the large positively charged arginine side chain are especially important structural features. Compare the structures of vasopressin and oxytocin and explain why their functions overlap. Can you suggest what will happen to the binding properties of vasopressin if the arginine at position 8 is replaced by lysine?

KEY CONCEPT

Although small in comparison to larger protein molecules, peptides have significant biological activity. They are involved in a variety of signal transduction processes.

COMPANION

GW

WEBSITE **Visit the companion website at www.oup.com/us/mckee to read the Biochemistry in Perspective box on protein poisons.**

5.3 PROTEINS

Of all the molecules encountered in living organisms, proteins have the most diverse functions, as the following list suggests.

1. **Catalysis**. *Enzymes* are proteins that direct and accelerate thousands of biochemical reactions in such processes as digestion, energy capture, and biosynthesis. These molecules have remarkable properties. For example, enzymes can increase reaction rates by factors of between 10^6 and 10^{12}. They can perform this feat under mild conditions of pH and temperature because they can induce or stabilize strained reaction intermediates. For example, ribulose bisphosphate carboxylase is an important enzyme in photosynthesis, and the protein complex nitrogenase is responsible for nitrogen fixation.

2. **Structure**. Some proteins provide structural support. Structural proteins often have very specialized properties. For example, collagen (the major components of connective tissues) and fibroin (silk protein) have significant mechanical strength. Elastin, the rubberlike protein found in elastic fibers, is found in several tissues in the body (e.g., blood vessels and skin) that must be elastic to function properly.

3. **Movement**. Proteins are involved in all cell movements. For example, actin, tubulin, and other proteins comprise the cytoskeleton. Cytoskeletal proteins are active in cell division, endocytosis, exocytosis, and the ameboid movement of white blood cells.

4. **Defense**. A wide variety of proteins are protective. In vertebrates, for example, keratin, a protein found in skin cells, aids in protecting the organism against mechanical and chemical injury. The blood-clotting proteins fibrinogen and thrombin prevent blood loss when blood vessels are damaged. The immunoglobulins (or antibodies) are produced by lymphocytes when foreign organisms such as bacteria invade an organism. Binding antibodies to an invading organism is the first step in its destruction. Many organisms protect themselves by producing toxic proteins that either kill or deter predators or competitors. Examples include the neurotoxin α-bungarotoxin produced by certain venomous snakes and ricin, a protein synthesis inhibitor found in the seeds of the castor bean plant.

5. **Regulation**. Binding a hormone molecule or a growth factor to cognate receptors on its target cell changes cellular function. For example, insulin and glucagon are peptide hormones that regulate blood glucose levels. Growth hormone stimulates cell growth and division. Growth factors are

polypeptides that control animal cell division and differentiation. Examples include platelet-derived growth factor (PDGF) and epidermal growth factor (EGF).

6. **Transport**. Many proteins function as carriers of molecules or ions across membranes or between cells. Examples of membrane proteins include the enzyme Na^+-K^+ ATPase and the glucose transporter. Other transport proteins include hemoglobin, which carries O_2 to the tissues from the lungs, and the lipoproteins LDL and HDL, which transport lipids from the liver and intestines to other organs. Transferrin and ceruloplasmin are serum proteins that transport iron and copper, respectively.

7. **Storage**. Certain proteins serve as a reservoir of essential nutrients. For example, ovalbumin in bird eggs and casein in mammalian milk are rich sources of organic nitrogen during development. Plant proteins such as zein perform a similar role in germinating seeds.

8. **Stress response**. The capacity of living organisms to survive a variety of abiotic stresses is mediated by certain proteins. Examples include cytochrome P_{450}, a diverse group of enzymes found in animals and plants that usually convert a variety of toxic organic contaminants into less toxic derivatives, and metallothionein, a cysteine-rich intracellular protein found in virtually all mammalian cells that binds to and sequesters toxic metals such as cadmium, mercury, and silver. Excessively high temperatures and other stresses result in the synthesis of a class of proteins called the **heat shock proteins** (hsps) that promote the correct refolding of damaged proteins. If such proteins are severely damaged, hsps promote their degradation. (Certain hsps function in the normal process of protein folding.) Cells are protected from radiation by DNA repair enzymes.

Protein research efforts in recent years have revealed that numerous proteins have multiple and often unrelated functions. Once thought to be a rare phenomenon, **multifunction proteins** (sometimes referred to as *moonlighting proteins*) are a diverse class of molecules. Prominent examples include glyceraldehyde-3-phosphate dehydrogenase (GAPD) and the crystallins. As the name suggests, GAPD (p. 273) is an enzyme that catalyzes the oxidation of glyceraldehyde-3-phosphate, an intermediate in glucose catabolism. The GAPD protein is now known to have roles in such diverse processes as DNA replication and repair, endocytosis, and membrane fusion events. The crystallins are the major water-soluble protein components of the transparent cells in the lens of the vertebrate eye. To perform their structural function, crystallin proteins must be soluble in the liquid-crystal cytoplasm and exceptionally stable. Most importantly, they must prevent the scattering of visible light, often by binding to small numbers of *chromophores*, substances that absorb light energy. Crystallins appear to have been genetically "recruited" from a variety of metabolic enzymes and molecular chaperones. They have retained some of their original functions, although often with some loss of activity and/or specificity. For example, the α-crystallins are heat shock proteins that protect cells from physiological stress. Another crystallin (ε-crystallin), found in the lens of ducks, is a lactate dehydrogenase (p. 277). η-Crystallin, found in elephant shrews, is also a retinal dehydrogenase, an essential enzyme in the metabolism of retinol (vitamin A).

In addition to their functional classifications, proteins are categorized on the basis of amino acid sequence similarities and overall three-dimensional shape. **Protein families** are composed of protein molecules that are related by amino acid sequence similarity. Such proteins share an obvious common ancestry. Classic protein families include the hemoglobins (blood oxygen transport proteins, pp. 168–171) and the immunoglobulins, the antibody proteins produced by the immune system in response to antigens (foreign substances). Proteins more distantly related are often classified into **superfamilies**. For example, the globin superfamily includes a variety of heme-containing proteins that serve in the binding and/or transport of oxygen. In addition to the hemoglobins and myoglobins (oxygen-binding proteins in muscle cells), the globin superfamily includes neuroglobin and cytoglobin (oxygen-binding proteins in brain and other tissues, respectively) and the leghemoglobins (oxygen-sequestering proteins in the root nodules of leguminous plants).

Because of their diversity, proteins are often classified in two additional ways: shape and composition. Proteins are classified into two major groups based on their shape. As the name suggests, **fibrous proteins** are long, rod-shaped molecules that are insoluble in water and physically tough. Fibrous proteins, such as the keratins found in skin, hair, and nails, have structural and protective functions. **Globular proteins** are compact spherical molecules that are usually water-soluble. Typically, globular proteins have dynamic functions. For example, nearly all enzymes have globular structures. Other examples include the immunoglobulins and the transport proteins hemoglobin and albumin (a carrier of fatty acids in blood).

On the basis of composition, proteins are classified as simple or conjugated. Simple proteins, such as serum albumin and keratin, contain only amino acids. In contrast, each **conjugated protein** consists of a simple protein combined with a nonprotein component. The nonprotein component is called a **prosthetic group**. (A protein without its prosthetic group is called an **apoprotein**. A protein molecule combined with its prosthetic group is referred to as a **holoprotein**.) Prosthetic groups typically play an important, even crucial, role in the function of proteins. Conjugated proteins are classified according to the nature of their prosthetic groups. For example, **glycoproteins** contain a carbohydrate component, **lipoproteins** contain lipid molecules, and **metalloproteins** contain metal ions. Similarly, **phosphoproteins** contain phosphate groups, and **hemoproteins** possess heme groups (p. 168).

Protein Structure

Proteins are extraordinarily complex molecules. Complete models depicting even the smallest of the polypeptide chains are almost impossible to comprehend. Simpler images that highlight specific features of a molecule are useful. Two methods of conveying structural information about proteins are presented in Figure 5.15. Another structural representation, referred to as a ball-and-stick model, is presented later (Figures 5.37 and 5.39).

Biochemists have distinguished several levels of the structural organization of proteins. **Primary structure**, the amino acid sequence, is specified by genetic information. As the polypeptide chain folds, it forms certain localized arrangements of adjacent (but not necessarily contiguous) amino acids that constitute **secondary structure**. The overall three-dimensional shape that a polypeptide assumes is called the **tertiary structure**. Proteins that consist of two or more polypeptide chains (or subunits) are said to have a **quaternary structure**.

PRIMARY STRUCTURE Every polypeptide has a specific amino acid sequence. The interactions between amino acid residues determine the protein's three-dimensional structure and its functional role and relationship to other proteins. Polypeptides that have similar amino acid sequences and have arisen from the same ancestral gene are said to be **homologous**. Sequence comparisons among homologous polypeptides have been used to trace the genetic relationships of different species. For example, the sequence homologies of the mitochondrial redox protein cytochrome c have been used extensively in the study of evolution of species. Sequence comparisons of cytochrome c, an essential molecule in energy production, among numerous species reveal a significant amount of sequence conservation. The amino acid residues that are identical in all homologues of a protein, referred to as *invariant*, are presumed to be essential for the protein's function. (In cytochrome c the invariant residues interact with heme, a prosthetic group, or certain other proteins involved in energy generation.)

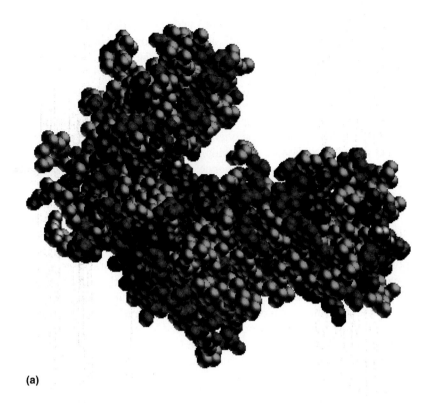

(a)

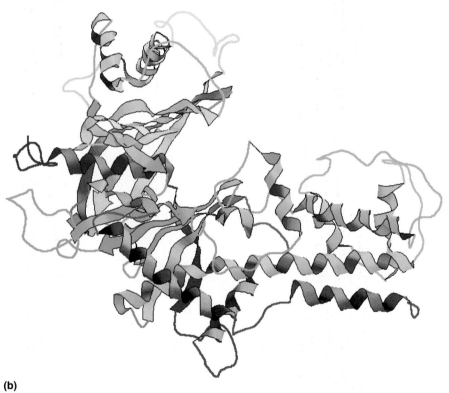

(b)

FIGURE 5.15

The Enzyme Adenylate Kinase

(a) This space-filling model illustrates the volume occupied by molecular components and overall shape. (b) In a ribbon model β-pleated segments are represented by flat arrows. The α-helices appear as spiral ribbons.

PRIMARY STRUCTURE, EVOLUTION, AND MOLECULAR DISEASES Each type of polypeptide folds into its own unique and functional conformation. Over time, however, as the result of evolutionary processes, the amino acid sequences of polypeptides change owing to the random and spontaneous alterations in DNA sequences called mutations. A significant number of primary sequence changes do not affect a polypeptide's function. Some of these substitutions are said to be *conservative* because an amino acid with a chemically similar side chain is substituted. For example, at certain sequence positions leucine and isoleucine, which both contain hydrophobic side chains, may be substituted for each other without affecting function. Some sequence positions are significantly less stringent. These residues, referred to as *variable*, apparently perform nonspecific roles in the polypeptide's function.

Substitutions at conservative and variable sites have been used to trace evolutionary relationships. These studies assume that the longer the time since two species diverged from each other, the larger the number of differences in a certain polypeptide's primary structure. For example, humans and chimpanzees are believed to have diverged relatively recently (perhaps only 4 million years ago). This presumption, based principally on fossil and anatomical evidence, is supported by cytochrome c primary sequence data indicating that the protein is identical in both species. Kangaroos, whales, and sheep, whose cytochrome c molecules each differ by 10 residues from the human protein, are believed to have evolved from a common ancestor that lived over 50 million years ago. It is

FIGURE 5.16

Segments of β-Chain in HbA and HbS

Individuals possessing the gene for sickle-cell hemoglobin produce β-chains with valine instead of glutamic acid at residue 6.

Hb A   Val—His—Leu—Thr—Pro—Glu—Glu—Lys—

Hb S   Val—His—Leu—Thr—Pro—Val—Glu—Lys—

         1   2   3   4   5   6   7   8

FIGURE 5.17

Sickle Cell Hemoglobin

HbS molecules aggregate into rodlike filaments because the hydrophobic side chain of valine, the substituted amino acid in the β-chain, interacts with a hydrophobic pocket in a second hemoglobin molecule.

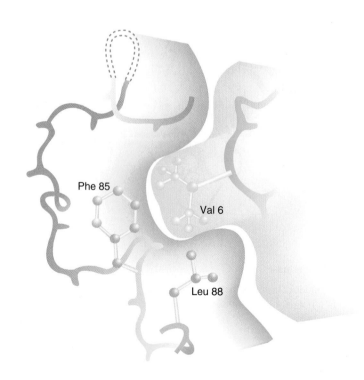

interesting to note that quite often the overall three-dimensional structure does not change despite numerous amino acid sequence changes. The shape of proteins coded for by genes that diverged millions of years ago may show a remarkable resemblance to each other.

Mutations, however, can also be deleterious. Such random changes in gene sequence can range from moderate to severe. Individual organisms with nonconservative amino acid substitutions at the invariant residues of cytochrome c, for example, are not viable. Mutations can also have a profound effect without being immediately lethal. Sickle-cell anemia, which is caused by mutant hemoglobin, is a classic example of a group of maladies that Linus Pauling and his colleagues referred to as **molecular diseases**. (Dr. Pauling first demonstrated that sickle-cell patients have a mutant hemoglobin through the use of electrophoresis.) Human adult hemoglobin (HbA) is composed of two identical α-chains and two identical β-chains. Sickle-cell anemia results from a single amino acid substitution in the β-chain of HbA. Analysis of the hemoglobin molecules of sickle-cell patients reveals that the only difference between HbA and sickle-cell hemoglobin (HbS) is at amino acid residue 6 in the β-chain (Figure 5.16). Because of the substitution of a hydrophobic valine for a negatively charged glutamic acid, HbS molecules aggregate to form rigid rodlike structures in the oxygen-free state (Figure 5.17). The patient's red blood cells become sickle shaped and are susceptible to hemolysis, resulting in severe anemia. These red blood cells have an abnormally low oxygen-binding capacity. Intermittent clogging of capillaries by sickled cells also causes tissues to be deprived of oxygen. Sickle-cell anemia is characterized by excruciating pain, eventual organ damage, and earlier death.

Until recently, because of the debilitating nature of sickle-cell disease, affected individuals rarely survived beyond childhood. Thus one might predict that the deleterious mutational change that causes this affliction would be rapidly eliminated from human populations. However, the sickle-cell gene is not as rare as would be expected. Sickle-cell disease occurs only in individuals who have inherited two copies of the sickle-cell gene. Such individuals, who are said to be *homozygous*, inherit one copy of the defective gene from each parent. Each of the parents is said to have the *sickle-cell trait*. Such people, referred to as *heterozygous* because they have one normal HbA gene and one defective HbS gene, are relatively sympton-free, even though about 40% of their hemoglobin is HbS. The incidence of sickle-cell trait is especially high in some regions of Africa. In these areas malaria, caused by the *Anopheles* mosquito–borne parasite *Plasmodium*, is a serious health problem. Individuals with the sickle-cell trait are less vulnerable to malaria because their red blood cells are a less favorable environment for the growth of the parasite than are normal cells. Because sickle-cell trait carriers are more likely to survive malaria than normal individuals, the incidence of the sickle-cell gene has remained high. (In some areas, the sickle-cell trait is present in as much as 40% of the native population.)

- The primary structure of a polypeptide is its amino acid sequence. The amino acids are connected by peptide bonds.

- Amino acid residues that are essential for the molecule's function are referred to as invariant.

- Proteins with similar amino acid sequences and functions and a common origin are said to be homologous.

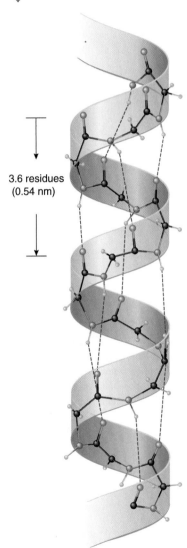

3.6 residues (0.54 nm)

FIGURE 5.18

The α-Helix

Hydrogen bonds form between carbonyl and N—H groups along the long axis of the α-helix. Note that there are 3.6 residues per turn of the helix, which has a pitch of 0.54 nm.

QUESTION 5.7

A genetic disease called *glucose-6-phosphate dehydrogenase deficiency* is inherited in a manner similar to that of sickle-cell anemia. The defective enzyme cannot keep erythrocytes supplied with sufficient amounts of the antioxidant molecule NADPH (Chapter 8). NADPH protects cell membranes and other cellular structures from oxidation. Describe in general terms the inheritance pattern of this molecular disease. Why do you think that the antimalarial drug primaquine, which stimulates peroxide formation, results in devastating cases of hemolytic anemia in carriers of the defective gene? Does it surprise you that this genetic anomaly is commonly found in African and Mediterranean populations?

SECONDARY STRUCTURE The secondary structure of polypeptides consists of several repeating patterns. The most commonly observed types of secondary structure are the α-helix and the β-pleated sheet. Both α-helix and β-pleated sheet patterns are stabilized by localized hydrogen bonding between the carbonyl and N—H groups in the polypeptide's backbone. These patterns occur when all the ϕ(phi) angles (rotation angle about C_α—N) in a polypeptide segment are equal and all the ψ (psi) angles (rotation angle about C_α—C) are equal (see Figure 5.12b). Because peptide bonds are rigid, the α-carbons are swivel points for the polypeptide chain. Several properties of the R groups (e.g., size and charge, if any) attached to the α-carbon influence the ϕ and ψ angles. Certain amino acids foster or inhibit specific secondary structural patterns. Many fibrous proteins are composed almost entirely of secondary structural patterns.

The *α-helix* is a rigid, rodlike structure that forms when a polypeptide chain twists into a right-handed helical conformation (Figure 5.18). Hydrogen bonds form between the N—H group of each amino acid and the carbonyl group of

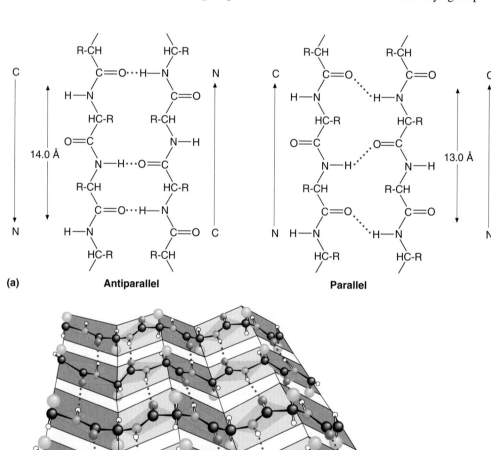

(a)   Antiparallel   Parallel

FIGURE 5.19

β-Pleated Sheet

(a) Two forms of β-pleated sheet: antiparallel and parallel. Hydrogen bonds are represented by dotted lines. (b) A more detailed view of antiparallel β-pleated sheet.

(b)

the amino acid four residues away. There are 3.6 amino acid residues per turn of the helix, and the pitch (the distance between corresponding points per turn) is 0.54 nm. Amino acid R groups extend outward from the helix. Because of several structural constraints (i.e., the rigidity of peptide bonds and the allowed limits on the values of the ϕ and ψ angles), certain amino acids do not foster α-helical formation. For example, glycine's R group (a hydrogen atom) is so small that the polypeptide chain may be too flexible. Proline, on the other hand, contains a rigid ring that prevents the N—C_α bond from rotating. In addition, proline has no N—H group available to form the intrachain hydrogen bonds that are crucial in α-helix structure. Amino acid sequences with large numbers of charged amino acids (e.g., glutamate and aspartate) and bulky R groups (e.g., tryptophan) are also incompatible with α-helix structures.

β-Pleated sheets form when two or more polypeptide chain segments line up side by side (Figure 5.19). Each individual segment is referred to as a β-strand. Rather than being coiled, each β-strand is fully extended. β-Pleated sheets are stabilized by hydrogen bonds that form between the polypeptide backbone N—H and carbonyl groups of adjacent chains. β-Pleated sheets are either parallel or antiparallel. In *parallel* β-pleated sheet structures, the hydrogen bonds in the polypeptide chains are arranged in the same direction; in antiparallel chains these bonds are arranged in opposite directions. Occasionally, mixed parallel-antiparallel β-sheets are observed.

Many globular proteins contain combinations of α-helix and β-pleated sheet secondary structures (Figure 5.20). These patterns are called **supersecondary structures** or **motifs**. In the *βαβ unit*, two parallel β-pleated sheets are connected by an α-helix segment. The structure of *βαβ* units is stabilized by hydrophobic interactions between nonpolar side chains projecting from the interacting surfaces of the β-strands and the α-helix. Glycine and proline residues often occur in β-turns. Glycine's lack of an organic side group permits a contiguous proline to assume a *cis* orientation (same side of the peptide plane), and a tight turn can form in a polypeptide strand. Proline is a helix-breaking residue that alters the direction of the polypeptide chain. The *β-turn* is common in proteins rich in α-helical segments.

In the *β-meander* pattern, two antiparallel β-sheets are connected by polar amino acids and glycines to effect a more abrupt change in direction called a reverse or *hairpin turn*. In *αα units* (or helix-loop-helix units), two α-helical regions separated by a nonhelical loop become aligned in a defined way because of interacting side chains. Several *β-barrel* arrangements are formed when various β-sheet configurations fold back on themselves. When an antiparallel β-sheet doubles back on itself in a pattern that resembles a common Greek pottery design the motif is called the *Greek key*.

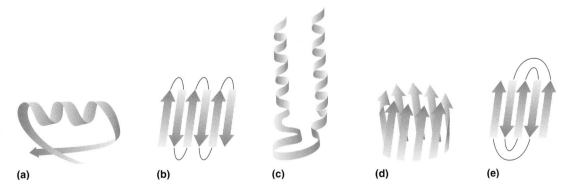

(a)   (b)   (c)   (d)   (e)

FIGURE 5.20

Selected Supersecondary Structures

(a) $\beta\alpha\beta$ units, (b) β-meander, (c) $\alpha\alpha$ unit, (d) β-barrel, and (e) Greek key.

TERTIARY STRUCTURE Although globular proteins often contain significant numbers of secondary structural elements, several other factors contribute to their structure. The term *tertiary structure* refers to the unique three-dimensional conformations that globular proteins assume as they fold into their native (biologically active) structures and prosthetic groups, if any, are inserted. **Protein folding**, a process in which an unorganized, *nascent* (newly synthesized) molecule acquires a highly organized structure, occurs as a consequence of the interactions between the side chains in their primary structure. Tertiary structure has several important features:

1. Many polypeptides fold in such a fashion that amino acid residues that are distant from each other in the primary structure come into close proximity.

2. Globular proteins are compact because of efficient packing as the polypeptide folds. During this process, most water molecules are excluded from the protein's interior making interactions between both polar and nonpolar groups possible.

3. Large globular proteins (i.e., those with more than 200 amino acid residues) often contain several compact units called domains. *Domains* (Figure 5.21) are typically structurally independent segments that have specific functions (e.g., binding an ion or small molecule). The core three-dimensional structure of a domain is called a **fold**. Well-known examples of folds include the nucleotide-binding Rossman fold and the globin fold. Domains are classified on the basis of their core motif structure. Examples include α, β, α/β, and $\alpha + \beta$. α-Domains are composed exclusively of α-helices, and β-domains consist of antiparallel β strands. α/β-Domains contain various combinations of an α-helix alternating with β-strands ($\beta\alpha\beta$ motifs). $\alpha + \beta$ Domains are primarily β-sheets with one or more outlying α-helices. Most proteins contain two or more domains.

4. A number of eukaryotic proteins, referred to as **modular** or **mosaic proteins**, contain numerous duplicate or imperfect copies of one or more domains that are linked in series. Fibronectin (Figure 5.22) contains three repeating domains: Fl, F2, and F3. All three domains, which are found in a variety of extracellular matrix (ECM) proteins, contain binding sites for other ECM molecules such as collagen (p. 165) and heparan sulfate (p. 250), as well as certain cell surface receptors. Domain modules are coded for by genetic sequences created by gene duplications and recombination events. Such sequences are used by living organisms to construct new proteins. For example, the immunoglobulin structural domain is found not only in antibodies, but also in a variety of cell surface proteins.

The following types of interactions stabilize tertiary structure (Figure 5.23):

1. **Hydrophobic interactions**. As a polypeptide folds, hydrophobic R groups are brought into close proximity because they are excluded from water. Then the highly ordered water molecules in solvation shells are released from the interior, increasing the disorder (entropy) of the water molecules. The favorable entropy change is a major driving force in protein folding. It should be noted that a few water molecules remain within the core of folded proteins, where each forms as many as 4 hydrogen bonds with the polypeptide backbone. The stabilization contributed by small "structural" water molecules may free the polypeptide from some of its internal interactions. The resulting increased flexibility of the polypeptide chain is believed to play a critical role in the binding of molecules called **ligands** to specific sites. Ligand binding is an important protein function.

2. **Electrostatic interactions**. The strongest electrostatic interaction in proteins occurs between ionic groups of opposite charge. Referred to as **salt bridges**, these noncovalent bonds are significant only in regions of the

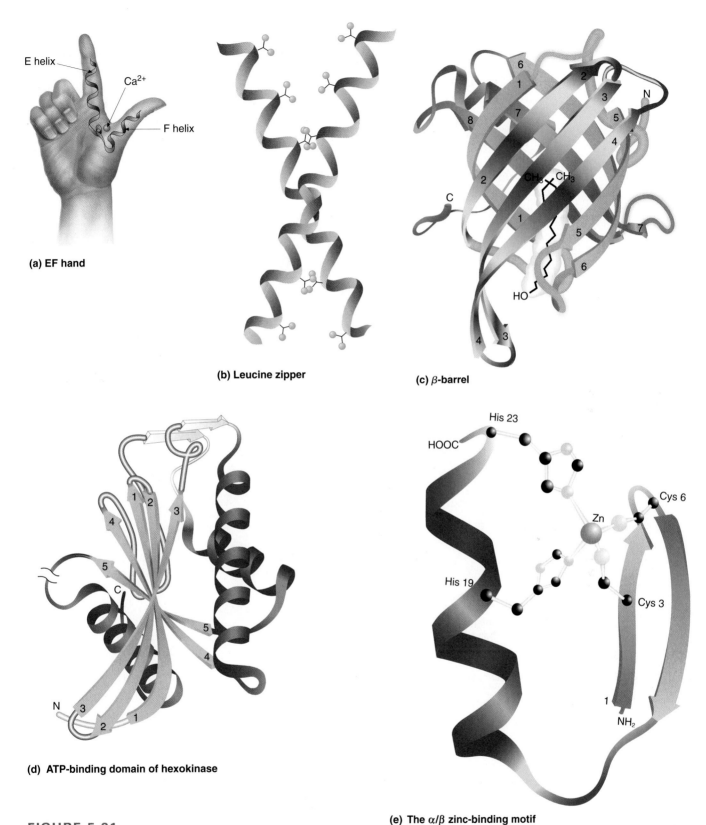

(a) EF hand

(b) Leucine zipper

(c) β-barrel

(d) ATP-binding domain of hexokinase

(e) The α/β zinc-binding motif

FIGURE 5.21

Selected Domains Found in Large Numbers of Proteins

(a) The EF hand, a helix-loop-helix that binds specifically to Ca^{2+}, and (b) the leucine zipper, a DNA-binding domain, are two examples of α-domains. (c) Human retinol-binding protein, a type of β-barrel domain (retinol, a visual pigment molecule, is shown in yellow). (d) The ATP-binding domain of hexokinase, a type of α/β-domain. (e) The α/β zinc-binding motif, a core feature of numerous DNA-binding domains.

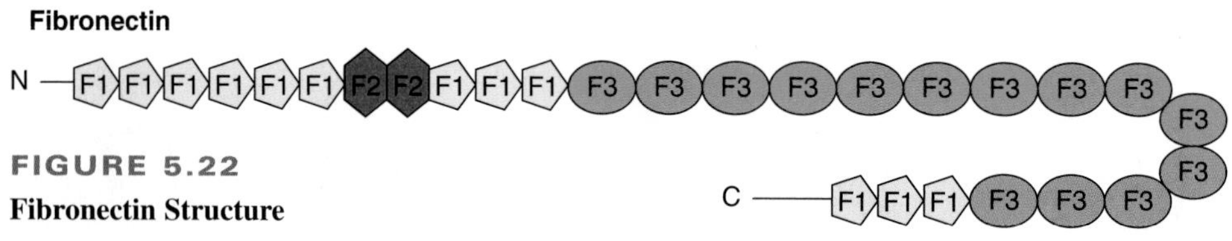

Fibronectin

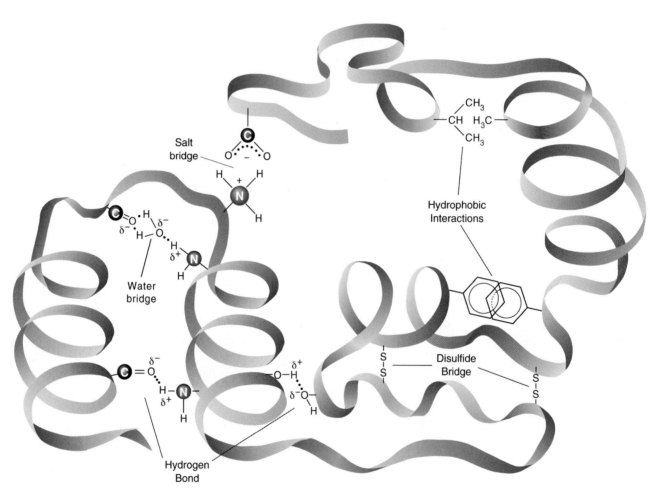

FIGURE 5.22

Fibronectin Structure

Fibronectin is a mosaic protein that is composed of multiple copies of F1, F2, and F3 modules.

FIGURE 5.23

Interactions That Maintain Tertiary Structure

protein where water is excluded because of the energy required to remove water molecules from ionic groups near the surface. Salt bridges have been observed to contribute to the interactions between adjacent subunits in complex proteins. The same is true for the weaker electrostatic interactions (ion-dipole, dipole-dipole, van der Waals). They are significant in the interior of the folded protein and between subunits or in protein-ligand interactions. (In proteins that consist of more than one polypeptide chain, each polypeptide is called a **subunit**.) Ligand-binding pockets are water-depleted regions of the protein.

3. **Hydrogen bonds**. A significant number of hydrogen bonds form within a protein's interior and on its surface. In addition to forming hydrogen bonds with one another, the polar amino acid side chains may interact with water or with the polypeptide backbone. Again, the presence of water precludes the formation of hydrogen bonds with other species.

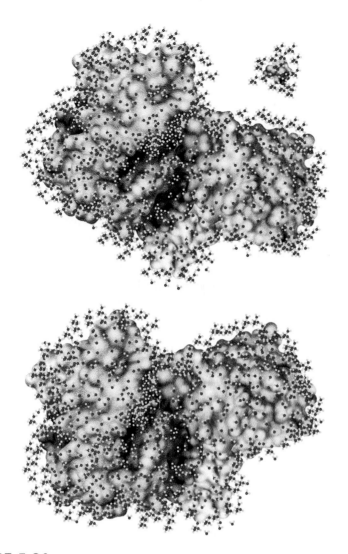

FIGURE 5.24

Hydration of a Protein

Three layers of structured water molecules surround a space-filling model of the enzyme hexokinase, before and after binding the sugar glucose. Hexokinase (p. 268) is an enzyme that catalyzes the nucleophilic attack of the carbon–6 hydroxyl group of glucose on the phosphorus in the terminal phosphate of ATP. As the hydrated glucose molecule enters its binding site in a cleft in the enzyme, it sheds its water molecules and displaces water molecules occupying the binding site. The water exclusion process promotes the conformation change that moves the domains together to create the catalytic site. Water exclusion from this site also prevents the unproductive hydrolysis of ATP.

4. **Covalent bonds**. Covalent linkages are created by chemical reactions that alter a polypeptide's structure during or after its synthesis. (Examples of these reactions, referred to as posttranslational modifications, are described in Section 19.2.) The most prominent covalent bonds in tertiary structure are the disulfide bridges found in many extracellular proteins. In extracellular environments, these strong linkages partly protect protein structure from adverse changes in pH or salt concentrations. Intracellular proteins do not contain disulfide bridges because of high cytoplasmic concentrations of reducing agents.

5. **Hydration**. As described previously (p. 80) structured water is an important stabilizing feature of protein structure. The dynamic hydration shell that forms around a protein (Figure 5.24) also contributes to the flexibility required for biological activity.

The precise nature of the forces that promote the folding of proteins (described on pp. 159–162) has not been completely resolved. It is clear, however, that protein folding is a thermodynamically favorable process with an overall negative free energy change. According to the free energy equation

$$\Delta G° = \Delta H° - T\Delta S°$$

a negative free energy change in a process is the result of a balance between favourable and unfavourable enthalpy and entropy changes. As a polypeptide folds, favorable (negative) ΔH values are the result in part of the sequestration of hydrophobic side chains within the interior of the molecule and the optimization of other noncovalent interactions. Opposing these factors is the unfavorable decrease in entropy that occurs as the disorganized polypeptide folds into its highly organized native state. The change in entropy of the water that surrounds the protein is positive because of the decreased organization of the water in going from the unfolded to the folded state of the protein. For most polypeptide molecules the net free energy change between the folded and unfolded state is relatively modest (the energy equivalent of several hydrogen bonds). The precarious balance between favorable and unfavorable forces allows proteins the flexibility they require for biological function.

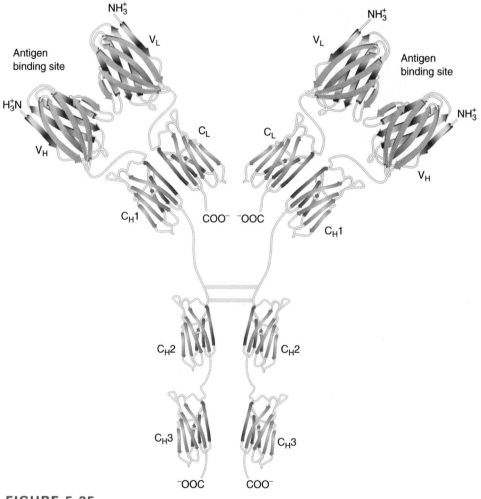

FIGURE 5.25

Structure of Immunoglobulin G

IgG is an antibody molecule composed of two heavy chains (H) and two light chains (L) that together form a Y-shaped molecule. Each of the heavy and light chains contains constant (C) and variable (V), β-barrel domains (the classic immunoglobulin fold). The chains are held together by disulfide bridges (yellow lines) and noncovalent interactions. The variable domains of the H and L chains form the site that binds to antigens (foreign molecules). Many antigenic proteins bind to the external surface of these sites. Note that disulfide bridges are also a structural feature within each constant domain.

QUATERNARY STRUCTURE Many proteins, especially those with high molecular weights, are composed of several polypeptide chains. As mentioned, each polypeptide component is called a subunit. Subunits in a protein complex may be identical or quite different. Multisubunit proteins in which some or all subunits are identical are referred to as **oligomers**. Oligomers are composed of **protomers**, which may consist of one or more subunits. A large number of oligomeric proteins contain two or four subunit protomers, referred to as dimers and tetramers, respectively. There appear to be several reasons for the common occurrence of multisubunit proteins:

1. Synthesis of separate subunits may be more efficient than substantially increasing the length of a single polypeptide chain.

2. In supramolecular complexes such as collagen fibers, replacement of smaller worn-out or damaged components can be managed more effectively.

3. The complex interactions of multiple subunits help regulate a protein's biological function.

Polypeptide subunits assemble and are held together by noncovalent interactions such as the hydrophobic effect, electrostatic interactions, and hydrogen bonds, as well as covalent cross-links. As with protein folding, the hydrophobic effect is clearly the most important because the structures of the complementary interfacing surfaces between subunits are similar to those observed in the interior of globular protein domains. Although they are less numerous, covalent cross-links significantly stabilize certain multisubunit proteins. Prominent examples include the disulfide bridges in the immunoglobulins (Figure 5.25), and the desmosine and lysinonorleucine linkages in certain connective tissue proteins. *Desmosine* (Figure 5.26) cross-links connect four polypeptide chains in the rubberlike connective tissue protein elastin. They are formed as a result of a series of reactions involving the oxidation and condensation of lysine side chains. A similar process results in the formation of *lysinonorleucine*, a cross-linking structure that is found in elastin and collagen.

Quite often the interactions between subunits are affected by the binding of ligands. In **allostery**, which is the control of protein function through ligand binding, binding a ligand to a specific site in a protein triggers a conformational change that alters its affinity for other ligands. Ligand-induced conformational changes in such proteins are called **allosteric transitions**, and the ligands that trigger them are called **effectors** or **modulators**. Allosteric effects can be positive or negative, depending on whether effector binding increases or decreases the protein's affinity for other ligands. One of the best understood examples of allosteric effects, the reversible binding of O_2 and other ligands to hemoglobin, is described on pp. 169–171. (Because allosteric enzymes play a key role in the control of metabolic processes, allostery is discussed further in Sections 6.3 and 6.5.)

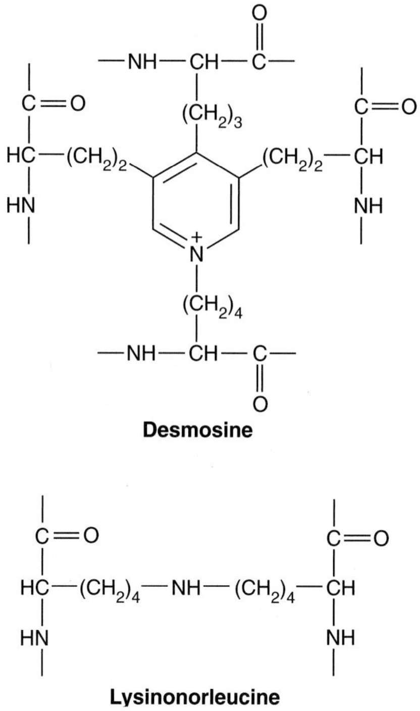

Desmosine

Lysinonorleucine

FIGURE 5.26

Desmosine and Lysinonorleucine Linkages

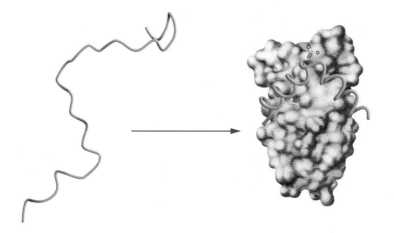

FIGURE 5.27

Disordered Protein Binding

The intrinsically disordered phosphorylated KID domain (pKID) (left) of the transcription regulatory protein CREB searches out and binds to the KIX domain of the transcription coactivator protein CBP (right). As pKID binds to KIX, it folds into a pair of helices.

Review the following illustrations of globular proteins. Identify examples of secondary and supersecondary structure.

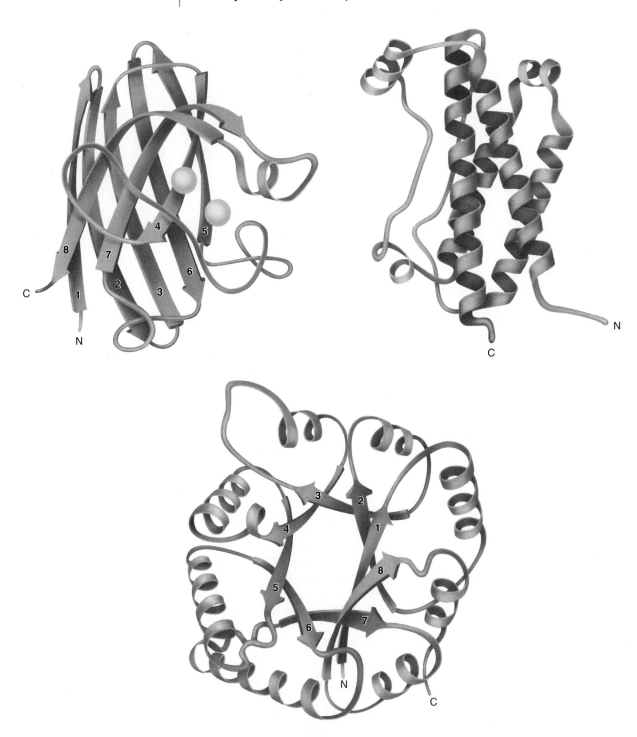

Illustrate the noncovalent interactions that can occur between the following side chain groups in folded polypeptides: (a) serine and glutamate, (b) arginine and aspartate, (c) threonine and serine, (d) glutamine and aspartate, (e) phenylalanine and tryptophan.

UNSTRUCTURED PROTEINS In the traditional view of proteins, a polypeptide's function is determined by its specific and relatively stable three-dimensional structure. In recent years, however, as a result of new genomic methodologies and new applications of various forms of spectroscopy, it has become apparent that many proteins are in fact partially or completely unstructured. Unstructured proteins are referred to as **IUPs (intrinsically unstructured proteins)**. If there is a complete lack of ordered structure, the term **natively unfolded proteins** is used. Most IUPs are eukaryotic. Amazingly, over 30% of eukaryotic proteins are partially or completely disordered, whereas only about 2 and 4% of archaean and bacterial proteins, respectively, can be described as unstructured. The folding of IUPs into stable three-dimensional conformations is prevented by biased amino acid sequences that contain high percentages of polar and charged amino acids (e.g., Ser, Gln, Lys, and Glu) and low quantities of hydrophobic amino acids (e.g., Leu, Val, Phe, and Trp).

IUPs have a diversity of functions. Many are involved in the regulation of such processes as signal transduction, transcription, translation, and cell proliferation. Highly extended and malleable disordered segments enable the molecule to "search" for binding partners. A representative example is provided by CREB, a transcription regulatory protein discussed later (Chapter 18) that binds to CRE, one type of DNA sequence called a *response element*. When the pKID (*k*inase *i*nducible *do*main) of CREB is phosphorylated by a kinase (an enzyme that attaches phosphate groups to specific amino acid side chains) it becomes unstructured. The unstructured pKID domain is then able to search out and bind to a domain of CREB-binding protein (CBP) called KIX (KID-binding domain) (Figure 5.27). As often happens with IUPs, the disordered pKID domain transitions into a more ordered conformation as it binds to the KIX domain of CBP. As a result of CREB-CBP binding, CREB forms a dimer that alters the expression of certain genes when it binds to its response element.

LOSS OF PROTEIN STRUCTURE Considering the small differences in the free energy of folded and unfolded proteins, it is not surprising that protein structure is especially sensitive to environmental factors. Many physical and chemical agents can disrupt a protein's native conformation. The process of structure disruption, which may or may not involve protein unfolding, is called **denaturation**. (Denaturation is not usually considered to include the breaking of peptide bonds.) Depending on the degree of denaturation, the molecule may partially or completely lose its biological activity. Denaturation often results in easily observable changes in the physical properties of proteins. For example, soluble and transparent egg albumin (egg white) becomes insoluble and opaque upon heating. Like many denaturations, cooking eggs is an irreversible process.

The following example of a reversible denaturation was demonstrated in the 1950s by Christian Anfinsen, who shared the Nobel Prize in Chemistry in 1972. Bovine pancreatic ribonuclease (a digestive enzyme from cattle that degrades RNA) is denatured when treated with β-mercaptoethanol and 8 M urea (Figure 5.28). During this process, ribonuclease, composed of a single polypeptide with four disulfide bridges, completely unfolds and loses all biological activity. Careful removal of the denaturing agents with dialysis results in a spontaneous and correct refolding of the polypeptide and re-formation of the disulfide bonds. Anfinsen's experimental treatment resulting in a full restoration of the enzyme's catalytic activity provided an important early insight into the roles of different forces and primary structure in protein folding. However, most proteins treated similarly do not renature.

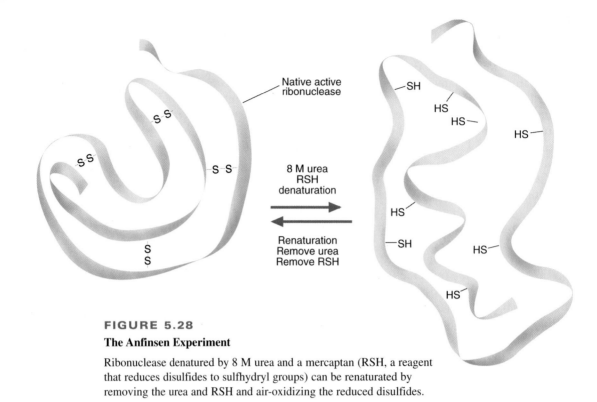

FIGURE 5.28

The Anfinsen Experiment

Ribonuclease denatured by 8 M urea and a mercaptan (RSH, a reagent that reduces disulfides to sulfhydryl groups) can be renatured by removing the urea and RSH and air-oxidizing the reduced disulfides.

- Biochemists distinguish four levels of the structural organization of proteins.
- In primary structure, the amino acid residues are connected by peptide bonds.
- The secondary structure of polypeptides is stabilized by hydrogen bonds. Prominent examples of secondary structure are α-helices and β-pleated sheets.
- Tertiary structure is the unique three-dimensional conformation that a protein assumes because of the interactions between amino acid side chains. Several types of interaction stabilize tertiary structure: the hydrophobic effect, electrostatic interactions, hydrogen bonds, and certain covalent bonds.
- Proteins that consist of several separate polypeptide subunits exhibit quaternary structure.
- Both noncovalent and covalent bonds hold the subunits together. Some proteins are partially or completely unstructured.

Denaturing conditions include the following:

1. **Strong acids or bases**. Changes in pH result in protonation of some protein side chain groups, which alters hydrogen bonding and salt bridge patterns. As a protein approaches its isoelectric point, it becomes less soluble and may precipitate from solution.

2. **Organic solvents**. Water-soluble organic solvents such as ethanol interfere with hydrophobic interactions because they interact with nonpolar R groups and form hydrogen bonds with water and polar protein groups. Nonpolar solvents also disrupt hydrophobic interactions.

3. **Detergents**. Detergents are substances that disrupt hydrophobic interactions, causing proteins to unfold into extended polypeptide chains. These molecules are called **amphipathic** because they contain both hydrophobic and hydrophilic components.

4. **Reducing agents**. In the presence of reagents such as urea, reducing agents (e.g., β-mercaptoethanol) convert disulfide bridges to sulfhydryl groups. Urea disrupts hydrogen bonds and hydrophobic interactions.

5. **Salt concentration**. When there is an increase in the salt concentration of an aqueous solution of protein, some of the water molecules that interact with the protein's ionizable groups are attracted to the salt ions. As the number of solvent molecules available to interact with these groups decreases, protein-protein interactions increase. If the salt concentration is high enough, there are so few water molecules available to interact with ionizable groups that the solvation spheres surrounding the protein's ionized groups are removed. The protein molecules aggregate and then precipitate. This process is referred to as *salting out*. Because salting out is usually reversible, it is often used as an early step in protein purification.

6. **Heavy metal ions**. Heavy metals such as mercury (Hg^{2+}) and lead (Pb^{2+}) affect protein structure in several ways. They may disrupt salt bridges by forming ionic bonds with negatively charged groups. Heavy metals also

(Continued on p. 159)

BIOCHEMISTRY IN PERSPECTIVE

Molecular Machines

How do living organisms utilize chemical bond energy to perform the thousands of tasks required to sustain life? Purposeful movement is the hallmark of living organisms. This behavior takes myriad forms that range from the record-setting 110 km/h chasing sprint of the cheetah to more subtle movements such as the migration of white blood cells in the animal body, cytoplasmic streaming in plant cells, intracellular transport of organelles, and the enzyme-catalyzed unwinding of DNA. The multisubunit proteins responsible for these phenomena (e.g., the muscle sarcomere and various other types of cytoskeletal components, and DNA polymerase) function as biological machines. Machines are defined as mechanical devices with moving parts that perform work (the product of force and distance). When machines are used correctly, they permit the accomplishment of tasks that would often be impossible without them. Although biological machines are composed of relatively fragile proteins that cannot withstand the physical conditions associated with human-made machines (e.g., heat and friction), the two types do share important features. In addition to having moving parts, all machines require energy-transducing mechanisms; that is, they convert energy into directed motion.

Despite the wide diversity of motion types in living organisms, in all cases, energy-driven changes in protein conformations result in the accomplishment of work. Protein conformation changes occur when a ligand is bound. When a specific ligand binds to one subunit of a multisubunit protein complex, the change in its conformation will affect the shapes of adjacent subunits. These changes are reversible; that is, ligand dissociation from a protein causes it to revert to its previous conformation. The work performed by complex biological machines requires that the conformational and, therefore, functional changes occur in an orderly and directed manner. In other words, an energy source (usually provided by the hydrolysis of ATP or GTP) drives a sequence of conformational changes of adjacent subunits in one functional direction. For example, the DNA replication machine DNA polymerase III incorporates nucleotides into a new DNA strand at the rate of 9000 nucleotides per minute. The directed functioning of this and other biological machines is possible because nucleotide hydrolysis is irreversible under physiological conditions.

Motor Proteins

Despite their functional diversity, all biological machines possess one or more protein components that bind nucleoside triphosphates (NTP). These subunits, called NTPases, function as mechanical transducers or **motor proteins**. The NTP hydrolysis–driven changes in the conformation of a motor protein trigger ordered conformational changes in adjacent subunits in the molecular machine. NTP-binding proteins perform a wide variety of functions in eukaryotes, most of which occur in one or more of the following categories.

1. **Classical motors.** Classical motor proteins are ATPases that move a load along a protein filament, as shown earlier (Figure 2.4). The best-known examples include the **myosins**, which move along actin filaments, and the kinesins and dyneins, which move vesicles and organelles along microtubules. **Kinesins** walk along the microtubules toward the (+) end, away from the centrosome (the microtubule organizing center). **Dyneins** walk along the microtubules toward the (−) end, toward the centrosome.

2. **Timing devices.** The function of certain NTP-binding proteins is to provide a delay period during a complex process that ensures accuracy. The prokaryotic protein synthesis protein EF-Tu (Biochemistry in Perpective Box EF-TU: A Motor Protein, Ch. 19) is a well-known example. The relatively slow rate of GTP hydrolysis by EF-Tu when it is bound to an aminoacyl-tRNA allows sufficient time for the dissociation of the complex from the ribosome if the tRNA-mRNA base sequence binding is not correct.

3. **Microprocessing switching devices.** A variety of GTP-binding proteins act as on-off molecular switches in signal transduction pathways. Examples include the β-subunits of the trimeric G proteins. Numerous intracellular signal control mechanisms are regulated by G proteins.

4. **Assembly and disassembly factors.** Numerous cellular processes require the rapid and reversible assembly of protein subunits into larger molecular complexes. Among the most dramatic examples of protein subunit polymerization are the assembly of tubulin and actin into microtubules and microfilaments, respectively. The slow hydrolysis of GTP by tubulin and ATP by actin monomers, after the incorporation of these molecules into their respective polymeric filaments, promotes subtle conformational changes that later allow disassembly.

The best-characterized motor protein is myosin. A brief overview of the structure and function of myosin in the molecular events in muscle contraction is provided.

Myosin

The myosins are a family of motor proteins that transduce ATP bond energy into unidirectional movement along actin filaments. Found in all eukaryotic cells, the myosins are involved in a wide variety of cellular movements. The role of myosin in movement

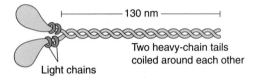

is best understood in skeletal muscle contraction. Skeletal muscle myosin, referred to as myosin II, consists of two heavy chains and two light chains (Figure 5A). The N-terminal globular head domain (Figure 5B) of the heavy chain possesses separate binding sites for ATP and actin. A long α-helix extending from the head domain forms the flexible neck region and the C-terminal tail. The actin-binding cleft and ATP-binding site are on opposite sides of the myosin head domain. These sites are connected by so-called switch helices. As a result, the ATP-hydrolyzing activity of the ATP-binding site is activated when the myosin head binds to actin. In the *swinging neck–lever model* of the actin-myosin crossbridge cycle, ATP-induced changes in myosin conformation cause a leverlike swinging motion in the neck relative to the catalytic domain. The relative motion of thick (myosin filaments) and thin filaments (actin polymers complexed with several other proteins) in the muscle sarcomere (the functional unit of skeletal muscle) is caused by the swinging stroke of the myosin neck domain (Figure 5C).

FIGURE 5A

Myosin II Structure

A myosin II molecule is composed of two heavy chains, each of which contains a globular head, a hinge region and a long rodlike tail, as well as four light chains. Two small light chains are wrapped around each myosin head.

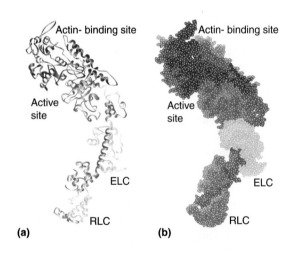

FIGURE 5B

Myosin II Head Domain Structure

(a) Ribbon model and (b) space-filling model: green, heavy-chain residues 4–204; red, heavy-chain residues 216–626; purple, heavy-chain residues 647–843; yellow, essential light chain (ELC); orange, regulatory light chain (RLC). The light chains encircle and stabilize the long α-helix formed by the heavy chain.

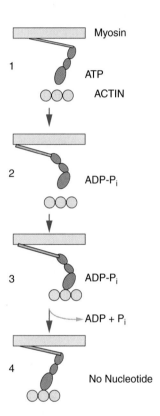

FIGURE 5C

The Actomyosin Cycle in Skeletal Muscle

(Step 1:) The myosin head has bound an ATP within the nucleotide-binding site and has detached from actin. (Step 2:) ATP hydrolysis causes the myosin head to become "cocked." (This is the energy-transducing event.) (Step 3:) Myosin binds weakly to actin. (Step 4:) Release of ADP and P_i causes the myosin head to bind tightly to actin, which is followed by the power stroke. The conformational change in the myosin head (the leverlike swinging motion) causes the myosin filament to move along the actin filament.

SUMMARY: Molecular machine function is made possible by conformational changes triggered by the hydrolysis of nucleotides bound to protein subunits called motor proteins.

bond with sulfhydryl groups, a process that may result in significant changes in protein structure and function. For example, Pb^{2+} binds to sulfhydryl groups in two enzymes in the heme synthetic pathway (Chapter 14). The resultant decrease in hemoglobin synthesis causes severe anemia. (In anemia the number of red blood cells or the hemoglobin concentration is lower than normal.) Anemia is one of the most easily measured symptoms of lead poisoning.

7. **Temperature changes**. As the temperature increases, the rate of molecular vibration increases. Eventually, weak interactions such as hydrogen bonds are disrupted and the protein unfolds. Some proteins are more resistant to heat denaturation, and this fact can be used in purification procedures.

8. **Mechanical stress**. Stirring and grinding actions disrupt the delicate balance of forces that maintain protein structure. For example, the foam formed when egg white is beaten vigorously contains denatured protein.

COMPANION

GW

W E B S I T E   Visit the companion website at www.oup.com/us/mckee to read the Biochemistry in Perspective box on lead poisoning.

The Folding Problem

The rapid and efficient folding of newly synthesized polypeptides is an essential phase of information transfer in living organisms. The direct relationship between a protein's primary sequence and its final three-dimensional conformation, and by extension its biological activity, is among the most important assumptions of modern biochemistry. One of the principal underpinnings of this paradigm has already been mentioned: the series of experiments reported by Christian Anfinsen in the late 1950s. Working with bovine pancreatic RNase, Anfinsen demonstrated that under favorable conditions a denatured protein could refold into its native and biologically active state (Figure 5.28). This discovery suggested that the three-dimensional structure of any protein could be predicted if the physical and chemical properties of the amino acids and the forces that drive the folding process (e.g., bond rotations, free energy considerations, and the behavior of amino acids in aqueous environments) were understood. Unfortunately, several decades of painstaking research with the most sophisticated tools available (e.g., X-ray crystallography and NMR in combination with site-directed mutagenesis and computer-based mathematical modeling) resulted in only limited progress. Nuclear magnetic resonance (NMR) spectrometry is an imaging technique that measures the absorption of electromagnetic radiation by atomic nuclei in the presence of a strong magnetic field. Molecular structure can be probed because the nuclei of atoms such as hydrogen in different positions in a molecule resonate at slightly different frequencies. **Site-directed mutagenesis** is a recombinant DNA technique in which specific sequence changes can be introduced into a predetermined position in cloned genes. Briefly, such work revealed that protein folding is a stepwise process in which secondary structure formation (i.e., α-helix and β-pleated sheet) is an early feature. Hydrophobic interactions appear to be an important force in folding. In addition, amino acid substitutions experimentally introduced into certain proteins reveal that changes in surface amino acids rarely affect the protein's structure. In contrast, substitutions of amino acids within the hydrophobic core often lead to serious structural changes in conformation.

The traditional protein-folding model is limited insofar as interactions between amino acid side chains alone force the molecule to fold into its final shape. The following considerations highlight additional problems associated with the model.

1. **Time constraints**. The time to synthesize proteins routinely ranges from a few seconds to no more than a few minutes. Yet, when even a relatively small number of possible bond rotations during the folding process are considered, the time required for folding is measured in years. Therefore, most

researchers have concluded that protein folding is not a random process based solely on primary sequence.

2. **Complexity**. The calculations required in the mathematical models of protein folding based on physical data (e.g., bond angles and degrees of rotation) are overwhelmingly complex. It is unlikely that they alone can resolve the principles of what appears in living organisms to be an astonishingly fast and elegant process.

In recent years important advances have been made by biochemists in protein-folding research by utilizing imaginative combinations of technologies such as multidimensional NMR and circular dichroism, to name a few. **Circular dichroism** (CD) is a type of spectroscopy in which the relationship between molecular motion and structure is probed with electromagnetic radiation.

By utilizing such techniques as multidimensional NMR and CD, protein-folding researchers have determined that the process does not consist, as was originally thought, of a single pathway. Instead, there are numerous routes that a polypeptide can take to fold into its native state. As illustrated in Figure 5.29a an energy landscape with a funnel shape appears to best describe how an unfolded polypeptide with its own unique set of constraints (e.g., its amino acid sequence and posttranslational modifications, and environmental features within the cell such as temperature, pH, and molecular crowding) negotiates its way to a low-energy folded state. Depending largely on its size, a polypeptide may or may

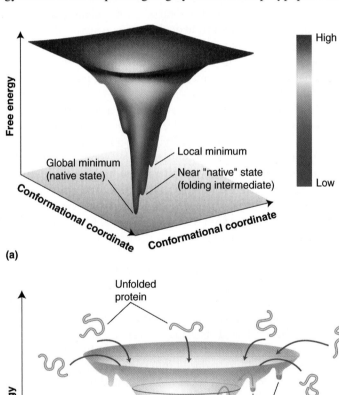

FIGURE 5.29

The Energy Landscape for Protein Folding

(a) Color is used to indicate the entropy level of the folding polypeptide. As folding progresses the polypeptide moves from a disordered stated (high entropy, red) toward a progressively more ordered conformation until its unique biologically active conformation is achieved (lower entropy, blue): (b) A more realistic view of the energy landscape: polypeptides can fold into their native states by several different pathways. Many molecules form transient intermediates, whereas others may become trapped in a misfolded state.

not form intermediates (species existing long enough to be detected) that are momentarily trapped in local energy wells (Figure 5.29b). Small molecules (fewer than 100 residues) often fold without intermediate formation (Figure 5.30a). As these molecules begin emerging from the ribosome, a rapid and cooperative folding process begins in which side chain interactions facilitate the formation and alignment of secondary structures. The folding of larger polypeptides typically involves the formation of several intermediates (Figure 5.30b, c). In many of these molecules or the domains within a molecule, the hydrophobically collapsed shape of the intermediate is referred to as a molten globule. The term **molten globule** refers to a partially organized globular state of a folding polypeptide that resembles the molecule's native state. Within the interior of a molten globule, tertiary interactions among amino acid side chains are fluctuating; that is, they have not yet stabilized.

It has also become increasingly clear that the folding and targeting of many proteins in living cells are aided by a group of molecules now referred to as the **molecular chaperones**. These molecules, most of which appear to be heat shock proteins (hsps), apparently occur in all organisms. Several classes of molecular chaperones have been found in organisms ranging from bacteria to the higher animals and plants. They are found in several eukaryotic organelles, such as mitochondria, chloroplasts, and ER. There is a high degree of sequence homology among the molecular chaperones of all species so far investigated. The properties of several of these important molecules are described next.

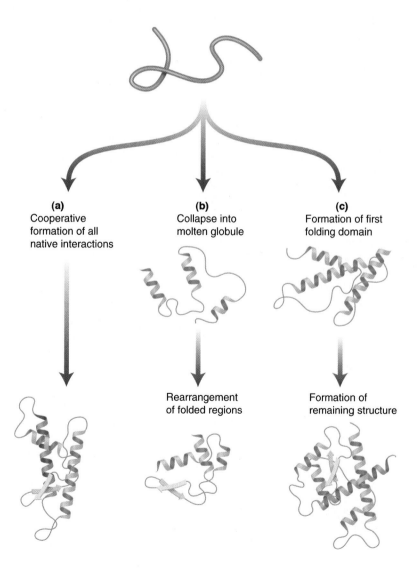

(a)
Cooperative formation of all native interactions

(b)
Collapse into molten globule

(c)
Formation of first folding domain

Rearrangement of folded regions

Formation of remaining structure

FIGURE 5.30

Protein Folding

(a) In many small proteins, folding is cooperative with no intermediates formed. (b) In some larger proteins, folding involves the initial formation of a molten globule followed by rearrangement into the native conformation. (c) Large proteins with multiple domains follow a more complex pathway, with each domain folding separately before the entire molecule progresses to its native conformation.

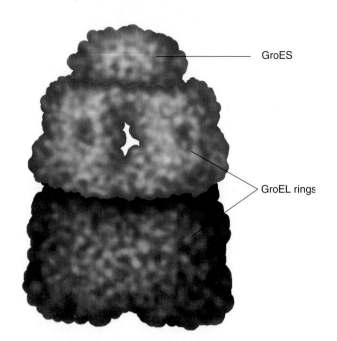

GroES

GroEL rings

FIGURE 5.31

Space-Filling Model of the *E. coli* Chaperonin called the GroES-GroEL Complex

GroES (a *co-chaperonin*, or *hsp10*) is a seven-subunit ring that sits on top of GroEL. GroEL (a *chaperonin*, or *hsp60*) is composed of two stacked, seven-subunit rings with a cavity in which ATP-dependent protein folding takes place.

MOLECULAR CHAPERONES Molecular chaperones apparently assist unfolded proteins in two ways. First, during a finite time between synthesis and folding, proteins must be protected from inappropriate protein-protein interactions. For example, certain mitochondrial and chloroplast proteins must remain unfolded until they are inserted in an organelle membrane. Second, proteins must fold rapidly and precisely into their correct conformations. Some must be assembled into multisubunit complexes. Investigations of protein folding in a variety of organisms reveal the existence of two major molecular chaperone classes in protein folding.

1. **Hsp70s.** The **hsp70s** are a family of molecular chaperones that bind to and stabilize proteins during the early stages of folding. Numerous hsp70 monomers bind to short hydrophobic segments in unfolded polypeptides, thereby preventing molten globule formation. Each type of hsp70 possesses two binding sites, one for an unfolded protein segment and another for ATP. Release of a polypeptide from an hsp70 involves ATP hydrolysis. Mitochondrial and ER-localized hsp70s are required for transmembrane translocation of some polypeptides.

2. **Hsp60s.** Once an unfolded polypeptide has been released by hsp70, it is passed on to a member of a family of molecular chaperones referred to as the **hsp60s** (also called the **chaperonins** or *Cpn60s*), which mediate protein folding. The hsp60s form a large structure composed of two stacked seven-subunit rings. The unfolded protein enters the hydrophobic cavity of the hsp60 complex (Figure 5.31). When folding is complete, ATP hydrolysis converts the cavity to a hydrophilic surface and the folded protein or domain moves out of the cavity. When the cavity is empty, it reverts to the hydrophobic surface in which form it can accept ATP and an unfolded protein for a repeat of the cycle.

In addition to promoting the folding of nascent protein, molecular chaperones direct the refolding of protein that was partially unfolded as a consequence of stressful conditions. If refolding is not possible, molecular chaperones promote protein degradation. A diagrammatic view of protein folding is presented in Figure 5.32.

Fibrous Proteins

Fibrous proteins typically contain high proportions of regular secondary structures, such as α-helices or β-pleated sheets. As a consequence of their rodlike or sheetlike shapes, many fibrous proteins have structural rather than dynamic

(Continued on p. 165)

KEY CONCEPTS

- All the information required for each newly synthesized polypeptide to fold into its biologically active conformation is encoded in the molecule's primary sequence.

- Some relatively simple polypeptides fold spontaneously into their native conformations.

- Other larger molecules require the assistance of proteins called molecular chaperones to ensure correct folding.

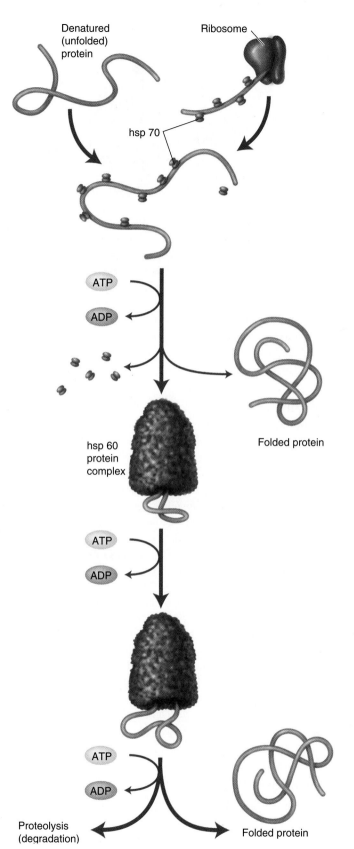

Denatured
(unfolded)
protein

Ribosome

hsp 70

ATP

ADP

Folded protein

hsp 60
protein
complex

ATP

ADP

ATP

ADP

Proteolysis
(degradation)

Folded protein

FIGURE 5.32

The Molecular Chaperones

Molecular chaperones bind transiently to both nascent proteins and
unfolded proteins (i.e., those denatured by stressful conditions).
The members of the hsp70 family stabilize nascent proteins and
reactivate some denatured proteins. Many proteins also require
hsp60 proteins to achieve their final conformations. If a protein
cannot be salvaged, the molecular chaperones help destroy it.

BIOCHEMISTRY IN PERSPECTIVE

Protein Folding and Human Disease

What are the effects of misfolded proteins on human health? The accumulation of insoluble, misfolded proteins is an important feature of several human neurodegenerative diseases. Alzheimer's and Huntington's diseases are prominent examples. Despite differences in their initiating events, specific brain areas affected, and symptoms, both diseases have in common cell-destroying dysfunctional processes set in motion by toxic proteins. A key feature of disease initiation is the conversion of normal protein structure, most commonly α-helices and random coils, into abnormal β-pleated sheet conformations. A brief overview of the molecular basis of each disease is provided.

Alzheimer's Disease

Alzheimer's disease (AD) is a progressive and ultimately fatal condition that is characterized by seriously impaired intellectual function. AD first manifests itself with short-term memory loss. Eventually, severe memory loss, disorientation, and agitation accompany a total loss of the patient's personality. Caused by neuronal death in brain regions related to memory and cognition, AD is a multifactorial disease. It is diagnosed at autopsy by the presence of insoluble aggregates of extracellular proteinaceous debris called **amyloid deposits** (or senile plaques) and intracellular accumulations of excessively phosphorylated versions of *tau*, an intrinsically unfolded microtubule-associated protein. Mutation of the tau gene, posttranslational modifications, and/or aberrant proteolysis of tau, all of which weaken its association with microtubules, increase the free concentration of tau in the cytoplasm. In its free form, tau aggregates into long filaments and/or becomes cleaved by the enzyme caspase-3, initiating apoptotic processes (p. 56) that lead to cell death. The link between this event and β-amyloid deposits is not clear, but *in vitro* studies show that β-amyloid fragments potentiate caspase-3/tau activity. The core of amyloid deposits is composed of Aβ40 and Aβ42 cleavage products of amyloid precursor protein (APP).

There are inherited and sporadic versions of AD. Most cases of inherited AD have an early onset (i.e., in middle age). Sporadic AD typically is diagnosed after age 65. The genes associated with familial (inherited) forms of AD code for mutant versions of APP, presenilin 1 (PS1), presenilin 2 (PS2), and apolipoprotein E4 (p. 392). APP is a transmembrane protein of unknown function with a large extracellular region that undergoes several proteolytic processing reactions. PS1 and PS2 are components of secretase γ, one of several proteases involved in APP processing. Mutations in the genes for APP, PS1, and PS2 have been associated with the release of the toxic Aβ40 and Aβ42 fragments. The mechanisms by which AD patients develop the sporadic form of the disease in the absence of known risk factors remain unresolved.

Huntington's Disease

Huntington's disease (HD) is one of a group of inherited neurodegenerative disorders called the polyglutamine diseases. The symptoms of HD (psychomotor skill loss, involuntary movements, and progressive dementia) are caused by neuron death in the frontal lobes and the basal ganglia of the brain. The HD gene codes for *huntingtin*, a large polypeptide of unknown function with a molecular mass of nearly 350 kD. A polyglutamine sequence containing between 6 and 34 glutamine residues occurs at the protein's N-terminal. In the mutant protein, the glutamine repeat sequence may contain as many as 150 glutamine residues, but as few as 37 residues can cause the disease. The age of symptom onset correlates with the length of the polyglutamine sequence. For example, early-onset (juvenile) HD occurs in individuals with over 55 repeats. It is believed that the initiating neurotoxic event is the association between polyglutamine sequences from nearby huntingtin molecules to form a β-sheet-like structure called a polar zipper. The subsequent formation of protein aggregates, which contain mutant huntingtin and other proteins, triggers a cascade of events that ends in apoptosis.

SUMMARY: The accumulation of misfolded proteins impedes cell function. Eventually protein aggregates cause cell death.

roles. Keratin (Figure 5.33) is a fibrous protein composed of bundles of α-helices, whereas the polypeptide chains of silk fibroin (Figure 5.34) are arranged in antiparallel β-pleated sheets. The structural features of collagen, the most abundant protein in vertebrates, are described in some detail.

COLLAGEN Collagen is synthesized by connective tissue cells and then secreted into the extracellular space to become part of the connective tissue matrix. The 20 major families of collagen molecules include many closely related proteins that have diverse functions. The genetically distinct collagen molecules in skin, bones, tendons, blood vessels, and corneas impart to these structures many of their special properties (e.g., the tensile strength of tendons and the transparency of corneas).

Collagen is composed of three left-handed polypeptide helices that are twisted around each other to form a right-handed triple helix (Figure 5.35). Type I collagen molecules, found in teeth, bone, skin, and tendons, are about 300 nm long and approximately 1.5 nm wide. Approximately 90% of the collagen found in humans is type I.

The amino acid composition of collagen is distinctive. Glycine constitutes approximately one-third of the amino acid residues. Proline and 4-hydroxyproline may account for as much as 30% of a collagen molecule's amino acid composition. Small amounts of 3-hydroxyproline and 5-hydroxylysine also occur. (Specific proline and lysine residues in collagen's primary sequence are hydroxylated within the rough ER after the polypeptides have been synthesized. These reactions, which are discussed in Chapter 19, require ascorbic acid (p. 763).

Collagen's amino acid sequence primarily consists of large numbers of repeating triplets with the sequence of Gly—X—Y, in which X and Y are often proline and hydroxyproline. Hydroxylysine is also found in the Y position. Simple carbohydrate groups are often attached to the hydroxyl group of hydroxylysine residues. It has been suggested that collagen's carbohydrate components are

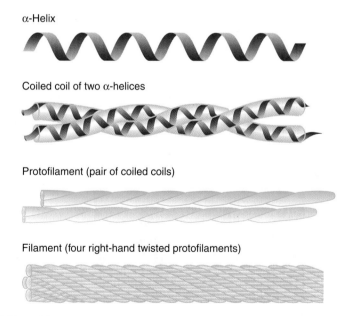

α-Helix

Coiled coil of two α-helices

Protofilament (pair of coiled coils)

Filament (four right-hand twisted protofilaments)

FIGURE 5.33

α-**Keratin**

The α-helical rodlike domains of two keratin polypeptides form a coiled coil. Two staggered antiparallel rows of these dimers form a supercoiled protofilament. Hydrogen bonds and disulfied bridges are the principal interactions between subunits. Hundreds of filaments, each containing four protofilaments, form a macrofibril. Each hair cell, also called a fiber, contains several macrofibrils. Each strand of hair consists of numerous dead cells packed with keratin molecules. In addition to hair, the keratins are also found in wool, skin, horns, and fingernails.

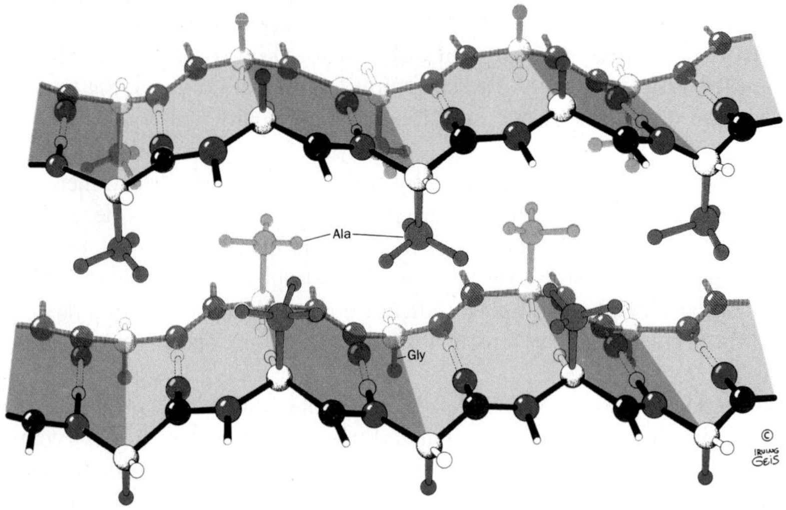

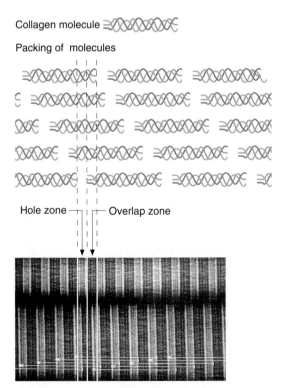

FIGURE 5.34

Molecular Model of Silk Fibroin

In fibroin, the silk fibrous protein, the polypeptide chains are arranged in fully extended antiparallel β-pleated sheet conformations. Note that the R groups of alanine on one side of each β-pleated sheet interdigitate with similar residues on the adjacent sheet. Silk fibers (fibroin embedded in an amorphous matrix) are flexible because the pleated sheets are loosely bonded to each other (primarily with weak van der Waals forces) and slide over each other easily.

FIGURE 5.35

Collagen Fibrils

The bands are formed by staggered collagen molecules. Cross-striations are about 680 Å apart. Each collagen molecule is about 3000 Å long.

Collagen molecule

Packing of molecules

Hole zone   Overlap zone

required for *fibrilogenesis*, the assembly of collagen fibers in their extracellular locations, such as tendons and bone.

The enzyme lysyl oxidase converts some of the lysine and hydroxylysine side groups to aldehydes through oxidative deamination, and this facilitates the spontaneous nonenzymatic formation of strengthening aldimine, and aldol cross-links. (An aldol cross-link is formed in a reaction, called an **aldol condensation**, in which two aldehydes form an α, β-unsaturated aldehyde linkage. In condensation reactions, a small molecule, in this case H_2O, is removed.) Cross-linkages between hydroxylysine-linked carbohydrates and the amino group of other lysine and hydroxylysine residues on adjacent molecules also occurs. Increased cross-linking with age leads to the brittleness and breakage of the collagen fibers that occur in older organisms.

Glycine is prominent in collagen sequences because the triple helix is formed by interchain hydrogen bonding involving the glycine residues. Therefore every third residue is in close contact with the other two chains. Glycine is the only amino acid with an R group sufficiently small for the space available. Larger R groups would destabilize the superhelix structure. The triple helix is further strengthened by hydrogen bonding between the polypeptides (caused principally by the large number of hydroxyproline residues) and lysinonorleucine linkages that stabilize the orderly arrays of triple helices in the final collagen fibril.

Globular Proteins

The biological functions of globular proteins usually involve the precise binding of small ligands or large macromolecules such as nucleic acids or other

---

QUESTION 5.10

Covalent cross-links contribute to the strength of collagen. The first reaction in cross-link formation is catalyzed by the copper-containing enzyme lysyl oxidase, which converts lysine residues to the aldehyde allysine:

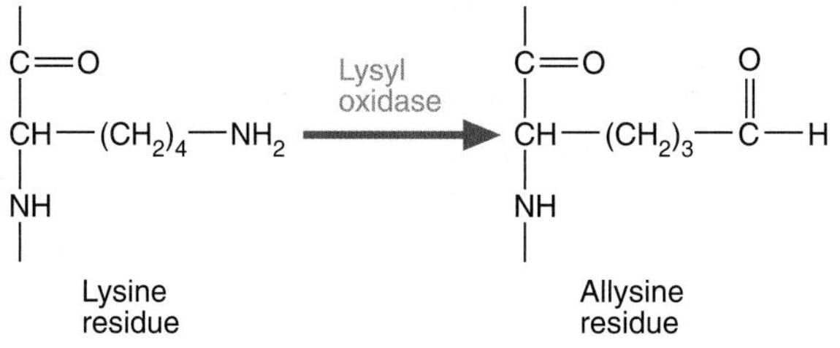

Allysine then reacts with other side chain aldehyde or amino groups to form cross-linkages. For example, two allysine residues react to form an aldol cross-linked product:

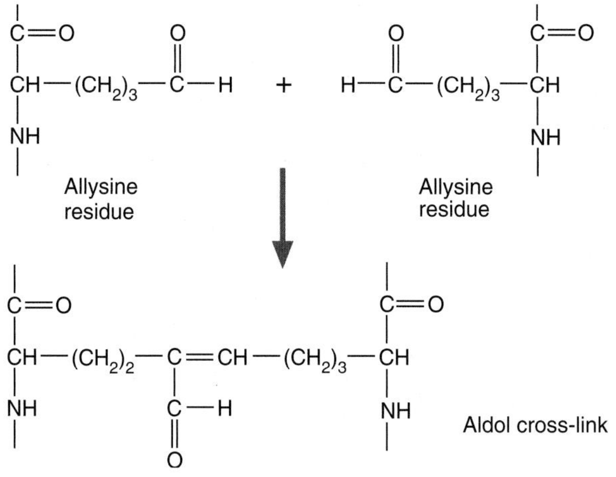

In a disease called *lathyrism*, which occurs in humans and several other animals, a toxin (β-aminopropionitrile) found in sweet peas *(Lathyrus odoratus)* inactivates lysyl oxidase. Consider the abundance of collagen in animal bodies and suggest some likely symptoms of this malady.

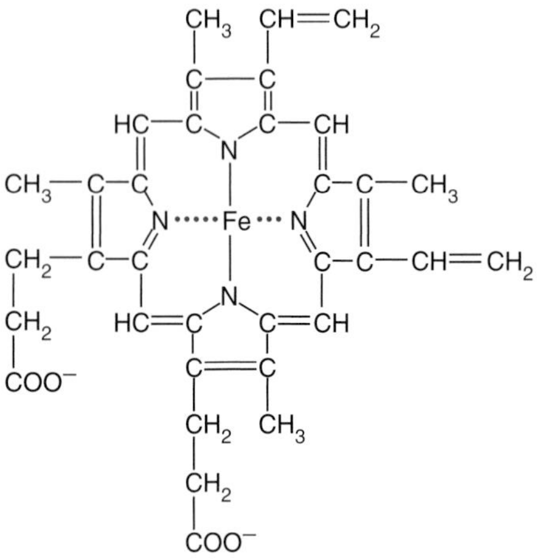

FIGURE 5.36

Heme

Heme consists of a porphyrin ring (composed of four pyrroles) with Fe^{2+} in the center.

proteins. Each protein possesses one or more unique cavities or clefts whose structure is complementary to a specific ligand. After ligand binding, a conformational change occurs in the protein that is linked to a biochemical event. For example, the binding of ATP to myosin in muscle cells is a critical event in muscle contraction.

The oxygen-binding proteins myoglobin and hemoglobin are interesting and well-researched examples of globular proteins. They are both members of the hemoproteins, a specialized group of proteins that contain the prosthetic group heme. Although the heme group (Figure 5.36) in both proteins is responsible for the reversible binding of molecular oxygen, the physiological roles of myoglobin and hemoglobin are significantly different. The chemical properties of heme are dependent on the Fe^{2+} ion in the center of the prosthetic group. Fe^{2+}, which forms six coordinate bonds, is bound to the four nitrogens in the center of the protoporphyrin ring. Two other coordinate bonds are available, one on each side of the planar heme structure. In myoglobin and hemoglobin, the fifth coordination bond is to the nitrogen atom in a histidine residue, and the sixth coordination bond is available for binding oxygen. In addition to serving as a reservoir for oxygen within muscle cells, myoglobin facilitates the diffusion of oxygen in metabolically active cells. The role of hemoglobin, the primary protein of red blood cells, is to deliver oxygen to cells throughout the body. A comparison of the structures of these two proteins illustrates several important principles of protein structure, function, and regulation.

MYOGLOBIN Myoglobin, found in high concentration in skeletal and cardiac muscle, gives these tissues their characteristic red color. The muscles of diving mammals such as whales, which remain submerged for long periods, have high myoglobin concentrations. Because of the extremely high concentrations of myoglobin, such muscles are typically brown. The protein component of myoglobin, called globin, is a single polypeptide chain that contains eight segments of α-helix (Figure 5.37). The folded globin chain forms a crevice that almost completely encloses a heme group. Free heme [Fe^{2+}] has a high affinity for O_2 and is irreversibly oxidized to form hematin [Fe^{3+}]. Hematin cannot bind O_2. Noncovalent interactions between amino acid side chains and the nonpolar porphyrin ring within the oxygen-binding crevice decrease heme's affinity for O_2. The decreased affinity protects Fe^{2+} from oxidation and allows for the reversible binding of O_2. All of the heme-interacting amino acids are nonpolar except for two histidines, one of which (the proximal histidine) binds directly to the heme iron atom (Figure 5.38). The other (the distal histidine) stabilizes the oxygen-binding site.

HEMOGLOBIN Hemoglobin is a roughly spherical molecule found in red blood cells, where its primary function is to transport oxygen from the lungs to every tissue in the body. Recall that HbA is composed of two α-chains and two β-chains (Figure 5.39). The HbA molecule is commonly designated $\alpha_2\beta_2$. [There is another type of adult hemoglobin, however: approximately 2% of human hemoglobin is HbA$_2$, which contains δ (delta)-chains instead of β-chains.] Before birth, several additional hemoglobin polypeptides are synthesized. The ε (epsilon)-chain, which appears in early embryonic life, and the γ-chain, found in the fetus, closely resemble the β-chain. Because both $\alpha_2\varepsilon_2$ and $\alpha_2\gamma_2$ hemoglobins have a greater affinity for oxygen than does $\alpha_2\beta_2$ (HbA), the fetus can preferentially absorb oxygen from the maternal bloodstream.

Although the three-dimensional configurations of myoglobin and the α- and β-chains of hemoglobin are very similar, their amino acid sequences have many differences. Comparison of these molecules from dozens of species has revealed nine invariant amino acid residues. Several invariant residues directly affect the oxygen-binding site, whereas others stabilize the α-helical peptide segments. The remaining residues may vary considerably. However, most

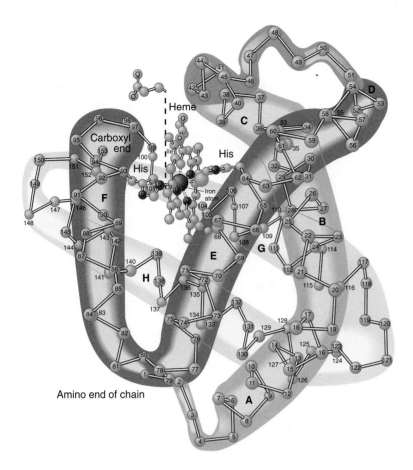

FIGURE 5.37

Myoglobin

With the exception of the side chain groups of two histidine residues, only the α-carbon atoms of the globin polypeptide are shown. Myoglobin's eight helices are designated A through H. The heme group has an iron atom that binds reversibly with oxygen. To improve clarity one of heme's propionic acid side chains has been displaced.

substitutions are conservative. For example, each polypeptide's interior remains nonpolar.

The four chains of hemoglobin are arranged in two identical dimers, designated as $\alpha_1\beta_1$, and $\alpha_2\beta_2$. Each globin polypeptide has a heme-binding unit similar to that described for myoglobin. Although both myoglobin and hemoglobin bind oxygen reversibly, the latter molecule has a complex structure and more complicated binding properties. The numerous noncovalent interactions (mostly hydrophobic) between the subunits in each αβ-dimer remain largely unchanged when hemoglobin interconverts between its oxygenated and deoxygenated forms. In contrast, the relatively small number of interactions between the two dimers change substantially during this transition. When hemoglobin is oxygenated, the salt bridges and certain hydrogen bonds are ruptured as the $\alpha_1\beta_1$ and $\alpha_2\beta_2$ dimers slide by each other and rotate 15° (Figure 5.40). The deoxygenated conformation of hemoglobin (deoxyHb) is often referred to as the T(taut) state and oxygenated hemoglobin (oxyHb) is said to be in the R (relaxed) state. The oxygen-induced readjustments in the interdimer contacts are almost simultaneous. In other words, a conformational change in one subunit is rapidly propagated to the other subunits. Consequently, hemoglobin alternates between two stable conformations, the T and R states.

Because of subunit interactions, the oxygen dissociation curve of hemoglobin has a sigmoidal shape (Figure 5.41). As the first O_2 binds to hemoglobin, the binding of additional O_2 to the same molecule is enhanced. This binding pattern, called **cooperative binding**, results from changes in hemoglobin's three-dimensional structure that are initiated when the first O_2 binds. The binding of the first O_2 facilitates the binding of the remaining three O_2 molecules to the tetrameric hemoglobin molecules. In the lungs, where O_2 tension is high, hemoglobin is quickly saturated (converted to the R state). In tissues depleted of O_2, hemoglobin gives up about half its oxygen. In contrast to hemoglobin, myoglobin's oxygen dissociation curve is hyperbolic. This simpler binding pattern,

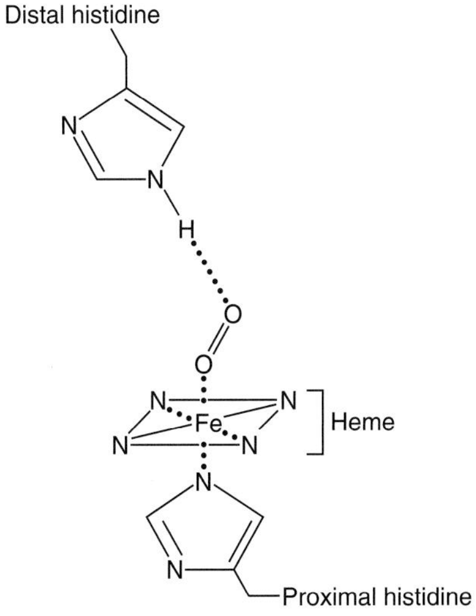

Distal histidine

FIGURE 5.38

The Oxygen-Binding Site of Heme Created by a Folded Globin Chain

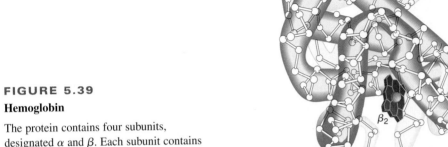

FIGURE 5.39

Hemoglobin

The protein contains four subunits, designated α and β. Each subunit contains a heme group that binds reversibly with oxygen.

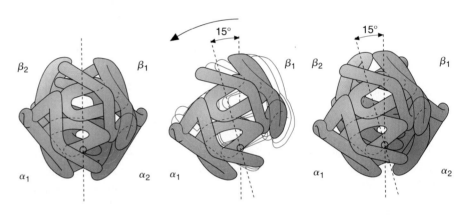

(a) Deoxyhemoglobin

(b) Oxyhemoglobin

FIGURE 5.40

The Hemoglobin Allosteric Transition

When hemoglobin is oxygenated, the $\alpha_1\beta_1$ and $\alpha_2\beta_2$ dimers slide by each other and rotate 15°.

a consequence of myoglobin's simpler structure, reflects several aspects of this protein's role in oxygen storage. Because its dissociation curve is well to the left of the hemoglobin curve, myoglobin gives up oxygen only when the muscle cell's oxygen concentration is very low (i.e., during strenuous exercise). In addition, because myoglobin has a greater affinity for oxygen than does hemoglobin, oxygen moves from blood to muscle.

The binding of ligands other than oxygen affects hemoglobin's oxygen-binding properties. For example, the dissociation of oxygen from hemoglobin is enhanced if pH decreases. By this mechanism, called the *Bohr effect*, oxygen is delivered to cells in proportion to their needs. Metabolically active cells, which require large amounts of oxygen for energy generation, also produce large amounts of the waste product CO_2. As CO_2 diffuses into blood, it reacts with water to form HCO_3^- and H^+. (The bicarbonate buffer was discussed on p. 96.) The subsequent binding of H^+ to several ionizable groups on hemoglobin molecules enhances the dissociation of O_2 by converting hemoglobin to its T state. (Hydrogen ions bind preferentially to deoxyHb. Any increase in H^+ concentration stabilizes the deoxy conformation of the protein and therefore shifts the equilibrium distribution between the T and R states.) When a small number of CO_2 molecules bind to terminal amino groups on hemoglobin (forming carbamate or $—NHCOO^-$ groups) the deoxy form (T state) of the protein is additionally stabilized.

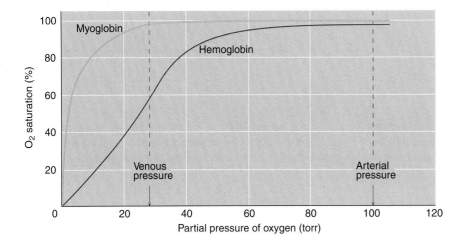

FIGURE 5.41

Equilibrium Curves Measure the Affinity of Hemoglobin and Myoglobin for Oxygen

2,3-Bisphosphoglycerate (BPG) (also called glycerate-2,3-bisphosphate) is also an important regulator of hemoglobin function. Although most cells contain only trace amounts of BPG, red blood cells contain a considerable amount. BPG is derived from glycerate-l,3-bisphosphate, an intermediate in the breakdown of the energy-rich compound glucose. In the absence of BPG, hemoglobin has a very high affinity for oxygen (Figure 5.42). As with H+ and CO_2, binding BPG stabilizes deoxyHb. A negatively charged BPG molecule binds in a central cavity within hemoglobin that is lined with positively charged amino acids.

In the lungs the process is reversed. A high oxygen concentration drives the conversion from the deoxyHb configuration to that of oxyHb. The change in the protein's three-dimensional structure initiated by the binding of the first oxygen molecule releases bound CO_2, H^+, and BPG. The H^+ recombines with HCO_3^- to form carbonic acid, which then dissociates to form CO_2 and H_2O. Afterward, CO_2 diffuses from the blood into the alveoli.

KEY CONCEPTS

- Globular protein function usually involves binding to small ligands or to other macromolecules.
- The oxygen-binding properties of myoglobin and hemoglobin are determined in part by the number of subunits they contain.

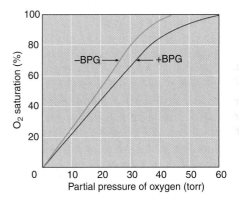

FIGURE 5.42

The Effect of 2,3-Bisphosphoglycerate (BPG) on the Affinity Between Oxygen and Hemoglobin

In the absence of BPG (–BPG), hemoglobin has a high affinity for O_2; where BPG is present and binds to hemoglobin (+BPG), its affinity for O_2 decreases.

QUESTION 5.11

Fetal hemoglobin (HbF) binds to BPG to a lesser extent than does HbA. Why do you think HbF has a greater affinity for oxygen than does maternal hemoglobin?

QUESTION 5.12

Myoglobin stores O_2 in muscle tissue to be used by the mitochondria only when the cell is in oxygen debt, while hemoglobin can effectively transport O_2 from the lungs and deliver it discriminately to cells in need of O_2. Describe the structural features that allow these two proteins to accomplish separate functions.

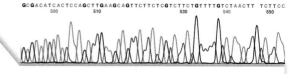

BIOCHEMISTRY IN THE LAB
Protein Technology

Living organisms produce a stunning variety of proteins. Consequently, it is not surprising that considerable time, effort, and funding have been devoted to investigating their properties. Since the amino acid sequence of bovine insulin was determined by Frederick Sanger in 1953, the structures of several thousand proteins have been elucidated.

In contrast to the 10 years required for insulin, current technologies allow protein sequence determination within a few days. In addition to the Edman degradation method and mass spectrometry, the amino acid sequence of a protein can be generated from its DNA or mRNA sequence if this information is available. After a brief review of protein purification methods, the Edman degradation method and mass spectrometry are described. Note that all the techniques for isolating, purifying, and characterizing proteins exploit differences in charge, molecular weight, and binding affinities. Many of these technologies apply to the investigation of other biomolecules.

Purification

Protein analysis begins with isolation and purification. Extraction of a protein requires cell disruption and homogenization (see Biochemistry in the Lab, Cell Technology, Chapter 2). This process is often followed by differential centrifugation and, if the protein is a component of an organelle, by density gradient centrifugation. After the protein-containing fraction has been obtained, several relatively crude methods may be used to enhance purification. **Salting out** is a technique in which high concentrations of salts such as ammonium sulfate $[(NH_4)_2SO_4]$ are used to precipitate proteins. Because each protein has a characteristic salting-out point, this technique removes many impurities. (Unwanted proteins that remain in solution are discarded when the liquid is decanted.) When proteins are tightly bound to membrane, organic solvents or detergents often aid in their extraction. Dialysis is routinely used to remove low-molecular-weight impurities such as salts, solvents, and detergents.

As a protein sample becomes progressively more pure, more sophisticated methods are used to achieve further purification. Among the most commonly used techniques are chromatography and electrophoresis.

Chromatography

Originally devised to separate low-molecular-weight substances such as sugars and amino acids, chromatography has become an invaluable tool in protein purification. A wide variety of chromatographic techniques are used to separate protein mixtures on the basis of molecular properties such as size, shape, and weight, or certain binding affinities. Often several techniques must be used sequentially to obtain a demonstrably pure protein.

In all chromatographic methods the protein mixture is dissolved in a liquid known as the **mobile phase**. As the protein molecules pass across the **stationary phase** (a solid matrix), they separate from each other because they are differently distributed between the two phases. The relative movement of each molecule results from its capacity to remain associated with the stationary phase while the mobile phase continues to flow.

Three chromatographic methods commonly used in protein purification are gel-filtration chromatography, ion-exchange chromatography, and affinity chromatography. **Gel-filtration chromatography** (Figure 5D) is a form of size-exclusion chromatography in which particles in an aqueous solution flow through a column (a hollow tube) filled with gel and are separated according to size. Molecules that are larger than the gel pores are excluded and therefore move through the column quickly. Molecules that are smaller than the gel pores diffuse in and out of the pores, so their movement through the column is retarded. Differences in the rates of particle movement separate the protein mixture into bands, which are then collected separately.

Ion-exchange chromatography separates proteins on the basis of their charge. Anion-exchange resins, which consist of positively charged materials, bind reversibly with a protein's negatively charged groups. Similarly, cation-exchange resins bind positively charged groups. After proteins that do not bind to the resin have been removed, the protein of interest is recovered by an appropriate change in the solvent pH and/or salt concentration. (A change in pH alters the protein's net charge.)

Affinity chromatography takes advantage of the unique biological properties of proteins. That is, it uses a special noncovalent binding affinity between the protein and a special molecule (the ligand). The ligand is covalently bound to an insoluble matrix, which is placed in a column. After nonbinding protein molecules have passed through the column, the protein of interest is removed by altering the conditions that affect binding (i.e., pH or salt concentration).

Electrophoresis

Because proteins are electrically charged, they move in an electric field. In this process, called **electrophoresis**, molecules separate from each other because of differences in their net charge. For example, molecules with a positive net charge migrate toward the negatively charged electrode (cathode). Molecules with a net negative charge will move toward the positively charged electrode (anode). Molecules with no net charge will not move at all.

Electrophoresis, one of the most widely used techniques in biochemistry, is nearly always carried out by using gels such as polyacrylamide or agarose. The gel, functioning much as it does in gel-filtration chromatography, also acts to separate proteins on the basis of their molecular weight and shape. Consequently, gel

BIOCHEMISTRY IN THE LAB cont

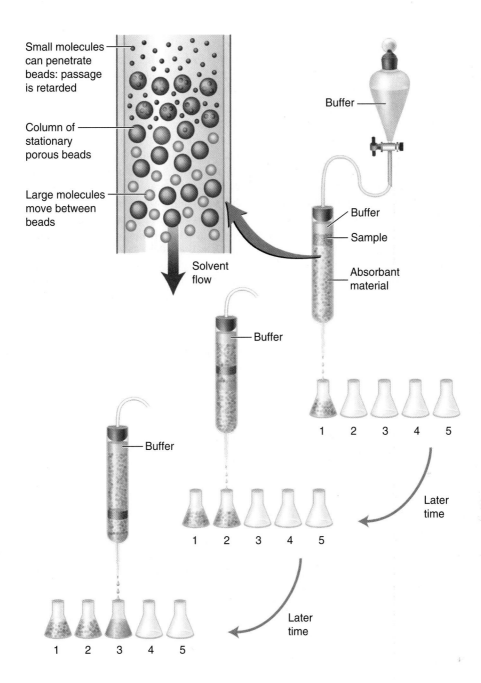

Small molecules can penetrate beads: passage is retarded

Column of stationary porous beads

Large molecules move between beads

Solvent flow

Buffer

Buffer

Sample

Absorbant material

Buffer

1 2 3 4 5

Later time

Buffer

1 2 3 4 5

Later time

1 2 3 4 5

FIGURE 5D

Gel-Filtration Chromatography

In gel-filtration chromatography the stationary phase is a gelatinous polymer with pore sizes selected by the experimenter to separate molecules according to their sizes. The sample is applied to the top of the column and is eluted with buffer (the mobile phase). As elution proceeds, larger molecules travel faster through the gel than smaller molecules, whose progress is slowed because they can enter the pores. If fractions are collected, the larger molecules appear in the earlier fractions and later fractions contain smaller molecules.

BIOCHEMISTRY IN THE LAB cont

electrophoresis is highly effective at separating complex mixtures of proteins or other molecules.

Bands resulting from a gel electrophoretic separation may be treated in several ways. Specific bands may be excised from the gel after visualization with ultraviolet light. Each protein-containing slice is then eluted with buffer and prepared for further analysis. Because of its high resolving power, gel electrophoresis is also used to assess the purity of protein samples. Staining gels with a dye such as Coomassie Brilliant Blue is a common method for quickly assessing the success of a purification step.

SDS–polyacrylamide gel electrophoresis (SDS-PAGE) is a widely used variation of electrophoresis that can be used to determine molecular weight (Figure 5E). SDS, a negatively charged detergent, binds to the hydrophobic regions of protein molecules, causing the proteins to denature and assume rodlike shapes. Because most molecules bind SDS in a ratio roughly proportional to their molecular weights, during electrophoresis SDS-treated proteins migrate toward the anode (+ pole) only in relation to their molecular weight.

Protein Sequence Analysis

The first step in protein sequence analysis is to determine how many of each type of amino acid residue are present in the molecule. The process for obtaining this information, referred to as the *amino acid composition*, begins with the complete hydrolysis of all peptide bonds. Hydrolysis is typically accomplished with 6 N HCl for 10 to 100 hours. Long reaction times are required because of difficulties in the hydrolysis of three aliphatic amino acids (Leu, Ile, and Val). Hydrolysis is followed by analysis of the resulting amino acid mixture, referred to as the *hydrolysate*. Because of the vigorous conditions of acid hydrolysis, several amino acids (Trp, Gln, Ser, Thr, Tyr, and cystine) are degraded. The concentrations of these molecules in the protein are determined by alternate means.

Currently, most protein hydrolysates are analyzed by automated high-pressure liquid chromatography (HPLC). In HPLC, after the hydrolysate has been treated with compounds such as Edman's reagent (described shortly), the products are forced at high pressure through a stainless steel column packed with a stationary phase. Each amino acid derivative is identified according to its retention time on the column. Amino acid analysis by HPLC takes about 1 hour.

Determining a protein's primary structure is similar to solving a complex puzzle. Several steps are involved in solving the amino acid sequence of any protein.

1. **Cleavage of all disulfide bonds.** Oxidation with performic acid is commonly used.
2. **Determination of the N-terminal and C-terminal amino acids.** Several methods are available to determine the N-terminal amino acid. In Sanger's method, the polypeptide chain is reacted with l-fluoro-2,4-dinitrobenzene. The

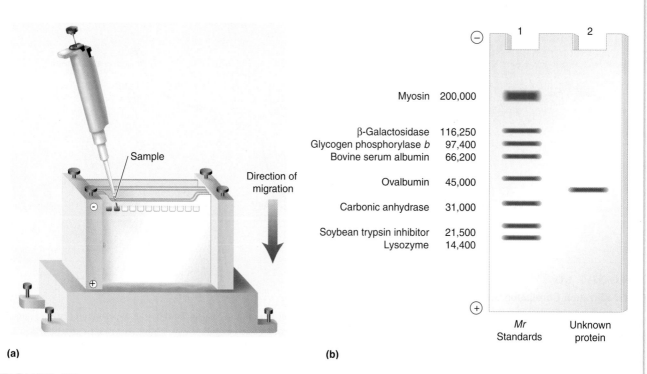

(a)   (b)

FIGURE 5E

Gel Electrophoresis

(a) Gel apparatus. The samples are loaded into wells. After an electric field is applied, the proteins move into the gel. (b) Molecules separate and move in the gel as a function of molecular weight and shape.

BIOCHEMISTRY **IN THE LAB** cont

dinitrophenyl (DNP) derivative of the N-terminal amino acid can then be isolated. Alternatively, dabsyl chloride is used to determine N-terminal amino acid residues. Dabsylation is a very sensitive technique because the dabsyl group, a fluorescent marker, is easily detected in small amounts. The N-terminal derivatives of both of these reagents can then be identified by HPLC analysis. A group of enzymes called the carboxypeptidases are used to identify the C-terminal residue. Because these enzymes sequentially cleave peptide bonds starting at the C-terminal residue, the first amino acid liberated is the C-terminal residue.

3. **Cleavage of the polypeptide into fragments.** The polypeptide is broken into smaller peptides because technical problems prevent the direct sequencing of long polypeptides. The use of several reagents, each of which cuts the chain at a different site, creates overlapping sets of fragments. After the amino acid sequence of each fragment has been determined, the investigator uses this information to work out the entire sequence of the polypeptide. Of all the enzymes commonly used, the pancreatic enzyme trypsin is the most reliable. It cleaves peptide bonds on the carboxy side of either lysine or arginine residues. The peptide fragments, referred to as *tryptic*

peptides, have lysine or arginine carboxy terminal residues. Chymotrypsin, another pancreatic enzyme, is also often used. It breaks peptide bonds on the carboxyl side of phenylalanine, tyrosine, leucine, methionine, or tryptophan. Treating the polypeptide with the reagent cyanogen bromide also generates peptide fragments. Cyanogen bromide specifically cleaves peptide bonds on the carboxyl side of methionine residues.

4. **Determination of the sequences of the peptide fragments.** Each fragment is sequenced through repeated cycles of a procedure called the *Edman degradation*. In this method phenylisothiocyanate (PITC), often referred to as Edman's reagent, reacts with the N-terminal residue of each fragment. Treatment of the product of this reaction with acid cleaves the N-terminal residue as a phenylthiohydantoin derivative. The derivative is then identified by comparing it with known standards, using electrophoresis or various chromatographic methods (most commonly HPLC). Because of the large number of steps involved in sequencing peptide fragments, Edman degradation is usually carried out by using a computer-programmed machine called a sequenator.

5. **Ordering the peptide fragments.** The amino acid sequence information derived from two or more sets of

DNP-amino acid

Dabsyl-amino acid

Edman degradation

PITC

Dilute | H⁺

Phenylthiohydantoin derivative of N-terminal amino acid

Peptide minus N-terminal residue

BIOCHEMISTRY **IN THE LAB** cont

polypeptide fragments is next examined for overlapping segments. Such segments make it possible to piece together the overall sequence.

A typical primary sequence determination problem is given, along with its solution.

Problem

Consider the following peptide:

Gly—Ile—Glu—Trp—Thr—Pro—Tyr—Gln—Phe—Arg—Lys

What amino acids and peptides are produced when this peptide is treated with each of the following reagents?

a. Carboxypeptidase   b. Chymotrypsin

c. Trypsin   d. DNFB (dinitrofluorobenzene)

Solution

a. Because carboxypeptidase cleaves at the carboxyl end of peptides, the products are

Gly—Ile—Glu—Trp—Thr—Pro—Tyr—Gln—Phe—Arg and Lys

b. Because chymotrypsin cleaves peptide bonds in which aromatic amino acids (i.e., Phe, Tyr, and Trp) contribute a carboxyl group, the products are

Gly—Ile—Glu—Trp, Thr—Pro—Tyr, Gin—Phe, and Arg—Lys

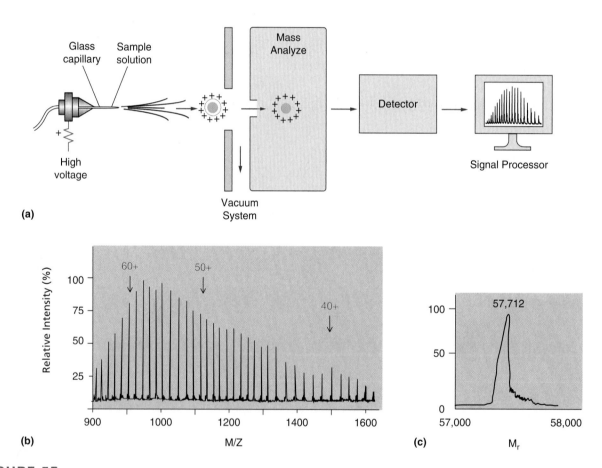

(a)

(b) M/Z

(c) M_r

FIGURE 5F

Mass Spectrometry

(a) The principal steps in electrospray ionization. The sample (a protein dissolved in a solvent) is injected via a glass capillary into the ionization chamber. The voltage difference between the electrospray needle and the injection port results in the creation of protein ions. The solvent evaporates during this phase. The ions enter the mass spectrometer, which then measures their m/z ratios. (b) An electrospray mass spectrum showing the m/z ratios for several peaks. (c) A computer analysis of the data showing the molecular mass of the sample protein (M_r = molecular weight).

BIOCHEMISTRY IN THE LAB cont

c. Trypsin cleaves at the carboxyl end of lysine and arginine. The products are

Gly—Ile—Glu—Trp—Thr—Pro—Tyr—Gln—Phe—Arg
and Lys

d. DNFB tags the amino-terminal amino acid. The product is

DNP—Gly—Ile—Glu—Trp—Thr—Pro—
Tyr—Gln—Phe—Arg—Lys

Hydrolysis then cleaves all the peptide bonds, and DNP—Gly can be identified by a chromatographic method.

Automation of the Edman degradation method has increased the speed and accuracy of the sequencing process. Computer-assisted devices called sequenators can determine the amino acid sequence of vanishingly small samples in a fraction of the time needed for the manual method.

Mass Spectrometry

In recent years biochemists have moved away from the Edman degradation method for sequencing proteins. Instead, many use **mass spectrometry** (MS), a powerful and sensitive technique for separating, identifying, and determining the mass of molecules by exploiting differences in their mass-to-charge (m/z) ratios. In a mass spectrometer, ionized molecules flow through a magnetic field (Figure 5F). The magnetic field force deflects the ions depending on their m/z ratios with lighter ions being more deflected from a straight-line path than heavier ions. A detector measures the deflection of each ion. In addition to protein identity and mass determinations, MS is also used to detect bound cofactors and protein modifications. Because MS analysis

involves the ionization and vaporization of the substances to be investigated, its use in the analysis of thermally unstable macromolecules such as proteins and nucleic acids did not become feasible until methods such as electrospray ionization and matrix-assisted laser desorption ionization (MALDI) had been developed. In electrospray ionization a solution containing the protein of interest is sprayed in the presence of a strong electrical field into a port in the spectrometer. As the protein droplets exit the injection device, typically an ultrafine glass tube, the protein molecules becomes charged. In MALDI, a laser pulse vaporizes the protein, which is embedded in a solid matrix. Once the sample has been ionized, its molecules, now in the gas phase, are separated according to their individual m/z ratios. A detector within the mass spectrometer produces a peak for each ion. In a computer-assisted process, information concerning each ion's mass is compared against data for ions of known structure and used to determine the sample's molecular identity.

Protein sequencing analysis makes use of tandem MS (two mass spectrometers linked in series, MS/MS). A protein of interest, often extracted from a band in a gel, is then digested by a proteolytic enzyme. Subsequently, the enzyme digest is injected into the first mass spectrometer, which separates the oligopeptides according to their m/z ratios. One by one, each oligopeptide ion is directed into a collision chamber, where it is fragmented by collisions with hot inert gas molecules. Product ions, peptides that differ from each other in size by one amino acid residue, are then sequentially directed into the second mass spectrometer. A computer identifies each peak and automatically determines the amino acid sequence of the peptides. The process is then repeated

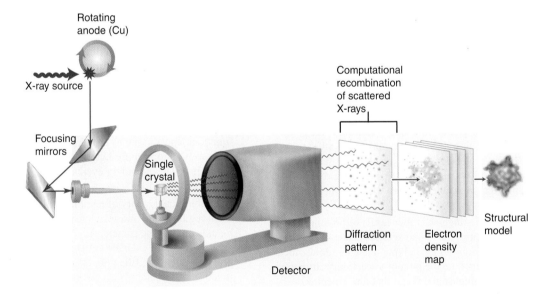

FIGURE 5G

Schematic Diagram of X-Ray Crystallography

X-rays are useful in the analysis of biomolecules because their wavelength range is quite similar to the magnitude of chemical bonds. Consequently, the resolving power of X-ray crystallography is equivalent to interatomic distances.

BIOCHEMISTRY IN THE LAB cont

with oligopeptides derived from digestion with another enzyme. The computer uses the sequence information derived from both digests to determine the amino acid sequence of the original polypeptide.

X-Ray Crystallography

Much of the three-dimensional structural information about proteins was obtained by X-ray crystallography. Because the bond distances in proteins are approximately 0.15 nm, the electromagnetic radiation used to resolve protein structure must have a short wavelength. Visible light wavelengths [$(\lambda) =$ 400–700 nm] clearly does not have sufficient resolving power for biomolecules. X-rays, however, have very short wavelengths (0.07–0.25 nm).

In X-ray crystallography, highly ordered crystalline specimens are exposed to an X-ray beam (Figure 5G). As the X-rays hit the crystal, they are scattered by the atoms in the crystal. The diffraction pattern that results is recorded on charge-coupled device (CCD) detectors. The diffraction patterns are used to construct an electron density map. Because there is no objective lens to recombine the scattered X-rays, the three-dimensional image is reconstructed mathematically. Computer programs now perform these extremely complex and laborious computations. ■

Chapter Summary

1. Polypeptides are amino acid polymers. Proteins may consist of one or more polypeptide chains.

2. Each amino acid contains a central carbon atom (the α-carbon) to which an amino group, a carboxylate group, a hydrogen atom, and an R group are attached. In addition to comprising protein, amino acids have several other biological roles. According to their capacity to interact with water, amino acids may be separated into four classes: nonpolar, polar, acidic, and basic.

3. Titration of amino acids and peptides illustrates the effect of pH on their structures. The pH at which a molecule has no net charge is called its isoelectric point.

4. Amino acids undergo several chemical reactions. Two reactions are especially important: peptide bond formation and cysteine oxidation.

5. Proteins have a vast array of functions in living organisms. In addition to serving as structural materials, proteins are involved in metabolic regulation, transport, defense, and catalysis. Some proteins are multifunctional; that is, they have two or more seemingly unrelated functions. Proteins can also be classified into families and superfamilies, according to their sequence similarities as well as their shapes and composition. Fibrous proteins (e.g., collagen) are long, rod-shaped molecules that are insoluble in water and physically tough. Globular proteins (e.g., hemoglobin) are compact, spherical molecules that are usually soluble in water.

6. Biochemists have distinguished four levels of protein structure. Primary structure, the amino acid sequence, is specified by genetic information. As the polypeptide chain folds, local folding patterns constitute the protein's secondary structure. The overall three-dimensional shape that a polypeptide assumes is called the tertiary structure. Proteins that consist of two or more polypeptides have quaternary structure. The functions of numerous proteins, especially molecules that participate in eukaryotic regulatory processes, are partially or completely unstructured. Many physical and chemical conditions disrupt protein structure. Denaturing agents include strong acids or bases, reducing agents, organic solvents, detergents, high salt concentrations, heavy metals, temperature changes, and mechanical stress.

7. One of the most important aspects of protein synthesis is the folding of polypeptides into their biologically active conformations. Despite decades of investigation into the physical and chemical properties of polypeptide chains, the mechanism by which a primary sequence dictates the molecule's final conformation is unresolved. Many proteins require molecular chaperones to fold into their final three-dimensional conformations. Protein misfolding is now known to be an important feature of several human diseases, including Alzheimer's disease and Huntington's disease.

8. Fibrous proteins (e.g., α-keratin and collagen), which contain high proportions of α-helices or β-pleated sheets, have structural rather than dynamic roles. Despite their varied functions, most globular proteins have features that allow them to bind to specific ligands or sites on certain macromolecules. These binding events involve conformational changes in the globular protein's structure.

9. The biological activity of complex multisubunit proteins is often regulated by allosteric interactions in which small ligands bind to the protein. Any change in the protein's activity is due to changes in the interactions among the protein's subunits. Effectors can increase or decrease the function of a protein.

Suggested**Readings**

Branden, C., and Tooze, J., *Introduction to Protein Structure*, 2nd ed., Garland, New York, 1999.

Bustamonte, C., Of Torques, Forces, and Protein Machines, *Protein Sci.* 13:3061–65, 2004.

Chothia, C., Gough, J., Vogel, C., and Teichmann, S. A., Evolution of the Protein Repertoire, *Science* 300:1701–1703, 2003.

Dyson, H. J., and Wright, P. E., Intrinsically Unstructured Proteins and Their Functions, *Nat. Rev. Mol. Cell Biol.* 6(3):197–208, 2005.

Fink, A. L., Natively Unfolded Proteins, *Curr. Opin. Struct Biol.* 15:35–41, 2005.

Lesk, A. M., *Introduction to Protein Science: Architecture, Function, and Genomics*, Oxford University Press, Oxford, 2004.

Lindorff, K., Rogen, P., Poci, E., Vendruscolo, M., and Dobson, M., Protein Folding and the Organization of the Protein Topology Universe, *Trends, Biochem. Sci.* 30(1):13–19, 2005.

Mattos, C., Protein-Water Interactions in a Dynamic World, *Trends Biochem. Sci.* 27(4):203–208, 2002.

Ponting, C. P., and Russell, R. R., The Natural History of Protein Domains, *Annu. Rev. Biophys. Biomed. Struct.* 31:45–71, 2002.

Tompa, P., Szasz, C., and Buday, L., Structural Disorder Throws New Light on Moonlighting. *Trends Biochem. Sci.* 30(9):484–489, 2005.

Key**Words**

affinity chromatography, *172*
aldimine, *138*
aldol condensation, *166*
aliphatic hydrocarbon, *128*
allosteric transition, *153*
allostery, *153*
Alzheimer's disease, *164*
amino acid residue, *124*
amphipathic molecule, *156*
amphoteric molecule, *127*
antigen, *141*
apoprotein, *142*
aromatic hydrocarbon, *128*
asymmetric carbon, *131*
chaperonins, *162*
chiral carbon, *131*
circular dichroism, *160*
conjugated protein, *142*
cooperative binding, *169*
denaturation, *155*
disulfide bridge, *136*
effector, *153*

electrophoresis, *172*
enantiomer, *131*
fibrous protein, *142*
fold, *136*
gel-filtration chromatography, *172*
globular protein, *142*
glycoprotein, *142*
heat shock protein, *141*
hemoprotein, *142*
holoprotein, *142*
homologous polypeptide, *142*
hormone, *130*
hsp60, *162*
hsp70, *162*
Huntington's disease, *164*
intrinsically unstructured protein, *155*
ion-exchange chromatography, *172*
isoelectric point, *133*
ligand, *148*

lipoprotein, *142*
metalloprotein, *142*
mobile phase, *172*
modular protein, *148*
modulator, *153*
molecular chaperone, *161*
molecular disease, *145*
molten globule, *161*
motif, *147*
motor protein, *157*
multifunction protein, *141*
natively unfolded protein, *155*
neurotransmitter, *130*
oligomer, *153*
optical isomer, *131*
peptide, *124*
peptide bond, *136*
phosphoprotein, *142*
polypeptide, *124*
primary structure, *142*
prosthetic group, *142*

protein, *124*
protein family, *141*
protein folding, *148*
protein superfamily, *141*
protomer, *153*
quaternary structure, *142*
response element, *155*
salt bridge, *172*
salting out, *172*
Schiff base, *138*
SDS-polyacrylamide gel electrophoresis, *174*
secondary structure, *142*
site-directed mutagenesis, *159*
stationary phase, *172*
stereoisomer, *131*
subunit, *150*
supersecondary structure, *147*
tertiary structure, *142*
zwitterion, *127*

Review**Questions**

These questions are designed to test your knowledge of the key concepts discussed in this chapter, before moving on to the next chapter. You may like to compare your answers to the solutions provided in the back of the book and in the accompanying Study Guide.

1. Distinguish between proteins, peptides, and polypeptides.
2. Indicate whether each of the following amino acids is polar, nonpolar, acidic, or basic:
   a. glycine
   b. tyrosine
   c. glutamic acid
   d. histidine
   e. proline
   f. lysine
   g. cysteine
   h. asparagine
   i. valine
   j. leucine
3. Arginine has the following pK_a values:

   $pK_1, = 2.17$, $pK_2 = 9.04$, $pK_R = 12.48$

   Give the structure and net charge of arginine at the following pH values: 1, 4, 7, 10, 12

4. Shown is the titration curve for histidine.
   a. What species are present at each plateau?
   b. Using the titration curve, determine the pK$_a$ of each ionization of histidine.
   c. What is the isoelectric point of histidine?

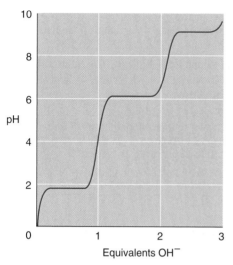

5. Consider the following molecule.

   a. Name it.
   b. Use the three-letter symbols for the amino acid to represent this molecule.

6. Rotation about the peptide bond in glycylglycine is hindered. Draw the resonance forms of the peptide bond and explain why.

7. List six functions of proteins in the body.

8. Differentiate the terms in each pair:
   a. globular and fibrous proteins
   b. simple and conjugated proteins
   c. apoprotein and holoprotein

9. Define the following terms:
   a. asymmetric carbon
   b. motor protein
   c. prosthetic group
   d. primary structure
   e. molten globule

10. Indicate the level(s) of protein structure to which each of the following contributes:
    a. amino acid sequence
    b. β-pleated sheet
    c. hydrogen bond
    d. disulfide bond

11. What type of secondary structure would the following amino acid sequence be *most* likely to have?
    a. polyproline

    b. polyglycine
    c. Ala—Val—Ala—Val—Ala—Val—
    d. Gly—Ser—Gly—Ala—Gly—Ala

12. List three factors that do not foster α-helix formation.

13. Denaturation is the loss of protein function from structural change or chemical reaction. At what level of protein structure or through what chemical reaction does each of the following denaturation agents act?
    a. heat
    b. strong acid
    c. saturated salt solution
    d. organic solvents (e.g., alcohol or chloroform)

14. A polypeptide has a high pI value. Suggest which amino acids might comprise it.

15. Outline the steps to isolate a typical protein. What is achieved at each step?

16. Outline the steps to purify a protein. What criteria are used to evaluate purity?

17. List the types of chromatography used to purify proteins. Describe how each separation method works.

18. In using carboxypeptidase to sequence a protein, the protein is first broken down into smaller fragments, which are separated from one another. Each fragment is then individually sequenced. If this initial fragmentation were not carried out, amino acid residues would build up in the reaction medium. How would their presence inhibit sequencing?

19. Define the following terms:
    a. mosaic protein
    b. homologous polypeptide
    c. cooperative binding
    d. aldol condensation
    e. globular protein

20. In an amino acid analysis, a large protein is broken down into overlapping fragments by using specific enzymes. Why must the sequences be overlapping?

21. Define the following terms:
    a. electrophoresis
    b. molecular disease
    c. α-carbon
    d. isoelectric point
    e. peptide bond

22. The following amino acid sequence represents bradykinin, a peptide released by certain organisms in response to wasp stings.

    Arg—Pro—Pro—Gly—Phe—Ser—Pro—Phe—Arg

    What amino acids or peptides are produced when bradykinin is treated with each of the following reagents?
    a. carboxypeptidase
    b. chymotrypsin
    c. trypsin
    d. DNFB

23. Describe the problems associated with using a polypeptide's primary sequence to determine its final three-dimensional shape.

24. Describe the forces involved in protein folding.

25. What are the characteristics of motor proteins? How do organisms use them?

26. Briefly outline the roles of molecular chaperones in protein folding.
27. Define the following terms:
    a. hydrophobic amino acid
    b. salt bridge
    c. dabsyl–amino acid
    d. site-directed mutagenesis
    e. proteomics

Thought Questions

These questions are designed to reinforce your understanding of all of the key concepts discussed in the book so far, including this chapter and all of the chapters before it. They may not have one right answer! The authors have provided possible solutions to these questions in the back of the book and in the accompanying Study Guide, for your reference.

28. Residues such as valine, leucine, isoleucine, methionine, and phenylalanine are often found in the interior of proteins, whereas arginine, lysine, aspartic acid, and glutamic acid are often found on the surface of proteins. Suggest a reason for this observation. Where would you expect to find glutamine, glycine, and alanine?

29. Proteins that are synthesized by living organisms adopt a biologically active conformation. Yet when such molecules are prepared in the laboratory, they usually fail to spontaneously adopt their active conformations. Can you suggest why?

30. The active site of an enzyme contains sequences that are conserved because they participate in the protein's catalytic activity. The bulk of an enzyme, however, is not part of the active site. Because a substantial amount of energy is required to assemble enzymes, why are they usually so large?

31. A structural protein may incorporate large amounts of immobilized water as part of its structure. Can you suggest how protein molecules "freeze" the water in place and make it part of the protein structure?

32. The peptide bond is a stronger bond than that of esters. What structural feature of the peptide bond gives it additional bond strength?

33. Because of their tendency to avoid water, nonpolar amino acids play an important role in forming and maintaining the three-dimensional structure of proteins. Can you suggest how these molecules accomplish this feat?

34. Hydrolysis of β-endorphin (a peptide containing 31 amino acid residues) produces the following amino acids:
    Tyr, Gly (3), Phe (2), Met, Thr (3), Ser (2), Lys (5), Gln (2), Pro, Leu (2), Val (2), Asn (2), Ala (2), Ile, His, and Glu
    Treatment with carboxypeptidase liberates Gln. Treatment with DNFB liberates DNP-Tyr. Treatment with trypsin produces the following peptides:
    Lys, Gly—Gln, Asn—Ala—Ile—Val—Lys,
    Tyr—Gly—Gly—Phe—Met—Thr—Ser—Glu—Lys,
    Asn—Ala—His—Lys, Ser—Gln—Thr—Pro—Leu—
    Val—Thr—Leu—Phe—Lys
    Treatment with chymotrypsin produces the following peptides:
    Lys—Asn—Ala—Ile—Val—Lys—Asn—Ala—
    His—Lys—Lys—Gly—Gln
    Tyr—Gly—Gly—Phe
    Met—Thr—Ser—Glu—Lys—Ser—Gln—Thr—Pro—
    Leu—Val—Thr—Leu—Phe
    What is the primary sequence of β-endorphin?.

35. Consider the following tripeptide:
    Gly—Ala—Val

    a. What is the approximate isoelectric point?
    b. In which direction will the tripeptide move if placed in an electric field at pH 1, 5, 10, and 12?

36. Chymotrypsin is an enzyme that cleaves other enzymes during sequencing. Why don't chymotrypsin molecules attack each other?

37. Most amino acids appear bluish purple when treated with ninhydrin reagent. Proline and hydroxyproline appear yellow. Suggest a reason for the difference.

38. When the multifunction protein glyceraldehyde-3-phosphate dehydrogenase (GAPD) catalyzes a key reaction in glycolysis (a metabolic pathway in cytoplasm), it does so as a homotetramer (four identical subunits). The GAPD monomer is a nuclear DNA repair enzyme. Describe in general terms what structural properties of multifunction proteins allow this phenomenon.

39. From the following analytical results, deduce the structure of a peptide isolated from the Alantian orchid, which contains 14 amino acids. Complete hydrolysis produces the following amino acids: Gly (3), Leu (3), Glu (2), Pro, Met, Lys (2), Thr, Phe. Treatment with carboxypeptidase releases glycine. Treatment with DNFB releases DNP-glycine. Treatment with a nonspecific proteolytic enzyme produces the following fragments:
    Gly—Leu—Glu, Gly—Pro—Met—Lys,
    Lys—Glu, Thr—Phe—Leu—Leu—Gly,
    Lys—Glu—Thr—Phe—Leu,
    Leu—Leu—Gly,
    Glu—Thr—Phe, Glu—Gly—Pro,
    Pro—Met—Lys—Lys,
    and Gly—Leu

40. Many proteins have several functions. Provide examples. What natural forces are responsible for this phenomenon?

41. Why are multifunctional proteins necessary and/or desirable?

42. Given the following decapeptide sequence, which amino acids would you expect to be on the surface of this molecule once it folds into its native conformation?
    Gly—Phe—Tyr—Asn—Tyr—Met—Ser—His—Val—Leu

43. What amino acid residues of the decapeptide in Question 42 would tend to be found on the interior of the molecule?

44. What would be the products of the acid hydrolysis for 3 hours of the decapeptide in Question 42?

45. A mutational change alters a polypeptide by substituting 3 adjacent prolines for 3 glycines. What possible effect will this event have on the protein's structure?

46. As a genetic engineer, you have been given the following task: alter a protein's structure by converting a specific amino acid sequence that forms an extended α-helix to one that forms a β-barrel. What types of amino acid are probably in the α-helix, and which ones would you need to substitute?

47. Of the naturally occurring amino acids Gly, Val, Phe, His, and Ser, which would be likely to form coordination compounds with metals?

48. β-Endorphin, an opiate peptide, is released by the anterior pituitary gland at the base of the mammalian brain in response to stress or pain. As with other signal molecules, the effects of β-endorphin on its target tissue (neurons) is triggered when it binds to its receptor. In general terms, outline the process by which the β-endorphin receptor would be isolated and its structure characterized.

49. Amino acids are the precursors of a vast number of biologically active nitrogen-containing molecules. Which amino acids are the precursors of the following molecules?

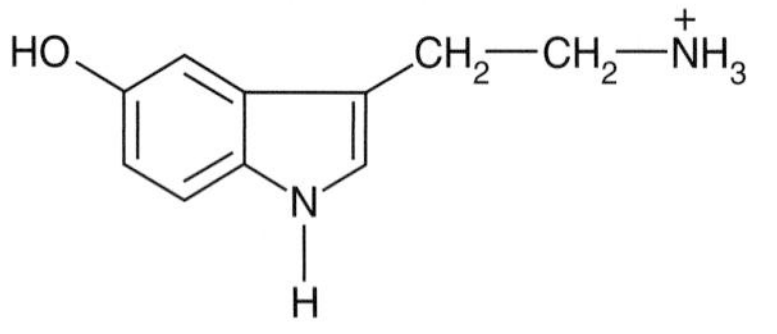

50. You are a materials engineer who decides that a synthetic fibrous protein might possess desirable properties for a new product. If technical constraints dictate that only two or three amino acids can be used, which ones would you choose?

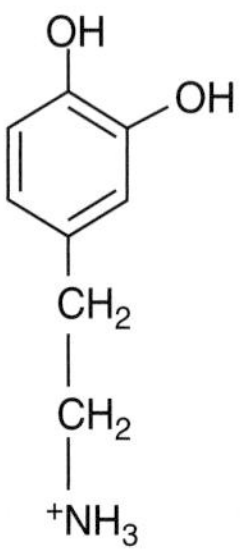

51. Suggest a protocol for separating oxytocin and vasopressin from an extract of the posterior pituitary gland.

52. The caging of water sequesters nonpolar amino acid residues to the interior of the folding protein. Can you explain how this phenomenon can have a significant impact on the formation and maintenance of protein native structure?

Serotonin

Dopamine

Enzymes

Space-Filling Model of Lysozyme
Lysozyme, an enzyme found in tears and saliva, destroys certain bacteria by hydrolyzing cell wall polysaccharides. In this space-filling model the polypeptide is shown with a bound segment of polysaccharide (green).

Overview

BIOCHEMISTS HAVE INVESTIGATED ENZYMES OR ENZYMATIC ACTIVITIES FOR OVER 120 YEARS. LONG BEFORE THEY HAD ANY REALISTIC UNDERSTANDING of the physical basis of the living state, biochemists instinctively appreciated the importance of enzymes. Using the technologies devised by biochemists, life scientists gradually determined the properties of biological systems. This work demonstrated that almost every event in living organisms occurs because of enzyme-catalyzed reactions. Until recently all known enzymes were proteins, but groundbreaking research led to the revelation that RNA molecules also have catalytic properties. This chapter is devoted to catalytic proteins, while the characteristics of catalytic RNA molecules are described in Chapter 18.

Living organisms can be viewed as biochemical factories composed of integrated constellations of efficiently operating molecular machines. Of all the classes of molecular machines, the enzymes are undoubtedly the most important. Without their catalytic capacities, most of the thousands of biochemical reactions that sustain living processes would occur at imperceptible rates. Recent determinations of uncatalyzed reaction rates in water range from 5 s for CO_2 hydration to 1.1 billion years for glycine decarboxylation. In contrast, enzyme-catalyzed reactions typically occur within time frames ranging from micro- to milliseconds. Enzymes are in fact the means by which living organisms channel the flow of energy and matter. Today, as a result of accumulating evidence derived from protein dynamics (conformational motion studies) and macromolecular crowding analysis, enzyme research is undergoing revolutionary changes. For example, according to long-held views, enzyme function depends almost entirely on the complementary shapes and catalytic interactions between reactant molecules and their more or less flexible binding sites. Recently, however, investigators have demonstrated that the catalytic function of certain enzymes can be linked to internal protein motions that extend throughout the protein molecule. Similarly, it is now recognized that enzymes function in conditions that are vastly different from those in which they have traditionally been studied (i.e., purified molecules in dilute concentration). Instead, the natural milieu of enzymes is a crowded gel-like environment. As a result of recent investigations, models of enzyme kinetics are evolving, and methods of experimentation, data collection, and simulation are becoming more sophisticated and closer to realistic *in vivo* conditions. This chapter reviews the structural and functional properties of enzymes.

6.1 PROPERTIES OF ENZYMES

How do enzymes work? The answer to this question requires a review of the role of catalysts. To proceed at a viable rate, most chemical reactions require an initial input of energy. At temperatures above absolute zero (0 K, or −273.1°C), all molecules possess vibrational energy, which increases as the molecules are heated. Consider the following reaction:

$$A + B \rightarrow C$$

As the temperature rises, vibrating molecules (A and B) are more likely to collide. A chemical reaction occurs when the colliding molecules possess a minimum amount of energy called the **activation energy** (E_a) or, more commonly in biochemistry, the free energy of activation ($\Delta G^{\ddagger}$). Not all collisions result in chemical reactions because only a fraction of the molecules have sufficient energy or the correct orientation to react (i.e., to break bonds or rearrange atoms into the product molecules).

Another way of increasing the likelihood of collisions, thereby increasing the formation of product, is to increase the temperature of the reaction or the concentrations of reactants. Yet another way is to provide a surface onto which reactant molecules can bind in favorable orientations (e.g., metal catalysts in laboratory and industrial reactions). In living systems, however, bringing about a temperature increase to facilitate reactions is unrealistic because the elevated temperature would damage the integrity of biomolecular structures. Living organisms use protein catalysts (enzymes) to circumvent this restriction of the kinetic energy of the system.

Enzymes are catalysts that have several remarkable properties. First, the rates of enzymatically catalyzed reactions are often phenomenally high. Rate increases of 10^7 to 10^{19} have been observed. Second, in marked contrast to inorganic catalysts, the enzymes are highly specific to the reactions they catalyze and side products are rarely formed. Finally, because of their relatively large and complex structures, enzymes can be regulated. This is an especially important consideration in living organisms, which must conserve energy and raw materials. By definition, a **catalyst** is a substance that enhances the rate of a chemical reaction but is not permanently altered by the reaction. Catalysts perform this feat because they decrease the activation energy required for a chemical reaction. In other words, catalysts provide an alternative reaction pathway that requires less energy (Figure 6.1). A **transition state** occurs at the apex of both reaction pathways in Figure 6.1. The free energy of activation, $\Delta G^{\ddagger}$, is defined as the amount of energy required to convert 1 mol of **substrate** (reactant) molecules from the ground state (the stable, low-energy form of a molecule) to the transition state. In the reaction in which ethanol is oxidized to form acetaldehyde

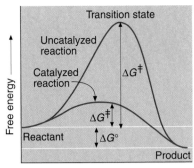

FIGURE 6.1

A Catalyst Reduces the Activation Energy of a Reaction

A catalyst alters the free energy of activation $\Delta G^{\ddagger}$, not the standard free energy ΔG° of the reaction.

this transition state might look like

Even with an inorganic catalyst, most laboratory reactions require an input of energy, usually in the form of heat. In addition, most inorganic catalysts are nonspecific; that is, they accelerate a wide variety of reactions. In contrast, enzymes perform their work at moderate temperatures and are quite specific in the reactions that each one catalyzes. The difference between inorganic catalysts and enzymes is directly related to their structures. In contrast to inorganic catalysts, each type of enzyme molecule contains a unique, intricately shaped binding surface called an **active site**. Substrates bind to the enzyme's active site, which is typically a small cleft or crevice on a large protein molecule. The active site is more than a binding site, however. Several of the amino acid side chains that line the active site actively participate in the catalytic process.

The shape and charge distribution of an enzyme's active site constrains the motions and allowed conformations of the substrate, forcing it to adopt a conformation more like that of the transition state. In other words, the structure of the active site is used to optimally orient the substrate. As a result the enzyme-substrate complex converts to product and free enzyme without the high-energy requirement of the constrained transition state. Consequently, the reaction rate increases significantly over that of the uncatalyzed reaction. Several other factors (described in Section 6.4) also contribute to rate enhancement.

Enzymes, like all catalysts, cannot alter the equilibrium of the reaction, but they can increase the rate toward equilibrium. Consider the following reversible reaction:

$$A \rightleftharpoons B$$

Without a catalyst, the reactant A is converted into the product B at a certain rate. Because this is a reversible reaction, B is also converted into A. The rate expression for the forward reaction is $k_F[A]^n$, and the rate expression for the reverse reaction is $k_R[B]^m$. The superscripts n and m represent the order of a reaction. Reaction order reflects the mechanism by which A is converted to B and vice versa. A reaction order of 2 for the conversion of A to B indicates that it is a bimolecular process and the molecules of A must collide for the reaction to occur (Section 6.3). At equilibrium, the rates for the forward and reverse reactions must be equal:

$$k_F[A]^n = k_R[B]^m \tag{1}$$

which rearranges to

$$\frac{k_F}{k_R} = \frac{[B]^m}{[A]^n} \tag{2}$$

The ratio of the forward and reverse constants is the equilibrium constant:

$$K_{eq} = \frac{[B]^m}{[A]^n} \tag{3}$$

For example, in Equation (3), if $m = n = 1$ and $k_F = 1 \times 10^{-3}$ s^{-1} and $k_R = 1 \times 10^{-6}$ s^{-1}, then

$$K_{eq} = \frac{10^{-3}}{10^{-6}} = 10^3$$

At equilibrium, therefore, the ratio of products to reactants is 1000 to 1.

In a catalyzed reaction, both the forward rate and the backward rate are increased, but the K_{eq} (in this case, 1000) remains unchanged. If the catalyst increases both the forward and reverse rates by a factor of 100, then the forward rate becomes 100,000 and the reverse rate becomes 100. Because of the dramatic increase in the rate of the forward reaction made possible by the catalyst, equilibrium is approached in seconds or minutes instead of hours or days.

It is important to recognize that the theory of chemical equilibrium assumes ideal conditions. Ideal solutions, for example, contain solutes in such low concentration that interactions such as steric repulsion or attractive forces are nonexistent. Most reactions that occur in solution, however, deviate from ideality. In such circumstances, equilibrium constants are based not on solute concentrations, but on activities, quantities called *effective concentrations* that take intermolecular interactions into account. The effective concentration or *activity* (a) of a solute is equal to

$$a = \gamma c \tag{4}$$

where γ is a correction factor called the *activity coefficient*, a factor dependent on the size and charge of the species and on the ionic strength of the solution in which the species is reacting, and c is the concentration in moles per liter. The impact of this phenomenon can be considerable. For example, the oxygen-binding capacity of hemoglobin, the predominant red blood cell protein, differs by several orders of magnitude depending on whether it is measured within red blood cells or in

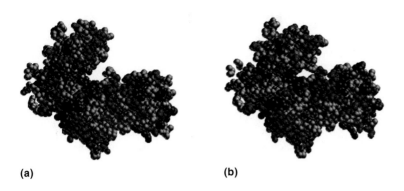

(a)   (b)

FIGURE 6.2

The Induced-Fit Model

Substrate binding causes enzymes to undergo conformational change. Hexokinase, a single polypeptide with two domains is shown (a) before and (b) after glucose binding. The domains move relative to each other to close around a glucose molecule (not shown).

dilute buffer. The equilibrium constant for a reaction in nonideal conditions is given by

$$K_{eq}^{\circ} = \gamma_B[B]/\gamma_A[A] = K_{eq}^{i}\Gamma \qquad (5)$$

where K_{eq}^{i} is the ideal constant and Γ is the nonideality factor, the ratio of the activity coefficients of the products and reactants. It should be noted that biochemists have traditionally minimized nonspecific interactions by carrying out investigations of enzymes, as well as other macromolecules, in dilute solutions. Over the past decade it has become increasingly apparent that the assumption of ideal conditions needs to be reevaluated. Consequently, many investigators now use high-molecular-weight "crowding agents" such as dextran (a glucose polymer produced by some bacteria) or serum albumin to simulate intracellular conditions in enzyme studies. Enzyme assays in the presence of crowding agents are closer to those of direct *in vivo* measurements, but the environment of the assay is still too homogeneous and differs significantly from the crowded heterogeneous conditions *in vivo*. It is a challenge for today's biochemists to construct assays and models that duplicate *in vivo* conditions.

Enzyme specificity is an enzyme property that is partially accounted for by the lock-and-key model, introduced by Emil Fischer in 1890. Each enzyme binds to a single type of substrate because the active site and the substrate have complementary structures. The substrate's overall shape and charge distribution allow it to enter and interact with the enzyme's active site. In a modern variation of the lock-and-key model, Daniel Koshland's *induced-fit model*, the flexible structure of proteins is taken into account (Figure 6.2). In this model, substrate does not fit precisely into a rigid active site. Instead, noncovalent interactions between the enzyme and substrate change the three-dimensional structure of the active site, conforming the shape of the active site to the shape of the substrate in its transition state conformation.

Although the catalytic activity of some enzymes depends only on interactions between active site amino acids and the substrate, other enzymes require nonprotein components for their activities. Enzyme **cofactors** may be ions, such as Mg^{2+} or Zn^{2+}, or complex organic molecules, referred to as **coenzymes**. The protein component of an enzyme that lacks an essential cofactor is called an **apoenzyme**. Intact enzymes with their bound cofactors are referred to as **holoenzymes**.

The activities of some enzymes can be regulated. Adjustments in the rates of enzyme-catalyzed reactions allow cells to respond effectively to environmental changes. Organisms may control enzyme activities directly, principally through the binding of activators or inhibitors, the covalent modification of enzyme molecules, or indirectly, by regulating enzyme synthesis. (Control of enzyme synthesis requires gene expression changes, a topic covered in Chapters 18 and 19.)

KEY CONCEPTS

- Enzymes are catalysts.
- Catalysts modify the rate of a reaction because they provide an alternative reaction pathway that requires less activation energy than the uncatalyzed reaction.
- Most enzymes are proteins.

QUESTION 6.1

The hexokinases are a class of enzymes that catalyze the ATP-dependent phosphorylation of hexoses (sugars with six carbons). The hexokinases will bind only D-hexose sugars and not their L-counterparts. In general terms, describe the features of enzyme structure that make this specificity possible.

6.2 CLASSIFICATION OF ENZYMES

In the early days of biochemistry, enzymes were named at the whim of their discoverers. Often enzyme names provided no clue to their function (e.g., trypsin), and sometimes several names were used for the same enzyme. Enzymes were often named by adding the suffix *-ase* to the name of the substrate. For example, urease catalyzes the hydrolysis of urea. To eliminate confusion, the International Union of Biochemistry (IUB) instituted a systematic naming scheme for enzymes. Each enzyme is now classified and named according to the type of chemical reaction it catalyzes. In this scheme, an enzyme is assigned a four-number classification and a two-part name called a *systematic name*. In addition, a shorter version of the systematic name, called the *recommended name*, is suggested by the IUB for everyday use. For example, alcohol:NAD$^+$ oxidoreductase (EC 1.1.1.1) is usually referred to as alcohol dehydrogenase. (The letters EC are an abbreviation for the Enzyme Commission of the IUB.) Because many enzymes were discovered before the institution of the systematic nomenclature, the old well-known names have been retained in quite a few cases.

The following are the six major enzyme categories:

1. **Oxidoreductases. Oxidoreductases** catalyze oxidation-reduction reactions in which the oxidation state of one or more atoms in a molecule is altered. Oxidation-reduction in biological systems involves one- or two-electron transfer reactions accompanied by the compensating change in the amount of hydrogen and oxygen in the molecule. Prominent examples include the redox reactions facilitated by the dehydrogenases and the reductases. For example, alcohol dehydrogenase catalyzes the oxidation of ethanol and other alcohols, and ribonucleotide reductase catalyzes the reduction of ribonucleotides to form deoxyribonucleotides. The oxygenases, oxidases, and peroxidases are among the enzymes that use O_2 as an electron acceptor.

2. **Transferases. Transferases** are enzymes that transfer molecular groups from a donor molecule to an acceptor molecule. Such groups include amino, carboxyl, carbonyl, methyl, phosphoryl, and acyl (RC=O). Common trivial names for the transferases often include the prefix *trans*; the transcarboxylases, transmethylases, and transaminases are examples.

3. **Hydrolases. Hydrolases** catalyze reactions in which the cleavage of bonds such as C—O, C—N, and O—P is accomplished by the addition of water. The hydrolases include esterases, phosphatases, and proteases.

4. **Lyases. Lyases** catalyze reactions in which groups (e.g., H_2O, CO_2, and NH_3) are removed by elimination to form a double bond or are added to a double bond. Decarboxylases, hydratases, dehydratases, deaminases, and synthases are examples of lyases.

5. **Isomerases.** A heterogeneous group of enzymes, the **isomerases** catalyze several types of intramolecular rearrangements. The epimerases catalyze the inversion of asymmetric carbon atoms and the mutases catalyze the intramolecular transfer of functional groups.

6. **Ligases. Ligases** catalyze bond formation between two substrate molecules. For example, DNA ligase links DNA strand fragments together. The names of many ligases include the term *synthetase*. Several other ligases are called carboxylases.

Table 6.1 presents an example from each enzyme class.

TABLE 6.1   Selected Examples of Enzymes

Enzyme Class	Example	Reaction Catalyzed
Oxidoreductase	Alcohol dehydrogenase	$CH_3-CH_2-OH + NAD^+ \longrightarrow CH_3-\overset{\displaystyle O}{\overset{\|}{C}}H + NADH + H^+$
Transferase	Hexokinase	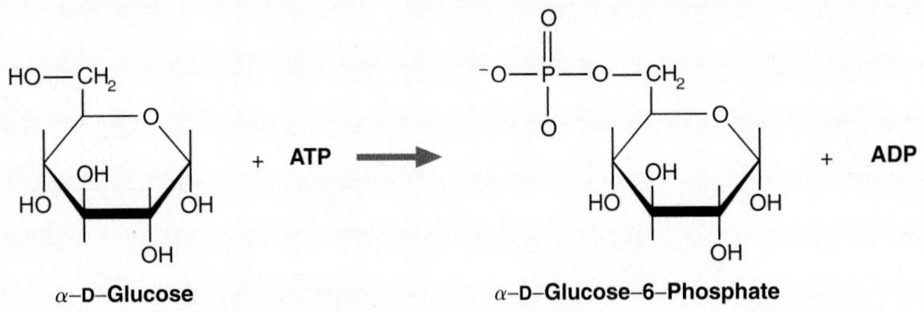 α–D–Glucose + ATP → α–D–Glucose–6–Phosphate + ADP
Hydrolase	Chymotrypsin	Polypeptide + H_2O ⟶ Peptides
Lyase	Pyruvate decarboxylase	$CH_3-\overset{\displaystyle O}{\overset{\|}{C}}-\overset{\displaystyle O}{\overset{\|}{C}}-O^- + H^+$ ⟶ $CH_3-\overset{\displaystyle O}{\overset{\|}{C}}-H$ (Acetaldehyde) + CO_2 (Carbon Dioxide)
Isomerase	Alanine racemase	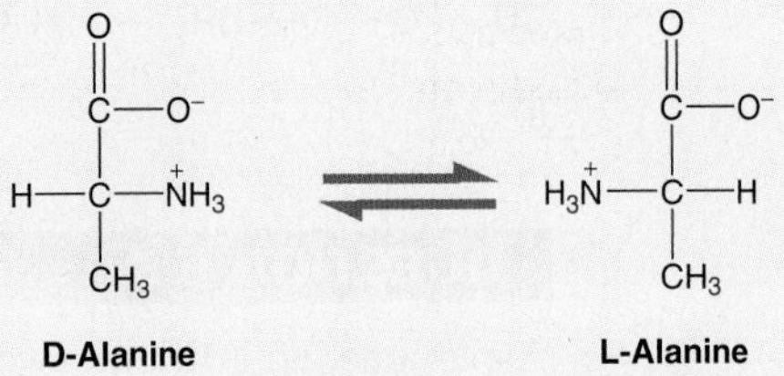 D-Alanine ⇌ L-Alanine
Ligase	Pyruvate carboxylase	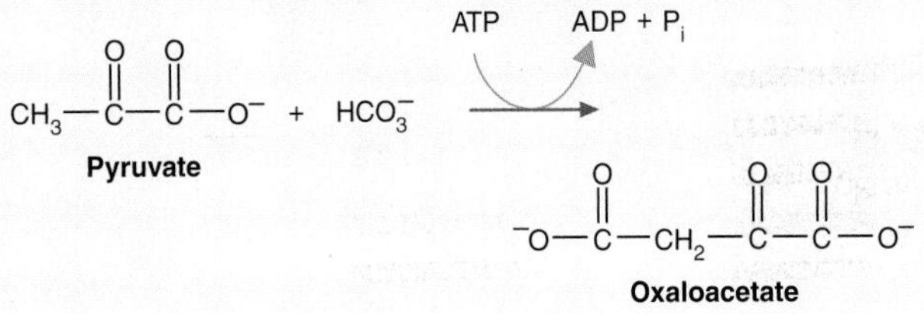 Pyruvate + HCO_3^- → Oxaloacetate (ATP → ADP + P_i)

QUESTION 6.2

Which type of enzyme catalyzes each of the following reactions?

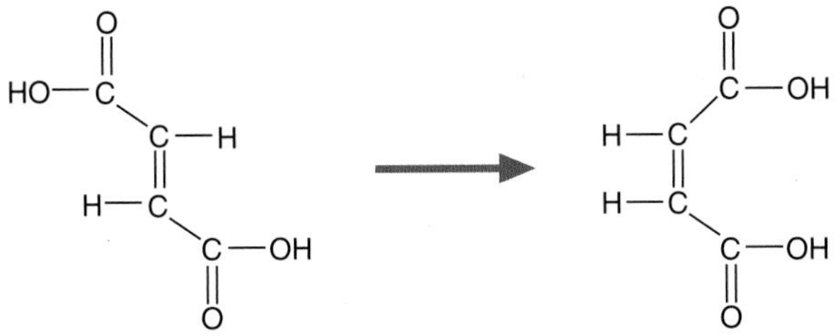

(a)

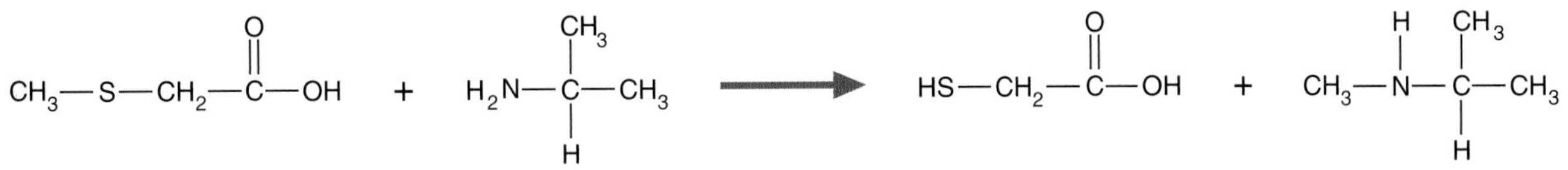

(b)

QUESTION 6.2 (CONT.)

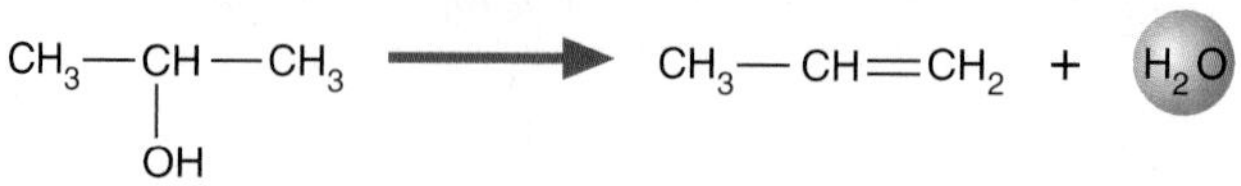

$CH_3-CH-CH_3 \longrightarrow CH_3-CH=CH_2 + H_2O$
$\quad\quad\;\; |$
$\quad\quad\; OH$

(c)

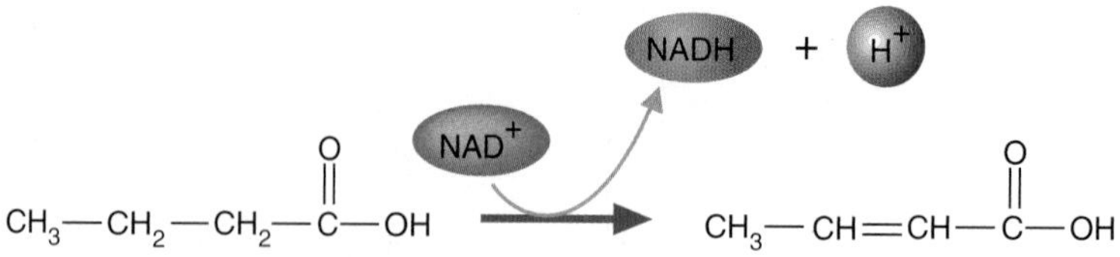

(d)

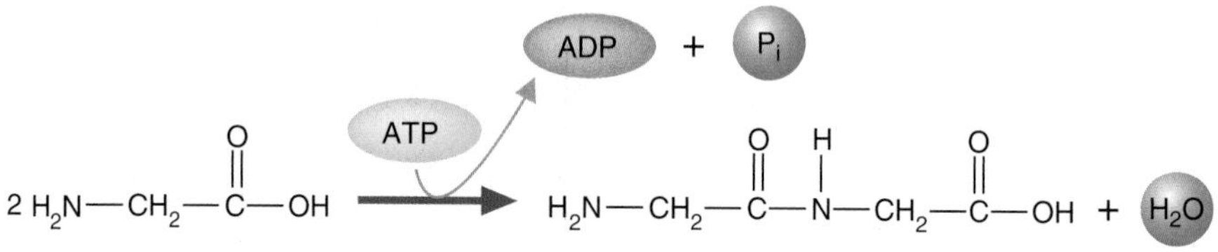

(e)

$CH_3-\overset{\overset{\textstyle O}{\|}}{C}-O-CH_3 + H_2O \longrightarrow CH_3-\overset{\overset{\textstyle O}{\|}}{C}-OH + CH_3-OH$

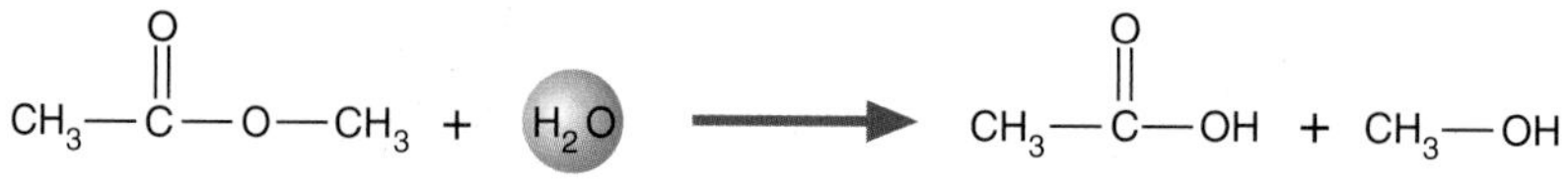

(f)

QUESTION 6.3

Aspartame, an artificial sweetener, has the following structure:

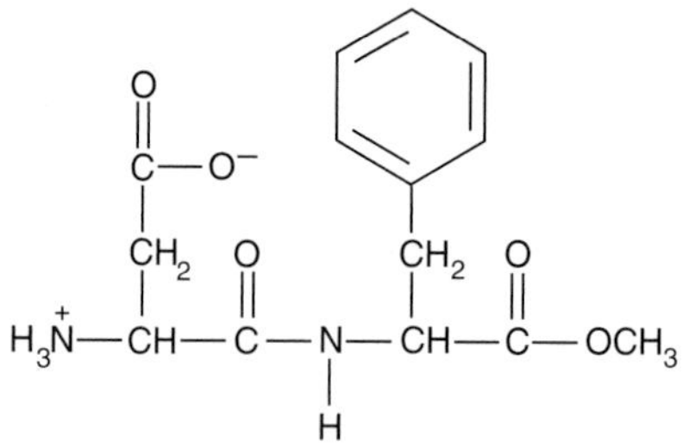

Once consumed in food or beverages, aspartame is degraded in the digestive tract to its component molecules. Predict what the products of this process are. What classes of enzymes are involved?

6.3 ENZYME KINETICS

Recall from Chapter 4 that the principles of thermodynamics can predict whether a reaction is spontaneous but cannot predict its rate. The rate or **velocity** of a biochemical reaction is defined as the change in the concentration of a reactant or product per unit time. The initial velocity v_0 of the reaction A → P, where A and P are substrate and product molecules, respectively, is

$$v_0 = \frac{-\Delta[A]}{\Delta t} = \frac{\Delta[P]}{\Delta t} \tag{6}$$

where [A] = concentration of substrate
$\quad\quad$ [P] = concentration of product
$\quad\quad\;\; t$ = time

Initial velocity (v_0) is the velocity of a reaction when [A] greatly exceeds the concentration of enzyme E, [E], and the reaction time is very short.

Measurements of v_0 are made immediately after the mixing of enzyme and substrate because it can be assumed that the reverse reaction (i.e., conversion of product into substrate) has not yet occurred to any appreciable extent.

The quantitative study of enzyme catalysis, referred to as **enzyme kinetics**, provides information about reaction rates. Kinetic studies also measure the affinity of enzymes for substrates and inhibitors and provide insight into reaction mechanisms. Enzyme kinetics, in turn, provides insight into the forces that regulate metabolic pathways. The rate of the reaction $A \rightarrow P$ is proportional to the frequency with which the reacting molecules form product. The reaction rate is

$$v_0 = k[A]^x \tag{7}$$

where v_0 = initial rate
k = a rate constant that depends on the reaction conditions (e.g., temperature, pH, and ionic strength)
x = the order of the reaction

Combining Equations (6) and (7), we have

$$\frac{\Delta[A]}{\Delta t} = k[A]^x \tag{8}$$

Order is a term that is useful in describing a reaction. Order is determined empirically, that is, by experimentation (Figure 6.3), and is defined as the sum of the exponents on the concentration terms in the rate expression. Determining the order of a reaction allows an experimenter to draw certain conclusions regarding the reaction's mechanism. A reaction is said to follow *first-order kinetics* if the rate depends on the first power of the concentration of a single reactant and suggests that the rate-limiting step is a unimolecular reaction (i.e., no molecular collisions are required). In the reaction $A \rightarrow P$ the experimental rate equation, it is assumed that

$$\text{Rate} = k[A]^1 \tag{9}$$

If [A] is doubled, the rate is observed to double. Reducing [A] by half results in halving the observed reaction rate. In first-order reactions the concentration of the reactant is a function of time, so k is expressed in units of s^{-1}. In any reaction the time required for one-half of the reactant molecules to be consumed is called a *half-life* ($t_{1/2}$).

In the reaction $A + B \rightarrow P$, if the order of A and B is 1 each, then the reaction is said to be *second-order*, and A and B must collide for product to form (a biomolecular reaction):

$$\text{Rate} = k[A]^1[B]^1 \tag{10}$$

In this circumstance the reaction rate depends on the concentrations of the two reactants. In other words, both A and B take part in the reaction's rate-determining step. Second-order rate constants are measured in units of $M^{-1}s^{-1}$.

Sometimes second-order reactions involve reactants such as water that are present in great excess:

$$A + H_2O \rightarrow P$$

The second-order rate expression is

$$\text{Rate} = k[A]^1 [H_2O]^1 \tag{11}$$

Because water is present in excess and $[H_2O]$ is essentially constant, however, the reaction appears to be first-order. Such reactions are said to be *pseudo-first-order*. Hydrolysis reactions in biochemical systems are assumed to be pseudo-first-order because of the ready availability of the second reactant, H_2O, in aqueous environments.

Another possibility is that only one of the two reactants is involved in the rate-determining step and it alone appears in the rate expression. For the above reaction, if rate = $k[A]^2$, then the rate-limiting step involves collisions between A molecules. The water is involved in a fast, non-rate-limiting step in the reaction mechanism.

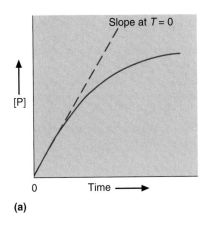

(a)

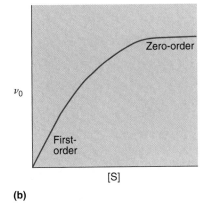

(b)

FIGURE 6.3
Enzyme Kinetic Studies

(a) Conversion of substrate to product per unit time. The slope of the curve at $t = 0$ equals the initial rate of the reaction. (b) Plot of initial velocity v versus substrate concentration [S]. The rate of the reaction is directly proportional to substrate concentration only when [S] is low. When [S] becomes high enough to saturate the enzyme, the rate of the reaction is zero-order with respect to substrate. At intermediate substrate concentrations, the reaction has a mixed order (i.e., the effect of substrate on reaction velocity is in transition).

When the addition of a reactant does not alter a reaction rate, the reaction is said to be *zero-order* for that reactant. For the reaction A → P, the experimentally determined rate expression under such conditions is

$$\text{Rate} = k[A]^0 = k \tag{12}$$

The rate is constant because the reactant concentration is high enough to saturate all the catalytic sites on the enzyme molecules. An example of an order determination is given in Problem 6.1.

Reaction order can also be characterized in another way. A theoretical term can be used to characterize simple reactions: *molecularity* is defined as the number of colliding molecules in a single-step reaction. A *unimolecular* reaction A → B has a molecularity of one, whereas the *bimolecular* reaction A + B → C + D has a molecularity of two.

WORKED PROBLEM 6.1

Consider the following reaction:

$$\underset{\text{H}_3\overset{+}{\text{N}}-\text{CH}_2-\overset{\overset{\text{O}}{\|}}{\text{C}}-\text{NH}-\text{CH}_2-\overset{\overset{\text{O}}{\|}}{\text{C}}-\text{O}^- + \text{H}_2\text{O}}{} \longrightarrow 2\ \text{H}_3\overset{+}{\text{N}}-\text{CH}_2-\overset{\overset{\text{O}}{\|}}{\text{C}}-\text{O}^-$$

Given the following rate data, determine the order in each reactant and the overall order of the reaction.

Initial Concentrations (mol/L)		
Glycylglycine	H₂O	Rate (mmol/s)
0.1	0.1	1×10^2
0.2	0.1	2×10^2
0.1	0.2	2×10^2
0.2	0.2	4×10^2

Solution

The overall initial rate expression is

$$\text{Rate} = k[\text{glycylglycine}]^x[\text{H}_2\text{O}]^y$$

To evaluate x and y, determine the effect on the rate of the reaction of increasing the concentration of one reactant while keeping the concentration of the other constant.

For this experiment, doubling the concentration of glycylglycine doubles the rate of the reaction; therefore, x is 1. Doubling the concentration of water doubles the rate of the reaction. So y is also 1. The rate expression then is

$$\text{Rate} = k[\text{glycylglycine}]^1[\text{H}_2\text{O}]^1$$

The reaction is first-order in both reactants and second-order overall. ∎

Michaelis-Menten Kinetics

One of the most useful models in the systematic investigation of enzyme rates was proposed by Leonor Michaelis and Maud Menten in 1913. The concept of the enzyme-substrate complex, first enunciated by Victor Henri in 1903, is central to Michaelis-Menten kinetics. When the substrate S binds in the active site of an enzyme E, an intermediate complex (ES) is formed. During the transition state, the substrate is converted into product. After a brief time, the product dissociates from the enzyme. This process can be summarized as follows:

$$\text{E} + \text{S} \underset{k_{-1}}{\overset{k_1}{\rightleftharpoons}} \text{ES} \overset{k_2}{\rightarrow} \text{E} + \text{P} \tag{13}$$

where k_1 = rate constant for ES formation

k_{-1} = rate constant for ES dissociation

k_{-2} = rate constant for product formation and release from the active site

Equation (13) ignores the reversibility of the step in which the ES complex is converted into enzyme and product. This simplifying assumption is allowed if the reaction rate is measured while [P] is still very low. Recall that initial velocities are measured in most kinetic studies.

According to the Michaelis-Menten model, as currently conceived, it is assumed that (1) k_{-1} is negligible compared with k_1 and (2) the rate of formation of ES is equal to the rate of its degradation over most of the course of the reaction. (The latter premise is referred to as the *steady state assumption*.)

$$\text{Rate} = \frac{\Delta P}{\Delta t} = k_2[\text{ES}] \tag{14}$$

To be useful, a reaction rate must be defined in terms of [S] and [E]. The rate of formation of ES is equal to $k_1[\text{E}][\text{S}]$, whereas the rate of ES dissociation is equal to $(k_{-1} + k_2)[\text{ES}]$. The steady state assumption equates these two rates:

$$k_1[\text{E}][\text{S}] = (k_{-1} + k_2)[\text{ES}] \tag{15}$$

$$[\text{ES}] = \frac{[\text{E}][\text{S}]}{(k_{-1} + k_2)/k_1} \tag{16}$$

Michaelis and Menten introduced a new constant, K_m (now referred to as the *Michaelis constant*):

$$K_m = \frac{k_{-1} + k_2}{k_1} \tag{17}$$

They also derived the equation

$$v = \frac{V_{\max}[\text{S}]}{[\text{S}] + K_m} \tag{18}$$

where $V_{\max}$ = maximum velocity that the reaction can attain. This equation, now referred to as the *Michaelis-Menten equation*, has proven to be very useful in defining certain aspects of enzyme behavior. For example, when [S] is equal to K_m, the denominator in Equation (18) is equal to 2[S], and v is equal to $V_{\max}/2$ (Figure 6.4). The experimentally determined value K_m is considered a constant that is characteristic of the enzyme and the substrate under specified conditions. It may reflect the affinity of the enzyme for its substrate. (If k_2 is much smaller than k_{-1}, that is, $k_2 \ll k_{-1}$, then the K_m value approximates k_1. In this circumstance, K_m is the dissociation constant for the ES complex.) The lower the value of K_m, the greater the affinity of the enzyme for ES complex formation.

An enzyme's kinetic properties can also be used to determine its catalytic efficiency. The **turnover number** (k_{cat}) of an enzyme is defined as

$$k_{cat} = \frac{V_{\max}}{[\text{E}_t]} \tag{19}$$

where k_{cat} = number of substrate molecules converted to product per unit time by an enzyme molecule under saturating conditions

$[\text{E}_t]$ = total enzyme concentration

Under physiological conditions, [S] is usually significantly lower than K_m. A more useful measure of catalytic efficiency is obtained by rearranging Equation (19) as

$$V_{\max} = k_{cat}[\text{E}_t] \tag{20}$$

and substituting this function in the Michaelis-Menten equation (Equation 18):

$$v = \frac{k_{cat}[\text{E}_t][\text{S}]}{K_m + [\text{S}]} \tag{21}$$

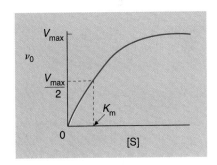

FIGURE 6.4

Initial Reaction Velocity v_0 and Substrate Concentration [S] for a Typical Enzyme-Catalyzed Reaction

The enzyme has half-maximal velocity at substrate concentration K_m.

When [S] is very low, $[E_t]$ is approximately equal to [E] and Equation (21) reduces to

$$v = (k_{cat}/K_m)[E][S] \tag{22}$$

In Equation (22) the term k_{cat}/K_m, also referred to as the *specificity constant*, is the second-order rate constant for a reaction in which $[S] \ll K_m$. In this reaction the [S] is sufficiently low that the value of k_{cat}/K_m reflects the relationship between catalytic rate and substrate binding affinity. A substrate with a high specificity constant will have a low K_m (high affinity) and a high kinetic efficiency (high turnover number). The specificity constant can be very useful when comparing different substrates for the same enzyme. When [S] is low and the $[E_t]$ can be accurately measured, Equation (22) holds. Examples of k_{cat}, K_m, and k_{cat}/K_m values of selected enzymes are provided in Table 6.2. It should be noted that the upper limit for an enzyme's k_{cat}/K_m value cannot exceed the maximal value of the rate at which the enzyme can bind to substrate molecules (k_1). This limit is imposed by the rate of diffusion of substrate into an enzyme's active site. The *diffusion control limit* on enzymatic reactions is approximately 10^8 to 10^9 M^{-1}s^{-1}. Several enzymes, for example, those listed in Table 6.2, have k_{cat}/K_m values that approach the diffusion control limit. Because such enzymes convert substrate to product virtually every time the substrate diffuses into the active site, they are said to have achieved

TABLE 6.2   The Values of k_{cat}, K_m, and k_{cat}/K_m for Selected Enzymes

Enzyme	Reaction Catalyzed	k_{cat} (s^{-1})	K_m (M)	k_{cat}/K_m (M^{-1}s^{-1})
Acetylcholinesterase	Acetylcholine + H_2O → Acetate + Choline + H^+	1.4×10^4	9×10^{-5}	1.6×10^8
Carbonic anhydrase	$HCO_3^- + H^+ \rightleftharpoons CO_2 + H_2O$	4×10^5	0.026	1.5×10^7
Catalase	$2 H_2O_2 \longrightarrow 2 H_2O + O_2$	4×10^7	1.1	4×10^7
Fumarase	Fumarate + $H_2O \rightleftharpoons$ Malate	8×10^2	5×10^{-6}	1.6×10^8
Triosephosphate isomerase	Glyceraldehyde–3–phosphate $\rightleftharpoons$ Dihydroxyacetone phosphate	4.3×10^3	4.7×10^{-4}	2.4×10^8

Source: Adapted from A. Fersht, Structure and Mechanism in Protein Science: A Guide to Enzyme Catalysis and Protein Folding, 2nd ed., W. H. Freeman, New York, 1999.

catalytic perfection. Living organisms overcome the diffusion control limit for the enzymes in biochemical pathways that do not achieve this high degree of catalytic efficiency by organizing them into multienzyme complexes. In these complexes, the active sites of the enzymes are in such close proximity to each other that diffusion is not a factor in the transfer of substrate and product molecules.

Enzyme activity is measured in *international units* (IU). One IU is defined as the amount of enzyme that produces 1 μmol of product per minute. An enzyme's *specific activity*, a quantity that is used to monitor enzyme purification, is defined as the number of international units per milligram of protein. [A new unit for measuring enzyme activity called the *katal* has recently been introduced. One katal (kat) indicates the amount of enzyme that transforms 1 mole of substrate per second. One katal is equal to 6×10^7 IU.]

An example of a kinetic determination is given in Problem 6.2.

KEY CONCEPTS

- The Michaelis-Menten kinetic model explains several aspects of the behavior of many enzymes.
- Each enzyme has a characteristic K_m for a particular substrate under specified conditions.

WORKED PROBLEM 6.2

Consider the Michaelis-Menten plot illustrated in Figure 6.5. Identify the following points on the curve:

a. V_{max}
b. K_m

Solution

a. The maximum rate the enzyme can attain is V_{max}. Further increases in substrate concentration do not increase the rate.

b. $K_m = $ [S] at $V_{max}/2$ ∎

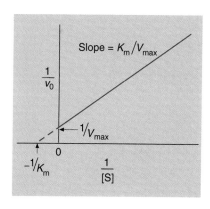

FIGURE 6.5

A Michaelis-Menten Plot

Lineweaver-Burk Plots

The K_m and V_{max} values for an enzyme are determined by measuring initial reaction velocities at various substrate concentrations. Approximate values of K_m and V_{max} can be obtained by constructing a graph, as shown in Figure 6.4. A more accurate determination of these values results from an algebraic transformation of the data. The Michaelis-Menten equation, whose graph is a hyperbola,

$$v = \frac{V_{max}[S]}{[S] + K_m}$$

can be rearranged by taking its reciprocal:

$$\frac{1}{v_0} = \frac{K_m}{V_{max}} \frac{1}{[S]} + \frac{1}{V_{max}}$$

The reciprocals of the initial velocities are plotted as functions of the reciprocals of substrate concentrations. In such a graph, referred to as a *Lineweaver-Burk double-reciprocal plot*, the straight line that is generated has the form $y = mx + b$, where y and x are variables (1/v and 1/[S], respectively) and m and b are constants (K_m/V_{max} and $1/V_{max}$, respectively). The slope of the straight line is K_m/V_{max} (Figure 6.6). As indicated in Figure 6.6, the intercept on the vertical axis is $1/V_{max}$. The intercept on the horizontal axis is $-1/K_m$.

Problem 6.3 is an example of a kinetics problem using the Lineweaver-Burk plot.

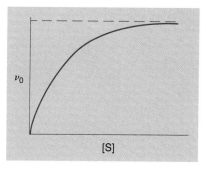

FIGURE 6.6

Lineweaver-Burk, or Double-Reciprocal, Plot

If an enzyme obeys Michaelis-Menten kinetics, a plot of the reciprocal of the reaction velocity $1/v_0$ as a function of the reciprocal of the substrate concentration $1/[S]$ will fit a straight line. The slope of the line is K_m/V_{max}. The intercept on the vertical axis is $1/V_{max}$. The intercept on the horizontal axis is $-1/K_m$.

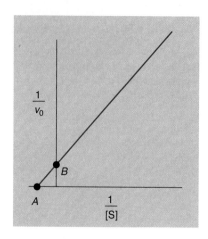

FIGURE 6.7

A Lineweaver-Burk Plot

WORKED PROBLEM 6.3

Consider the Lineweaver-Burk plot in Figure 6.7. Identify:
a. $-1/K_m$
b. $1/V_{max}$
c. $K/_m/V_{max}$

Solution
a. $A = -1/K_m$
b. $B = 1/V_{max}$
c. $K_m/V_{max} = $ slope

∎

Multisubstrate Reactions

Most biochemical reactions involve two or more substrates. The most common multisubstrate reaction, the bisubstrate reaction, is represented as

$$A + B \rightleftharpoons C + D$$

In most of these reactions there is a transfer of a specific functional group from one substrate to the other (e.g., phosphate or methyl groups), or oxidation-reduction reactions in which redox coenzymes (e.g., $NAD^+/NADH$ or $FAD/FADH_2$) are substrates. The kinetic analysis of bisubstrate reactions is, by necessity, more complicated than that of one-substrate reactions. Often, however, the determination of the K_m for each substrate (when the other substrate is present in a saturating amount) is sufficient for most purposes. On the basis of kinetic analysis, multisubstrate reactions can be divided into two classes: sequential and double displacement.

SEQUENTIAL REACTIONS A sequential reaction cannot proceed until all substrates have been bound in the enzyme's active site. There are two possible sequential mechanisms: ordered and random. In a bisubstrate ordered mechanism, the first substrate must bind to the enzyme before the second for the reaction to proceed to product formation. The notation for such a reaction, introduced by W. W. Cleland, is

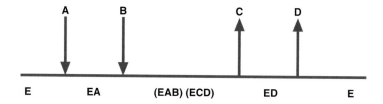

In a random sequential mechanism, the substrates can bind in any order.

DOUBLE-DISPLACEMENT REACTIONS In a double displacement, or "ping-pong" mechanism, the first product is released before the second substrate binds.

where E* = modified enzyme

In such a reaction, the enzyme is altered by the first phase of the reaction. The enzyme is then restored to its original form during the second phase, in which the second substrate is converted to product.

Enzyme Inhibition

The activity of enzymes can be inhibited. Molecules that reduce an enzyme's activity, called **inhibitors**, include many drugs, antibiotics, food preservatives, and poisons. The investigations of enzyme inhibition and inhibitors carried out by biochemists are important for several reasons. First and foremost, in living systems enzyme inhibition is an important means by which metabolic pathways are regulated. Numerous small biomolecules are used routinely to modulate the rates of specific enzymatic reactions so that the needs of the organism are consistently met. Second, numerous clinical therapies are based on enzyme inhibition. For example, many antibiotics and other drugs reduce or eliminate the activity of specific enzymes. The most effective AIDS treatment is a multidrug therapy that includes protease inhibitors, molecules that disable a viral enzyme required to make new virus. Finally, investigations of enzyme inhibition have enabled biochemists to develop techniques for probing the physical and chemical architecture, as well as the functional properties, of enzymes.

Enzyme inhibition can be reversible or irreversible. **Reversible inhibition** occurs when the inhibitory effect of a compound can be counteracted by increasing substrate levels or removing the inhibitor compound while the enzyme remains intact. Reversible inhibition will be **competitive** if the inhibitor binds to the free enzyme and competes with the substrate for occupation of the active site, **noncompetitive** if the inhibitor binds to both the free enzyme and the enzyme-substrate complex, and **uncompetitive** if the inhibitor binds to only the enzyme-substrate complex. **Irreversible inhibition** occurs when inhibitor binding permanently impairs the enzyme, usually through a covalent reaction that chemically modifies the enzyme.

COMPETITIVE INHIBITORS Competitive inhibitors bind reversibly to free enzyme, not the ES complex, to form an enzyme-inhibitor (EI) complex.

The substrate and the inhibitor compete for the same site on the enzyme. As soon as the inhibitor binds, substrate access to the active site is blocked. The concentration of EI complex depends on the concentration of free inhibitor and on the dissociation constant K_I:

$$K_I = \frac{[E][I]}{[EI]} = \frac{k_{I-1}}{k_{I1}}$$

Because the EI complex readily dissociates, the enzyme is again available for substrate binding. The enzyme's activity declines with an increased K_m and

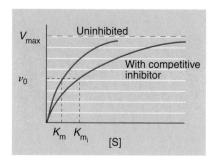

FIGURE 6.8

Michaelis-Menten Plot of Uninhibited Enzyme Activity Versus Competitive Inhibition

Initial velocity v_0 is plotted against substrate concentration [S]. With competitive inhibition, V_{max} stays constant and K_m increases.

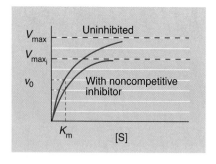

FIGURE 6.9

Malonate

unchanged V_{max} (Figure 6.8) in the presence of a competitive inhibitor because the EI complex does not participate in the catalytic process. Although V_{max} is unchanged, the effect of a competitive inhibitor on activity is reversed by increasing the concentration of substrate. At high [S], all the active sites are filled with substrate, and reaction velocity reaches the value observed without an inhibitor.

Substances that behave as competitive inhibitors (i.e., reduce an enzyme's apparent affinity for substrate) are often similar in structure to the substrate. Such molecules include reaction products or unmetabolizable analogues or derivatives of substrate molecules. Succinate dehydrogenase, an enzyme in the Krebs citric acid cycle (Chapter 9), catalyzes a redox reaction that converts succinate to fumarate.

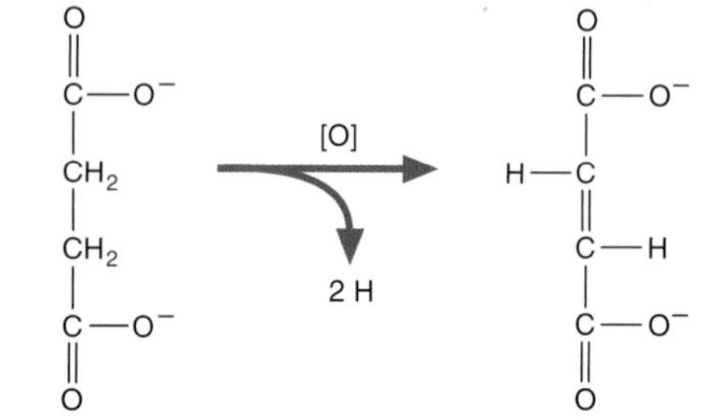

This reaction is inhibited by malonate (Figure 6.9). Malonate binds to the enzyme's active site but cannot be converted to product.

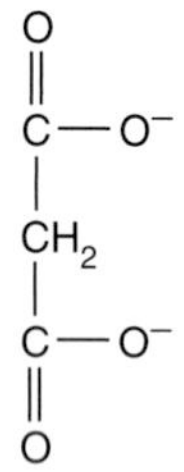

NONCOMPETITIVE INHIBITORS In some enzyme-catalyzed reactions an inhibitor can bind to both the enzyme and the enzyme-substrate complex:

$$E + S \rightleftharpoons ES \longrightarrow P + E$$

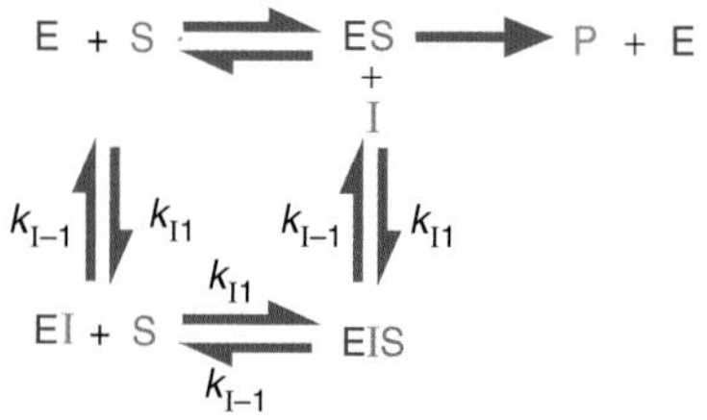

In such circumstances, referred to as **noncompetitive** inhibition, the inhibitor binds to a site other than the active site. Inhibitor binding results in a modification of the enzyme's conformation that prevents product formation (Figure 6.10). Noncompetitive inhibitors have little or no structural resemblance to substrate, but they may influence substrate binding if their binding sites are in close proximity to the substrate binding site. In some cases, noncompetitive inhibition may be partially reversed by increasing the substrate concentration. Analysis of reactions inhibited by noncompetitive inhibitors is often complex because usually two or more substrates are involved. Thus the characteristics of the inhibition that are observed may depend in part on factors such as the order in which the different substrates bind. In addition, there are two determinations of K_I for the binding of noncompetitive inhibitors

$$K_I = \frac{[E][I]}{[EI]}$$

and

$$K_I = \frac{[E][S][I]}{[EIS]}$$

Depending on the inhibitor being considered, the values of these dissociation constants may or may not be equivalent. There are two forms of noncompetitive inhibition: pure and mixed. In pure noncompetitive inhibition, a rare phenomenon, both K_I values are equivalent. Mixed noncompetitive inhibition is typically more complicated because the K_I values are different.

UNCOMPETITIVE INHIBITORS In uncompetitive inhibition, which is considered a type of noncompetitive inhibition, the inhibitor binds only to the enzyme-substrate complex, not to the free enzyme. As a consequence, the inhibitor

FIGURE 6.10

Michaelis-Menten Plot of Uninhibited Enzyme Activity Versus Noncompetitive Inhibition

Initial velocity v_0 is plotted against substrate concentration [S]. With noncompetitive inhibition V_{max} decreases and K_m stays constant.

is ineffective at low substrate concentrations because very little of the ES complex is present.

$$E + S \rightleftharpoons ES \longrightarrow P + E$$

$$+$$
$$I$$

$$k_{I-1} \Big\updownarrow k_{I1}$$

$$EIS$$

The dissociation constant for the binding step of an uncompetitive inhibitor to an enzyme is

$$K_I = \frac{[ES][I]}{[EIS]}$$

When I binds to ES, the substrate is not free to dissociate from the enzyme and the K_m decreases, giving the appearance of higher substrate affinity. At high [S], the inhibitor is in its most effective ES-bound form, and the V_{max} of the system decreases significantly. The K_m and V_{max} may or may not decrease to the same degree. Uncompetitive inhibition is most commonly observed for enzymes that bind more than one substrate.

KINETIC ANALYSIS OF ENZYME INHIBITION Competitive, noncompetitive, and uncompetitive inhibition can be distinguished with double-reciprocal plots (Figure 6.11 a–e). In two sets of rate determinations, enzyme concentration is held constant. The first experiment establishes the velocity and kinetic parameters (K_m and V_{max}) of the uninhibited enzyme. In the second experiment, a constant amount

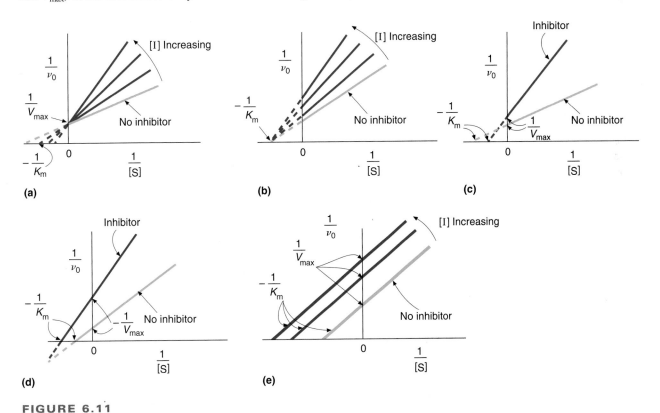

FIGURE 6.11

Kinetic Analysis of Enzyme Inhibition

(a) Competitive inhibition. Plots of $1/v$ versus $1/[S]$ in the presence of several concentrations of the inhibitor result in no change in the vertical intercept (i.e., $1/V_{max}$, so V_{max} is unchanged) and a decrease in the vertical intercept (i.e., $-1/K_m$, so K_m is increased). (b) Noncompetitive inhibition. Plots of $1/v$ versus $1/[S]$ in the presence of several concentrations of the inhibitor intersect at the same point on the **horizontal** axis, $-1/K_m$. In noncompetitive inhibition the constants for ES and EIS are assumed to stay the same. In mixed noncompetitive inhibition the plots of $1/v$ versus $1/[S]$ will intersect the vertical axis above the horizontal axis (c) if K_I' is greater than K_I or below the horizontal axis (d) if K_I' is less than K_I. (e) Uncompetitive inhibition. Plots of $1/v$ versus $1/[S]$ in the presence of several concentrations of the inhibitor result in a decrease in K_m and a decrease in V_{max} such that the slope of the line is unchanged but shifts up and to the left.

of inhibitor is included in each enzyme assay. Figure 6.11 illustrates the different effects that inhibitors have on enzyme activity. Competitive inhibition increases the K_m of the enzyme but the V_{max} is unchanged. (This is shown in the double-reciprocal plot as a shift in the horizontal intercept.) In pure noncompetitive inhibition, V_{max} is lowered (i.e., the vertical intercept is shifted); the K_m is unchanged in this rare case because the k values for the binding to E and ES are the same. In mixed noncompetitive inhibition these k values are different. Consequently, both V_{max} and K_m change. In such cases, the plots of $1/v$ versus $1/[S]$ will intersect to the left of the vertical axis either above the horizontal axis if K_I' is greater than K_I, or below the axis if K_I' is less than K_I. In uncompetitive inhibition both K_m and V_{max} are changed, although their ratio (i.e., the slope K_m/V_{max}) remains the same.

IRREVERSIBLE INHIBITION In reversible inhibition the inhibitor can dissociate from the enzyme because it binds through noncovalent bonds. Irreversible inhibitors usually bond covalently to the enzyme, often to a side chain group in the active site. For example, enzymes containing free sulfhydryl groups can react with alkylating agents such as iodoacetate:

$$\text{Enzyme}-CH_2-SH \;+\; I-CH_2-\overset{\overset{\displaystyle O}{\|}}{C}-O^- \;\longrightarrow\; \text{Enzyme}-CH_2-S-CH_2-\overset{\overset{\displaystyle O}{\|}}{C}-O^- \;+\; HI$$

Glyceraldehyde-3-phosphate dehydrogenase, an enzyme in the glycolytic pathway (Chapter 8), is inactivated by alkylation with iodoacetate. Enzymes that use sulfhydryl groups to form covalent bonds with metal cofactors are often irreversibly inhibited by heavy metals (e.g., mercury and lead). The anemia that is symptomatic of lead poisoning is caused in part because lead binds to a sulfhydryl group of ferrochelatase. Ferrochelatase catalyzes the insertion of Fe^{2+} into the heme prosthetic group of hemoglobin.

Problem 6.4 is concerned with enzyme inhibition.

WORKED PROBLEM 6.4

Consider the Lineweaver-Burk plot illustrated in Figure 6.12.

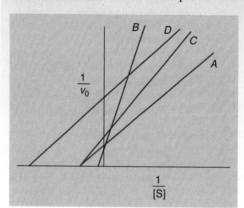

FIGURE 6.12
A Lineweaver-Burk Plot

Line A = normal enzyme-catalyzed reaction
Line B = compound B added
Line C = compound C added
Line D = compound D added

Identify the type of inhibitory action shown by compounds B, C, and D.

Solution

Compound B is a competitive inhibitor because the K_m only has changed. Compound C is a pure noncompetitive inhibitor because the V_{max} only has changed. Compound D is an uncompetitive inhibitor because both K_m and V_{max} have changed. ∎

QUESTION 6.4

Iodoacetamide is an irreversible inhibitor of several enzymes that have a cysteine residue in their active sites. After examining its structure, predict the products of the reaction of iodoacetamide with such an enzyme.

$$I-CH_2-\overset{\overset{\displaystyle O}{\|}}{C}-NH_2$$

ALLOSTERIC ENZYMES Although the Michaelis-Menten model is an invaluable tool, it does not explain the kinetic properties of many enzymes. For example, plots of reaction velocity versus substrate concentration for many enzymes with multiple subunits are often sigmoidal rather than hyperbolic, as predicted by the Michaelis-Menten model (Figure 6.13). Such effects are seen in an important group of enzymes called the **allosteric enzymes**. Note that the substrate-binding curve in Figure 6.13 resembles the oxygen-binding curve of hemoglobin.

The properties of allosteric enzymes are discussed on pp. 216–218.

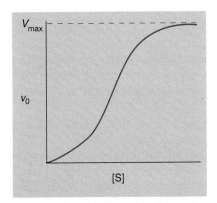

FIGURE 6.13

The Kinetic Profile of an Allosteric Enzyme

The sigmoidal binding curve displayed by many allosteric enzymes resembles the curve for the cooperative binding of O_2 to hemoglobin.

QUESTION 6.5

Drinking methanol can cause blindness in humans, as well as a severe acidosis that may be life-threatening. Methanol is toxic because it is converted in the liver to formaldehyde and formic acid by the enzymes alcohol dehydrogenase and aldehyde dehydrogenase. Methanol poisoning is treated with dialysis and infusions of bicarbonate and ethanol. Explain why each treatment is used.

Enzyme Kinetics, Metabolism, and Macromolecular Crowding

The ultimate goal of enzyme kinetic investigations is the development of a realistic understanding of metabolism in living organisms. Despite a wealth of knowledge, gathered over many decades, about biochemical pathways and the *in vitro* ("in glass," i.e., measured in the laboratory) kinetic properties of enzymes, insight into metabolic processes in "live" cells is elusive. An important reason for this lack of success is now believed to lie in the assumptions of the law of mass action. According to this rule, when an equilibrium constant for a reaction such as

$$A + B \underset{k_R}{\overset{k_F}{\rightleftharpoons}} C$$

is calculated, it is assumed that the forward rate is *linear* (i.e., is directly proportional) to the concentrations of A and B, and the rate of the reverse reaction (k_R) is linear with respect to the concentration of C. For the K_{eq} value derived for this or any reaction to be valid, it is further assumed that (1) the system in which the reaction occurs is homogeneous (i.e., there is uniform mixing of its contents) and (2) interacting molecules move randomly and independently of each other. It is now understood that under crowded *in vivo* ("in life") conditions, neither of these assumptions appears to hold true. The crowded interior of cells is highly heterogeneous (i.e., an enormous variety of different molecular species are present) and filled with macromolecules, membranes, cytoskeletal components, and other obstacles that impede molecular movement. This phenomenon is called **macromolecular crowding**. Among its observed effects within cells are increased values for the effective concentrations of macromolecules, enhanced protein folding and binding affinities, changes in reaction rates and equilibrium constants, and decreased diffusion rates. These effects are *nonlinear*; that is, they are difficult to predict. For example, the *in vivo* kinetic values (i.e., K_m, V_{max}, and K_{eq}) of each biochemical reaction depend on the activity coefficient of its enzyme and the diffusion rates of substrate and effector molecules, neither of which are easy to describe in the heterogeneous environment of the cell. Slow substrate diffusion can be overcome by the segregation of reactions into microcompartments where substrate concentration is relatively high, or the recruitment of some or all of a pathway's enzymes into complexes called *metabolons* in which pathway intermediates are directly transferred, or "channeled," from one active site to the next. These microenvironments have their own unique ionic composition that promotes the efficient operation of the resident enzymes. It is a significant challenge to biochemists to identify these local conditions and duplicate them *in vitro* to facilitate the investigation of specific enzymes and determine realistic kinetic parameters. Discrepancies between results derived *in vivo*, *in vitro*, and in silico (by computer simulation) suggest that an enzyme isolated from its metabolic pathway and other intersecting pathways does not function with the same kinetic parameters.

Despite the daunting complexities of living cells, systems biologists have endeavored to develop new strategies to investigate metabolic processes. This work is often done by metabolic engineers, industrial scientists who improve the metabolic capacities of microorganisms used to produce commercial products. Among the most important tools used by metabolic engineers are dynamic computer simulations of metabolic processes that are based on both *in vivo* and *in vitro* data. The mathematical model upon which the computer simulation is based is constructed from kinetic data and equations that define *metabolic flux* (the rate of flow of metabolites, such as substrates, products, and intermediates along biochemical pathways). (Despite the limitations of *in vitro* kinetic values just described, they are recognized as invaluable baseline data during the early phases of model development.) The ultimate goal of a computer simulation is to approximate the *in vivo* steady state behavior of metabolic processes and predict how they will respond to parameter changes such as nutrient availability or mutant forms of enzymes. Insights already gained from computer simulations (e.g., see Biochemistry in Perspective, Glycolysis and Jet Engines, in Chapter 8) guarantee that in silico modeling and pathway simulation will form an increasingly significant part of future enzyme kinetics investigations.

6.4 CATALYSIS

However valuable kinetic studies are, they reveal little about how enzymes catalyze biochemical reactions. Biochemists use other techniques to investigate the catalytic mechanisms of enzymes. [A *mechanism* describes how bonds are broken and new ones formed in the conversion of substrate(s) to product(s).] Enzyme mechanism investigations seek to relate enzyme activity to the structure and function of the active site. Scientists who study these mechanisms use such methods as X-ray crystallography, chemical inactivation of active site side chains, and modeling with simple model compounds as substrates and as inhibitors.

Organic Reactions and the Transition State

Almost all the chemical reactions in living organisms involve organic molecules. Although there are significant differences in context, biochemical reactions follow the same set of rules as the reactions studied by organic chemists. The essential features of both are the reaction between electron-deficient atoms (electrophiles) and electron-rich atoms (nucleophiles), and the formation of transition states. Each is discussed briefly.

Chemical bonds form when a nucleophile donates an electron pair to an electrophile. For example, in the following reaction

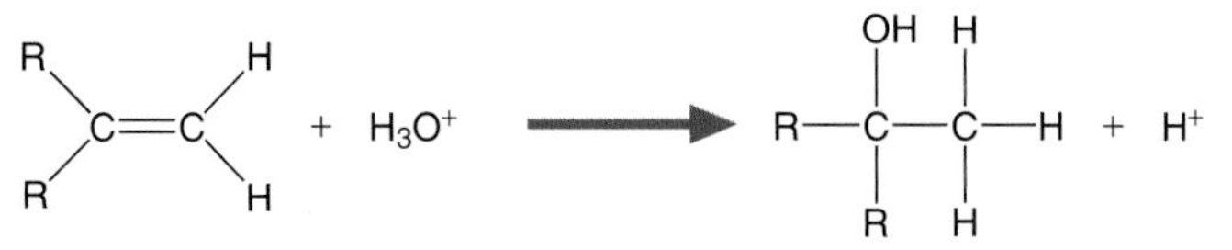

the π electrons of the nucleophilic double bond will react with a partially positive hydrogen atom of an electrophilic hydronium ion. As in all organic reactions, product formation takes place in several steps.

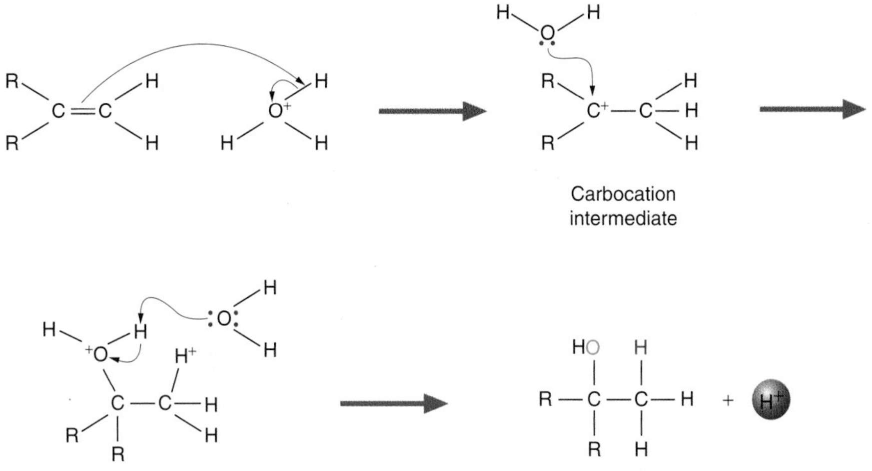

Carbocation intermediate

This step-by-step description is referred to as a **reaction mechanism**. The curved arrows illustrate the flow of electrons away from a nucleophile and toward an electrophile. During the course of the reaction, one or more intermediates form. An **intermediate** is a species that exists for a finite length of time (10^{-13} s or less) and then is transformed into product. In the example reaction, a carbocation intermediate forms as π electrons in the double bond attack an electrophilic hydronium ion. (A **carbocation**, or carbonium ion, contains an electron-deficient, positively charged carbon atom.) This type of intermediate is stabilized by the R groups that decrease the positive charge on the carbon atom. In the reaction's next step, an electron pair of the oxygen atom in a water molecule forms a σ bond with the positively charged carbon atom. In the final step the alcohol product results from a proton transfer to another water molecule. Other examples of reactive intermediates observed in biochemical reactions include **carbanions** (nucleophilic carbon anions with three bonds and an unshared electron pair) and **free radicals** (highly reactive species with at least one unpaired electron).

In any chemical reaction, only molecules that reach the activated condition known as the transition state can convert into product molecules. The conversion of a stable reactant molecule into an activated one is analogous to the energy-requiring process of rolling a boulder up a hill. Once the boulder has reached the top, only a slight push will cause it to slide down the other side, releasing energy as it does so. According to transition state theory, the rate of reaction is determined by the relative number of reactant molecules that possess sufficient energy to overcome the activation energy barrier (E_a) (refer to Figure 6.1). Reaction rates increase if E_a can be lowered. Enzymes do so primarily by stabilizing the transition state. Several factors that enzymes use to perform this feat are described in the next section. It should be noted, however, that an important feature of transition state stabilization is the relative affinity of active sites for substrate, product, and the

FIGURE 6.14

Energy Profile for a Two-Step Reaction

The energy maxima are the unstable transient transition states $TS_1^{\ddagger}$, and $TS_2^{\ddagger}$, and the energy minimum is the intermediate (I).

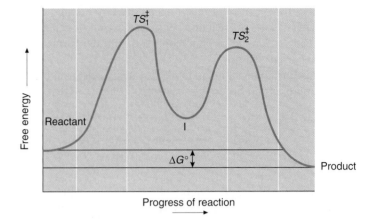

transition state. Investigations of transition state analogues, stable molecules that closely resemble a transition state, have revealed that enzyme active sites bind the transition state more tightly than they bind substrate or product molecules. The energy diagram for a two-step reaction in Figure 6.14 illustrates the difference between an intermediate (a reactive species with a finite lifetime) and a transition state (an unstable species with maximum free energy).

Catalytic Mechanisms

Despite extensive research, the mechanisms of only a few enzymes are known in significant detail. It is known that enzymes achieve significantly higher catalytic rates than other catalysts because their active sites possess structures that are uniquely suited to promote catalysis. Several factors contribute to enzyme catalysis. The most important of these are proximity and strain effects, electrostatic effects, acid-base catalysis, covalent catalysis, and quantum tunneling. It should be noted that none of these catalytic factors are mutually exclusive. In varying degrees, they all participate in each type of catalytic mechanism.

PROXIMITY AND STRAIN EFFECTS For a biochemical reaction to occur, the substrate must come into close proximity to catalytic functional groups (side chain groups involved in a catalytic mechanism) within the active site. In addition, the substrate must be precisely oriented to the catalytic groups. When the substrate is correctly positioned, a change in the enzyme's conformation may result in a strained enzyme-substrate complex. This strain helps to bring the enzyme-substrate complex into the transition state.

ELECTROSTATIC EFFECTS Recall that the strength of electrostatic interactions is inversely related to the hydration of participating species (Chapter 3). Hydration shells increase the distance between charge centers and reduce electrostatic attraction. Because water is largely excluded from the active site as the substrate binds, the local dielectric constant is often low. The charge distribution in the relatively anhydrous active site may facilitate the optimum positioning of substrate molecules as well as influence their chemical reactivity. In addition, weak electrostatic interactions, such as those between permanent and induced dipoles in both the active site and the substrate, are believed to contribute to catalysis.

ACID-BASE CATALYSIS Acid-base catalysis (proton transfer) is an important factor in chemical reactions. For example, consider the hydrolysis of an ester:

$$R-\overset{\overset{\displaystyle O}{\|}}{C}-O-R' \quad + \quad H_2O \quad \longrightarrow \quad R-\overset{\overset{\displaystyle O}{\|}}{C}-OH \quad + \quad R'-OH$$

Because water is a weak nucleophile, ester hydrolysis is relatively slow in neutral solution. Ester hydrolysis takes place much more rapidly if the pH is raised. As the hydroxide ion attacks the polarized carbon atom of the carbonyl

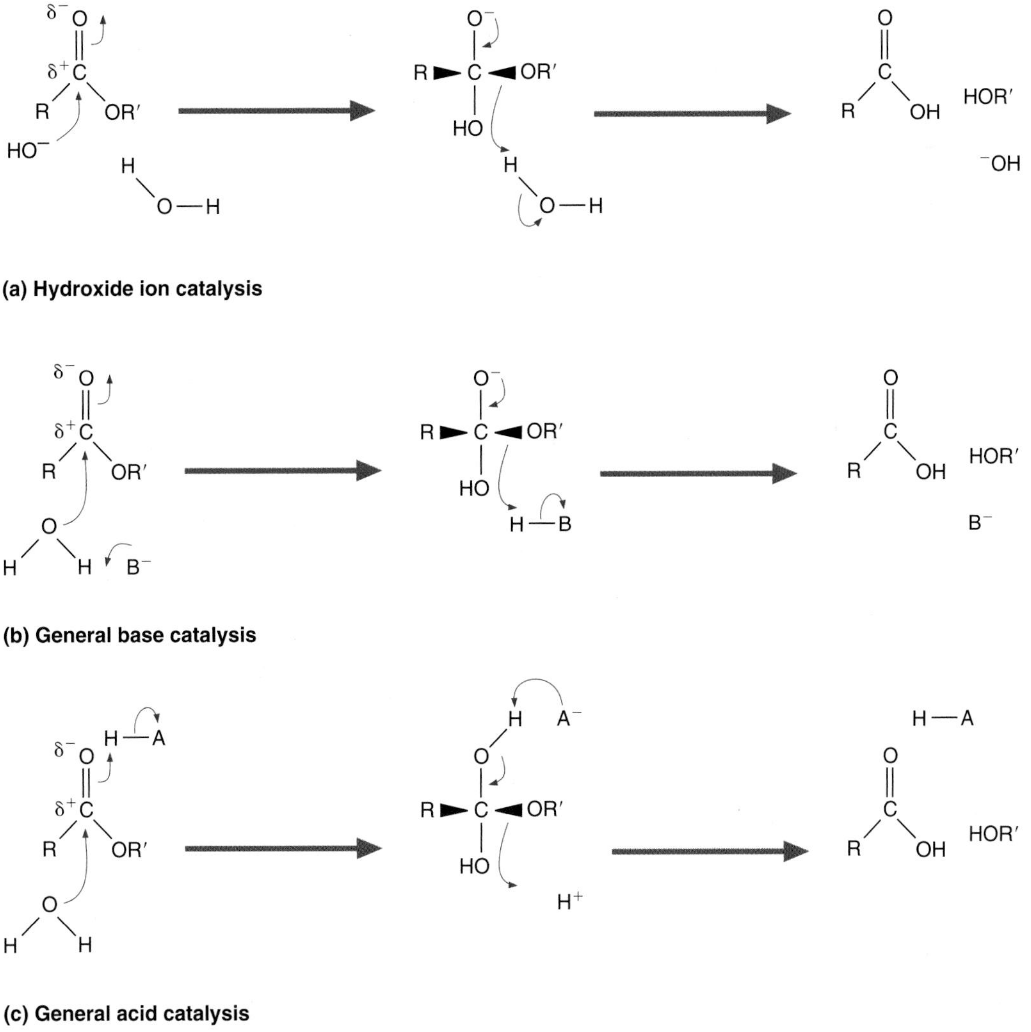

(a) Hydroxide ion catalysis

(b) General base catalysis

(c) General acid catalysis

FIGURE 6.15

Ester Hydrolysis

Esters can be hydrolyzed in several ways: (a) catalysis by free hydroxide ion, (b) general base catalysis, and (c) general acid catalysis. A colored arrow represents the movement of an electron pair during each mechanism.

group (Figure 6.15a), a tetrahedral intermediate is formed. As the intermediate breaks down, a proton is transferred from a nearby water molecule. The reaction is complete when the alcohol is released. However, hydroxide ion catalysis is not practical in living systems. Enzymes use several functional groups that behave as general bases to transfer protons efficiently. Such groups can be precisely positioned in relation to the substrate (Figure 6.15b). Ester hydrolysis can also be catalyzed by a general acid (Figure 6.15c). As the oxygen of the ester's carbonyl group binds to the proton, the carbon atom becomes more electrophilic. The ester then becomes more susceptible to the nucleophilic attack of a water molecule.

Within enzyme active sites, the functional groups on the side chains of histidine (imidazole), aspartate and glutamate (carboxylate), tyrosine (hydroxyl), cysteine (sulfhydryl), and lysine (amine) can act either as proton donors (called general acids) or proton acceptors (called general bases) depending on their state of protonation. Each of these side chain groups has a characteristic pK_a value that may be influenced by nearby charged or polar atoms. For example, the side chain of histidine often participates in concerted acid-base catalysis because its pK_a

range is close to physiological pH. The protonated imidazole ring can serve as a general acid, and the deprotonated imidazole ring can serve as a general base:

General acid   General base

Histidine

Within the active site of the serine proteases (a class of proteolytic enzymes such as trypsin and chymotrypsin, see pp. 212–213), the close proximity of an aspartate carboxylate group to histidine raises the latter's pK_a. As a consequence, histidine, acting as a general base, abstracts a proton from a nearby serine side chain, thus converting the oxygen of serine into a better nucleophile.

COVALENT CATALYSIS In some enzymes a nucleophilic side chain group forms an unstable covalent bond with an electrophilic group on the substrate. As just described, the serine proteases use the —CH_2—OH group of serine as a nucleophile to hydrolyze peptide bonds. During the first step, the nucleophile (i.e., the oxygen of serine) attacks the carbonyl group of the peptide substrate. As the ester bond is formed, the peptide bond is broken. The resulting acyl-enzyme intermediate is hydrolyzed by water in a second reaction:

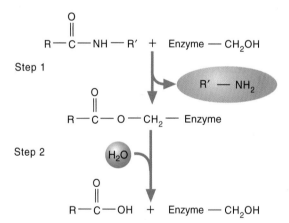

Several other amino acid side chains may act as nucleophiles. The sulfhydryl group of cysteine, and the carboxylate groups of aspartate and glutamate, can play this role.

Quantum Tunneling and Enzyme Catalysis

Investigations with increasingly sophisticated technologies have revealed that transition state theory does not fully account for the catalytic power of the several enzymes thus far studied. It is now believed that when hydrogen is transferred in the form of hydrogen atoms (H·), hydride ions (H:⁻) or protons (H⁺), a phenomenon called tunneling occurs. *Tunneling* is a quantum mechanical process in which a particle passes through an energy barrier instead of over it. *Quantum mechanics* is the science that deals with the properties of matter and electromagnetic radiation on the atomic and subatomic scales. One of the most remarkable contributions of quantum theory is particle-wave duality, the recognition that particles of matter have wavelike properties and radiation has particle-like properties. For example, light (pp. 476–479) is composed of photons (packets of energy) that propagate through space as a wave. In the quantum world, a particle such as a proton can move through an energy barrier rather than over it, because it behaves like a wave. Such an event is forbidden in classical (Newtonian) physics. Quantum tunneling is possible only for light particles such as electrons and hydrogen atoms because a particle's wavelength is inversely proportional to the square root of the particle's mass. According to quantum mechanical calculations, a proton can tunnel over a distance of 0.58 Å. Among the best-documented cases of hydrogen tunneling is the hydride transfer reaction catalyzed by dihydrofolate reductase (DHFR). DHFR, the enzyme that reduces 7,8-dihydrofolate to 5,6,7,8-tetrahydrofolate (the biologically active form of folic acid, see pp. 525–526), uses the coenzyme NADPH as a hydride donor. Compelling evidence indicates that vibrational energy resulting from protein conformation changes during the reaction reduces the tunneling distance, thereby compressing the activation barrier and increasing the probability of hydride transfer between NADPH and the substrate. The progress of a reaction *through* rather than *over* an energy barrier (therefore, requiring more kinetic energy) would lead to an acceleration of reaction rate comparable to that seen in many enzyme-catalyzed reactions. Macromolecular dynamics and its link to the phenomenon of quantum tunneling is an important area for future research and appears to provide an explanation for the high rates of enzymatic reactions.

QUESTION 6.6

Quantum tunneling appears to play a significant role in the facilitation of efficient enzyme catalysis, particularly in reactions involving small-species transfer (H^+, $H^{:-}$, etc). Provide an energy diagram for a hypothetical reaction in the presence and absence of an enzyme that clearly illustrates the quantum tunneling process. [*Hint:* Refer to the progress of reaction diagram in Figure 6.1.]

The Roles of Amino Acids in Enzyme Catalysis

The active sites of enzymes are lined with amino acid side chains that together create a microenvironment that is conducive to catalysis. The functions of these residues fall into two major categories: catalytic and noncatalytic. Catalytic residues directly participate in the catalytic mechanism, while noncatalytic residues have support functions. Of the 20 amino acids found in proteins (Figure 5.2), only those with polar and charged side chains actually participate in catalysis. These amino acids (and their side chains groups) are as follows: serine, threonine, and tyrosine (hydroxyl); cysteine (thiol), glutamine and asparagine (amide); glutamate and aspartate (carboxylate); lysine (amine); arginine (guanidinium); and histidine (imidazole).

Extensive research has revealed that catalytic mechanisms require the precise positioning of one or more catalytic units that are composed of either two or three amino acid side chains called dyads or triads, respectively. Although the number of enzymes is vast, the catalytic units are composed of relatively few combinations of amino acids. Commonly observed examples include the arginine-arginine, carboxylate-carboxylate, and carboxylate-histidine dyads. An arginine-arginine dyad is a catalytic unit in adenylate kinase, an enzyme that catalyzes the transfer of phosphoryl groups from ATP to other nucleotides. The polarizing effect of the two arginines on the phosphate group's oxygens has the effect of converting phosphate into a good leaving group. Carboxylate-carboxylate dyads occur in the active sites of the aspartic proteases, a family of proteolytic enzymes such as pepsin, which animals use to digest dietary protein. The close proximity of the two negatively charged aspartate carboxyl groups raises the pK_a of one of the aspartates making it less acidic and more basic. Its enhanced capacity to accept a proton initiates an acid-base hydrolytic mechanism. The functional properties of aspartate-histidine dyads result from a polarized imidazolium ring caused by the close proximity of the aspartate's negatively charged carboxylate group. In the active site of aconitase, the enzyme that catalyzes the isomerization of citrate to form isocitrate (p. 327), histidine acts as a general acid and protonates the –OH group of the citrate, making it a better leaving group.

The resulting HOH is held in the active site by histidine long enough to be able to attack the intermediate and generate the isomer of citrate (isocitrate). An aspartate-histidine dyad is also, most notably, a component of the well-researched serine protease triad (Asp-His-Ser) (p. 212). The close proximity of the aspartate carboxylate group to the imidazole group of histidine raises the latter group's pK_a, thus facilitating its ability to remove a proton from serine. The deprotonated serine is thus converted into a better nucleophile.

The functions of noncatalytic side groups, which include substrate orientation and transition state stabilization, are more subtle than those of catalytic residues. For example, the substrate specificity of chymotrypsin (p. 212–213), manifested in the cleavage of peptide bonds on the C-terminal side of the aromatic amino acids tryptophan, tyrosine, and phenylalanine, is made possible by the relatively large size of a hydrophobic pocket within the active site that both accommodates and orients the aromatic side chains. The oxyanion intermediate (p. 213) that forms during the catalytic mechanism is stabilized by interactions between the substrate's peptide bond carbonyl group and the backbone amide hydrogens of serine and glycine residues.

The Role of Cofactors in Enzyme Catalysis

In addition to active site amino acid side chains, many enzymes require nonprotein cofactors, that is, metal cations and the coenzymes. The structural properties and chemical reactivities of each group are briefly discussed.

METALS The important metals in living organisms fall into two classes: the alkali and alkaline earth metals (e.g., Na^+, K^+, Mg^{2+}, and Ca^{2+}) and the transition metals (e.g., Zn^{2+}, Fe^{2+}, and Cu^{2+}). The alkali and alkaline earth metals in enzymes are loosely bound and usually have structural roles. In contrast, the transition metals play key roles in catalysis either bound to functional groups such as carboxylate, imidazole, or hydroxyl groups, or as components of prosthetic groups such as Fe^{2+} in heme.

Several properties of transition metals make them useful in catalysis. Metal ions provide a high concentration of positive charge that is especially useful in binding small molecules. Because transition metals act as *Lewis acids* (electron pair acceptors), they are effective electrophiles. (Amino acid side chains are poor electrophiles because they cannot accept unshared pairs of electrons.) Because the metal's directed d shell valences allow them to interact with two or more ligands, metal ions help orient the substrate within the active site. As a consequence, the substrate–metal ion complex polarizes the substrate and promotes catalysis. For example, carbonic anhydrase is the enzyme that catalyzes the reversible hydration of CO_2 to form bicarbonate (HCO_3^-). Its active site contains a zinc (Zn^{2+}) cofactor that is coordinated with three histidine side chains. The zinc ion polarizes a water molecule resulting in a Zn^{2+}-bound OH group. The OH group (acting as a nucleophile) attacks CO_2, converting it into HCO_3^-:

Finally, because transition metals have two or more valence states, they can mediate oxidation-reduction reactions by reversibly gaining or losing electrons. For example, the reversible oxidation of Fe^{2+} to form Fe^{3+} is important in the function of cytochrome P_{450}, enzyme that processes toxic substances in animals. [See Biochemistry in Perspective boxes in Chapters 10 (Ischemia and Reperfusion) and 12 (Biotransformation).][30]

QUESTION 6.7

Copper is a cofactor in several enzymes, including lysyl oxidase and superoxide dismutase. Ceruloplasmin, a deep-blue glycoprotein, is the principal copper-containing protein in blood. It is used to transport Cu^{2+} and maintain appropriate levels of Cu^{2+} in the body's tissues. Ceruloplasmin also catalyzes the oxidation of Fe^{2+} to Fe^{3+}, an important reaction in iron metabolism. Because the metal is widely found in foods, copper deficiency is rare in humans. Deficiency symptoms include anemia, leukopenia (reduction in blood levels of white blood cells), bone defects, and weakened arterial walls. The body is partially protected from exposure to excessive copper (and several other metals) by metallothionein, a small, metal-binding protein that possesses a large proportion of cysteine residues. Certain metals (most notably zinc and cadmium) induce the synthesis of metallothionein in the intestine and liver.

In *Menkes' syndrome* intestinal absorption of copper is defective. How can affected infants be treated to avoid the symptoms of the disorder, which include seizures, retarded growth, and brittle hair?

QUESTION 6.8

In another rare inherited disorder, called *Wilson's disease*, excessive amounts of copper accumulate in liver and brain tissue. A prominent symptom of the disease is the deposition of copper in greenish-brown layers surrounding the cornea, called Kayser-Fleischer rings. Wilson's disease is now known to be caused by a defective ATP-dependent protein that transports copper across cell membranes. Apparently, the copper transport protein is required to incorporate copper into ceruloplasmin and to excrete excess copper. In addition to a diet low in copper, Wilson's disease is treated with zinc sulfate and the chelating agent penicillamine (p. 137). Describe how these treatments work. [*Hint:* Metallothionein has a greater affinity for copper than for zinc.]

COENZYMES The coenzymes are a group of organic molecules that provide enzymes with chemical versatility because they either possess reactive groups not found on amino acid side chains or can act as carriers for substrate molecules. Some coenzymes are only transiently bound to the enzyme and are essentially cosubstrates, whereas others are tightly bound by covalent or noncovalent bonds. Unlike ordinary catalysts, coenzyme structures are changed by the reactions in which they participate. Their catalytically active forms must be regenerated before another catalytic cycle can occur. Transiently bound coenzymes are usually regenerated by a reaction catalyzed by another enzyme, whereas tightly bound coenzymes are regenerated during a step in the catalytic cycle. Most coenzymes are derived from vitamins. **Vitamins** (organic nutrients required in small amounts in the human diet) are divided into two classes: water-soluble and lipid-soluble. In addition, there are certain vitamin-like substances that organisms can synthesize in sufficient amounts to facilitate enzyme-catalyzed reactions. Examples include lipoic acid, carnitine, coenzyme Q, biopterin, *S*-adenosylmethionine, and *p*-aminobenzoic acid.

Coenzymes can be classified according to function into three groups: electron transfer, group transfer, and high-energy transfer potential. Coenzymes involved in redox (electron or hydrogen transfer) reactions include nicotinamide adenine dinucleotide (NAD^+), nicotinamide adenine dinucleotide phosphate ($NADP^+$), flavin adenine dinucleotide (FAD), flavin mononucleotide (FMN), coenzyme Q (CoQ), and tetrahydrobiopterin (BH_4). Coenzymes such as thiamine pyrophosphate (TPP), coenzyme A (CoASH), and pyridoxal phosphate are involved in the transfer of aldehyde, acyl, and amino groups, respectively. One-carbon transfers, a diverse set of reactions in which carbon atoms are transferred in various oxidation states between substrates, require biotin, tetrahydrofolate (TH_4) or *S*-adenosylmethionine (SAM). Nucleotides, known for their high-energy transfer potential, function as coenzymes in that they activate metabolic intermediates, and/or serve as phosphate donors or as carriers for small molecules. Examples of the latter include UDP-glucose, an intermediate in glycogen synthesis (p. 296), and CDP (cytidine dephosphate)-ethanolamine, an intermediate in the synthesis of certain lipids (p. 444). In such cases the nucleotide serves as a molecular carrier. The numerous noncovalent bonds that form when it binds within the enzyme's active site assure that the attached metabolite is correctly positioned. Note that the structures of coenzymes such as NAD^+, $NADP^+$, FAD, FMN, and CoASH also contain adenine nucleotide components. Table 6.3 lists the vitamins and vitamin-like substances, their coenzyme forms, and the reactions they facilitate, and gives the pages on which their structural and functional properties are described.

KEY CONCEPTS

- The amino acid side chains in the active site of enzymes catalyze proton transfers and nucleophilic substitutions. Other reactions require nonprotein cofactors, that is, metal cations and the coenzymes.

- Metal ions are effective electrophiles, and they help orient the substrate within the active site. In addition, certain metal cations mediate redox reactions.

- Coenzymes are organic molecules that have a variety of functions in enzyme catalysis.

QUESTION 6.9

Identify each of the following as a cofactor, coenzyme, apoenzyme, holoenzyme, or none of these. Explain each answer.
- a. Zn^{2+}
- b. active alcohol dehydrogenase
- c. alcohol dehydrogenase lacking Zn^{2+}
- d. FMN
- e. NAD^+
- f. CDP-ethanolamine
- g. biotin

TABLE 6.3   Vitamins, Vitamin-like Molecules, and Their Coenzyme Forms

Molecules	Coenzyme Form	Reaction or Process Promoted	See Page
Water-Soluble Vitamins			
Thiamine (B_1)	Thiamine pyrophosphate	Decarboxylation, aldehyde group transfer	319
Riboflavin (B_2)	FAD and FMN	Redox	314
Pyridoxine (B_6)	Pyridoxal phosphate	Amino group transfer	509
Nicotinic acid (niacin)	NAD and NADP	Redox	313
Pantothenic acid	Coenzyme A	Acyl transfer	318
Biotin	Biotin	Carboxylation	432
Folic acid	Tetrahydrofolic acid	One-carbon group transfer	525
Vitamin B_{12}	Deoxyadenosylcobalamin, methylcobalamin	Intramolecular rearrangements	526
Ascorbic acid (vitamin C)	Unknown	Hydroxylation	365
Lipid-Soluble Vitamins			
Vitamin A	Retinal	Vision, growth, and reproduction	365
Vitamin D	1, 25-Dihydroxycholecalciferol	Calcium and phosphate metabolism	390
Vitamin E	Unknown	Lipid antioxidant	365
Vitamin K	Unknown	Blood clotting	388
Vitamin-like Molecules			
Coenzyme Q	Coenzyme Q	Redox	341
Biopterin	Tetrahydrobiopterin	Redox	534
S-Adenosylmethionine	S-Adenosylmethionine	Methylation	526
Lipoic acid	Lipoamide	Redox	319

Effects of Temperature and pH on Enzyme-Catalyzed Reactions

Any environmental factor that disturbs protein structure may change enzymatic activity. Enzymes are especially sensitive to changes in temperature and pH.

TEMPERATURE   All chemical reactions are affected by temperature. In general, the higher the temperature, the higher the reaction rate; that is, the greater the increase in the number of collisions. The reaction velocity increases because more molecules have sufficient energy to enter into the transition state. The rates of enzyme-catalyzed reactions also increase with increasing temperature. However, enzymes are proteins that become denatured at high temperatures. An enzyme's *optimum temperature*, the temperature at which it operates at maximum efficiency (Figure 6.16), is determined in the laboratory under specified conditions of pH, ionic strength, and solute concentrations. In living organisms there is no single definable optimal temperature for enzymes.

pH   Hydrogen ion concentration, expressed as a pH value, affects enzymes in several ways. First, catalytic activity is related to the ionic state of the active site. Changes in hydrogen ion concentration can affect the ionization of active site groups. For example, the catalytic activity of a certain enzyme requires the protonated form of a side chain amino group. If the pH becomes so alkaline that the group loses its proton, the enzyme's activity may be depressed. In addition, substrates may be affected. If a substrate contains an ionizable group, a change in pH may alter its capacity to bind to the active site. Second, changes in ionizable groups may change the tertiary structure of the enzyme. Drastic changes in pH often lead to denaturation.

Although a few enzymes tolerate large changes in pH, most enzymes are active only within a narrow pH range. For this reason, living organisms employ buffers to closely regulate pH. The pH value at which an enzyme's activity is maximal

FIGURE 6.16

Effect of Temperature on Enzyme Activity

Modest increases in temperature increase the rate of enzyme-catalyzed reactions because of an increase in the number of collisions between enzyme and substrate. Eventually, increasing the temperature decreases the reaction velocity. Catalytic activity is lost because heat denatures the enzyme.

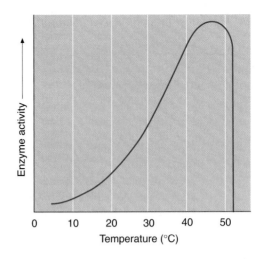

FIGURE 6.17

Effect of pH on Two Enzymes

Each enzyme has a certain pH at which it is most active. A change in pH can alter the ionizable groups within the active site or affect the enzyme's conformation.

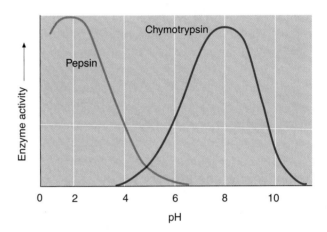

is called the **pH optimum** (Figure 6.17). The pH optima of enzymes, determined in the laboratory, vary considerably. For example, the optimum pH of pepsin, a proteolytic enzyme produced in the stomach, is approximately 2. For chymotrypsin, which digests protein in the small intestine the optimum pH is approximately 8.

Detailed Mechanisms of Enzyme Catalysis

Each of the thousands of enzymes that have been investigated has a unique structure, substrate specificity, and reaction mechanism. The mechanisms of a variety of enzymes have been investigated intensively over the past several decades. The subsections that follow describe the catalytic mechanisms of two well-characterized enzymes.

CHYMOTRYPSIN Chymotrypsin is a 27,000 D protein that belongs to the serine proteases. The active sites of all serine proteases contain a characteristic set of amino acid residues, often referred to as the serine protease triad (p. 208). In the chymotrypsin numbering system these are Asp 102, His 57, and Ser 195. Studies of crystallized enzyme bound to substrate analogues reveal that these residues are close to each other in the active site. The active site serine residue plays an especially important role in the catalytic mechanisms of this group of enzymes. Serine proteases are irreversibly inhibited by diisopropylfluorophosphate (DFP). In DFP-inhibited enzymes, the inhibitor is covalently bound only to Ser

195 and not to any of the other 29 serines. The special reactivity of Ser 195 is attributed to the proximity of His 57 and Asp 102. The imidazole ring of His 57 lies between the carboxyl group of Asp 102 and the —CH₂OH group of Ser 195. The carboxyl group of Asp 102 polarizes His 57, thus allowing it to act as a general base (i.e., the abstraction of a proton by the imidazole group is facilitated):

$$-CH_2-\overset{\overset{\displaystyle O}{\|}}{C}-O^- \cdots HN \diagdown N: \quad HO-CH_2-$$

Removing the proton from the serine OH group converts it into a more effective nucleophile.

Chymotrypsin catalyzes the hydrolysis of peptide bonds adjacent to aromatic amino acids. The probable mechanism for this reaction is illustrated in Figure 6.18. Step (a) of the figure shows the initial enzyme-substrate complex. The alignment of the amino acid residues within the active site, including the catalytic triad of Asp 102, His 57, and Ser 195, creates a temporarily unoccupied position called the oxyanion hole. In addition, the active site provides a hydrophobic binding pocket for the substrate's phenylalanine side chain. The nucleophilic hydroxyl oxygen of Ser 195 launches a nucleophilic attack on the carbonyl carbon of the substrate. The tetrahedral intermediate that forms (step b) is sufficiently distorted that the newly created oxyanion is stabilized by hydrogen bonds to the amide hydrogens of Ser 195 and Gly 193, and enters the oxyanion hole. It is believed that the formation of the tetrahedral intermediate, which resembles the transition state, and its preferential binding by the enzyme, is responsible for the catalytic efficiency of the serine proteases.

The tetrahedral intermediate subsequently decomposes to form the covalently bound acyl-enzyme intermediate (step c). The residue His 57, acting as a general acid, is believed to facilitate this decomposition. In steps (d) and (e), the two previous steps are reversed. With water acting as an attacking nucleophile, a tetrahedral (oxyanion) intermediate is formed. By step (f) (the final enzyme-product complex), the bond between the serine oxygen and the carbonyl carbon has been broken. Serine is again hydrogen-bonded to His 57.

ALCOHOL DEHYDROGENASE Recall that alcohol dehydrogenase, an enzyme found in many eukaryotic cells (e.g., animal liver, plant leaves, and yeast), catalyzes the reversible oxidation of an alcohol to form an aldehyde:

$$CH_3-CH_2-OH \quad + \quad \boxed{NAD^+} \quad \rightleftharpoons \quad CH_3-\overset{\overset{\displaystyle O}{\|}}{C}-H \quad + \quad \boxed{NADH} \quad + \quad \boxed{H^+}$$

In this reaction, which involves the removal of two electrons and two protons, the coenzyme NAD acts as a hydride ion (H:⁻) acceptor.

The active site of alcohol dehydrogenase contains two cysteine residues (Cys 48 and Cys 174) and a histidine residue (His 67), all of which are coordinated to a zinc ion (Figure 6.19a). After NAD⁺ binds to the active site, the substrate ethanol enters and binds to the Zn²⁺ as the alcoholate anion (Figure 6.19b). The electrostatic effect of Zn²⁺ stabilizes the transition state. As the intermediate decomposes, the hydride ion is transferred from the substrate to the nicotinamide ring of NAD⁺. After the aldehyde product is released from the active site, NADH also dissociates.

KEY CONCEPTS

- Each enzyme has a unique structure, substrate specificity, and reaction mechanism.
- Each mechanism is affected by catalysis-promoting factors that are determined by the structure of the substrate and the enzyme's active site.

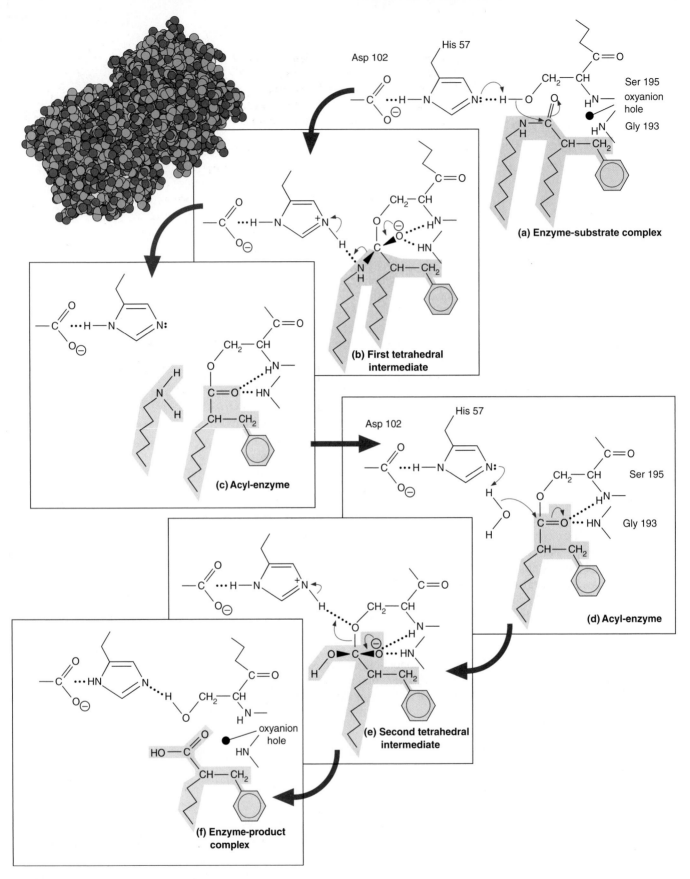

FIGURE 6.18

The Probable Mechanism of Action of Chymotrypsin

There is a fast acylation step during which the carbonyl end of the target peptide bond of the substrate is transferred to the enzyme to form an acyl-enzyme adduct and the amino end of the substrate leaves the active site. A second slow deacylation of the enzyme follows, releasing the carbonyl end of the substrate.

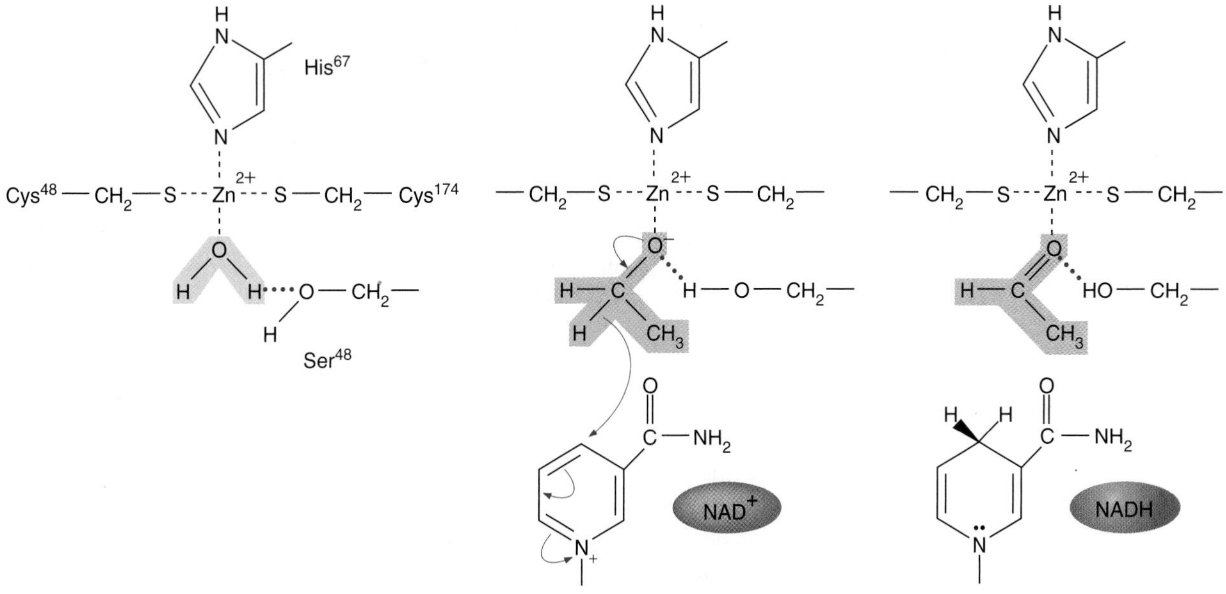

(a) Free enzyme

(b) Enzyme-ethanol
complex

(c) Enzyme-acetaldehyde
complex

FIGURE 6.19

Functional Groups of the Active Site of Alcohol Dehydrogenase

(a) Without a substrate, a molecule of water is one of the ligands of the Zn^{2+} ion. (b) The substrate ethanol probably binds to the Zn^{2+} as the alcoholate anion, displacing the water molecule. (c) NAD^+ accepts a hydride ion from the substrate and the aldehyde product is formed.

6.5 ENZYME REGULATION

Living organisms have sophisticated mechanisms for regulating their vast networks of biochemical pathways. Regulation is essential for several reasons:

1. **Maintenance of an ordered state**. Regulation of each pathway results in the production of the substances required to maintain cell structure and function in a timely fashion and without wasting resources.

2. **Conservation of energy**. Cells ensure that they consume just enough nutrients to meet their energy requirements by constantly controlling energy-generating reactions.

3. **Responsiveness to environmental changes**. Cells can make relatively rapid adjustments to changes in temperature, pH, ionic strength, and nutrient concentrations because they can increase or decrease the rates of specific reactions.

The regulation of biochemical pathways is complex. It is achieved primarily by adjusting the concentrations and activities of certain enzymes. Control is accomplished by (1) genetic control, (2) covalent modification, (3) allosteric regulation, and (4) compartmentation.

Genetic Control

The synthesis of enzymes in response to changing metabolic needs, a process referred to as **enzyme induction**, allows cells to respond efficiently to changes in their environment. For example, *E. coli* cells grown without the sugar lactose initially cannot metabolize this nutrient when it is introduced into the bacterium's growth medium. The introduction of lactose in the absence of glucose, however, activates the genes that code for enzymes needed to utilize lactose as an energy source. After all the lactose has been consumed, synthesis of these enzymes is terminated.

The synthesis of certain enzymes may also be specifically inhibited. In a process called *repression*, the end product of a biochemical pathway may inhibit the synthesis of a key enzyme in the pathway. For example, in *E. coli* the products of some amino acid synthetic pathways regulate the synthesis of key enzymes. The mechanism of this form of metabolic control is discussed in Chapter 18.

Covalent Modification

Some enzymes are regulated by the reversible interconversion between their active and inactive forms. Several covalent modifications of enzyme structure cause these changes in function. Many such enzymes have specific residues that may be phosphorylated and dephosphorylated. For example, glycogen phosphorylase (Chapter 8) catalyzes the first reaction in the degradation of glycogen, a carbohydrate energy storage molecule. In a process controlled by hormones, the inactive form of the enzyme (glycogen phosphorylase b) is converted to the active form (glycogen phosphorylase a) by adding a phosphate group to a specific serine residue. Other types of reversible covalent modification include methylation, acetylation, and nucleotidylation (the covalent addition of a nucleotide).

Several enzymes are produced and stored as inactive precursors called **proenzymes** or **zymogens**. Zymogens are converted into active enzymes by the irreversible cleavage of one or more peptide bonds. For example, chymotrypsinogen is produced in the pancreas. After chymotrypsinogen is secreted into the small intestine, it is converted to its active form that is responsible for digesting dietary protein in several steps (Figure 6.20).

Allosteric Regulation

In each biochemical pathway there are one or more enzymes whose catalytic activity can be modulated (i.e., increased or decreased) by the binding of effector

FIGURE 6.20

The Activation of Chymotrypsinogen

The inactive zymogen chymotrypsinogen is activated in several steps. After its secretion into the small intestine, chymotrypsinogen is converted into π-chymotrypsin when trypsin, another proteolytic enzyme, cleaves the peptide bond between Arg 15 and Ile 16. Later chymotrypsin cleaves several other peptide bonds, and conformational changes cause the formation of α-chymotrypsin.

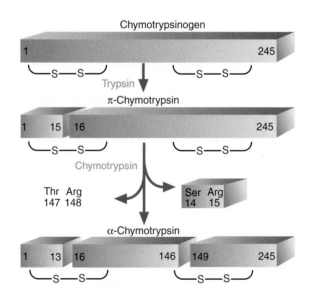

molecules. The binding of ligands to *allosteric sites* on such an enzyme triggers rapid conformational changes that can increase or decrease its rate of substrate binding. A plot of a reaction catalyzed by an allosteric enzyme differs from those of enzymes that observe Michaelis-Menten kinetics. Instead the rate curve is sigmoidal (Figure 6.21). If the ligand is the same as the substrate (i.e., if the binding of a substrate influences the binding of additional substrate), the allosteric effects are referred to as *homotropic. Heterotropic effects* involve modulating ligands, which are different from the substrate. Most allosteric enzymes are multisubunit enzymes with multiple substrate and effector binding sites.

Two theoretical models that attempt to explain the behavior of allosteric enzymes are the concerted model and the sequential model (Figure 6.22). In the *concerted* (or *symmetry*) model, it is assumed that the enzyme exists in only two states: T(aut) and R(elaxed). Activators bind to and stabilize the R conformation, whereas inhibitors bind to and stabilize the T conformation. The term *concerted* is applied to this model because the conformations of all the protein's subunits are believed to change simultaneously when the first effector binds. (This rapid concerted change in conformation maintains the protein's overall symmetry.) The binding of an activator shifts the equilibrium in favor of the R form. An inhibitor shifts the equilibrium toward the T conformation.

The concerted model explains some properties of allosteric enzymes, but not others. For example, the concerted model accounts for **positive cooperativity**, in which the first ligand increases subsequent ligand binding, but not **negative cooperativity**, a phenomenon observed in enzymes in which the binding of the first ligand reduces the affinity of the enzyme for similar ligands. In addition, the concerted model makes no allowances for hybrid conformations. Examples of enzymes whose behavior appears to be consistent with the concerted model are aspartate transcarbamoylase (ATCase) in *E. coli* and phosphofructokinase (PFK).

ATCase catalyzes the first step in a reaction pathway that leads to the synthesis of the pyrimidine nucleotide cytidine triphosphate (CTP) (Figure 6.23). CTP acts as an inhibitor of ATCase activity. It shifts the rate curve to the right, indicating an increase in the apparent K_m of ATCase that corresponds to a lower

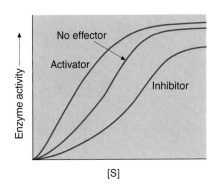

FIGURE 6.21

The Rate of an Enzyme-Catalyzed Reaction as a Function of Substrate Concentration

The activity of allosteric enzymes is affected by positive effectors (activators) and negative effectors (inhibitors).

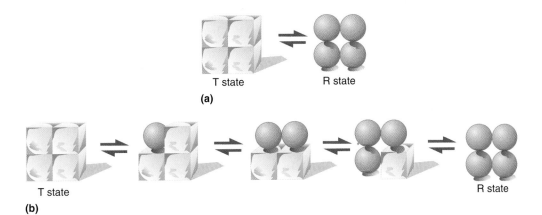

FIGURE 6.22

Allosteric Interaction Models

(a) In the concerted model, the enzyme exists in only two conformations. Substrates and activators have a greater affinity for the R state. Inhibitors favor the T state. (b) In the sequential model, one subunit assumes an R conformation as it binds to substrate. As the first subunit changes its conformation, the affinity of nearby subunits for ligand is affected.

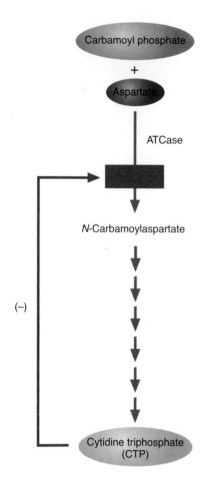

FIGURE 6.23

Feedback Inhibition

ATCase (aspartate transcarbamoylase) catalyzes the committed step in the synthesis of CTP. The binding of CTP, the product of the pathway, to ATCase inhibits the enzyme.

velocity of enzyme activity at a given concentration of substrate. CTP inhibition of ATCase is an example of negative feedback inhibition (p. 26), the process in which the product of a pathway inhibits the activity of an enzyme at or near the beginning of a biochemical pathway or at a branch point. The purine nucleotide ATP acts as an activator. ATP activation of ATCase makes sense because nucleic acid biosynthesis requires relatively equal amounts of purine and pyrimidine nucleotides. When ATP concentration is higher than that of CTP, ATCase is activated. When ATP concentration is lower than that of CTP, the net effect on ATCase is inhibitory.

Phosphofructokinase catalyzes an important reaction in glycolysis (p. 265–276): the transfer of a phosphate group from ATP to the OH group on C-1 of fructose-6-phosphate.

$$\text{Fructose-6-phosphate} + \text{ATP} \rightarrow \text{fructose-1,6-bisphosphate} + \text{ADP}$$

Phosphofructokinase is composed of four identical subunits, each of which has an active site and a number of allosteric sites. The R state of the enzyme is stabilized by ADP and AMP (a sensitive indicator of ATP depletion and the cell's need for energy). The T-state is stabilized by molecules such as ATP and other molecules that are indicators that cellular energy is high.

In the sequential model, first proposed by Daniel Koshland and his associates, the binding of a ligand to one flexible subunit in a multisubunit protein triggers a conformational change that is sequentially transmitted to adjacent subunits. The more sophisticated sequential model allows for the intermediate conformations that may be more realistic representations of the operation of some enzymes. The model can also accommodate negative cooperativity (ligand binding to one subunit induces conformational changes in adjacent subunits that make ligand binding less likely). Consequently, the sequential model can be considered to be a general model that accounts for all allosteric possibilities. In this view the concerted model is a simple example of the sequential model.

There are two important features of allosteric regulation that should be noted. First, the concerted and sequential models are theoretical models; that is, the behavior of many allosteric proteins appears to be more complex than can be accounted for by either model. For example, the cooperative binding of O_2 by hemoglobin (the most thoroughly researched allosteric protein) appears to exhibit features of both models. The binding of the first O_2 initiates a concerted $T \rightarrow R$ transition that involves small changes in the conformation of each subunit (a feature of the sequential model; see p. 217). In addition, hemoglobin species with only one or two bound O_2 have been observed. A second and more important issue is that there are no simple rules that explain metabolic regulation. For example, attempts to increase the metabolic flux of pathways in organisms such as yeast by engineering increased synthesis of allosteric enzymes failed. (*Flux* is the rate of turnover of molecules through a pathway.) Flux increases were observed only when all of the enzymes in the pathway were increased. It appears that pathway regulation is the result of regulatory contributions to a greater or lesser degree by all or most of the enzymes.

Compartmentation

Compartmentation, created by cellular infrastructure, is a significant means for regulating biochemical reactions because physical separation makes separate control possible. The intricate internal architecture of cells contains compartments (e.g., eukaryotic organelles) and microcompartments of diverse types (e.g., individual enzymes or multiprotein complexes attached to membranes or cytoskeletal filaments). Cellular compartmentation solves several interrelated problems:

1. **Divide and control**. The physical separation of competing reactions (i.e., those that would undo the other, such as kinases and phosphatases) allows for coordinated regulation that prevents the wasteful dissipation of resources.

2. **Diffusion barriers**. Within crowded cells, diffusion of substrate molecules is a potentially limiting factor in reaction rates. Cells circumvent this problem by means of creating microenvironments in which enzymes and their substrates are concentrated, as well as by metabolite channeling, the transfer of product molecules from one enzyme to the next in a multienzyme complex.

3. **Specialized reaction conditions**. Certain reactions require an environment with unique properties. For example, the low pH within lysosomes facilitates hydrolytic reactions.

4. **Damage control**. The segregation of potentially toxic reaction products protects other cellular components.

   Overall metabolic control requires the integration of all the cell's biochemical pathways, which is accomplished, in part, by transport mechanisms that transfer metabolites and signal molecules between compartments.

KEY CONCEPTS

- All biochemical pathways are regulated to maintain the ordered state of living cells.
- Regulation is accomplished by genetic control, covalent modification of enzymes, allosteric regulation, and cell compartmentation.

QUESTION 6.10

Certain enzymes are involved in the metabolism of the medicinal chemicals that alter or enhance physiological processes. For example, aspirin suppresses pain, and antibiotics kill infectious organisms. Once such a drug has been consumed, it is absorbed and distributed to the tissues, where it performs its function. Eventually, drug molecules are processed (primarily in the liver) and excreted. The dosage of each drug that physicians prescribe is based on the amount required to achieve a therapeutic effect and the drug's average rate of excretion from the body. Various reactions prepare drug molecules for excretion. Examples include oxidation, reduction, and conjugation reactions. (In conjugation reactions, small polar or ionizable groups are attached to a drug molecule to improve its solubility.) The amount of certain enzymes directly affects a patient's ability to metabolize a specific drug. For example, isoniazid is used to treat tuberculosis, a highly infectious, chronic, debilitating disease caused by *Mycobacterium tuberculosis*. This drug is metabolized by N-acetylation, in which an amide bond forms between the substrate and an acetyl group. The rate at which isoniazid is acetylated determines its clinical effectiveness.

Two tuberculosis patients with similar body weights and symptoms are given the same dose of isoniazid. Although both take the drug as prescribed, one patient fails to show a significant clinical improvement. The other patient is cured. Genetic factors appear to be responsible for the differences in drug metabolism. Can you suggest a reason why the two patients reacted so differently to isoniazid? How can physicians improve the percentage of patients who are cured?

BIOCHEMISTRY IN PERSPECTIVE

Enzymes and Clinical Medicine

How are enzymes used to promote human health? Enzymes are useful in modern medical practice for several reasons. Enzyme assays provide important information concerning the presence and severity of disease. In addition, enzymes often provide a means of monitoring a patient's response to therapy.

In the following discussion, the use of enzymes in the diagnosis of myocardial infarction and as therapeutic agents is described.

Diagnostic Enzymes

Myocardial infarction, the interruption of the heart's blood supply, leads to the death of cardiac muscle cells. The symptoms of myocardial infarction include pain in the left side of the chest that may radiate to the neck, left shoulder and arm, and irregular breathing. Several clinical tests for myocardial infarction are based on detection of enzymes and other proteins that leak from damaged tissue into the blood. Among the first enzymes used to detect heart damage was lactate dehydrogenase (LDH). Later, creatine kinase (CK) assays were developed. Both LDH and CK are produced by various cells in slightly different forms called **isozymes**. Isozymes can be distinguished from each other because they migrate differently during electrophoresis. LDH is a tetramer composed of two subunits: heart type (H) and muscle type (M). There are five different LDH isozymes. LDH 1 (H_4) and LDH 2 (H_3M) are found only in heart muscle and red blood cells. LDH 5 (M_4) occurs in both liver and skeletal muscle. LDH_3 and LDH_4 are found in other organs. The differences in the migration patterns of LDH isozymes of a normal individual and a myocardial infarction patient are illustrated in Figure 6A. CK, which occurs as a dimer, has two types of subunit, muscle type (M) and brain type (B). Heart muscle contains CK_2 (MB) and CK_3 (MM). Only the MB isozyme is found exclusively in heart muscle. Its concentration in blood reaches a maximum within a day following the infarction. CK_3 (MM), which is also found in other tissues, peaks a day after CK_2 (Figure 6B). Of the two enzymes, CK is the first enzyme detected during a myocardial infarction. CK's serum concentration rises and falls so rapidly that it is of little clinical use after several days. LDH, whose serum concentration rises later, is used to monitor the later stages of heart damage.

In recent years the strategy of measuring LDH has been replaced with tests for the cardiac variant of troponin I (cTnI). (The troponin complex, composed of three different subunits, regulates striated muscle contraction.) Troponin I is highly specific for myocardial injury and can be detected earlier than CK-MB and its serum levels remain elevated longer than those of LDH.

Therapeutic Enzymes

The use of enzymes in medical therapy has been limited. When administered to patients, enzymes are often rapidly inactivated or degraded. The large amounts of enzyme that are often required to sustain a therapy may provoke allergic reactions. However, there have been several successful enzyme therapies, including the use of streptokinase and asparaginase. Streptokinase, a noncatalytic protein produced by *Streptococcus pyogenes*, is

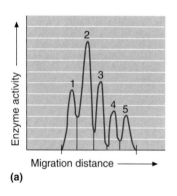
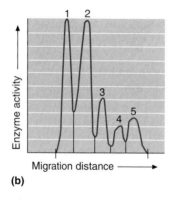

(a)   **(b)**

FIGURE 6A

Electrophoresis Pattern of Lactate Dehydrogenase Isozymes

(a) LDH isozymes from a normal individual. (b) LDH isozymes from a myocardial infarction patient. In recent years the use of LDH in the diagnosis of myocardial infarction has been replaced with test for cTnI.

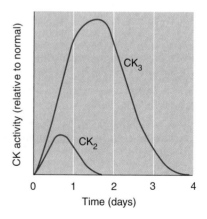

FIGURE 6B

Characteristic Pattern of Serum Creatine Kinase Concentrations Following a Myocardial Infarction

BIOCHEMISTRY IN PERSPECTIVE cont

used with great success in the treatment of myocardial infarction. Once injected into the bloodstream, streptokinase activates plasminogen, the inactive precursor of the proteolytic enzyme plasmin. Plasmin (a trypsinlike enzyme) digests fibrin, the primary component of blood clots. If administered soon after the beginning of a heart attack streptokinase can often prevent or significantly reduce further damage to the heart because it can dissolve blood clots that occlude the coronary arteries. Human tissue plasminogen activator (tPA), a product of recombinant DNA technology (Biochemistry in the Lab Nucleic Acid Methods I) that acts in a similar fashion, is also used to treat myocardial infarction.

The enzyme asparaginase is used to treat several types of cancer. It catalyzes the following reaction:

L-Asparagine + H_2O → L-aspartate + NH_3

Asparaginase occurs in plants, vertebrates, and bacteria but not in human blood. Unlike most normal cells, the cells in tumors of certain kinds such as those in several forms of adult leukemia,

cannot synthesize asparagine. The infusion of asparaginase reduces the blood's concentration of asparagine and often causes tumor regression. Asparaginase therapy has several serious side effects, including allergic reactions and liver damage. Unfortunately, after several asparaginase treatments some patients develop *resistance*; that is, their tumor cells grow despite the presence of the drug that previously had caused tumor regression. Apparently, some tumor cells can induce the synthesis of asparagine synthetase, an enzyme that converts aspartate to asparagine.

In some diseases the body's capacity to produce certain digestive enzymes is compromised. In cystic fibrosis, for example, the absence of a functional chloride channel results in malabsorption of nutrients in the digestive tract. Inadequate chloride transport in the pancreas results in the failure to synthesize enzymes such as chymotrypsin and trypsin that digest food molecules. For this reason CF patients must consume digestive enzymes prepared by pharmaceutical companies during every meal.

 **SUMMARY:** Specific enzymes are used in the diagnosis or treatment of a variety of human diseases.

Chapter**Summary**

1. Enzymes are biological catalysts. They enhance reaction rates because they provide an alternative reaction pathway that requires less energy than an uncatalyzed reaction. In contrast to some inorganic catalysts, most enzymes catalyze reactions at mild temperatures. In addition, enzymes are specific to the types of reaction they catalyze. Each type of enzyme has a unique, intricately shaped binding surface called an active site. Substrate binds to the enzyme's active site, which is a small cleft or crevice in a large protein molecule. In the lock-and-key model of enzyme action, the structures of the enzyme's active site and the substrate transition state are complementary. In the induced-fit model, the protein molecule is assumed to be flexible.

2. Enzymes are classified and named according to the type of reaction each one catalyzes. There are six major enzyme categories: oxidoreductases, transferases, hydrolases, lyases, isomerases, and ligases.

3. Enzyme kinetics is the quantitative study of enzyme catalysis. According to the Michaelis-Menten model, when the substrate S binds in the active site of an enzyme E, an ES transition state complex is formed. During the transition state, the substrate is converted into product. After a time the product dissociates from the enzyme. The symbol V_{max}

represents the maximal velocity for the reaction, and K_m is a rate constant called the Michaelis constant. Experimental determinations of K_m and V_{max} are made with Lineweaver-Burk double-reciprocal plots.

4. The turnover number (k_{cat}) is a measure of the number of substrate molecules converted to product per unit time by an enzyme when it is saturated with substrate. Because [S] is relatively low under physiological conditions ([S] << K_m), the specificity constant k_{cat}/K_m is a more reliable gauge of the catalytic efficiency of enzymes.

5. Enzyme inhibition may be reversible or irreversible. Irreversible inhibitors usually bind covalently to enzymes. In reversible inhibition, the inhibitor can dissociate from the enzyme. The most common types of reversible inhibition are competitive, noncompetitive, and uncompetitive. The assumptions of the law of mass action upon which *in vitro* enzyme analysis are based do not appear to hold true in the crowded and heterogeneous interior space of living organisms.

6. The kinetic properties of allosteric enzymes are not explained by the Michaelis-Menten model. Most allosteric enzymes are multisubunit proteins. The binding of substrate or effector to one subunit affects the binding properties of other subunits.

7. Enzymes use the same catalytic mechanisms as nonenzymatic catalysts. Several factors contribute to enzyme catalysis: proximity and strain effects, electrostatic effects, acid-base catalysis, covalent catalysis, and quantum tunneling. Combinations of these factors affect enzyme mechanisms.

8. Active site amino acid side chains can facilitate proton transfer and nucleophilic substitutions as well as stabilize the transition state. Many enzymes use nonprotein cofactors (metals and coenzymes) to facilitate reactions.

9. Enzymes are sensitive to environmental factors such as temperature and pH. Each enzyme has an optimum temperature and an optimum pH.

10. The chemical reactions in living cells are organized into a series of biochemical pathways. The pathways are controlled primarily by adjusting the concentrations and activities of enzymes through genetic control, covalent modification, allosteric regulation, and compartmentation.

 Take your learning further by visiting the **companion website** for Biochemistry at **www.oup.com/us/mckee** where you can complete a multiple-choice quiz on enzymes to help you prepare for exams.

Suggested Readings

Agarwal, P. K., Enzymes: An Integrated View of Structural Dynamics and Function, *Microbial Cell Factories* 5:2–14, 2006 [http:www.microbialcellfactories.com/contents/5/1/2].

Fersht, A., *Structure and Mechanism in Protein Science: A Guide to Enzyme Catalysis and Protein Folding*, W. H. Freeman, New York, 1999.

Garcia-Viloca, M., Gao, J., Karplus, M., and Truhlar, D. G., How Enzymes Work: Analysis by Modern Rate Theory and Computer Simulations, *Science* 303:186–195, 2004.

Gutteridge, A., and Thornton, J. M., Understanding Nature's Toolkit, *Trends Biochem. Sci.* 30(11):622–629, 2005.

Kraut, J., How Do Enzymes Work? *Science* 242:533–540, 1998.

Miller, J. A., Women in Chemistry, in Kass-Simon, G., and Fannes, P. (eds.), *Women in Science: Righting the Record*, pp. 300–334, Bloomington, Indiana University Press, 1990.

Palson, B., The Challenges of In Silico Biology, *Nature Biotechnol.* 18:1147–1150, 2000.

Schnell, S., and Turner, T. E., Reaction Kinetics in Intracellular Environments with Macromolecular Crowding: Simulations and Rate Laws, *Prog. Biophys. Mol. Biol.* 85:234–260, 2004.

Storey, K. B. (ed.), *Functional Metabolism: Regulation and Adaptation*, Wiley-Liss, Hoboken, New Jersey, 2004.

Sutcliffe, M. J., and Scrutton, N. S., Enzyme Catalysis: Over the Barrier or Through the Barrier? *Trends Biochem. Sci.* 25(9):405–408, 2000.

Teusink, B. et al., Can Yeast Glycolysis Be Understood in Terms of *In Vitro* Kinetics of the Constituent Enzymes? Testing Biochemistry, *Eur. J. Biochem.* 267:5313–5329, 2000.

Wolfenden, R., and Snider, M. J., The Depth of Chemical Time and the Power of Enzymes as Catalysts, *Acc. Chem. Res.* 34(12):938–945, 2001.

Key Words

activation energy, *187*
active site, *187*
activity coefficient, *186*
allosteric enzyme, *203*
apoenzyme, *189*
carbanion, *205*
carbocation, *205*
catalyst, *187*
coenzyme, *189*
cofactor, *189*
competitive inhibition, *199*

enzyme induction, *217*
enzyme kinetics, *192*
free radical, *205*
holoenzyme, *189*
hydrolase, *190*
inhibitor, *199*
intermediate, *205*
irreversible inhibition, *199*
isomerase, *190*
isozyme, *222*
ligase, *191*

lyase, *190*
macromolecular crowding, *204*
negative cooperativity, *219*
noncompetitive inhibition, *199*
oxidoreductase, *190*
oxyanion, *208*
pH optimum, *214*
positive cooperativity, *219*
proenzyme, *218*

reaction mechanism, *205*
reversible inhibition, *199*
specificity constant, *194*
substrate, *187*
transferase, *190*
transition state, *187*
turnover number, *195*
uncompetitive inhibition, *199*
velocity, *192*
vitamin, *212*
zymogen, *218*

Review Questions

These questions are designed to test your knowledge of the key concepts discussed in this chapter, before moving on to the next chapter. You may like to compare your answers to the solutions provided in the back of the book and in the accompanying Study Guide.

1. Clearly define the following terms:
   a. activation energy
   b. catalyst
   c. active site
   d. coenzyme
   e. velocity of a chemical reaction

2. What are four important properties of enzymes?

3. Living things must regulate the rate of catalytic processes. Explain how the cell regulates enzymatic reactions.

4. Determine the class of enzyme that is most likely to catalyze each of the following reactions.

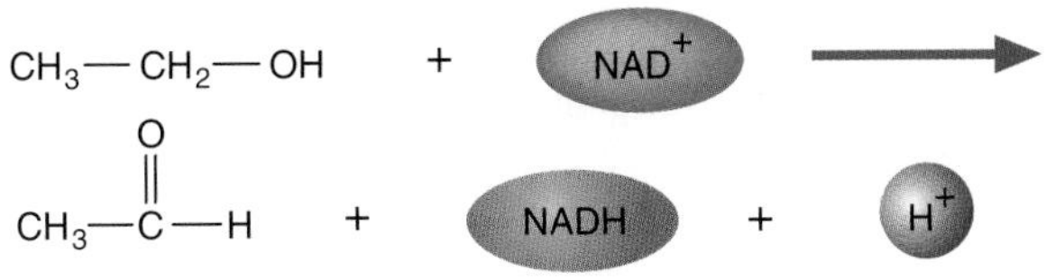

CH_3—CH_2—OH + NAD$^+$ →

CH_3—$\overset{\overset{O}{\|}}{C}$—H + NADH + H$^+$

(a)

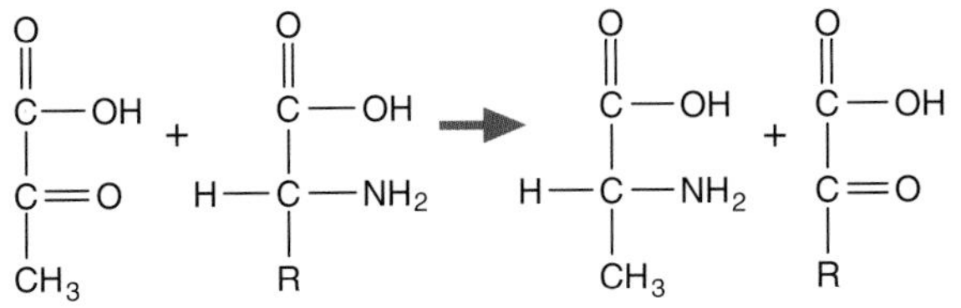

$\overset{\overset{O}{\|}}{C}$—OH + $\overset{\overset{O}{\|}}{C}$—OH → $\overset{\overset{O}{\|}}{C}$—OH + $\overset{\overset{O}{\|}}{C}$—OH

C=O + H—C—NH$_2$ → H—C—NH$_2$ + C=O

CH$_3$ + R → CH$_3$ + R

(b)

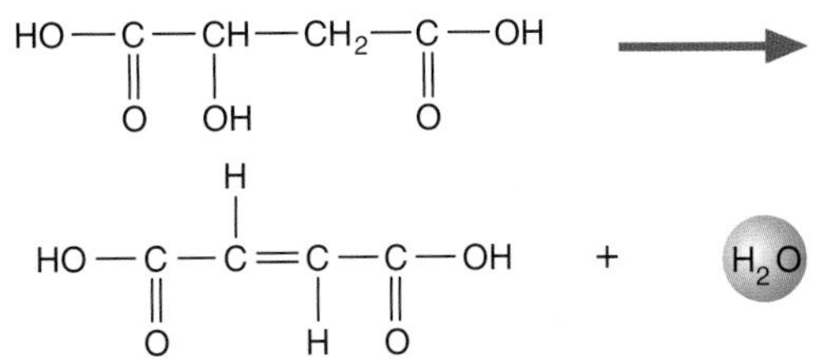

HO—C—CH—CH$_2$—C—OH →

with O, OH, O substituents

HO—C—C=C—C—OH + H$_2$O

with O, H, O substituents

(c)

$\overset{\overset{O}{\|}}{C}$—H → $\overset{\overset{O}{\|}}{C}$—H

HO—C—H → H—C—H

HO—C—H → HO—C—H

H—C—OH → H—C—OH

H—C—OH → H—C—OH

CH$_2$OH → CH$_2$OH

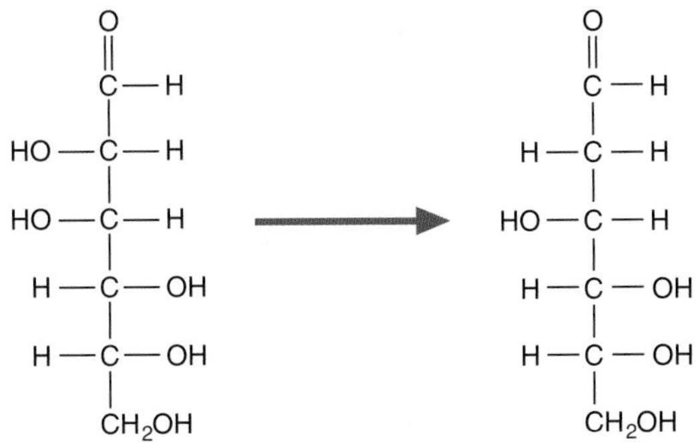

(d)

CO$_2$ + CH$_3$—C—C—OH + ATP →

with O, O substituents

HO—C—CH$_2$—C—C—OH + ADP + P$_i$

with O, O, O substituents

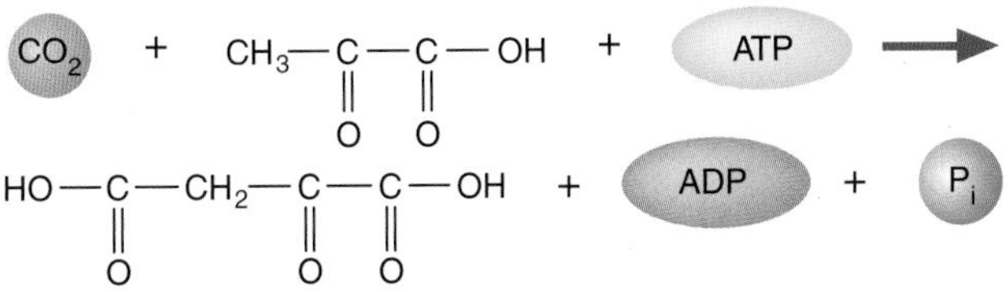

(e)

5. Define the following terms:
   a. reaction mechanism
   b. carbanion
   c. free radical
   d. quantum tunneling
   e. reaction order

6. Several factors contribute to enzyme catalysis. What are they? Briefly explain the effect of each.

7. Why is the regulation of biochemical processes important? List three reasons.

8. Describe negative feedback inhibition.

9. Define the following terms:
   a. oxidoreductase
   b. lyase
   c. ligase
   d. transferase
   e. hydrolase
   f. isomerase

10. Describe the two models that explain the binding of allosteric enzymes. Use either model to explain the binding of oxygen to hemoglobin.

11. Define the following terms:
    a. vitamin
    b. cofactor
    c. transition state
    d. inhibitor
    e. allosteric enzyme

12. What are the major coenzymes? Briefly describe the function of each.

13. What properties of transition metals make them useful as enzyme cofactors?

14. Define the following terms:
    a. cooperativity
    b. oxyanion
    c. apoenzyme
    d. holoenzyme
    e. isozyme

15. The change in enthalpy ΔH for the following reaction is −28.2 kJ/mol:

$$C_6H_{12}O_6 + 6O_2 \rightarrow 6CO_2 + 6H_2O$$

Explain why glucose ($C_6H_{12}O_6$) is stable in an oxygen atmosphere for appreciable periods of time.

16. Enzymes act by reducing the activation energy of a reaction. Describe several ways in which this is accomplished.

17. Define the following terms:
    a. half-life
    b. repression
    c. turnover number
    d. katal
    e. noncompetitive inhibitor

18. In enzyme kinetics, why are measurements made at the start of a reaction?

19. Histidine is frequently used as a general acid or general base in enzyme catalysis. Consider the pK_a values of the side groups of the amino acids listed in Table 5.2 to suggest a reason why this is so.

20. Enzymes are stereochemically specific; that is, they often convert only one stereoisomeric form of substrate into product. Why is such specificity inherent in their structure?

21. The uncatalyzed reaction rate for the conversion of substrate X to product Y is one year. The enzyme-catalyzed rate is one millisecond. Describe the features of the

enzyme that are probably responsible for this rate difference.

22. Describe the relationship between the law of mass action, solute concentration, and effective concentration.

23. Define the following terms.
    a. oxygenase
    b. epimerase
    c. protease
    d. hydroxylase
    e. oxidase

24. Define the following terms:
    a. metabolon
    b. *in vivo*
    c. *in vitro*
    d. in silico
    e. metabolic flux

25. Draw the structures of the amino acids whose side chains most commonly participate in catalytic mechanisms when they occur in the active sites of enzymes.

26. Use specific examples to describe the roles played in enzyme catalysis by the amino acids side chains from Question 25.

27. Describe compartmentation within eukaryotic cells. What problems does this phenomenon solve for living organisms?

28. Explain the role that compartmentation plays in the regulation of metabolic pathways. Provide several examples.

29. What are the two types of enzyme inhibitors? Give an example of each.

30. Review the structure of the standard amino acids and list those that are capable of acting as acids or bases in enzyme catalysis. Are there any that can function as either acids or bases?

31. In any given enzyme, the active site is only a small portion of the entire molecule. Synthesis of such a relatively large molecular machine requires an enormous amount of cellular energy. Explain why this inefficiency is tolerated.

ThoughtQuestions

These questions are designed to reinforce your understanding of all of the key concepts discussed in the book so far, including this chapter and all of the chapters before it. They may not have one right answer! The authors have provided possible solutions to these questions in the back of the book and in the accompanying Study Guide, for your reference.

32. Consider the following reaction:

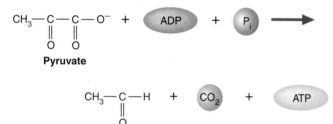

Pyruvate

Using the following data, determine the order of the reaction for each substrate and the overall order of the reaction.

	Concentration (mol/L)			Rate
Experiment	Pyruvate	ADP	P_1	$(mol\ L^{-1}\ s^{-1})$
1	0.1	0.1	0.1	8×10^{-4}
2	0.2	0.1	0.1	1.6×10^{-3}
3	0.2	0.2	0.1	3.2×10^{-3}
4	0.1	0.1	0.2	3.2×10^{-3}

33. Consider the following data for an enzyme-catalyzed hydrolysis reaction in the presence and absence of inhibitor I:

[Substrate] (M)	v_0 (μmol/min)	v_{0I} (μmol/min)
6×10^{-6}	20.8	4.2
1×10^{-5}	29	5.8
2×10^{-5}	45	9
6×10^{-5}	67.6	13.6
1.8×10^{-4}	87	16.2

Using a Michaelis-Menten plot, determine K_m for the uninhibited reaction and the inhibited reaction.

34. Use the data in Question 2 to do the following.
    a. Generate Lineweaver-Burk plots for the data.
    b. Explain the significance of the horizontal intercept, the vertical intercept, and the slope.
    c. Identify the type of inhibition being measured.

35. Two experiments were performed with the enzyme ribonuclease. In experiment 1 the effect of increasing substrate concentration on reaction velocity was measured. In experiment 2 the reaction mixtures were identical to those in experiment 1 except that 0.1 mg of an unknown compound was added to each tube. Plot the data according to the Lineweaver-Burk method. Determine the effect of the unknown compound on the enzyme's activity. (Substrate concentration is measured in millimoles per liter. Velocity is measured in the change in optical density per hour.)

Experiment 1		Experiment 2	
[S]	v	[S]	v
0.5	0.81	0.5	0.42
0.67	0.95	0.67	0.53
1	1.25	1	0.71
2	1.61	2	1.08

36. Describe the mechanism for chymotrypsin. Show how the amino acid residues of the active site participate in the reaction.

37. Alcohol dehydrogenase (ADH) is inhibited by numerous alcohols. Using the data given in the following table,

calculate the k_{cat}/K_m values for each of the alcohols. Which of the listed alcohols is most easily metabolized by alcohol dehydrogenase?

Kinetic parameters for hamster testes ADH.

Substrate	K_m (μM)	k_{cat}(min^{-1})
Ethanol	960	480
1-Butanol	440	450
1-Hexanol	69	182
12-Hydroxydodecanoate	50	146
All-*trans*-retinol	20	78
Benzyl alcohol	410	82
2-Butanol	250,000	285
Cyclohexanol	31,000	122

38. 4-Methyl pyrazole (4-MP) has been developed as a long-acting and less toxic alternative to ethanol in the treatment of ethylene glycol poisoning. Shown is a Lineweaver-Burk plot of the inhibition of alcohol dehydrogenase by various concentrations of 4-methyl pyrazole. What type of inhibition appears to be exhibited by this molecule?

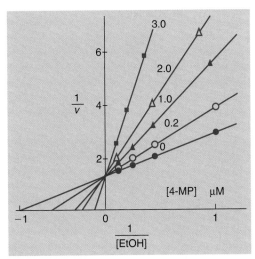

39. The table presents the rates of reaction at specific substrate concentrations for an enzyme that displays classical Michaelis-Menten kinetics. Two sets of inhibitor data are also included. Determine the K_m and V_{max} for the uninhibited enzyme.

[S] (mM)	Without inhibitor	v_o (μM/s) With inhibitor A	With inhibitor B
1.3	2.50	1.17	0.62
2.6	4.00	2.10	1.42
6.5	6.30	4.00	2.65
13.0	7.60	5.70	3.12
26.0	9.00	7.20	3.58

40. Determine the type of inhibition exhibited by each inhibitor in Question 39. What kind of affinity would these inhibitors have for ES vs E?

41. Catalase has a K_m of 25 mM and a k_{cat} of 4.0×10^7 s^{-1} with H_2O_2 as a substrate. Carbonic anhydrase has a K_m of 26 mM and a k_{cat} of 4.0×10^5 s^{-1}. What do these data tell you about these two enzymes?

42. Fumarase is an enzyme of the citric acid cycle that catalyzes the following reaction.

Fumarate

Malate

To what class of enzyme does fumarase belong?

43. The K_m, and k_{cat} for fumarase with fumarate as a substrate are 5×10^{-6} M and 8×10^2 s^{-1}, respectively. When malate is the substrate the K_m and k_{cat} are 2.5×10^{-5} M and 9×10^2 s^{-1}, respectively. What do the data tell you about the operation of this enzyme in the citric acid cycle?

44. What possible effect would macromolecular crowding have on activity coefficients? Explain how macromolecular crowding can increase the effective concentration of a substance.

45. Consider the following reaction, and determine the class of enzymes that would catalyze it.

46. Consider the following reaction along with its rate information.

A + B → C

[A] (mM)	[B] (mM)	Rate (mM/s)
0.05	0.05	2×10^7
0.10	0.05	4×10^7
0.05	0.1	4×10^7
0.1	0.1	8×10^7

What is the overall rate expression for this reaction? What is the order of the reaction?

47. Deuterium ($_1^2$H or D) is a naturally occurring but rare isotope of hydrogen ($_1^1$H) that has a neutron in its nucleus. Deuterium's atomic structure results in several differences in its chemical and physical properties in comparison to hydrogen. For example, the normal boiling points of H_2 and D_2 are −252.8°C (20.4 K) and −250.7°C (22.5 K), respectively. Compared to the pH of pure H_2O of 7.00, the pD for D_2O is 7.35. Considering the quantum tunneling properties of hydrogen within the active site of DHFR, will NADPH react faster or slower than NADD (nicotine adenine dinucleotide deuteride). Explain your answer.

48. In the serine protease triad, the proximity of an aspartate carboxylate group to the imidazole group of histidine raises the latter's pK_a. Explain.

49. Review the electronic configuration of the magnesium ion and explain why Mg^{2+} can function as a Lewis acid.

50. Provide the reaction between the residue Ser 195 of chymotrypsin and diisopropylfluorophosphate. Why is this particular serine so reactive?

51. Ethylene glycol ($HO—CH_2—CH_2—OH$) is frequently used as antifreeze in automobile engines. Every year children and pets are poisoned because they tasted this sweet-tasting material. Ethylene glycol is metabolized in the liver by alcohol dehydrogenase. Suggest a possible medical treatment for ethylene glycol intoxication.

52. Given the following rate expression, complete the following table.

$$Rate = k[A]^2[B]$$

[A] (mM)	[B] (mM)	Rate (mM/s)
0.1	0.01	1×10^6
0.1	0.02	_____
0.2	0.01	_____
0.2	0.02	_____

53. Consider the following graph and illustrate how the K_m value would be determined. Why is the Lineweaver-Burk plot method used instead?

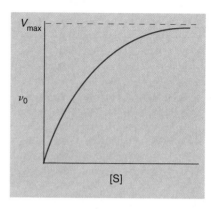

54. Write out a mechanism for the serine triad hydrolysis of glycylglycine.

55. When free water molecules are excluded from an enzyme's catalytic pocket, is a catalytic —OH group a stronger or weaker nucleophile? Explain.

Carbohydrates

Sugars and Cells Sugar molecules shape the molecular landscape of living organisms. Carbohydrates linked to membrane proteins and lipids are especially prominent on the external surface of cells.

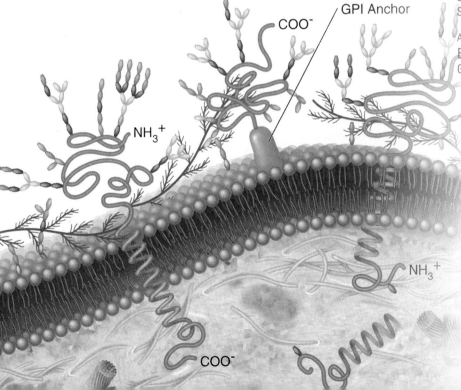

COO⁻

GPI Anchor

NH₃⁺

NH₃⁺

COO⁻

Overview

CARBOHYDRATES ARE NOT JUST AN IMPORTANT SOURCE OF RAPID ENERGY PRODUCTION FOR LIVING CELLS. THEY ARE ALSO STRUCTURAL BUILDING blocks of cells and components of numerous metabolic pathways. Sugar polymers linked to proteins and lipids are now recognized as a high-density coding system. Their vast structural diversity is exploited by living organisms to produce the informational capacity required for living processes. Chapter 7 describes the structures and chemistry of typical carbohydrate molecules found in living organisms, and provides an introduction to glycomics, the investigation of the sugar code.

Carbohydrates, the most abundant biomolecules in nature, are a direct link between solar energy and the chemical bond energy of living organisms. (More than half of all "organic" carbon is found in carbohydrates.) They are formed during *photosynthesis* (Chapter 13), a biochemical process in which light energy is captured and used to drive the biosynthesis of energy-rich organic molecules from the energy-poor inorganic molecules CO_2 and H_2O. Most carbohydrates contain carbon, hydrogen, and oxygen in the ratio $(CH_2O)_n$, hence the name "hydrate of carbon." They have been adapted for a wide variety of biological functions, which include energy sources (e.g., glucose), structural elements (e.g., cellulose and chitin in plants and insects, respectively), cellular communication and identity, and precursors in the production of other biomolecules (e.g., amino acids, lipids, purines, and pyrimidines). Carbohydrates are classified as monosaccharides, disaccharides, oligosaccharides, and polysaccharides according to the number of simple sugar units they contain. Carbohydrate moieties also occur as components of other biomolecules. A vast array of *glycoconjugates* (protein and lipid molecules with covalently linked carbohydrate groups) are distributed among all living species, most notably among the eukaryotes. Certain carbohydrate molecules (the sugars ribose and deoxyribose) are structural elements of nucleotides and nucleic acids.

In recent years it has become increasingly apparent that carbohydrate provides living organisms with enormous informational capacities. Investigations of biological processes such as signal transduction, cell-cell interactions, and endocytosis have revealed that they typically involve the binding of glycoconjugates such as glycoproteins and glycolipids or free carbohydrate molecules with complementary receptors. Chapter 7 provides a foundation for understanding the complex processes in living organisms by reviewing the structure and function of the most common carbohydrates and glycoconjugates. The chapter ends with a discussion of the *sugar code*, the mechanism by which carbohydrate structure is used to encode biological information.

7.1 MONOSACCHARIDES

Monosaccharides, or simple sugars, are polyhydroxy aldehydes or ketones. Recall from Chapter 1 that monosaccharides with an aldehyde functional group are called **aldoses**, whereas those with a ketone group are called *ketoses* (Figure 7.1). The simplest aldose and ketose are glyceraldehyde and dihydroxyacetone, respectively (Figure 7.2). Sugars are also classified according to the number of carbon atoms they contain. For example, the smallest sugars, called *trioses*, contain three carbon atoms. Four-, five-, and six-carbon sugars are called *tetroses,*

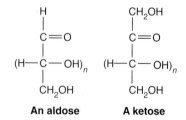

FIGURE 7.1

General Formulas for the Aldose and Ketose Forms of Monosaccharide

pentoses, and *hexoses*, respectively. The most abundant monosaccharides found in living cells are the pentoses and hexoses. Often, class names such as aldohexoses and ketopentoses, which combine information about carbon number and functional groups, describe monosaccharides. For example, glucose, a six-carbon, aldehyde-containing sugar, is referred to as an aldohexose.

The sugar structures shown in Figures 7.1 and 7.2 are known as Fischer projections (in honor of the great Nobel Prize–winning German chemist Emil Fischer). In these structures the carbohydrate backbone is drawn vertically with the most highly oxidized carbon usually shown at the top. The horizontal lines are understood to project toward the viewer; the vertical lines recede from the viewer.

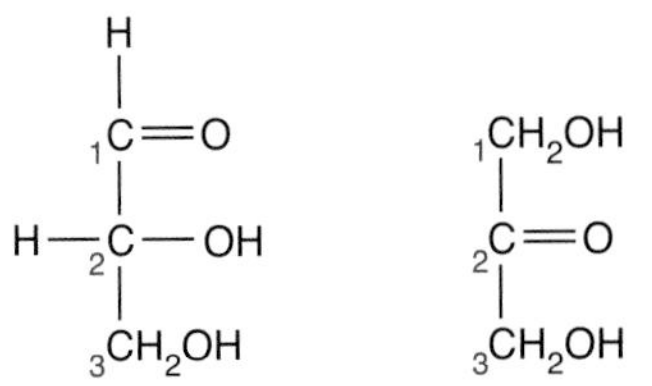

Glyceraldehyde   **Dihydroxyacetone**

FIGURE 7.2

Glyceraldehyde (an Aldotriose) and Dihydroxyacetone (a Ketotriose)

QUESTION 7.1

Identify the class of each of the following sugars. For example, glucose is an aldohexose.

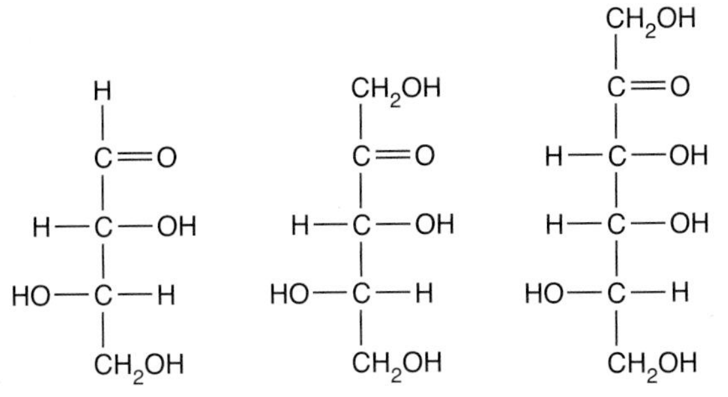

(a)   (b)   (c)

Monosaccharide Stereoisomers

When the number of chiral carbon atoms increases in optically active compounds, the number of possible optical isomers also increases. The total number of possible isomers can be determined by using van't Hoff's rule: A compound with n chiral carbon atoms has a maximum of 2^n possible stereoisomers. For example, when n is 4, there are 2^4 or 16 stereoisomers (8 D-stereoisomers and 8 L-stereoisomers).

In optical isomers the reference carbon is the asymmetric carbon that is most remote from the carbonyl carbon. Its configuration is similar to that of the asymmetric carbon in either D- or L-glyceraldehyde. Almost all naturally occurring sugars have the D-configuration. They can be considered to be derived from either the triose D-glyceraldehyde (the aldoses) or the nonchiral triose dihydroxyacetone (the ketoses). (Note that although dihydroxyacetone does not have an asymmetric carbon, it clearly is the parent compound for the ketoses.) In the D-aldose family of sugars (Figure 7.3), which contains most biologically important monosaccharides, the hydroxyl group is to the right on the chiral carbon atom in the Fischer model farthest from the most oxidized carbon (in this case the aldehyde group) in the molecule (e.g., carbon 5 in a six-carbon sugar).

Stereoisomers that are not enantiomers (mirror-image isomers) are called **diastereomers**. For example, the aldopentoses D-ribose and L-ribose are enantiomers, as are D-arabinose and L-arabinose (Figure 7.4). The sugars D-ribose and D-arabinose, which are isomers but not mirror images, are diastereomers. Diastereomers that differ in the configuration at a single asymmetric carbon atom are called **epimers**. For example, D-glucose and D-galactose are epimers because their structures differ only in the configuration of the OH group at carbon 4 (Figure 7.3). D-Mannose and D-galactose are not epimers because their configurations differ at more than one carbon.

FIGURE 7.3

The D Family of Aldoses

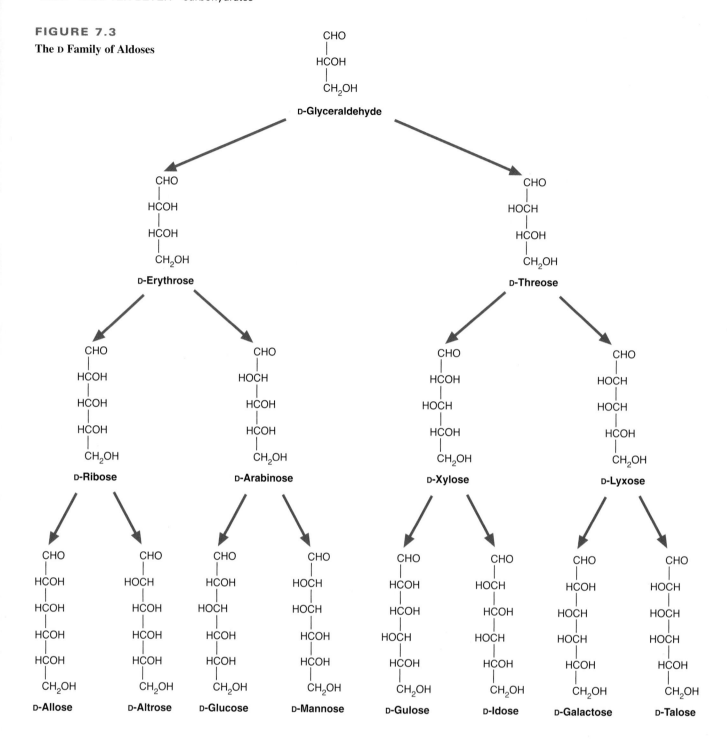

QUESTION 7.2

When viewed in two dimensions (as on the printed page), the structural differences between optical isomers may appear trivial. However, many biomolecules are optically active, and the capacity of enzymes to distinguish between D- and L-substrate molecules is an important feature of the chemistry of living cells. For example, most of the enzymes that break down and use dietary carbohydrates can bind to D-sugars but not to their L-isomers. Can you convince yourself that the D- and L-isomers of an optically active molecule are indeed different in three-dimensional space? Make models of D- and L-glyceraldehyde with an organic chemistry model kit or with colored styrofoam balls and toothpicks.

Cyclic Structure of Monosaccharides

Sugars that contain four or more carbons exist primarily in cyclic forms. Ring formation occurs in aqueous solution because aldehyde and ketone groups react reversibly with hydroxyl groups present in the sugar to form cyclic **hemiacetals** and **hemiketals**, respectively. Ordinary hemiacetals and hemiketals, which form when molecules containing an aldehyde or ketone functional group react with an alcohol, are unstable and easily revert to the aldehyde or ketone forms (Figure 7.5). When the aldehyde or ketone group and the alcohol functional group are part of the same molecule, however, an intramolecular cyclization reaction occurs that can form stable products. The most stable cyclic hemiacetal and hemiketal rings contain five or six atoms. As cyclization occurs, the carbonyl carbon becomes a new chiral center. This carbon is called the *anomeric carbon atom*. The two possible diastereomers that may form during the cyclization reaction are called **anomers**.

In aldose sugars the hydroxyl group of the newly formed hemiacetal occurs on carbon 1 (the anomeric carbon) and may occur either above the ring (in the "up" position) or below the ring (in the "down" position). For D-sugars, when the hydroxyl is down, the structure is in the α-anomeric form. If the hydroxyl is up, the structure is in the β-anomeric form. In Fischer projections, the α-anomeric hydroxyl occurs on the right and the β-hydroxyl occurs on the left (Figure 7.6).

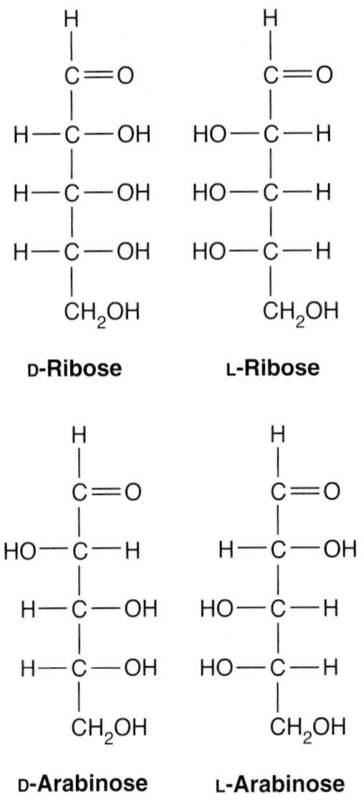

FIGURE 7.4

The Optical Isomers D- and L-Ribose and D- and L-Arabinose

D-Ribose and D-arabinose are diastereomers; that is, they are not mirror images.

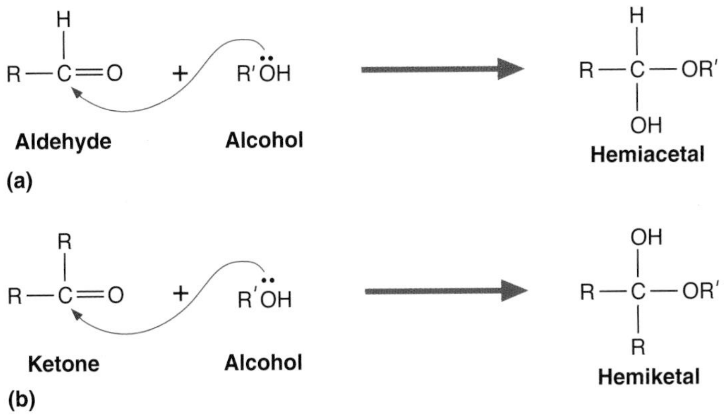

FIGURE 7.5

Formation of Hemiacetals and Hemiketals

(a) From an aldehyde. (b) From a ketone.

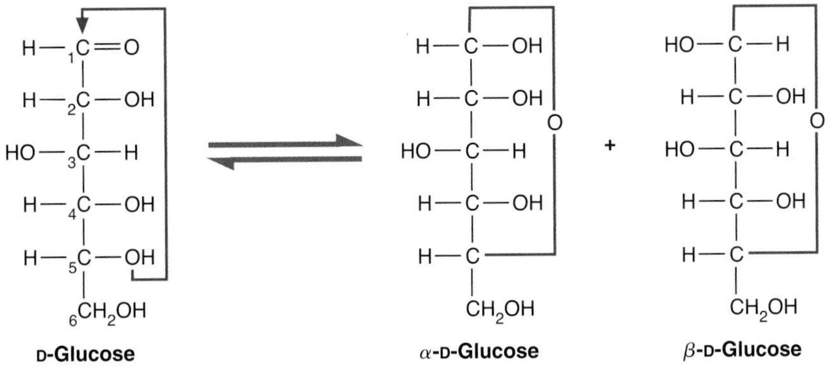

FIGURE 7.6

Monosaccharide Structure

Formation of the hemiacetal structure of glucose. Both the α- and β-anomers of glucose form. Note that the right angles in the hemiacetal linkage in Fischer models of monosaccharides do not represent methylene groups.

It is important to note that the anomers are defined relative to the D- and L-classification of sugars. These rules apply only to D-sugars, the most common ones found in nature. In the L-sugars the α-anomeric OH group is above the ring. The cyclization of sugars is more easily visualized by using Haworth structures.

HAWORTH STRUCTURES Fischer representations of cyclic sugar molecules use a long bond to indicate ring structure. A more accurate picture of carbohydrate structure was developed by the English chemist W. N. Haworth (Figure 7.7). Haworth structures more closely depict proper bond angles and lengths than do Fischer representations.

Five-membered hemiacetal rings are called *furanoses* because of their structural similarity to furan (Figure 7.8). For example, the cyclic form of fructose depicted in Figure 7.9 is called fructofuranose. Six-membered rings are called *pyranoses* because of their similarity to pyran. Glucose, in the pyranose form, is called glucopyranose.

CONFORMATIONAL STRUCTURES Although Haworth projection formulas are often used to represent carbohydrate structure, they are oversimplifications. Bond angle analysis and X-ray analysis demonstrate that *conformational formulas* are more accurate representations of monosaccharide structure (Figure 7.10). Conformational structures are more accurate because they illustrate the puckered nature of sugar rings.

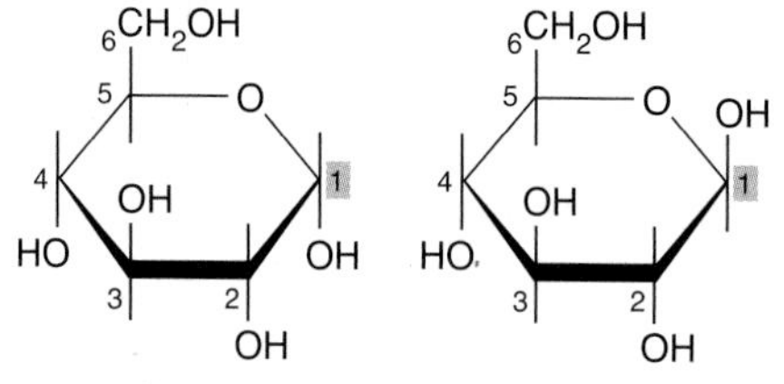

KEY CONCEPTS

- Monosaccharides, which may be polyhydroxy aldehydes or ketones, are either aldoses or ketoses.
- Sugars that contain four or more carbons primarily have cyclic forms.
- Cyclic aldoses or ketoses are hemiacetals and hemiketals, respectively.

FIGURE 7.7

Haworth Structures of the Anomers of D-Glucose

(a) α-D-Glucose. (b) β-D-Glucose. Note that in carbohydrate chemistry, the hydrogens bonded to carbons in sugar rings can be represented by single lines.

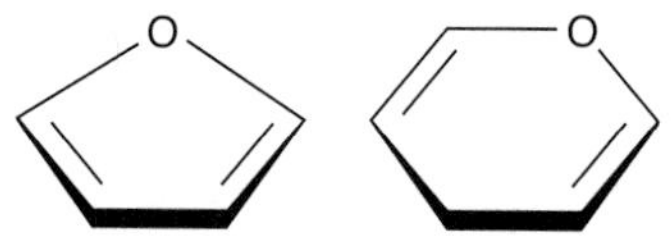

Furan   Pyran

FIGURE 7.8

Furan and Pyran

FIGURE 7.9

Fischer and Haworth Representations of D-Fructose

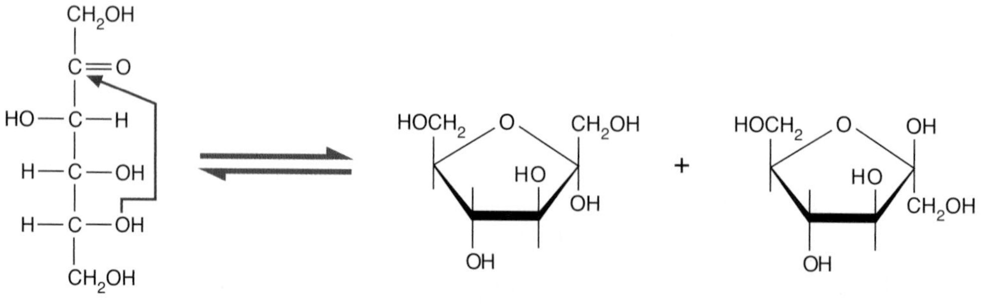

D-Fructose α-D-Fructofuranose β-D-Fructofuranose

Space-filling models, whose dimensions are proportional to the van der Waals radius of the atoms, also give useful structural information (see later: Figures 7.21, 7.22, and 7.23).

Monosaccharides undergo most of the reactions that are typical of aldehydes, ketones, and alcohols. The most important of these reactions in living organisms are described.

MUTAROTATION The α- and β-forms of monosaccharides are readily interconverted when dissolved in water. This spontaneous process, called **mutarotation**, produces an equilibrium mixture of α- and β-forms in both furanose and pyranose ring structures. The proportion of each form differs according to sugar type. Glucose, for example, exists primarily as a mixture of α- (38%) and β- (62%) pyranose forms (Figure 7.11). Fructose is predominantly found in the α- and β-furanose forms. The open chain formed during mutarotation can participate in oxidation-reduction reactions.

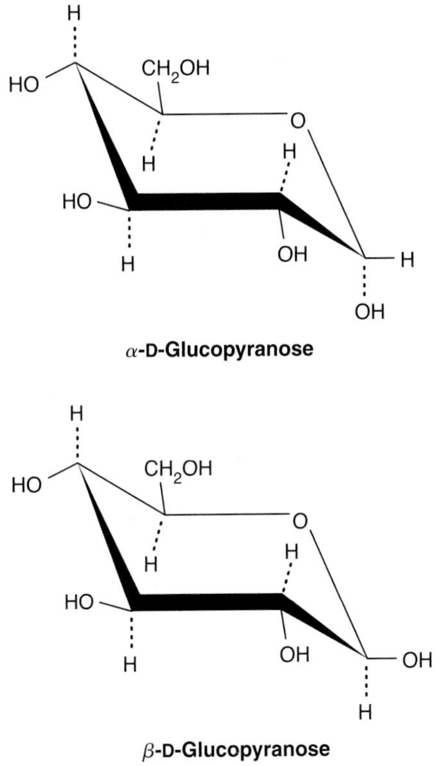

α-D-Glucopyranose

β-D-Glucopyranose

FIGURE 7.10

α- and β-D-Glucose

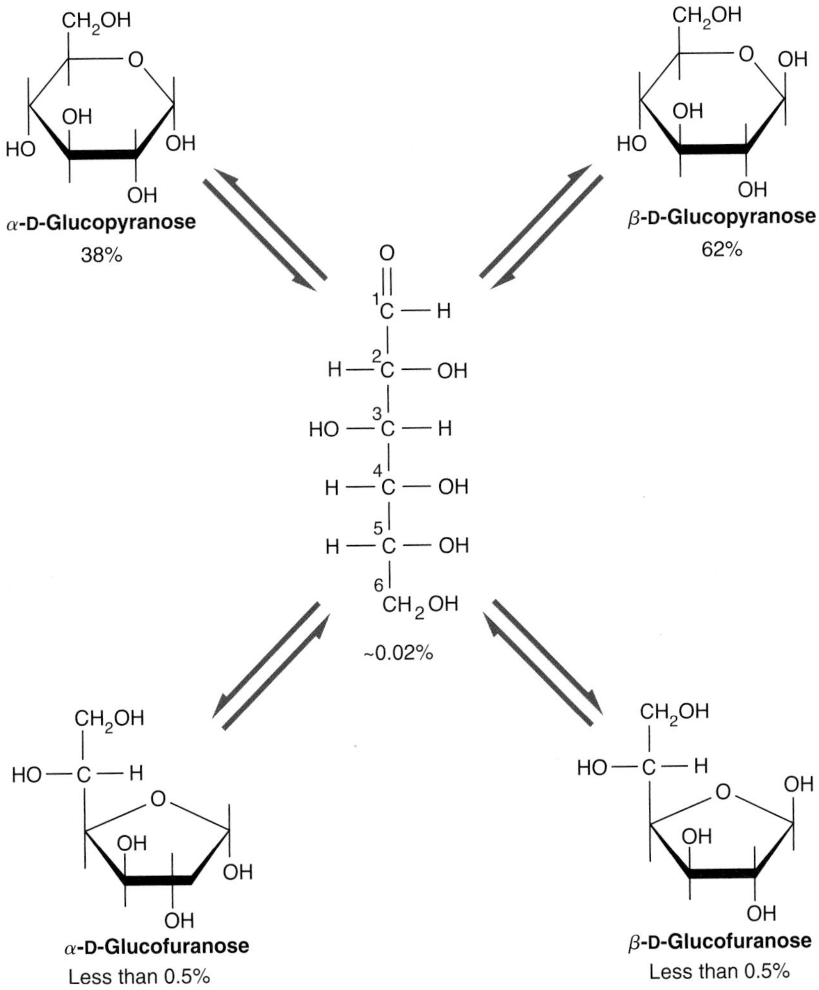

FIGURE 7.11

Equilibrium Mixture of D-Glucose

When glucose is dissolved in water at 25°C, the anomeric forms of the sugar undergo very rapid interconversions. When equilibrium is reached (i.e., there is no net change in the occurrence of each form), the glucose solution contains the percentages shown.

Reactions of Monosaccharides

The carbonyl and hydroxyl groups of sugars can undergo several chemical reactions. Among the most important are oxidation, reduction, isomerization, esterification, glycoside formation, and glycosylation reactions.

OXIDATION In the presence of oxidizing agents, metal ions such as Cu^{2+}, and certain enzymes, monosaccharides readily undergo several oxidation reactions. Oxidation of an aldehyde group yields an **aldonic acid**, whereas oxidation of a terminal CH_2OH group (but not the aldehyde group) gives a **uronic acid**. Oxidation of both the aldehyde and CH_2OH gives an **aldaric acid** (Figure 7.12).

The carbonyl groups in both aldonic and uronic acids can react with an OH group in the same molecule to form a cyclic ester known as a **lactone**:

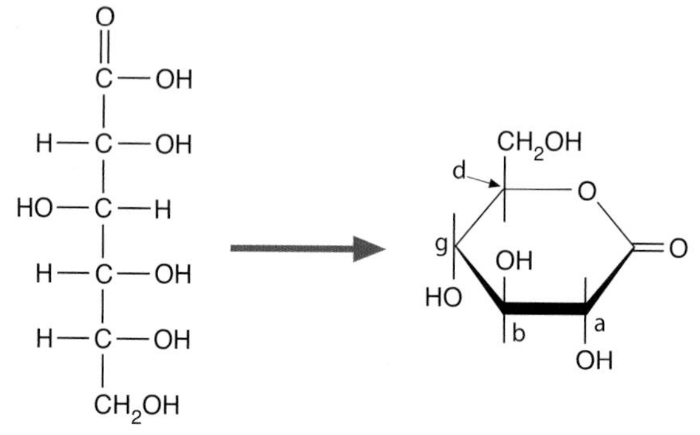

D-Gluconic acid
(An aldonic acid)

D-Glucono-δ-lactone

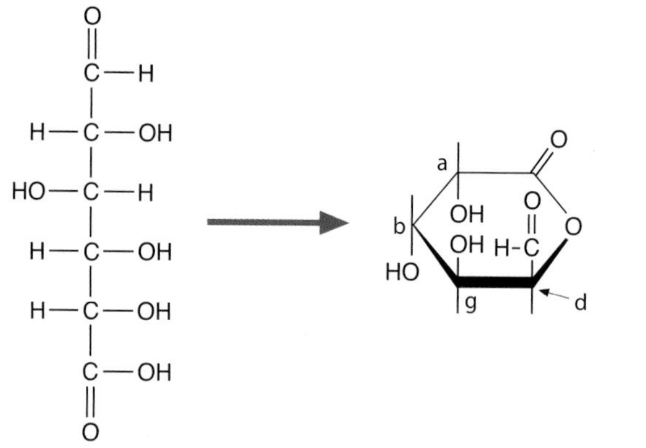

D-Glucuronic acid
(A uronic acid)

D-Glucurono-δ-lactone

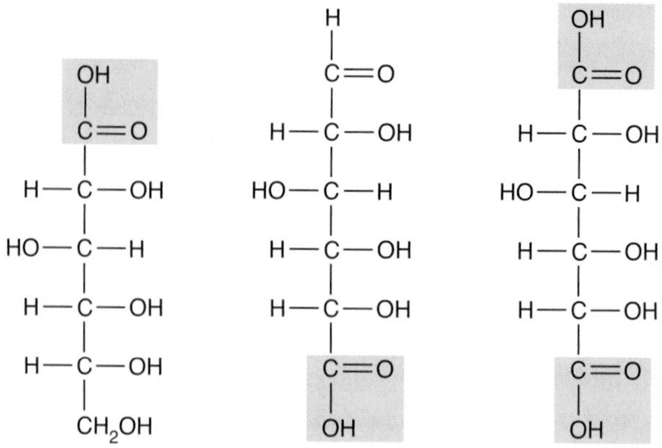

FIGURE 7.12

Oxidation Products of Glucose

The newly oxidized groups are highlighted.

D-Gluconic acid

D-Glucuronic acid

D-Glucaric acid

Lactones are commonly found in nature. For example, L-ascorbic acid, also known as vitamin C (Figure 7.13), is a lactone derivative of D-glucuronic acid. It is synthesized by all mammals except guinea pigs, apes, fruit-eating bats, and, of course, humans. These species must obtain ascorbic acid in their diet, hence the name vitamin C. Ascorbic acid is a powerful reducing agent; that is, it protects cells from highly reactive oxygen and nitrogen species (see pp. 365–366). In addition, it is required in the hydroxylation reactions of collagen.

Sugars that can be oxidized by weak oxidizing agents such as Benedict's reagent are called **reducing sugars** (Figure 7.14). Because the reaction occurs only with sugars that can revert to the open chain form, all monosaccharides are reducing sugars.

REDUCTION Reduction of the aldehyde and ketone groups of monosaccharides yields the sugar alcohols (**alditols**). Reduction of D-glucose, for example, yields D-glucitol, also known as D-sorbitol (Figure 7.15). Sugar alcohols are used commercially in processing foods and pharmaceuticals. Sorbitol, for example, improves the shelf life of candy because it helps prevent moisture loss. Adding sorbitol syrup to artificially sweetened canned fruit reduces the unpleasant aftertaste of the artificial sweetener saccharin. Sorbitol is converted into fructose in the liver.

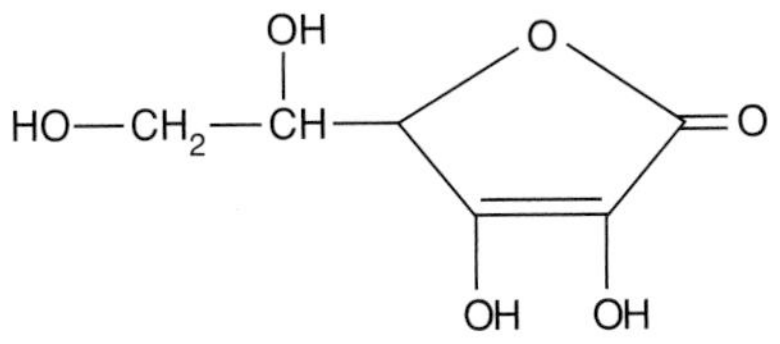

ISOMERIZATION Monosaccharides undergo several types of isomerization. For example, after several hours an alkaline solution of D-glucose also contains D-mannose and D-fructose. Both isomerizations involve an intramolecular shift of a hydrogen atom and a relocation of a double bond (Figure 7.16). The intermediate formed is called an **enediol**. The reversible transformation of glucose to fructose is an aldose-ketose interconversion. Because the configuration at a single asymmetric carbon changes, the conversion of glucose to mannose is

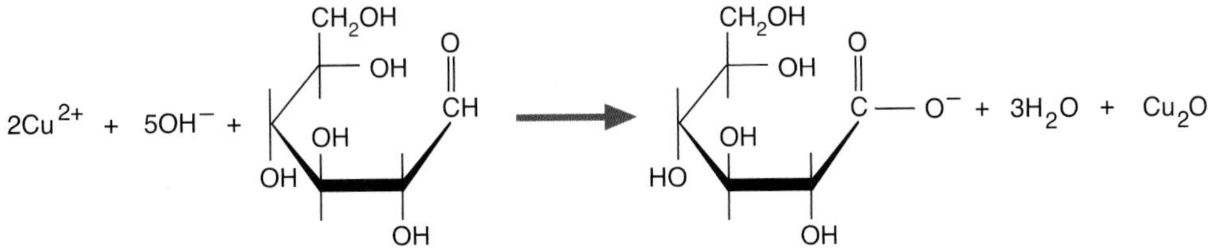

COMPANION

WEBSITE Visit the companion website at www.oup.com/us/mckee to read the Biochemistry in Perspective box on scurvy and ascorbic acid.

Ascorbic acid

FIGURE 7.13

Structure of Ascorbic Acid

Humans and guinea pigs cannot synthesize ascorbic acid because they lack gluconolactone oxidase, one of the three enzymes required to synthesize the acid from its precursor glucuronate.

$$2Cu^{2+} + 5OH^- + \quad \longrightarrow \quad + 3H_2O + Cu_2O$$

FIGURE 7.14

Reaction of Glucose with Benedict's Reagent

Benedict's reagent, copper(II) sulfate in a solution of sodium carbonate and sodium citrate, is reduced by the monosaccharide glucose. Glucose is oxidized to form the salt of gluconic acid. The reaction also forms the reddish-brown precipitate Cu_2O and other oxidation products (not shown).

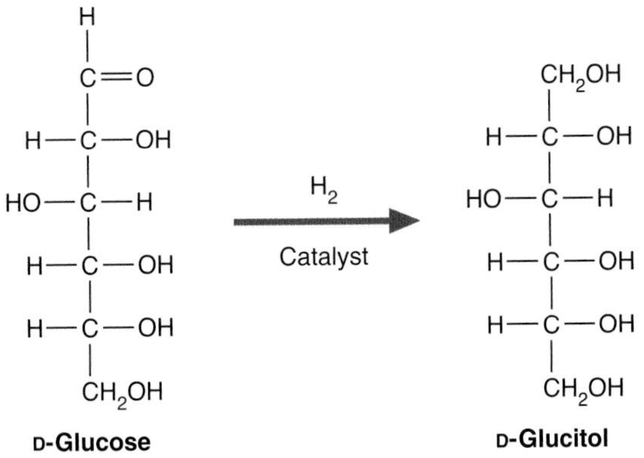

D-Glucose → **D-Glucitol**

FIGURE 7.15

Laboratory Reduction of Glucose to Form D-Glucitol (Sorbitol)

FIGURE 7.16

Isomerization of D-Glucose to Form D-Mannose and D-Fructose

An enediol intermediate is formed in this process.

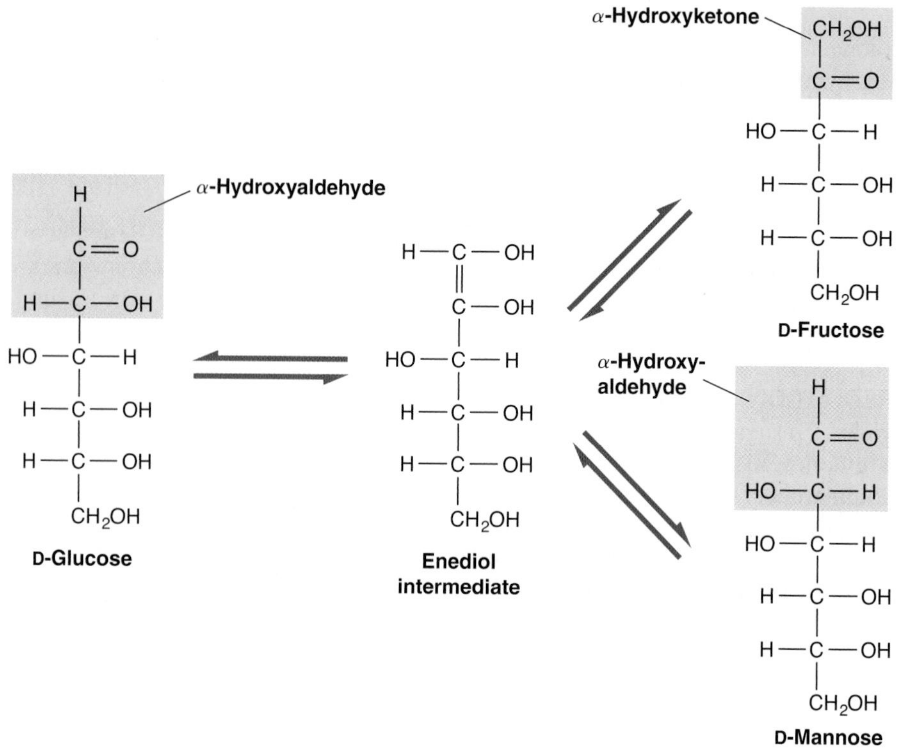

referred to as an **epimerization**. Several enzyme-catalyzed reactions involving enediols occur in carbohydrate metabolism (Chapter 8).

ESTERIFICATION Like all free OH groups, those of carbohydrates can be converted to esters by reactions with acids. Esterification often dramatically changes a sugar's chemical and physical properties. Phosphate and sulfate esters of carbohydrate molecules are among the most common ones found in nature.

Phosphorylated derivatives of certain monosaccharides are metabolic components of living cells that are frequently formed during reactions with ATP. They are important because many biochemical transformations use nucleophilic substitution reactions. Such reactions require a leaving group. In a carbohydrate molecule this group is most likely to be an OH group. However, because OH groups are poor leaving groups, any substitution reaction is unlikely. The problem is solved by converting an appropriate OH group to a phosphate ester, which can then be displaced by an incoming nucleophile. As a consequence, a slow reaction now occurs much more rapidly.

Sulfate esters of carbohydrate molecules are found predominantly in the proteoglycan components of connective tissue. Sulfate esters are charged so they bind large amounts of water and small ions. They also participate in forming salt bridges between carbohydrate chains.

QUESTION 7.3

Draw the following compounds:
  a. α- and β-anomers of D-galactose
  b. aldonic acid, uronic acid, and aldaric acid derivatives of galactose
  c. galactitol
  d. δ-lactone of galactonic acid

GLYCOSIDE FORMATION Hemiacetals and hemiketals react with alcohols to form the corresponding **acetal** or **ketal** (Figure 7.17). When the cyclic hemiacetal or hemiketal form of the monosaccharide reacts with an alcohol, the new linkage is called a **glycosidic linkage**, and the compound is called a **glycoside**. The name of the glycoside specifies the sugar component. For example, the acetals of glucose and the ketals of fructose are called *glucoside* and *fructoside*, respectively. Additionally, glycosides derived from sugars with five-membered rings are called *furanosides*; those from six-membered rings are called *pyranosides*. A relatively simple example shown in Figure 7.18 illustrates the reaction of glucose with methanol to form two anomeric types of methyl glucosides. Because glycosides are acetals, they are stable in basic solutions. Carbohydrate molecules that contain only acetal groups do not test positive with Benedict's reagent. (Acetal formation "locks" a ring so it cannot undergo oxidation or mutarotation.) Only hemiacetals act as reducing agents.

If an acetal linkage is formed between the hemiacetal hydroxyl group of one monosaccharide and a hydroxyl group of another monosaccharide, the resulting glycoside is called a **disaccharide**. A molecule containing a large number of monosaccharides linked by glycosidic linkages is called a **polysaccharide**.

QUESTION 7.4

Draw the structure of a D-glucosamine molecule linked to threonine via a β-glycosidic linkage.

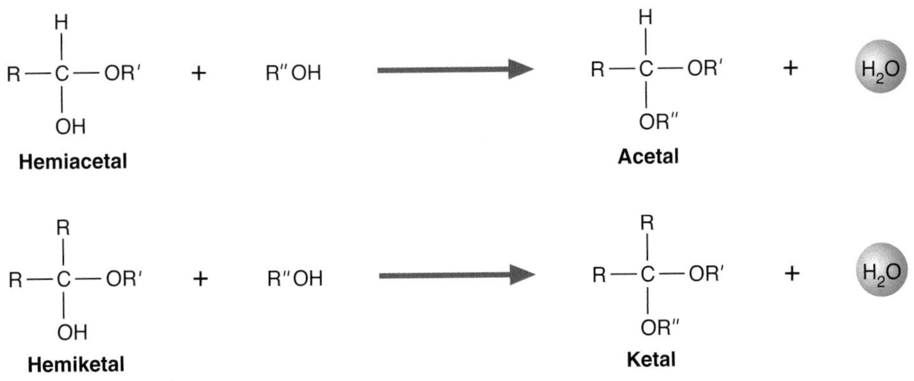

FIGURE 7.17

Formation of Acetals and Ketals

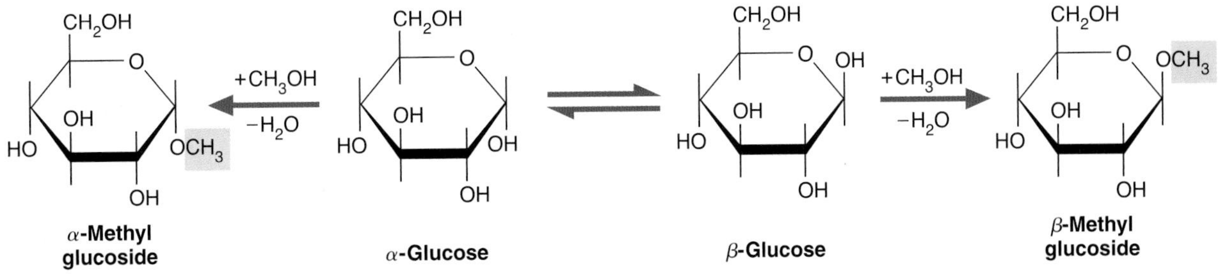

FIGURE 7.18

Methyl Glucoside Formation

Noncarbohydrate components of glycosides are called aglycones. The highlighted methyl groups are aglycones.

QUESTION 7.5

Glycosides are commonly found in nature. One example is salicin (Figure 7.19), a compound found in willow tree bark that has antipyretic (fever-reducing) and analgesic properties. Can you identify the carbohydrate and aglycone (noncarbohydrate) components of salicin?

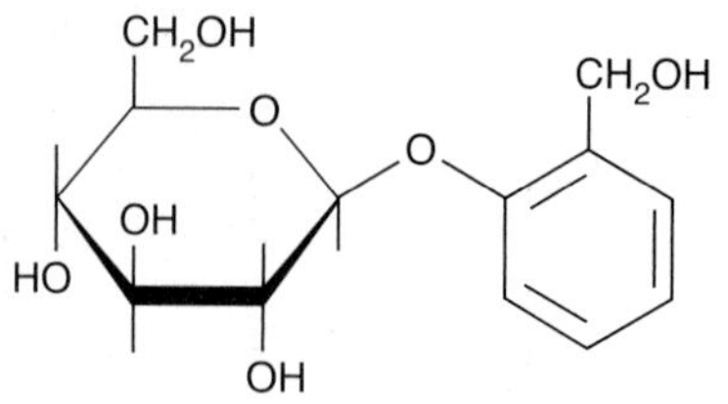

FIGURE 7.19
Salicin

GLYCOSYLATION REACTIONS Glycosylation reactions attach sugars or glycans (sugar polymers) to proteins or lipids. Analogous to glycoside formation between sugar molecules, the glycosylation reactions, catalyzed by the glycosyl transferases, form glycosidic bonds between anomeric carbons in certain glycans and nitrogen or oxygen atoms in other types of molecules. For example, both N- and O-glycosidic bonds are prominent structural features of proteins. N-Glycosidic linkages form between oligosaccharides and the side chain amide nitrogen of certain asparagine residues, whereas O-glycosidic linkages attach glycans to the side chain hydroxyl oxygens of serine or threonine residues (p. 244) or the hydroxyl oxygens of membrane lipids.

Reducing sugars can also react with nucleophilic nitrogen atoms in nonenzymatic reactions. These so-called *glycation* reactions occur rapidly in the presence of heat (e.g., during the cooking or baking of sugar-containing food) or slowly within the body when excess sugar molecules are present. The best-researched example is the reaction of glucose with the side chain amino nitrogen of lysine residues in proteins. The nonenzymatic glycation of protein, called the Maillard reaction (named for the French chemist who discovered it in 1912), begins with the nucleophilic attack of the amino nitrogen on the anomeric carbon of the reducing sugar (Figure 7.20). The Schiff base that forms rearranges to form a stable ketoamine called the *Amadori product*. Both the protein-bound Schiff base and the Amadori product can undergo further reactions (e.g., oxidations, rearrangements, and dehydrations) to produce additional protein-bound products, referred to collectively as *advanced glycation end products* (AGEs). Reactive carbonyl-containing products such as the dicarbonyl compound glyoxal (CHOCHO) cause rapid protein cross-linkage and adduct formation. (An **adduct** is the product of an addition reaction, that is, the reactions of two molecules to form a third molecule.) Consequently, glycation alters the structural and functional properties of proteins. For example, the glycation of long-lived proteins such as collagen and elastin disrupts the structure of vascular and connective tissues. In addition, AGEs trigger the production of molecules such as the cytokines that promote inflammatory processes. The accumulation of AGEs has been linked to such age-related conditions as vascular and neurodegenerative diseases, and arthritis. In one vascular disease, *atherosclerosis*, cells lining the arterial blood vessels are damaged by AGE formation. When the body initiates a repair process involving macrophages and growth factors, inflammation results, leading to the formation of artery-clogging deposits called *plaque*. The capacity of affected blood vessels to nourish nearby tissue is eventually compromised. The excessively high blood glucose levels that occur in diabetes mellitus (see Biochemistry in Perspective box on Diabetes Mellitus in Chapter 16) cause an accelerated form of atherosclerosis as well as numerous other AGE-related pathological changes.

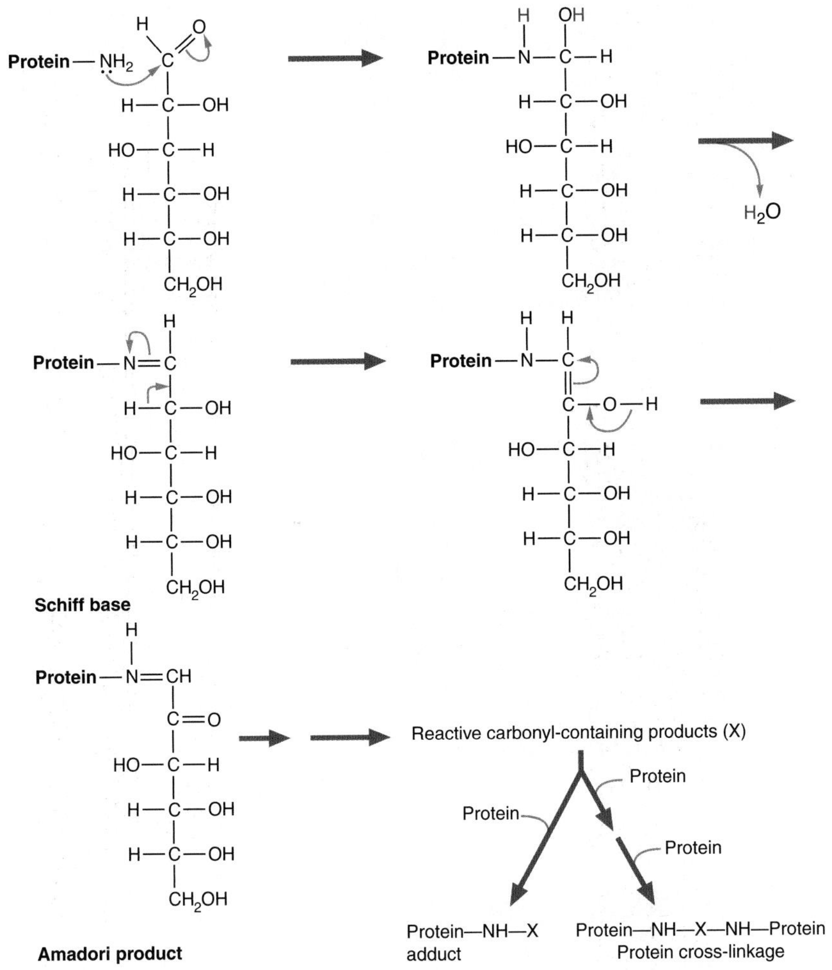

FIGURE 7.20

The Maillard Reaction

Any molecule that contains an amino group can undergo the Maillard reaction, so nucleotides and amines also react with glucose molecules. Since proteins have greater exposure to elevated circulating simple sugars, they are more compromised by the process.

Important Monosaccharides

Among the most important monosaccharides found in living organisms are glucose, fructose, and galactose. The principal functional roles of these molecules are briefly described.

GLUCOSE D-Glucose, originally called dextrose, is found in large quantities throughout the living world (Figure 7.21). It is the primary fuel for living cells. In animals, glucose is the preferred energy source of brain cells and cells that have few or no mitochondria, such as erythrocytes. Cells that have a limited oxygen supply, such as those in the eyeball, also use large amounts of glucose to generate energy. Dietary sources include plant starch and the disaccharides lactose, maltose, and sucrose.

FRUCTOSE D-Fructose, originally called levulose, is often referred to as fruit sugar because of its high content in fruit. It is also found in some vegetables and in honey (Figure 7.22). This molecule is an important member of the ketose family of sugars. On a per-gram basis, fructose is twice as sweet as sucrose. It can therefore be used in smaller amounts. For this reason, fructose is often used as a sweetening agent in processed food products. Large amounts of fructose are

FIGURE 7.21

α-D-**Glucopyranose**

Compare the information provided by these two representations. (a) The space-filling model, with carbon, oxygen, and hydrogen atoms in green, red, and white, respectively and (b) the Haworth structure.

(a)   (b)

FIGURE 7.22

β-D-**Fructofuranose**

(a) Space-filling model and (b) Haworth structure.

(a)   (b)

FIGURE 7.23

α-D-**Galactopyranose**

(a) Space-filling model and (b) Haworth structure.

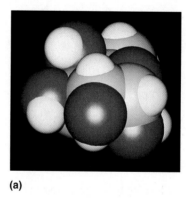

(a)   (b)

used in the male reproductive tract. It is synthesized in the seminal vesicles and then incorporated into semen. Sperm use the sugar as an energy source.

GALACTOSE Galactose is necessary to synthesize a variety of biomolecules (Figure 7.23). These include lactose (in lactating mammary glands), glycolipids, certain phospholipids, proteoglycans, and glycoproteins. Synthesis of these substances is not diminished by diets that lack galactose or the disaccharide lactose (the principal dietary source of galactose) because the sugar is readily synthesized from glucose-1-phosphate. As mentioned previously, galactose and glucose are epimers at carbon 4. The interconversion of galactose and glucose is catalyzed by an enzyme called an epimerase.

In *galactosemia*, a genetic disorder, an enzyme required to metabolize galactose is missing. Galactose, galactose-1-phosphate, and galactitol (a sugar alcohol derivative) accumulate and cause liver damage, cataracts, and severe mental retardation. The only effective treatment is early diagnosis and a diet free of galactose.

KEY CONCEPT

Glucose, fructose, and galactose are among the most important monosaccharides in living organisms.

Monosaccharide Derivatives

Simple sugars may be converted to closely related chemical compounds. Several of these are important metabolic and structural components of living organisms.

URONIC ACIDS Recall that uronic acids are formed when the terminal CH_2OH group of a monosaccharide is oxidized. Two uronic acids are important in animals: D-glucuronic acid and its epimer, L-iduronic acid (α-D-glucuronate and β-L-iduronate in Figure 7.24). In liver cells, glucuronic acid is combined with molecules such as steroids, certain drugs, and bilirubin (a degradation product of heme in the oxygen-carrying protein hemoglobin) to improve water solubility. This process helps remove waste products from the body. Both D-glucuronic acid and L-iduronic acid are abundant in connective tissue carbohydrate components.

AMINO SUGARS In amino sugars a hydroxyl group (most commonly on carbon 2) is replaced by an amino group (Figure 7.25). These compounds are common constituents of the complex carbohydrate molecules found attached to cellular proteins and lipids. The most common amino sugars of animal cells are D-glucosamine and D-galactosamine. Amino sugars are often acetylated. One such molecule is *N*-acetylglucosamine. *N*-Acetylneuraminic acid (the most common form of sialic acid) is a condensation product of D-mannosamine and pyruvic acid, a 2-ketocarboxylic acid. Sialic acids are ketoses containing nine carbon atoms that may be amidated with acetic or glycolic acid (hydroxyacetic acid). They are common components of glycoproteins and glycolipids.

DEOXY SUGARS Monosaccharides in which an —H has replaced an —OH group are known as *deoxy sugars*. Two important deoxy sugars found in cells are L-fucose (formed from D-mannose by reduction reactions) and 2-deoxy-D-ribose (Figure 7.26). Fucose is often found among the carbohydrate components of glycoproteins, such as those of the ABO blood group determinantes on the surface of red blood cells. 2-Deoxyribose, the pentose sugar component of DNA, was shown earlier (Figure 1.9).

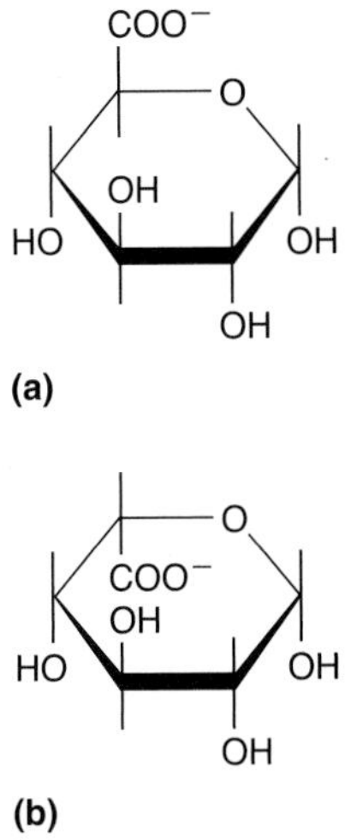

FIGURE 7.24
Uronic Acids

(a) α-D-Glucuronate and (b) β-L-Iduronate.

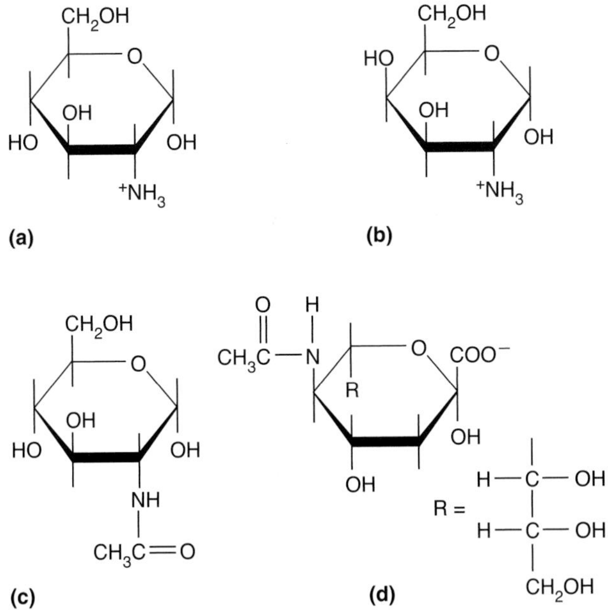

FIGURE 7.25
Amino Sugars

(a) α-D-Glucosamine, (b) α-D-galactosamine, (c) *N*-acetyl-α-D-glucosamine, and (d) *N*-acetylneuraminic acid (sialic acid).

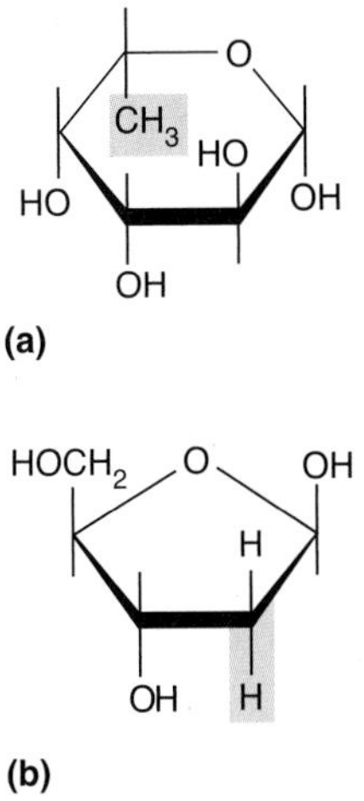

FIGURE 7.26
Deoxy Sugars

(a) β-L-Fucose (6-deoxygalactose) and (b) 2-deoxy-β-D-ribose. The carbon atoms that have —OH goups replaced by —H are highlighted.

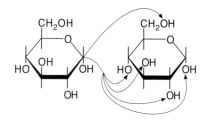

FIGURE 7.27
Glycosidic Bonds

Several types of glycosidic bonds can form between monosaccharides. The sugar α-D-glucopyranose (left) can theoretically form glycosidic linkages with any of the alcoholic functional groups of another monosaccharide, in this case another molecule of α-D-glucopyranose.

α-Lactose

β-Lactose

FIGURE 7.28
α- and β-Lactose

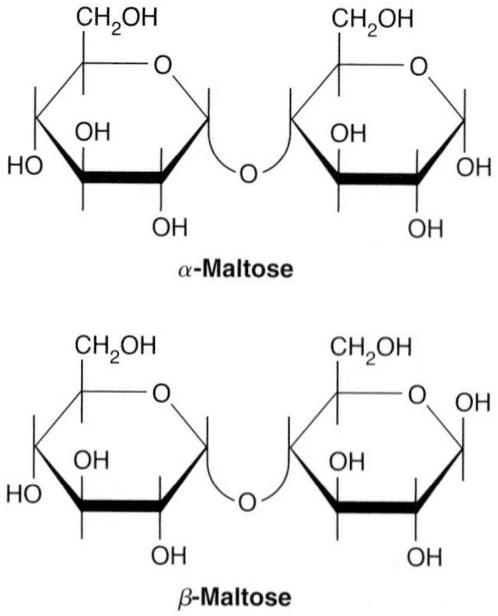

7.2 DISACCHARIDES

Disaccharides are molecules composed of two monosaccharides that are linked by a glycosidic bond. If one monosaccharide molecule is linked through its anomeric carbon atom to the hydroxyl group on carbon 4 of another monosaccharide, the glycosidic linkage is designated as 1,4. Because the anomeric hydroxyl group may be in either the α- or β-configuration, two possible disaccharides may form when two sugar molecules are linked: α(1,4) or β(1,4). Other varieties of glycosidic linkages [i.e., α or β(1,1), (1,2), (1,3), and (1,6) linkages] also occur (Figure 7.27).

Digestion of disaccharides and other carbohydrates is mediated by enzymes synthesized by cells lining the small intestine. Deficiency of any of these enzymes causes unpleasant symptoms when the undigestible disaccharide sugar is ingested. Because carbohydrates are absorbed principally as monosaccharides, any undigested disaccharide molecules pass into the large intestine, where osmotic pressure draws water from the surrounding tissues (diarrhea). Bacteria in the colon digest the disaccharides (fermentation), thus producing gas (bloating and cramps). The most commonly known deficiency is *lactose intolerance*, which may occur in most human adults except those with ancestors from northern Europe and/or certain African groups. Caused by the greatly reduced synthesis of the enzyme lactase following childhood, lactose intolerance is treated by eliminating the sugar from the diet or (in some cases) by treating food with the enzyme lactase.

Lactose (milk sugar) is a disaccharide found in milk. It is composed of one molecule of galactose linked through the hydroxyl group on carbon 1 in a β-glycosidic linkage to the hydroxyl group of carbon 4 of a molecule of glucose (Figure 7.28). Because the anomeric carbon of galactose is in the β-configuration, the linkage between the two monosaccharides is designated as β(1,4). The inability to hydrolyze the β(1,4) linkage (caused by lactase deficiency) is common among animals, which are unable to digest carbohydrates with such linkages (such as cellulose). Because the glucose component contains a hemiacetal group, lactose is a reducing sugar.

Maltose, also known as malt sugar, is an intermediate product of starch hydrolysis and does not appear to exist freely in nature. Maltose is a disaccharide with an α(1,4) glycosidic linkage between two D-glucose molecules. In solution the free anomeric carbon undergoes mutarotation, which results in an equilibrium mixture of α- and β-maltoses (Figure 7.29).

Cellobiose, a degradation product of cellulose, contains two molecules of glucose linked by a β(1,4) glycosidic bond (Figure 7.30). Like maltose, whose structure is identical except for the direction of the glycosidic bond, cellobiose does not occur freely in nature.

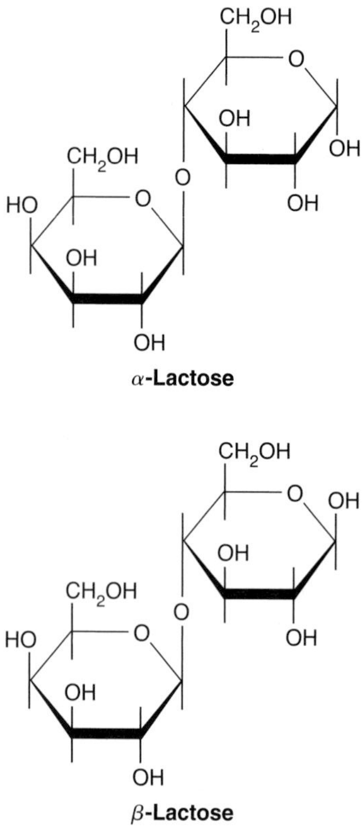

α-Maltose

β-Maltose

FIGURE 7.29
α- and β-Maltose

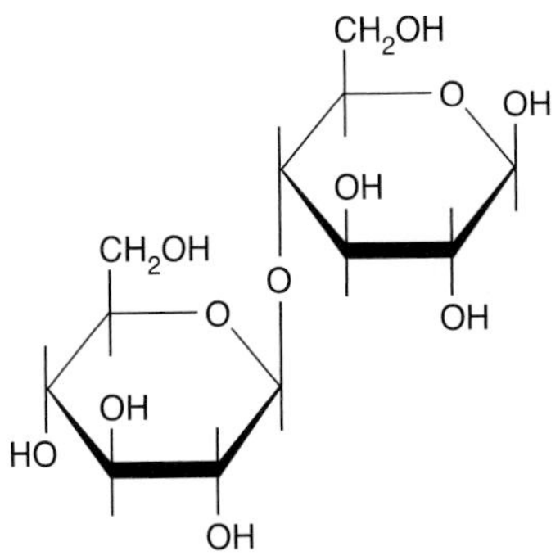

FIGURE 7.30

β-Cellobiose

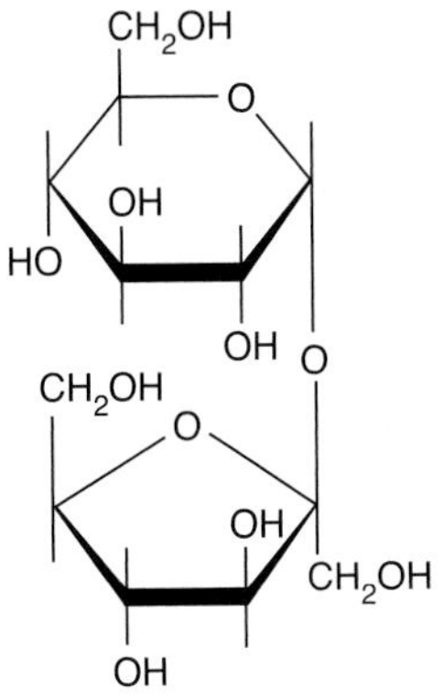

FIGURE 7.31

Sucrose

The glucose and fructose residues are linked by an α, β(1,2) glycosidic bond.

Sucrose (common table sugar: cane sugar or beet sugar) is produced in the leaves and stems of plants. It is a transportable energy source throughout the entire plant. Containing both α-glucose and β-fructose residues, sucrose differs from the previously described disaccharides in that the monosaccharides are linked through a glycosidic bond between both anomeric carbons (Figure 7.31). Because neither monosaccharide ring can revert to the open-chain form, sucrose is a nonreducing sugar.

QUESTION 7.6

Which of the following sugars or sugar derivatives are reducing sugars?
  a. glucose
  b. fructose
  c. α-methyl-D-glucoside
  d. sucrose
Which of these compounds are capable of mutarotation?

KEY CONCEPTS

- Disaccharides are glycosides composed of two monosaccharide units.
- Maltose, lactose, cellobiose, and sucrose are disaccharides.

7.3 POLYSACCHARIDES

Polysaccharides, also referred to as **glycans**, are composed of large numbers of monosaccharide units connected by glycosidic linkages. Smaller glycans, called oligosaccharides, are polymers containing up to about 10 or 15 monomers, most often attached to polypeptides in glycoproteins (p. 252) and some glycolipids (p. 383). Among the best-characterized oligosaccharide groups are those attached to membrane and secretory proteins. There are two broad classes of oligosaccharides: N-linked and O-linked. The N-linked oligosaccharides are attached to polypeptides by an N-glycosidic bond with the side chain amide group of the amino acid asparagine. There are three major types of asparagine-linked oligosaccharide: high-mannose, hybrid, and complex (Figure 7.32). The O-Linked oligosaccharides are attached to polypeptides by the side chain hydroxyl group of the amino acids serine or threonine in polypeptide chains or the hydroxyl group of membrane lipids. Larger glycans may contain from hundreds to thousands of sugar units. These molecules may have a linear structure or they may have branched shapes. Polysaccharides may be divided into two classes: **homoglycans**, which are composed of one type of monosaccharide, and **heteroglycans**, which contain two or more types of monosaccharides.

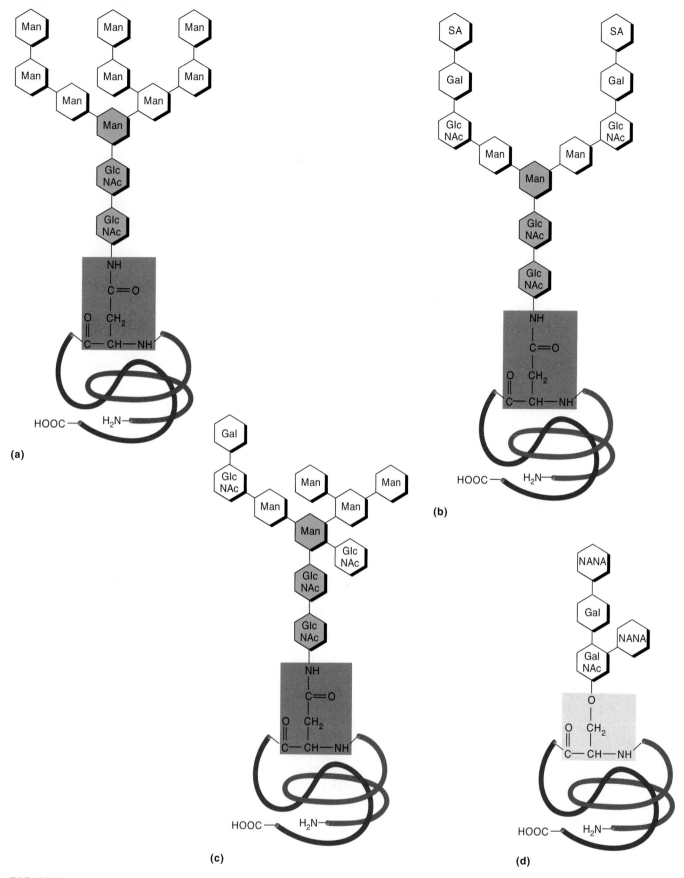

FIGURE 7.32

Oligosaccharides Linked to Polypeptides

(a) High-mannose, (b) complex, and (c) hybrid oligosaccharides are N-linked. (d) Typical O-linked oligosaccharides. The asparagine links (a, b, and c) and serine links (d) are highlighted. (SA = sialic acid, NANA = *N*-acetylneuraminic acid, Man = mannose, Gal = galactose, GlcNAc = *N*-acetylglucosamine). Note that N-linked oligosaccharides share a core structure (highlighted in green).

Homoglycans

The **homoglycans** found in abundance in nature are starch, glycogen, cellulose, and chitin. Starch, glycogen, and cellulose all yield D-glucose when they are hydrolyzed. Starch and glycogen are energy storage molecules in plants and animals, respectively. Cellulose is the primary structural component of plant cells. **Chitin**, the principal structural component of the exoskeletons of arthropods such as insects and crustaceans and the cell walls of many fungi, yields the glucose derivative *N*-acetylglucosamine when it is hydrolyzed. It should be noted that polysaccharides such as starch and glycogen, unlike proteins and nucleic acids, have no fixed molecular weights. The size of such molecules is a reflection of the metabolic state of the cell producing them. For example, when blood sugar levels are high (e.g., after a meal), the liver synthesizes glycogen. Glycogen molecules in a well-fed animal may have molecular weights as high as 2×10^7 D. When blood sugar levels fall, the liver enzymes begin breaking down the glycogen molecules, releasing glucose into the bloodstream. If the animal continues to fast, the process continues until glycogen reserves are almost used up.

STARCH Starch, the energy reservoir of plant cells, is a significant source of carbohydrate in the human diet. Much of the nutritional value of the world's major foodstuffs (e.g., potatoes, rice, corn, and wheat) comes from starch. Two polysaccharides occur together in starch: amylose and amylopectin.

Amylose is composed of long, unbranched chains of D-glucose residues that are linked with $\alpha(1,4)$ glycosidic bonds (Figure 7.33). A number of polysaccharides, including both types of starch, have one *reducing end* in which the ring can open to form a free aldehyde group with reducing properties. The internal anomeric carbons in these molecules are involved in acetal linkages and are not free to act as reducing agents.

Amylose molecules, which typically contain several thousand glucose residues, vary in molecular weight from 150,000 to 600,000 D. Because the linear amylose molecule forms long, tight helices, its compact shape is ideal for its storage function. The common iodine test for starch works because molecular iodine inserts itself into these helices. (The intense blue color of a positive test comes from electronic interactions between iodine molecules and the helically arranged glucose residues of the amylose.)

The other form of starch, **amylopectin**, is a branched polymer containing both $\alpha(1,4)$ and $\alpha(1,6)$ glycosidic linkages. The $\alpha(1,6)$ branch points may occur every

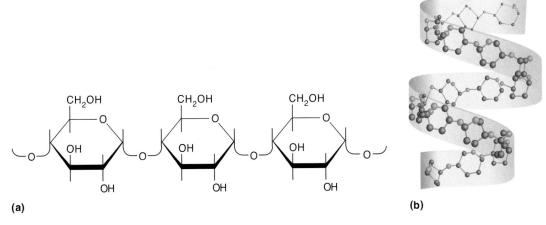

(a) (b)

FIGURE 7.33

Amylose

(a) The D-glucose residues of amylose are linked through $\alpha(1,4)$ glycosidic bonds.

(b) The amylose polymer forms a left-handed helix.

20 to 25 glucose residues and prevent helix formation (Figure 7.34a). The number of glucose units in amylopectin may vary from a few thousand to a million.

Starch digestion begins in the mouth, where the salivary enzyme α-amylase initiates hydrolysis of the glycosidic linkages. Digestion continues in the small intestine, where pancreatic α-amylase randomly hydrolyzes all the $\alpha(1,4)$ glycosidic bonds except those next to the branch points. The products of α-amylase are maltose, the trisaccharide maltotriose, and the α-limit dextrins [oligosaccharides that typically contain eight glucose units with one or more $\alpha(1,6)$ branch points]. Several enzymes secreted by cells that line the small intestine convert these intermediate products into glucose. Glucose molecules are then absorbed into intestinal cells. After passage into the bloodstream, they are transported to the liver and then to the rest of the body.

GLYCOGEN **Glycogen** is the carbohydrate storage molecule in vertebrates. It is found in greatest abundance in liver and muscle cells. (Glycogen may make

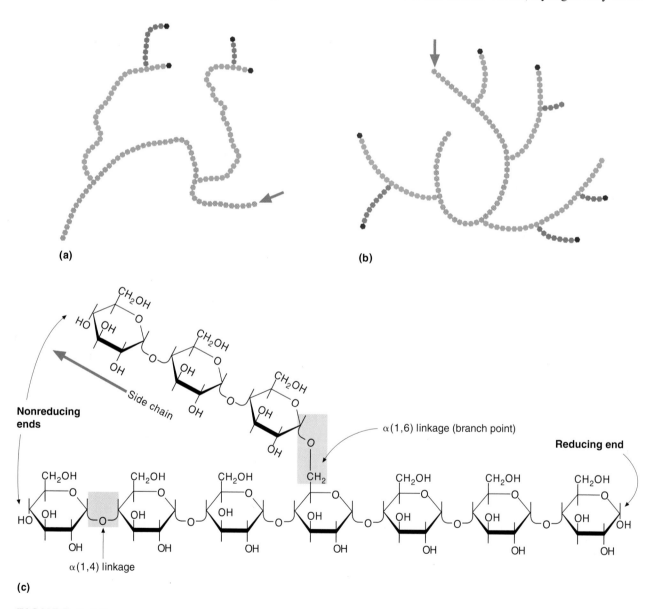

(a)

(b)

Nonreducing ends

Side chain

$\alpha(1,6)$ linkage (branch point)

Reducing end

$\alpha(1,4)$ linkage

(c)

FIGURE 7.34

(a) Amylopectin and (b) Glycogen

Each hexagon represents a glucose molecule. Notice that each molecule has only one reducing end (arrow) and numerous nonreducing ends. (c) Detail from (a) or (b).

up as much as 8–10% of the wet weight of liver cells and 2–3% of that of muscle cells.) Glycogen (Figure 7.34b) is similar in structure to amylopectin except that it has more branch points, possibly at every fourth glucose residue in the core of the molecule. In the outer regions of glycogen molecules, branch points are not so close together (approximately every 8–12 residues). Because the molecule is more compact than other polysaccharides, it takes up little space, an important consideration in mobile animal bodies. The many nonreducing ends in glycogen molecules allow for stored glucose to be rapidly mobilized when the animal demands much energy.

QUESTION 7.7

It has been estimated that two high-energy phosphate bonds must be expended to incorporate one glucose molecule into glycogen. Why is glucose stored in muscle and liver in the form of glycogen, not as individual glucose molecules? In other words, why is it advantageous for a cell to expend metabolic energy to polymerize glucose molecules? [*Hint*: Besides the reasons given in Section 7.3, refer to Chapter 3 for another problem that glucose polymerization solves.]

CELLULOSE **Cellulose** is a polymer composed of D-glucopyranose residues linked by $\beta(1,4)$ glycosidic bonds (Figure 7.35). It is the most important structural polysaccharide of plants. Because cellulose comprises about one-third of plant biomass, it is the most abundant organic substance on earth. Approximately 100 trillion kg of cellulose are produced each year.

Pairs of unbranched cellulose molecules, which may contain as many as 12,000 glucose units each, are held together by hydrogen bonding to form sheet-like strips called *microfibrils* (Figure 7.36). Each bundle of microfibrils contains

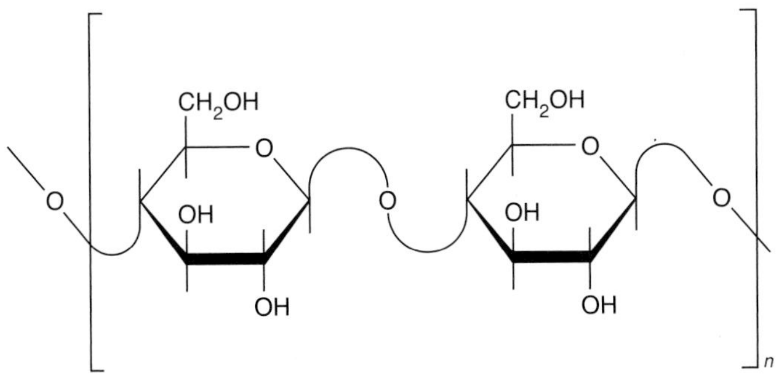

FIGURE 7.35
Cellulose

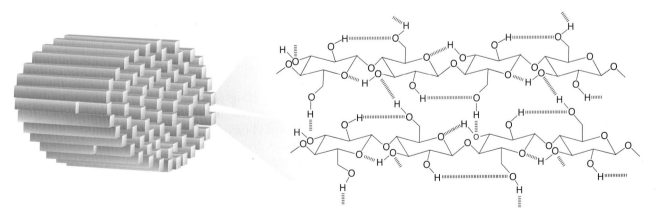

FIGURE 7.36

Cellulose Microfibrils

Intermolecular hydrogen bonds between adjacent cellulose molecules are largely responsible for the great strength of cellulose.

approximately 40 of these pairs. These structures are found in both plant primary and secondary cell walls, where they provide a structural framework that both protects and supports cells.

The ability to digest cellulose is found only in microorganisms that possess the enzyme cellulase. Certain animal species (e.g., termites and cows) use such organisms in their digestive tracts to digest cellulose. The breakdown of the cellulose makes glucose available to both the microorganisms and their host. Although many animals cannot digest cellulose-containing plant materials, these substances play a vital role in nutrition. Cellulose is one of several plant products that make up the dietary fiber that is now believed to be important for good health.

Because of its structural properties, cellulose has enormous economic importance. Products such as wood, paper, and textiles (e.g., cotton, linen, and ramie) owe many of their unique characteristics to their cellulose content.

Heteroglycans

Heteroglycans are high-molecular-weight carbohydrate polymers that contain more than one kind of monosaccharide. The major classes of heteroglycans found in mammals are N- and O-linked heteropolysaccharides (N- and O-glycans) attached to proteins, the glycosaminoglycans of the extracellular matrix, and the glycan components of glycolipids and GPI (glycosylphosphatidylinositol) anchors. The structure and properties of the N- and O-glycans and the glycosaminoglycans are described next. Discussion of glycolipids and GPI anchors, a means for attaching peripheral proteins to membrane, is deferred to Chapter 11.

N- AND O-GLYCANS Many proteins have both N- and O-linked oligosaccharides that can comprise a significant proportion of the molecule's molecular weight. The N-linked oligosaccharides (**N-glycans**) are linked via a β-glycosidic bond between the core N-acetylglucosamine anomeric carbon and a side chain amide nitrogen of an asparagine residue. The three major types, high mannose, complex, and hybrid, were shown in Figure 7.32. In addition to N-acetylglucosamine, the most commonly observed sugars among the N-glycans include mannose, galactose, N-acetylneuraminic acid, glucose, and fucose. Note that all N-glycans have the same core structure:

$$^+NH_3$$

Man——β 1,4——GlcNac——β 1,4——GlcNAc——A$_{SN}$

N-Glycan Core Structure

$$C=O$$
$$|$$
$$O^-$$

The O-linked oligosaccharides (**O-glycans**) have a disaccharide core of galactosyl-β-(1,3)-N-acetylgalactosamine linked to the protein via an α-glycosidic bond to the hydroxyl oxygen of serine or threonine residues. In the collagens, the core β-linked disaccharide may be Gal-Gal or Glc-Gal and is linked to the side chain hydroxyl oxygen of 5-hydroxylysine. Other sugars found in O-glycans are N-acetylneuraminic acid, xylose, sialic acid, and fucose.

GLYCOSAMINOGLYCANS Glycosaminoglycans (GAGs) are linear polymers with disaccharide repeating units (Table 7.1). Many of the sugar residues are amino derivatives. The repeating units contain a hexuronic acid (a uronic acid

TABLE 7.1   Disaccharide Repeating Units in Selected Glycosaminoglycans

Name	Repeating Unit	Molecular Weight (D)	Comments
Chondroitin sulfate	(1,4)-*O*-β-D-Glucopyranosyluronic acid-(1,3)-2-acetamido-2-deoxy-6-*O*-sulfo-β-D-galactopyranose	5,000–50,000	May also have sulfate on carbon 6. Important component of cartilage.
Dermatan sulfate	(1,4)t-*O*-α-L-Idopyranosyluronic acid-(1,3)-2-acetamido-2-deoxy-4-*O*-sulfo-β-D-galactopyranose	15,000–40,000	Varying amounts of D-glucuronic acid may be present. Concentration increases during aging process
Heparin	(1,4)-*O*-α-D-Glucopyranosyluronic acid-2-sulfo-(1,4)-2-sulfamido-2-deoxy-6-*O*-sulfo-α-D-glucopyranose	6,000–25,000	Anticoagulant activity. Found in mast cells. Also contains D-glucuronic acid
Keratan sulfate	(1,3)-*O*-β-D-Galactopyranose-(1,4)-2-acetamido-2-deoxy-6-*O*-sulfo-β-D-glucopyranose	4,000–19,000	Minor constituent of proteoglycans. Found in cornea, cartilage, and intervertebral disks.
Hyaluronic acid	(1,4)-*O*-β-D-Glucopyranosyluronic acid-(1,3)-2-acetamido-2-deoxy-β-D-glucopyranose	4,000	Most abundant GAG in the vitreous humor of the eye and the synovial fluid of joints.

containing six carbon atoms), except for keratan sulfate, which contains galactose. Usually an *N*-acetylhexosamine sulfate is also present, except in hyaluronic acid, which contains *N*-acetylglucosamine. Many disaccharide units contain both carboxyl and sulfate groups. GAGs are classified according to their sugar residues, the linkages between these residues, and the presence and location of sulfate groups. Five classes are distinguished: *hyaluronic acid, chondroitin sulfate, dermatan sulfate, heparin and heparan sulfate, and keratan sulfate.*

GAGs have many negative charges at physiological pH. The charge repulsion keeps GAGs separated from each other. Additionally, the relatively inflexible polysaccharide chains are strongly hydrophilic. GAGs occupy a huge volume relative to their mass because they attract large volumes of water. For example, hydrated hyaluronic acid may occupy a volume 1000 times greater than its dry state.

7.4 GLYCOCONJUGATES

The compounds that result from the covalent linkages of carbohydrate molecules to both proteins and lipids are collectively known as the **glycoconjugates**. These substances have profound effects on the function of individual cells, as well as the cell-cell interactions of multicellular organisms. There are two classes of carbohydrate-protein conjugate: proteoglycans and glycoproteins. Although both molecular types contain carbohydrate and protein, their structures and functions appear, in general, to be substantially different. The *glycolipids* (Chapter 11), which are oligosaccharide-containing lipid molecules, are found predominantly on the outer surface of plasma membranes.

Proteoglycans

Proteoglycans are distinguished from the more common glycoproteins by their extremely high carbohydrate content, which may constitute as much as 95% of the dry weight of such molecules. These molecules occur on cell surfaces or are secreted into the extracellular matrix. All proteoglycans contain GAG chains that are linked to protein molecules (known as *core proteins*) by N- and O-glycosidic linkages. The diversity of proteoglycans results from both the number of different core proteins and the large variety of classes and lengths of the carbohydrate chains. Examples include the syndecans, glypicans, and aggrecan. The syndecans are a class of heparan sulfate and chondroitin sulfate containing proteoglycans in which the core protein is a transmembrane protein. The glypicans are proteoglycans that contain heparan sulfate and are linked to cell membrane by GPI anchors (p. 400). Aggrecan, a proteoglycan found in abundance in cartilage, consists of a core protein to which are attached over 100 chondroitin sulfate and about 40 keratan sulfate chains. Up to 100 aggrecan monomers are in turn attached to hyaluronic acid to form a proteoglycan aggregate (Figure 7.37).

Proteoglycans are multifunctional owing to their physical and biochemical properties and to their abundance. In addition to their roles in organizing extracellular matrices, proteoglycans participate in all cellular processes that involve events at cell surfaces. For example, the membrane-bound syndecans and glypicans that bind to specific signal molecules (e.g., growth factors) are components in several signal transduction pathways that regulate the cell cycle. Because of their vast numbers of polyionic GAG chains, the aggrecans trap large volumes of water. Consequently, these molecules occupy thousands of times as much space as a densely packed molecule of the same mass. The strength, flexibility, and resilience of cartilage are made possible by the combination of the compressive stiffness contributed by repulsion between the negatively charged GAGs and the tensile strength of collagen fibers.

KEY CONCEPT

Polysaccharide molecules, composed of large numbers of monosaccharide units, are used in energy storage and as structural materials.

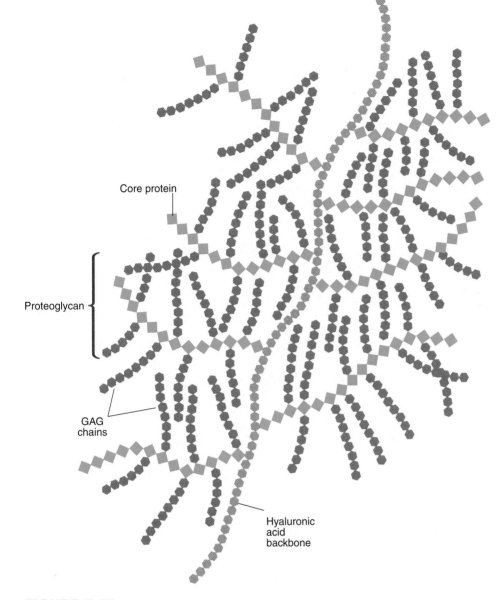

FIGURE 7.37

Proteoglycan Aggregate

Proteoglycan aggregates are typically found in the extracellular matrix of connective tissue. The noncovalent attachment of each aggrecan monomer to hyaluronic acid via the core protein is mediated by two linker proteins (not shown). Proteoglycans interact with numerous fibrous proteins in the extracellular matrix such as collagen, elastin, and fibronectin (a glycoprotein involved in cell adhesion).

A number of genetic diseases associated with proteoglycan metabolism, known as *mucopolysaccharidoses*, have been identified. Because proteoglycans are constantly being synthesized and degraded, their excessive accumulation (caused by missing or defective lysosomal enzymes) has very serious consequences. For example, in *Hurler's syndrome*, deficiency of a specific enzyme causes dermatan sulfate to accumulate. Symptoms include mental retardation, skeletal deformity, and death in early childhood. Like Tay-Sachs disease, Hurler's syndrome is an autosomal recessive disorder (i.e., one copy of the defective gene is inherited from each parent).

Glycoproteins

Glycoproteins are commonly defined as proteins that are covalently linked to carbohydrate through N- or O-linkages. The carbohydrate composition of glycoprotein varies from 1% to more than 85% of total weight. The carbohydrates found include monosaccharides and disaccharides such as those attached to the structural protein collagen and the branched oligosaccharides on plasma glycoproteins. Although the glycoproteins are sometimes considered to include the proteoglycans, for structural reasons they are examined separately. There is a relative absence in glycoproteins of uronic acids, sulfate groups, and the disaccharide repeating units that are typical of proteoglycans.

As described previously, the carbohydrate groups of glycoproteins are linked to the polypeptide by either an N-glycosidic linkage between *N*-acetylglucosamine (GlcNAc) and the amino acid asparagine (Asn) or an O-glycosidic linkage between *N*-acetylgalactosamine (GalNAc) and the hydroxyl group of an amino acid, serine (Ser) or threonine (Thr). The former glycoprotein class is sometimes referred to as *asparagine-linked*; the latter is often called *mucin-type*. All three forms of asparagine-linked oligosaccharides [i.e., high mannose, complex, and hybrid N-glycans (Figure 7.32)] are constructed on a membrane-bound lipid molecule and covalently linked to asparagine residues during ongoing protein synthesis (Chapter 19). Several additional reactions, in the lumen of the endoplasmic reticulum and the Golgi complex, form the final N-linked oligosaccharide structures. Examples of proteins with asparagine-linked oligosaccharides include the iron transport protein transferrin and ovalbumin, a nutritional storage protein in chicken eggs. Mucin-type carbohydrate (O-glycan) units vary considerably in size and structure, from disaccharides such as Gal-1,3-GalNAc, found in the antifreeze glycoprotein of antarctic fish (Figure 7.38), to the complex oligosaccharides of blood groups such as those of the ABO system. O-Glycan synthesis is initiated later than that of N-glycans, most probably in the Golgi complex. The covalent reaction of *N*-acetylgalactosamine with the hydroxyl oxygen of serine or threonine side chains is followed by the addition of monosaccharides such as galactose and sialic acid.

GLYCOPROTEIN FUNCTIONS Glycoproteins are a diverse group of molecules that are ubiquitous constituents of most living organisms (Table 7.2). They occur in cells, in both soluble and membrane-bound form, and in extracellular fluids. Vertebrate animals are particularly rich in glycoproteins.

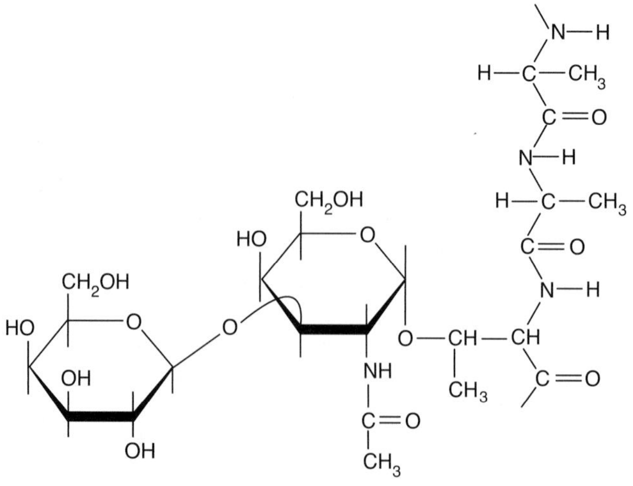

FIGURE 7.38

Antifreeze Glycoprotein

This segment is a recurring glycotripeptide unit. Each disaccharide unit, composed of β-1,3-linked residues of galactose and *N*-acetylgalactosamine, is attached to the polypeptide chain by a glycosidic linkage to a threonine residue.

TABLE 7.2 Glycoproteins

Type	Example	Source	Molecular Weight (D)
Enzyme	Ribonuclease B	Bovine	14,700
Immunoglobulin	Immunoglobulin A	Human	160,000
	Immunoglobulin M	Human	950,000
Hormone	Chorionic gonadotropin	Human placenta	38,000
	Follicle-stimulating hormone	Human	34,000
Membrane protein	Glycophorin	Human red blood cells	31,000
Lectin	Potato lectin	Potato	50,000
(carbohydrate-	Soybean agglutinin	Soybean	120,000
binding proteins)	Ricinus lectin	Castor bean	120,000

Examples include the metal-transport proteins transferrin and ceruloplasmin, the blood-clotting factors, and many of the components of complement (proteins involved in cell destruction during immune reactions). A number of hormones are glycoproteins. Consider, for example, follicle-stimulating hormone (FSH), produced by the anterior pituitary gland. FSH stimulates the development of both eggs and sperm. Additionally, many enzymes are glycoproteins. Ribonuclease (RNase), the enzyme that degrades ribonucleic acid, is a well-researched example. Other glycoproteins are integral membrane proteins (Chapter 11). Of these, Na^+-K^+-ATPase (an ion pump found in the plasma membrane of animal cells) and the major histocompatibility antigens (cell surface markers used to cross-match organ donors and recipients) are especially interesting examples.

Recent research has focused on how carbohydrate stabilizes protein molecules and functions in recognition processes in multicellular organisms. The presence of carbohydrate on protein molecules protects them from denaturation. For example, bovine RNase A is more susceptible to heat denaturation than its glycosylated counterpart RNase B. Several other studies have shown that sugar-rich glycoproteins are relatively resistant to proteolysis (splitting of polypeptides by enzyme-catalyzed hydrolytic reactions). Because the carbohydrate is on the molecule's surface, it may shield the polypeptide chain from proteolytic enzymes.

The carbohydrates in glycoproteins also affect biological function. In some glycoproteins this contribution is more easily discerned than in others. For example, a large content of sialic acid residues is responsible for the high viscosity of salivary mucins (the lubricating glycoproteins of saliva). Another interesting example is the antifreeze glycoproteins of antarctic fish. Their disaccharide residues form hydrogen bonds with water molecules. This process retards the growth of ice crystals.

Glycoproteins, as components of the glycocalyx (p. 47), are now known to be important in complex recognition phenomena. A prime example is the insulin receptor, whose binding to insulin facilitates the transport of glucose into numerous cell types. It does so, in part, by recruiting glucose transporters to the plasma membrane. A variety of cell surface glycoproteins are also involved in cellular adhesion, a critical event in the cell-cell interactions of growth and differentiation (Figure 7.39). The best-characterized of these substances are called cell adhesion molecules (CAMs). Examples include the *selectins* (transient cell-cell interactions), the *integrins* (cell attachment to the components of the extracellular matrix), and the *cadherins* (Ca^{2+}-dependent binding of cells to each other within a tissue). The roles of glycoconjugates in living processes are explored further in the next section.

KEY CONCEPTS

- Glycoconjugates are biomolecules in which carbohydrate is covalently linked to either proteins or lipids.

- Proteoglycans are composed of relatively large amounts of carbohydrate (GAG units) covalently linked to small polypeptide components.

- Glycoproteins are proteins covalently linked to carbohydrate through N- or O-linkages.

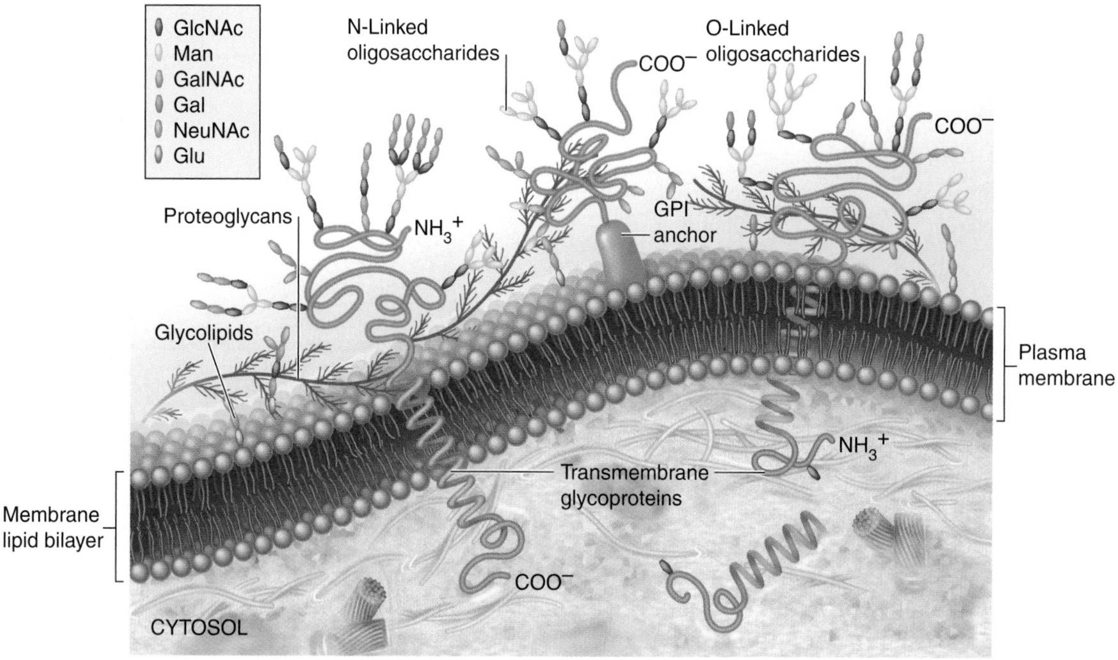

FIGURE 7.39

The Glycocalyx

A GPI (glycosylphosphatidylinositol) anchor is a specialized structure that attaches several diverse types of oligosaccharide to the plasma membrane of some eukaryotic cells.

7.5 THE SUGAR CODE

Living organisms require extraordinarily large, coding capacities because each information transfer event, whether it is the conversion of a substrate to product within an enzyme's active site, the transduction of a hormonal signal, or the engulfment of a bacterial cell by a macrophage, is initiated by the specific binding of one unique molecule by another that has been selected from millions of other nearby molecules. In other words, the functioning of systems as profoundly complicated as living organisms requires a correspondingly large repertoire of molecular codes. To succeed as a coding mechanism, a class of molecules must provide a large capacity for variations in shape because the number of different messages that must be quickly and unambiguously deciphered is tremendous.

For over 50 years the focus of research efforts to understand information flow in biosystems was focused primarily on the nucleic acids DNA and RNA. As a result of this monumental work, life scientists fully expected to find that approximately 100,000 genes existed to code for proteins in humans. Instead, analysis of the data generated by the Human Genome Project (HGP) (pp. 704–706) produced a much lower number: estimates vary (depending on methodology and computer software) between 20,000 and 40,000 protein-coding genes, with most estimates around 30,000.

Whatever the reasons for this surprisingly low number, it is known that living organisms have two strategies for expanding the coding capacity of their genes: alternative splicing and covalent modification. *Alternative splicing* (described in Chapter 18) is a mechanism whereby eukaryotes produce several polypeptides from the same gene by cutting mRNA transcripts and then splicing together various combinations of the RNA fragments. Each type of spliced mRNA product is translated into a unique polypeptide. *Posttranslational modifications* (described in Chapter 19) are enzyme-catalyzed changes in a protein's structure that occur after its synthesis.

Of all the types of posttranslational modification (e.g., phosphorylation, acetylation, and proteolytic cleavage), glycosylation is the most important in terms of coding capacity, as the following examples illustrate. Recall that just 20 amino acids account for the enormous diversity of proteins observed in living organisms. The total number of hexapeptides that can be synthesized from these amino acids is an impressive 20^6 (6.4×10^7). Carbohydrates have structural properties (e.g., glycosidic linkage variations, branching, and anomeric isomers) that provide them with significant coding capacity. In contrast to the peptide linkages that form exclusively between the amino and carboxyl groups in amino acid residues to create a linear peptide molecule, the glycosidic linkages between monosaccharides can be considerably more variable. Consequently, the potential number of permutations in oligosaccharides is substantially higher than that predicted for peptides. For example, the total number of possible linear and branched hexasaccharides that can form from 20 simple or modified monosaccharides is 1.44×10^{15}. In addition to their immense combinatorial possibilities, oligosaccharides, whether they are attached to proteins or lipids, have one other property: their relative inflexibility (in comparison to peptides), which allows them to bind more precisely with ligands.

Lectins: Translators of the Sugar Code

Once information has been encoded, it must be translated. The translation of the sugar code is accomplished by lectins. **Lectins** are carbohydrate-binding proteins (CBPs) that are not antibodies and have no enzymatic activity. Originally discovered in plants, they are now known to exist in all organisms. Lectins, which usually consist of two or four subunits, possess recognition domains that bind to specific carbohydrate groups via hydrogen bonds, van der Waals forces, and hydrophobic interactions. Biological processes that involve lectin binding include an array of cell-cell interactions (Figure 7.40). Prominent examples include infections by microorganisms, the mechanisms of many toxins, and physiological processes such as leukocyte rolling. The essential features of each are briefly described.

Infection by many bacteria is initiated when the microorganisms become firmly attached to host cells. Often, attachment is mediated by the binding of bacterial lectins to oligosaccharides on the cell's surface. *Helicobacter pylori*, the causative agent of gastritis and stomach ulcers, possesses several lectins that allow it to establish a chronic infection in the mucous lining of the stomach. One of these lectins binds with high affinity to a portion of the type O blood group determinant, an oligosaccharide. This circumstance explains the observation that humans with type O blood are at considerably greater risk of developing ulcers than those with other blood types. Individuals with type A or B blood are not immune to infection, however, since the bacterium can also use other lectins to achieve adhesion.

The damaging effects of many bacterial toxins occur only after endocytosis into the host cell, a process that is initiated by lectin-ligand binding. The binding of the B subunit of cholera toxin to a glycolipid on the surface of intestinal cells results in the uptake of the toxic A subunit. Once internalized, the A subunit proceeds to disrupt the mechanism that regulates chloride transport, a process that results in a life-threatening diarrhea.

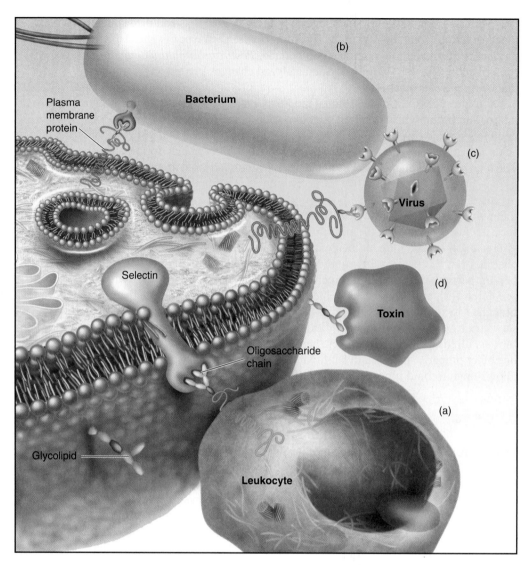

FIGURE 7.40

Role of Oligosaccharides in Biological Recognition

The specific binding of lectins (carbohydrate-binding proteins) to the oligosaccharide groups of glyco-conjugate molecules is an essential feature of many biological phenomena. (a) Cell-cell interactions (e.g., leukocyte rolling), (b and c) cell-pathogen infections, and (d) the binding of toxins (e.g., cholera toxin) to cells.

Leukocyte rolling is a well-known example of cell-cell interaction mediated by lectin binding. When a tissue becomes damaged in an animal either by infection with a pathogenic organism or by physical trauma, it emits signal molecules that trigger an inflammatory process. In response to certain of these molecules, the endothelial cells that line nearby blood vessels produce and insert into their plasma membranes a protein called selectin. The selectins, a family of lectins that act as cell adhesion molecules, are displayed on the surface of the endothelial cells. They bind transiently to the *selectin* ligand (an oligosaccharide) on white blood cells such as neutrophils. These relatively weak binding events can serve to slow the rapid motion of the neutrophils as they flow in blood, causing them to appear to be rolling along the luminal surface of the blood vessel. Once rolling has been initiated, and white blood cells approach the inflammation site, they encounter other signal molecules that cause them to express another lectin called *integrin* on their surfaces. The binding of integrin with its oligosaccharide ligand on the endothelial surface of the blood vessel causes the neutrophils to stop rolling. Subsequently, the neutrophils undergo changes that allow them to squeeze between the cells of the endothelium and migrate to the infected site, where they proceed to consume and degrade bacteria and cellular debris.

The Glycome

The term **glycome**, derived from *glyco* (sweet) and *-ome* (as in genome), was created to describe the total set of sugars and glycans that a cell or organism produces. Glycomes are constantly in flux because cells responding to environmental signals fine-tune biological responses by altering the glycan structures attached to proteins and lipids. This capacity exists in large part because there is no template for glycan biosynthesis. In contrast to nucleic acid and protein biosynthesis, which are template-driven processes (i.e., multiple identical copies are produced using a nucleotide base sequence), glycans are constructed stepwise, on an assembly line within the ER and Golgi complex. Because of factors such as variations in sugar nucleotide precursor concentrations and the localization of glycan-processing enzymes, the glycan components of each type of glycoprotein may be produced in a series of slightly different forms called **glycoforms**. It has been suggested that this phenomenon, referred to as **microheterogeneity**, may be a means by which cells can generate cell- or tissue-specific signal transduction ligands and/or a mechanism whereby cells elude pathogens whose binding to certain glycan structures initiates an infective process.

QUESTION 7.8

The sugar code, with its diverse and subtle nontemplate mechanism for encoding information, has been described as an "analog" system, whereas genetic information processing (DNA- and RNA-directed protein synthesis) is considered to be "digital." Explain.

COMPANION

WEBSITE Visit the companion website at www.oup.com/us/mckee to read the Biochemistry in the Lab box on glycomics.

KEY CONCEPTS

- The covalent modification of biomolecules such as proteins and lipids provides living organisms with enormous coding capacity.

- The entire set of sugars and glycans produced by a cell or organism is referred to as the glycome.

- Glycoproteins are often produced in slightly different versions called glycoforms.

BIOCHEMISTRY IN PERSPECTIVE

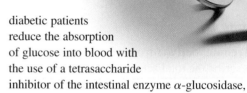

Sweet Medicine

How does carbohydrate biochemistry impact human health? The importance of carbohydrates in human biological processes is emphasized by the following statistics: at least 200 genes code for glycan-processing enzymes, and at least 50% of all proteins are covalently linked to one or more carbohydrate groups. Thus it is not surprising that defects in glycan synthesis and degradation can have very serious health consequences. For example, there are at least 30 rare genetic diseases called *congenital disorders of glycosylation* (CDGs), in which various steps in N-glycan processing are defective. Other rare carbohydrate-related diseases include several forms of muscular dystrophy, as well as Tay-Sachs (p. 385) and Gaucher's diseases. Carbohydrates are now known to play roles in several more common illnesses.

For example, changes in the galactose content of the antibody class IgG have recently been shown to be directly related to the severity (i.e., the degree of inflammation) of juvenile arthritis. Additionally, recent evidence indicates that changes in glycosylation patterns accompany changes in the behavior of cancer cells. Infectious diseases are often initiated by the binding of pathogen CBPs to glycoconjugate receptors on host cell surfaces. Glycan research has increased the accessibility of these disorders to investigation.

Biological processes with prominent glycan involvement are potential targets for therapeutic intervention. With several notable exceptions, however, until the recent introduction of innovations that improved glycan analysis and synthesis technologies, attempts to develop carbohydrate-based therapies have failed. The most prominent examples of successful attempts are the DNA recombinant glycoprotein products erythropoietin (which stimulates red blood cell production) and tissue plasminogen activator (a blood-clot-dissolving serine protease, see Biochemistry in Perspective box on Enzymes and Clinical Medicine in Chapter 6), and heparin, a GAG that prevents blood clot formation. Currently, numerous carbohydrate-based drugs are either in use or in clinical trials. Among these are the following.

1. **Enzyme inhibitors**. Certain glycan molecules can be used to inhibit specific enzyme activities. For example, some diabetic patients reduce the absorption of glucose into blood with the use of a tetrasaccharide inhibitor of the intestinal enzyme α-glucosidase,

2. **Enzyme replacement**. The therapeutic replacement of missing enzymes can be effective only if the new enzymes can be targeted to their site of action within cells. The missing enzyme in Gaucher's disease (β-glucosidase) has been successfully engineered by altering its oligosaccharide group so that it binds to the intracellular mannose receptor that ensures delivery to lysosomes.

3. **Vaccines**. Vaccines, substances that protect against disease by triggering antibody production, are often composed of antigenic proteins. Despite current problems associated with large-scale synthesis, glycans are potentially useful in vaccine production because they have a lower risk of allergic reactions. In addition, they are more stable and therefore easier to convert into therapeutic agents. Because of their low immunogenic potential, glycan vaccines are usually composed of a glycan linked to certain proteins known to boost immune responsiveness. A glycan-based vaccine called Hib protects against infection by the bacterium *Hemophilus influenzae* type b, the leading cause of meningitis in children. Since its introduction, Hib has significantly reduced the occurrence of this potentially lethal disease.

4. **Redesigned drugs**. Careful analysis of the function and physiological processing of existing glycan-based drugs can result in the production of redesigned (or "second-generation") drugs. For example, the effectiveness of recombinant erythropoietin was increased by the attachment of additional N-linked glycans that both improved its biological activity and reduced its rate of excretion.

SUMMARY: Carbohydrate biochemistry provides insight into several human diseases. Carbohydrate biotechnology is being used to develop carbohydrate-based drugs and vaccines.

Chapter**Summary**

1. Carbohydrates, the most abundant organic molecules in nature, are classified as monosaccharides, disaccharides, oligosaccharides, and polysaccharides according to the number of simple sugar units they contain. Carbohydrate moieties also occur as components of other biomolecules. Glycoconjugates are protein and lipid molecules with covalently linked carbohydrate groups. They include proteoglycans, glycoproteins, and glycolipids.

2. Monosaccharides with an aldehyde functional group are called aldoses; those with a ketone group are known as ketoses. Simple sugars belong to either the D or L family, according to whether the configuration of the asymmetric carbon farthest from the aldehyde or ketone group resembles the D- or L-isomer of glyceraldehyde. The D family aldoses contains most biologically important sugars.

3. Sugars containing five or six carbons exist in cyclic forms that result from reactions between hydroxyl groups and either aldehyde (hemiacetal product) or ketone groups (hemiketal product). In both five-membered rings (furanoses) and six-membered rings (pyranoses), the hydroxyl group attached to the anomeric carbon lies either below (α) or above (β) the plane of the ring. The spontaneous interconversion between α- and β-forms is called mutarotation.

4. Simple sugars undergo a variety of chemical reactions. Derivatives of these molecules, such as uronic acids, amino sugars, deoxy sugars, and phosphorylated sugars, have important roles in cellular metabolism. Glycosylation reactions attach sugars to proteins or lipids. Glycation reactions are nonenzymatic reactions in which reducing sugars react with nucleophilic nitrogens.

5. Hemiacetals and hemiketals react with alcohols to form acetals and ketals, respectively. When the cyclic hemiacetal or hemiketal form of a monosaccharide reacts with an alcohol, the new linkage is called a glycosidic linkage, and the compound is called a glycoside.

6. Glycosidic bonds form between the anomeric carbon of one monosaccharide and one of the free hydroxyl groups of another monosaccharide. Disaccharides are carbohydrates composed of two monosaccharides. Oligosaccharides, carbohydrates that typically contain as many as 10 monosaccharide units, are often attached to proteins and lipids. Polysaccharide molecules, which are composed of large numbers of monosaccharide units, may have a linear structure like cellulose and amylose or a branched structure like glycogen and amylopectin. Oligosaccharides and polysaccharides are now referred to as glycans. Glycans may consist of only one sugar type (homoglycans) or multiple types (heteroglycans).

7. The three most common homoglycans found in nature (starch, glycogen, and cellulose) all yield D-glucose when hydrolyzed. Cellulose is a plant structural material; starch and glycogen are storage forms of glucose in plant and animal cells, respectively. Chitin, the principal structural material in insect exoskeletons, is composed of N-acetylglucosamine residues linked in unbranched chains. The major classes of heteroglycans, carbohydrate polymers that contain more than one kind of monosaccharide, are N- and O-glycans, glycosaminoglycans, and the glycan components of glycolipids and GPI-anchors.

8. The enormous heterogeneity of proteoglycans, which are found predominantly in the extracellular matrix of tissues, allows them to play diverse, but as yet poorly understood, roles in living organisms. Glycoproteins occur in cells, in both soluble and membrane-bound forms, and in extracellular fluids. The diverse structures of the glycoconjugates, which include proteoglycans, glycoproteins, and glycolipids, allow them to play important roles in information transfer in living organisms. The glycome is the total set of sugars and glycans that a cell or organism produces.

 Take your learning further by visiting the **companion website** for Biochemistry at **www.oup.com/us/mckee** where you can complete a multiple-choice quiz on carbohydrates to help you prepare for exams.

Suggested**Readings**

Bouvet, V., and Ben, R. N., Antifreeze Glycoproteins: Structure, Conformation, and Biological Applications, *Cell Biochem. Biophys.* 39(2):133–144, 2003.

Feizi, T., and Chai, W., Oligosaccharide Microarrays to Decipher the Glyco Code, *Nat. Rev. Mol. Cell Biol.* 5(7):582–588, 2004.

Freeze, H., Genetic Defects in the Human Glycome, *Nat. Rev. Gene.* 7(7):537–551, 2006.

Gabius, H.-J., Biological Information Transfer Beyond the Genetic Code: The Sugar Code, *Naturwissenschaften* 87(3):108–121, 2000.

Gabius, H.-J., Siebert, H.-C., Andre, S., Jimenez-Barbero, J., and Rudiger, H., Chemical Biology of the Sugar Code, *ChemBioChem* 5(6):740–764, 2004.

Maeder, T., Sweet Medicines, *Sci. Am.* 287(1):40–47, 2002.

Melton, L., AGE Breakers: Rupturing the Body's Sugar-Protein Bonds Might Turn Back the Clock, *Sci. Am.* 283(1):16, 2000.

Merry, A. H., and Merry, C. L. R., Glycoscience Finally Comes of Age, *EMBO Rep.* 6(10):900–903, 2005.

Sasisekaran, R., and Myette, J. R., The Sweet Science of Glycobiology, *Am. Sci.* 91:432–441, 2003.

Seeberger, P. H., Exploring Life's Sweet Spot, *Nature* 437:1239, 2005.

Shriver, Z., Raguram, S., and Sasisekharan, R., Glycomics: A Pathway to a Class of New and Improved Therapeutics, *Nat. Rev. Drug Discovery* 3(10):863–873, 2004.

Raman, R., Raguram, S., Venkataraman, G., Paulson, J. C., and Sasisekharan, R., Glycomics: An Integrated Systems Approach to Structure-Function Relationships of Glycans, *Nat. Methods* 2(11):817–824, 2005.

Varki, A., Cummings, R., Esko, J., Freeze, H., Hart, G., and Marth, J. (eds.), *Essentials of Glycobiology*, Cold Spring Harbor Laboratory Press, New York, 1999.

Key Words

acetal, *237*
adduct, *238*
aldaric acid, *234*
alditol, *235*
aldonic acid, *234*
aldose, *228*
amylopectin, *245*
amylose, *245*
anomer, *231*
cellobiose, *242*
cellulose, *247*
chitin, *245*

diastereomer, *229*
disaccharide, *237*
enediol, *235*
epimer, *229*
epimerization, *236*
glycan, *243*
glycoconjugate, *250*
glycoform, *257*
glycogen, *246*
glycome, *257*
glycomics, *257*
glycosaminoglycan, *248*

glycoside, *237*
glycosidic linkage, *237*
hemiacetal, *231*
hemiketal, *231*
heteroglycan, *243*
homoglycan, *243*
ketal, *237*
lactone, *234*
lactose, *242*
lectin, *255*
maltose, *242*

microheterogeneity, *257*
monosaccharide, *228*
mutarotation, *233*
N-glycan, *248*
O-glycan, *248*
oligosaccharide, *243*
polysaccharide, *237*
proteoglycan, *250*
reducing sugars, *235*
sucrose, *243*
uronic acid, *234*

Review Questions

These questions are designed to test your knowledge of the key concepts discussed in this chapter, before moving on to the next chapter. You may like to compare your answers to the solutions provided in the back of the book and in the accompanying Study Guide.

1. Define the following terms:
   a. hemiacetal
   b. space-filling model
   c. uronic acid
   d. aldaric acid
   e. lactone

2. Give an example of each of the following:
   a. epimer
   b. glycosidic linkage
   c. reducing sugar
   d. monosaccharide
   e. anomer
   f. diastereomer

3. What structural relationship is indicated by the term D-sugar? Why are (+) glucose (shifts polarized light to the right) and (−) fructose (shifts polarized light to the left) both classified as D-sugars?

4. Name an example of each of the following classes of compounds:
   a. glycoprotein
   b. proteoglycan
   c. disaccharide
   d. glycosaminoglycan (GAG)

5. Define the following terms:
   a. alditol
   b. epimerization
   c. enediol
   d. acetal
   e. glycoside

6. What is the difference between a heteroglycan and a homoglycan? Give examples.

7. Which of the following carbohydrates are reducing and which nonreducing?
   a. starch
   b. cellulose
   c. fructose
   d. sucrose
   e. ribose

8. Define the following terms:
   a. glycosylation
   b. glycation
   c. atherosclerosis
   d. AGE
   e. deoxy sugar

9. What structural differences characterize starch, cellulose, and glycogen?

10. Draw the structure of a disaccharide unit in a polysaccharide composed of D-glucose linked α(1,4) to D-galactosamine.

11. Determine the number of possible stereoisomers for the following compounds:

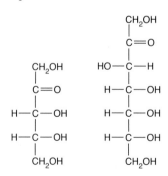

Ribulose   Sedoheptulose

Use asterisks to indicate the chiral carbons in each molecule.

12. Define the following terms:
    a. glycan
    b. heteroglycan
    c. homoglycan
    d. GAG
    e. syndecan

13. Raffinose is the most abundant trisaccharide found in nature.
    a. Provide the systematic name for this trisaccharide.
    b. Is raffinose a reducing or nonreducing sugar?
    c. Is raffinose capable of mutarotation?

14. Give at least one function of each of the following:
    a. glycogen
    b. glycosaminoglycans
    c. glycoconjugates
    d. proteoglycans
    e. glycoproteins
    f. polysaccharides

15. The polymer chains of glycosaminoglycans are widely spread apart and bind large amounts of water.
    a. What two functional groups of the polymer make this binding of water possible?
    b. What type of bonding is involved?

16. In glycoproteins, what are the three amino acids to which the carbohydrate groups are most frequently linked? To what functional group is the glycan linked in each case?

17. Chondroitin sulfate chains have been likened to a large fishnet, passing small molecules through their matrix but excluding large ones. Use the structure of chondroitin sulfate and proteoglycans to explain this analogy.

18. Define the term *reducing sugar*. What structural feature does a reducing sugar have?

19. Define the following terms:
    a. Hurler's syndrome
    b. heparin
    c. lectin
    d. N-glycan
    e. O-glycan

20. Compare the structures of proteoglycans and glycoproteins. How are structural differences related to their functions?

21. What role is carbohydrate thought to play in maintaining glycoprotein stability?

22. How does the structure of cellulose differ from starch and glycogen?

23. Classify each of the following sugar pairs as enantiomers, diastereomers, epimers, or an aldose-ketose pair.
    a. D-erythrose and D-threose
    b. D-glucose and D-mannose
    c. D-ribose and L-ribose
    d. D-allose and D-galactose
    e. D-glyceraldehyde and dihydroxyacetone

24. Define the following terms:
    a. sugar code
    b. glycome
    c. glycoforms
    d. microheterogeneity
    e. glycomics

ThoughtQuestions

These questions are designed to reinforce your understanding of all of the key concepts discussed in the book so far, including this chapter and all of the chapters before it. They may not have one right answer! The authors have provided possible solutions to these questions in the back of the book and in the accompanying Study Guide, for your reference.

25. β-Galactosidase is an enzyme that hydrolyses only β(1,4) linkages of lactose. An unknown trisaccharide is converted by β-galactosidase into maltose and galactose. Draw the structure of the trisaccharide.

26. Steroids are large, polycyclic complex, lipid-soluble molecules that are very insoluble in water. Reaction with glucuronic acid makes a steroid much more water soluble and enables transport through the blood. What structural feature of the glucuronic acid increases the solubility?

27. Many bacteria are surrounded by a proteoglycan coat. Use your knowledge of the properties of this substance to suggest a function for such a coat.

28. It has long been recognized that breast milk protects infants from infectious diseases, especially those that affect the digestive tract. The main reason for this protection appears to be a large group of oligosaccharides that are components of human milk. Suggest a rationale for the protective effect of these oligosaccharides.

29. Describe the steps in the conversion of the Fischer formula of a sugar to a Haworth formula: for a furanose or pyranose sugar, draw a five- or six-membered ring, respectively, with oxygen placed as shown.

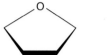

Five-membered ring   **Six-membered ring**

Place the hydroxyl groups above the plane of the ring if they are pointing to the left in the Fischer formula and below the ring if they are pointing to the right. In D-sugars, the last carbon position (e.g., C-6 in glucose) is always pointing up. Draw the furanose forms of the following sugars:

(a)   (b)   (c)   (d)

30. Draw the pyranose forms of the sugars in Question 29.

31. Provide the complete names for the sugars in Questions 29 and 30.

32. A polysaccharide is found in the shells of arthropods (e.g., lobsters and grasshoppers) and of mollusks (e.g., oysters and snails). It can be obtained from these sources by soaking the shells in cold dilute hydrochloric acid to dissolve the calcium carbonate. The threadlike substance formed is composed of linear long-chain molecules. Hydrolysis with boiling acid gives D-glucosamine and acetic acid in equimolar amounts. Milder enzymatic hydrolysis gives *N*-acetyl-D-glucosamine as the sole product. The polysaccharide's linkages are identical to those of cellulose. What is the structure of this polymer?

33. Proteoglycan aggregates in tissues form hydrated, viscous gels. Can you think of any obvious mechanical reason why their capacity to form gels is important to cell function? [*Hint:* Liquid water is virtually incompressible.]

34. Alginic acid, isolated from seaweed and used as a thickening agent for ice cream and other foods, is a polymer of D-mannuronic acid with $\beta(1,4)$ glycosidic linkages.
    a. Draw the structure of alginic acid.
    b. Why does this substance act as a thickening agent?

D-Mannuronic acid

35. What is the maximum number of stereoisomers for mannuronic acid?

36. The ABO blood group antigens are the terminal sugars covalently linked to the end of glycolipid in the red blood cell membrane. The H antigen is the precursor of the A and B antigens. Individuals with type A blood produce a gene that codes for an enzyme that adds *N*-acetylgalactosamine in an $\alpha(1,3)$ linkage to the Gal* residue in the H antigen. Type B

blood requires that an enzyme add a D-α galactose in an α(1,3) linkage to the Gal*. Draw the structures of the A and B antigens.

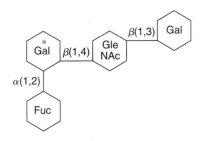

H Antigen

37. What would happen to the H antigen precursor (see Question 36) if an individual has both A and B genes? Keep in mind that the substrate of the respective enzymes is the H antigen.

38. Phosphate esters can form at positions 2 to 6 of an aldohexose but not at position 1. Explain.

39. When glucose is reduced, only one alditol is produced. When fructose undergoes the same reaction, however, two diasteriomeric sugars are produced. Draw their structures.

40. In the liver, the water solubility of some hydrophobic molecules (e.g., drug molecules and steroid hormones) is improved by converting them into sulfate derivatives. Once such molecules are water soluble, they can be easily excreted. How does sulfate improve solubility?

41. Sucrose does not undergo mutarotation. Explain.

42. An oligosaccharide isolated from an organism is found to contain two glucose residues and one galactose residue.

Exhaustive methylation followed by hydrolysis produced two glucoses with methoxy groups at positions 2, 3, and 6 and galactose with methoxy groups at positions 2, 3, 4, and 6. What is the structure of the original oligosaccharide?

43. A newly isolated aldohexose is oxidized to produce the corresponding aldaric acid that has an internal plane of symmetry; that is, it is a symmetrical molecule. What is the structure of the original aldohexose?

44. What sugar is produced by the epimerization of galactose?

45. Before a sugar can be analyzed by gas chromatography (GC) or GC/MS, it must be converted to a volatile derivative. Why can't sugars be analyzed directly?

46. The carbohydrate molecule 3-ketoglucose can exist in several ring forms. Draw them and determine which is the most stable.

47. Suggest a reason why sorbitol prevents moisture loss in candy.

48. Glyceraldehyde, the simplest aldose, does not have a cyclic form when tetroses with only one more carbon can form a ring readily. Can you suggest a reason for this phenomenon?

49. Olestra has been used in certain snack foods as an alternative to fats and oils. Its structure consists of a sucrose molecule in which all free hydroxyl groups have formed esters with oleic acid (an 18-carbon monounsaturated fatty acid). Olestra molecules contain no calories because they are exceptionally large and cannot be digested. Draw the structure of olestra. Use R—COOH as an abbreviation for oleic acid.

50. For a sugar to behave as a reducing sugar, it must have a free aldehyde group. Fructose is a ketose, yet it behaves like a reducing sugar. Explain.

Carbohydrate Metabolism

Wine: A Product of Fermentation Humans use microorganisms, in this case yeast, to metabolize sugar in the absence of oxygen. The aging of wine in oak barrels improves its taste and aroma.

Overview

▼

CARBOHYDRATES PLAY SEVERAL CRUCIAL ROLES IN THE METABOLIC PROCESSES OF LIVING ORGANISMS. THEY SERVE AS ENERGY SOURCES AND as structural elements in living cells. Chapter 8 focuses on the role of carbohydrates in energy production. Because the monosaccharide glucose is a prominent energy source in almost all living cells, major emphasis is placed on its synthesis, degradation, and storage. The use of other sugars is also discussed.

L iving cells are in a state of ceaseless activity. To maintain its "life" each cell depends on highly coordinated biochemical reactions. Carbohydrates are an important source of the energy that drives these reactions. In Chapter 8, the energy-generating pathways of carbohydrate metabolism are discussed. During **glycolysis**, an ancient pathway found in almost all organisms, a small amount of energy is captured as a glucose molecule is converted to two molecules of pyruvate. Glycogen, a storage form of glucose in vertebrates, is synthesized by **glycogenesis** when glucose levels are high and degraded by **glycogenolysis** when glucose is in short supply. Glucose can also be synthesized from noncarbohydrate precursors by reactions referred to as **gluconeogenesis**. The **pentose phosphate pathway** enables cells to convert glucose-6-phosphate, a derivative of glucose, to ribose-5-phosphate (the sugar used to synthesize nucleotides and nucleic acids) and other types of monosaccharides. NADPH, an important cellular reducing agent, is also produced by this pathway. In Chapter 9, the *glyoxylate cycle*, used by some organisms (primarily plants) to manufacture carbohydrate from fatty acids, is considered. *Photosynthesis*, a process in which light energy is captured to drive carbohydrate synthesis, is described in Chapter 13.

Any discussion of carbohydrate metabolism focuses on the synthesis and usage of glucose, the major fuel of most organisms. In vertebrates, glucose is transported throughout the body in the blood. If cellular energy reserves are low, glucose is degraded by the glycolytic pathway. Glucose molecules not required for immediate energy production are stored as glycogen in liver and muscle. The energy requirements of many tissues (e.g., brain, red blood cells, and exercising skeletal muscle cells) depend on an uninterrupted flow of glucose. Depending on a cell's metabolic requirements, glucose can also be used to synthesize, for example, other monosaccharides, fatty acids, and certain amino acids. Figure 8.1 summarizes the major pathways of carbohydrate metabolism in animals.

8.1 GLYCOLYSIS

Glycolysis, a series of reactions that occur, at least in part, in almost every living cell, is believed to be among the oldest of all the biochemical pathways. Both the enzymes and the number and mechanisms of the steps in the pathway are highly conserved in prokaryotes and eukaryotes. Also, glycolysis is an anaerobic process, which would have been necessary in the oxygen-poor atmosphere of pre-eukaryotic Earth.

In glycolysis, also referred to as the *Embden-Meyerhof-Parnas pathway*, each glucose molecule is split and converted to two three-carbon units (pyruvate). During this process several carbon atoms are oxidized. The small amount of energy captured during glycolytic reactions (about 5% of the total available) is stored temporarily in two molecules each of ATP and NADH. The subsequent metabolic fate of pyruvate depends on the organism being considered and its

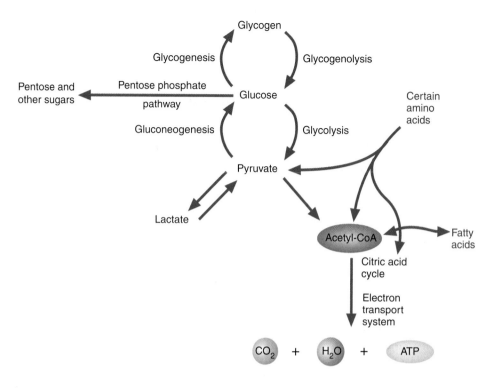

FIGURE 8.1

Major Pathways in Carbohydrate Metabolism

In animals, excess glucose is converted to its storage form, glycogen, by glycogenesis. When glucose is needed as a source of energy or as a precursor molecule in biosynthetic processes, glycogen is degraded by glycogenolysis. In some cells glucose is converted to ribose-5-phosphate (a component of nucleotides) and NADPH (a powerful reducing agent) by means of the pentose phosphate pathway. Glucose is oxidized by glycolysis, an energy-generating pathway that converts it to pyruvate. In the absence of oxygen, pyruvate is converted to lactate. When oxygen is present, pyruvate is further degraded to form acetyl-CoA. Significant amounts of energy in the form of ATP can be extracted from acetyl-CoA by the citric acid cycle and the electron transport system. Note that carbohydrate metabolism is inextricably linked to the metabolism of other nutrients. For example, acetyl-CoA is also generated from the breakdown of fatty acids and certain amino acids. When acetyl-CoA is present in excess, a different pathway converts it into fatty acids.

metabolic circumstances. In **anaerobic organisms** (those that do not use oxygen to generate energy), pyruvate may be converted to waste products such as ethanol, lactic acid, acetic acid, and similar molecules. Using oxygen as a terminal electron acceptor, aerobic organisms such as animals and plants completely oxidize pyruvate to form CO_2 and H_2O in an elaborate stepwise mechanism known as **aerobic respiration**.

Glycolysis (Figure 8.2), which consists of 10 reactions, occurs in two stages:

1. Glucose is phosphorylated twice and cleaved to form two molecules of glyceraldehyde-3-phosphate (G-3-P). The two ATP molecules consumed during this stage are like an investment, because this stage creates the actual substrates for oxidation in a form that is trapped inside the cell.

2. Glyceraldehyde-3-phosphate is converted to pyruvate. Four ATP molecules and two NADH are produced. Because two ATP were consumed in stage 1, the net production of ATP per glucose molecule is 2.

The glycolytic pathway can be summed up in the following equation:

$$\text{D-Glucose} + 2\,\text{ADP} + 2\,P_i + 2\,\text{NAD}^+ \rightarrow 2\,\text{pyruvate} + 2\,\text{ATP} + 2\,\text{NADH} + 2\text{H}^+ + 2\text{H}_2\text{O}$$

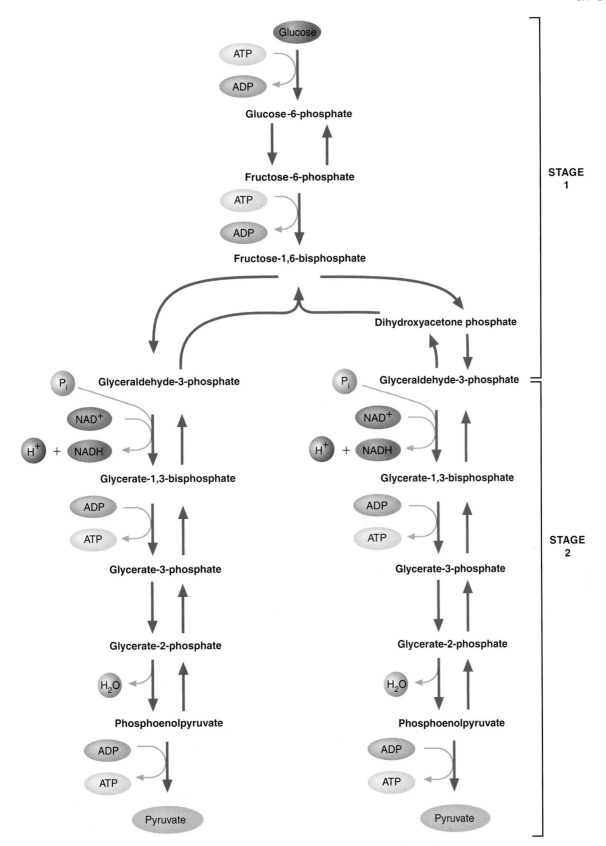

FIGURE 8.2

The Glycolytic Pathway

In glycolysis each glucose molecule is converted into two pyruvate molecules. In addition, two molecules each of ATP and NADH are produced. Reactions with double arrows are reversible reactions, and those with single arrows are irreversible reactions that serve as control points in the pathway.

The Reactions of the Glycolytic Pathway

Glycolysis is summarized in Figures 8.3 and 8.4. The 10 reactions of the gly-colytic pathway are as follows.

1. **Synthesis of glucose-6-phosphate**. Immediately after entering a cell, glucose and other sugar molecules are phosphorylated. Phosphorylation prevents transport of glucose out of the cell and increases the reactivity of the oxygen in the resulting phosphate ester. Several enzymes, called the hexokinases, catalyze the phosphorylation of hexoses in all cells in the body. ATP, a cosubstrate in the reaction, is complexed with Mg^{2+}. (ATP-Mg^{2+} complexes are common in kinase-catalyzed reactions.) Under intracellular conditions the reaction is irreversible; that is, the enzyme has no ability to retain or accommodate the product of the reaction in its active site, regardless of the concentration of G-6-P.

Glucose + ATP → (Hexokinase, Mg^{2+}) → **Glucose-6-phosphate** + ADP

The animal liver has four hexokinases. Three of these enzymes (hexokinases I, II, and III) are found in varying concentrations in other body tissues, where they bind reversibly to an anion channel (called a *porin*) in the outer membrane of mitochondria. As a result, ATP is readily available. These isozymes have high affinities for glucose relative to its concentration in blood; that is, they are half-saturated at concentrations of less than 0.1 mM, although blood glucose levels are approximately 4 to 5 mM. In addition, hexokinases I, II, and III are inhibited from phosphorylating glucose molecules by glucose-6-phosphate, the product of the reaction. When blood glucose levels are low, these properties allow cells such as those in brain and muscle to obtain sufficient glucose. When blood glucose levels are high, cells do not phosphorylate more glucose molecules than required to meet their immediate needs. The fourth enzyme, called hexokinase IV (or glucokinase), catalyzes the same reaction but has significantly different kinetic properties. Gluco-kinase (GK), found in liver as well as in certain cells in pancreas, intestine, and brain, requires much higher glucose concentrations for optimal activity (about 10 mM), and it is not inhibited by glucose-6-phosphate. In liver, GK diverts glucose into storage as glycogen. This capacity provides the resources used to maintain blood glucose levels, a major role of the liver. Consequently, after a carbohydrate meal the liver does not remove large quantities of glucose from the blood for glycogen synthesis until other tissues have satisfied their requirements for this molecule. In cell types where it occurs, GK is believed to be a *glucose sensor*. Because glucokinase does not usually work at maximum velocity, it is highly sensitive to small changes in blood glucose levels. Its activity is linked to a signal transduction pathway. For example, the release of insulin (the hormone that promotes the uptake of glucose into muscle and adipose tissue cells) by pancreatic β-cells in response to rising blood levels of glucose is initiated by GK. GK regulation involves its binding to GK regulator protein (GKRP), a process that is triggered by high fructose-6-phosphate levels. GKRP/GK then moves into the nucleus. When blood glucose levels rise after a meal, GKRP releases GK (caused by exchange with fructose-1-phosphate), and GK moves back through the nuclear pores and can again phosphorylate glucose.

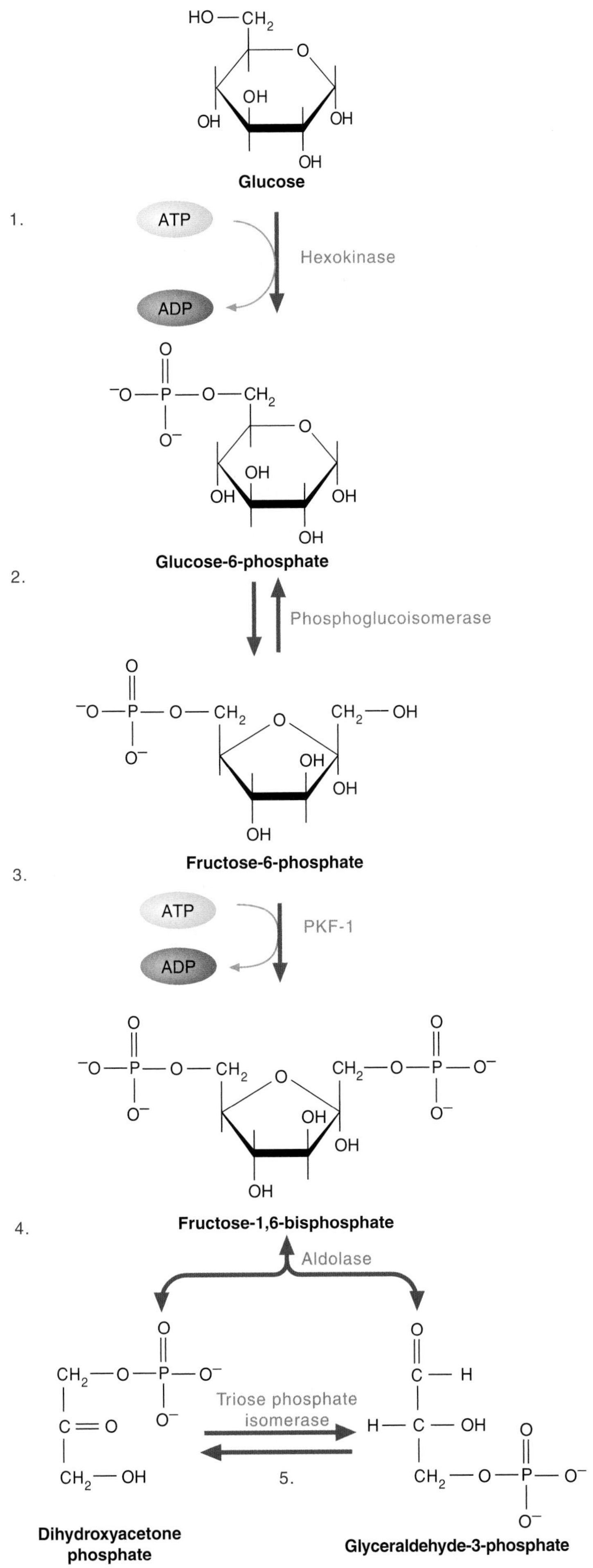

FIGURE 8.3

Stage 1: Reactions of Glycolysis

In all there are 10 reactions in the glycolytic pathway. The number of each reaction is highlighted in red.

FIGURE 8.4
Stage 2 Reactions of Glycolysis

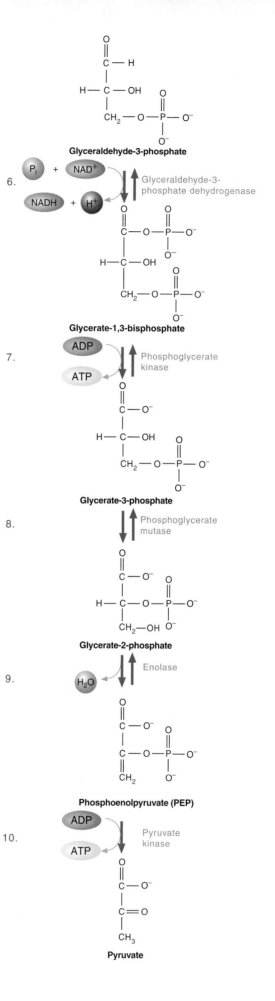

2. **Conversion of glucose-6-phosphate to fructose-6-phosphate**. During reaction 2 of glycolysis, the aldose glucose-6-phosphate is converted to the ketose fructose-6-phosphate by phosphoglucose isomerase (PGI) in a readily reversible reaction:

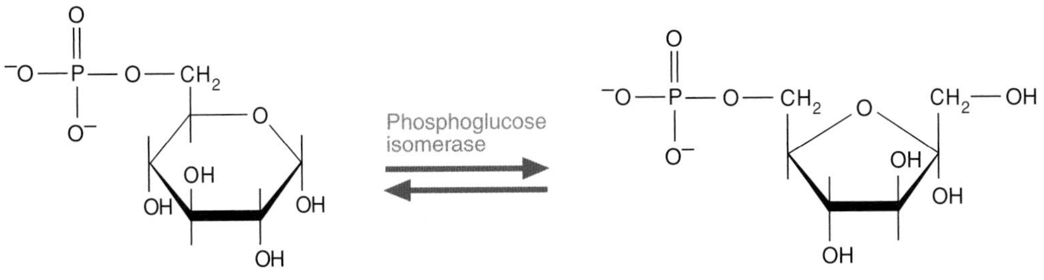

Glucose-6-phosphate　　　　　　**Fructose-6-phosphate**

Recall that the isomerization reaction of glucose and fructose involves an enediol intermediate (Figure 7.16). This transformation makes C-1 of the fructose product available for phosphorylation.

3. **The phosphorylation of fructose-6-phosphate**. Phosphofructokinase-1 (PFK-1) irreversibly catalyzes the phosphorylation of fructose-6-phosphate to form fructose-1,6-bisphosphate:

Fructose-6-phosphate　　　　　　**Fructose-1,6-bisphosphate**

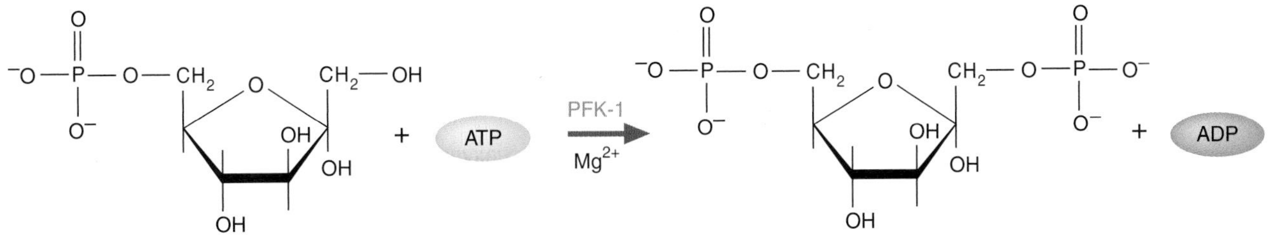

Investing a second molecule of ATP serves several purposes. First of all, because ATP is used as the phosphorylating agent, the reaction proceeds with a large decrease in free energy. After fructose-1,6-bisphosphate has been synthesized, the cell is committed to glycolysis. Because fructose-1,6-bisphosphate eventually splits into two trioses, another purpose for phosphorylation is to prevent any later product from diffusing out of the cell. PFK-1 is a major regulatory enzyme in glycolysis. Its activity is allosterically inhibited by high levels of ATP and citrate, which are indicators that the cell's energy charge is high and that the citric acid cycle, a major component of the cell's energy-generating capacity, has slowed down. AMP is an allosteric activator of PFK-1. AMP levels, which increase when the energy charge of the cell is low, are a better predictor of energy deficit than ADP levels. Fructose-2,6-bisphosphate is an allosteric activator of PFK-1 activity in the liver and is synthesized by phosphofructokinase-2 (PFK-2) in response to hormonal signals correlated to blood glucose levels. When serum glucose levels are high, hormone-stimulated increase in fructose-2,6-bisphosphate coordinately increases the activity of PFK-1 (activates glycolysis) and decreases the activity of the enzyme that catalyzes the reverse reaction, fructose-1,6-bisphosphatase (inhibits gluconeogenesis, Section 8.2). AMP is an allosteric inhibitor of fructose-1,6-bisphosphatase. PFK-2 is a bifunctional enzyme that behaves as a phosphatase when phosphorylated in response to the hormone glucagon (released into blood in reponse to low blood sugar, see p. 279). It functions as a kinase when dephosphorylated in response to the hormone insulin (high blood sugar).

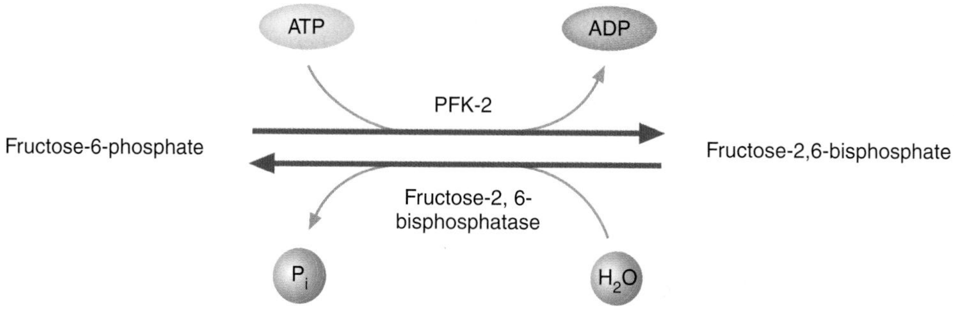

4. **Cleavage of fructose-1,6-bisphosphate**. Stage 1 of glycolysis ends by cleaving fructose-1,6-bisphosphate into two three-carbon molecules: glyceraldehyde-3-phosphate (G-3-P) and dihydroxyacetone phosphate (DHAP). This reaction is an **aldol cleavage**, hence the name of the enzyme: aldolase. Aldol cleavages are the reverse of aldol condensations, described on p. 166. In aldol cleavages an aldehyde and a ketone are products.

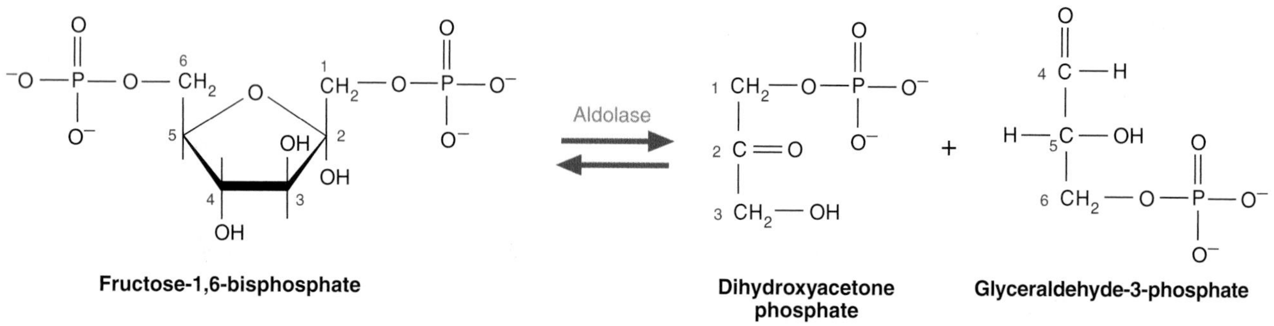

Fructose-1,6-bisphosphate   **Dihydroxyacetone phosphate**   **Glyceraldehyde-3-phosphate**

Although the cleavage of fructose-1,6-bisphosphate is thermodynamically unfavorable ($\Delta G^{\circ\prime} = +23.8$ kJ/mol), the reaction proceeds because the products are rapidly removed.

5. **The interconversion of glyceraldehyde-3-phosphate and dihydroxyacetone phosphate**. Of the two products of the aldolase reaction, only G-3-P serves as a substrate for the next reaction in glycolysis. To prevent the loss of the other three-carbon unit from the glycolytic pathway, triose phosphate isomerase catalyzes the reversible conversion of DHAP to G-3-P:

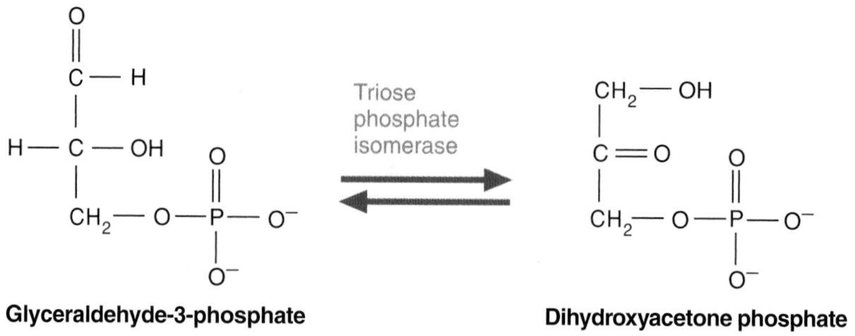

Glyceraldehyde-3-phosphate   **Dihydroxyacetone phosphate**

After this reaction, the original molecule of glucose has been converted to two molecules of G-3-P.

6. **Oxidation of glyceraldehyde-3-phosphate**. During reaction 6 of glycolysis, G-3-P undergoes oxidation and phosphorylation. The product,

glycerate-1,3-bisphosphate, contains a high-energy phosphoanhydride bond, which may be used in the next reaction to generate ATP:

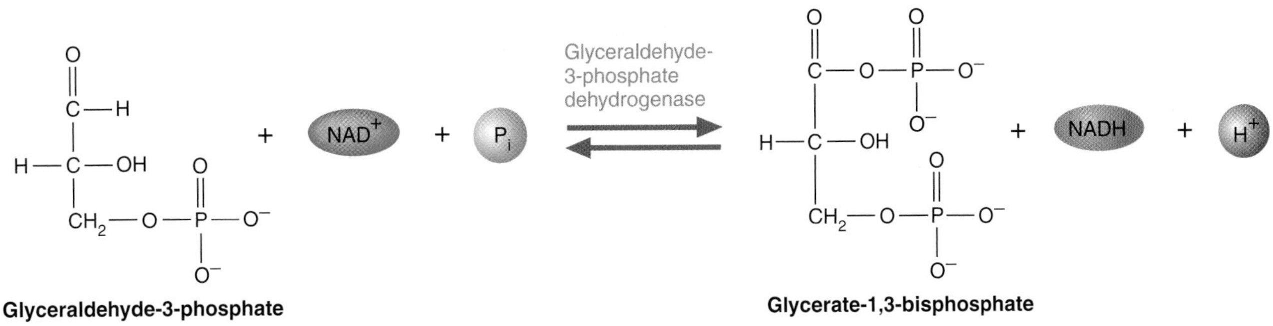

Glyceraldehyde-3-phosphate **Glycerate-1,3-bisphosphate**

This complex process is catalyzed by glyceraldehyde-3-phosphate dehydrogenase, a tetramer composed of four identical subunits. Each subunit contains one binding site for G-3-P and another for NAD^+. As the enzyme forms a covalent thioester bond with the substrate (Figure 8.5), a hydride ion ($H:^-$) is transferred to NAD^+ in the active site. The NADH then leaves the active site and is replaced by NAD^+. The acyl enzyme adduct is attacked by inorganic phosphate and the product leaves the active site.

7. **Phosphoryl group transfer**. In this reaction ATP is synthesized as phosphoglycerate kinase catalyzes the transfer of the high-energy phosphoryl group of glycerate-1,3-bisphosphate to ADP:

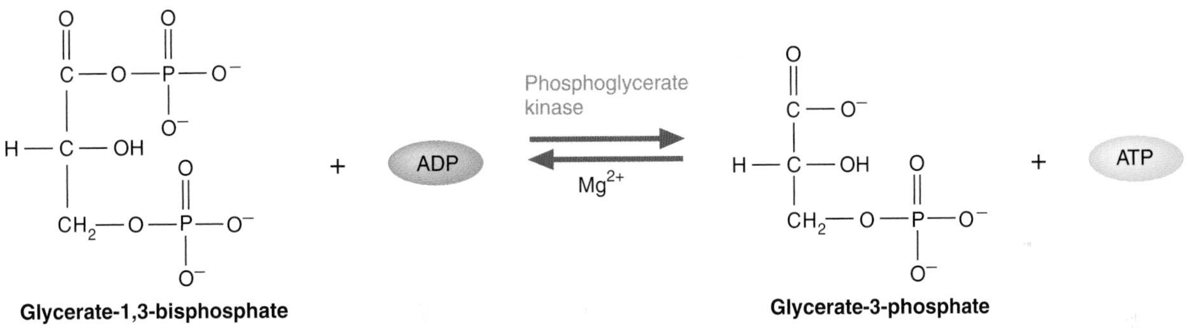

Glycerate-1,3-bisphosphate **Glycerate-3-phosphate**

Reaction 7 is an example of a substrate-level phosphorylation. Because the synthesis of ATP is endergonic, it requires an energy source. In **substrate-level phosphorylations**, ATP is produced by the transfer of a phosphoryl group from a substrate with a high phosphoryl transfer potential (glycerate-1,3-bisphosphate) (refer to Table 4.1) to produce a compound with a lower transfer potential (ATP) and therefore $\Delta G < 0$. Because two molecules of glycerate-1,3-bisphosphate are formed for every glucose molecule, this reaction produces two ATP molecules, and the investment of phosphate bond energy is recovered. ATP synthesis later in the pathway represents a net gain.

8. **The interconversion of 3-phosphoglycerate and 2-phosphoglycerate**. Glycerate-3-phosphate has a low phosphoryl group transfer potential. As such, it is a poor candidate for further ATP synthesis ($\Delta G^{\circ\prime}$ for ATP synthesis is −30.5 kJ/mol). Cells convert glycerate-3-phosphate with its energy-poor phosphate ester to phosphoenolpyruvate (PEP), which has an exceptionally high phosphoryl group transfer potential. (The standard free energies of hydrolysis of glycerate-3-phosphate and PEP are −12.6 and −61.9 kJ/mol, respectively.) In the first step in this conversion (reaction 8), phosphoglycerate mutase catalyzes the conversion of a C-3 phosphorylated compound to a C-2 phosphorylated compound through a two-step addition/elimination cycle.

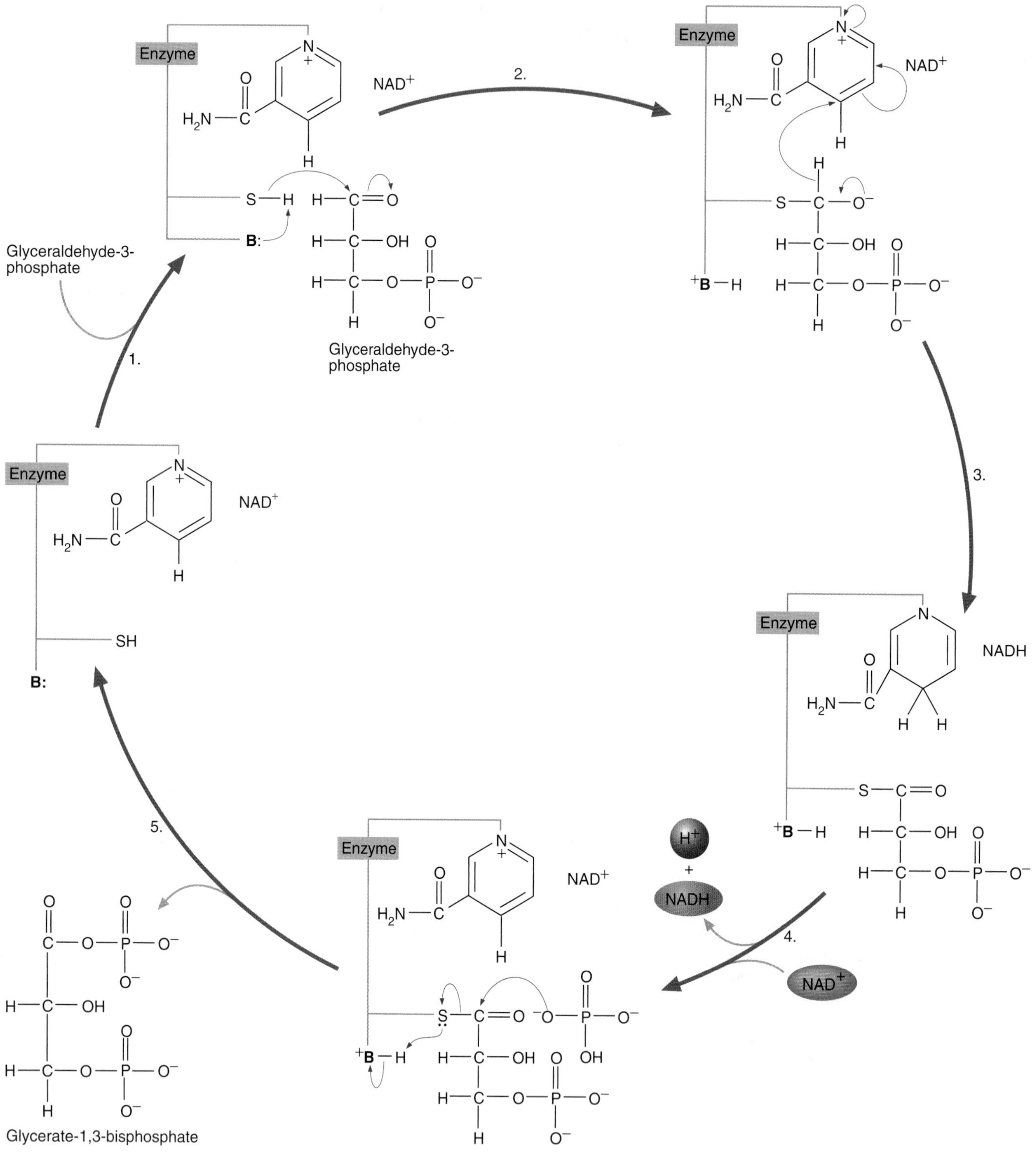

FIGURE 8.5

Glyceraldehyde-3-Phosphate Dehydrogenase Reaction

In the first step the substrate, glyceraldehyde-3-phosphate, enters the active site. As the enzyme catalyzes the reaction of the substrate with a sulfhydryl group within the active site (step 2), the substrate is oxidized (step 3). The bound NADH is exchanged for a cytoplasmic NAD^+ (step 4). Displacement of the enzyme by inorganic phosphate (step 5) liberates the product, glycerate-1, 3-bisphosphate, thus returning the enzyme to its original form.

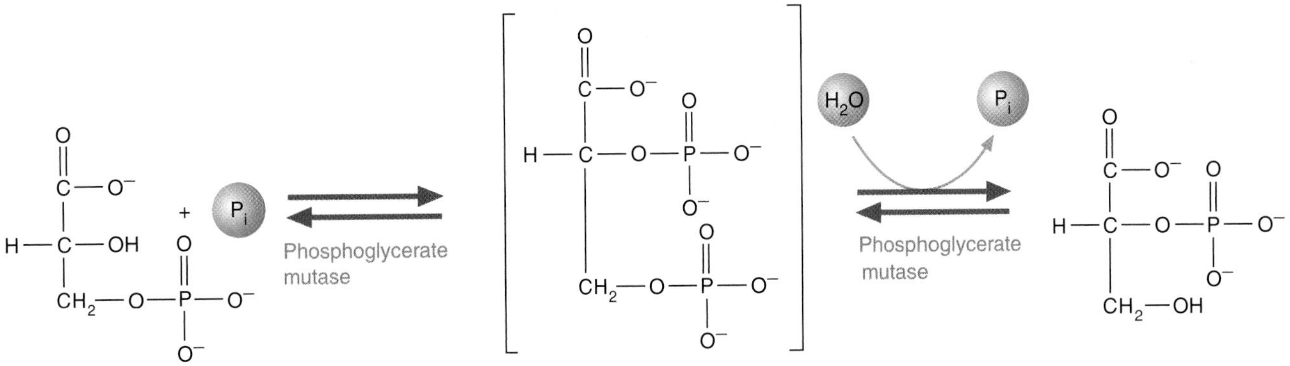

Glycerate-3-phosphate   **Glycerate-2,3-Bisphosphate**   **Glycerate-2-phosphate**

9. **Dehydration of 2-phosphoglycerate.** Enolase catalyzes the dehydration of glycerate-2-phosphate to form PEP:

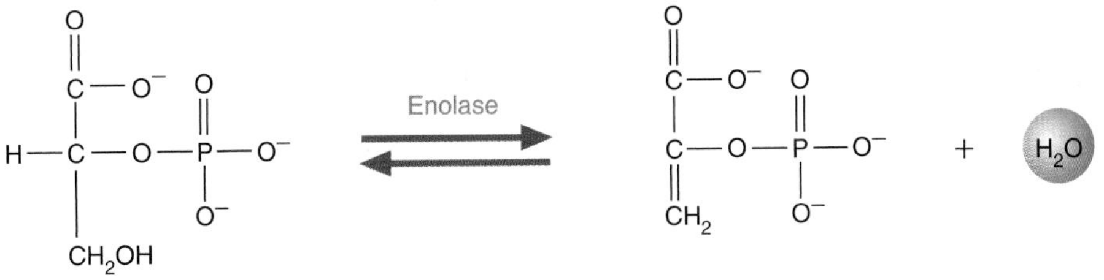

Glycerate-2-phosphate   **Phosphoenolpyruvate (PEP)**

PEP has a higher phosphoryl group transfer potential than does glycerate-2-phosphate because it contains an enol-phosphate group instead of a simple phosphate ester. The reason for this difference is made apparent in the next reaction. Aldehydes and ketones have two isomeric forms. The *enol* form contains a carbon-carbon double bond and a hydroxyl group. Enols exist in equilibrium with the more stable carbonyl-containing *keto* form. The interconversion of keto and enol forms, also called **tautomers**, is referred to as **tautomerization**:

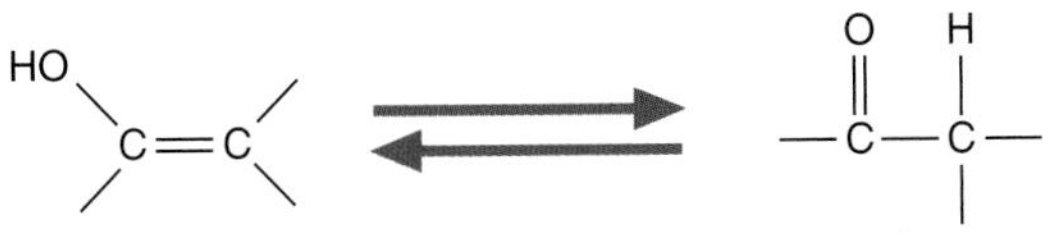

Enol form   **Keto form**

This tautomerization is restricted by the presence of the phosphate group, as is the resonance stabilization of the free phosphate ion. As a result, phosphoryl transfer to ADP in reaction 10 is highly favored.

10. **Synthesis of pyruvate.** In the final reaction of glycolysis, pyruvate kinase catalyzes the transfer of a phosphoryl group from PEP to ADP. Two molecules of ATP are formed for each molecule of glucose.

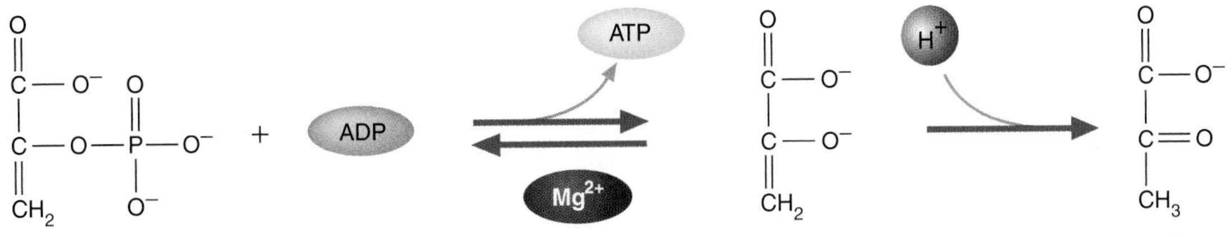

PEP   **Pyruvate (enol form)**   **Pyruvate (keto form)**

PEP is irreversibly converted to pyruvate because in this reaction the transfer of a phosphoryl group from a molecule with a high transfer potential to one with a lower transfer potential there is an exceptionally large free energy loss (refer to Table 4.1). This energy loss is associated with the spontaneous conversion (tautomerization) of the enol form of pyruvate to the more stable keto form.

The Fates of Pyruvate

In terms of energy, the result of glycolysis is the production of two ATPs and two NADHs per molecule of glucose. Pyruvate, the other product of glycolysis, is still an energy-rich molecule, which can produce a substantial amount of ATP. Whether or not further energy can be produced, however, depends on the cell type and the availability of oxygen. Under aerobic conditions, most cells in the body convert pyruvate into acetyl-CoA, the entry-level substrate for the **citric acid cycle**, an amphibolic pathway that completely oxidizes two carbons to form CO_2, NADH, and $FADH_2$. (An **amphibolic pathway** functions in both anabolic and catabolic processes.) The **electron transport system**, a series of oxidation-reduction reactions, transfers electrons from NADH and $FADH_2$ to O_2 to form water. The energy that is released during electron transport is coupled to a mechanism that synthesizes ATP. Under anaerobic conditions, further oxidation of pyruvate is impeded. A number of cells and organisms compensate by converting this molecule to a more reduced organic compound and regenerating the NAD^+ required for glycolysis to continue (Figure 8.6). This process of NAD^+ regeneration

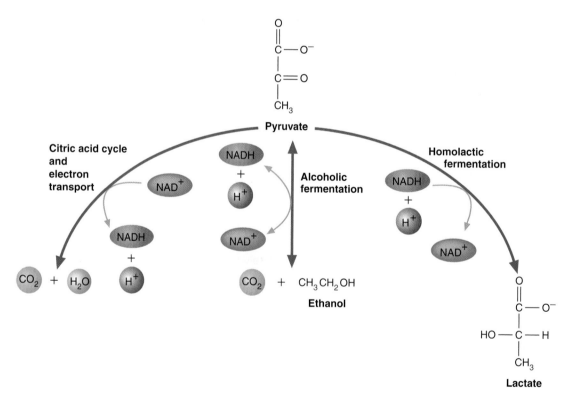

FIGURE 8.6

The Fates of Pyruvate

When oxygen is available (*left*), aerobic organisms completely oxidize pyruvate to CO_2 and H_2O. In the absence of oxygen, pyruvate can be converted to several types of reduced molecules. In some cells (e.g., yeast), ethanol and CO_2 are produced (*middle*). In others (e.g., muscle cells), homolactic fermentation occurs in which lactate is the only organic product (*right*). Some microorganisms use heterolactic fermentation reactions (not shown) that produce other acids or alcohols in addition to lactate. In all fermentation processes, the principal purpose is to regenerate NAD^+ so that glycolysis can continue.

is referred to as **fermentation**. Muscle cells and certain bacterial species (e.g., *Lactobacillus*) produce NAD$^+$ by transforming pyruvate into lactate:

$$CH_3-CO-COO^- + NADH + H^+ \underset{\text{Lactate dehydrogenase}}{\rightleftharpoons} CH_3-CHOH-COO^- + NAD^+$$

Pyruvate   **Lactate**

In rapidly contracting muscle cells, the demand for energy is high. After the O$_2$ supply is depleted, *lactic acid fermentation* provides sufficient NAD$^+$ to allow glycolysis (with its low level of ATP production) to continue for a short time (Figure 8.7).

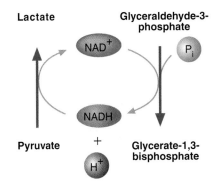

FIGURE 8.7

Recycling of NADH During Anaerobic Glycolysis

The NADH produced during the conversion of glyceraldehyde-3-phosphate to glycerate-1,3-bisphosphate is oxidized when pyruvate is converted to lactate. This process allows the cell to continue producing ATP for a short time until O$_2$ is again available.

QUESTION 8.1

Most molecules of ethanol are detoxified in the liver by two reactions. In the first, ethanol is oxidized to form acetaldehyde. This reaction, catalyzed by alcohol dehydrogenase, produces large amounts of NADH:

$$CH_3-CH_2-OH + NAD^+ \xrightarrow{ADH} CH_3\overset{O}{\underset{\|}{C}}-H + NADH + H^+$$

Soon after its production, acetaldehyde is converted to acetate by aldehyde dehydrogenase, which catalyzes a reaction that also produces NADH:

$$CH_3-\overset{O}{\underset{\|}{C}}-H + NAD^+ + H_2O \xrightarrow{\text{Aldehyde dehydrogenase}}$$

$$CH_3-\overset{O}{\underset{\|}{C}}-O^- + NADH + 2 H^+$$

One common effect of alcohol intoxication is the accumulation of lactate in the blood. Can you explain why this effect occurs?

In yeast and certain bacterial species, pyruvate is decarboxylated to form acetaldehyde, which is then reduced by NADH to form ethanol. (In a **decarboxylation** reaction, an organic acid loses a carboxyl group as CO$_2$.)

$$CH_3-CO-COO^- \xrightarrow[CO_2]{\text{Pyruvate decarboxylase}} CH_3-\overset{O}{\underset{\|}{C}}-H$$

Pyruvate   **Acetaldehyde**

$$\xrightarrow[NADH \quad NAD^+]{\text{Alcohol dehydrogenase}} CH_3-CH_2-OH$$

+ H$^+$

Ethanol

COMPANION

WEBSITE Visit the companion website at www.oup.com/us/mckee to read the Biochemistry in Perspective box on fermentation.

KEY CONCEPTS

- During glycolysis, glucose is converted to two molecules of pyruvate. A small amount of energy is captured in two molecules each of ATP and NADH.

- In anaerobic organisms, pyruvate is converted to waste products in a process called fermentation.

- In the presence of oxygen the cells of aerobic organisms convert pyruvate into CO_2 and H_2O.

This process, called alcoholic fermentation, is used commercially to produce wine, beer, and bread. Certain bacterial species produce organic molecules other than ethanol. For example, *Clostridium acetobutylicum*, an organism related to the causative agents of botulism and tetanus, produces butanol. Until recently, this organism was used commercially to synthesize butanol, an alcohol used to produce detergents and synthetic fibers. A petroleum-based synthetic process has now replaced microbial fermentation.

The Energetics of Glycolysis

During glycolysis, the energy released as glucose is broken down to pyruvate is coupled to the phosphorylation of ADP with a net yield of 2 ATP. However, evaluation of the standard free energy changes of the individual reactions (Figure 8.8) does not explain the efficiency of this pathway. A more useful method for evaluating free energy changes takes into account the conditions (e.g., pH and metabolite concentrations) under which cells actually operate. As illustrated in Figure 8.8, free energy changes measured in red blood cells indicate that only three reactions (1, 3, and 10, see pp. 269–270) have significantly negative ΔG values. These reactions, catalyzed by hexokinase, PFK-1, and pyruvate kinase, respectively, are for all practical purposes irreversible; that is, each goes to completion as written.

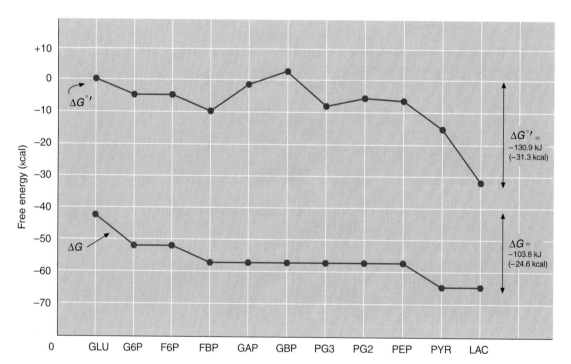

FIGURE 8.8

Free Energy Changes During Glycolysis in Red Blood Cells

Note that the standard free energy changes ($\Delta G^{\circ\prime}$) for the reactions in glycolysis show no consistent pattern (upper plot). In contrast, actual free energy values (ΔG) based on metabolite concentrations measured in red blood cells (lower plot) clearly illustrate why reactions 1, 3, and 10 (the conversions of glucose to glucose-6-phosphate, fructose-6-phosphate to fructose-1,6-bisphosphate, and phosphoenolpyruvate to pyruvate, respectively) are irreversible. The ready reversibility of the remaining reactions is indicated by their near-zero ΔG values. (GLU = glucose, G6P = glucose-6-phosphate, F6P = fructose-6-phosphate, FBP = fructose-1,6-bisphosphate, GAP = glyceraldehyde phosphate, PG3 = glycerate-3-phosphate, PG2 = glycerate-2-phosphate, PEP = phosphoenolpyruvate, PYR = pyruvate, LAC = lactate) Note that the conversion of DHAP to GAP is not counted in this list, since FBP is broken into GAP and DHAP, which is reconverted into GAP.

The values for the remaining reactions (2, 4–9) are so close to zero that they operate near equilibrium. Consequently, these latter reactions are easily reversible; small changes in substrate or product concentrations can alter the direction of each reaction. Not surprisingly, in gluconeogenesis (Section 8.2), the pathway by which glucose can be generated from pyruvate and certain other substrates, all of the glycolytic enzymes are involved except for those that catalyze reactions 1, 3, and 10. Gluconeogenesis uses different enzymes to bypass the irreversible steps of glycolysis.

Regulation of Glycolysis

The rate at which the glycolytic pathway operates is directly controlled primarily by allosteric regulation of three enzymes: hexokinase, PFK-1, and pyruvate kinase. The reactions catalyzed by these enzymes are irreversible and can be switched on and off by allosteric effectors. In general, allosteric effectors are molecules whose cellular concentrations are sensitive indicators of a cell's metabolic state. Some allosteric effectors are product molecules. For example, hexokinase is inhibited by excess glucose-6-phosphate. Several energy-related molecules also act as allosteric effectors. For example, a high AMP concentration (an indicator of low energy production) activates PFK-1 and pyruvate kinase. In contrast, a high ATP concentration (an indicator that the cell's energy requirements are being met) inhibits both enzymes. Citrate and acetyl-CoA, which accumulate when ATP is in rich supply, inhibit PFK-1 and pyruvate kinase, respectively. Fructose-2,6-bisphosphate, produced via hormone-induced covalent modification of PFK-2, is an indicator of high levels of available glucose and allosterically activates PFK-1. Accumulated fructose-1,6-bisphosphate activates pyruvate kinase, providing a feed-forward mechanism of control (i.e., fructose-1,6-bisphosphate is an allosteric activator). The allosteric regulation of glycolysis is summarized in Table 8.1.

Glycolysis is also regulated by the peptide hormones glucagon and insulin. **Glucagon**, released by pancreatic α-cells when blood glucose is low, activates the phosphatase function of PFK-2, thereby reducing the level of fructose-2,6-bisphosphate in the cell. As a result, PFK-1 activity and flux through glycolysis are decreased. In liver, glucagon also inactivates pyruvate kinase. Glucagon's effects, triggered by binding to its receptor on target cell surfaces, are mediated by cyclic AMP. Cyclic AMP (cAMP) is a second messenger molecule produced from ATP in a reaction catalyzed by adenylate cyclase, a plasma membrane protein. Once synthesized, cAMP binds to and activates protein kinase A (PKA). PKA then initiates a cascade of phosphorylation/dephosphorylation reactions that alter the activities of a diverse set of enzymes and transcription factors. **Transcription factors** are proteins that regulate or initiate RNA synthesis by binding to specific DNA sequences called **response elements.**

Insulin is a peptide hormone released from pancreatic β-cells when blood glucose levels are high. The effects of insulin on glycolysis include activation of the kinase function of PFK-2, which increases the level of fructose-2,6-bisphosphate in the cell, in turn increasing glycolytic flux. In cells containing

TABLE 8.1   Allosteric Regulation of Glycolysis

Enzyme	Activator	Inhibitor
Hexokinase		Glucose-6-phosphate, ATP
PFK-1	Fructose-2,6-bisphosphate, AMP	Citrate, ATP
Pyruvate kinase	Fructose-1,6-bisphosphate, AMP	Acetyl-CoA, ATP

insulin-sensitive glucose transporters (muscle and adipose tissue but not liver or brain) insulin promotes the translocation of glucose transporters to the cell surface. When insulin binds to its cell-surface receptor, the receptor protein undergoes several autophosphorylation reactions, which trigger numerous intracellular signal cascades that involve phosphorylation and dephosphorylation of target enzymes and transcription factors. Many of insulin's effects on gene expression are mediated by SREBP1c, a sterol regulatory element binding protein (p. 442). As a result of SREBP1c activation, there is increased synthesis of glucokinase and pyruvate kinase.

AMP-activated protein kinase (AMPK), an enzyme first discovered as a regulator of lipid metabolism, is now known to affect glucose metabolism as well. Once AMPK is activated, as a result of an increase in a cell's AMP:ATP ratio, it phosphorylates target proteins (enzymes and transcription factors). AMPK switches off anabolic pathways (e.g., protein and lipid synthesis) and switches on catabolic pathways (e.g., glycolysis and fatty acid oxidation). AMPK's regulatory effects on glycolysis include the following. In cardiac and skeletal muscle, AMPK promotes glycolysis by facilitating the stress- or exercise-induced recruitment of glucose transporters to the plasma membrane. In cardiac cells, AMPK stimulates glycolysis by activating PFK-2. AMPK structure and its functional properties are described in Chapter 12.

QUESTION 8.2

Insulin is a hormone secreted by the pancreas when blood sugar increases. Its most easily observable function is to reduce the blood sugar level to normal. The binding of insulin to most body cells promotes the transport of glucose across the plasma membrane. The capacity of an individual to respond to a carbohydrate meal by reducing blood glucose concentration quickly is referred to as *glucose tolerance*. Chromium-deficient animals show a decreased glucose tolerance; that is, they cannot remove glucose from blood quickly enough. The metal is believed to facilitate the binding of insulin to cells. Do you think the chromium is acting as an allosteric activator or cofactor?

QUESTION 8.3

Louis Pasteur, the great nineteenth-century French chemist and microbiologist, was the first scientist to observe that cells that can oxidize glucose completely to CO_2 and H_2O use glucose more rapidly in the absence of O_2 than in its presence. The oxygen molecule seems to inhibit glucose consumption. Explain in general terms the significance of this finding, now referred to as the **Pasteur effect**.

BIOCHEMISTRY IN PERSPECTIVE

Glycolysis and Jet Engines

Can systems biology improve our understanding of biochemical pathways such as glycolysis?

Modern species are the result of billions of years of rigorous natural selection that has adapted organisms to their various environments. This selection process has also occurred for the metabolic pathways that manage the biochemical transformations that sustain life. As systems biologists analyzed metabolic processes in their search for design principles, it became apparent that evolution, operating under thermodynamic and kinetic constraints, has converged again and again into a relatively small set of metabolic designs. Catabolic pathways, those that degrade organic molecules and release energy, provide an important example. Catabolic pathways typically have two characteristics: optimal ATP production and kinetic efficiency (i.e., minimal response time to changes in cellular metabolic requirements). Living organisms have optimized catabolic pathways, in part, by utilizing highly exergonic reactions at the beginning of a pathway. The early "activation" of nutrient molecules thus makes subsequent ATP-producing reactions (usually near the end of the pathway) a thermodynamically downhill process. As a result, the pathway can produce ATP under varying substrate and product concentrations. The term "turbo design," inspired by the turbo engines in jet aircraft, is now used to describe this phenomenon.

A jet engine creates propulsion by mixing air with fuel to create hot, expanding, fast-moving exhaust gases that are blasted out the back. Some of the air drawn in at the front of the engine is diverted into compressors, where its pressure is substantially increased before flowing into combustion chambers and mixing with fuel molecules. As the burning fuel molecules expand, they

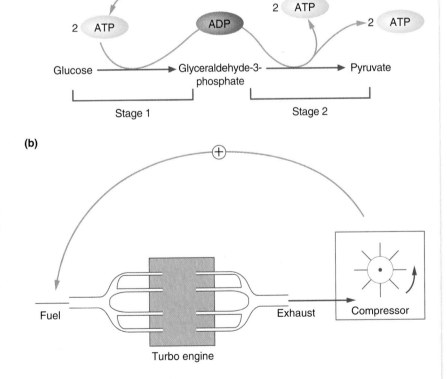

(a)

Stage 1   Stage 2

(b)

FIGURE 8A

Comparison of Glycolysis and the Turbo Jet Engine

(a) Glycolysis is a catabolic pathway in which two of the four ATPs produced in stage 2 are used to activate an incoming glucose molecule. The ADPs used in substrate-level phosphorylation reactions in stage 2 are generated from the two ATPs used in stage 1 and in ATP-requiring reactions throughout the cell. (b) The schematic representation of a turbo jet engine illustrates how energy generated by an engine can be used to improve efficiency. Before exiting the engine, hot exhaust gases are diverted around the compressor turbines, raising the temperature and increasing the efficiency of fuel combustion.

BIOCHEMISTRY IN PERSPECTIVE cont

flow through turbines fitted with fan blades that drive the compressors. An important feature of this process is that hot exhaust gases are also fed back into the engine to accelerate the fuel input step (Figure 8A).

Catabolic pathways with a turbo design, such as glycolysis, are optimized and efficient. However, the early phases of such pathways must be negatively regulated to prevent buildup of intermediates and overuse of fuel. As described, two of the four ATPs produced from each glucose molecule are fed back to the fuel input stage of the pathway to drive the pathway forward. As indicated by its use by most modern living organisms, glycolysis has been a tremendously successful energy-generating strategy. It is not perfect, however. Under certain circumstances, the turbo design of glycolysis makes some cells vulnerable to a phenomenon called "substrate-accelerated death" as the following example illustrates.

Certain types of mutant yeast cells are unable to grow anaerobically on glucose despite having a completely functional glycolytic pathway. These mutants die when exposed to large concentrations of glucose. Amazingly, research efforts have revealed that defects in *TPS1*, the gene that codes for the catalytic subunit of trehalose-6-phosphate synthase, are responsible. Trehalose-6-phosphate (Tre-6-P), an α-(1,1)-disaccharide of glucose, is a compatible solute (see the Biochemistry in Perspective box in Chapter 3 entitled Water, Abiotic Stress, and Compatible Solutes) used by yeast and various other organisms to resist several forms of abiotic stress. Apparently, Tre-6-P is a normal inhibitor of HK (and possibly a glucose transporter). In the absence of a functional TPS1 protein and when glucose becomes available, glycolytic flux in the mutant cells rapidly accelerates. In a relatively short time, as a result of the turbo design of the pathway, most available phosphate has been incorporated into glycolytic intermediates, and the cell's ATP level is too low to sustain cellular processes. This and other similar examples of substrate accelerated cell death in other species provide insight into the importance of the intricate regulatory mechanisms observed in living organisms.

SUMMARY: Certain catabolic pathways are optimized by the use of highly exergonic reactions in an early phase. The ATP product of the pathway is used to drive the pathway forward.

8.2 GLUCONEOGENESIS

Gluconeogenesis, the formation of new glucose molecules from noncarbohydrate precursors, occurs primarily in the liver. These precursors include lactate, pyruvate, glycerol, and certain α-keto acids (molecules derived from amino acids). Under certain conditions (i.e., metabolic acidosis or starvation) the kidney can make small amounts of new glucose. Between meals adequate blood glucose levels are maintained by the hydrolysis of liver glycogen. When liver glycogen is depleted (e.g., owing to prolonged fasting or vigorous exercise), the gluconeogenesis pathway provides the body with adequate glucose. Brain and red blood cells rely exclusively on glucose as their energy source.

Gluconeogenesis Reactions

The reaction sequence in gluconeogenesis is largely the reverse of glycolysis. Recall, however, that three glycolytic reactions (the reactions catalyzed by hexokinase, PFK-1, and pyruvate kinase) are irreversible. In gluconeogenesis, alternate reactions catalyzed by different enzymes are used to bypass these obstacles. The reactions unique to gluconeogenesis are listed next. The entire

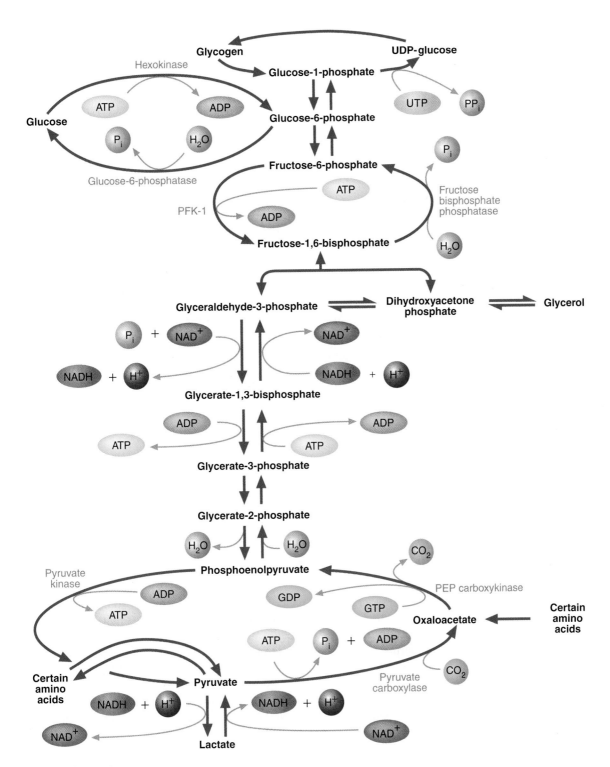

FIGURE 8.9

Carbohydrate Metabolism: Gluconeogenesis and Glycolysis

In gluconeogenesis, which occurs when blood sugar levels are low and liver glycogen is depleted, 7 of the 10 reactions of glycolysis are reversed. Three irreversible glycolytic reactions are bypassed by alternative reactions. The major substrates for gluconeogenesis are certain amino acids (derived from muscle), lactate (formed in muscle and red blood cells), and glycerol (produced from the degradation of triacylglycerols). In contrast to the reactions of glycolysis, which occur only in cytoplasm, the gluconeogenesis reactions catalyzed by pyruvate carboxylase and, in some species, PEP carboxykinase occur within the mitochondria, and the reaction catalyzed by glucose-6-phosphatase takes place in the endoplasmic reticulum.

gluconeogenic pathway and its relationship to glycolysis are illustrated in Figure 8.9. The bypass reactions of gluconeogenesis are as follows:

1. **Synthesis of PEP**. PEP synthesis from pyruvate requires two enzymes: pyruvate carboxylase and PEP carboxykinase. Pyruvate carboxylase, found within mitochondria, converts pyruvate to oxaloacetate (OAA):

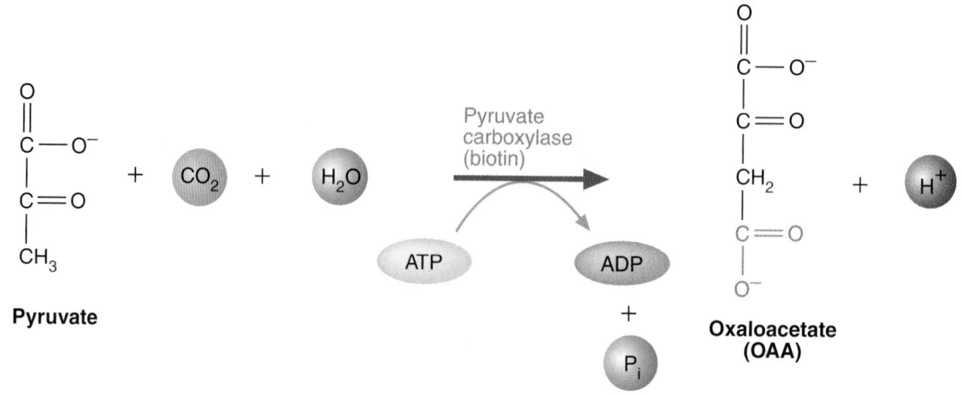

The coenzyme *biotin* (p. 432), which functions as a CO_2 carrier, is covalently bound to the enzyme through the side chain amino group of a lysine residue. OAA is decarboxylated and phosphorylated by PEP carboxykinase in a reaction driven by the hydrolysis of guanosine triphosphate (GTP):

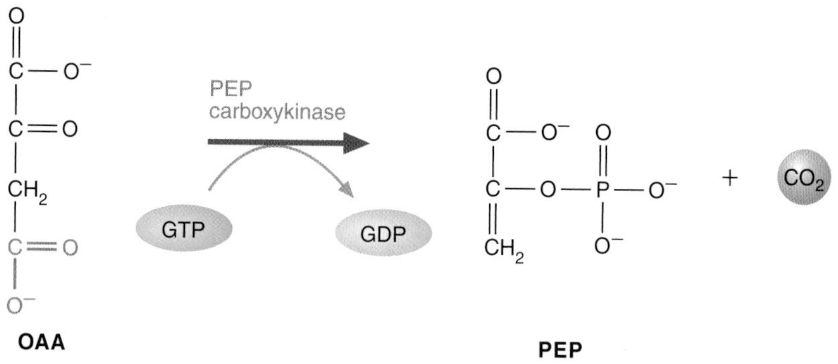

PEP carboxykinase is found within the mitochondria of some species and in the cytoplasm of others. In humans this enzymatic activity is found in both compartments. Because the inner mitochondrial membrane is impermeable to OAA, cells that lack mitochondrial PEP carboxykinase transfer OAA into the cytoplasm by using, for example, the **malate shuttle**. In this process, OAA is converted into malate by mitochondrial malate dehydrogenase. After the transport of malate across mitochondrial membrane, the reverse reaction (to form OAA) is catalyzed by cytoplasmic malate dehydrogenase.

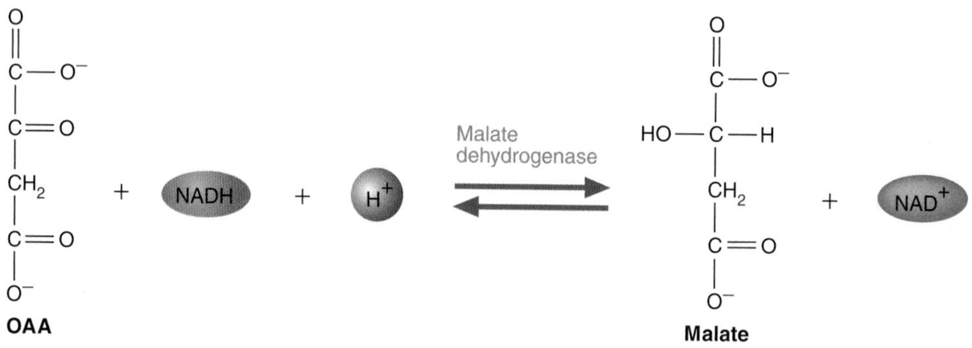

2. **Conversion of fructose-1,6-bisphosphate to fructose-6-phosphate.**
The irreversible PFK-1–catalyzed reaction in glycolysis is bypassed by
fructose-1,6-bisphosphatase:

Fructose-1,6-bisphosphate + H_2O → (Fructose-1,6-bisphosphatase) → **Fructose-6-phosphate** + P_i

This exergonic reaction ($\Delta G^{\circ\prime} = -16.7$ kJ/mol) is also irreversible under
cellular conditions. ATP is not regenerated, and inorganic phosphate (P_i)
is also produced. Fructose-1,6-bisphosphatase is an allosteric enzyme.
Its activity is stimulated by citrate and inhibited by AMP and fructose-
2,6-bisphosphate.

3. **Formation of glucose from glucose-6-phosphate.** Glucose-6-phosphatase,
found only in liver and kidney, catalyzes the irreversible hydrolysis of glu-
cose-6-phosphate to form glucose and P_i. Glucose is subsequently released
into the blood.

As stated, each of the foregoing reactions is matched by an opposing
irreversible reaction in glycolysis. Each set of such paired reactions is
referred to as a *substrate cycle*. Because they are coordinately regulated (an
activator of the enzyme catalyzing the forward reaction serves as an
inhibitor of the enzyme catalyzing the reverse reaction), very little energy
is wasted, even though both enzymes may be operating at some level at the
same time. *Flux control* (regulation of the flow of substrate and removal of
product) is more effective if transient accumulation of product is funneled
back through the cycle. The catalytic velocity of the forward enzyme
will remain high if the concentration of the substrate is maximized. The
gain in catalytic efficiency more than makes up for the small energy loss
in recycling the product.

Gluconeogenesis is an energy-consuming process. Instead of generating
ATP (as in glycolysis), gluconeogenesis requires the hydrolysis of six high-
energy phosphate bonds.

QUESTION 8.4

Malignant hyperthermia is a rare, inherited disorder triggered during surgery
by certain anesthetics. A dramatic (and dangerous) rise in body temperature (as
high as 112°F) is accompanied by muscle rigidity and acidosis. The excessive
muscle contraction is initiated by a large release of calcium from the sarcoplasmic reticulum, a calcium-storing organelle in muscle cells. Acidosis results from
excessive lactic acid production. Prompt treatment to reduce body temperature
and to counteract the acidosis is essential to save the patient's life. A probable
contributing factor to this disorder is wasteful cycling between glycolysis and gluconeogenesis. Explain why this is a reasonable explanation.

QUESTION 8.5

After examining the gluconeogenic pathway reaction summary illustrated here,
account for each component in the equation. [*Hint:* The hydrolysis of each
nucleotide releases a proton.]

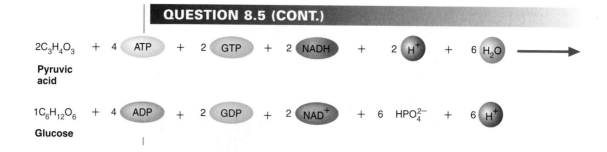

QUESTION 8.5 (CONT.)

$2C_3H_4O_3$ + 4 ATP + 2 GTP + 2 NADH + 2 H^+ + 6 H_2O ⟶

Pyruvic acid

$1C_6H_{12}O_6$ + 4 ADP + 2 GDP + 2 NAD^+ + 6 HPO_4^{2-} + 6 H^+

Glucose

QUESTION 8.6

Patients with *von Gierke's disease* (a glycogen storage disease) lack glucose-6-phosphatase activity. Two prominent symptoms of this disorder are fasting hypoglycemia and lactic acidosis. Can you explain why these symptoms occur?

Gluconeogenesis Substrates

As previously mentioned, several metabolites are gluconeogenic precursors. Three of the most important substrates are described briefly.

FIGURE 8.10

The Cori Cycle

During strenuous exercise, lactate is produced anaerobically in muscle cells. After passing through blood to the liver, lactate is converted to glucose by gluconeogenesis.

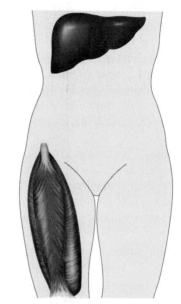

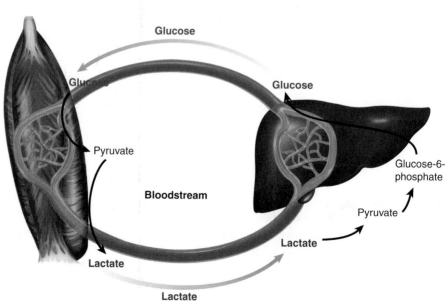

Lactate is released by red blood cells and other cells that lack mitochondria or have low oxygen concentrations. In the **Cori cycle**, lactate is released by skeletal muscle during exercise (Figure 8.10). After lactate is transferred to the liver, it is reconverted to pyruvate by lactate dehydrogenase and then to glucose by gluconeogenesis.

Glycerol, a product of fat metabolism in adipose tissue, is transported to the liver in the blood and then converted to glycerol-3-phosphate by glycerol kinase. Oxidation of glycerol-3-phosphate to form DHAP occurs when cytoplasm NAD^+ concentration is relatively high.

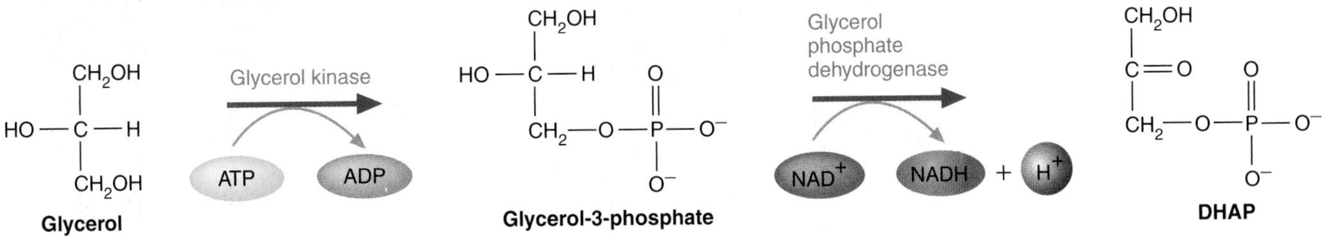

Of all the amino acids that can be converted to glycolytic intermediates (molecules referred to as *glucogenic*), alanine is perhaps the most important. (The metabolism of the glucogenic amino acids is described in Chapter 15.) When exercising muscle produces large quantities of pyruvate, some of these molecules are converted to alanine by a transamination reaction involving glutamate:

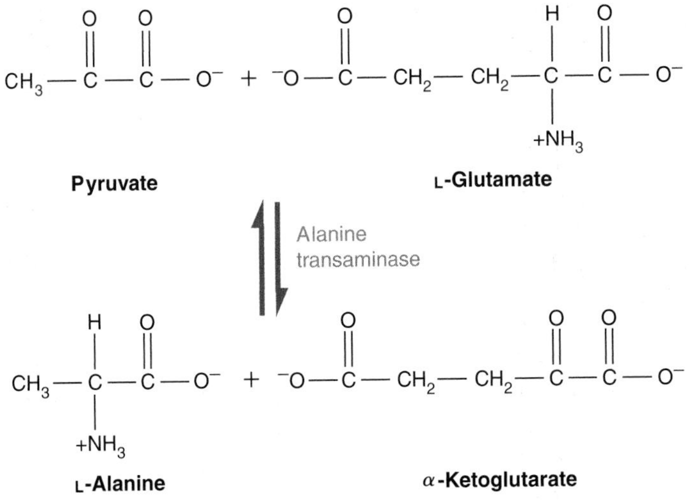

After it has been transported to the liver, alanine is reconverted to pyruvate and then to glucose. The **glucose-alanine cycle** (Figure 8.11) serves several purposes. In addition to its role in recycling α-keto acids between muscle and liver, the glucose-alanine cycle is a mechanism for transporting amino nitrogen to the liver. In α-keto acids, sometimes referred to as carbon skeletons, a carbonyl group is directly attached to the carboxyl group. Once alanine reaches the liver it is reconverted to pyruvate and the amino nitrogen is then incorparated into urea (Chapter 15).

Gluconeogenesis Regulation

As with other metabolic pathways, the rate of gluconeogenesis is affected primarily by substrate availability, allosteric effectors, and hormones. Not surprisingly, gluconeogenesis is stimulated by high concentrations of lactate, glycerol, and amino acids. A high-fat diet, starvation, and prolonged fasting make large quantities of these molecules available.

The four key enzymes in gluconeogenesis (pyruvate carboxylase, PEP carboxykinase, fructose-1,6-bisphosphatase, and glucose-6-phosphatase) are affected

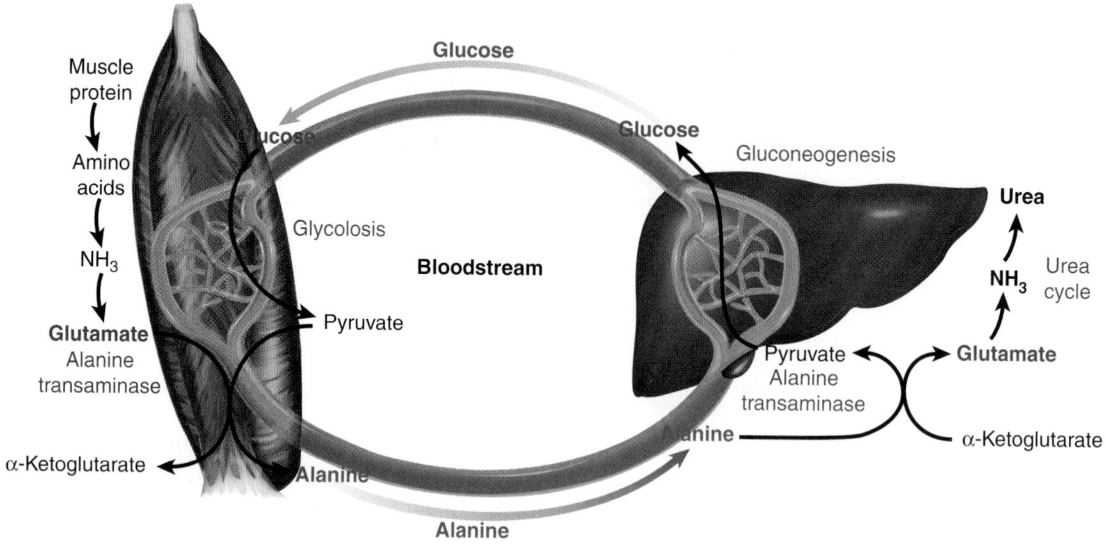

FIGURE 8.11

The Glucose-Alanine Cycle

Alanine is formed from pyruvate in muscle. After it has been transported to the liver, alanine is reconverted to pyruvate by alanine transaminase. Eventually pyruvate is used in the synthesis of new glucose. Because muscle cannot synthesize urea from amino nitrogen, the glucose-alanine cycle is used to transfer amino nitrogen to the liver.

to varying degrees by allosteric modulators. For example, fructose-1,6-bisphosphatase is activated by ATP and inhibited by AMP and fructose-2,6-bisphosphate. Acetyl-CoA activates pyruvate carboxylase. (The concentration of acetyl-CoA, a product of fatty acid degradation, is especially high during starvation.) An overview of the allosteric regulation of glycolysis and gluconeogenesis is provided in Figure 8.12.

As with other biochemical pathways, hormones affect gluconeogenesis by altering the concentrations of allosteric effectors and the key rate-determining enzymes. As mentioned previously, glucagon depresses the synthesis of fructose-2,6-bisphosphate, which releases the inhibition of fructose-1,6-bisphosphatase, and inactivates the glycolytic enzyme pyruvate kinase. Hormones also influence gluconeogenesis by altering enzyme synthesis. For example, the synthesis of gluconeogenic enzymes is stimulated by cortisol, a steroid hormone produced in the cortex of the adrenal gland that facilitates the body's adaptation to stressful situations. Finally, insulin action leads to the synthesis of new molecules of glucokinase, PFK-1 (SREBP1c-induced), and PFK-2 (glycolysis favored). Insulin also depresses the synthesis (also via SREBP1c) the synthesis of glucose-6-phosphatase, fructose-1,6-bisphosphatase, and PEP carboxykinase. Glucagon action leads to the synthesis of additional molecules of PEP carboxykinase, fructose-1,6-bisphosphatase, and glucose-6-phosphatase (gluconeogenesis favored).

These hormones accomplish this feat by altering the phosphorylation state of certain target proteins in the liver cell, which in turn modifies gene expression. The key point to remember is that insulin and glucagon have opposing effects on carbohydrate metabolism. The direction of flux, either glycolysis or gluconeogenesis, is largely determined by the ratio of insulin to glucagon. After a carbohydrate meal, the insulin/glucagon ratio is high and glycolysis in the liver predominates over gluconeogenesis. After a period of fasting or following a high-fat, low-carbohydrate meal, the insulin/glucagon ratio is low and gluconeogenesis in the liver predominates over glycolysis. The availability of ATP is the second

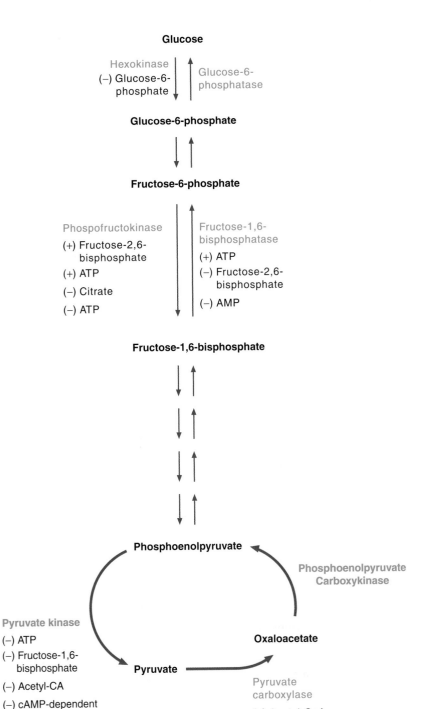

Glucose

Hexokinase | Glucose-6-
(−) Glucose-6- | phosphatase
phosphate

Glucose-6-phosphate

Fructose-6-phosphate

Phospofructokinase | Fructose-1,6-
| bisphosphatase
(+) Fructose-2,6- | (+) ATP
bisphosphate | (−) Fructose-2,6-
(+) ATP | bisphosphate
(−) Citrate | (−) AMP
(−) ATP |

Fructose-1,6-bisphosphate

Phosphoenolpyruvate

Phosphoenolpyruvate
Carboxykinase

Pyruvate kinase

(−) ATP

(−) Fructose-1,6-
bisphosphate

(−) Acetyl-CA

(−) cAMP-dependent
phosphorylation

Oxaloacetate

Pyruvate

Pyruvate
carboxylase

(+) Acetyl-CoA

FIGURE 8.12

**Allosteric Regulation of Glycolysis and
Gluconeogenesis**

The key enzymes in glycolysis and
gluconeogenesis are regulated by allosteric
effectors. Activator, +; inhibitor, −.

important regulator in the reciprocal control of glycolysis and gluconeogenesis in
that high levels of AMP, the low-energy hydrolysis product of ATP, increase the flux
through glycolysis at the expense of gluconeogenesis, and low levels of AMP
increase the flux through gluconeogenesis at the expense of glycolysis. Although
control at the PFK-1/fructose-1,6-bisphosphatase cycle would appear to be
sufficient for this pathway, control at the pyruvate kinase step is key because it
permits the maximal retention of PEP, a molecule with a very high phosphate transfer potential.

KEY CONCEPTS

- Gluconeogenesis, the synthesis of new
  glucose molecules from noncarbohydrate
  precursors, occurs primarily in the liver.

- The reaction sequence is the reverse of
  glycolysis except for three reactions that
  bypass irreversible steps in glycolysis.

8.3 THE PENTOSE PHOSPHATE PATHWAY

The pentose phosphate pathway is an alternative metabolic pathway for glucose oxidation in which no ATP is generated. Its principal products are NADPH, a reducing agent required in several anabolic processes, and ribose-5-phosphate, a structural component of nucleotides and nucleic acids. The pentose phosphate pathway occurs in the cytoplasm in two phases: oxidative and nonoxidative. In the oxidative phase of the pathway, the conversion of glucose-6-phosphate to ribulose-5-phosphate is accompanied by the production of two molecules of NADPH. The nonoxidative phase involves the isomerization and condensation of a number of different sugar molecules. Three intermediates in this process that are useful in other pathways are ribose-5-phosphate, fructose-6-phosphate, and glyceraldehyde-3-phosphate.

The oxidative phase of the pentose phosphate pathway consists of three reactions (Figure 8.13a). In the first reaction, glucose-6-phosphate dehydrogenase (G-6-PD) catalyzes the oxidation of glucose-6-phosphate. 6-Phosphogluconolactone and NADPH are products in this reaction. 6-Phospho-D-glucono-δ-lactone is then hydrolyzed to produce 6-phospho-D-gluconate. A second molecule of NADPH is produced during the oxidative decarboxylation of 6-phosphogluconate, a reaction that yields ribulose-5-phosphate.

A substantial amount of the NADPH required for reductive processes (i.e., lipid biosynthesis) is supplied by these reactions. For this reason this pathway is most active in cells in which relatively large amounts of lipids are synthesized, (e.g., adipose tissue, adrenal cortex, mammary glands, and the liver). NADPH is also a powerful **antioxidant**. (**Antioxidants** are substances that prevent the oxidation of other molecules. Their roles in living processes are described in Chapter 10.) Consequently, the oxidative phase of the pentose phosphate pathway is also quite active in cells that are at high risk for oxidative damage, such as red blood cells.

The nonoxidative phase of the pathway begins with the conversion of ribulose-5-phosphate to ribose-5-phosphate by ribulose-5-phosphate isomerase or to xylulose-5-phosphate by ribulose-5-phosphate epimerase. During the remaining reactions of the pathway (Figure 8.13b), transketolase and transaldolase catalyze the interconversions of trioses, pentoses, and hexoses. Transketolase is a TPP-requiring enzyme that transfers two-carbon units from a ketose to an aldose. (TPP, thiamine pyrophosphate, is the coenzyme form of thiamine, also known as vitamin B_1.) Two reactions are catalyzed by transketolase. In the first reaction, the enzyme transfers a two-carbon unit from xylulose-5-phosphate to ribose-5-phosphate, yielding glyceraldehyde-3-phosphate and sedoheptulose-7-phosphate. In the second transketolase-catalyzed reaction, a two-carbon unit from another xylulose-5-phosphate molecule is transferred to erythrose-4-phosphate to form a second molecule of glyceraldehyde-3-phosphate and fructose-6-phosphate. (Erythrose-4-phosphate is used by some organisms to synthesize aromatic amino acids.) Transaldolase transfers three-carbon units from a ketose to an aldose. In the reaction catalyzed by transaldolase, a three-carbon unit is transferred from sedoheptulose-7-phosphate to glyceraldehyde-3-phosphate. The products formed are fructose-6-phosphate and erythrose-4-phosphate. The result of the nonoxidative phase of the pathway is the synthesis of ribose-5-phosphate and the glycolytic intermediates glyceraldehyde-3-phosphate and fructose-6-phosphate.

When pentose sugars are not required for biosynthetic reactions, the metabolites in the nonoxidative portion of the pathway are converted into glycolytic intermediates that can then be further degraded to generate energy or converted into precursor molecules for biosynthetic processes (Figure 8.14). For this reason the pentose phosphate pathway is also referred to as the *hexose monophosphate shunt*. In plants, the pentose phosphate pathway is involved in the synthesis of glucose during the dark reactions of photosynthesis (Chapter 13).

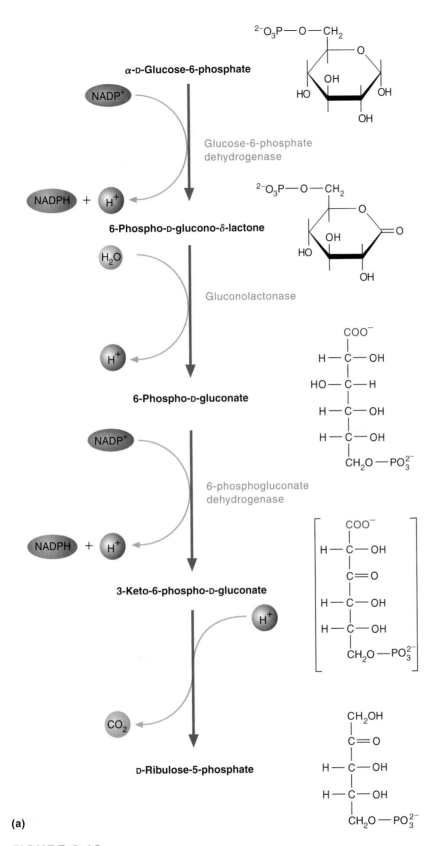

(a)

FIGURE 8.13a

The Pentose Phosphate Pathway

(a) The oxidative phase. NADPH is an important product of these reactions.

FIGURE 8.13b

**The Pentose Phosphate
Pathway**

(b) The nonoxidative phase.
When cells require more
NADPH than pentose phos-
phates, the enzymes in the
nonoxidative phase convert
ribose-5-phosphate into the
glycolytic intermediates
fructose-6-phosphate and
glyceraldehyde-3-phosphate.

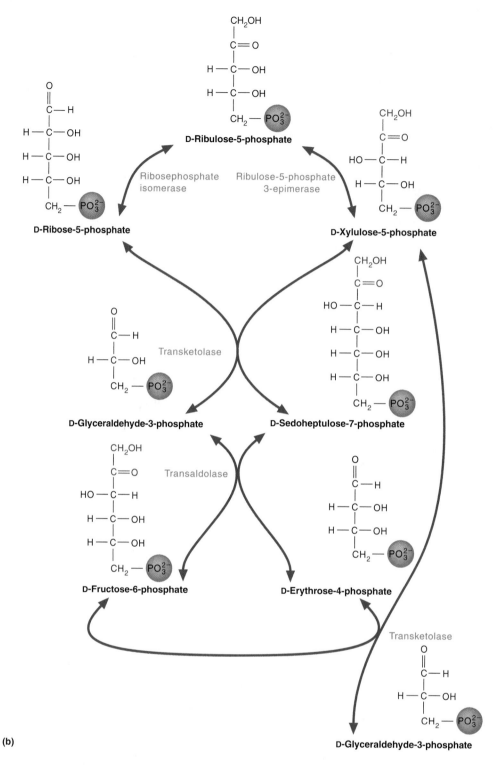

(b)

KEY CONCEPT

The pentose phosphate pathway produces
NADPH, ribose-5-phosphate, and the glycolytic intermediates fructose-6-phosphate
and glyceraldehyde-3-phosphate.

The pentose phosphate pathway is regulated to meet the cell's moment-by-
moment requirements for NADPH and ribose-5-phosphate. The oxidative phase
is very active in cells such as red blood cells or hepatocytes in which demand
for NADPH is high. In contrast, the oxidative phase is virtually absent in cells
(e.g., muscle cells) that synthesize little or no lipid. G-6-PD catalyzes a key
regulatory step in the pentose phosphate pathway. Its activity is inhibited by
NADPH and stimulated by GSSG, the oxidized form of glutathione, an important cellular antioxidant (Chapter 10) and glucose-6-phosphate. In addition, diets
high in carbohydrate increase the synthesis of both G-6-PD and phosphogluconate
dehydrogenase.

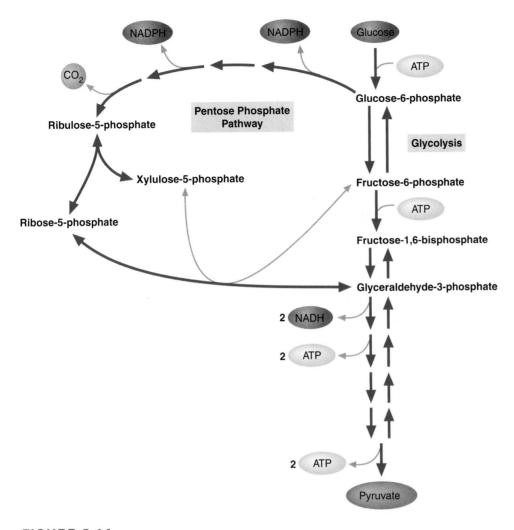

FIGURE 8.14

Carbohydrate Metabolism: Glycolysis and the Pentose Phosphate Pathway

If the cell requires more NADPH than ribose molecules, it can channel the products of the nonoxidative phase of the pentose phosphate pathway into glycolysis. As this overview of the two pathways illustrates, excess ribose-5-phosphate can be converted into the glycolytic intermediates fructose-6-phosphate and glyceraldehyde-3-phosphate.

8.4 METABOLISM OF OTHER IMPORTANT SUGARS

Several sugars other than glucose are important in vertebrates. The most notable of these are fructose, galactose, and mannose. Besides glucose, these molecules are the most common sugars found in oligosaccharides and polysaccharides. They are also energy sources. The reactions by which these sugars are converted into glycolytic intermediates are illustrated in Figure 8.15. The metabolism of fructose, an important component of the human diet, is discussed.

Fructose Metabolism

Dietary sources of fructose include fruit, honey, sucrose, and high-fructose corn syrup, an inexpensive sweetener used in a wide variety of processed foods and beverages.

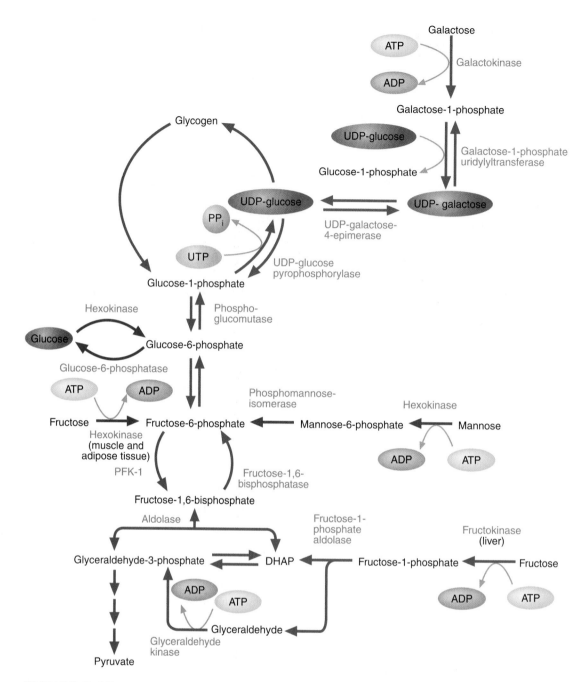

FIGURE 8.15

Carbohydrate Metabolism: Other Important Sugars

Fructose enters the glycolytic pathway by two routes. Fructokinase in liver cells converts fructose to fructose-1-phosphate that is then split into DHAP and glyceraldehyde. In muscle and adipose tissue, fructose is phosphorylated by hexokinase to form the glycolytic intermediate fructose-6-phosphate. Galactose is converted into galactose-1-phosphate, which then reacts with UDP-glucose to form UDP-galactose. UDP-galactose is converted to its epimer, UDP-glucose, the substrate for glycogen synthesis. Mannose is phosphorylated by hexokinase to form mannose-6-phosphate, which is then isomerized to fructose-6-phosphate.

Fructose, second only to glucose as a source of carbohydrate in the human diet, can enter the glycolytic pathway by two routes. In the liver, fructose is converted to fructose-1-phosphate by fructokinase:

Fructose → **Fructose-1-phosphate** (Fructokinase, ATP → ADP)

When fructose-1-phosphate enters the glycolytic pathway, it is first split into dihydroxyacetone phosphate (DHAP) and glyceraldehyde by fructose-1-phosphate aldolase. DHAP is then converted to glyceraldehyde-3-phosphate by triose phosphate isomerase. Glyceraldehyde-3-phosphate is generated from glyceraldehyde and ATP by glyceraldehyde kinase.

The conversion of fructose-1-phosphate into glycolytic intermediates bypasses two regulatory steps (the reactions catalyzed by hexokinase and PFK-1); thus in comparison to glucose, the entrance of fructose into the glycolytic pathway is essentially unregulated.

In muscle and adipose tissue, fructose is converted to the glycolytic intermediate fructose-6-phosphate by hexokinase. Because the hexokinases have a low affinity for fructose, this reaction is of minor importance unless fructose consumption is exceptionally high.

Fructose → **Fructose-6-phosphate** (Hexokinase, ATP → ADP)

8.5 GLYCOGEN METABOLISM

Glycogen is the storage form of glucose. The synthesis and degradation of glycogen are carefully regulated so that sufficient glucose is available for the body's energy needs. Both glycogenesis and glycogenolysis are controlled primarily by three hormones: insulin, glucagon, and epinephrine.

Glycogenesis

Glycogen synthesis occurs after a meal, when blood glucose levels are high. It has long been recognized that the consumption of a carbohydrate meal is followed promptly by liver glycogenesis. The synthesis of glycogen from glucose-6-phosphate involves the following set of reactions.

1. **Synthesis of glucose-1-phosphate**. Glucose-6-phosphate is reversibly converted to glucose-1-phosphate by phosphoglucomutase, an enzyme that contains a phosphoryl group attached to a reactive serine residue:

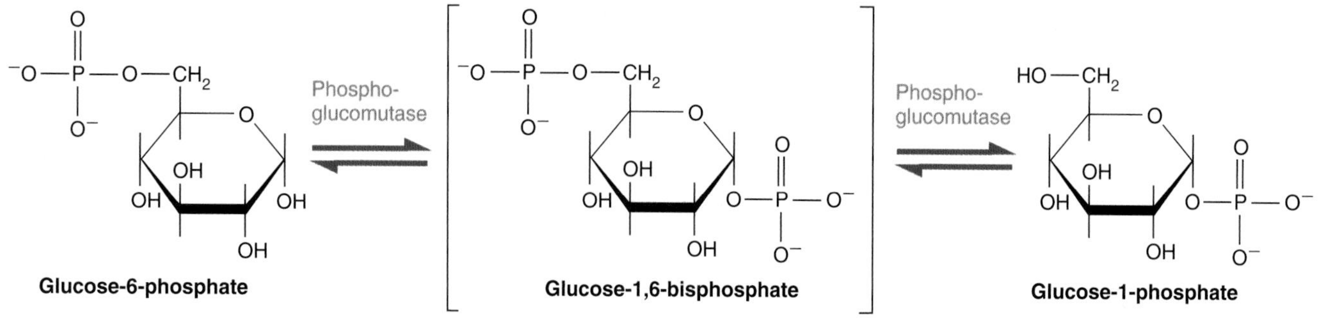

Glucose-6-phosphate   **Glucose-1,6-bisphosphate**   **Glucose-1-phosphate**

The enzyme's phosphoryl group is transferred to glucose-6-phosphate, forming glucose-1,6-bisphosphate. As glucose-1-phosphate forms, the phosphoryl group attached to C-6 is transferred to the enzyme's serine residue.

2. **Synthesis of UDP-glucose**. Glycosidic bond formation is an endergonic process. Derivatizing the sugar with a good leaving group provides the driving force for most sugar transfer reactions. For this reason, sugar-nucleotide synthesis is a common reaction preceding sugar transfer and polymerization processes. Uridine diphosphate glucose (UDP-glucose) is more reactive than glucose and is held more securely in the active site of the enzymes catalyzing transfer reactions (referred to as a group as glycosyl transferases). Because UDP-glucose contains two phosphoryl bonds, it is a highly reactive molecule. Formation of UDP-glucose, whose $\Delta G^{\circ\prime}$ value is near zero, is a reversible reaction catalyzed by UDP-glucose pyrophosphorylase:

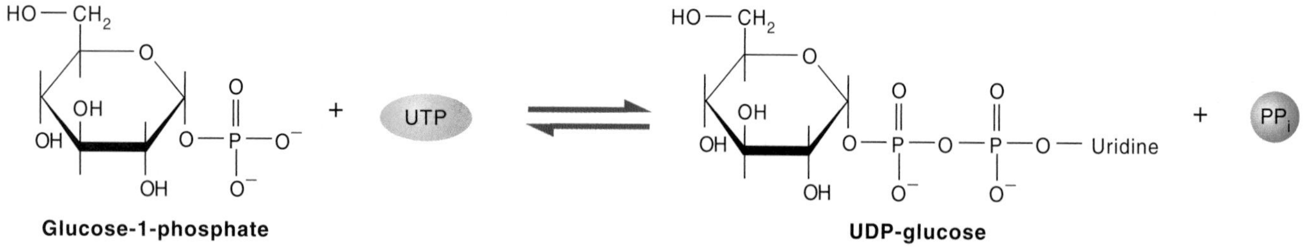

Glucose-1-phosphate   **UDP-glucose**

However, the reaction is driven to completion because pyrophosphate (PP_i) is immediately and irreversibly hydrolyzed by pyrophosphatase with a large loss of free energy ($\Delta G^{\circ\prime} = -33.5$ kJ/mol):

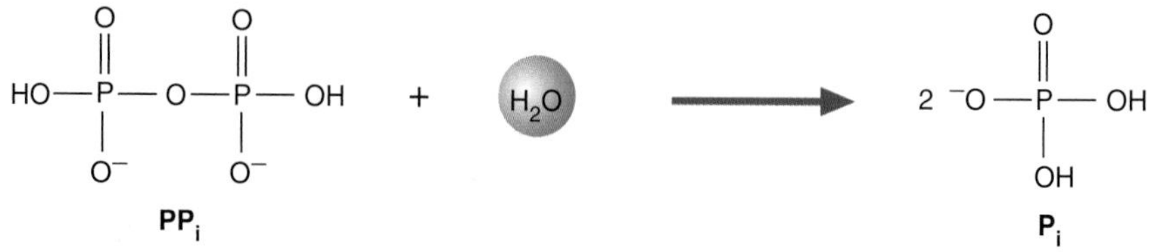

PP_i   **P_i**

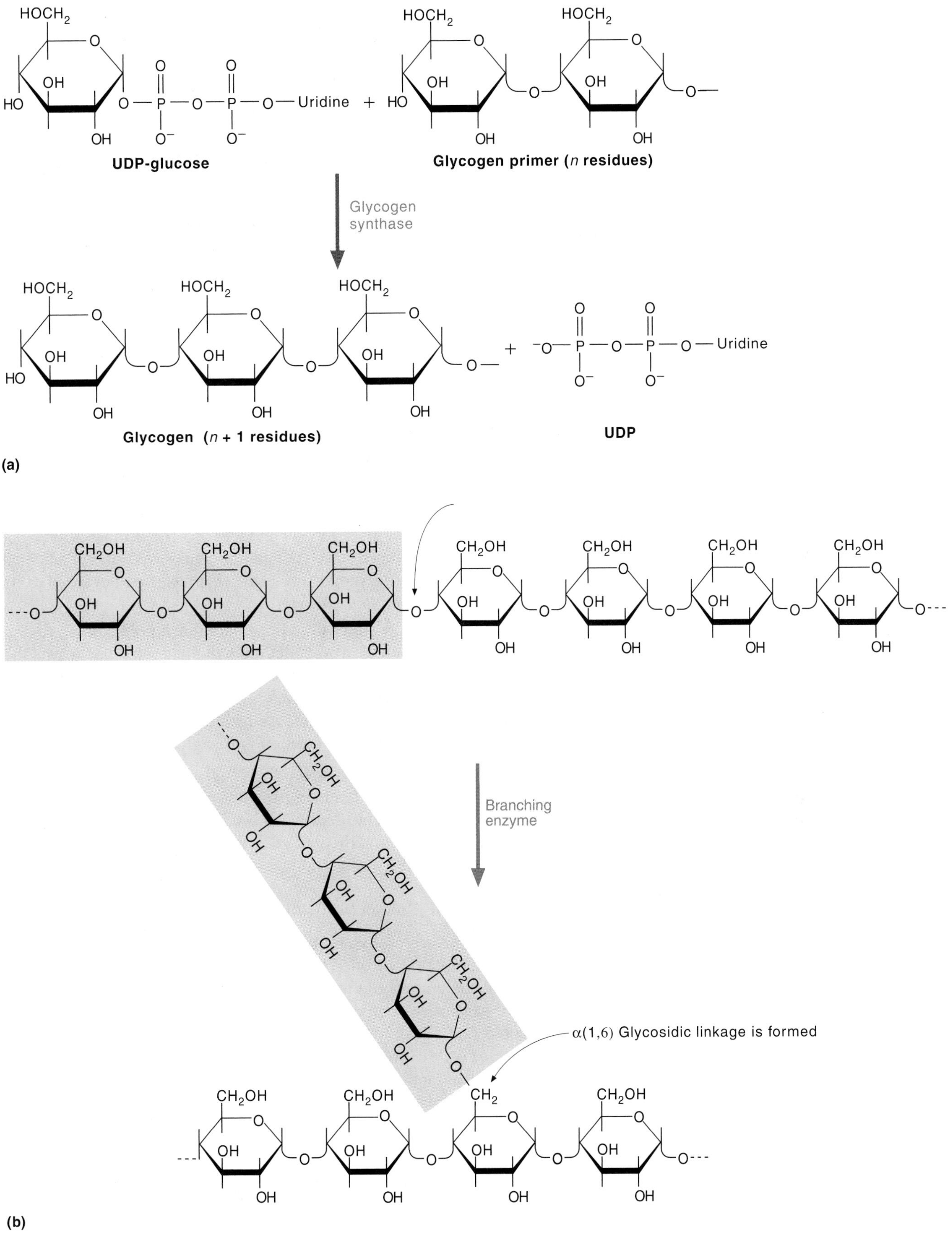

FIGURE 8.16

Glycogen Synthesis

(a) The enzyme glycogen synthase breaks the ester linkage of UDP-glucose and forms an $\alpha(1,4)$ glycosidic bond between glucose and the growing glycogen chain. (b) Branching enzyme is responsible for the synthesis of $\alpha(1,6)$ linkages in glycogen.

(Recall that removing product shifts the reaction equilibrium to the right. This cellular strategy is common.)

3. **Synthesis of glycogen from UDP-glucose**. The formation of glycogen from UDP-glucose requires two enzymes: (a) glycogen synthase, which catalyzes the transfer of the glucosyl group of UDP-glucose to the nonreducing ends of glycogen (Figure 8.16a), and (b) amylo-$\alpha(1,4 \rightarrow 1,6)$-glucosyl transferase (branching enzyme), which creates the $\alpha(1,6)$ linkages for branches in the molecule (Figure 8.16b).

Glycogen synthesis requires a preexisting tetrasaccharide composed of four $\alpha(1,4)$-linked glucosyl residues. The first of these residues is linked to a specific tyrosine residue in a "primer" protein called *glycogenin*. The glycogen chain is then extended by glycogen synthase and branching enzyme. Large glycogen granules, each consisting of a single highly branched glycogen molecule, can be observed in the cytoplasm of liver and muscle cells of well-fed animals. The enzymes responsible for glycogen synthesis and degradation coat each granule's surface.

Glycogenolysis

Glycogen degradation requires the following two reactions.

1. Removal of glucose from the nonreducing ends of glycogen. Glycogen phosphorylase uses inorganic phosphate (P_i) to cleave the $\alpha(1,4)$ linkages on the outer branches of glycogen to yield glucose-1-phosphate. Glycogen phosphorylase stops when it comes within four glucose residues of a branch point (Figure 8.17). (A glycogen molecule that has been degraded to its branch points is called a *limit dextrin*.)

2. Hydrolysis of the $\alpha(1,6)$ glycosidic bonds at branch points of glycogen. Amylo-$\alpha(1,6)$-glucosidase, also called debranching enzyme, begins the removal of $\alpha(1,6)$ branch points by transferring the outer three of the four glucose residues attached to the branch point to a nearby nonreducing end. It then removes the single glucose residue attached at each branch point. The product of this latter reaction is free glucose (Figure 8.18).

Glucose-1-phosphate, the major product of glycogenolysis, is diverted to glycolysis in muscle cells to generate energy for muscle contraction. In hepatocytes, glucose-1-phosphate is converted to glucose, by phosphoglucomutase and glucose-6-phosphatase, which is then released into the blood. A summary of glycogenolysis is shown in Figure 8.19.

Regulation of Glycogen Metabolism

Glycogen metabolism is carefully regulated to avoid wasting energy. Both synthesis and degradation are controlled through a complex mechanism involving insulin, glucagon, and epinephrine, as well as allosteric regulators. Glucagon is released from the pancreas when blood glucose levels drop in the hours after a meal. It binds to receptors on hepatocytes and initiates a signal transduction process that elevates intracellular cAMP levels. The second messenger cAMP amplifies the original glucagon signal and initiates a phosphorylation cascade that leads to the activation of glycogen phosphorylase along with a number of other proteins. Within seconds, glycogenolysis leads to the release of glucose into the bloodstream.

When occupied, the insulin receptor becomes an active tyrosine kinase enzyme that causes a phosphorylation cascade that ultimately has the opposite effect of the glucagon/cAMP system: the enzymes of glycogenolysis are inhibited and the enzymes of glycogenesis are activated. Insulin also increases the rate of glucose uptake into several types of target cells, but not liver or brain cells.

Emotional or physical stress releases epinephrine from the adrenal medulla. Epinephrine promotes glycogenolysis and inhibits glycogenesis. In emergency situations, when epinephrine is released in relatively large quantities, massive

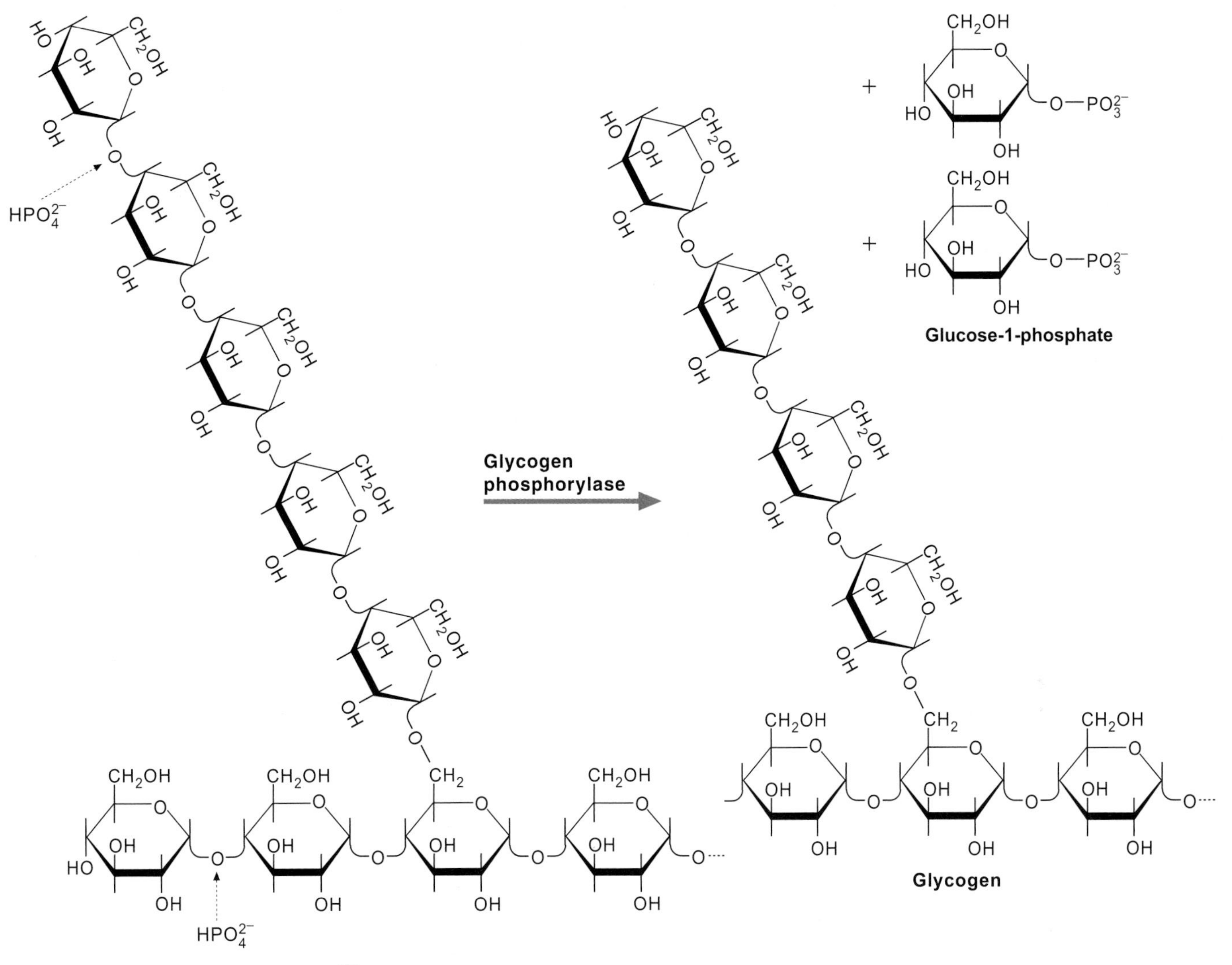

FIGURE 8.17

Glycogen Degradation

Glycogen phosphorylase catalyzes the removal of glucose residues from the nonreducing ends of a glycogen chain to yield glucose-1-phosphate. In this illustration one glucose residue is removed from each of two nonreducing ends. Removal of glucose residues continues until there are four residues at a branch point.

production of glucose provides the energy required to manage the situation. This effect is referred to as the flight-or-fight response. Epinephrine initiates the process by activating adenylate cyclase in liver and muscle cells. Two other second messengers, calcium ions and inositol trisphosphate (Chapter 16), are also believed to be involved in epinephrine's action.

Glycogen synthase (GS) and glycogen phosphorylase have both active and inactive conformations that are interconverted by covalent modification. The active form of glycogen synthase, known as the I (independent) form, is converted to the inactive or D (dependent) form by phosphorylation. The activity of GS can be finely modulated in response to a range of signal intensities because it is inactivated by phosphorylation reactions catalyzed by a large number of kinases. Physiologically, the most important kinases are glycogen synthase kinase 3 (GSK3) and casein kinase 1 (CS1). In contrast to GS, the inactive form of glycogen phosphorylase (phosphorylase b) is converted to the active form

(Continued on p. 303)

FIGURE 8.18

Glycogen Degradation via Debranching Enzyme

Branch points in glycogen are removed by the debranching enzyme amylo-α(1,6)-glucosidase. After transferring the three-residue unit that precedes the branch point to a nearby nonreducing end of the glycogen molecule, the enzyme cleaves the α(1,6) linkage, releasing a glucose molecule.

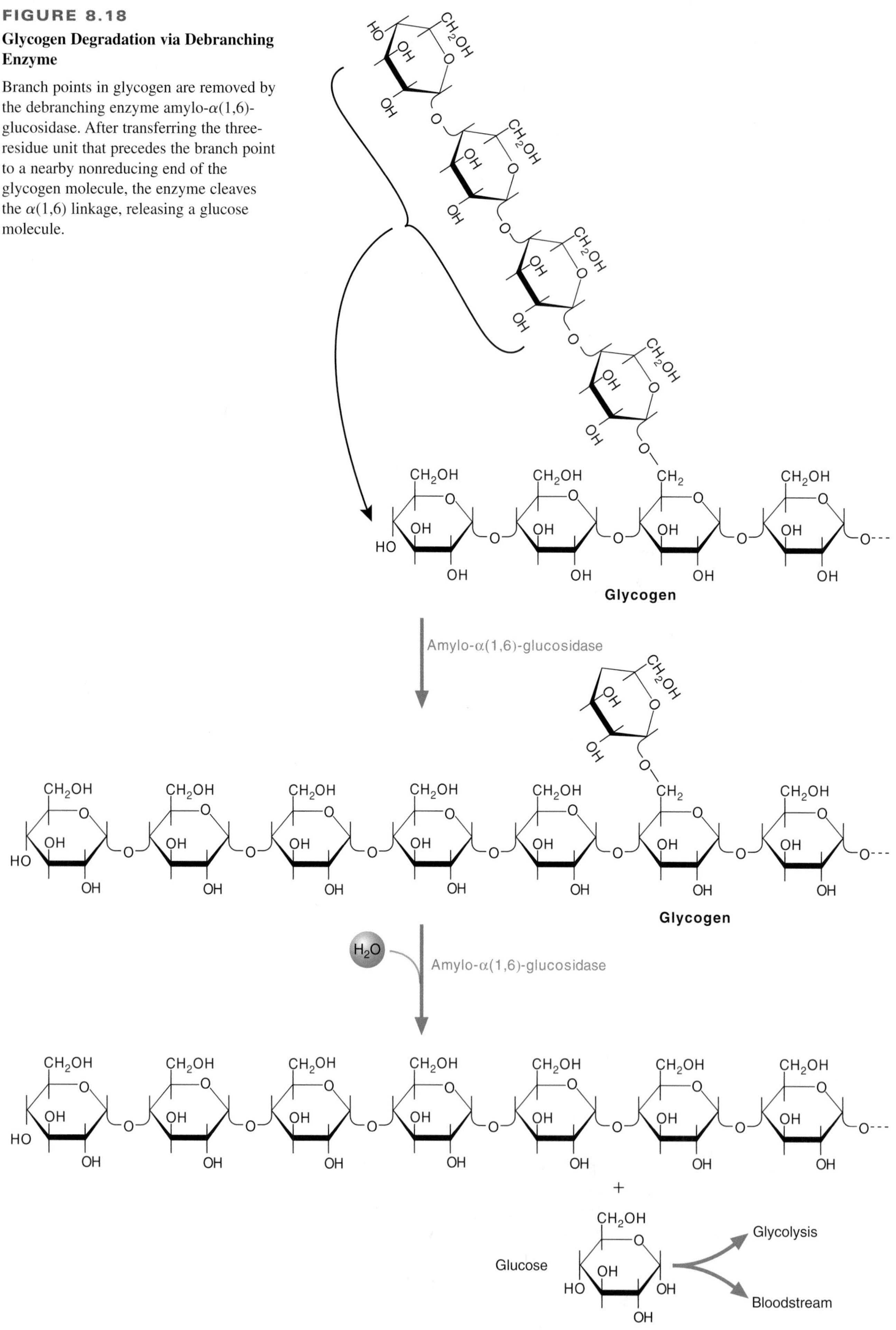

Glycogen

Amylo-α(1,6)-glucosidase

Glycogen

H_2O

Amylo-α(1,6)-glucosidase

Glucose

Glycolysis

Bloodstream

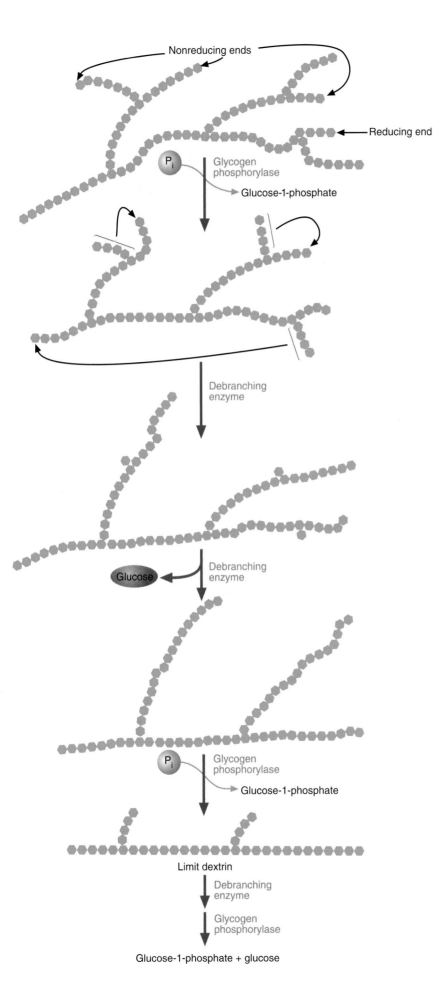

FIGURE 8.19
Glycogen Degradation: Summary

Glycogen phosphorylase cleaves the $\alpha(1,4)$ linkages of glycogen to yield glucose-1-phosphate until it comes within four glucose residues of a branch point. Debranching enzyme transfers three of these residues to a nearby nonreducing end and releases the fourth residue as free glucose. The repeated actions of both enzymes can lead to the complete degradation of glycogen.

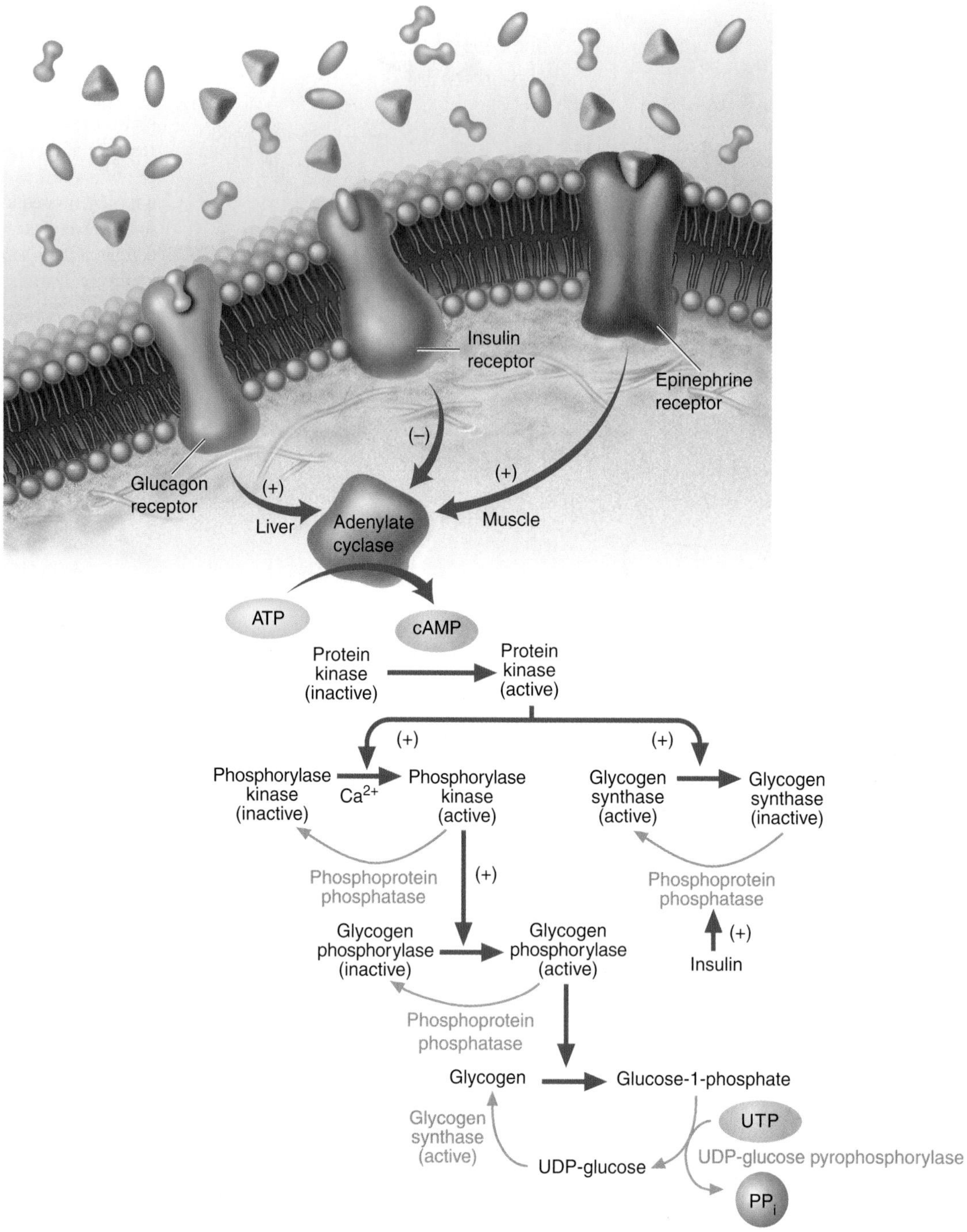

FIGURE 8.20

Major Factors Affecting Glycogen Metabolism

The binding of glucagon (released from the pancreas in response to low blood sugar) and/or epinephrine (released from the adrenal glands in response to stress) to their cognate receptors on the surface of target cells initiates a reaction cascade that converts glycogen to glucose-1-phosphate and inhibits glycogenesis. Insulin inhibits glycogenolysis and stimulates glycogenesis in part by decreasing the synthesis of cAMP and activating phosphoprotein phosphatase.

(phosphorylase a) by the phosphorylation of a specific serine residue. The phosphorylating enzyme is called phosphorylase kinase. Phosphorylation of both glycogen synthase (inactivating) and phosphorylase kinase (activating) is catalyzed by a protein kinase, which is activated by cAMP. Glycogen synthesis occurs when glycogen synthase and glycogen phosphorylase have been dephosphorylated. This conversion is catalyzed by phosphoprotein phosphatase 1 (PP1), which also inactivates phosphorylase kinase. It is noteworthy that PP1 is linked to both glycogen synthase and glycogen phosphorylase by an anchor protein (p. 66) called PTG (protein targeting to glycogen).

Several allosteric regulators also regulate glycogen metabolism. In muscle cells, both calcium ions released during muscle contraction and AMP bind to sites on glycogen phosphorylase b and promote its conversion to phosphorylase a. The reverse process, the conversion of glycogen phosphorylase a to phosphorylase b, is promoted by high levels of ATP and glucose-6-phosphate. Glycogen synthase activity is stimulated by glucose-6-phosphate. In hepatocytes, glucose is an allosteric regulator that promotes the inhibition of glycogen phosphorylase. The major factors in glycogen metabolic regulation are summarized in Figure 8.20.

KEY CONCEPTS

- During glycogenesis, glycogen synthase catalyzes the transfer of the glucosyl group of UDP-glucose to the nonreducing ends of glycogen, and glycogen branching enzyme catalyzes the formation of branch points.

- Glycogenolysis requires glycogen phosphorylase and debranching enzyme. Glycogen metabolism is regulated by the actions of three hormones: glucagon, insulin, and epinephrine and several allosteric regulators.

QUESTION 8.7

Glycogen storage diseases are caused by inherited defects of one or more enzymes involved in glycogen synthesis or degradation. Patients with *Cori's disease*, caused by a deficiency of debranching enzyme, have enlarged livers (*hepatomegaly*) and low blood sugar concentrations (**hypoglycemia**). Can you suggest what causes these symptoms?

Chapter**Summary**

1. The metabolism of carbohydrates is dominated by glucose because this sugar is an important fuel molecule in most organisms. If cellular energy reserves are low, glucose is degraded by the glycolytic pathway. Glucose molecules that are not required for immediate energy production are stored as either glycogen (in animals) or starch (in plants).

2. During glycolysis, glucose is phosphorylated and cleaved to form two molecules of glyceraldehyde-3-phosphate. Each glyceraldehyde-3-phosphate is then converted to a molecule of pyruvate. A small amount of energy is captured in two molecules each of ATP and NADH. In anaerobic organisms, pyruvate is converted to waste products. During this process, NAD^+ is regenerated so that glycolysis can continue. In the presence of O_2, aerobic organisms convert pyruvate to acetyl-CoA and then to CO_2 and H_2O. Glycolysis is controlled primarily by allosteric regulation of three enzymes—hexokinase, PFK-1, and pyruvate kinase—and by the hormones glucagon and insulin.

3. During gluconeogenesis, molecules of glucose are synthesized from noncarbohydrate precursors (lactate, pyruvate, glycerol, and certain amino acids). The reaction sequence in gluconeogenesis is largely the reverse of glycolysis. The three irreversible glycolytic reactions (the synthesis of pyruvate, the conversion of fructose-1,6-bisphosphate to fructose-6-phosphate, and the formation of glucose from

glucose-6-phosphate) are bypassed by alternate energetically favorable reactions.

4. The pentose phosphate pathway, in which glucose-6-phosphate is oxidized, occurs in two phases. In the oxidative phase, two molecules of NADPH are produced as glucose-6-phosphate is converted to ribulose-5-phosphate. In the nonoxidative phase, ribose-5-phosphate and other sugars are synthesized. If cells need more NADPH than ribose-5-phosphate, a component of nucleotides and the nucleic acids, then metabolites of the nonoxidative phase are converted into glycolytic intermediates.

5. Several sugars other than glucose are important in vertebrate carbohydrate metabolism. These include fructose, galactose, and mannose.

6. The substrate for glycogen synthesis is UDP-glucose, an activated form of the sugar. UDP-glucose pyrophosphorylase catalyzes the formation of UDP-glucose from glucose-1-phosphate and UTP. Glucose-6-phosphate is converted to glucose-1-phosphate by phosphoglucomutase. Glycogen synthesis requires two enzymes: glycogen synthase and branching enzyme. Glycogen degradation requires glycogen phosphorylase and debranching enzyme. The balance between glycogenesis (glycogen synthesis) and glycogenolysis (glycogen breakdown) is carefully regulated by several hormones (insulin, glucagon, and epinephrine) and allosteric regulators.

COMPANION Take your learning further by visiting the **companion website** for Biochemistry at **www.oup.com/us/mckee** where you can complete a multiple-choice quiz **WEBSITE** on carbohydrate metabolism to help you prepare for exams.

Suggested Readings

Bakker, B. M., Mensonides, F. I. C., Teusink, B., van Hoek, P., Michels, P. A. M., and Westerhoff, H. V., Compartmentation Protects Trypanosomes from the Dangerous Design of Glycolysis, *Proc. Nat. Acad. Sci. USA* 97(5):2087–2092, 2000.

Eldar-Finkelman, H., Glycogen Synthase Kinase 3: An Emerging Therapeutic Target, *Trends Mol. Med.* 8(3):126–132, 2002.

Fothergill-Gilmore, L. A., and Michels, P. A., Evolution of Glycolysis, *Prog. Biophys. Mol. Biol.* 59:105–135, 1993.

Frommer, W. B., Schulze, W. X., and LaLonde, S., Hexokinase, Jack-of-All-Trades, *Science* 300:261, 263, 2003.

Hallfrisch, J., Metabolic Effects of Dietary Fructose, *FASEB J.* 4:2652–2660, 1990.

Malik, V. S., Schulze, M. B., and Hu, F. B., Intake of Sugar-Sweetened Beverages and Weight Gain: A Systematic Review, *Am. J. Clin. Nutr.* 84:274–288, 2006.

Melendez-Hevia, E., Waddell, T. G., and Shelton, E. D., Optimization of Molecular Design in the Evolution of Metabolism: The Glycogen Molecule, *Biochem. J.* 295:477–483, 1993.

Melendez-Hevia, E., Waddell, T. G., Heinrich, R., and Montero, F., Theoretical Approaches to the Evolutionary Optimization of Glycolysis: Chemical Analysis, *Eur. J. Biochem.* 244:527–543, 1997.

Payne, V. A., Arden, C., Wu, C., Lange, A. J., and Agius, 1., Dual Role of Phosphofructokinase-2/Fructose Bisphosphatase-2 in Regulating the Compartmentation and Expression of Glucokinase in Hepatocytes, *Diabetes* 54:1949–1957, 2005.

Teusink, B., Walsh, M. C., van Dam, K., and Westerhoff, H. V., The Danger of Metabolic Pathways with Turbo Design, *Trends Biochem. Sci.* 23(5):162–169, 1998.

Wilson, J. E., Isozymes of Mammalian Hexokinase: Structure, Subcellular Localization and Metabolic Function, *J. Exp. Biol.* 206:2049–2057, 2003.

Key Words

aerobic respiration, *266*	decarboxylation, *277*	glycogenesis, *265*	pentose phosphate pathway, *265*
aldol cleavage, *272*	electron transport system, *276*	glycogenolysis, *265*	response element, *273*
amphibolic pathway, *276*	epinephrine, *298*	glycolysis, *265*	substrate-level phosphorylation, *279*
anaerobic organisms, *266*	fermentation, *277*	hypoglycemia, *303*	
antioxidant, *290*	glucagon, *279*	insulin, *279*	tautomer, *275*
citric acid cycle, *276*	gluconeogenesis, *265*	malate shuttle, *284*	tautomerization, *275*
Cori cycle, *286*	glucose-alanine cycle, *287*	Pasteur effect, *280*	transcription factor, *279*

Review Questions

These questions are designed to test your knowledge of the key concepts discussed in this chapter, before moving on to the next chapter. You may like to compare your answers to the solutions provided in the back of the book and in the accompanying Study Guide.

1. Upon entering a cell, glucose is phosphorylated. Give two reasons why this reaction is required.

2. Describe the functions of the following molecules:
   a. insulin
   b. glucagon
   c. fructose-2,6-bisphosphate
   d. UDP-glucose
   e. cAMP
   f. GSSG
   g. NADPH

3. Describe the structural differences between ribose-5-phosphate and ribulose-5-phosphate.

4. In which locations in the eukaryotic cell do the following processes occur?
   a. gluconeogenesis
   b. glycolysis
   c. pentose phosphate pathway

5. Compare the entry-level substrates, products, and metabolic purposes of glycolysis and gluconeogenesis.

6. Define substrate-level phosphorylation. Which two reactions in glycolysis are in this category?

7. What is the principal reason that organisms such as yeast produce alcohol?

8. Why is pyruvate not oxidized to CO_2 and H_2O under anaerobic conditions?

9. Describe how epinephrine promotes the conversion of glycogen to glucose.

10. Glycolysis occurs in two stages. Describe what is accomplished in each stage.

11. What effects do the following molecules have on gluconeogenesis?
    a. lactate
    b. ATP
    c. pyruvate
    d. glycerol
    e. AMP
    f. acetyl-CoA

12. Describe the physiological conditions that activate gluconeogenesis.

13. The following two reactions constitute a wasteful cycle:

$$\text{Glucose} + \text{ATP} \rightarrow \text{glucose-6-phosphate}$$
$$\text{Glucose-6-phosphate} + H_2O \rightarrow \text{glucose} + P_i$$

Suggest how such wasteful cycles are prevented or controlled.

14. Define the following terms:
    a. fermentation

b. citric acid cycle
c. electron transport system
d. tautomerization
e. aldol cleavage

15. Define the following terms:
    a. anaerobic
    b. aerobic
    c. antioxidants
    d. glucose-alanine cycle
    e. malate shuttle

16. Describe the central role of glucose in carbohydrate metabolism.

17. Draw the structure of glucose with the carbons numbered 1 through 6. Number these carbons again as they appear in the two molecules of pyruvate formed during glycolysis.

18. Draw the reactions that convert glucose into ethanol.

19. After reviewing Figure 8.15, draw the reactions that convert galactose into a glycolytic intermediate. Include the Haworth formulas in your work.

20. Draw the reactions that convert fructose into glycolytic intermediates.

21. Why is severe hypoglycemia so dangerous?

22. Which reaction in glycolysis involves a redox reaction?

23. Explain why ATP hydrolysis occurs so early in glycolysis, an ATP-producing pathway.

24. In which reaction in glycolysis does a dehydration occur?

25. What would be the net result of the conversion of a molecule of sucrose to pyruvate?

26. Describe the Cori cycle. What is its physiological function?

27. Describe the effects of insulin and glucagon on glycogen metabolism.

28. Describe the effects of insulin and glucagon on blood glucose.

29. What cells produce insulin, glucagon, epinephrine, and cortisol?

30. Describe the different functions of glycogen in liver and muscle.

31. Describe the fate of pyruvate under anaerobic and aerobic conditions.

Thought Questions

These questions are designed to reinforce your understanding of all of the key concepts discussed in the book so far, including this chapter and all of the chapters before it. They may not have one right answer! The authors have provided possible solutions to these questions in the back of the book and in the accompanying Study Guide, for your reference.

32. An individual has a genetic deficiency that prevents the production of glucokinase. Following a carbohydrate meal, do you expect blood glucose levels to be high, low, or about normal? What organ accumulates glycogen under these circumstances?

33. Glycogen synthesis requires a short primer chain. Explain how new glycogen molecules are synthesized given this limitation.

34. Why is fructose metabolized more rapidly than glucose?

35. What is the difference between an enol-phosphate ester and a normal phosphate ester that gives PEP such a high phosphoryl group transfer potential?

36. In aerobic oxidation, oxygen is the ultimate oxidizing agent (electron acceptor). Name two common oxidizing agents in anaerobic fermentation.

37. Why is it important that gluconeogenesis is not the exact reverse of glycolysis?

38. Compare the structural formulas of ethanol, acetate, and acetaldehyde. Which molecule is the most oxidized? Which is the most reduced? Explain your answers.

39. *Trypanosoma brucei* is a parasitic protozoan that causes sleeping sickness in humans, Transmitted by the tsetse fly, sleeping sickness is a fatal disease characterized by fever, anemia, inflammation, lethargy, headache, and convulsions. When trypanosomes are present in the human bloodstream, they depend on glycolysis entirely for energy generation. The first seven glycolytic enzymes in these organisms are localized in peroxisome-like organelles called glycosomes, which are only regulated weakly by allosteric regulator molecules. Glycosomes take up glucose and export glycerate-3-phosphate. There are two pools of ADP and ATP (cytoplasmic and glycosomal), and the glycosomal membrane is impermeable to both nucleotides as well as most other glycolytic intermediates. If the glycosomal membrane is compromised, the concentration of phosphoylated glycolytic intermediates rises and the cells die. Explain.

40. The consumption of large amounts of soft drink beverages and processed foods sweetened with high-fructose corn syrup has been linked to obesity. After reviewing Figures 8.1 and 8.15, suggest a likely reason for this phenomenon.

41. How does phosphorylation increase the reactivity of glucose?

42. Examine the structure of phosphoenolpyruvate and explain why it has such a high phosphoryl group transfer potential.

43. Both glycogen and triacylglycerols are energy sources used by the body. Suggest a reason why both are required.

44. Severe dieting results in both the reduction of fat stores and the loss of muscle mass. Use biochemical reactions to trace the conversion of muscle protein to glucose production.

45. Suggest a reason why glycolysis produces NADH and the pentose phosphate pathway produces NADPH.

46. Transamination reactions are used to interconvert certain types of amino acids. For example alanine transaminase reversibly converts L-glutamate to L-alanine. Glutamate can also be converted to aspartate using oxaloacetate as the α-keto acid. Write this reaction.

47. Cells in culture are fed glucose molecules labeled with ^{14}C at carbon 2. Trace the radioactive label through one pass through the pentose phosphate pathway.

48. Ethanol is especially toxic in children for several reasons. For example, ethanol consumption results in elevated levels of NADH in the liver. Suggest a mechanism that explains this phenomenon.

Aerobic Metabolism I: The Citric Acid Cycle

CHAPTER 9

OUTLINE

9.1 OXIDATION-REDUCTION REACTIONS
Redox Coenzymes
Aerobic Metabolism

9.2 CITRIC ACID CYCLE
Conversion of Pyruvate to Acetyl-CoA
Reactions of the Citric Acid Cycle
Fate of Carbon Atoms in the Citric
　　Acid Cycle
The Amphibolic Citric Acid Cycle
Citric Acid Cycle Regulation
The Glyoxylate Cycle

BIOCHEMISTRY IN PERSPECTIVE
The Evolutionary History of the Citric
Acid Cycle

Available online:
BIOCHEMISTRY IN PERSPECTIVE
Hans Krebs and the Citric Acid Cycle

In aerobic cells most energy is generated within the mitochondrion. Dioxygen (O_2) is the final electron acceptor in the oxidation of nutrient molecules.

Overview

OVER TWO BILLION YEARS AGO, PROKARYOTES SUCH AS THE CYANOBAC-
TERIA BEGAN CREATING AN OXYGENATED ATMOSPHERE. THE DIOXYGEN
(O_2, often referred to as oxygen) these primitive organisms produced as a
waste product of photosynthesis triggered a revolution in the living world. As O_2
accumulated in the atmosphere, organisms developed aerobic respiration, a powerful molecular mechanism that exploits oxygen as a means of generating energy.
The resulting vast increase in energy-generating capacity was a key factor in the
evolution of complex eukaryotic life forms. Modern aerobic organisms transduce
the chemical bond energy of food molecules into the bond energy of ATP by
using oxygen as the terminal acceptor of the electrons extracted from food molecules. The capacity to use oxygen to oxidize nutrients such as glucose and fatty
acids yields a substantially greater amount of energy than does fermentation.

As the first primordial life forms emerged on Earth, they were sustained
by preformed, simple organic molecules such as carboxylic and amino
acids. The source of these substances, which primitive living systems used
as building block and fuel molecules, is believed to have been chemical reactions
driven by electrical discharges, solar radiation, and thermal forces deep within
the planet. In addition, untold tons of chemical matter arrived from outer space
as Earth was bombarded by meteorites and other cosmic debris. Early organisms
(primordial prokaryotic cells) eventually became so abundant that they consumed
organic molecules faster than those molecules were formed by natural forces. As
supplies of preformed molecules dwindled, some organisms evolved new mechanisms for obtaining food. The most important innovation was the capacity to
synthesize photosensitive pigments that captured light energy and converted it
to chemical bond energy. In this mechanism, called photosynthesis, sugar molecules are synthesized from CO_2 and hydrogen atoms. In early forms of photosynthesis hydrogen was supplied by hydrogen gas (H_2), small organic
molecules or H_2S, all of which were available in limited quantities. Eventually
cyanobacteria developed a photosynthetic mechanism that utilized water, a far
more abundant substance, as the hydrogen source. Dioxygen (O_2) is a by-product of this process. As oxygenic photosynthesis occurred on an ever increasing
scale, the oxygen content of the atmosphere increased. Because O_2 combines
with other molecules (e.g., $4NH_3 + 3O_2 \rightarrow 2N_2 + 6H_2O$), the earth's atmosphere
was converted, over a billion-year time span, to one consisting principally of
dinitrogen, water vapor, carbon dioxide, and oxygen.

It has long been believed that the appearance of oxygen in the atmosphere,
beginning about 2.7 billion years ago, resulted in mass extinctions of organisms.
In this view most of the organisms that arose under the reducing (oxygen-free) conditions of the primitive Earth, which were devoid of O_2, were unprepared for living in an oxidizing atmosphere. However, recent reevaluations of available
evidence suggest that the organisms that existed when molecular oxygen levels
increased were already adapted to the toxic effects of the gas. The lethal effects
of oxygen toxicity and radiation are caused by the same mechanism: the production
of toxic reactive intermediates called *reactive oxygen species* (ROS). During the
early phase of the planet's existence, Earth was bombarded by high-intensity ultraviolet (UV) radiation that damages organic molecules by splitting water molecules
and thereby generating ROS. When O_2 began to accumulate, ozone (O_3) was
formed in the upper stratosphere as oxygen molecules came in contact with UV

light or electrical discharges. It was the ozone, acting as a shield, that eventually lowered radiation levels by about 30-fold. Before the formation of the ozone layer, primitive cells protected themselves from the damaging effects of radiation by evading it (e.g., living in the oceans below the depth where light penetrates) or by producing protective molecules such as the antioxidant enzyme catalase. These same tactics were used to protect against free radicals derived from molecular oxygen. Eventually, some organisms developed the means for utilizing oxygen to generate significantly more energy than can be obtained by means of fermentation. The various strategies used by modern organisms to protect themselves from oxygen toxicity and/or to exploit oxygen to generate energy are as follows.

Obligate anaerobes, organisms that grow only in the absence of oxygen, avoid the gas by living in highly reduced environments such as soil. They use fermentative processes to satisfy their energy requirements. **Aerotolerant anaerobes**, which also depend on fermentation for their energy needs, possess detoxifying enzymes and antioxidant molecules that protect against oxygen's toxic products. **Facultative anaerobes** not only possess the mechanisms needed for detoxifying oxygen metabolites, they can also generate energy by using oxygen as an electron acceptor when the gas is present. Finally, **obligate aerobes** are highly dependent on oxygen for energy production. They protect themselves from the potentially dangerous consequences of exposure to oxygen with elaborate mechanisms composed of enzymes and antioxidant molecules.

Facultative anaerobes and obligate aerobes, which use oxygen to generate energy, employ the following biochemical processes: the citric acid cycle, the electron transport pathway, and oxidative phosphorylation (Figure 9.1). In

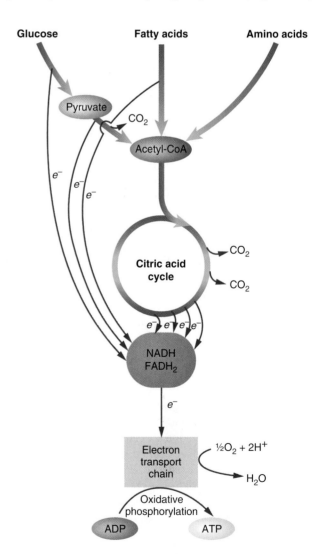

FIGURE 9.1

Overview of Aerobic Metabolism

In aerobic metabolism, the nutrient molecules glucose, fatty acids, and some amino acids are degraded to form acetyl-CoA. Acetyl-CoA then enters the citric acid cycle. Electron carriers (e^-), produced by glucose and fatty acid degradation and several citric acid cycle reactions, donate electrons to the electron transport chain via NADH and $FADH_2$. Energy captured by the electron transport chain is then used to synthesize ATP in a process referred to as oxidative phosphorylation. Note that O_2, the terminal electron acceptor in aerobic metabolism, combines with protons to form water molecules.

eukaryotes these processes occur within the mitochondrion (Figure 9.2). The citric acid cycle is a metabolic pathway in which two-carbon fragments derived from organic fuel molecules are oxidized to form CO_2 and the coenzymes NAD^+ and FAD are reduced to form NADH and $FADH_2$, which act as electron carriers. The electron transport pathway, also referred to as the electron transport chain (ETC), is a mechanism by which electrons are transferred from reduced coenzymes to an acceptor (usually O_2). In oxidative phosphorylation, the energy released by electron transport is captured in the form of a proton gradient that drives the synthesis of ATP, the energy currency of living organisms.

Chapter 9 begins with a review of oxidation-reduction reactions and the relationship between electron flow and energy transduction. This is followed by a detailed discussion of the citric acid cycle, the central pathway in aerobic metabolism, and its roles in energy generation and biosynthesis. In Chapter 10, the discussion of aerobic metabolism continues with an examination of electron transport and oxidative phosphorylation, the means by which aerobic organisms use oxygen to generate significant amounts of ATP. It ends with a review of *oxidative stress*, a series of reactions in which toxic oxygen species are created and subsequently damage cell structure and function. The principal mechanisms used by living organisms to protect themselves from oxidative stress are also described.

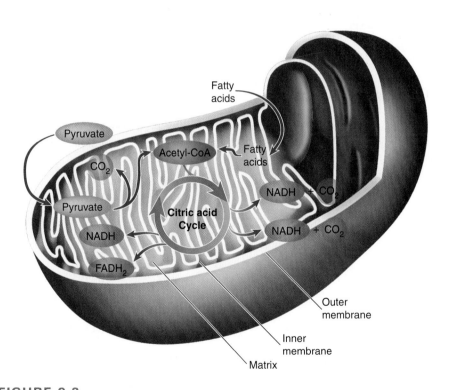

FIGURE 9.2

Aerobic Metabolism in the Mitochondrion

In eukaryotic cells aerobic metabolism occurs within the mitochondrion. Acetyl-CoA, the oxidation product of pyruvate, fatty acids, and certain amino acids (not shown), is oxidized by the reactions of the citric acid cycle within the mitochondrial matrix. The principal products of the cycle are the reduced coenzymes NADH and $FADH_2$, and CO_2. The high-energy electrons of NADH and $FADH_2$ are subsequently donated to the electron transport chain (ETC), a series of electron carriers in the inner membrane. The terminal electron acceptor for the ETC is O_2. The energy derived from the electron transport mechanism drives ATP synthesis by creating a proton gradient across the inner membrane. The large folded surface of the inner membrane is studded with ETC complexes, transport proteins of numerous types, and ATP synthase, the enzyme complex responsible for ATP synthesis.

9.1 OXIDATION-REDUCTION REACTIONS

In living organisms, both energy-capturing and energy-releasing processes consist largely of redox reactions. Recall that redox reactions occur when electrons are transferred between an electron donor (reducing agent) and an electron acceptor (oxidizing agent). In some redox reactions, only electrons are transferred. For example, in the reaction

$$Cu^+ + Fe^{3+} \rightleftharpoons Cu^{2+} + Fe^{2+}$$

an electron is transferred from Cu^+ to Fe^{3+}. Cu^+, the reducing agent, is oxidized to form Cu^{2+}. Meanwhile, Fe^{3+} is reduced to Fe^{2+}. In many reactions, however, both electrons and protons are transferred. For example, the reaction catalyzed by lactate dehydrogenase begins with the transfer of a hydride ion, that is, a hydrogen nucleus and two electrons ($H{:}^-$). Then, as pyruvate is reduced to form lactate and NAD^+, a proton (H^+) is gained from the environment (Figure 9.3).

Redox reactions are more easily understood if they are separated into half-reactions. In the reaction between copper and iron the Cu^+ ion loses an electron to become Cu^{2+}:

$$Cu^+ \rightleftharpoons Cu^{2+} + e^-$$

This equation indicates that Cu^+ is the electron donor. (Together Cu^+ and Cu^{2+} constitute a **conjugate redox pair**.) As Cu^+ loses an electron, Fe^{3+} gains an electron to form Fe^{2+}:

$$Fe^{3+} + e^- \rightleftharpoons Fe^{2+}$$

In this half-reaction, Fe^{3+} is an electron acceptor. The separation of redox reactions emphasizes that electrons are always the common intermediates between half-reactions.

The constituents of half-reactions may be observed in an electrochemical cell (Figure 9.4). Each half-reaction takes place in a separate container or *half-cell*.

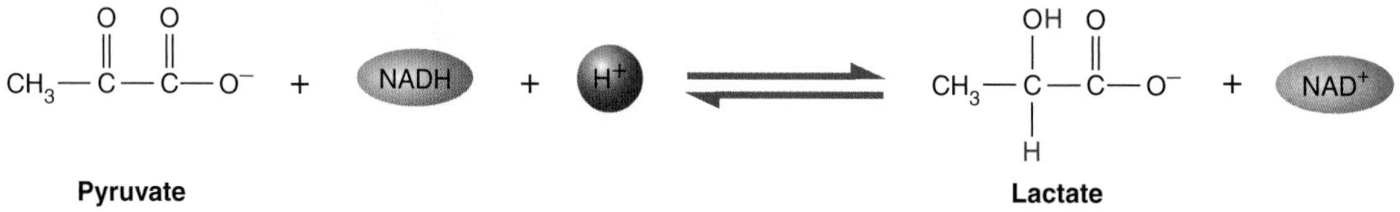

FIGURE 9.3

Reduction of Pyruvate by NADH

In this redox reaction, a hydride ion ($H{:}^-$) is transferred from NADH to pyruvate and the product is protonated from the surrounding medium to form lactate.

FIGURE 9.4

An Electrochemical Cell

Electrons flow from the copper electrode through the voltmeter to the iron electrode. The salt bridge containing KCl completes the electrical circuit. The voltmeter measures the electrical potential, which drives electrons from one half-cell to the other.

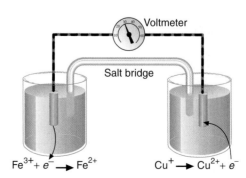

The movement of electrons generated in the half-cell undergoing oxidation (e.g., $Cu^+ \rightarrow Cu^{2+} + e^-$) generates a voltage (or potential difference) between the two half-cells. The sign of the voltage (measured by a voltmeter) is positive or negative according to the direction of the electron flow. The magnitude of the potential difference is a measure of the energy that drives the reaction.

The tendency for a specific substance to gain electrons is called its **reduction potential**. The **standard reduction potential** of a substance is measured in a galvanic cell relative to a standard hydrogen electrode. A standard cell has all solutes at 1.0 M concentration, all gases at 1 atm pressure, and the temperature at 25°C. The reduction potential for the half-reaction, $2H^+ + 2e^- \rightarrow H_2(g)$ against the standard hydrogen electrode is set at 0.00 V.

In biochemistry the reference half-reaction is

$$2H^+ + 2\,e^- \rightleftharpoons H_2$$

when pH $= 7$
temperature $= 25°C$
pressure $= 1$ atm

Under these conditions the reduction potential of the hydrogen electrode is −0.42 V when measured against the standard hydrogen electrode in which the hydrogen ion concentration is 1 M. Substances with reduction potentials lower than −0.42 V (i.e., those with more negative values) have a lower affinity for electrons than does H^+. Substances with higher reduction potentials (i.e., those with more positive values) have a greater affinity for electrons (Table 9.1). The pH in the test electrode is 7.0 for each of the redox half-reactions, and the pH of the reference standard electrode is 0 or the $[H^+]$ is 1.0 M.

TABLE 9.1   Standard Reduction Potentials*

Redox Half-Reaction	Standard Reduction Potentials ($E^{\circ\prime}$) (V)
$2H^+ + 2e^- \rightarrow H_2$	−0.42
α-Ketoglutarate $+ CO_2 + 2H^+ + 2e^- \rightarrow$ isocitrate	−0.38
$NADP^+ + 2H^+ + 2e^- \rightarrow NADPH$	−0.324
$NAD^+ + H^+ + 2e^- \rightarrow NADH$	−0.32
$S + 2H^+ + 2e^- \rightarrow H_2S$	−0.23
$FAD + 2H^+ + 2e^- \rightarrow FADH_2$	−0.22
Acetaldehyde $+ 2H^+ + 2e^- \rightarrow$ ethanol	−0.20
Pyruvate $+ 2H^+ + 2e^- \rightarrow$ lactate	−0.19
Oxaloacetate $+ 2H^+ + 2e^- \rightarrow$ malate	−0.166
$Cu^{2+} + e^- \rightarrow Cu^+$	+0.16
Fumarate $+ 2H^+ + 2e^- \rightarrow$ succinate	+0.031
Cytochrome b $(Fe^{3+}) + e^- \rightarrow$ cytochrome b (Fe^{2+})	+0.075
Cytochrome c_1 $(Fe^{3+}) + e^- \rightarrow$ cytochrome c_1 (Fe^{2+})	+0.22
Cytochrome c $(Fe^{3+}) + e^- \rightarrow$ cytochrome c (Fe^{2+})	+0.235
Cytochrome a $(Fe^{3+}) + e^- \rightarrow$ cytochrome a (Fe^{2+})	+0.29
$NO_3^- + 2H^+ + 2e^- \rightarrow NO_2^- + H_2O$	+0.42
$NO_2^- + 8H^+ + 6e^- \rightarrow NH_4^+ + 2H_2O$	+0.44
$Fe^{3+} + e^- \rightarrow Fe^{2+}$	+0.77
$\frac{1}{2}O_2 + 2H^+ + 2e^- \rightarrow H_2O$	+0.82

* By convention, redox reactions are written with the reducing agent to the right of the oxidizing agent and the number of electrons transferred. In this table the redox pairs are listed in order of increasing $E^{\circ\prime}$ values. The more negative the $E^{\circ\prime}$ value is for a redox pair, the lower the affinity of the oxidized component for electrons. The more positive the $E^{\circ\prime}$ value is, the greater the affinity of the oxidized component of the redox pair for electrons. Under appropriate conditions, a redox half-reaction reduces any of the half-reactions below it in the table.

A substance with a more negative (less positive) reduction potential will receive electrons from a substance with a more positive reduction potential, and the overall cell potential ($\Delta E°'$) will be positive. The relationship between $\Delta E°'$ and $\Delta G°'$ is

$$\Delta G°' = -nF \, \Delta E°'$$

where $\Delta G°'$ = the standard free energy
n = the number of electrons transferred
F = the Faraday constant (96,485 J/V·mol)
$\Delta E°'$ = the difference in reduction potential between the electron donor and the electron acceptor under standard conditions

Living organisms utilize redox coenzymes as high-energy electron carriers. The most prominent examples are described next.

Redox Coenzymes

The coenzyme forms of the vitamin molecules nicotinic acid and riboflavin are universal electron carriers. Their structural and functional properties are as follows.

NICOTINIC ACID There are two coenzyme forms of nicotinic acid: **nicotinamide adenine dinucleotide (NAD)** and **nicotinamide adenine dinucleotide phosphate (NADP)**. These coenzymes occur in oxidized forms (NAD^+ and $NADP^+$) and reduced forms (NADH and NADPH). The structures of NAD^+ and $NADP^+$ both contain adenosine and the N-ribosyl derivative of nicotinamide, which are linked together through a pyrophosphate group (Figure 9.5a). $NADP^+$ has an additional phosphate attached to the 2′ OH group of adenosine. (The ring atoms of the sugar in a nucleotide are designated with a prime to distinguish them from atoms in the base.) Both NAD^+ and $NADP^+$ carry electrons for several enzymes in a group known as the dehydrogenases. (Dehydrogenases catalyze hydride transfer reactions. Many dehydrogenases that catalyze reactions involved in energy generation use the coenzyme NADH. The enzymes that require NADPH usually catalyze biosynthetic reactions. A small number of dehydrogenases can use either NADH or NADPH.)

Alcohol dehydrogenase catalyzes the reversible oxidation of ethanol to form acetaldehyde:

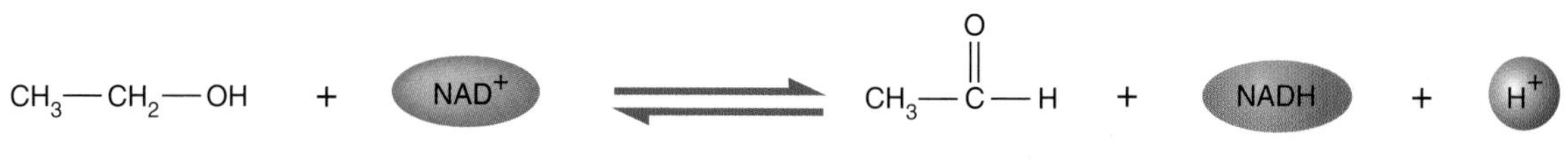

During this reaction NAD^+ accepts a hydride ion from ethanol, the substrate molecule undergoing oxidation. The product deprotonates to form the acetaldehyde molecule. The reversible reduction of NAD^+ is illustrated in Figure 9.5b.

In most reactions catalyzed by dehydrogenases, the NAD^+ (or $NADP^+$) is bound only transiently to the enzyme. After the reduced version of the coenzyme is released from the enzyme, it donates the hydride ion to another molecule, called an *electron acceptor*, with a more positive reduction potential than NADH.

RIBOFLAVIN Riboflavin (vitamin B_2) is a component of two coenzymes: **flavin mononucleotide (FMN)** and **flavin adenine dinucleotide (FAD)** (Figure 9.6). FMN and FAD function as tightly bound prosthetic groups in a class of enzymes known as the **flavoproteins**. Flavoproteins are a diverse group

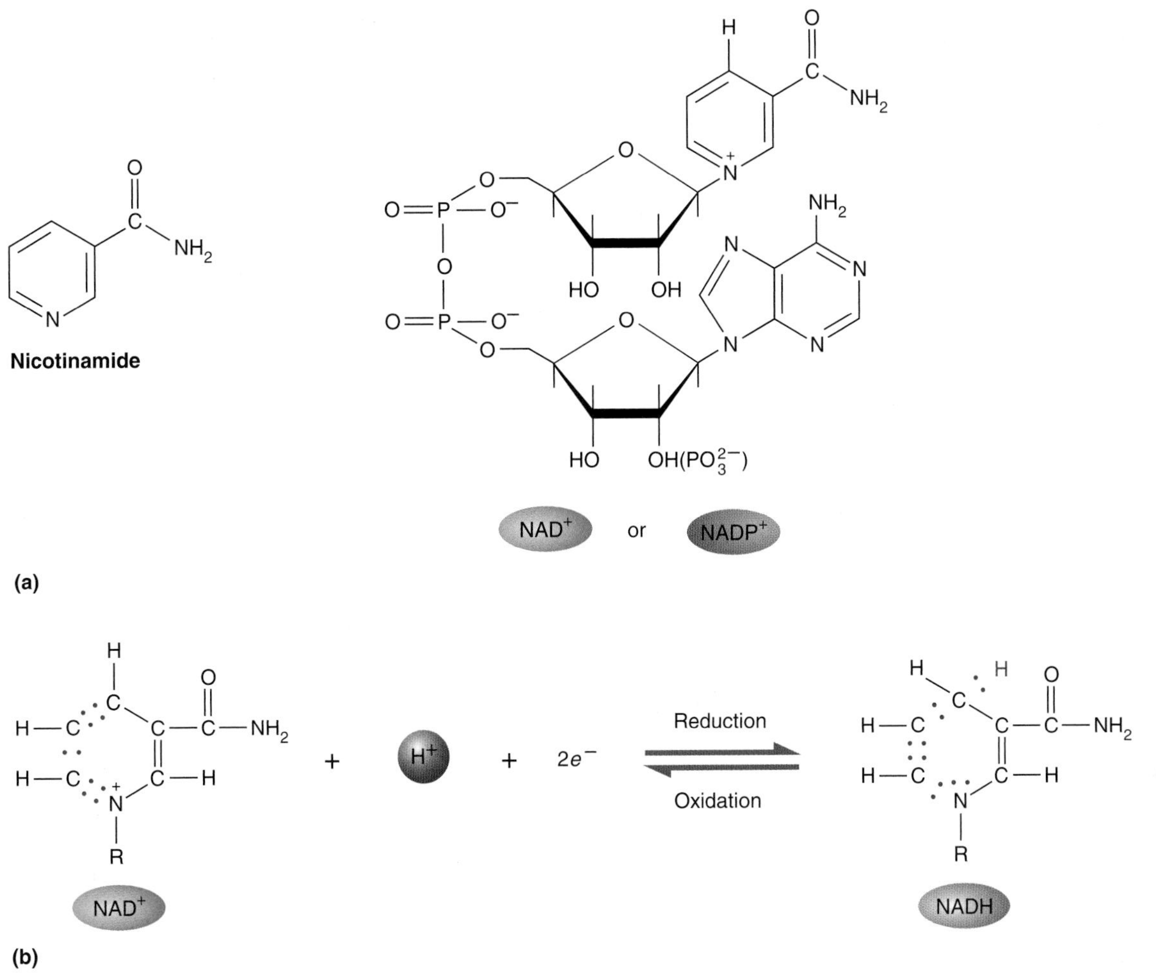

(a)

(b)

FIGURE 9.5

Nicotinamide Adenine Dinucleotide (NAD)

(a) Nicotinamide and NAD(P)+. (b) Reversible reduction of NAD+ to NADH. To simplify the equation, only the nicotinamide ring is shown. The rest of the molecule is designated R.

of catalysts; they function as dehydrogenases, oxidases, and hydroxylases. These enzymes, which catalyze oxidation-reduction reactions, use the isoalloxazine group of FAD or FMN as a donor or acceptor of two hydrogen atoms. FMN plays a key role in the link between two-electron transfer reactions in the mitochondrial matrix and the one-electron transfer reactions of the electron transport chain, as it can transfer one hydrogen atom at a time. Succinate dehydrogenase is a prominent example of a flavoprotein. It catalyzes the oxidation of succinate to form fumarate, an important reaction in energy production.

QUESTION 9.1

Use Table 9.1 to determine which of the following reactions will proceed as written:

$$CH_3CH_2OH + 2 \text{ cyt b (Fe}^{3+}) \rightarrow CH_3CHO + 2 \text{ cyt b (Fe}^{2+}) + 2H^+$$

$$NO_2^- + H_2O + 2 \text{ cyt b (Fe}^{3+}) \rightarrow 2 \text{ cyt b (Fe}^{2+}) + NO_3^- + 2H^+$$

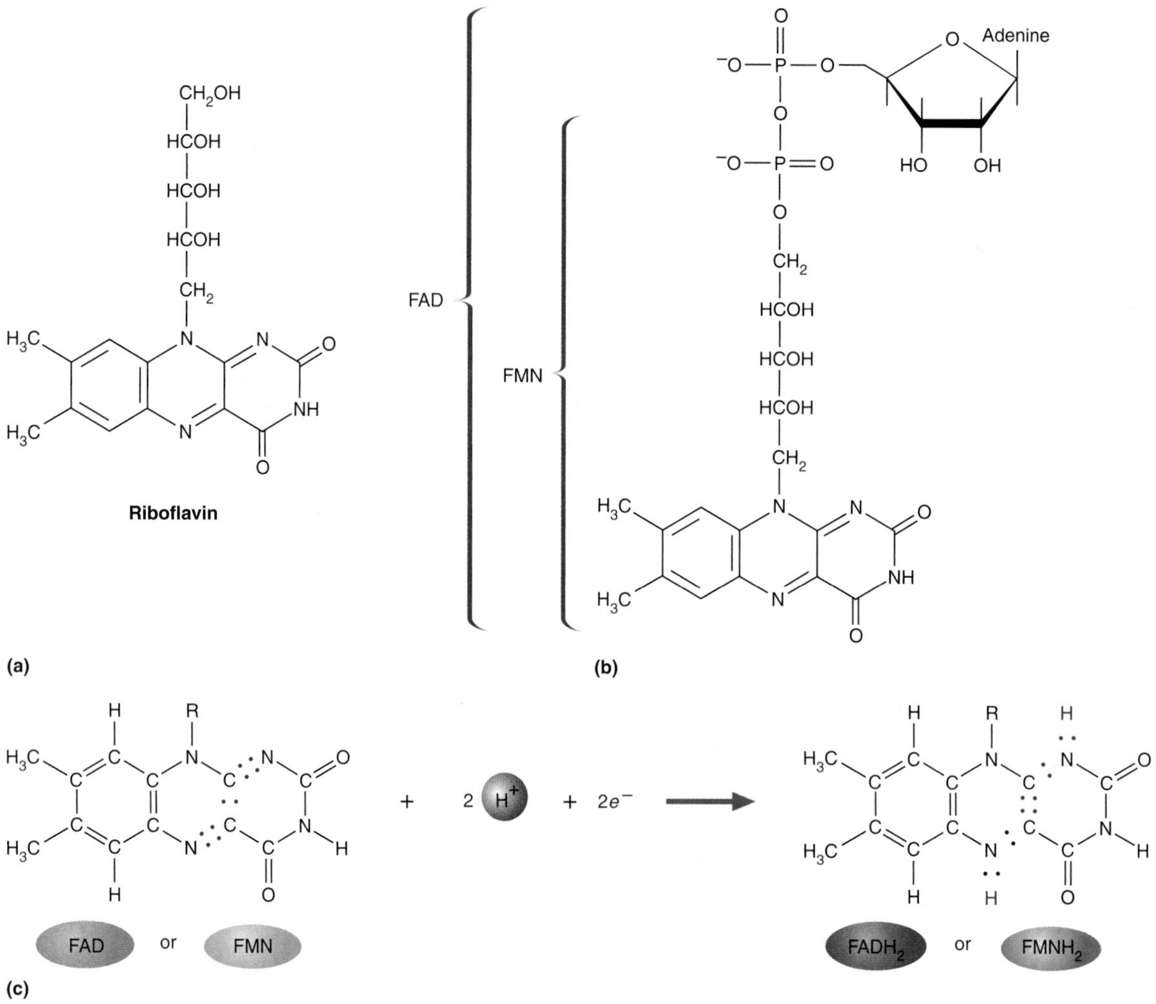

FIGURE 9.6

Flavin Coenzymes

(a) The vitamin riboflavin. (b) FAD and FMN. (c) Reversible reduction of flavin coenzymes: to simplify the equation, only the isoalloxazine ring system is shown. The rest of the coenzyme is designated R.

QUESTION 9.2

Which of the following reactions are redox reactions? For each redox reaction identify the oxidizing and reducing agents.

1. Glucose + ATP → glucose-1-phosphate + ADP
2.

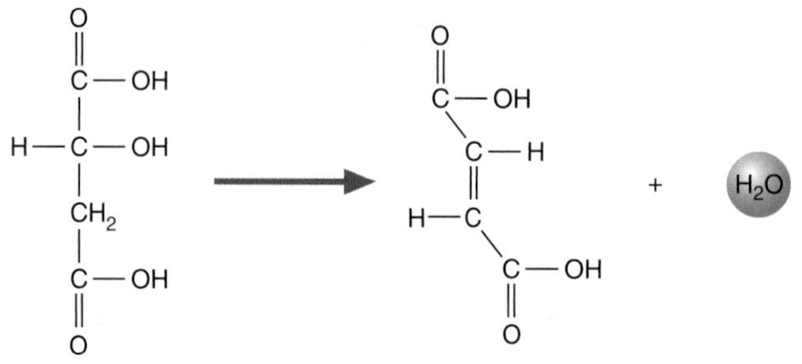

3. Lactate + NAD^+ → pyruvate + NADH + H^+
4. $NO_2^- + 8H^+ + 6$ cyt b (Fe^{2+}) → $NH_4^+ + 2H_2O + 6$ cyt b (Fe^{3+})
5. $CH_3CHO + NADH + H^+$ → $CH_3CH_2OH + NAD^+$

WORKED PROBLEM 9.1

Use the following half-cell potentials to calculate (a) the overall cell potential and (b) $\Delta G^{\circ\prime}$.

Succinate + $\frac{1}{2}O_2 \rightarrow$ fumarate + H_2O

The half-reactions are

Fumarate + $2H^+ + 2e^- \rightarrow$ succinate	$(E^{\circ\prime} = -0.031\ V)$
$\frac{1}{2}O_2 + 2H^+ + 2e^- \rightarrow H_2O$	$(E^{\circ\prime} = +0.82\ V)$

Solution

Write the fumarate reaction as an oxidation (lower reduction potential), balance if the number of electrons transferred differs between the two half-reactions (not necessary here), and add the two reactions to get the net reaction.

Succinate + $\frac{1}{2}O_2 \rightarrow$ fumarate + H_2O

a. The overall potential is defined by the following:

$$\Delta E^{\circ\prime} = E^{\circ\prime}\text{ (electron acceptor)} - E^{\circ\prime}\text{ (electron donor)}$$
$$\Delta E^{\circ\prime} = (+0.82\ V) - (-0.031\ V)$$
$$\Delta E^{\circ\prime} = +0.85\ V$$

b. Use the formula to find $\Delta G^{\circ\prime}$.

$$\begin{aligned}\Delta G^{\circ\prime} &= -nF\Delta E^{\circ\prime}\\ &= -(2)(96.5\ \text{kJ/V·mol})(0.85\ V)\\ &= -164.05\ \text{kJ/mol}\\ &= -164\ \text{kJ/mol}\end{aligned}$$

■

QUESTION 9.3

Because redox reactions play an important role in living processes, biochemists need to determine the oxidation state of the atoms in a molecule. In one method, the oxidation state of an atom is determined by assigning numbers to carbon atoms based on the type of groups attached to them. For example, a bond to a hydrogen is assigned the value −1. A bond to another carbon atom is valued at 0, and a bond to an electronegative atom such as oxygen or nitrogen is valued at +1. The values of a single carbon atom in a molecule may range from −4 (e.g., CH_4) to +4 (CO_2). Note that methane is a high-energy molecule and carbon dioxide is a low-energy molecule. As carbon changes its oxidation state from −4 (methane) to +4 (carbon dioxide), a large amount of energy is released. This process is therefore highly exothermic.

Ethanol is degraded in the liver by a series of redox reactions. Identify the oxidation state of the indicated carbon atom in each molecule in the following reaction sequence:

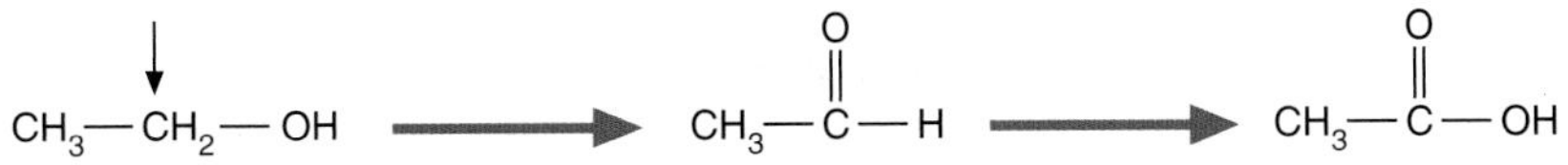

Ethanol Acetaldehyde Acetic acid

QUESTION 9.4

As CO_2 is incorporated into organic molecules during photosynthesis, is it being oxidized or reduced?

FIGURE 9.7

Electron Flow and Energy

Electron flow may be used to generate and capture energy in aerobic respiration. Energy may also be used to drive electron flow in photosynthesis. ($NADP^+$ is a more phosphorylated version of NAD^+.)

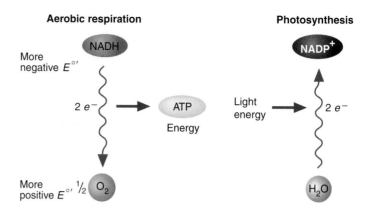

Aerobic Metabolism

Most of the aerobic cell's free energy is captured by the mitochondrial electron transport chain (ETC). During this process, electrons are transferred from a redox pair with a more negative reduction potential ($NADH/NAD^+$) to those with more positive reduction potentials. The last component in the system is the $H_2O/\frac{1}{2}O_2$ pair:

$$\frac{1}{2}O_2 + NADH + H^+ \rightarrow H_2O + NAD^+$$

The free energy released as a pair of electrons passes from NADH to O_2 under standard conditions is calculated as follows:

$$\begin{aligned}
\Delta G^{\circ\prime} &= -nF\Delta E^{\circ\prime} \\
&= -2(96.5 \text{ kJ/V·mol})[0.815 - (-0.32)] \\
&= -220 \text{ kJ/mol}
\end{aligned}$$

A significant portion of the free energy generated as electrons move from NADH to O_2 in the ETC is used to drive ATP synthesis.

It should be noted that in several metabolic processes, electrons move from redox pairs with more positive reduction potentials to those with more negative reduction potentials. Of course, energy is required. The most prominent example of this phenomenon is photosynthesis (Chapter 13). Photosynthetic organisms use captured light energy to drive electrons from electron donors, such as water, to electron acceptors with more negative reduction potentials (Figure 9.7). The energized electrons eventually flow back to acceptors with more positive reduction potentials, thereby providing energy for ATP synthesis and CO_2 reduction to form carbohydrate.

In Section 9.2 the citric acid cycle is examined. In this pathway, which is the first phase of aerobic metabolism, the energy released by the oxidation of two-carbon fragments derived from glucose, fatty acids, and some amino acids is captured by and carried in the reduced coenzymes NADH and $FADH_2$.

9.2 CITRIC ACID CYCLE

The citric acid cycle (Figure 9.8) is a series of biochemical reactions aerobic organisms use to release chemical energy stored in the two-carbon acetyl group in acetyl-CoA. Acetyl-CoA is composed of an acetyl group derived from the breakdown of carbohydrates, lipids, and some amino acids that is linked to the acyl carrier molecule **coenzyme A** (Figure 9.9). Acetyl-CoA is synthesized from pyruvate (a partially oxidized product of the degradation of sugars and certain amino acids) in a series of reactions. Acetyl-CoA is also the product of fatty acid catabolism (described in Chapter 11) and certain reactions in amino acid metabolism (Chapter 15). In the citric acid cycle, the acetyl group's carbon atoms are oxidized to CO_2 and the electrons are transferred as hydride ions to NAD^+.

KEY CONCEPTS

- In living organisms, both energy-capturing and energy-releasing processes consist primarily of redox reactions.
- In redox reactions, electrons move between an electron donor and an electron acceptor.
- In many reactions, both electrons and protons are transferred.
- In biological systems, most redox reactions involve hydride ion transfer ($NADH/NAD^+$) or hydrogen atom transfer ($FADH_2/FAD$).

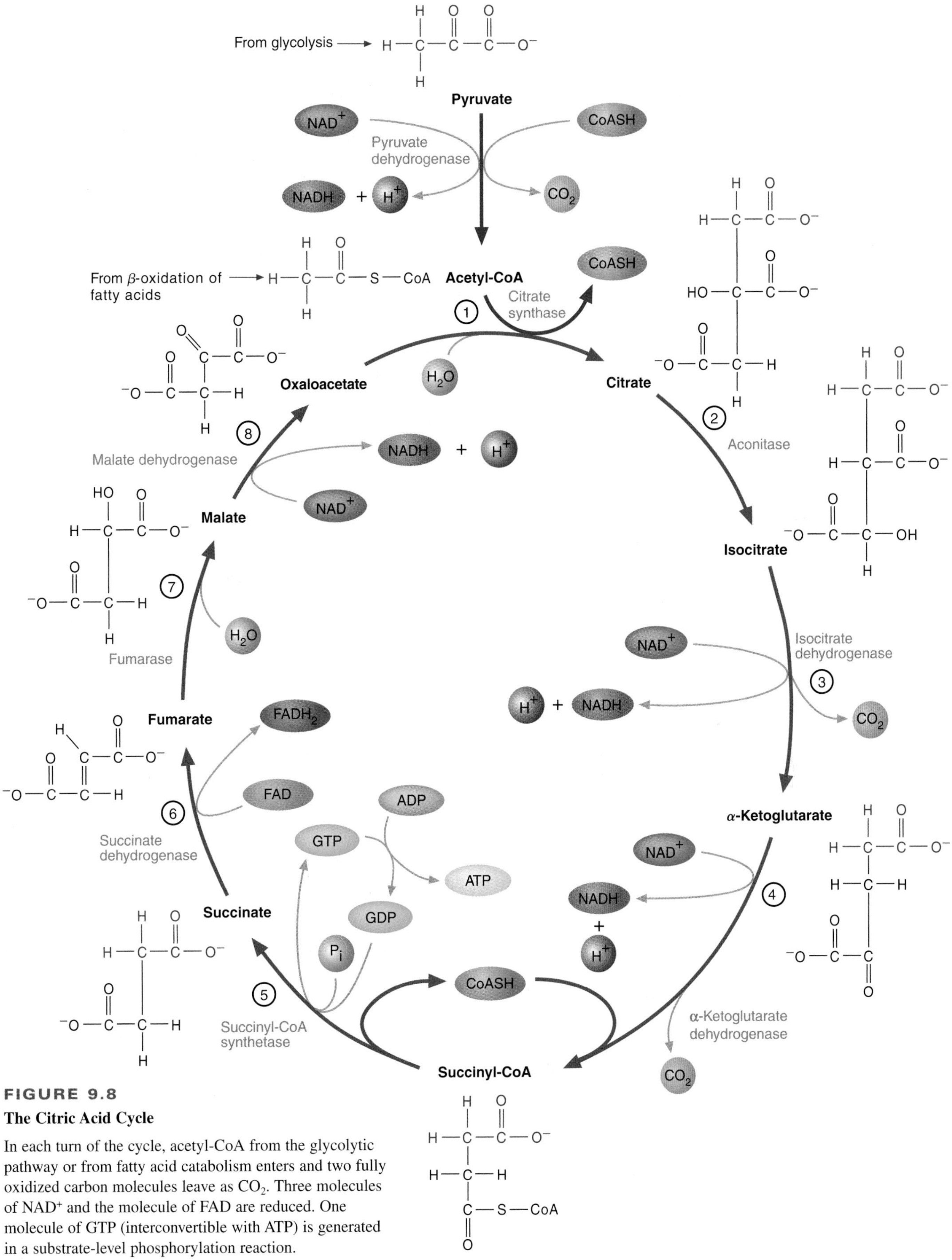

FIGURE 9.8

The Citric Acid Cycle

In each turn of the cycle, acetyl-CoA from the glycolytic pathway or from fatty acid catabolism enters and two fully oxidized carbon molecules leave as CO_2. Three molecules of NAD^+ and the molecule of FAD are reduced. One molecule of GTP (interconvertible with ATP) is generated in a substrate-level phosphorylation reaction.

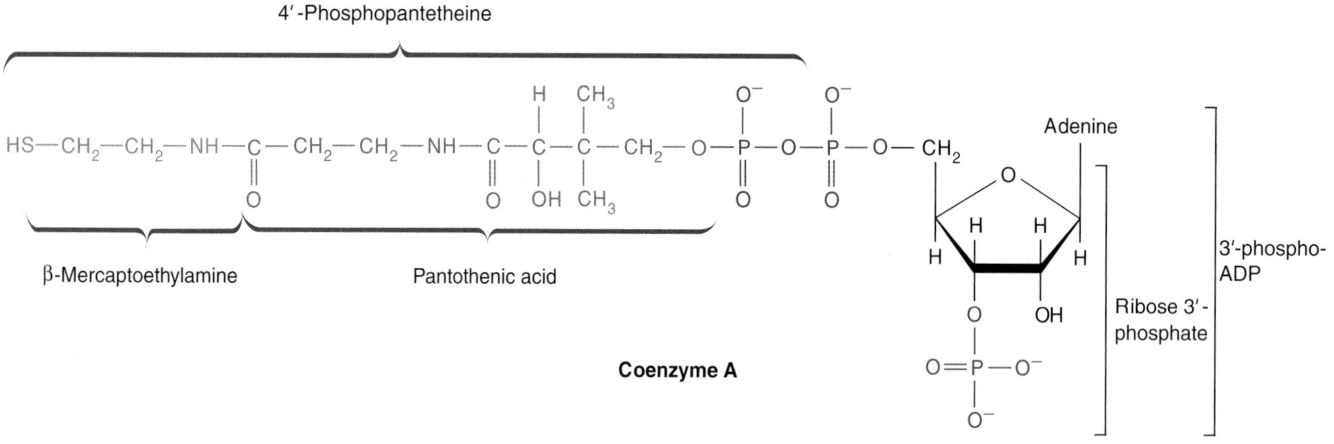

FIGURE 9.9

Coenzyme A

In coenzyme A a 3′ phosphate derivative of ADP is linked to pantothenic acid via a phosphate ester bond. The β-mercaptoethylamine group of coenzyme A is attached to pantothenic acid by an amide bond. Coenzyme A is a carrier of acyl groups that range in size from the acetyl group to long-chain fatty acids. Because the reactive SH group forms a thioester bond with acyl groups, coenzyme A is often abbreviated as CoASH. The carbon-sulfur bond of a thioester is more easily cleaved than the carbon-oxygen bond of an ester.

TABLE 9.2 Summary of the Coenzymes in the Citric Acid Cycle

Coenzyme	Functions
Thiamine pyrophosphate (TPP)	Decarboxylation and aldehyde group transfer
Lipoic acid	Carrier of hydrogens or acetyl groups
NADH	Electron carrier
FADH$_2$	Electron carrier
Coenzyme A (CoASH)	Acetyl group carrier

Subsequently, the conversion of succinate to oxaloacetate involves the transfer of both hydrogen atoms to FAD and hydride ions to NAD$^+$ to form FADH$_2$ and NADH, respectively.

In the first reaction of the citric acid cycle, a two-carbon acetyl group condenses with a four-carbon molecule (oxaloacetate) to form a six-carbon molecule (citrate). During the subsequent seven reactions, in which two CO$_2$ molecules are produced and four pairs of electrons are removed from carbon compounds, citrate is reconverted to oxaloacetate. During one step in the cycle, the high-energy molecule guanosine triphosphate (GTP) is produced during a substrate-level phosphorylation. The net reaction for the citric acid cycle is as follows:

$$\text{Acetyl-CoA} + 3\text{ NAD}^+ + \text{FAD} + \text{GDP} + P_i + 2H_2O \rightarrow$$
$$2CO_2 + 3\text{ NADH} + \text{FADH}_2 + \text{CoASH} + \text{GTP} + 2H^+$$

In addition to its role in energy production, the citric acid cycle plays another important role in metabolism. Cycle intermediates are substrates in a variety of biosynthetic reactions. Table 9.2 provides a summary of the roles of coenzymes in the citric acid cycle.

Conversion of Pyruvate to Acetyl-CoA

After its transport into the mitochondrial matrix, pyruvate is converted to acetyl-CoA in a series of reactions catalyzed by the enzymes in the pyruvate dehydrogenase complex. The net reaction, an oxidative decarboxylation, is as follows:

$$Pyruvate + NAD^+ + CoASH \rightarrow Acetyl\text{-}CoA + NADH + CO_2 + H^+$$

Despite the apparent simplicity of this highly exergonic reaction ($\Delta G^{o\prime} = -33.5$ kJ/mol), its mechanism is one of the most complex known. The pyruvate dehydrogenase complex is a large multienzyme structure that contains three enzyme activities: pyruvate dehydrogenase (E_1), also known as pyruvate decarboxylase, dihydrolipoyl transacetylase (E_2), and dihydrolipoyl dehydrogenase (E_3). Each enzyme activity is present in multiple copies. Table 9.3 summarizes the number of copies of each enzyme and the required coenzymes of *E. coli* pyruvate dehydrogenase.

In the first step, pyruvate dehydrogenase catalyzes the decarboxylation of pyruvate. A nucleophile is formed when a basic residue of the enzyme extracts a proton from the thiazole ring of **thiamine pyrophosphate** (TPP). (TPP is the coenzyme form of thiamine, also called vitamin B_1.) The intermediate, hydroxyethyl-TPP (HETPP), forms after the nucleophilic thiazole ring has attacked the carbonyl group of pyruvate with the resulting loss of CO_2 (Figure 9.10).

In the next several steps, the hydroxyethyl group of HETPP is converted to acetyl-CoA by dihydrolipoyl transacetylase. **Lipoic acid** (Figure 9.11) an acyl transfer coenzyme that contains two thiol groups that can be reversibly oxidized, plays a crucial role in this transformation. Lipoic acid is bound to the enzyme through an amide linkage with the ε-amino group of a lysine residue. It reacts with HETPP to form an acetylated lipoic acid and free TPP. The acetyl group is then transferred to the sulfhydryl group of coenzyme A. Subsequently, the reduced lipoic acid is reoxidized by dihydrolipoyl dehydrogenase. The $FADH_2$ is reoxidized by NAD^+ (with its more negative reduction potential) to form the FAD required for the oxidation of the next reduced lipoic acid residue. The mobile NADH can deliver its electrons to the ETC and is replaced in the enzyme by another NAD^+ molecule so that the cycle can begin again.

The activity of pyruvate dehydrogenase is regulated by two mechanisms: product inhibition and covalent modification (Section 6.5). The enzyme complex is allosterically activated by NAD^+, CoASH, and AMP. It is inhibited by high concentrations of ATP and the reaction products acetyl-CoA and NADH. In vertebrates these molecules also activate a kinase, which converts the active pyruvate dehydrogenase complex to an inactive phosphorylated form. High concentrations of the substrates pyruvate, CoASH, and NAD^+ inhibit the activity of the kinase. The pyruvate dehydrogenase complex is reactivated by a dephosphorylation reaction catalyzed by a phosphoprotein phosphatase. The phosphoprotein phosphatase is activated when the mitochondrial ATP concentration is low.

KEY CONCEPT

Pyruvate is converted to acetyl-CoA by the enzymes in the pyruvate dehydrogenase complex. TPP, FAD, NAD^+, Coenzyme A, and lipoic acid are required coenzymes.

TABLE 9.3 *E. coli* Pyruvate Dehydrogenase Complex

Enzyme Activity	Function	Copies per Complex*	Coenzymes
Pyruvate dehydrogenase (E_1)	Decarboxylates pyruvate	24 (20–30)	TPP
Dihydrolipoyl transacetylase (E_2)	Catalyzes transfer of acetyl group to CoASH	24 (60)	Lipoic acid, CoASH
Dihydrolipoyl dehydrogenase (E_3)	Reoxidizes dihydrolipoamide	12 (20–30)	NAD^+, FAD

* The number of molecules of each enzyme activity found in mammalian pyruvate dehydrogenase is shown in parentheses.

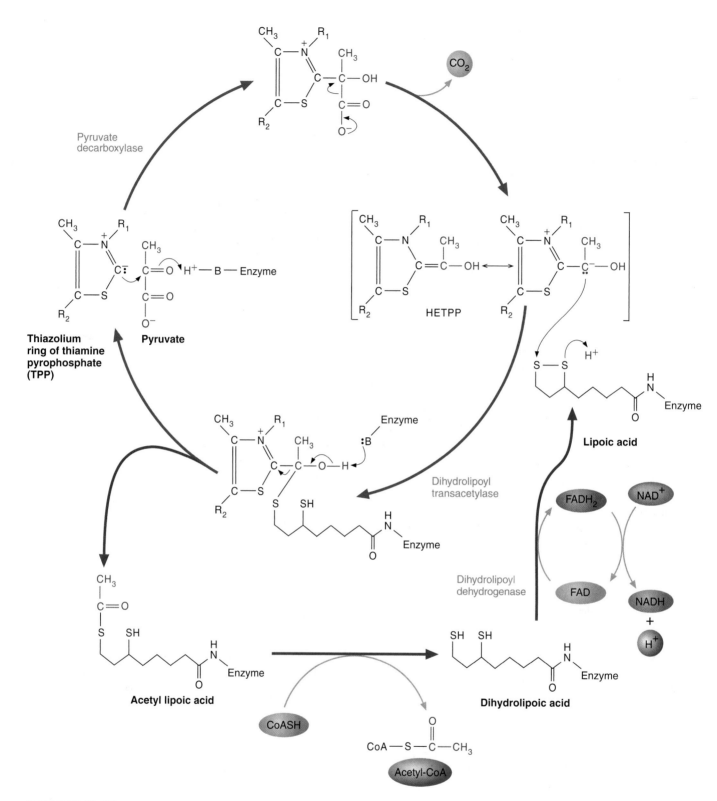

FIGURE 9.10

The Reactions Catalyzed by the Pyruvate Dehydrogenase Complex

Pyruvate decarboxylase, which contains TPP, catalyzes the formation of the HETPP. Dihydrolipoyl transacetylase uses lipoic acid as a cofactor to convert the hydroxyethyl group of HETPP to acetyl-CoA. Dihydrolipoyl dehydrogenase reoxidizes the reduced lipoic acid. (Refer to Figure 9.11 for the structure of lipoic acid.)

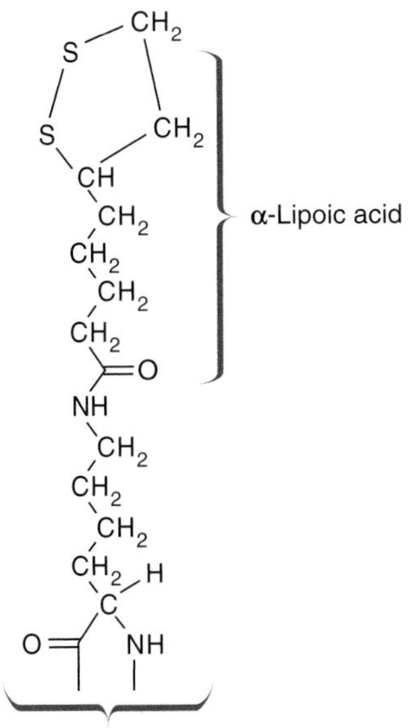

FIGURE 9.11

Lipoamide

Lipoic acid is covalently bonded to the enzyme through an amide linkage with the ε-amino group of a lysine residue.

Reactions of the Citric Acid Cycle

The citric acid cycle is composed of eight reactions that occur in two stages:

1. The two-carbon acetyl group of acetyl-CoA enters the cycle by reacting with the four-carbon compound oxaloacetate and two molecules of CO_2 are subsequently liberated (reactions 1–4).

2. Oxaloacetate is regenerated so it can react with another acetyl-CoA (reactions 5–8).

The reactions of the citric acid cycle are as follows.

1. **Introduction of two carbons as acetyl-CoA**. The citric acid cycle begins with the condensation of acetyl-CoA with oxaloacetate to form citrate:

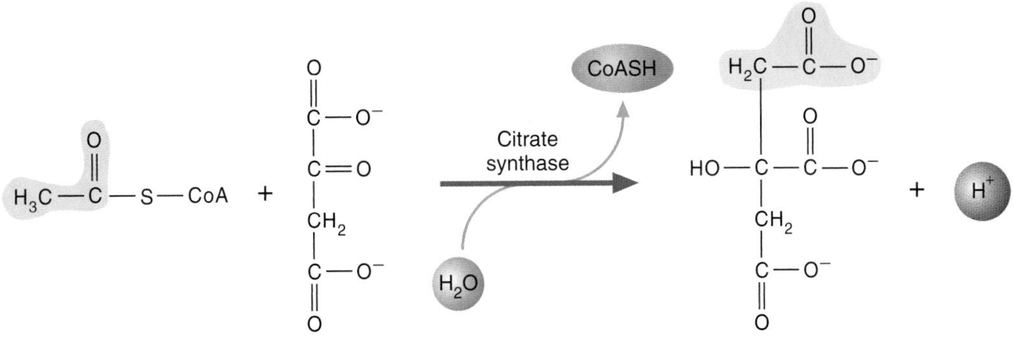

Note that this reaction is an aldol condensation. In this reaction (Figure 9.12) the enzyme removes a proton from the methyl group of acetyl-CoA, thereby converting it to an enol. The enol subsequently attacks the carbonyl carbon of oxaloacetate. The product, citroyl-CoA, rapidly hydrolyzes to form citrate and CoASH. Because of the hydrolysis of the high-energy thioester bond, the overall standard free energy change is –33.5 kJ/mol, and citrate formation is highly exergonic.

2. **Citrate is isomerized to form a secondary alcohol that can be easily oxidized**. In the next reaction of the cycle, citrate, which contains a tertiary alcohol, is reversibly converted to isocitrate by aconitase. During this isomerization reaction, an intermediate called *cis*-aconitate is formed

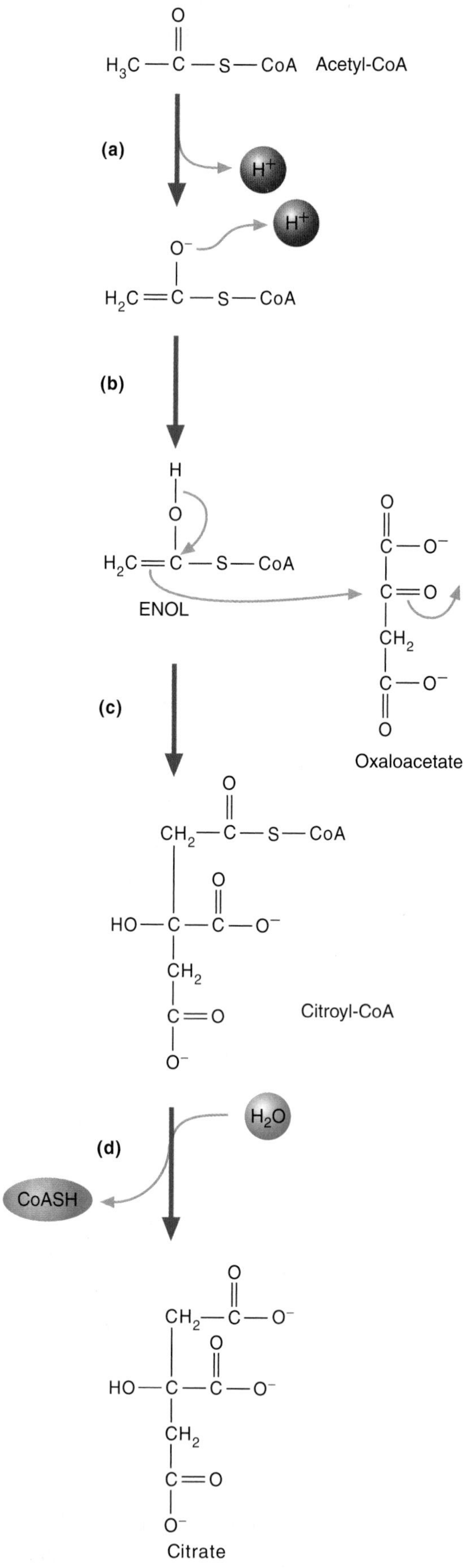

FIGURE 9.12

Citrate Synthesis

(a) A side chain carboxylate group of the enzyme citrate synthase removes a proton from the methyl group of acetyl-CoA.
(b) Simultaneously, a side chain NH group of a histidine residue donates a proton to the carbonyl oxygen, thus generating an enol intermediate. (c) The same histidine side chain then deprotonates the enol to produce an enolate anion that attacks the carbonyl carbon of oxaloacetate. (d) The product, citroyl-CoA, is then hydrolyzed in a nucleophilic acyl substitution reaction to yield citrate and CoASH.

by dehydration. The carbon-carbon double bond of *cis*-aconitate is then rehydrated in a nucleophilic addition reaction to form the more reactive secondary alcohol, isocitrate. Although the standard free energy change of citrate isomerization is positive ($\Delta G^{\circ\prime} = 13.3$ kJ), the reaction is pulled forward by the rapid removal of isocitrate by the next reaction.

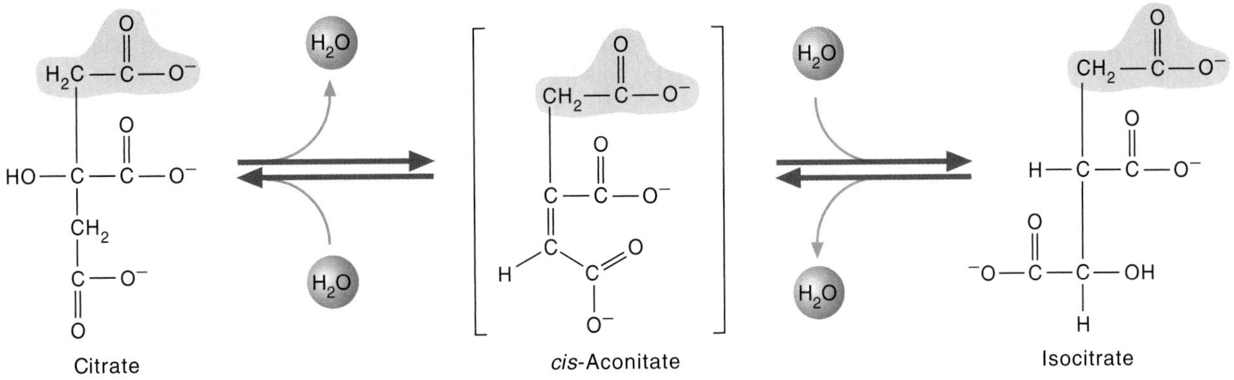

Citrate   *cis*-Aconitate   Isocitrate

3. **Isocitrate is oxidized to form NADH and CO$_2$.** The oxidative decarboxylation of isocitrate, catalyzed by isocitrate dehydrogenase, occurs in three steps. First, isocitrate is oxidized to form a transient ketone-group-containing intermediate called oxalosuccinate. A manganese (Mn^{2+}) cofactor facilitates the polarization of the newly formed ketone group. Immediate decarboxylation of oxalosuccinate results in the formation of an enol intermediate that rearranges to form α-ketoglutarate, an α-keto acid.

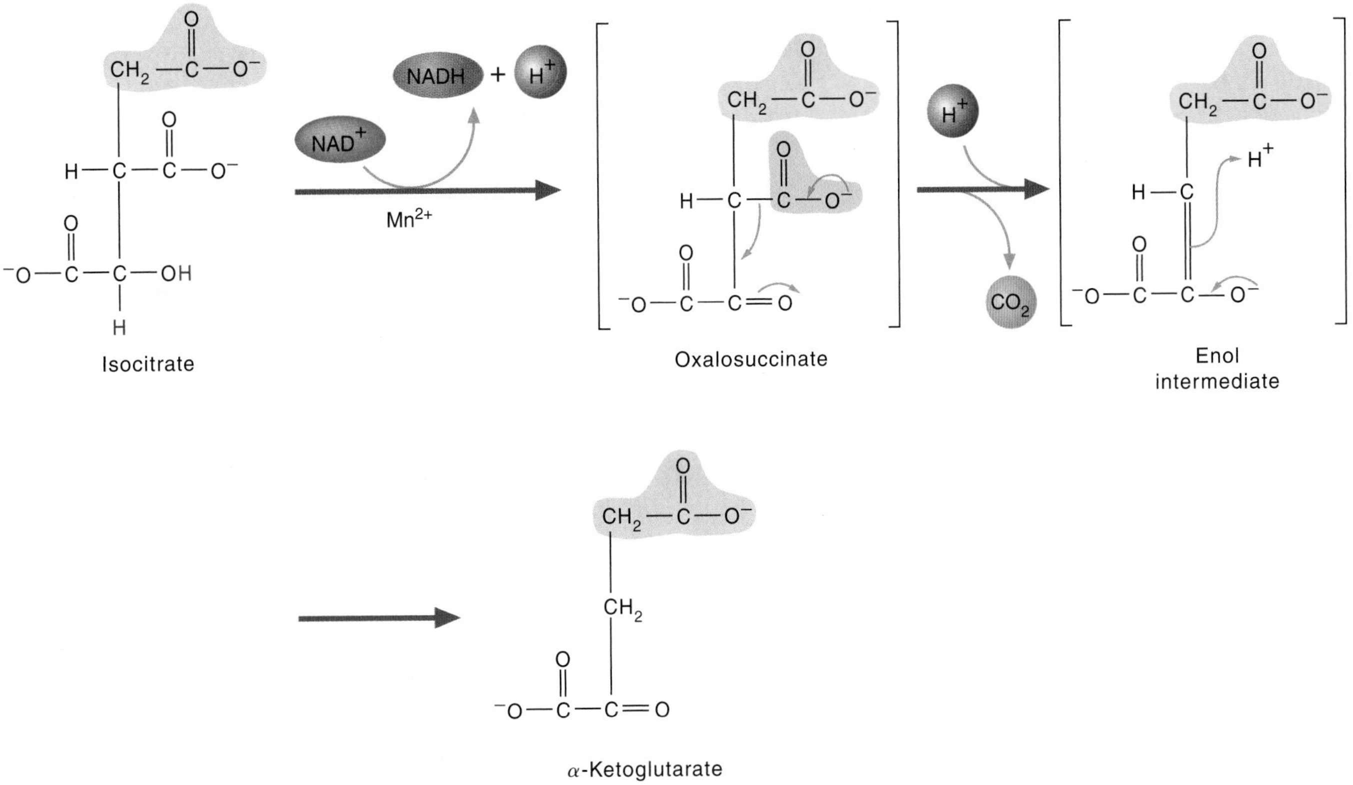

Isocitrate   Oxalosuccinate   Enol intermediate

α-Ketoglutarate

There are two forms of isocitrate dehydrogenase in mammals. The NAD$^+$-requiring isozyme is found only within mitochondria. The other isozyme, which requires NADP$^+$, is found in both the mitochondrial matrix and the cytoplasm. In some circumstances the latter enzyme is used in both compartments to generate NADPH, which is required in biosynthetic processes. Note that the NADH produced in the conversion of isocitrate to α-ketoglutarate is the first link between the citric acid cycle and the ETC and oxidative phosphorylation.

4. **α-Ketoglutarate is oxidized to form a second molecule each of NADH and CO_2.** The conversion of α-ketoglutarate to succinyl-CoA is catalyzed by the enzyme activities in the α-ketoglutarate dehydrogenase complex: α-ketoglutarate dehydrogenase, dihydrolipoyl transsuccinylase, and dihydrolipoyl dehydrogenase.

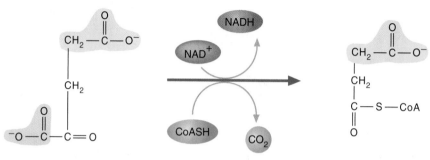

This highly exergonic reaction ($\Delta G^{\circ\prime} = -33.5$ kJ/mol), an oxidative decarboxylation, is analogous to the conversion of pyruvate to acetyl-CoA catalyzed by pyruvate dehydrogenase. Both reactions have energy-rich thioester molecules as products, that is, acetyl-CoA and succinyl-CoA. Other similarities between the two multienzyme complexes are that the same cofactors (TPP, CoASH, lipoic acid, NAD^+, and FAD) are required, and the same or similar allosteric effectors are inhibitors. α-Ketoglutarate dehydrogenase is inhibited by succinyl-CoA, NADH, ATP, and GTP. An important difference between the two complexes is that the control mechanism of the α-ketoglutarate dehydrogenase complex does not involve covalent modification.

5. **The cleavage of succinyl-CoA is coupled to a substrate-level phosphorylation.** The cleavage of the high-energy thioester bond of succinyl-CoA to form succinate is a reversible reaction catalyzed by succinyl-CoA synthetase (also called succinate thiokinase). In mammals the reaction is coupled to the substrate-level phosphorylation of ADP or GDP. There are two forms of succinyl-CoA synthetase: one is specific for ATP and the other for GTP. In many tissues both enzymes are produced, although their relative amounts vary.

The direction of the reaction depends on the relative concentrations of nucleoside diphosphates (ADP and/or GDP) and nucleotide triphosphates (ATP and/or GTP). The phosphoryl group of GTP can be donated to ADP in a reversible reaction catalyzed by nucleoside diphosphate kinase.

6. **The four-carbon molecule succinate is oxidized to form fumarate and FADH$_2$.** Succinate dehydrogenase catalyzes the oxidation of succinate to form fumarate:

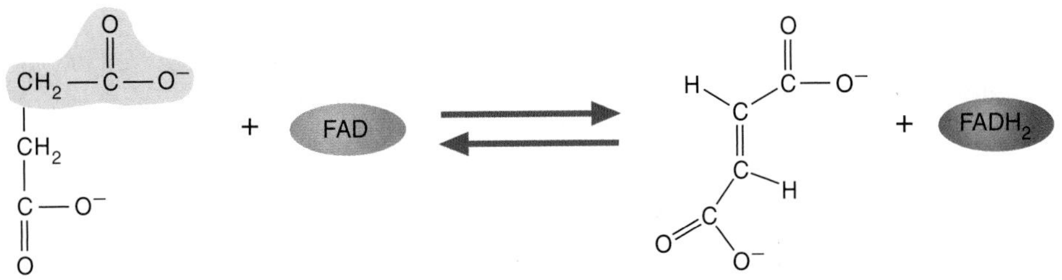

Succinate   Fumarate

Unlike the other citric acid cycle enzymes, succinate dehydrogenase is not found within the mitochondrial matrix. Instead, it is tightly bound to the inner mitochondrial membrane. The oxidation of an alkane requires a stronger oxidizing agent than NAD$^+$. Succinate dehydrogenase is a flavoprotein that uses FAD to drive the oxidation of succinate to fumarate. (In the complete reaction the electrons donated to the FAD covalently bound to succinate dehydrogenase are subsequently passed to coenzyme Q, a component of the ETC. The $\Delta G^{\circ\prime}$ for this process is -5.6 kJ/mol.) Succinate dehydrogenase is activated by high concentrations of succinate, ADP, and P$_i$ and inhibited by oxaloacetate. The enzyme is also inhibited by malonate, a structural analogue of succinate, whose accumulation redirects energy to fatty acid synthesis. This inhibitor was used by Hans Krebs in his pioneering work on the citric acid cycle.

COMPANION

WEBSITE Visit the companion website at www.oup.com/us/mckee to read the Biochemistry in Perspective box on Hans Krebs and the citric acid cycle.

7. **Fumarate is hydrated.** Fumarate is converted to L-malate in a reversible stereospecific hydration reaction catalyzed by fumarase (also referred to as fumarate hydratase):

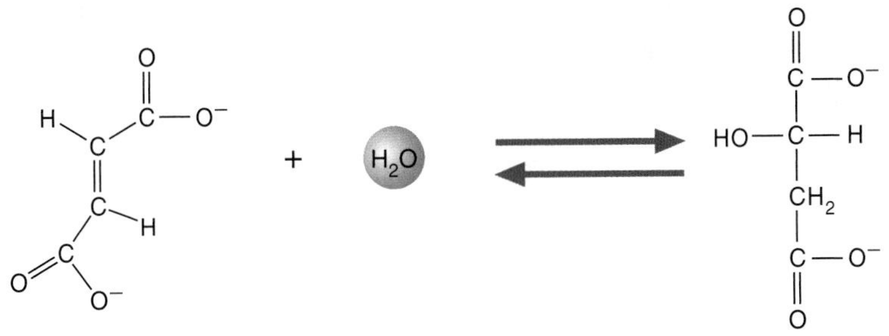

Fumarate   L-Malate

8. **Malate is oxidized to form oxaloacetate and a third NADH.** Finally, oxaloacetate is regenerated with the oxidation of L-malate:

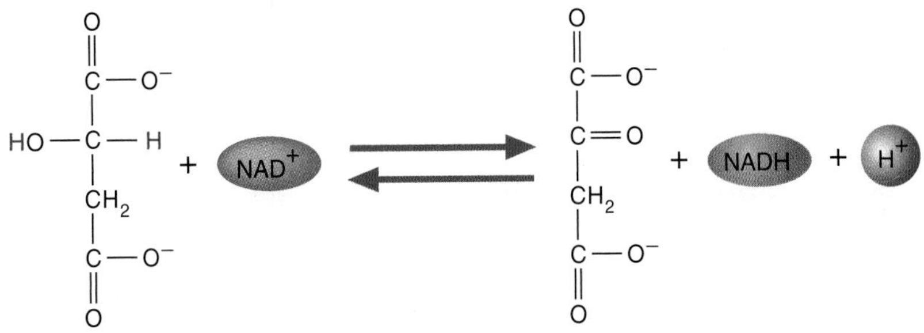

L-Malate   L-Oxaloacetate

Malate dehydrogenase uses NAD$^+$ as the oxidizing agent in a highly endergonic reaction ($\Delta G^{\circ\prime} = +29$ kJ/mol). The reaction is pulled to completion because of the removal of oxaloacetate in the next round of the cycle.

- The citric acid cycle begins with the condensation of a molecule of acetyl-CoA with oxaloacetate to form citrate, which is eventually reconverted to oxaloacetate.
- During this process, two molecules of CO_2, three molecules of NADH, one molecule of $FADH_2$, and one molecule of GTP are produced.

Fate of Carbon Atoms in the Citric Acid Cycle

In each turn of the citric acid cycle, two carbon atoms enter as the acetyl group of acetyl-CoA and two molecules of CO_2 are released. A careful review of Figure 9.8 reveals that the two carbon atoms released as CO_2 molecules are not the same two carbons that just entered the cycle. Instead, the released carbon atoms are derived from oxaloacetate that reacted with the incoming acetyl-CoA. The incoming carbon atoms subsequently form one-half of succinate. Because of the symmetric structure of succinate, the carbon atoms derived from the incoming acetyl group become evenly distributed in all of the molecules derived from succinate. Consequently, incoming carbon atoms are released as CO_2 only after two or more turns of the cycle.

QUESTION 9.5

Trace the labeled carbon in $CH_3{}^{14}\overset{\displaystyle O}{\overset{\|}{C}}$—SCoA through one round of the citric acid cycle. After examining Figure 9.8, show why more than two turns of the cycle are required before all the labeled carbon atoms are released as $^{14}CO_2$.

The Amphibolic Citric Acid Cycle

Amphibolic pathways can function in both anabolic and catabolic processes. The citric acid cycle is obviously catabolic: acetyl groups are oxidized to form CO_2, and energy is conserved in reduced coenzyme molecules. The citric acid cycle is also anabolic, since several citric acid cycle intermediates are precursors in biosynthetic pathways (Figure 9.13). For example, oxaloacetate is a gluconeogenesis substrate (Chapter 8) and a precursor in the synthesis of the amino acids lysine, threonine, isoleucine, and methionine (Chapter 14). α-Ketoglutarate also plays an important role in amino acid synthesis as a precursor of glutamate, glutamine, proline, and arginine. The synthesis of porphyrins such as heme requires succinyl-CoA (Chapter 14). Finally, excess citrate molecules are transported into the cytoplasm, where they are cleaved to form oxaloacetate and acetyl-CoA. The latter molecule is used to synthesize fatty acids and steroid molecules such as cholesterol (Chapter 12).

Anabolic processes drain the citric acid cycle of the molecules required to sustain its role in energy generation. Several reactions, referred to as **anaplerotic** reactions, replenish them. One of the most important anaplerotic reactions is catalyzed by pyruvate carboxylase (p. 284). A high concentration of acetyl-CoA, an indicator of an insufficient oxaloacetate concentration, activates pyruvate carboxylase. As a result, oxaloacetate concentration increases. Other anaplerotic reactions include the synthesis of succinyl-CoA from certain fatty acids (Chapter 12) and the α-keto acids α-ketoglutarate and oxaloacetate from the amino acids glutamate and aspartate, respectively, via transamination reactions (Chapter 14).

- The citric acid cycle is an amphibolic pathway; that is, it plays a role in both anabolism and catabolism.
- The citric acid cycle intermediates used in anabolic processes are replenished by several anaplerotic reactions.

QUESTION 9.6

Pyruvate carboxylase deficiency, a disease that is usually fatal, is caused when the enzyme that converts pyruvate to oxaloacetate is missing or defective. It is characterized by varying degrees of mental retardation and disturbances in several metabolic pathways, especially those involving amino acids and their degradation products. A prominent symptom of this malady is *lactic aciduria* (lactic acid in the urine). After reviewing the function of pyruvate carboxylase, explain why this symptom occurs.

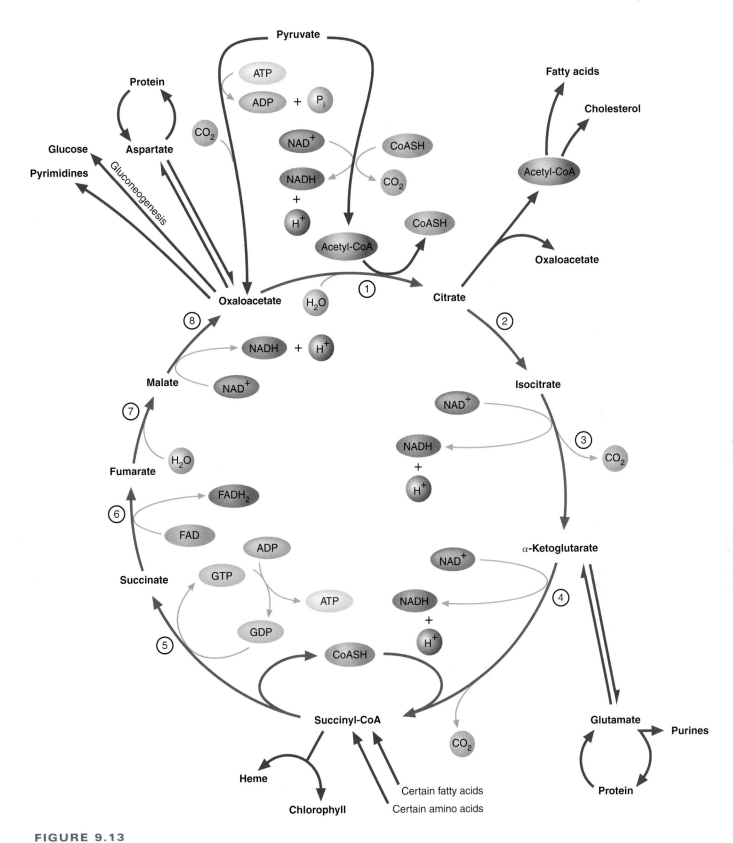

FIGURE 9.13

The Amphibolic Citric Acid Cycle

The citric acid cycle operates in both anabolic processes (e.g., the synthesis of fatty acids, cholesterol, heme, and glucose) and catabolic processes (e.g., amino acid degradation and energy production).

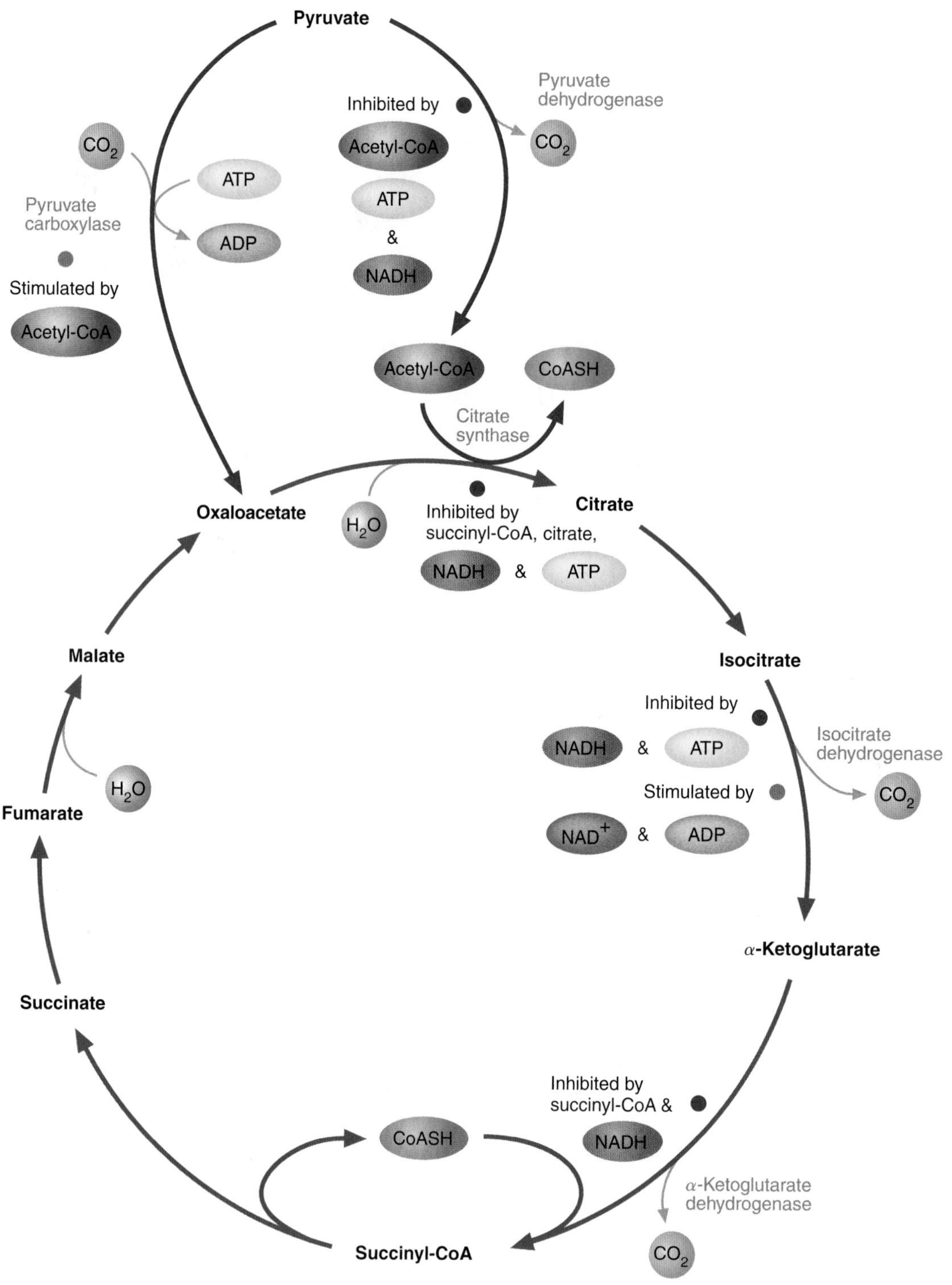

FIGURE 9.14

Control of the Citric Acid Cycle

The major regulatory sites of the cycle are indicated. Activators and inhibitors of regulated enzymes are shown in color.

Citric Acid Cycle Regulation

The citric acid cycle is precisely regulated so that the cell's energy and biosynthetic requirements are constantly met (Figure 9.14). Regulation is achieved via control of three irreversible enzymes within the cycle: citrate synthase, isocitrate dehydrogenase, and α-ketoglutarate dehydrogenase. These three enzymes operate far from equilibrium (i.e., with highly negative $\Delta G^{\circ\prime}$ values), and they also catalyze reactions that represent important metabolic branch points. Strategies of control include substrate availability, product inhibition, and competitive feedback inhibition.

Citrate synthase, the first enzyme in the cycle, catalyzes the formation of citrate from acetyl-CoA and oxaloacetate. Because the concentrations of acetyl-CoA and oxaloacetate are low in mitochondria in relation to the amount of the enzyme, any increase in substrate availability stimulates citrate synthesis. Succinyl-CoA (a "downstream" intermediate product of the cycle) is a competitive inhibitor of the enzyme, occupying the acetyl-CoA site when in excess, and citrate inhibits the enzyme by acting as an allosteric inhibitor. Other allosteric regulators of this reaction are NADH and ATP, whose concentrations reflect the cell's current energy status. A resting cell has high NADH/NAD$^+$ and ATP/ADP ratios. As a cell becomes metabolically active, NADH and ATP concentrations decrease. Consequently, key enzymes such as citrate synthase become more active.

Isocitrate dehydrogenase catalyzes the second closely regulated reaction in the cycle. Its activity is stimulated by relatively high concentrations of ADP and NAD$^+$ and inhibited by ATP and NADH. Isocitrate dehydrogenase is closely regulated because of its important role in citrate metabolism (Figure 9.15). As described earlier, the conversion of citrate to isocitrate is reversible. An equilibrium mixture of the two molecules consists largely of citrate. (The reaction is driven forward because isocitrate is rapidly transformed to α-ketoglutarate.) Of the two molecules, only citrate can penetrate the mitochondrial inner membrane. When cellular energy demands are met, excess citrate molecules are transported out of mitochondria. Once in the cytoplasm, citrate is cleaved by citrate lyase to yield acetyl-CoA and oxaloacetate. The acetyl-CoA formed is used in several biosynthetic processes, such as fatty acid synthesis. (Citrate transport is used to transfer acetyl-CoA out of mitochondria because acetyl-CoA cannot penetrate the inner mitochondrial membrane.) Oxaloacetate is used in biosynthetic reactions, or it can be converted to malate. Malate either reenters the mitochondrion, where it is reconverted to oxaloacetate, or is converted in the cytoplasm to pyruvate by malic enzyme. Pyruvate then reenters the mitochondrion. In addition to being a precursor of acetyl-CoA and oxaloacetate in the cytoplasm, citrate also acts directly to regulate several cytoplasmic processes. Citrate is an allosteric activator of the first reaction of fatty acid synthesis. In addition, citrate metabolism provides some of the NADPH used in fatty acid synthesis. Finally, because citrate is an inhibitor of PFK-1 (phosphofructokinase-1), it inhibits glycolysis.

The activity of α-ketoglutarate dehydrogenase is strictly regulated because of the important role of α-ketoglutarate in several metabolic processes (e.g., amino acid metabolism). When a cell's energy stores are low, α-ketoglutarate dehydrogenase is activated and α-ketoglutarate is retained within the cycle at the expense of biosynthetic processes. As the cell's supply of NADH rises, the enzyme is inhibited, and α-ketoglutarate molecules become available for biosynthetic reactions. It is also inhibited by its product succinyl-CoA and is activated by AMP, a critical indicator of low energy charge.

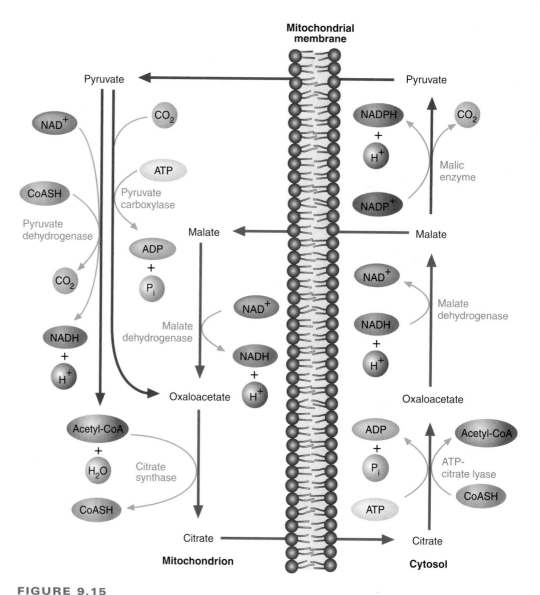

FIGURE 9.15

Citrate Metabolism

When citrate, a citric acid cycle intermediate, moves from the mitochondrial matrix into the cytoplasm, it is cleaved to form acetyl-CoA and oxaloacetate by citrate lyase. The citrate lyase reaction is driven by ATP hydrolysis. Most of the oxaloacetate is reduced to malate by malate dehydrogenase. Acetyl-CoA molecules can be used in biosynthetic pathways such as fatty acid synthesis. Malate may then be oxidized to pyruvate and CO_2 by malic enzyme. The NADPH produced in this reaction is used in cytoplasmic biosynthetic processes, such as fatty acid synthesis. Pyruvate enters the mitochondria, where it may be converted to oxaloacetate or acetyl-CoA. Malate may also reenter the mitochondria, where it is reoxidized to form oxaloacetate.

KEY CONCEPTS

- The citric acid cycle is closely regulated, thus ensuring that the cell's energy and biosynthetic needs are met.
- Allosteric effectors and substrate availability primarily regulate the enzymes citrate synthase, isocitrate dehydrogenase, α-ketoglutarate dehydrogenase, pyruvate dehydrogenase, and pyruvate carboxylase.

Two enzymes outside the citric acid cycle profoundly affect its regulation. The relative activities of pyruvate dehydrogenase and pyruvate carboxylase determine the degree to which pyruvate is used to generate energy and biosynthetic precursors. For example, if a cell is using a cycle intermediate such as α-ketoglutarate in biosynthesis, the concentration of oxaloacetate falls and acetyl-CoA accumulates. Because acetyl-CoA is an activator of pyruvate carboxylase (and an inhibitor of pyruvate dehydrogenase), more oxaloacetate is produced from pyruvate, thus replenishing the cycle.

Fluoroacetate, $F-CH_2-COO^-$, is a toxic substance found in certain plants that grow in Australia and South Africa. Animals that ingest these plants die. Research indicates, however, that fluoroacetate is not poisonous by itself. Rather, once consumed, fluoroacetate is converted into a toxic metabolite, fluorocitrate. In affected cells, citrate accumulates. Can you suggest how fluoroacetate is converted to fluorocitrate? Why are animals killed by fluoroacetate, whereas the plants are unaffected?

The Glyoxylate Cycle

Plants and some fungi, algae, protozoans, and bacteria can grow by using two-carbon compounds. (Molecules such as ethanol, acetate, and acetyl-CoA, derived from fatty acids, are the most common substrates.) The series of reactions responsible for this capability, referred to as the **glyoxylate cycle**, is a modified version of the citric acid cycle. In plants the glyoxylate cycle occurs in organelles called glyoxysomes (p. 58). In the absence of photosynthesis, for example, growth in germinating seed is supported by the conversion of oil reserves (triacylglycerol) to carbohydrate. In other eukaryotic organisms and in bacteria, glyoxylate enzymes occur in cytoplasm.

The glyoxylate cycle (Figure 9.16) consists of five reactions. The first two reactions (the synthesis of citrate and isocitrate) are familiar because they also occur in the citric acid cycle. However, the formation of citrate from oxaloacetate and acetyl-CoA and the isomerization of citrate to form isocitrate are catalyzed by glyoxysome-specific isozymes. The next two reactions are unique to the glyoxylate cycle. Isocitrate is split into two molecules (succinate and glyoxylate) by isocitrate lyase. (This reaction is an aldol cleavage.) Succinate, a four-carbon molecule, is eventually converted to malate by mitochondrial

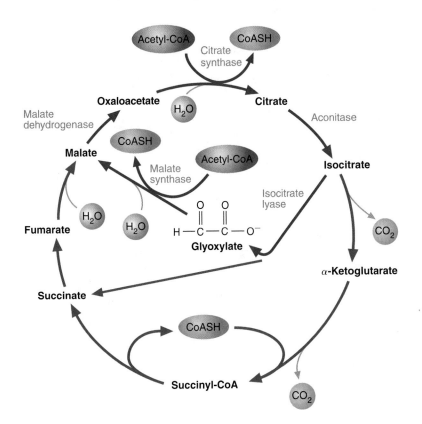

FIGURE 9.16

The Glyoxylate Cycle

By using some of the enzymes of the citric acid cycle, the glyoxylate cycle converts two molecules of acetyl-CoA to one molecule of oxaloacetate. Both decarboxylation reactions of the citric acid cycle are bypassed.

enzymes (Figure 9.17). The two-carbon molecule glyoxylate reacts with a second molecule of acetyl-CoA to form malate in a reaction catalyzed by malate synthase. The cycle is completed as malate is converted to oxaloacetate by malate dehydrogenase.

The glyoxylate cycle allows for the net synthesis of larger molecules from two-carbon molecules for the following reason. The decarboxylation reactions of the citric acid cycle, in which two molecules of CO_2 are lost, are bypassed. By using two molecules of acetyl-CoA, the glyoxylate cycle produces one molecule each of succinate and oxaloacetate. The succinate product is used in the synthesis of metabolically important molecules, the most notable of which is glucose. (In organisms, such as animals, that do not possess isocitrate lyase and malate synthase, gluconeogenesis substrates are always molecules with at least three carbon atoms. In these organisms there is no net synthesis of glucose from fatty acids.) The oxaloacetate product is used to sustain the glyoxylate cycle.

KEY CONCEPTS

- Organisms in which the glyoxylate cycle occurs can use two-carbon molecules to sustain growth.
- In plants, the glyoxylate cycle is found in organelles called glyoxysomes.

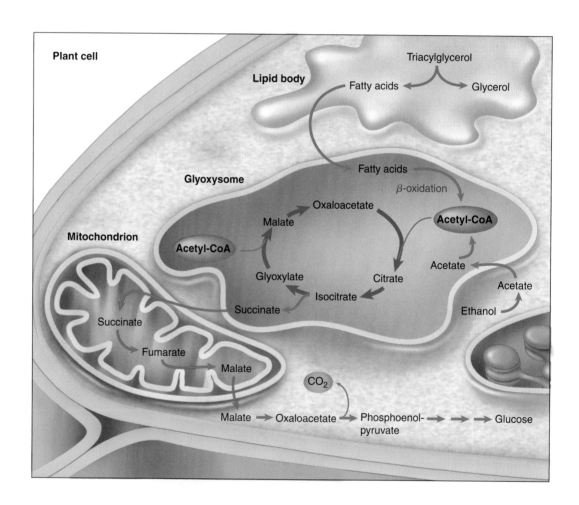

FIGURE 9.17

Role of the Glyoxylate Cycle in Gluconeogenesis

The acetyl-CoA used in the glyoxylate cycle is derived from the breakdown of fatty acids (β-oxidation, see Chapter 12). In organisms with the appropriate enzymes, glucose can be produced from two-carbon compounds such as ethanol and acetate. In plants, the reactions are localized in lipid bodies, glyoxysomes, mitochondria, and the cytoplasm.

BIOCHEMISTRY IN PERSPECTIVE

The Evolutionary History of the Citric Acid Cycle

How and why did the citric acid cycle originate?

Biological evolution is, in general, a conservative process: metabolites and enzymes that perform vital functions are retained. Over time, as environmental conditions change, organisms often recruit preexisting components for new functions. The citric acid cycle provides an interesting example. As a result of the genomic and biochemical analysis of a vast number of prokaryotic and eukaryotic

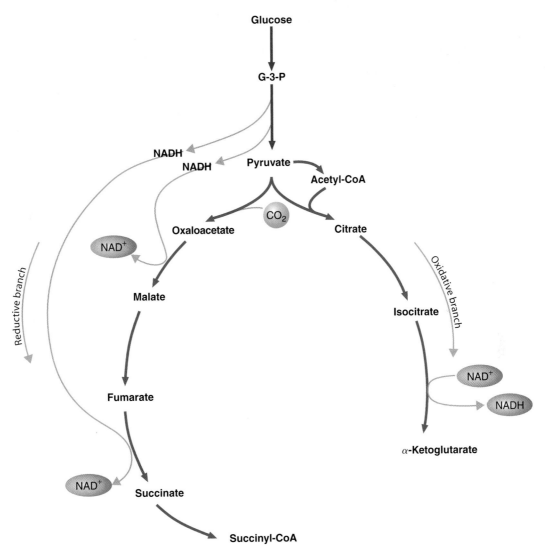

FIGURE 9A

The Incomplete Citric Acid Cycle

Under the anaerobic conditions, primitive prokaryotes developed two branches of what would eventually become the modern citric acid cycle. In the reductive branch (pyruvate → succinate or, in this illustration, succinyl-CoA), a portion of the NADH molecules generated from hexose molecules were reoxidized so that glycolysis could continue to function, thus sparing pyruvate for use in biosynthetic reactions. The oxidative branch (pyruvate → α-ketoglutarate, in this illustration, or succinyl-CoA) was used to generate NADH to be used to generate energy (probably using sulfur or sulfur compounds as electron acceptors) and biosynthetic precursor molecules.

BIOCHEMISTRY IN PERSPECTIVE cont

species, life scientists have begun to piece together the origins of the cycle. A major feature of this work has been the examination of organisms that use fragments of the cycle, use it in reverse, or do not use it at all.

It is almost certainly true that the citric acid cycle, the hub of metabolic processes in most modern organisms, did not emerge in its modern form as soon as O_2 began to accumulate in Earth's atmosphere. There is significant evidence to suggest that the citric acid cycle originally evolved in anaerobic prokaryotes as two separate pathways: a reductive branch (oxaloacetate → succinate or succinyl-CoA) and an oxidative branch (oxaloacetate → α-ketoglutarate or succinyl-CoA) (Figure 9A).

The reductive branch solved several problems of primordial organisms. Among these are providing a source of electron acceptors and biosynthetic precursors. Recall that the continued function of the glycolytic pathway requires that pyruvate molecules be reduced so that NADH can be reoxidized. This "redox balance," however, prevents the use of pyruvate molecules as metabolic precursors. This problem was apparently solved by two reactions in the reductive branch of the pathway that reoxidize NADH. The first is the conversion of oxaloacetate to malate. This reaction is catalyzed by malate dehydrogenase, an enzyme that is believed to have originated from a duplication of the lactate dehydrogenase gene. An extension of the reductive branch was made possible by the addition of fumarase and fumarate reductase, which catalyze the reversible conversion of malate to fumarate, and the reduction of fumarate to form succinate, respectively. Organisms that developed this mechanism for reoxidizing NADH would have had a selective advantage because of increased biosynthetic potential. In addition to utilizing some pyruvate molecules as biosynthetic precursors (e.g., certain amino acids and acetyl-CoA), these organisms were also able to exploit the intermediates in the reductive branch for biosynthesis. Examples include the synthesis of aspartate from oxaloacetate and porphyrins from succinyl-CoA.

The oxidative branch of an incomplete citric acid cycle generates α-ketoglutarate (or possibly succinyl-CoA) from citrate, the product of the reaction of oxaloacetate and acetyl-CoA. Both citrate and α-ketoglutarate are biosynthetic precursor molecules (e.g., fatty acids and glutamate, respectively). The maintenance of redox balance of these reactions required a terminal electron acceptor for the NADH produced by the oxidation of isocitrate to form α-ketoglutarate. In primitive anaerobic organisms, sulfur is believed to have served this function.

As atmospheric oxygen levels rose, the exploitation of O_2 as an electron acceptor was made possible by a complete citric acid cycle. Two possible mechanisms have been proposed. According to one scenario, the two branches (pyruvate → α-ketoglutarate and pyruvate → succinyl-CoA) were linked by α-ketoglutarate dehydrogenase, the product of the duplication of the pyruvate dehydrogenase gene. Alternatively, succinyl-CoA synthetase was the enzyme used to link the branches (pyruvate → succinyl-CoA and pyruvate → succinate), thus completing the cycle.

SUMMARY: The citric acid cycle probably developed in primordial cells as two separate pathways: the reductive branch, which provided a means of reoxidizing NADH, and an oxidative branch, which produced the biosynthetic precursor molecules citrate and α-ketoglutarate.

Chapter Summary

1. Aerobic organisms have an enormous advantage over anaerobic organisms, that is, a greater capacity to obtain energy from organic food molecules. To use oxygen to generate energy requires the following biochemical pathways: the citric acid cycle, the electron transport pathway, and oxidative phosphorylation.

2. Most reactions that capture or release energy are redox reactions. In these reactions, electrons are transferred between an electron donor (reducing agent) and an electron acceptor (oxidizing agent). In most biochemical reactions, hydride ions are transferred to NAD^+ or $NADP^+$ or hydrogen atoms are transferred to FAD or FMN. The tendency for a substance to gain electron(s) is called its reduction potential. Electrons flow spontaneously from a substance with a less positive (more negative) reduction potential to a substance with a more positive (less negative) reduction potential. In favorable redox reactions $\Delta E^{\circ\prime}$ is positive and $\Delta G^{\circ\prime}$ is negative.

3. The citric acid cycle is a series of biochemical reactions that eventually completely oxidize organic substrates, such as

glucose and fatty acids, to form CO_2, H_2O, and the reduced coenzymes NADH and $FADH_2$. Pyruvate, the product of the glycolytic pathway, is converted to acetyl-CoA, the citric acid cycle substrate.

4. In addition to its role in energy generation, the citric acid cycle plays an important role in several biosynthetic

processes, such as gluconeogenesis, amino acid synthesis, and porphyrin synthesis.

5. The glyoxylate cycle, found in plants and some fungi, algae, protozoans, and bacteria, is a modified version of the citric acid cycle in which two-carbon molecules, such as acetate, are converted to precursors of glucose.

 Take your learning further by visiting the **companion website** for Biochemistry at **www.oup.com/us/mckee** where you can complete a multiple-choice quiz on the citric acid cycle to help you prepare for exams.

Suggested Readings

Bilgen, T., Metabolic Evolution and the Origin of Life, in Storey, K. B. (ed.), *Functional Metabolism: Regulation and Adaptation*, pp. 557–582, Wiley-Liss, Hoboken, New Jersey, 2004.

Graham, T. E., and Gibala, M. J., Anaplerosis of the Tricarboxylic Acid Cycle in Human Skeletal Muscle During Exercise: Magnitude, Sources, and Potential Physiological Significance, *Adv. Exp. Med. Biol.* 441:271–286, 1998.

Huynen, M. A., Dandekar, T., and Bork, P., Variation and Evolution of the Citric Acid Cycle: A Genomic Perspective, *Trends Microbiol.* 7:281–291, 1999.

Kornberg, H., Krebs and His Trinity of Cycles, *Nat. Rev. Mol. Cell Biol.* 1(3):225–227, 2000.

Krebs, H. A., The History of the Tricarboxylic Cycle, *Perspect. Biol. Med.* 14:154–170, 1970.

Lane, N., *Oxygen: The Molecule That Made the World*, Oxford University Press, Oxford, 2002.

Lorenz, M. C., and Fink, G. R., Life and Death in a Macrophage: Role of the Glyoxylate Cycle in Virulence, *Eukaryotic Cell* 1(5):657–662, 2002.

Melendez-Hevia, E., Waddell, T. G., and Cascante, M., The Puzzle of the Krebs Citric Acid Cycle: Assembling the Pieces of Chemically Feasible Reactions and Opportunism in the Design of Metabolic Enzymes During Evolution, *J. Mol. Evol.* 43(3):293–303, 1996.

Scheffler, I. E., *Mitochondria*, Wiley-Liss, Hoboken, New Jersey, 1999.

Smith, E., and Morowitz, H. J., Universality in Intermediary Metabolism, *Proc. Natl. Acad. Sci. USA* 101(36): 13168–13173, 2004.

Zubay, G., *Origins of Life on the Earth and in the Cosmos*, 2nd ed., Academic Press, San Diego, 2000.

Key Words

aerotolerant anaerobe, *308*

amphibolic pathway, *326*

anaplerotic, *326*

coenzyme A, *316*

conjugate redox pair, *310*

facultative anaerobe, *308*

flavoprotein, *312*

flavin adenine dinucleotide, *312*

flavin mononucleotide, *312*

glyoxylate cycle, *331*

lipoic acid, *319*

nicotinamide adenine dinucleotide, *312*

nicotinamide adenine dinucleotide phosphate, *312*

obligate aerobe, *308*

obligate anaerobe, *308*

redox potential, *311*

reduction potential, *311*

thiamine pyrophosphate, *319*

Review Questions

These questions are designed to test your knowledge of the key concepts discussed in this chapter, before moving on to the next chapter. You may like to compare your answers to the solutions provided in the back of the book and in the accompanying Study Guide.

1. Define the following terms:
   a. anaerobes
   b. anaplerotic reactions
   c. glyoxysomes
   d. reduction potential
   e. conjugate redox pair

2. Describe in general terms how the appearance of molecular oxygen in Earth's atmosphere about 3 billion years ago affected the history of living organisms.

3. Describe the lifestyles of the following organisms:
   a. obligate anaerobes
   b. aerotolerant anaerobes
   c. facultative anaerobes
   d. obligate aerobes

4. A runner needs a tremendous amount of energy during a race. Explain how the use of ATP by contracting muscle affects the citric acid cycle.

5. Describe two important roles of the citric acid cycle.

6. Acetyl-CoA is manufactured in the mitochondria and used in the cytoplasm to synthesize fatty acids. However, acetyl-CoA cannot penetrate the mitochondrial membrane. How is this problem solved?

7. Describe the glyoxylate cycle. How is it used to synthesize complex molecules from two-carbon molecules?

8. Which of the following conditions indicates a low cell energy status? What impact does each of the following conditions have on flux through the citric acid cycle?
   a. high NADH/NAD$^+$ ratio
   b. high ATP/ADP ratio
   c. high acetyl-CoA concentration
   d. low citrate concentration
   e. high succinyl-CoA concentration

9. Define the following terms:
   a. coenzyme A
   b. amphibolic
   c. reactive oxygen species
   d. aerobes
   e. NAD(P)$^+$

10. Review the structures shown and identify the components of each molecule. What additional covalently attached molecules will be needed to convert these structures into redox coenzymes?

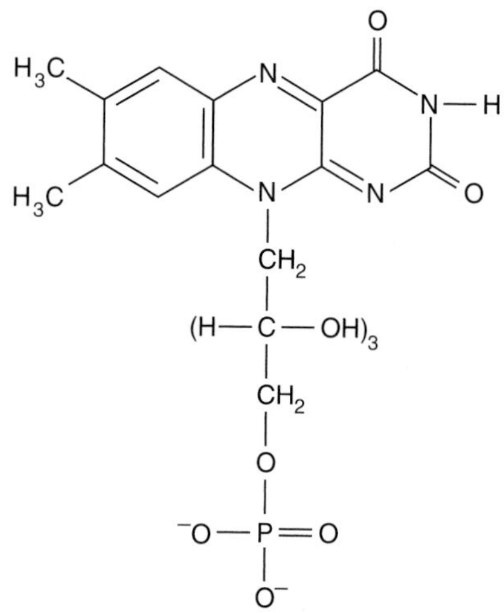

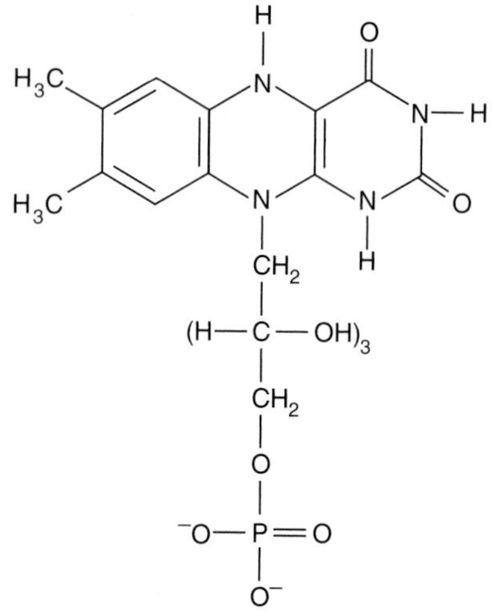

11. Write balanced equations for each of the reactions of the citric acid cycle.

12. What are the coenzymes required in each of the reactions in Question 11?

13. Discuss the mechanisms of control of the irreversible steps in the citric acid cycle.

14. The citric acid cycle operates only when O_2 is present, yet O_2 is not a substrate for the cycle. Explain.

15. If a small amount of [1-^{14}C]glucose is added to an aerobic yeast culture, where will the ^{14}C label initially appear in citrate molecules?

16. Define the following terms:
   a. flavoprotein
   b. FAD
   c. FMN
   d. lipoic acid
   e. TPP

17. Describe in detail the structure of the pyruvate dehydrogenase complex.

18. Describe the role played by each enzyme, cofactor, and coenzyme of the pyruvate dehydrogenase complex.

19. What are the two stages of the citric acid cycle? List the products of each stage.

20. Describe the role of citrate in fatty acid biosynthesis.

21. A diet that contains excess carbohydrate results in the biosynthesis of triacylglycerols (fatty acid esters of glycerol). Explain.

22. Explain why animals cannot produce glucose from two-carbon molecules such as acetate or ethanol.

23. Write the net equation for the citric acid cycle.

24. What steps in the citric acid cycle are regulated? Why are they regulated?

25. Provide examples of biosynthetic pathways that utilize citric acid cycle intermediates as precursor molecules.

ThoughtQuestions

These questions are designed to reinforce your understanding of all of the key concepts discussed in the book so far, including this chapter and all of the chapters before it. They may not have one right answer! The authors have provided possible solutions to these questions in the back of the book and in the accompanying Study Guide, for your reference.

26. What is the significance of substrate-level phosphorylation reactions? Which of the reactions in the citric acid cycle involve a substrate-level phosphorylation? Name another example from a biochemical pathway with which you are familiar.

27. You have just consumed a piece of fruit. Trace the carbon atoms in the fructose in the fruit through the biochemical pathways between their uptake into tissue cells and their conversion to CO_2.

28. Outline the biosynthesis of the amino acid glutamate from pyruvate.

29. Outline the degradation of glutamate to form CO_2.

30. Determine the standard free energy ($\Delta G^{\circ\prime}$) for the following reactions:
    a. $NADH + H^+ + \frac{1}{2}O_2 \rightarrow NAD^+ + H_2O$
    b. Cytochrome c (Fe^{2+}) $+ \frac{1}{2}O_2 \rightarrow$ cytochrome c (Fe^{3+}) $+ H_2O$

31. Both facultative aerobes and obligate aerobes possess fermentative and aerobic energy-generating pathways. As their name suggests, obligate aerobes cannot live without oxygen for even a few minutes. Suggest a reason why the aerobic lifestyle is so vulnerable to oxygen deprivation.

32. One of the many effects of chronic alcohol abuse is thiamine deficiency, caused by impaired absorption of the vitamin through the intestinal wall and diminished storage in a damaged liver. When thiamine levels are inadequate, cellular energy generation is diminished. What are the three enzymes involved in cellular metabolism that require thiamine? Describe the metabolic consequences of inadequate thiamine levels.

33. Despite the absence of the glyoxylate cycle in animals, when ^{14}C-labeled acetate is fed to lab animals, small amounts of the radioactive label later appear in glycogen stores. Explain.

34. Malonate (p. 198), poisons the citric acid cycle because it inhibits succinate dehydrogenase. After reviewing its structure, describe how the inhibitory effect of malonate can be overcome.

35. Fatty acid degradation stimulates the citric acid cycle through the activation of pyruvate carboxylase by acetyl-CoA. Why would the activation of pyruvate carboxylase increase energy generation from fatty acids?

36. Shock (failure of the circulatory system) is an abnormal condition in which blood flow is inadequate. The most common causes of shock are massive blood loss and obstruction of blood flow. As a result of shock there is a failure to provide cells with oxygen and nutrients. In this circumstance cells swell and lysosomal membranes rupture, among other effects. Describe how energy is generated during shock and why cell structure becomes destabilized.

37. Lactic acidosis occurs as a result of shock. Explain why low oxygen levels promote lactate production.

38. Because dichloroacetate inhibits the enzyme pyruvate dehydrogenase kinase, this compound has been used, with limited results, to treat lactic acidosis. The phosphorylation of the α-subunit of the pyruvate dehydrogenase component of the pyruvate dehydrogenase complex by pyruvate dehydrogenase kinase causes complete loss of enzymatic activity. Describe the theory behind the clinical use of dichloroacetate.

39. Pyruvate dehydrogenase deficiency is a fatal disease usually diagnosed in children. Symptoms include severe neurological damage. Elevated blood levels of lactate, pyruvate, and alanine are also seen. Explain how the deficiency of pyruvate dehydrogenase causes these elevated values.

40. The large amount of energy used during aerobic exercise (e.g., running) requires large amounts of oxaloacetate. Explain why acetyl-CoA cannot be used to produce oxaloacetate in this circumstance. What is the source of oxaloacetate molecules during aerobic activity?

41. Using the data in Table 9.1 and the equation given here, calculate the free energy (ΔG°) produced by the reduction of sulfur to hydrogen sulfide and oxygen to water by NADH. How much more free energy is produced by the reduction of oxygen compared to that of sulfur?

    $$\Delta G^{\circ} = -nF\Delta E^{\circ}$$

42. Systemic fungal infections (e.g., *Candida* and *Cryptococcus*) and tuberculosis (*Mycobacterium*) are on the rise. One of the reasons these organisms have high virulence is that they fare better following macrophage phagocytosis than other microorganisms. Phagocytosis activates the glyoxylate cycle in fungi and mycobacteria, allowing them to use two-carbon substrates to sustain growth in an otherwise nutrient-poor environment of the phagolysosome. Suggest some likely molecules that could be processed through the glyoxylate cycle in this circumstance.

Aerobic Metabolism II: Electron Transport and Oxidative Phosphorylation

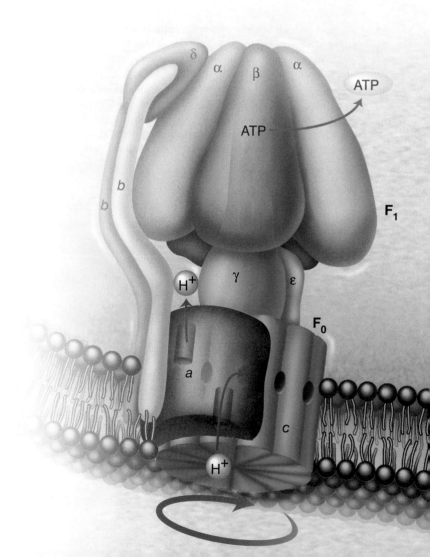

The ATP Synthase The ATP synthase is a rotating molecular machine that synthesizes ATP. This multiprotein complex is composed of two principal domains: the membrane-spanning F_0 component and the ATP-synthesizing F_1 component. The flow of protons through F_0, made possible by a gradient created by electron transport, generates a torque that forces the shaft (the γ subunit) to rotate. Rotational force within F_1 then triggers conformational changes that result in ATP synthesis.

Overview

THE USE OF OXYGEN BY AEROBIC ORGANISMS PROVIDES ENORMOUS
ADVANTAGES IN COMPARISON TO AN ANAEROBIC LIFESTYLE BECAUSE
the aerobic oxidation of nutrients such as glucose and fatty acids yields a
substantially greater amount of energy than does fermentation occurring in the
absence of O_2. Oxygen also facilitates reactions such as hydroxylations.
Because oxygen is a highly reactive molecule, however, an inescapable price is
paid for its use: toxic intermediate metabolites are formed. Research efforts
reveal that aerobic cells have an array of mechanisms that protect organisms
against oxygen's deleterious effects. Numerous enzymes and antioxidant
molecules usually prevent oxidative cell damage. Despite this high level of
protection, however, oxygen metabolites sometimes cause serious damage to
living cells. Abnormally stressful conditions can overwhelm an organism's
antioxidant mechanisms. Disorders such as cancer, heart disease, and the nerve
damage following spinal cord injury are now known to be caused in part by
oxygen metabolites.

The complex structure and functioning of aerobic organisms is sustained by
the extraordinarily large amounts of ATP they can generate. This capacity is made possible by their ability to use O_2 as the terminal acceptor of
the electrons extracted from fuel molecules. Other organisms can generate
energy with anaerobic electron transport pathways (e.g., the anaerobic
chemoorganotrophs synthesize ATP by using terminal electron acceptors such
as sulfur, sulfate, nitrate, ferric iron, or CO_2). However, anaerobes are relatively
simple in structure and occupy only isolated ecological niches. In contrast, aerobic organisms are found in great abundance throughout the biosphere.

The vast difference in energy-generating capacity between aerobic and anaerobic organisms is directly related to the chemical and physical properties of
oxygen. Oxygen has several properties that in combination have made possible
a highly favorable mechanism for extracting energy from organic molecules. First
of all, oxygen is found almost everywhere on the earth's surface. In contrast, most
other electron acceptors are relatively rare. Second, oxygen diffuses easily across
cell membranes. This is not true of several other electron acceptors. For example,
charged species such as sulfate and nitrate do not readily diffuse across cell membranes. Finally, under certain circumstances oxygen is highly reactive, so it
can also readily accept electrons. This capacity is also responsible for another
property of oxygen, its tendency to form highly destructive metabolites called
reactive oxygen species (ROS).

Chapter 10 describes the basic principles of oxidative phosphorylation, the
complex mechanism by which modern aerobic cells manufacture ATP. The discussion begins with a review of the electron transport system in which electrons
are donated by reduced coenzymes to the electron transport chain (ETC). The
ETC is a series of electron carriers in the inner membrane of the mitochondria
of eukaryotes and the plasma membrane of aerobic prokaryotes. The chapter's
next topic is chemiosmosis, the means by which the energy extracted from
electron flow is captured and used to synthesize ATP. Chapter 10 ends with a
discussion of the formation of toxic oxygen products and the strategies that cells
use to protect themselves.

10.1 ELECTRON TRANSPORT

The mitochondrial electron transport chain (ETC), also referred to as the electron transport system, is a series of electron carriers arranged in the inner membrane in order of increasing electron affinity; it is these molecules that transfer the electrons derived from reduced coenzymes to oxygen. During this transfer, a decrease in reduction potential ($\Delta E^{\circ\prime}$) occurs. When NADH is the electron donor and oxygen is the electron acceptor, the change in standard reduction potential is +1.14 V (i.e., +0.82V − (−0.32V), see Table 9.1). This process, in which oxygen is used to generate energy from food molecules, is sometimes referred to as **aerobic respiration**. The energy released during electron transfer is coupled to several endergonic processes, the most prominent of which is ATP synthesis. Other processes driven by electron transport pump Ca^{2+} into the mitochondrial matrix and generate heat in brown adipose tissue (described on p. 356). Reduced coenzymes, derived from glycolysis, the citric acid cycle, and fatty acid oxidation, are the principal sources of electrons.

Electron Transport and Its Components

The components of the ETC in eukaryotes are located in the inner mitochondrial membrane (Figure 10.1). Most ETC components are organized into four complexes. Each of these complexes consists of several proteins and prosthetic groups. The structural and functional properties of each complex are briefly

FIGURE 10.1

The Electron Transport Chain

Complexes I and II transfer electrons from NADH and succinate, respectively, to UQ. Complex III transfers electrons from UQH_2 to cytochrome c. Complex IV transfers electrons from cytochrome c to O_2. The arrows represent the flow of electrons.

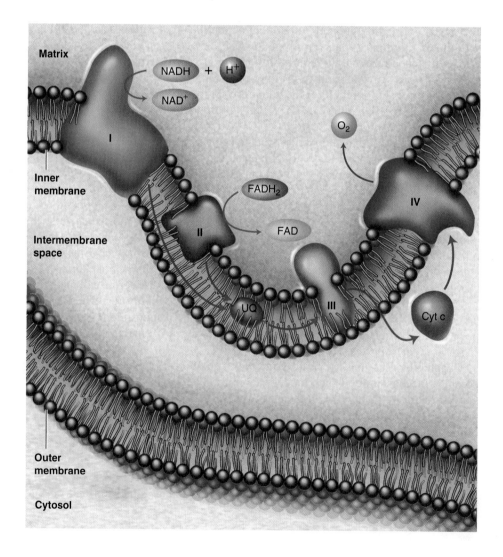

described. The roles of coenzyme Q (ubiquinone, UQ) and the cytochromes are also described. Figure 10.1 provides an overview of electron flow in the ETC.

Complex I, also referred to as the NADH dehydrogenase complex, catalyzes the transfer of electrons from NADH to UQ. The major sources of NADH include several reactions of the citric acid cycle (see pp. 316–325), and fatty acid oxidation (Chapter 12). Composed of at least 25 different polypeptides, complex I is the largest protein component in the inner membrane. In addition to one molecule of flavin mononucleotide (FMN), the complex contains seven iron-sulfur centers (Figure 10.2). Iron-sulfur centers, which may consist of two or four iron atoms complexed with an equal number of sulfide ions, mediate one-electron transfer reactions. Proteins that contain iron-sulfur centers are often referred to as *nonheme iron proteins*. Although the structure and function of complex I are still poorly understood, it is believed that NADH reduces FMN to $FMNH_2$. Electrons are then transferred from $FMNH_2$ to an iron-sulfur center, 1 electron at a time. After transfer from one iron-sulfur center to another, the electrons are eventually donated to UQ, a lipid-soluble, mobile electron carrier capable of accepting/donating electrons, one at a time (Figure 10.3).

Figure 10.4 illustrates the transfer of electrons through complex I. Electron transport is accompanied by the net movement of protons from the matrix across the inner membrane and into the intermembrane space. The significance of this phenomenon for ATP synthesis will be discussed.

The succinate dehydrogenase complex (complex II) consists primarily of the citric acid cycle enzyme succinate dehydrogenase and two iron-sulfur proteins. Complex II mediates the transfer of electrons from succinate to UQ. The oxidation site for succinate is located on the larger of the iron-sulfur proteins. This molecule also contains a covalently bound FAD. Other flavoproteins also donate electrons to UQ (Figure 10.5). In some cell types glycerol-3-phosphate dehydrogenase, an enzyme located on the outer face of the inner mitochondrial membrane, transfers electrons from cytoplasmic NADH to the ETC (see p. 287). Acyl-CoA dehydrogenase, the first enzyme in fatty acid oxidation (Chapter 12), transfers electrons to UQ from the matrix side of the inner membrane.

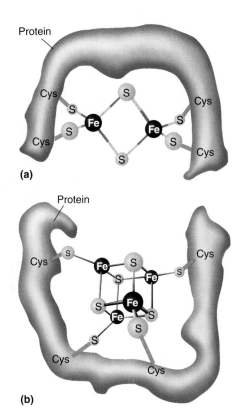

(a)

(b)

FIGURE 10.2

Two Iron-Sulfur Centers

(a) Protein whose iron-sulfur center contains 2 irons and 2 sulfurs. (b) A 4 Fe-4S iron-sulfur center. In both cases, the cysteine residues are part of a polypeptide.

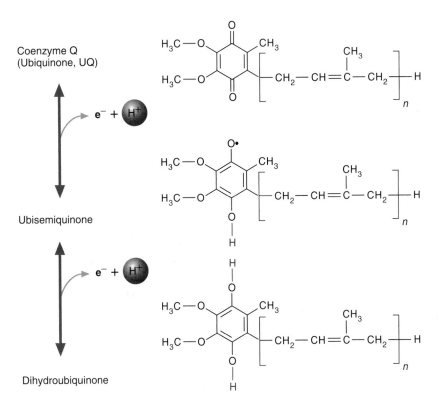

Coenzyme Q
(Ubiquinone, UQ)

Ubisemiquinone

Dihydroubiquinone

FIGURE 10.3

Structure and Oxidation States of Coenzyme Q

The length of the side chain varies among species. For example, some bacteria have six isoprene units. For mammals, however, $n = 10$, and such molecules are referred to as Q_{10}.

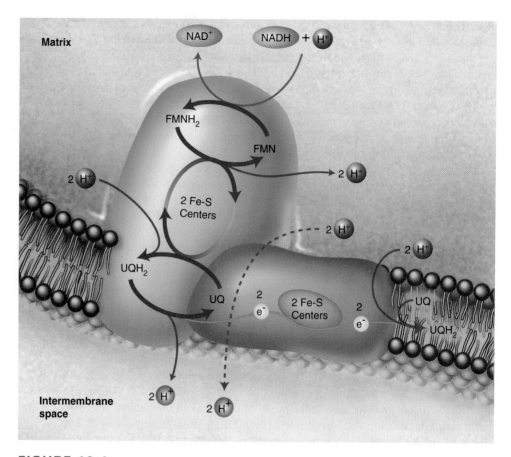

FIGURE 10.4

Transfer of Electrons Through Complex I of the Mitochondrial Electron Transport Chain

Electron transfer begins with the reduction of FMN by NADH, a process that requires one proton from the matrix. FMNH$_2$ subsequently transfers two electrons, one at a time, to six to eight Fe-S centers. (Because the path of the electrons is not known, only four Fe-S centers are shown.) The sequential transfer of the two electrons to the first Fe-S center eventually releases four protons into the intermembrane space. The mechanism by which these protons are transferred across the membrane is still unclear. However, an internal UQ is believed to be involved in the transfer of two of the protons. The second pair of protons is transferred to an external UQ as the two electrons are sequentially transferred from the internal UQ through a series of Fe-S centers.

Complex III, also referred to as cytochrome b_1 complex, contains as many as 11 subunits. Among these are three cytochromes (cyt b_L, cyt b_H, and cyt c_1) and one iron-sulfur center. The *cytochromes* are a series of electron transport proteins that contain a heme prosthetic group similar to those found in hemoglobin and myoglobin. Electrons are transferred by cytochromes one at a time in association with a reversible change in the oxidation state of a heme iron (i.e., between a reduced Fe^{2+} and an oxidized Fe^{3+}). The function of complex III is the transfer of electrons from reduced coenzyme Q (UQH$_2$) to a protein called cytochrome c (cyt c). Cytochrome c (Figure 10.6) is a mobile electron carrier that is loosely associated with the outer face of the inner mitochondrial membrane. The passage of electrons through complex III (Figure 10.7), referred to as the **Q cycle**, is complex. However, the overall reaction for the process whereby each UQH$_2$ donates two electrons to cytochrome c (cyt c) is straightforward:

$$UQH_2 + 2 \text{ cyt } c_{ox}(Fe^{3+}) + 2H^+_{matrix} \longrightarrow UQ + 2 \text{ cyt } c_{red}(Fe^{2+}) + 4H^+_{cytosol}$$

During the Q cycle, coenzyme Q molecules diffuse within the inner membrane between the electron donors in complex I or II and the electron acceptor in complex III (the iron-sulfur protein). UQH$_2$ donates its two electrons one at a time.

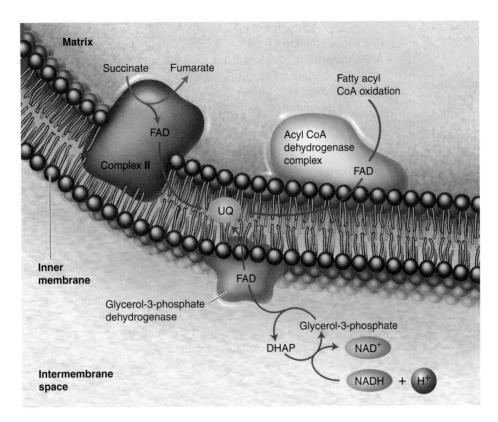

FIGURE 10.5

Path of Electrons from Succinate, Glycerol-3-Phosphate, and Fatty Acids to UQ

Electrons from succinate are transferred to FAD in complex II and several Fe-S centers and then to UQ. Electrons from cytoplasmic NADH are transferred to UQ via a pathway involving glycerol-3-phosphate and the flavoprotein glycerol-3-phosphate dehydrogenase (see p. 287). Fatty acids are oxidized as coenzyme A derivatives. Acyl-CoA dehydrogenase, one of several enzymes in fatty acid oxidation, transfers two electrons to FAD. They are then donated to UQ.

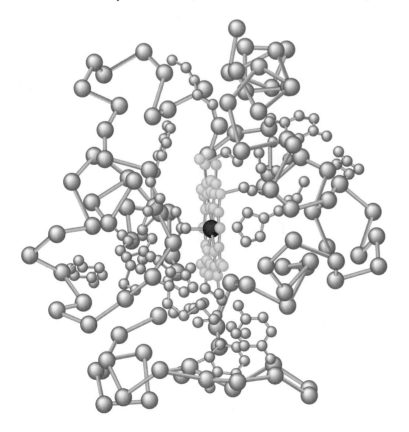

FIGURE 10.6

Structure of Cytochrome c

Cytochrome c is a member of the class of small proteins called cytochromes, each of which possesses a heme prosthetic group. During the electron transport process, the iron in the heme is alternately oxidized and reduced.

One electron flows to the Fe-S protein, which then transfers it to cyt c_1, after which it is donated to cyt c (Figure 10.7a). The second electron flows to cyt b_L and then to cyt b_H. Cyt b_H transfers this electron to an oxidized CoQ molecule (UQ) to generate UQ semiquinone (UQ^-). Note that the protons released by the oxidation of UQH_2 are released into the inner membrane space. An electron from a second molecule of UQH_2 then flows to the Fe-S center of complex III and then to cyt c_1 and cyt c (Figure 10.7b). The other electron (from the second UQH_2) is transferred to cyt b_L and b_H and is then donated to UQ semiquinone. As UQ^- accepts this electron, it also binds to two protons from the matrix to form UQH_2. For each pair of electrons that are transported through complex III, two molecules of cyt c are reduced, four protons are released into the inner membrane space, and a molecule of UQH_2 is regenerated from UQ^-.

Cytochrome oxidase (complex IV) is a protein complex that catalyzes the four-electron reduction of O_2 to form H_2O. The membrane-spanning complex (Figure 10.8) in mammals may contain between six and thirteen subunits, depending on species. Complex IV contains cytochromes a and a_3 and three copper ions. Two copper ions form Cu_A/Cu_A, a binuclear Cu-Cu center, and heme a_3 and Cu_B form a binuclear Fe-Cu center. Both centers accept electrons, one at a time. Electrons flow from cytochrome c to Cu_A/Cu_A to cytochrome a and then to a_3-Cu_B and, finally, to O_2. Four protons and four electrons are shuttled through complex IV from the outer face of the inner mitochondrial membrane to the matrix for delivery to the cytochrome a_3-Fe(II)-bound dioxygen. Two water molecules are formed and leave the site:

$$O_2 + 4H^+ + 4\,e^- \rightarrow 2H_2O$$

There is an ATP-binding regulatory site on complex IV. When ATP concentrations are high, ATP, acting as an allosteric inhibitor, binds to the site and causes a decrease in the activity of cytochrome oxidase.

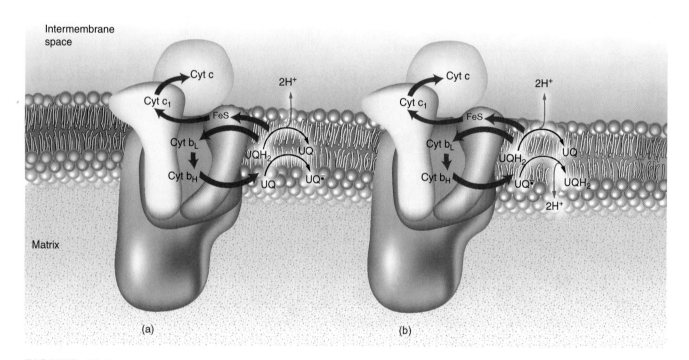

FIGURE 10.7

Electron Transport Through Complex III

The blue arrows represent the flow of electrons through complex III. Two molecules of UQH_2 are oxidized in two steps. The first electron from each UQH_2 is transferred to the Fe-S protein and then to cyt c_1 and cyt c. The second electron is transferred sequentially to cyt b_L and cyt b_H. (a) Cyt b_H then transfers the electron to UQ to generate UQ^- (UQ in its semiquinone anion form). (b) UQ^- is reduced by an electron from the second UQH_2 molecule, along with two additional protons taken up from the matrix to form UQH_2.

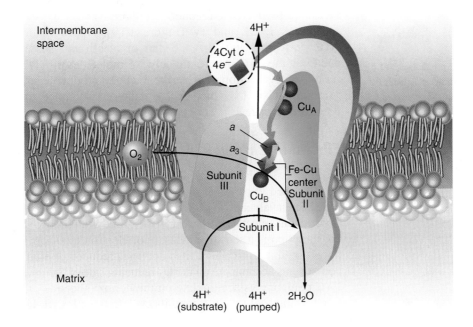

FIGURE 10.8

Electron Transport Through Complex IV

Each of two reduced cyt c molecules donates two electrons, one at a time, to Cu_A on subunit II. The electrons are then transferred to cyt a, cyt a_3, and Cu_B on subunit I. The O_2 binds to the heme group in cyt a and is converted to the protonated form of O^{-2}, its peroxy derivative, by two electrons transferred from the Fe-Cu center (cyt a_3 and Cu_B) in subunit I. Transfer of two additional electrons from cyt c and four protons from the matrix results in the formation of two water molecules. Simultaneously, four more protons are pumped from the matrix into the inner membrane space.

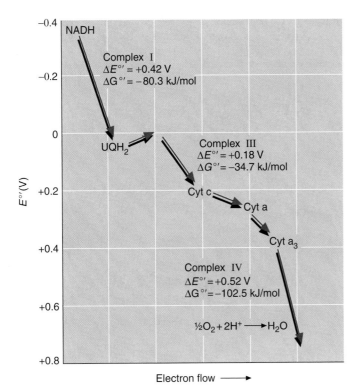

FIGURE 10.9

The Energy Relationships in the Mitochondrial Electron Transport Chain

Relatively large decreases in free energy occur in three steps. During each of these steps, sufficient energy is released to account for the synthesis of ATP.

KEY CONCEPT

The electron transport chain is a series of complexes consisting of molecular electron carriers located in the inner mitochondrial membrane of eukaryotic cells.

NADH oxidation results in the release of a substantial amount of energy, measured by decreased reduction potentials ($\Delta E^{\circ\prime}$), as electrons flow through complexes I, III, and IV (Figure 10.9). Approximately 2.5 molecules of ATP are synthesized for each pair of electrons transferred between NADH and O_2 in the ETC. Approximately 1.5 molecules of ATP result from the transfer of each pair donated by the $FADH_2$ produced by succinate oxidation. The mechanism by which ATP synthesis is coupled to electron transport is described on page 347. The components of the ETC are summarized in Table 10.1.

Electron Transport Inhibitors

Several molecules specifically inhibit the electron transport process (Figure 10.10). Used in conjunction with reduction potential measurements, inhibitors have been invaluable in determining the correct order of ETC components. In such experiments, electron transport is measured with an oxygen electrode. (Oxygen consumption is a sensitive measure of electron transport.) When electron transport is inhibited, oxygen consumption is reduced or eliminated. Oxidized ETC components accumulate on the O_2-reducing side of the site of inhibition. Reduced ETC components accumulate on the nonoxygen side of the site of inhibition. For example, antimycin A inhibits cyt b in complex III. If this inhibitor is added to a suspension of mitochondria, NAD^+, the flavins, and cyt b molecules become more reduced. The cytochromes c_1, c, and a become more oxidized. Other prominent examples of ETC inhibitors include rotenone and amytal, which inhibit NADH dehydrogenase (complex I). Carbon monoxide (CO), azide (N_3^-), and cyanide (CN^-) inhibit cytochrome oxidase.

TABLE 10.1 Supramolecular Components of the Electron Transport Chain

Enzyme Complex	Prosthetic Groups
Complex I (NADH dehydrogenase)	FMN, FeS
Complex II (succinate dehydrogenase)	FAD, FeS
Complex III (cytochrome bc_1 complex)	Hemes, FeS
Cytochrome c	Heme
Complex IV (cytochrome oxidase)	Hemes, Fe, Cu

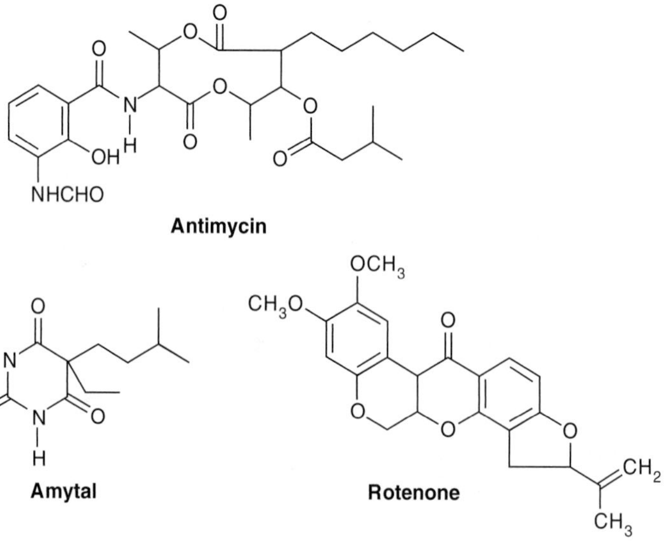

FIGURE 10.10
Several Inhibitors of the Mitochondrial Electron Transport Chain

Antimycin blocks the transfer of electrons from the b cytochromes. Amytal and rotenone block NADH dehydrogenase.

QUESTION 10.1

Which compound in each of the following pairs is the better reducing agent?

a. $NADH/H_2O$

b. $UQH_2/FADH_2$

c. Cyt c (reduced)/cyt b (reduced)

d. $FADH_2/NADH$

e. $NADH/FMNH_2$

10.2 OXIDATIVE PHOSPHORYLATION

Oxidative phosphorylation, the process whereby the energy generated by the ETC is conserved by the phosphorylation of ADP to yield ATP, has been studied since the 1940s. The only type of phosphorylation reaction with which biochemists were then familiar was substrate-level phosphorylation (e.g., the oxidation and subsequent phosphorylation of glyceraldehyde-3-phosphate to form bisphosphoglycerate, p. 272). However, the high-energy intermediate that was believed to couple electron transport and ATP synthesis was never identified. In 1961, British biochemist Peter Mitchell proposed a mechanism by which the free energy released during electron transport drives ATP synthesis. Now widely accepted, Mitchell's model is referred to as the **chemiosmotic coupling theory**.

The Chemiosmotic Theory

The chemiosmotic coupling theory has the following features:

1. As electrons pass through the ETC, protons are transported from the matrix and released into the intermembrane space. As a result, an electrical potential Ψ and a proton gradient ΔpH arise across the inner membrane. The electrochemical proton gradient is sometimes referred to as the **protonmotive force** Δp.

2. Protons, which are present in the intermembrane space in great excess as a result of the electron transport process, can pass through the inner membrane and back into the matrix down their concentration gradient only through special channels. (The inner membrane itself is impermeable to ions such as protons.) As the thermodynamically favorable flow of protons occurs through a channel, each of which contains an ATP synthase activity, ATP synthesis occurs.

Mitchell suggested that the free energy release associated with electron transport and ATP synthesis is coupled by the protonmotive force created by the ETC. (The term *chemiosmotic* emphasizes that chemical reactions can be coupled to osmotic gradients.) An overview of the chemiosmotic model as it operates in the mitochrondion is illustrated in Figure 10.11.

Examples of the evidence that supports the chemiosmotic theory include the following:

1. Actively respiring mitochondria expel protons. The pH of a weakly buffered suspension of mitochondria measured by an electrode drops when O_2 is added. The typical pH gradient across the inner membrane is approximately 0.05 pH unit.

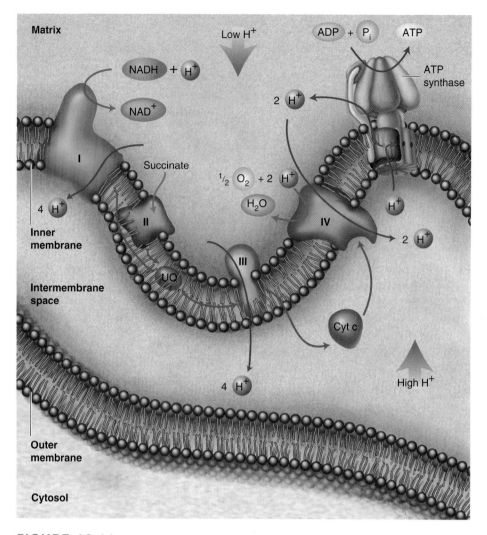

FIGURE 10.11

Overview of the Chemiosmotic Model

In Mitchell's model, protons are driven from the mitochondrial matrix across the inner membrane and into the intermembrane space by the electron transport mechanism. The energy captured from electron transport is used to create an electrical potential and a proton gradient. Because the inner membrane is impermeable to protons, they can traverse the membrane only by flowing through specific proton channels. The flow of protons through the ATP synthase drives the synthesis of ATP. (See Figure 10.1 for brief descriptions of the roles of complexes I, II, III, and IV in electron transport.)

2. ATP synthesis stops when the inner membrane is disrupted. For example, although electron transport continues, ATP synthesis stops in mitochondria placed in a hypertonic solution. Mitochondrial swelling results in proton leakage across the inner membrane.

3. A variety of molecules that inhibit ATP synthesis are now known to specifically dissipate the proton gradient (Figure 10.12). According to the chemiosmotic theory, a disrupted proton gradient dissipates the energy derived from food molecules as heat. **Uncouplers** such as dinitrophenol collapse the proton gradient by equalizing the proton concentration on both sides of membranes. (As they diffuse across the membrane, uncouplers pick up protons from one side and release them on the other.) **Ionophores** are hydrophobic molecules that dissipate osmotic gradients by inserting themselves into a membrane and forming a channel. For example, gramicidin is an antibiotic that forms a channel in membranes that allows the passage of H^+, K^+, and Na^+.

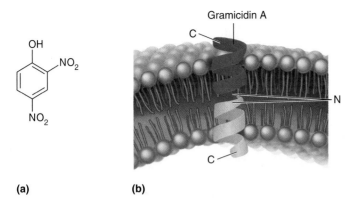

(a)   **(b)**

FIGURE 10.12

Uncouplers

(a) Dinitrophenol, which diffuses across the membrane, picks up protons on one side and releases them on the other. (b) Gramicidin A, an 11-residue peptide, forms an end-to-end dimer that creates a proton permeable pore in the membrane: C = carboxy terminus; N = amino terminus.

The proton gradients generated by electron transport systems are dissipated for two general purposes: ATP is synthesized as protons flow through the ATP synthase and regulated proton leakage is used to drive several other types of biological work. Examples include heat generation and transport of substances such as phosphate. In the next section, descriptions of ATP synthesis and its regulation within mitochondria are followed by a brief overview of energy generation from glucose. Section 10.2 ends with a discussion of nonshivering thermogenesis, the mechanism by which the mitochondrial proton gradient in certain animal cells regulates body temperature.

QUESTION 10.2

Dinitrophenol (DNP) is an uncoupler that was used as a diet aid in the 1920s, until several people who had been taking it died. Suggest why DNP consumption results in weight loss. The deaths caused by DNP were a result of liver failure. Explain. [*Hint:* Liver cells contain an extraordinarily large number of mitochondria.]

ATP Synthesis

Early electron microscopic studies of mitochondria showed numerous lollipop-shaped structures studding the inner membrane on its inner surface (Figure 10.13). Experiments begun in the early 1960s revealed that each lollipop is a proton translocating ATP synthase. These investigations made use of *submitochondrial particles* (SMP), which are small membranous vesicles formed when mitochondria are subjected to sonication. Further work showed that the ATP synthase (Figure 10.14) consists of two major components. The F_1 unit, the active ATPase, possesses five different subunits present in the ratio α_3: β_3: γ: δ: ε. There are three nucleotide-binding catalytic sites on F_1. The F_0 unit, a transmembrane channel for protons, has three subunits present in the ratio a: b_2: c_{12}.

The ATP synthase consists of two rotors (rotary motors) linked together by a strong, flexible *stator* (a stationary component of a motor). In respiring organisms the F_0 motor converts the protonmotive force into a rotational force that drives ATP synthesis catalyzed by the F_1 unit. The revolving component, the c ring

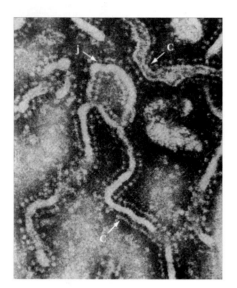

FIGURE 10.13

The ATP Synthase

An early electron micrograph of submito-chondrial particles revealing the "lollipop"-like structures that would eventually be identified as ATP synthase.

(formed from the c subunits), which is attached to a central shaft composed of the ε and γ subunits, rotates within the α,β hexamer of the F_1 unit. The stator (the b and δ subunits) prevents the α,β hexamer from rotating.

The synthesis of each ATP requires the translocation of three protons through the ATP synthase. (The transfer of an additional proton is required for the transport of ATP and OH^- out of the matrix in exchange for ADP and P_i.) As protons flow through F_0, the rotation of the proton channel (c_{12}, also referred to as the c ring) is transmitted to the γ subunit that projects into the core of the F_1 unit. The rotation of the central shaft puts it in three possible positions relative to each α,β dimer. In effect, the protonmotive force induces three sequential 120° rotations of the α,β hexamer. As rotation proceeds, each of three nucleotide-binding sites undergoes a series of conformational changes that result in ATP synthesis. In certain circumstances (e.g., *E. coli* in anaerobic conditions and fermentative lactic acid bacteria) the F_1 unit, acting as a motor, works in the opposite direction so that ATP is hydrolyzed. As a result, protons are pumped outward. The outward-moving proton gradient is then used to perform cellular work such as flagella rotation and nutrient transport.

The mechanism of ATP synthesis by the ATP synthase is as follows. In the F_0 unit each c subunit in the c ring consists of two transmembrane antiparallel helices. The C-terminal helix of c subunits contains an essential aspartate residue that upon protonation causes a swiveling motion that in turn triggers the rotation

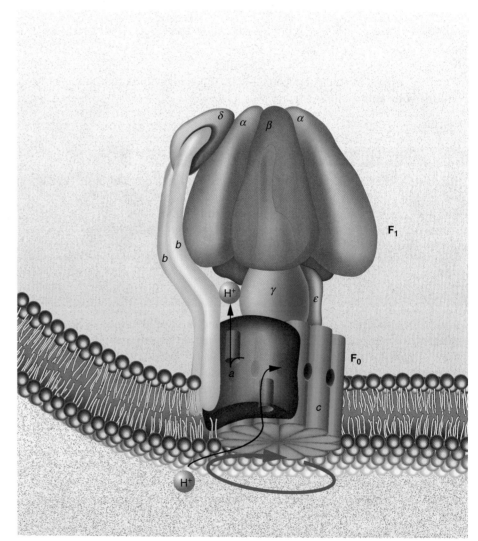

FIGURE 10.14

ATP Synthase from *Escherichia coli*

The rotor consists of ε, γ, and c_{12} subunits. The stator consists of a, b_2, δ, α_3, and β_3 subunits. The molecular components of ATP synthase are well conserved among bacteria, plants, and animals.

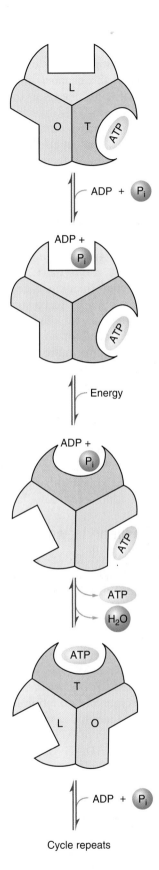

FIGURE 10.15

ATP Synthesis Model

The catalytic sites in the β subunits of ATP synthase are believed to have three conformations: open (O: inactive with low affinity for adenine nucleotides), tight (T: active with high affinity), and loose (L: also inactive). ATP synthesis begins with the binding of ADP and P_i to an L site. A conformational change driven by the rotation of the central shaft within the α,β hexamer converts the L site into a T conformation. ATP is then synthesized. As the shaft continues to rotate, the T site is converted to an O site that releases the ATP molecule. ATP cannot be released from the O site unless ADP and P_i are bound to the adjacent β-subunit site in the T conformation.

of the entire subunit. Deprotonation of this residue causes the C-terminal helix to return to its original conformation. Protons enter the c ring through a channel in the *a* subunit. At the end of this channel, at the interface of the *a* subunit and the proximal *c* subunit, a basic residue (Arg) transfers the incoming proton to the Asp residue in the *c* subunit. As a result of the rotation of the c subunit, it is displaced from the a subunit. As the next proton moves through the *a* subunit channel, the process repeats itself. The net effect is the rotation of the c ring in a counterclockwise direction (as viewed from the membrane). As each *c* subunit reaches a channel on the *a* subunit, deprotonation occurs, and the proton exits into the mitochondrial matrix. The *torque* (twisting force) generated by the c subunit rotation causes the asymmetric central shaft (composed of the ε and γ subunits) to rotate within a sleeve inside the α,β hexamer. The catalytic sites in the α,β hexamer occur on the β subunits. They occur in three conformations in terms of affinity for adenine nucleotide ligands: *open* (O) (inactive with low affinity), *tight* (T) (active with high affinity), and *loose* (L) (also inactive) (Figure 10.15). Interconversion between these conformations is caused by interaction with the rotating γ subunit. There are essentially three steps in the ATP synthesizing process: (1) ADP and P_i bind to an L site, (2) ATP is synthesized when the L conformation is transformed to a T conformation, and (3) ATP is released as the T conformation converts to an O conformation. As the γ subunit rotates and sequentially interacts with each β subunit, each active site is forced through the O, T, and L conformations.

KEY CONCEPTS

- In aerobic organisms, the energy used to drive the synthesis of most ATP molecules is the protonmotive force.
- The protonmotive force is generated as free energy is released when electrons flow through the electron transport chain.

QUESTION 10.3

A suspension of inside-out submitochondrial particles is placed in a solution that contains ADP, P_i, and NADH. Will increasing the proton concentration of the solution result in ATP synthesis? Explain.

Control of Oxidative Phosphorylation

Control of oxidative phosphorylation allows cells to adapt to ever-fluctuating energy requirements. Recall that under normal circumstances, electron transport and ATP synthesis are tightly coupled. The value of the *P/O ratio* (the number of moles of P_i consumed for each oxygen atom reduced to H_2O) reflects the degree of coupling observed between the protonmotive force, created by electron transport, and ATP synthesis. The measured maximum ratio for the oxidation of NADH is 2.5. The maximum P/O ratio for $FADH_2$ is 1.5. If isolated mitochondria are provided with an oxidizable substrate (e.g., succinate), all of the ADP is eventually converted to ATP. At this point, oxygen consumption becomes greatly depressed. Oxygen consumption increases dramatically when ADP is supplied. The control of aerobic respiration by ADP is referred to as **respiratory control**. The formation of ATP appears to be strongly related to the ATP mass action ratio ([ATP]/[ADP][P_i]). In other words, ATP synthase is inhibited by a high concentration of its product (ATP) and activated when ADP and P_i concentrations are high. The relative amounts of ATP and ADP within mitochondria are controlled largely by two transport proteins in the inner membrane: the ADP-ATP translocator and the phosphate carrier. The *ADP-ATP translocator* (Figure 10.16) is a dimeric protein responsible for the 1:1 exchange of intramitochondrial ATP for the ADP produced in the cytoplasm. As previously described, there is a potential difference across the inner mitochondrial membrane (negative inside). Because ATP molecules have one more negative charge than ADP molecules, the outward transport of ATP and inward transport of ADP are favored. The transport of $H_2PO_4^-$ along with a proton is mediated by the phosphate translocase, also referred to as the $H_2PO_4^-/H^+$ symporter.

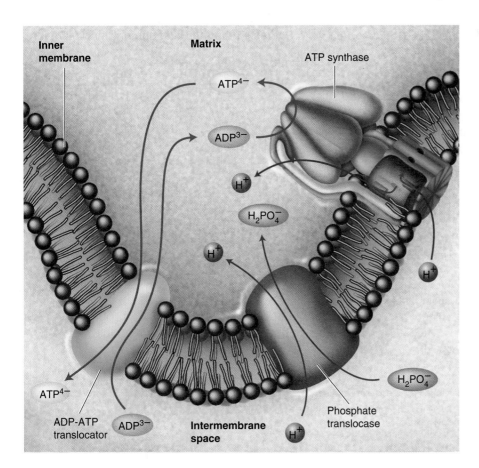

FIGURE 10.16

The ADP-ATP Translocator and the Phosphate Translocase

The transport of $H_2PO_4^-$ across the inner mitochondrial membrane by the phosphate translocase is driven by the proton gradient. For every four protons that are transported out of the matrix, three drive the ATP synthase rotor and one drives the inward transport of phosphate. The simultaneous exchange of ADP^{3-} and ATP^{4-}, required for continuing ATP synthesis and mediated by the ADP-ATP translocator, is driven by the potential difference across the inner membrane.

(*Symporters* are transmembrane transport proteins that move solutes across a membrane in the same direction. See Section 11.2 for a discussion of membrane transport mechanisms.) The inward transport of 4 protons is required for the synthesis of each ATP molecule: 3 to drive the ATP synthase rotor and 1 to drive the inward transport of phosphate.

The Complete Oxidation of Glucose

A summary of the sources of ATP produced from one molecule of glucose is provided in Table 10.2. ATP production from fatty acids, the other important energy source, is discussed in Chapter 12. Several aspects of this summary require further discussion. Recall that two molecules of NADH are produced during glycolysis. When oxygen is available, the oxidation of this NADH by the ETC is preferable (in terms of energy production) to lactate formation. The inner mitochondrial membrane, however, is impermeable to NADH. Animal cells have evolved several shuttle mechanisms to transfer electrons from cytoplasmic NADH to the mitochrondrial ETC. The most prominent examples are the glycerol phosphate shuttle and the malate-aspartate shuttle.

In the **glycerol phosphate shuttle** (Figure 10.17a), DHAP, a glycolytic intermediate, is reduced by NADH to form glycerol-3-phosphate. This reaction is followed by the oxidation of glycerol-3-phosphate by mitochondrial glycerol-3-phosphate dehydrogenase. (The mitochondrial enzyme uses FAD as an electron acceptor.) Because glycerol-3-phosphate interacts with the mitochondrial enzyme on the outer face of the inner membrane, the substrate does not actually enter the matrix. The $FADH_2$ produced in this reaction is then oxidized by the

TABLE 10.2 Summary of ATP Synthesis from the Oxidation of One Molecule of Glucose

	NADH	FADH$_2$	ATP
Glycolysis (cytoplasm)			
Glucose → glucose-6-phosphate			−1
Fructose-6-phosphate → fructose-1,6-bisphosphate			−1
Glyceraldehyde-3-phosphate → glycerate-1,3-bisphosphate	+2		
Glycerate-1,3-bisphosphate → glycerate-3-phosphate			+2
Phosphoenolpyruvate → pyruvate			+2
Mitochondrial Reactions			
Pyruvate → acetyl-CoA	+2		
Citric acid cycle			
Oxidation of isocitrate, α-ketoglutarate, and malate	+6		
Oxidation of succinate		+2	
GDP → GTP			+1.5*
Oxidative Phosphorylation			
2 Glycolytic NADH			+4.5† (3)‡
2 NADH (pyruvate to acetyl-CoA)			+5
6 NADH (citric acid cycle)			+15
2 FADH$_2$ (citric acid cycle)			+3
			31 (29.5)

* This number reflects the price of transport into the cytoplasm.
† Assumes the malate-aspartate shuttle.
‡ Assumes the glycerol phosphate shuttle.

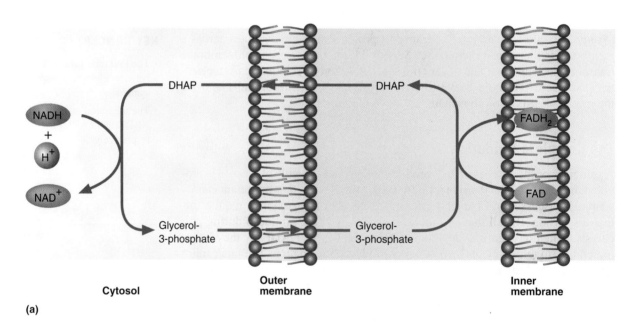

(a)

FIGURE 10.17

Shuttle Mechanisms That Transfer Electrons from Cytoplasmic NADH to the Respiratory Chain

(a) The glycerol-3-phosphate shuttle. Dihydroxyacetone phosphate (DHAP) is reduced to form glycerol-3-phosphate. Glycerol-3-phosphate (G-3-P) is reoxidized by mitochondrial glycerol-3-phosphate dehydrogenase and FAD is reduced to FADH$_2$. (b) The aspartate-malate shuttle. Oxaloacetate is reduced by NADH to form malate. Malate is transported into the mitochondrial matrix, where it is reoxidized to form oxaloacetate and NADH. Because oxaloacetate cannot penetrate the inner membrane, it is converted to aspartate in a transamination involving glutamate. Two inner membrane carriers are required for this shuttle mechanism: the glutamate-aspartate transport protein and the malate–α-ketoglutarate transport protein.

ETC. FAD as an electron acceptor produces only 1.5 ATP per molecule of cytoplasmic NADH.

Although the **malate-aspartate shuttle** (Figure 10.17b) is a more complicated mechanism than the glycerol phosphate shuttle, it is more energy efficient. The shuttle begins with the reduction of cytoplasmic oxaloacetate to malate by NADH. After its transport into the mitochondrial matrix, malate is reoxidized. The NADH produced is then oxidized by the ETC. For the shuttle to continue, oxaloacetate must be returned to the cytoplasm. Because the inner membrane is impermeable to oxaloacetate, it is converted to aspartate in a transamination reaction (Chapter 14) involving glutamate.

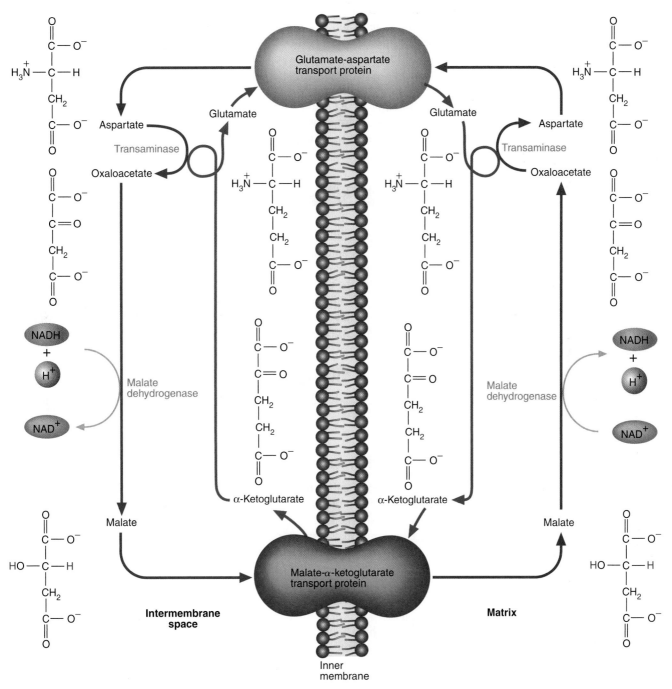

(b)

The aspartate is transported to the cytoplasm in exchange for glutamate (via the glutamate-aspartate transport protein), where it can be converted to oxaloacetate. The α-ketoglutarate is transported to the cytoplasm in exchange for malate (via the malate-α-ketoglutarate transport protein), where it can be converted to glutamate. The glutamate-aspartate transporter requires moving a proton into the matrix. Therefore, the net ATP synthesis using this mechanism is somewhat reduced. Instead of generating 2.5 molecules of ATP for each NADH molecule, the yield is approximately 2.25 molecules of ATP.

One final issue concerned with ATP synthesis from glucose remains. Recall that two molecules of ATP are produced in the citric acid cycle (from GTP). The price for their transport into the cytoplasm, where they will be used, is the uptake of two protons into the matrix. Therefore the total amount of ATP produced from a molecule of glucose is reduced by about half a molecule of ATP.

Depending on the shuttle used, the total number of molecules of ATP produced per molecule of glucose varies (approximately) from 29.5 to 31. Assuming that the average amount of ATP produced is 30 molecules, the net reaction for the complete oxidation of glucose is as follows:

$$C_6H_{12}O_6 + 6O_2 + 30\ ADP + 30\ P_i \rightarrow 6CO_2 + 6H_2O + 30\ ATP$$

The number of ATP molecules generated during the complete oxidation of glucose is in sharp contrast to the two molecules of ATP formed by glycolysis. Quite obviously, organisms that use oxygen to oxidize glucose have a substantial advantage.

KEY CONCEPT

The aerobic oxidation of glucose yields between 29.5 and 31 ATP molecules.

QUESTION 10.4

Traditionally, the oxidation of each NADH and FADH$_2$ by the ETC was believed to result in the synthesis of three molecules of ATP and two molecules of ATP, respectively. As noted, recent measurements, which have considered such factors as proton leakage across the inner membrane, have reduced these values somewhat. Use the earlier values to calculate the number of ATP molecules generated by the aerobic oxidation of a glucose molecule. First assume that the glycerol phosphate shuttle is operating; then assume that the malate-aspartate shuttle is transferring reducing equivalents into the mitochondrion.

QUESTION 10.5

Calculate the maximum number of ATP that can be generated from a mole of sucrose.

Uncoupled Electron Transport

Certain proteins, called **uncoupling proteins**, partially dissipate oxidative energy by translocating protons across the mitochondrial inner membrane without involving ATP synthesis. Uncoupling protein 1 (UCP1), the best-characterized example, is a dimer that forms a proton channel. UCP1, or *thermogenin*, is found exclusively in the mitochondria of brown fat, a specialized form of adipose tissue present in newborn babies (to help normalize body temperature) and hibernating mammals (to warm their bodies at the end of winter). (The characteristic color of brown adipose tissue results from the large number of mitochondria it contains.)

UCP1, which constitutes about 10% of the protein in the mitochondrial inner membrane, is activated when it is bound to fatty acids. During the reduction of the proton gradient by UCP1, the energy captured during electron transport is dissipated as heat. The entire process of heat generation from brown fat, called *non-shivering thermogenesis*, is regulated by norepinephrine (p. 534). (In shivering thermogenesis, heat is produced by nonvoluntary muscle contraction.) Norepinephrine, a neurotransmitter released from specialized neurons that terminate in brown adipose tissue, initiates a cascade mechanism that ultimately hydrolyzes fat molecules. The fatty acid products of fat hydrolysis activate the uncoupler protein. Fatty acid oxidation continues until the norepinephrine signal is terminated or the cell's fat reserves are depleted.

Two other uncoupling proteins have been identified. UCP2, found in most mammalian tissues, is believed to be linked to body weight regulation. The thermogenic effect of thyroid hormone (see later; Table 16.1) has been linked to UCP3, which has been detected in skeletal muscle as well as brown adipose tissue.

10.3 OXYGEN, CELL FUNCTION, AND OXIDATIVE STRESS

All living processes take place within a redox environment that may be defined as the sum of the reduction potential and the reducing capacity of linked redox pairs such as $NAD(P)H/NAD(P)^+$ and GSH/GSSG (reduced and oxidized forms, respectively, of glutathione, a key cellular reducing agent, see p. 362). Each cell's "redox state" is regulated within a narrow range because of the redox-sensitive nature of many metabolic and signaling pathways. These processes contain numerous proteins whose functional properties (activation or inactivation) are altered when the redox status of critical thiols (–SH groups) is altered. For example, the oxidation of sulfhydryl groups in proteins to form sulfenic (R-SOH), sulfinic ($R-SO_2H$), and sulfonic ($R-SO_3H$) acids can change the functional properties of these molecules. Limited redox changes do occur during certain normal cellular processes. For example, the cytoplasm of cells entering into cell division becomes more reduced, whereas that of differentiated cells becomes relatively more oxidized. Intracellular compartments also have distinctive redox conditions. The nucleus and mitochondria are more reduced than the cytoplasm (GSH/GSSG is high) and the ER is more oxidized (GSH/GSSG is low).

Redox regulation is of paramount importance because of the nature of molecular oxygen. Oxygen is used by the vast majority of living organisms to extract energy from organic molecules because of the large amounts of energy that can be generated, its ready availability, and its ease of distribution (especially in multicellular organisms). As previously mentioned (see p. 339), however, the advantages of using oxygen are linked to a dangerous property: oxygen can accept single electrons to form unstable derivatives, referred to as **reactive oxygen species (ROS)**. Examples of ROS include the superoxide radical, hydrogen peroxide, the hydroxyl radical, and singlet oxygen. Because ROS are so reactive, they can seriously damage living cells if formed in significant amounts. In living organisms, ROS formation is usually kept to a minimum by antioxidant defense mechanisms. **Antioxidants** are substances that react more easily with ROS than with critical biomolecules and, therefore, mitigate the tissue-damaging effects of these highly reactive metabolic by-products. Despite their potentially toxic properties, ROS function in small amounts as cell signaling devices by altering the redox status of target proteins such as metabolic enzymes, cytoskeletal components, cell cycle regulator proteins, transcription factors, and translation regulators. Their small size, easy diffusibility, and short half-lives allow ROS, in controlled amounts, to act as important integrators of cellular function.

Under certain conditions, referred to collectively as **oxidative stress**, antioxidant mechanisms are overwhelmed, ROS levels rise, and some damage may occur. Damage results primarily from enzyme inactivation, polysaccharide depolymerization, DNA breakage, and membrane destruction. Examples of circumstances that may cause serious oxidative damage include infection, inflammation, certain metabolic abnormalities, the overconsumption of certain drugs or exposure to intense radiation, and repeated contact with certain environmental contaminants (e.g., tobacco smoke).

In addition to contributing to the aging process, oxidative damage has been linked to at least 100 human diseases. Examples include cancer, cardiovascular disorders such as atherosclerosis, myocardial infarction, and hypertension, and neurological disorders such as amyotrophic lateral sclerosis (ALS, or Lou Gehrig's disease), Parkinson's disease, and Alzheimer's disease. Several types of cells are now known to deliberately produce large quantities of ROS. For example, in animal bodies scavenger cells such as macrophages and neutrophils continuously search for microorganisms and damaged cells. In an oxygen-consuming process referred to as the **respiratory burst**, ROS are generated and used to kill and dismantle these cells.

Reactive Oxygen Species

The properties of oxygen are of course directly related to its molecular structure. The diatomic oxygen molecule is a diradical. A **radical** is an atom or group of atoms that contains one or more unpaired electrons. Dioxygen is a diradical because it possesses two unpaired electrons. For this and other reasons, when it reacts, dioxygen can accept only one electron at a time.

Recall that during mitochondrial electron transport, H_2O is formed as a consequence of the sequential transfer of four electrons to O_2. During this process, several ROS are formed. Cytochrome oxidase (like other oxygen-activating proteins) traps these reactive intermediates within its active site until all four electrons have been transferred to oxygen. However, electrons may leak out of the electron transport pathway and react with O_2 to form ROS (Figure 10.18).

Under normal circumstances, cellular antioxidant defense mechanisms minimize any damage. ROS are also formed during nonenzymatic processes. For example, exposure to UV light and ionizing radiation causes ROS formation.

The first ROS formed during the reduction of oxygen is the superoxide radical $O_2^{\overline{\cdot}}$. Most superoxide radicals are produced by electrons derived from the Q cycle in complex III and by the flavoprotein NADH dehydrogenase (complex I). However, $O_2^{\overline{\cdot}}$ acts as a nucleophile and (under specific circumstances) as either an oxidizing agent or a reducing agent. Because of its solubility properties, $O_2^{\overline{\cdot}}$ causes considerable damage to the phospholipid components of membranes. When it is generated in an aqueous environment, $O_2^{\overline{\cdot}}$ reacts with itself to produce O_2 and hydrogen peroxide (H_2O_2):

$$2H^+ + 2O_2^{\overline{\cdot}} \longrightarrow O_2 + H_2O_2$$

Since H_2O_2 does not have any unpaired electrons, it is not a radical. The limited reactivity of H_2O_2 allows it to cross membranes and become widely dispersed. The subsequent reaction of H_2O_2 with Fe^{2+} (or other transition metals) results in the production of the hydroxyl radical (•OH), a highly reactive species:

$$Fe^{2+} + H_2O_2 \longrightarrow Fe^{3+} + \text{•OH} + OH^-$$

The hydroxyl radical diffuses only a short distance before it reacts with whatever biomolecule it collides with. Radicals such as the hydroxyl radical are especially dangerous because they can initiate an autocatalytic radical chain reaction (Figure 10.19). Singlet oxygen ($^1O_2$), an excited state of dioxygen in which the unpaired electrons have become paired, can form from superoxide:

$$2O_2^{\overline{\cdot}} + 2H^+ \longrightarrow H_2O_2 + {}^1O_2$$

or from peroxides:

$$2ROOH \longrightarrow 2ROH + {}^1O_2$$

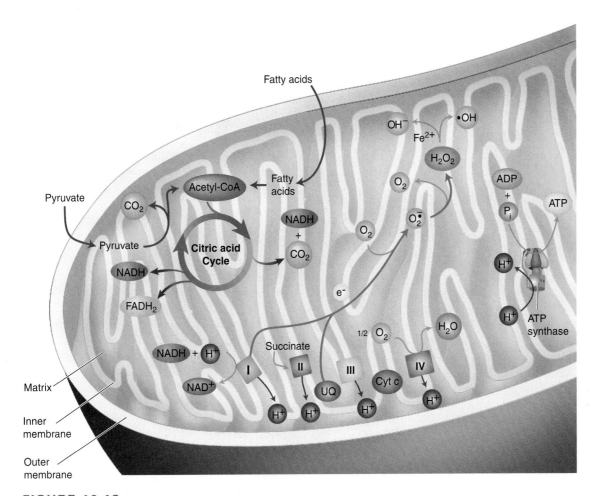

FIGURE 10.18

Overview of Oxidative Phosphorylation and ROS Formation in the Mitochondrion

Oxidative phosphorylation involves five multiprotein complexes: complexes I, II, III, and IV (the principal components of the ETC) and the ATP synthase. Pyruvate and fatty acids, the major fuel molecules, are transported into the mitochondrion, where they are oxidized by the citric acid cycle. The hydrogen atoms liberated during this process are carried by NADH and $FADH_2$ to the ETC. The energy that is released by the electron transport system is used to pump protons from the matrix through the inner membrane into the intermembrane space. The electrochemical gradient across the inner membrane created by proton pumping is used to synthesize ATP as protons flow through the ATP synthase. However, no system is perfect. Electrons leak from the ETC and react with O_2 to form superoxide ($O^{\bullet}$). In the presence of Fe^{2+}, superoxide is converted into the hydroxyl radical ($\bullet OH$). Superoxide is also converted to hydrogen peroxide.

Singlet oxygen formed from certain reactions of H_2O_2 and during photosynthetic light harvesting can react with double bonds in biomolecules. It is particularly damaging to aromatics and conjugated alkenes.

As mentioned (see p. 357), ROS are generated during several other cellular activities besides the reduction of O_2 to form H_2O. These include the biotransformation of xenobiotics and the respiratory burst (Figure 10.20) in white blood cells. In addition, electrons carried on ROS often leak from the electron transport pathways in the endoplasmic reticulum (e.g., the cytochrome P_{450} electron transport system) to form superoxide by combining with O_2. Reactions involving ROS include hydroxylation and peroxidation. Another such reaction, carbonylation, is a nonenzymatic protein modification resulting from the oxidation of amino acid side chains (i.e., Thr, Lys, Arg, or Pro) or the reaction of Cys, Lys, or His side chains with reactive carbonyl radicals.

There are also several nitrogen-containing radicals. Because their synthesis is often linked to that of ROS, **reactive nitrogen species** (RNS) are often classified

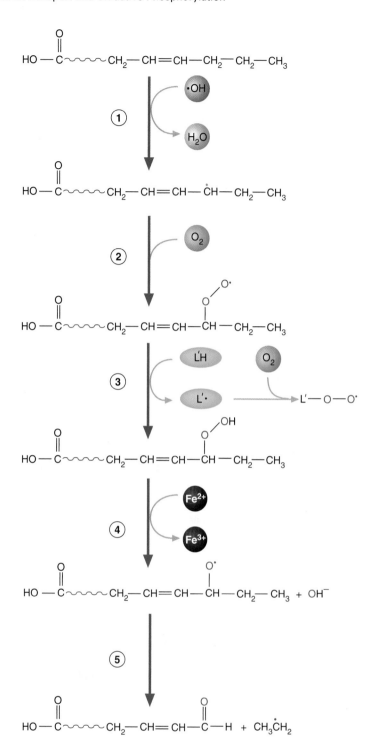

FIGURE 10.19

Radical Chain Reaction

Step 1: Lipid peroxidation reactions begin after the extraction of a hydrogen atom from an unsaturated fatty acid (LH → L•). Step 2: The lipid radical (L•) then reacts with O_2 to form a peroxyl radical (L• + O_2 → L—O—O•). Step 3: The radical chain reaction begins when the peroxyl radical extracts a hydrogen atom from another fatty acid molecule (L—O—O• + L′H → L—O—OH + L′•). Step 4: The presence of a transition metal such as Fe^{2+} initiates further radical formation (L—O—O—H + Fe^{2+} → LO• + HO^- + Fe^{3+}). Step 5: One of the most serious consequences of lipid peroxidation is the formation of α,β-unsaturated aldehydes, which involves a radical cleavage reaction. The chain reaction continues as the free radical product then reacts with a nearby molecule. Reactive carbonyl products are also products of this process.

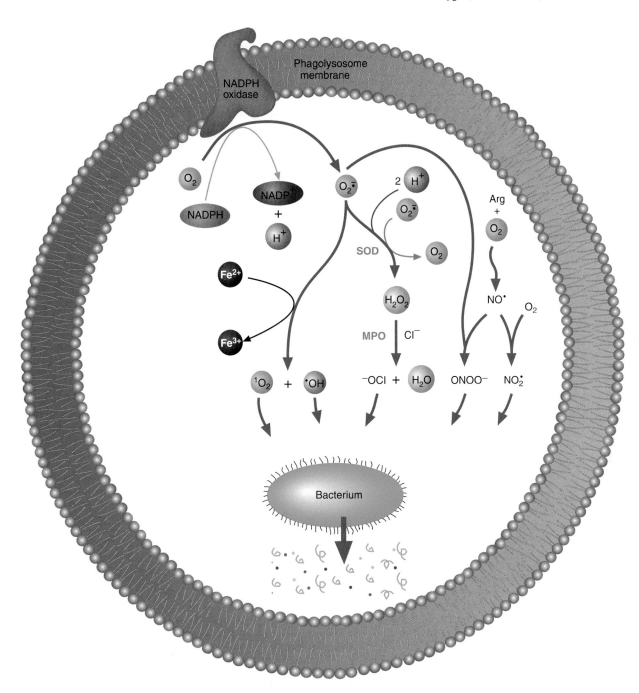

FIGURE 10.20

The Respiratory Burst

The respiratory burst provides a dramatic example of the destructiveness of ROS. Within seconds after a phagocytic cell binds to a bacterium (or other foreign structure), its oxygen consumption increases nearly 100-fold. During endocytosis the bacterium is incorporated into a large vesicle called a phagosome. Phagosomes then fuse with lysosomes to form phagolysosomes. Two destructive processes then ensue: the respiratory burst and digestion by lysosomal enzymes. The respiratory burst is initiated when NADPH oxidase converts O_2 to $O_2^{\bullet-}$. Two molecules of $O_2^{\bullet-}$ combine in a reaction catalyzed by SOD (superoxide dismutase) to form H_2O_2. H_2O_2 is next converted to several types of bactericidal (bacteria-killing) molecules by myeloperoxidase (MPO), an enzyme found in abundance in phagocytes. For example, MPO catalyzes the oxygenation of halide ions (e.g., Cl^-) to form hypohalides. Hypochlorite (the active ingredient in household bleach) is extremely bactericidal. In the presence of Fe^{2+}, $O_2^{\bullet-}$ and H_2O_2 react to form $\bullet OH$ and $^1O_2$ (singlet oxygen), both of which are extremely reactive. Nitric oxide, synthesized by nitric oxide synthase from arginine and O_2, reacts with superoxide to form peroxynitrite and with molecular oxygen to form nitrogen dioxide. In addition to the damage inflicted by various types of ROS, these species, along with MPO products, activate proteases, which degrade microbial proteins. Proteases are themselves protected from oxidative injury by MPO, which also possesses a catalase activity. After the disintegration of the bacterial cell, lysosomal enzymes digest the fragments that remain.

as ROS. Among the most important examples are nitric oxide ($^{\bullet}$NO), nitrogen dioxide ($^{\bullet}$NO$_2$), and peroxynitrite (ONOO$^-$). Nitric oxide ($^{\bullet}$NO) is a highly reactive gas. Because of its free radical structure $^{\bullet}$NO was regarded, until recently, primarily as a contributing factor in the destruction of the ozone layer in Earth's atmosphere and as a precursor of acid rain. Recent research has revealed, however, that $^{\bullet}$NO is an important signal molecule that is produced throughout the mammalian body. Especially high concentrations are found in the central nervous system. Physiological functions in which $^{\bullet}$NO is now believed to play a role include the regulation of blood pressure, the inhibition of blood clotting, and the destruction of foreign, damaged, or cancerous cells by macrophages. The disruption of the normally precise regulation of $^{\bullet}$NO synthesis has been linked to numerous pathological conditions that include stroke, migraine headache, male impotence, septic shock, and several neurodegenerative diseases such as Parkinson's disease. $^{\bullet}$NO can damage proteins with sulfhydryl groups, such as glyceraldehyde-3-phosphate dehydrogenase (p. 273), by converting SH groups into nitrosothiol (—SNO) derivatives. $^{\bullet}$NO also damages iron-sulfur proteins. Some of the damage attributed to $^{\bullet}$NO is, in fact, caused by its oxidation products, $^{\bullet}$NO$_2$ (2 $^{\bullet}$NO + O$_2$ → 2 $^{\bullet}$NO$_2$) and peroxynitrite ($^{\bullet}$NO + O$_2$ → ONOO$^-$).

Antioxidant Enzyme Systems

To protect themselves from oxidative stress, living organisms have developed several antioxidant defense mechanisms. These mechanisms employ several metalloenzymes and antioxidant molecules.

The major enzymatic defenses against oxidative stress are provided by four enzymes: superoxide dismutase, glutathione peroxidase, peroxiredoxin, and catalase. The wide distribution of these enzymatic activities underscores the ever-present problem of oxidative damage.

The superoxide dismutases (SOD) are a class of enzymes that catalyze the formation of H$_2$O$_2$ and O$_2$ from the superoxide radical:

$$2 O_2^{\bullet-} + 2H^+ \longrightarrow H_2O_2 + O_2$$

There are two major forms of SOD. In humans the Cu-Zn isoenzyme occurs in cytoplasm. A manganese-containing isozyme is found in the mitochondrial matrix. Lou Gehrig's disease, a fatal degenerative condition in which motor neurons are destroyed, is now known to be caused by a mutation in the gene that codes for the cytosolic Cu-Zn isozyme of SOD.

Glutathione peroxidase, a selenium-containing enzyme, is a key component in an enzymatic system most responsible for controlling cellular peroxide levels. Recall that this enzyme catalyzes the reduction of a variety of substances by the reducing agent GSH (Table 5.3). In addition to reducing H$_2$O$_2$ to form water, glutathione peroxidase transforms organic peroxides into alcohols:

$$2 \text{ GSH} + \text{R—O—O—H} \longrightarrow \text{G—S—S—G} + \text{R—OH} + \text{H}_2\text{O}$$

Several ancillary enzymes support glutathione peroxidase function (Figure 10.21). GSH is regenerated from GSSG by glutathione reductase:

$$\text{G—S—S—G} + \text{NADPH} + \text{H}^+ \longrightarrow 2 \text{ GSH} + \text{NADP}^+$$

The NADPH required in the reaction is provided primarily by several reactions of the pentose phosphate pathway (Chapter 8). Recall that NADPH is also produced by the reactions catalyzed by isocitrate dehydrogenase (p. 323) and malic enzyme (p. 330).

The *peroxiredoxins* (PRX) are a class of enzymes that detoxify peroxides. Their catalytic mechanism involves the oxidation of a redox-active cysteine side chain sulfhydryl group by the peroxide substrate to form a sulfinate. The sulfinate is subsequently reduced by a thiol-containing protein such as thioredoxin.

KEY CONCEPTS

- Reactive oxygen species form because oxygen is reduced by accepting one electron at a time.
- ROS formation is a normal by-product of metabolism and the result of conditions such as exposure to radiation.
- Reactive nitrogen species are often classified as ROS because the synthesis of these species is often linked.

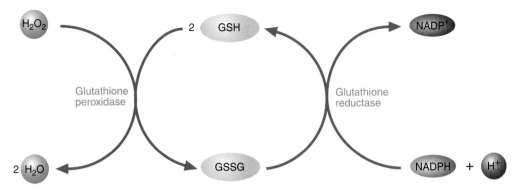

FIGURE 10.21

The Glutathione-Centered System

Glutathione peroxidase utilizes GSH to reduce the peroxides generated by cellular aerobic metabolism. GSH is regenerated from its oxidized form, GSSG, by glutathione reductase. NADPH, the reducing agent in this reaction, is supplied by the pentose phosphate pathway and several other reactions.

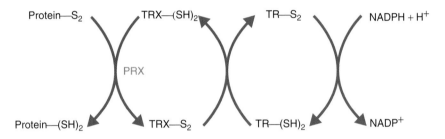

FIGURE 10.22

The Thioredoxin-Centered System

A substrate protein that contains function-altering disulfides is reduced by electron transfer from thioredoxin [TRX-(SH)$_2$] in a reaction mediated by the peroxiredoxin enzyme (PRX). The oxidized thioredoxin (TRX-S$_2$) is returned to its reduced form by thioredoxin reductase (TR). The electrons required to restore the TR to its reduced form come from NADPH via an enzyme-bound FAD.

Thioredoxin (TRX) is involved in redox reactions mediated by the peroxiredoxin/thioredoxin reductase (TR) system (sometimes referred to as the TRX-centered system) (Figure 10.22). TR and TRX also return many oxidized cellular proteins, including many transcription factors, to their functional reduced sulfhydryl form; this change is accomplished by means of an enzyme-catalyzed shift of electrons from reduced thioredoxin [TRX-(SH)$_2$] to the target protein. Thioredoxin reductase reduces oxidized thioredoxin [TRX(S$_2$)] with electrons delivered to it through a mobile NADPH and a bound FADH$_2$. TRX also serves as an electron shuttle for other nonantioxidant enzyme systems such as that of ribonucleotide reductase (p. 550).

Catalase is an enzyme whose primary role is to catalyze the conversion of H$_2$O$_2$ to water and dioxygen. (For every two molecules of H$_2$O$_2$ that are detoxified, one is oxidized to form O$_2$ and the other is reduced to H$_2$O.) Catalase has a heme-Fe(III) prosthetic group with the following mechanism:

$$\text{Step 1: } H_2O_2 + Fe(III)\text{—enzyme} \longrightarrow H_2O + O{=}Fe(IV)\text{—enzyme}$$

$$\text{Step 2: } H_2O_2 + O{=}Fe(IV)\text{—enzyme} \longrightarrow O_2 + H_2O + Fe(III)\text{—enzyme}$$

Two isozymes of catalase have been identified: HPI and HPII. HPI is a bifunctional enzyme in that the second H$_2$O$_2$ can be replaced by an oxygen-containing organic group (phenols, aldehydes, acids, and alcohols). This peroxidative reaction converts the organic molecule to an often less toxic form:

$$R—CH_2—OH + O=Fe(IV)—enzyme \longrightarrow R—CH_3 + O_2 + Fe(III)—enzyme$$

The respiratory burst of macrophages, for example, generates H_2O_2 as a microbicide primarily through the incomplete oxidation of fatty acids. Unlike HPI, HPII has only one substrate, H_2O_2. It is found in high quantity in erythrocytes and peroxisomes of phagocytic white blood cells. Outside the peroxisome, the primary H_2O_2 generator is SOD.

QUESTION 10.6

Selenium is generally considered to be a toxic element. (It is the active component of loco weed.) However, there is growing evidence that selenium is also an essential trace element. Because glutathione peroxidase activity is essential to protect red blood cells against oxidative stress, a selenium deficiency can damage red blood cells. Although sulfur is in the same chemical family as selenium, it cannot be substituted. Can you explain why? [*Hint*: Selenium is more easily oxidized than sulfur.] Is sulfur or selenium a better scavenger for oxygen when this gas is present in trace amounts?

QUESTION 10.7

Ionizing radiation is believed to damage living tissue by producing hydroxyl radicals. Drugs that protect organisms from radiation damage, which must be taken *before* radiation exposure usually have —SH groups. How do such drugs protect against radiation damage? Can you suggest any type of nonsulfhydryl group–containing molecule that would protect against hydroxyl radical–induced damage?

QUESTION 10.8

In some regions where malaria is endemic (e.g., the Middle East), fava beans are a staple food. Fava beans are now known to contain two β-glycosides called vicine and convicine:

Vicine   Convicine

It is believed that the aglycone components of these substances, called divicine and isouramil, respectively, can oxidize GSH. Individuals who eat fresh fava beans are protected to a certain extent from malaria. A condition known as *favism* results when some glucose-6-phosphate dehydrogenase–deficient individuals develop a severe hemolytic anemia after eating the beans. Explain why.

Antioxidant Molecules

Living organisms use antioxidant molecules to protect themselves from radicals. Some prominent antioxidants include GSH, α-tocopherol (vitamin E), ascorbic acid (vitamin C), and β-carotene (Figure 10.23).

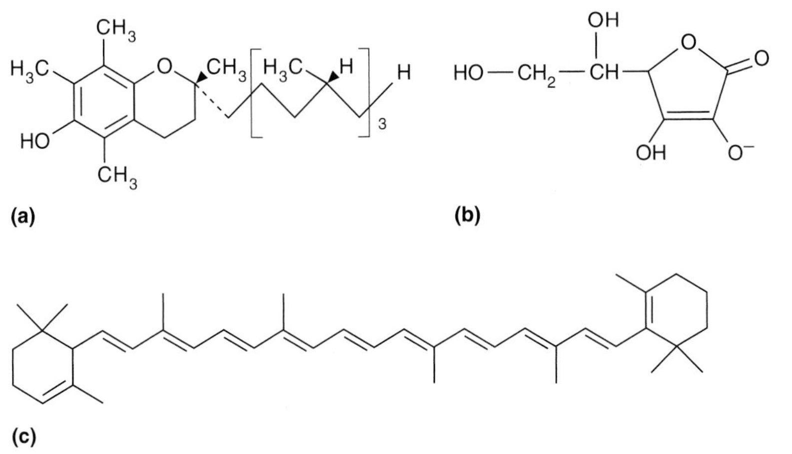

FIGURE 10.23
Selected Antioxidant Molecules
(a) α-Tocopherol (vitamin E). (b) ascorbate (vitamin C). (c) β-carotene.

α-Tocopherol, a potent radical scavenger, belongs to a class of compounds referred to as *phenolic antioxidants*. Phenols are effective antioxidants because the radical products of these molecules are resonance stabilized and thus relatively stable:

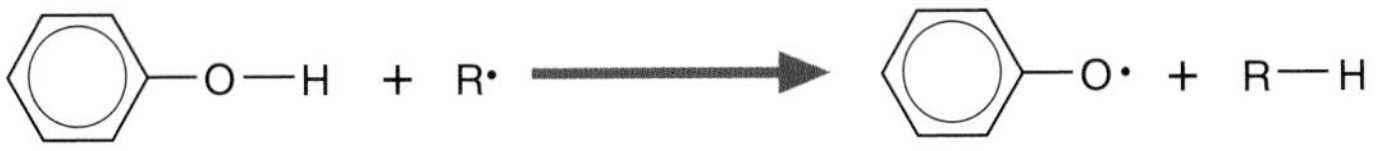

Phenol

Resonance - stabilized phenolic radical

Because vitamin E (found in vegetable and seed oils, whole grains, and green, leafy vegetables) is lipid soluble, it plays an important role in protecting membranes from lipid peroxyl radicals.

β-Carotene, found in yellow-orange and dark green fruits and vegetables such as carrots, sweet potatoes, broccoli, and apricots, is a member of a class of plant pigment molecules referred to as the *carotenoids*. In plant tissue the carotenoids absorb some of the light energy used to drive photosynthesis and protect against the ROS that form at high light intensities. In animals, β-carotene is a precursor of retinol (vitamin A) and an important antioxidant in membranes. (Retinol is a precursor of retinal, the light-absorbing pigment found in the rod cells of the retina.)

Ascorbic acid has been shown to be an efficient antioxidant. Present largely as ascorbate, this water-soluble molecule scavenges a variety of ROS within the aqueous compartments of cells and in extracellular fluids. Ascorbate is reversibly oxidized as shown:

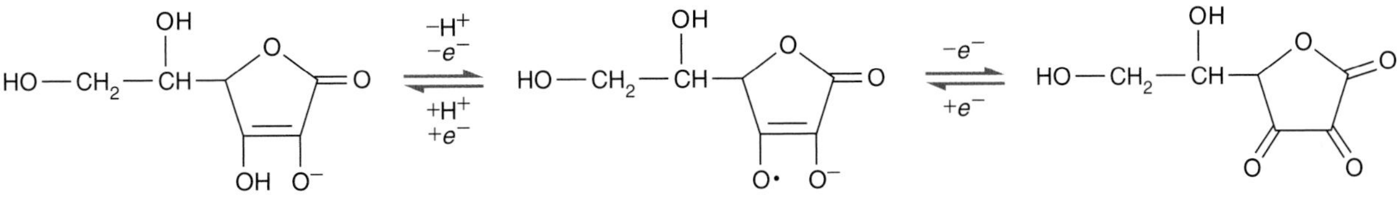

L- Ascorbate

L- Ascorbyl radical

Dehydro-L-ascorbic acid

Ascorbate protects membranes through two mechanisms. First, ascorbate prevents lipid peroxidation by reacting with peroxyl radicals formed in the cytoplasm before they can reach the membrane. Second, ascorbate enhances the antioxidant activity of vitamin E by regenerating reduced α-tocopherol from the α-tocopheroxyl radical (Figure 10.24). Ascorbate is then regenerated by reacting with GSH.

It is noteworthy that in well-nourished individuals, the consumption of excessive amounts of antioxidant supplements renders the body's cells more vulnerable to oxidative stress. In small quantities, ROS act as signal molecules. When

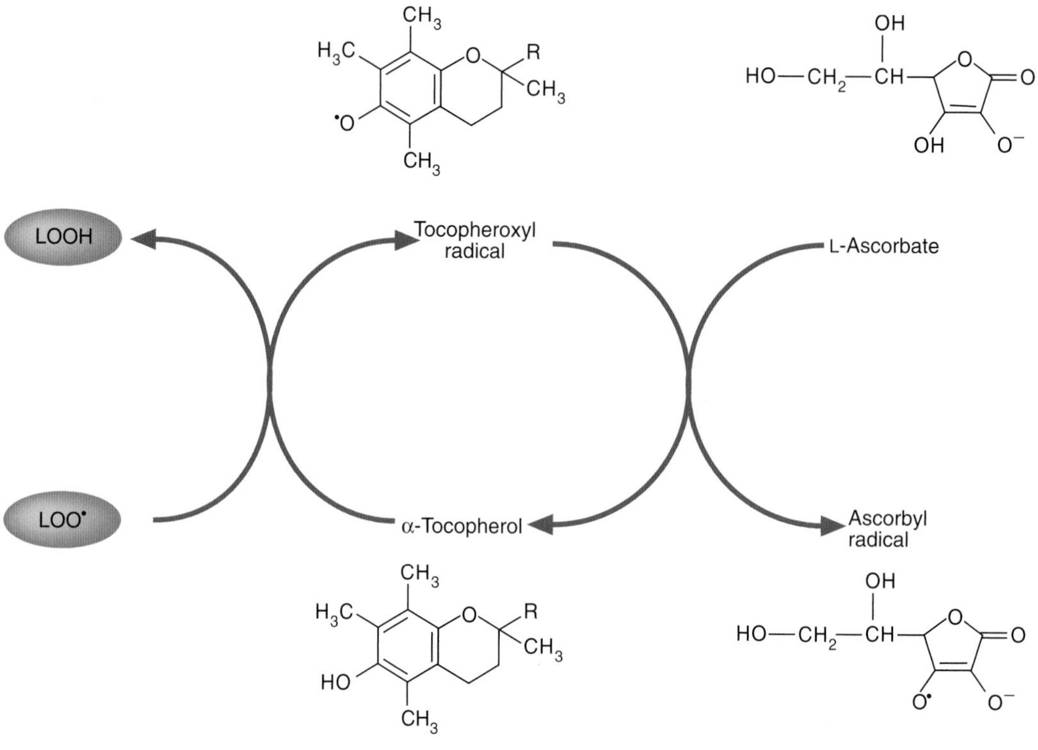

FIGURE 10.24

Regeneration of α-Tocopherol by L-Ascorbate

L-Ascorbate, a water-soluble molecule, protects membranes from oxidative damage by regenerating α-tocopherol from α-tocopheroxyl radical. The ascorbyl radical formed in this process is reconverted to L-ascorbate during a reaction with GSH.

KEY CONCEPTS

- Antioxidant molecules protect cell components from oxidative damage.
- Prominent antioxidants include GSH and the dietary components α-tocopherol, β-carotene, and ascorbic acid.

cells are experiencing oxidative stress (e.g., infection or inflammation), ROS levels begin to rise. Early in this process, ROS oxidize and/or covalently modify the sulfur groups of transcription factors, thereby triggering the expression of dozens of genes that strengthen the antioxidant defenses of the cell. In addition to increased concentrations of catalase, SOD, and other antioxidant enzymes, other stress proteins are produced. If cells contain excessive amounts of antioxidant molecules, obtained from dietary supplements, ROS-triggered defense mechanisms are compromised.

QUESTION 10.9

The antioxidant BHT (butylated hydroxytoluene) is widely used as a food preservative. Quercitin is a member of a large group of potent antioxidants found in fruits and vegetables called the flavonoids.

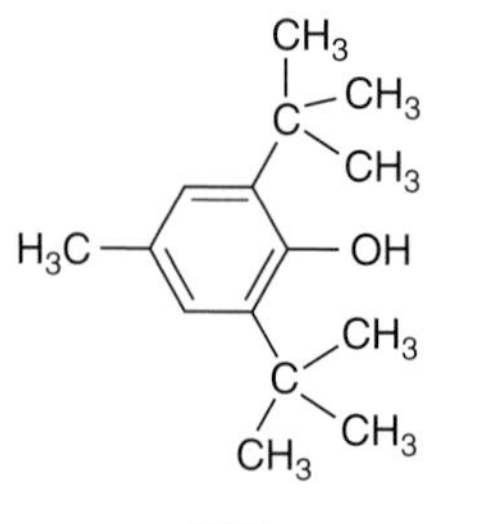

BHT

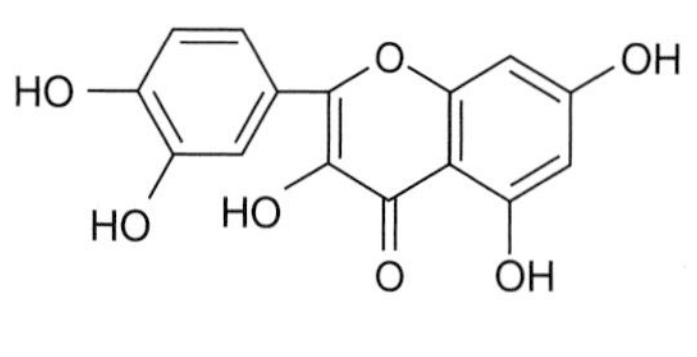

Quercitin

What structural characteristic of these molecules is responsible for their antioxidant properties?

BIOCHEMISTRY IN PERSPECTIVE

Ischemia and Reperfusion

How are heart and brain cells damaged by the inadequate nutrient and oxygen flow caused by blood clots, and why does the reintroduction of O_2 cause further damage?   The tissue damage that occurs during a myocardial infarction (heart attack) or a cerebrovascular accident (stroke) is caused by *ischemia*, a condition in which there is inadequate blood flow. Heart attacks and strokes are usually caused by atherosclerosis accompanied by blood clot formation in an essential artery. In atherosclerosis, soft masses of fatty material called *plaques* are formed in the linings of blood vessels. Unlike skeletal muscle, which is fairly resistant to ischemic injury, heart and brain are extremely sensitive to hypoxic (low-oxygen) conditions. For example, significant brain damage occurs if the brain is deprived of oxygen for more than a few minutes. The stimulation of anaerobic glycolysis, which leads to lactate production and acidosis, is an early response of cells to ischemia. Because energy production by glycolysis is inefficient, ATP levels begin to fall. As they do so, adenine nucleotides are degraded to form hypoxanthine (Chapter 15).

Without sufficient ATP, cells cannot maintain appropriate intracellular ion concentrations. For example, cytoplasmic calcium levels rise. One of the consequences of this circumstance is the activation of calcium-dependent enzymes such as proteases and phospholipases (enzymes that degrade the phospholipids in membranes). As osmotic pressure increases, affected cells swell and leak their contents. (Recall from Chapter 6 that the leakage of specific enzymes into blood is used to diagnose heart and liver damage, as detailed in the Biochemistry in Perspective box entitled Enzymes and Clinical Medicine.) The blood supply is further compromised as neutrophils, attracted to the damaged site via chemotaxis, clog the blood vessels. Eventually, lysosomal enzymes begin to leak from lysosomes. Because lysosomal enzymes are active only at low pH values, their presence in an increasingly acidic cytoplasm eventually results in the hydrolysis of cell components. ER stress is another important feature of hypoxic tissue. Under normal circumstances, the ER is an oxidizing environment that promotes protein folding and disulfide bond formation. Under hypoxic conditions protein folding is compromised and the unfolded protein response (p. 49) is triggered. If an oxygen supply is not soon reestablished, affected cells are irreversibly damaged.

The reoxygenation of an ischemic tissue, a process referred to as *reperfusion*, can be a life-saving therapy. For example, the use of streptokinase to digest artery-occluding clots in heart attack patients, accompanied by administration of oxygen, has saved many lives. However, depending on the duration of the hypoxic episode, the reintroduction of oxygen to ischemic tissue may also result in further damage.

Recent research with antioxidants reveals that ROS are largely responsible for reperfusion-initiated cell damage. The exact mechanism by which reperfusion causes ROS production is still unclear. However, there are several likely possibilities. For example, the leakage of electrons from swollen mitochondria may result in ROS formation. In addition, the release of iron from cell components such as myoglobin, which can result from ROS-inflicted damage, can cause additional production of •OH. Other reperfusion-caused damage is the result of the conversion of hypoxanthine to uric acid (Chapter 15), which involves the formation of O_2^- and •OH, and ROS synthesis in neutrophils. Finally, the acidosis caused by lactate accumulation in compromised heart muscle cells unloads abnormally high amounts of oxygen from hemoglobin. This latter condition greatly facilitates further ROS synthesis. Reperfusion also promotes the synthesis of •NO, which modifies oxidizable cysteine residues in the Ca^{2+} channels of the endoplasmic reticulum, another factor contributing to ER stress. Currently, ROS-quenching antioxidant molecules are being investigated for clinical use.

SUMMARY: Damage to heart and brain cells due to oxygen deprivation originates with inefficient energy production, followed by osmotic pressure increases, lysosomal breakage, and ER stress. In the absence of preventive measures, the reperfusion of damaged cells with O_2 leads to ROS formation, causing further damage.

Chapter**Summary**

1. Dioxygen (O_2), generally referred to as oxygen, is used by aerobic organisms as a terminal electron acceptor in energy generation. Several physical and chemical properties of oxygen make it suitable for this role. In addition to its ready availability (it occurs almost everywhere on the earth's surface), oxygen diffuses easily across cell membranes. Moreover, it is highly reactive, hence readily accepts electrons, and it is an oxidizing agent.

2. The NADH and $FADH_2$ molecules produced in glycolysis, the β-oxidation pathway, and the citric acid cycle generate usable energy in the electron transport pathway. The pathway consists of a series of redox carriers that receive electrons from NADH and $FADH_2$. At the end of the pathway the electrons, along with protons, are donated to oxygen to form H_2O.

3. During the oxidation of NADH, there are three steps in which the energy loss is sufficient to account for ATP synthesis. These steps occur within complexes I, III, and IV of the ETC.

4. Oxidative phosphorylation is the mechanism by which electron transport is coupled to the synthesis of ATP. According to the chemiosmotic theory, the creation of a proton gradient that accompanies electron transport is coupled to ATP synthesis.

5. The complete oxidation of a molecule of glucose results in the synthesis of 29.5 to 31 molecules of ATP, depending on whether the glycerol phosphate shuttle or the malate-aspartate shuttle transfers electrons from cytoplasmic NADH to the mitochondrial ETC.

6. The use of oxygen by aerobic organisms is linked to the production of ROS. These species form because the diradical oxygen molecule accepts electrons one at a time. Examples of ROS include the superoxide radical, hydrogen peroxide, the hydroxyl radical, and singlet oxygen. Prominent RNS include nitric oxide, nitrogen dioxide, and peroxynitrite.

Take your learning further by visiting the **companion website** for Biochemistry at www.oup.com/us/mckee where you can complete a multiple-choice quiz on electron transport and oxidative phosphorylation to help you prepare for exams.

Suggested**Readings**

Buetler, T. M., Krauskopf, A., and Ruegg, U. T., Role of Superoxide as a Signaling Molecule, *News Physiol. Sci.* 19:120–123, 2004.

Burch, P. M., and Heintz, N. H., Redox Regulation of Cell-Cycle Reentry: Cyclin D1 as a Primary Target for the Mitogenic Effects of Reactive Oxygen and Nitrogen Species, *Antioxidants Redox Signaling* 7(5–6):741–751, 2005.

Cooke, M. S., and Evans, M. D., Reactive Oxygen Species: From DNA Damage to Disease, *Sci. Med.*, 10(2):98–111, 2005.

Das, J., The Role of Mitochondrial Respiration in Physiological and Evolutionary Adaptation, *BioEssays* 28(9):890–901, 2006.

de Magalhaes, J. P., and Church, G. M., Cells Discover Fire: Employing Reactive Oxygen Species in Development and Consequences for Aging, *Exp. Gerontol.* 41:1–10, 2006.

Krauss, S., Zhang, C.-Y., and Lowell, B. B., The Mitochondrial Uncoupling–Protein Homologues, *Nat. Rev. Mol. Cell Biol.* 6:248–261, 2005.

Menon, S. G., and Goswami, P. C., A Redox Cycle Within the Cell Cycle: Ring in the Old with the New. *Oncogene* 26(8):1101–1109, 2007.

Nathan, C., Specificity of a Third Kind: Reactive Oxygen and Nitrogen Intermediates in Cell Signaling, *J. Clin. Invest.* 111(6):769–778, 2003.

Nicholls, D. G., and Ferguson, S. J., *Bioenergetics 3*, Academic Press, London, 2002.

Rothstein, E. C., and Lucchesi, P. A., Redox Control of the Cell Cycle: A Radical Encounter, *Antioxidants Radical Signaling* 7(5–6):701–703, 2005.

Tielens, A. G. M., Rotte, C. van Hellemond, J. J., and Martin, W., Mitochondria as We Don't Know Them, *Trends Biochem. Sci.* 27(11):564–572, 2002.

Valko, M., et al., Free Radicals and Antioxidants in Normal Physiological Functions and Human Disease, *Int. J. Biochem. Cell Biol.* doi:10.1016/j.biocel.2006.07.001

Winyard, P. G., Moody, C. J., and Jacob, C., Oxidative Activation of Antioxidant Defense, *Trends Biochem. Sci.* 30(8):453–461, 2005.

Key**Words**

aerobic respiration, *340*

antioxidant, *357*

β-carotene, *365*

chemiosmotic coupling theory, *347*

glycerol phosphate shuttle, *353*

ionophore, *348*

malate-aspartate shuttle, *365*

oxidative phosphorylation, *347*

oxidative stress, *357*

protonmotive force, *347*

Q cycle, *342*

radical, *358*

reactive nitrogen species, *359*

reactive oxygen species (ROS), *357*

respiratory burst, *358*

respiratory control, *352*

α-tocopherol, *365*

uncoupler, *348*

uncoupling protein, *356*

Review**Questions**

These questions are designed to test your knowledge of the key concepts discussed in this chapter, before moving on to the next chapter. You may like to compare your answers to the solutions provided in the back of the book and in the accompanying Study Guide.

1. Define the following terms:
   a. chemical coupling hypothesis
   b. chemiosmotic coupling theory
   c. ionophore
   d. respiratory control
   e. ischemia
2. What are the principal sources of electrons for the electron transport pathway?
3. Describe the processes that are believed to be driven by mitochondrial electron transport.
4. Describe the principal features of the chemiosmotic theory.
5. The chemical coupling hypothesis failed to explain why mitochondrial membrane must be intact during ATP synthesis. How does the chemiosmotic theory account for this phenomenon?
6. How does dinitrophenol inhibit ATP synthesis?
7. Four protons are required to drive the phosphorylation of ADP. Account for the function of each proton in this process.
8. List several reasons why oxygen is widely used in energy production.
9. Define the following terms:
   a. thioredoxin
   b. catalase
   c. β-carotene
   d. α-tocopherol
   e. Q cycle
10. Define the following terms:
    a. rotor
    b. stator
    c. uncoupler
    d. malate-aspartate shuttle
    e. ROS
11. Which of the following are reactive oxygen species? Why is each ROS dangerous?
    a. O_2
    b. OH^-
    c. $RO\bullet$
    d. O_2^-
    e. CH_3OH
    f. $^1O_2$
12. Define the following terms:
    a. RNS
    b. glutathione
    c. submitochondrial particle

d. dinitrophenol
   e. torque
13. Describe the types of cellular damage produced by ROS.
14. Define the following terms:
    a. radical
    b. protonmotive force
    c. antioxidant
    d. aerobic respiration
    e. respiratory burst
15. Describe the enzymatic activities used by cells to protect themselves from oxidative damage.
16. Explain how a defect in the gene for glucose-6-phosphate dehydrogenase can provide a survival advantage. [*Hint*: Visit the companion website at www.oup.com/us/mckee to read the Biochemistry in Perspective box for Chapter 10 entitled Glucose-6-Phosphate Dehydrogenase Deficiency.]
17. What would be the end products when the following substances are final electron acceptors in an electron transport system: nitrate, ferric ion, carbon dioxide, sulfate, and sulfur?
18. What advantage does dioxygen have over the oxidizing agents in Question 17?
19. Describe the effect of dinitrophenol on flagellar movement in respiring bacteria.
20. Provide the reaction equations that illustrate the production of ROS from electrons leaking from the electron transport system.
21. Valinomycin is an ionophore antibiotic that renders biological membranes permeable to K^+. Its side effects in patients with bacterial infections include a rise in body temperature and sweating. Explain.
22. Describe the mechanism whereby uncoupling agents disrupt phosphorylation.
23. The electron transport system consists of a series of oxidations rather than one reaction. Why is this an important feature of energy capture?
24. What metabolites accumulate when azide is added to actively respiring mitochondria?
25. What metabolites accumulate when cyanide is added to actively respiring mitochondria?
26. What is the minimum voltage drop for individual electron transfer events in the mitochondrial electron transport systems that is necessary for ATP synthesis?
27. When rotenone is added to actively respiring mitochondria, the ratio of NADH to NAD^+ increases, but the $FADH_2/FAD$ ratio remains unchanged. What step in the system is being inhibited?

ThoughtQuestions

These questions are designed to reinforce your understanding of all of the key concepts discussed in the book so far, including this chapter and all of the chapters before it. They may not have one right answer! The authors have provided possible solutions to these questions in the back of the book and in the accompanying Study Guide, for your reference.

28. During an experiment, $^{14}CH_3$—COOH is fed to microorganisms. Trace the ^{14}C label through the citric acid cycle. How many ATP molecules can be generated from 1 mol of this substance? (The conversion of acetate to acetyl-CoA requires the consumption of 2 ATP.)

29. Ethanol is oxidized in the liver to form acetate, which is converted to acetyl-CoA. Determine how many molecules of ATP are produced from 1 mol of ethanol. Note that 2 mol of NADH are produced when ethanol is oxidized to form acetate.

30. Glutamine is degraded to form NH_4^+, CO_2, and H_2O. How many molecules of ATP can be generated from 1 mol of this amino acid?

31. Consumption of dinitrophenol by animals results in an immediate increase in body temperature. Explain the phenomenon. Why should this decoupler not be used as a diet aid?

32. According to the chemiosmotic theory, what would be the effect on oxidative phosphorylation of allowing other positive ions to diffuse across the inner mitochondrial membrane?

33. Cyanide causes an irreversible inhibition of electron transport that prevents ATP synthesis, whereas the inhibitory effect of small amounts of dinitrophenol on ATP synthesis is reversible. Explain the difference.

34. The reduction potentials of the iron in each of the cytochromes in the electron transport system vary from –0.1 V to –0.39 V. Explain why these different values are necessary for the operation of the system.

35. Explain why an inhibitor of complex I will not only cause an increase in the ratio of NADH to NAD but also an increase in the UQ/UQH_2 ratio.

36. Cyanide binds effectively to ferrous ion. While there are several iron-containing complexes in the electron transport system, only the final step is inhibited. Explain.

37. Suppose that the cytochrome complexes were not embedded within the mitochondrial inner membrane. According to the chemiosmotic theory, what would be the consequences?

38. Explain why rotenone inhibits oxidative phosphorylation when the substrate is pyruvate, but not when succinate is used.

39. If nitrate is used as the terminal electron acceptor in the electron transport system, how many ATPs could be synthesized? [*Hint*: Refer to the reduction potential differences between NADH and nitrate (Table 9.1).]

40. How would you determine the difference between the effects of uncouplers and electron transport inhibitors on ATP synthesis?

41. Dehydroascorbate is unstable at pH values greater than 6 and decomposes to form tartrate and oxalate. Cells use GSH to reduce the loss of ascorbate. What is the reaction pathway for the regeneration of ascorbate?

42. ATP synthesis within mitochondria is regulated by several mechanisms. According to a recent hypothesis, aconitase and superoxide contribute to redox regulation of energy metabolism. A rise in superoxide levels inactivates aconitase by converting the enzyme's iron center of 4FE-4S to 3Fe-4S. Subsequently, when superoxide levels are lower, the active iron center is reassembled. Describe the immediate metabolic effects of aconitase inactivation.

43. Referring to Question 42, describe the general mechanism whereby aconitase is reactivated.

44. Mice in which the MnSOD gene has been inactivated die prematurely. Among their symptoms are massive accumulations of lipid in liver and skeletal muscle. Explain. [*Hint*: Refer to Question 42.]

45. Among the many destructive consequences of oxidative stress are reactions of •OH with polypeptide backbone atoms. The process begins with the abstraction of α-hydrogen atoms to form carbon radicals.

$$-NH-\underset{\bullet}{\overset{\overset{\displaystyle R}{|}}{C}}-\overset{\overset{\displaystyle O}{||}}{C}-$$

Subsequently, such radicals react with O_2 to form alkylperoxyl radicals (ROO•). Starting with the undamaged backbone atoms and O_2, describe the pathway that results in the formation of an alkylperoxyl radical.

46. Polypeptides under oxidative attack can form intra- and intermolecular cross-links. Referring to Question 45, suggest one mechanism whereby a cross-link could form.

Lipids and Membranes

Biological Membrane
A biological membrane is a dynamic compartmental barrier composed of a lipid bilayer noncovalently complexed with proteins, glycoproteins, glycolipids, and cholesterol.

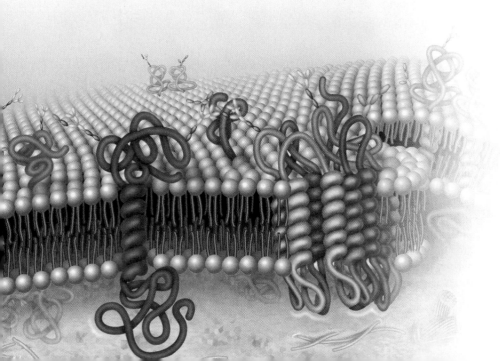

Overview

LIPIDS ARE NATURALLY OCCURRING SUBSTANCES THAT DISSOLVE IN HYDROCARBONS AND ORGANIC SOLVENTS, BUT NOT IN WATER. THEY perform a stunning array of functions in living organisms. Some lipids are vital energy reserves. Others are the primary structural components of biological membranes. Still other lipid molecules act as hormones, antioxidants, pigments, or vital growth factors and vitamins. This chapter describes the structures and properties of the major lipid classes found in living organisms as well as the structural and functional properties of biomembranes.

Lipids are a diverse group of biomolecules. Molecules such as fats and oils, phospholipids, steroids, and the carotenoids, which differ widely in both structure and function, are all considered lipids. Because of this diversity, the term **lipid** has an operational rather than a structural definition. Lipids are defined as substances from living organisms that dissolve in nonpolar solvents such as ether, chloroform, and acetone but not appreciably in water. The functions of lipids are also diverse. Several types of lipid molecules, (e.g., phospholipids and sphingolipids) are important structural components in cell membranes. Another type, the fats and oils (both of which are triacylglycerols), store energy efficiently. Other lipid molecules are chemical signals, vitamins, or pigments. Finally, some lipid molecules that occur in the outer surfaces of various organisms have protective or waterproofing functions.

Chapter 11 describes the structure and function of each major type of lipid and discusses the lipoproteins, complexes of protein and lipid that transport lipids in animals. Chapter 11 ends with an overview of membrane structure and function. In Chapter 12 the metabolism of several major lipids is described.

11.1 LIPID CLASSES

Lipids may be classified in many different ways. For this discussion, lipids can be subdivided into the following classes:

1. Fatty acids
2. Triacylglycerols
3. Wax esters
4. Phospholipids (phosphoglycerides and sphingomyelin)
5. Sphingolipids (molecules other than sphingomyelin that contain the amino alcohol sphingosine)
6. Isoprenoids (molecules made up of repeating isoprene units, a branched five-carbon hydrocarbon)

Each class is discussed.

Fatty Acids

Fatty acids are monocarboxylic acids that typically contain hydrocarbon chains of variable lengths (between 12 and 20 or more carbons) (Figure 11.1). Fatty acids are numbered from the carboxylate end. Greek letters are used to designate certain carbon atoms. The α-carbon in a fatty acid is adjacent to the carboxylate group, the β-carbon is two atoms removed from the carboxylate group, and so forth.

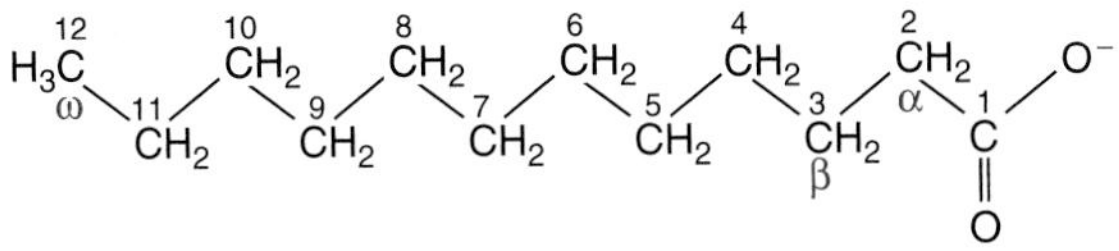

FIGURE 11.1

Fatty Acid Structure

Fatty acids consist of a long-chain hydrocarbon covalently bonded to a carboxylate group. The lipid shown is dodecanoic acid (common name, lauric acid), a 12-carbon saturated fatty acid (12:0).

TABLE 11.1 Examples of Fatty Acids

Common Name	Structure	Abbreviation
Saturated Fatty Acids		
Myristic acid	$CH_3(CH_2)_{12}COOH$	14:0
Palmitic acid	$CH_3(CH_2)_{12}CH_2CH_2COOH$	16:0
Stearic acid	$CH_3(CH_2)_{12}CH_2CH_2CH_2CH_2COOH$	18:0
Arachidic acid	$CH_3(CH_2)_{12}CH_2CH_2CH_2CH_2CH_2CH_2COOH$	20:0
Lignoceric acid	$CH_3(CH_2)_{12}CH_2CH_2CH_2CH_2CH_2CH_2CH_2CH_2CH_2CH_2COOH$	24:0
Cerotic acid	$CH_3(CH_2)_{12}CH_2CH_2CH_2CH_2CH_2CH_2CH_2CH_2CH_2CH_2CH_2CH_2COOH$	26:0
Unsaturated Fatty Acids		
Palmitoleic acid	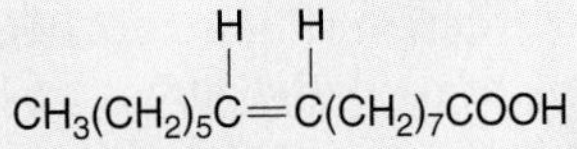	$16:1^{\Delta 9}$
Oleic acid	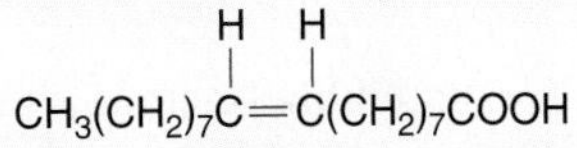	$18:1^{\Delta 9}$
Linoleic acid	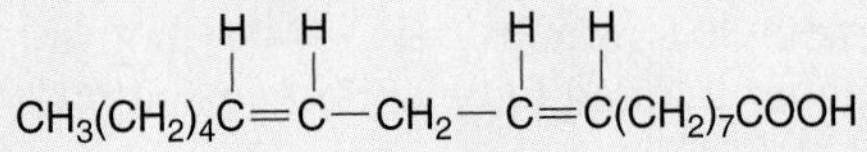	$18:2^{\Delta 9,12}$
α-Linolenic acid	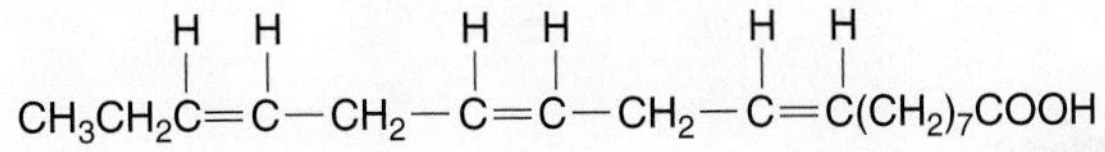	$18:3^{\Delta 9,12,15}$
γ-Linolenic acid	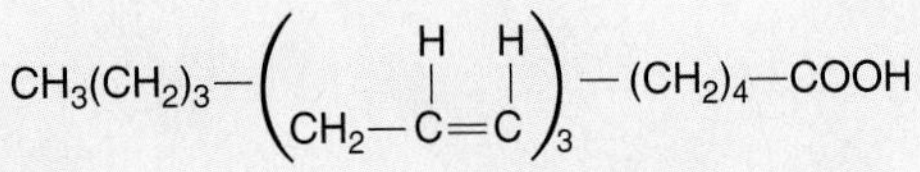	$18:3^{\Delta 6,9,12}$
Arachidonic acid	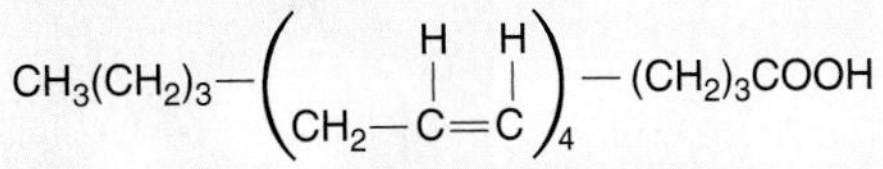	$20:4^{\Delta 5,8,11,14}$

The terminal methyl carbon atom is designated the omega (ω) carbon. Table 11.1 gives the structures, names, and standard abbreviations of several common fatty acids. Fatty acids are important components of several types of lipid molecules. They occur primarily in triacylglycerols and several types of membrane-bound lipid molecules.

Most naturally occurring fatty acids have an even number of carbon atoms that form an unbranched chain. (Unusual fatty acids with branched or ring-containing chains are found in some species.) Fatty acid chains that contain only carbon-carbon single bonds are referred to as *saturated*. Those molecules that contain one or more double bonds are said to be *unsaturated*. Double bonds are rigid structures. Consequently, molecules that contain them can occur in two isomeric forms: *cis* and *trans*. In *cis* **isomers**, similar or identical groups are on the same side of

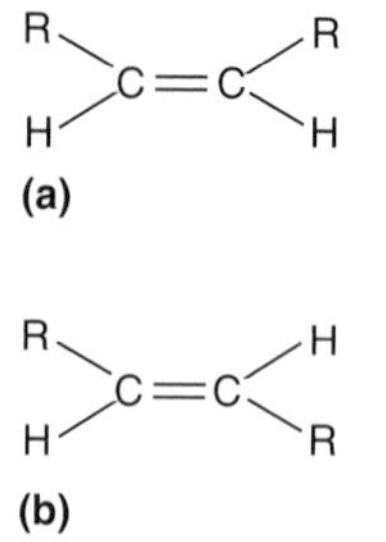

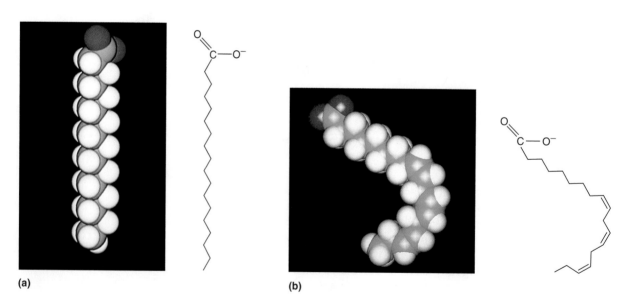

FIGURE 11.2

Isomeric Forms of Unsaturated Molecules

In *cis* isomers (a) both R groups are on the same side of the carbon-carbon double bond. In trans isomers (b), the R groups are on different sides.

a double bond (Figure 11.2a). When such groups are on opposite sides of a double bond, the molecule is said to be a *trans*-**isomer** (Figure 11.2b). The double bonds in most naturally occurring fatty acids are in a *cis* configuration. The presence of a *cis* double bond causes an inflexible "kink" in a fatty acid chain (Figure 11.3). Because of this structural feature, unsaturated fatty acids do not pack as closely together as saturated fatty acids. Less energy is required to disrupt the intermolecular forces between unsaturated fatty acids. Therefore, they have lower melting points and are liquids at room temperature. For example, a sample of palmitic acid (16:0), a saturated fatty acid, melts at 63°C, whereas palmitoleic acid ($16:1^{\Delta 9}$) melts at 0°C. Note that in the abbreviations for specific fatty acids, the number to the left of the colon is the total number of carbon atoms, and the number to the right is the number of double bonds. A superscript denotes the placement of a double bond. For example, $\Delta 9$ signifies that there are eight carbons between the carboxyl group and the double bond; that is, the double bond occurs between carbons 9 and 10.

Unsaturated fatty acids are also classified according to the location of the first double bond relative to the terminal methyl (omega, ω) end of the molecule. For example, linoleic acid and α-linolenic acid can be designated as $18:2\omega\text{-}6$ (equivalent to $18:2^{\Delta 9,12}$) and $18:3\omega\text{-}3$ (equivalent to $18:3^{\Delta 9,12,15}$), respectively. (The number to the right of ω designates the carbon at which the first double bond occurs, counting from the methyl end of the fatty acid. Sequential double bonds are always three carbons apart.) It is noteworthy that fatty acids with *trans* double bonds have three-dimensional structures similar to those of unsaturated fatty acids. In addition, the presence of one or more double bonds in a fatty acid makes it susceptible to oxidation (Figure 10.19). Consequences of this phenomenon include the effects of oxidative stress on cell membranes and the tendency of oils to become rancid (i.e., unpleasant smelling or tasting short-chain organic acids).

Fatty acids with one double bond are referred to as **monounsaturated** molecules. When two or more double bonds occur in fatty acids, usually separated by methylene groups (—CH$_2$—), they are referred to as **polyunsaturated**. The monounsaturated fatty acid oleic acid ($18:1^{\Delta 9}$) and the polyunsaturated linoleic acid ($18:2^{\Delta 9,12}$) are among the most abundant fatty acids in living organisms.

(a)   **(b)**

FIGURE 11.3

Space-Filling and Conformational Models

(a) A saturated fatty acid (stearic acid) and (b) an unsaturated fatty acid (α-linolenic acid). (Green spheres = carbon atoms; white spheres = hydrogen atoms; red spheres = oxygen atoms)

Organisms such as plants and bacteria can synthesize all the fatty acids they require from acetyl-CoA (Chapter 12). Mammals obtain most of their fatty acids from dietary sources. However, mammals can synthesize saturated fatty acids and some monounsaturated fatty acids. They can also modify some dietary fatty acids by adding two-carbon units and introducing some double bonds. Fatty acids that can be synthesized are called **nonessential fatty acids**. Because mammals do not possess the enzymes required to synthesize linoleic ($18:2^{\Delta 9,12}$) and α-linolenic ($18:2^{\Delta 9,12,15}$) acids, these **essential fatty acids** must be obtained from the diet.

Linoleic acid ($18:2\omega$-6) is the precursor for numerous derivatives, formed by elongation and/or desaturation reactions. Prominent examples include γ-linolenic acid ($18:3\omega$-6), arachidonic acid ($20:4\omega$-6), and docosapentaenoic acid ($22:4\omega$-6) (DPA). Together, linoleic acid and its derivatives are referred to as the **omega-6 fatty acids**. Food sources include various vegetable oils (e.g., sunflower and soybean oils), eggs, and poultry. α-Linolenic acid ($18:3\omega$-3) and its derivatives such as eicosapentaenoic acid (EPA) ($20:5\omega$-3) and docosahexaenoic acid (DHA) ($22:6\omega$-3) are referred to as the **omega-3 fatty acids**. Sources of α-linolenic include flaxseed and soybeans oils, and walnuts. EPA and DHA, also found in fish and fish oils (e.g., salmon, tuna, and sardines), are now believed to promote cardiovascular health. Effects associated with diets with adequate amounts of these two fatty acids include lower blood levels of triacylglycerols, lower blood pressure, and decreased platelet aggregation. Essential fatty acids are used as structural components (e.g., phospholipids in membranes) and as precursors for several important metabolites. Prominent examples of the latter include the eicosanoids and anandamine. The eicosanoids (p. 376) are hormone molecules derived from arachidonic acid. In general, omega-6-derived eicosanoids promote inflammation, whereas omega-3 derivatives have anti-inflammatory properties. Anandamine (*N*-arachidonyl ethanolamine), a derivative of arachidonic acid, is an *endocannabinoid*, a substance produced in the body that binds to the same receptor as tetrahydrocannabinol, a psychoactive drug. Anandamine acts as a neurotransmitter in the central and peripheral nervous systems, where it affects eating and sleep behavior, short-term memory, and pain relief.

Dermatitis (scaly skin) is an early symptom in individuals on low-fat diets that are deficient in essential fatty acids. Other signs of this deficiency include poor wound healing, reduced resistance to infection, alopecia (hair loss), and thrombocytopenia (reduction in the number of platelets, the blood component involved in the clotting process).

Fatty acids have several important chemical properties. The reactions that they undergo are typical of short-chain carboxylic acids. For example, fatty acids react with alcohols to form esters:

$$
\underset{\substack{\text{O}\\\|\\ \text{R}-\text{C}-\text{OH}}}{} + \text{R}'-\text{OH} \rightleftharpoons \underset{\substack{\text{O}\\\|\\ \text{R}-\text{C}-\text{O}-\text{R}'}}{} + \text{H}_2\text{O}
$$

This reaction is reversible; that is, under appropriate conditions a fatty acid ester can react with water to produce a fatty acid and an alcohol. Unsaturated fatty acids with double bonds can undergo hydrogenation reactions to form saturated fatty acids. Finally, unsaturated fatty acids are susceptible to oxidation. (This feature of fatty acid chemistry is described in Chapter 12.)

Certain fatty acids are covalently attached to a wide variety of eukaryotic proteins. Such proteins are referred to as *acylated* proteins. Fatty acid groups (called **acyl groups**) clearly facilitate the interactions between membrane proteins and their hydrophobic environment. Myristoylation and palmitoylation, the most common forms of protein acylation, are now known to influence a variety of structural and functional properties of proteins. Promotion of protein binding to membranes is a prominent example. In addition, hydrophobic fatty acid

KEY CONCEPT

Fatty acids are monocarboxylic acids, most of which are found in triacylglycerol molecules, several types of membrane-bound lipid molecules, or acylated membrane proteins.

molecules are transported from fat cells to body cells by means of the acylation of water-soluble serum proteins.

The Eicosanoids

The **eicosanoids** are a diverse group of extremely powerful hormonelike molecules produced in most mammalian tissues. They include the prostaglandins, thromboxanes, and leukotrienes (Figure 11.4). Together the eicosanoids mediate a wide variety of physiological processes, including smooth muscle contraction, inflammation, pain perception, and blood flow regulation. Eicosanoids are also implicated in several diseases such as myocardial infarction and rheumatoid arthritis.

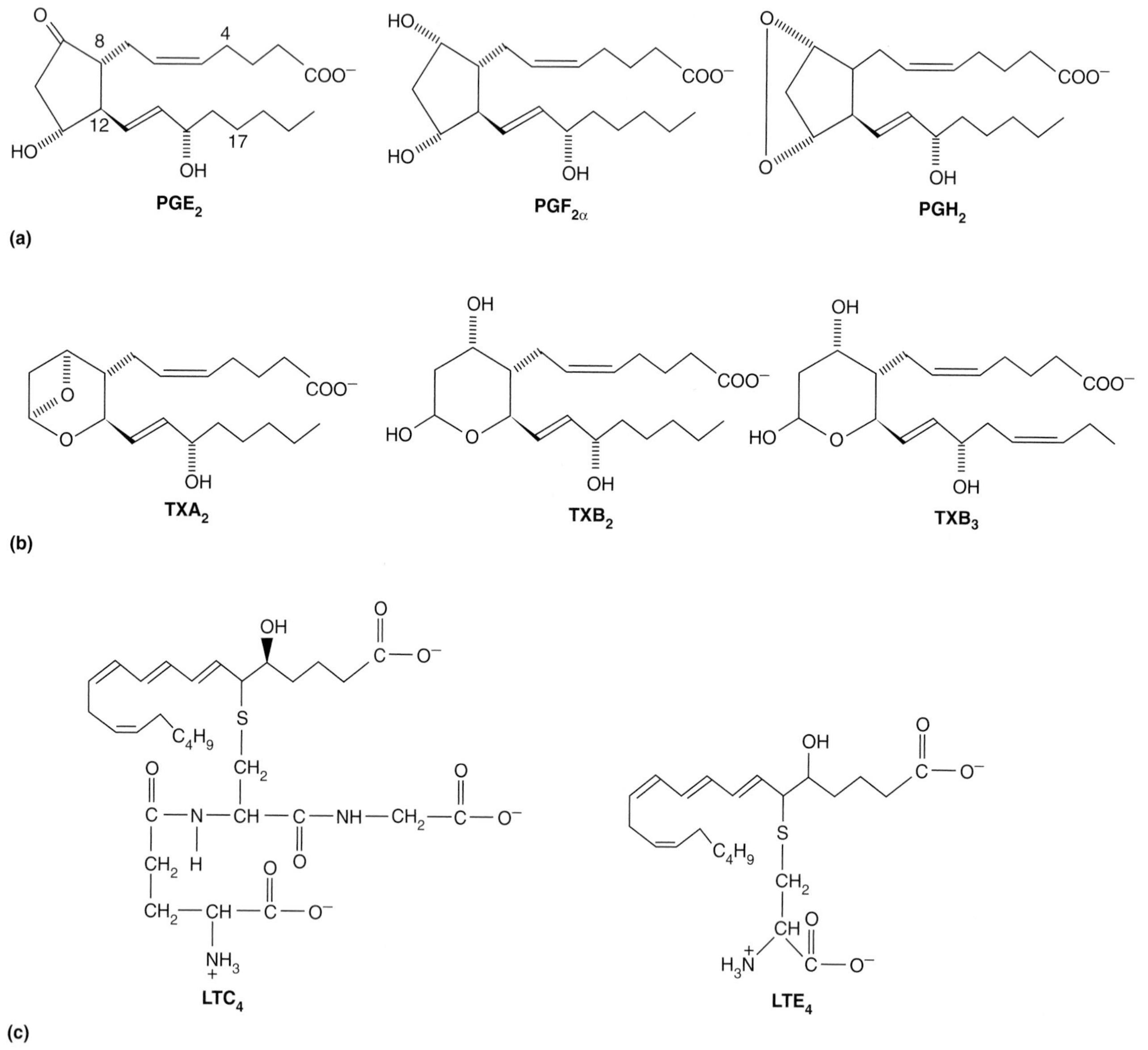

FIGURE 11.4

Eicosanoids

(a) Prostaglandins E_2, $F_{2\alpha}$, and H_2. (b) Thromboxanes A_2, B_2, and B_3. (c) Leukotrienes C_4 and E_4. Note that LTC_4 and LTE_4 both have four double bonds. The difference between the two molecules is the type of substituent attached via a thioether linkage. The substituents in LTC_4 and LTE_4 are glutathione and cysteine, respectively. In one metabolite of LTC_4 (LTD_4, not shown), the glutamate group is removed from the glutathione substituent.

Because they are generally active within the cell in which they are produced, the eicosanoids are called **autocrine** regulators instead of hormones.

Eicosanoids, which are usually designated by their abbreviations, are named according to the following system. The first two letters indicate the type of eicosanoid (PG = prostaglandin, TX = thromboxane, LT = leukotriene). The third letter identifies the type of modification made to the parent compound of the eicosanoid (e.g., A = hydroxyl group and an ether ring, B = two hydroxyl groups). The number in an eicosanoid name indicates the number of double bonds in the molecule. Eicosanoids are extremely difficult to study because they are active for short periods (often measured in seconds or minutes). In addition, they are produced only in small amounts.

Eicosanoids are derived from either arachidonic acid or EPA. Production of eicosanoids begins after arachidonic acid or EPA is released from membrane phospholipid molecules by the enzyme phospholipase A_2. A brief overview of each class of eicosanoid is provided next.

Prostaglandins contain a cyclopentane ring and hydroxy groups at C-11 and C-15. Molecules belonging to the E series of prostaglandins have a carbonyl group at C-9, whereas F series molecules have an OH group at the same position. The 2 series, derived from arachidonic acid, appears to be the most important group of prostaglandins in humans. EPA is the precursor of the 3 series of prostaglandins. Prostaglandins are involved in a wide range of regulatory functions. For example, prostaglandins promote inflammation, an infection-fighting process that produces pain and fever. They are also involved in reproduction (e.g., ovulation and uterine contractions during conception and labor) and digestion (e.g., inhibition of gastric secretion). Prostaglandin metabolism is complex for the following reasons:

1. There are many types of prostaglandin.

2. The types and amounts are different in each tissue or organ.

3. Certain prostaglandins have opposite effects in different organs, that is, their receptors are tissue specific. For example, several E-series prostaglandins cause smooth muscle relaxation in organs such as the intestine and uterus. The same molecules promote contraction of the smooth muscle in the cardiovascular system.

The **thromboxanes** are also derivatives of arachidonic acid or EPA. They differ from other eicosanoids in that their structures have a cyclic ether. TXA_2 and TXB_2 are produced from arachidonic acid primarily by platelets. Once TXA_2 has been synthesized, it is converted to TXB_2 by an isomerase. TXB_3 is synthesized from EPA and is a hydrolysis product of TXA_3. It is synthesized by polymorphonuclear leukocytes and plays a role in platelet aggregation and vasoconstriction following tissue injury.

The **leukotrienes** are linear (noncyclic) molecules whose synthesis is initiated by a peroxidation reaction catalyzed by lipoxygenase. The leukotrienes differ in the position of this peroxidation step and the nature of the thioether group attached near the site of peroxidation. The name *leukotrienes* stems from their early discovery in white blood cells (leukocytes) and the presence of a *triene* (three conjugated double bonds) in their structures. Leukotrienes LTC_4, LTD_4, and LTE_4 have been identified as components of slow-reacting substance of anaphylaxis (SRS-A). *Anaphylaxis* is an unusually severe allergic reaction that results in respiratory distress, low blood pressure, and shock. During inflammation (a normal response to tissue damage) SRS-A molecules increase fluid leakage from blood vessels into affected areas. LTB_4, a potent chemotactic agent, attracts infection-fighting white blood cells to damaged tissue. (Chemotactic agents are also referred to as chemoattractants.) Other effects of leukotrienes include vasoconstriction and bronchoconstriction (both caused by the contraction of smooth muscle in blood vessels and the air passages in the lungs) and edema (increased capillary permeability that causes fluid to leak out of blood vessels).

Triacylglycerols

Triacylglycerols are esters of glycerol with three fatty acid molecules (Figure 11.5). Glycerides with one or two fatty acid groups, called monoacylglycerols and diacylglycerols, respectively, are metabolic intermediates. They are normally present in small amounts. Because triacylglycerols have no charge (i.e., the carboxyl group of each fatty acid is joined to glycerol through a covalent bond), they are sometimes referred to as **neutral fats**. Most triacylglycerol molecules contain fatty acids of varying lengths; the acids themselves may be unsaturated, saturated, or a combination (Figure 11.6). Depending on their fatty acid compositions, triacylglycerol mixtures are referred to as fats or oils. *Fats*, which are solid at room temperature, contain a large proportion of saturated fatty acids. *Oils* are liquid at room temperature because of their relatively high unsaturated fatty acid content. (Recall that unsaturated fatty acids do not pack together as closely as do saturated fatty acids.)

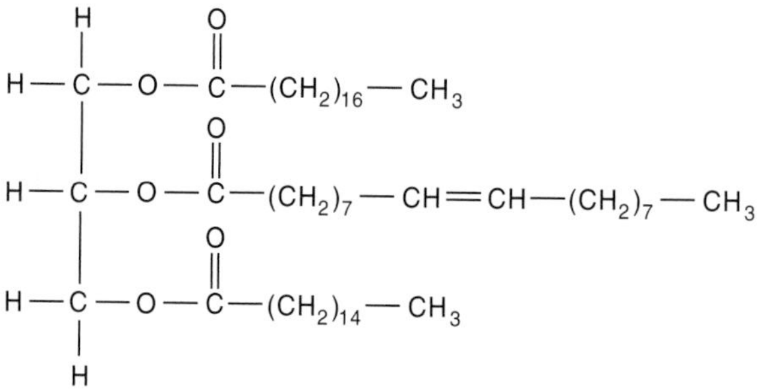

$$H-\overset{\overset{\displaystyle H}{|}}{C}-O-\overset{\overset{\displaystyle O}{\|}}{C}-(CH_2)_{16}-CH_3$$

$$H-\overset{|}{C}-O-\overset{\overset{\displaystyle O}{\|}}{C}-(CH_2)_7-CH=CH-(CH_2)_7-CH_3$$

$$H-\overset{|}{C}-O-\overset{\overset{\displaystyle O}{\|}}{C}-(CH_2)_{14}-CH_3$$

FIGURE 11.5

Triacylglycerol

Each triacylglycerol molecule is composed of glycerol esterified to three (usually different) fatty acids.

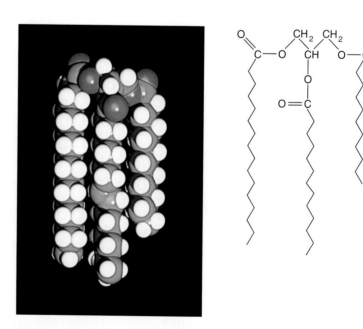

FIGURE 11.6

Space-Filling and Conformational Models of a Triacylglycerol

Triacylglycerols are molecules that serve as a rich source of chemical bond energy.

In animals, triacylglycerols (usually referred to as *fat*) have several roles. First, they are the major storage and transport form of fatty acids. Triacylglycerol molecules store energy more efficiently than glycogen for several reasons:

1. Triacylglycerols are hydrophobic, and therefore they coalesce into compact, anhydrous droplets within cells. A specialized type of cell called the *adipocyte*, found in adipose tissue, stores triacylglycerols. The anhydrous triacylglycerols store an equivalent amount of energy in about one-eighth of the volume of glycogen (the other major energy storage molecule), which binds a substantial amount of water.

2. Triacylglycerol molecules are more reduced and can thus release more electrons per molecule when oxidized than carbohydrate molecules. Therefore, triacylglycerols release more energy (38.9 kJ/g of fat compared with 17.2 kJ/g of carbohydrate) when they are degraded.

A second important function of fat is to provide insulation in low temperatures. Fat, a poor conductor of heat, prevents heat loss. Adipose tissue, with its high triacylglycerol content, is found throughout the body (especially underneath the skin). Finally, in some animals fat molecules secreted by specialized glands make fur or feathers water-repellent.

In plants, triacylglycerols constitute an important energy reserve in fruits and seeds. Because these molecules contain relatively large amounts of unsaturated fatty acids (e.g., oleic and linoleic), they are referred to as plant oils. Seeds rich in oil include peanut, corn, palm, safflower, soybean, and flax. Avocados and olives are fruits with a high oil content.

KEY CONCEPTS

- Triacylglycerols are molecules consisting of glycerol esterified to three fatty acids.
- In both animals and plants they are a rich energy source.

QUESTION 11.1

Oils can be converted to fats in a commercial nickel-catalyzed process referred to as *partial hydrogenation*. Under relatively mild conditions (180°C and pressures of about 1013 torr or 1.33 atm) enough double bonds are hydrogenated for liquid oils to solidify. This solid material, oleomargarine, has a consistency like butter. However, oils are not completely hydrogenated during commercial hydrogenation processes. Propose a practical reason for this.

QUESTION 11.2

Soapmaking is an ancient process. The Phoenicians, a seafaring people who dominated trade in the Mediterranean area about 3000 years ago, are believed to have been the first to manufacture soap. Traditionally, soap has been made by heating animal fat with potash. (Potash is a mixture of potassium hydroxide (KOH) and potassium carbonate (K_2CO_3) obtained by mixing wood ash with water.) Currently, soap is made by heating beef tallow or coconut oil with sodium or potassium hydroxide. During this reaction, which is a *saponification* (the reverse of esterification), triacylglycerol molecules are hydrolyzed to give glycerol and the sodium or potassium salts of fatty acids:

$$
\begin{array}{ccc}
\text{Triacylglycerol} + 3\ \text{NaOH} & \longrightarrow & \text{Soap} + \text{Glycerol}
\end{array}
$$

Triacylglycerol   **Soap**   **Glycerol**

QUESTION 11.2 (CONT.)

Fatty acid salts (soaps) are amphipathic molecules (i.e., they possess polar and nonpolar domains) that spontaneously form into micelles (Figure 3.11). Soap micelles have negatively charged surfaces that repel each other. Soap is used to remove dirt mixed with grease because it is an *emulsifying agent*; that is, it promotes the dispersion of one substance in another. The mixing of soap and grease results in an emulsion—specifically, a system in which the soap molecules are dispersed in the oil droplets. Complete the following diagram, and explain how this process occurs. [*Hint*: Recall that "like dissolves like."]

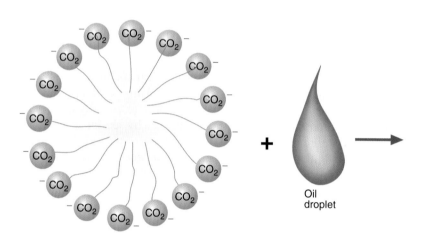

Oil droplet

$$CH_3—(CH_2)_{24}—\overset{\displaystyle O}{\overset{\displaystyle \|}{C}}—O—(CH_2)_{29}—CH_3$$

FIGURE 11.7

The Wax Ester Melissyl Cerotate

Found in carnauba wax, melissyl cerotate is an ester formed from melissyl alcohol and cerotic acid.

Wax Esters

Waxes are complex mixtures of nonpolar lipids. They are protective coatings on the leaves, stems, and fruits of plants and on the skin and fur of animals. Esters composed of long-chain fatty acids and long-chain alcohols are prominent constituents of most waxes. Well-known waxes include carnauba wax, produced by the leaves of the Brazilian wax palm, and beeswax. The predominant constituent of carnauba wax is the wax ester melissyl cerotate (Figure 11.7). Triacontyl hexadecanoate is one of several important wax esters in beeswax. Waxes also contain hydrocarbons, alcohols, fatty acids, aldehydes, and sterols (steroid alcohols).

Phospholipids

Phospholipids have several roles in living organisms. They are first and foremost structural components of membranes. In addition, several phospholipids are emulsifying agents and surface active agents. (A *surface active agent* is a substance that lowers the surface tension of a liquid, usually water, so that it spreads out over a surface.) Phospholipids are suited to these roles because, like fatty acid salts, they are amphipathic molecules. The hydrophobic domain of a phospholipid is composed largely of the hydrocarbon chains of fatty acids; the hydrophilic domain, called a **polar head group**, contains phosphate and other charged or polar groups.

When phospholipids are suspended in water, they spontaneously rearrange into ordered structures (Figure 11.8). As these structures form, phospholipid hydrophobic groups are buried in the interior to exclude water. Simultaneously, hydrophilic polar head groups are oriented so that they are exposed to water. When phospholipid molecules are present in sufficient concentration, they form bimolecular

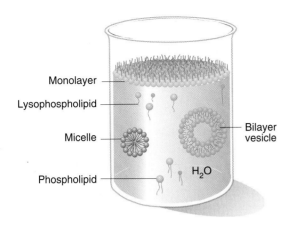

FIGURE 11.8

Phospholipid Molecules in Aqueous Solution

Each molecule is represented as a polar head group attached to one or two fatty acyl chains. (Lysophospholipid molecules possess only one fatty acyl chain.) The monolayer on the surface of the water forms first. As the phospholipid concentration increases, bilayer vesicles begin to form. Because of their wedge shape (compared to the cylindrical shape of phospholipids that contain two fatty acid chains), lysophospholipid molecules form micelles.

layers. This property of phospholipids (and other amphipathic lipid molecules) is the basis of membrane structure (see pp. 394–400).

There are two types of phospholipid: phosphoglycerides and sphingomyelins. **Phosphoglycerides** are molecules that contain glycerol, fatty acids, phosphate, and an alcohol (e.g., choline). **Sphingomyelins** differ from phosphoglycerides in that they contain sphingosine instead of glycerol. Because sphingomyelins are also classified as sphingolipids, their structures and properties are discussed separately.

Phosphoglycerides are the most numerous phospholipid molecules found in cell membranes. The simplest phosphoglyceride, phosphatidic acid, is the precursor for all other phosphoglyceride molecules. Phosphatidic acid is composed of glycerol-3-phosphate that is esterified with two fatty acids. Phosphoglyceride molecules are classified according to which alcohol becomes esterified to the phosphate group. For example, if the alcohol is choline, the molecule is called phosphatidylcholine (PC) (also referred to as *lecithin*). Other types of phosphoglyceride include phosphatidylethanolamine (PE), phosphatidylserine (PS), diphosphatidylglycerol (dPG), and phosphatidylinositol (PI). (Refer to Table 11.2 for the structures of the common types of phosphoglyceride.) The most common fatty acids in the phosphoglycerides have between 16 and 20 carbons. Saturated fatty acids usually occur at C-1 of glycerol. The fatty acid substituent at C-2 is usually unsaturated. A derivative of phosphatidylinositol, namely, phosphatidylinositol-4,5-bisphosphate (PIP_2), is found in only small amounts in plasma membranes. PIP_2 is now recognized as an important component of intracellular signal transduction. The *phosphatidylinositol cycle*, initiated when certain hormones bind to membrane receptors, is described in Section 16.2.

Phosphatidylinositol is also a prominent structural component of glycosyl phosphatidylinositol (GPI) anchors. **GPI anchors** (Figure 11.9), which are also composed of a trimannosylglucosamine group and phosphoethanolamine, attach certain proteins to the external surface of the plasma membrane. Proteins are attached to the anchor molecule via an amide linkage between the carboxyl terminal of the protein and the amino nitrogen of ethanolamine. The two fatty acids of the phosphatidylinositol component are embedded in the plasma membrane.

KEY CONCEPTS

- Phospholipids are amphipathic molecules that play important roles in living organisms as membrane components, emulsifying agents, and surface active agents.

- There are two types of phospholipid: phosphoglycerides and sphingomyelins.

TABLE 11.2 Major Classes of Phosphoglycerides

$$\text{R}_2\text{CO}\overset{\displaystyle O}{\underset{\displaystyle}{\|}}\text{—}\underset{\displaystyle}{\text{CH}}\begin{matrix}\text{CH}_2\text{O}\overset{\displaystyle O}{\overset{\|}{\text{—C}}}\text{—R}_1\\ \\ \text{CH}_2\text{O}\overset{\displaystyle O}{\underset{\underset{\displaystyle \text{O}^-}{|}}{\overset{\|}{\text{—P}}}}\text{—O—X}\end{matrix}$$

	X Substituent	
Name of X-OH	**Formula of X**	**Name of Phospholipid**
Water	—H	Phosphatidic acid
Choline	—CH$_2$CH$_2\overset{+}{\text{N}}$(CH$_3$)$_3$	Phosphatidylcholine (lecithin)
Ethanolamine	—CH$_2$CH$_2\overset{+}{\text{N}}$H$_3$	Phosphatidylethanolamine (cephalin)
Serine	—CH$_2$—CH$\overset{\overset{+}{\text{NH}_3}}{\underset{\text{COO}^-}{\|}}$	Phosphatidylserine
Glycerol	—CH$_2$CHCH$_2$OH$\underset{\text{OH}}{|}$	Phosphatidylglycerol
Phosphatidylglycerol	—CH$_2$CH—CH$_2$—O—P—O—CH$_2$ (with OH, O$^-$, RCOCH, CH$_2$OCR groups)	Diphosphatidylglycerol (cardiolipin)
Inositol	(inositol ring with OH groups)	Phosphatidylinositol

QUESTION 11.3

Dipalmitoylphosphatidylcholine is the major component of *surfactant*, or surface active agent (an amphipathic molecule), that is secreted into lung alveoli to reduce the surface tension of the primarily aqueous extracellular fluid of the alveolar epithelia. Alveoli, also referred to as alveolar sacs, are the functional units of respiration. Oxygen and carbon dioxide diffuse across the walls of alveolar sacs, which are one cell thick.

The water on alveolar surfaces has a high surface tension because of the attractive forces between the molecules. If the water's surface tension is not reduced, the alveolar sac tends to collapse, making breathing extremely difficult. If premature infants lack sufficient surfactant, they are likely to die of suffocation. This condition is called *respiratory distress syndrome*. Draw the structure of dipalmitoylphosphatidylcholine. Considering the general structural features of phospholipids, propose a reason why surfactant is effective in reducing surface tension.

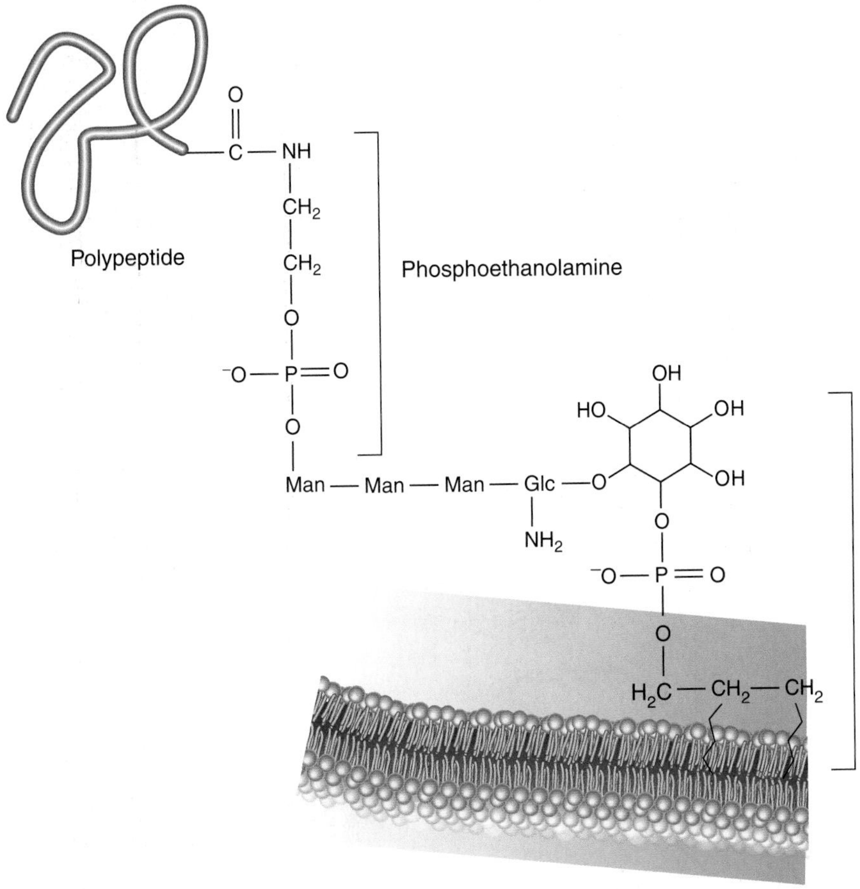

FIGURE 11.9

GPI Anchor

GPI-anchored proteins are attached to the external surface of the membrane through a linker element, phosphoethanolamine-Man$_3$-GlcNH$_2$, connecting the polypeptide at its carboxy terminus via an amide bond to a membrane phosphatidylinositol via an ether bond. Note that there are variations of this structure. For example, GlcNH$_2$ can be acetylated, and the phosphate of phosphatidic acid can be linked to either C-2 or C-3 of inositol.

Sphingolipids

Sphingolipids are important components of animal and plant membranes. All sphingolipid molecules contain a long-chain amino alcohol. In animals this alcohol is primarily sphingosine (Figure 11.10). Phytosphingosine is found in plant sphingolipids. The core of each type of sphingolipid is *ceramide*, a fatty acid amide derivative of sphingosine. In *sphingomyelin*, the 1-hydroxyl group of ceramide is esterified to the phosphate group of phosphorylcholine or phosphorylethanolamine (Figure 11.11). Sphingomyelin is found in most animal cell membranes. However, as its name suggests, sphingomyelin is found in greatest abundance in the myelin sheath of nerve cells. The myelin sheath is formed by successive wrappings of the cell membrane of a specialized myelinating cell around a nerve cell axon. Its insulating properties facilitate the rapid transmission of nerve impulses.

The ceramides are also precursors for the **glycolipids**, sometimes referred to as the *glycosphingolipids* (Figure 11.12). In glycolipids a monosaccharide, disaccharide, or oligosaccharide is attached to a ceramide through an O-glycosidic linkage. Glycolipids also differ from sphingomyelin in that they contain no

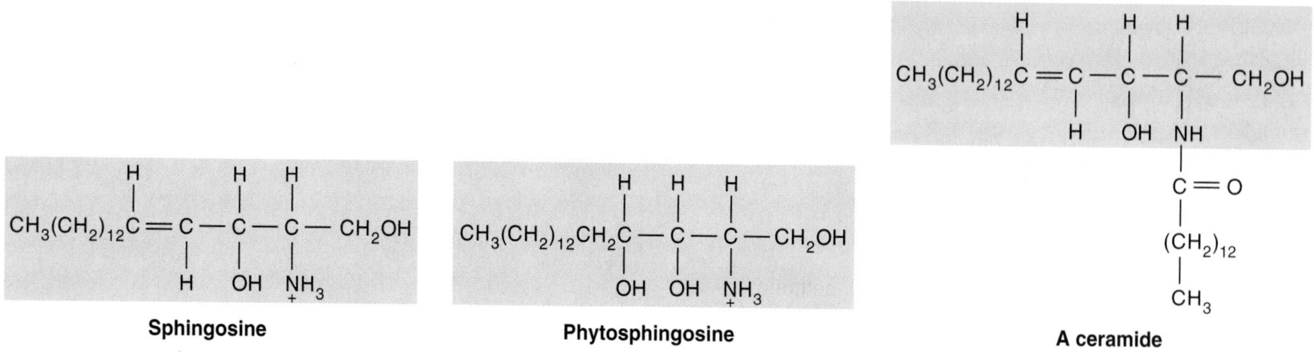

$$CH_3(CH_2)_{12}\overset{\overset{\displaystyle H}{|}}{C}=C-\overset{\overset{\displaystyle H}{|}}{\underset{\underset{\displaystyle OH}{|}}{C}}-\overset{\overset{\displaystyle H}{|}}{\underset{\underset{\displaystyle \overset{+}{N}H_3}{|}}{C}}-CH_2OH$$

Sphingosine

$$CH_3(CH_2)_{12}CH_2\overset{\overset{\displaystyle H}{|}}{\underset{\underset{\displaystyle OH}{|}}{C}}-\overset{\overset{\displaystyle H}{|}}{\underset{\underset{\displaystyle OH}{|}}{C}}-\overset{\overset{\displaystyle H}{|}}{\underset{\underset{\displaystyle \overset{+}{N}H_3}{|}}{C}}-CH_2OH$$

Phytosphingosine

$$CH_3(CH_2)_{12}\overset{\overset{\displaystyle H}{|}}{C}=C-\overset{\overset{\displaystyle H}{|}}{\underset{\underset{\displaystyle OH}{|}}{C}}-\overset{\overset{\displaystyle H}{|}}{\underset{\underset{\displaystyle NH}{|}}{C}}-CH_2OH$$

$$\underset{\displaystyle CH_3}{\overset{\displaystyle C=O}{\underset{\displaystyle (CH_2)_{12}}{|}}}$$

A ceramide

FIGURE 11.10

Sphingolipid Components

Note that the *trans* isomer of sphingosine occurs in sphingolipids.

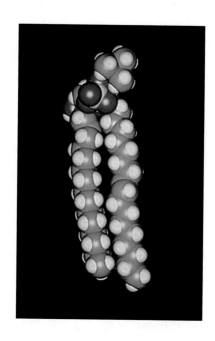

FIGURE 11.11

Conformational and Space-Filling Models of Sphingomyelin

The fatty acid component of sphingomyelins can be saturated or monounsaturated and 16 to 24 carbons in length, depending on the species and tissue of origin. The sphingosine base can be replaced by sphinganine (no double bond) and other C-20 homologues, although sphingosine is by far the most abundant.

phosphate. The most important glycolipid classes are the cerebrosides, the sulfatides, and the gangliosides.

Cerebrosides are sphingolipids in which the head group is a monosaccharide. (These molecules, unlike phospholipids, are nonionic.) Galactocerebrosides, the most common example of this class, are found almost entirely in the cell membranes of the brain. If a cerebroside is sulfated, it is referred to as a *sulfatide*. Sulfatides are negatively charged at physiological pH.

Sphingolipids that possess oligosaccharide groups with one or more sialic acid residues are called *gangliosides*. Although gangliosides were first isolated from nerve tissue, they also occur in most other animal tissues. The names of gangliosides include subscript letter and numbers. The letters M, D, and T indicate whether

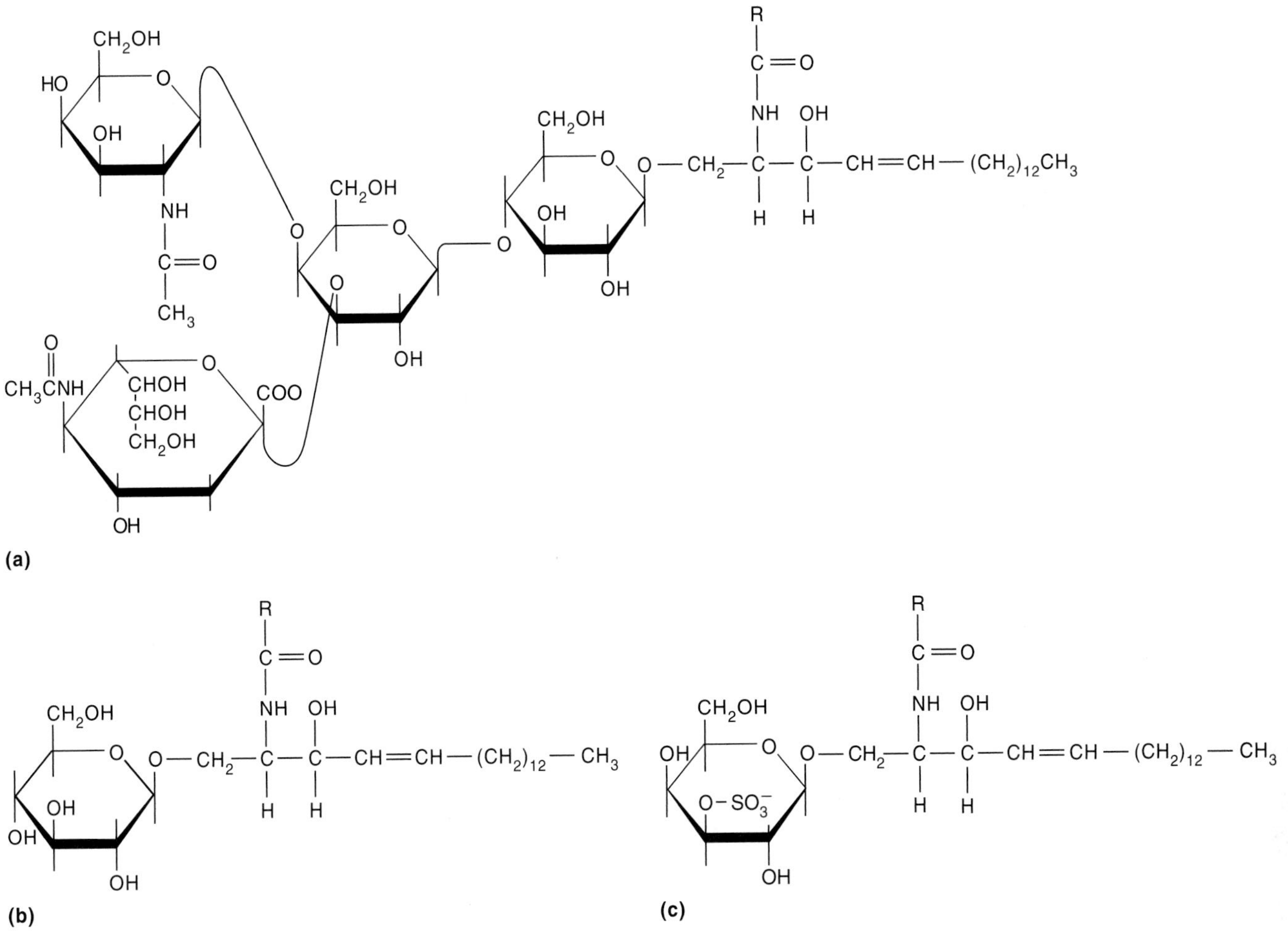

(a)

(b)   **(c)**

FIGURE 11.12

Selected Glycolipids

(a) Tay-Sachs ganglioside (GM$_2$), (b) glucocerebroside, and (c) galactocerebroside sulfate (a sulfatide).

the molecule contains one, two, or three sialic acid residues (see Fig. 7.25d), respectively. The numbers designate the sequence of sugars that are attached to ceramide. The Tay-Sachs ganglioside G$_{M2}$ is illustrated in Figure 11.12.

The normal role of glycolipids is still unclear. Certain glycolipid molecules may bind bacterial toxins, as well as bacterial cells, to animal cell membranes. For example, the toxins that cause cholera, tetanus, and botulism bind to glycolipid cell membrane receptors. Bacteria that have been shown to bind to glycolipid receptors include *E. coli*, *Streptococcus pneumoniae*, and *Neisseria gonorrhoeae*, the causative agents of urinary tract infections, pneumonia, and gonorrhea, respectively.

Sphingolipid Storage Diseases

Each lysosomal storage disease (see p. 53) is caused by hereditary deficiency of an enzyme required for the degradation of a specific metabolite. Several lysosomal storage diseases are associated with sphingolipid metabolism. Most of these diseases, also referred to as the *sphingolipidoses*, are fatal. The most common sphingolipid storage disease, Tay-Sachs disease, is caused by a deficiency of β-hexosaminidase A, the enzyme that degrades the ganglioside G$_{M2}$. As cells accumulate this molecule, they swell and eventually die. Tay-Sachs symptoms (blindness, muscle weakness, seizures, and mental retardation) usually appear

KEY CONCEPTS

- Sphingolipids, important membrane components of animals and plants, contain a complex long-chain amino alcohol (either sphingosine or phytosphingosine).

- The core of each sphingolipid is ceramide, a fatty acid amide derivative of the alcohol molecule. Glycolipids are derivatives of ceramide that possess a carbohydrate component.

several months after birth. Because there is currently no therapy for Tay-Sachs disease or for any other of the sphingolipidoses, the condition is always fatal (usually by age 3). Examples of the sphingolipidoses are summarized in Table 11.3.

Isoprenoids

The **isoprenoids** are a vast array of biomolecules that contain repeating five-carbon structural units known as *isoprene units* (Figure 11.13). Isoprenoids are not synthesized from isoprene (methylbutadiene). Instead, their biosynthetic pathways all begin with the formation of isopentenyl pyrophosphate from acetyl-CoA (Chapter 12).

The isoprenoids consist of terpenes and steroids. **Terpenes** are an enormous group of molecules that are found largely in the essential oils of plants. Steroids are derivatives of the hydrocarbon ring system of cholesterol.

TERPENES The terpenes are classified according to the number of isoprene residues they contain (Table 11.4). *Monoterpenes* are composed of two isoprene units (10 carbon atoms). Geraniol is a monoterpene found in the *essential oils*, volatile hydrophobic liquid mixtures extracted from plants, fruits, or flowers (e.g., roses, lemon, and geranium). Each essential oil has a characteristic odor, and some are used to make perfumes.

Terpenes that contain three isoprenes (15 carbons) are referred to as *sesquiterpenes*. Farnesene, an important constituent of oil of citronella, which is used in

TABLE 11.3 Selected Sphingolipid Storage Diseases*

Disease	Symptom	Accumulating Sphingolipid	Enzyme Deficiency
Tay-Sachs disease	Blindness, muscle weakness, seizures, mental retardation	Ganglioside G_{M2}	β-Hexosaminidase A
Gaucher's disease	Mental retardation, liver and spleen enlargement, erosion of long bones	Glucocerebroside	β-Glucosidase
Krabbe's disease	Demyelination, mental retardation	Galactocerebroside	β-Galactosidase
Niemann-Pick disease	Mental retardation	Sphingomyelin	Sphingomyelinase

*Many diseases are named for the physicians who first described them. Tay-Sachs disease was reported by Warren Tay (1843–1927), a British ophthalmologist, and Bernard Sachs (1858–1944), a New York neurologist. Phillipe Gaucher (1854–1918), a French physician, and Knud Krabbe (1885–1961), a Danish neurologist, first described Gaucher's disease and Krabbe's disease, respectively. Niemann-Pick disease was first characterized by the German physicians Albert Niemann (1880–1921) and Ludwig Pick (1868–1944).

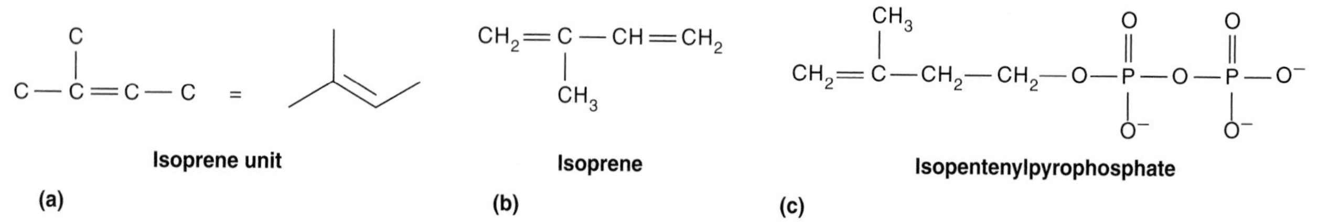

FIGURE 11.13

Isoprene

(a) Basic isoprene structure. (b) The organic molecule isoprene. (c) Isopentenylpyrophosphate.

TABLE 11.4 Examples of Terpenes

Type	Number of Isoprene Units	Example Name	Example Structure
Monoterpene	2	Geraniol	
Sesquiterpene	3	Farnesene	
Diterpene	4	Phytol	
Triterpene	6	Squalene	
Tetraterpene	8	β-Carotene	
Polyterpene	9–24	Dolichol	
	Thousands	Rubber	

soap and perfumes, is a sesquiterpene. Phytol, a plant alcohol, is a *diterpene*, a molecule composed of four isoprene units. Squalene is a prominent example of the *triterpenes*; this intermediate in the synthesis of the steroids is found in large quantities in shark liver oil, olive oil, and *yeast*. **Carotenoids**, the orange pigments found in most plants, are the only *tetraterpenes* (molecules composed of eight isoprene units). The *carotenes* are hydrocarbon members of this group. The *xanthophylls* are oxygenated derivatives of the carotenes. *Polyterpenes* are high-molecular-weight molecules composed of up to thousands of isoprene units. Natural rubber is a polyterpene composed of between 3000 and 6000 isoprene units. *Dolichols* are polyisoprenoid alcohols (16–19 isoprene units) that function as sugar carriers in glycoprotein synthesis.

Several important biomolecules are composed of nonterpene components attached to isoprenoid groups (often referred to as *prenyl* or *isoprenyl* groups). Examples of these biomolecules, referred to as **mixed terpenoids**, include vitamin E (α-tocopherol) (Figure 10.23a), ubiquinone (Figure 10.3), vitamin K, and some cytokinins (plant hormones) (Figure 11.14).

A variety of proteins in eukaryotic cells are now known to be covalently attached to prenyl groups after their biosynthesis on ribosomes. The prenyl groups most often involved in this process, referred to as **prenylation**, are farnesyl and geranylgeranyl groups (Figure 11.15). The function of protein prenylation is not clear. There is some evidence that it plays a role in the control of cell growth. For example, *Ras proteins*, a group of cell growth regulators, are activated by prenylation reactions.

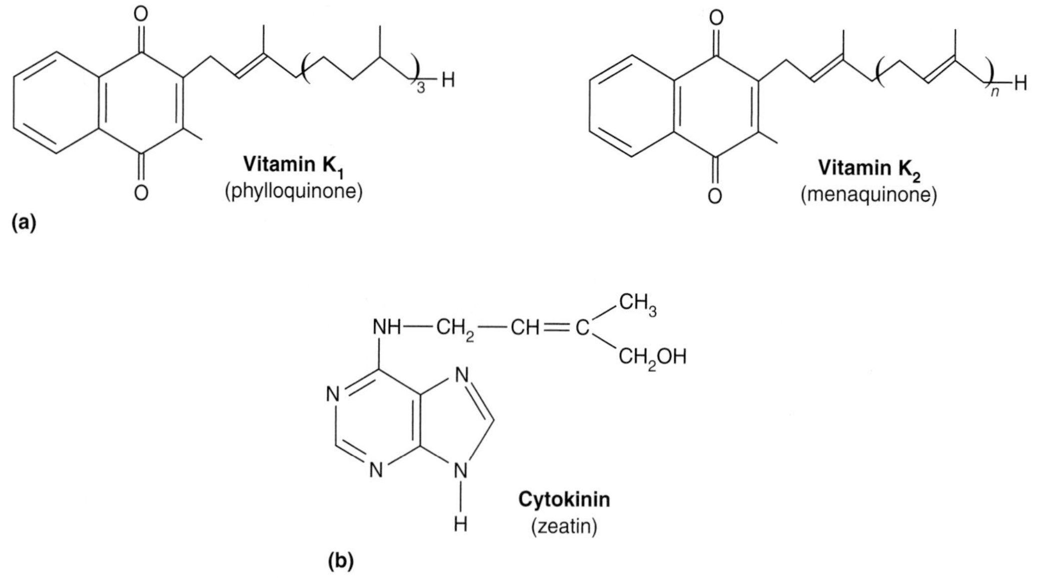

(a)

(b)

FIGURE 11.14

Selected Mixed Terpenoids

(a) Vitamin K_1 (phylloquinone) is found in plants, where it acts as an electron carrier in photosynthesis. Vitamin K_2 (menaquinone) is synthesized by intestinal bacteria and plays an important role in blood coagulation. (b) Cytokinins are cell-division-promoting substances in plants. Some cytokinins are mixed terpenoids. In combination with auxin (a plant growth regulator), zeatin, originally found in immature maize seeds, stimulates mature plant cells to undergo cell division.

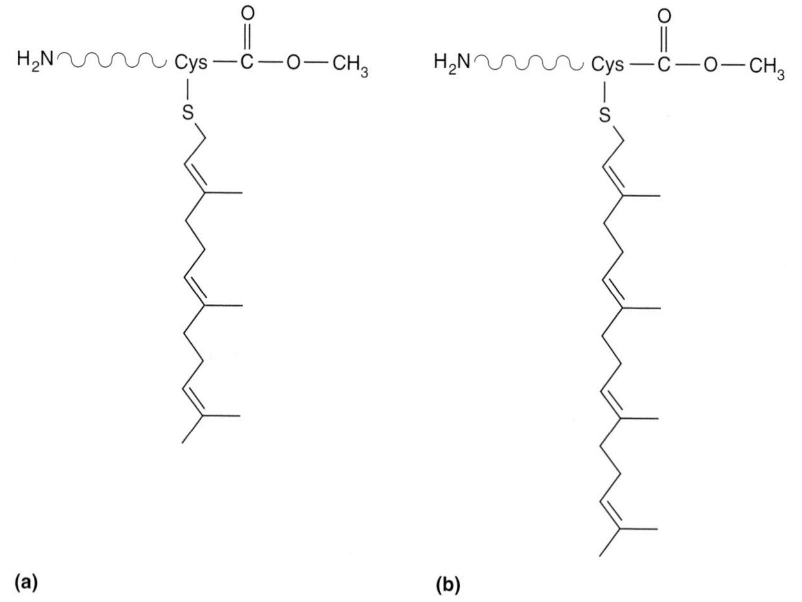

FIGURE 11.15

Prenylated Proteins

Prenyl groups are covalently attached at the SH group of C-terminal cysteine residues. Many prenylated proteins are also methylated at this residue. (a) Farnesylated protein. (b) Geranylgeranylated protein.

(a)

(b)

QUESTION 11.4

The majority of terpenes contain one or more ring structures. Consider the following examples. Determine which terpene class they belong to, and outline the positions of the isoprene units.

Carvone
(spearmint oil)

Camphor

Abscisic acid
(plant growth regulator)

STEROIDS **Steroids** are derivatives of triterpenes with four fused rings. They are found in all eukaryotes and a small number of bacteria. Steroids are distinguished from each other by the placement of carbon-carbon double bonds and various substituents (e.g., hydroxyl, carbonyl, and alkyl groups).

Cholesterol, an important molecule in animals, is an example of a steroid (Figure 11.16). In addition to being an essential component in animal cell

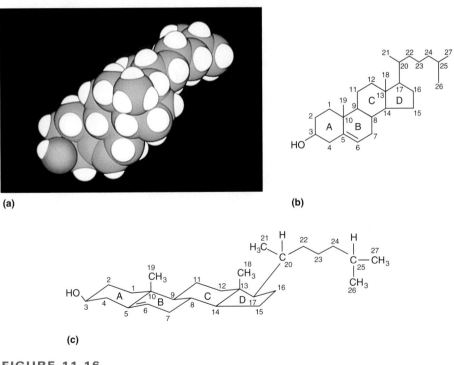

(a)

(b)

(c)

FIGURE 11.16

Structure of Cholesterol

(a) Space-filling model, (b) conventional view, and (c) conformational model. Space-filling models and conformational models (see p. 232) represent molecular structure more accurately than the conventional view.

membranes, cholesterol is a precursor in the biosynthesis of all steroid hormones, vitamin D, and bile salts (Figure 11.17). Cholesterol (C-27) is formed from the linear triterpene squalene (C-30) by intramolecular ring closure, oxidation, and cleavage. The only double bond retained migrates to the Δ5 position, and C-3 is oxidized to a hydroxyl group, which justifies its classification as a *sterol*. (Although the term *steroid* is most properly used to designate molecules that contain one or more carbonyl or carboxyl groups, it is often used to describe all derivatives of the steroid ring structure.) Cholesterol is usually stored within cells as a fatty acid ester. The esterification reaction is catalyzed by the enzyme *acyl-CoA:cholesterol acyltransferase* (ACAT), located on the cytoplasmic face of the endoplasmic reticulum.

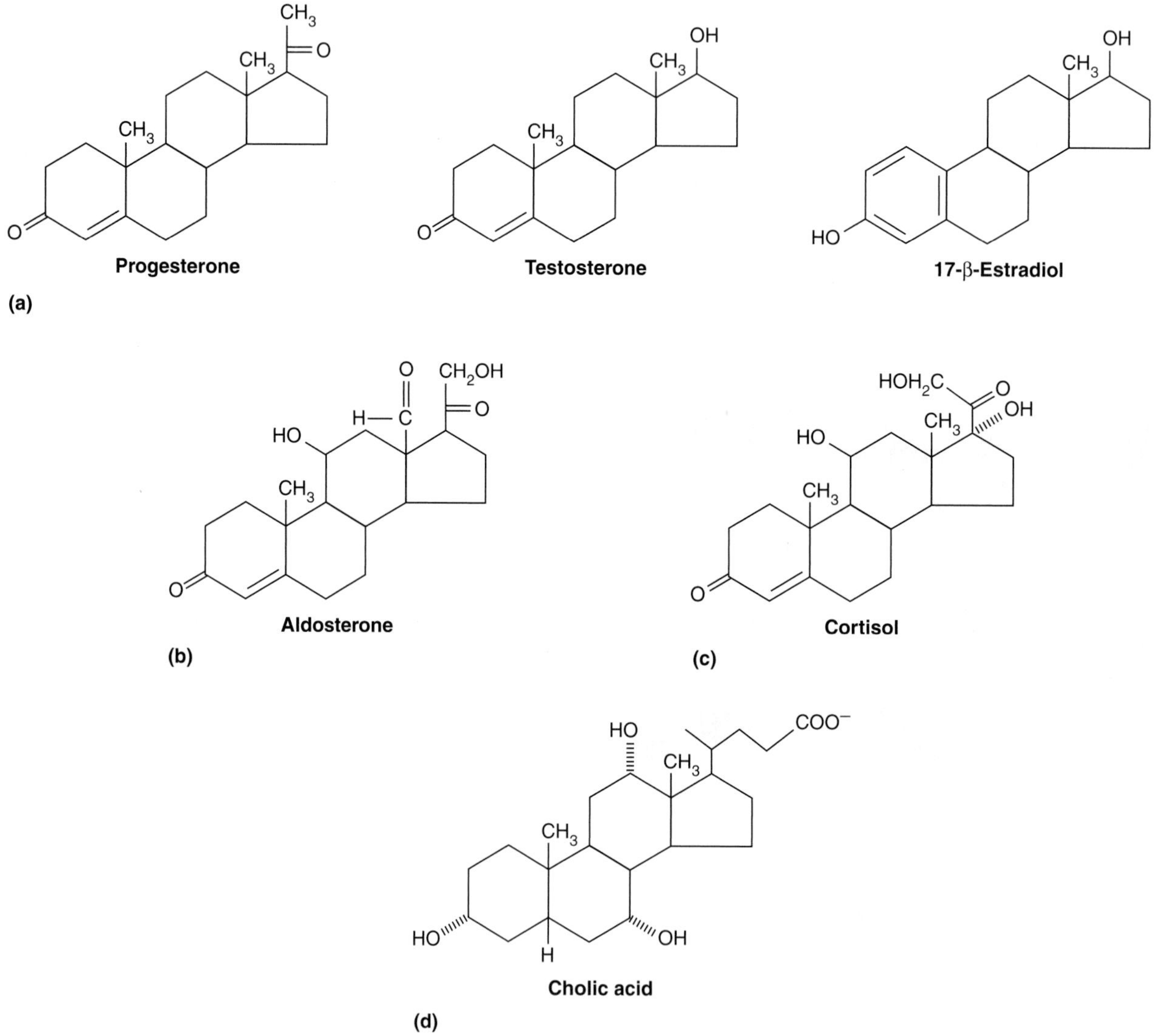

FIGURE 11.17

Animal Steroids

(a) Sex hormones (molecules that regulate the development of sexual structures and various reproductive behaviors).
(b) A mineralocorticoid (a molecule produced in the adrenal cortex that regulates plasma concentrations of several ions, especially sodium). (c) A glucocorticoid (a molecule that regulates the metabolism of carbohydrates, fats, and proteins).
(d) A bile acid (a molecule produced in the liver that aids the absorption of dietary fats and fat-soluble vitamins in the intestine).

QUESTION 11.5

Bile salts are emulsifying agents; that is, they promote the formation of mixtures of hydrophobic substances and water. Produced in the liver, bile salts assist in the digestion of fats in the small intestine. They are formed by linking bile acids to hydrophilic substances such as the amino acid glycine. After reviewing the structure of cholic acid in Figure 11.17, suggest how the structural features of bile salts contribute to their function.

Practically all plant steroid molecules are sterols. The function of plant sterols is still relatively unclear. They undoubtedly play an important role in membrane structure and function. Certain sterol derivatives, such as the cardiac glycosides, are known to protect plants that produce them from predators. Most plant sterols possess a one- or two-carbon substituent attached to C-24. The most abundant sterols in green algae and higher plants are β-sitosterol and stigmasterol (Figure 11.18).

Cardiac glycosides, molecules that increase the force of cardiac muscle contraction, are among the most interesting steroid derivatives. Glycosides are carbohydrate-containing acetals (see p. 237). Although several cardiac glycosides are extremely toxic (e.g., *ouabain*, obtained from the seeds of the plant *Strophanthus gratus*), others have valuable medicinal properties (Figure 11.19). For example, *digitalis*, an extract of the dried leaves of *Digitalis purpurea* (the foxglove plant), is a time-honored stimulator of cardiac muscle contraction. *Digitoxin*, the major "cardiotonic" glycoside in digitalis, is used to treat congestive heart failure, an illness in which the heart is so damaged by disease processes (e.g., myocardial infarcts) that pumping is impaired. In higher than therapeutic doses, digitoxin is extremely toxic. Both ouabain and digitoxin inhibit Na^+–K^+ ATPase (see p. 402).

KEY CONCEPTS

- Isoprenoids are a large group of biomolecules with repeating units derived from isopentenyl pyrophosphate.
- There are two types of isoprenoids: terpenes and steroids.

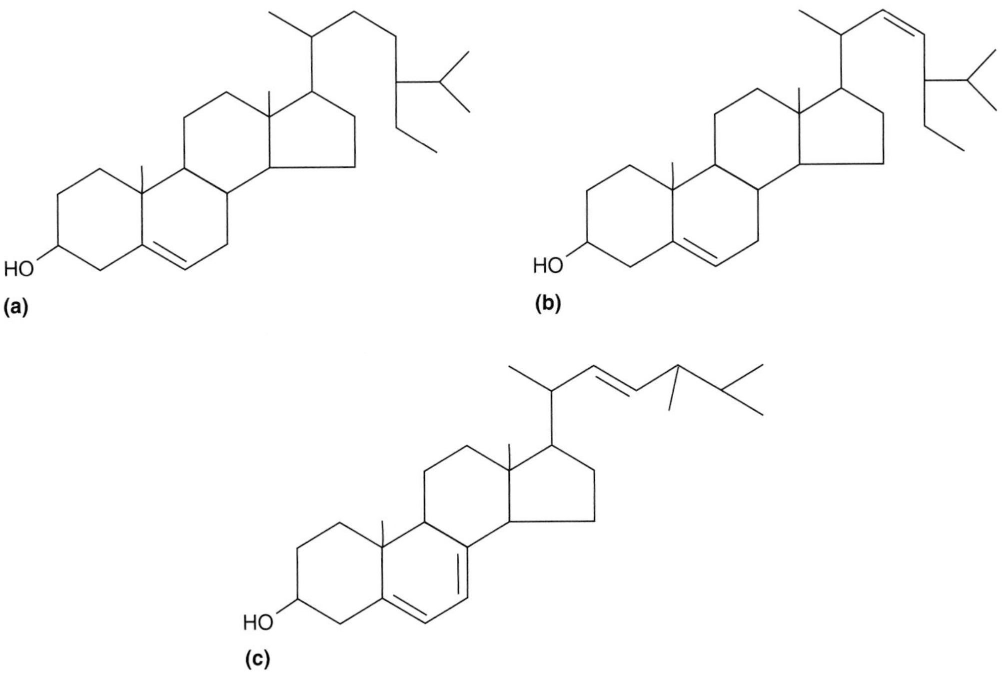

(a)

(b)

(c)

FIGURE 11.18

Plant Steroids

(a) β-Sitosterol, (b) stigmasterol, and (c) ergosterol (found in fungi). One of the most important roles of plant sterols is the stabilization of cell membranes.

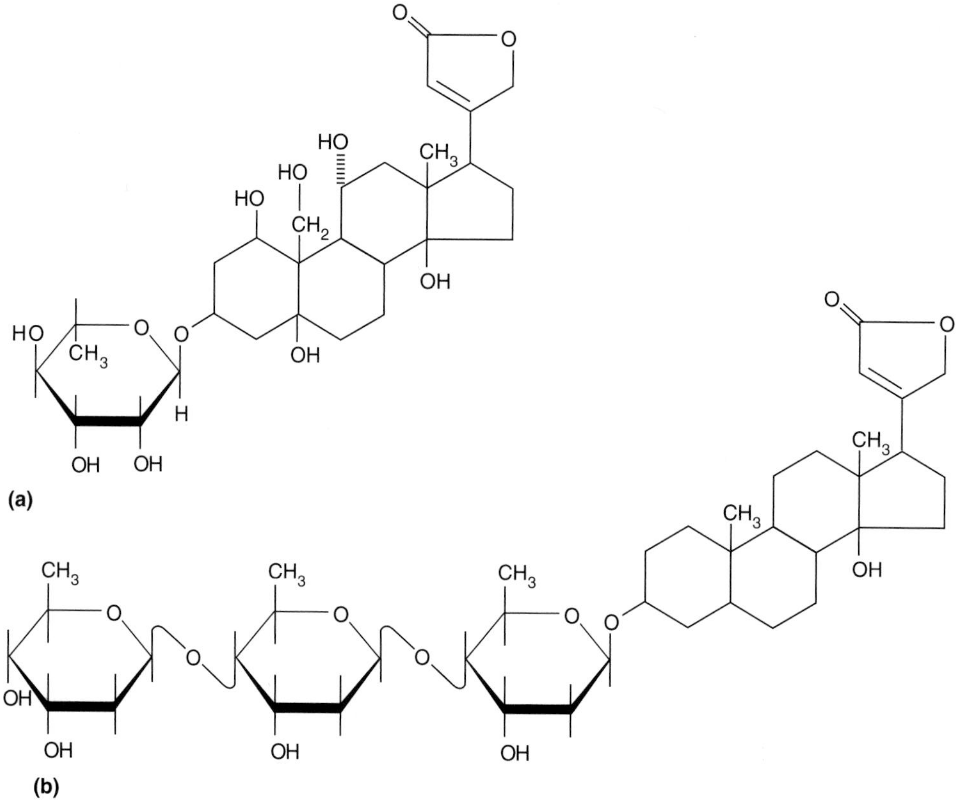

FIGURE 11.19

Cardiac Glycosides

Each cardiac glycoside possesses a glycone (carbohydrate) and an aglycone component. (a) In ouabain the glycone is one rhamnose residue. The steroid aglycone of ouabain is called ouabagenin. (b) The glycone of digitoxin is composed of three digitoxose residues. The aglycone of digitoxin is called digitoxigenin.

Lipoproteins

Although the term *lipoprotein* can describe any protein that is covalently linked to lipid groups (e.g., fatty acids or prenyl groups), it is most often applied to a group of molecular complexes found in the blood plasma of mammals (especially humans). Plasma lipoproteins transport lipid molecules (triacylglycerols, phospholipids, and cholesterol) through the bloodstream from one organ to another. Lipoproteins also contain several types of lipid-soluble antioxidant molecules (e.g., α-tocopherol and several carotenoids). (The function of *antioxidants*, substances that protect biomolecules from free radicals, is described in Chapter 10.) The protein components of lipoproteins, called *apolipoproteins* or *apoproteins*, are synthesized in the liver. There are five major classes of apolipoproteins: A, B, C, D, and E. A generalized lipoprotein is shown in Figure 11.20. The relative amounts of lipid and protein components of the major types of lipoprotein are summarized in Figure 11.21.

Lipoproteins are classified according to their density. **Chylomicrons**, which are large lipoproteins ($D \leq 1000$ nm) of extremely low density (<0.95 g/cm^3), transport dietary triacylglycerols and cholesteryl esters from the intestine to muscle and adipose tissues. Chylomicron remnants are then taken up by the liver via endocytosis. **Very low density lipoproteins** (VLDL) (0.98 g/cm^3, $D = 30$–90 nm), synthesized in the liver, transport lipids to tissues. As VLDLs are depleted of triacylglycerol and some apolipoprotein and phospholipids, they shrink in size, become more dense, and are referred to as **intermediate-density lipoproteins** (IDL)

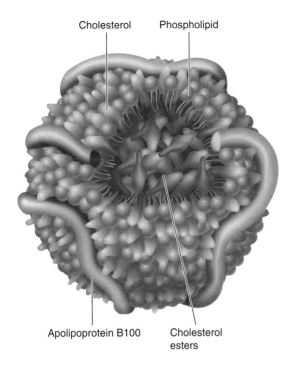

Cholesterol   Phospholipid

Apolipoprotein B100   Cholesterol esters

FIGURE 11.20

Plasma Lipoproteins

Lipoproteins vary in diameter from 5 to 1000 nm. Each type of lipoprotein contains a neutral lipid core composed of cholesteryl esters and/or triacylglycerols. This core is surrounded by a layer of phospholipid, cholesterol, and protein. Charged and polar residues on the surface of a lipoprotein enable it to dissolve in blood. In an LDL (low-density lipoprotein), as illustrated in this figure, each particle is composed of a core of cholesteryl esters surrounded by a monolayer that consists of hundreds of cholesterol and phospholipid molecules and several apolipoproteins, including apolipoprotein B-100, the ligand for the LDL receptor (p. 408).

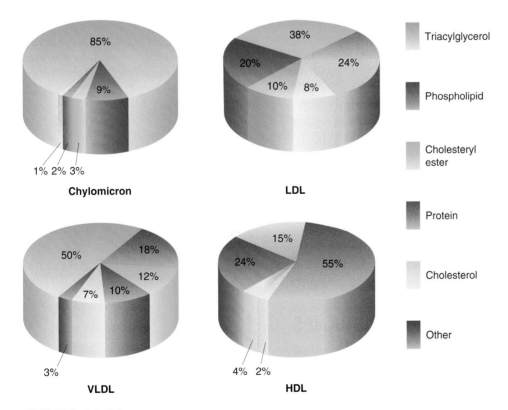

FIGURE 11.21

Proportional (Relative) Number of Cholesterol, Cholesteryl Ester, Phospholipid, and Protein Molecules in Four Major Classes of Plasma Lipoproteins

Chylomicrons are the largest plasma lipoproteins because they contain the largest percentage of triacylglycerols. In contrast, HDL are smaller particles with higher percentages of protein and low lipid content (mostly cholesterol).

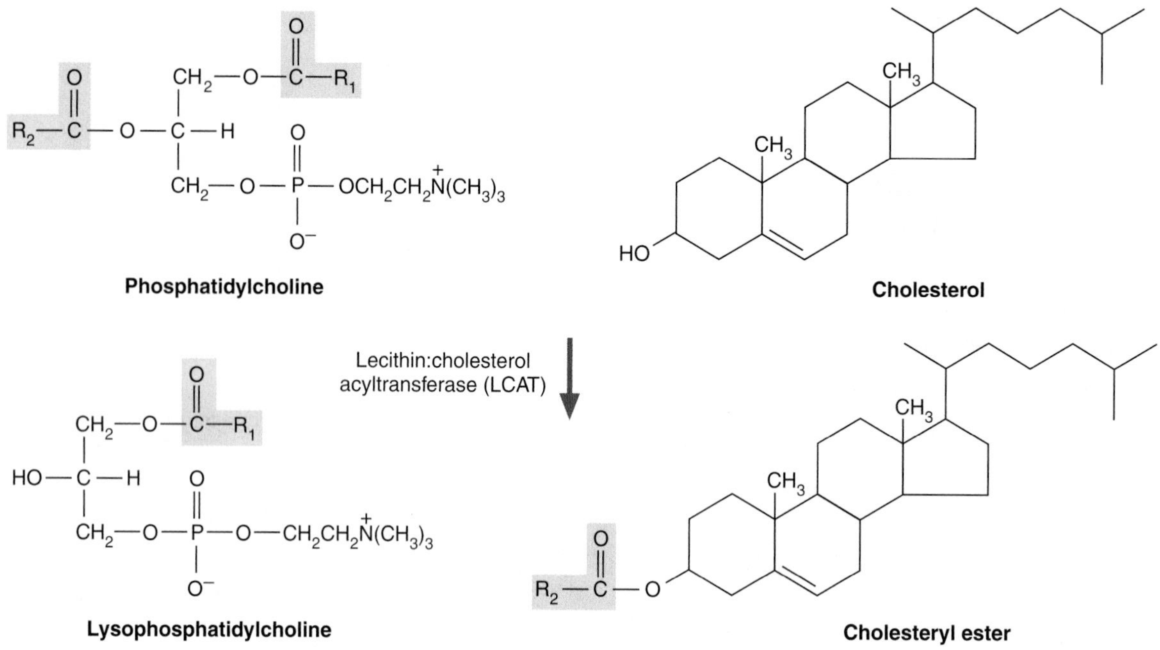

Phosphatidylcholine

Cholesterol

Lecithin:cholesterol
acyltransferase (LCAT)

Lysophosphatidylcholine

Cholesteryl ester

FIGURE 11.22

Reaction Catalyzed by Lecithin:Cholesterol Acyltransferase (LCAT)

Cholesteryl ester transfer protein, a protein associated with the LCAT-HDL complex, transfers cholesteryl esters from HDL to VLDL and LDL. The acyl groups are highlighted in color.

KEY CONCEPTS

- Plasma lipoproteins transport lipids through the bloodstream.

- On the basis of density, lipoproteins are classified into five major classes: chylomicrons, VLDL, IDL, LDL, and HDL.

(1 g/cm^3, D = 40 nm). IDLs may continue to lose triacylglycerol to form a higher density lipoprotein called low-density lipoprotein, or they may be removed from the bloodstream by the liver. **Low-density lipoproteins** (LDL) (1.04 g/cm^3, D = 20 nm) are the principal transporters of cholesterol and cholesteryl esters to tissues. In an intricate process (pp. 408–409) elucidated by Michael Brown and Joseph Goldstein (recipients of the 1985 Nobel Prize in Physiology or Medicine), LDL particles bind to LDL receptors and then are engulfed by cells. The role of **high-density lipoprotein** (HDL) (1.2 g/cm^3, D = 9 nm), a protein-rich particle produced in the liver and the intestine, appears to be the scavenging of excessive cholesterol from cell membranes and cholesteryl esters from VLDL and LDL. Cholesteryl esters are formed when the plasma enzyme lecithin:cholesterol acyltransferase (LCAT) transfers a fatty acid residue from lecithin to cholesterol (Figure 11.22). HDL transports these cholesteryl esters to the liver. The liver, the only organ that can dispose of excess cholesterol, converts most of it to bile acids (Chapter 12). The role of lipoproteins in atherosclerosis, a chronic disease of the cardiovascular system, is discussed in Chapter 12.

11.2 MEMBRANES

Most of the properties attributed to living organisms (e.g., movement, growth, reproduction, and metabolism) depend, either directly or indirectly, on membranes. All biological membranes have the same general structure. As previously mentioned (Chapter 2), membranes contain lipid and protein molecules. In the currently accepted concept of membranes, referred to as the **fluid mosaic model**, or the Singer-Nicholson model, a membrane is a noncovalent heteropolymer of a **lipid bilayer** and associated proteins. The nature of these molecules determines each membrane's biological functions and mechanical properties. Because of the importance of membranes in biochemical processes, the remainder of this chapter is devoted to a discussion of their structure and functions.

Membrane Structure

Each type of living cell has a unique set of functions. Consequently, the membranes of each cell type have unique features. Not surprisingly, the proportions of lipid and protein vary considerably among cell types and among organelles within each cell (Table 11.5). The types of lipid and protein found in each membrane also vary.

MEMBRANE LIPIDS When amphipathic molecules are suspended in water, they spontaneously rearrange into ordered structures (Figure 11.8). As these structures form, hydrophobic groups become buried in the water-depleted interior. Simultaneously, hydrophilic groups become oriented so that they are exposed to water. Phospholipids form into bimolecular layers at relatively low concentration. This property of phospholipids (and other amphipathic lipid molecules) is the basis of membrane structure. Membrane lipids are largely responsible for several other important features of biological membranes, as well.

Membrane Fluidity The term *fluidity* describes the degree of resistance of membrane components to movement. Rapid lateral movement (Figure 11.23) of lipid molecules is apparently responsible for the proper functioning of many membrane proteins. The movement of lipid molecules from one side of a lipid bilayer to the other occurs only during membrane synthesis or under conditions

TABLE 11.5 Chemical Composition of Some Cell Membranes

Membrane	Protein (%)	Lipid (%)	Carbohydrate (%)
Human erythrocyte plasma membrane	49	43	8
Mouse liver cell plasma membrane	46	54	2–4
Amoeba plasma membrane	54	42	4
Mitochondrial inner membrane	76	24	1–2
Spinach chloroplast lamellar membrane	70	30	6
Halobacterium purple membrane	75	25	0

Source: G. Guidotti, Membrane Proteins, *Annu. Rev. Biochem.* 41:731, 1972.

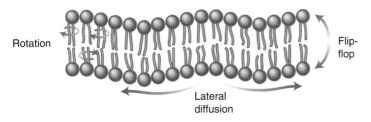

FIGURE 11.23

Lateral Diffusion in Biological Membranes

Lateral movement of phospholipid molecules is usually relatively rapid. "Flip-flop," the transfer of a lipid molecule from one side of a lipid bilayer to the other, occurs during new-membrane synthesis and membrane remodeling. Rotation of phospholipids within cell membranes is very rapid.

of lipid imbalance and requires the presence of a mediator protein, *flippase*, in the process called facilitated diffusion. A membrane's fluidity is largely determined by the percentage of unsaturated fatty acids in its phospholipid molecules and its composition and distribution of cholesterol. (Recall that unsaturated hydrocarbon chains pack less densely than saturated ones; see p. 374) Cholesterol contributes stability to the membrane because of its rigid ring system and ability to form van der Waals interactions with contiguous hydrocarbon chains. However, fluidity remains high because of its incomplete penetration into the membrane and flexible hydrocarbon tail (Figure 11.24). One measure of membrane fluidity, the ability of membrane components to diffuse laterally, can be demonstrated when cells from two different species are fused to form a **heterokaryon** (Figure 11.25). (Certain viruses or chemicals are used to promote cell-cell fusion.) The plasma membrane proteins of each cell type can be tracked when they are labeled with different fluorescent markers. Initially, the proteins are confined to their own side of the heterokaryon membrane. As time passes, the two fluorescent markers intermix, indicating that proteins move freely in the lipid bilayer.

FIGURE 11.24

Diagrammatic View of a Lipid Bilayer

The flexible hydrocarbon chains in the hydrophobic core (lightly shaded area in the middle) make the membrane fluid. The phospholipids in the membrane have different levels of unsaturation and vary in the nature of the polar head group. Cholesterol's compact and rigid ring system creates structural stability in the outer region of each leaflet. Red blood cells and other cells that are subjected to mechanical stress have a high content of cholesterol and cardiolipin (two phospholipids linked by a glycerol). Cell membranes are about 7 to 9 nm thick. (Nitrogen atoms are red, oxygen atoms are blue, and phosphorus atoms are orange.)

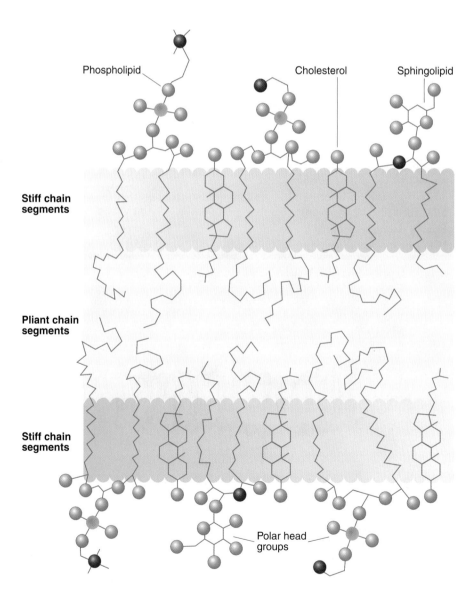

Phospholipid   Cholesterol   Sphingolipid

Stiff chain segments

Pliant chain segments

Stiff chain segments

Polar head groups

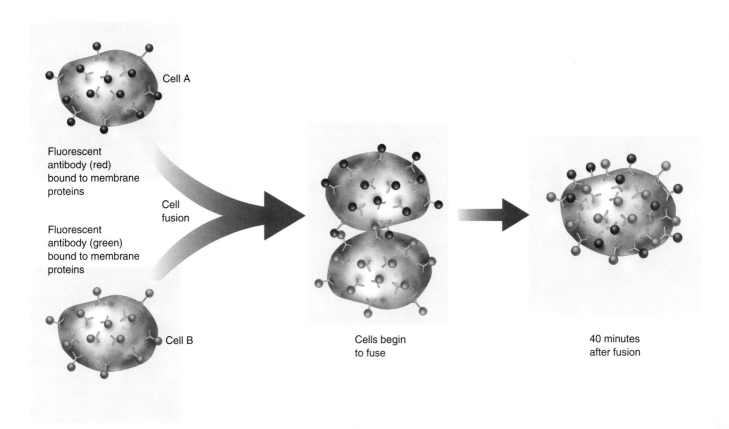

FIGURE 11.25

Fusion of Two Different Fluorescence-Labeled Cells to Form a Heterokaryon

This experiment demonstrates that membranes are fluid and that proteins can move freely within the lipid bilayer. Metabolic inhibitors do not slow down protein movement, but lowering the temperature below 15°C does.

Selective Permeability Hydrophobic hydrocarbon chains in lipid bilayers provide a virtually impenetrable barrier to the transport of ionic and polar substances. Specific membrane proteins regulate the movement of such substances into and out of cells. To cross a lipid bilayer, a polar substance must shed some or all of its hydration sphere and bind to a carrier protein for membrane translocation or pass through an aqueous protein channel. Both methods shield the hydrophilic molecule from the hydrophobic core of the membrane. Water crosses the membrane through protein channels called *aquaporins*, which exhibit cell type variation in their permeability to both water and accompanying ions (see Biochemistry in Perspective box entitled Aquaporins, p. 406). Nonpolar substances simply diffuse through the lipid bilayer down their concentration gradients. Each membrane exhibits its own transport capability or selectivity based on its protein component.

Self-Sealing When lipid bilayers are disrupted, they soon reseal. Small plasma membrane breaks spontaneously seal via the lateral flow of lipid molecules. Repair of larger lesions caused by mechanical stress, however, is an energy-requiring, Ca^{2+}-dependent process. A tear in the plasma membrane results in the inward flow of Ca^{2+} down its concentration gradient. Calcium ions trigger the movement of nearby endomembrane-derived vesicles to the lesion site. In an exocytosis-like process that involves cytoskeletal rearrangements, motor proteins such as dynein and kinesin, and membrane fusion proteins, the vesicles fuse with the plasma membrane to form a membrane patch. The repair process is rapid, usually occurring within a few seconds of the traumatic event.

COMPANION

Gw

W E B S I T E **Visit the companion website at www.oup.com/us/mckee to read the Biochemistry in Perspective box on membrane fusion and botulism.**

Asymmetry Biological membranes are asymmetric; that is, the lipid composition of each half of a bilayer is different. For example, the human red blood cell membrane possesses substantially more phosphatidylcholine and sphingomyelin on its outside surface. Most of the membrane's phosphatidylserine and phosphatidylethanolamine are on the inner side. New membrane is synthesized by insertion of additional phospholipid molecules from the cytoplasmic face of existing membranes. Flippase proteins specific to each type of phospholipid transfer lipid molecules to the opposite leaflet until membrane stability has been attained. Because the two faces of the resulting membrane are not chemically equivalent, the resulting leaflets are not identical in lipid composition. The protein components of membranes (discussed next) also exhibit considerable asymmetry, with distinctly different functional domains within membrane and on the cytoplasmic and extracellular faces of membrane.

MEMBRANE PROTEINS Most of the functions associated with biological membranes require protein molecules. Membrane proteins are often classified by the function they perform; structure, transport, catalysis, signal transduction, or immunological identity. Membrane proteins are also classified according to their structural relationship to membrane. Proteins that are embedded in and/or extend through a membrane are referred to as *integral proteins* (Figure 11.26). Such molecules can be extracted only by disrupting the membrane with organic solvents or detergents. *Peripheral proteins* are bound to membrane primarily through noncovalent interactions with integral membrane proteins or covalent bonds to myristic, palmitic, or prenyl groups. GPI anchors link a wide variety of cell-surface proteins (e.g., lipoprotein lipase, folate receptor, alkaline phosphatase, and the core proteins of glypicans) to plasma membranes. Some peripheral proteins interact directly with the lipid bilayer.

The anion channel protein (Figure 11.27), found in red blood cell membrane, is a well-researched example of integral membrane protein. *Anion channel protein* (also referred to as band 3) is composed of two identical subunits, each consisting of 929 amino acids. This protein channel plays an important role in CO_2 transport in blood. The HCO_3^- ion formed from CO_2 with the aid of carbonic anhydrase diffuses into and out of the red blood cell through the anion channel in exchange for chloride ion Cl^-. (The exchange of Cl^- for HCO_3^-, called

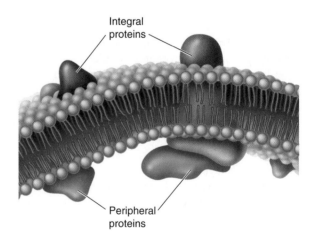

FIGURE 11.26

Integral and Peripheral Membrane Proteins

Integral membrane proteins are released only if the membrane is disrupted with detergents. Many peripheral membrane proteins can be removed with mild reagents, such as high salt concentration.

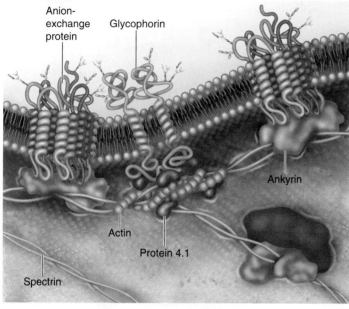

FIGURE 11.27

Red Blood Cell Integral Membrane Proteins

The integral membrane proteins glycophorin and anion-exchange protein are components in a network of linkages that connect the plasma membrane to structural elements of the cytoskeleton (e.g., actin, spectrin, protein band 4.1, and ankyrin). Note that the oligosaccharides on glycophorin are the ABO and MN blood group antigens.

the *chloride shift*, preserves the electrical potential of the red blood cell membrane.)

Red blood cell membrane peripheral proteins, composed largely of spectrin, ankyrin, and band 4.1, are primarily involved in preserving the cell's unique biconcave shape, which maximizes the surface area/volume ratio and exposure of diffusing O_2 to intracellular hemoglobin. *Spectrin* is a tetramer, composed of two $\alpha\beta$ dimers, that binds to ankyrin and band 4.1. *Ankyrin* is a large globular polypeptide (215 kD) that links spectrin to the anion channel protein. (This is a connecting link between the red blood cell's cytoskeleton and its plasma membrane.) *Band 4.1* binds to both spectrin and *actin filaments* (a cytoskeletal component found in many cell types). Because band 4.1 also binds to glycophorin, it too links the cytoskeleton and the membrane.

MEMBRANE MICRODOMAINS The lipids and proteins in membranes are not distributed uniformly. A notable example is provided by "lipid rafts," specialized microdomains in the external leaflet of eukaryotic plasma membranes (Figure 11.28). The components of lipid rafts are primarily cholesterol and sphingolipids, and certain membrane proteins. The rigid fused rings of cholesterol pack tightly alongside the more saturated acyl chains of sphingolipid molecules. Consequently, the lipid molecules in these microdomains are more ordered (i.e., less fluid) than those in nonraft regions, where unsaturated acyl chains

KEY CONCEPTS

- The basic structural feature of membrane structure is the lipid bilayer, which consists of phospholipids and other amphipathic lipid molecules.
- Membrane proteins, embedded in or associated with the phospholipid bilayer, contribute specialized functions to the membrane depending on its cell type and its role in biological processes.

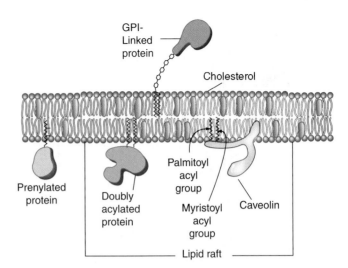

FIGURE 11.28

Lipid Rafts

As a result of stable associations between cholesterol and sphingolipids, slightly thicker membrane microdomains called lipid rafts form. Lipid rafts are also enriched in specific types of membrane protein. (Caveolin is a membrane protein found in caveolae, curved lipid rafts involved in clathrin-independent endocytosis and other processes.)

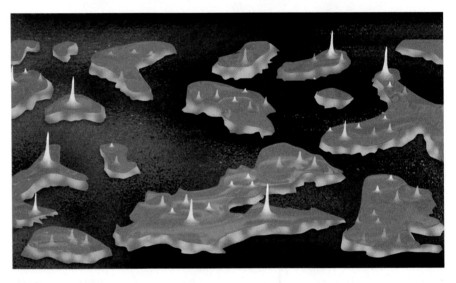

FIGURE 11.29

The Lipid Raft Environment

Atomic force microscopy provides a view of thick lipid rafts surrounded by a more fluid bilayer. The peaks in this micrograph represent GPI-linked proteins.

(in phospholipids) are more common. As their name suggests, lipid rafts seemingly float in a sea of more loosely packed membrane lipids (Figure 11.29).

Lipid rafts are enriched in some classes of proteins and they exclude others. Lipid-raft-associated proteins include GPI-anchored proteins, doubly acetylated tyrosine kinases, and certain transmembrane proteins. Some proteins are always present in lipid rafts, whereas others enter lipid rafts only as a result of an activation process. Lipid rafts have been implicated in a variety of cellular processes. Examples include exocytosis, endocytosis, and signal transduction. Lipid rafts are believed to function as platforms where the molecules that drive these processes are spatially organized.

Membrane Function

Among the vast array of membrane functions are the transport of polar and charged substances into and out of cells and organelles and the relay of signals that initiate change in metabolic and developmental aspects of cell function. Each of these topics is discussed briefly. A description of membrane receptors follows.

MEMBRANE TRANSPORT Membrane transport mechanisms are vital to living organisms. Ions and molecules constantly move across cell plasma membranes and across the membranes of organelles. This flux must be carefully regulated to meet each cell's metabolic needs. For example, a cell's plasma membrane regulates the entrance of nutrient molecules and the exit of waste products. Additionally, it regulates intracellular ion concentrations. Because lipid bilayers are generally impenetrable to ions and polar substances, specific transport components must be inserted into cellular membranes. Several examples of these structures, referred to as transport proteins or permeases, are discussed.

Biological transport mechanisms are classified according to whether they require energy. Major types of biological transport are illustrated in Figure 11.30. In **passive transport**, there is no direct input of energy. In contrast, **active transport** requires energy to transport molecules against a concentration gradient.

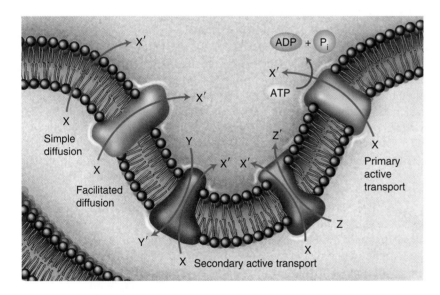

FIGURE 11.30

Transport Across Membranes

The major transport processes are simple and facilitated diffusion and primary and secondary active transport. In simple diffusion the spontaneous transport of a specific solute is driven by its concentration gradient. Facilitated diffusion, the movement of a solute down its concentration gradient across a membrane, occurs through protein channels or carriers. Both primary and secondary active transport require energy to move solutes across a membrane against their concentration gradients. In primary active transport, this energy is usually provided directly by ATP hydrolysis. In secondary active transport, solutes are moved across a membrane by energy stored in a concentration gradient of a second substance that has been created by ATP hydrolysis or another energy-generating mechanism.

In **simple diffusion**, each solute, propelled by random molecular motion, moves down its concentration gradient (i.e., from an area of high concentration to an area of low concentration). In this spontaneous process, there is a net movement of solute until an equilibrium is reached. A system reaching equilibrium becomes more disordered: that is, entropy increases. Because there is no input of energy, transport occurs with a negative change in free energy. In general, the higher the concentration gradient, the faster the rate of solute diffusion. The diffusion of gases such as O_2 and CO_2 across membranes is proportional to their concentration gradients. The diffusion of nonpolar organic molecules (such as steroid hormones) also depends on molecular weight and lipid solubility.

In **facilitated diffusion**, the second type of passive transport, certain large or charged molecules are moved through special channels or carriers. *Channels* are tunnel-like transmembrane proteins. Each type usually transports a specific solute. Many channels are regulated. Chemically regulated channels open or close in response to a specific chemical signal. For example, a *chemically gated Na^+ channel* in the nicotinic acetylcholine receptor complex (found in muscle cell plasma membranes) opens when the the neurotransmitter molecule acetylcholine binds. Then Na^+ rushes into the cell, and the membrane potential falls. Because membrane potential is an electrical gradient across the membrane (see p. 88), a decrease in membrane potential is membrane *depolarization*. Local depolarization caused by acetylcholine leads to the opening of nearby Na^+ channels (these are referred to as *voltage-gated Na^+ channels*). *Repolarization*, the reestablishment of the membrane potential, begins with the diffusion of K^+ ions out of the cell through *voltage-gated K^+ channels*. (The diffusion of K^+ ions out of the cell makes the inside less positive, that is, more negative.)

Another form of facilitated diffusion involves membrane proteins called *carriers* (sometimes referred to as *passive transporters*). In carrier-mediated transport, a specific solute binds to the carrier on one side of a membrane and causes a conformational change in the carrier. The solute is then translocated across the membrane and released. The red blood cell *glucose transporter* is the best characterized example of passive transporters. It allows D-glucose to diffuse across the red blood cell membrane with its concentration gradient for use in glycolysis and the pentose phosphate pathway.

The two forms of active transport are primary and secondary. In *primary active transport*, energy is provided by ATP. Transmembrane ATP-hydrolyzing enzymes use the energy derived from ATP to drive the transport of ions or molecules. The Na^+-K^+ pump (also referred to as the Na^+-K^+ ATPase) is a prominent example of a primary transporter. (The Na^+ and K^+ gradients are required for maintaining normal cell volume and membrane potential. Refer to the Biochemistry in Perspective box entitled Cell Volume Regulation in Chapter 3 on the companion website). In *secondary active transport*, concentration gradients generated by primary active transport are harnessed to move substances across membranes. For example, the Na^+ gradient created by the Na^+-K^+ ATPase pump is used in kidney tubule cells and intestinal cells to transport D-glucose against its concentration gradient (Figure 11.31).

Impaired membrane transport mechanisms can have very serious consequences. One of the best understood examples of dysfunctional transport occurs in cystic fibrosis. **Cystic fibrosis** (CF), a fatal autosomal recessive disease, is caused by a missing or defective plasma membrane glycoprotein called **cystic fibrosis transmembrane conductance regulator (CFTR)**. CFTR (Figure 11.32), which functions as a chloride channel in epithelial cells, is a member of a family of proteins referred to as ABC transporters. (ABC transporters are so named because each contains a polypeptide segment called an *ATP-binding cassette*.) CFTR contains five domains. Two domains, each containing six membrane-spanning

COMPANION

GW

WEBSITE Visit the companion website at www.oup.com/us/mckee to read the Biochemistry in Perspective box on cell volume regulation in Chapter 3.

KEY CONCEPTS

- Membrane transport mechanisms are classified as passive or active according to whether they require energy.

- In passive transport, solutes moving across membranes move down their concentration gradient.

- In active transport, energy derived directly or indirectly from ATP hydrolysis or another energy source is required to move an ion or molecule against its concentration gradient.

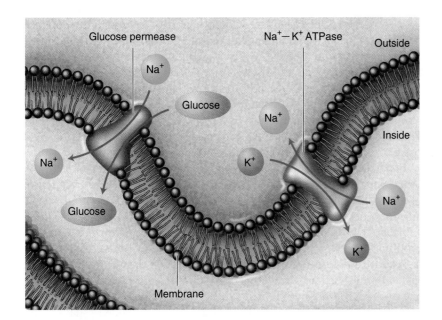

FIGURE 11.31

The Na⁺-K⁺ ATPase and Glucose Transport

The Na⁺-K⁺ ATPase preserves the Na⁺ gradient essential to maintain membrane potential. In certain cells, glucose transport depends on the Na⁺ gradient. Glucose permease transports both Na⁺ and glucose. Only when both substrates are bound does the protein change its conformation, thus initiating transport.

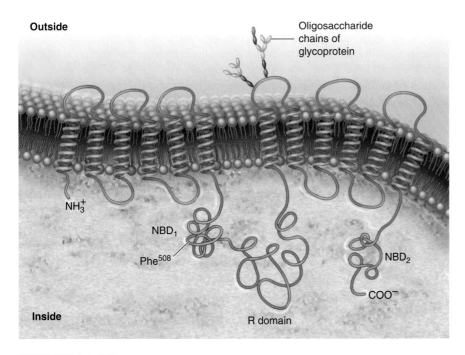

FIGURE 11.32

The Cystic Fibrosis Transmembrane Regulator (CFTR)

CFTR is a chloride channel composed of two domains (each with six membrane-spanning helices) that constitute the Cl^- pore, two nucleotide-binding domains (NBD), and a regulatory (R) domain. Transport of Cl^- through the pore, driven by ATP hydrolysis, occurs when specific amino acid residues on the R domains are phosphorylated. The most commonly observed CF-causing mutation is a deletion of Phe^{508} in NBD_1, which prevents proper targeting of CFTR-containing vesicles to the plasma membrane. The precise structural relationships among the pore-forming helices remain unclear.

helices, form the Cl⁻ channel pore. Chloride transport through the pore is controlled by the other three domains (all of which occur on the cytoplasmic side of the plasma membrane). Two are nucleotide-binding domains (NBD_1 and NBD_2) that bind and hydrolyze ATP and use the released energy to drive conformational changes in the pore. The regulatory (R) domain contains several amino acid residues that must be phosphorylated by cAMP-dependent protein kinase for chloride transport to occur.

The chloride channel is vital for proper absorption of salt (NaCl) and water across the apical (top) membrane surface of epithelial cells that line ducts and tubes in tissues such as lungs, liver, small intestine, and sweat glands. Chloride channel opening occurs in response to a signal molecule, PKA (protein kinase A). PKA phosphorylates specific residues in the R domain, causing a change in its conformation that triggers the binding of ATP molecules to NBD_1 and NBD_2. The two nucleotide-binding domains then form a head-to-tail heterodimer-like structure with the ATP-binding sites on the inside surfaces. As a result of these intramolecular rearrangements, the chloride channel gate opens and chloride ions flow down their concentration gradient. Hydrolysis of one of the NBD-bound ATP molecules causes dimer disruption that results in channel closing. The NBD dimer acts as a timing device in that the rate of ATP hydrolysis determines the length of time that the channel is open.

In cystic fibrosis, the failure of CFTR channels results in the retention of Cl⁻ within the cells. A thick mucus or other secretion forms because osmotic pressure causes the excessive uptake of water from the mucus. The most obvious features of CF are lung disease (obstructed air flow and chronic bacterial infections) and pancreatic insufficiency (impaired production of digestive enzymes that can result in severe nutritional deficits). In the majority of CF patients, CFTR is defective because of a deletion mutation at Phe^{508}, which causes protein misfolding that prevents the processing and insertion of the mutant protein into the plasma membrane. Less common causes of CF (over 100 have been described) include defective formation of CFTR mRNA molecules, mutations in the nucleotide-binding domains that result in ineffective binding or hydrolysis of ATP, and mutations in the pore-forming domains that cause reduced chloride transport.

Before the development of modern therapies, CF patients rarely survived childhood. It is only because of antibiotics (used principally in the treatment of lung infections) and commercially available digestive enzymes (replacing the enzymes normally produced in the pancreas) that many CF patients can now expect to live into their 30s. Yet as with the sickle-cell gene (see p. 145), defective CF genes are not rare. With an approximate incidence of 1 in every 2500 Caucasians, CF is the most common fatal genetic disorder in this population. Recent experiments with "knockout" mice indicate that carriers of the mutant gene are protected from diseases that kill because of diarrhea. (Knockout animals are bred to contain one copy of a defective gene in all of their cells.) The experimental animals used in CF research lose significantly less body fluid because they have a reduced number of functional chloride channels. It is suspected that CF carriers (individuals having only one copy of a defective CF gene) are also less susceptible to fatal diarrhea (e.g., in cholera) for the same reason. The CF gene did not spread beyond western Europe (the incidence among East Asians is approximately 1 in 100,000) because CF carriers secrete slightly more salt in their sweat than noncarriers and the epithelial cells that line sweat gland ducts cannot reabsorb chloride efficiently. In warmer climates, where sweating is a common feature of daily life, chronic excessive salt loss is far more dangerous than intermittent exposure to diarrhea-causing microorganisms.

QUESTION 11.6

Suggest the mechanism(s) by which each of the following substances is (are) transported across cell membranes:

    a. CO_2

    b. Glucose

    c. Cl^-

    d. K^+

    e. Fat molecules

    f. α-Tocopherol

QUESTION 11.7

Describe the types of noncovalent interaction that promote the stability and the functional properties of biological membranes.

QUESTION 11.8

Transport mechanisms are often categorized according to the number of transported solutes and the direction of solute transport.

    1. *Uniporters* transport one solute.

    2. *Symporters* transport two different solutes simultaneously in the same direction.

    3. *Antiporters* transport two different solutes simultaneously in opposite directions.

After examining the examples of transport discussed in this chapter, determine the category to which each belongs.

MEMBRANE RECEPTORS Membrane receptors provide mechanisms by which cells monitor and respond to changes in their environment. In multicellular organisms the binding to membrane receptors of chemical signals, such as the hormones and neurotransmitters of animals, is a vital link in intracellular communication. Other receptors are engaged in cell-cell recognition or adhesion. For example, lymphocytes perform a critical function of the immune system by transiently binding to the cell surface of virus-infected cells. This binding event triggers the lymphocyte-induced death of the infected cell. Similarly, the capacity of cells to recognize and adhere to other appropriate cells in a tissue is of crucial importance in many organismal processes, such as embryonic and fetal development.

The binding of a ligand to a membrane receptor results in a conformational change, which then causes a specific programmed response. Sometimes, receptor responses appear to be relatively straightforward. For example, the binding of acetylcholine to an acetylcholine receptor opens a cation channel. However, most responses are complex. The most intensively researched example of membrane receptor function is LDL receptor–mediated endocytosis, which is discussed next.

The low-density lipoprotein receptor is responsible for the uptake of cholesterol-containing lipoproteins into cells. The LDL receptor is a glycoprotein

(Continued on p. 408)

BIOCHEMISTRY IN PERSPECTIVE

The Aquaporins

How do water molecules flow so rapidly across hydrophobic cell membranes? A basic characteristic of living cells is the capacity to rapidly move water across cell membranes in response to changes in osmotic pressure. For years, many researchers assumed that simple diffusion was responsible for most water flow. In a wide variety of cell types such as red blood cells and certain kidney cells, it became apparent that water flow is extraordinarily rapid. In the early 1990s, investigators characterized the first of a series of water channel proteins, now called the aquaporins. Found initially in red blood cell membrane and then in kidney tubule cells, **aquaporin** 1 (AQP-1) is an intrinsic membrane protein complex that facilitates water flow, about 3×10^9 water molecules/s/channel. Aquaporins have been found in almost all living organisms, with at least 10 different forms in mammals, with different water and ion permeability characteristics.

Recent experimental evidence suggests that water flow through aquaporin channels is regulated. For example, three mammalian aquaporins appear to be regulated by pH. Others are regulated by phosphorylation reactions or by the binding of specific signal molecules. In 1993, the cause of a rare inherited form of **nephrogenic diabetes insipidis** (a disease in which the kidneys of affected individuals cannot produce concentrated urine) was discovered to be a mutation in the gene for AQP-2. The mutated AQP-2 does not respond to the antidiuretic hormone vasopressin (Table 16.1, p. 596).

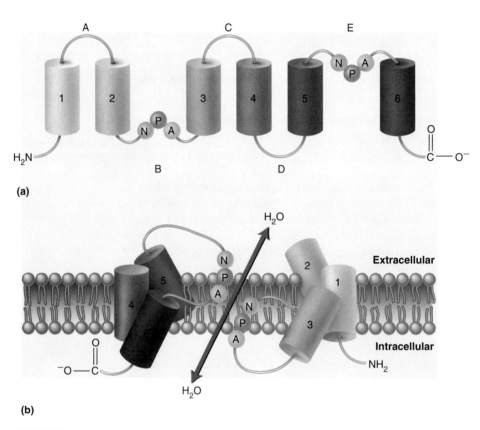

(a)

(b)

FIGURE 11A

Diagrammatic View of the Aquaporin Monomer

Each AQP-1 monomer possesses six membrane-spanning α-helices that are connected by five loops (a). In the functional monomer, the two loops containing the NPA sequences (asn-pro-ala) form the water-selective site (b).

Of all the aquaporins, AQP-1, a homotetramer that has high permeability to water only, is the best characterized. Each subunit is a polypeptide containing 269 amino acid residues that form a water-transporting pore with six α-helical membrane-spanning domains connected by five loops (Figure 11A). Although each monomer is an independent water channel, the formation of the tetramer is required for full function. In the functional monomer, the two loops that both possess an Asn-Pro-Ala (NPA) sequence meet in the middle to form the water-binding site. The pore, which has been measured at 3 Å, is only slightly larger than a water molecule (2.8 Å). As illustrated in Figure 11A, the NPA sequences are juxtaposed in the channel. The movement through the channel of water molecules only, and not smaller species such as H^+, is believed to be made possible by the formation of hydrogen bonds between water molecules and the Asn residues of the two NPA sequences (Figure 11B). The hydrophobic environment created by the amino acid residues on the other helices that comprise the pore causes the hydrogen bonds between water molecules to break as they move in single file toward the narrowest portion of the pore. It also forces the oxygen atom of each water molecule to orient itself toward the Asn residues. When the water molecule approaches the 3 Å constriction in the pore, its oxygen atom sequentially forms and breaks hydrogen bonds with the side chains of the two Asn residues. The absence of other hydrogen bonding partners prevents the ionization of H_2O and the generation of protons. Peter Agre received the Nobel Prize in Chemistry in 2003 for his discovery and investigation of aquaporins.

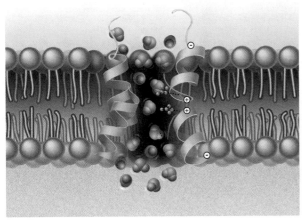

(a)

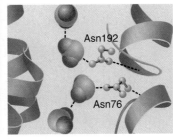

Asn192

Asn76

(b)

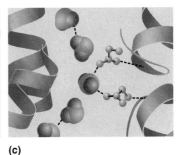

(c)

FIGURE 11B

Water Transport Through the AQP-1 Monomer

Water molecules move through the pore in single file. As each molecule approaches the constriction in the pore, it is forced to orient its oxygen atom so it can then form and break hydrogen bonds with the side chains of the two Asn residues. (a) Within the pore of the aquaporin monomer, the B and E loops (see Figure 11A) create a positive electrostatic environment in which the oxygen atom of each water molecule is oriented toward the two Asn residues. (b) and (c) The sequential formation and breakage of hydrogen bonds between the oxygen of water molecules and the two Asn residue side chains mediate the movement of water through the pore.

SUMMARY: The aquaporins are a class of membrane channel proteins that are responsible for water flow into and out of cells in response to osmotic pressure changes. The water pore within AQP-1, the best characterized aquaporin, has a structure that allows the movement across the plasma membrane of water molecules, but not those of other small species.

The LDL Receptor

The cytoplasmic domain of the LDL receptor plays a critical role in forming coated pits, an important feature of receptor-mediated endocytosis. Once LDL has bound to a receptor, both are rapidly internalized.

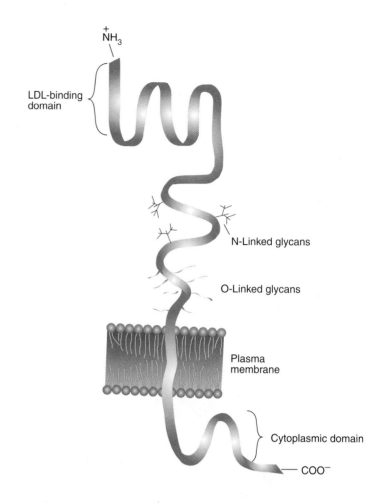

(Figure 11.33) found on the surface of many cells. When cells need cholesterol for the synthesis of membrane or steroid hormones, they produce LDLs receptors and insert them into discrete regions of plasma membrane. (These membrane regions usually constitute about 2% of a cell's surface. The protein *clathrin*, which has a unique structure referred to as *triskelion*, is the major protein component. It forms a latticelike polymer on the cytosolic side of the membrane during the initial stages of endocytosis.) The number of receptors per cell varies from 15,000 to 70,000, depending on cell type and cholesterol requirements.

The process of LDL receptor–mediated endocytosis occurs in several steps (Figure 2.23a). It begins within several minutes after LDLs have bound to LDL receptors. The region surrounding the bound receptor, referred to as a *coated pit*, pinches off and becomes a *coated vesicle*. Subsequently, *uncoated vesicles* are formed as clathrin depolymerizes. Before uncoated vesicles fuse with lysosomes, LDLs are uncoupled from LDL receptors as the pH changes from 7 to 5. (This change is created by ATP-driven proton pumps in the vesicle membrane.) LDL receptors are recycled to the plasma membrane, and LDL-containing vesicles fuse with lysosomes. Subsequently, proteins associated with the LDL particles are degraded to amino acids, and cholesteryl esters are hydrolyzed to cholesterol and fatty acids. Under normal circumstances, LDL receptor–mediated endocytosis is a highly regulated process. In liver cells, transcription factors called SREBPs (sterol regulatory element binding proteins) have been characterized. SREBP precursors are membrane-bound ER proteins. When liver cell cholesterol levels are low, the precursor proteins are transported to the Golgi complex, where they are cleaved to form the active transcription factors. The SREBPs then migrate into the nucleus and bind to sterol regulatory elements (SRE); afterward, they together activate up to 30 genes involved in lipid metabolism, including the gene for the LDL receptor. Cellular cholesterol levels then rise in response to a combination of LDL uptake and increased endogenous cholesterol synthesis. High-fat diets block LDL receptor synthesis because the accumulation of ingested cholesterol in ER membranes prevents SREBP processing reactions.

The LDL receptor was discovered during an investigation of an inherited disease, in this case, *familial hypercholesterolemia* (FH). The LDL receptor was discovered as Brown and Goldstein were investigating the uptake of LDLs into the fibroblasts of FH patients. The biochemical defect that causes FH was then identified as mutations in the LDL receptor gene.

Patients with FH have elevated levels of plasma cholesterol because they have missing or defective LDL receptors. (Recall that LDLs transport cholesterol to tissues.) Heterozygous individuals (also referred to as *heterozygotes*) inherit one defective LDL receptor gene. Consequently, they possess half the number of functional LDL receptors. With blood cholesterol values of 300 to 600 mg/100 mL, it is not surprising that heterozygotes have heart attacks as early as the age of 40. They also develop disfiguring *xanthomas* (cholesterol deposits in the skin) in their 30s. With a population frequency of 1 in 500, heterozygous FH is one of the most common human genetic anomalies. In contrast, *homozygotes* (individuals who have inherited a defective LDL receptor gene from both parents) are rare (approximately one in one million). These patients have plasma cholesterol values of 650 to 1200 mg/100 mL. Both xanthomas and heart attacks occur during childhood or early adolescence. Death usually occurs before the age of 20. The genetic defects that cause FH prevent affected cells from obtaining sufficient cholesterol from LDLs. The most common defect is failure to synthesize the receptor. Other defects include ineffective intracellular processing of newly synthesized receptor, defects in the receptor's binding of LDL, and the inability of receptors to cluster in coated pits.

Chapter**Summary**

1. Lipids are a diverse group of biomolecules that dissolve in non-polar solvents. They can be separated into the following classes: fatty acids and their derivatives, triacylglycerols, wax esters, phospholipids, lipoproteins, sphingolipids, and the isoprenoids.

2. Fatty acids are monocarboxylic acids that occur primarily in triacylglycerols, phospholipids, and sphingolipids. The eicosanoids are a group of powerful hormonelike molecules derived from long-chain fatty acids. The eicosanoids include the prostaglandins, the thromboxanes, and the leukotrienes.

3. Triacylglycerols are esters of glycerol with three fatty acid molecules. Triacylglycerols that are solid at room temperature (i.e., possess mostly saturated fatty acids) are called fats. Those that are liquid at room temperature (i.e., possess a high unsaturated fatty acid content) are referred to as oils. Triacylglycerols, the major storage and transport form of fatty acids, are an important energy storage form in animals. In plants they store energy in fruits and seeds.

4. Phospholipids are structural components of membranes. There are two types of phospholipid: phosphoglycerides and sphingomyelins.

5. Sphingolipids are also important components of animal and plant membranes. Like the sphingomyelins they contain a ceramide base (*N*-acylsphingosine) but do not contain phosphate. Their polar head group is one or more sugar residues.

6. Isoprenoids are molecules that contain repeating five-carbon isoprene units. The isoprenoids consist of the terpenes and the steroids.

7. Plasma lipoproteins transport lipid molecules through the bloodstream from one organ to another. They are classified according to their density. Chylomicrons are large lipoproteins of extremely low density that transport dietary triacylglycerols and cholesteryl esters from the intestine to adipose tissue and skeletal muscle. Very low density lipoproteins, which are synthesized in the liver, transport lipids to tissues. As VLDLs unload some of their lipid molecules, they are converted to LDLs. LDLs bind to LDL receptors on the plasma membrane and then are engulfed by cells. HDLs, also produced in the liver, scavenge cholesterol from cell membranes and other lipoprotein particles. LDLs play an important role in the development of atherosclerosis.

8. According to the fluid mosaic model, the basic structure of membranes is a lipid bilayer in which proteins float. Membrane lipids (the majority of which are phospholipids) are primarily responsible for the fluidity, regional partitioning (lipid rafts), and sealing and fusion properties of membranes.

Membrane proteins usually define the biological functions of specific membranes. Depending on their location, membrane proteins can be classified as integral or peripheral. Membranes are involved in transport and in the binding of hormones and other extracellular metabolic signals.

9. Membranes exhibit heterogeneity with respect to their lipid, protein, and carbohydrate components. New lipid bilayer is synthesized from the cytoplasmic face, and lipids are distributed throughout the bilayer by specific flippase mediators. The compartments on either side of the membrane are chemically different, and the membrane surfaces reflect that difference.

10. The movement of substances across the cell membrane can be accomplished by passive diffusion (small nonpolar molecules), facilitated diffusion (polar or charged substances requiring a carrier or channel protein), primary active transport (ATP energy used to concentrate a substance on one side of the membrane), secondary active transport (ion gradient generated by primary active transport used to concentrate a substance on one side of the membrane), or receptor-mediated transport (receptor and ligand in coated pits engulfed by endocytosis). Some transport channels are gated; that is, they open only in the presence of a certain gating substance (neurotransmitter, ions, etc.) or membrane condition (voltage, pH).

COMPANION WEBSITE Take your learning further by visiting the **companion website** for Biochemistry at **www.oup.com/us/mckee** where you can complete a multiple-choice quiz on lipids and membranes to help you prepare for exams.

Suggested**Readings**

Edidin, M., The State of Lipid Rafts: From Model Membranes to Cells, *Annu. Rev. Biophys. Biomol. Struct.* 32:257–283, 2003.

Gadsby, D. C., Vergani, P., and Csanady, L., The ABC Protein Turned Chloride Channel Whose Failure Causes Cystic Fibrosis, *Nature* 440:477–483, 2006.

Jahn, R., and Scheller, R. H., SNAREs: Enzymes for Membrane Fusion, *Nat. Rev. Mol. Cell Biol.* 7:631–643, 2006.

Janmey, P. A., and Kinnunen, P. K. J., Biophysical Properties of Lipids and Dynamic Membranes, *Trends Cell Biol.* 16(10):538–546, 2006.

Jeon, H., and Blocklow, S. C., Structure and Physiological Function of the Low Density Lipoprotein Receptor, *Annu. Rev. Biochem.* 74:535–562, 2005.

King, L. S., Kozono, D., and Agre, P., From Structure to Disease: The Evolving Tale of Aquaporin Biology, *Nat. Rev. Mol. Cell Biol.* 5:687–698, 2004.

Mayor, S., and Riezman, H., Sorting GPI-Anchored Proteins, *Nat. Rev. Mol. Cell Biol.* 5:110–120, 2004.

Parker, M. W., and Fell, S. C., Pore-Forming Protein Toxins: From Structure to Function, *Prog. Biophys. Mol. Biol.* 88:91–142, 2005.

Welsh, M. J., and Smith, A. E., Cystic Fibrosis, *Sci. Am.* 273(6):52–59, 1995.

Wine, J. T., Cystic Fibrosis Lung Disease, *Sci. Med.* 6(3):34–43, 1999.

Key**Words**

active transport, *401*

acyl group, *375*

aquaporin, *406*

autocrine, *377*

carotenoid, *387*

chylomicron, *392*

cis isomer, *373*

cystic fibrosis, *402*

cystic fibrosis transmembrane conductance regulator, *402*

eicosanoid, *376*

essential fatty acid, *375*

facilitated diffusion, *402*

fluid mosaic model, *394*

glycolipid, *383*

GPI anchor, *381*

heterokaryon, *396*

high-density lipoprotein, *394*

integral protein, *398*

intermediate-density lipoprotein, *392*

isoprenoid, *386*

leukotriene, *377*

lipid, *372*

lipid bilayer, *394*

low-density lipoprotein, *394*

mixed terpenoid, *387*

monounsaturated, *374*

nephrogenic diabetes insipidis, *406*

neutral fat, *378*

nonessential fatty acid, *375*

omega-3 fatty acid, *375*

omega-6 fatty acid, *375*

passive transport, *401*

peripheral protein, *398*

phosphoglyceride, *381*

phospholipid, *380*

polar head group, *380*

polyunsaturated, *374*

prenylation, *387*

prostaglandin, *377*

simple diffusion, *402*

sphingolipid, *385*

sphingomyelin, *381*

steroid, *389*

terpene, *386*

thromboxane, *377*

trans isomer, 374

very low density lipoprotein, *392*

wax, *380*

Review**Questions**

These questions are designed to test your knowledge of the key concepts discussed in this chapter, before moving on to the next chapter. You may like to compare your answers to the solutions provided in the back of the book and in the accompanying Study Guide.

1. Clearly define the following terms:
   a. lipid
   b. autocrine regulator
   c. amphipathic
   d. sesquiterpene
   e. lipid bilayer

2. Define the following terms:
   a. prenylation
   b. membrane fluidity
   c. chylomicron
   d. voltage-gated channel
   e. terpene

3. Define the following terms:
   a. peripheral protein
   b. isoprenoid
   c. omega-3 fatty acid
   d. endocannabinoid
   e. neutral fat

4. Define the following terms:
   a. CFTR
   b. aquaporin
   c. GPI anchor
   d. essential oil
   e. mixed terpenoid

5. Define the following terms:
   a. intermediate-density lipoprotein
   b. fluid mosaic model
   c. lipid raft
   d. SREBP
   e. trans fatty acid

6. List a major function of each of the following classes of lipid:
   a. phospholipids
   b. sphingolipids
   c. oils
   d. waxes
   e. steroids
   f. carotenoids

7. To which lipid class does each of the following molecules belong?

b.

$CH_3(CH_2)_{10}CH_2-O-\overset{\overset{O}{\|}}{C}-(CH_2)_{10}CH_3$

c.

$CH_3(CH_2)_7CH=CH(CH_2)_7COOH$

d.

e.

f.

a.

8. What role do plasma lipoproteins play in the human body?

9. Why do plasma lipoproteins require a protein component to accomplish their role?

10. Describe several factors that influence membrane fluidity.

11. Which of the following statements concerning ionophores are true?
    a. Ionophores form channels through which ions flow.
    b. They require energy.

c. Ions may diffuse in either direction.

d. Ionophores may cause voltage gates.

e. They transport all ions with equal ease.

12. Explain the differences in the ease of lateral movement and bilayer translocation movement of phospholipids.

13. Explain how potassium moves across a nerve cell membrane during both depolarization and repolarizations.

14. From what fatty acid are most of the eicosanoids derived? List several medical conditions in which the suppression of eicosanoid synthesis may appear to be advantageous.

15. In which of the following processes do the prostaglandins not have a major recognized role?
    a. reproduction
    b. blood clotting
    c. respiration
    d. inflammation
    e. blood pressure modification

16. Classify each of the following as a monoterpene, diterpene, triterpene, tetraterpene, sesquiterpene, or polyterpene:

a.

b.

c.

d.

e.

f.

17. Discuss the functions of triacylglycerols.

18. Sphingomyelins are amphipathic molecules. Review the structure of a typical sphingomyelin. Which regions are hydrophilic and which are hydrophobic?

19. What changes might you make in the structure of a cell membrane to increase the cell's resistance to mechanical stress?

20. How does the function of HDL promote the reduction of risk of coronary artery disease?

21. Identify the membrane components responsible for each of the following tasks. Explain how each component accomplishes its task.
    a. selective permeability
    b. self-sealing capability
    c. fluidity
    d. asymmetry
    e. active transport of ions

22. Describe how glucose is transported across membranes in the kidney. What type of transport is involved?

23. Describe and give a specific example of a facilitated diffusion process.

24. Compare and contrast the following processes. Give an example of each to illustrate.
    a. primary vs secondary active transport
    b. passive vs facilitated diffusion
    c. carrier-mediated vs channel-mediated transport

25. How do the following substances move across the plasma membranes of animal cells?
    a. CO_2
    b. H_2O
    c. glucose
    d. Cl^-
    e. Na^+

26. How do lipoproteins transport water-insoluble lipid molecules in the bloodstream?

27. Suggest a reason why trans fatty acids have melting points similar to analogous saturated fatty acids.

28. Detergents are synthetic soaplike substances that are used to disrupt membranes and extract membrane proteins. Explain how this process works.

29. Explain why triacylglycerols are not components of lipid bilayers.

30. What happens to the structure of a phospholipid bilayer in a nonpolar solvent such as hexane?

31. What happens to the melting point of a fatty acid mixture when branched-chain fatty acids (found in some bacteria) are added?

32. Eggs, which contain lecithin in addition to ovalbumin and other proteins, are used in the preparation of mayonnaise because they prevent the separation of oil and water mixtures. Explain.

Thought**Questions**

These questions are designed to reinforce your understanding of all of the key concepts discussed in the book so far, including this chapter and all of the chapters before it. They may not have one right answer! The authors have provided possible solutions to these questions in the back of the book and in the accompanying Study Guide, for your reference.

33. Animal cells are enclosed in a cell membrane. According to the fluid mosaic model, this membrane is held together by hydrophobic interactions. Consider the shear forces involved. Why does this membrane not break every time an animal moves?

34. Species-specific antigens are located on the surfaces of human and canine cells. If a heterokaryon is formed from red blood cell membrane from both species, what will happen to each set of antigens? What does this suggest about the nature of membranes?

35. Suggest a reason why elevated LDL levels are a risk factor for coronary artery disease.

36. Glycolipids are nonionic lipids that can orient themselves into bilayers as phospholipids do. They accomplish this feat although they lack an ionic group like that of the phospholipids. Suggest a reason why this is possible.

37. Explain why spontaneous phospholipid translocation (the movement of a molecule from one side of a bilayer to the other) is so slow.

38. Mammals in the Arctic (e.g., reindeer) have higher levels of unsaturated fatty acids in their legs than in the rest of their bodies. Suggest a reason for this phenomenon. Does it have a survival advantage?

39. Explain why entropy increases when a lipid bilayer forms from phospholipid molecules.

40. The fluid mosaic model of membrane structure has been very useful in explaining membrane behavior. However, the description of membrane as proteins floating in a phospholipid sea is oversimplified. Describe some components of membrane that are restricted in their lateral motion.

41. Plants often produce waxes on the surface of their leaves to prevent dehydration and protect against insects. What structural feature of waxes make them more suitable for this task than carbohydrates or proteins?

42. Changes in temperature affect membrane properties. How would you expect the lipid composition of thermophile membranes to differ from those of prokaryotes that live at more normal temperatures?

43. Men on calorie-restricted diets frequently experience an intensified growth of facial hair, which correlates with a rise in blood steroid levels. How would you explain this phenomenon?

44. Boric acid is a potent insecticide that dissipates the wax coat of insects. Explain how this substance kills insects.

45. The myelin sheath insulates the axons of certain neurons in the body. What structural feature of this covering makes it a good insulator?

Lipid Metabolism

Near the Finish Line
The bodies of runners are physiologically adapted to rapidly transform the chemical bond energy in fatty acids, a compact and efficient energy source, into ATP, the molecule that drives muscle contraction.

Overview

THE ROLES LIPIDS PLAY IN LIVING ORGANISMS ARE LARGELY DUE TO THEIR HYDROPHOBIC STRUCTURES. AS PROMINENT COMPONENTS OF CELL membranes, lipids are primarily responsible for the integrity of individual cells and the intracellular compartments that are the hallmark of eukaryotic organisms. The hydrophobic and highly reduced structure of triacylglycerols makes them a compact and rich source of energy for cellular processes. Chapter 12 focuses on the metabolism of the major classes of lipids, that is, how they are synthesized and degraded and how these processes are regulated. Major emphasis is placed on the central metabolite in lipid metabolism: acetyl-coenzyme A. Because of its prominent role in cardiovascular disease, the metabolism of cholesterol is also discussed.

Acetyl-CoA, the energy-rich molecule composed of coenzyme A and a two-carbon acyl group, plays a preeminent role in the metabolism of lipids. In most lipid-related metabolic processes, acetyl-CoA is either a substrate or a product. For example, acetyl-CoA that is not immediately required by a cell in energy production is used in fatty acid synthesis. When fatty acids are degraded to generate energy, acetyl-CoA is produced. Similarly, three acetyl-CoA molecules are combined to form isopentenyl pyrophosphate, the precursor in the synthesis of isoprenoids such as cholesterol and vitamin E. Therefore, molecules as diverse as the terpenes and steroids found in animals and plants are all synthesized from acetyl-CoA. In chapter 12 the metabolism of the major classes of lipid is discussed: fatty acids, triacylglycerols, phospholipids, sphingolipids, and isoprenoids. In addition, the metabolism of some important fatty acid metabolites, the eicosanoids and the ketone bodies, is reviewed. Several metabolic control mechanisms are discussed throughout the chapter.

12.1 FATTY ACIDS AND TRIACYLGLYCEROLS

Fatty acids are an important and efficient energy source for many living cells. In animals, most fatty acids are obtained in the diet. For example, in the average U.S. diet, between 30 and 40% of calories ingested are provided by fat. Triacylglycerol molecules are digested within the lumen of the small intestine (Figure 12.1). After being mixed with **bile salts**, amphipathic molecules with detergent properties that are produced in the liver and temporarily stored in the gallbladder (see p. 391), triacylglycerol molecules are digested by pancreatic lipase to form fatty acids and monoacylglycerol. These latter molecules are then transported across the plasma membrane of intestinal wall cells (enterocytes), where they are reconverted to triacylglycerols. Enterocytes combine triacylglycerols with dietary cholesterol and newly synthesized phospholipids and protein to form chylomicrons (large, low-density lipoproteins). After their secretion into the lymph (tissue fluid derived from blood), chylomicrons pass from the lymph into the bloodstream at the thoracic duct.

Most of the triacylglycerol content of circulating chylomicrons is removed from blood by muscle and the adipose tissue cells (adipocytes), which comprise the body's primary lipid storage depot. Lipoprotein lipase, synthesized by cardiac and skeletal muscle, lactating mammary gland, and adipose tissue, is transferred to the endothelial surface of the capillaries, where it converts the

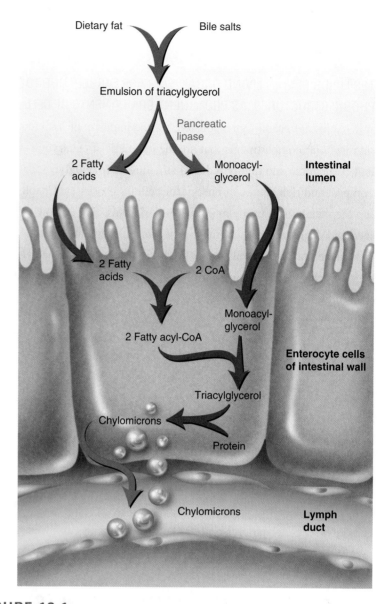

FIGURE 12.1

Digestion and Absorption of Triacylglycerols in the Small Intestine

After triacylglycerols have been emulsified (solubilized) by mixing with bile salts, they are digested by intestinal lipases, the most important of which is pancreatic lipase. The products, fatty acids and monoacylglycerol, are transported into enterocytes and resynthesized to form triacylglycerol. The triacylglycerol molecules, along with newly synthesized phospholipid and protein, are then incorporated into chylomicrons. After the chylomicrons have been transported into lymph, via exocytosis, and then into blood, the triacylglycerols are drawn off by muscle and fat cells. Chylomicron remnants are removed from the blood by the liver.

triacylglycerol in chylomicrons into fatty acids and glycerol. Lipoprotein lipase is activated when it binds to the apoprotein component of chylomicrons known as complex II (C-II). The triacylglycerols in VLDL are also degraded by lipoprotein lipase.

The fatty acids are taken up by cells, whereas glycerol is carried in the blood to the liver, where the enzyme glycerol kinase converts it to glycerol-3-phosphate. Glycerol-3-phosphate is then used in the synthesis of triacylglycerols, phospholipids, or glucose. When lipoprotein lipase has removed about 90% of the triacylglycerols in chylomicrons, the **chylomicron remnants** are removed from the blood by liver cells via receptor-mediated endocytosis.

Depending on an animal's current metabolic needs, fatty acids may be converted to triacylglycerols, degraded to generate energy, or used in membrane synthesis. Immediately after a meal, for example, insulin is released in response to high blood glucose levels. Insulin promotes triacylglycerol storage by inactivating *hormone-sensitive lipase* (an enzyme that hydrolyzes the ester bonds of fat molecules) and promoting triacylglycerol synthesis in fat and muscle cells. Insulin also stimulates the release of VLDL from the liver and activates lipoprotein lipase synthesis and transport to the endothelial cells serving fat and muscle tissue. As a result, the uptake of fatty acids and storage as triacylglycerols is increased. When blood glucose falls after a meal, insulin levels decrease and glucagon levels increase, and the process reverses. Triacylglycerols are degraded to form glycerol and fatty acids. As discussed, glycerol is a substrate for gluconeogenesis in liver. Fatty acids are degraded by the body's cells to generate energy.

Triacylglycerol synthesis (referred to as **lipogenesis**) is illustrated in Figure 12.2. Glycerol-3-phosphate or dihydroxyacetone phosphate reacts sequentially with three molecules of acyl-CoA (fatty acid esters of CoASH). Acyl-CoA molecules are produced in the following reaction:

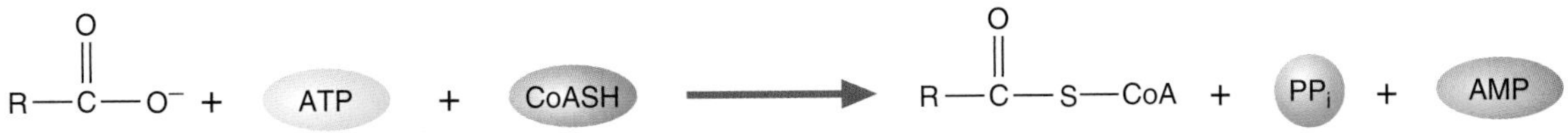

$$R-\overset{\overset{\text{O}}{\|}}{C}-O^- + \text{ATP} + \text{CoASH} \longrightarrow R-\overset{\overset{\text{O}}{\|}}{C}-S-\text{CoA} + \text{PP}_i + \text{AMP}$$

Note that the reaction is driven to completion by the hydrolysis of pyrophosphate by pyrophosphatase.

In the synthesis of triacylglycerols, phosphatidic acid is formed by two sequential acylations of glycerol-3-phosphate or by a pathway involving the direct acylation of dihydroxyacetone phosphate. In the latter pathway, acyldihydroxyacetone phosphate is later reduced to form lysophosphatidic acid. Depending on the pathway used, lysophosphatidic acid synthesis utilizes either an NADH or an NADPH cofactor. Phosphatidic acid is produced when lysophosphatidic acid reacts with a second acyl-CoA. Once formed, phosphatidic acid is converted to diacylglycerol by phosphatidic acid phosphatase. A third acylation reaction forms triacylglycerol. Fatty acids derived from both the diet and *de novo* synthesis are incorporated into triacylglycerols. (The term *de novo* is used by biochemists to indicate new synthesis.) *De novo* synthesis of fatty acids is discussed on p. 429.

When energy reserves are low, the body's fat stores are mobilized in a process referred to as **lipolysis** (Figure 12.3). Lipolysis occurs during fasting, during vigorous exercise, and in response to stress. Several hormones (e.g., glucagon and epinephrine) bind to specific adipocyte plasma membrane receptors and begin a reaction sequence similar to the activation of glycogen phosphorylase. Hormone binding to the receptor elevates cytosolic cAMP levels, which, in turn, activates hormone-sensitive triacyglycerol lipase. Both products of lipolysis (i.e., fatty acids and glycerol) are released into the blood. As stated previously (see p. 287), glycerol is transported to the liver, where it can be used in lipid or glucose synthesis. After their transport across the adipocyte plasma membrane, fatty acids become bound to serum albumin. (Serum albumin is an abundant protein that transports numerous substances in blood. Other examples of hydrophobic albumin ligands include steroids and eicosanoids.) The albumin-bound fatty acids are carried to tissues throughout the body, where they are released from albumin molecules, taken up by cells and used in new-membrane synthesis and energy generation, or stored for use later. Fatty acids are transported into cells by a protein in the plasma membrane. This process is linked to the active transport of sodium. The amount of fatty acid that is transported depends on its concentration in blood and the relative activity of the fatty acid transport mechanism. Cells vary widely in their capacity to transport and use fatty acids. Some cells (e.g., brain and red blood cells) cannot use fatty acids as fuel, although others (e.g., cardiac muscle)

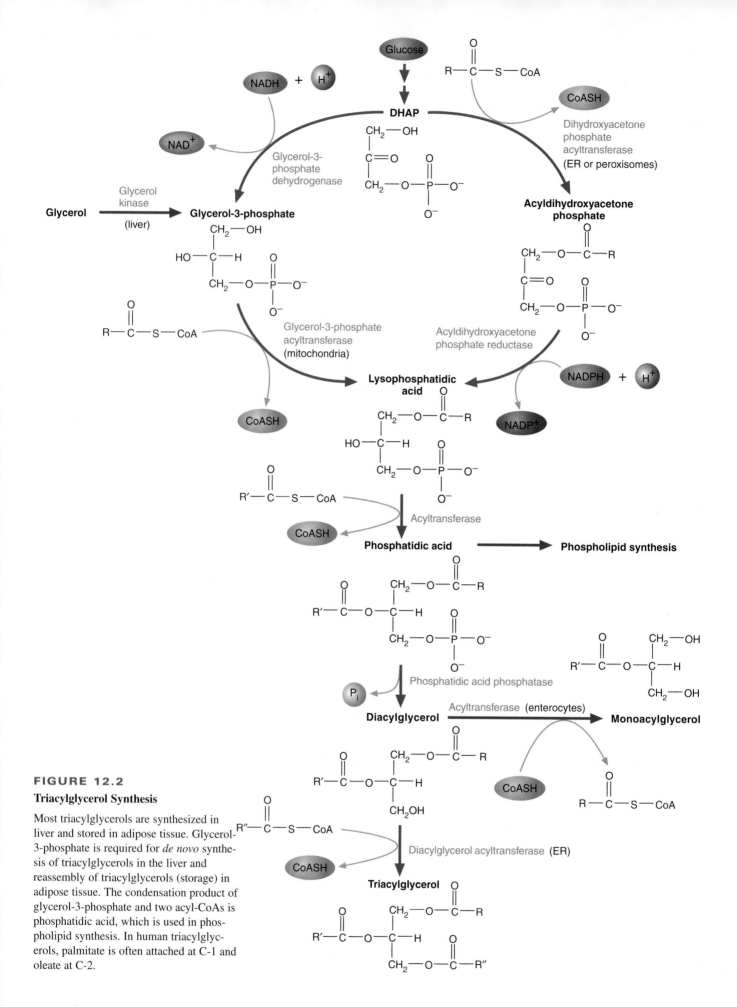

FIGURE 12.2

Triacylglycerol Synthesis

Most triacylglycerols are synthesized in liver and stored in adipose tissue. Glycerol-3-phosphate is required for *de novo* synthesis of triacylglycerols in the liver and reassembly of triacylglycerols (storage) in adipose tissue. The condensation product of glycerol-3-phosphate and two acyl-CoAs is phosphatidic acid, which is used in phospholipid synthesis. In human triacylglycerols, palmitate is often attached at C-1 and oleate at C-2.

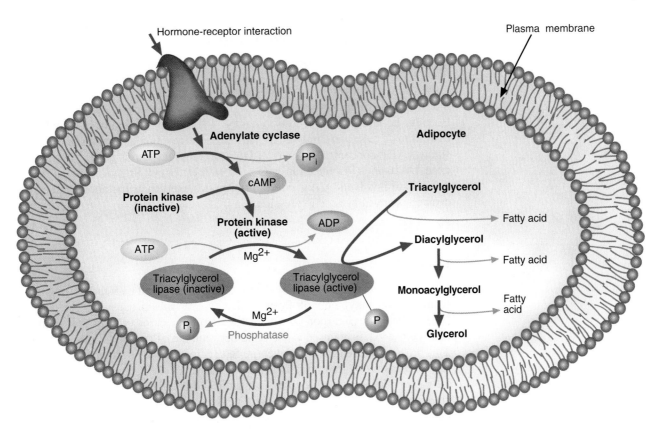

FIGURE 12.3

Diagrammatic View of Lipolysis in Adipocytes

When certain hormones bind to their receptors in adipose tissue, a cascade mechanism releases fatty acids and glycerol from triacylglycerol molecules. Hormone-receptor interaction activates adenylate cyclase. Elevated cAMP levels initiate a phosphorylation cascade that activates HSL (hormone-sensitive lipase or triacylglycerol lipase) via phosphorylation. After their transport across the plasma membrane, fatty acids are transported in blood to other organs bound to serum albumen.

rely on them for a significant portion of their energy requirements. Once they enter a cell, fatty acids must be transported to their destinations (mitochondria, endoplasmic reticulum, and other organelles). Several **fatty acid–binding proteins** (water-soluble proteins whose sole function is to bind and transport hydrophobic fatty acids) may be responsible for this transport.

Most fatty acids are degraded to form acetyl-CoA within mitochondria in a process referred to as β-oxidation. β-Oxidation also occurs in peroxisomes. Other oxidative mechanisms are also available to degrade certain nonstandard fatty acids.

Fatty acids are synthesized when an organism has met its energy needs and nutrient levels are high. (Glucose and several amino acids are substrates for fatty acid synthesis.) Fatty acids are synthesized from acetyl-CoA in a process that is similar to the reverse of β-oxidation. Although most fatty acids are supplied in the diet, most animal tissues can synthesize some saturated and unsaturated fatty acids. In addition, animals can elongate and desaturate dietary fatty acids. For example, arachidonic acid is produced by adding a two-carbon unit and introducing two double bonds to linoleic acid.

KEY CONCEPTS

- When energy reserves are high, triacylglycerols are stored in the process called lipogenesis.

- When energy reserves are low, triacylglycerols are degraded to form fatty acids and glycerol. This process is called lipolysis.

QUESTION 12.1

You have just consumed a cheeseburger. Trace the fat molecules (triacylglycerol) from the cheeseburger to your adipocytes (fat cells).

VLDL secretion by liver cells depends directly on the intracellular concentration of fatty acids. A cytoplasmic fatty acid–binding protein (FABP) may be responsible for the transport of fatty acids to the SER (smooth endoplasmic reticulum), the site of lipid synthesis (triacylglycerols, membrane lipids, steroids, etc.). Because hepatic VLDL secretion has been found to be greater in female rats than in male rats, there may be a connection between an animal's sex hormone status, FABP concentration, and VLDL secretion rate. If this is so, what do you expect to happen if a male rat is injected with estrogen? In general, what mechanisms are involved? [*Hint*: Steroid hormones exert their metabolic effects by causing changes in gene expression.]

Fatty Acid Degradation

Most fatty acids are degraded by the sequential removal of two-carbon fragments from the carboxyl end of fatty acids. During this process, referred to as **β-oxidation**, the β-carbon (second carbon from the carboxyl group) is oxidized and acetyl-CoA is released as the bond between the α- and β-carbons is cleaved. This process is repeated until the entire fatty acid chain has been processed. Other mechanisms for degrading fatty acids are known. Most of these degrade unusual fatty acids; for example, odd-chain or branched-chain molecules usually require an α-oxidation step in which the fatty acid chain is shortened by one carbon by means of a stepwise oxidative decarboxylation. In some organisms, the carbon farthest from the carboxyl group may be oxidized by a process called ω-oxidation. The subsequent β-oxidation generates a short-chain dicarboxylic acid. Very little is known about the structure, mechanism, or relevance of ω-oxidizing enzymes. Once formed, acetyl-CoA and other short-chain products can be used to generate energy or provide metabolic intermediates. β-Oxidation is discussed next. The degradation of odd-chain, branched-chain, and unsaturated fatty acids is also described.

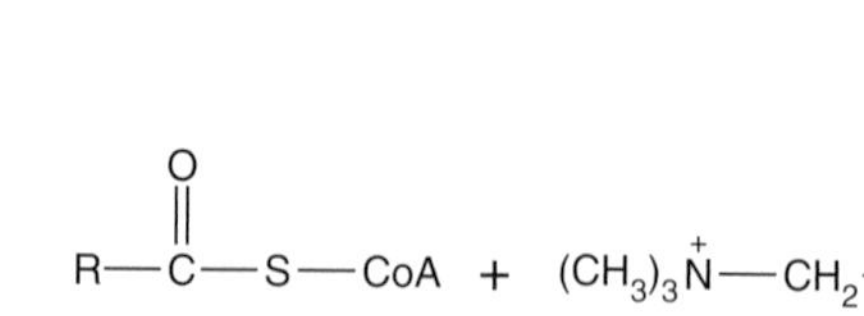

β-Oxidation occurs primarily within mitochondria. Before β-oxidation begins, each fatty acid is activated in a reaction with ATP and CoASH (see p. 417). The enzyme that catalyzes this reaction, acyl-CoA synthetase, is found in the outer mitochondrial membrane. Because the mitochondrial inner membrane is impermeable to most acyl-CoA molecules, a special carrier called *carnitine* is used to transport acyl groups into the mitochondrion (Figure 12.4). Carnitine-mediated transfer of acyl groups into the mitochondrial matrix is accomplished through the following mechanism (Figure 12.5):

FIGURE 12.4

Structure of Carnitine

1. Each acyl-CoA molecule is converted to an acylcarnitine derivative:

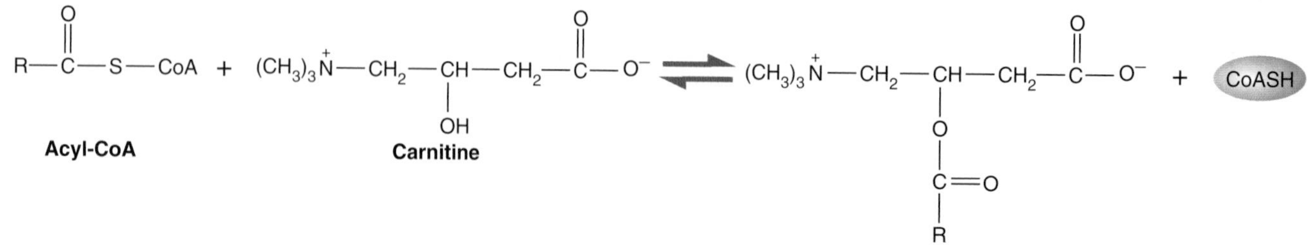

Acyl-CoA **Carnitine** **Acylcarnitine**

This reaction is catalyzed by carnitine acyltransferase I (CAT-I).

2. A carrier protein within the mitochondrial inner membrane transfers acylcarnitine into the mitochondrial matrix.

3. Acyl-CoA is regenerated by carnitine acyltransferase II (CAT-II).

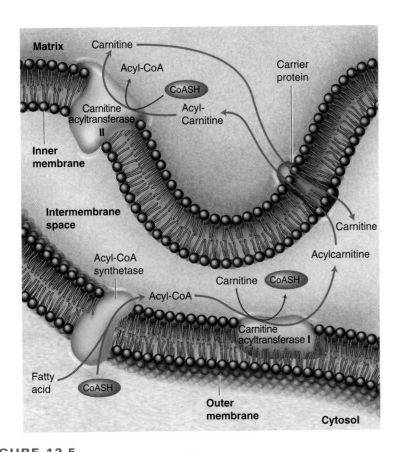

FIGURE 12.5

Fatty Acid Transport into the Mitochondrion

Fatty acids are activated to form acyl-CoA by acyl-CoA synthetase, an enzyme in the outer mitochrondrial membrane. Acyl-CoA then reacts with carnitine to form an acylcarnitine derivative. Carnitine acyltransferase I catalyzes this reaction. Acylcarnitine is transported across the inner membrane by a carrier protein and then reconverted to carnitine and acyl-CoA by carnitine acyltransferase II.

4. Carnitine is transported back into the intermembrane space by the carrier protein. It then reacts with another acyl-CoA.

A summary of the reactions of the β-oxidation of saturated fatty acids is shown in Figure 12.6. The pathway begins with an oxidation-reduction reaction, catalyzed by acyl-CoA dehydrogenase (an inner mitochondrial membrane flavoprotein), in which one hydrogen atom each is removed from the α- and β-carbons and transferred to the enzyme-bound FAD:

The FADH$_2$ produced in this reaction then donates two electrons to the mitochondrial electron transport chain (ETC). There are several isozymes of acyl-CoA dehydrogenase, each specific to a different fatty acid chain length. The product of this reaction is *trans-α,β-enoyl-CoA*.

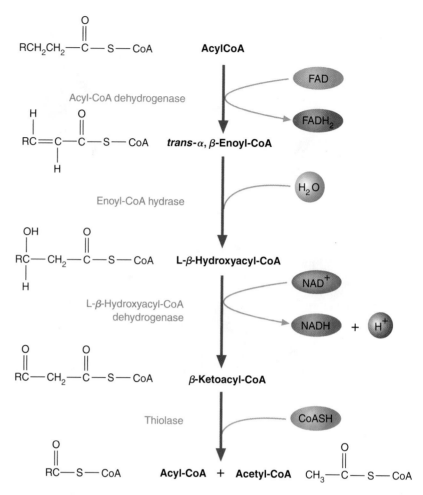

FIGURE 12.6

β-Oxidation of Acyl-CoA

The β-oxidation of acyl-CoA molecules consists of four reactions that occur in the mito-chondrial matrix. Each cycle of reactions forms acetyl-CoA and an acyl-CoA that is shorter by two carbons.

The second reaction, catalyzed by enoyl-CoA hydrase, involves a hydration of the double bond between the α- and β-carbons:

The β-carbon is now hydroxylated. In the next reaction this hydroxyl group is oxidized. The production of a β-ketoacyl-CoA is catalyzed by β-hydroxyacyl-CoA dehydrogenase:

The electrons transferred to NAD$^+$ are later donated to complex I of the ETC. Finally, thiolase (sometimes referred to as β-ketoacyl-CoA thiolase) catalyzes a C_α-C_β cleavage:

$$R-\underset{\beta}{\overset{O}{\underset{\|}{C}}}-\underset{\alpha}{CH_2}-\overset{O}{\overset{\|}{C}}-S-CoA \;+\; CoASH \longrightarrow R-\overset{O}{\overset{\|}{C}}-S-CoA \;+\; CH_3-\overset{O}{\overset{\|}{C}}-S-CoA$$

β-Ketoacyl-CoA **Acyl-CoA** **Acetyl-CoA**

In this reaction, sometimes called a **thiolytic cleavage**, an acetyl-CoA molecule is released. The other product, an acyl-CoA, now contains two fewer C atoms.

The four steps just outlined constitute one cycle of β-oxidation. During each later cycle, a two-carbon fragment is removed. This process, sometimes called the *β-oxidation spiral*, continues until, in the last cycle, a four-carbon acyl-CoA is cleaved to form two molecules of acetyl-CoA.

In muscle, the rate of β-oxidation depends on the availability of its substrate (i.e., the concentration of fatty acids in blood) and the tissue's current energy requirements. When NADH/NAD$^+$ ratios are high, β-hydroxyacyl–CoA dehydrogenase is inhibited. High acetyl-CoA levels depress the activity of thiolase. In liver, where fatty acids are also used in the synthesis of triacylglycerols and phospholipids, the rate of β-oxidation depends on how quickly these molecules are transported into mitochondria. When glucose levels are high and excess glucose molecules are being converted into fatty acids, malonyl-CoA, the product of the first committed step in fatty acid synthesis, prevents a futile cycle by inhibiting CAT-I.

The following equation summarizes the oxidation of palmitoyl-CoA:

$$CH_3(CH_2)_{14}\overset{O}{\overset{\|}{C}}-S-CoA \;+\; 7\;\boxed{FAD} \;+\; 7\;\boxed{NAD^+} \;+\; 7\;\boxed{CoASH} \;+\; 7\;\boxed{H_2O} \longrightarrow$$

$$8\;CH_3\overset{O}{\overset{\|}{C}}-S-CoA \;+\; 7\;\boxed{FADH_2} \;+\; 7\;\boxed{NADH} \;+\; 7\;\boxed{H^+}$$

The acetyl-CoA molecules produced by fatty acid oxidation are converted via the *citric acid cycle* to CO_2 and H_2O as additional NADH and FADH$_2$ are formed. A portion of the energy released as NADH and FADH$_2$ are oxidized by the ETC is later captured in ATP synthesis via *oxidative phosphorylation*. The complete oxidation of acetyl-CoA is discussed in Chapter 10. The calculation of total number of ATP that can be generated from palmitoyl is reviewed next.

QUESTION 12.3

Identify each of the following biomolecules:

$$R'-\overset{O}{\overset{\|}{C}}-O-\underset{\underset{CH_2-O-\underset{\underset{O^-}{|}}{\overset{\overset{O}{\|}}{P}}-O^-}{\overset{\overset{CH_2-O-\overset{O}{\overset{\|}{C}}-R}{|}}{C}}-H$$

(a)

$$R-\overset{O}{\overset{\|}{C}}-S-CoA$$

(b)

$$(CH_3)_3\overset{+}{N}-CH_2-CH-CH_2-\overset{O}{\overset{\|}{C}}-O^-$$

(c)

QUESTION 12.4

In the absence of oxygen, cells can produce small amounts of ATP from the anaerobic oxidation of glucose. This is not true for fatty acid oxidation. Explain.

The Complete Oxidation of a Fatty Acid

The aerobic oxidation of a fatty acid generates a large number of ATP molecules. As previously described (see p. 352), the oxidation of each $FADH_2$ during electron transport and oxidative phosphorylation yields approximately 1.5 molecules of ATP. Similarly, the oxidation of each NADH yields approximately 2.5 molecules of ATP. The yield of ATP from the oxidation of palmitoyl-CoA, which generates 7 $FADH_2$ 7 NADH, and 8 acetyl-CoA molecules to form CO_2 and H_2O, is calculated as follows:

$$7\ FADH_2 \times 1.5\ ATP/FADH_2 = 10.5\ ATP$$

$$7\ NADH \times 2.5\ ATP/NADH = 17.5\ ATP$$

$$8\ Acetyl\text{-}CoA \times 10\ ATP/acetyl\text{-}CoA = \frac{80\quad ATP}{108\quad ATP}$$

The formation of palmitoyl-CoA from palmitic acid uses two ATP equivalents. The net synthesis of ATP per molecule of palmitoyl-CoA is therefore 106 molecules of ATP.

The yield of ATP from the oxidations of palmitic acid and glucose can be compared. Recall that the total number of ATP molecules produced per glucose molecule is approximately 31. If glucose and palmitic acid molecules are compared in terms of the number of ATP molecules produced per carbon atom, palmitic acid is a superior energy source. The ratio for glucose is 31/6 or 5.2 ATP molecules per carbon atom. Palmitic acid yields 106/16 or 6.6 ATP molecules per carbon atom. The oxidation of palmitic acid generates more energy than that of glucose because palmitic acid is a more reduced molecule. (Glucose with its six oxygen atoms is a partially oxidized molecule.)

QUESTION 12.5

Determine the number of moles of NADH, $FADH_2$, and ATP molecules that can be synthesized from 1 mol of stearic acid.

β-OXIDATION IN PEROXISOMES β-Oxidation of fatty acids also occurs within peroxisomes. In animals, peroxisomal β-oxidation appears to shorten very long-chain fatty acids. The resulting medium-chain fatty acids are further degraded within mitochondria. Peroxisomal membrane possesses an acyl-CoA ligase activity that is specific for very long chain fatty acids. Mitochondria apparently cannot activate long-chain fatty acids such as tetracosanoic (24:0) and hexacosanoic (26:0) acids. Peroxisomal carnitine acyltransferases catalyze the transfer of these molecules into peroxisomes, where they are oxidized to form acetyl-CoA and medium-chain acyl-CoA molecules (i.e., possessing between 6 and 12 carbons). Medium-chain acyl-CoAs are further degraded via β-oxidation within mitochondria.

Although the reactions of peroxisomal β-oxidation are similar to those in mitochondria, there are some notable differences. First, the initial reaction in the peroxisomal pathway is catalyzed by a different enzyme. This reaction is a dehydration catalyzed by an acyl-CoA oxidase. The reduced coenzyme $FADH_2$ then donates its electrons directly to O_2 instead of UQ (coenzyme Q). This feature of peroxisomal β-oxidation is in sharp contrast to the mitochondrial pathway, which synthesizes ATP. The H_2O_2 produced when $FADH_2$ is oxidized is converted to H_2O by catalase. Second, the next two reactions in peroxisomal β-oxidation are catalyzed by two enzyme activities (enoyl-CoA hydrase and 3-hydroxyacyl CoA dehydrogenase) found on the same protein molecule. Finally, the last enzyme in the pathway (β-ketoacyl-CoA thiolase) and its mitochondrial version have different substrate specificities. The former does not efficiently bind medium-chain acyl-CoAs.

THE KETONE BODIES Most of the acetyl-CoA produced during fatty acid oxidation is used by the citric acid cycle or in isoprenoid synthesis (Section 12.3). Under normal conditions, fatty acid metabolism is so carefully regulated that only small amounts of excess acetyl-CoA are produced. In a process called **ketogenesis**, excess acetyl-CoA molecules are converted to acetoacetate, β-hydroxybutyrate, and acetone, a group of molecules called the **ketone bodies** (Figure 12.7).

Ketone body formation, which occurs within the matrix of liver mitochondria, begins with the condensation of two acetyl-CoAs to form acetoacetyl-CoA. Then acetoacetyl-CoA condenses with another acetyl-CoA to form β-hydroxy-β-methylglutaryl-CoA (HMG-CoA). In the next reaction, HMG-CoA is cleaved to form acetoacetate and acetyl-CoA. Acetoacetate is then reduced to form β-hydroxybutyrate. Acetone is formed by the spontaneous decarboxylation of acetoacetate when the latter molecule's concentration is high. (This condition, referred to as **ketosis**, occurs during starvation and in uncontrolled diabetes,

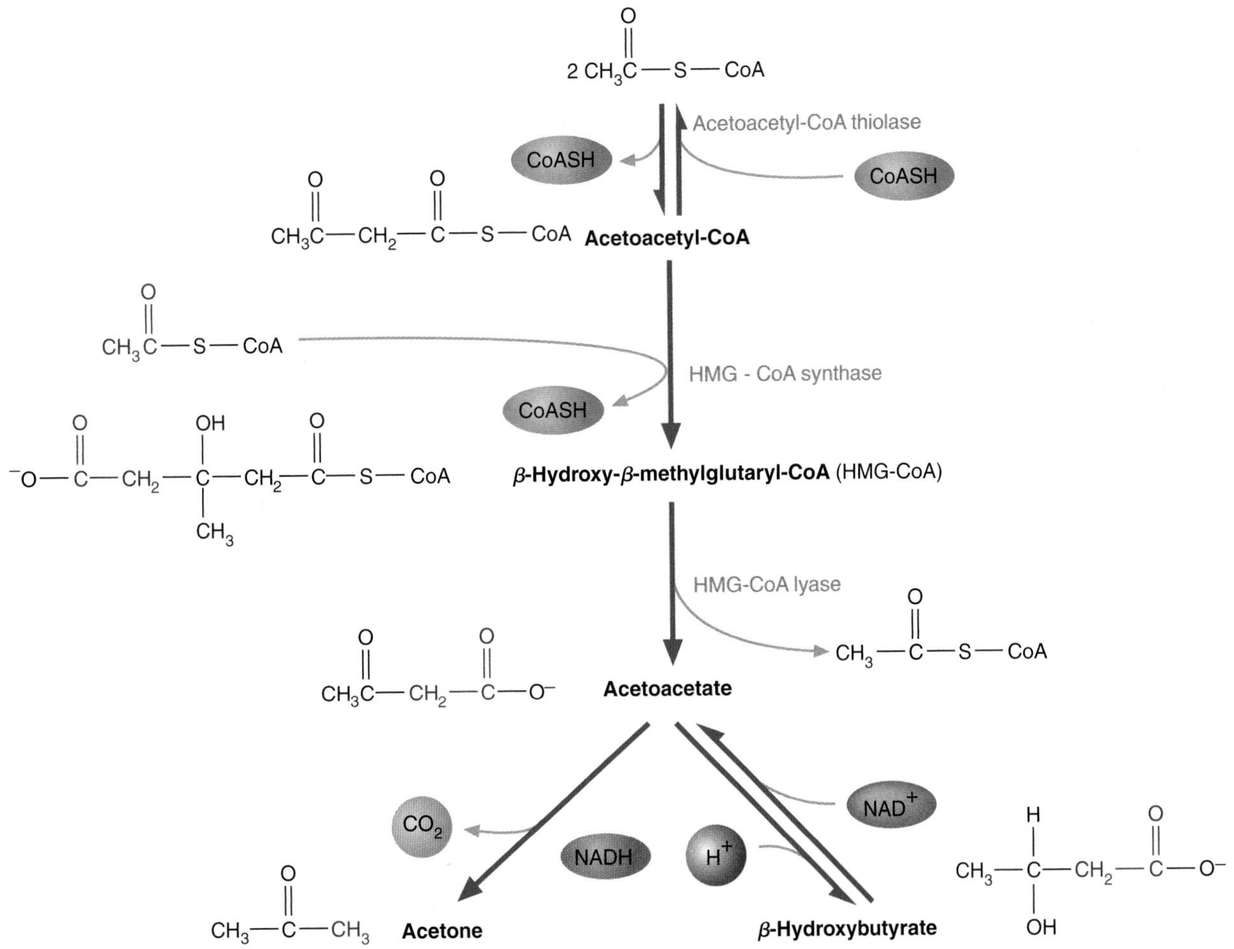

FIGURE 12.7

Ketone Body Formation

Ketone bodies (acetoacetate, acetone, and β-hydroxybutyrate) are produced within the mitochondria when excess acetyl-CoA is available. Under normal circumstances, only small amounts of ketone bodies are produced.

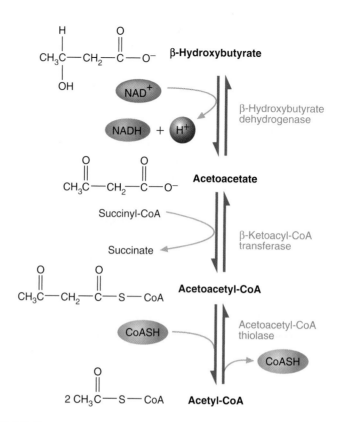

FIGURE 12.8

Conversion of Ketone Bodies to Acetyl-CoA

Some organs (e.g., heart and skeletal muscle) can use ketone bodies (β-hydroxybutyrate and acetoacetate) as an energy source under normal conditions. During starvation, however, ketone bodies become an important fuel source for the brain. Because liver does not have β-ketoacid-CoA transferase, it cannot use ketone bodies as an energy source. These reactions are reversible.

KEY CONCEPTS

- In β-oxidation, fatty acids are degraded by breaking the bond between the α- and β-carbon atoms.
- The ketone bodies are produced from excess molecules of acetyl-CoA.

a metabolic disease discussed in Chapter 16: see the Biochemistry in Perspective box entitled Diabetes Mellitus. In both these conditions there is a heavy reliance on fat stores and β-oxidation of fatty acids to supply energy.)

Several tissues, most notably cardiac and skeletal muscle, use ketone bodies to generate energy. During prolonged starvation (i.e., in the absence of sufficient glucose) the brain uses ketone bodies as an energy source. The mechanism by which acetoacetate and β-hydroxybutyrate are converted to acetyl-CoA is illustrated in Figure 12.8.

Fatty Acid Oxidation: Double Bonds and Odd Chains

The β-oxidation pathway degrades saturated fatty acids with an even number of carbon atoms. Certain additional reactions are required to degrade unsaturated, odd-chain, and branched-chain fatty acids.

UNSATURATED FATTY ACID OXIDATION The oxidation of unsaturated fatty acids such as oleic acid requires additional enzymes. They are needed because, unlike the *trans* double bonds introduced during β-oxidation, the double bonds of most naturally occurring unsaturated fatty acids have a *cis* configuration. The enzyme enoyl-CoA isomerase converts the *cis-β,γ* double bond to a *trans-α,β* double bond. The β-oxidation of oleic acid is illustrated in Figure 12.9.

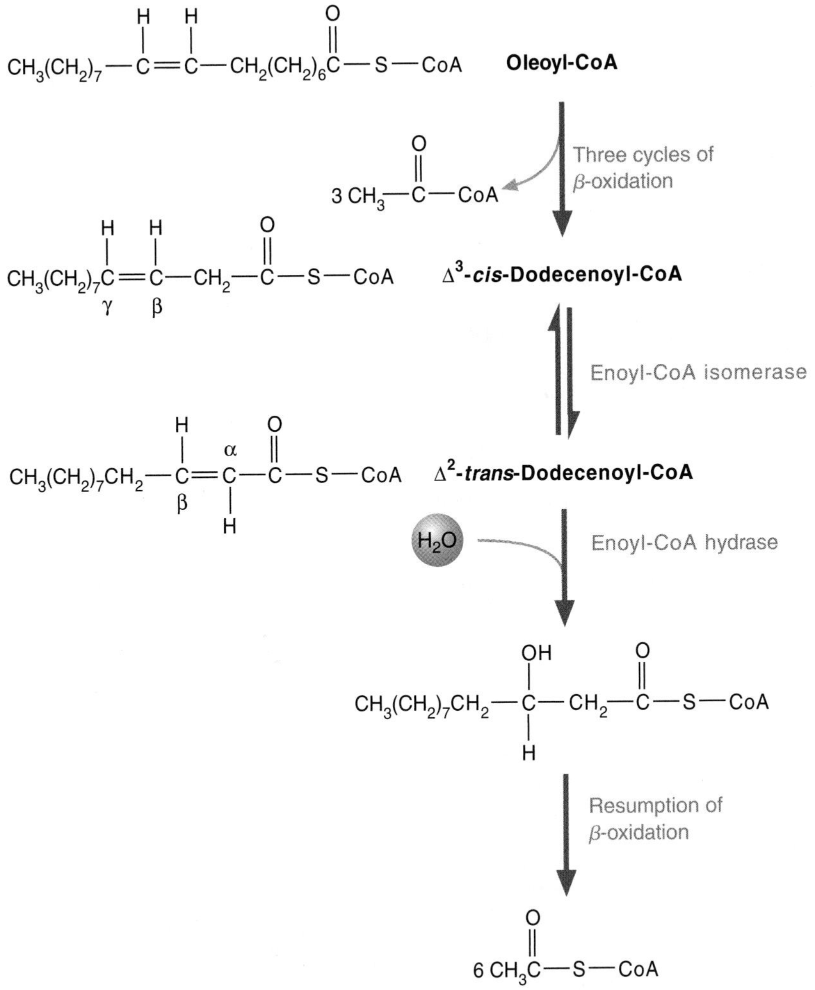

FIGURE 12.9

β-Oxidation of Oleoyl-CoA

β-Oxidation of the CoA derivative of oleic acid progresses until Δ^3-*cis*-dodecenoyl-CoA is produced. This molecule is not a suitable substrate for β-oxidation because it contains a *cis* double bond. After conversion of the β,γ-*cis* double bond to an α,β-*trans* double bond, β-oxidation recommences.

ODD-CHAIN FATTY ACID OXIDATION Although most fatty acids contain an even number of carbon atoms, certain organisms (e.g., some plants and microorganisms) produce odd-chain fatty acid molecules. β-Oxidation of such fatty acids proceeds normally until the last β-oxidation cycle, which yields one acetyl-CoA molecule and one propionyl-CoA molecule. Propionyl-CoA is then converted to succinyl-CoA, a citric acid cycle intermediate (Figure 12.10). Ruminant animals such as cattle and sheep derive a substantial amount of energy from the oxidation of odd-chain fatty acids. These molecules are produced by microbial fermentation processes in the rumen (stomach).

α-OXIDATION *α-Oxidation* is a mechanism for degrading branched-chain fatty acid molecules such as phytanic acid. Phytanic acid, a 20-carbon fatty acid, is an oxidation product of phytol, a diterpene alcohol esterified to chlorophyll, the photosynthetic pigment. Phytol, found in green vegetables, is converted to phytanic acid after ingestion. Phytanic acid is a component of dairy products and other foods derived from herbivorous (plant-eating) animals. In humans,

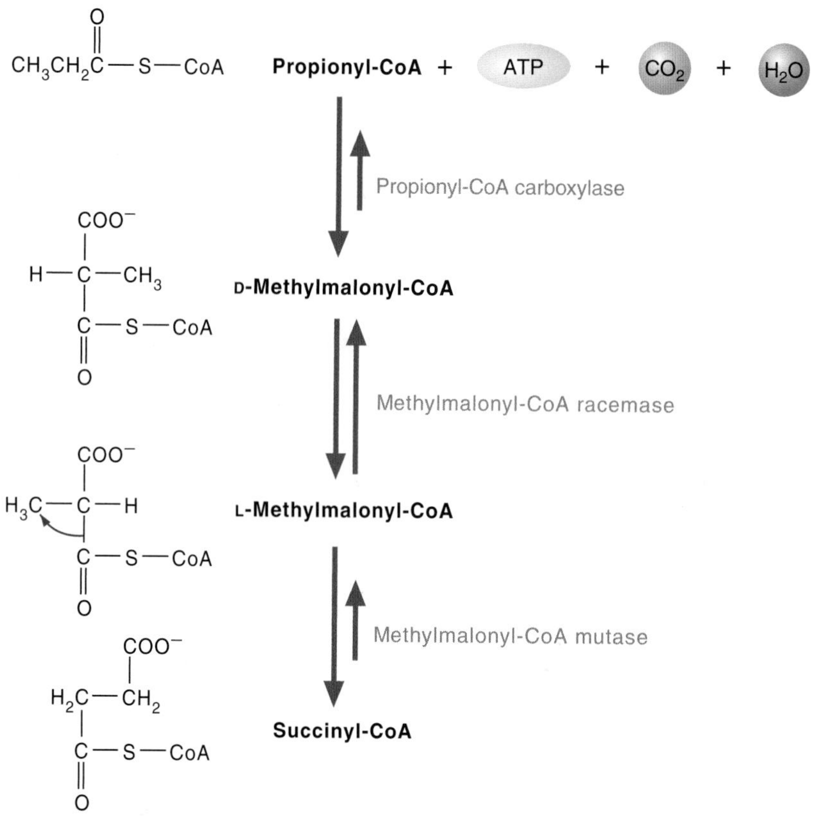

$$CH_3CH_2\overset{\overset{\displaystyle O}{\|}}{C}\!\!-\!\!S\!\!-\!\!CoA \qquad \textbf{Propionyl-CoA} \;+\; \boxed{ATP} \;+\; \bigcirc\!\!CO_2 \;+\; \bigcirc\!\!H_2O$$

Propionyl-CoA carboxylase

$$\begin{array}{c} COO^- \\ | \\ H\!\!-\!\!C\!\!-\!\!CH_3 \\ | \\ C\!\!-\!\!S\!\!-\!\!CoA \\ \| \\ O \end{array} \qquad \textbf{D-Methylmalonyl-CoA}$$

Methylmalonyl-CoA racemase

$$\begin{array}{c} COO^- \\ | \\ H_3C\!\!-\!\!C\!\!-\!\!H \\ | \\ C\!\!-\!\!S\!\!-\!\!CoA \\ \| \\ O \end{array} \qquad \textbf{L-Methylmalonyl-CoA}$$

Methylmalonyl-CoA mutase

$$\begin{array}{c} COO^- \\ | \\ H_2C\!\!-\!\!CH_2 \\ | \\ C\!\!-\!\!S\!\!-\!\!CoA \\ \| \\ O \end{array} \qquad \textbf{Succinyl-CoA}$$

FIGURE 12.10

Conversion of Propionyl-CoA to Succinyl-CoA

In the first step, propionyl-CoA is carboxylated by propionyl-CoA carboxylase, an enzyme with a biotin cofactor. The product, D-methylmalonyl-CoA, is isomerized by methylmalonyl-CoA racemase to form L-methylmalonyl-CoA. In the last step, a hydrogen atom and the carbonyl-CoA group exchange positions. This unusual reaction is catalyzed by methylmalonyl-CoA mutase, an enzyme that require 5′-deoxyadenosylcobalamin, usually designated as vitamin B_{12}.

KEY CONCEPT

Several reactions in addition to β-oxidation are required to degrade unsaturated, odd-chain, and branched-chain fatty acids.

α-oxidation takes place in peroxisomes. In some plant tissues (e.g., leaves and seeds), α-oxidation is a major mechanism in the degradation of long-chain fatty acids.

β-Oxidation of phytanic acid is blocked by the methyl group substituent on C-3 (the β-position). Consequently, the first step in phytanic acid catabolism is an α-oxidation in which the molecule is converted to a α-hydroxy fatty acid. This reaction is followed by the removal of the carboxyl group (Figure 12.11). After activation to a CoA derivative, the product, pristanic acid, can be further degraded by β-oxidation. All subsequent side chain methyl groups will now be in the α-position, which is not a problem for β-oxidation enzymes. The ability to oxidize phytanic acid is critical because large quantities of this molecule are found in the diet. In *Refsum's disease* (also referred to as *phytanic acid storage syndrome*) a buildup of phytanic acid causes very serious neurological problems. In this rare autosomal recessive condition, nerve damage is due to a missing or defective gene that codes for phytanoyl-CoA hydroxylase. The mechanism by which phytanic acid accumulation causes nerve damage is unknown. Eating less phytanic acid–containing foods (i.e., dairy products) can significantly reduce nerve damage.

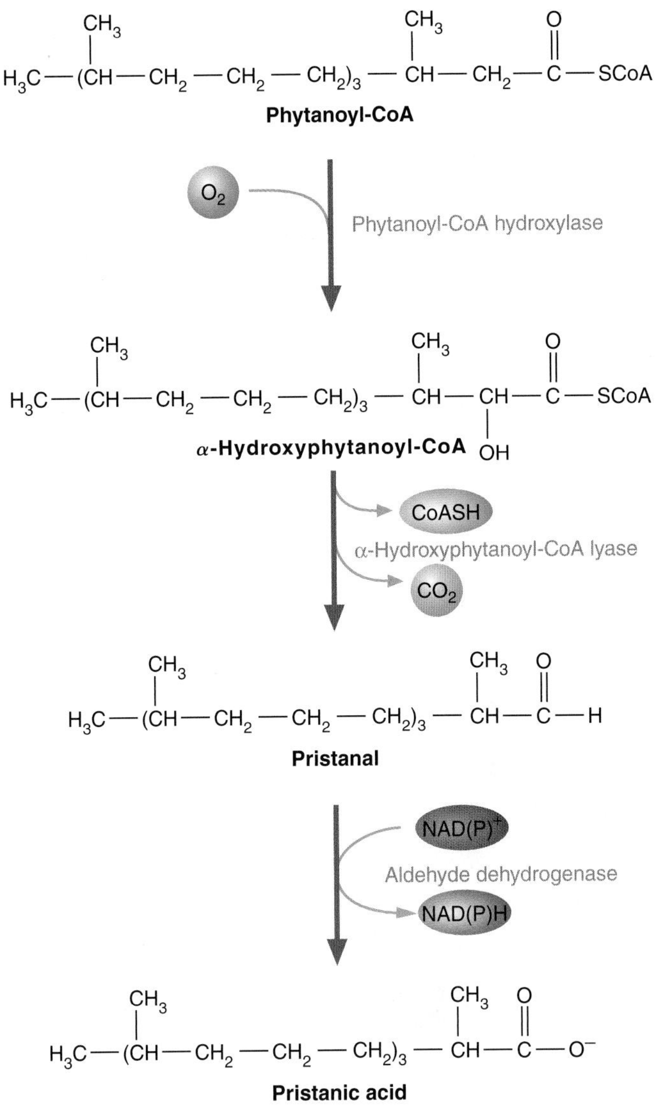

FIGURE 12.11

α-Oxidation of Phytanic Acid in Peroxisomes

In α-oxidation, phytanoyl-CoA is converted to α-hydroxyphytanoyl-CoA in an O_2-requiring reaction. α-Hydroxyphytanoyl-CoA is converted to pristanal in a TPP-requiring decarboxylation reaction in which the α-carbon is oxidized. The thioester bond of the other product, formyl-CoA, is subsequently cleaved to form CoASH and formic acid (HCOOH), which is then oxidized to form CO_2. Pristanal is oxidized in an NAD(P)$^+$-requiring reaction to form pristanic acid. Pristanic acid is then further degraded by β-oxidation after esterification with CoASH. Products of this process are three acetyl-CoAs, three propionyl-CoAs, and one isobutyl-CoA.

QUESTION 12.6

In the past, mammals were believed to be unable to use fatty acids in gluconeogenesis. (Acetyl-CoA cannot be converted to pyruvate because the reaction catalyzed by pyruvate dehydrogenase is irreversible.) Recent experimental evidence indicates that certain unusual fatty acids (i.e., those with odd chains or two carboxylic acid groups) can be converted to glucose in small but measurable quantities. One molecule of propionyl-CoA is produced when an odd-carbon chain fatty acid is oxidized. Describe a possible biochemical pathway by which a liver cell might synthesize glucose from propionyl-CoA. [*Hint*: Refer to Figure 12.10.] One of the products of the β-oxidation of dicarboxylic acids is succinyl-CoA. Propose a biochemical pathway for the conversion of the molecule illustrated in Figure 12.12 to glucose.

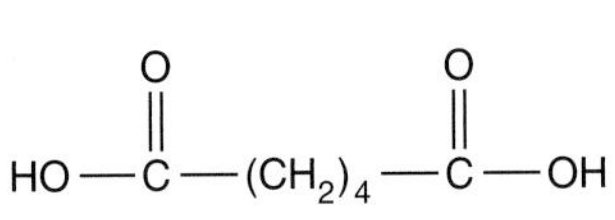

FIGURE 12.12

Adipic Acid

Fatty Acid Biosynthesis

Although fatty acid synthesis occurs within the cytoplasm of most animal cells, liver is the major site for this process. (Recall, for example, that liver produces VLDL. See p. 392.) Fatty acids are synthesized when the diet is low in fat and/or high in carbohydrate or protein. Most fatty acids are synthesized from dietary

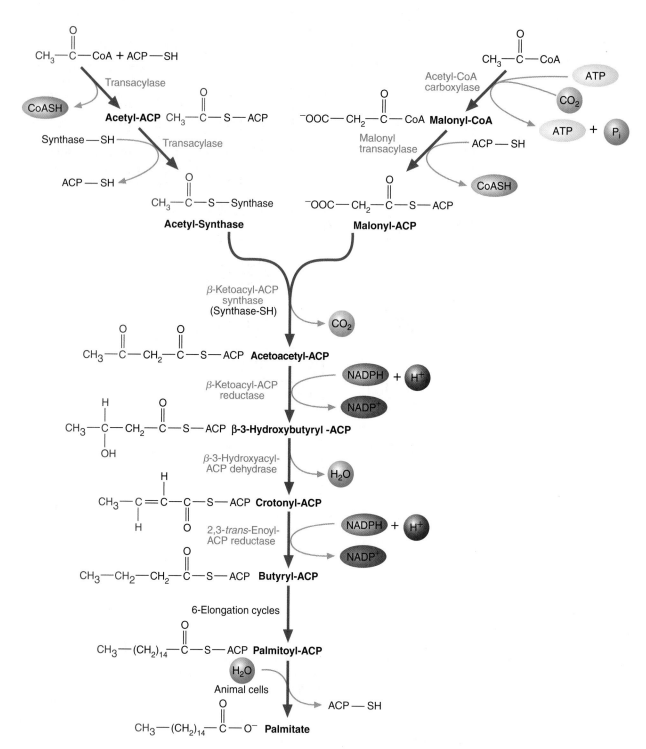

FIGURE 12.13

Fatty Acid Biosynthesis

The substrates for fatty acid synthesis are acetyl-ACP (formed from acetyl-CoA and ACP by a transacetylase) and malonyl-ACP (formed from acetyl-CoA, CO_2, and ACP by ACC and malonyl transacetylase). During each elongation cycle, the nascent fatty acid molecule grows by two carbons when a malonyl-ACP condenses with the growing acyl-ACP chain in a reaction catalyzed by β-ketoacyl-ACP synthase. This is followed by a reduction of the new carbonyl group by β-ketoacyl-ACP reductase to form an alcohol, and the removal of water to form a carbon-carbon double bond. Each elongation cycle ends with the reduction of the carbon-carbon double bond. In animal cells, the process ends with the release of palmitate from ACP.

glucose. As discussed, glucose is converted to pyruvate in the cytoplasm. After entering the mitochondrion, pyruvate is converted to acetyl-CoA, which condenses with oxaloacetate, a citric acid cycle intermediate, to form citrate. When mitochondrial citrate levels are sufficiently high (i.e., cellular energy requirements are low), citrate enters the cytoplasm, where it is cleaved to form acetyl-CoA and oxaloacetate. The net reaction for the synthesis of palmitic acid from acetyl-CoA is as follows:

8 Acetyl-CoA + 14 NADPH + 14 H^+ + 7 ATP $\longrightarrow$

Palmitate + 14 $NADP^+$ + 7 ADP + 7 P_i + 8 CoASH + 6 H_2O

A relatively large quantity of NADPH is required in fatty acid synthesis. A substantial amount of NADPH is provided by the pentose phosphate pathway (see p. 290). Reactions catalyzed by isocitrate dehydrogenase (see p. 323) and malic enzyme (see p. 330) provide smaller amounts.

The biosynthesis of fatty acids is outlined in Figure 12.13. At first glance, fatty acid synthesis appears to be the reverse of the β-oxidation pathway. For example, fatty acids are constructed by the sequential addition of two-carbon groups supplied by acetyl-CoA. Additionally, the same intermediates (i.e., β-ketoacyl-, β-hydroxyacyl-, and α,β-unsaturated acyl groups) are found in both pathways. A closer examination, however, reveals several notable differences between fatty acid synthesis and β-oxidation.

1. **Location**. Fatty acid synthesis occurs predominantly in the cytoplasm. (Recall that β-oxidation occurs within mitochondria and peroxisomes.)

2. **Enzymes**. The enzymes that catalyze fatty acid synthesis are significantly different in structure from those in β-oxidation. In eukaryotes, most of these enzymes are components of a multienzyme complex referred to as fatty acid synthase.

3. **Thioester linkage**. The intermediates in fatty acid synthesis are linked through a thioester linkage to **acyl carrier protein** (ACP), a component of fatty acid synthase. (Recall that acyl groups are attached to CoASH through a thioester linkage during β-oxidation.) Note that acyl groups are attached to both ACP and CoASH via a phosphopantetheine prosthetic group (Figure 12.14).

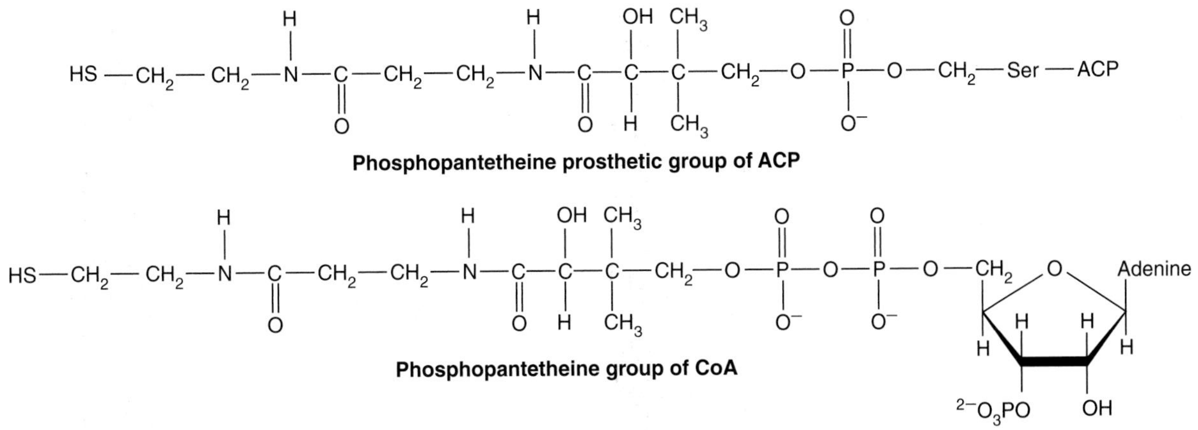

Phosphopantetheine prosthetic group of ACP

Phosphopantetheine group of CoA

FIGURE 12.14

Comparison of the Phosphopantetheine Group in Acyl Carrier Protein (ACP) and in Coenzyme A (CoASH)

Fatty acids are attached to this prosthetic group on ACP during fatty acid biosynthesis and on CoASH during β-oxidation.

4. Electron carriers. In contrast to β-oxidation, which produces NADH and FADH$_2$, fatty acid synthesis consumes NADPH.

Fatty acid synthesis begins with the carboxylation of acetyl-CoA to form malonyl-CoA, an irreversible reaction that is catalyzed by acetyl-CoA carboxylase (ACC) (Figure 12.15). The first phase of this reaction is ATP-dependent carboxylation of biotin to give carboxybiotin. Subsequent decarboxylation results in the transfer of an activated CO$_2$ from biotin to acetyl-CoA. Acetyl-CoA carboxylation, the rate-limiting step in fatty acid synthesis, is an activating reaction that is necessary because the next reaction, a carbon-carbon condensation, is thermodynamically unfavorable.

Acetyl-CoA carboxylase is found in most organisms. In eukaryotes, ACC contains three domains: BCCP (biotin carboxyl carrier protein), BC (biotin carboxylase), and CT (carboxyltransferase). Biotin is bound to BCCP via an amide linkage to the side chain of a lysine residue. The flexible lysine side chain is, in effect, a swinging arm that transfers the newly carboxylated biotin from the active site of the BC domain to the active site of the CT domain (a 7 Å distance). CT then catalyzes the transfer of the carboxyl group from biotin to acetyl-CoA to form the product malonyl-CoA. In mammals, there are two forms of ACC. A cytoplasmic enzyme, ACC1, is expressed in lipogenic tissues such as liver, adipose tissue, and lactating mammary gland. A mitochondrial enzyme, ACC2, occurs in oxidative tissues such as cardiac and skeletal muscle, where its product, malonyl-CoA, acts as a strong inhibitor of carnitine acyltransferase I.

Mammalian ACC contains two subunits, each with a bound biotin cofactor. ACC becomes active when ACC dimers aggregate to form high-molecular-weight polymers (4 million to 8 million daltons) composed of 10 to 20 dimers. ACC, a key enzyme in fatty acid metabolism, is highly regulated by allosteric modulators and phosphorylation reactions (Figure 12.16). The allosteric effects of citrate, a feed-forward activator that promotes polymerization, and palmitoyl-CoA, an end product inhibitor that causes depolymerization, are dependent on the phosphorylation state of the enzyme. ACC is phosphorylated and therefore,

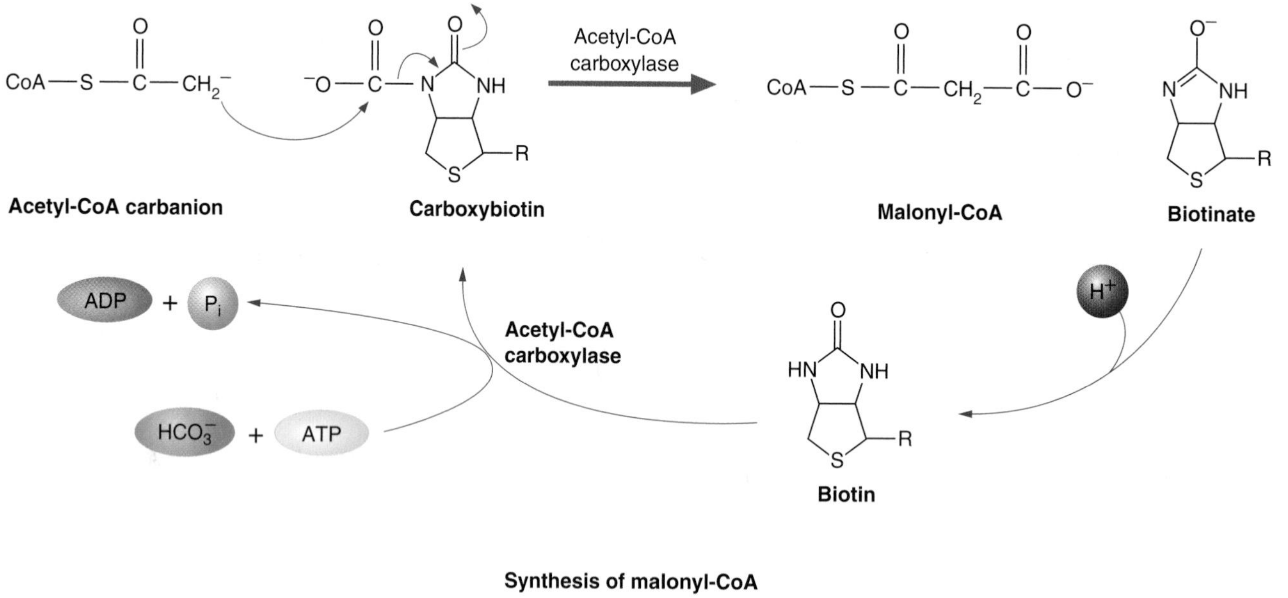

Synthesis of malonyl-CoA

FIGURE 12.15

Synthesis of Malonyl-CoA

The reaction begins with the ATP-dependent carboxylation of the biotin cofactor of the enzyme. The carboxylase abstracts a proton from the α-carbon of the acetyl-CoA to generate a reactive carbanion. The carbanion attacks the carbon of carboxybiotin to yield malonyl-CoA and biotinate. The biotinate is protonated by the enzyme to regenerate its biotin form.

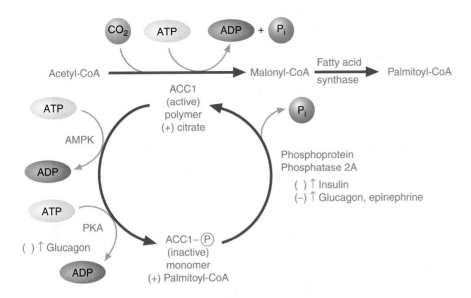

FIGURE 12.16

Regulation of Acetyl-CoA Carboxylase 1

ACC1 is inactivated (depolymerized) by phosphorylation reactions catalyzed by AMPK (as a result of high AMP levels), PKA (stimulated by glucagon), and palmitoyl-CoA accumulation. ACC1 is activated (polymerized) by phosphoprotein phosphatase 2A, which is activated by high insulin levels (glucose readily available) and deactivated by high glucagon levels (low blood glucose), and/or the presence of epinephrine (stress demand for energy mobilization).

inhibited (depolymerized) by AMP-activated protein kinase (AMPK), an important regulatory enzyme in energy metabolism. (Note that AMP is a more sensitive indicator of a cell's energy charge than either ADP or ATP.) Phosphorylation by PKA, stimulated by glucagon, also plays a role in ACC inhibition. Depolymerization is also favored by the presence of palmitoyl-CoA, which binds to and stabilizes the dimeric form of the enzyme. Glucagon and epinephrine help to maintain the phosphorylated inactive form of ACC by inactivating phosphoprotein phosphatase-2A (PP-2A), the enzyme that mediates the dephosphorylation of a number of target proteins including ACC. Insulin activates PP-2A, which dephosphorylates ACC and permits its polymerization and, therefore, activation. The polymerized form of ACC is stabilized by binding to citrate, which accumulates when acetyl-CoA levels are high.

The remaining reactions in fatty acid synthesis take place on the fatty acid synthase multienzyme complex (Figure 12.17). This complex, the site of seven enzyme activities and ACP, is a 500 kD dimer. Because the enormous polypeptides in the dimer are arranged in a head-to-tail configuration, two fatty acids can be constructed simultaneously.

During the first reaction on fatty acid synthase (Figure 12.18), acetyl transacylase (ATase) catalyzes the transfer of the acetyl group from an acetyl-CoA molecule to the SH group of a cysteinyl residue of β-ketoacyl-ACP synthase (KSase). Malonyl-ACP is formed when malonyl transacylase (MTase) transfers a malonyl group from malonyl-CoA to the SH group of the pantetheine prosthetic group of ACP (reaction 2). KSase then catalyzes a condensation reaction (reaction 3) in which acetoacetyl-ACP is formed (Figure 12.19).

During the next three steps, consisting of two reductions and a dehydration, the acetoacetyl group is converted to a butyryl group. (The flexible

(Continued on p. 436)

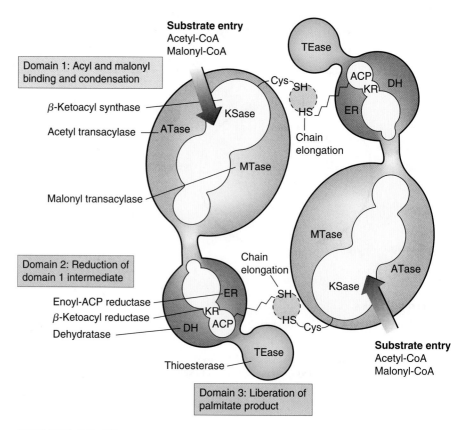

FIGURE 12.17

Fatty Acid Synthase Structure

In animals, fatty acid synthase consists of a head-to-tail dimer of identical polypeptides. Each polypeptide has seven enzyme activities. The initial stages of the cyclic pathway involve acetylating KSase (β-ketoacyl synthase, acetyl-CoA substrate) and charging ACP (malonyl-CoA substrate). Note that, because of the head-to-tail orientation of the dimer, domain 1 of one monomer is next to domain 2 of the other monomer. This allows MTase (malonyl transacylase) to charge the ACP of the other monomer and for the acetyl-ACP generated to serve as a substrate for KSase on the neighboring polypeptide. Condensation of the two entities generates acetoacetyl-ACP located close to KR (β-ketoacyl reductase), DH (β-hydroxyacyl-dehydratase), and ER (enoyl-ACP reductase) activities. The product of these enzymes, butyryl-ACP, is a substrate for ATase (acetyl transacetylase), which transfers the butyryl group to KSase, leaving the ACP free to react with another malonyl-CoA via MTase. The cycle repeats until the ACP carries a palmitoyl group, at which time TEase (thioesterase) cleaves the thioester bond, releasing palmitate from the synthase complex.

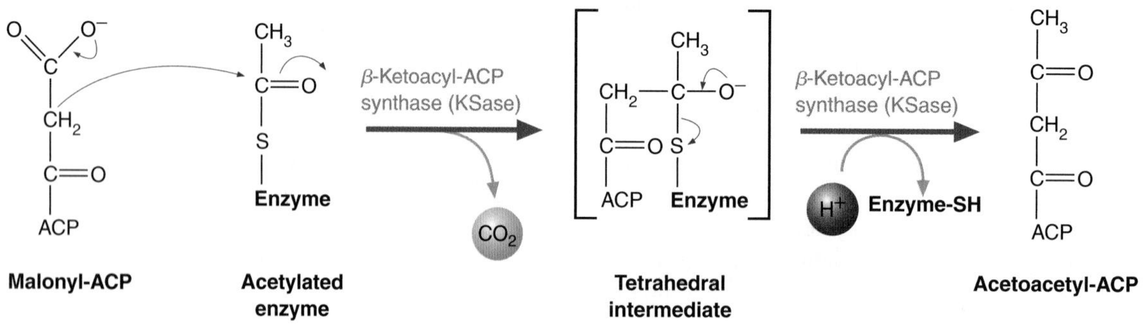

FIGURE 12.18

Formation of Acetoacetyl-ACP

The acetyl group is shown bound to the enzyme β-ketoacyl-ACP synthase through a cysteine residue. The carbonyl group of the acetyl group is attacked by the central carbon on the malonyl group attached to ACP. Acetoacetyl-ACP is generated as the C—S bond is broken.

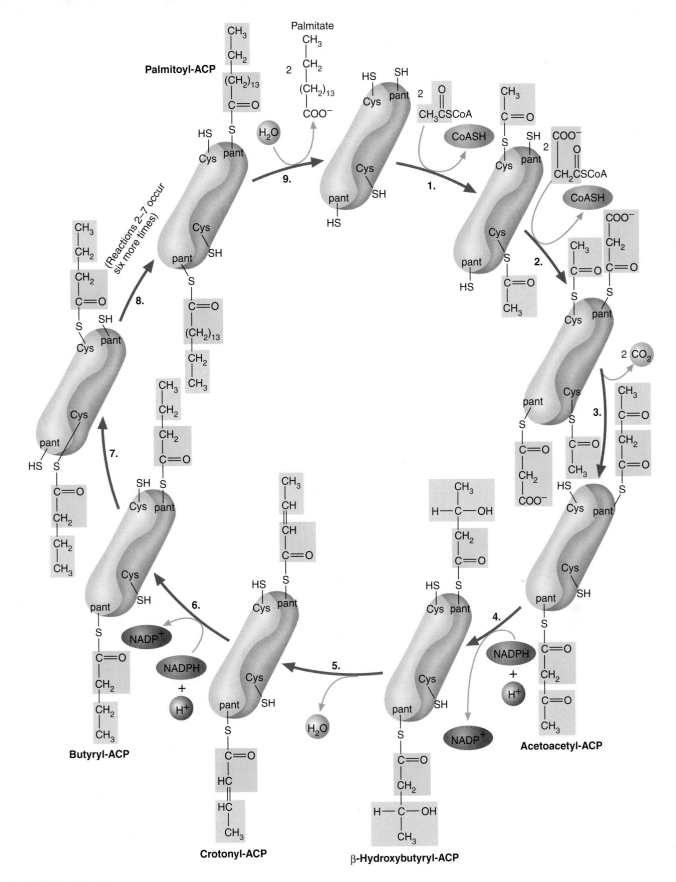

FIGURE 12.19

Fatty Acid Biosynthesis

The bicolored structures represent the dimer of fatty acid synthase. Each component in the dimer possesses a Cys-SH residue belonging to β-ketoacyl ACP synthase and pant-SH, the pantetheine sulfhydryl group of ACP. (Note that fatty acids are attached via a thioester linkage to the thiol terminal on ACP during fatty acid biosynthesis and on CoA during β-oxidation.) The enzymes that catalyze the reactions in steps 1 through 6 are (1) acetyl-CoA-ACP transacylase, (2) malonyl-CoA-ACP transacylase, (3) β-ketoacyl-ACP synthase, (4) β-ketoacyl-ACP reductase, (5) β-hydroxyacyl-ACP dehydratase, and (6) 2,3-*trans*-enoyl-ACP reductase. In step 7 the butyryl group is transferred from ACP to the cysteine-SH group of β-ketoacyl-ACP synthase by acetyl-CoA-ACP transacylase. Step 8 represents the repeated cycles of condensation and reduction needed to produce palmitoyl-ACP. Step 9, the release of the product of the process, palmitic acid, from the enzyme complex, is catalyzed by a thioesterase.

phosphopantetheine arm acts as a tether so that the substrate does not diffuse away between steps in the cycle. This greatly increases the speed and efficiency of the process.) β-Ketoacyl-ACP reductase (KRase) catalyzes the reduction of acetoacetyl-ACP to form β-hydroxybutyryl-ACP. β-Hydroxyacyl-ACP dehydratase (DH) later catalyzes a dehydration, thus forming crotonyl-ACP. Butyryl-ACP is produced when 2,3-*trans*-enoyl-ACP reductase (ERase) reduces the double bond in crotonyl-ACP. In the last step of the first cycle of fatty acid synthesis, the butyryl group is transferred from the pantetheine group to the cysteine residue of KSase via the action of ATase. The newly freed ACP-SH group now binds to another malonyl group. The process is then repeated. Eventually, palmitoyl-ACP is synthesized. Now the palmitoyl group is released from fatty acid synthase when thioesterase (TEase) cleaves the thioester bond. Depending on cellular conditions, palmitate can be used directly in the synthesis of several types of lipid (e.g., triacylglycerol or phospholipids), or it can enter the mitochondrion, where several enzymes catalyze elongating and desaturating reactions. Endoplasmic reticulum (ER) possesses similar enzymes.

FATTY ACID ELONGATION AND DESATURATION Elongation and desaturation of fatty acids synthesized in cytoplasm or obtained from the diet are accomplished primarily by ER enzymes. (These processes occur only when the diet provides an inadequate supply of appropriate fatty acids.) Fatty acid elongation and desaturation (the formation of double bonds) are especially important in the regulation of membrane fluidity and the synthesis of the precursors for a variety of fatty acid derivatives, such as the eicosanoids. For example, myelination, the formation of a myelin sheath around certain nerve cells, depends especially on the ER fatty acid synthetic reactions. Very long chain saturated and monounsaturated fatty acids are important constituents of the cerebrosides and sulfatides found in myelin. Cells apparently regulate membrane fluidity by adjusting the types of fatty acids that are incorporated into membrane lipids. For example, in cold weather, more unsaturated fatty acids are incorporated. (Recall that unsaturated fatty acids have a lower freezing point than do saturated fatty acids. See p. 374.) If the diet does not provide a sufficient number of these molecules, fatty acid synthetic pathways are activated. Although elongation and desaturation are closely integrated processes, for the sake of clarity they are discussed separately.

ER fatty acid chain elongation, which uses two-carbon units provided by malonyl-CoA, is a cycle of condensation, reduction, dehydration, and reduction reactions similar to those observed in cytoplasmic fatty acid synthesis. In contrast to the cytoplasmic process, the intermediates in the ER elongation process are CoA esters. These reactions can lengthen both saturated and unsaturated fatty acids. Reducing equivalents are provided by NADPH.

Acyl-CoA molecules are desaturated in ER membrane in the presence of NADH and O_2. All components of the desaturase system are integral membrane proteins that are apparently randomly distributed on the cytoplasmic surface of the ER. The association of cytochrome b_5 reductase (a flavoprotein), cytochrome b_5, and oxygen-dependent desaturases constitutes an electron transport system. This system efficiently introduces double bonds into long-chain fatty acids (Figure 12.20). Both the flavoprotein and cytochrome b_5 (found in a ratio of approximately 1:30) have hydrophobic peptides that anchor the proteins into the microsomal membrane. Animals typically have Δ^9, Δ^6, and Δ^5 desaturases that use electrons supplied by NADH via the electron transport system to activate the oxygen needed to create the double bond. Plants contain additional desaturases for the Δ^{12} and Δ^{15} positions.

Because elongation and desaturation systems are in close proximity to each other in microsomal membrane, a variety of long-chain polyunsaturated acids are typically produced. A prominent example of this interaction is the synthesis of arachidonic acid ($20:4^{\Delta5,8,11,14}$) from linoleic acid ($18:2^{\Delta9,12}$).

KEY CONCEPTS

- In animals, fatty acids are synthesized in the cytoplasm from acetyl-CoA and malonyl-CoA.

- Mitochondrial and ER enzymes elongate and desaturate newly synthesized fatty acids as well as those obtained in the diet.

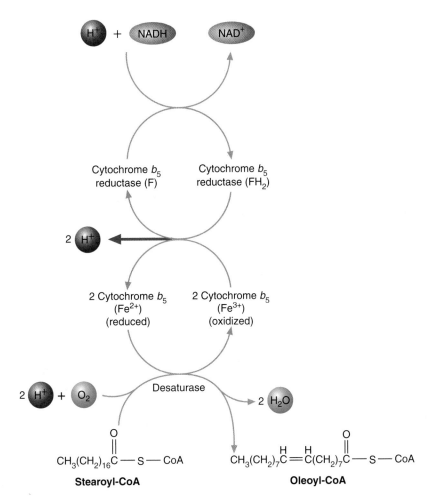

FIGURE 12.20

Desaturation of Stearoyl-CoA

The desaturase uses electrons provided by an electron transport system composed of cytochrome b_5 reductase and cytochrome b_5 to activate the oxygen (not shown) needed to create the double bond. NADH is the electron donor.

Eicosanoid Metabolism

As previously described (p. 376) the eicosanoids are hormone-like substances that are derived from either arachidonic acid or EPA (eicosapentaenoic acid). Of all the biosynthetic pathways of the eicosanoids, those that originate with arachidonic acid are the best understood. Almost all cellular arachidonic acid is stored in cell membranes as esters at C-2 of glycerol in phosphoglycerides. Release of arachidonic acid from membrane, considered to be the rate-limiting step in eicosanoid synthesis (Figure 12.21), results from binding an appropriate chemical signal to its receptor on a target cell plasma membrane. For example, the release of arachidonic acid in platelets is caused by binding thrombin, an enzyme that plays an important role in blood clotting. (Thrombin is a proteolytic enzyme that converts the soluble plasma protein fibrinogen into fibrin, which then forms an insoluble meshwork.) Platelet aggregation, triggered by the eicosanoid TXA, is an early critical step in the blood-clotting process.

Most often, the release of arachidonic acid is catalyzed by phospholipase A_2. Certain steroids that suppress inflammation inhibit phospholipase A_2. Phospholipase A_2 cleaves acyl groups from C-2 of a phosphoglyceride, thus forming a fatty acid and lysophosphoglyceride. Once released, arachidonic acid molecules may be converted (depending on both cell type and intracellular conditions) into a variety of eicosanoid molecules. Prostaglandin synthesis begins when

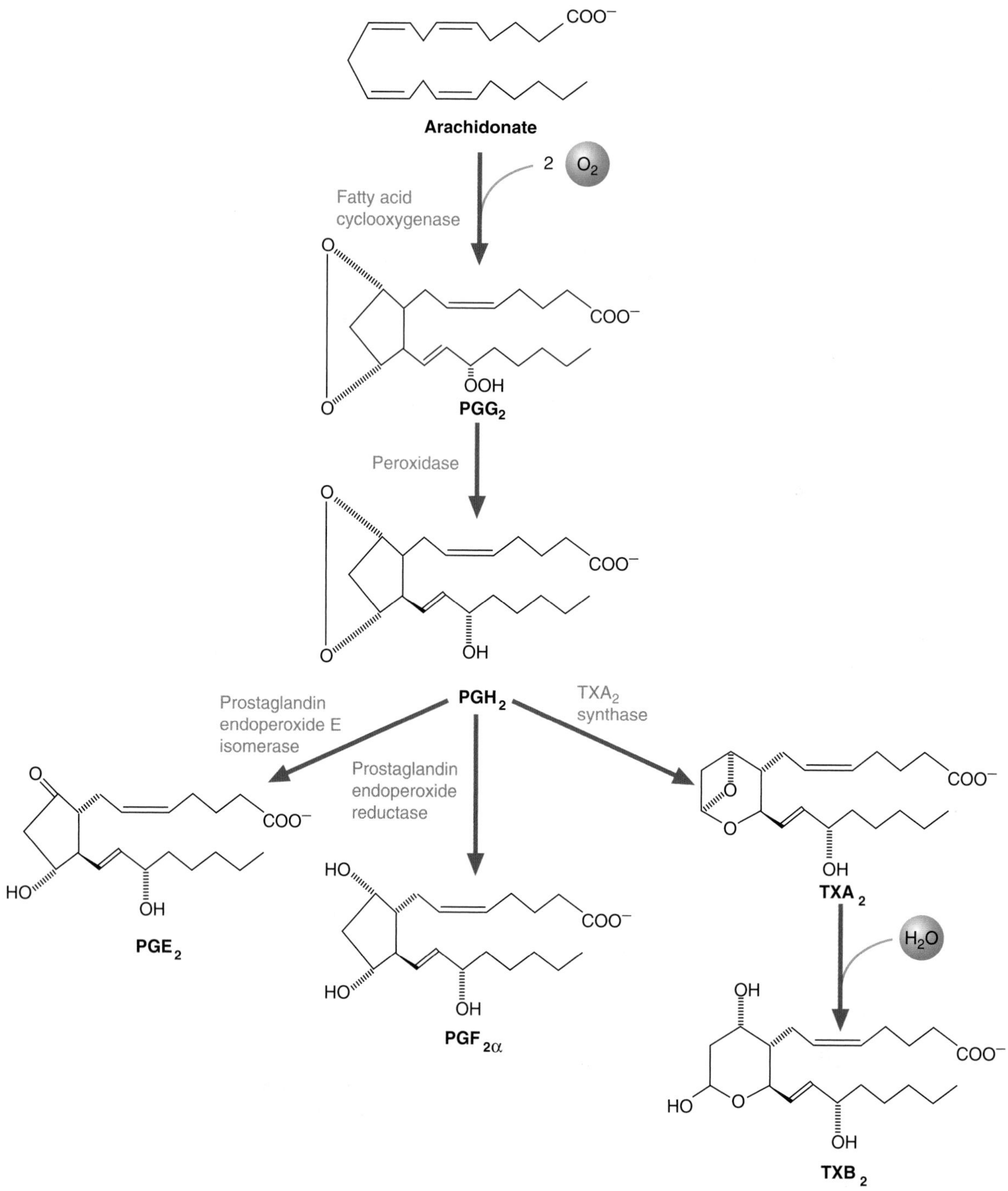

FIGURE 12.21

Synthesis of Selected Prostaglandins and Thromboxanes

Each step is catalyzed by a cell-specific enzyme. Note that at physiological pH, fatty acids are ionized.

cyclooxygenase converts arachidonic acid into PGG_2. (Aspirin inactivates cyclooxygenase by acetylating a critical serine residue in the enzyme.) Then the formation of PGH_2 from PGG_2 is catalyzed by peroxidase. (Prostaglandin endoperoxidase synthase, an ER enzyme, possesses both the cyclooxygenase and the peroxidase activities.) PGH_2 is a precursor of several eicosanoids. For example, PGE_2 and PGF_2 are formed from PGH_2 by the actions of prostaglandin endoperoxide isomerase and prostaglandin endoperoxide reductase, respectively. In platelets and lung cells, TXA_2 synthase catalyzes the conversion of PGH_2 to TXA_2. Within seconds, TXA_2 is nonenzymatically hydrolyzed to the inactive molecule TXB_2.

In a separate pathway, arachidonic acid is converted into the leukotrienes. Several enzymes, referred to as lipoxygenases, catalyze the addition of hydroperoxy groups to arachidonic acid. Mammalian cells that produce leukotrienes (Figure 11.4c) include a variety of white blood cells. Various leukotriene molecules are released during allergic reactions and inflammatory processes. For example, LTC_4, LTD_4, and LTE_4 are components of slow-reacting substance of anaphylaxis. These molecules are derived from LTA_4, which is produced when 5-lipooxygenase introduces a peroxide group into arachidonic acid. The product of this reaction is then converted by LTA_4 synthase into an epoxide, a three-membered ring that contains an ether functional group.

QUESTION 12.7

Rheumatoid arthritis is an autoimmune disease in which the joints are chronically inflamed. In **autoimmune diseases** the immune system fails to distinguish between self and nonself. For reasons that are not understood, specific lymphocytes are stimulated to produce antibodies, referred to as *autoantibodies*. These molecules bind to surface antigens on the patient's own cells as if they were foreign. In rheumatoid arthritis, the binding of an autoantibody called rheumatoid factor (RF) to fragments of immunoglobulin G (IgG) promotes inflammation because it stimulates the infiltration of joint tissue by several types of white blood cells. The leakage of lysosomal enzymes from actively phagocytosing cells (neutrophils and macrophages) leads to further tissue damage. The inflammatory response is perpetuated by the release of several eicosanoids. For example, macrophages are known to produce PGE_4, TXA_2, and several leukotrienes.

Currently, the treatment of rheumatoid arthritis consists of suppressing pain and inflammation. Despite treatment, however, the disease continues to progress. Aspirin plays an important role in the treatment of rheumatoid arthritis and other types of inflammation because of its low cost and relative safety. Certain steroids are more potent than aspirin in reducing inflammation; that is, they immediately and dramatically reduce painful symptoms. However, steroids have serious side effects. For example, prednisone may depress the immune system, cause fat redistribution to the neck ("buffalo hump"), and cause serious behavioral changes. For these and other reasons, prednisone is used to treat rheumatoid arthritis only when a patient does not respond to aspirin or similar drugs.

Review the effects of aspirin and steroids on eicosanoid metabolism and suggest a reason why this information is relevant to the treatment of rheumatoid arthritis. Does it explain the difference between the effectiveness of aspirin and steroids in treating inflammation?

QUESTION 12.8

Excessive consumption of fructose has been linked to obesity and to the condition referred to as *hypertriglyceridemia* (high blood levels of triacylglycerols). Common sources of fructose for most Americans are sucrose and high-fructose corn syrup. Over the past several decades, high-fructose corn syrup has replaced sucrose in many processed foods and beverages because it is inexpensive in comparison to sucrose. It now comprises 40% of caloric sweeteners. (The fructose content of fresh fruits and vegetables is so low that it would be difficult to consume sufficient quantities to induce hypertriglyceridemia.) Sucrose is digested in the small intestine by the enzyme sucrase to yield one molecule each of fructose and glucose. Digestion is so rapid that the blood concentrations of these sugars become quite high. Whatever its source, once fructose reaches the

QUESTION 12.8 (CONT.)

liver, it is converted to fructose-1-phosphate (see p. 294). It is now believed that fructose-1-phosphate promotes hexokinase D activity. (Apparently, fructose-1-phosphate binds to and inactivates a protein that depresses hexokinase D activity.) In addition, whereas high blood levels of glucose trigger the release of insulin and *leptin* (a hormone secreted by adipose tissue), both of which curb appetite; this does not happen with fructose. After reviewing fructose metabolism and fatty acid and triacylglycerol synthesis, suggest how hypertriglyceridemia and obesity might result from a diet that is rich in sucrose and high-fructose corn syrup.

Regulation of Fatty Acid Metabolism in Mammals

Animals have such varying requirements for energy that the metabolism of fatty acids (*the* major energy source in animals) is carefully regulated. Both short- and long-term regulatory mechanisms are used. In short-term regulation (measured in minutes) the activities of existing molecules of key regulatory enzymes are modified by allosteric regulators (Figure 12.22), covalent modification, and hormones. When energy levels are high, β-oxidation is depressed by the binding of the allosteric modulators NADH and acetyl-CoA to β-hydroxyacyl–CoA dehydrogenase and thiolase, respectively. Similarly, malonyl-CoA, the product of ACC, is an allosteric regulator of CAT-I. In liver when insulin/glucagon ratios are high, malonyl-CoA levels rise and cause the inhibition of β-oxidation, thus preventing a futile cycle. High cellular concentrations of long-chain fatty acyl-CoA esters inhibit ACC by promoting its depolymerization.

Fatty acid synthesis and β-oxidation are also rapidly regulated by changes in energy demand by 5′-AMP-activated protein kinase (AMPK). **AMPK** is a trimeric enzyme composed of an α subunit (catalytic) and β and γ subunits (regulatory). As AMP/ATP ratios begin to rise, AMPK is activated by upstream AMPK kinases and by its allosteric modulator, AMP. In addition to acting as an allosteric activator, AMP promotes the activating phosphorylation reactions and inhibits dephosphorylation by protein phosphatases. AMP levels are a sensitive indicator of cellular energy status, and they rise in response to stresses that deplete cellular ATP levels (e.g., nutrient deprivation, hypoxia, heat shock, prolonged exercise). Once activated, AMPK switches off anabolic pathways (e.g., fatty acid and triacylglycerol synthesis by phosphorylating ACC1 (p. 432) and glycerol-3-phosphate acyltransferase, respectively) and switches on catabolic pathways (e.g., β-oxidation by activating ACC2 and malonyl-CoA decarboxylase (MCD), the enzyme that decreases the concentration of malonyl-CoA).

Hormones play an important role in both short- and long-term regulation of fatty acid metabolism. Short-term effects of insulin that promote fat synthesis are caused by rapid signal transduction mechanisms. Insulin activates phosphoprotein phosphatase 2A, which dephosphorylates and activates ACC1, in combination with allosteric regulators and covalent modifications (p. 432). Insulin also activates ATP-citrate lyase (p. 330) and pyruvate dehydrogenase (in adipocytes). Insulin promotes fat synthesis in adipocytes by triggering the movement of the transporter GLUT-4 to the cell surface, thus facilitating the entry into the cell of glucose (the precursor of glycerol-3-phosphate and fatty acids). Insulin simultaneously depresses fat mobilization in adipocytes by stimulating the phosphorylation of hormone-sensitive lipase. Glucagon, epinephrine, and cortisol increase lipolysis by stimulating the dephosphorylation and inactivation of hormone-sensitive lipase via signal transduction–mediated phosphorylation reactions.

Changes in the long-term regulation of fatty acid metabolism, which occur in response to fluctuating nutrient availability and energy demand, are effected

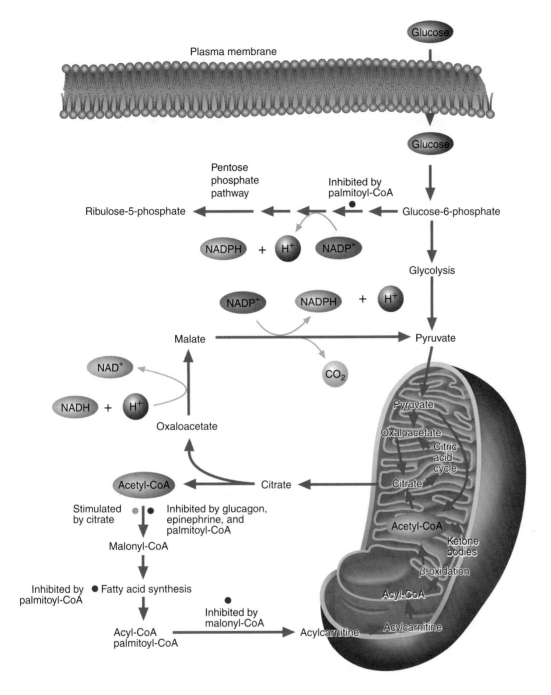

FIGURE 12.22

Regulation of Intracellular Fatty Acid Metabolism

Fatty acids are synthesized in cytoplasm from acetyl-CoA, which is formed within the mitochondrion. Because the inner membrane is impermeable to acetyl-CoA, it is transferred out as citrate. Citrate is produced from acetyl-CoA and oxaloacetate in the citric acid cycle, a reaction pathway in the mitochondrial matrix. Citrate is transferred to cytoplasm when β-oxidation is suppressed, that is, when the cell needs little energy. Later it is cleaved to form oxaloacetate and acetyl-CoA. When the cell needs more energy, fatty acids are transported into the mitochondrion as acylcarnitine derivatives. Then acyl-CoA is degraded to acetyl-CoA via β-oxidation. (The further oxidation of acetyl-CoA to generate ATP is described in Chapter 9.) Note that the hormones glucagon and epinephrine and the substrates citrate, malonyl-CoA, and palmitoyl-CoA are important regulators of fatty acid metabolism. Fatty acid metabolism and carbohydrate metabolism are interrelated. Pyruvate, the precursor of acetyl-CoA, is a product of glycolysis. A portion of the NADPH, the reducing agent required in fatty acid synthesis, is generated by several reactions in the pentose phosphate pathway. NADPH is also produced by converting malate, formed by the reduction of oxaloacetate, to pyruvate.

by alterations in gene expression. Two classes of transcription factors are prominent components of an intricate regulatory process: the SREBPs and PPARs. Each type of transcription factor, when activated, binds to a regulatory element near target genes, a process that triggers the binding of coactivator molecules and, subsequently, transcription.

The **SREBPs** (sterol regulatory element binding proteins) are a group of three proteins, SREBP1a, SREBP lc, and SREBP 2, which are coded for by two genes. SREBPla and (more prominently) SREBPlc regulate the expression of genes involved in fatty acid metabolism. (SREBPla, which activates all SREBP-responsive genes, is continuously expressed at low levels in most animal tissues.) SREBP2 regulates genes in cholesterol metabolism (see p. 456). In liver and adipose tissue, the activation of SREBP1c in response to insulin upregulates the transcription of the genes that code for enzymes in fatty acid synthetic pathway and NADPH synthesis. Glucagon and high levels of long-chain fatty acids inhibit SREBPlc. **PPARs** (peroxisome proliferator–activated receptors), named for the ability of certain synthetic compounds to cause the proliferation of liver cell peroxisomes, are ligand-activated transcription factors that bind to PPAR response elements associated with target genes. PPARα, a transcription factor originally discovered because of the peroxisome proliferation phenomenon, controls the expression of several genes in lipid metabolism. In liver and adipose tissues, under fasting conditions, it stimulates fatty acid catabolism and ketogenesis. PPARγ, primarily expressed in adipose tissue, in combination with insulin and SREBP1, stimulates glucose uptake and fatty acid and triacylglycerol synthesis. The activity of PPARs is stimulated by the binding of several lipid molecules (e.g., saturated and unsaturated fatty acids, prostaglandins and leukotrienes).

KEY CONCEPTS

- The metabolism of fatty acids, the major energy source in animals, is regulated in the short term by allosteric modulators, covalent modification, and hormones.
- Long-term regulation, which occurs in response to fluctuating nutrient availability and energy demand, is effected by changes in gene expression.

QUESTION 12.9

Identify each of the following biomolecules:

(a) $CH_3-\overset{H}{\underset{OH}{C}}-CH_2-\overset{O}{C}-O^-$

(b) $CoA-S-\overset{O}{C}-CH_2-\overset{O}{C}-O^-$

(c) [biotin structure with HN—NH, S, R]

(d) $CH_3-\overset{O}{C}-S-ACP$

What is the function of each?

BIOCHEMISTRY IN PERSPECTIVE

Atherosclerosis

What is the biochemical basis of arterial damage in the disease process called atherosclerosis? Atherosclerosis is a chronic disease in which soft masses called atheromas accumulate within arterial walls, eventually compromising their functional structure. Normal arterial walls are strong and flexible. They consist of three well-defined layers: the *intima* (a single layer of endothelial cells attached to an underlying extracellular matrix), the *media* (layers of smooth muscle cells embedded in an extracellular matrix consisting of elastic fibers, collagen, and proteoglycan), and the *adventitia* (the outermost layer, which consists of fibroblasts, smooth muscle cells, collagen, and elastin). Sandwiched between these concentric layers are additional elastic fibers.

Once believed to be little more than passive conduits of blood, blood vessels are now known to be physiologically active. The endothelial cells that line arteries, for example, perform several vital functions. Among these are providing a barrier that prevents toxic substances from penetrating into the vessel wall, and regulating the response of arteries to shear stress (the fluctuating mechanical force created by blood flow). Nitric oxide (NO), the vasodilator produced by endothelial NO synthase, relaxes smooth muscle cells. Endothelial cells also produce molecules that provide arterial linings with smooth Teflon-like surfaces that prevent white blood cells from adhering. As with all such cells, arterial endothelial cells are continuously replaced. In the course of aging, the endothelial barrier becomes leaky and vulnerable, a result that is accelerated by poor diet, smoking, and a sedentary lifestyle.

The atherosclerotic process is initiated by the accumulation of LDL within the intima, where they become ensnared by interactions with proteoglycans. Here both LDL phospholipid and protein components undergo oxidation reactions. For example, aldehydes formed by the peroxidation of phospholipids react with lysine residues of apolipoprotein B. Aldehyde-altered apo B, as well as other oxidation products and glycated proteins (p. 238), are biomarkers recognized by certain white blood cells. These cells begin accumulating within the subendothelial space. This inflammatory process accelerates as endothelial and nearby smooth muscle cells release signal molecules called *chemokines,* which attract fully activated macrophages to the injured site. The attempt of macrophages to clear the site of oxidized LDL is overwhelmed, and the phagocytes become so filled with lipoprotein that they become "foam cells." Together, damaged endothelial cells, macrophages, and smooth muscle cells release molecules that promote cell division and cell migration, which has the effect of reorganizing the vessel wall architecture. In effect, the attempt to heal the injury results in the formation of a fibrous cap that walls off the damaged tissue.

It is important to note that the atherosclerotic process usually causes the formation of atheromas that extend sideways, and not outward, which would cut off blood flow. Atherosclerotic lesions (also called *plaque*) usually do not cause obvious problems for long periods of time, perhaps decades. Eventually however, the inflammatory process weakens the fibrous cap, which may suddenly rupture. It is the subsequent formation of a thrombus (blood clot) that prevents blood flow, especially in smaller vessels such as the coronary arteries. Sudden death, the most common symptom of coronary artery damage, results from such an event. In a small number of myocardial infarctions (heart attacks), plaque formation grows outward, slowly occluding blood flow. In these cases, decreased blood flow through one or more coronary arteries causes angina pectoris (pain and tightness in the chest) by preventing O_2 and nutrients from reaching the myocardial cells. Prompt medical attention to life-threatening occlusions can prevent death.

SUMMARY: Atherosclerosis, which may lead to myocardial infarction, is initiated by damage to the endothelial cells that line arteries. The formation of atherosclerotic lesions begins with the accumulation of LDL and progresses to an inflammatory process that degrades arterial structure and function.

12.2 MEMBRANE LIPID METABOLISM

The lipid bilayer of cell membranes is composed primarily of phospholipids and sphingolipids. The metabolism of these lipid classes is briefly described.

Phospholipid Metabolism

Most of the reactions involved in lipid biosynthesis appear to be located in the smooth endoplasmic reticulum (SER), although several enzyme activities have also been detected in the Golgi complex. The biosynthesis of phospholipid occurs at the interface of ER membrane and cytoplasm because each enzyme is a membrane-associated protein with its active site facing the cytoplasm. The fatty acid composition of phospholipids changes somewhat after their synthesis. (Typically, unsaturated fatty acids replace the original saturated fatty acids.) Most of this remodeling is accomplished by certain phospholipases and acyltransferases. Presumably, the process allows a cell to adjust the fluidity of its membranes.

The syntheses of phosphatidylethanolamine and phosphatidylcholine are similar (Figure 12.23). Phosphatidylethanolamine synthesis begins in the cytoplasm when ethanolamine enters the cell and is immediately phosphorylated. Subsequently, phosphoethanolamine reacts with CTP (cytidine triphosphate) to form the activated intermediate CDP-ethanolamine. Several nucleotides serve as high-energy carriers of specific molecules. CDP derivatives have an important role in the transfer of polar head groups in phosphoglyceride synthesis. (Recall that UDP plays a similar role in glycogen synthesis; see p. 296). CDP-ethanolamine is converted to phosphatidylethanolamine when it reacts with diacylglycerol. This reaction is catalyzed by an enzyme on the endoplasmic reticulum. As noted, the biosynthesis of phosphatidylcholine is similar to that of phosphatidylethanolamine. The choline required in this pathway is obtained in the diet. However, phosphatidylcholine is also synthesized in the liver from phosphatidylethanolamine (Figure 12.24). Phosphatidylethanolamine is methylated in three steps by the enzyme phosphatidylethanolamine-*N*-methyltransferase to form the trimethylated product phosphatidylcholine. *S-Adenosylmethionine* (SAM) is the methyl donor in this set of reactions. (The role of SAM in cellular methylation processes is discussed in Chapter 14.)

In animals, phosphatidylserine is generated by polar head group exchange with phosphatidylethanolamine mediated by phosphatidylethanolamine–serine transferase (Figure 12.25), a reversible ER enzyme.

Phospholipid turnover is rapid. (**Turnover** is the rate at which all molecules in a structure are degraded and replaced with newly synthesized molecules.) For example, in animal cells, approximately two cell divisions are required for the replacement of one-half of the total number of phospholipid molecules. Phosphoglycerides are degraded by the phospholipases. Each phospholipase, which catalyzes the cleavage of a specific bond in phosphoglyceride molecules, is named according to the bond cleaved. Phospholipases A_1 and A_2, which hydrolyze the ester bonds of phosphoglycerides at C-1 and C-2, respectively, contribute to the phospholipid remodeling described.

KEY CONCEPTS

- Phospholipid synthesis occurs in the membrane of the SER. After phospholipids have been synthesized, they are remodeled by altering their fatty acid composition.
- Phospholipid degradation is catalyzed by several phospholipases.

FIGURE 12.23

Phospholipid Synthesis

After ethanolamine or choline has entered a cell, it is phosphorylated and converted to a CDP derivative. Diacylglycerol then reacts with the CDP derivative, and phosphatidylethanolamine or phosphatidylcholine is formed. Triacylglycerol is produced if diacylglycerol reacts with acyl-CoA. CDP-diacylglycerol, formed from phosphatidic acid and CTP, is a precursor of several phospholipids (e.g., phosphatidylglycerol and phosphatidylinositol).

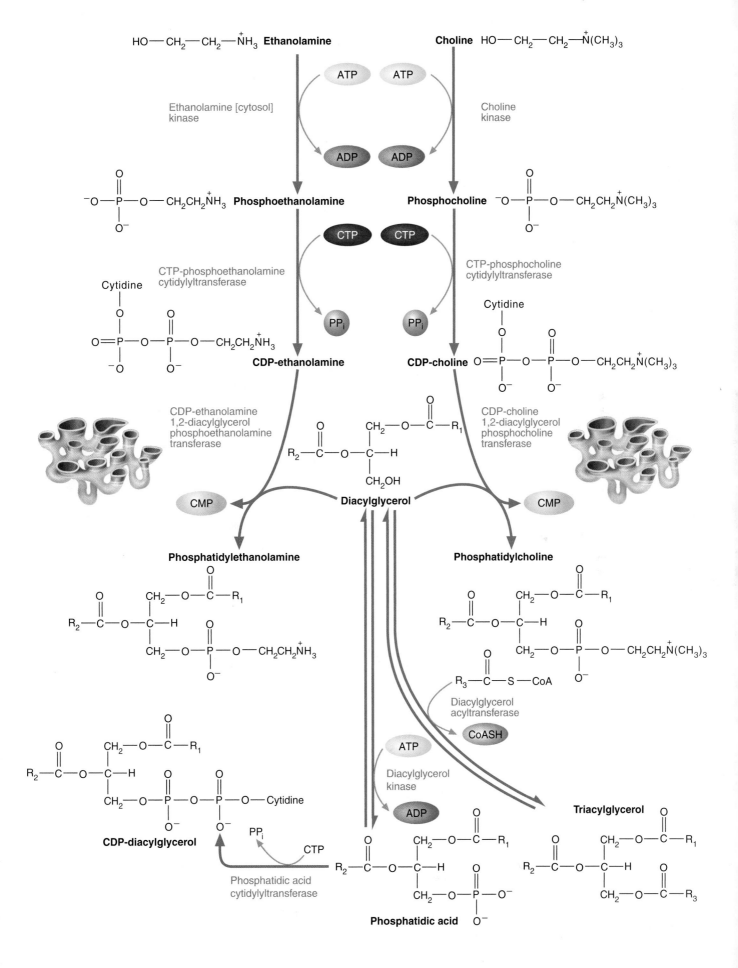

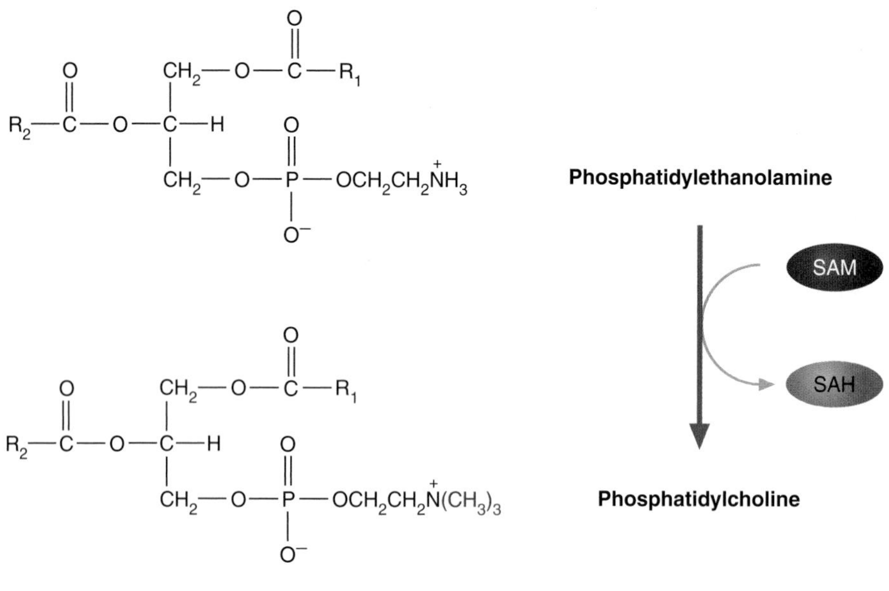

FIGURE 12.24

Conversion of Phosphatidylethanolamine to Phosphatidylcholine by Phosphatidyl-ethanolamine *N*-methyltransferase (PEMT)

Methylation reactions that use 5-adenosyl-methionine (SAM) are discussed in Chapter 19. (SAH = *S*-adenosylhomocysteine.)

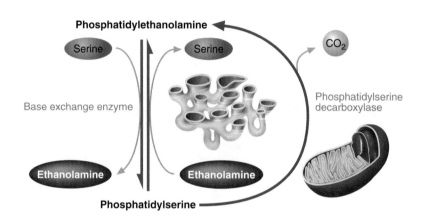

Phosphatidylethanolamine

SAM

SAH

Phosphatidylcholine

FIGURE 12.25

Phosphatidylserine Synthesis

Phosphatidylserine can be synthesized from phosphatidylethanolamine in a reaction in which the polar head groups are exchanged. Phosphatidylethanolamine can also be synthesized from phosphatidylserine in a decarboxylation reaction. This reaction is an important source of ethanolamine in many eukaryotes.

Phosphatidylethanolamine

Serine — Serine — CO_2

Base exchange enzyme

Phosphatidylserine decarboxylase

Ethanolamine — Ethanolamine

Phosphatidylserine

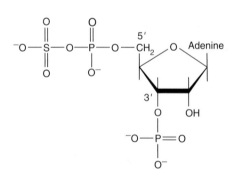

FIGURE 12.26

3′-Phosphoadenosine-5′-Phosphosulfate (PAPS)

PAPS is a high-energy sulfate donor.

Sphingolipid Metabolism

Recall that sphingolipids in animals possess ceramide, a derivative of the amino alcohol sphingosine. The synthesis of ceramide begins with the condensation of palmitoyl-CoA with serine to form 3-ketosphinganine. This reaction is catalyzed by 3-ketosphinganine synthase, a pyridoxal-5′-phosphate–dependent enzyme. (Because pyridoxal-5′-phosphate plays an important role in amino acid metabolism, the biochemical function of this coenzyme is discussed in Chapter 14.) 3-Ketosphinganine is subsequently reduced by NADPH to form sphinganine. In a two-step process involving acyl-CoA and $FADH_2$, sphinganine is converted to ceramide. Sphingomyelin is formed when ceramide reacts with phosphatidylcholine. (In an alternative reaction, CDP-choline is used in place of phosphatidylcholine.) When ceramide reacts with UDP-glucose, glucosylceramide (a common cerebroside, sometimes referred to as glucosylcerebroside) is produced. Galactocerebroside, a precursor of other glycolipids, is synthesized when ceramide reacts with UDP-galactose. The sulfatides are synthesized when the galactocerebrosides react with the sulfate donor molecule **3′-phosphoadenosine-5′-phosphosulfate** (PAPS) (Figure 12.26). The transfer of sulfate groups is catalyzed by the microsomal enzyme sulfotransferase. (*Microsomes* are small vesicles derived from fragmented endoplasmic reticulum when cells are homogenized in the laboratory.) Sphingolipids are degraded within lysosomes. Recall that specific diseases, called the sphingolipidoses (p. 385), result when enzymes

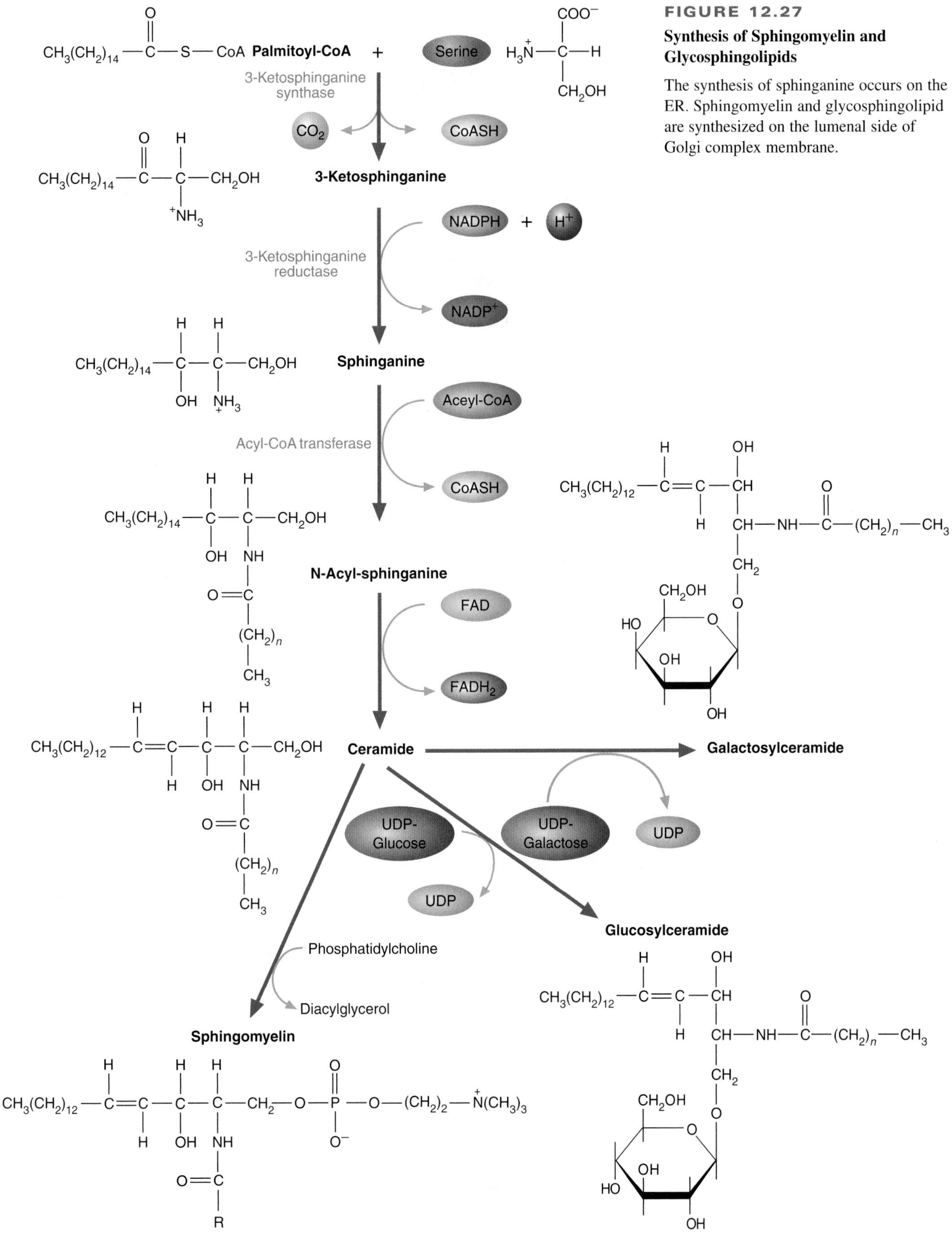

FIGURE 12.27

Synthesis of Sphingomyelin and Glycosphingolipids

The synthesis of sphinganine occurs on the ER. Sphingomyelin and glycosphingolipid are synthesized on the lumenal side of Golgi complex membrane.

required for degrading these molecules are missing or defective. The synthesis of sphingomyelin and glycosphingolipids is shown in Figure 12.27).

12.3 ISOPRENOID METABOLISM

Isoprenoids occur in all eukaryotes. Despite the astonishing diversity of isoprenoid molecules, the mechanisms by which different species synthesize them are similar. In fact, the initial phase of isoprenoid synthesis (the synthesis of isopentenyl pyrophosphate) appears to be identical in all of the species in which this process has been investigated. Figure 12.28 illustrates the relationships among the isoprenoid classes.

Because of its importance in human biology, cholesterol has received enormous attention from researchers. For this reason the metabolism of cholesterol is better understood than that of any other isoprenoid molecule.

Cholesterol Metabolism

The cholesterol that is used throughout the body is derived from two sources: diet and *de novo* synthesis. When the diet provides sufficient cholesterol, the synthesis of this molecule is depressed. In normal individuals cholesterol delivered by LDL suppresses cholesterol synthesis. Cholesterol biosynthesis is stimulated when the diet is low in cholesterol. As described previously, cholesterol is a vital cell

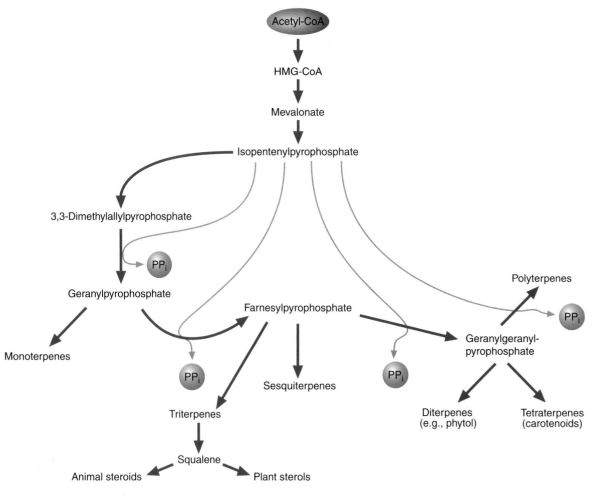

FIGURE 12.28

Isoprenoid Biosynthesis

Isoprenoid biosynthetic pathways produce an astonishing variety of products in different cell types and in different species. Despite this diversity, the beginning of isoprenoid biosynthesis appears to be identical in most of the species investigated (e.g., yeast, mammals, and plants). (HMG-CoA = β-hydroxy-β-methylglutaryl-CoA)

membrane component and a precursor in the synthesis of important metabolites. An important mechanism for disposing of cholesterol is conversion to bile acids.

CHOLESTEROL SYNTHESIS Although all tissues (e.g., adrenal glands, ovaries, testes, skin, and intestine) can make cholesterol, most cholesterol molecules are synthesized in the liver. Cholesterol synthesis can be divided into three phases:

1. formation of HMG-CoA (β-hydroxy-β-methylglutaryl-CoA) from acetyl-CoA
2. conversion of HMG-CoA to squalene
3. conversion of squalene to cholesterol

The first phase of cholesterol synthesis is a cytoplasmic process (Figure 12.29). (Recall that the initial substrate, acetyl-CoA, is produced in mitochondria from fatty acids or pyruvate. Also observe the similarity of the first phase of cholesterol synthesis to ketone body synthesis. Refer to Figure 12.7.) The condensation of two acetyl-CoA molecules to form β-ketobutyryl-CoA (also referred to as acetoacetyl-CoA) is catalyzed by thiolase.

$$2\ CH_3-\overset{\overset{\displaystyle O}{\|}}{C}-S-CoA \rightleftharpoons CH_3-\overset{\overset{\displaystyle O}{\|}}{C}-CH_2-\overset{\overset{\displaystyle O}{\|}}{C}-S-CoA\ +\ \text{CoASH}$$

      Acetyl-CoA   **β-Ketobutyryl-CoA**

In the next reaction, β-ketobutyryl-CoA condenses with another acetyl-CoA to form β-hydroxy-β-methylglutaryl-CoA (HMG-CoA). This reaction is catalyzed by β-hydroxy-β-methylglutaryl-CoA synthase (HMG-CoA synthase).

$$CH_3-\overset{\overset{\displaystyle O}{\|}}{C}-CH_2-\overset{\overset{\displaystyle O}{\|}}{C}-S-CoA\ +\ CH_3-\overset{\overset{\displaystyle O}{\|}}{C}-S-CoA \longrightarrow {}^-O-\overset{\overset{\displaystyle O}{\|}}{C}-CH_2-\overset{\overset{\displaystyle CH_3}{\underset{\displaystyle OH}{|}}}{C}-CH_2-\overset{\overset{\displaystyle O}{\|}}{C}-S-CoA\ +\ \text{CoASH}$$

   β-Ketobutyryl-CoA   **Acetyl-CoA**   **HMG-CoA**

The second phase of cholesterol synthesis begins with the reduction of HMG-CoA to form mevalonate. This reaction is catalyzed by HMG-CoA reductase (HMGR), the rate-limiting enzyme in cholesterol synthesis. NADPH is the reducing agent.

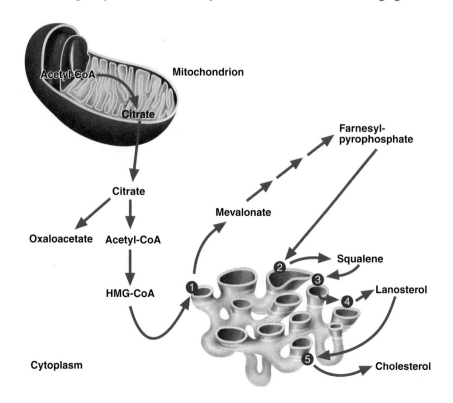

FIGURE 12.29

Cholesterol Synthesis

Several reactions occur in the cytoplasm, but most enzymes involved in cholesterol biosynthesis occur within ER membrane. The enzymes are indicated by the following numbers: 1 = HMG-CoA reductase, 2 = squalene synthase, 3 = squalene monooxygenase, 4 = 2,3-oxidosqualene lanosterol cyclase, 5 = enzymes catalyzing 20 separate reactions. Note that squalene and lanosterol are acted upon by ER membrane enzymes while they are bound to carrier proteins in the cytoplasm.

The HMGR polypeptide consists of three major domains: an N-terminal transmembrane anchor domain, a catalytic domain, and a linker that connects the membrane and catalytic domains. The enzyme, located on the cytoplasmic face of the ER, consists of two HMGR dimers that associate to form a tetramer. Each dimer has an active site at the interface between the two dimers. The reaction (Figure 12.30) begins with the nucleophilic acyl substitution in which there is a hydride transfer from NADPH to the thioester carbonyl group of HMG-CoA. This transfer is assisted by a hydrogen bond between a lysine and the thioester carbonyl oxygen. The C—S bond in the product mevaloyl-CoA is hydrolyzed to form mevaldehyde. The CoA-thiolate anion product is protonated by a histidine residue and then released. Protonation of the carbonyl oxygen of mevaldehyde by the lysine residue facilitates a second hydride transfer from NADPH to form mevalonate.

In a series of cytoplasmic reactions, mevalonate is converted to farnesylpyrophosphate. Mevalonate kinase catalyzes the synthesis of phosphomevalonate. A second phosphorylation reaction catalyzed by phosphomevalonate kinase produces 5-pyrophosphomevalonate.

(Phosphorylation reactions significantly increase the solubility of these hydrocarbon molecules in the cytoplasm.) 5-Pyrophosphomevalonate is converted to isopentenylpyrophosphate in a process involving a decarboxylation and a dehydration:

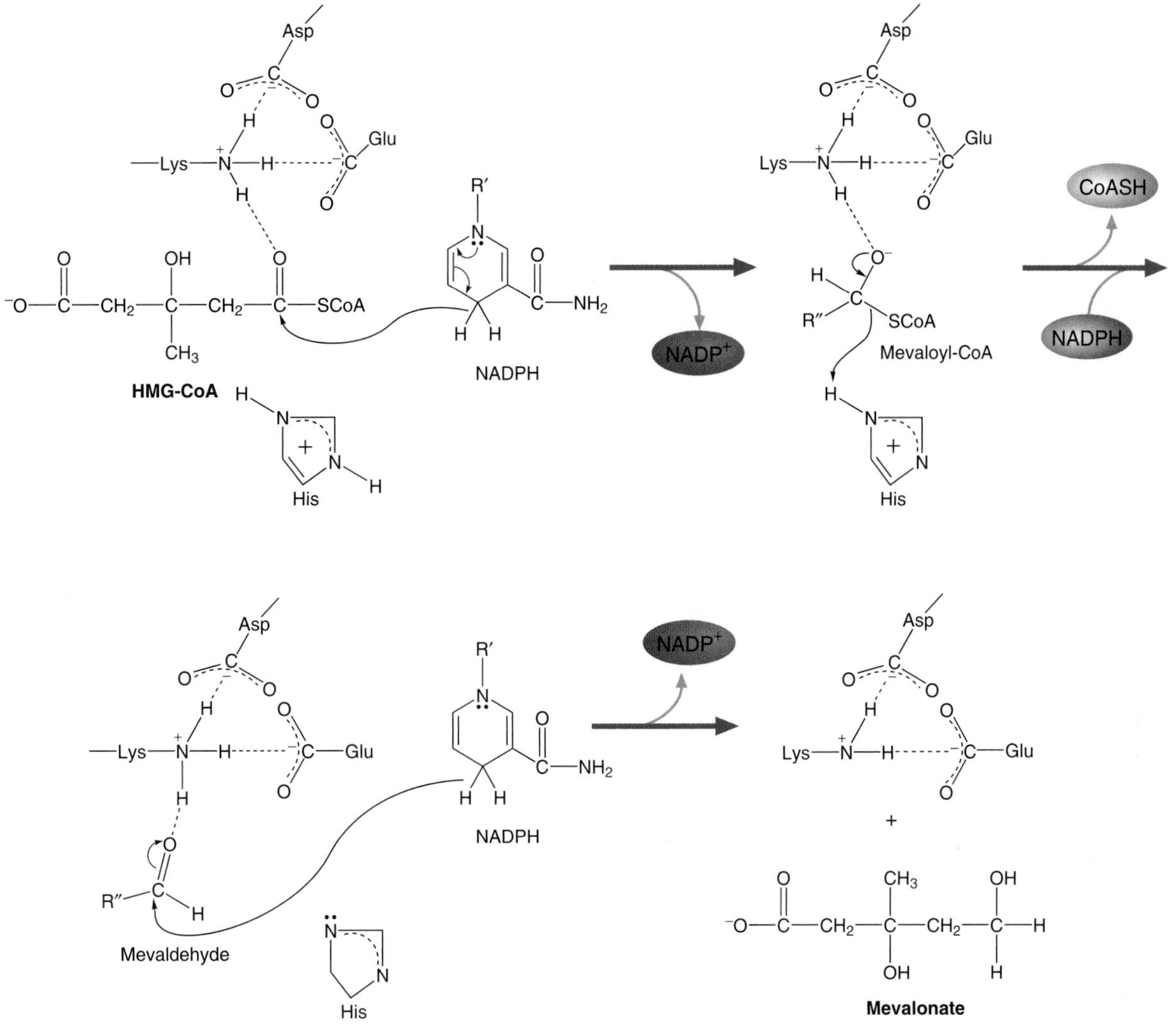

FIGURE 12.30

The HMGR-Catalyzed Reaction

In the HMGR-catalyzed reaction, there is an initial hydride transfer from NADPH to the thioester carbonyl of the substrate HMG-CoA. Subsequently, the C—S bond of mevaloyl-CoA is hydrolyzed to form mevaldehyde, and a histidine residue protonates the CoA-thiolate anion. A second hydride transfer from NADPH, facilitated by the protonation of the carbonyl oxygen of mevaldehyde, results in the formation of mevalonate, the product of the reaction. Note that carboxylate side chain groups of Asp and Glu residues orient the Lys side chain amino group within the active site.

Isopentenyl pyrophosphate is next converted to its isomer dimethylallylpyrophosphate by isopentenylpyrophosphate isomerase. (A CH_2=CH—CH_2— group on an organic molecule is sometimes referred to as an *allyl group*.)

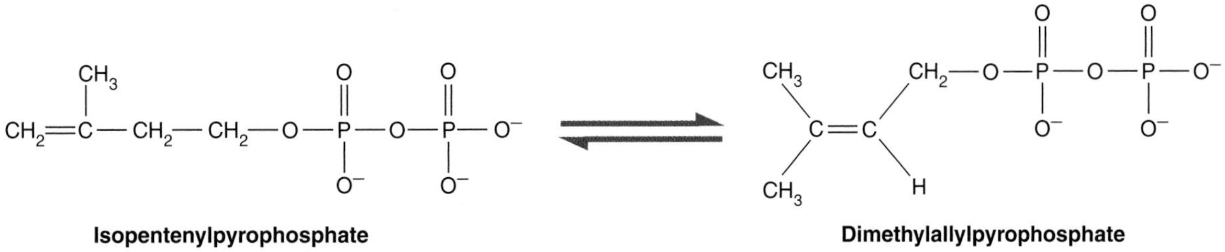

Geranylpyrophosphate is generated during a condensation reaction between isopentenylpyrophosphate and dimethylallylpyrophosphate (Figure 12.31). Pyrophosphate is also a product of this reaction and two subsequent reactions.

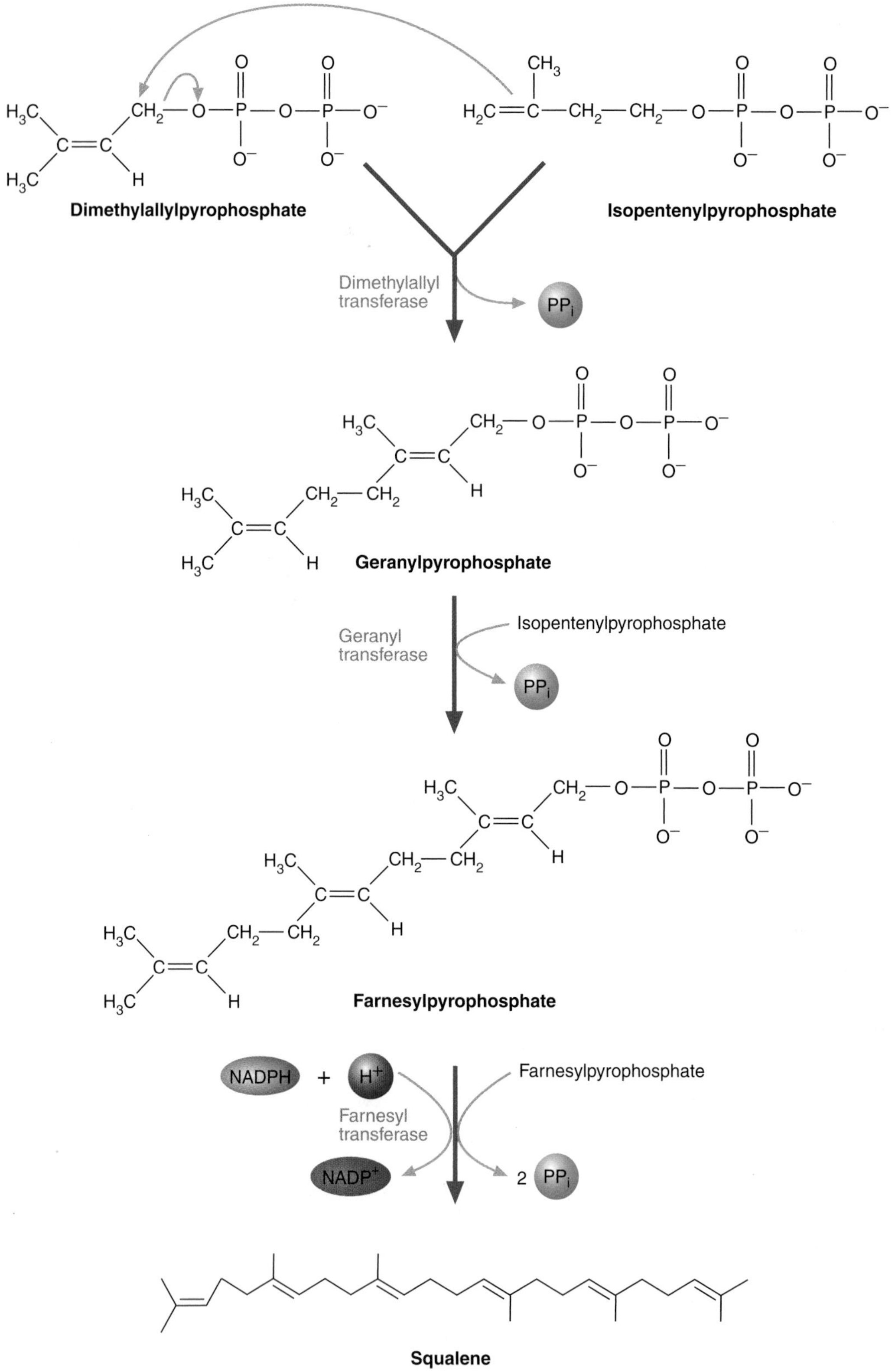

FIGURE 12.31

Synthesis of Squalene

The head-to-tail condensation of dimethylallylpyrophosphate and isopentenylpyrophosphate generates the terpene geranylpyrophosphate. A subsequent head-to-tail condensation with another isopentenylpyrophosphate generates the C_{15} farnesylpyrophosphate. Head-to-head condensation of two molecules of farnesylpyrophosphate produces the C_{30} triterpene, squalene.

(Recall that reactions in which pyrophosphate is released are irreversible because of subsequent pyrophosphate hydrolysis.) Geranyl transferase catalyzes the condensation reaction between geranylpyrophosphate and isopentenylpyrophosphate that forms farnesylpyrophosphate. Squalene is synthesized when farnesyl transferase (a microsomal enzyme) catalyzes the condensation of two farnesylpyrophosphate molecules. (Farnesyl transferase is sometimes referred to as squalene synthase.) This reaction requires NADPH as an electron donor.

The last phase of the cholesterol biosynthetic pathway (Figure 12.32) begins by binding squalene to a specific cytoplasmic protein carrier called **sterol carrier protein**. The conversion of squalene to lanosterol occurs while the intermediates

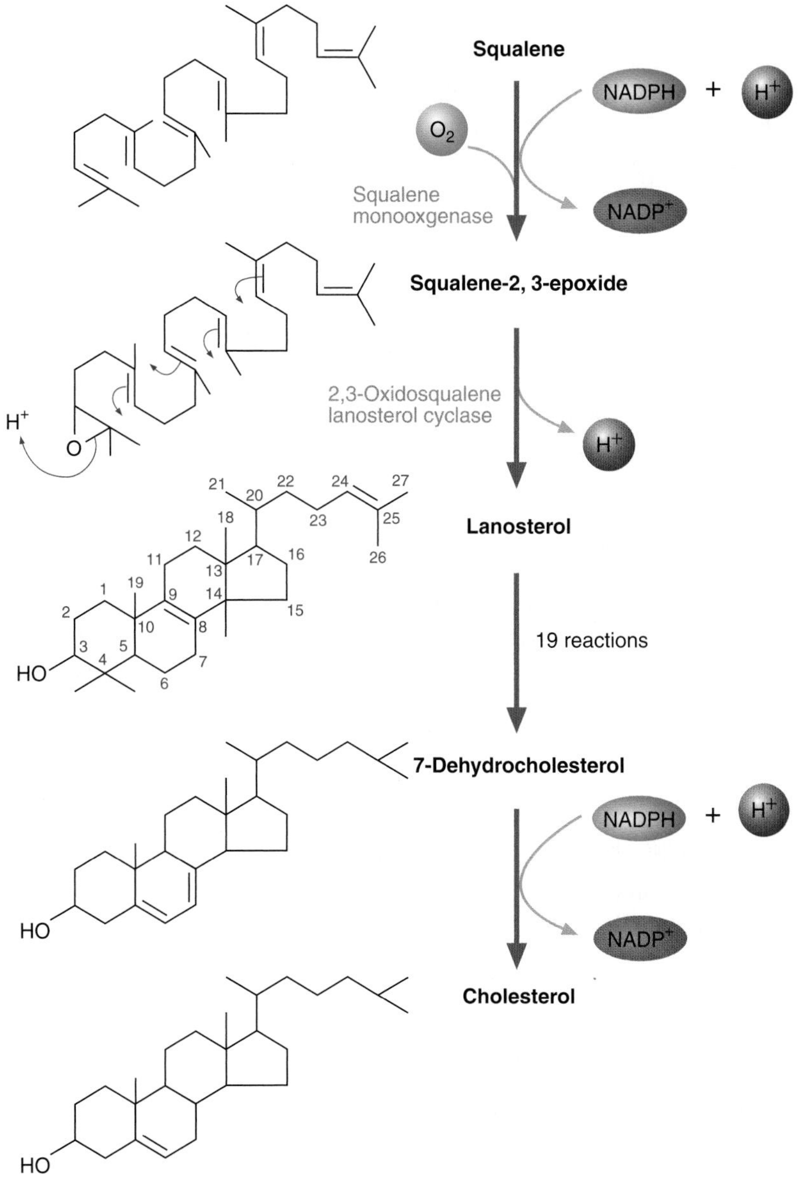

FIGURE 12.32

Synthesis of Cholesterol from Squalene

This is the major route in mammals. In an alternative minor route, squalene is converted to desmosterol, which is then reduced to form cholesterol. The details of these and many reactions in the major route are poorly understood. (Desmosterol differs from cholesterol because of a C=C bond between C-24 and C-25.)

are bound to this protein. The enzyme activities required for the oxygen-dependent epoxide formation (squalene monooxygenase) and subsequent cyclization (2,3-oxidosqualene lanosterol cyclase) that result in lanosterol synthesis have been localized in microsomes. Squalene monooxygenase requires NADPH and FAD for activity. After its synthesis, lanosterol binds to a second carrier protein, to which it remains attached during the remaining reactions. All of the enzyme activities that catalyze the remaining 20 reactions needed to convert lanosterol to cholesterol are embedded in ER membranes. In a series of transformations involving NADPH and oxygen, lanosterol is converted to 7-dehydrocholesterol. This product is then reduced by NADPH to form cholesterol.

CHOLESTEROL DEGRADATION Unlike many other types of biomolecules, cholesterol and other steroids cannot be degraded to smaller molecules. Instead, they are converted to derivatives whose improved solubility properties allow their excretion. The most important mechanism for degrading and eliminating cholesterol is the synthesis of the bile acids. Bile acid synthesis, which occurs in the liver, is outlined in Figure 12.33. The conversion of cholesterol to 7-α-hydroxycholesterol, catalyzed by cholesterol-7-hydroxylase (an ER enzyme), is the rate-limiting reaction in bile acid synthesis. Cholesterol-7-hydroxylase is a cytochrome P_{450} enzyme (see the Biochemistry in Perspective box entitled Biotransformation, pp. 461–463). In later reactions, the double bond at C-5 is rearranged and reduced, and an additional hydroxyl group is introduced. The products of this process, cholic acid and deoxycholic acid, are converted to bile salts by ER enzymes that catalyze conjugation reactions. (In **conjugation reactions** a molecule's solubility is increased by converting it into a derivative that contains a water-soluble group. Amides and esters are common examples of these conjugated derivatives.) Most bile acids are conjugated with glycine or taurine (Figure 12.34).

The bile salts are important components of *bile*, a yellowish green liquid produced by hepatocytes that aids in the digestion of lipids. In addition to bile salts, bile contains cholesterol, phospholipids, and bile pigments (bilirubin and biliverdin). The bile pigments are degradation products of heme. After it is secreted into the bile ducts and stored in the gallbladder, bile is used in the small intestine to enhance the absorption of dietary fat. Bile acts as an emulsifying agent; that is, it promotes the breakup of large fat droplets into smaller ones. Bile salts are also involved in the formation of so-called biliary micelles, which aid in absorbing fat and the fat-soluble vitamins (A, D, E, and K). Most bile salts are reabsorbed in the distal ilium (near the end of the small intestine). They enter the blood and are transported back to the liver, where they are resecreted into the bile ducts with other bile components. The biological significance of bile acid conjugation reactions appears to be that the conjugation process prevents premature absorption of bile acids in the biliary tract (the duct system and gallbladder) and small intestine. The reabsorption of bile salts in the distal ilium of the small intestine (necessary for effective recycling) is apparently triggered by the glycine or taurine signal. (It has been estimated that bile salt molecules are recycled 18 times before they are finally eliminated.)

QUESTION 12.10

The formation in the gallbladder or bile ducts of gallstones (crystals usually composed of cholesterol and inorganic salts) afflicts millions of people. Predisposing factors for this excruciatingly painful malady include obesity and infection of the gallbladder (*cholecystitis*). Because cholesterol is virtually insoluble in water, it is solubilized in bile by its incorporation into micelles composed of bile salts and phospholipids. Gallstones tend to form when cholesterol is secreted into bile in excessive quantities. Suggest a reason why an obese person is prone to gallstone formation. [*Hint*: HMG-CoA reductase activity is higher in obese individuals.]

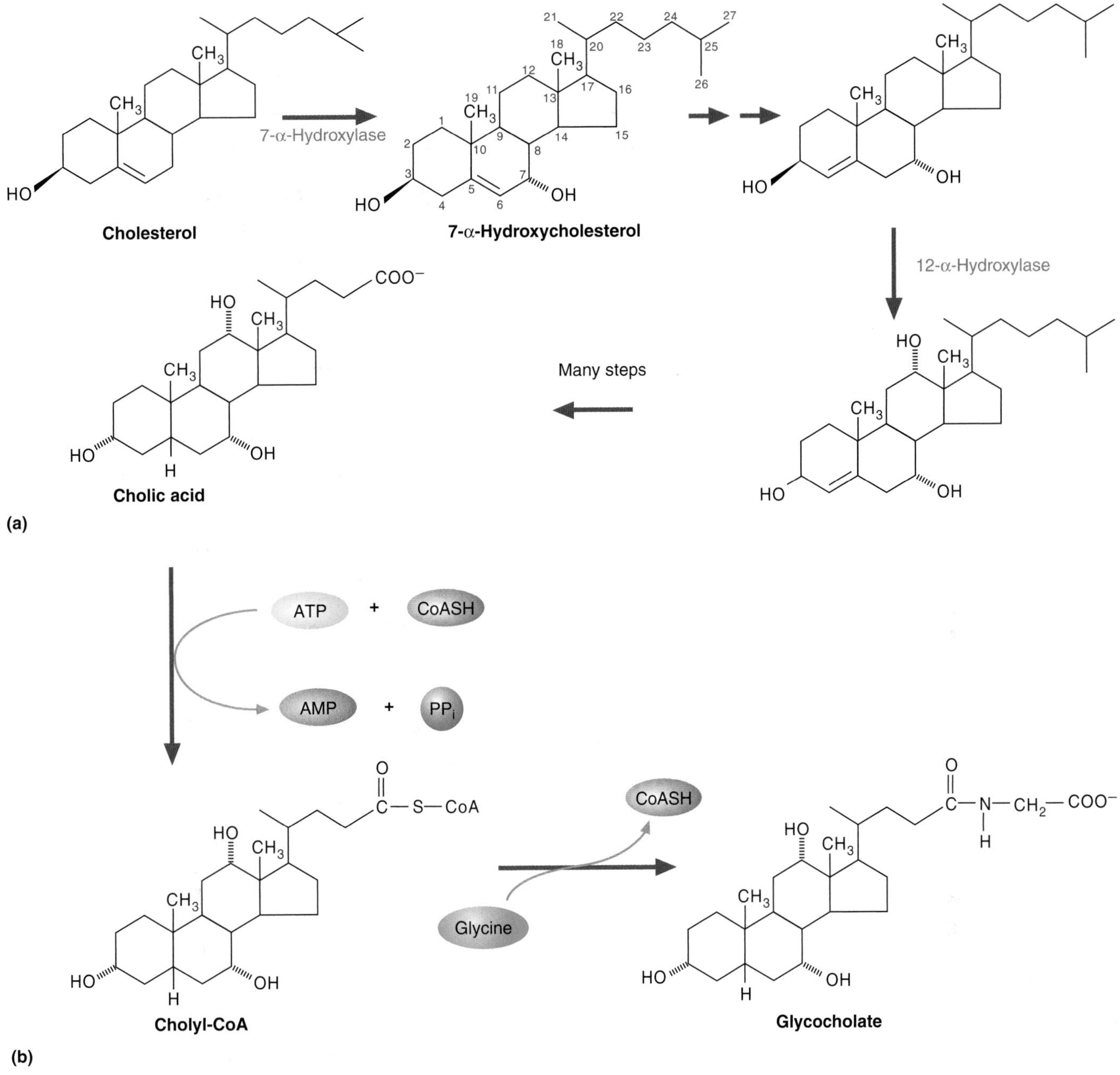

(a)

(b)

FIGURE 12.33

Bile Acid Synthesis

Because of their higher solubility, bile acids primarily act as detergents during dietary fat digestion. (a) Synthesis of cholic acid. (b) Synthesis of glycocholate, a bile salt.

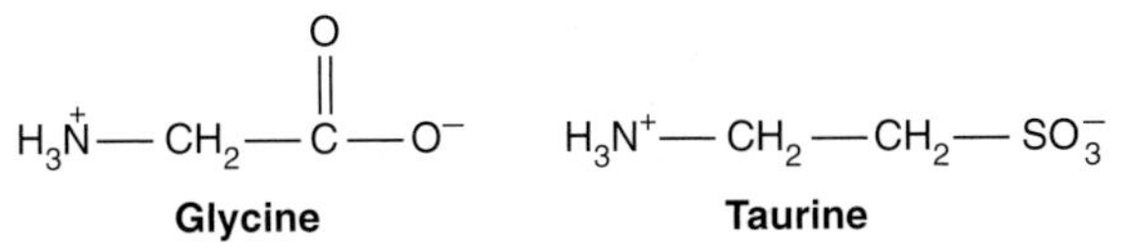

FIGURE 12.34

Structure of Glycine and Taurine

Most bile acids are conjugated in the liver with glycine or taurine.

CHOLESTEROL HOMEOSTASIS The critical roles of cholesterol in animal bodies combined with the potentially toxic properties it exhibits when present in excessive amounts, require that its concentration be maintained within normal limits. Cholesterol homeostasis is achieved through intricate mechanisms that regulate its biosynthetic pathway, LDL receptor activity, and bile acid biosynthesis. The regulation of cholesterol biosynthesis is primarily accomplished by modulation of existing HMGR molecules, gene expression changes, and enzyme degradation. The principal means by which the activity of existing HMGR molecules are regulated is downregulation via phosphorylation reactions (Figure 12.35). HMGR activity is depressed by phosphorylation by AMPK in response to high cellular concentrations of AMP, which have the effect of integrating cholesterol biosynthesis, a metabolically expensive process, into the cell's energy metabolism. cAMP, which is regulated by hormones such as glucagon and epinephrine, also depresses HMGR by activating phosphoprotein phosphatase inhibitor 1 (PPI-1) by means of a phosphorylation reaction catalyzed by PKA (protein kinase A). Activated PPI-1 inhibits several phosphatases that can increase HMGR activity by removing phosphate groups. Insulin increases HMGR activity, in part, by inhibiting cAMP synthesis. HMGR is also regulated by a negative feedback mechanism involving various sterols, including cholesterol derived from the endocytosis of LDL receptors, and nonsterol derivatives of mevalonate.

Sterol-mediated changes in gene expression are a major feature of cholesterol homeostasis. The ER membrane protein SREBP2 is the predominant regulator of cholesterol biosynthesis. In addition to stimulating the expression of cholesterol biosynthesis genes, SREBP2 activates the LDL receptor gene, and

FIGURE 12.35

Regulation of HMGR by Covalent Modification

HMGR is inactivated by phosphorylation reactions catalyzed by AMPK, in response to rising AMP levels. Activation of HMGR is effected by phosphoprotein phosphatase. When cAMP levels are high PPI-1, activated by a phosphorylation reaction, inhibits phosphoprotein phosphatase. An inhibited phosphoprotein phosphatase insures the inactive status of HMGR.

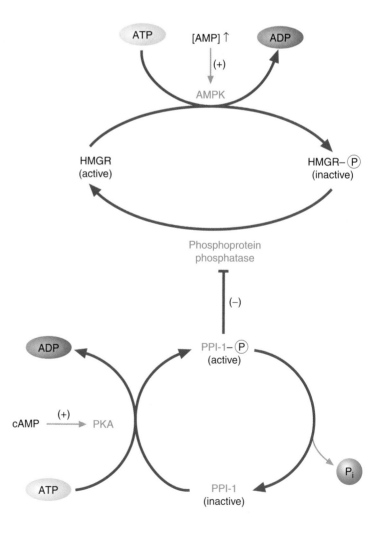

three genes required for NADPH synthesis (G-6-PD, 6-phosphogluconate-dehydrogenase, and malic enzyme). SREBP2 is bound to SREBP cleavage-activating protein (SCAP). When cells have sufficient cholesterol, the sterol-sensing domain (SSD) of SCAP binds cholesterol and an ER retention protein called Insig (*ins*ulin-*i*nduced *g*ene). In cholesterol-depleted cells, Insig no longer binds to SCAP, which now escorts SREBP-2 from the ER to the Golgi (Figure 12.36). When the SREBP-SCAP complex reaches the Golgi, two proteases release the N-terminal domain of SREBP2, the active transcription factor, from the membrane. The activated SREBP2 product then translocates to the nucleus, where it binds to the SREs (sterol regulatory elements) of the target genes. In addition to SREBP, the expression of some target genes requires the binding of coregulatory transcription factors. For example, transcription of the genes for HMGR

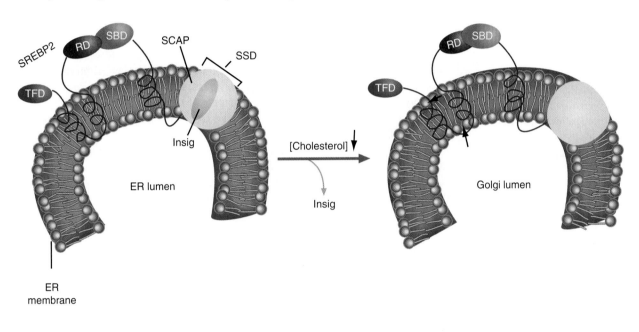

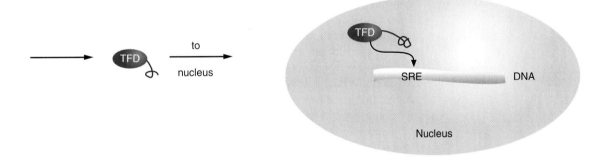

FIGURE 12.36

SREBP2 Regulation

SREBP2, an ER protein, is complexed to SCAP (SREBP cleavage-activating protein) via the binding of the regulatory domain (RD) of SREBP2 with the SREBP-binding domain (SBD) of SCAP. The SSD (sterol-sensing domain) of SCAP binds cholesterol. When cholesterol levels are high and cholesterol is bound to SSD, the ER retention protein Insig is also bound. When cholesterol levels are depleted and cholesterol is no longer bound to SSD, Insig is released, and the SREBP/SCAP complex is transferred to the Golgi complex. When SREBP2 is present in the Golgi complex, two proteases cleave it at two sites (arrows) to release the now active TFD (transcription factor domain). The SREBP transcription factor then moves into the nucleus, where it binds to SREs (sterol regulatory elements) that are associated with sterol-related genes.

and HMG-CoA synthase also requires the binding of NF-1 (nuclear factor-1) and CREB (cAMP response element binding protein).

As cellular cholesterol levels begin to rise as a result of newly synthesized enzymes and the internalization of LDL from the bloodstream via LDL receptors, a concentration threshold is reached and a few molecules bind to the sterol-sensing domains of SCAP. As the result of a conformation change in the SCAP-SREBP complex, the ER protein Insig replaces cholesterol in the sterol-sensing domain of SCAP and the complex is retained in the ER, thus ending SREBP2 processing and the transcription of target genes. High levels of cholesterol and other metabolites of mevalonate also depress further cholesterol synthesis by inhibiting the translation of existing HMGR mRNA. In liver, excess cholesterol activates acyl-CoA–cholesterol acyltransferase (ACAT), the enzyme that catalyzes the transfer of a fatty acid from a fatty acyl–CoA molecule to the hydroxyl group of cholesterol to generate a cholesterol ester storage molecule.

HMGR degradation, another means by which cholesterol levels are controlled, is mediated by Insig. When cholesterol levels are high, sterols bind to the enzyme's sterol-sensing N-terminal domain. Subsequently HMGR binds to Insig, which in turn is associated with ubiquitin ligase, an enzyme that initiates a major proteolytic mechanism (described in Chapter 15).

High cholesterol concentrations in the liver also trigger bile acid biosynthesis. When cholesterol begins to accumulate, some molecules become oxidized to form oxysterols (e.g., 25-hydroxycholesterol). Oxysterols bind to and activate LXR (liver X receptor), which then forms an active heterodimer transcription factor with RXR (retinol X receptor). The heterodimer then causes the transcription of 7-α-hydroxylase, the rate-limiting enzyme in bile synthesis.

Steroid Hormone Synthesis

Recall that cholesterol is a precursor for all steroid hormones. These syntheses are briefly outlined now. The metabolism of steroids is extremely complex and as yet incompletely understood. A variety of cells and organelles figure prominently in the production and processing of these powerful substances. The initial reaction in steroid hormone synthesis, the conversion of cholesterol to pregnenolone (Figure 12.37), is catalyzed by desmolase, a mitochondrial enzyme. Desmolase is an enzyme complex composed of two hydroxylases, one of which is a cytochrome P_{450} enzyme. Cytochrome P_{450} is involved in reactions of steroid and xenobiotic metabolism (see Biotransformation, the Biochemistry in Perspective box that concludes this chapter). After its synthesis, pregnenolone is transported to the ER, where it is converted to progesterone. Pregnenolone and progesterone are precursors for all other steroid hormones (Figure 12.38). In addition to its precursor role, progesterone also acts as a hormone. Its primary hormonal role is the regulation of several physiological changes in the uterus. During the menstrual cycle, progesterone is produced by specialized cells within the ovary. During pregnancy the progesterone that is produced in large quantities by the placenta prevents uterine smooth muscle contractions.

QUESTION 12.11

During an experiment cholesterol is synthesized using

as the substrate. Which atoms in the cholesterol that is recovered will be labeled?

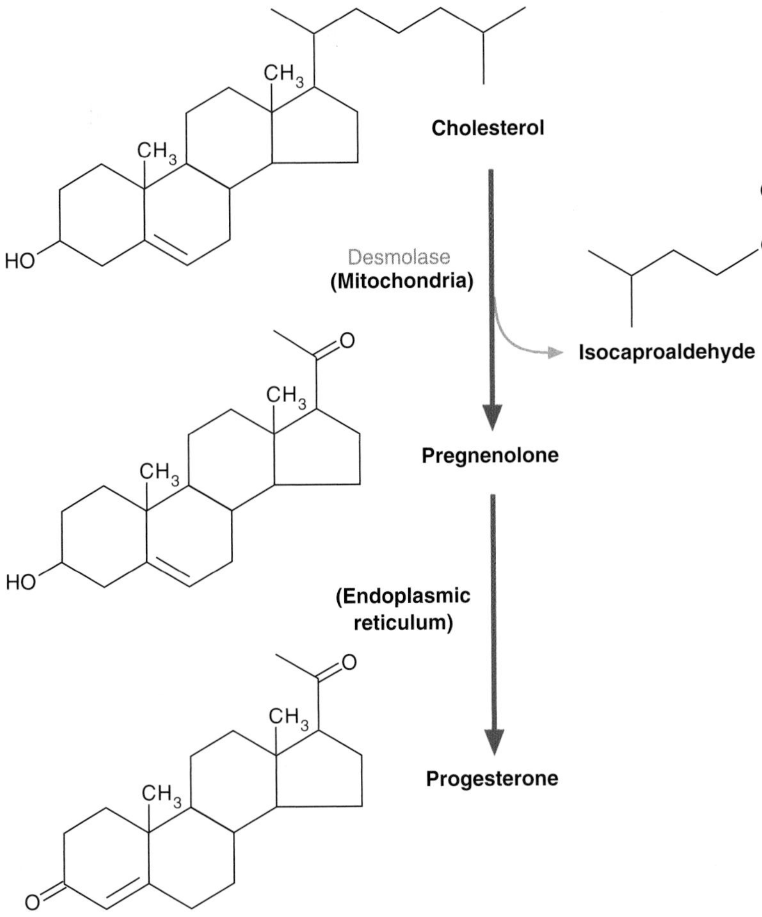

Cholesterol

Desmolase
(Mitochondria)

Isocaproaldehyde

Pregnenolone

**(Endoplasmic
reticulum)**

Progesterone

FIGURE 12.37

Progesterone Synthesis

Pregnenolone is synthesized in the mitochondria. It is transported to the ER, where
it is converted to progesterone. This latter
process oxidizes the hydroxyl group and
isomerizes a C=C bond.

QUESTION 12.12

Cortisol (also referred to as *hydrocortisone*) is a potent glucocorticoid.
Glucocorticoids, hormones that promote carbohydrate, protein, and fat metabolism, stimulate such physiological processes as gluconeogenesis, lipolysis,
and an increased uptake of amino acids by the liver. Cortisol also possesses
a small amount of mineralocorticoid activity. **Mineralocorticoids** regulate Na^+
and K^+ metabolism. For example, aldosterone, the most important mineralocorticoid in humans, induces the reabsorption of Na^+ from urine. It also promotes the secretion of K^+ and H^+ into urine. For steroids to have either
glucocorticoid or mineralocorticoid activity, they must possess a hydroxy
group at C-11.

In *Addison's disease* an inadequate secretion of glucocorticoids and
mineralocorticoids results in hypoglycemia, an imbalance in the body's Na^+ and
K^+ concentrations, dehydration, and low blood pressure. In the past, individuals who had undiagnosed Addison's disease found that consumption of
large amounts of licorice provided some relief from their symptoms. (Licorice,
an extract of the plant *Glycyrrhiza glabra*, is used as a flavoring agent in
candy and some medicines.) In these patients, licorice consumption resulted
in sodium retention, hypokalemia (low blood K^+ concentration), and a rise
in blood pressure. This effect was more pronounced when cortisol was
administered.

Recently, it has been discovered that the active ingredient in licorice, called
glycyrrhizic acid, inhibits 11-β-hydroxysteroid dehydrogenase, the enzyme that
reversibly converts cortisol to cortisone, its inactive metabolite. Refer to the

QUESTION 12.12 (CONT.)

structure of cortisol (Figure 12.38) and deduce the structure of cortisone. Cortisol is regarded as a glucocorticoid. Why, then, does glycyrrhizic acid consumption appear to affect mineral metabolism? Cortisone is often used to treat Addison's disease even though it is physiologically inactive. Suggest a reason to justify its use.

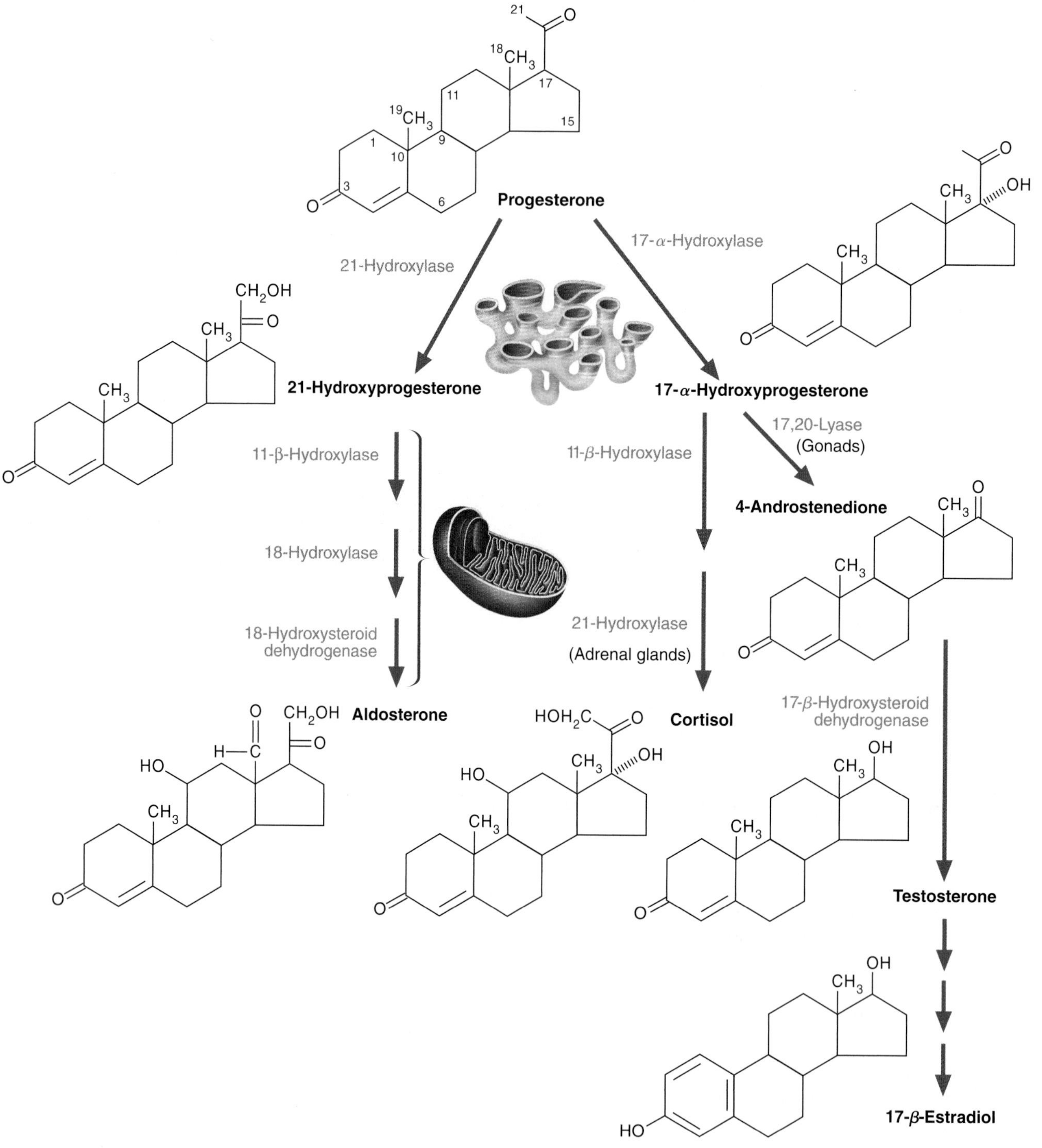

FIGURE 12.38

Synthesis of Selected Steroids

The enzyme 17-α-hydroxylase is found in all steroid-producing cells. Most other enzymes are tissue specific.

BIOCHEMISTRY IN PERSPECTIVE

Biotransformation

How are potentially toxic hydrophobic molecules metabolized by the body? In **biotransformation**, a series of enzyme-catalyzed processes in which toxic substances are converted into less toxic metabolites, the enzymes generally possess broad specificities. In mammals, biotransformation is used principally to convert toxic molecules, which are usually hydrophobic, into water-soluble derivatives so that they may be more easily excreted. The enzymes that catalyze the biotransformation of xenobiotics (foreign molecules) are similar to several of the enzymes that dispose of hydrophobic endogenous molecules. Although biotransformation reactions occur in several locations within the cell (e.g., the cytoplasm and mitochondria), most occur within the ER. Cell types also differ in their biotransforming potential. In general, cells located near the major points of xenobiotic entry into the body (e.g., liver, lung, and intestine) possess greater concentrations of biotransforming enzymes than others.

Biotransformation processes have been differentiated into two major types. During **phase I**, reactions involving oxidoreductases and hydrolases convert hydrophobic substances into more polar molecules. **Phase II** consists of reactions in which metabolites containing appropriate functional groups are conjugated with substances such as glucuronate, glutamate, sulfate, or glutathione. In general, conjugation dramatically improves solubility, which then promotes rapid excretion. (In animals, excretion of biotransformed molecules is sometimes referred to as phase III.) Although many substances undergo these phases sequentially, a significant number do not. For example, some molecules are excreted as phase I metabolites, while others undergo only phase II reactions. Moreover, variations in enzyme concentrations, availability of cosubstrates, and the order in which the reactions occur may cause certain substances to be converted into more than one end product. However, despite these and other complications, basic biotransformation patterns have emerged. Several well-researched examples of phase I and phase II reactions are described below. In the following discussions, the term **detoxication** refers to the process by which a toxic molecule is converted to a more soluble (and usually less toxic) product. The more familiar term **detoxification** implies the correction of a state of toxicity, that is, the chemical reactions that produce sobriety in an intoxicated person.

Phase I reactions usually convert substrates to more polar forms by introducing or unmasking a functional group (e.g., —OH, —NH$_2$, or —SH). Many phase I enzymes are located in the ER membrane, but others such as the dehydrogenases (e.g., alcohol dehydrogenases and peroxidases) occur in the cytoplasm, while still others (e.g., monoamine oxidase) are localized in mitochondria. The predominant enzymes of ER oxidative metabolism are the monooxygenases, sometimes referred to as mixed function oxidases. They are so named because in a typical reaction, one molecule of oxygen is consumed (reduced) per substrate molecule: one oxygen atom appearing in the product and the other in a molecule of water. Monooxygenases can carry out an immense variety of chemical reactions. Some of these reactions form highly unstable (and therefore toxic) intermediates.

There are two major types of microsomal monooxygenase, both of which require NADPH as an external reductant: the cytochrome P$_{450}$ (cyt P$_{450}$) system and flavin-containing monooxygenases. The **cytochrome P$_{450}$ system**, which consists of two enzymes (NADPH–cytochrome P$_{450}$ reductase and cytochrome P$_{450}$) is involved in the oxidative metabolism of many endogenous substances (e.g., steroids and bile acids), as well as the detoxication of a wide variety of xenobiotics. **Flavin-containing monooxygenases** catalyze an NADPH- and an oxygen-requiring oxidation of substances (primarily xenobiotics) bearing functional groups containing nitrogen, sulfur, or phosphorus. The properties of the cyt P$_{450}$ electron transport systems are described.

Cytochrome P450 Electron Transport Systems

In cytochrome P$_{450}$ electron transport systems, found in microsomal ER and inner mitochondrial membranes, two electrons are transferred one at a time from NADPH to a cytochrome P$_{450}$ protein by NADPH–cytochrome P$_{450}$ reductase (Figure 12 A). The latter enzyme is a flavoprotein that contains both FAD and FMN in a ratio of 1:1 per mole of enzyme. In addition to its role in the cytochrome P$_{450}$ system, the reductase is also believed to be involved in the function of heme oxygenase (p. 585).

The hemoproteins referred to as cytochrome P$_{450}$ are so named because of the complexes they form with carbon monoxide. In the presence of the gas, light is strongly absorbed at a wavelength of 450 nm. Over 6000 cyt P$_{450}$ genes have been identified in species as diverse as mammals and bacteria. In humans, 63 cyt P$_{450}$ genes, classified into 18 families, have been identified. Each gene codes for a protein with a unique specificity range. Cytochrome P$_{450}$ proteins found in the liver have broad and overlapping specificities. For example, molecules as diverse as alkanes, aromatics, ethers, and sulfides are routinely oxidized. In contrast, cytochrome P$_{450}$ proteins in the adrenal glands, ovaries, and testes that add hydroxyl groups to steroid

BIOCHEMISTRY IN PERSPECTIVE cont

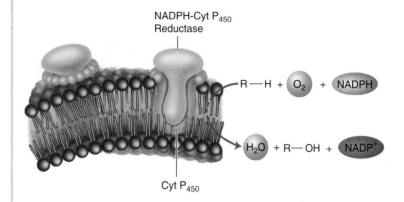

FIGURE 12A

The Cytochrome P_{450} Electron Transport System

Cytochrome P_{450} and cytochrome P_{450} reductase are components of an electron transport system used to oxidize both endogenous and exogenous molecules.

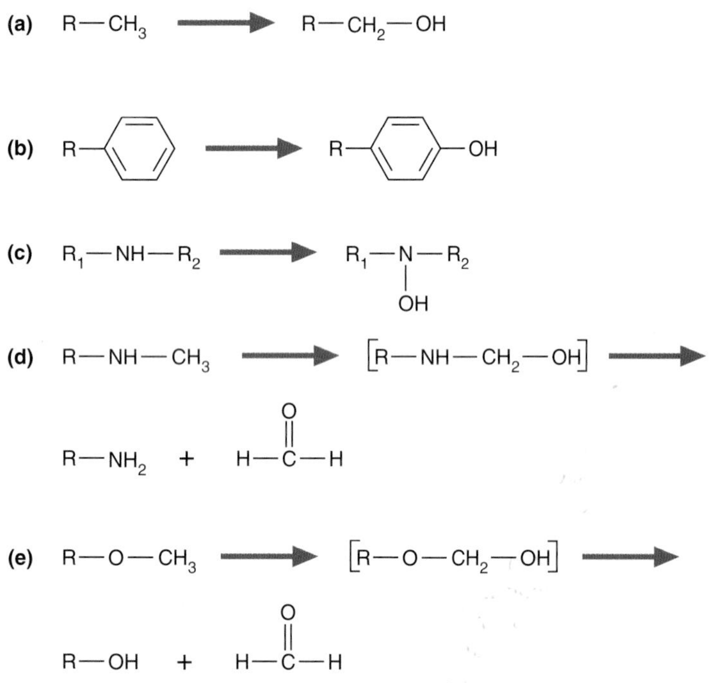

(a) $R—CH_3 \longrightarrow R—CH_2—OH$

(b) $R—\langle\bigcirc\rangle \longrightarrow R—\langle\bigcirc\rangle—OH$

(c) $R_1—NH—R_2 \longrightarrow R_1—\underset{\underset{OH}{|}}{N}—R_2$

(d) $R—NH—CH_3 \longrightarrow [R—NH—CH_2—OH] \longrightarrow$

$R—NH_2 \quad + \quad H—\overset{\overset{O}{\|}}{C}—H$

(e) $R—O—CH_3 \longrightarrow [R—O—CH_2—OH] \longrightarrow$

$R—OH \quad + \quad H—\overset{\overset{O}{\|}}{C}—H$

FIGURE 12B

Diverse Substrates Oxidized by Cytochrome P_{450} Isozymes

Among the reactions catalyzed by cytochrome P_{450} are (a) aliphatic oxidation, (b) aromatic hydroxylation, (c) N-hydroxylation, (d) N-dealkylation, and (e) O-dealkylation.

molecules have narrow specificities. Despite this diversity, all cytochrome P_{450} isozymes contain one molecule of heme. In addition, the cytochrome P_{450}s are similar in their physical properties and catalytic mechanisms.

Despite an enormous variety of substrates, all of the oxidative reactions catalyzed by cytochrome P_{450} may be viewed as hydroxylation reactions (i.e., an OH group appears in each reaction) (Figure 12B). The general reaction is as follows:

$$R—H + O_2 + NADPH + H^+ \longrightarrow ROH + H_2O + NADP^+$$

where R—H is the substrate.

The oxygenation reaction is initiated when the substrate binds to oxidized cytochrome P_{450} (Fe^{3+}). This binding promotes a reduction of the enzyme substrate complex by an electron

transferred from NADPH via cytochrome P_{450} reductase (Fe^{2+}—substrate). After reduction, cytochrome P_{450} can bind O_2. Then the electron from heme iron is transferred to the bound O_2, thus forming a transient $Fe^{3+}—O_2^-$—substrate species. (If the bound substrate is easily oxidized, it can be converted into a peroxy radical.) A second electron transferred from the flavoprotein results in the generation of a $Fe^{3+}—O_2^{2-}$—substrate complex. This brief association ends when the oxygen-oxygen bond is broken. One atom is released in a water molecule, while the other remains bound to heme. After a hydrogen atom or electron has been abstracted from the substrate, the oxygen species (now a powerful oxidant) is transferred to the substrate. The cycle ends with release of the product from the active site. Depending on the nature of the substrate, the product is either an **epoxide**

BIOCHEMISTRY IN PERSPECTIVE cont

(a highly reactive ether in which the oxygen is incorporated into a three-membered ring) or an alcohol. The function of conjugation (phase II) reactions is to inactivate biologically active substances and/or to form more polar (and therefore more easily excretable) derivatives. During this process, lipophilic metabolites, bearing functional groups that can act as acceptors, undergo enzyme-catalyzed reactions along with second (or donor) substrates. Among the most frequently used donor substrates are glucuronic acid, glutathione (see later, Section 14.3), sulfate, and amino acids.

SUMMARY: Biotransformation, the enzyme-catalyzed process in which toxic, hydrophobic molecules are converted to less toxic, water-soluble molecules, consists of two types of biochemical reaction class: phase I and phase II. Phase I reactions introduce or unmask polar functional groups in hydrophobic molecules. In phase II the water solubility of substrate molecules is substantially improved by the conjugation of functional groups with substances such as glucuronic acid.

Chapter**Summary**

1. Acetyl-CoA plays a central role in most lipid-related metabolic processes. For example, acetyl-CoA is used in the synthesis of fatty acids. When fatty acids are degraded to generate energy, acetyl-CoA is the product.

2. Depending on the body's current energy requirements, newly digested fat molecules are used to generate energy or are stored within adipocytes. When the body's energy reserves are low, fat stores are mobilized in a process referred to as lipolysis. In lipolysis, triacylglycerols are hydrolyzed to fatty acids and glycerol. Glycerol is transported to the liver, where it can be used in lipid or glucose synthesis. Most fatty acids are degraded to form acetyl-CoA within mitochondria in a process referred to as β-oxidation. In the β-oxidation pathway, acetyl-CoA is formed from fatty acids in a series of four reactions in which the β-carbon is oxidized, followed by the breakage of the bond between the α- and β-carbons. Peroxisomal β-oxidation appears to shorten very long fatty acids. Additional reactions are required to degrade odd-chain and unsaturated fatty acids and dicarboxylic acids. When the acetyl-CoA product is present in excess, ketone bodies are produced and are used as an energy source in some tissues.

3. The first step in eicosanoid synthesis is the release of arachidonic acid from C-2 of glycerol in membrane phosphoglyceride molecules. Cyclooxygenase converts arachidonic acid into PGG_2, which is a precursor of the prostaglandins and the thromboxanes. The lipoxygenases convert arachidonic acid to the precursors of the leukotrienes.

4. Fatty acid synthesis begins with the carboxylation of acetyl-CoA to form malonyl-CoA. ACC, a key enzyme in fatty acid metabolism, is regulated by allosteric modulators and phosphorylation reactions. The remaining reactions of fatty acid synthesis take place on the fatty acid synthase multienzyme complex. Several enzymes are available to elongate and desaturate dietary and newly synthesized fatty acids.

5. Both short- and long-term regulatory mechanisms are used to control fatty acid metabolism. Short-term regulation includes the use of malonyl-CoA as an inhibitor of CAT-I, and AMPK-catalyzed phosphorylation of ACC1 and glycerol-3-phosphate acyltransferase. Hormones such as insulin, glucagon, epinephrine, and cortisol have roles in short- and long-term regulation. Long-term regulation of fatty acid metabolism involves changes in gene expression triggered by transcription factors. Two prominent examples are the SREBPs and the PPARs.

6. After phospholipids have been synthesized at the interface of the SER and the cytoplasm, they are often "remodeled"; that is, their fatty acid composition is adjusted. The turnover (i.e., the degradation and replacement) of phospholipids, mediated by the phospholipases, is rapid.

7. Synthesis of the ceramide component of sphingolipids begins with the condensation of palmitoyl-CoA with serine to form 3-ketosphinganine. In a two-step process involving acyl-CoA and $FADH_2$, sphinganine (formed when 3-ketosphinganine is reduced by NADPH) is converted to ceramide. Sphingolipids are degraded within lysosomes.

8. Cholesterol synthesis can be divided into three phases: formation of HMG-CoA from acetyl-CoA, conversion of HMG-CoA to squalene, and conversion of squalene to cholesterol. Cholesterol is the precursor for all steroid hormones and the bile salts. Bile salts are used to emulsify dietary fat. They are the primary means by which the body can rid itself of cholesterol. Cholesterol homeostasis is achieved through intricate mechanisms that regulate cholesterol biosynthesis, LDL receptor activity, and bile acid biosynthesis.

Take your learning further by visiting the **companion website** for Biochemistry at **www.oup.com/us/mckee** where you can complete a multiple-choice quiz on lipid metabolism to help you prepare for exams.

Suggested Readings

Hardie, D. G., Minireview: The AMP-Activated Protein Kinase Cascade: The Key Sensor of Cellular Energy Status, *Endocrinology* 144(12):5179–5183, 2003.

Feige, J. N., Gelman, L., Michalik, L., Desvergne, B., and Wahli, W., From Molecular Action to Physiological Outputs: Peroxisome Proliferator–Activated Receptors Are Nuclear Receptors at the Crossroads of Cellular Functions, *Prog. Lipid Res.* 45:120–159, 2006.

Ferre, P., The Biology of Peroxisome Proliferator–Activated Receptors: Relationship with Lipid Metabolism and Insulin Sensitivity, *Diabetes* 53(suppl 1):S43–S50, 2004.

Gurr, M. I., Frayn, K. N., and Harwood, J. L., *Lipid Biochemistry: An Introduction*, 5th ed., Blackwell Scientific, Oxford, 2001.

Munday, M. R., Regulation of Mammalian Acetyl-CoA Carboxylase, *Biochem. Soc. Trans.* 30:1059–1064, 2002.

Neels, J. G., and Olefsky, J. M., A New Way to Burn Fat, *Science* 312:1756–1758, 2006.

Rawson, R. B., The SREBP Pathway: Insights from Insigs and Insects, *Nat. Rev. Mol. Cell. Biol.* 4:631–640, 2003.

Sampath, H., and Ntambi, J. M., Regulation of Gene Expression by Polyunsaturated Fatty Acids, *Heart Metab.* 32:32–35, 2006.

Santamarina-Fojo, S., Gonzalez-Navarro, H., Freeman, L., Wagner, E., and Nong, Z., Hepatic Lipase, Lipoprotein Metabolism, and Atherosclerosis, *Arterioscler. Thromb. Vasc. Biol.* 24:1750–1754, 2004.

Steinberg, G. R., Macaulay, S. L., Febbraio, M. A., and Kemp, B. E., AMP-Activated Protein Kinase: The Fat Controller of the Energy Railroad, *Can. J. Physiol. Pharmacol.* 84(7):655–665, 2006.

Stocker, R., and Keaney, J. F., Role of Oxidative Modifications in Atherosclerosis, *Physiol. Rev.* 84:1381–1478, 2003.

Vance, D. E., and Vance, J. E., *Biochemistry of Lipids, Lipoproteins, and Membranes*, 4th ed., Elsevier, Amsterdam, 2002.

Key Words

acyl carrier protein, *431*
AMPK, *440*
autoimmune disease, *439*
bile salts, *415*
biotransformation, *461*
chylomicron remnants, *416*
conjugation reaction, *454*

cytochrome P$_{450}$ system, *461*
detoxication, *461*
detoxification, *461*
epoxide, *462*
fatty acid–binding protein, *419*
glucocorticoid, *459*

ketogenesis, *425*
ketone body, *425*
ketosis, *425*
lipogenesis, *417*
lipolysis, *417*
mineralocorticoid, *459*
β-oxidation, *420*
phase I reaction, *461*

phase II reaction, *461*
3′-phosphoadenosine-5′-phosphosulfate, *446*
PPAR, *442*
SREBP, *442*
sterol carrier protein, *453*
thiolytic cleavage, *423*
turnover, *444*

Review Questions

These questions are designed to test your knowledge of the key concepts discussed in this chapter, before moving on to the next chapter. You may like to compare your answers to the solutions provided in the back of the book and in the accompanying Study Guide.

1. Define the following terms:
   a. *de novo*
   b. oil bodies
   c. β-oxidation
   d. turnover
   e. thiolytic cleavage

2. What is the function of each of the following substances?
   a. carnitine
   b. sterol carrier protein
   c. thrombin

   d. thiolase
   e. desmolase

3. What are the differences between β-oxidation in mitochondria and in peroxisomes? What similarities are there between these processes?

4. List three differences between fatty acid synthesis and β-oxidation.

5. Define the following terms:
   a. autoantibodies
   b. ketone bodies

c. conjugation reaction

d. ACP

e. AMPK

6. Explain how hormones act to modify the metabolism of fatty acids in both the short and long term. Give examples.

7. What is the difference between a steroid and a sterol?

8. Define the following terms:

a. hypertriglyceridemia

b. SREBP1

c. SREBP2

d. chemokine

e. atheroma

9. Show how the following fatty acid is oxidized:

$$CH_3CH_2CH_2\underset{\underset{CH_3}{|}}{CH}-CH_2-\underset{\underset{}{\overset{O}{\|}}}{C}-OH$$

10. What are the products of the oxidation of the molecule in Question 9?

11. Insulin is released after carbohydrate intake. Describe two ways insulin acts to influence fatty acid metabolism.

12. β-Oxidation of naturally occurring monounsaturated fatty acids requires an additional enzyme. What is this enzyme, and how does it accomplish its task?

13. Define the following terms:

a. plaque

b. SAM

c. PAPS

d. HMGR

e. allyl group

14. Identify the hydrophobic and hydrophilic regions in the following molecule. How does it orient itself in a membrane?

$$CH_3-(CH_2)_{15}CH_2-\overset{\overset{O}{\|}}{C}-O-\underset{\underset{CH_2-O-\overset{\overset{O}{\|}}{P}-O-CH_2CH_2N(CH_3)_3}{|}}{\overset{\overset{CH_2-O-\overset{\overset{O}{\|}}{C}-CH_2(CH_2)_{15}-CH_3}{|}}{C}}-H$$

15. What is the function of each of the following?

a. autoantibodies

b. ketone bodies

c. biotransformation

d. phase I reaction

e. ACP

16. Gaucher's disease is an inherited deficiency of β-glucocerebrosidase. Glucocerebroside is deposited in macrophages that die, releasing their contents into the tissues. Some affected individuals experience neurologic disorders while quite young; others do not show ill effects until much later in life. The disease may be detected by assaying white blood cells for the ability to hydrolyze the β-glycosidic bond of artificial substrates. Examine the following glucocerebroside and determine which bond is cleaved by glucocerebrosidase.

17. Determine the number of moles of ATP that can be generated from the fatty acids in 1 mol of tristearin, a triacylglycerol composed of glycerol esterified to three stearic acid molecules. What is the fate of the glycerol?

18. How would you describe the three phases of biotransformation?

19. Define the following terms:

a. detoxication

b. detoxification

c. epoxide

d. monooxygenase

e. cytochrome P_{450}

20. Outline the biosynthesis of bile salts. What are the functions of these substances?

21. How are lipid molecules such as animal steroid molecules like estrogen and β-carotene related to each other? What biosynthetic reactions do these specific molecules have in common?

22. What is the function of each of the following substances?

a. phospholipid exchange protein

b. sterol carrier protein

c. pregnenolone

d. glucocorticoid

e. hormone sensitive lipase

23. List and describe the components of the cytochrome P_{450} electron transport system. What is the role of each component?

24. What is the function of conjugation reactions in the biotransformation process?

25. The absorption of triacylglycerols in the small intestine is an energy-requiring process that involves hydrolytic reactions to yield glycerol and fatty acids. After their transport into enterocytes, fatty acids and glycerol are reconverted into triacylglycerols. Suggest a reason why triacylglycerols are not absorbed directly without the hydrolytic reactions.

26. The peroxisomal enzyme β-ketoacyl–CoA-thiolase does not bind medium-chain acyl-CoA, in contrast to the analogous mitochondrial enzyme. Explain why this phenomenon is an advantage to the cell.

27. Under severe starvation conditions, people develop "acetone breath." Explain.

28. Provide an explanation for the intracellular separation of fatty acid metabolic processes (i.e., fatty acid biosynthesis in cytoplasm and degradation in mitochondria and peroxisomes).

29. Describe the mechanism in which NADH is involved in the activation of oxygen molecules in fatty acid desaturation processes.

30. Describe the role of insulin in lipid metabolism.

31. Membranes that contain a significant proportion of *cis* unsaturated fatty acids are more fluid than similar membranes with higher levels of saturated fatty acids. Explain.

32. Describe why cholesterol is used as a stiffening agent in animal membranes.

33. When $^{14}CO_2$ is used in the synthesis of malonyl-CoA from acetyl-CoA, no label appears in the eventual fatty acid products. Explain.

34. Physical exercise is known to lower cholesterol levels. Explain.

35. Draw a cholesterol molecule and indicate the mevalonate units.

36. Determine how many ATP equivalents are obtained by the oxidation of oleic acid.

37. Describe the fate of glycerol generated from triacylglycerol hydrolysis in adipocytes.

38. Many processed and fast foods contain *trans* fatty acids. Explain why these molecules are a problem for the body.

39. Before β-oxidation, fatty acids are converted to their CoASH derivatives. Explain why this reaction is necessary.

40. Review the steps in β-oxidation and determine which ones are actually oxidation reactions.

ThoughtQuestions

These questions are designed to reinforce your understanding of all of the key concepts discussed in the book so far, including this chapter and all of the chapters before it. They may not have one right answer! The authors have provided possible solutions to these questions in the back of the book and in the accompanying Study Guide, for your reference.

41. Pharmaceuticals in a class called the statins inhibit the enzyme HMG-CoA reductase. What is the primary effect of this drug on patients?

42. What are the potential consequences of ineffective regulation of the opposing processes of β-oxidation and fatty acid synthesis?

43. Describe the possible effects of low levels of carnitine on a person's metabolism.

44. Phospholipases show an enhanced activity for a substrate above the critical micelle concentration. (The critical micelle concentration, or cmc, is that concentration of a lipid above which micelles begin to form.)
    a. What type of noncovalent interactions are possible between the lipid and the enzyme at this stage?
    b. What do these interactions suggest about the structure of phospholipases?

45. When the production of acetyl-CoA exceeds the body's capacity to oxidize it, acetoacetate, β-hydroxybutyrate, and acetone accumulate. When generated in large amounts, these substances can exceed the blood's buffering capacity. As the blood pH falls, the ability of red blood cells to carry oxygen is affected. Subsequently, the brain can be starved for oxygen, and a fatal coma can result. Explain how severe dieting can produce this condition.

46. The acyl-CoA dehydrogenase deficiency diseases are a group of inherited defects that impair the β-oxidation of fatty acids. Symptoms of the disease range from nausea and vomiting to frequent comas. Symptoms may be alleviated by eating regularly and avoiding periods of starvation (12 hours or more). Why does this simple procedure alleviate the symptoms?

47. There is an unusually high concentration of phosphatidylcholine on the lumenal side of the ER. What structural feature of phosphatidylcholine is responsible for this? Explain how this structural feature produces this effect.

48. During periods of stress or fasting, blood glucose levels fall. In response, fatty acids are released by adipocytes. Explain how the drop in blood glucose triggers fatty acid release.

49. Butyric acid, a simple four-carbon fatty acid, is oxidized by β-oxidation. Calculate the number of $FADH_2$ and NADH molecules produced in this oxidation. How many acetyl-CoA molecules are also produced?

50. The adaptations of desert animals to their environment include water conservation mechanisms. A number of these organisms conserve water so successfully that they never actually drink it. They depend instead on water generated during metabolism. Determine how much water can be obtained by the oxidation of one mole of palmitic acid.

51. Consider the structure of the following fatty acid and determine the reactions by which it is oxidized.

$$H_3C-(CH-CH_2-CH_2-CH_2)_3-CH-CH_2-C-O^-$$

(with CH_3 substituents and a terminal O carbonyl)

52. Cholestyramine is an anion-exchange resin used by clinicians to lower serum cholesterol in patients. This oral medication forms insoluble complexes with bile salts in the intestine, thus preventing their reabsorption. Explain the mechanism by which cholestyramine lowers serum cholesterol levels.

CHOLESTYRAMINE
RESIN

53. Provide an explanation for the fact that most fatty acids are 16 or 18 carbons long.

54. An experimenter using acetyl-CoA with a ^{14}C label on the carbonyl group traces the label in a cell synthesizing cholesterol. On what atoms of mevalonate will the label appear?

55. Determine the position of ^{14}C (see Question 54) in isopentenylpyrophosphate.

13

Photosynthesis

A Light in the Forest
Light has a profound impact on most living organisms on Earth. Its energy, captured by a molecular mechanism called photosynthesis, is used to manufacture organic biomolecules.

Overview

WITHOUT QUESTION, PHOTOSYNTHESIS IS THE MOST IMPORTANT BIOCHEM-ICAL PROCESS ON EARTH. WITH A FEW MINOR EXCEPTIONS, PHOTOSYNTHESIS is the only mechanism by which an external abiotic source of energy is harnessed by the living world. As with other energy-yielding processes, photosynthesis involves oxidation-reduction reactions. Water is the source of electrons and protons that reduce CO_2 to form organic compounds. Chapter 13, which is devoted to a discussion of the principles of photosynthetic processes, emphasizes the relationship between photosynthetic reactions and the structure of chloroplasts and the relevant properties of light.

As living organisms became abundant on the primitive Earth, their consumption of the organic nutrients produced by geochemical processes outpaced production. The abundant CO_2 in the planet's early atmosphere, a natural carbon source for organic synthesis, was largely of volcanic origin or had been generated during the anaerobic degradation of organic nutrients by living organisms. However, CO_2 is an oxidized, low-energy molecule. For this reason the processes by which CO_2 is incorporated into organic molecules require energy and reducing power. (The formation of carbon-carbon bonds requires free energy now provided by ATP hydrolysis. Reducing power is necessary because a strong electron donor must provide high-energy electrons to convert CO_2 to a CH_2O unit once the former has been incorporated into an organic molecule.)

The evolution of photosynthetic mechanisms (referred to as **photosystems**) provided both the energy and the reducing power for organic synthesis. The organisms so equipped had a definite survival advantage because they no longer needed to depend on an uncertain supply of preformed organic nutrients. These primitive organisms are believed to have been similar to modern green sulfur bacteria, which have a photosystem that uses light energy to drive a relatively simple electron transport process. As a pigment molecule in the photosystem absorbs light energy, an electron is energized and then donated to the first of several electron acceptors. Eventually, two light-excited electrons are donated to NAD^+, thus forming the reducing agent NADH. (Modern green sulfur bacteria may also use NADPH as a reducing agent in photosynthesis. In more advanced species, NADPH is used exclusively as the reducing agent.) The membrane component that mediates the conversion of light energy into chemical energy is a pigment-protein complex referred to as a **reaction center**. The electrons removed from the reaction center are replaced when oxidized components of the reaction center strip electrons from H_2S, thus generating elemental sulfur (S). As electrons flow through the photosystem, protons are pumped across the membrane, creating an electrochemical gradient. ATP is synthesized as protons move back into the cell through an ATP synthase. This photosystem therefore provides the bacterial cell with the NADH and ATP required to drive the incorporation of CO_2 into organic molecules. Today, the availability of H_2S is restricted to narrow ecological niches such as the hydrothemal vents described on page 2.

The next critical step in the evolution of life was a photosystem that removed electrons from H_2O with the resulting release of O_2. Photosynthetic organisms penetrated and occupied vast new areas on the planet. Water-based photosynthesis also had a profound impact on other organisms. As photosynthetic organisms proliferated, they provided a new and richer supply of organic molecules for other life-forms. Their most significant contribution, however, was the accumulation of gaseous oxygen in the atmosphere. Eventually, organisms began to use oxygen to generate energy.

The reduction potential of H_2O is significantly more positive than that of H_2S. A larger input of energy is therefore required to drive the transfer of electrons from H_2O to electron acceptors with more negative reduction potentials (see p. 316). Consequently, the mechanisms for converting water's low-energy electrons into the high-energy electrons needed in ATP and NADPH synthesis are more elaborate and sophisticated. Chapter 13 describes the principles of this process, that is, photosynthesis in plants and algae. The discussion begins with a detailed view of chloroplast structure. After a brief review of the relevant properties of light, the reactions that constitute modern photosynthesis will be described. These include the light reactions and the light-independent reactions. During the light reactions, electrons are energized and eventually used in the synthesis of both ATP and NADPH. These molecules are then used in the light-independent reactions (often referred to as the dark reactions) to drive the synthesis of carbohydrate. Several photosynthetic variations, referred to as C4 metabolism and crassulacean acid metabolism, are also discussed. Chapter 13 ends with a discussion of several mechanisms that control photosynthesis in plants.

KEY CONCEPTS

- Incorporating CO_2 into organic molecules requires energy and reducing power.

- In photosynthesis, both these requirements are provided by a complex process driven by light energy.

13.1 CHLOROPHYLL AND CHLOROPLASTS

The essential feature of photosynthesis is the absorption of light energy by specialized pigment molecules (Figure 13.1). The **chlorophylls** are green pigment molecules that resemble heme. *Chlorophyll a* plays the principal role in eukaryotic photosynthesis, because its absorption of light energy directly drives photochemical events. *Chlorophyll b* acts as a light-harvesting pigment by absorbing light energy and passing it on to chlorophyll a. The orange-colored **carotenoids** are isoprenoid molecules that either function as light-harvesting pigments (e.g., lutein, a xanthophyll, see p. 387) or protect against ROS (e.g., β-carotene).

In plants and algae, photosynthesis takes place within specialized organelles called chloroplasts (Chapter 2, p. 59). Chloroplasts resemble mitochondria in several respects. First, both organelles have an outer and an inner membrane with different permeability characteristics (Figure 13.2). The outer membrane of each organelle is highly permeable, whereas the inner membrane possesses specialized carrier molecules that regulate molecular traffic. Second, the chloroplast inner membrane encloses an inner space, referred to as the **stroma**, that resembles the mitochondrial matrix. The stroma possesses a variety of enzymes (e.g., those that catalyze the light-independent reactions and starch synthesis), DNA, and ribosomes.

There are also notable differences between the organelles. For example, chloroplasts are substantially larger than mitochondria. Although their shapes and sizes vary, many plant mitochondria are rod-shaped structures approximately 1500 nm long and 500 nm wide. Many chloroplasts are spheroidal, from 4000 to 6000 nm long and approximately 2000 nm wide. In addition, chloroplasts possess a distinct third membrane, referred to as the **thylakoid membrane**, that forms an intricate set of flattened vesicles. Thylakoid membrane is folded into a series of the disklike vesicular structures called **grana**. Each *granum* is a stack of several flattened vesicles. The internal compartment created by the formation of grana is referred to as the **thylakoid lumen** (or *space*). The thylakoid membranes that interconnect the grana are called **stromal lamellae**. Adjacent layers of membrane that fit closely together within each granum are said to be appressed. The stromal lamellae are nonappressed.

The pigments and proteins responsible for the light-dependent reactions of photosynthesis are found within thylakoid membrane (Figure 13.3). Most of these molecules are organized into the working units of photosynthesis.

1. **Photosystem I**. Photosystem I (PSI) (Figure 13.4), which energizes and transfers the electrons that eventually are donated to $NADP^+$, is a large

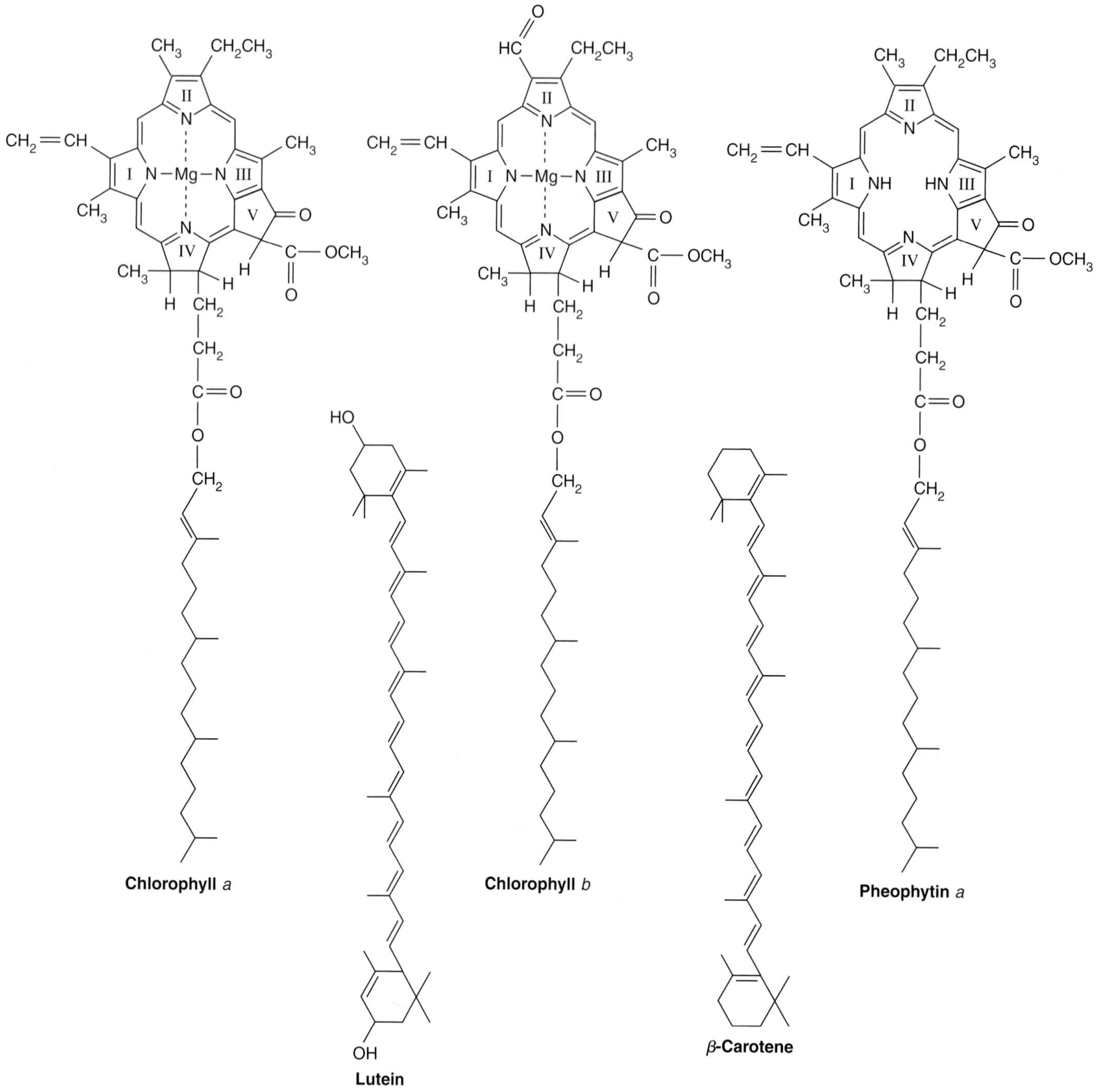

FIGURE 13.1

Pigment Molecules Used in Photosynthesis

Chlorophylls a and b are found in almost all photosynthesizing organisms. They possess a complex cyclic structure (called a porphyrin) with a magnesium atom at its center. Chlorophyll a possesses a methyl group attached to ring II of the porphyrin, whereas chlorophyll b has an aldehyde group attached to the same site. Pheophytin a is similar in its structure to chlorophyll a. The magnesium atom is replaced by two protons. Chlorophylls a and b and pheophytin a all possess a phytol chain esterified to the porphyrin. The phytol chain extends into and anchors the molecule to the membrane. Lutein and β-carotene are the most abundant carotenoids in thylakoid membranes.

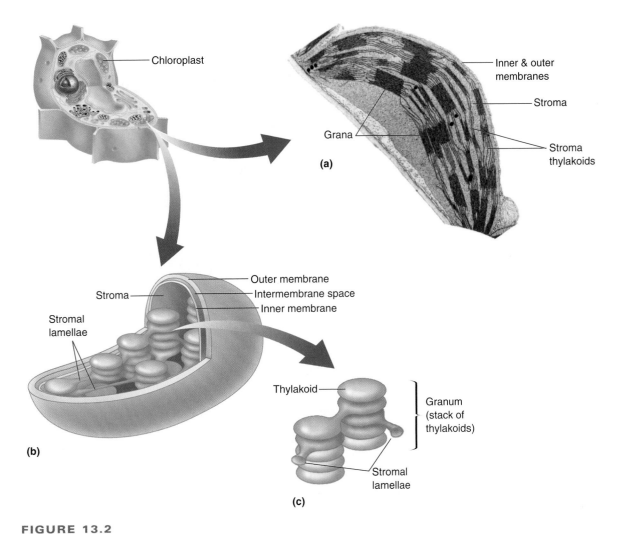

FIGURE 13.2

Chloroplast Structure

Chloroplasts have inner and outer membranes. A third membrane forms within the aqueous, enzyme-rich stroma into flattened sacs called thylakoids. A stack of thylakoids is called a granum. Unstacked, connecting thylakoid membrane is referred to as stromal lamellae. (a) Electron micrograph of a chloroplast. (b) Diagrammatic view of a chloroplast. (c) Cutaway view of a granum.

membrane-spanning, multisubunit complex. The largest polypeptides are two nearly identical 83 kD subunits designated PsaA and PsaB, each of which contains 11 transmembrane helices. They serve to anchor and organize the photosynthetic and accessory pigments in the reaction center. Although PSI possesses over 200 chlorophyll a molecules, its essential role (the donation of energized electrons to a series of electron carriers within thylakoid membrane) is performed by two special chlorophyll a molecules that reside within the reaction center. These molecules, referred to as a special pair, are located in the core complex of PSI, the PsaAB dimer. The special pair in photosystem I is sometimes referred to as P700, because they absorb light at 700 nm. In addition to the special pair the PsaAB dimer contains a series of single electron carriers: A_0, A_1, and F_x. A_0 is a specific chlorophyll a molecule that accepts an energized electron

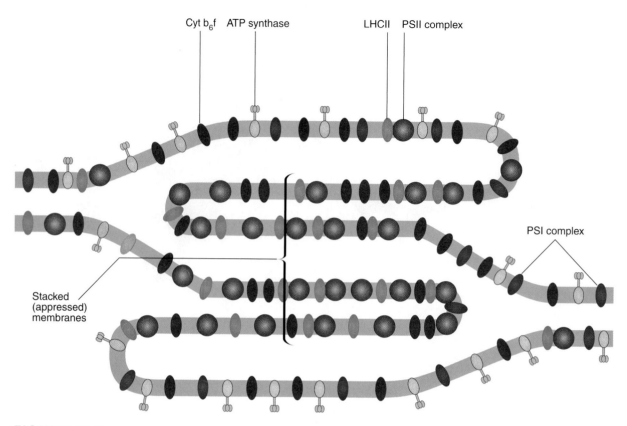

FIGURE 13.3

The Working Units of Photosynthesis

PSI complexes are most abundant in the unstacked stromal lamellae. In contrast, PSII complexes are located primarily in the stacked regions of thylakoid membrane. Cytochrome b_6f is found in both areas of thylakoid membrane. The ATP synthase is found only in thylakoid membrane that is directly in contact with the stroma.

FIGURE 13.4

Structure of Photosystem I

PSI is composed of a reaction center surrounded by the peripheral light-harvesting complex (LHCI), shown here as green ribbon structures. Two other electron transport proteins are also shown: plastocyanin (orange ribbon structure) and ferredoxin (dark pink ribbon structure).

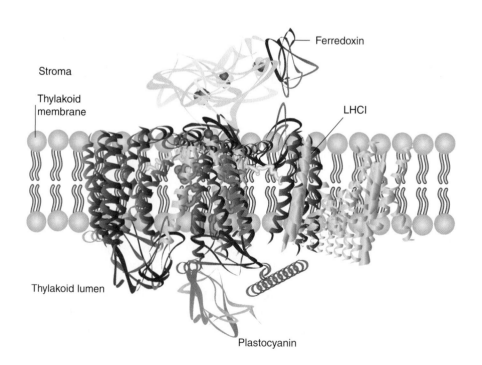

from P700 and transfers it to A_1. A_1, which has been identified as phylloquinone (vitamin K_1), is a molecule similar in structure to ubiquinone (Figure 10.3) and is sometimes abbreviated as Q. The electron is then transferred from A_1 to F_x, a 4Fe-4S center. Subsequently, the electron is donated to F_A and F_B, two 4Fe-4S centers in an adjacent 9 kD protein. Chlorophyll a molecules other than the special pair, as well as small amounts of chlorophyll b and carotenoids that occur in photosystem I, act as antenna pigments. **Antenna pigments** absorb light energy and transfer it to the reaction center. Additional antenna pigment molecules in a peripheral light-harvesting complex (LHCI) associated with PSI also contribute to efficient absorption of light energy. This phenomenon is described more fully in Section 13.2. Most PSI complexes are located in nonappressed thylakoid membrane, that is, membrane that is directly exposed to the stroma.

2. **Photosystem II**. The function of photosystem II is to oxidize water molecules and donate energized electrons to electron carriers that eventually reduce photosystem I. Photosystem II is a large membrane-spanning protein-pigment complex now believed to possess more than 25 proteins (Figure 13.5). The most active form of PSII, a dimer, is primarily located in appressed grana membrane. The most prominent component of PSII is the reaction center, a protein-pigment complex composed of two polypeptide subunits known as D_1 (33 kD) and D_2 (31 kD) (the D_1/D_2 dimer), cytochrome b_{559}, and a special pair of chlorophyll a molecules (referred to as P680) that absorb light at 680 nm. The oxygen-evolving component of PSII consists of manganese-stabilizing protein (MSP), a

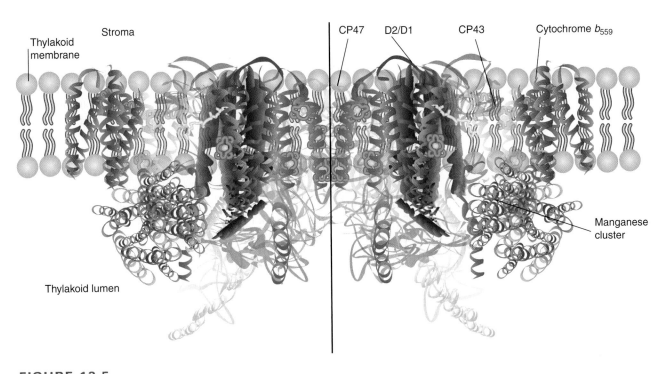

FIGURE 13.5

Structure of Photosystem II

In this view of a PSII dimer, the D_1 and D_2 subunits are shown as dark blue ribbon structures, and the intrinsic light-harvesting proteins CP43 and CP47 are shown as purple ribbon structures. (CP43 and CP47 bind 14 and 16 chlorophyll a molecules, respectively.) Chlorophyll molecules here are dark green, with their magnesium ions shown as yellow spheres. Cytochrome b_{599} subunits are illustrated as light blue ribbon structures. The manganese ions and calcium ion of the oxygen-evolving complex are shown as light blue and dark pink spheres, respectively.

membrane-associated complex that contains a cubelike Mn_3CaO_4 cluster linked to a fourth Mn by a mono-μ-oxo bridge, and a tyrosine residue (Tyr[161]), often referred to as Y_z, located on D_1. Also associated with the D_1/D_2 dimer are several electron acceptors. Pheophytin a is a chlorophyll-like pigment that accepts an electron from P680. This electron is donated in turn to two forms of plastoquinone (PQ), a molecule similar to ubiquinone. One plastoquinone, Q_A, is permanently bound to D_2, whereas Q_B is reversibly bound to D_1. Several hundred antenna pigment molecules are also associated with the reaction center. A fraction of these light-harvesting molecules are associated with 43 kD (CP43) and 47 kD (CP47) proteins that are components of the core structure of PSII. The preponderance of accessory pigment molecules and several proteins, however, belong to a detachable unit referred to as *light-harvesting-complex II (LHCII)*, a major component of thylakoid membrane. LHCII is a trimer of the light-harvesting proteins Lhcb1, Lhcb2, and Lhcb3, each of which binds 12 to 14 chlorophyll a and b molecules as well as several carotenoid molecules. Energy is transferred from LHCII to CP43 and CP47 or D1/D2.

PSI (P700) and PSII (P680) have absorption properties that overlap and, if they were physically close to each other, most of the light energy would be preferentially transferred to PSI, which has a lower energy absorption requirement (longer wavelength). Balanced absorption is maintained through a reduced plastoquinone (PQH_2)-dependent phosphorylation of LHCII. When PSII absorption exceeds that of PSI, PQH_2 accumulates and LHCII becomes phosphorylated. A conformational change in LHCII releases it from its association with PSII and confinement to the appressed lamellae. Movement of LHCII to regions of the membrane containing PSI leads to an increased oxidation of PQH_2 or its broken association with PSII reduces the production of PQH_2. Regardless of which event predominates, the end result is that balance is restored. This is because when PQH_2 levels drop, the LHCII is dephosphorylated and locks its association with PSII and the appressed lamellae. Physical separation of PSI and PSII is required for this excitation balancing function to work efficiently and is absolutely necessary because both the intensity and energy composition of light vary significantly throughout the day and from sunny to shady conditions. The transfer of electrons from PSII to PSI is not compromised by this physical separation because the transfer is mediated through mobile electron carriers. Most PSII units are found in thylakoid membrane within grana, that is, in appressed membrane, not exposed to the stroma.

3. **Cytochrome b_6f complex**. Cytochrome b_6f complex, found throughout the thylakoid membrane, is similar in structure and function to the cytochrome bc_1 complex in mitochondrial inner membrane. (Recall that cytochrome bc_1 complex is involved in transferring electrons from UQ to cytochrome c in the mitochondrial ETC and pumping protons across the inner membrane.) The cytochrome b_6f complex plays a critical role in the transfer of photoexcited electrons from PSII to PSI. An iron-sulfur site on the complex accepts electrons from the membrane-soluble electron carrier plastoquinone and donates them to a small water-soluble, copper-containing protein called plastocyanin. The mechanism that transports electrons from PQH_2 through the cytochrome b_6f complex appears to be similar to the Q cycle in mitochondria (Figure 10.7).

4. **ATP synthase**. The chloroplast ATP synthase (Figure 13.6), also referred to as CF_0CF_1ATP *synthase*, is structurally similar to the mitochondrial ATP synthase. The CF_0 component is a membrane-spanning protein complex that contains a proton-conducting channel. The CF_1 head piece, which projects into the stroma, possesses an ATP-synthesizing activity. A transmembrane proton gradient produced during light-driven electron transport

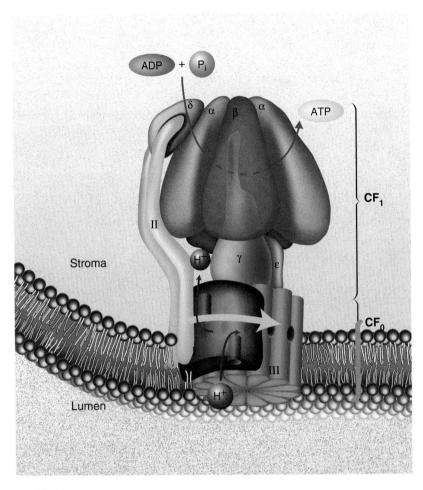

FIGURE 13.6

Diagrammatic View of the Chloroplast ATP Synthase

The ATP synthase is composed of two components: an integral membrane protein complex (CF_0) that contains a proton pore and an extrinsic protein complex (CF_1) that synthesizes ATP. CF_0 contains four different types of subunits: I, II, III, and IV. The proton pore is composed of 12–14 copies of subunit III. Subunit IV (not shown) binds CF_0 to CF_1. CF_1 consists of five different subunits: α, β, γ, δ, and ϵ.

drives ADP phosphorylation. The synthesis of each ATP molecule is believed to require pumping approximately three protons across the membrane into the thylakoid space. Thylakoid membrane that is directly in contact with the stroma contains the ATP synthase.

KEY CONCEPTS

- In chloroplasts a double membrane encloses an inner space called the stroma. The stroma contains the enzymes that catalyze the light-independent reactions of photosynthesis. The third membrane forms into flattened sacs called thylakoids.

- Thylakoid membrane contains the pigments and proteins of the light-dependent reactions.

QUESTION 13.1

Describe where each of the following molecules, molecular complexes, or processes is localized in the chloroplast. Explain their functions.

a. LHCII

b. lutein

c. PSII

d. MSP

e. CF_0CF_1

f. P700

g. P680

h. O_2 generation

13.2 LIGHT

The sun emits energy in the form of electromagnetic radiation, which propagates through space as waves, some of which impinge on Earth. Visible light, the energy source that drives photosynthesis, occupies a small part of the electromagnetic radiation spectrum (Figure 13.7). Many of the properties of light are explained by its wave behavior (Figure 13.8). Energy waves are described by the following terms:

1. **Wavelength**. Wavelength (λ) is the distance from the crest of one wave to the crest of the next wave.

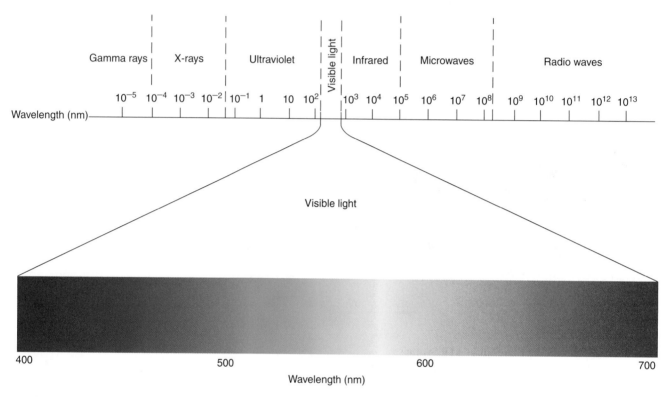

FIGURE 13.7

The Electromagnetic Spectrum

Gamma rays, which have short wavelengths, have high energy. At the other end of the spectrum, the radio waves (long wavelengths) have low energy. Visible light is the portion of the spectrum to which the visual pigments in the retina of the eye are sensitive. Pigment molecules in chloroplasts are also sensitive to portions of the visible spectrum.

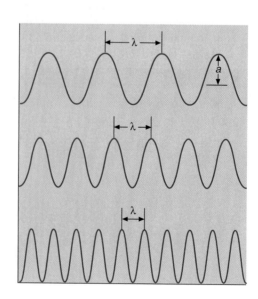

FIGURE 13.8

Properties of Waves

A wavelength (λ) is the distance between two consecutive peaks in a wave. The amplitude (a) or height of a wave is related to the intensity of electromagnetic radiation. Frequency is the number of waves that pass a point in space per second. Radiation with the shortest wavelength has the highest frequency.

2. **Amplitude**. Amplitude (a) is the height of a wave. The intensity of electromagnetic radiation (e.g., the brightness of light) is proportional to a^2.

3. **Frequency**. Frequency (v) is the number of waves that pass a point in space per second.

For each type of radiation the wavelength multiplied by the frequency equals the velocity (c) of the radiation:

$$\lambda v = c$$

This equation rearranges to

$$\lambda = c/v$$

The wavelength therefore depends on both the frequency and the velocity of the wave.

The wavelengths of visible light range from 400 nm (violet light) to 700 nm (red light). In comparison, highly energetic X-rays and γ-rays have wavelengths that are 10^4 to 10^7 times shorter. On the other end of the spectrum are low-energy radio waves; these have wavelengths on the order of meters to kilometers.

QUESTION 13.2

Why do green light waves have less energy than blue light waves?

In addition to behaving like a wave, visible light (and other types of electromagnetic radiation) exhibits the properties of particles such as mass and acceleration. Einstein's observation that energy has mass, or $E = mc^2$, applies to the photon. When light interacts with matter, it does so in discrete packets of energy called photons. The energy (E) of a photon is proportional to the frequency of the radiation:

$$E = hv$$

where h is Planck's constant (6.63×10^{-34} J $\bullet$ s).

According to quantum theory, radiant energy can be absorbed or emitted only in specific quantities called quanta. When a molecule absorbs a quantum of energy, an electron is promoted from its ground state orbital (lowest energy level) to a higher energy state. For absorption to occur, the energy difference between the two energy states must exactly equal the energy of the absorbed photon. Complex molecules often absorb at several wavelengths. For example, chlorophyll produces an absorption spectrum with broad and multiple peaks (blue-violet region and red region). Both of these facts suggest that chlorophyll absorbs photons of many different energies with varying probabilities. The wavelengths that are not absorbed are visible to us, and so a chlorophyll solution (or a leaf) appears green. Molecules that absorb electromagnetic energy have structural components called chromophores. Electrons in **chromophores** move easily to higher energy levels when energy is absorbed. Visible chromophores typically possess extended chains of conjugated double bonds and aromatic rings. They undergo electronic transitions wherein an electron moves from a ground state occupied orbital to a higher-energy unoccupied orbital (Figure 13.9). Molecules with a small number of conjugated double bonds or isolated double bonds absorb energy in the ultraviolet portion of the electromagnetic spectrum. The extensive conjugation of the chromophores of photosynthetic and accessory pigments allows electronic transitions to occur at longer wavelengths (lower energy) across the visible spectrum. The π electrons of these systems require less energy to make the transition into a higher-energy orbital.

Once excited, an electron can return to its ground state in several ways:

1. **Fluorescence**. In **fluorescence** a molecule's excited state decays as it emits a photon. Because the excited electron loses some energy initially by relaxing

COMPANION

Gw

WEBSITE Visit the companion website at www.oup.com/us/mckee to read the Biochemistry in the Lab box on photosynthetic studies for a discussion of the absorption spectrum of chlorophyll and several other pigment molecules.

FIGURE 13.9

An Electron Absorbing Visible Light

If a molecule absorbs a photon of visible light, an electron becomes excited and moves to a higher orbital. There is usually no change in the spin of the excited electron. As long as the spins of the two unpaired electrons remain antiparallel, the molecule is said to be in an excited singlet state. An excited molecule can return to its ground state by releasing energy as fluorescence or heat. In addition, the energy may be transferred to another molecule, or the energized electron itself may be donated to another molecule.

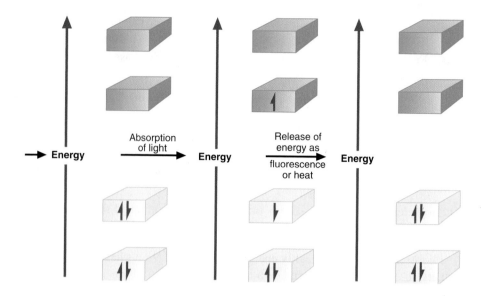

to a lower vibrational (energy) state, a transition resulting in the emission of a photon has lower energy than the photon originally absorbed. Fluorescent decay can occur as quickly as 10^{-15} s. (Although various chlorophylls absorb light energy throughout the visible spectrum, they emit only photons with low energy at or beyond the red end of the visible spectrum.)

2. **Resonance energy transfer**. In **resonance energy transfer**, the excitation energy is transferred to a neighboring chromophore through interaction between adjacent molecular orbitals. A chromophore whose absorption spectrum overlaps the emission spectrum of the target chromophore can absorb photons released when that chromophore returns to its ground state.

3. **Oxidation-reduction**. An excited electron is transferred to a neighboring molecule. An excited electron occupies a normally unoccupied orbital and is bound less tightly than when it occupies a ground state orbital. A molecule with an excited electron is a strong reducing agent. It returns to its ground state by reducing another molecule.

4. **Radiationless decay**. The excited molecule decays to its ground state by converting the excitation energy into heat.

Of all these responses to energy absorption, the most important in photosynthesis are resonance energy transfer and oxidation-reduction. Resonance energy transfer plays a critical role in the harvesting of light energy by accessory pigment molecules (Figure 13.10). Eventually, the energy absorbed and transmitted by light-harvesting complexes reaches the reaction center chlorophyll molecules. When these molecules become excited, they can lose an electron to a specific acceptor molecule (Figure 13.11). For example, P700 passes electrons to ferredoxin (an iron-sulfur protein), and P680 passes them to pheophytin a. (Once oxidized, P700 and P680 are referred to as P700* and P680*.) The electron hole left in the reaction center chlorophyll molecules is filled by an electron from a donor molecule. Plastocyanin and water play this role in PSI and PSII, respectively. Fluorescence also plays a role in photosynthesis when light absorption exceeds the capacity of the photosystems to transfer energy. Then photons are reemitted by a protective mechanism.

KEY CONCEPTS

- The light energy absorbed by chromophores causes electrons to move to higher energy levels.

- In photosynthesis, it is energy absorption that drives electron flow.

QUESTION 13.3

Explain the observation that the different absorption spectra of antenna pigment molecules are different from those of P680 and P700.

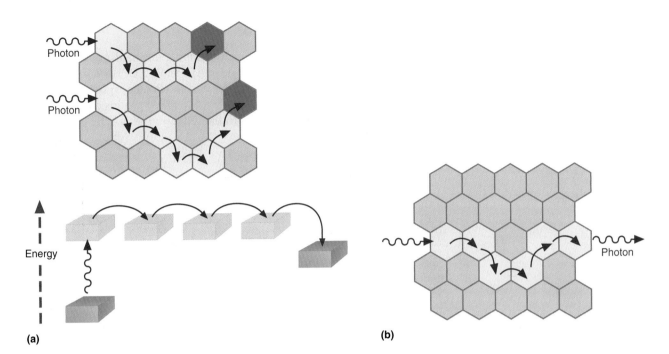

FIGURE 13.10

Resonance Energy Transfer

Energy flows through a light-harvesting complex. (a) A photon that has been absorbed by a molecule in a light-harvesting complex migrates randomly through the complex by resonance energy transfer. The energy is donated from one antenna molecule to another (yellow hexagons) until it is trapped by a reaction center (dark green hexagons) or is reemitted (b). The reaction center traps the excitation energy because its lowest excited state has lower energy than that of the antenna molecules.

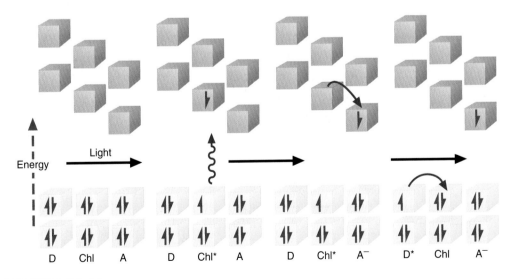

FIGURE 13.11

Electron Transfer

Electron transfer initiates photosynthesis. When light energy is absorbed by a chlorophyll complex (Chl), an electron is transferred to an acceptor (A). The oxidized chlorophyll complex (Chl*) extracts an electron from a donor (D).

13.3 LIGHT REACTIONS

As described previously, the **light reactions** are a mechanism by which electrons are energized and subsequently used in ATP and NADPH synthesis. Species that evolve O_2 require both photosystems I and II. (In other species only photosystem I is used.) During photosynthesis, the two photosystems couple the light-driven oxidation of water molecules to the reduction of $NADP^+$. The overall reaction is

$$2\,NADP^+ + 2H_2O \rightleftharpoons 2\,NADPH + O_2 + 2H^+$$

The standard reduction potentials for the half-reactions are

$$O_2 + 4e^- + 4H^+ \rightleftharpoons 2H_2O \qquad E^{\circ\prime} = +0.816\,V$$

and

$$NADP^+ + H^+ + 2e^- \rightleftharpoons NADPH \quad E^{\circ\prime} = -0.320\,V$$

Therefore, the coupled process has a standard redox potential of -1.136 V. The minimum free energy change for this process (calculated from $\Delta G'^{\circ} = -nF\,\Delta E^{\circ\prime}$; Section 9.1) is approximately 438 kJ (104.7 kcal) per mole of O_2 generated. In comparison, a mole of photons of 700 nm light provides approximately 170 kJ (40.6 kcal). Experimental observations have revealed that the absorption of 8 or more photons (i.e., 2 photons per electron) is required for each O_2 generated. Consequently, a total of 1360 kJ (325 kcal) (i.e., 8 times 170 kJ) is absorbed for each mole of O_2 produced. This energy is more than sufficient to account for reducing $NADP^+$ and to establish the proton gradient for ATP synthesis.

The process of light-driven photosynthesis begins with the excitation of PSII by light energy. One electron at a time is transferred to a chain of electron carriers that connects the two photosystems. As electrons are transferred from PSII to PSI, protons are pumped across the thylakoid membrane from the stroma into the thylakoid space. ATP is synthesized as protons flow back into the stroma through the ATP synthase. When P700 absorbs an additional photon, it releases an energized electron. This electron is immediately replaced by an electron provided by PSII. The newly energized PSI electron is passed through a series of iron-sulfur proteins and a flavoprotein to $NADP^+$, the final electron acceptor. This entire sequence, referred to as the **Z scheme**, is outlined in Figure 13.12.

Photosystem II and Oxygen Generation

When LHCII absorbs a photon, its energy is transferred to P680 in PS II; the newly energized electron is ejected and subsequently donated to *pheophytin a* (Figure 13.13), a molecule similar to chlorophyll in its structure. Reduced pheophytin a passes this electron to Q_A (plastoquinone). When a second electron is transferred from P680 and two protons are transferred from the stroma, reduced plastoquinone (Q_AH_2), also referred to as plastoquinol, is formed:

Plastoquinone (Q_A)　　　　　Plastoquinol (Q_AH_2)

One at a time, Q_AH_2 donates its two electrons to Q_B. Two additional stromal protons are used in the reduction of Q_B. Reduced Q_B (Q_BH_2) donates its electrons, in turn, to the cytochrome b_6f complex. Finally, the cytochrome b_6f complex donates its electrons to plastocyanin (PC), a mobile peripheral membrane protein. PC, a single-electron carrier, then transfers these electrons to P700 in PSI.

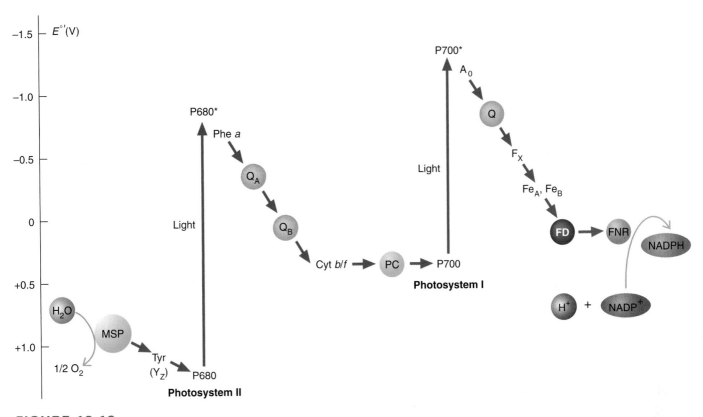

FIGURE 13.12

More Details of the Z Scheme

The flow of electrons from photosystem II to photosystem I drives the transport of protons into the thylakoid lumen. Electron transfer through the iron-sulfur proteins Fe_A and Fe_B is not understood. The $E^{\circ\prime}$ values are approximate. (MSP = Manganese-stabilizing protein (MSP) contains a manganese cluster; PC = plastocyanin, FD = ferredoxin, FNR = ferredoxin-NADP oxidoreductase.)

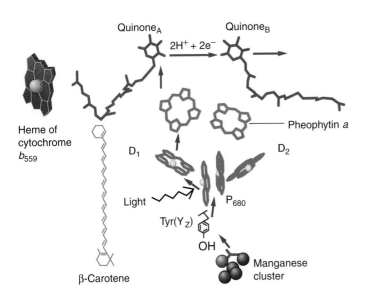

FIGURE 13.13

Photosystem II Electron Transport

The cofactors in PSII are shown in the same positions as in the left monomer in the PSII dimer shown in Figure 13.5. The arrows indicate the electron transport pathway. A photon of light energizes an electron in P_{680}. P_{680} donates this electron to pheophytin a, which is eventually transferred to the quinones. Note that the manganese cluster and the Tyr(Y_Z) side chain replace the electron in oxidized P_{680}.

Recall that electrons transferred from P680 are replaced when H_2O is oxidized. An *oxygen-evolving complex*, composed in part of the $MnCaO_3$ cluster in MSP and the tyrosine residue located on D_1, is responsible for the transfer of electrons from H_2O to P680*. (It should be noted that P680* has a very high redox potential, measured at over 1.25 V.) As P680 electrons absorb light energy and translocate, P680* oxidizes a tyrosine (referred to as Y_Z) in D_1. Tyrosine is effective in electron transfer because the tyrosyl radical formed is resonance stabilized.

The evolution of one O_2 requires splitting two H_2O molecules, which releases 4 protons and 4 electrons. Experimental evidence indicates that H_2O is converted to O_2 by a mechanism referred to as the *water-oxidizing clock* (Figure 13.14). The O_2-evolving complex has five oxidation states: S_0, S_1, S_2, S_3, and S_4. In this complex, S_0 is the most reduced state and S_4 is the most oxidized state. It is now believed that the Mn_4CaO_3 cluster, near the PSII reaction center, is responsible for these transitions. Oxygen–oxygen bond formation is the rate-limiting step in water oxidation. The oxygen-evolving complex also abstracts protons from H_2O as it cycles through the oxidation states. The protons are released into the thylakoid lumen, where they contribute to the pH gradient that drives ATP synthesis.

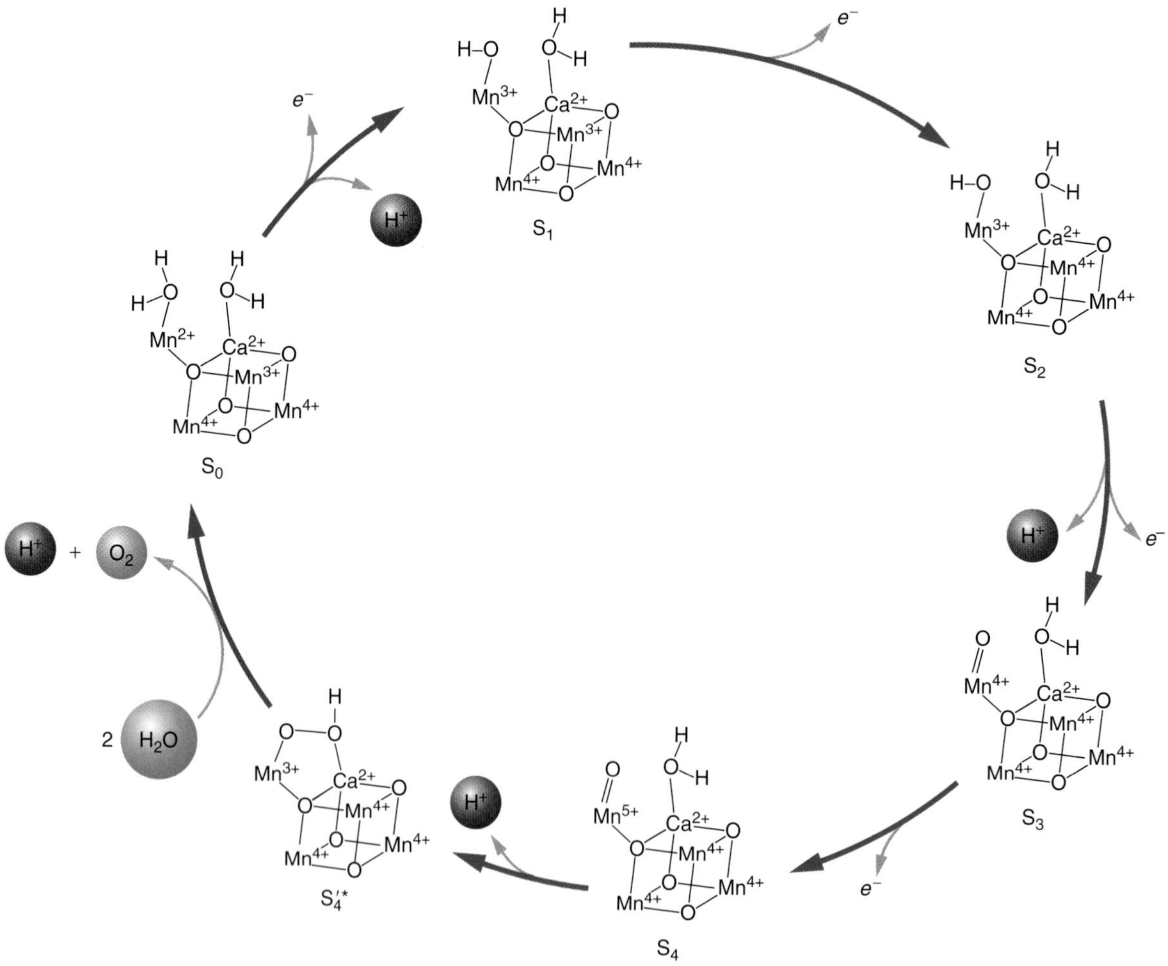

FIGURE 13.14

The Water-Oxidizing Clock

Changes in the oxidation states of manganese atoms in the Mn_4CaO_3 cluster during the electron transport process. The O_2-evolving complex has five oxidation states. Four electrons are removed for every O_2 that is evolved. At S_4, the unshared electrons of the water molecule coordinated with the Ca^{2+} form a bond with the manganyl oxygen as the manganese is being reduced from +5 to +3. The S_4' phase may represent a transition state between S_4 and S_0. Subsequently, O_2 is released. Protons are also released during the cycle.

QUESTION 13.4

Excessive amounts of light can depress photosynthesis. Recent research indicates that PSII is extremely vulnerable to light damage. Plants often survive this damage because they possess efficient repair systems. It now appears that cells delete and resynthesize damaged components and recycle undamaged ones. For example, the D_1 polypeptide, apparently the most vulnerable component of PSII is rapidly replaced after it is damaged. Review the role of PSII, and suggest the proximate cause of light-induced damage of D_1. [*Hint*: The D_1/D_2 dimer binds two molecules of β-carotene.]

Photosystem I and NADPH Synthesis

The absorption of a photon by P700 leads to the release of an energized electron that is passed through a series of electron carriers (Figure 13.15). The first electron carrier is a chlorophyll a molecule (A_0). As the electron is donated sequentially to phylloquinone (Q) and to several iron-sulfur proteins (the last of which is ferredoxin), it is moved from the lumenal surface of the thylakoid membrane to its stromal surface. Ferredoxin, a mobile, water-soluble protein, then donates each electron to a flavoprotein called ferredoxin-NADP oxidoreductase (FNR). The flavoprotein uses a total of two electrons and a stromal proton to reduce $NADP^+$ to NADPH. The transfer of electrons from ferredoxin to $NADP^+$ is referred to as the *noncyclic electron transport pathway* (Figure 13.16). In this pathway the absorption of eight photons yields an ATP/NADPH ratio of 3:2. In *cyclic electron transport* (Figure 13.17), reduced ferredoxin donates its

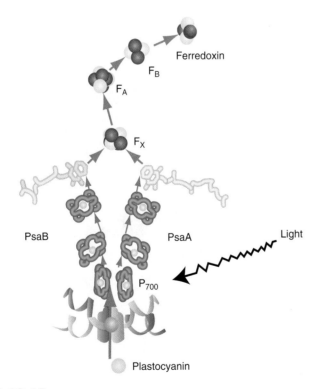

FIGURE 13.15

Photosystem I Electron Transport

These cofactors are in the same positions as in Figure 13.4. A photon of light energizes an electron in P_{700}, which is then sequentially transferred to the iron-sulfur centers F_X, F_A, and F_B and finally to ferredoxin. The missing electron in oxidized P_{700} is transferred from plastocyanin. Two tryptophan residues in the two polypeptide segments shown in this illustration (as pink and blue ribbon structures) are believed to assist in the transfer of electrons from plastocyanin.

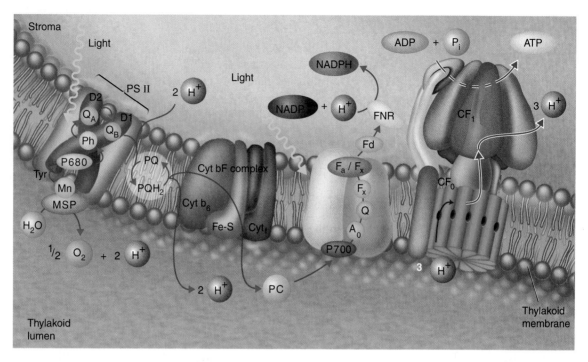

FIGURE 13.16

Membrane Organization of the Light Reactions in Chloroplasts: The Noncyclic Electron Transport Chain and the ATP Synthase Complex

As two electrons move from each water molecule to $NADP^+$ (blue arrows), about two H^+ are pumped from the stroma into the thylakoid lumen. Two additional H^+ are generated within the lumen by the oxygen-evolving complex. The flow of protons through the proton pore in CF_0 drives the synthesis of ATP in CF_1. (MSP = manganese-stabilizing protein; Ph = pheophytin; Fd = ferredoxin; FNR = the flavoprotein ferredoxin-NADP oxidoreductase)

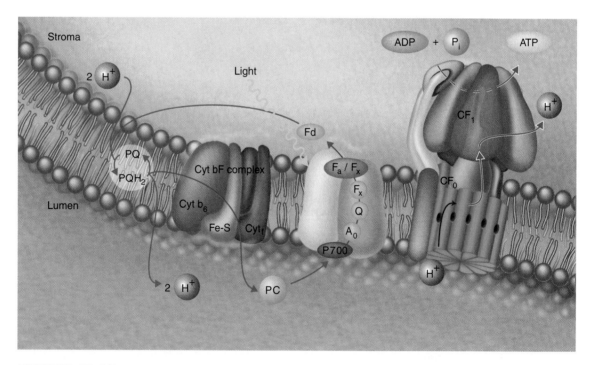

FIGURE 13.17

The Cyclic Electron Transport Pathway

A Q cycle similar to that observed in the electron transport pathway that links PSI and PSII is believed to be responsible for pumping two H^+ across the thylakoid membrane for each electron transported. The proton flow drives ATP synthesis. No NADPH is produced.

electrons to plastoquinone, which then passes them to the cyt b$_f$ complex, plastocyanin, and eventually P700. In this process, which typically occurs when a chloroplast has a high NADPH/NADP$^+$ ratio, no NADPH is produced. Instead, electron transport results in the transfer of additional protons across the thylakoid membrane. As a result, additional molecules of ATP are synthesized.

The Calvin cycle (p. 486) requires an ATP/NADPH ratio of 3:2. However, ATP is also used for processes other than carbohydrate synthesis. Consequently, both noncyclic and cyclic photophosphorylation pathways are required for sufficient ATP synthesis during photosynthesis.

QUESTION 13.5

Describe the role of each of the following molecules in photosynthesis:

a. plastocyanin

b. β-carotene

c. ferredoxin

d. plastoquinone

e. pheophytin a

f. lutein

QUESTION 13.6

Because PSII and PSI operate in series, photosynthesis is efficient when the two systems receive light at equal rates; recent research indicates that phosphorylation helps balance their activities. After reviewing the structural and functional properties of PSI, PSII, and LHCII, describe the regulatory roles of plastoquinone and the kinase and phosphatase enzymatic activities that make the balanced regulation of PSI and PSII possible. Review the effects of covalent modification of proteins (Chapter 5), and suggest why phosphorylation has this effect on photosynthesis.

Photophosphorylation

During photosynthesis light energy captured by an organism's photosystems is transduced into ATP phosphate bond energy. This conversion is referred to as **photophosphorylation**. It is apparent from the preceding discussions that there are many similarities between mitochondrial and chloroplast ATP synthesis. For example, many of the same molecules and terms that are encountered in aerobic respiration (Chapter 10) are also relevant to discussions of photosynthesis. Additionally, in both organelles, electron transport is used to induce a proton gradient, which in turn drives ATP synthesis. Although there are a variety of differences between aerobic respiration and photosynthesis, the essential difference between the two processes is the conversion of light energy into redox energy by chloroplasts. (Recall that mitochondria produce redox energy by extracting high-energy electrons from food molecules.) Another critical difference involves the permeability characteristics of mitochondrial inner membrane and thylakoid membrane. In contrast to the inner membrane, the thylakoid membrane is permeable to Mg^{2+} and Cl$^-$. Therefore, Mg^{2+} and Cl$^-$ move across the thylakoid membrane as electrons and protons are transported during the light reaction. The electrochemical gradient across the thylakoid membrane that drives ATP synthesis therefore consists mainly of a proton gradient that may be as great as 3.5 pH units.

Experimental measurements of H$^+$/ATP ratios indicate that the movement across the thylakoid membrane of about 12 protons in noncyclic photophosphorylation yields three molecules of ATP. The synthesis of these ATPs is made possible by the absorption of eight photons, one for each of the electrons from two water molecules. Proton transport occurs as these electrons are transported down the noncyclic electron transport system (Figure 13.16). In cyclic photophosphorylation, the pumping of eight protons by the cyt b_6f complex, as the result of the absorption of four photons, yields two molecules of ATP.

QUESTION 13.7

Describe the effect of uncoupler molecules on ATP synthesis in chloroplasts. Compare it with that observed in mitochondria.

QUESTION 13.8

When chloroplasts are first soaked in an acidic solution (pH 4) and then transferred to a basic solution (pH 8), there is a quick and brief burst of ATP synthesis. Explain.

QUESTION 13.9

A variety of herbicides kill plants by inhibiting photosynthetic electron transport. Atrazine, a triazine herbicide, blocks electron transport between Q_A and Q_B in PSII. The compound 3-(3,4-dichlorophenyl)-1,1-dimethylurea (DCMU) also blocks electron flow between the two molecules of plastoquinone. Paraquat is a member of a family of compounds called bipyridylium herbicides. Paraquat is reduced by PSI but is easily reoxidized by O_2 in a process that produces superoxide and hydroxyl radicals. Plants die because their cell membranes are destroyed by radicals. Of the herbicides just discussed, determine which, if any, are most likely to be toxic to humans and other animals. What specific damage may occur?

13.4 THE LIGHT-INDEPENDENT REACTIONS

The incorporation of CO_2 into carbohydrate by eukaryotic photosynthesizing organisms, a process that occurs within chloroplast stroma, is often referred to as the **Calvin cycle**. Because the reactions of the Calvin cycle can occur without light if sufficient ATP and NADPH are supplied, they have often been called the *dark reactions*. The term is somewhat misleading, however. The Calvin cycle reactions typically occur only when the plant is illuminated, because ATP and NADPH are produced by the light reactions. Therefore **light-independent reactions** is a more appropriate term. Because of the types of reaction that occur in the Calvin cycle, it is also referred to as the *reductive pentose phosphate cycle* (RPP cycle) and the *photosynthetic carbon reduction cycle* (PCR cycle).

The Calvin Cycle

The net equation for the Calvin cycle (Figure 13.18) is

$$3CO_2 + 6\,NADPH + 9\,ATP \rightarrow$$
$$\text{glyceraldehyde-3-phosphate} + 6\,NADP^+ + 9\,ADP + 8\,P_i$$

For every three molecules of CO_2 that are incorporated into carbohydrate molecules, there is a net gain of one molecule of glyceraldehyde-3-phosphate. The fixation of six CO_2 into one hexose molecule occurs at the expense of 12 NADPH and 18 ATP. The reactions of the cycle can be divided into three phases.

1. **Carbon fixation. Carbon fixation**, the mechanism by which inorganic CO_2 is incorporated into organic molecules, consists of a single reaction. Ribulose-1,5-bisphosphate carboxylase catalyzes the carboxylation of ribulose-1,5-bisphosphate to form two molecules of glycerate-3-phosphate.

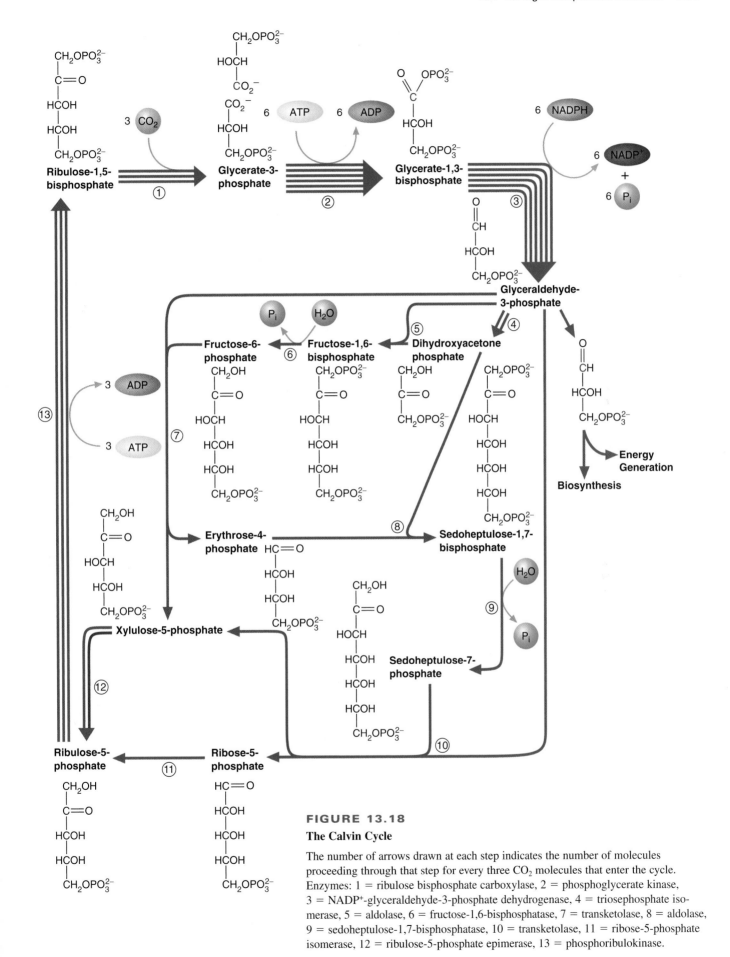

FIGURE 13.18

The Calvin Cycle

The number of arrows drawn at each step indicates the number of molecules proceeding through that step for every three CO_2 molecules that enter the cycle. Enzymes: 1 = ribulose bisphosphate carboxylase, 2 = phosphoglycerate kinase, 3 = $NADP^+$-glyceraldehyde-3-phosphate dehydrogenase, 4 = triosephosphate isomerase, 5 = aldolase, 6 = fructose-1,6-bisphosphatase, 7 = transketolase, 8 = aldolase, 9 = sedoheptulose-1,7-bisphosphatase, 10 = transketolase, 11 = ribose-5-phosphate isomerase, 12 = ribulose-5-phosphate epimerase, 13 = phosphoribulokinase.

Plants that produce glycerate-3-phosphate as the first stable product of photosynthesis are referred to as **C3 plants**. (Notable exceptions are described in the box on pages 491–493.) Ribulose-1,5-bisphosphate carboxylase, a complex molecule composed of eight large (L 56 kD) subunits and eight small (S 14 kD) subunits, is the pacemaker enzyme of the Calvin cycle. Its activity is regulated by CO_2, O_2, Mg^{2+}, and pH, as well as other metabolites. Each L subunit contains an active site that binds substrate. The catalytic activity of the L subunits is enhanced by the S subunits. Because the CO_2 fixation reaction is extremely slow, plants compensate by producing a large number of copies of the enzyme, which often constitutes approximately half of a leaf's soluble protein. For this reason, ribulose-1,5-bisphosphate carboxylase is often described as the world's most abundant enzyme.

2. **Reduction.** The next phase of the cycle consists of two reactions. Six molecules of glycerate-3-phosphate are phosphorylated at the expense of six ATP molecules to form glycerate-1,3-bisphosphate. The latter molecules are then reduced by $NADP^+$-glyceraldehyde-3-phosphate dehydrogenase to form six molecules of glyceraldehyde-3-phosphate. These reactions are similar to reactions encountered in gluconeogenesis. Unlike the dehydrogenase in gluconeogenesis, the Calvin cycle enzyme uses NADPH as a reducing agent.

3. **Regeneration.** As noted previously, the net production of fixed carbon in the Calvin cycle is one molecule of glyceraldehyde-3-phosphate. The other five glyceraldehyde-3-phosphate molecules are processed in the remainder of the Calvin cycle reactions to regenerate three molecules of ribulose-1,5-bisphosphate. Two molecules of glyceraldehyde-3-phosphate are isomerized to form dihydroxyacetone phosphate. One dihydroxyacetone molecule combines with a third glyceraldehyde-3-phosphate molecule to form fructose-1,6-bisphosphate. The latter molecule is then hydrolyzed to fructose-6-phosphate. Fructose-6-phosphate subsequently combines with a fourth molecule of glyceraldehyde-3-phosphate to form xylulose-5-phosphate and erythrose-4-phosphate. Erythrose-4-phosphate combines with dihydroxyacetone phosphate to form sedoheptulose-1,7-bisphosphate, which is then hydrolyzed to form sedoheptulose-7-phosphate. The fifth molecule of glyceraldehyde-3-phosphate combines with sedoheptulose-7-phosphate to form ribose-5-phosphate and a second molecule of xylulose-5-phosphate. Ribose-5-phosphate and both molecules of xylulose-5-phosphate are separately isomerized to ribulose-5-phosphate. In the last step, three molecules of ribulose-5-phosphate are phosphorylated at the expense of three ATP molecules to form three molecules of ribulose-1,5-bisphosphate. The remaining molecule of glyceraldehyde-3-phosphate is either used within the chloroplast in starch synthesis or exported to the cytoplasm, where it may serve in the synthesis of sucrose or other metabolites.

COMPANION

Gw

WEBSITE Visit the companion website at www.oup.com/us/mckee to read the Biochemistry in Perspective box on starch and sucrose metabolism.

QUESTION 13.10

Many of the reactions in the regeneration phase of the Calvin cycle are similar to reactions encountered in previous chapters in this textbook. Review the reactions in this phase of the Calvin cycle and determine which reactions they resemble. [*Hint*: Review Chapter 8.]

QUESTION 13.11

When plant cells are illuminated, their cytoplasmic ATP/ADP and $NADH/NAD^+$ ratios rise significantly. The following shuttle mechanism is believed to contribute to the transfer of ATP and reducing equivalents from the chloroplast into the cytoplasm. Once dihydroxyacetone phosphate has been transported from the stroma

QUESTION 13.11 (CONT.)

into the cytoplasm, it is converted to glyceraldehyde-3-phosphate and then to glycerate-1,3-bisphosphate. (This reaction is the reverse of the reaction in which glyceraldehyde-3-phosphate is formed during carbon fixation.) In the cytoplasmic reaction, the reducing equivalents are donated to NAD^+ to form NADH. In a later reaction, glycerate-1,3-bisphosphate is converted to glycerate-3-phosphate with the concomitant production of one molecule of ATP. Glycerate-3-phosphate is then transported back into the chloroplast, where it is reconverted to glyceraldehyde-3-phosphate.

This shuttle somewhat depresses mitochondrial respiration processes. Review the regulation of aerobic respiration (Chapter 9) and suggest how photosynthesis suppresses this aspect of mitochondrial function.

Photorespiration

Photorespiration is perhaps the most curious feature of photosynthesis. In this light-dependent process, oxygen is consumed and CO_2 is liberated by plant cells that are actively engaged in photosynthesis. Photorespiration is a multistep mechanism initiated by ribulose bisphosphate carboxylase. In addition to its carboxylation function, this enzyme possesses an oxygenase activity. (For this reason the name *ribulose-1, 5-bisphosphate carboxylase-oxygenase*, or *rubisco*, is sometimes used.) Because the enzyme's active site binds to both CO_2 and O_2, these substrates compete.

In the oxygenation reaction, ribulose-1,5-bisphosphate is converted to glycolate-2-phosphate and glycerate-3-phosphate (Figure 13.19). In a complex series of reactions, glycolate-2-phosphate is oxidized by O_2. Ultimately, the glycerate-3-phosphate produced by this pathway is used to produce (via the Calvin cycle) ribulose-1,5-bisphosphate. Photorespiration is a wasteful process. It loses fixed carbon (as CO_2), and consumes both ATP and NADH.

The rate of photorespiration depends on several parameters. These include the concentrations of CO_2 and O_2 to which photosynthesizing cells are exposed. Photorespiration is depressed by CO_2 concentrations above 0.2%. (Because photorespiration and photosynthesis occur concurrently, CO_2 is released during CO_2 fixation. When the rates of CO_2 release and fixation are equal, the *CO_2 compensation point* has been reached. The lower the CO_2 compensation point, the less photorespiration takes place. Many C3 plants have CO_2 compensation points between 0.02 and 0.03% of CO_2 in the air near photosynthesizing cells.) In contrast, high O_2 concentrations and high temperatures promote photorespiration. Consequently, this process is favored when plants are exposed to high temperatures and any condition that causes low CO_2 and/or high O_2 concentrations. For example, photorespiration is a serious problem for C3 plants in hot, dry environments. To conserve water, these plants close their stomata, thus reducing the CO_2 concentration within leaf tissue. (*Stomata* are pores on the surface of leaves. When they are open, CO_2, O_2, and H_2O vapor can readily diffuse down the concentration gradients between the leaf's interior and the external environment.) In addition, as photosynthesis continues, O_2 levels increase. Depending on the severity of the circumstances, from 30 to 50% of a plant's yield of fixed carbon may be lost. This effect can be serious because several C3 plant's (e.g., soybeans and oats) are major food crops.

Photorespiration is an artifact of the evolutionary history of photosynthesis. In the early atmosphere in which the first photosystem evolved, oxygen levels were very low. Thus, over the long time period before oxygen levels became a problem, there was no selection pressure to improve the capacity of the rubisco active site to distinguish between CO_2 and O_2. Selection pressure could occur only when oxygen levels increased significantly. It is noteworthy that selection

KEY CONCEPTS

- The Calvin cycle is a series of light-independent reactions in which CO_2 is incorporated into organic molecules.
- The Calvin cycle reactions occur in three phases: carbon fixation, reduction, and regeneration.
- Photorespiration is a wasteful process in which photosynthesizing cells evolve CO_2.

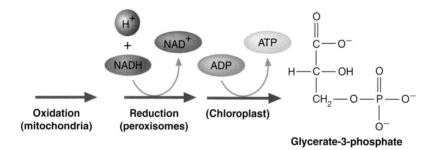

FIGURE 13.19

Photorespiration

Photorespiration is a complex multistep process catalyzed by enzymes in several cellular compartments. (a) The synthesis of glycolate occurs within the stroma. (b) Glycolate is converted to glyoxylate within peroxisomes. In a complex series of reactions that occur in peroxisomes, mitochondria, and chloroplasts, glyoxylate is converted to glycerate-3-phosphate and CO_2. Note that because of CO_2 release during the photorespiration pathway, two molecules of glycolate-2-phosphate are required to eventually form one molecule of glycerate-3-phosphate

for CO_2 over O_2 is higher in modern green plants than in bacteria. The pathway that evolved to convert glycolate-3-phosphate to glycerate-3-phosphate, although costly in ATP and NADH consumption, is viewed as a salvage operation that recovers previously fixed and partly reduced carbon. C4 plants, which have developed an elaborate mechanism to suppress photorespiration, are described next.

Alternatives to C3 Metabolism

In addition to C3 photosynthesis, which is used by most plants, there are two other mechanisms for fixing CO_2: C4 metabolism and crassulacean acid metabolism. Both improve the efficiency of photosynthesis in climates where temperatures are high and water is scarce.

C4 METABOLISM C4 plants include sugarcane and maize (corn); they are found primarily in the tropics, and can thrive under conditions of drought and high temperatures. The name *C4 plants* indicates the prominent role of a four-carbon molecule (oxaloacetate) in a biochemical pathway that avoids photorespiration. This pathway is called **C4 metabolism**, the *C4 pathway*, or the *Hatch-Slack pathway* (after its discoverers).

The leaves of C4 plants possess two types of photosynthesizing cells: mesophyll cells and bundle sheath cells. (In C3 plants, photosynthesis occurs in mesophyll cells.) Most mesophyll cells in both plant types are positioned so that they are in direct contact with air when the leaf's stomata are open. In C4 plants, CO_2 is captured in specialized mesophyll cells that incorporate it into oxaloacetate (Figure 13.20). Phosphoenolpyruvate carboxylase (PEP carboxylase) catalyzes this reaction, which is an indirect means of carbon fixation because CO_2 must first be converted to bicarbonate. Since PEP carboxylase has a very low K_m for CO_2 (i.e., a higher affinity) than rubisco, and O_2 is a poor substrate, C4 plants are more effective at capturing CO_2 than are C3 plants. Once formed, oxaloacetate is reduced to malate, which then diffuses into bundle sheath cells. As the name implies, bundle sheath cells form a layer around vascular bundles, which contain phloem and xylem vessels. Unlike C3 plants, the bundle sheath cells of most C4 plants possess chloroplasts.

In the bundle sheath cells, malate is decarboxylated to pyruvate in a reaction that reduces $NADP^+$ to NADPH. The pyruvate product of this latter reaction diffuses back to a mesophyll cell, where it can be reconverted to PEP. Although this reaction is driven by the hydrolysis of one molecule of ATP, there is a net cost of two ATP molecules. An additional ATP molecule is required to convert the AMP product to ADP so that it can be rephosphorylated during photosynthesis. This circuitous process delivers CO_2 and NADPH to the chloroplasts of bundle sheath cells, where ribulose-1,5-bisphosphate carboxylase and the other enzymes of the Calvin cycle use them to synthesize triose phosphates. The concentrations of CO_2 available to rubisco in the bundle sheath cells of C4 plants are significantly higher (10–20 times as great) than in C3 plants. C4 plants also use water more efficiently than C3 plants because they can close stomata when ambient temperature is high, thereby reducing transpiration.

CRASSULACEAN ACID METABOLISM Crassulacean acid metabolism (CAM) is a mechanism that conserves water in plants that live in deserts and other regions with high light intensity and very limited water supply. (The Crassulaceae are a family of plants in which the CAM pathway was first investigated.) CAM plants, most of which are succulents (e.g., cacti), open their stomata only at night after the air temperature has decreased and the risk of water loss is low. The CO_2 enters through the Stomata, where it is immediately incorporated into an oxaloacetate molecule via the carboxylation of phosphoenolpyruvate catalyzed by PEP carboxylase (Figure 13.21). Oxaloacetate is then reduced to malate, which is stored overnight in the vacuoles of mesophyll cells. In the daytime, malate molecules are broken down to pyruvate and the rubisco substrate CO_2. The temporal separation of carbon fixation and the Calvin cycle allows CAM plants to close their stomata during daytime, thus minimizing water loss through transpiration.

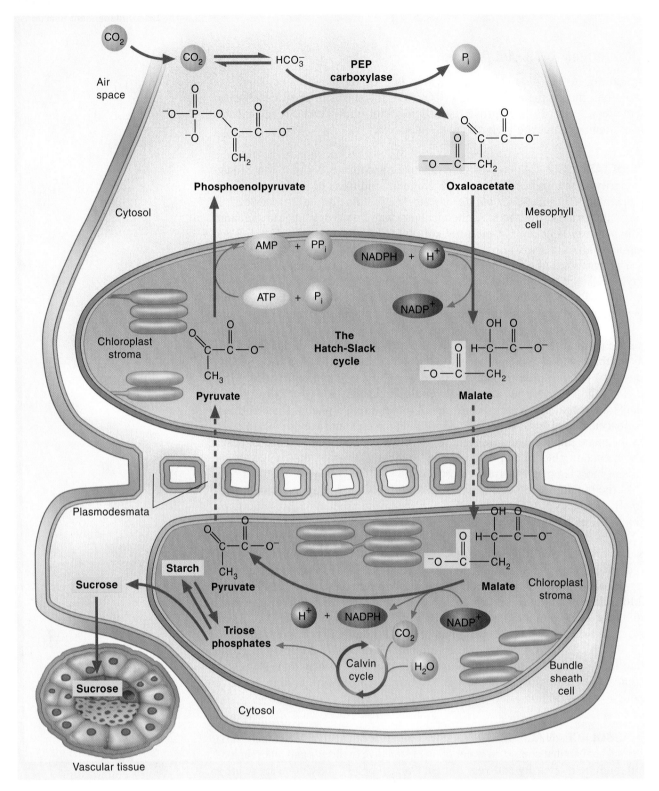

FIGURE 13.20

C4 Metabolism

In the C4 pathway mesophyll cells, which are in direct contact with the air space in the leaf, take up CO_2 and use it to synthesize oxaloacetate, which is then reduced to malate. (Some C4 plants synthesize aspartate instead of malate.) Malate then diffuses to bundle sheath cells, where it is reconverted to pyruvate. The CO_2 released in this reaction is used in the Calvin cycle, eventually yielding triose phosphate molecules. Triose phosphate is subsequently converted to starch or sucrose. Pyruvate returns to the mesophyll. Starch is synthesized from glucose-1-phosphate, which is subsequently converted to ADP-glucose by ADP-glucose pyrophosphorylase. ADP glucose molecules are then incorporated into a preexisting polysaccharide chain by starch synthase. Sucrose-6-phosphate is synthesized from UDP-glucose and fructose-6-phosphate by sucrose phosphate synthase. Sucrose phosphatase catalyzes the hydrolysis of sucrose-6-phosphate to form sucrose and P_i. Note that Glyceraldehyde-3-phosphate and dihydroxyacetone phosphate are referred to as triose phosphates.

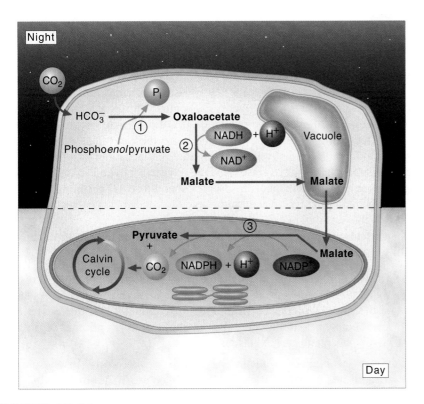

FIGURE 13.21

Crassulacean Acid Metabolism

At night the stomata of CAM plants open to allow CO_2 to enter. Within mesophyll cells, PEP carboxylase (1) incorporates CO_2 (as HCO_3^-) into oxaloacetate. Afterward, oxaloacetate is reduced by malate dehydrogenase (2) to form malate. Malate is stored in the cell's vacuole until daylight. Light stimulates the decarboxylation of malate by malic enzyme (3) to form pyruvate and CO_2. As a result of this temporal separation of reactions, CO_2 can be incorporated into sugar molecules via the Calvin cycle during the day, when the plant's stomata are closed to avoid water loss.

QUESTION 13.12

Briefly explain the significance of photorespiration and the strategies that C4 plants use to avoid it.

13.5 REGULATION OF PHOTOSYNTHESIS

Because plants must adapt to a wide variety of environmental conditions, the regulation of photosynthesis is complex. Although the control of most photosynthetic processes is far from being completely understood, several control features are well established. Most of these processes are directly or indirectly controlled by light. After a brief description of general light-related effects, the control of the activity of ribulose-1, 5-bisphosphate carboxylase, the key regulatory enzyme in photosynthesis, is discussed.

Light Control of Photosynthesis

Investigations of photosynthesis are complicated by several factors. The most prominent of these is that the photosynthetic rate depends on temperature and cellular CO_2 concentration, as well as on light. Nevertheless, numerous investigations have firmly established light as an important regulator of most aspects of photosynthesis. This is not surprising, considering light's role in driving photosynthesis.

Many of the effects of light on plants are mediated by changes in the activities of key enzymes. Because plant cells possess enzymes that operate in several competing pathways (i.e., glycolysis, pentose phosphate pathway, and the Calvin cycle), careful metabolic regulation is critical. Light assists in this regulation by activating certain photosynthetic enzymes and deactivating several enzymes in degradative pathways. Among the light-activated enzymes are ribulose-1,5-bisphosphate carboxylase, $NADP^+$-glyceraldehyde-3-phosphate dehydrogenase, fructose-1,6-bisphosphatase, sedoheptulose-1,7-bisphosphatase, and phosphoribulokinase. Light-inactivated enzymes include phosphofructokinase and glucose-6-phosphate dehydrogenase.

Light affects enzymes by indirect mechanisms. Among the best researched are the following.

1. **pH**. Recall that during the light reactions, protons are pumped across the thylakoid membrane from the stroma into the thylakoid lumen. As the pH of the stroma increases from 7 to approximately 8, the activities of several enzymes are affected. For example, the pH optimum of ribulose-1,5-bisphosphate carboxylase is 8.

2. Mg^{2+}. Several photosynthetic enzymes (e.g., fructose-1,6-bisphosphatase) are activated by Mg^{2+}. Light induces an increase in the stromal Mg^{2+} concentration from 1 to 3 mM to about 3 to 6 mM. (Recall that Mg^{2+} moves across thylakoid membrane into the stroma during the light reactions.)

3. **The ferredoxin-thioredoxin system**. Thioredoxins are small proteins that transfer electrons from reduced ferredoxin to certain enzymes (Figure 13.22). (Recall that ferredoxin is an electron donor in PSI.) When exposed to light, PSI reduces ferredoxin, which then reduces ferredoxin-thioredoxin reductase (FTR), an iron-sulfur protein that mediates the transfer of electrons between ferredoxin and thioredoxin.

FIGURE 13.22

The Ferredoxin-Thioredoxin System

Using the light energy captured by PSI, energized electrons are donated to ferredoxin. Electrons donated by ferredoxin to FTR (ferredoxin-thioredoxin reductase) are used to reduce the disulfide bridge of thioredoxin. Thioredoxin then reduces the disulfide bridges of susceptible enzymes. Some enzymes are activated by this process, whereas others are inactivated.

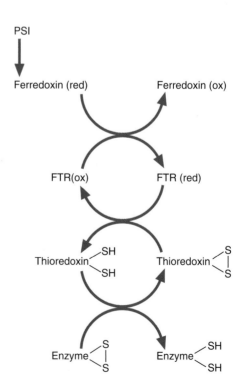

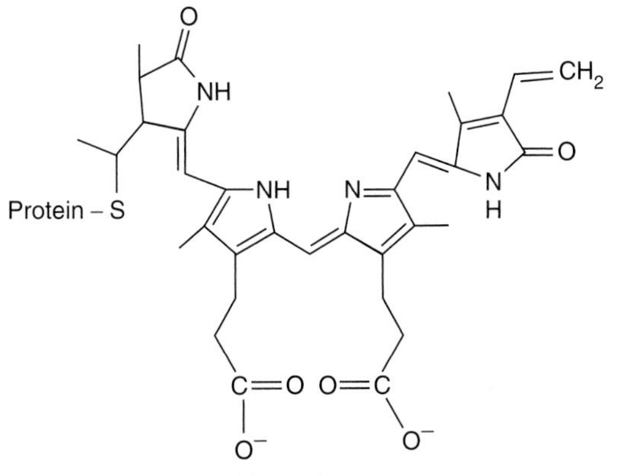

FIGURE 13.23

Phytochrome

The absorption of light changes the arrangement of conjugated double bonds in the molecule.

Reduced thioredoxins activate several enzymes (e.g., fructose-1,6-bisphosphatase, sedoheptulose-1,7-bisphosphatase, and $NADP^+$-glyceraldehyde-3-phosphate dehydrogenase.

4. **Phytochrome**. Phytochrome is a 120 kD protein that possesses a red light-sensitive chromophore (Figure 13.23) that exists in two forms: P_r and P_{fr}. The inactive blue form, P_r, absorbs red light (670 nm). The absorption of longer wavelengths (i.e., far red light, 720 nm) converts P_r to P_{fr}, the active green form. In the dark, P_{fr} decays back to P_r. Phytochrome activation triggers several signal transduction pathways that mediate hundreds of plant responses to light. Phytochrome-mediated processes include seed germination, stem elongation, flower production, and the differentiation of chloroplasts from proplastids. (Several of these processes are also promoted by certain plant hormones.) Phytochrome has specific effects on photosynthetic processes. These include controlling the rate of synthesis of the small subunit of ribulose-1,5-bisphosphate carboxylase, regulation of light absorption by LHCIIb, a component of LHCII, and positioning chloroplasts within photosynthesizing cells. Glucose-6-phosphate dehydrogenase, a NADPH-producing enzyme, is inactivated by light-driven reduction of a disulfide bond by ferredoxin. The enzyme is active at night when NADPH cannot be provided by photosynthesis.

Control of Ribulose-1,5-Bisphosphate Carboxylase

The genes that code for ribulose-1,5-bisphosphate carboxylase (rubisco) are found within the chloroplast (the L subunit) and the nucleus (the S subunit). The activation of these genes is mediated by an increase in light intensity (illumination). Phytochrome also appears to play a role in this activating process. Once the S subunit has been transported from the cytoplasm into the chloroplast, both subunits assemble to form the L_8S_8 holoenzyme. A protein called the *large subunit-binding protein* appears to assist in the assembly of the holoenzyme. When illumination is low, the synthesis of both subunits is rapidly depressed.

The activity of rubisco is modified by a number of metabolic signals. When photosynthesis is active, the pH in the stroma increases (protons are being pumped out of the stroma into the thylakoid lumen) and the Mg^{2+} concentration increases (Mg^{2+} moves into the stroma as H^+ moves out). Both changes increase the activity of rubisco. In other words, when the photosynthetic rate is high, the rate of CO_2 fixation is high. An important consideration in this process is whether the stomata are open or closed (see the discussion of photorespiration on p. 489). Although CO_2 is the preferred substrate for rubisco, under physiological conditions both the carboxylase activity and the oxidase activity compete significantly with each other. If the stomata are closed, as they would be on a hot, dry day, O_2 accumulation in the leaf tissue greatly decreases the proportional participation of the carboxylase activity of rubisco. Recall that C4 plants diminish this competition by trapping the CO_2 in a four-carbon intermediate and delivering the CO_2 via decarboxylation directly to a rubisco molecule that is protected from exposure to O_2.

Rubisco is subject to covalent modification as a means of regulation. The active site of the L subunit must be carbamoylated at a specific lysine residue to be active. Carbamoylation is the nonenzymatic carboxylation of a free primary amino group, in this case the ε-amino group of a certain lysine in the active site of rubisco. The rate of carbamoylation is dependent on the CO_2 concentration and an alkaline pH. The ribulose-1,5-bisphosphate can and does bind to the active site in both its modified and unmodified forms, but catalysis can occur only when rubisco is carbamoylated. The level of activation is cooperative and increases as more of the eight subunits are modified. A specific protein called rubisco activase mediates an ATP-dependent removal of ribulose-1,5-bisphosphate from the active site so that carbamoylation can occur, followed by enzyme activation. In the absence of light, photosynthesis is depressed and the ATP required for this activation process is greatly reduced, as is the NADPH required for the Calvin cycle. In the dark, some plants manufacture the substrate analogue carboxyarabinitol-1-phosphate (CA1P), which serves as an effective competitive inhibitor of the activation of rubisco. (A substrate analogue is a molecule that has a structure similar to the transition state intermediate of the enzyme; however, the enzyme cannot act on the analogue catalytically.)

KEY CONCEPT

Light is the principal regulator of photosynthesis. In photosynthetic processes, light affects the activities of regulatory enzymes such as rubisco by means of indirect mechanisms, which include changes in pH, Mg^{2+} concentration, the ferredoxin-thioredoxin system, and phytochrome. Rubisco is also regulated by covalent modification in that carbamoylation of an active site lysine residue is required for optimal activity.

COMPANION

WEBSITE **Visit the companion website at www.oup.com/us/mckee to read the Biochemistry in the Lab box on photosynthetic studies for a brief discussion of the principles of these technologies and examples of their use.**

D-Carboxyarabinitol-1-phosphate   Rubisco transition state

This regulatory mechanism ensures that CO_2 fixation occurs at an appreciable rate only when the concentration of CO_2 and available energy are high.

Despite the complexity of photosynthesis, many aspects of this life-sustaining process are now understood. Technologies whose use contributed (and continues to contribute) to this research effort include spectroscopy, photochemistry, X-ray crystallography, and radioactive tracers.

BIOCHEMISTRY IN PERSPECTIVE

Photosynthesis in the Deep

How can light-dependent photosynthesis occur at the bottom of the ocean, where there is no sunlight? Sunlight penetrates ocean water to a depth of 200 m (about 660 ft). Accordingly, it has long been assumed that photosynthetic organisms do not exist below this depth. It came as a surprise, therefore, when in 2003 an intriguing exception was discovered more than a mile below the surface of the Pacific Ocean off the coast of Mexico. Investigators isolated a new species of green sulfur bacteria (referred to as GSB1) in close proximity to the extremely hot water (measured at 350°C) of hydrothermal vents (p. 2). In this nearly pitch-black environment, GSB1, an obligate anaerobic, photosynthesizing organism, apparently harnesses blackbody radiation emitted by the hot water (Figure 13A). A *blackbody* is a substance that absorbs radiant energy, in this case geothermal energy, and emits light energy. The hotter the substance becomes, the shorter the wavelength of light that it emits. The elements in an electric stove, for example, begin to appear dull red as the oven temperature reaches a certain level. It should be noted that some researchers believe that chemical reactions may also contribute to the dim light emitted from hydrothermal vents.

GSB1 is an autotroph that requires only elemental sulfur ($S°$), H_2S, CO_2, and light energy to maintain its existence. It can absorb sufficient amounts of low-energy light because it possesses an extremely efficient light-harvesting antenna called a chlorosome. *Chlorosomes* are membrane-bound vesicles attached to the plasma membrane that collect low-energy photons of light, no matter how rare, and transfer them to the photosynthetic reaction centers in the plasma membrane.

In addition to aggregates of bacteriochlorophyll c (a molecule similar to chlorophyll, but with an absorption maxima near the infrared), chlorosomes contain smaller amounts of bacteriochlorophyll a, carotenoids, and quinones. They also contain small amounts of lipid and protein molecules. However

amazing GSB1 is, it is instructive to put its lifestyle in context. It survives in a very hostile and low-light environment, but its growth (as measured by cell division time) is measured in years instead of the minutes or hours that are characteristic of photosynthetic bacteria in more temperate environments.

FIGURE 13A

Light Emitted at a Hydrothermal Vent Heated by Geothermal Energy.

SUMMARY: Dim light, sufficient for a slow-growing type of photosynthetic green sulfur bacterium, is used to fix carbon near hydrothermal vents. This light is emitted by hydrothermal vents, which are hot enough to cause "blackbody" radiation.

Chapter**Summary**

1. In plants, photosynthesis takes place in chloroplasts. Chloroplasts possess three membranes. The outer membrane is highly permeable, whereas the inner membrane possesses a variety of carrier molecules that regulate

molecular traffic into and out of the chloroplast. A third membrane, called the thylakoid membrane, forms an intricate series of flattened vesicles called grana.

2. Photosynthesis consists of two major phases: the light reactions and the light-independent reactions. During the light reactions, water is oxidized, O_2 is evolved, and the ATP and NADPH required to drive carbon fixation are produced. The major working units of the light reactions are photosystems I and II, the cytochrome b_6f complex, and the ATP synthase. In noncyclic electron transport, electrons from water molecules are transferred from photosystem II to photosystem I to $NADP^+$ with the production of O_2 and NADPH. Cyclic electron transport involves only PSI and generates additional ATP but no NADPH. During the light-independent reactions, CO_2 is incorporated into organic molecules. The first stable product of carbon fixation is glycerate-3-phosphate. The Calvin cycle is composed of three phases: carbon fixation, reduction, and regeneration.

3. Most of the carbon incorporated during the Calvin cycle is used initially to synthesize starch and sucrose, both of which are important energy sources. Sucrose is also important because it is used to translocate fixed carbon throughout the plant.

4. Photorespiration is an apparently wasteful process whereby O_2 is consumed and CO_2 is released from plants. Its role in plant metabolism is not understood. C4 plants, which grow in relatively stringent environments, have developed biochemical and anatomical mechanisms for suppressing photorespiration.

5. Because plants must adapt to ever-changing environmental conditions, the regulation of photosynthesis is complex. Several features of this control are now well established. Light is an important regulator of most aspects of photosynthesis. Many of the effects of light are mediated by changes in the activities of key enzymes. The mechanisms by which light effects these changes include changes in pH, Mg^{2+} concentration, the ferredoxin-thioredoxin system, and phytochrome. The most important enzyme in photosynthesis is ribulose-1,5-bisphosphate carboxylase. Its activity is highly regulated. Light activates the synthesis of both types of the enzyme's subunits. Its activity is affected by allosteric effectors as well as covalent modification. Carbamoylation of an active site lysine residue is required for activation of rubisco.

COMPANION WEBSITE Take your learning further by visiting the **companion website** for Biochemistry at **www.oup.com/us/mckee** where you can complete a multiple-choice quiz on photosynthesis to help you prepare for exams.

Suggested Readings

Allen, J. F., Photosynthesis of ATP—Electrons, Proton Pumps, Rotors, and Poise, *Cell* 110:273–276, 2002.

Buchanan, B. B., and Luan, S., Redox Regulation in the Chloroplast Thylakoid Lumen: A New Frontier in Photosynthesis Research, *J. Exp. Bot.* 56(416):1439–1447, 2005.

Danielsson, R., et al., Dimeric and Monomeric Organization of Photosystem II: Distribution of Five Distinct Complexes in the Different Domains of the Thylakoid Membrane, *J. Biol. Chem.* 281(20):14241–14249, 2006.

Ensmenger, P. A., *Life Under the Sun*, Yale University Press, New Haven, Connecticut, and London, 2001.

Gross, M., *Light and Life*, Oxford University Press, Oxford, 2002.

Heathcote, P., Fyfe, P. K., and Jones, M. R., Reaction Centres: The Structure and Evolution of Biological Solar Power, *Trends Biochem. Sci.* 27(2):79–87, 2002.

Horton, P., and Ruban, A., Molecular Design of the Photosystem II Light-Harvesting Antenna: Photosynthesis and Photoprotection, *J. Exp. Bot.* 56(411):365–373, 2005.

Jones, M. R., and Fyfe, P. K., Photosynthesis: A New Step in Oxygen Evolution, *Current Biol.* 14:R320–R322, 2004.

Melkozernov, A. N., Barber, J., and Blankenship, R. E., Light Harvesting in Photosystem I Supercomplexes, *Biochemistry* 45(2):331–345, 2006.

Merchant, S., and Sawaya, M. R., The Light Reactions: A Guide to Recent Acquisitions for the Picture Gallery, *Plant Cell*, 17:648–663, 2005.

Nelson, N., and Ben-Shem, A., The Structure of Photosystem I and Evolution of Photosynthesis, *BioEssays* 27:914–922, 2005.

Nugent, J. H. A., and Evans, M. C. W., Structure of Biological Solar Energy Converters—Further Revelations, *Trends Plant Sci.* 9(8):368–370, 2004.

Key Words

antenna pigment, *473*

C3 plants, *488*

C4 metabolism, *491*

C4 plants, *490*

Calvin cycle, *486*

Carbon fixation, *486*

carotenoid, *469*

chlorophyll, *469*

chromophore, *474*

crassulacean acid metabolism, *491*

fluorescence, *477*

granum, *469*

light-independent reaction, *486*

light reaction, *480*

photophosphorylation, *485*

photorespiration, *489*

photosystem, *478*

reaction center, *468*

resonance energy transfer, *478*

stroma, *469*

stromal lamella, *469*

thylakoid membrane, *469*

thylakoid lumen, *469*

Z scheme, *480*

Review**Questions**

These questions are designed to test your knowledge of the key concepts discussed in this chapter, before moving on to the next chapter. You may like to compare your answers to the solutions provided in the back of the book and in the accompanying Study Guide.

1. Define the following terms:
   a. photosystem
   b. reaction center
   c. light reaction
   d. dark reaction

2. What was the most significant contribution of early photosynthetic organisms to Earth's environment?

3. List the three primary photosynthetic pigments and describe the role each plays in photosynthesis.

4. List five ways in which chloroplasts resemble mitochondria.

5. Define the following terms:
   a. chloroplast
   b. photorespiration
   c. chlorosome
   d. stromal lamellae
   e. granum

6. Excited molecules can return to the ground state by several means. Describe each briefly. Which of these processes are important in photosynthesis? Describe how they function in a living organism.

7. What is the final electron acceptor in photosynthesis when the $NADPH/NADP^+$ ratio is low? Does your answer change if the $NADPH/NADP^+$ ratio is high?

8. What reactions occur during the light reactions of photosynthesis?

9. What reactions occur during the light-independent reactions of photosynthesis?

10. Define the following terms:
    a. D_1/D_2 dimer
    b. CF_0
    c. wavelength
    d. chromophore
    e. fluorescence

11. The overall equation for photosynthesis is

$$6CO_2 + 6H_2O \xrightarrow{\text{light}} C_6H_{12}O_6 + 6O_2$$

    From which of the two substrates are the atoms of molecular oxygen derived?

12. Why is the oxygen-evolving system referred to as a clock?

13. If the rate of photosynthesis is plotted versus the incident wavelength of light, an action spectrum is obtained. How can the action spectrum provide information about the nature of the light-absorbing pigments involved in photosynthesis? [*Hint*: Refer to the Biochemistry in Perspective box entitled Photosynthetic Studies at the companion website, www.oup.com/us/mckee]

14. Define the following terms:
    a. radiationless decay
    b. resonance energy transfer
    c. Z scheme
    d. PSI
    e. PSII

15. Using the action spectrum for photosynthesis determine what wavelengths of light appear to be optimal for photosynthesis. [*Hint*: Refer to the Biochemistry in Perspective box entitled Photosynthetic Studies at the companion website, www.oup.com/us/mckee]

16. List the types of metal that are components of the photosynthesis mechanism. What functions do they serve?

17. Explain the following observation. When a photosynthetic system is exposed to a brief flash of light, no oxygen is evolved. Only after several bursts of light is oxygen evolved.

18. Define the following terms:
    a. carbon fixation
    b. C3 metabolism
    c. C4 metabolism
    d. CAM
    e. phytochrome

19. Describe the Z scheme of photosynthesis. How are the products of this reaction used to fix carbon dioxide?

20. Where does carbon dioxide fixation take place in the cell with reference to the light-dependent reaction?

21. The chloroplast has a highly organized structure. How does this structure help make photosynthesis possible?

22. Visit the companion website at www.oup.com/us/mckee to read the Biochemistry in Perspective box entitled Photosynthetic Studies. Then describe the Emerson enhancement effect. How was it used to demonstrate the existence of two different photosystems?

23. Why does CO_2 depress photorespiration?

24. After one turn of the Calvin cycle in a C3 plant, where would you expect to find the radioactive label of $^{14}CO_2$?

25. What are triose phosphates? Where are they formed in plant carbohydrate metabolism?

26. A C4 plant is exposed to $^{15}O_2$. Where will this radioactive label first appear?

27. Review the net reaction for photosynthesis and explain the origin of the oxygen atoms in glucose molecules.

28. Can plants be grown in an oxygen-free atmosphere?

29. If corn, a C4 plant, is exposed to $^{14}CO_2$, in what molecule will it initially appear? Indicate the portion of the molecule in which the label will be found.

30. What is the end product of photosynthesis when H_2S is the source of hydrogen atoms?

31. What effect would you expect dinitrophenol to have on photosynthesis?

Thought Questions

These questions are designed to reinforce your understanding of all of the key concepts discussed in the book so far, including this chapter and all of the chapters before it. They may not have one right answer! The authors have provided possible solutions to these questions in the back of the book and in the accompanying Study Guide, for your reference.

32. Without carbon dioxide, chlorophyll fluoresces. How does carbon dioxide prevent this fluorescence?

33. The statement has been made that the more extensively conjugated a chromophore, the less energy a photon needs to excite it. What is conjugation, and how does it contribute to this phenomenon?

34. Increasing the intensity of the incident light but not its energy increases the rate of photosynthesis. Why is this so?

35. Both oxidative phosphorylation and photophosphorylation trap energy in high-energy bonds. How are these processes different? How are they the same?

36. In C3 plants, high concentrations of oxygen inhibit photosynthesis. Why is this so?

37. Generally, increasing the concentration of carbon dioxide increases the rate of photosynthesis. What conditions could prevent this effect?

38. It has been suggested that chloroplasts, like mitochondria, evolved from living organisms. What features of the chloroplast suggest that this is true?

39. Explain why photorespiration is repressed by high concentrations of carbon dioxide.

40. Why does exposing C3 plants to high temperatures raise the carbon dioxide compensation point?

41. Herbicides that act by promoting photorespiration are lethal to C3 plants but do not affect C4 plants. Why is this so?

42. Corn, a grain of major economic importance, is a C4 plant, and many weeds in temperate climates are C3 plants. Therefore the herbicides described in Question 41 are widely used. What effect is likely if these toxic materials are not degraded before they wash into the ocean?

43. Although both organelles originated as free-living prokaryotes, mitochondria are significantly smaller than chloroplasts. Suggest a reason for this discrepancy.

    [*Hint*: Consider the energy sources for both organelles.]

44. Suggest a reason why primitive photosynthetic organisms used H_2S rather than H_2O as the proton donor.

45. Suggest a reason why photosynthetic pigments absorb in the blue region of the visible spectrum, but not in the ultraviolet.

46. If a C3 plant and a C4 plant are placed in separate sealed containers and provided with adequate light and water, both will survive. If both plants are placed in a single sealed container, the C3 plant will die. Explain.

47. Which plants expend more energy per molecule of glucose produced, C3 or C4?

48. Increased industrial production has raised the levels of CO_2 in the atmosphere. What type of plant (C3 or C4) would you expect to benefit from this change?

49. Triazine herbicides are effective C3 plant toxins that are used to suppress weed growth. They apparently bind to a plastocyanin-binding protein. Suggest a mechanism whereby triazines undermine C3 growth.

50. Photosynthesizing organisms of the deep ocean capture long-wave radiation. Determine how many quanta of light at 1000 nm are required to provide as much energy as plants that absorb at 700 nm. [*Hint*: Recall that $\Delta E = hc/\lambda$, where h is Planck's constant (1.58×10^{-37} kcal/s), c is the speed of light (3×10^8 m/s), and λ is wavelength.]

Nitrogen Metabolism I: Synthesis

Nitrogen Metabolites
Glutamatin, glutamine, and ammonia are among the most important molecules in nitrogen metabolism.

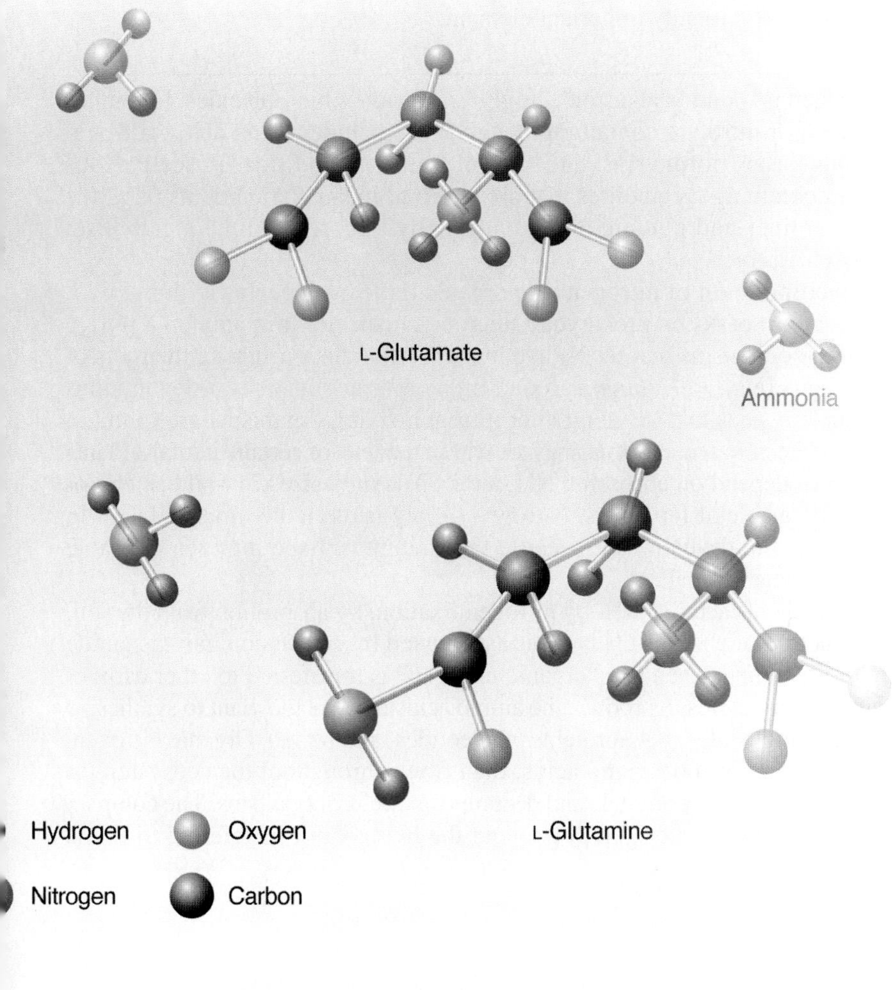

L-Glutamate

Ammonia

L-Glutamine

Hydrogen
Oxygen
Nitrogen
Carbon

Overview

NITROGEN IS AN ESSENTIAL ELEMENT FOUND IN PROTEINS, NUCLEIC ACIDS, AND MYRIAD OTHER BIOMOLECULES. DESPITE THE IMPORTANT role that it plays in living organisms, biologically useful nitrogen is scarce. Although nitrogen gas (N_2) is plentiful in the atmosphere, it is almost chemically inert. Therefore, converting N_2 to a useful form requires a major expenditure of energy. Only a few organisms can perform this function, referred to as nitrogen fixation. Certain microorganisms (all of which are bacteria or cyanobacteria) can reduce N_2 to form NH_3 (ammonia). Plants and microorganisms can absorb NH_3 and NO_3^- (nitrate), the oxidation product of ammonia. Both molecules are then used to synthesize nitrogen-containing biomolecules. (The conversion of NH_3 to NO_3^- in a process referred to as nitrification is also performed by certain microorganisms.) Animals cannot synthesize the nitrogen-containing molecules they require from NH_3 and NO_3^-. Instead, they must acquire "organic nitrogen," primarily amino acids, from dietary sources (i.e., plants and plant-eating animals). In a complex series of reaction pathways, animals then use amino acid nitrogen to synthesize important metabolites. Chapter 14 describes the synthesis of the major nitrogen-containing molecules (e.g., amino acids and nucleotides) and a small selection of molecules that represent the rich diversity of metabolites that contain this critically important element.

Nitrogen is found in an astonishingly vast array of biomolecules. Examples of major nitrogen-containing metabolites include amino acids, nitrogenous bases, porphyrins, and several lipids. In addition, miscellaneous nitrogen-containing metabolites that are required in smaller amounts (e.g., the biogenic amines and glutathione) are critically important in the metabolism of many eukaryotes.

The incorporation of nitrogen into organic molecules begins with the fixation (reduction) of N_2 by prokaryotic microorganisms to form ammonia (NH_3). Some of the organisms that fix N_2 live in symbiotic relationships in the roots of certain plants (e.g., *Rhizobium* is found in the root nodule cells of leguminous plants such as peas and beans). Other nitrogen-fixing organisms are found in marine and fresh water, in hot springs, or within the guts of certain animals. Plants such as corn depend on absorbing NH_3 and NO_3^- synthesized by soil bacteria or provided by artificial fertilizers. Nitrogen supply is often the limiting factor in plant growth and development because the amount of fixed nitrogen available to plants is usually small.

Whether plants acquire NH_3 by nitrogen fixation, by absorption from the soil, or by reduction of absorbed NO_3^-, it is assimilated by conversion into the amide group of glutamine. Then this "organic nitrogen" is transferred to other carbon-containing compounds to produce the amino acids used by the plant to synthesize nitrogenous molecules (e.g., proteins, nucleotides, and heme). Organic nitrogen, primarily in the form of amino acids, then flows throughout the ecosystem as plants are consumed by animals and decomposing microorganisms. The complex process that transfers nitrogen throughout the living world is referred to as the *nitrogen cycle*.

Organisms other than plants vary widely in their capacity to synthesize amino acids from metabolic intermediates and fixed nitrogen. For example, many microorganisms can produce all the amino acids they need. In contrast, animals can synthesize only about half the amino acids they require. The **nonessential amino acids (NAA)** are synthesized from readily available metabolites. The amino acids that must be provided in the diet to ensure proper nitrogen balance and adequate growth are referred to as **essential amino acids (EAA)**.

Following the digestion of dietary protein in the body's digestive tract, free amino acids are transported across intestinal enterocytes and into the blood. Most diets do not provide amino acids in the proportions that the body requires. Their concentrations, therefore, must be adjusted by metabolic mechanisms. The amino acids released to the blood in the intestine already show some changes in their relative concentrations. The intestinal mucosa is a very active and constantly replaced tissue that sustains its structure and function by means of incoming nutrients other than glucose. The amino acid glutamine, for example, is a primary energy source for enterocytes. The blood from the gastrointestinal (GI) tract goes first to the liver, another active tissue. The liver synthesizes the serum proteins, among others, and draws amino acids from the blood for this purpose. It also preferentially uses amino acids (especially alanine and serine) to synthesize glucose for export. The blood that leaves the liver to nourish the rest of the body has a much higher concentration of **branched-chain amino acids (BCAA:** leucine, isoleucine, and valine) than that leaving the GI tract. The BCAA are essential amino acids and provide critical hydrophobic side chains in protein structure (e.g., leucine zipper motifs in DNA-binding proteins). BCAA also represent a major transport form of amino nitrogen from the liver to other tissues, where they are used in the synthesis of the nonessential amino acids required for protein synthesis, as well as various amino acid derivatives.

Transamination reactions dominate amino acid metabolism. In these reactions, catalyzed by a group of enzymes referred to as the *aminotransferases* or *transaminases*, α-amino groups are transferred from an α-amino acid to an α-keto acid:

Acceptor keto acid   Donor amino acid   New keto acid   New amino acid

(Recall that in *α-keto acids* such as α-ketoglutarate and pyruvate, a carbonyl group is directly adjacent to the carboxyl group.) Transamination reactions, which are readily reversible, play important roles in both the synthesis and degradation of the amino acids.

After a discussion of nitrogen fixation, the essential features of amino acid biosynthesis are described. This is followed by descriptions of the biosynthesis of selected nitrogen-containing molecules. A special emphasis is placed on the anabolic pathways of the nucleotides. Chapter 15 traces the flow of nitrogen atoms through several catabolic pathways to the nitrogenous waste products excreted by animals.

14.1 NITROGEN FIXATION

Several circumstances limit the amount of usable nitrogen available in the biosphere. The most notable are the limited number of species that can convert N_2 into NH_3, a more chemically reactive molecule, and the high energy requirements of this process, which is referred to as **nitrogen fixation**. Among the most prominent nitrogen-fixing species are free-living bacteria (e.g., *Azotobacter vinelandii* and *Clostridium pasteurianum*), the cyanobacteria (e.g., *Nostoc muscorum* and *Anabaena azollae*), and symbiotic bacteria (e.g., several species of *Rhizobium*). Symbiotic organisms form mutualistic, that is, mutually beneficial, relationships with host plants or animals. *Rhizobium* species, for example, infect the roots of leguminous plants such as soybeans and alfalfa.

Nitrogen fixation requires a large energy input because reduction of N_2 to form NH_3 involves the breaking of the nonpolar triple bond of atmospheric dinitrogen gas. In commercial nitrogen fixation, NH_3 is the product of the Haber-Bosch reaction, where H_2 and N_2 are heated at 400 to 650°C under 200 to 400 atm pressure in the presence of an iron catalyst. Unlike the Haber-Bosch process, nitrogen-fixing species convert N_2 to NH_3 at ambient temperature and atmospheric pressure. The energy requirements for the biological process, however, are also high, with a minimum of 16 ATP required to reduce one N_2 to form two ammonia molecules.

The Nitrogen Fixation Reaction

All species that can fix nitrogen possess the *nitrogenase complex*. Its structure, similar in all species so far investigated, consists of two proteins called dinitrogenase and dinitrogenase reductase (Figure 14.1). Dinitrogenase (240 kD), also referred to as MoFe *protein*, is an $\alpha_2\beta_2$ heterotetramer. Each $\alpha\beta$ dimer is a catalytic unit that contains two unique metal prosthetic groups: a *P cluster* [8Fe-7S] and *molybdenum-iron cofactor* (MoFe cofactor), which is linked

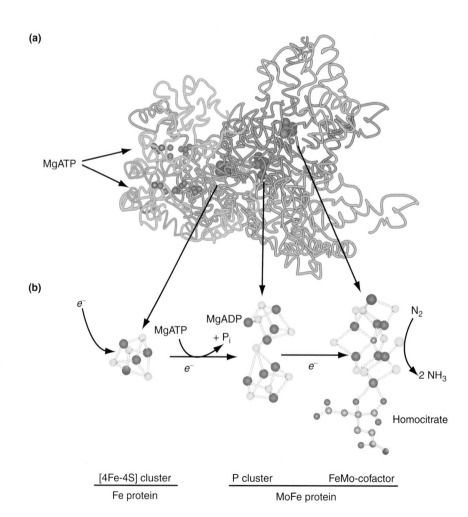

FIGURE 14.1

Nitrogenase Complex Structure

(a) Left: Fe protein dimer (pink and red); right: an $\alpha\beta$ dimer of the MoFe protein, with the α subunit in blue and the β subunit in green. (b) The metal clusters of the Fe protein [4Fe-4S] and the P cluster and FeMo cofactor of the FeMo protein are illustrated as ball-and-stick models. Atom colors: carbon, gray; nitrogen, blue; oxygen, red; phosphorus, dark green; sulfur, yellow; magnesium, orange; iron, green; molybdenum, pink.

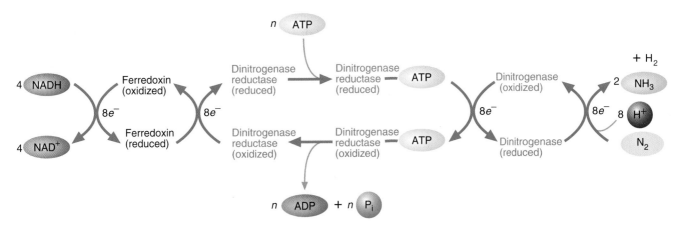

FIGURE 14.2

Schematic Diagram of the Nitrogenase Complex Illustrating the Flow of Electrons and Energy in Enzymatic Nitrogen Fixation

The high energy of activation of nitrogen fixation is overcome by a large number of ATP molecules (about 16 ATP per N_2 molecule). Both the binding of ATP to dinitrogenase reductase and its subsequent hydrolysis cause conformational changes in the protein that facilitate the transfer of electrons to dinitrogenase.

to homocitrate (an isomer of citrate). The MoFe protein catalyzes the reaction $N_2 + 8H^+ + 8e^- \rightarrow 2NH_3 + H_2$. Dinitrogenase reductase (60 kD) (also referred to as *Fe protein*) is a dimer containing identical subunits, each of which has a MgATP-binding site. A 4Fe-4S cluster is bound at the interface of the two subunits, 15 Å from the MgATP-binding site and close to the docking site for the dinitrogenase tetramer (MoFe protein). The Fe protein transfers electrons, ultimately derived from NAD(P)H, one at a time to the Mo-Fe protein. Both proteins in the nitrogenase complex are irreversibly inactivated by O_2.

The first step in nitrogen fixation (Figure 14.2) is the transfer of electrons from NAD(P)H to ferredoxin, a powerful reducing agent that in turn donates the electrons to the Fe protein FeS cluster, one at a time. Each electron transfer from the Fe protein to the P cluster of the MoFe protein involves the docking of the MgATP-bound, electron-charged Fe protein followed by electron transfer requiring a MgATP hydrolysis-dependent conformational change. The Fe protein (MgADP) then dissociates from the MoFe protein and exchanges its MgADP for MgATP. The electron transfer process is then repeated until eight electrons have been delivered to the MoFe cofactor. (In its oxidized electron-deficient form, the Fe protein has a higher affinity for MgATP.) The reduction potentials for the Fe protein and the P cluster have been recorded in the –400 and –300 mV ranges, respectively. The transfer of the first two electrons leads to the reduction of H^+ (the MoFeH$_2$ state of the enzyme). This stage of the process occurs with or without the presence of N_2. The incoming N_2 exchanges with H_2 in the active site (between the molybdenum ion and four coordinating Fe centers) to form a stable intermediate (MoFeN$_2$H). Subsequently, six electrons and six protons are transferred to the active site to form the diimine (HN $=$ NH, two electrons added), then the hydrazine (H$_2$N—NH$_2$, four electrons added) and, finally, two NH$_3$ products (six electrons added). Therefore, two electrons are used to reduce 2H$^+$ to H$_2$ (essential to the catalytic process) and six electrons are used to reduce N_2 in three steps to yield two NH$_3$ molecules.

In addition to the large quantities of ATP that are required to drive the reduction of N_2 (a minimum of 16 ATP), a substantial number of copies of nitrogenase complex proteins must be synthesized because of the enzyme's slow turnover time: about 6 NH$_3$ produced per second per molecule of enzyme. (Nitrogenase complex protein may constitute as much as 20% of cellular protein in diazotrophs, the bacteria that fix nitrogen.) As a result, the regulation of nitrogen fixation by diazotrophs, which must respond to numerous environmental variations (e.g., fixed nitrogen and oxygen levels and carbon source availability), is both intricate and stringent. The major form of regulation is transcriptional control of the approximately 20 nitrogen fixation (*nif*) genes. In addition to coding for dinitrogenase reductase (Fe protein) and the α and β subunits of dinitrogenase (MoFe protein), *nif* genes code for a variety of enzymes that synthesize components of the nitrogen fixation process such as metal clusters, homocysteine, or ferredoxin, and regulatory proteins that mediate responses to environmental cues.

Nitrogen Assimilation

Nitrogen assimilation is the incorporation of inorganic nitrogen compounds into organic molecules. In plants, nitrogen assimilation begins in the roots, whether by the transfer of NH_4^+ (ammonium ions) from symbiotic bacteria in the root nodules of leguminous plants or by the absorption of NH_4^+ or NO_3^- (nitrate) from soil. Nitrate is produced by nitrifying soil bacteria. Organisms such as *Nitrosomonas* oxidize NH_4^+ to form NO_2^- (nitrite), which is then further oxidized to NO_3^- by bacteria such as *Nitrobacter*. The assimilation of inorganic nitrogen into biomolecules in plants is effected by the incorporation of ammonium nitrogen into amino acids. Glutamine synthetase, the most important enzyme in nitrogen assimilation, catalyzes the ATP-dependent reaction of glutamate with NH_4^+ to form glutamine. In the next step, glutamine reacts with α-ketoglutarate to form glutamate. The synthesis of other amino acids via the transfer of the amino group of glutamate is described in Section 14.2. When nitrate is the nitrogen source, it must first be converted to NH_4^+ in a two-step process. After nitrate has been converted to nitrite by nitrate reductase, ammonia is produced by the reduction of nitrite, which is catalyzed by nitrite reductase.

QUESTION 14.1

The nitrogenase complex can reduce molecules other than N_2. Provide the structures for the products for each of the following substrates (real and hypothetical) that contain triple bonds:

a. hydrogen cyanide

b. dinitrogen

c. acetylene

QUESTION 14.2

What are the substrates and enzyme components in nitrogen fixation? Arrange each set in the order in which electrons and protons are transferred to dinitrogen.

QUESTION 14.3

A red heme-containing protein called leghemoglobin is found in the nitrogen-fixing root nodules of leguminous plants. The protein component is produced by the plant, whereas bacterial cells produce heme. Can you deduce the function of leghemoglobin? [*Hint*: Review the function of the hemoglobins.]

14.2 AMINO ACID BIOSYNTHESIS

Living organisms differ in their capacity to synthesize the amino acids required for protein synthesis. Although plants and many microorganisms can produce all their amino acids from readily available precursors, other organisms must obtain some preformed amino acids from their environment. For example, mammalian tissues can synthesize NAA (Table 14.1) by relatively simple reaction pathways. In contrast, EAA must be obtained from the diet because mammals lack the long and complex reaction pathways required for their synthesis.

Amino Acid Metabolism Overview

Amino acids serve a number of functions. In addition to their most important role in the synthesis of proteins, amino acids are the principal source of the nitrogen atoms required in various synthetic reaction pathways. The nonnitrogen components of amino acids (referred to as carbon skeletons) are a source of energy, as well as precursors in several reaction pathways. Therefore an adequate intake of amino acids, in the form of dietary protein, is essential for an animal's proper growth and development.

Dietary protein sources differ widely in their proportions of the EAA. In general, complete proteins (those containing sufficient quantities of EAA) are of animal origin (e.g., meat, milk, and eggs). Plant proteins often lack one or more EAA. For example, gliadin (wheat protein) has insufficient amounts of lysine, and zein (corn protein) is low in both lysine and tryptophan. Because plant proteins differ in their amino acid compositions, plant foods can provide a high-quality source of essential amino acids only if they are eaten in appropriate combinations. One such combination consists of beans (low in methionine) and cereal grains (low in lysine).

The amino acid molecules that are immediately available for use in metabolic processes are referred to as the **amino acid pool**. In animals, amino acids in the pool are derived from the breakdown of both dietary and tissue proteins. Amino acid metabolism is a complex series of reactions in which the amino acid molecules required for the syntheses of proteins and metabolites are continuously being synthesized and degraded. Depending on current metabolic requirements, certain amino acids are synthesized or interconverted and then transported to tissue, where they are used.

TABLE 14.1   The Essential and Nonessential Amino Acids in Humans

Essential	Nonessential
Isoleucine	Alanine
Leucine	Arginine*
Lysine	Asparagine
Methionine	Aspartate
Phenylalanine	Cysteine
Threonine	Glutamate
Tryptophan	Glutamine
Valine	Glycine
	Histidine*
	Proline
	Serine
	Tyrosine

*Essential for infants.

When nitrogen intake (primarily amino acids) equals nitrogen loss, the body is said to be in *nitrogen balance*. This is the condition of healthy adults. In *positive nitrogen balance*, a condition that is characteristic of growing children, pregnant women, and recuperating patients, nitrogen intake exceeds nitrogen loss. The excess of nitrogen is retained because the amount of tissue proteins being synthesized exceeds the amount being degraded. *Negative nitrogen balance* exists when an individual cannot replace nitrogen losses with dietary sources. *Kwashiorkor* ("the disease the first child gets when the second is on the way") is a form of malnutrition caused by a prolonged insufficient intake of protein. Its symptoms include growth failure, apathy, ulcers, liver enlargement, and diarrhea, as well as decreased mass and function of the heart and kidneys. Prevalent in Africa, Asia, and Central and South America, kwashiorkor can be treated by feeding the children high-quality, protein-rich foods such as milk, eggs, and meat.

Transport of amino acids into cells is mediated by specific membrane-bound transport proteins, several of which have been identified in mammalian cells. They differ in their specificity for the types of amino acid transported and in whether the transport process is linked to the movement of Na^+ across the plasma membrane. (Recall that the gradient created by the active transport of Na^+ can move molecules across the membrane. Na^+-dependent amino acid transport is similar to that observed in the glucose transport process illustrated in Figure 11.31.) For example, several Na^+-dependent transport systems have been identified in the lumenal plasma membrane of enterocytes. Na^+-independent transport systems are responsible for transporting amino acids across the portion of enterocyte plasma membrane in contact with blood vessels. The γ-glutamyl cycle (see later: Figure 14.20) is believed to assist in transporting some amino acids into specific tissues (i.e., brain, intestine, and kidney).

Reactions of Amino Groups

Once amino acid molecules have entered cells, the amino groups are available for numerous synthetic reactions. This metabolic flexibility is effected primarily by transamination reactions in which amino groups are transferred from an α-amino acid to an α-keto acid. However, another class of reactions, in which NH_4^+ or the amide nitrogen of glutamine is used to supply the amino group or the amide nitrogen of certain amino acids, also occurs. These reaction types are discussed next.

TRANSAMINATION Eukaryotic cells possess a large variety of aminotransferases. Found within both the cytoplasm and mitochondria, these enzymes possess two types of specificity: (1) the type of α-ammo acid that donates the α-amino group and (2) the α-keto acid that accepts the α-amino group. Although the aminotransferases vary widely in the type of amino acids they bind, most of them use glutamate as the amino group donor:

Acceptor α-keto acid   Glutamate   New amino acid α-Ketoglutarate

Because glutamate is produced when α-ketoglutarate (a citric acid cycle intermediate) accepts an amino group, these two molecules, the *α-ketoglutarate/ glutamate pair*, have a strategically important role in both amino acid metabolism and metabolism in general. Two other such pairs have important functions in metabolism. In addition to its role in transamination reactions, the *oxaloacetate/aspartate pair* is involved in the disposal of nitrogen in the urea

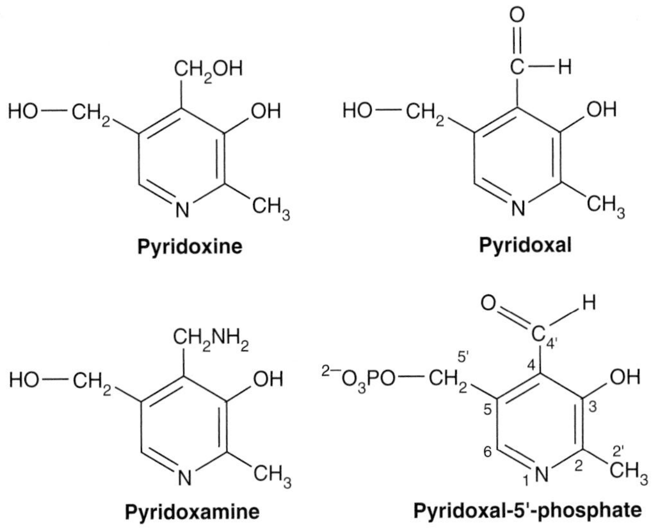

FIGURE 14.3

Vitamin B$_6$

Vitamin B$_6$ includes (a) pyridoxine, (b) pyridoxal, and (c) pyridoxamine. (Pyridoxine is found in leafy green vegetables. Pyridoxal and pyridoxamine are found in animal foods such as fish, poultry, and red meat.) The biologically active form of vitamin B$_6$ is (d) pyridoxal-5'-phosphate.

cycle (Chapter 15). One of the most important functions of the *pyruvate/alanine pair* is in the alanine cycle (Figure 8.11). α-Ketoglutarate and oxaloacetate are citric acid cycle intermediates. Consequently, transamination reactions often represent an important mechanism for meeting the energy requirements of cells.

Transamination reactions require the coenzyme pyridoxal-5'-phosphate (PLP), which is derived from pyridoxine (vitamin B$_6$). PLP is also required in numerous other reactions of amino acids. Examples include racemizations, decarboxylations, and several side chain modifications. (**Racemizations** are reactions in which mixtures of L- and D-amino acids are formed.) The structures of the vitamin and its coenzyme form are illustrated in Figure 14.3.

PLP is bound in the enzyme active site by noncovalent interactions and a Schiff base (R'—CH=N—R, an aldimine) formed by the condensation of the aldehyde group of PLP and the ε-amino group of a lysine residue.

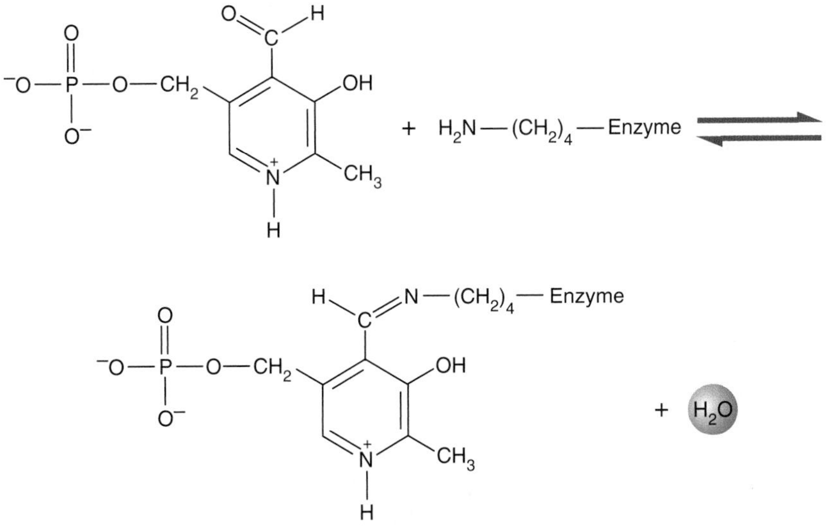

Additional stabilizing forces include ionic interactions between amino acid side chains and PLP's pyridinium ring and phosphate group. The positively charged pyridinium ring also functions as an electron sink, stabilizing negatively charged reaction intermediates.

Amino acid substrates become bound to PLP via the α-amino group in an imine exchange reaction. Then one of three bonds in the substrate is selectively broken in the active sites in each type of PLP-dependent enzyme.

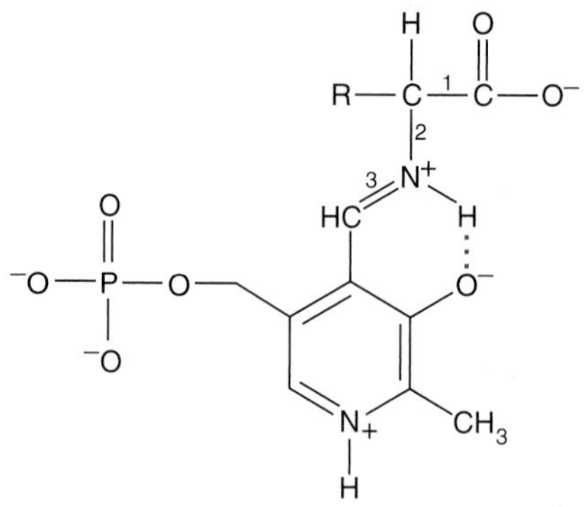

This selectivity is dependent upon the presence or absence of a nearby base catalyst and the orientation of the amino acid in the active site. If an initial deprotonation of the α-carbon of the amino group donor occurs, then transamination (bond 2 broken) or racemization or elimination (bond 3 broken) may occur. If the initial deprotonation does not occur, then decarboxylation results (bond 1 broken).

Despite the apparent simplicity of the transamination reaction, the mechanism is quite complex. The reaction begins with the formation of a Schiff base between PLP and the α-amino group of an α-amino acid (Figure 14.4). When the α-hydrogen atom is removed by a general base in the enzyme active site, a resonance-stabilized intermediate forms. With the donation of a proton from a general acid and a subsequent hydrolysis, the newly formed α-keto acid is released from the enzyme. A second α-keto acid then enters the active site and is converted into an α-amino acid in a reversal of the reaction process that has just been described. Transamination reactions are examples of a reaction mechanism referred to as a double displacement, or *ping-pong reaction* (p. 196). The mechanism is so named because the first substrate must leave the active site before the second one can enter.

QUESTION 14.4

Provide the structure of the α-keto acid product of the transamination of each of the following molecules:

   a. glutamine

   b. isoleucine

   c. phenylalanine

   d. aspartate

   e. cysteine

Transamination reactions are reversible. It is, therefore, theoretically possible for all amino acids to be synthesized by transamination. However, experimental evidence indicates that there is no net synthesis of an amino acid if its α-keto acid precursor is not independently synthesized by the organism. For example, alanine, aspartate, and glutamate are nonessential for animals because their α-keto acid precursors (i.e., pyruvate, oxaloacetate, and α-ketoglutarate) are readily available metabolic intermediates. Because the reaction pathways for synthesizing molecules such as phenylpyruvate, α-keto-β-hydroxybutyrate, and imidazolepyruvate do not occur in animal cells, phenylalanine, threonine, and histidine must be provided in the diet. (Reaction pathways that synthesize amino

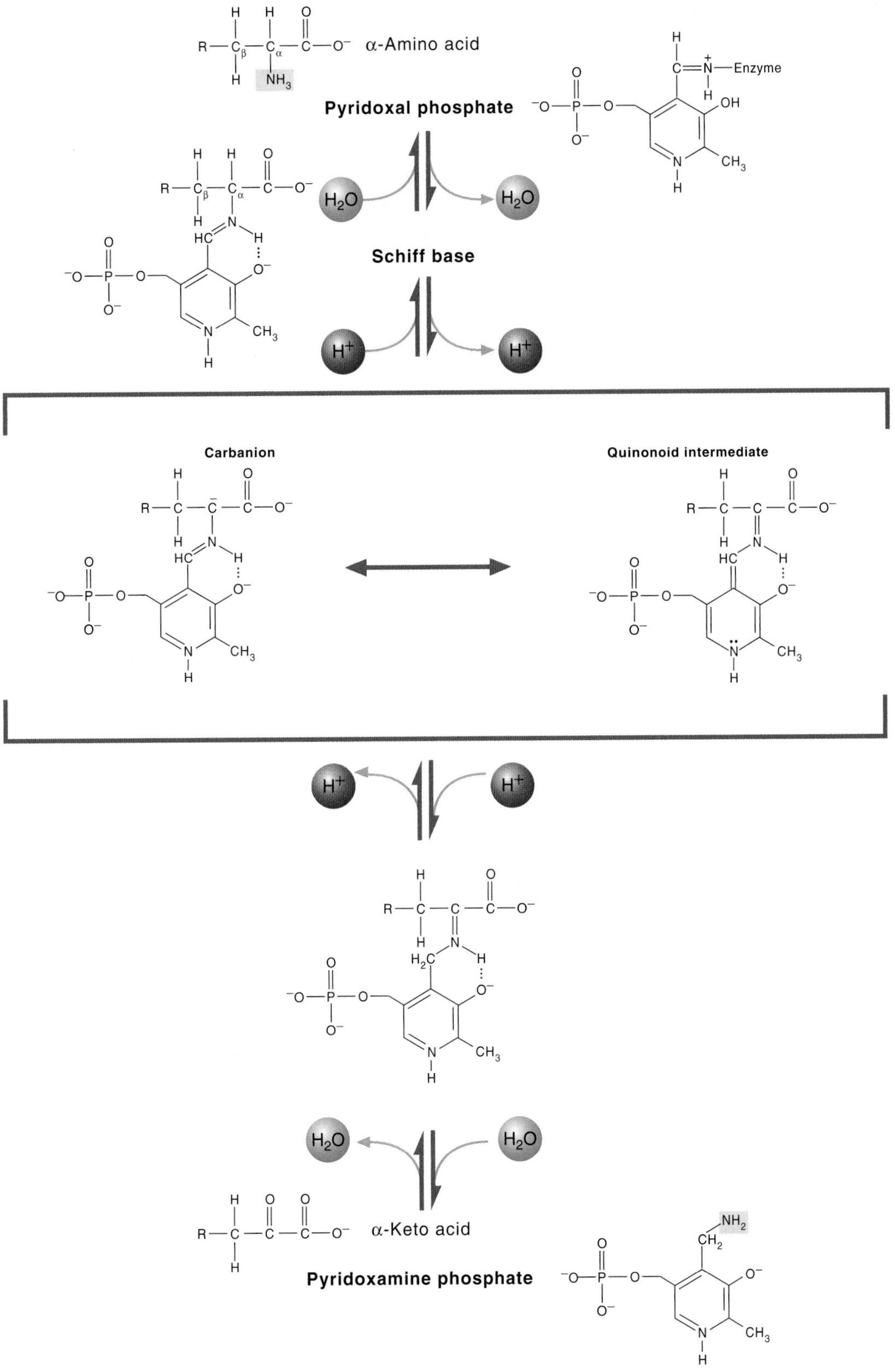

FIGURE 14.4

The Transamination Mechanism

The donor amino acid forms a Schiff base with pyridoxal phosphate within the enzyme's active site. A proton is lost, and a carbanion forms and is resonance-stabilized by interconversion to a quinonoid intermediate. After an enzyme-catalyzed proton transfer and a hydrolysis, the α-keto product is released. A second α-keto acid then enters the active site. This acceptor α-keto acid is converted to an α-amino acid product as the mechanism just described is reversed. Note that the chirality of the donor amino acid is preserved in the α-amino acid product. Within the active site, the orientation of the quinonoid intermediate allows the proton to be added in a manner that confers on the resulting product, the Schiff base, an L-configuration.

acids from metabolic intermediates, and not by transamination, are referred to as de novo pathways.)

DIRECT INCORPORATION OF AMMONIUM IONS INTO ORGANIC MOLECULES There are two principal means by which ammonium ions are incorporated into amino acids and eventually other metabolites: (1) reductive amination of α-keto acids and (2) formation of the amides of aspartic and glutamic acid with subsequent transfer of the amide nitrogen to form other amino acids.

Glutamate dehydrogenase, an enzyme found in both the mitochondria and cytoplasm of eukaryotic cells and in some bacterial cells, catalyzes the direct amination of α-ketoglutarate:

$$^-O-\overset{\overset{O}{\parallel}}{C}-CH_2-CH_2-\overset{\overset{O}{\parallel}}{C}-\overset{\overset{O}{\parallel}}{C}-O^- \;+\; NH_4^+ \;+\; NADH \;+\; H^+ \;\rightleftharpoons$$

α-Ketoglutarate

$$^-O-\overset{\overset{O}{\parallel}}{C}-CH_2-CH_2-\underset{\underset{+NH_3}{|}}{\overset{\overset{H}{|}}{C}}-\overset{\overset{O}{\parallel}}{C}-O^- \;+\; NAD^+ \;+\; H_2O$$

Glutamate

The primary function of this enzyme in eukaryotes appears to be catabolic (i.e., a means of producing NH_4^+ in preparation for nitrogen excretion). However, the reaction is reversible. When excess ammonia is present, the reaction is driven toward glutamate synthesis.

Ammonium ions are also incorporated into cell metabolites by the formation of glutamine, the amide of glutamate:

$$^-O-\overset{\overset{O}{\parallel}}{C}-CH_2-CH_2-\underset{\underset{+NH_3}{|}}{\overset{\overset{H}{|}}{C}}-\overset{\overset{O}{\parallel}}{C}-O^- \;+\; ATP \;+\; NH_4^+ \;\longrightarrow$$

Glutamate

$$H_2N-\overset{\overset{O}{\parallel}}{C}-CH_2-CH_2-\underset{\underset{+NH_3}{|}}{\overset{\overset{H}{|}}{C}}-\overset{\overset{O}{\parallel}}{C}-O^- \;+\; ADP \;+\; P_i$$

Glutamine

KEY CONCEPTS

- In transamination reactions, amino groups are transferred from one carbon skeleton to another.

- In reductive amination, amino acids are synthesized by the incorporation of free NH_4^+ or the amide nitrogen of glutamine or asparagine into α-keto acids.

- Ammonium ions are also incorporated into cellular metabolites by the amination of glutamate to form glutamine.

The brain, a rich source of the enzyme glutamine synthetase, is especially sensitive to the toxic effects of NH_4^+. Brain cells convert NH_4^+ and glutamate to glutamine, a neutral, nontoxic molecule. Glutamine is then transported to the liver, where the nitrogenous waste is produced.

Synthesis of the Amino Acids

The amino acids differ from other classes of biomolecules in that each member of this class is synthesized by a unique pathway. Despite the tremendous diversity of amino acid synthetic pathways, they have one common feature. The carbon skeleton of each amino acid is derived from a commonly available metabolic intermediate. Thus in animals, all NAA molecules are derivatives of either glycerate-3-phosphate, pyruvate, α-ketoglutarate, or oxaloacetate. Tyrosine, synthesized from the essential amino acid phenylalanine, is an exception to this rule.

On the basis of the similarities in their synthetic pathways, the amino acids can be grouped into six families: glutamate, serine, aspartate, pyruvate, the aromatics, and histidine (Figure 14.5). The amino acids in each family are ultimately derived from one precursor molecule. In the discussions of amino acid synthesis that follow, the intimate relationship between amino acid metabolism and several other metabolic pathways is apparent. Amino acid biosynthesis is outlined in Figure 14.6.

THE GLUTAMATE FAMILY The glutamate family includes—in addition to glutamate—glutamine, proline, and arginine. As described, α-ketoglutarate may be converted to glutamate by reductive amination and by transamination reactions

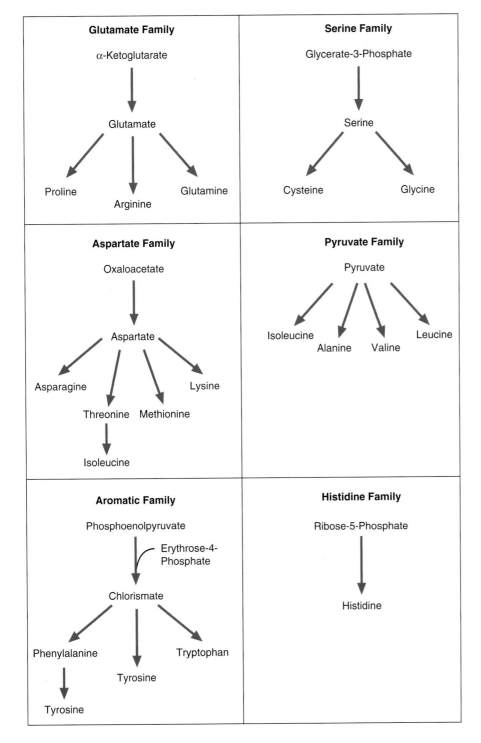

FIGURE 14.5

The Amino Acid Biosynthetic Families

Each family of amino acids is derived from a common precursor molecule.

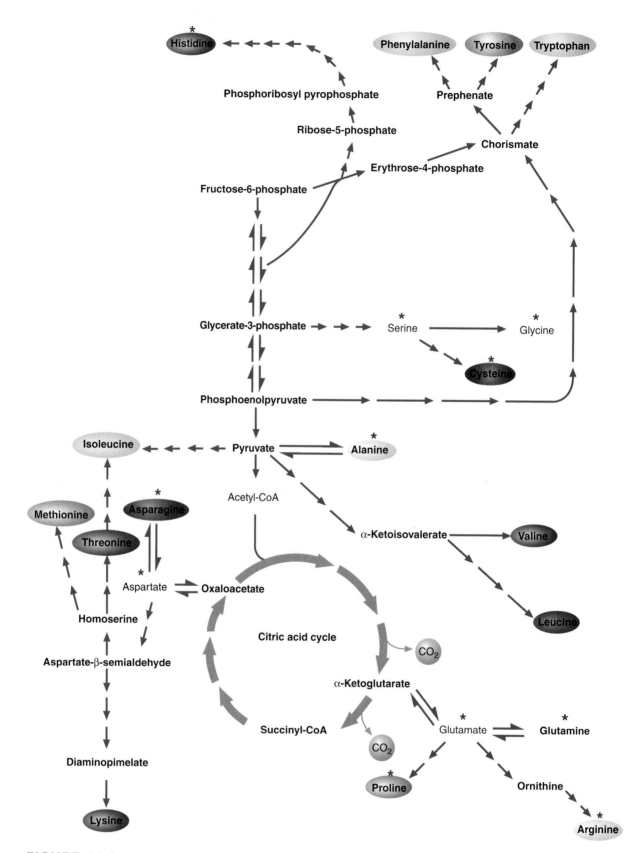

FIGURE 14.6

Biosynthesis of the Amino Acids

Intermediates in the central metabolic pathways provide the carbon skeleton precursor molecules required for the synthesis of each amino acid. The number of reactions in each pathway is indicated. The nonessential amino acids for mammals are indicated by asterisks. (In mammals, tyrosine can be synthesized from phenylalanine.)

involving a number of amino acids. Although the relative contribution of these reactions to glutamate synthesis varies with cell type and metabolic circumstances, transamination appears to play a major role in the synthesis of most glutamate molecules in eukaryotic cells.

The conversion of glutamate to glutamine, catalyzed by glutamine synthetase, takes place in a number of mammalian tissues (liver, brain, kidney, muscle, and intestine). BCAA (branched chain amino acids) are an important source of amino groups in glutamine synthesis. As mentioned, blood that leaves the liver is selectively enriched in BCAA. Many more BCAA are taken up by peripheral tissues than are needed for protein synthesis. The amino groups of the BCAA may be used primarily for the synthesis of nonessential amino acids. In addition to its role in protein synthesis, glutamine is the amino group donor in numerous biosynthetic reactions (e.g., purine, pyrimidine, and amino sugar syntheses) and, as previously mentioned, as a safe storage and transport form of NH_4^+. Glutamine is therefore a major metabolite in living organisms. Other functions of glutamine vary, depending on the cell type being considered. For example, in the kidney and small intestine, glutamine is a major source of energy. In the small intestine, approximately 55% of glutamine carbon is oxidized to CO_2.

Proline is a cyclized derivative of glutamate. As shown in Figure 14.7, a γ-glutamyl phosphate intermediate is reduced to glutamate-γ-semialdehyde. The enzyme catalyzing the phosphorylation of glutamate (γ-glutamyl kinase) is regulated by negative feedback inhibition by proline. Glutamate-γ-semialdehyde cyclizes spontaneously to form Δ^1-pyrroline-5-carboxylate. Then Δ^1-pyrroline-5-carboxylate reductase catalyzes the reduction of Δ^1-pyrroline-5-carboxylate to form proline. The interconversion of Δ^1-pyrroline-5-carboxylate and proline may act as a shuttle mechanism to transfer reducing equivalents derived from the pentose phosphate pathway into mitochondria. This process may partially explain the high turnover of proline in many cell types. Proline can also be synthesized from ornithine, a urea cycle intermediate. (The urea cycle, discussed in Chapter 15, is a pathway in which urea, the principal mammalian nitrogenous waste product, is produced.) The enzyme catalyzing ornithine's conversion to glutamate-γ-semialdehyde, ornithine aminotransferase, is found in relatively high concentration in fibroblasts where the demand for proline incorporation into collagen is high.

Glutamate is also a precursor of arginine. Arginine synthesis begins with the acetylation of the α-amino group of glutamate. *N*-Acetylglutamate is then converted to ornithine in a series of reactions that include a phosphorylation, a reduction, a transamination, and a deacetylation (removal of an acetyl group). The subsequent reactions in which ornithine is converted to arginine are part of the urea cycle. In infants, in whom the urea cycle is insufficiently functional, arginine is an essential amino acid.

THE SERINE FAMILY The members of the serine family—serine, glycine, and cysteine—derive their carbon skeletons from the glycolytic intermediate glycerate-3-phosphate. The members of this group play important roles in numerous anabolic pathways. Serine is a precursor of ethanolamine and sphingosine. Glycine is used in the purine, porphyrin, and glutathione synthetic pathways. Together, serine and glycine contribute to a series of biosynthetic pathways that are referred to collectively as one-carbon metabolism (discussed in Section 14.3). Cysteine plays a significant role in sulfur metabolism (Chapter 15).

Serine is synthesized in a direct pathway from glycerate-3-phosphate that involves dehydrogenation, transamination, and hydrolysis by a phosphatase (Figure 14.8). Cellular serine concentration controls the pathway through feedback inhibition of phosphoglycerate dehydrogenase and phosphoserine phosphatase. The latter enzyme catalyzes the only irreversible step in the pathway.

The conversion of serine to glycine consists of a single complex reaction catalyzed by serine hydroxymethyltransferase, a pyridoxal phosphate–requiring enzyme. During the reaction, which is an aldol cleavage, serine binds to pyridoxal

(Continued on p. 518)

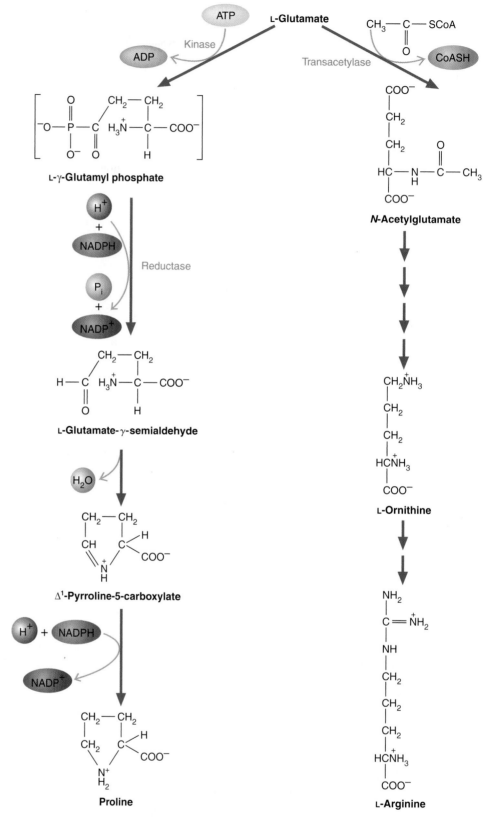

FIGURE 14.7

Biosynthesis of Proline and Arginine from Glutamate

Proline is synthesized from glutamate in three steps. The second step is a spontaneous cyclization reaction. In arginine synthesis the acetylation of glutamate prevents the cyclization reaction. In mammals the reactions that convert ornithine to arginine are part of the urea cycle.

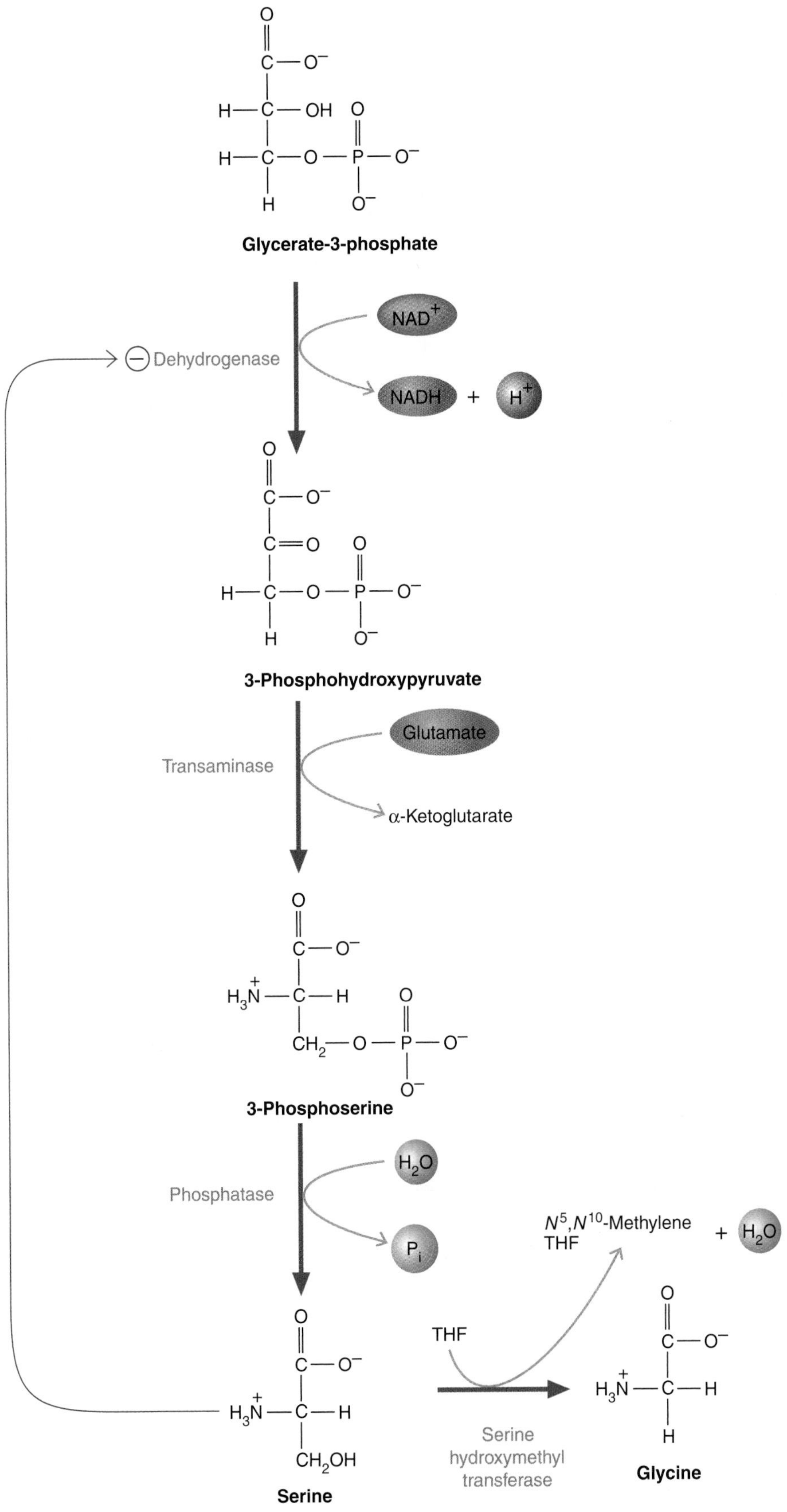

FIGURE 14.8
Biosynthesis of Serine and Glycine
Serine inhibits glycerate-3-phosphate-dehydrogenase, the first reaction in the pathway.

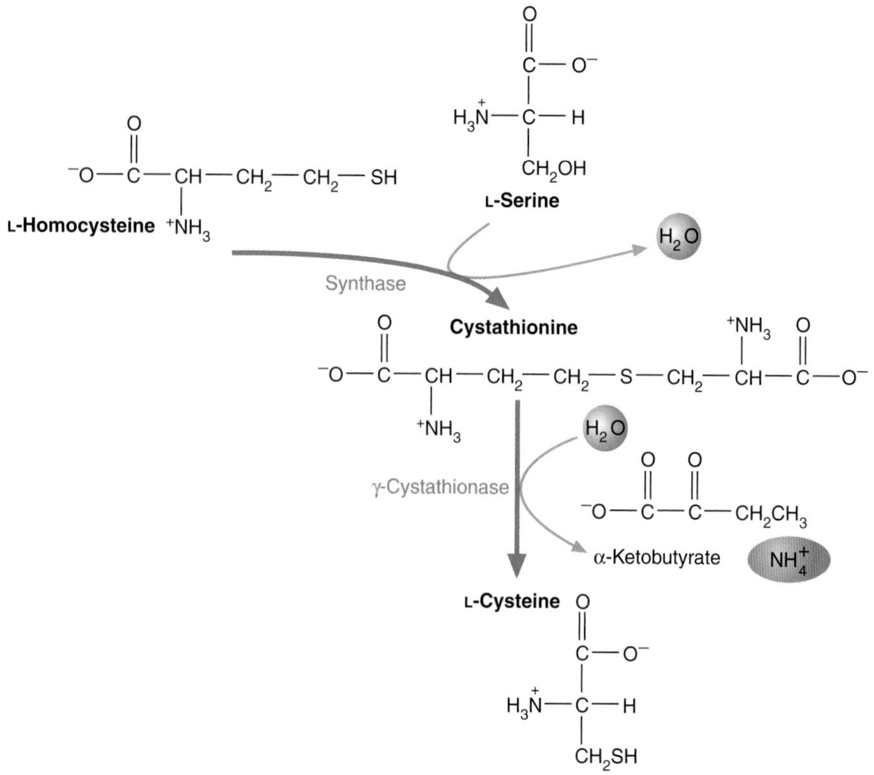

FIGURE 14.9

Biosynthesis of Cysteine

In animals, serine condenses with homocysteine (derived from methionine) to form cystathionine in a reaction catalyzed by cystathionine β-synthase (CBS). Cystathionine γ-lyase (CGL) catalyzes the cleavage of cystathionine to yield cysteine, α-ketobutyrate, and NH_4^+.

phosphate. The reaction yields glycine and a chemically reactive formaldehyde group that is transferred to tetrahydrofolate (THF) to form N^5,N^{10}-methylene tetrahydrofolate. (The coenzyme tetrahydrofolate is discussed in Section 14.3.) Serine is the major source of glycine. Smaller amounts of glycine can be derived from choline when the latter molecule is present in excess. The synthesis of glycine from choline consists of two dehydrogenations and a series of demethylations.

Cysteine synthesis is a primary component of sulfur metabolism. The carbon skeleton of cysteine is derived from serine (Figure 14.9). In animals the sulfhydryl group is transferred from methionine by way of the intermediate molecule homocysteine. Both enzymes involved in the conversion of serine to cysteine (cystathionine β-synthase, or CBS, and cystathionine γ-lyase, or CGL) require pyridoxal phosphate.

THE ASPARTATE FAMILY Aspartate, the first member of the aspartate family of amino acids, is derived from oxaloacetate in a transamination reaction:

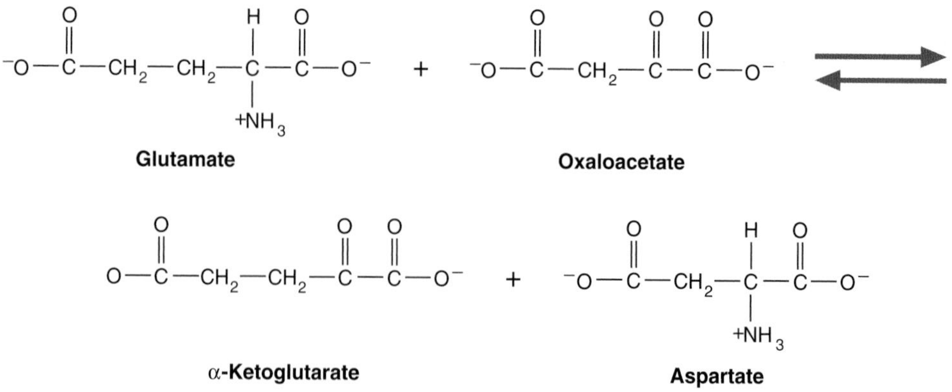

Aspartate transaminase (AST) (also known as glutamic oxaloacetic transaminase, or GOT), the most active of the aminotransferases, is found in most cells. AST isozymes occur in both mitochondria and the cytoplasm, and the reaction that it catalyzes is reversible. This enzymatic activity, therefore, significantly influences the flow of carbon and nitrogen within the cell. For example, excess glutamate is converted via AST to aspartate. Aspartate is then used as a source of both nitrogen (for urea formation) and the citric acid cycle intermediate oxaloacetate. Aspartate is also an important precursor in nucleotide synthesis.

The aspartate family also contains asparagine, lysine, methionine, and threonine. Threonine contributes to the reaction pathway in which isoleucine is synthesized. The synthesis of isoleucine, often considered to be a member of the pyruvate family, is discussed on p. 521.

Asparagine, the amide of aspartate, is not formed directly from aspartate and NH_4^+. Instead, the amide group of glutamine is transferred by amide group transfer during an ATP-requiring reaction catalyzed by asparagine synthase:

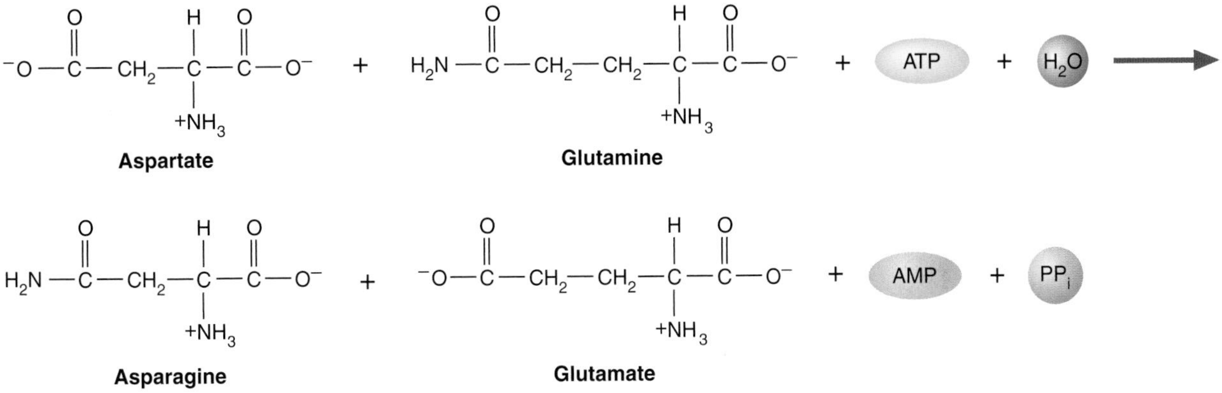

The synthesis of the other members of the aspartate family (Figure 14.10) is initiated by aspartate kinase (often referred to as aspartokinase) in an ATP-requiring reaction in which the side chain carboxyl group is phosphorylated. Aspartate β-semialdehyde, produced by the NADPH-dependent reduction of β-aspartylphosphate, represents an important branch point in plant and bacterial amino acid synthesis. The semialdehyde can either react with pyruvate to form dihydropicolinic acid (a precursor of lysine) or be reduced to homoserine. Lysine is synthesized from dihydropicolinic acid in a series of reactions that is still poorly characterized. Homoserine also occurs at a branch point. It is the precursor in the synthesis of both methionine and threonine.

THE PYRUVATE FAMILY The pyruvate family consists of alanine, valine, leucine, and isoleucine. Alanine is synthesized from pyruvate in a single step:

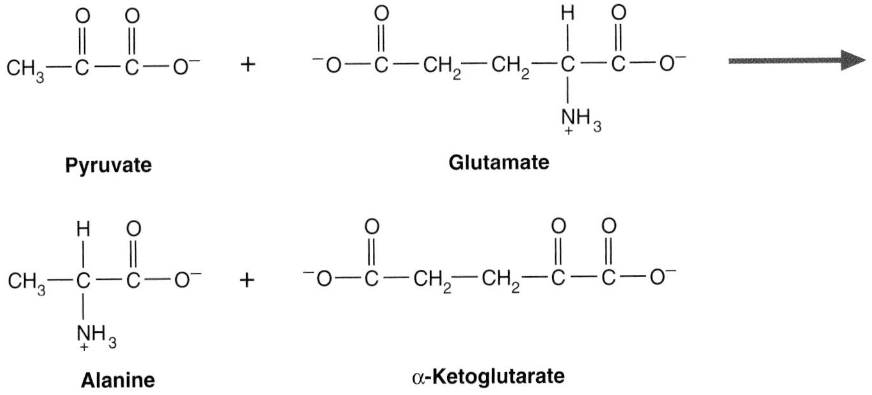

Although the enzyme that catalyzes this reaction, alanine aminotransferase, has cytoplasmic and mitochondrial forms, the majority of its activity has been found in the cytoplasm. Recall that the alanine cycle (Chapter 8) contributes to the maintenance of blood glucose. BCAA are the ultimate source of many of the amino groups transferred from glutamate in the alanine cycle (Figure 8.11).

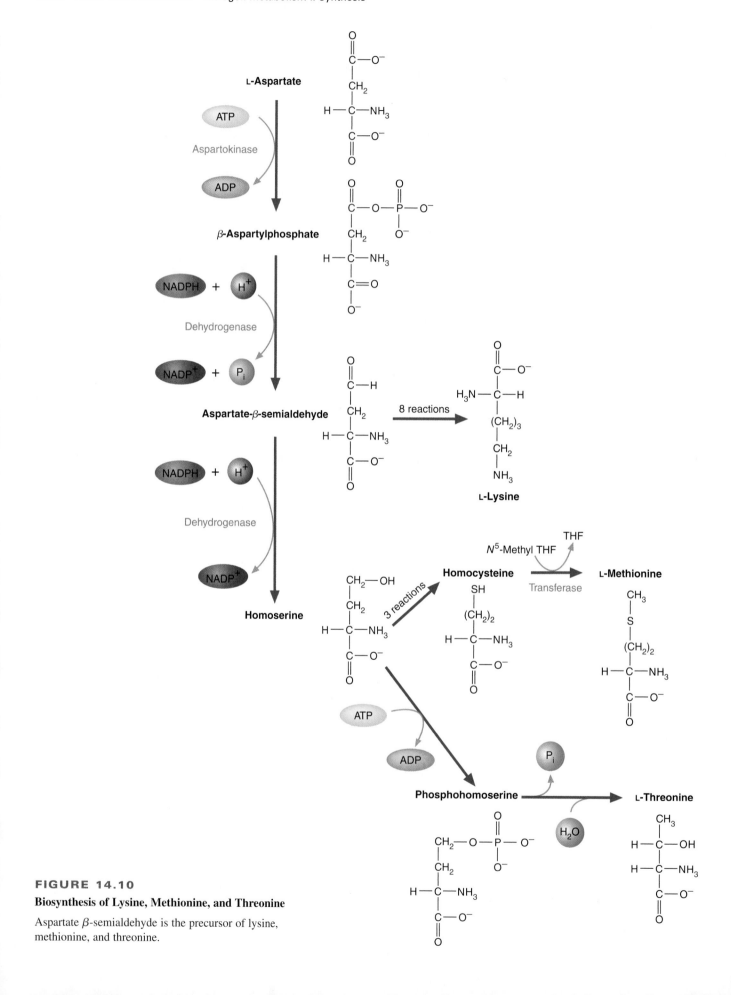

FIGURE 14.10

Biosynthesis of Lysine, Methionine, and Threonine

Aspartate β-semialdehyde is the precursor of lysine, methionine, and threonine.

The syntheses of valine, leucine, and isoleucine from pyruvate are illustrated in Figure 14.11. Valine and isoleucine are synthesized in parallel pathways with the same four enzymes. Valine synthesis begins with the condensation of pyruvate with hydroxyethyl-TPP (a decarboxylation product of a pyruvate-thiamine pyrophosphate intermediate) catalyzed by acetohydroxy acid synthase. The α-acetolactate product is then reduced to form α,β-dihydroxyisovalerate followed by a dehydration to α-ketoisovalerate. Valine is produced in a subsequent transamination reaction from α-ketoisovalerate which is also a precursor of leucine.

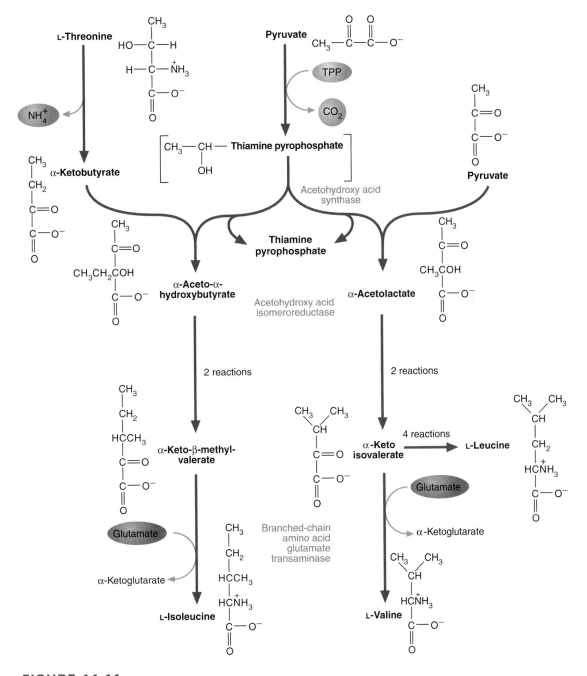

FIGURE 14.11

Biosynthesis of Valine, Leucine, and Isoleucine

The valine and isoleucine biosynthetic pathways share four enzymes. Isoleucine synthesis begins with the reaction of α-ketobutyrate (a derivative of threonine) with pyruvate. In valine synthesis the condensation of two pyruvate molecules is the first step. Leucine is produced by a series of reactions that begins with α-ketoisovalerate, an intermediate in valine synthesis.

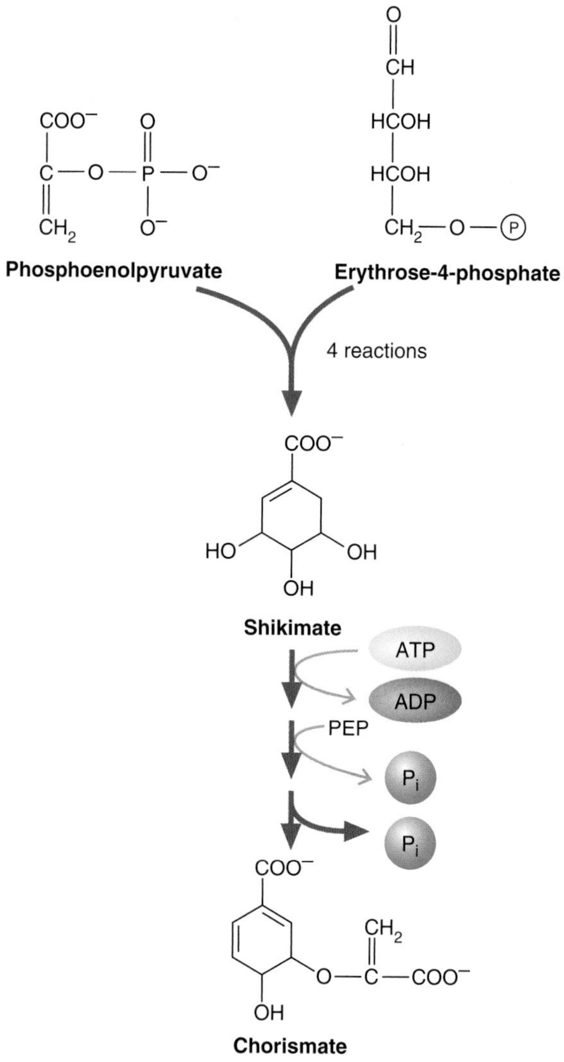

FIGURE 14.12

Chorismate Biosynthesis

Chorismate is an intermediate in the shikimate pathway. The formation of chorismate involves the ring closure of an intermediate (not shown) and the subsequent creation of two double bonds. The side chain of chorismate is derived from phosphoenolpyruvate (PEP).

Isoleucine synthesis also involves hydroxyethyl-TPP, which condenses with α-ketobutyrate to form α-aceto-α-hydroxybutyrate. α-Ketobutyrate is derived from L-threonine in a deamination reaction catalyzed by threonine deaminase. α-Aceto-α-hydroxybutyrate undergoes a reduction reaction and subsequently loses an H_2O molecule, thus forming α-keto-β-methylvalerate. Isoleucine is then produced during a transamination reaction. In the first step of leucine biosynthesis from α-ketoisovalerate, acetyl-CoA donates a two-carbon unit. Leucine is formed after isomerization, reduction, and transamination reactions.

THE AROMATIC FAMILY The aromatic family of amino acids includes phenylalanine, tyrosine, and tryptophan. Of these, only tyrosine is considered to be nonessential in mammals. Either phenylalanine or tyrosine is required for the synthesis of dopamine, epinephrine, and norepinephrine, an important class of biologically potent molecules referred to as the *catecholamines* (pp. 534–535). Tryptophan is a precursor in the synthesis of NAD^+, $NADP^+$, and the neurotransmitter serotonin.

The benzene ring of the aromatic amino acids is formed by the *shikimate pathway* (Figure 14.12). The carbons in the benzene ring are derived from erythrose-4-phosphate and phosphoenolpyruvate. These two molecules condense to form a molecule that is subsequently converted to chorismate. Chorismate is the branch point in the syntheses of various aromatic compounds.

Figure 14.13 illustrates the syntheses of phenylalanine, tyrosine, and tryptophan from chorismate. (Chorismate is also a precursor in the synthesis of the

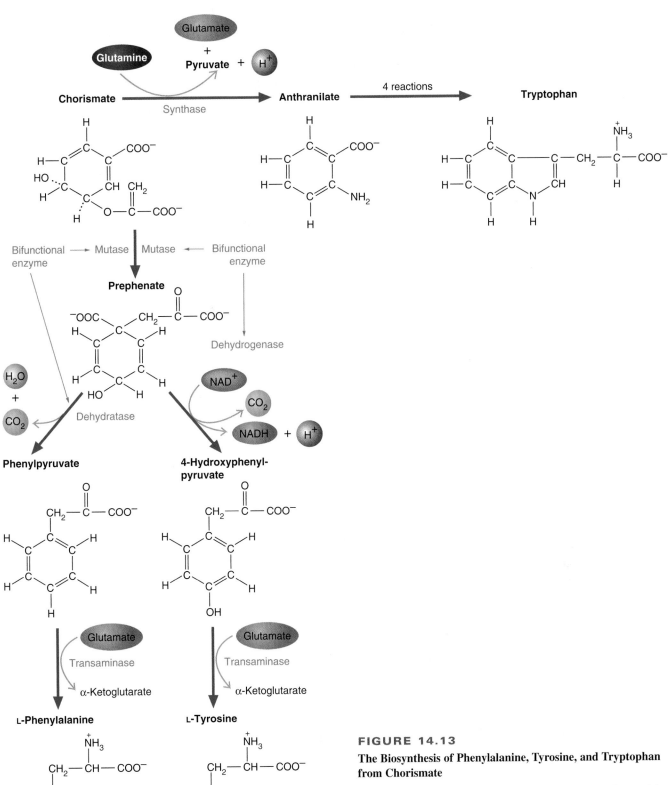

FIGURE 14.13

The Biosynthesis of Phenylalanine, Tyrosine, and Tryptophan from Chorismate

Chorismate is converted to prephenate (the precursor of phenylalanine and tyrosine) and anthranilate (the precursor of tryptophan). Chorismate can also be converted to 4-hydroxybenzoic acid, the precursor of the ubiquinones. 4-Hydroxyphenylpyruvate is also a precursor in the synthesis of plastoquinone and various tocopherols.

aromatic rings in the mixed terpenoids, e.g., the tocopherols, the ubiquinones, and plastoquinone.)

Tyrosine is not an essential amino acid in animals because it is synthesized from phenylalanine in a hydroxylation reaction. The enzyme involved, phenyl-alanine-4-monooxygenase, requires the coenzyme tetrahydrobiopterin (Section 14.3), a folic acid–like molecule derived from GTP. Because this reaction also is a first step in phenylalanine catabolism, it is discussed further in Chapter 15.

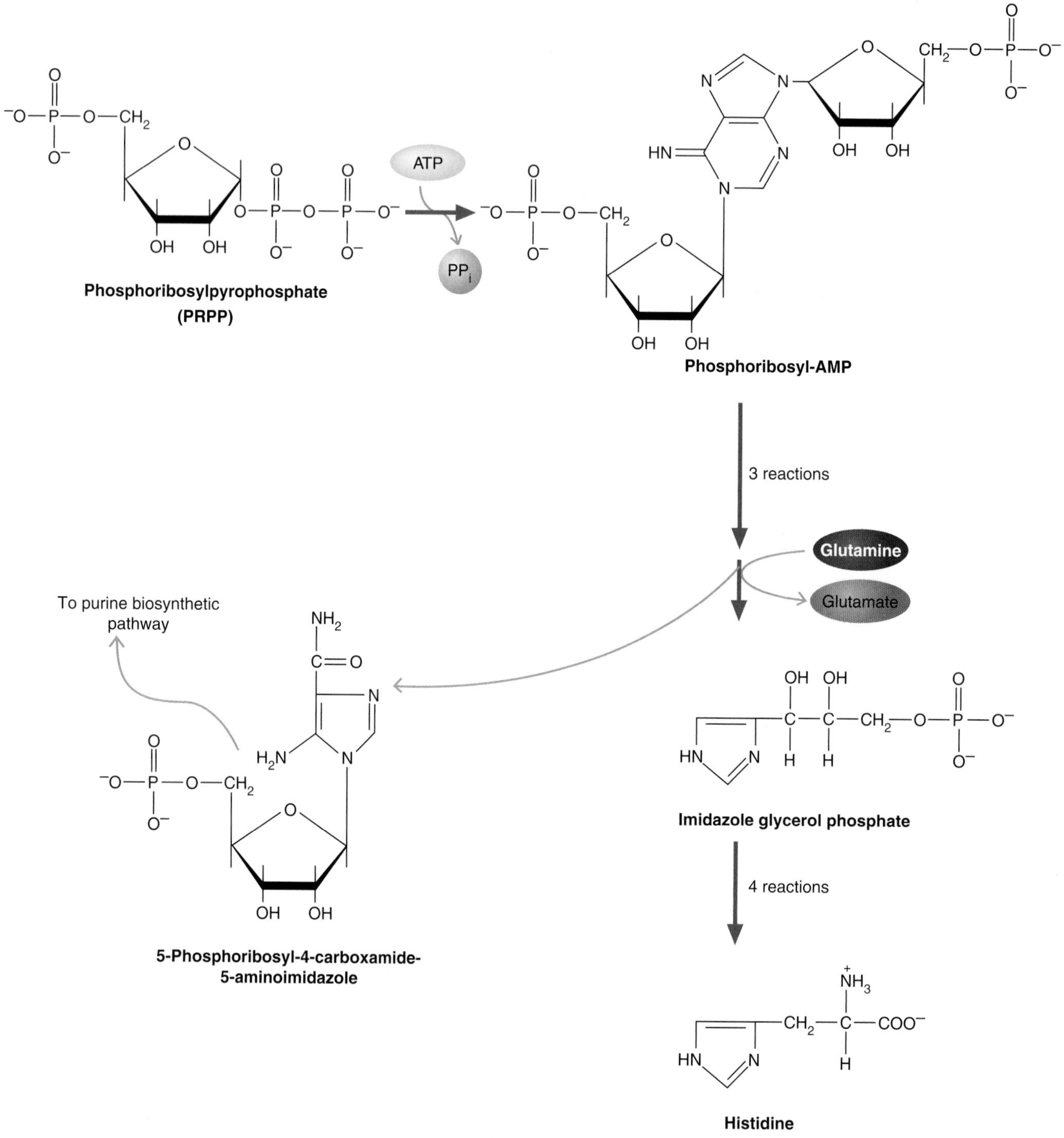

FIGURE 14.14

Histidine Biosynthesis

Histidine is derived from three biomolecules: PRPP (five carbons), the adenine ring from ATP (one nitrogen and one carbon), and glutamine (one nitrogen). The ATP used in the first reaction in the pathway is regenerated when 5-phosphoribosyl-4-carboxamide-5-aminoimidazole (released in a subsequent reaction) is diverted into the purine nucleotide biosynthetic pathway.

HISTIDINE Histidine is considered to be nonessential in healthy human adults. In human infants and many animals, histidine must be provided by the diet. Histidine contributes substantially to protein structure and function because of its unique chemical properties. Recall, for example, that histidine residues bind heme prosthetic groups in hemoglobin. In addition, histidine often acts as a general acid during enzyme-catalyzed reactions. Of all the amino acids, histidine's biosynthesis is the most unusual. Histidine is synthesized from phosphoribosylpyrophosphate (PRPP), ATP, and glutamine (Figure 14.14). Synthesis begins with the condensation of PRPP with ATP to form phosphoribosyl-ATP. Phosphoribosyl-ATP is then hydrolyzed by phosphoribosyl-ATP pyrophosphorylase to phosphoribosyl-AMP. Subsequently, a hydrolysis reaction opens the adenine ring. After an isomerization and the transfer of an amino group from glutamine, imidazole glycerol phosphate is synthesized. (The other product of the latter reaction, 5'-phosphoribosyl-4-carboxamide-5-aminoimidazole, is used in the synthesis of purine nucleotides. See Section 14.3.) Histidine is produced from imidazole glycerol phosphate in a series of reactions that include a dehydration, a transamination, a phosphorolysis, and an oxidation.

KEY CONCEPTS

- There are six families of amino acids: glutamate, serine, aspartate, pyruvate, the aromatics, and histidine.
- The nonessential amino acids are derived from precursor molecules available in many organisms.
- The essential amino acids are synthesized from metabolites produced only in plants and some microorganisms.

14.3 BIOSYNTHETIC REACTIONS INVOLVING AMINO ACIDS

As described, amino acids are precursors of many physiologically important nitrogen-containing molecules, in addition to serving as building blocks for polypeptides. In the following discussion the syntheses of several examples of these molecules (neurotransmitters, glutathione, alkaloids, nucleotides, and heme) are described. Because many of these processes involve the transfer of carbon groups, this section begins with a brief description of one-carbon metabolism.

One-Carbon Metabolism

Carbon atoms have several oxidation states. Those of biological interest are found in methanol ($+1$), formaldehyde ($+2$), and formate ($+3$). Table 14.2 lists the equivalent one-carbon groups that are actually involved in synthetic reactions.

The most important carriers of one-carbon groups in biosynthetic pathways are folic acid and S-adenosylmethionine. The metabolism of each is described briefly. (The function of biotin, a carrier of CO_2 groups, is discussed in Section 8.2.)

FOLIC ACID Folic acid, also known as folate or folacin, is a B vitamin. Its structure consists of a pteridine nucleus and *para*-aminobenzoic acid, linked to one or more glutamic acid residues (Figure 14.15). Once absorbed by the body, folic acid is converted by dihydrofolate reductase to the biologically active form, tetrahydrofolic acid (THF). The carbon units carried by THF (i.e., methyl, methylene, methenyl, and formyl groups) are bound to N^5 of the pteridine ring and/or N^{10} of the para-aminobenzoate ring. Figure 14.16 illustrates the interconversions of the one-carbon units carried by THF, as well as their origin and metabolic fate. A substantial number of one-carbon units enter the THF pool as N^5,N^{10}-methylene THF, produced during the conversion of serine to glycine and the cleavage of glycine (catalyzed by glycine synthase).

TABLE 14.2 One-Carbon Groups

Oxidation Level	Methanol (most reduced)	Formaldehyde	Formate (most oxidized)
One-carbon group	Methyl (—CH$_3$)	Methylene (—CH$_2$—)	Formyl (—CHO)
			Methenyl (—CH=)

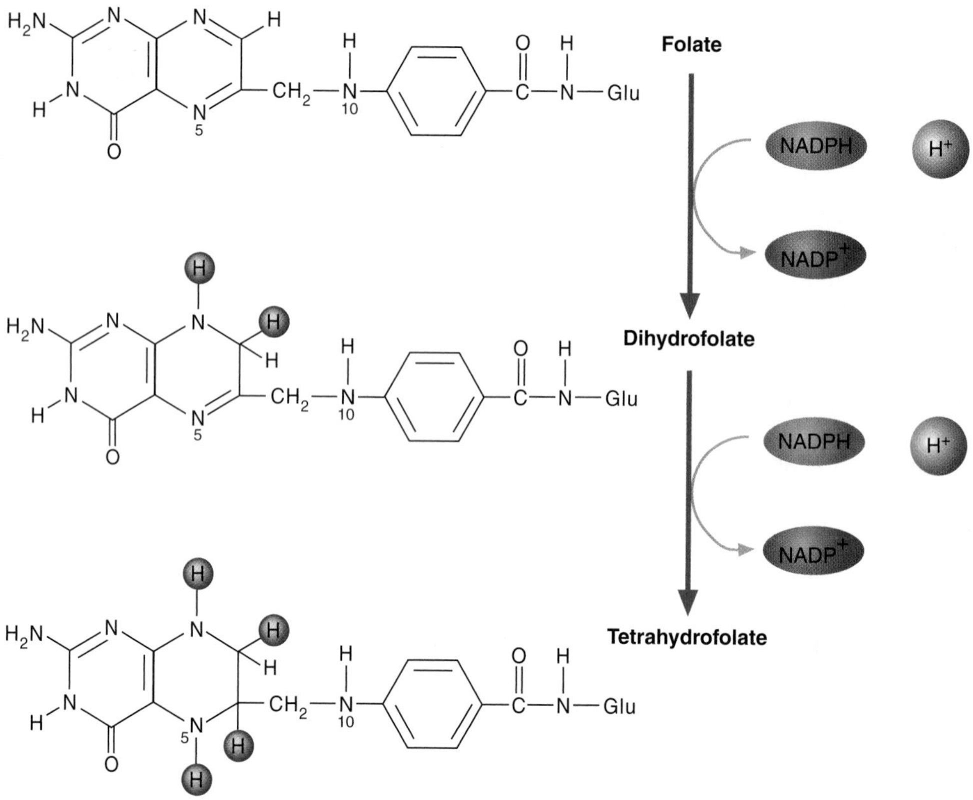

Folate

NADPH   H⁺

NADP⁺

Dihydrofolate

NADPH   H⁺

NADP⁺

Tetrahydrofolate

FIGURE 14.15

Biosynthesis of Tetrahydrofolate (THF)

The vitamin folic acid (folate) is converted to its biologically active form by two successive reductions of the pteridine ring. Both reactions are catalyzed by dihydrofolate reductase.

Methylcobalamin, a coenzyme form of vitamin B_{12}, is required for the N^5-methyl THF–dependent conversion of homocysteine to methionine (Figure 14.16). **Vitamin B_{12}** (cobalamin) is a complex, cobalt-containing molecule synthesized only by microorganisms (Figure 14.17). (During the purification of cobalamin, a cyanide group attaches to cobalt.) Another coenzyme form of vitamin B_{12}, 5′-deoxyadenosylcobalamin, is required for the isomerization of methylmalonyl-CoA to succinyl-CoA, which is catalyzed by methylmalonyl-CoA mutase (refer to Figures 12.10 and 15.10). Animals obtain cobalamin from intestinal flora and by consuming foods derived from other animals (e.g., liver, eggs, shrimp, chicken, and pork). A deficiency of vitamin B_{12} results in **pernicious anemia**. In addition to low red blood cell counts, the symptoms of this malady include weakness and various neurological disturbances. Pernicious anemia is most often caused by decreased secretion of intrinsic factor, a glycoprotein secreted by stomach cells, which is required for the absorption of the vitamin in the intestine. Vitamin B_{12} absorption can also be inhibited by several gastrointestinal disorders, such as celiac disease and tropical sprue, both of which damage the lining of the intestine. A reduction in vitamin B_{12} absorption has also been observed in the presence of intestinal overgrowths of microorganisms induced by antibiotic treatments.

S-ADENOSYLMETHIONINE S-Adenosylmethionine (SAM) is the major methyl group donor in one-carbon metabolism. Formed from methionine and ATP (Figure 14.18), SAM contains an "activated" methyl thioether group, which can be transferred to a variety of acceptor molecules (Table 14.3). S-Adenosylhomocysteine (SAH) is a product in these reactions. The loss of free energy that accompanies S-adenosylhomocysteine formation makes the methyl

(Continued on p. 529)

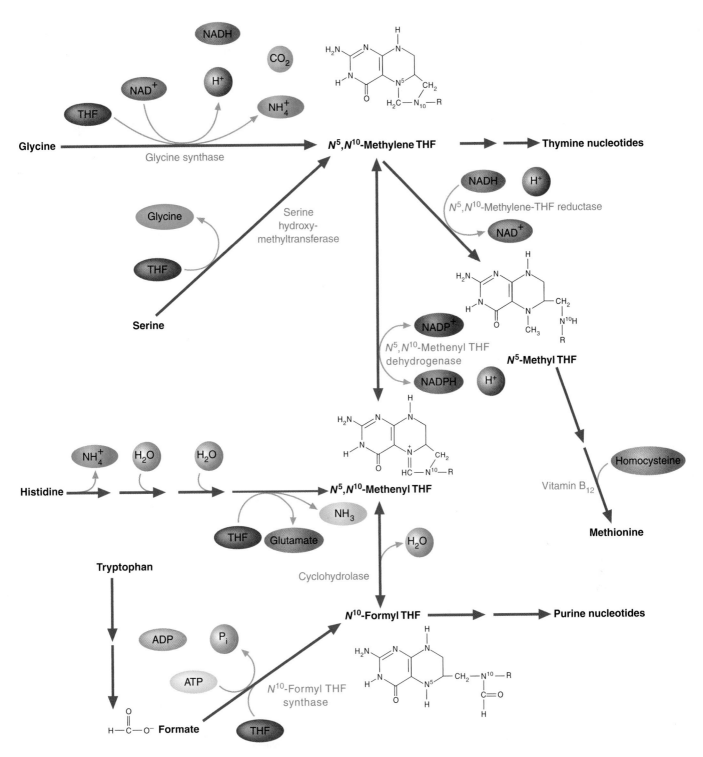

FIGURE 14.16

Structures and Enzymatic Interconversions of THF Coenzymes

The THF coenzymes play a critical role in one-carbon metabolism. The interconversions of the coenzymes are reversible except for the conversion of N^5,N^{10}-methylene THF to N^5-methyl THF.

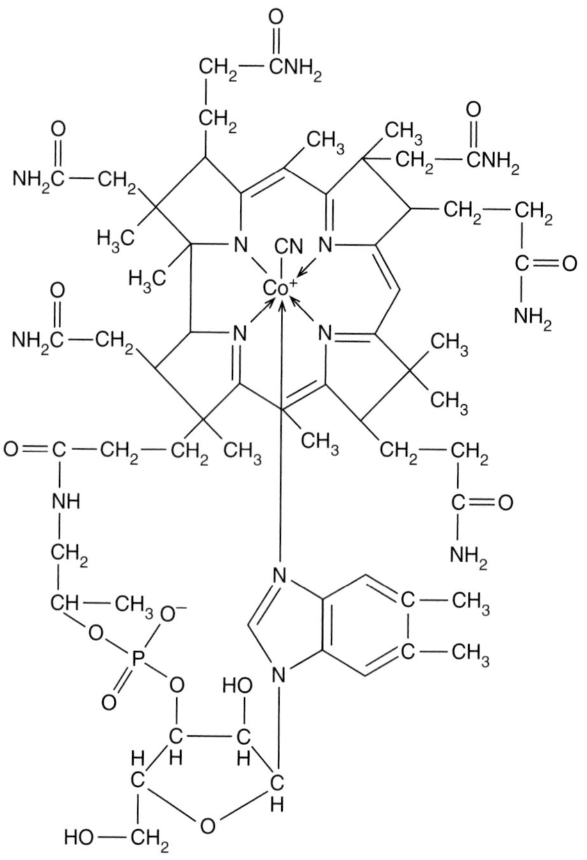

FIGURE 14.17

Structure of Cyanocobalamin, a Derivative of Vitamin B$_{12}$

During the purification of cobalamin, a cyanide group attaches to cobalt.

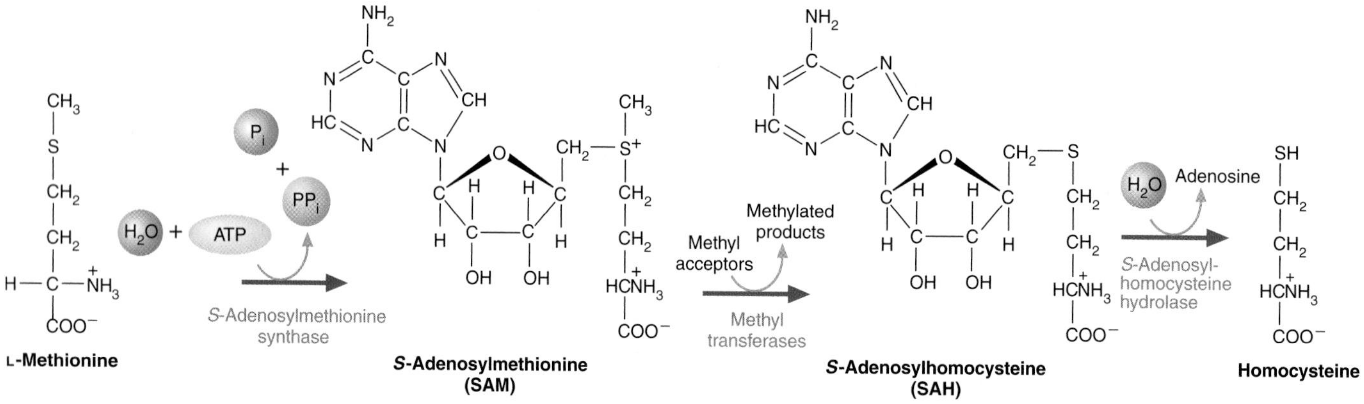

FIGURE 14.18

The Formation of S-Adenosylmethionine

One of the principal functions of SAM is to serve as a methylating agent. The S-adenosylmethionine product of these reactions is then hydrolyzed to form homocysteine.

TABLE 14.3 Examples of Transmethylation Acceptors and Products

Methyl Acceptors	Methylated Product
Phosphatidylethanolamine	Phosphatidylcholine (p. 381)
Norepinephrine	Epinephrine (p. 535)
Guanidinoacetate	Creatine (p. 531)
γ-Aminobutyric acid	Carnitine (p. 420)

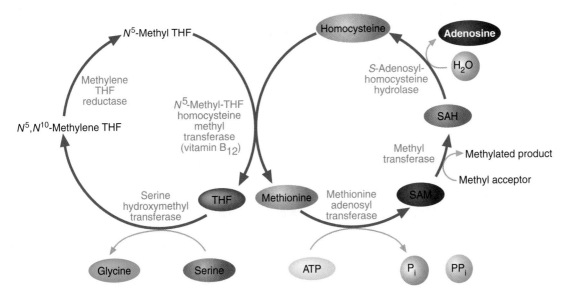

FIGURE 14.19

The Tetrahydrofolate and *S*-Adenosylmethionine Pathways

The THF and SAM pathways intersect at the reaction, catalyzed by N^5-methyl THF homocysteine methyltransferase, in which homocysteine is converted to methionine.

transfer irreversible. SAM acts as a methyl donor in at least 115 transmethylation reactions, some of which occur in the synthesis of phospholipids, several neurotransmitters, and glutathione. *DNA* methylation reactions play a significant role in gene expression.

The importance of SAM in metabolism is reflected in the several mechanisms that provide for the synthesis of sufficient amounts of its precursor, methionine, when the latter molecule is temporarily absent from the diet. For example, choline is used as source of methyl groups to convert homocysteine into methionine. Homocysteine can also be methylated in a reaction utilizing N^5-methyl THF. This latter reaction is a bridge between the THF and SAM pathways (Figure 14.19).

COMPANION

GW

W E B S I T E Visit the companion website at www.oup.com/us/mckee to read the Biochemistry in Perspective box on epigenetics and the epigenome in Chapter 17.

QUESTION 14.5

The following substances are derived from amino acids. For each, provide the name of the amino acid precursor.

$H_3\overset{+}{N}-CH_2-CH_2-CH_2-\overset{\displaystyle O}{\overset{\|}{C}}-O^-$

(a)

$H_3\overset{+}{N}-CH_2-CH_2-$ (indole ring with OH)

(b)

(imidazole ring)
CH_2
CH_2
$+NH_3$

(c)

$H_2N-CH_2-CH_2-$ (benzene ring with OH, OH)

(d)

QUESTION 14.5 (CONT.)

(e)

QUESTION 14.6

Amethopterin, also referred to as **methotrexate**, is a structural analogue of folate. (**Analogues** are compounds that closely resemble other molecules.) Methotrexate has been used to treat several types of cancer and autoimmune diseases. It has been especially successful in childhood leukemia. Autoimmune diseases treated with methotrexate include rheumatoid arthritis, Crohn's disease, and psoriasis (excessive skin cell production and inflammation).

Amethopterin (methotrexate)

Using your knowledge of cell biology and biochemistry, suggest a biochemical mechanism that explains why amethopterin is effective against cancer. [*Hints*: Compare the structures of folate and methotrexate. Review Figure 14.15.] Explain why methotrexate treatment for cancer causes temporary baldness.

QUESTION 14.7

Melatonin is a hormone derived from serotonin. It is produced in the brain's light-sensitive pineal gland. The pineal's secretion of melatonin is depressed by nerve impulses that originate in the retina of the eye and other light-sensitive tissue in the body in response to light. Pineal function is involved in *circadian rhythms*, patterns of activity associated with light and dark, such as sleep/wake cycles. In many mammals the functioning of the pineal gland also regulates seasonal cycles of fertility and infertility. (Melatonin inhibits the secretion of certain hormones from the hypothalamus and pituitary that stimulate the functioning of the ovaries and testes.) For example, in some species (e.g., deer) the males are fertile only in early spring, ensuring that newborn animals will be mature enough to survive the next winter. Melatonin is also a powerful antioxidant, especially in the central nervous system.

After serotonin is produced in the pineal gland, it is converted to 5-hydroxy-*N*-acetyltryptamine by *N*-acetyltransferase. 5-Hydroxy-*N*-acetyltryptamine is then methylated by *O*-methyltransferase. SAM is the methylating agent. With this information, draw the synthetic pathway of melatonin.

Creatine is a nitrogen-containing organic acid found primarily in muscle and brain. Both of these tissues experience large and fluctuating energy demands. Phosphocreatine, the product of the reaction of creatine and ATP, serves as a short-term storage form of high-energy phosphate. (Refer to Table 4.1.) When energy demands are high and available ATP molecules are hydrolyzed, phosphocreatine donates its phosphoryl group to ADP to yield ATP. Most creatine molecules are synthesized in the body in a two-reaction pathway. In the first step arginine and glycine are converted in the kidney to guanidinoacetate and ornithine (p. 516) by L-arginine:glycine amidinotransferase (AGAT).

$$H_2\overset{+}{N}=C\underset{NH-CH_2-\overset{\displaystyle O}{\overset{\|}{C}}-O^-}{\overset{NH_2}{\big<}}$$

Guanidinoacetate

In liver, guanidinoacetate reacts with SAM to form creatine and SAH in a reaction catalyzed by *S*-adenosyl-L-methionine:*N*-guanidinoacetate methyltransferase (GAMT). With the information provided, write out the creatine biosynthetic pathway.

KEY CONCEPT

Tetrahydrofolate, the biologically active form of folic acid, and *S*-adenosylmethionine are important carriers of single carbon atoms in a variety of synthetic reactions.

Glutathione

The nitrogen-containing molecule glutathione (γ-glutamylcysteinylglycine) is the most common intracellular thiol. (Its concentration in mammalian cells varies from 0.5 to 10 M.) The functions of glutathione (GSH) include involvement in DNA and RNA synthesis and in the synthesis of certain eicosanoids and other biomolecules. (In many of these processes, GSH acts as a reducing agent. As such, it maintains the sulfhydryl groups of enzymes and other molecules in a reduced state.) In addition to protecting cells from radiation, oxygen toxicity, and environmental toxins, GSH also promotes amino acid transport. After a brief discussion of its synthesis, the transport role of GSH is described. This is followed by a discussion of an interesting class of enzymes called the glutathione-*S*-transferases.

GSH is synthesized in a pathway composed of two reactions. In the first reaction, γ-glutamylcysteine synthase catalyzes the condensation of glutamate with cysteine (Figure 14.20). γ-Glutamylcysteine, the product of this reaction, then combines with glycine to form GSH in a reaction catalyzed by glutathione synthase.

TRANSPORT Transport of GSH out of cells appears to serve several functions. Among these are (1) transfer of the sulfur atoms of cysteine between cells, (2) protection of the plasma membrane from oxidative damage, and (3) transfer to membrane-bound γ-glutamyl transpeptidase leading to the formation of γ-glutamyl amino acid derivatives. The latter process, which initiates the γ-glutamyl cycle, occurs in brain, intestine, pancreas, liver, and kidney cells.

In addition to maintaining cellular GSH levels, the γ-glutamyl cycle has been observed to stimulate the Na^+-dependent transport of amino acids, especially in kidney cells and lactating mammary gland cells. It also contributes to the regulation of amino acid transport across the blood-brain barrier. (The *blood-brain barrier*, which consists of connective tissue and specialized capillary endothelial cells, protects the brain from toxic substances by regulating the passage of substances from the blood into the central nervous system.)

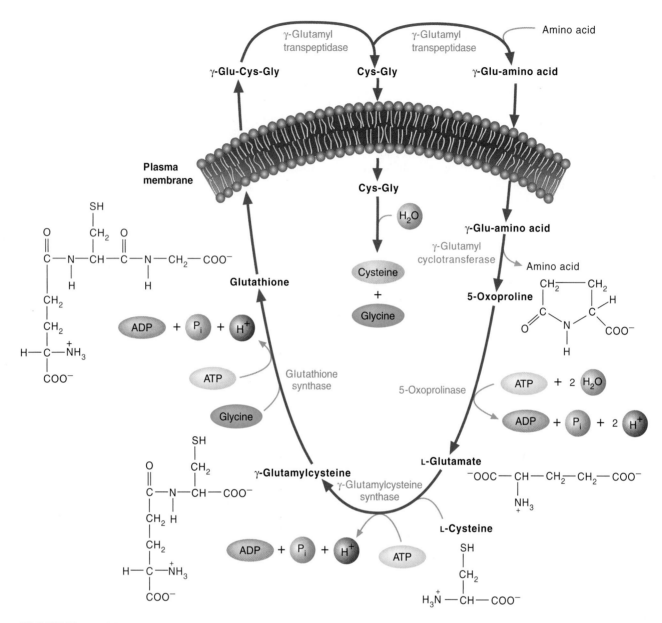

FIGURE 14.20

The γ-Glutamyl Cycle

The steps in the γ-glutamyl cycle are as follows. (1) Glutathione is excreted from the cell. γ-Glutamyltranspeptidase converts GSH to γ-Glu-amino acid derivative and Cys-Gly. (2) The γ-Glu-amino acid is transported into the cell, where it is converted to 5-oxoproline and the free amino acid. 5-Oxoproline is eventually reconverted to glutamate, a component of GSH. (3) Cys-Gly is transported into the cell, where (4) it is hydrolyzed to cysteine and glycine. The functions of the γ-glutamyl cycle are described in the text. (5) GSH is produced in two reactions catalyzed by γ-glutamyl cysteine synthase and glutathione synthase.

KEY CONCEPTS

- Glutathione (GSH), the most common intercellular thiol, is involved in many cellular activities.
- In addition to reducing sulfhydryl groups, GSH protects cells against toxins and promotes the transport of some amino acids.

GLUTATHIONE-S-TRANSFERASES As mentioned, GSH contributes to the protection of cells from environmental toxins. GSH does so by reacting with a large variety of foreign molecules to form GSH conjugates (Figure 14.21). The bonding of these substances with GSH, which prepares them for excretion, may be spontaneous, or it may be catalyzed by the GSH-S-transferases (also known as the *ligandins*). Before their excretion in urine, GSH conjugates are usually converted to mercapturic acids by a series of reactions initiated by γ-glutamyltransferase.

Neurotransmitters

More than 30 different substances, including several amino acids, have been proven or proposed to act as neurotransmitters. **Neurotransmitters**, signal molecules released from neurons, are either excitatory or inhibitory. Excitatory neurotransmitters (e.g., glutamate and acetylcholine) open sodium channels and promote the depolarization of the membrane in another cell (either another neuron or an effector cell, such as a muscle cell). If the second (postsynaptic) cell is a neuron, the wave of depolarization (referred to as an action potential) triggers the release of neurotransmitter molecules as it reaches the end of the axon. (Most neurotransmitter molecules are stored in numerous membrane-enclosed *synaptic vesicles*.) When the action potential reaches the nerve ending, the neurotransmitter molecules are released by exocytosis into the synapse. If the postsynaptic cell is a muscle cell, sufficient release of excitatory neurotransmitter molecules results in muscle contraction. Inhibitory neurotransmitters (e.g., glycine) open chloride channels and make the membrane potential in the postsynaptic cell even more negative, that is, they inhibit the formation of an action potential.

A significant percentage of neurotransmitter molecules are either amino acids or amino acid derivatives (Table 14.4). The latter class, referred to as the **biogenic amines**, includes γ-aminobutyric acid, the catecholamines, serotonin, and histamine.

γ-AMINOBUTYRIC ACID γ-*Aminobutyric acid* (GABA) acts as an inhibitory neurotransmitter in the central nervous system. The binding of GABA to its receptor opens certain channels, which results in the inward flow of Cl^- or the outward flow of K^+. (The benzodiazepines, a class of tranquilizers that alleviate anxiety and aggressive behavior, have been shown to enhance GABA's ability to increase membrane conductance of chloride.)

GABA is produced by the decarboxylation of glutamate. The reaction is catalyzed by glutamate decarboxylase, which is a pyridoxal phosphate–requiring enzyme:

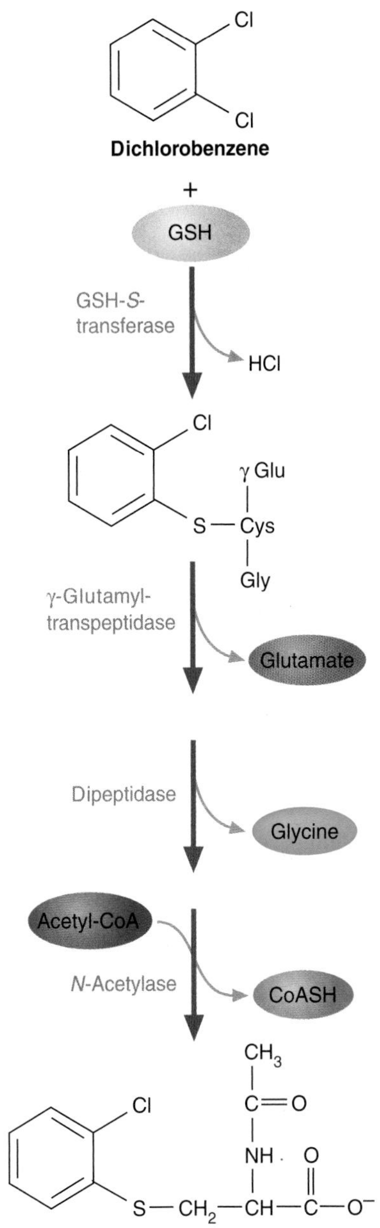

FIGURE 14.21

Formation of a Mercapturic Acid Derivative of a Typical Organic Contaminant

GSH-*S*-transferase catalyzes the synthesis of a GSH derivative of dichlorobenzene.

TABLE 14.4 Amino Acid and Amine Neurotransmitters

Amino Acids	Amines
Glycine	Norepinephrine*
Glutamate	Epinephrine*
γ-Aminobutyric acid (GABA)	Dopamine*
	Serotonin
	Histamine

*A catecholamine.

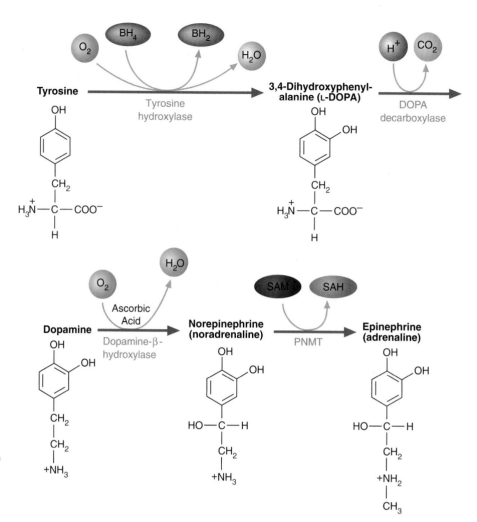

FIGURE 14.22

Biosynthesis of the Catecholamines

Dopamine, norepinephrine, and epinephrine act as neurotransmitters and/or hormones. (PNMT = phenylethanolamine-N-methyl-transferase)

THE CATECHOLAMINES The *catecholamines* (dopamine, norepinephrine, and epinephrine) are derivatives of tyrosine. Dopamine (D) and norepinephrine (NE) are used in the brain as excitatory neurotransmitters. Outside the central nervous system, NE and epinephrine (E) are released primarily from the adrenal medulla, as well as the peripheral nervous system. Because both NE and E regulate aspects of metabolism, they are often considered to be hormones.

The first, and rate-limiting, step in catecholamine synthesis is the hydroxylation of tyrosine to form 3,4-dihydroxyphenylalanine (DOPA) (Figure 14.22). Tyrosine hydroxylase, the mitochondrial enzyme that catalyzes the reaction, requires a cofactor known as *tetrahydrobiopterin* (BH_4). A folic acid–like molecule, BH_4 is an essential cofactor in the hydroxylation of aromatic amino acids; it is regenerated from its oxidized metabolite, BH_2, by reduction with NADPH.)

COMPANION

WEBSITE **Visit the companion website at www.oup.com/us/mckee to read the Biochemistry in Perspective box on Parkinson's disease and dopamine.**

Tyrosine hydroxylase uses BH_4 to activate O_2. One oxygen atom is attached to tyrosine's aromatic ring, while the other atom oxidizes the coenzyme. DOPA, the product of the reaction, is used in the synthesis of the other catecholamines.

DOPA decarboxylase, a pyridoxal phosphate–requiring enzyme, catalyzes the synthesis of dopamine from DOPA. Dopamine is produced in neurons found in certain structures in the brain. It is believed to exert an inhibitory action in the central nervous system. Deficiency in dopamine production has been found to be associated with Parkinson's disease, a serious degenerative neurological disorder. The precursor L-DOPA is used to alleviate the symptoms of Parkinson's disease because dopamine cannot penetrate the blood-brain barrier. (Although most lipid-soluble substances readily pass across the blood-brain barrier, many polar molecules and ions cannot move from blood capillaries.) Once L-DOPA has been transported into appropriate nerve cells, it is converted to dopamine.

Norepinephrine is synthesized from tyrosine in the chromaffin cells of the adrenal medulla in response to fright, cold, and exercise, as well as low levels of blood glucose. NE acts to stimulate the degradation of triacylglycerol and glycogen. It also increases cardiac output and blood pressure. The hydroxylation of dopamine to produce NE is catalyzed by the copper-containing enzyme dopamine-β-hydroxylase, an oxidase that requires ascorbic acid, acting as a reducing agent, for full activity.

As described, the secretion of epinephrine in response to stress, trauma, extreme exercise, or hypoglycemia causes a rapid mobilization of energy stores, that is, glucose from the liver and fatty acids from adipose tissue. The reaction in which NE is methylated to form E is catalyzed by the enzyme phenylethanolamine-*N*-methyltransferase (PNMT). Although the enzyme occurs predominantly in the chromaffin cells of the adrenal medulla, it is also found in certain portions of the brain, where E functions as a neurotransmitter. Recent evidence indicates that both E and NE are present in several other organs (e.g., liver, heart, and lung). Bovine PNMT is a monomeric protein (30 kD) that uses SAM as a source of methyl groups.

SEROTONIN Found in various cells in the central nervous system, serotonin inhibits feeding. This biogenic amine has been implicated in human eating disorders such as anorexia nervosa, bulimia, and the carbohydrate craving associated with seasonal affective disorder (SAD). SAD is a clinical depression triggered by the decreased daylight in autumn and winter. Additionally, serotonin appears to affect mood, temperature regulation, pain perception, and sleep. The hallucinogenic drug LSD (lysergic acid diethylamide) apparently competes with serotonin for specific brain cell receptors. Serotonin is also found in the gastrointestinal tract, blood platelets, and mast cells.

Tryptophan hydroxylase uses O_2 and the electron donor BH_4 to hydroxylate C-5 of tryptophan. The product, called 5-hydroxytryptophan, then undergoes a decarboxylation catalyzed by 5-hydroxytryptophan decarboxylase, a pyridoxal phosphate–requiring enzyme. Serotonin, often referred to as 5-hydroxytryptamine, is the product of this reaction.

Tryptophan   **5-Hydroxytryptophan**   **5-Hydroxytryptamine (serotonin)**

(Continued on p. 538)

BIOCHEMISTRY IN PERSPECTIVE

Gasotransmitters

How do gas molecules, previously thought to be toxic at any concentration, act as signal molecules?   **Gasotransmitters** are endogenous gaseous molecules that have been recently recognized as a class of signal molecules. These molecules, which are synthesized by highly regulated enzymes, have several characteristics in common: (1) they are lipid-soluble gases that can diffuse through cellular membranes; (2) their biological effects at physiological concentrations are mediated by second messenger molecules and/or ion channels; and (3) signals are not terminated by enzymes but largely by discontinued synthesis and diffusion away from cellular targets. Gasotransmitters include nitric oxide (NO), carbon monoxide (CO), hydrogen sulfide (H_2S), and certain ROS. It is noteworthy that each of these substances is extremely toxic and/or potentially lethal at high concentrations. Their biologically relevant actions occur at very low (usually μmolar) concentrations. A brief overview of the synthesis and functions of NO, CO, and H_2S is provided. ROS signals are described on p. 365.

Nitric Oxide

Nitric oxide (NO) is a highly reactive gas. Because of its free radical structure NO (symbolized as NO') was regarded until recently primarily as a contributing factor in the destruction of the ozone layer in the Earth's atmosphere and as a precursor of acid rain. Recent research has revealed, however, that NO' is an important signal molecule, produced throughout the mammalian body. Physiological functions such as the regulation of blood pressure, the inhibition of blood clotting, and the destruction of foreign, damaged, or cancerous cells by macrophages are triggered when NO' binds to the heme-containing domain of guanylate cyclase. The product of guanylate cyclase is the second messenger molecule cGMP (cyclic GMP, see p. 594). NO' is also a neurotransmitter and is produced in many areas of the brain where its formation has been linked to glutamate, which also acts as a neurotransmitter. When glutamate is released from a neuron and binds to a certain class of glutamate receptor, a transient flow of Ca^{2+} through the postsynaptic membrane is triggered that stimulates NO' synthesis. Once synthesized, NO' diffuses back to the presynaptic cell, where it signals further release of glutamate. Thus NO' acts as a retrograde neurotransmitter; that is, it promotes a cycle in which glutamate is released from the presynaptic neuron and then binds to and promotes action potentials in the postsynaptic neuron. This potentiating mechanism is now believed to play a role in learning and memory formation, as well as other functions in mammalian brain. The disruption of the normally precise regulation of NO' synthesis has been linked to numerous pathological conditions that include stroke, migraine headache, male impotence, septic shock, and several neurodegenerative diseases such as Parkinson's disease.

NO' is synthesized by NO synthase (NOS), a heme-containing metalloenzyme that catalyzes a two-step oxidation of L-arginine to L-citrulline (Figure 14A). In this complex reaction electrons are transferred from NADPH to O_2 by an electron transport chain that involves several redox components. The functional enzyme is a

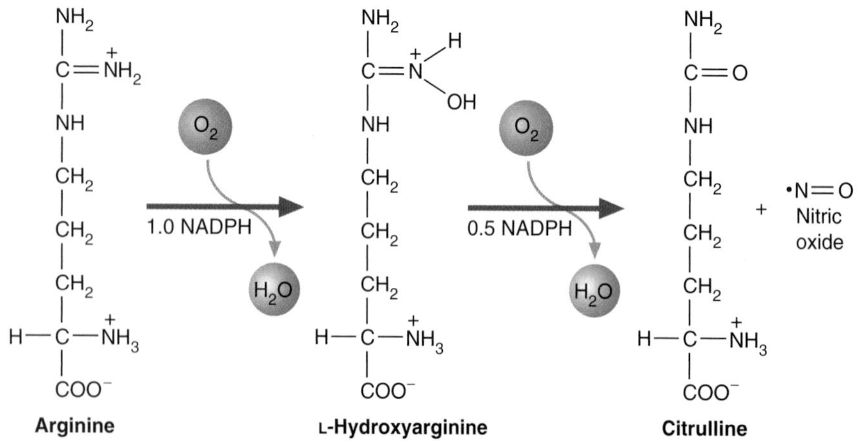

FIGURE 14A

The NOS-Catalyzed Reaction

The biosynthesis of NO' is a two-step oxidation of arginine to form citrulline. During the reaction, 2 mol of O_2 and 1.5 mol of NADPH are consumed per mol of citrulline formed.

BIOCHEMISTRY IN PERSPECTIVE cont

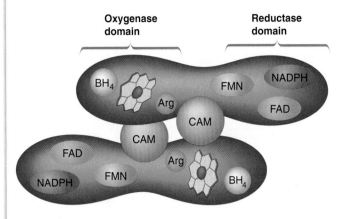

FIGURE 14B

Diagrammatic Structure of NOS

The catalytically active NOS is a homodimer. Each monomer binds NADPH, FAD, FMN, BH$_4$, and CAM in addition to the substrates arginine and O$_2$.

homodimer (Figure 14B). Each monomer has two major domains. The reductase domain possesses binding sites for NADPH, FAD, and FMN. The oxygenase domain, binds BH$_4$ and the substrates arginine and O$_2$. Between the two major domains is the binding site for calmodulin (CAM), a small calcium-binding protein that regulates a variety of enzymes. During NO$^{\bullet}$ synthesis CAM accelerates the rate of electron transfer from the reductase domain to the heme group.

The biosynthesis of NO$^{\bullet}$ begins with the hydroxylation of L-arginine. The electron source for this reaction is NADPH, which donates two electrons to FAD, which in turn reduces FMN. BH$_4$ is essential for the activation of O$_2$ by the electrons donated by NADPH. The product of this reaction, L-hydroxyarginine, remains bound to NOS. The steps in the subsequent reaction have not yet been resolved. It is believed that L-hydroxyarginine reacts with a heme-peroxy complex (R—O—OH) to give citrulline and NO$^{\bullet}$.

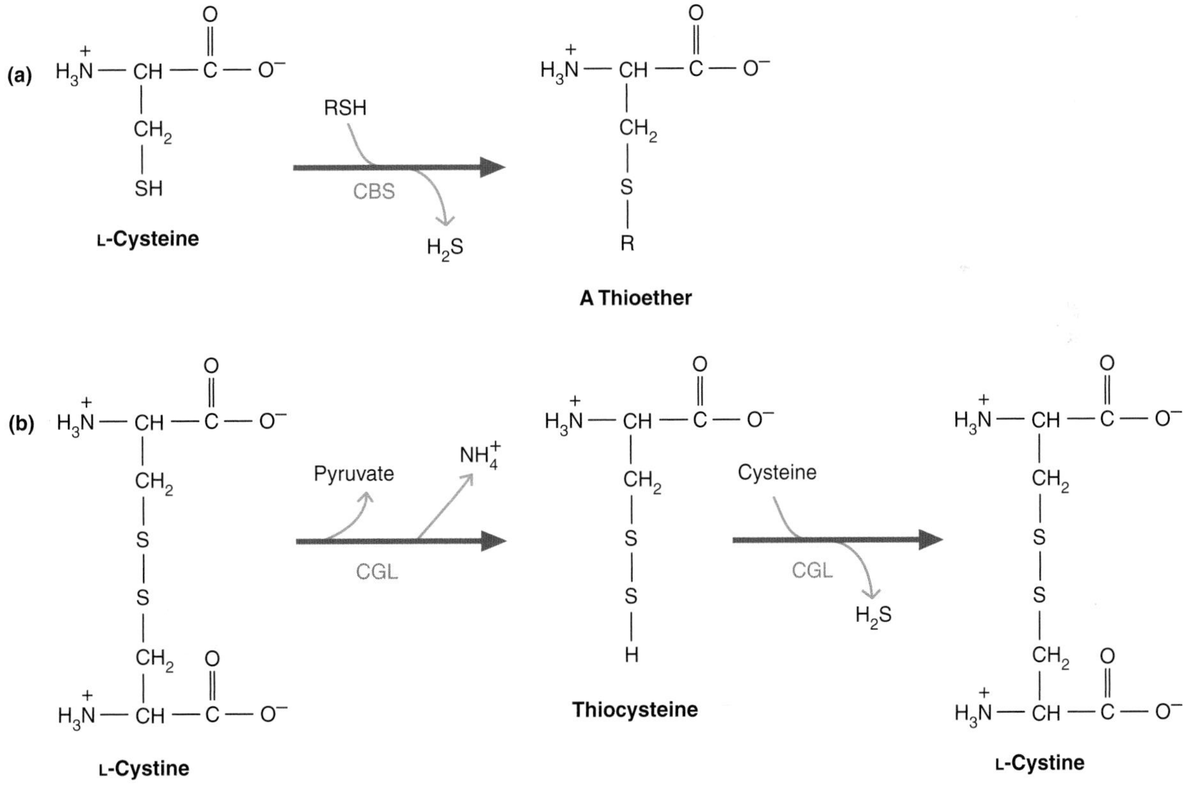

FIGURE 14C

H$_2$S Biosynthesis

Several enzymes catalyze H$_2$S synthesis in animals. The most important of these are cystathionine β-synthase (CBS) and cystathionine γ-lyase (CGL), both of which require pyridoxal 5′-phosphate as a coenzyme. (a) CBS in brain catalyzes the formation of H$_2$S by the substitution of the thiol group of L-cysteine with one of several thiol molecules (RSH) to yield a thioether (R—S—R′). (b) CGL catalyzes a β-desulhydration of cystine that yields pyruvate, NH$_4^+$, and thiocysteine. The reaction of thiocysteine with cysteine yields cystine and H$_2$S.

BIOCHEMISTRY **IN PERSPECTIVE** cont

Carbon Monoxide

The most important physiological roles played by CO, a color-less and odorless gas, include neuromodulation (i.e., regulation of neurotransmitter release and other neural activities related to learning, memory, and thermal regulation), cardiac protection during hypoxia against ischemia-reperfusion damage (see the Biochemistry in Perspective box in Chapter 10 entitled Ischemia and Reperfusion, p. 367), and vascular relaxation (i.e. promotion of smooth muscle relaxation and, therefore, blood vessel dilation). CO is produced by the catabolism of heme in a reaction (Figure 15.18 shown later) catalyzed by heme oxygenase (HO), an ER enzyme. In addition to CO, the products of this reaction are biliverdin and free iron atoms. The catabolism of heme is an ongoing activity in all tissues that is particularly persistent in spleen (hemoglobin breakdown) and liver (cytochrome P_{450} turnover). Heme itself and its oxidized form, hemin, are strong oxidants and have the potential to confer significant tissue damage. Biliverdin and its reduced form, bilirubin, have much lower oxidant potential than the parent heme. All cells must have a pool of available free heme to serve as a resource for the synthesis of heme-containing proteins, but that level is kept in a safe range by the action of HO. The free iron is sequestered by ferritin to prevent redox damage. Many actions of CO are mediated by its binding to heme-containing proteins or activation of K^+ channels. CO, at physiologically safe levels, has also been shown to reduce inflammation and suppress apoptosis (cell death).

Hydrogen Sulfide

H_2S is a toxic gas with an unpleasant, foul odor. Within the body, where it is produced in small quantities, H_2S is known to affect numerous physiological processes. The most prominent of these are its functions as a neuromodulator in brain and as a vascular relaxant. H_2S is synthesized from cysteine by several enzymes (Figure 14C). The most important of these are the pyridoxal phosphate–requiring enzymes cystathionine γ-lyase (CGL) and cystathione β-synthetase (CBS), both of which catalyze other reactions in the transulfuration pathway (see later: Section 15.1). With cysteine used as an alternate substrate, both enzymes catalyze a hydrolytic reaction to generate H_2S. CGL is the dominant H_2S-producing enzyme in the liver and neuronal tissue. H_2S dilates arterioles by triggering the opening of ATP-sensitive K^+ channels, a process that results in the hyperpolarization of smooth muscle cell membrane. H_2S exerts its neuromodulatory effects by enhancing the activation of receptors for NMDA (*N*-methyl D-aspartate). *NMDA receptors* are a type of glutamate receptor; when activated, they open an ion channel that allows the inward flow of Ca^{2+} and other ions. NMDA receptors are believed to play an important role in *synaptic plasticity*, the capacity of a synapse (the connection between two neurons) to change in chemical strength. Changes in synaptic strength are believed to be the basis for memory formation.

SUMMARY: At very low concentrations, NO, CO, and H_2S are signal molecules that diffuse easily through cell membranes and whose synthesis is rigorously controlled.

HISTAMINE An amine produced in numerous tissues throughout the body, histamine has complex physiological effects. It is a mediator of allergic and inflammatory reactions, a stimulator of gastric acid production, and a neurotransmitter in several areas of the brain. Histamine is formed by the decarboxylation of L-histidine in a reaction catalyzed by histidine decarboxylase, a pyridoxal phosphate–requiring enzyme.

Histidine → (Histidine decarboxylase, releasing H^+ and CO_2) → Histamine

Histidine   **Histamine**

Alkaloids

The **alkaloids** are a large, heterogeneous group of basic nitrogen-containing molecules produced in the leaves, seeds, or bark of some plants. They are derived from α-amino acids (or closely related molecules) in complex and poorly understood pathways. Although many alkaloids have profound physiological properties when consumed by animals, their roles in plants are relatively obscure. Because they often have bitter tastes or are poisonous, alkaloids may protect plants against herbivores, insects, and microbes. Several examples of this group of natural products are illustrated in Figure 14.23.

Alkaloids are classified according to their heterocyclic rings. For example, cocaine, a central nervous system stimulant, and atropine, a muscle relaxant, are examples of the *tropane alkaloids*, in which a nitrogen appears in a bridge of a seven-membered ring structure. The addictive and toxic molecule called nicotine is an example of the *pyridine alkaloids* in which nitrogen appears as a member of a six-atom aromatic ring. The tobacco plant takes advantage of nicotine's potent insecticidal properties. The addictive components of opium (codeine and morphine) are examples of the *isoquinoline alkaloids*, in which a nitrogen appears in one ring of a multiring system.

Nucleotides

Nucleotides are complex nitrogen-containing molecules required for cell growth and development. Not only are nucleotides the building blocks of the nucleic acids, they also play several essential roles in energy transformation and regulate many metabolic pathways. As described, each nucleotide is composed of three parts: a nitrogenous base, a pentose sugar, and one or more phosphate groups.

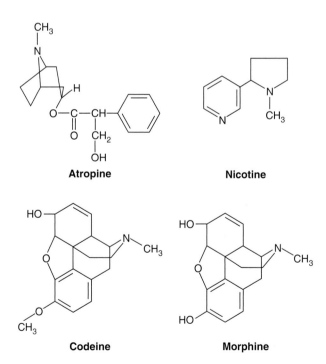

FIGURE 14.23

Structures of Several Alkaloids

More than 5000 alkaloids have been isolated from plants. Their roles in plants are usually unknown. Alkaloids are often physiologically potent molecules in animals.

The nitrogenous bases are derivatives of either purine or pyrimidine, which are planar heterocyclic aromatic compounds.

Purine   **Pyrimidine**

Common naturally occurring **purines** include adenine, guanine, xanthine, and hypoxanthine; thymine, cytosine, and uracil are common **pyrimidines** (Figure 14.24). Because of their aromatic structures, the purines and pyrimidines absorb UV light. At pH 7, this absorption is especially strong at 260 nm. Purine and pyrimidine bases have tautomeric forms; that is, they undergo spontaneous shifts in the relative position of a hydrogen atom and a double bond in a three-atom sequence involving heteroatoms. This property is especially important because the precise location of hydrogen atoms on the oxygen and nitrogen atoms affects the interaction of bases in nucleic acid molecules. Adenine and cytosine have both amino and imino forms; guanine, thymine, and uracil have both keto (lactam) and enol (lactim) forms (Figure 14.25). At physiological pH the amino and keto forms are the most stable.

When a purine or pyrimidine base is linked through a β-N-glycosidic linkage to C-1 of a pentose sugar, the molecule is called a **nucleoside**, and it contains one of two types of sugar: ribose or deoxyribose. Ribose-containing nucleosides with adenine, guanine, cytosine, and uracil are referred to as adenosine, guanosine, cytidine, and uridine, respectively. When the sugar component is deoxyribose, the prefix *deoxy* is used. For example, the deoxy nucleoside with adenine is called deoxyadenosine. Deoxythymidine is called thymidine because the base thymine usually occurs only in deoxyribonucleosides. Possible confusion in the identification of atoms in the base and sugar

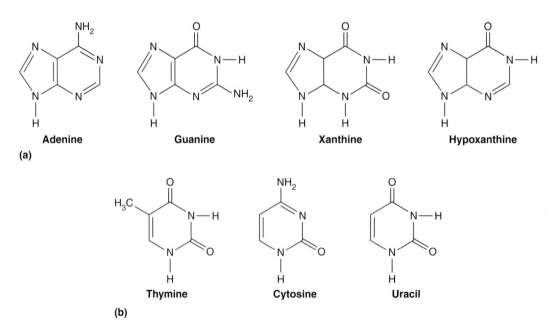

Adenine   **Guanine**   **Xanthine**   **Hypoxanthine**

(a)

Thymine   **Cytosine**   **Uracil**

(b)

FIGURE 14.24

The Most Common Naturally Occurring Purines (a) and Pyrimidines (b)

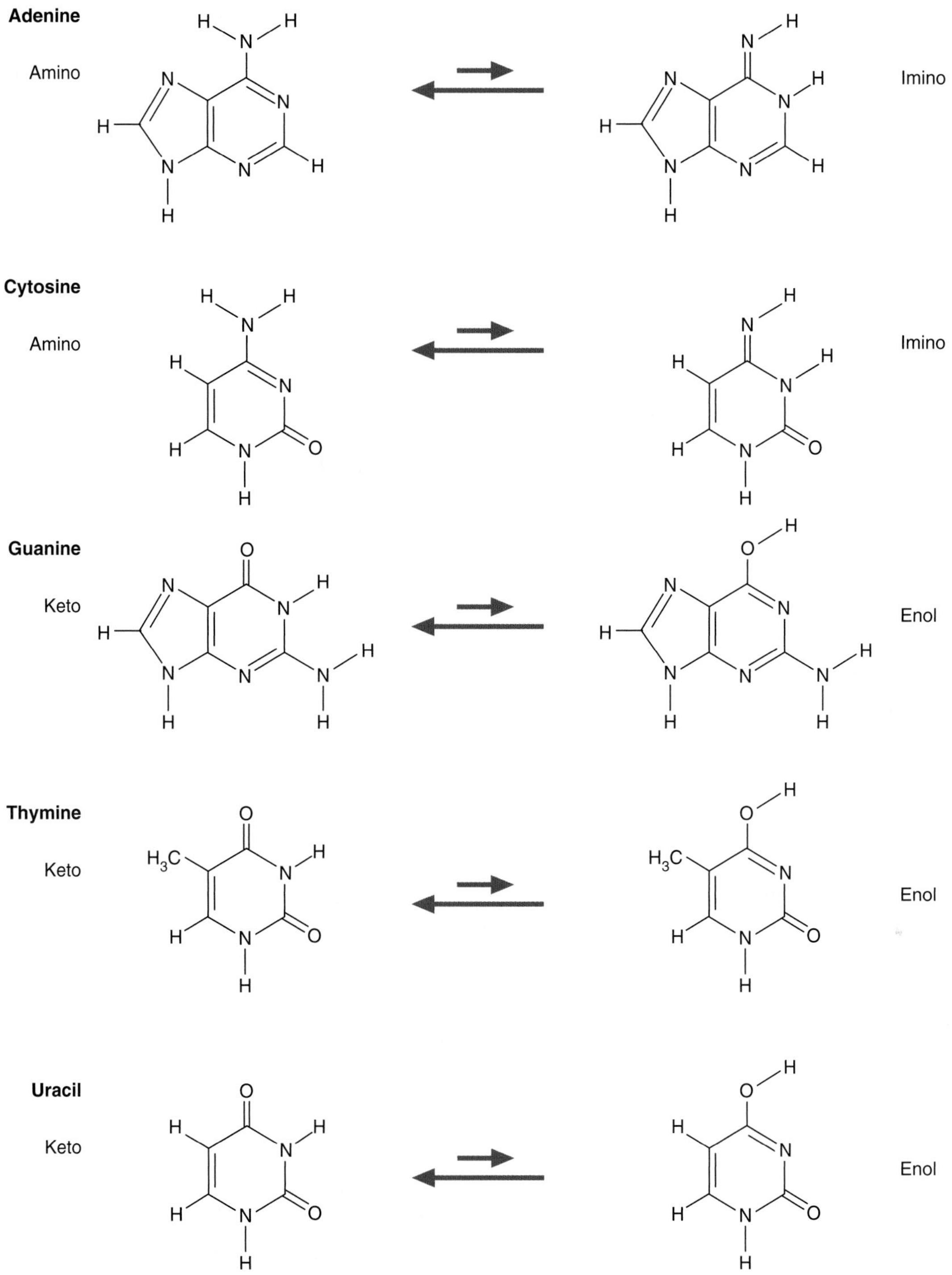

FIGURE 14.25

Tautomers of Adenine, Cytosine, Guanine, Thymine, and Uracil

At physiological pH, the amino and keto tautomers of nitrogenous bases are the predominant forms.

components of nucleosides is avoided by using a superscript prime to denote the atoms in the sugar.

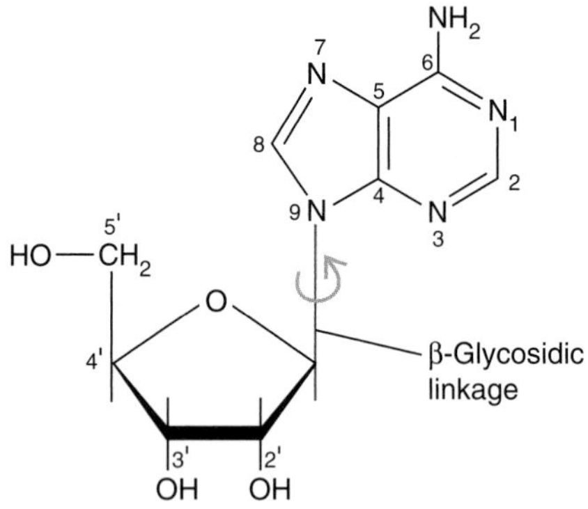

Rotation around the N-glycosidic bond of nucleosides creates two conformations: *syn* and *anti*. Purine nucleosides occur as either *syn* or *anti* forms. In pyrimidine nucleosides the *anti* conformation predominates because of steric hinderance between the pentose sugar and the carbonyl oxygen at C-2.

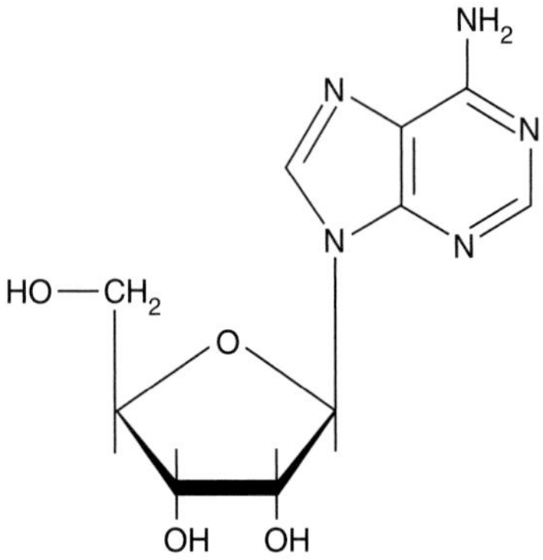

anti-Adenosine 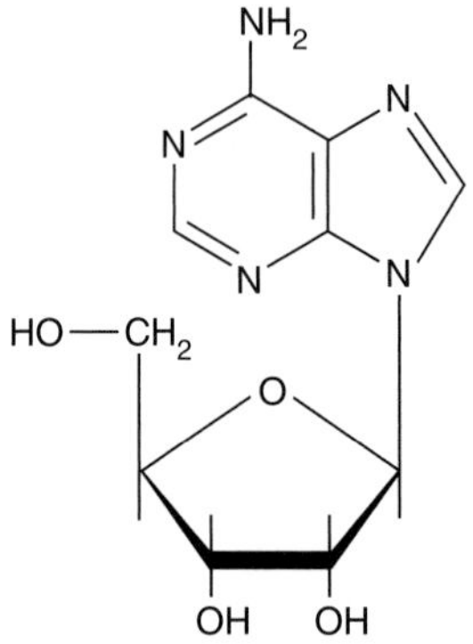   *syn*-Adenosine 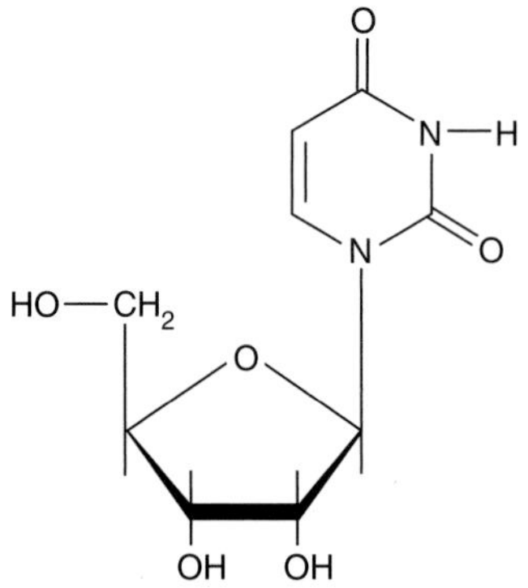   *anti*-Uridine

Nucleotides are nucleosides in which one or more phosphate groups are bound to the sugar (Figure 14.26). Most naturally occurring nucleotides are 5′-phosphate esters. If one phosphate group is attached at the 5′-carbon of the sugar, the molecule is named as a nucleoside monophosphate (e.g., adenosine-5′-monophosphate: AMP). Nucleoside di- and triphosphates contain two and three phosphate groups, respectively. Phosphate groups make nucleotides strongly acidic. (Protons dissociate from the phosphate groups at physiological pH.) Because of their acidic nature, nucleotides may also be named as acids. For example, AMP is often referred to as adenylic acid or adenylate. Nucleoside di- and triphosphates form complexes with Mg^{2+}. In nucleoside triphosphates such as ATP, Mg^{2+} can form α,β (shown) and β,γ complexes:

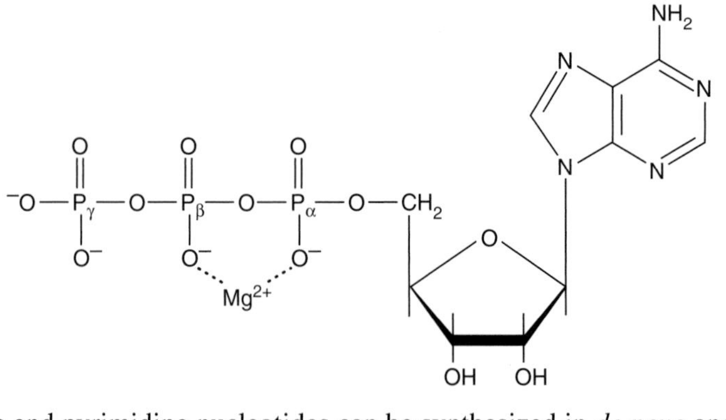

Purine and pyrimidine nucleotides can be synthesized in *de novo* and salvage pathways. Both types of pathway are described.

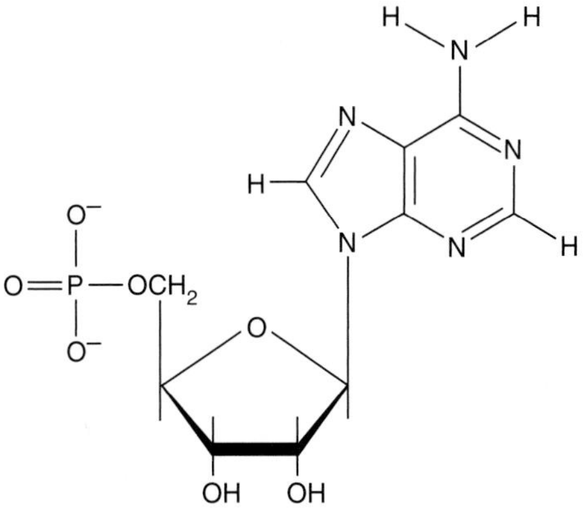

Adenosine-5'-monophosphate (AMP)

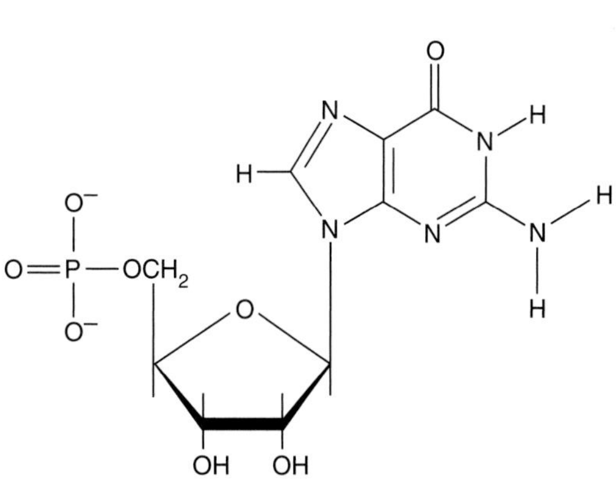

Guanosine-5'-monophosphate (GMP)

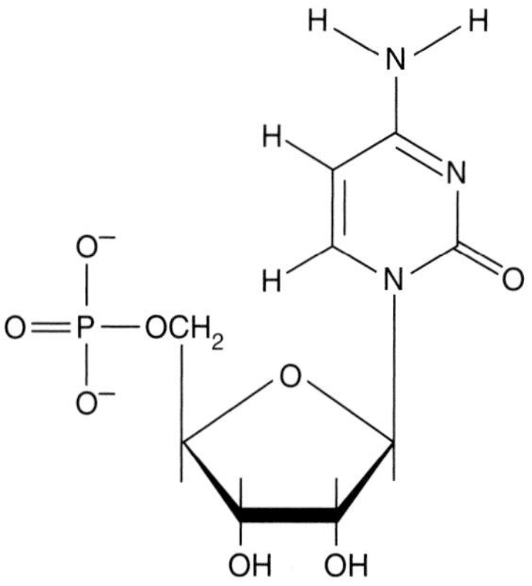

Cytidine-5'-monophosphate (CMP)

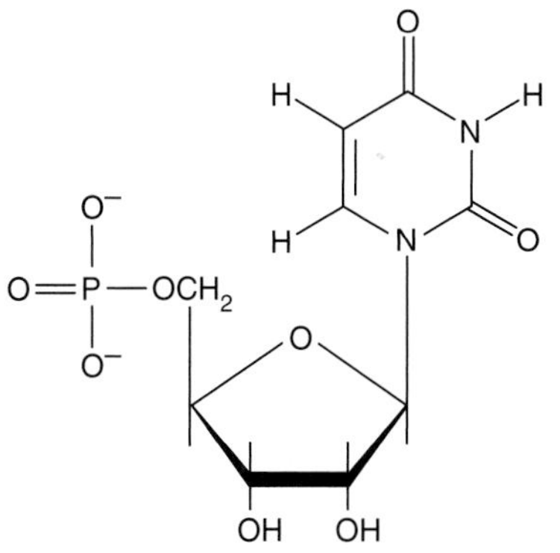

Uridine-5'-monophosphate (UMP)

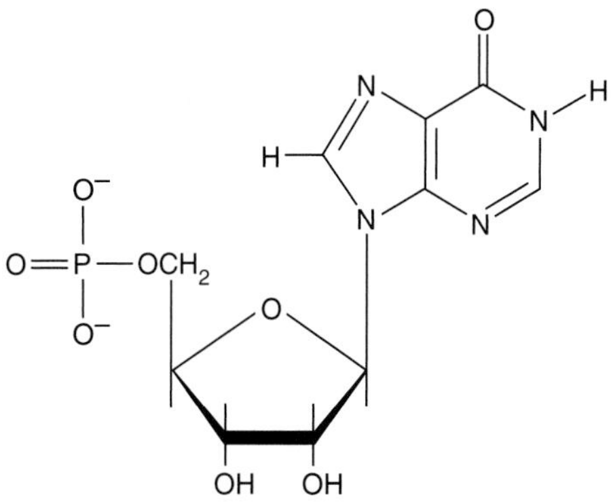

Inosine-5'-monophosphate (IMP)

FIGURE 14.26

Common Ribonucleotides

Ribonucleotides contain ribose. When nucleotides contain deoxyribose instead of ribose, their names include the prefix *deoxy*. Inosine-5′-monophosphate (IMP) is an intermediate in purine nucleotide synthesis. The base component of IMP is hypoxanthine.

PURINE NUCLEOTIDES The *de novo* synthesis of purine nucleotides begins with the formation of 5-phospho-α-D-ribosyl-1-pyrophosphate (PRPP) catalyzed by ribose-5-phosphate pyrophosphokinase (PRPP synthetase).

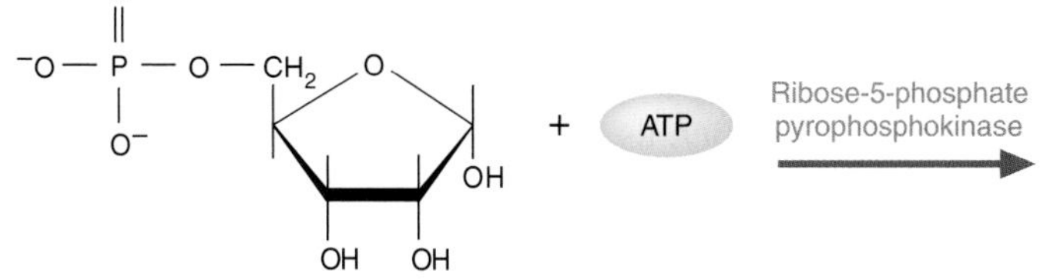

α-D-ribose-5-phosphate

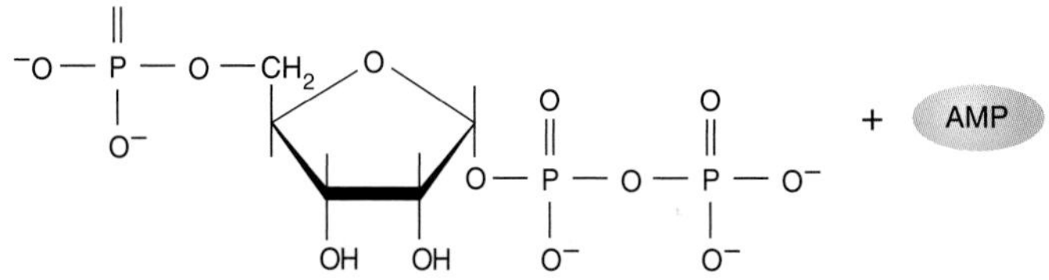

5-Phospho-α-D-ribosyl-1-pyrophosphate (PRPP)

(The substrate for this reaction, α-D-ribose-5-phosphate, is a product of the pentose phosphate pathway.) Figure 14.27 illustrates the pathway by which PRPP is converted to inosine monophosphate (inosinate), the first purine nucleotide. The process begins with the displacement of the pyrophosphate group of PRPP by the amide nitrogen of glutamine in a reaction catalyzed by glutamine PRPP amidotransferase. This reaction is the committed step in purine synthesis. The product formed is 5-phospho-β-D-ribosylamine.

Once 5-phospho-β-D-ribosylamine has formed, the building of the purine ring structure begins. Phosphoribosylglycinamide synthase catalyzes the formation of an amide bond between the carboxyl group of glycine and the amino group of 5-phospho-β-D-ribosylamine. In eight subsequent reactions, the first purine nucleotide IMP is formed. Other precursors of the base component of IMP (hypoxanthine) include CO_2, aspartate, and N^{10}-formyl THF. This pathway requires the hydrolysis of four ATP molecules.

The conversion of IMP to either adenosine monophosphate (AMP or adenylate) or guanosine monophosphate (GMP or guanylate) requires two reactions (Figure 14.28).

AMP differs from IMP in only one respect: an amino group replaces a keto oxygen at position 6 of the purine base. The amino nitrogen provided by aspartate becomes linked to IMP in a GTP-requiring reaction catalyzed by adenylosuccinate synthetase. In this step the product adenylosuccinate eliminates fumarate to form AMP. (The enzyme that catalyzes this reaction also catalyzes a similar step in IMP synthesis.) The conversion of IMP to GMP begins with a dehydrogenation utilizing NAD^+ which is catalyzed by IMP dehydrogenase. The product, referred to as xanthosine monophosphate (XMP), is then converted to GMP by the donation of an amino nitrogen by glutamine in an ATP-requiring reaction catalyzed by GMP synthase.

Nucleoside triphosphates are the most common nucleotide used in metabolism. They are formed in the following manner. Recall that ATP is synthesized from ADP and P_i during certain reactions in glycolysis and aerobic metabolism. ADP is synthesized from AMP and ATP in a reaction catalyzed by adenylate kinase:

$$AMP + ATP \rightarrow 2\,ADP$$

Other nucleoside triphosphates are synthesized in ATP-requiring reactions catalyzed by a series of nucleoside monophosphate kinases:

$$NMP + ATP \rightleftharpoons NDP + ADP$$

Nucleoside diphosphate kinase catalyzes the formation of nucleoside triphosphates

$$N_1DP + N_2TP \rightleftharpoons N_1TP + N_2DP$$

where N_1 and N_2 are purine or pyrimidine bases.

In the purine salvage pathway, purine bases obtained from the normal turnover of cellular nucleic acids or (to a lesser extent) from the diet are reconverted into nucleotides. Because the *de novo* synthesis of nucleotides is metabolically expensive (i.e., relatively large amounts of phosphoryl bond energy are used), many cells have mechanisms to retrieve purine bases. Hypoxanthine-guaninephosphoribosyltransferase (HGPRT) catalyzes nucleotide synthesis using PRPP and either hypoxanthine or guanine. The hydrolysis of pyrophosphate makes these reactions irreversible.

Hypoxanthine + PRPP $\longrightarrow$ IMP + PP$_i$

Guanine + PRPP $\longrightarrow$ GMP + PP$_i$

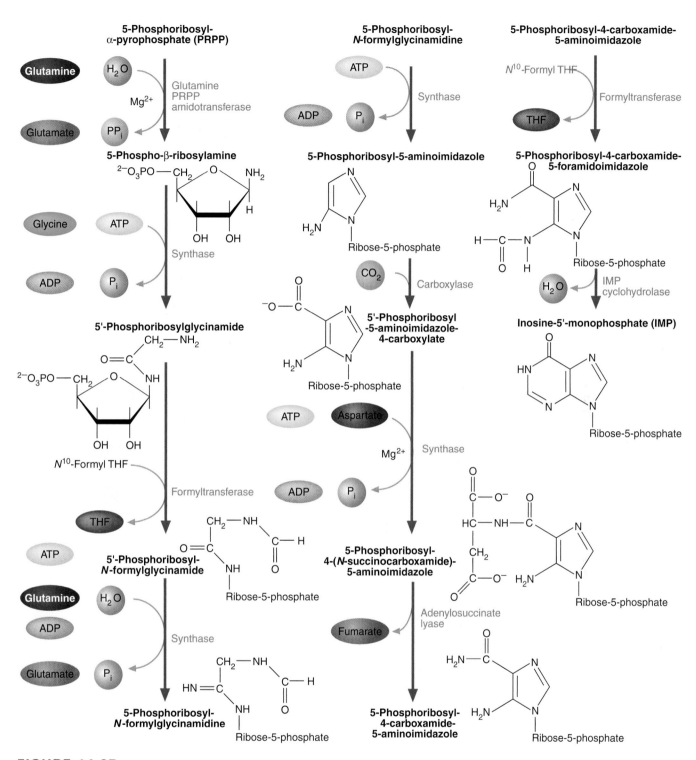

FIGURE 14.27

Synthesis of Inosine-5′-Monophosphate

The biosynthesis of IMP begins with the reaction between an amino group of glutamine with C-1 of PRPP. The product, 5-phospho-β-ribosyl-amine, subsequently undergoes a series of reactions in which the purine ring is constructed using carbon atoms from formate (via N^{10}-formyl THF) and CO_2, and nitrogen atoms from glycine, glutamine, and aspartate.

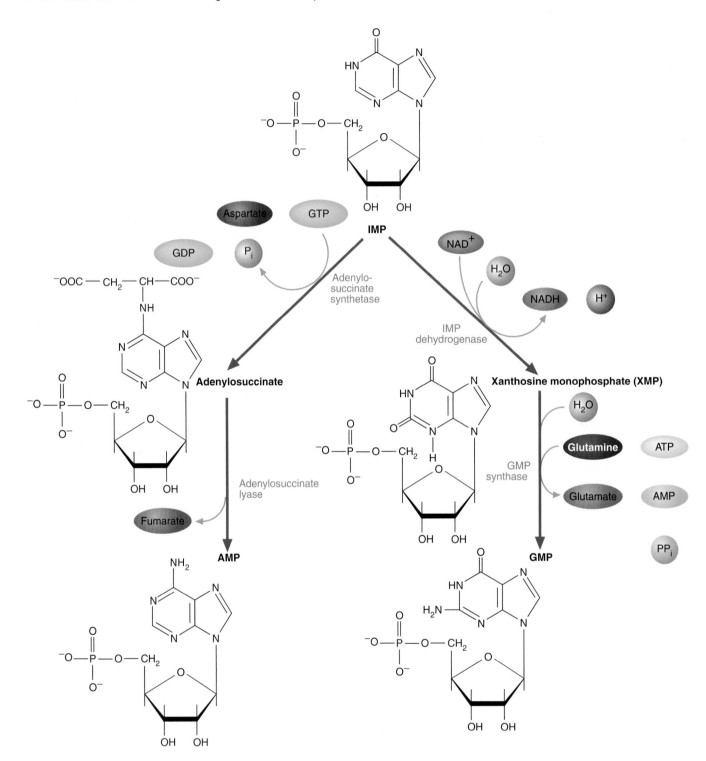

FIGURE 14.28

Biosynthesis of AMP and GMP from IMP

In the first step of AMP synthesis, the C-6 keto oxygen of the hypoxanthine base moiety of IMP is replaced by the amino group of aspartate. In the second step the product of the first reaction, adenylosuccinate, is hydrolyzed to form AMP and fumarate. GMP synthesis begins with the oxidation of IMP to form XMP. GMP is produced as the amide nitrogen of glutamine replaces the C-2 keto oxygen of XMP. Note that AMP formation requires GTP and GMP formation requires ATP.

Deficiency of HGPRT causes *Lesch-Nyhan syndrome*, a devastating X-linked disease that occurs primarily in males. It is characterized by excessive production of uric acid, the degradation product of purine nucleotides (Section 15.3), and certain neurological symptoms (self-mutilation, involuntary movements, and mental retardation). Although a powerful antioxidant, uric acid can act as a prooxidant when present in large amounts. Oxidative stress is now believed to contribute to the symptoms of Lesch-Nyhan syndrome. Affected children appear normal at birth but begin to deteriorate at about 3 to 4 months of age. Death, usually caused by renal failure, occurs in childhood. A partially defective HGPRT enzyme causes one form of *gout* (a condition in which high blood uric acid concentrations result in the accumulation of sodium urate crystals in joints, especially those in feet).

Adenine phosphoribosyltransferase (ARPT) catalyzes the transfer of adenine to PRPP, thus forming AMP:

The relative importance of the *de novo* and salvage pathways is unclear. However, the severe symptoms of hereditary HGPRT deficiency indicate that the purine salvage pathway is vitally important. In addition, investigations of purine nucleotide synthesis inhibitors for treating cancer indicate that both pathways must be inhibited for significant tumor growth suppression.

The regulation of purine nucleotide biosynthesis is summarized in Figure 14.29. The pathway is controlled to a considerable degree by PRPP availability. Several products of the pathway inhibit both ribose-5-phosphate pyrophosphokinase and glutamine-PRPP amidotransferase. The combined inhibitory effect of the end products is synergistic (i.e., the net inhibition is greater than the inhibition of each nucleotide acting alone). At the IMP branch point, both AMP and

COMPANION

Visit the companion website at www.oup.com/us/mckee to read the Biochemistry in Perspective box on gout in Chapter 15.

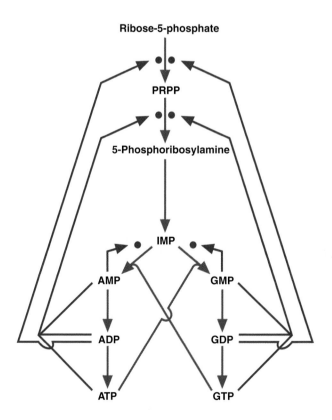

FIGURE 14.29

Purine Nucleotide Biosynthesis Regulation

Feedback inhibition is indicated by red arrows. The stimulation of AMP synthesis by GTP and GMP synthesis by ATP ensures a balanced synthesis of both purine nucleotide families.

GMP regulate their own syntheses by feedback inhibition of adenylosuccinate synthetase and IMP dehydrogenase. The hydrolysis of GTP drives the synthesis of adenylosuccinate, whereas ATP drives XMP synthesis. This reciprocal arrangement is believed to facilitate the maintenance of appropriate cellular concentrations of adenine and guanine nucleotides.

PYRIMIDINE NUCLEOTIDES In pyrimidine nucleotide synthesis the pyrimidine ring is assembled first and then linked to ribose phosphate. The carbon and nitrogen atoms in the pyrimidine ring are derived from bicarbonate, aspartate, and glutamine. Synthesis begins with the formation of carbamoyl phosphate in an ATP-requiring reaction catalyzed by the cytoplasmic enzyme carbamoyl phosphate synthetase II (Figure 14.30). (Carbamoyl phosphate synthetase I is a

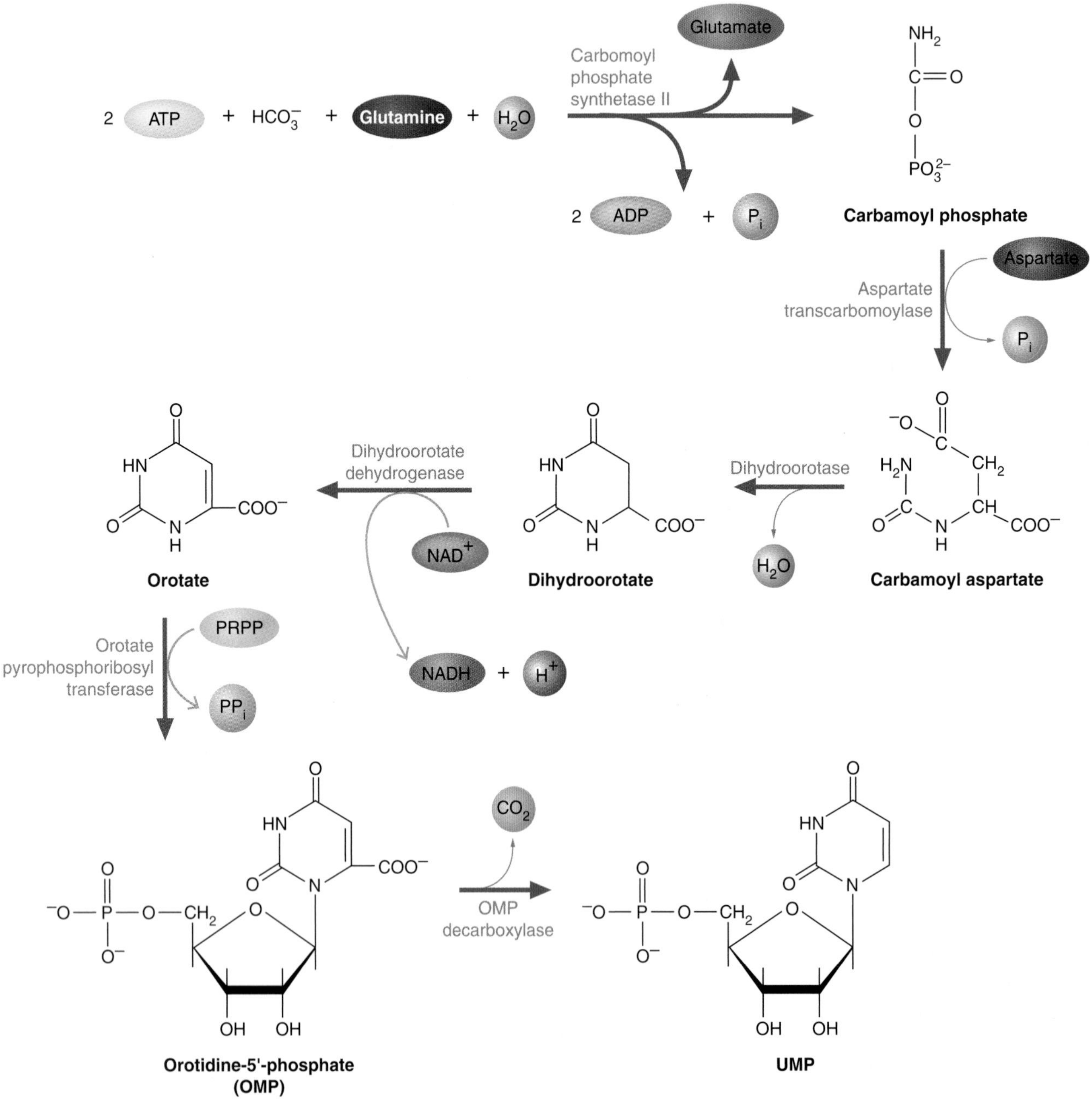

FIGURE 14.30

Pyrimidine Nucleotide Synthesis

The metabolic pathway in which UMP is synthesized is composed of six enzyme-catalyzed reactions.

mitochondrial enzyme involved in the urea cycle, described in Chapter 15.) One molecule of ATP provides a phosphate group, while the hydrolysis of another ATP drives the reaction. Carbamoyl phosphate next reacts with aspartate to form carbamoyl aspartate. The closure of the pyrimidine ring is then catalyzed by dihydroorotase. The product, dihydroorotate, is then oxidized to form orotate. Dihydroorotate dehydrogenase, the enzyme that catalyzes this reaction, is a flavoprotein associated with the inner mitochondrial membrane. (The NADH produced in this reaction donates its electrons to the electron transport complex.) Once synthesized on the cytoplasmic face of the inner mitochondrial membrane, orotate is converted by orotate pyrophosphoribosyl transferase to orotidine-5′-phosphate (OMP), the first nucleotide in the pathway, by reacting with PRPP. Uridine-5′-phosphate (UMP) is produced when OMP is decarboxylated in a reaction catalyzed by OMP decarboxylase. Both orotate pyrophosphoribosyl transferase and OMP decarboxylase activities occur on a protein referred to as UMP synthase. (In a rare genetic disease called *orotic aciduria*, there is excessive urinary excretion of orotic acid because UMP synthase is defective. Symptoms include anemia and retardation in growth. Treatment with a combination of pyrimidine nucleotides, which inhibit the production of orotate and provide the building blocks for nucleic acid synthesis, reverses the disease process.) UMP is a precursor for the other pyrimidine nucleotides. Two sequential phosphorylation reactions form UTP, which then accepts an amide nitrogen from glutamine to form CTP.

DEOXYRIBONUCLEOTIDES All the nucleotides discussed so far are ribonucleotides, molecules that are principally used as the building blocks of RNA, as nucleotide derivatives of molecules such as sugars, or as energy sources. The nucleotides required for DNA synthesis, the 2′-deoxyribonucleotides, are produced by reducing ribonucleoside diphosphates in a reaction catalyzed by ribonucleotide reductase (Figure 14.31). The electrons used in the synthesis of 2′-deoxyribonucleotides are ultimately donated by NADPH. Thioredoxin mediates the transfer of hydrogen atoms from NADPH to ribonucleotide reductase. The regeneration of reduced thioredoxin is catalyzed by thioredoxin reductase.

Ribonuclease reductase I (RNRI), found in mammals, is a tetramer of two different subunits. Subunit 1 has a number of reactive thiols required for catalysis plus the allosteric sites involved in regulation. Subunit 2 possesses a critical binuclear Fe(III) center that generates and stabilizes a tyrosyl radical essential for enzyme function. The interface of the four subunits forms the active site. The tyrosyl radical initiates the radical-mediated reduction of substrate NDPs by abstracting an H atom from one of the thiols in subunit 1, generating a transient thiyl radical. The thiyl radical abstracts an H atom from C3′ of the substrate, generating a radical. A nearby thiol protonates the C2′—OH group and it departs as H_2O, leaving behind a carbocation. Hydride ion shift from an active site thiol resolves the carbocation and a disulfide bridge forms in the active site. The H atom abstracted by the thiyl radical is returned to C3′, and tyrosine transfers an H atom to resolve the thiyl radical. The product dNDP leaves the active site, and the enzyme is returned to its reduced free thiol state by electron transfer from NADPH mediated through thioredoxin (Figure 14.32).

Regulation of ribonucleotide reductase is intricate. The binding of dATP (deoxyadenosine triphosphate) to a regulatory site on the enzyme decreases catalytic activity. The binding of deoxyribonucleoside triphosphates to several other enzyme sites alters substrate specificity so that there are differential increases in the concentrations of each of the deoxyribonucleotides. This latter process balances the production of the 2′-deoxyribonucleotides required for cellular processes, especially that of DNA synthesis.

The deoxyuridylate (dUMP) produced by dephosphorylation of the dUDP product of ribonucleotide reductase is not a component of DNA, but its

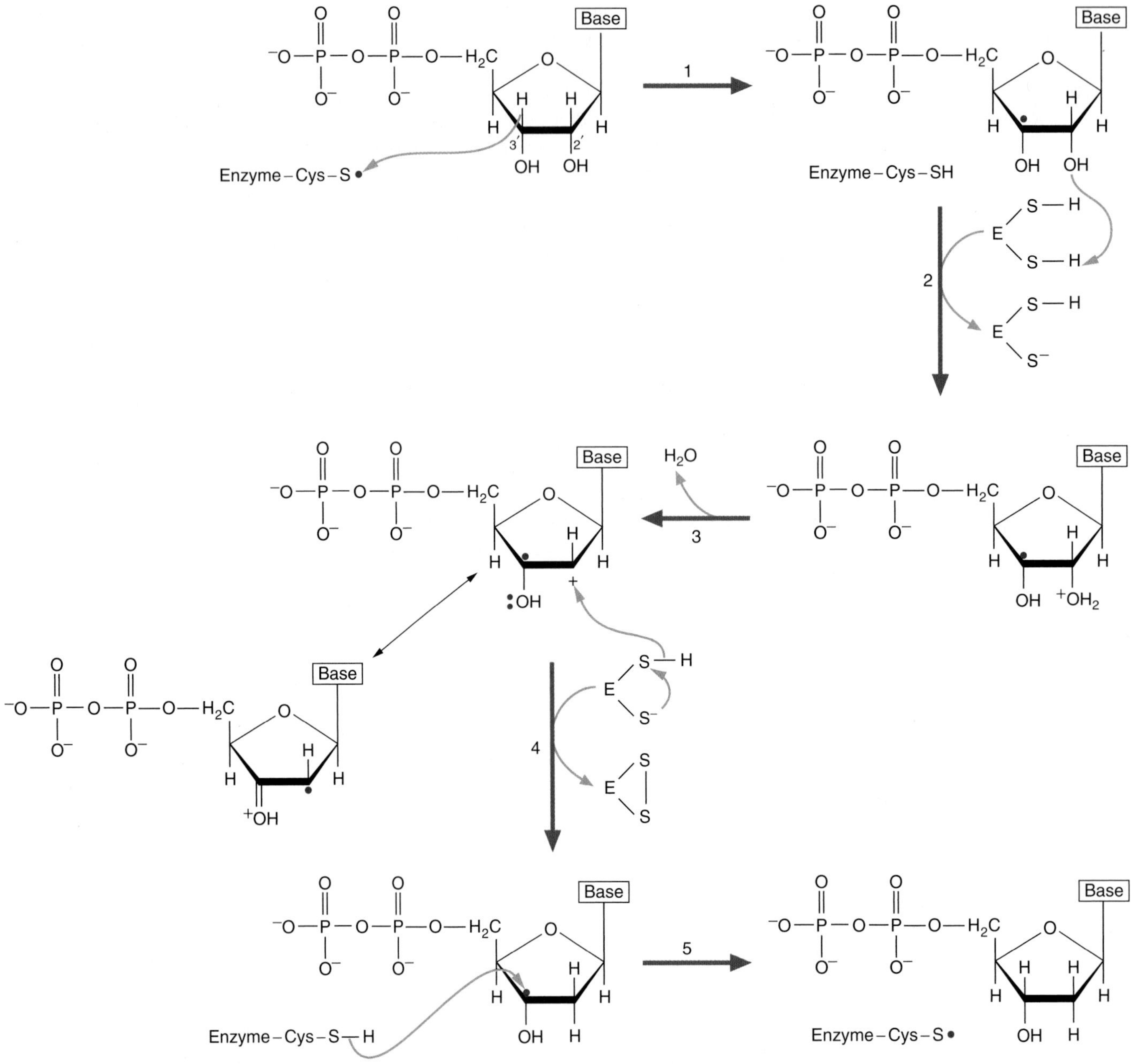

FIGURE 14.31

Mechanism of Ribonucleotide Reductase

The reaction begins with the tyrosyl radical–induced formation of a transient thiyl radical and the binding of NDP. (1) The thiyl radical abstracts a H atom from C3'. The C2'—OH (2) is protonated by a reactive thiol and H_2O is eliminated (3) to generate a carbocation. A dithiol reduces the cation radical (4) and (5) an H atom is transferred from the initiating thiyl S to the C3' and the product dNDP leaves the active site. The subsequent reduction of a disulfide–mediated by thioredoxin/NADPH and regeneration of the tyrosyl radical returns the enzyme to its ready state to receive new substrate.

methylated derivative deoxythymidylate (dTMP) is. The methylation of dUMP is catalyzed by thymidylate synthase, which utilizes N^5, N^{10}-methylene THF. As the methylene group is transferred, it is reduced to a methyl group, while the folate coenzyme is oxidized to form dihydrofolate. THF is regenerated from dihydrofolate by dihydrofolate reductase and NADPH. (This reaction is the site of action of some anticancer drugs, such as methotrexate.) Deoxyuridylate can also be synthesized from dCMP by deoxycytidylate deaminase.

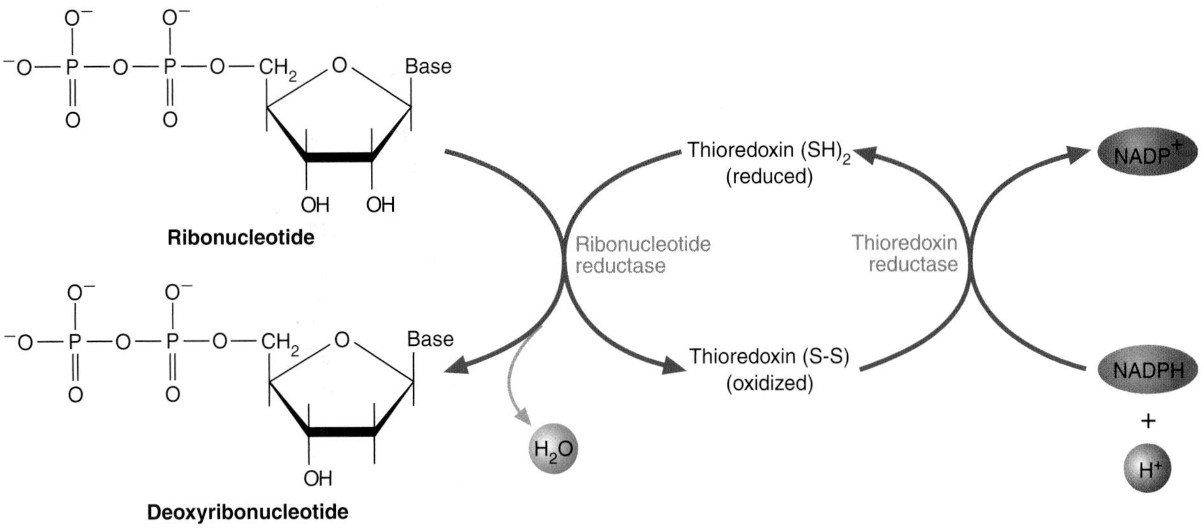

Ribonucleotide

Deoxyribonucleotide

Thioredoxin (SH)$_2$
(reduced)

Thioredoxin (S-S)
(oxidized)

Ribonucleotide
reductase

Thioredoxin
reductase

NADP$^+$

NADPH
+
H$^+$

H$_2$O

Base

Base

FIGURE 14.32

Deoxyribonucleotide Biosynthesis

Electrons for the reduction of ribonucleotides ultimately come from NADPH. Thioredoxin, a small protein with two thiol groups, mediates the transfer of electrons from NADPH to ribonucleotide reductase.

In mammals, carbamoyl phosphate synthetase II is the key regulatory enzyme in the biosynthesis of pyrimidine nucleotides. The enzyme is inhibited by UTP, the product of the pathway, and stimulated by purine nucleotides. In many bacteria, aspartate carbamoyl transferase is the key regulatory enzyme. It is inhibited by CTP and stimulated by ATP. The pyrimidine salvage pathway, which uses preformed pyrimidine bases from dietary sources or from nucleotide turnover, is of minor importance in mammals.

Heme

Heme, one of the most complex molecules synthesized by mammalian cells, has an iron-containing porphyrin ring. Heme is an essential structural component of hemoglobin, myoglobin, peroxidase, and the cytochromes. Although it occurs in almost all aerobic cells, the heme biosynthetic pathway is especially prominent in liver and reticulocytes (the nucleus-containing precursor cells of red blood cells in bone marrow). Heme is synthesized from the relatively simple components glycine and succinyl-CoA.

In the first step of heme synthesis, glycine and succinyl-CoA condense to form δ-aminolevulinate (ALA) (Figure 14.33). This reaction is catalyzed by ALA synthase (ALAS), a pyridoxal phosphate–requiring mitochondrial enzyme. Mammals have two nuclear ALAS genes: *ALAS1*, which is expressed in all nucleated cells, and *ALAS2*, which is expressed only in reticulocytes. *ALAS1* activity is regulated by free heme molecules and hematin (an oxidized derivative of heme), which prevent the transcription of the *ALAS1* gene and the translation of *ALAS1* mRNA. Heme also prevents the transport of ALAS into mitochondria.

In the next step of porphyrin synthesis, two molecules of ALA condense to form porphobilinogen. Porphobilinogen synthase, which catalyzes this reaction, is a zinc-containing enzyme that is extremely sensitive to heavy-metal poisoning. Uroporphyrinogen I synthase catalyzes the symmetric condensation of four porphobilinogen molecules. An additional protein is also required in this reaction. Uroporphyrinogen III cosynthase alters the specificity of uroporphyrinogen

KEY CONCEPTS

- Nucleotides are the building blocks of the nucleic acids. They also regulate metabolism and transfer energy.
- The purine and pyrimidine nucleotides are synthesized in both de novo and salvage pathways.

COMPANION

W E B S I T E Visit the companion website at www.oup.com/us/mckee to read this chapter's Biochemistry in Perspective box on lead poisoning.

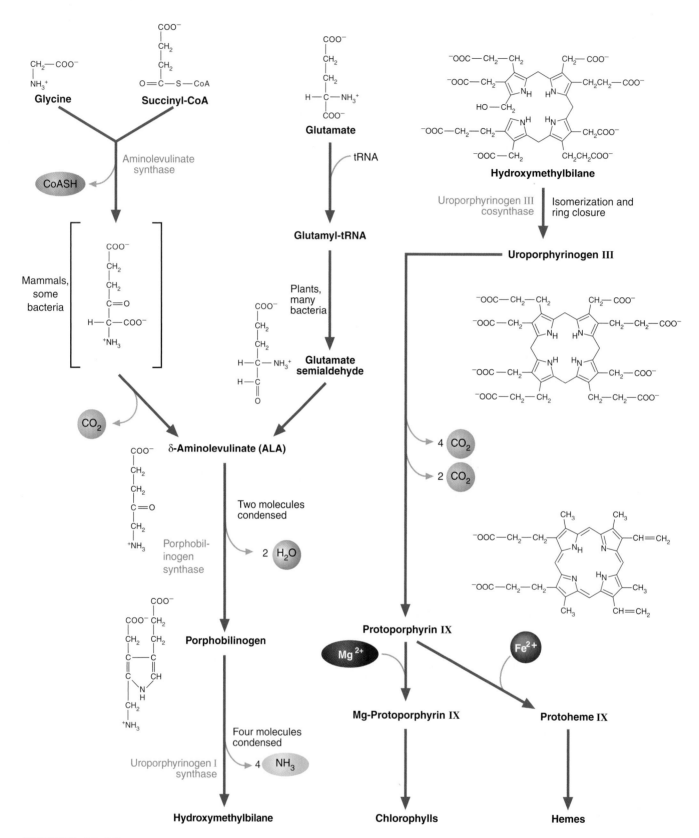

FIGURE 14.33

Heme and Chlorophyll Biosynthesis

In animals, the first reaction and the last three reactions occur within mitochondria, whereas the remaining four reactions occur in cytoplasm. In plants, the entire chlorophyll biosynthetic pathway occurs within chloroplasts. In plants and some bacterial species, ALA is synthesized from glutamate in a process involving glutamyl tRNA. Refer to Chapter 17 for a discussion of tRNA (transfer RNA).

synthase so that the asymmetrical molecule uroporphyrinogen III is produced. When four CO_2 molecules are removed, catalyzed by uroporphyrinogen decarboxylase, coproporphyrinogen is synthesized. This reaction is followed by the removal of two additional CO_2 molecules, thus forming protoporphyrinogen IX. Oxidation of the porphyrin ring's methylene groups forms protoporphyrin IX, the direct precursor of heme. The final step in the synthesis of heme (also called *protoheme IX*) is the insertion of Fe^{2+}, a reaction that occurs spontaneously but is accelerated by ferrochelatase.

The *porphyrias*, a group of rare inherited metabolic disorders, are characterized by toxic accumulations of porphyrins and various precursor molecules. Each type of porphyria is caused by the deficient activity of a specific enzyme (with the exception of ALAS) in the heme biosynthetic pathway. The major symptoms of the porphyrias may be neurological (e.g., pain, agitation, hallucinations, and/or convulsions) and gastrointestinal (e.g., abdominal pain or vomiting). Photosensitivity, caused by high blood levels of heme precursors, may cause skin blisters. The absorption of wavelengths of UV and visible light by heme precursors generates cytotoxic ROS.

Protoporphyrin IX is also a precursor of the chlorophylls (Figure 14.33). After magnesium (Mg^{2+}) has been incorporated, the enzyme Mg-protoporphyrin methylesterase catalyzes the addition of a methyl group to form Mg-protophorphyrin IX monomethylester. This molecule is then converted to chlorophyll in several light-induced reactions.

KEY CONCEPTS

- Heme is an iron- and nitrogen-containing porphyrin ring system synthesized from glycine and succinyl-CoA.

- Protoporphyrin IX, the precursor of heme, is also a precursor of the chlorophylls.

QUESTION 14.9

Identify each of the following biomolecules. Explain the function of each in biochemical processes.

(a)   (b)   (c)   (d)

Chapter**Summary**

1. Nitrogen, found in proteins, nucleic acids, and myriad other biomolecules, is an essential element in living systems. Biologically useful nitrogen, a scarce resource, is produced in a process referred to as nitrogen fixation. The nitrogenase enzyme complex that converts N_2 to NH_3 resides in some free soil bacteria, cyanobacteria, and symbiotic root nodule bacteria.

2. Organisms vary widely in their ability to synthesize amino acids. Some organisms (e.g., plants and some microorganisms) can produce all required amino acid molecules from fixed nitrogen. Animals can produce only some amino acids.

Nonessential amino acids are produced from readily available precursor molecules, whereas essential amino acids must be acquired in the diet.

3. Two types of reaction play prominent roles in amino acid metabolism. In transamination reactions, new amino acids are produced when α-amino groups are transferred from donor α-amino acids to acceptor α-keto acids. Because transamination reactions are reversible, they play an important role in both amino acid synthesis and degradation. Ammonium ions or the amide nitrogen of glutamine can also be directly incorporated into amino acids and eventually other metabolites.

4. On the basis of the biochemical pathways in which they are synthesized, the amino acids can be divided into six families: glutamate, serine, aspartate, pyruvate, aromatics, and histidine.

5. Amino acids are precursors of many physiologically important biomolecules. Many of the processes that synthesize these molecules involve the transfer of carbon groups. Because many of these transfers involve one-carbon groups (e.g., methyl, methylene, methenyl, and formyl), the overall process is referred to as one-carbon metabolism. *S*-Adenosylmethionine (SAM) and tetrahydrofolate (THF) are the most important carriers of one-carbon groups.

6. Molecules derived from amino acids include several neurotransmitters (e.g., GABA, the catecholamines, serotonin, and histamine) and hormones (e.g., melatonin). Glutathione is an amino acid derivative that plays an essential role in cells. Alkaloids are a diverse group of basic nitrogen-containing molecules. The role of alkaloids in the plants that produce them is poorly understood. Several alkaloid molecules have profound physiological effects on animals. The nucleotides, molecules that serve as the building blocks of the nucleic acids (as well as energy sources and metabolic regulators), possess heterocyclic nitrogenous bases as part of their structures. These bases, called the purines and the pyrimidines, are derived from various amino acid molecules. Heme is an example of a complex heterocyclic ring system that is derived from glycine and succinyl-CoA. The biosynthetic pathway that produces heme is similar to the one that produces chlorophyll in plants.

 Take your learning further by visiting the **companion website** for Biochemistry at **www.oup.com/us/mckee** where you can complete a multiple-choice quiz on synthesis to help you prepare for exams.

Suggested**Readings**

Bellinger, D. C., Lead, *Pediatrics* 113:1016–1022, 2004.

Christensen, K., and MacKenzie, R. E., Mitochondrial One-Carbon Metabolism Is Adapted to the Specific Needs of Yeast, Plants, and Mammals, *BioEssays* 28:595–605, 2006.

Gotoh, T., and Mori, M., Nitric Oxide and Endoplasmic Reticulum Stress, *Anterioscler. Thromb. Vasc. Biol.* 26:1439–1446, 2006.

Hawkins, R. A., O'Kane, R. L., Simpson, I. A., and Vina, J. R., Structure of the Blood-Brain Barrier and Its Role in the Transport of Amino Acids, *J. Nutr.* 136:218S–226S, 2006.

Igarshi, R. Y., and Seefeldt, L. C., Nitrogen Fixation: The Mechanism of the Mo-Dependent Nitrogenase, *Crit. Rev. Biochem. Mol. Biol.* 38:351–384, 2003.

Kauppinen, R., Porphyrias, *Lancet* 365:241–252, 2005.

Lefler, C. W., Parfenova, H., Jagger, J. H., and Wang, R., Carbon Monoxide and Hydrogen Sulfide: Gaseous Messengers in Cerebrovascular Circulation, *J. Appl. Physiol.* 100:1065–1076, 2006.

Magni, G., Amici, A., Emanuelli, M., Orsomando, G., Raffaelli, N., and Ruggieri, S., Enzymology of NAD$^+$ Homeostasis in Man, *Cell Mol. Life Sci.* 61:19–34, 2004.

Wu, G., Fang, Y.-Z., Yang, S., Lupton, J. R., and Turner, N. D., Glutathione Metabolism and Its Implications for Health, *J. Nutr.* 134:489–492, 2004.

Wu, L., and Wang, R., Carbon Monoxide: Endogenous Production, Physiological Functions, and Pharmacological Applications, *Pharmacol. Rev.* 57:585–630, 2005.

Key**Words**

alkaloid, *539*

amethopterin, *530*

amino acid pool, *507*

analogue, *530*

biogenic amine, *533*

branched-chain amino acid, *503*

catecholamine, *522*

essential amino acid, *503*

methotrexate, *530*

neurotransmitter, *533*

nitrogen fixation, *504*

nonessential amino acid, *503*

nucleoside, *540*

pernicious anemia, *526*

purine, *540*

pyrimidine, *540*

racemization, *509*

transamination, *503*

vitamin B$_{12}$, *526*

Review**Questions**

These questions are designed to test your knowledge of the key concepts discussed in this chapter, before moving on to the next chapter. You may like to compare your answers to the solutions provided in the back of the book and in the accompanying Study Guide.

1. Define the following terms:
   a. essential amino acid
   b. nitrogen balance
   c. *de novo* pathway
   d. biogenic amine
   e. vitamin B$_{12}$

2. Why are transamination reactions important in both the synthesis and degradation of amino acids?

3. Why are nitrogen compounds limited in the biosphere? Give two reasons.

4. Nitrogenase complexes are irreversibly inactivated by oxygen. Explain how nitrogen-fixing bacteria solve this problem.

5. Define the following terms:
   a. analogue
   b. pernicious anemia
   c. excitatory neurotransmitter
   d. inhibitory neurotransmitter
   e. retrograde neurotransmitter

6. Use reaction equations to illustrate how α-ketoglutarate is converted to glutamate. Name the enzymes and cofactors required.

7. In PLP-catalyzed reactions, the pyridinium ring acts as an electron sink. Describe this process.

8. The concentrations of the following amino acids differ significantly in blood flowing from the liver to the rest of the body and in the nutrient pool entering the enterocyte. Explain why the concentration changes in each case.
   a. isoleucine
   b. glutamine
   c. serine
   d. valine
   e. alanine

9. Define the following terms:
   a. blood-brain barrier
   b. catecholamine
   c. tetrahydrobiopterin
   d. L-DOPA
   e. seasonal affective disorder

10. What are the two major classes of neurotransmitter? How do their modes of action differ? Give an example of each type of neurotransmitter.

11. Illustrate the pathways to synthesize the following amino acids:
    a. glutamine
    b. methionine
    c. threonine
    d. glycine
    e. cysteine

12. Determine the synthetic family to which each of the following amino acids belongs:
    a. alanine
    b. phenylalanine
    c. methionine
    d. tryptophan
    e. histidine
    f. serine

13. Define the following terms:
    a. LSD
    b. alkaloid
    c. *anti*-adenosine
    d. PRPP
    e. Lesch-Nyhan syndrome

14. What are the two most important carriers in one-carbon metabolism? Give two examples of processes in which each one participates.

15. Glutathione is an important intracellular thiol. List five functions of glutathione in the body.

16. List ten essential amino acids in humans. Why are they essential?

17. Indicate which of the following compounds are nucleosides, nucleotides, or purine or pyrimidine bases, providing a correct name in each case.

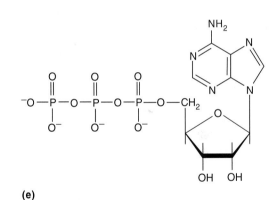

(a)   (b)   (c)   (d)   (e)

18. Define the following terms:
    a. heme oxygenase
    b. orotic aciduria
    c. thioredoxin
    d. *ALAS*
    e. creatine

19. In pyrimidine nucleosides the anti conformation predominates because of steric interactions with pentose. Do the purine nucleosides have similar interactions?

20. Describe the steric interactions that determine the conformations that pyrimidine nucleosides assume.

21. Outline the reactions involved during the assembly of the purine ring. Include structures in your answer.

22. Referring to Question 21, calculate the number of ATP molecules that are required to synthesize a purine. Referring to the purine salvage pathway, calculate the number of ATP molecules that are required to prepare the same molecule. How many ATP molecules are saved by the purine salvage pathway?

23. Explain how the γ-glutamyl cycle acts to transport amino acids across a membrane. How does the location of the γ-glutamyl transpeptidase help drive this process?

24. The amino acids glutamine and glutamate are central to amino acid metabolism. Explain.

25. A mutant bacterium is unable to synthesize glycine. What intermediate in purine biosynthesis will accumulate?

26. Transamination reactions have been described as ping-pong reactions. Use the reaction of alanine with α-ketoglutarate to indicate how this ping-pong reaction works.

27. Pyridoxal phosphate acts as an intermediate carrier of amino groups during transamination reactions. Write a series of reactions to show the role of pyridoxal phosphate in the reaction of alanine and α-ketoglutarate.

28. What is the biologically active form of folic acid? How is it formed?

29. Identify the following biomolecules. Describe the metabolic role of each.

(a) (b) (c)

(d) (e)

30. Both oxygen and nitrogen are present in the atmosphere as gases. Oxygen is reactive and nitrogen is relatively inert. What features of the two molecules account for this difference?

31. Long before living organisms developed the nitrogenase system, there was ammonia in Earth's atmosphere. Suggest a naturally occurring method of fixing nitrogen.

32. What carbon in uracil is derived from carbon dioxide?

33. What is the function of ATP in the conversion of glutamate to glutamine?

34. Individuals with inadequate tyrosine metabolism are often light-sensitive and easily develop severe cases of sunburn. Explain. [*Hint*: The skin pigment melanin (p. 53) is derived from L-DOPA.]

35. In the nitrogen reductase system, for every molecule of ammonia released, one molecule of hydrogen gas is also produced. What is the source of the hydrogen gas?

ThoughtQuestions

These questions are designed to reinforce your understanding of all of the key concepts discussed in the book so far, including this chapter and all of the chapters before it. They may not have one right answer! The authors have provided possible solutions to these questions in the back of the book and in the accompanying Study Guide, for your reference.

36. Trace the following radiolabeled molecule through the biosynthetic route to AMP synthesis.

H_3N^+—$^{14}CH_2$—COO^-

37. Although they are consumed by animals in the diet, the purine and pyrimidine bases (unlike fatty acids and sugars) are not used to generate energy. Explain.

38. Why do marathon runners prefer beverages with sugar instead of amino acids during a long run?

39. When susceptible people consume monosodium glutamate, they experience several extremely unpleasant symptoms, such as increased blood pressure and body temperature. Use your knowledge of glutamate activity to explain these symptoms.

40. Erythropoietic protoporphyria is a form of porphyria in which ferrochelatase levels are abnormally low. As a result of this deficiency, protoporphyrin accumulates in bone marrow, blood plasma, red blood cells, skin, and liver. Treatment with β-carotene provides some protection from sun exposure. Liver transplantation remains the only treatment for liver damage. When such patients undergo this surgery, their skin is covered and the lights in the operating theater are dimmed. Explain.

41. Radiation exerts part of its damaging effect by causing the formation of hydroxyl radicals. Write a reaction equation to explain how glutathione acts to protect against this form of radiation damage.

42. In pyrimidine nucleotide synthesis, the carbon and nitrogen atoms are derived from bicarbonate, aspartate, and glutamine. Devise a simple experiment to prove the source of the nitrogen atoms. Do not forget to take into account the nitrogen exchange in amino acids.

43. Most amino acids can be readily interconverted with the corresponding α-keto acid. This is not true of lysine. In the multistep process to form the corresponding α-keto acid of lysine, an early step leads to the formation of a Schiff base. Explain how this product might be formed.

44. By definition, essential amino acids are not synthesized by an organism. Arginine is classified as an essential amino acid in children even though it is part of the urea cycle. Explain.

45. If an organism is incubated with $^{14}CH_2(OH)CH(NH_2)COOH$, what positions in the purine ring will be labeled?

46. In PLP-catalyzed reactions, the bond broken in the substrate molecule must be perpendicular to the plane of the pyridinium ring. Considering the bonds present in this ring, describe why this arrangement stabilizes the carbanion.

47. During your experimental investigation of the conversion of glutamine to proline, you have labeled the γ-carbonyl group with ^{14}C. What carbon in the proline product will be labeled?

48. In the biosynthesis of lysine, aspartate β-semialdehyde reacts with pyruvate to produce dihydropicolinate. Indicate which carbon of the pyruvate corresponds to each carbon of the product.

49. Pyruvate is a precursor of isoleucine. Which carbon of isoleucine corresponds to C-2 of pyruvate?

50. In water, cytosine will gradually convert to uracil. Using reactions you have learned in this and previous chapters, show how the conversion takes place.

Nitrogen Metabolism II: Degradation

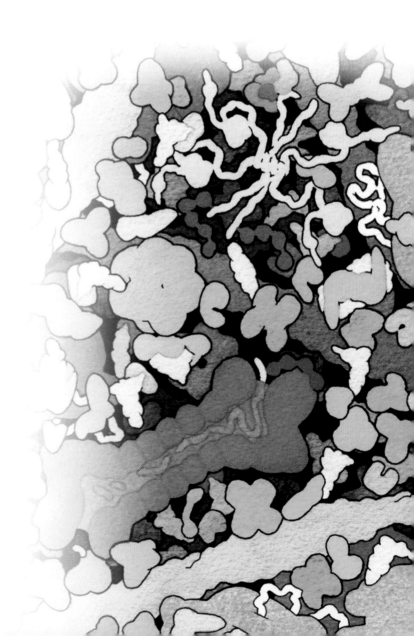

The Proteasome The proteasome is a large multisubunit molecular machine that contains several proteolytic activities and ATPases. It destroys worn-out or abnormal proteins, and short-lived signal proteins. In this illustration p53 (shown in yellow), a cell cycle regulator protein, is the target protein to be degraded. Before its destruction, p53 is tagged with several ubiquitin molecules (shown in red). Ubiquitin chains are required for a protein's entry into the proteasome.

Overview

▼

THE METABOLISM OF NITROGEN-CONTAINING MOLECULES SUCH AS
PROTEINS AND NUCLEIC ACIDS DIFFERS SIGNIFICANTLY FROM THAT OF
carbohydrates and lipids. Whereas the latter molecules can be stored and
mobilized as needed for biosynthetic reactions or for energy generation, there
is no nitrogen-storing molecule. (One exception to this rule is storage protein
in seeds.) Organisms must constantly replenish their supply of usable nitrogen
to replace organic nitrogen that is lost in catabolism. For example, animals
must have a steady supply of amino acids in their diets to replace the nitrogen
excreted as urea, uric acid, and other nitrogenous waste products.

Despite their apparent stability, most living cells are constantly undergoing
renovation. One of the most obvious aspects of cellular renovation is the
turnover of protein and nucleic acids, a process that results in the continuous flow of nitrogen atoms through living organisms. Considering the difficulty and expense associated with the fixing of nitrogen (and its subsequent
scarcity), it is not surprising that living organisms recycle organic nitrogen
into a variety of metabolites before the element is reconverted to its inorganic
form. In some organisms (e.g., plants and microorganisms) this process does
not begin until death. Then *decomposers*, microorganisms that inhabit soil and
water, convert the organic nitrogen of all dead organisms to ammonia.
Ammonia, or its oxidized products nitrate and nitrite, may subsequently be
absorbed and used by nearby organisms. Alternatively, nitrate may be converted
to atmospheric nitrogen, a process referred to as *denitrification*.

Animals, which typically have a more energetic and aggressive lifestyle,
appear to be more wasteful of organic nitrogen. To maintain metabolic flexibility, animals have had to develop mechanisms for disposing of excess and toxic
nitrogen-containing molecules (i.e., amino acids and nucleotides) that are not
immediately required in cellular processes. Such molecules are converted into
nitrogenous waste. Although many variations are observed among species, the
following generalizations can be made. The nitrogen in amino acids is removed
by deamination reactions and converted to ammonia. The toxic nature of this
molecule requires that it be detoxified and/or excreted as fast as it is generated.
Many aquatic animals can excrete ammonia itself, which dissolves in the surrounding water and is quickly diluted. (These organisms are referred to as
ammonotelic.) Terrestrial animals, which must conserve body water, convert
ammonia to molecules that can be excreted without a large loss of water.
Mammals, for example, convert ammonia to urea. (Urea-producing organisms
are referred to as *ureotelic*.) Other animals, such as birds, certain reptiles, and
insects, which have even more stringent water conservation problems, are called
uricotelic because they convert ammonia to uric acid. In many animals (e.g.,
humans and birds), uric acid is also the nitrogenous waste product of purine
nucleotide catabolism.

Nitrogen catabolic pathways are similar in many organisms, and most research
efforts in nitrogen catabolism have concentrated on mammals. The mammalian
pathways are, therefore, the focus of this chapter. Chapter 15 begins by tracing
the nitrogen atoms in proteins to amino acids and their degradation products.
A brief review of the degradation of several neurotransmitters follows. The
chapter ends with descriptions of the catabolic pathways of the nucleotides and
the porphyrin heme.

15.1 PROTEIN TURNOVER

The cellular concentration of each type of protein is a consequence of a balance between its synthesis and its degradation. Although it appears to be wasteful, the continuous degradation and resynthesis of proteins, a process referred to as **protein turnover**, serves several purposes. First of all, metabolic flexibility is afforded by relatively quick changes in the concentrations of key regulatory enzymes, peptide hormones, and receptor molecules. Protein turnover also protects cells from the accumulation of abnormal proteins. Finally, numerous physiological processes are just as dependent on timely degradative reactions as they are on synthetic ones. For example, the progression of eukaryotic cells through the phases of the cell cycle (Section 18.1) is regulated by the precisely timed synthesis and degradation of a class of proteins called the *cyclins*.

Proteins differ significantly in their turnover rates, which are measured in half-lives. (A *half-life* is the time required for 50% of a specified amount of a protein to be degraded.) Proteins that play structural roles typically have long half-lives. For example, some connective tissue proteins (e.g., the collagens) often have half-lives that are measured in years. In contrast, the half-lives of regulatory enzymes are typically measured in minutes. Several selected examples are listed in Table 15.1.

Although some proteins are degraded by proteolytic enzymes in cytoplasm (e.g., Ca^{2+}-activated calpains) and lysosomes (e.g., cathepsins), most cellular proteins are degraded by a rapid, efficient, and elaborate mechanism referred to as the **ubiquitin proteasomal system** (UPS). UPS-mediated protein degradation, which occurs both in the cytoplasm and in the nucleus, is initiated with a covalent modification referred to as ubiquination.

The mechanisms that target protein for destruction by ubiquination or by other degradative processes are not fully understood. However, the following features of proteins appear to mark them for destruction:

1. **N-terminal residues**. Very short-lived proteins often have very basic or bulky hydrophobic N-terminal residues. More stable proteins characteristically have sulfur-containing, hydroxyl-containing, or nonbulky hydrophobic amino acids at the N-terminus.

2. **Peptide motifs**. Proteins with certain homologous sequences are rapidly degraded. For example, proteins that have extended sequences containing proline, glutamate, serine, and threonine have half-lives of less than 2 hours. (PEST sequences are named for the one-letter abbreviations for these amino acids. See Table 5.1.) Ensuring rapid ubiquination is the *cyclin destruction box*, a set of homologous nine-residue sequences near the N-terminus of cyclins.

3. **Oxidized residues**. Oxidized amino acid residues (i.e., residues that are altered by oxidases or by attack by ROS) promote protein degradation.

TABLE 15.1 Human Protein Half-Lives

Protein	Approximate Value of Half-Life (h)
Ornithine decarboxylase	0.5
Tyrosine aminotransferase	2
Tryptophan oxygenase	2
PEP carboxykinase	5
Arginase	96
Aldolase	118
Glyceraldehyde-3-phosphate dehydrogenase	130
Cytochrome c	150
Hemoglobin	2880

Once ubiquinated, the proteins are transferred to massive proteolytic molecular machines called **proteasomes** that cleave them into peptide fragments with an average of seven to eight amino acid residues. Such fragments are further degraded by cytoplasmic proteases to amino acids that can be recycled into new protein molecules. **Ubiquination** (Figure 15.1), the attachment of a small, highly conserved 76-residue protein called ubiquitin to worn-out or damaged proteins, or short-lived regulatory proteins, occurs in several stages and involves three enzyme classes: E1, E2, and E3. In the first step, E1 (*ubiquitin-activating enzyme*) activates a ubiquitin molecule via adenylation and transfers it to an active site thiol of E1 to form a high-energy thioester. The C-terminal glycine carboxyl group of the ubiquitin molecules participates in this reaction. Ubiquitin is then transferred to an active site thiol of E2 (*ubiquitin-conjugating enzyme*) via a transthiolation reaction. There is only one E1, but as many as 25 E2 enzymes in mammalian cells. The E2 enzymes vary in their specificity for association with E3 (*ubiquitin ligase*). The E3 enzymes interact with E2 and the target protein and transfer ubiquitin

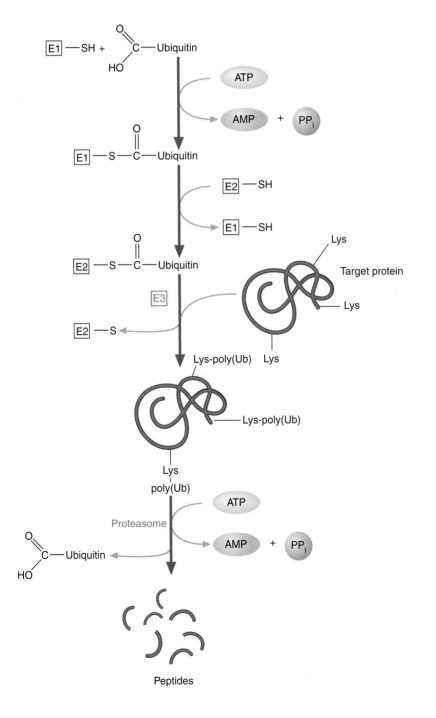

FIGURE 15.1

Ubiquination of Protein

Three enzymes are involved in preparing ubiquitin for its role in protein degradation. In the first step, an activating enzyme E1 forms a thiol ester with ubiquitin. (This reaction is driven by the hydrolysis of ATP to AMP.) Ubiquitin is then transferred from E1 to E2 (ubiquitin-conjugating enzyme). Next, E3 (ubiquitin ligase) binds both E2 and the target protein and mediates the transfer of ubiquitin from E2 to specific lysine side chains of the target protein. Enzyme E3 is also responsible for generating the polyubiquitin chains that are usually required for proteasome targeting. Most cells possess a single type of E1 and numerous families of E2 and E3 to accomplish high-specificity protein turnover.

FIGURE 15.2

The Proteasome

The 20S proteasome (core particle) (a) is a barrel-shaped structure that contains 28 proteins: two α rings with seven subunits each and two β rings with seven subunits each. A cutaway view (b) reveals the inner catalytic chamber.

KEY CONCEPTS

- Protein turnover, the continuous synthesis and degradation of proteins, provides living organisms with metabolic flexibility and protects cells from the accumulation of abnormal proteins.

- Most cellular proteins are degraded by the ubiquitin proteasomal system to yield short peptides.

- The amino acid products of the peptides that are cleaved by cytoplasmic proteases enter the amino acid pool and are available for incorporation into new protein molecules.

to a specific internal lysine side chain of the target protein via a thioester-to-amide transition. The high specificity required for regulated proteolysis is provided by E3. In mammals, about 1000 E3 genes have been identified. Chains with a minimum of four ubiquitin monomers are required for transfer of the target protein to the proteasome. The specificity of ubiquination is derived in part from the substrate specificities of a large number of E2s and E3s.

There are also temporal and spatial regulatory mechanisms. For example, the ubiquination of some proteins (e.g., signal transduction molecules) is limited to those that have already undergone certain covalent modifications such as phosphorylation. Spatial regulation is accomplished by the subcellular localization of specific E2s and E3s and modifying enzymes such as specific kinases. It is noteworthy that the covalent modification of certain proteins with ubiquitin or ubiquitin-like proteins promotes functions other than proteolysis. Examples include transcription, DNA repair, and membrane transport.

The proteasome (Figure 15.2) is a large (2000 kD) multisubunit complex in the form of a hollow cylinder with dimensions of 1500 nm by 1150 nm. Referred to as the 26S proteasome, this structure consists of a 20S core particle and two 19S regulatory particles. [The Svedberg unit (S) is a measure of the rate at which a particle sediments in an ultracentrifuge. Since S values are related to both the mass and the shape of particles, they are not additive.]

The 20S particle consists of four heptameric protein rings ($\alpha_7\beta_7\beta_7\alpha_7$). The two inner β rings possess three different types of proteolytic activity that face the inner chamber. The α-ring components have an N-terminal segment that limits access to the proteolytic chamber to unfolded peptide chains. The 19S particles consist of a nine-subunit *lid* and a ten-subunit *base*. Six of the base subunits possess ATPase activity and are thought to facilitate the protein unfolding that is critical for threading substrate polypeptides into the catalytic chamber. The lid subunits participate in substrate selection and ubiquitin processing. This gate-keeping function ensures that only properly polyubiquinated target proteins are translocated into the catalytic chamber for degradation. One protein is degraded at a time, and the six- to ten-residue peptide products are released from the proteasome for hydrolysis to free amino acids by cytoplasmic proteases.

15.2 AMINO ACID CATABOLISM

Despite the complexity of amino acid degradative pathways, the following generalizations can be made. The catabolism of the amino acids usually begins by removing the amino group. Amino groups can then be disposed of in urea synthesis. The carbon skeletons produced from the standard amino acids are then degraded to form seven metabolic products: acetyl-CoA, acetoacetyl-CoA, pyruvate, α-ketoglutarate, succinyl-CoA, fumarate, and oxaloacetate. Depending on the animal's current metabolic requirements, these molecules are used to synthesize fatty acids or glucose or to generate energy. Amino acids degraded to form acetyl-CoA or acetoacetyl-CoA are referred to as **ketogenic** because they can be converted to either fatty acids or ketone bodies. The carbon skeletons of the **glucogenic** amino acids, which are degraded to pyruvate or a citric acid cycle intermediate, can then be used in gluconeogenesis. Discussions of deamination pathways and urea synthesis are followed by descriptions of the pathways that degrade carbon skeletons.

Deamination

The removal of the α-amino group from amino acids involves two types of biochemical reaction: transamination and oxidative deamination. Both reactions have been described (Section 14.2). (Recall that transamination reactions occupy important positions in nonessential amino acid synthesis.) Because these reactions are reversible, amino groups are easily shifted from abundant amino acids and used to synthesize those that are scarce. Amino groups become available for urea synthesis when amino acids are in excess. Urea is synthesized in especially large amounts when the diet is high in protein or when there is massive breakdown of protein, for example, during starvation.

In muscle, excess amino groups are transferred to α-ketoglutarate to form glutamate:

$$\alpha\text{-Ketoglutarate} + \text{L-Amino acid} \rightleftharpoons \text{L-Glutamate} + \alpha\text{-Keto acid}$$

The amino groups of glutamate molecules are transported in blood to the liver by the alanine cycle (Figure 8.11):

$$\text{Pyruvate} + \text{L-Glutamate} \rightleftharpoons \text{L-Alanine} + \alpha\text{-Ketoglutarate}$$

In the liver, glutamate is formed as the reaction catalyzed by alanine transaminase is reversed. The oxidative deamination of glutamate yields α-ketoglutarate and NH_4^+.

In most extrahepatic tissues, the amino group of glutamate is released via oxidative deamination as NH_4^+. Ammonia is carried to the liver as the amide group of glutamine. The ATP-requiring reaction in which glutamate is converted to glutamine is catalyzed by glutamine synthetase:

$$\text{L-Glutamate} + NH_4^+ + \text{ATP} \longrightarrow \text{L-Glutamine} + \text{ADP} + P_i$$

After its transport to the liver, glutamine is hydrolyzed by glutaminase to form glutamate and NH_4^+. An additional NH_4^+ is generated as glutamate dehydrogenase converts glutamate to α-ketoglutarate:

$$\text{L-Glutamine} + H_2O \longrightarrow \text{L-Glutamate} + NH_4^+$$

$$\text{L-Glutamate} + H_2O + NAD^+ \longrightarrow \alpha\text{-Ketoglutarate} + \text{NADH} + H^+ + NH_4^+$$

Most of the ammonia generated in amino acid degradation is produced by the oxidative deamination of glutamate. Additional ammonia is produced in several other reactions catalyzed by the following enzymes.

1. **L-Amino acid oxidases**. Small amounts of ammonia are generated by various L-amino acid oxidases, found in liver and kidney, that require a flavin mononucleotide (FMN) coenzyme. FMN is regenerated from $FMNH_2$ by reacting with O_2 to form H_2O_2.

2. **Serine and threonine dehydrases**. Serine and threonine are not substrates in transamination reactions. Their amino groups are removed by the pyridoxal phosphate–requiring hepatic enzymes serine dehydratase and threonine dehydratase. The carbon skeleton products of these reactions are pyruvate and α-ketobutyrate, respectively.

3. **Bacterial urease**. A major source of ammonia in liver (approximately 25%) is produced by the action of certain bacteria in the intestine that possess the

enzyme urease. Urea present in the blood circulating through the lower digestive tract diffuses across cell membranes and into the intestinal lumen. Once urea has been hydrolyzed by bacterial urease to form ammonia, the latter substance diffuses back into the blood, which transports it to the liver.

4. **Adenosine deaminase.** Adenosine deaminase is an enzyme in nucleotide metabolism. It catalyzes the conversion of AMP to IMP, a reaction that releases the C-6 amino group of adenine as NH_4^+.

Urea Synthesis

In ureotelic organisms the urea cycle disposes of approximately 90% of surplus nitrogen. As shown in Figure 15.3, urea is formed from ammonia, CO_2, and aspartate in a cyclic pathway referred to as the **urea cycle**. Because the urea cycle was discovered by Hans Krebs and Kurt Henseleit, it is often referred to as the **Krebs urea cycle** or the *Krebs-Henseleit cycle*. The overall equation for urea synthesis is

$$CO_2 + NH_4^+ + Aspartate + 3\ ATP + 2\ H_2O \longrightarrow$$
$$Urea + Fumarate + 2\ ADP + 2\ P_i + AMP + PP_i + 5\ H^+$$

Urea synthesis, which occurs in hepatocytes, begins with the formation of carbamoyl phosphate in the matrix of mitochondria. The substrates for this reaction, catalyzed by carbamoyl phosphate synthetase I, are NH_4^+ and HCO_3^-.

Carbamate

Carbamoyl Phosphate

Carbamoyl phosphate synthesis is essentially irreversible because two molecules of ATP are consumed. (One is used to activate HCO_3^-. The second molecule is used to phosphorylate carbamate.) Carbamoyl phosphate subsequently reacts with ornithine to form citrulline. Citrulline is synthesized in a nucleophilic acyl substitution reaction in which the side chain amino group of ornithine is the nucleophile and phosphate is the leaving group.

Ornithine   **Carbamoyl Phosphate**

Citrulline

This reaction, catalyzed by ornithine transcarbamoylase, is driven to completion because phosphate is released from carbamoyl phosphate. (Recall from Table 4.1 that carbamoyl phosphate has a high phosphoryl group transfer potential.) Once formed, citrulline is transported to the cytoplasm, where it reacts with aspartate to form argininosuccinate. (The α-amino group of aspartate, formed

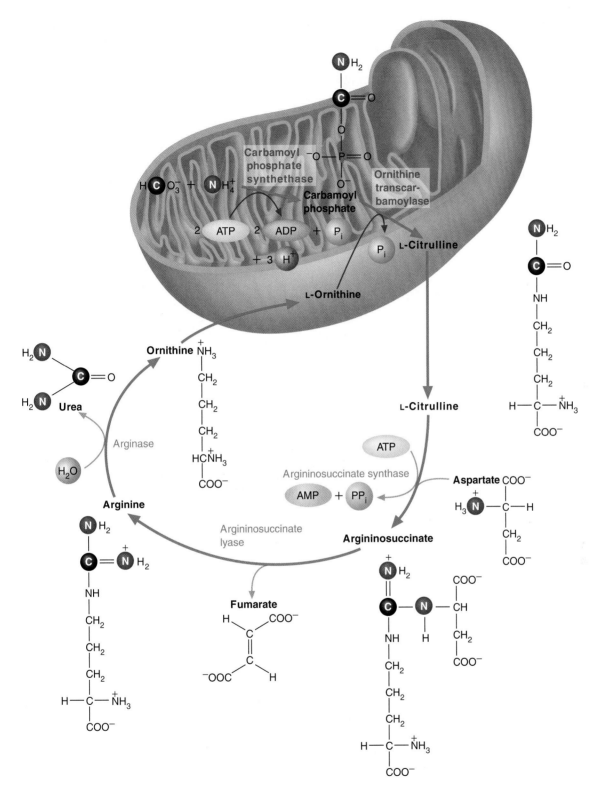

FIGURE 15.3

The Urea Cycle

The urea cycle converts NH_4^+ to urea, a less toxic molecule. The sources of the atoms in urea are shown in color. Citrulline is transported across the inner membrane by a carrier for neutral amino acids. Ornithine is transported in exchange for H^+ or citrulline. Fumarate is transported back into the mitochondrial matrix (for reconversion to malate) by carriers for α-ketoglutarate or tricarboxylic acids.

from oxaloacetate by transamination reactions in the liver, provides the second nitrogen that is ultimately incorporated into urea.) In this reaction, which is catalyzed by argininosuccinate synthase, citrulline is activated by reacting with ATP to form an citrulline-AMP intermediate and pyrophosphate. The amino nitrogen of aspartate, acting as a nucleophile, adds to the C＝N bond of the citrulline-AMP intermediate.

Citrulline-AMP

Aspartate

Arginosuccinate

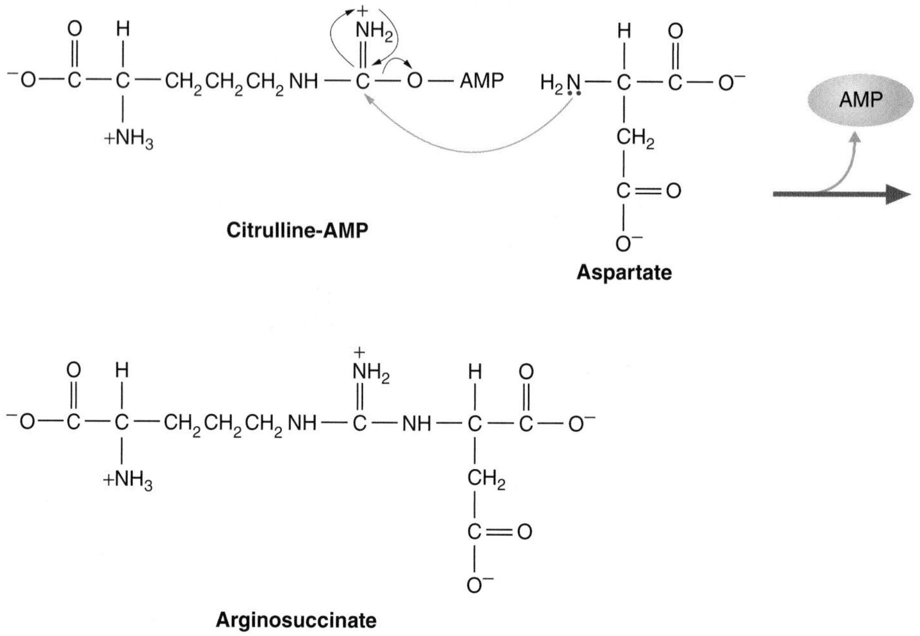

This reaction, a nucleophilic acyl substitution reaction that is pulled forward by the cleavage of pyrophosphate by pyrophosphatase, yields arginosuccinate. AMP is the leaving group. Argininosuccinate lyase subsequently cleaves argininosuccinate to form arginine (the immediate precursor of urea) and fumarate.

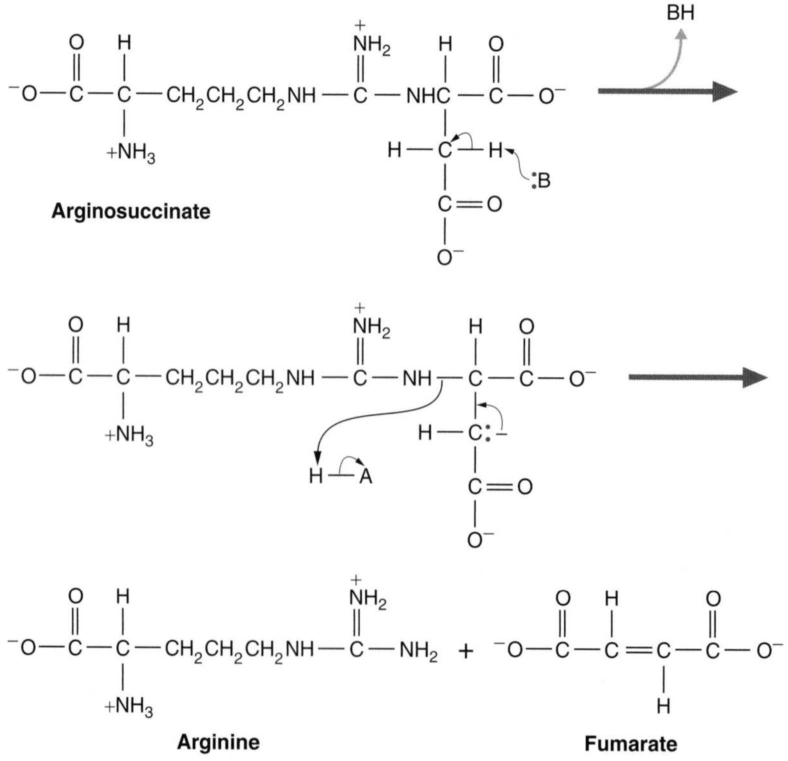

Arginosuccinate

Arginine

Fumarate

A histidine residue within the enzyme's active site, acting as a base (B:), removes a proton from the substrate to form a carbanion. The carbanion then expels the nitrogen to form a $C = C$ bond. The nitrogen accepts a proton from a proton donor (HA) (perhaps the protonated histidine).

In the final reaction of the urea cycle, arginase catalyzes the hydrolysis of arginine to form ornithine and urea.

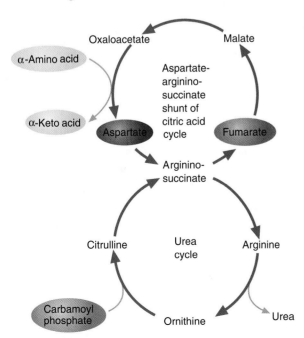

Arginine

Ornithine

Urea

Once it forms, urea diffuses out of the hepatocytes and into the bloodstream. It is ultimately eliminated in urine by the kidney. Ornithine returns to the mitochondria for condensation with carbamoyl phosphate to begin the cycle again. Because arginase is found in significant amounts only in the ureotelic animal liver, urea is produced only in this organ.

After its transport back into the mitochondrial matrix, fumarate is hydrated to form malate, a component of the citric acid cycle. The oxaloacetate product of the citric acid cycle can be used in energy generation, or it can be converted to glucose or aspartate. Four high-energy phosphates are consumed in the synthesis of one molecule of urea. Two molecules each of ATP are required to regenerate ATP from AMP and two ATPs from two ADPs. Energy is generated from the conversion of fumarate to oxaloacetate in the citric acid cycle. The relationship between the urea cycle and the citric acid cycle, often referred to as the **Krebs bicycle**, is outlined in Figure 15.4.

FIGURE 15.4

The Krebs Bicycle

The aspartate used in urea synthesis is generated from oxaloacetate, a citric acid cycle intermediate.

Hyperammonemia, a potentially fatal condition in which blood levels of NH_4^+ become excessive when the liver's capacity to synthesize urea is compromised, is discussed in the Biochemistry in Perspective box on that topic (see companion website at www.oup.com/us/mckee).

Control of the Urea Cycle

Urea is a toxic molecule. Its synthesis is, therefore, stringently regulated. There are long- and short-term regulatory mechanisms. The levels of all five urea cycle enzymes are altered by variations in dietary protein consumption. Within several days after a significant dietary change, there are twofold to threefold changes in enzyme levels. Several hormones are involved in the altered rates of enzyme synthesis. Glucagon and the glucocorticoids activate the transcription of urea cycle enzymes, whereas insulin represses their synthesis.

The urea cycle enzymes are controlled in the short term by the concentrations of their substrates. Carbamoyl phosphate synthetase I is also allosterically activated by *N-acetylglutamate*. This latter molecule is a sensitive indicator of the cell's glutamate concentration. (Recall that a significant amount of NH_4^+ is derived from glutamate.) *N*-Acetylglutamate is produced from glutamate and acetyl-CoA in a reaction catalyzed by *N*-acetylglutamate synthetase, which is allosterically activated by arginine.

KEY CONCEPTS

- Urea is synthesized from ammonia, CO_2, and aspartate.
- The urea cycle is carefully regulated.

QUESTION 15.1

Although arginine is an intermediate in the urea cycle, it is an essential amino acid in young animals. Suggest a reason for this phenomenon.

QUESTION 15.2

In some clinical circumstances, patients with hyperammonemia are treated with antibiotics. Suggest a rational basis for this therapy.

Catabolism of Amino Acid Carbon Skeletons

The α-amino acids can be grouped into classes according to their end products: acetyl-CoA, acetoacetyl-CoA, pyruvate, and several citric acid cycle intermediates. Each group is briefly discussed. The degradation pathways for the 20 α-amino acids found in proteins are outlined in Figure 15.5.

AMINO ACIDS FORMING ACETYL-CoA In all, 10 α-amino acids yield acetyl-CoA. This group is further divided according to whether pyruvate is an intermediate in acetyl-CoA formation. (Recall that pyruvate is converted to acetyl-CoA by the pyruvate dehydrogenase complex.) The amino acids whose degradation involves pyruvate are alanine, serine, glycine, cysteine, and threonine. The other five amino acids converted to acetyl-CoA by pathways not involving pyruvate are lysine, tryptophan, tyrosine, phenylalanine, and leucine. The two reaction sequences are outlined in Figures 15.6 and 15.7.

The individual catabolic pathways for these molecules are as follows:

1. **Alanine.** Recall that the reversible transamination reaction involving alanine and pyruvate is an important component of the alanine cycle discussed previously (Section 8.2).

2. **Serine.** As described, serine is converted to pyruvate by serine dehydratase.

3. **Glycine.** Glycine can be converted to serine by serine hydroxymethyltransferase. (The hydroxymethyl group is donated by N^5, N^{10}-methylene

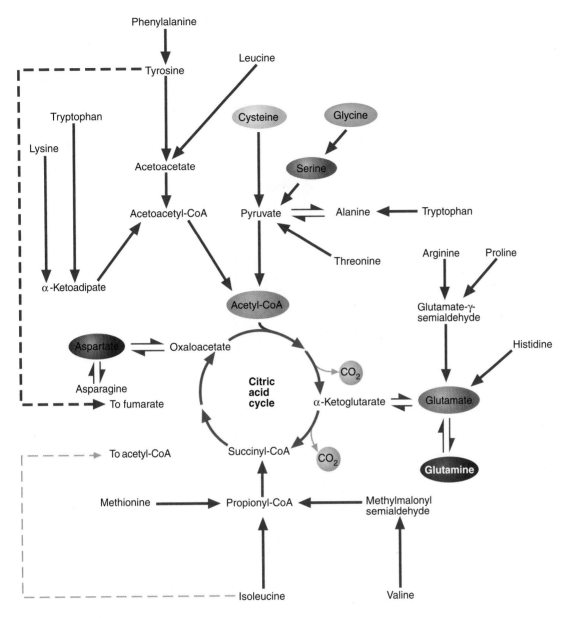

FIGURE 15.5

Degradation of the 20 α-Amino Acids Found in Proteins

The α-amino groups are removed early in the catabolic pathways. Carbon skeletons are converted to common metabolic intermediates.

THF as described in Section 14.3.) Then serine is converted to pyruvate, as previously described. Most glycine molecules, however, are degraded to CO_2, NH_4^+, and a methylene group that is removed by THF. The enzyme involved is glycine synthase (also referred to as glycine cleavage enzyme), which requires NAD^+.

4. **Cysteine**. In animals, cysteine is converted to pyruvate by several pathways. In the principal pathway, the conversion occurs in three steps. Initially, cysteine is oxidized to cysteine sulfate. Pyruvate is produced after a transamination and a desulfuration reaction.

5. **Threonine**. In the major degradative pathway, threonine is oxidized by threonine dehydrogenase to form α-amino-β-ketobutyrate. The latter molecule is metabolized further to form pyruvate, or it can be cleaved by α-amino-β-ketobutyrate lyase to form acetyl-CoA and glycine.

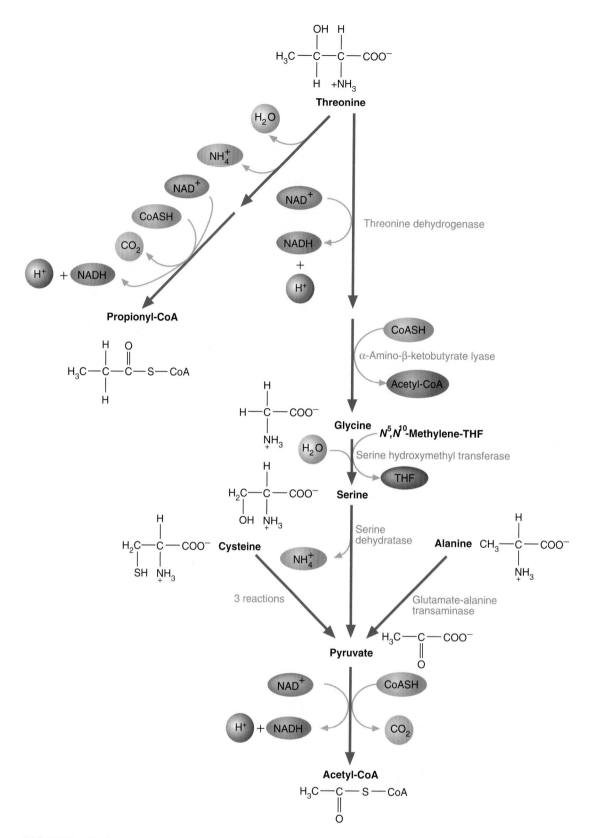

FIGURE 15.6

The Catabolic Pathways of Threonine, Glycine, Serine, Cysteine, and Alanine

Pyruvate is an intermediate in the conversion of these amino acids to acetyl-CoA. Note that glycine is also degraded by glycine synthase to form CO_2, NH_4^+ and N^5,N^{10}-methylene-THF in an NAD^+-requiring reaction. In primates, most threonine molecules are degraded to propionyl-CoA.

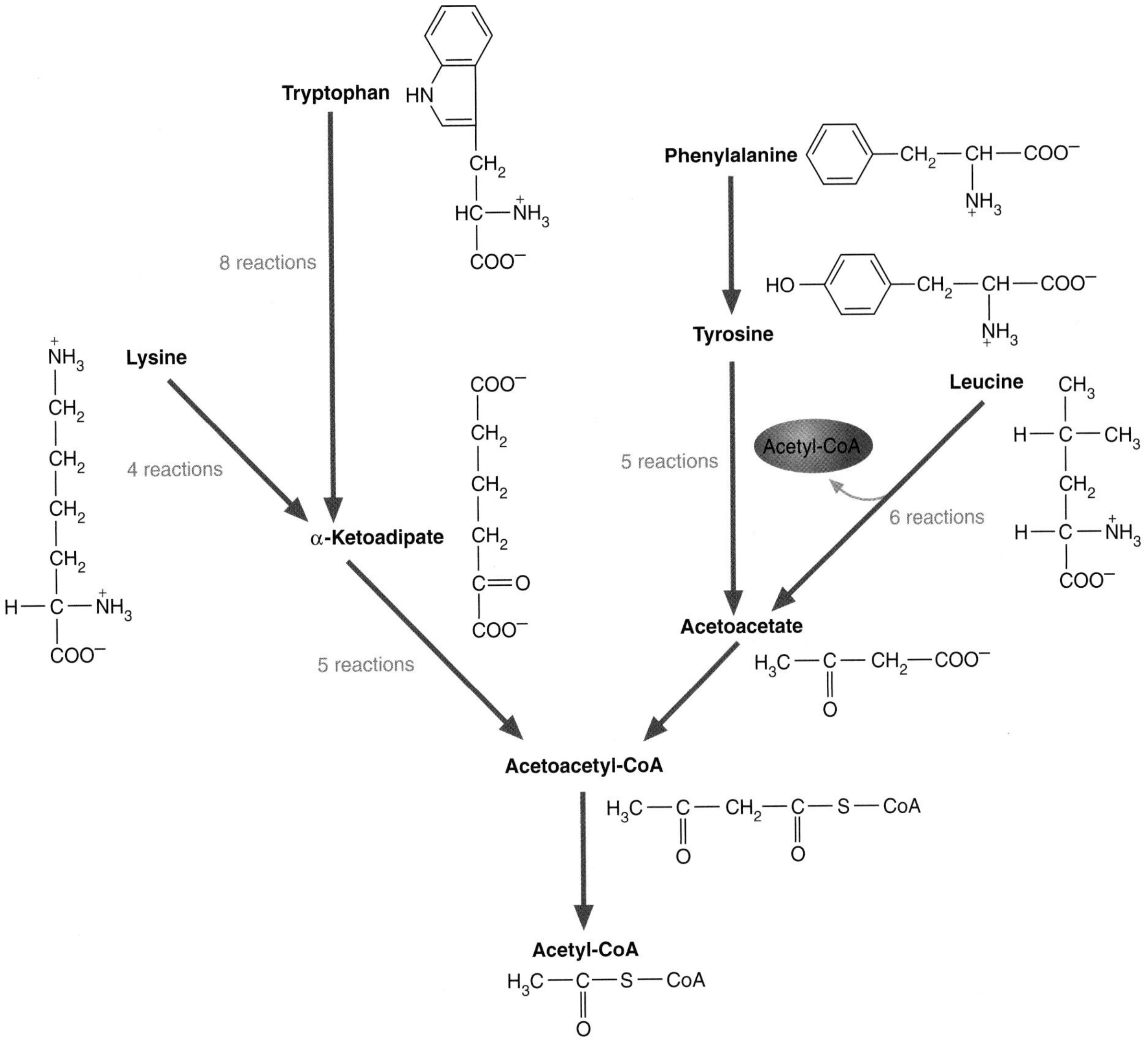

FIGURE 15.7

The Catabolic Pathways of Lysine, Tryptophan, Phenylalanine, Tyrosine, and Leucine

These pathways are long and complex. The number of reactions in each segment is indicated.

As previously discussed, glycine is converted to acetyl-CoA via pyruvate. Alternatively, threonine can be degraded to α-ketobutyrate by threonine dehydratase and subsequently to propionyl-CoA. Propionyl-CoA is then converted to succinyl-CoA (see p. 427).

6. **Lysine.** Lysine is converted to α-ketoadipate in a series of reactions that include two oxidations, removal of the side chain amino group, and a transamination. Acetoacetyl-CoA is produced in a further series of reactions that involve several oxidations, a decarboxylation, and a hydration. Acetoacetyl-CoA can be converted to acetyl-CoA in a reaction that is the reverse of a step in ketone body formation.

7. **Tryptophan.** Tryptophan is converted to α-ketoadipate in a long, complex series of eight reactions, which also yield formate and alanine. Acetyl-CoA is synthesized from α-ketoadipate as described for lysine. The alanine produced in this pathway is converted to acetyl-CoA via pyruvate.

8. **Tyrosine.** Tyrosine catabolism begins with a transamination and a dehydroxylation. Homogentisate is synthesized in the latter reaction, catalyzed

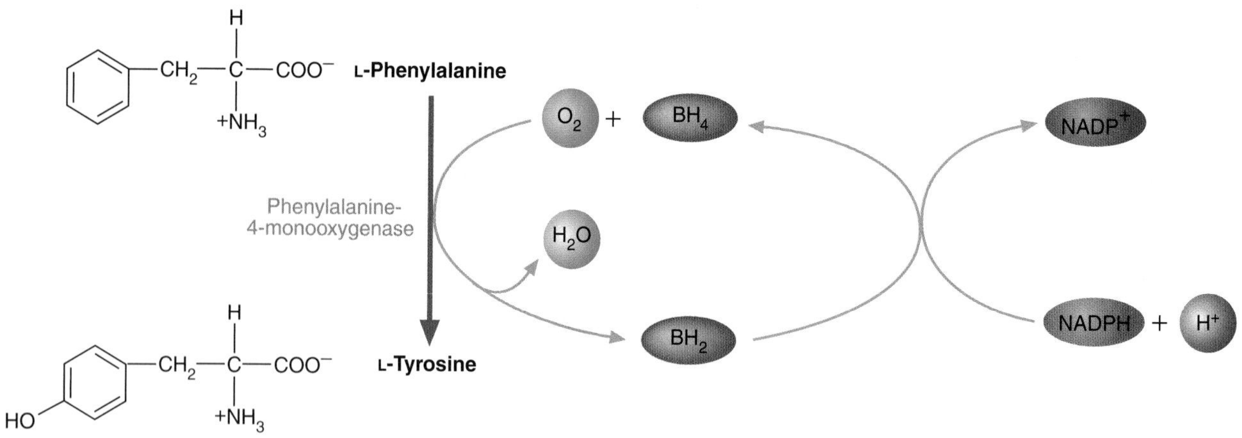

FIGURE 15.8

The Conversion of Phenylalanine to Tyrosine

The reaction catalyzed by phenylalanine-4-monooxygenase is irreversible. The electrons required for the hydroxylation of phenylalanine are carried to O_2 from NADPH by tetrahydrobiopterin.

by the ascorbate-requiring enzyme parahydroxyphenylpyruvate dioxygenase. Homogentisate is converted to maleylacetoacetate by homogentisate oxidase. Acetoacetate and fumarate are then generated in isomerization and hydration reactions.

9. **Phenylalanine**. Phenylalanine is converted to tyrosine by phenylalanine-4-monooxygenase in a reaction illustrated in Figure 15.8. Tyrosine is degraded to form acetoacetate and fumarate.

10. **Leucine**. Leucine, one of the branched-chain amino acids, is converted to HMG-CoA in a series of reactions that include a transamination, two oxidations, a carboxylation, and a hydration. HMG-CoA is then converted to acetyl-CoA and acetoacetate by HMG-CoA lyase.

AMINO ACIDS FORMING α-KETOGLUTARATE Five amino acids (glutamate, glutamine, arginine, proline, and histidine) are degraded to α-ketoglutarate. An outline of their catabolism is illustrated in Figure 15.9. Each pathway is briefly described.

1. **Glutamate and glutamine**. Glutamine is converted to glutamate by glutaminase. As described previously, glutamate is converted to α-ketoglutarate by glutamate dehydrogenase or by transamination.

2. **Arginine**. Recall that arginine is cleaved by arginase to form ornithine and urea. In a subsequent transamination reaction, ornithine is converted to glutamate-γ-semialdehyde. Glutamate is then produced as glutamate-γ-semialdehyde is hydrated and oxidized. α-Ketoglutarate is produced by a transamination reaction or by oxidative deamination.

3. **Proline**. Proline catabolism begins with an oxidation reaction that produces Δ^1-pyrroline. The latter molecule is converted to glutamate-γ-semialdehyde by a hydration reaction. Glutamate is then formed by another oxidation reaction.

4. **Histidine**. Histidine is converted to glutamate in four reactions: a nonoxidative deamination, two hydrations, and the removal of a formamino group (NH = CH—) by THF.

AMINO ACIDS FORMING SUCCINYL-CoA Succinyl-CoA is formed from the carbon skeletons of methionine, isoleucine, valine, and threonine (as already

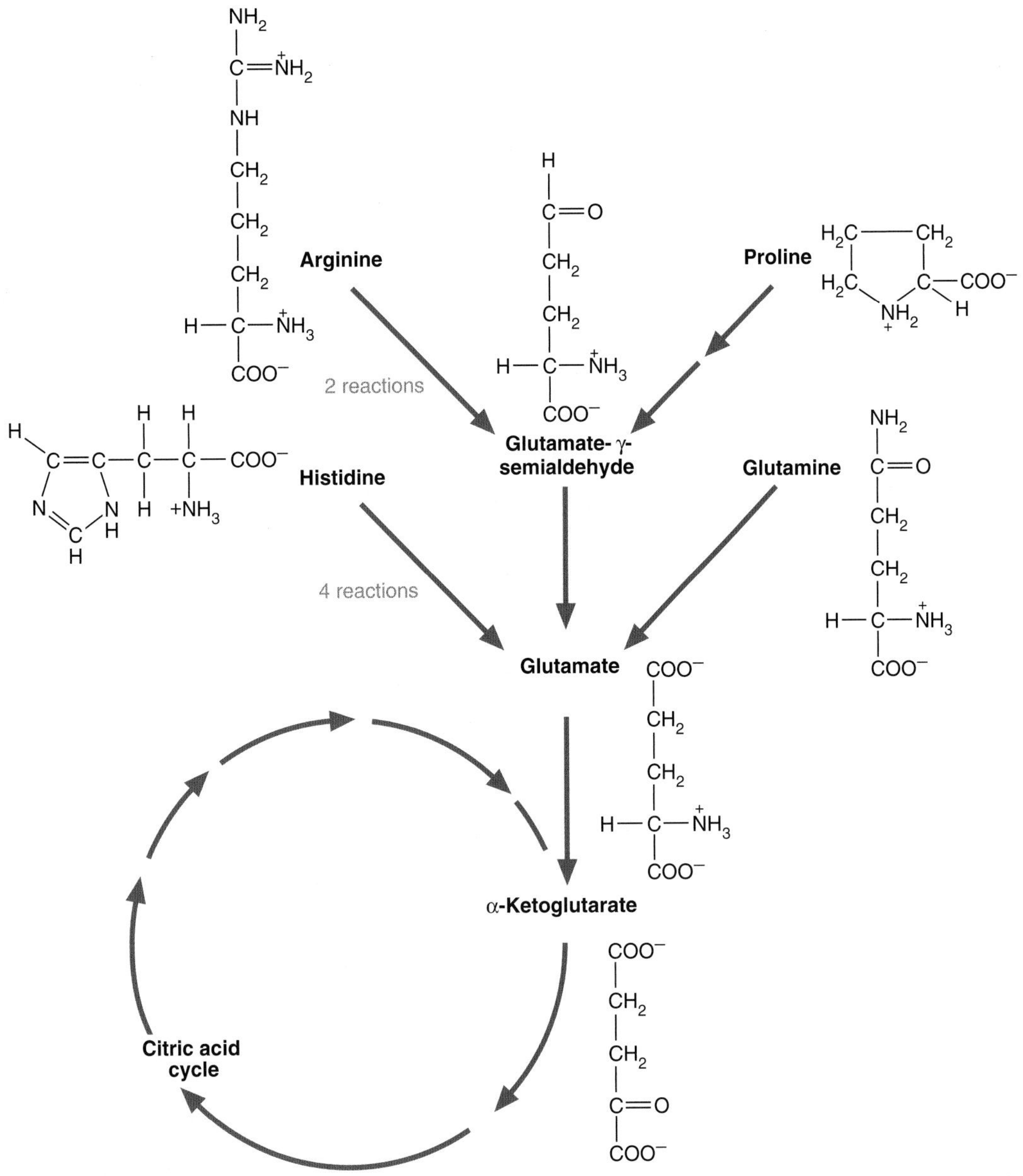

FIGURE 15.9

The Catabolic Pathways of Glutamate, Glutamine, Arginine, Proline, and Histidine

All these amino acids are eventually converted to α-ketoglutarate.

discussed). An outline of the reactions that degrade the first three of these amino acids is illustrated in Figure 15.10.

1. **Methionine**. Methionine degradation begins with the formation of
   S-adenosylmethionine, which is followed by a demethylation reaction,
   as described (Figure 14.18). S-Adenosylhomocysteine, the product of
   the latter reaction, is hydrolyzed to adenosine and homocysteine. Homocysteine then combines with serine to yield cystathionine. Cysteine,
   α-ketobutyrate, and NH_4^+ result from the cleavage of cystathionine.
   α-Ketobutyrate is then converted to propionyl-CoA by α-keto acid dehydrogenase. Propionyl-CoA is converted to succinyl-CoA in three steps.
   The enzyme that catalyzes the last of these, methylmalonyl-CoA mutase,
   requires methylcobalamin. The conversion of methionine to cysteine is

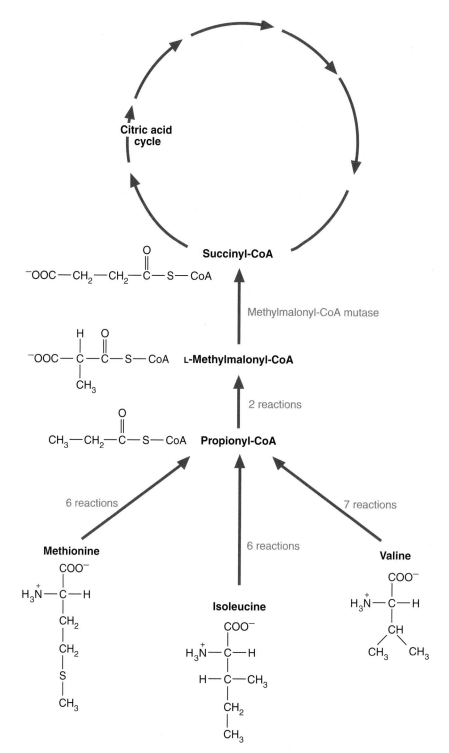

FIGURE 15.10

The Catabolic Pathways of Methionine, Isoleucine, and Valine

Propionyl-CoA and L-methylmalonyl-CoA are intermediates in the conversion of these amino acids to succinyl-CoA. Methylmalonyl-CoA mutase is a vitamin B_{12}–requiring enzyme. Note that threonine is also degraded via the propionyl-CoA/succinyl-CoA pathway (see Figure 15.6).

sometimes referred to as the **transsulfuration pathway** (Figure 15.11). A substantial amount of the sulfate produced from cysteine degradation is excreted in urine. Sulfate in the form of PAPS (p. 446) is also used in the synthesis of sulfatides and proteoglycans. Additionally, molecules such as steroids and certain drugs are excreted as sulfate esters. Also recall that the gasotransmitter H_2S is synthesized from cysteine in reactions catalyzed by CBS or CGL.

2. **Isoleucine and valine.** The first four reactions in the degradation of isoleucine and valine are identical. Initially, both amino acids undergo transamination reactions to form α-keto-β-methylvalerate and α-ketoisovalerate, respectively. This is followed by the formation of CoA derivatives, and oxidative decarboxylation, oxidation, and hydration reactions. The product of the isoleucine pathway is then hydrated, dehydrogenated, and cleaved to form acetyl-CoA and propionyl-CoA. In the valine degradative pathway the α-keto acid intermediate is converted into propionyl-CoA after a double bond is hydrated and CoA is removed by hydrolysis. After the formation of an aldehyde by the oxidation of the hydroxyl group,

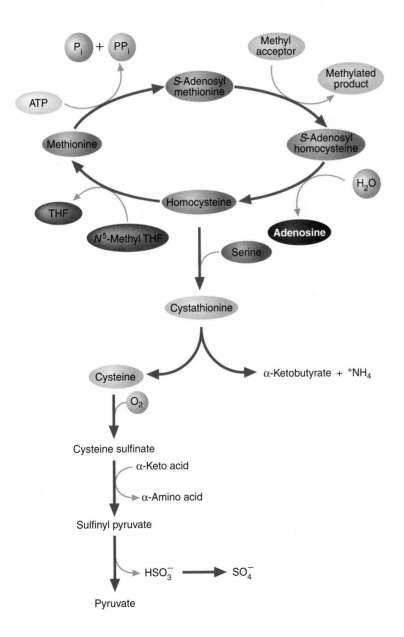

FIGURE 15.11

The Transsulfuration Pathway

The sulfur atom of methionine becomes the sulfur atom of cysteine in two reactions. Cystathionine β-synthase (CBS) converts homocysteine and serine into cystathionine. The latter molecule is then converted to cysteine, α-ketobutyrate, and NH_4^+ by cystathionine γ-lyase (CGL). Cysteine may then be incorporated into glutathione, coenzyme A, or proteins; or it may be oxidized by cysteine dioxygenase to form cysteine sulfinate. Cysteine sulfinate can undergo a transamination reaction followed by desulfuration to yield pyruvate and sulfite (HSO_3^-). Sulfite is subsequently converted to sulfate (SO_4^-) by sulfite oxidase. The sulfate generated in cysteine catabolism is excreted or used in several biosynthetic or catabolic pathways. Note that the transsulfuration and methylation pathways are intimately related.

KEY CONCEPT

Amino acid carbon skeletons can be degraded into one or more of several metabolites. These include acetyl-CoA, acetoacetyl-CoA, α-ketoglutarate, succinyl-CoA, and oxaloacetate.

propionyl-CoA is produced as a new thioester is formed during an oxidative decarboxylation.

AMINO ACIDS FORMING OXALOACETATE Both aspartate and asparagine are degraded to form oxaloacetate. Aspartate is converted to oxaloacetate with a single transamination reaction. Asparagine is initially hydrolyzed to yield aspartate and NH_4^+ by asparaginase.

QUESTION 15.3

Taurine is a sulfur-containing amine synthesized from cysteine. Although taurine is present in high concentrations in mammalian cells, except for its incorporation in bile salts, the physiological role of this amine is still poorly understood. However, several pieces of information suggest that taurine is an important metabolite. For example, taurine is found in brain tissue in large amounts. In addition, domestic cats have been observed to develop congestive heart failure if fed a taurine-free diet. (Cats cannot synthesize taurine. For this reason they must consume meat in their diet. Cats that are fed vegetarian diets soon become listless and will die prematurely.) In most animals, taurine is synthesized from cysteine sulfinate (the oxidation product of cysteine) in two reactions: a decarboxylation followed by an oxidation of the sulfinate group ($-SO_2^-$) to form sulfonate ($-SO_3^-$). With this information, determine the biosynthetic pathway for taurine. [*Hint*: The structure of taurine is illustrated in Chapter 12 on p. 455. Also refer to Figure 15.11.]

QUESTION 15.4

Taurine is the most abundant amino acid in white blood cells, where it reacts with HOCl produced during the respiratory burst catalyzed by myeloperoxidase. The product of this reaction, taurine monochloramine (Tau-Cl), is relatively nontoxic and stable compared with HOCl. Tau-Cl modulates the inflammation process by downregulating the production of NO$^\bullet$ and proinflammatory proteins such as tumor necrosis factor α (TNF-α). Provide the reactions in which HOCl and Tau-Cl are produced. [*Hint*: Review the respiratory burst.]

BIOCHEMISTRY IN PERSPECTIVE

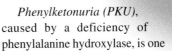

Disorders of Amino Acid Catabolism

What are the effects on human health of deficiency of a single enzyme in amino acid metabolism? Defects in amino acid catabolism were among the first genetic diseases to be recognized and investigated by medical scientists. These "inborn errors of metabolism" result from **mutations** (permanent changes in genetic information, i.e., DNA structure). Most commonly, in the genetic diseases related to amino acid metabolism, the defective gene codes for an enzyme. The metabolic blockage that results from such a deficit disrupts what are ordinarily highly coordinated cellular and organismal processes, producing abnormal amounts and/or types of metabolites. Because these metabolites (or their heightened concentrations) are often toxic, permanent damage or death ensues. Several of the most commonly observed inborn errors of amino acid metabolism are discussed below.

Alkaptonuria, caused by a deficiency of homogentisate oxidase, was the first disease to be linked to genetic inheritance involving a single enzyme. In 1902 Archibald Garrod proposed that a single inheritable unit (later called a gene) was responsible for the urine in alkaptonuric patients turning black. Large quantities of homogentisate, the substrate for the defective enzyme, are excreted in urine. Homogentisate turns black when it is oxidized as the urine is exposed to air. Although black urine appears to be an essentially benign (if somewhat disconcerting) condition, alkaptonuria is not innocuous, because alkaptonuric patients develop arthritis in later life. In addition, pigment accumulates gradually and unevenly darkens the skin.

Albinism is an example of a genetic defect with serious consequences. The enzyme tyrosinase is deficient. Consequently, *melanin*, a black pigment found in skin, hair, and eyes, is not produced. It is formed from tyrosine in several cell types, for example, the melanocytes in skin. In such cells, tyrosinase converts tyrosine to DOPA and DOPA to dopaquinone. A large number of molecules of the latter product, which is highly reactive, condense to form melanin. Because of the lack of pigment, affected individuals (called albinos) are extremely sensitive to sunlight. In addition to their susceptibility to skin cancer and sunburn, they often have poor eyesight.

Phenylketonuria (PKU), caused by a deficiency of phenylalanine hydroxylase, is one of the most common genetic diseases associated with amino acid metabolism. If this condition is not identified and treated immediately after birth, mental retardation and other forms of irreversible brain damage occur. This damage results mostly from the accumulation of phenylalanine. High phenylalanine blood levels result in the saturation of the transport mechanism for large neutral amino acids across the blood-brain barrier. (The other members of this group are Val, Met, Ile, Leu, Tyr, Trp, and His.) Brain damage results from decreased levels of protein and neurotransmitter synthesis. When present in excess, phenylalanine undergoes transamination to form phenylpyruvate, which is also converted to phenyllactate and phenylacetate. Large amounts of these molecules are excreted in the urine. Phenylacetate gives the urine its characteristic musty odor. PKU is treated with a low-phenylalanine diet.

In *maple syrup urine disease*, also called *branched-chain ketoaciduria*, the α-keto acids derived from leucine, isoleucine, and valine accumulate in large quantities in blood. Their presence in urine imparts a characteristic odor that gives the malady its name. All three α-keto acids accumulate because of a deficient branched-chain α-keto acid dehydrogenase complex. (This enzymatic activity is responsible for the conversion of the α-keto acids to their acyl-CoA derivatives.) If left untreated, affected individuals experience vomiting, convulsions, severe brain damage, and mental retardation. They often die before 1 year of age. As with phenylketonuria, treatment consists of rigid dietary control.

Deficiency of methylmalonyl-CoA mutase results in *methylmalonic acidemia*, a condition in which methylmalonate accumulates in blood. The symptoms are similar to those of maple syrup urine disease. Methylmalonate may also accumulate because of a deficiency of adenosylcobalamin or weak binding of this coenzyme by a defective enzyme. Some affected individuals respond to injections of large daily doses of vitamin B_{12}.

SUMMARY: The deficiency in humans of a single enzyme in amino acid metabolism has widespread effects that typically include brain damage.

15.3 DEGRADATION OF SELECTED NEUROTRANSMITTERS

The previous discussion of amino acid catabolic disorders indicates that catabolic processes are just as important for the proper functioning of cells and organisms as are anabolic processes. This is no less true for molecules that act as neurotransmitters. To maintain precise information transfer, neurotransmitters are usually quickly degraded or removed from the synaptic cleft. An extreme example of enzyme inhibition illustrates the importance of neurotransmitter degradation. Recall that acetylcholine is the neurotransmitter that initiates muscle contraction. Shortly afterward, the action of acetylcholine is terminated by the enzyme acetylcholinesterase. (Acetylcholine must be destroyed rapidly so that muscle can relax before the next contraction.) Acetylcholinesterase is a serine esterase that hydrolyzes acetylcholine to acetate and choline. Serine esterases have catalytic mechanisms similar to those of the serine proteases (Section 6.4). Both types of enzyme are irreversibly inhibited by DFP (diisopropylfluorophosphate). Exposure to DFP causes muscle paralysis because acetylcholinesterase is irreversibly inhibited. With each nerve impulse, more acetylcholine molecules enter the neuromuscular synaptic cleft. The accumulating acetylcholine molecules repetitively bind to acetylcholine receptors. The overstimulated muscle cells soon become paralyzed (nonfunctional). Affected individuals suffocate because of paralyzed respiratory muscles.

The catecholamines epinephrine, norepinephrine, and dopamine are inactivated by oxidation reactions catalyzed by monoamine oxidase (MAO) (Figure 15.12).

FIGURE 15.12

Inactivation of the Catecholamines

Monoamine oxidase is a flavoprotein that catalyzes the oxidative deamination of amines to form the corresponding aldehydes; O_2 is the electron acceptor, and NH_3 and H_2O_2 are the other products. (PNMT = phenylethanolamine-N-methyltransferase)

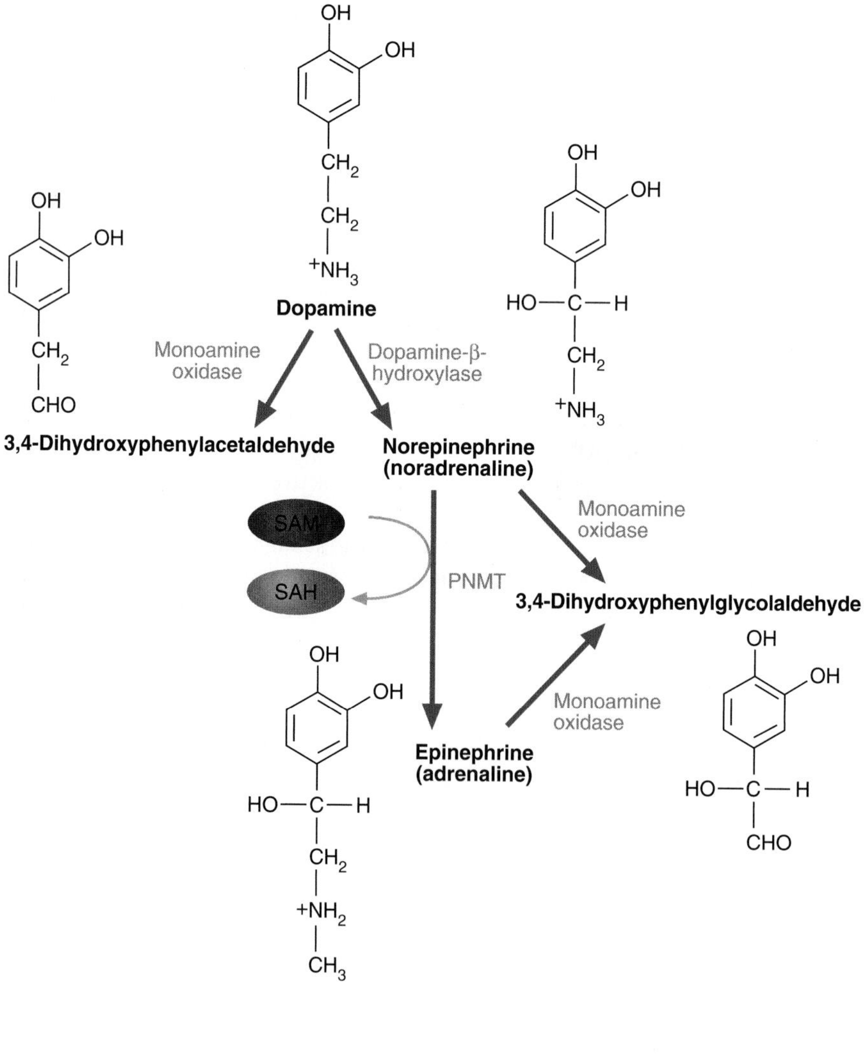

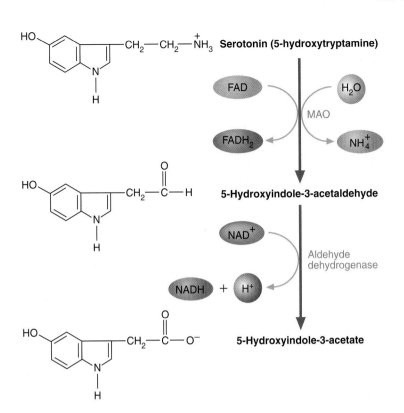

FIGURE 15.13

Degradation of Serotonin

In the major catabolic pathway, serotonin is deaminated and oxidized to form 5-hydroxyindole-3-acetaldehyde. The latter molecule is then further oxidized to form 5-hydroxyindole-3-acetate.

Catecholamines must be transported out of the synaptic cleft before inactivation because MAO is located in nerve endings. (The process by which neurotransmitters are transported back into nerve cells so that they can be reused or degraded is referred to as *reuptake*.) Epinephrine, released as a hormone from the adrenal gland, is carried in the blood and is catabolized in nonneural tissue (perhaps the kidney). Catecholamines are also inactivated in methylation reactions catalyzed by catechol-*O*-methyltransferase (COMT). These two enzymes (MAO and COMT) work together to produce a large variety of oxidized and methylated metabolites of the catecholamines.

After its reuptake into nerve cells, serotonin is degraded in a two-step pathway (Figure 15.13). In the first reaction, serotonin is oxidized by MAO. The product, 5-hydroxyindole-3-acetaldehyde, is then further oxidized by aldehyde dehydrogenase to form 5-hydroxyindole-3-acetate.

KEY CONCEPT

Information transfer in animals requires that after their release, neurotransmitters be quickly degraded or removed from the synaptic cleft.

QUESTION 15.5

Identify each of the following neurotransmitters. Explain how each is synthesized and inactivated.

(a) (b) (c) (d)

Myasthenia gravis is an autoimmune disease in which autoantibodies bind to and initiate the destruction of the acetylcholine receptor in skeletal muscle cell membranes. It is treated with drugs that inhibit acetylcholinesterase, the enzyme that degrades acetylcholine. As the disease progresses, the number of functional acetylcholine receptors is reduced. This condition is characterized by muscle weakness and fatigability. Eventually, patients develop difficulty in speaking and swallowing. However, a short time after consuming reversible cholinesterase inhibitors (e.g., neostigmine or physostigmine), patients experience significant improvement in their symptoms. Based on your knowledge of the action of acetylcholine, can you suggest how anticholinesterase drugs achieve this short-term clinical improvement? [*Hint*: For a muscle cell to contract, a threshold number of acetylcholine receptors must bind acetylcholine. In normal individuals, the receptor occupancy required to initiate muscle contraction is quite low. Also note that productive binding and unbinding of a neurotransmitter to its receptor are often rapid.]

15.4 NUCLEOTIDE DEGRADATION

In most living organisms, purine and pyrimidine nucleotides are constantly degraded and/or recycled. During digestion, nucleic acids are hydrolyzed to oligonucleotides by enzymes called **nucleases**. (Short nucleic acid segments containing fewer than 50 nucleotides are called **oligonucleotides**.) Enzymes that are specific for breaking internucleotide bonds in DNA are called *deoxyribonucleases* (DNases); those that degrade RNA are called *ribonucleases* (RNases). Once formed, oligonucleotides are further hydrolyzed by various *phosphodiesterases*, a process that produces a mixture of mononucleotides. *Nucleotidases* remove phosphate groups from nucleotides, yielding nucleosides. These latter molecules are hydrolyzed by *nucleosidases* to free bases and ribose or deoxyribose, which are then absorbed.

Generally speaking, dietary purine and pyrimidine bases are not used in significant amounts to synthesize cellular nucleic acids. Instead, they are degraded within enterocytes. Purines are degraded to uric acid in humans and birds. Pyrimidines are degraded to β-alanine or β-aminoisobutyric acid, as well as NH_3 and CO_2. In contrast to the catabolic processes for other major classes of biomolecules (e.g., sugars, fatty acids, and amino acids), purine and pyrimidine catabolism does not result in ATP synthesis. The major pathways for the degradation of purine and pyrimidine bases are described next.

Purine Catabolism

Purine nucleotide catabolism is outlined in Figure 15.14. There is some variation in the specific pathways used by different organisms or tissues to degrade AMP. In most tissues, AMP is hydrolyzed by 5′-nucleotidase to form adenosine. Adenosine is then deaminated by adenosine deaminase (also called adenosine aminohydrolase) to form inosine. In muscle, AMP is initially converted to IMP by AMP deaminase (also referred to as adenylate aminohydrolase). IMP is subsequently hydrolyzed to inosine by 5′-nucleotidase. The AMP deaminase reaction is also a component of the purine nucleotide cycle (Figure 15.15). In this pathway the product of the AMP deaminase-catalyzed reaction, IMP, reacts with aspartate to yield adenylosuccinate. This GTP-requiring reaction is catalyzed by adenylosuccinate synthetase. Adenylosuccinate is then converted by adenylosuccinase to AMP and fumarate. The purine nucleotide cycle is a means of converting amino acids (via aspartate) to citric acid cycle intermediates

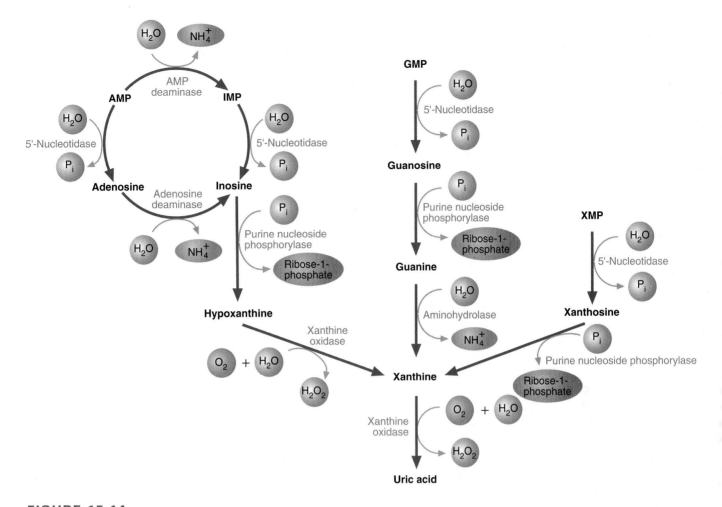

FIGURE 15.14

Purine Nucleotide Catabolism

Ribose-1-phosphate is released in AMP, GMP, and XMP catabolism. Xanthine oxidase–catalyzed reactions generate O_2^-, as well as H_2O_2.

(via fumarate). In skeletal muscle, AMP deaminase activity is exceptionally high. In contrast to other tissues, skeletal muscle can replenish citric acid cycle intermediates only with the fumarate produced by the purine nucleotide cycle. Activation of muscle AMP deaminase (called myoadenylate deaminase) and increased flux through the purine nucleotide cycle (Figure 15.15) occur during intense exercise.

Purine nucleoside phosphorylase converts inosine, guanosine, and xanthosine to hypoxanthine, guanine, and xanthine, respectively. (The ribose-1-phosphate, formed during these reactions is reconverted to PRPP by ribose-5-phosphate pyrophosphokinase.) Hypoxanthine is oxidized to xanthine by xanthine oxidase, an enzyme that contains molybdenum, FAD, and two different Fe-S centers. (Xanthine oxidase–catalyzed reactions produce O_2^- in addition to forming H_2O_2.) Guanine is deaminated to xanthine by guanine deaminase (also called guanine aminohydrolase). Xanthine molecules are further oxidized to uric acid by xanthine oxidase.

Many animals degrade uric acid further (Figure 15.16). Urate oxidase converts uric acid to allantoin, an excretory product in many mammals. Allantoinase catalyzes the hydration of allantoin to form allantoate, which is excreted by bony fish. Other fish, as well as amphibians, produce allantoicase, which splits

FIGURE 15.15

The Purine Nucleotide Cycle

In skeletal muscle, the purine nucleotide cycle is an anaplerotic process that replenishes citric acid cycle intermediates by producing fumarate from aspartate.

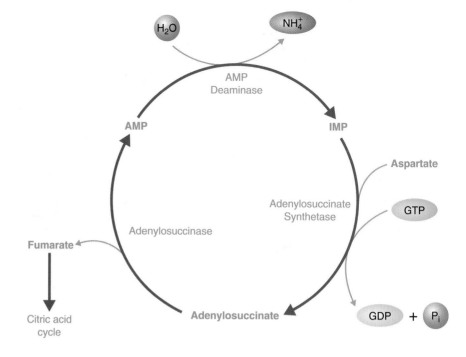

FIGURE 15.16

Uric Acid Catabolism

Many animals possess enzymes that allow them to convert uric acid to other excretory products. The final excretory products of specific animal groups are indicated.

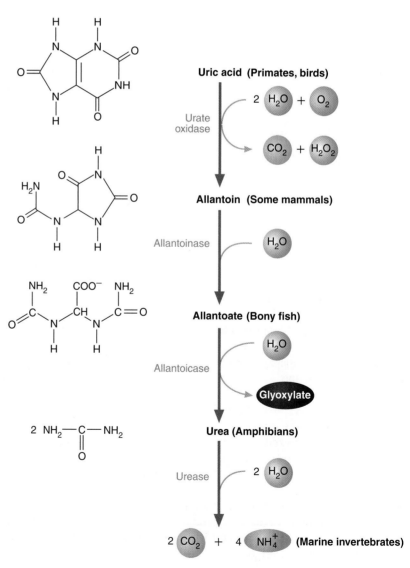

allantoic acid into glyoxylate and urea. Finally, marine invertebrates degrade urea to NH_4^+ and CO_2 in a reaction catalyzed by urease.

Several diseases result from defects in purine catabolic pathways. *Gout*, which is often characterized by high blood levels of uric acid and recurrent attacks of arthritis, is caused by several metabolic abnormalities. Two different immun-odeficiency diseases are now known to result from defects in purine catabolic reactions. In *adenosine deaminase deficiency*, large concentrations of dATP inhibit ribonucleotide reductase. Consequently, DNA synthesis is depressed. For reasons that are not yet clear, this metabolic distortion is observed primarily in the T and B lymphocytes. (*T lymphocytes*, or **T cells**, bear antibody-like molecules on their surfaces. They bind to and destroy foreign cells in a process referred to as **cellular immunity**. *B lymphocytes*, or **B cells**, produce antibodies that bind to foreign substances, thereby initiating their destruction by other immune system cells. The production of antibodies by B cells is referred to as the **humoral immune response**.) Children with adenosine deaminase deficiency usually die before the age of 2 because of massive infections. In *purine nucleoside phosphorylase deficiency*, levels of purine nucleotides are high and synthesis of uric acid decreases. High levels of dGTP are apparently responsible for the impairment of T cells that is characteristic of this malady. Individuals with *myoadenylate deaminase deficiency* exhibit exercise-induced muscle fatigue.

COMPANION

W E B S I T E Visit the companion website at www.oup.com/us/mckee to read the **Biochemistry in Perspective box on gout.**

QUESTION 15.7

Unlike primates and birds, many animals possess the enzyme urate oxidase. Suggest a reason why these organisms do not suffer from gout.

QUESTION 15.8

One of the more fascinating aspects of biochemistry is that living organisms use the same molecule for different purposes. Two interesting examples are allantoin and allantoate, which serve as nitrogenous waste in several animal groups. Certain legumes, such as soybeans and snapbeans, begin to synthesize allantoin and allantoate when they become infected by nitrogen-fixing bacteria. Both molecules, referred to as the *ureides*, are nitrogen transport compounds. (Other legumes, such as peas and alfalfa, use asparagine for nitrogen transport whether they are infected or not.) Once synthesized, the ureides are transported through the xylem to the leaves. In leaves, the nitrogen is released and used primarily in amino acid synthesis. Allantoate is degraded to glyoxylate, four molecules of NH_4^+, and two molecules of CO_2 by three reactions that are not yet completely characterized. Based on the information provided in Chapter 14 and in this chapter, trace the transport of NH_4^+ in root nodules to its incorporation into amino acids in leaves. Assume that the same or similar enzymes are used to synthesize allantoin and allantoate as observed in animals.

Pyrimidine Catabolism

In humans the purine ring cannot be degraded. This is not true for the pyrimidine ring. An outline of the pathway for pyrimidine nucleotide catabolism is illustrated in Figure 15.17.

Before they can be degraded, cytidine and deoxycytidine are converted to uridine and deoxyuridine, respectively, by deamination reactions catalyzed by cytidine deaminase. Similarly, deoxycytidylate (dCMP) is deaminated to form deoxyuridylate (dUMP). The latter molecule is then converted to deoxyuridine

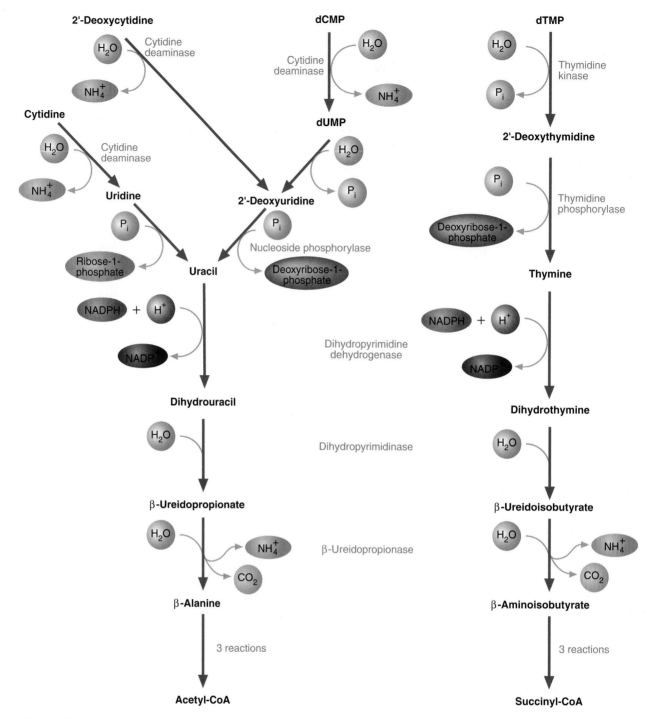

FIGURE 15.17

Degradation of Pyrimidine Bases

Uracil and thymine are degraded to β-alanine and β-aminoisobutyrate, respectively, in parallel pathways. The entire pathway is present in mammalian liver.

by 5'-nucleotidase. Uridine and deoxyuridine are then further degraded by nucleoside phosphorylase to form uracil. Thymine is formed from thymidylate (dTMP) by the sequential actions of thymidine kinase and thymidine phosphorylase.

Uracil and thymine are converted to their end products, β-alanine and β-aminoisobutyrate, respectively, in parallel pathways. In the first step, uracil and thymine are reduced by dihydropyrimidine dehydrogenase to their corresponding dihydro derivatives. As these latter molecules are hydrolyzed, the rings open, yielding β-ureidopropionate and β-ureidoisobutyrate, respectively. Finally, β-alanine and β-aminoisobutyrate are produced in deamination reactions catalyzed by β-ureidopropionase.

In several conditions, β-aminoisobutyrate is produced in such large quantities that it appears in urine. Among these are a genetic predisposition for slow β-aminoisobutyrate conversion to succinyl-CoA and diseases that cause massive cell destruction, such as leukemia. Because it is soluble, excess β-aminoisobutyrate does not cause problems comparable to those observed in gout.

QUESTION 15.9

Identify each of the following biomolecules. Explain how they are produced.

(a) $NH_2-C(=O)-NH_2$

(b) (purine/xanthine-type ring structure)

(c) $H_3\overset{+}{N}-CH_2-CH_2-C(=O)-O^-$

QUESTION 15.10

The products of pyrimidine base catabolism, β-alanine and β-aminoisobutyrate, can be further degraded to acetyl-CoA and succinyl-CoA, respectively. Can you suggest the types of reaction required to accomplish these transformations?

15.5 HEME BIOTRANSFORMATION

In the body, the majority of heme molecules (80%) occur in red blood cells as oxygen-binding prosthetic groups in hemoglobin. Senescent red blood cells (life span 120 days) are removed from the bloodstream by reticuloendothelial cells, the phagocytic cells in liver, spleen, bone marrow, lung, and lymph nodes that ingest worn-out or abnormal body cells. Heme is released from hemoglobin, and globin is converted to amino acids. Heme is then rapidly degraded because of its cytotoxic properties (due to ROS generation). As described previously (p. 538), heme is oxidized by heme oxygenase (HO), an ER enzyme. HO is a component of an electron transport system similar to that of cytochrome P_{450}. The products of this reaction are the dark blue-green pigment biliverdin, iron, and carbon monoxide (CO).

The reaction (Figure 15.18) begins with the O_2 and NADPH-dependent hydroxylation of the α-methene carbon. The subsequent reaction of the intermediate, α-hydroxyhemin, with O_2 (probably a nonenzymatic step) yields biliverdin, CO (the α-methene carbon), and a ferric iron ion. CO diffuses out of the cell, and is transported in the blood to the lungs; it leaves the body on exhalation. NADPH converts Fe^{3+} to Fe^{2+} before its release. Biliverdin is subsequently converted into bilirubin in a reaction catalyzed by the cytoplasmic enzyme biliverdin reductase.

Bilirubin is a very toxic molecule. It is known to inhibit RNA and protein synthesis and carbohydrate metabolism in the brain. Mitochondria appear to be especially sensitive to its effects. Bilirubin is also a metabolically expensive

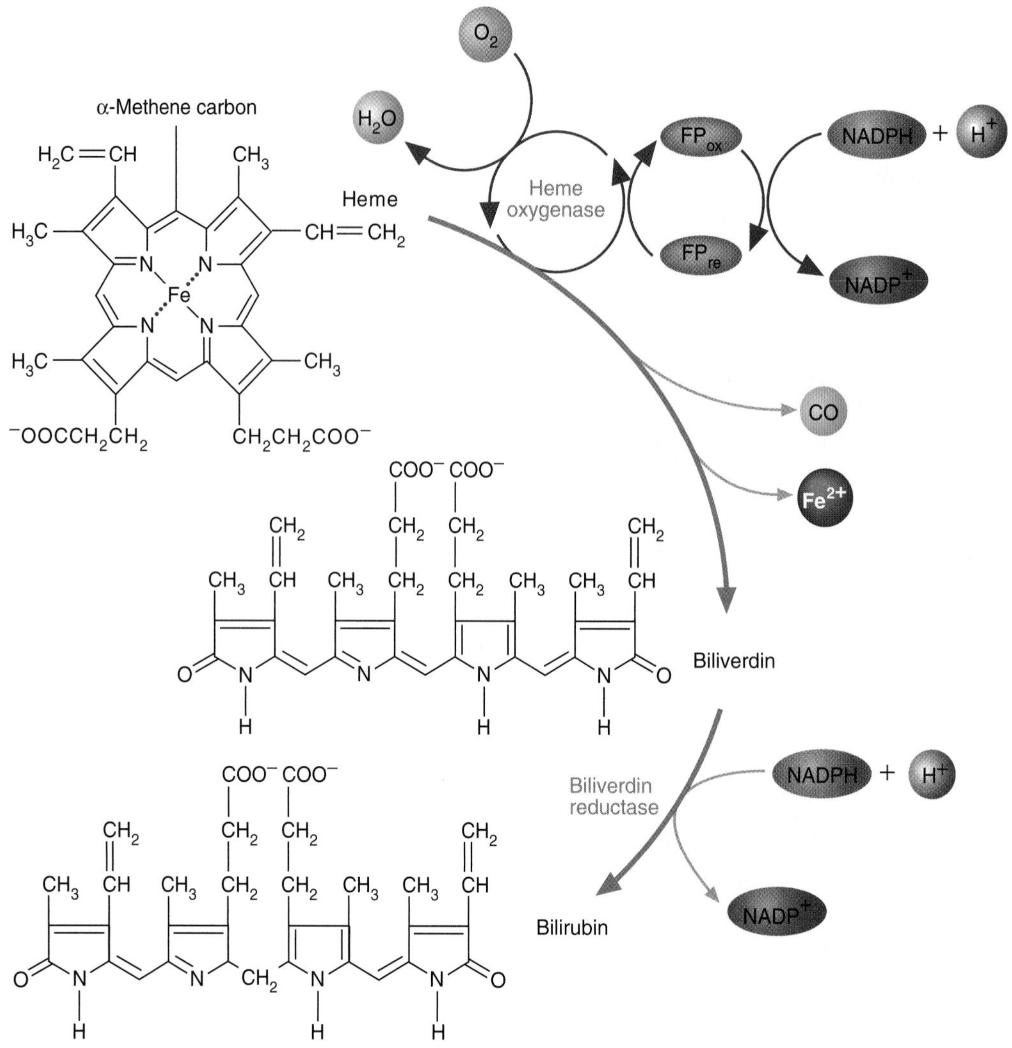

FIGURE 15.18

Bilirubin Synthesis

Heme oxygenase, which catalyzes the conversion of free heme groups to biliverdin and CO, functions as part of a microsomal electron transport system similar to that of cytochrome P_{450} (FP = NADPH-cytochrome P_{450} reductase). Heme oxygenase requires three molecules of O_2 and five of NADPH. Biliverdin reductase can use NADPH or NADH as a reductant.

molecule to produce. For example, bilirubin is virtually insoluble in water, because of intramolecular hydrogen bonding. Therefore, sophisticated transport mechanisms and conjugation reactions in the liver (Figure 15.19) are required for excretion as components of bile in the gastrointestinal tract. Because bilirubin creates so many problems, considerable effort has been devoted to elucidating its purpose. (Amphibians, reptiles, and birds excrete the water-soluble precursor biliverdin.) Despite its toxic properties (at higher than normal levels) in humans, bilirubin is now recognized as a powerful antioxidant. During bilirubin transport in blood, the radical-scavenging pigment is distributed throughout the circulatory system. (The association of bilirubin with the plasma protein albumin protects cells from the molecule's toxic effects.) Plasma levels of bilirubin have been linked inversely to risk of atherosclerosis-related diseases. Free heme, oxidized LDL, and other risk factors for cardiovascular disease induce increased expression of HO1, one of two isozymes with heme oxygenase activity. The subsequent generation of CO, biliverdin, and bilirubin provides an adaptive

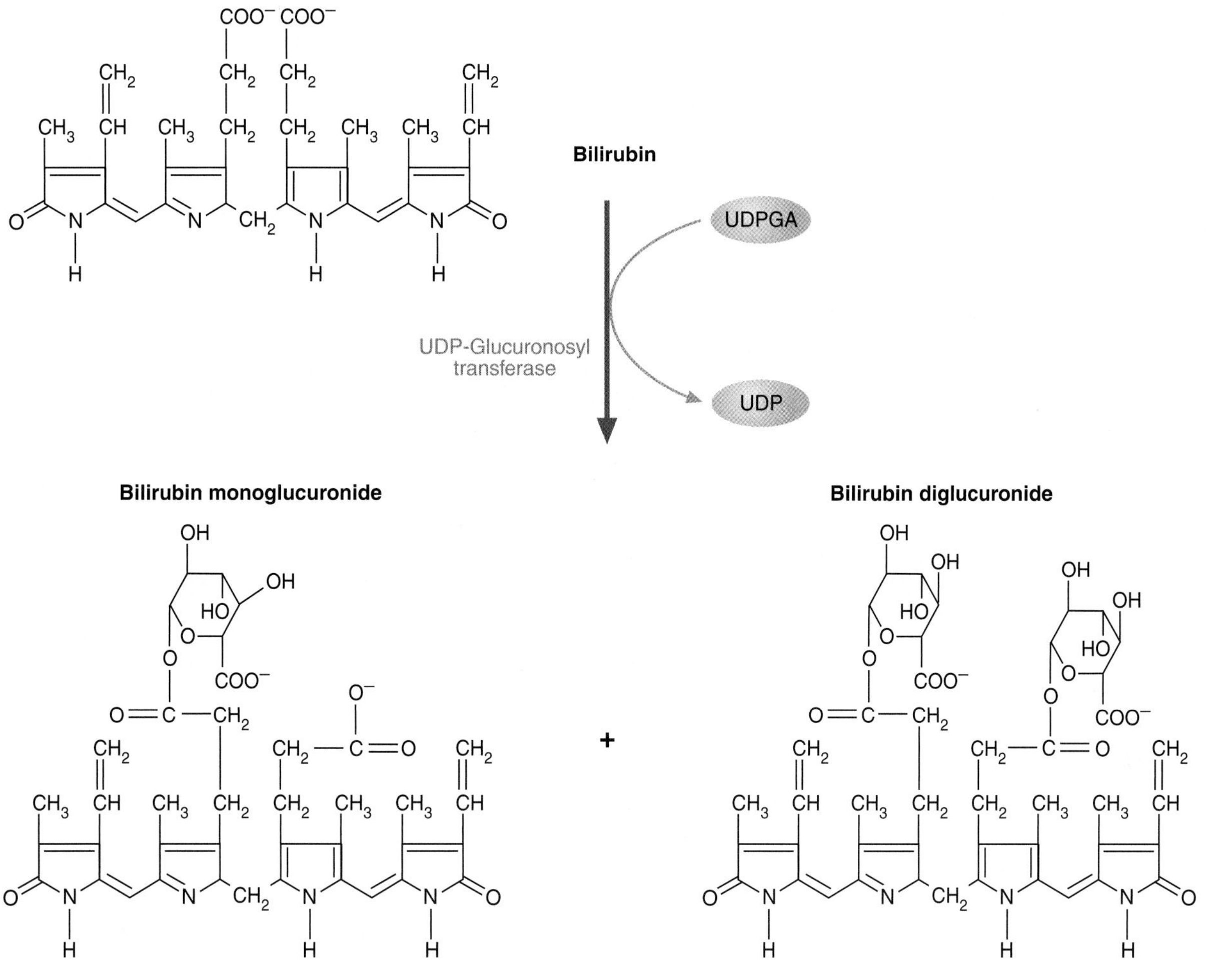

FIGURE 15.19

Bilirubin Conjugation

Before bilirubin is excreted in bile, its propionyl carboxyl groups are esterified with glucuronic acid to form both monoglucuronides and diglucuronides. (UDPGA = UDP-glucuronic acid.) The diglucuronide is the major form produced in many animals. In a number of species, especially mammals, bilirubin conjugation is required for efficient secretion into bile.

response to injury associated with ROS generation. The antioxidant capacity of bilirubin, in particular, provides resistance to the maturation of atherosclerotic changes in the vasculature. It also inhibits the white blood cell adhesion processes associated with inflammation and plaque development. The CO product of HO1 action inhibits the production of a number of growth factors associated with smooth muscle proliferation, a key component of plaque development.

KEY CONCEPTS

- The heme group of hemoproteins is first converted to biliverdin and then to bilirubin.
- After undergoing conjugation reactions in the liver, bilirubin is excreted in bile.
- Bilirubin is a potent antioxidant.

Chapter**Summary**

1. Animals are constantly synthesizing and degrading nitrogen-containing molecules such as proteins and nucleic acids. Protein turnover is believed to provide cells with metabolic flexibility, protection from accumulations of abnormal proteins, and the timely destruction of proteins during developmental processes. Most cellular proteins are degraded by the

ubiquitin proteasomal system. The process begins with the covalent modification of target proteins called ubiquination. Ubiquitin, a small, highly conserved protein is linked to worn-out or damaged proteins, or short-lived regulatory proteins.

2. In general, amino acid degradation begins with deamination. Most deamination is accomplished by transamination

reactions, which are followed by oxidative deaminations that produce ammonia. Although most deaminations are catalyzed by glutamate dehydrogenase, other enzymes also contribute to ammonia formation. Ammonia is prepared for excretion by the enzymes of the urea cycle. Aspartate and CO_2 also contribute atoms to urea.

3. Amino acids are classified as ketogenic or glucogenic on the basis of whether their carbon skeletons are converted to fatty acids or to glucose. Several amino acids can be classified as both ketogenic and glucogenic because their carbon skeletons are precursors for both fat and carbohydrates.

4. The degradation of neurotransmitters is critical to the proper functioning of information transfer in animals. The amine neurotransmitters such as acetylcholine, the catecholamines, and serotonin are among the best-researched examples.

5. The turnover of nucleic acids is accomplished by enzymes of several types. The nucleases degrade the nucleic acids to oligonucleotides. (The deoxyribonucleases degrade DNA; the ribonucleases degrade RNA.) The phosphodiesterases convert the oligonucleotides to mononucleotides. By removing phosphate groups, the nucleotidases convert nucleotides to nucleosides. The nucleosidases hydrolyze nucleosides to form free bases and ribose or deoxyribose. The nucleoside phosphorylases convert ribonucleosides to free bases and ribose-1-phosphate. Dietary nucleic acids are generally degraded in the intestine and are not used in salvage pathways. Cellular purines are converted to uric acid. Many animals degrade uric acid further because they produce enzymes that are not present in primates. Pyrimidine bases are degraded to either β-alanine (UMP, CMP, dCMP) or β-aminoisobutyrate (dTMP).

6. The porphyrin heme is degraded to form the excretory product bilirubin in a biotransformation process that involves the enzymes heme oxygenase and biliverdin reductase and UDP-glucuronosyltransferase. After undergoing a conjugation reaction, bilirubin is excreted as a component of bile.

7. The heme oxygenase-1 gene is inducible by a number of cellular stressors (free heme, LDL_{ox}, UV radiation, ROS). The increase in CO, biliverdin, and bilirubin in the vasculature provides protection against the cytotoxic effects of oxidative stress.

COMPANION GW WEBSITE Take your learning further by visiting the **companion website** for Biochemistry at **www.oup.com/us/mckee** where you can complete a multiple-choice quiz on degradation to help you prepare for exams.

Suggested Readings

Ajoka, R. S., Phillips, J. D., and Kushner, J. P., Biosynthesis of Heme in Mammals, *Biochim. Biophys. Acta* 1763:723–736, 2006.

Ciechanover, A., The Ubiquitin Proteolytic System: From a Vague Idea, Through Basic Mechanisms, and on to Human Diseases and Drug Targeting, *Neurology* 66 (Suppl. 1):S7–S19, 2006.

Finkelstein, J. D., Inborn Errors of Sulfur-Containing Amino Acid Metabolism, *J. Nutr.* 136:1750S–1754S, 2006.

Hague, S. M., Klaffke, S., and Bandmann, O., Neurodegenerative Disorders: Parkinson's Disease and Huntington's Disease, *J. Neurol. Neurosurg Psychiatry* 76:1058–1063, 2005.

Mani, A., and Gelmann, E. P., The Ubiquitin-Proteasome Pathway and Its Role in Cancer, *J. Clin. Oncol.* 23(21):4776–4789, 2005.

Morita, T., Heme Oxygenase and Atherosclerosis, *Arterioscler. Thromb. Vasc. Biol.* 25:1786–1795, 2005.

Neuwelt, E. A., Mechanisms of Disease: The Blood-Brain Barrier, *Neurosurgery* 54(1):131–140, 2004,

Rigato, I., Ostrow, J. D., and Tiribelli, C., Bilirubin and the Risk of Non-hepatic Diseases, *Trends Mol. Med.* 11(6): 277–283, 2005.

Ryter, S. W., and Tyrell, R. M., The Heme Synthesis and Degradative Pathways: Role in Oxidant Sensitivity: Heme Oxygenase Has Both Pro- and Antioxidant Properties, *Free Radical Biol. Med.* 28(2):289–309, 2000.

Schuller-Lewis, G. B., and Park, E., Taurine: New Implications for an Old Amino Acid, *FEMS Microbiol. Lett.* 226:195–202, 2003.

Key Words

B cell, 583

cellular immunity, 583

glucogenic, 563

humoral immune response, 583

hyperammonemia, 568

ketogenic, 563

Krebs bicycle, 567

Krebs urea cycle, 564

mutation, 577

nuclease, 580

oligonucleotide, 580

protein turnover, 560

proteasome, 561

T cell, 583

transsulfuration pathway, 575

ubiquination, 561

ubiquitin, 561

ubiquitin proteasomal system, 560

urea cycle, 564

Review**Questions**

These questions are designed to test your knowledge of the key concepts discussed in this chapter, before moving on to the next chapter. You may like to compare your answers to the solutions provided in the back of the book and in the accompanying Study Guide.

1. Define the following terms:
   a. denitrification
   b. ammonotelic
   c. protein turnover
   d. ubiquination
   e. humoral immune response

2. What are the major molecules that serve in the excretion of nitrogen?

3. What are three purposes served by protein turnover?

4. What are the structural features of proteins that mark them for destruction?

5. Define the following terms:
   a. glucogenic
   b. ubiquitin
   c. mutation
   d. Krebs bicycle
   e. transsulfuration

6. What are the seven metabolic products produced by the degradation of amino acids?

7. Describe how aspartate is degraded.

8. In humans the purine ring cannot be degraded. How is it excreted? What reactions are involved?

9. Define the following terms:
   a. oligonucleotide
   b. nuclease
   c. urea cycle
   d. ketogenic
   e. ubiquitin proteasomal system

10. Indicate which of the following amino acids are ketogenic and which are glucogenic:
    a. tyrosine
    b. lysine
    c. glycine
    d. alanine
    e. valine
    f. threonine

11. Describe how glutamate is degraded.

12. The urea cycle occurs partially in the cytosol and partially in the mitochondria. Discuss the urea cycle reactions with reference to their cellular locations.

13. Describe how the glucose-alanine cycle acts to transport ammonia to the liver.

14. Describe how alanine is degraded.

15. In individuals with PKU, is tyrosine an essential amino acid?

16. Urea formation is energetically expensive, requiring the expenditure of 4 mol of ATP per mole of urea formed. However, NADH is produced when fumarate is reconverted to aspartate. How many ATP molecules are produced by the mitochondrial oxidation of the NADH? What is the net ATP requirement for urea synthesis?

17. Define the following terms:
    a. porphyrin
    b. alkaptonuria
    c. L-DOPA
    d. cellular immunity
    e biliverdin

18. Describe the Krebs bicycle. What compound links the citric acid and urea cycles?

19. Most amino acids are degraded in the liver. This is not true of the branched-chain amino acids, most of which are degraded in extrahepatic tissues with high protein turnover. Suggest some examples of these tissues.

20. Describe how a protein is targeted for degradation.

21. Describe how lysine is degraded.

22. Provide the names of the organisms that form the following substances as nitrogenous waste molecules:
    a. uric acid
    b. urea
    c. allantoate
    d. NH_4^+
    e. allantoin

23. Define the following terms:
    a. α-methene carbon
    b. allantoate
    c. allantoin
    d. MAO
    e. maple syrup urine disease

24. Which of the following molecules yields uric acid when degraded?
    a. DNA
    b. FAD
    c. CTP
    d. PRPP
    e. β-alanine
    f. urea
    g. NAD^+

25. Describe how tyrosine is degraded.

26. Explain how a defective enzyme in the urea cycle can produce high levels of ammonia.

27. Explain why providing domestic cats with a vegetarian diet is a bad idea.

28. Trace the origin of the nitrogen atoms in urea.

29. Fish excrete nitrogen as ammonia, whereas mammals excrete urea, and birds and desert reptiles excrete uric acid. Explain this set of observations in terms of each animal's water economy.

30. Explain why ketogenic amino acids yielding acetylCoA only cannot be converted into glucose.

31. Describe how the urea cycle is activated after a high-protein meal.

32. Foods high in purines can trigger a gout attack. Provide several examples of these foods and explain how they contribute to this malady. (Visit the companion website at www.oup.com/us/mckee to read the Biochemistry in Perspective box on gout.)

COMPANION
Gw
WEBSITE

33. Which nitrogen atom in uracil eventually ends up in urea molecules?

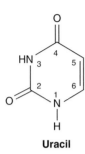

Uracil

34. Explain why the amino acid tryptophan is both ketogenic and glucogenic.

35. Describe the symptoms of maple syrup urine disease and the metabolic basis of this malady.

36. What molecules cause the urine odor that is characteristic of the disease mentioned in Question 35?

37. Alkaptonuria is a disease in which the urine turns black. Explain the cause of this symptom.

38. What amino acids are involved in alkaptonuria?

39. Why can't humans simply excrete waste nitrogen atoms as ammonia rather than utilize the energetically expensive process of urea synthesis?

Thought Questions

These questions are designed to reinforce your understanding of all of the key concepts discussed in the book so far, including this chapter and all of the chapters before it. They may not have one right answer! The authors have provided possible solutions to these questions in the back of the book and in the accompanying Study Guide, for your reference.

40. Mammals excrete most nitrogen atoms as urea. The urea cycle itself is costly, requiring considerable amounts of ATP energy. What mechanism does the cell have to compensate for that energy input?

41. Create a diagram that illustrates how the purine nucleotide cycle contributes to energy metabolism in skeletal muscle. Include glycolysis and the citric acid cycle in your drawing. What enzymatic activities would you expect to be absent in skeletal muscle?

42. Describe how increasing concentrations of ammonia stimulate the formation of *N*-acetylglutamate and turn on the urea cycle.

43. Phenylketonuria can be caused by deficiencies in phenylalanine hydroxylase and by enzymes catalyzing the formation and regeneration of 5,6,7,8-tetrahydrobiopterin. How can this second defect cause the symptoms of PKU?

44. Individuals who cannot produce 5,6,7,8-tetrahydrobiopterin must be supplied with L-dopa and 5-hydroxytryptophan, metabolic precursors to norepinephrine and serotonin. Why does supplying 5,6,7,8-tetrahydrobiopterin have no effect?

45. In their *in vitro* studies using liver slices, Krebs and Henseleit observed that urea formation was stimulated by the addition of ornithine, citrulline, and arginine. Other amino acids had no effect. Explain these observations.

46. Specify the type of carbon unit that is transferred by each of the following compounds:
    a. N^5,N^{10}-methylene THF
    b. serine
    c. choline
    d. *S*-adenosylmethionine

47. Caffeine, a methylated xanthine found in chocolate, coffee, and tea, is excreted as uric acid. Use your knowledge of the metabolism of other purine compounds to suggest how caffeine is metabolized.

48. Some animals living in a fluid medium excrete nitrogen as ammonia. Land animals, which conserve water, excrete urea and uric acid. Why does the excretion of these molecules aid in water conservation?

49. Parkinson's disease is a devastating, progressive disorder of the central nervous system. Damage to dopaminergic nerve tracts in the brain causes tremors and impaired muscle coordination. An early-onset (inherited) form of the disease has been linked to a defective protein, now called parkin. Parkin has E3 activity. Describe, in general terms, how neurons with this defective protein are damaged.

50. Cardiovascular disease is caused by various forms of oxidative stress to the vasculature. Explain the cytoprotective function of *HO1* in the development of atherosclerosis.

51. Dihydrouracil and β-uredidopropionate (*N*-carbamoyl-β-alanine) are intermediates in the conversion of uracil to β-alanine. Provide the structures of the molecules in this pathway. (Refer to the numbered illustration of uracil in Question 33.)

52. Refer to Question 33 and outline the pathway by which the nitrogen of uracil is used in the synthesis of urea.

53. Why do primates suffer from gout while most other animals do not?

54. Diabetes is a complex set of metabolic diseases with the common symptom of an inability to transport glucose into target cells (muscle cells and adipocytes). The body compensates in part by degrading muscle protein to generate energy. Explain how this process works.

55. Individuals with hyperammonemia are given α-keto acids as a treatment. Explain. (Visit the companion website at www.oup.com/us/mckee to read the Biochemistry in Perspective box on hyperammonemia.)

COMPAN

WEBS

Integration of Metabolism

A Grizzly Bear Feeding on Jumping Salmon The food that animals consume supplies their bodies with the nourishment required to sustain living processes. Complex regulatory mechanisms ensure that the demands of each cell for energy and metabolites are consistently met.

Overview

▼

PREVIOUS CHAPTERS DEAL WITH THE METABOLISM OF CARBOHYDRATES, LIPIDS, AND OTHER MOLECULES. HOWEVER, THE WHOLE IS NOT JUST THE sum of its parts. Multicellular organisms are extraordinarily complex, more so than their components would suggest. Chapter 16 takes a wider view of functioning of the mammalian body. A review of the mechanisms of action of hormones and other signal molecules that make the sophisticated regulation possible, is followed by a consideration of the contributions of major organs in mammals. The chapter also describes the feeding-fasting cycle, a physiological process that ensures that adequate energy resources are available.

The most distinctive characteristic of living organisms is their capacity to sustain adequate, if not always optimal, operating conditions despite changes in their internal and external environments. And if this feat is not amazing enough, consider that they must also simultaneously repair damaged components and, when possible, undergo cell divisions and other forms of growth. To accomplish these functions, the anabolic and catabolic reaction pathways that use carbohydrates, lipids, and proteins as energy sources and biosynthetic precursors must be precisely regulated. For multicellular organisms, this endeavor is astonishingly complicated. However, the capacity to sustain life and efficiently exploit energy resources in diverse environments is made possible by a division of labor among their constituent cells, tissues, and organ systems. Mammals, the most carefully investigated group of multicellular organisms, have a sophisticated and mutually beneficial division of labor. Each organ performs specific functions to serve both the short- and long-term interests of the body.

The operation of such a complex system as the body is maintained by a continuous flow of information among its parts. A simple system for information transfer is composed of a primary signal (e.g., a hormone), a target (a specific receptor), and a transducer system (that converts the signal to a cellular response). Physiological systems, however, require finely modulated responses to complex stimuli. In addition, for coordinated functioning, each body part must receive information about events in other parts. Considering the complexities of multicellular organisms, the need for a large number of primary signals, specific receptors, and transducer systems is not surprising. In the mammalian body, much information transfer is accomplished by hormones. These messenger molecules are arranged in complex hierarchies that allow for a high degree of sophisticated regulation.

In Chapter 16 the focus of the discussion is the integration of the major metabolic processes in mammals. The chapter begins with an overview of metabolic processes and descriptions of the major classes of signal molecules and their mechanisms of action. This is followed by a discussion of the metabolic contributions of several major organs. The chapter ends with an overview of the feeding-fasting cycle, which has great physiological importance because of its role in energy acquisition. The chapter also provides descriptions of two disorders of metabolic regulation: diabetes and obesity.

16.1 OVERVIEW OF METABOLISM

The central metabolic pathways are common to most organisms. Throughout the life of an organism, a precise balance is struck between anabolic (synthetic) and catabolic (degradative) processes. An overview of the principal anabolic and catabolic pathways in heterotrophs such as animals is illustrated in Figure 16.1. As a young animal grows and matures, the rate of anabolic processes is greater than that of catabolic processes. As healthy adulthood is reached, anabolic processes slow, and growth essentially stops. Except during illness or pregnancy, the animal's tissues exist in a metabolic steady state throughout the remainder of its life. In a **steady state**, the rate of anabolic processes is approximately

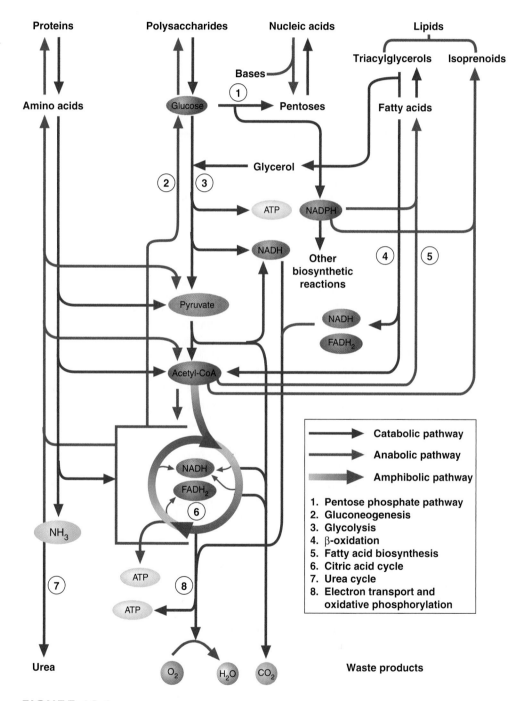

FIGURE 16.1

Overview of Metabolism

This simplified overview of metabolism illustrates the anabolic and catabolic pathways of the major biomolecules in heterotrophs (i.e., the biochemical pathways that synthesize, degrade, or interconvert important biomolecules and generate energy).

equal to that of catabolic processes. Consequently, the appearance and functioning of the animal change little from one day to the next. Only over long periods do the inevitable signs of aging appear.

How are animals (or other multicellular organisms) able to maintain a balance between anabolic and catabolic processes as they respond and adapt to changes in their environment? The answer to this question is not fully understood. However, various forms of intercellular communication are believed to play an important role. Most intercellular communication occurs by means of chemical signals. Once released into the extracellular environment, each chemical signal is recognized by specific cells (called **target cells**), which then respond in a specific manner. Most chemical signals are modified amino acids, fatty acid derivatives, peptides, proteins, or steroids.

In animals the nervous and endocrine systems are primarily responsible for coordinating metabolism. The nervous system provides a rapid and efficient mechanism for acquiring and processing environmental information. Nerve cells, called neurons, release neurotransmitters (Section 14.3) at the end of long cell extensions called axons into tiny intercellular spaces called synapses. The neurotransmitter molecules bind to nearby cells, evoking specific responses from those cells.

Metabolic regulation by the endocrine system is achieved by secretion of chemical signals called hormones directly into the blood. The endocrine system is composed of specialized cells, many of which are found in glands. After these hormone molecules, referred to as **endocrine** hormones, have been secreted, they travel through the blood until they reach a target cell. Some hormones exert very specific effects on one type of target cell; other hormones act on a variety of target cells. For example, thyroid-stimulating hormone (TSH) stimulates follicular cells in the thyroid gland to release T_3 (triiodothyronine) and T_4 (thyroxine) (Figure 16.2). In contrast, T_3 (the most active form of thyroid hormone) and T_4 stimulate a variety of cellular reactions in numerous cell types (e.g., T_3 stimulates glycogenolysis in liver cells and glucose absorption in the small intestine).

Most hormone-induced changes in cell function result from alterations in the activity or concentration of enzymes. Hormones interact with cells by binding to specific receptor molecules. The receptors for most water-soluble hormones (e.g., polypeptides and epinephrine) are located on the surface of target cells. The binding of these hormones to membrane-bound receptors triggers an intracellular response. The intracellular actions of many hormones are mediated by a group of molecules referred to as **second messengers**. (The hormone molecule is the first messenger.) Several second messengers have been identified. These include the nucleotides cyclic AMP (cAMP) and cyclic GMP (cGMP), calcium ions, and the inositol-phospholipid system. Most second messengers act to modulate enzymes, often by a powerful amplification device called an enzyme cascade. In an *enzyme cascade* (Figure 16.3), enzymes undergo conformational transitions that switch the enzymes from their inactive forms to their active forms, or vice versa, in a sequentially expanding array leading to a substantial amplification of the original signal. This process is often initiated when a second messenger binds to a specific enzyme. For example, the binding of cAMP to inactive protein kinase A converts it to active protein kinase A which, in turn, modifies the activity of many target enzymes through phosphorylation. The original signal generates an amplified and diversified response, via a second messenger (at the signal level) in some cases and an enzyme cascade (at the catalytic level) in most cases. A cAMP system accomplishes amplification at both levels. Table 16.1 lists the most important mammalian hormones. Note that the hormones in Table 16.1 fall into three categories: peptide or polypeptide, steroid, and amino acid derivatives.

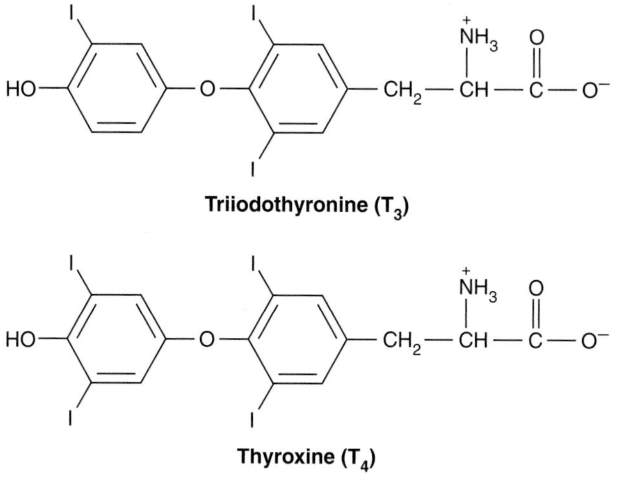

FIGURE 16.2

Structure of the Thyroid Hormones T$_3$ and T$_4$.

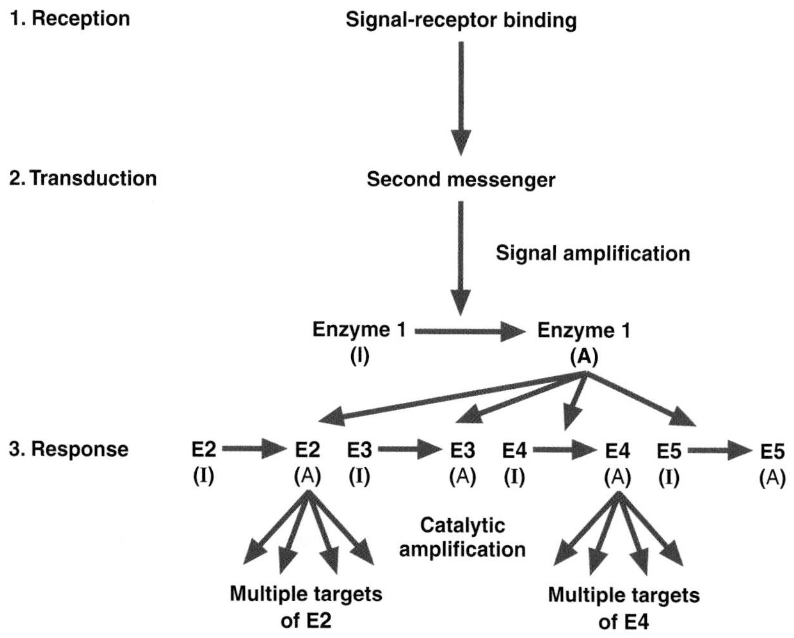

Triiodothyronine (T$_3$)

Thyroxine (T$_4$)

FIGURE 16.3

Signal Transduction

Signal transduction mechanisms occur in three phases. (1) Reception: a signal molecule binds to its receptor. (2) Transduction: as a result of receptor binding, an enzyme cascade is initiated, typically by a second messenger molecule. The enzyme activated by the second messenger modifies multiple copies of a number of different target enzymes. Target enzymes that are activated may also modify multiple copies of a second set of target proteins. (3) Response: cellular functions are altered by changes in the activities of existing enzymes, rearrangements of the cytoskeleton, and/or altered gene expression. Signal amplification occurs at each step of the cascade. (I = inactive, A = active)

TABLE 16.1 Selected Mammalian Hormones

Source	Hormone	Function
Hypothalamus	Gonadotropin-releasing hormone* (GnRH)	Stimulates LH and FSH secretion
	Corticotropin-releasing hormone* (CRH)	Stimulates ACTH secretion
	Growth hormone-releasing hormone* (GHRH)	Stimulates GH secretion
	Somatostatin*	Inhibits GH and TSH secretion
	Thyrotropin-releasing hormone* (TRH)	Stimulates TSH and prolactin secretion
Pituitary	Luteinizing hormone* (LH)	Stimulates cell development and synthesis of sex hormones in ovaries and testes
	Follicle-stimulating hormone* (FSH)	Promotes ovulation and estrogen synthesis in ovaries and sperm development in testes
	Corticotropin* (ACTH) (adrenocorticotropic hormone)	Stimulates steroid synthesis in adrenal cortex
	Growth hormone* (GH)	Exerts general anabolic effects in many tissues
	Thyrotropin* (TSH) (thyroid-stimulating hormone)	Stimulates thyroid hormone synthesis
	Prolactin*	Stimulates milk production in mammary glands and assists in the regulation of the male reproductive system
	Oxytocin*	Stimulates uterine contractions and milk ejection
	Vasopressin*	Regulates blood pressure and water balance
Gonads	Estrogens† (estradiol)	Regulates maturation and function of reproductive system in females
	Progestins† (progesterone)	Facilitates implantation of fertilized eggs and maintenance of pregnancy
	Androgens† (testosterone)	Regulate maturation and function of reproductive system in males
Adrenal cortex	Glucocorticoids† (cortisol, corticosterone)	Exert diverse metabolic effects, as well as inhibiting the inflammatory response
	Mineralocorticoids† (aldosterone)	Regulate mineral metabolism
Thyroid	Triiodothyronine‡ (T_3)	Stimulates many cellular reactions
	Thyroxine‡ (T_4) (after conversion to T_3)	
Gastrointestinal tract	Gastrin*	Stimulates secretion of stomach acid and pancreatic enzymes
	Secretin*	Regulates pancreatic exocrine secretions
	Cholecystokinin*	Stimulates secretion of digestive enzymes and bile
	Somatostatin*	Inhibits secretion of gastrin and glucagon
Pancreas	Insulin*	Exerts general anabolic effects including glucose uptake and lipogenesis
	Glucagon*	Promote glycogenolysis and lipolysis
	Somatostatin*	Inhibits the secretion of glucagon

*Peptide or polypeptide.

†Steroid.

‡Amino acid derivative.

16.2 HORMONES AND INTERCELLULAR COMMUNICATION

After a review of peptide and steroid hormones and their mechanisms of action, a brief overview of growth factors (hormonelike molecules that regulate cell growth and cell division) is provided.

Peptide Hormones

In mammals, a vast number of metabolic activities are controlled to one degree or another by peptide hormones. Synthesis and secretion of many of these hormones are regulated by a complex cascade mechanism and ultimately controlled by the central nervous system. In this system, outlined in Figure 16.4, sensory signals are received by the hypothalamus, an area in the brain that integrates the nervous and endocrine systems. Once it is appropriately stimulated, the hypothalamus synthesizes and releases specific peptide-releasing hormones. These molecules enter a specialized capillary bed surrounding the pituitary. The pituitary gland, which is attached to the hypothalamus by the pituitary stalk, consists of two distinct lobes: the anterior lobe, which receives hormonal signals from the hypothalamus via a capillary system, and the posterior lobe, which receives neurosecretory signals directly from the hypothalamus. The hypothalamus-releasing hormones target specific clusters of cells in the anterior lobe and stimulate the secretion of trophic ("turning" or "changing") hormones. Once secreted into the bloodstream, these trophic hormones stimulate the synthesis and release of specific hormones from other endocrine glands. For example, the tripeptide thyrotropin-releasing hormone (TRH) stimulates the secretion of TSH. TSH in turn stimulates the thyroid gland to release the thyroid hormones T_3 and T_4.

In anatomy and function, the posterior pituitary differs from the anterior lobe. The hormones secreted (i.e., the peptides vasopressin and oxytocin) are actually synthesized in neurons of separate types that originate in the hypothalamus. After their synthesis, both hormones are packaged into secretory granules. The granules then migrate down the axons into the posterior lobe. They are secreted into the bloodstream when an action potential reaches the nerve endings and initiates exocytosis. Oxytocin is secreted as a response to nerve impulses initiated by the stretching of the uterus late in pregnancy and suckling during breast feeding. Vasopressin (antidiuretic hormone) is secreted in response to

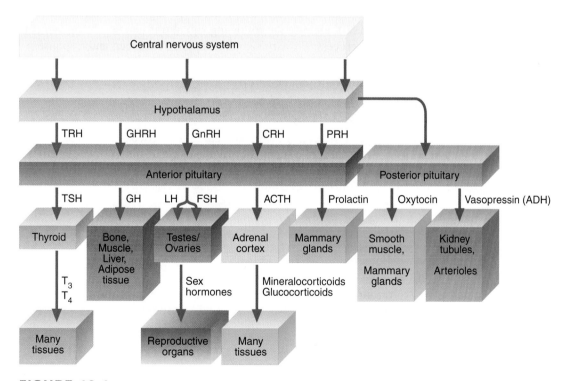

FIGURE 16.4

The Endocrine Hormone Cascade System

Hormone synthesis and secretion are ultimately controlled by the central nervous system. Neurons in the hypothalamus either stimulate or inhibit hormone synthesis and/or secretion in the pituitary. Pituitary hormones initiate metabolic activities in their target cells.

neural signals from specialized cells called osmoreceptors, which are sensitive to changes in blood osmolality.

Animals employ several mechanisms to prevent excessive hormone synthesis and release. The most prominent of these is feedback inhibition. The hypothalamus and anterior pituitary are controlled by the target cells they regulate. For example, TSH release by the anterior pituitary is inhibited when blood levels of T_3 and T_4 rise. The thyroid hormones inhibit the responsiveness of TSH-synthesizing cells to TRH. In addition, several tropic hormones inhibit the synthesis of their releasing factors.

Target cells also possess mechanisms that protect against overstimulation by hormones. In a process referred to as **desensitization**, target cells adjust to changes in stimulation levels by decreasing the number of cell-surface receptors or by inactivating those receptors. The reduction in cell-surface receptors in response to stimulation by specific hormone molecules is called **downregulation**. In downregulation, receptors are internalized by endocytosis. Depending on cell type and several metabolic factors, the receptors may eventually be recycled to the cell surface or be degraded. If degraded, new receptor proteins must be synthesized to replace old receptors. Some disease states are caused by or associated with target cell insensitivity to specific hormones. For example, some cases of diabetes are associated with **insulin resistance**, caused by a decrease in functional insulin receptors. (Diabetes is discussed in a Biochemistry in Perspective box later in this chapter: pp. 606–608.)

Sensitive techniques are now available to detect and measure hormones. The most common of these are enzyme-linked immunosorbent assays (ELISA).

KEY CONCEPT

Many of the hormones in the mammalian body are controlled by a complex cascade mechanism and ultimately regulated by the central nervous system.

COMPANION

GW

WEBSITE Visit the companion website at www.oup.com/us/mckee to read the Biochemistry in the Lab box on hormone methods.

QUESTION 16.1

Review the epinephrine-stimulated activation of glycogen breakdown presented earlier (Figure 8.20). Identify the following signal transduction components in this biochemical process: Primary signal, receptor, transducer, and response.

QUESTION 16.2

Cortisone is one of several synthetic adrenal glucocorticosteroids that are prescribed for the treatment of inflammatory and allergic disorders. It can be administered orally. In contrast, the hormone insulin, used in the treatment of diabetes mellitus, must be injected. Explain.

QUESTION 16.3

When a hormone molecule binds reversibly to its receptor in or on a target cell, it does so through noncovalent interactions. Describe the types of interaction that are most likely to be involved in this binding process.

QUESTION 16.4

The thymus gland is a bilobed organ found just above the heart. It promotes the differentiation of certain lymphocytes to form T cells, which confer cellular immunity. In addition, the thymus gland secretes several thymic hormones that stimulate T-cell function after these cells have left the thymus gland.

QUESTION 16.4 (CONT.)

Stress, both physical and emotional, is now known to depress cellular immunity. For example, the risk of developing cancer and severe infections increases with the duration of chronic stress. Elevated blood levels of glucocorticoids (e.g., cortisol) are believed to play an important role in the mechanism by which stress induces this change in function. Thymic hormone secretion is depressed when glucocorticoid levels are high. In addition, circadian rhythms alter thymic hormone and glucocorticoid levels. (Thymic hormone levels are higher and glucocorticoid levels are lower at night than during the day.) Assuming that glucocorticoids depress thymic hormone secretion, outline (in general terms) how thymic hormone is affected by stressful situations and by light/dark cycles. What other hormones are involved in both of these processes? [*Hint*: Review melatonin synthesis in Chapter 14, Question 14.7.]

Water-soluble hormone molecules (e.g., polypeptides, proteins, and epinephrine) initiate their actions by binding to receptor molecules on the outer surface of the target cell's plasma membrane. There are two major types of cell-surface receptor: G-protein-coupled receptors and receptor tyrosine kinases. The functional properties of each type are described.

G-protein-coupled receptors (GPCRs), the largest known protein receptor family, are composed of seven membrane spanning helices with an extracellular N-terminal segment that forms part of the ligand-binding site and an intracellular C-terminal segment that interacts with G-proteins. GPCRs transduce a wide variety of stimuli into intracellular signals. In addition to hormones such as glucagon, TSH, the catecholamines, and the endocannabinoids (e.g., the arachidonic acid derivative anandamide), GPCRs also respond to neurotransmitters (e.g., glutamate, dopamine, and GABA), neuropeptides (e.g., vasopressin and oxytocin), odorants and tastants (molecules that stimulate the senses of odor and taste, respectively), and light (rhodopsin).

G proteins, also known as heterotrimeric GTP-binding proteins, are the molecular switches that transduce ligand binding to GPCRs into intracellular signals (Figure 16.5). G proteins are composed of α, β, and γ subunits. The α subunit binds GTP and GDP.

Of the 20 known α subunits, the best researched examples include those of G_S and G_i, which stimulate and inhibit adenylate cyclase, respectively, The $\beta\gamma$ complex, composed of β and γ subunits, binds to and inhibits the α subunit. G proteins are anchored to the plasma membrane by myristoyl or palmitoyl groups attached to the α subunits and farnesyl or geranylgeranyl groups attached to γ subunits. The $\beta\gamma$ dimer promotes the association of the α subunit to the GPCR and, in the absence of receptor activation prevents GDP/GTP exchange (discussed shortly). It also facilitates the anchoring of the α subunit to the membrane and plays a role in effector signaling downstream (postreceptor activation).

G-protein activation occurs when ligand binds to the GPCR. A conformational change in the transmembrane region of the receptor leads to GDP/GTP exchange mediated by a **guanine nucleotide exchange factor (GEF)** followed by GTP-α subunit dissociation. The activated α subunit moves over the cytoplasmic surface of the membrane to activate a second messenger–generating enzyme. The GTP-α_s subunit activates adenylate cyclase and increases the intracellular synthesis of cAMP, a second messenger in the signal transduction cascade that follows. GTP-α_i inhibits adenylate cyclase, reducing the intracellular concentration of cAMP. Other molecules in addition to cAMP are involved in the transduction of some GPCR signaling mechanisms. These include components of the phosphatidylinositol cycle and calcium ions. cGMP, a second messenger molecule synthesized by guanylate cylase is also discussed.

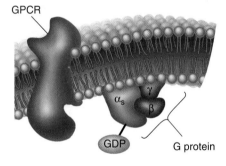

FIGURE 16.5

G-Protein-Coupled Receptor and G Protein

When a ligand binds to a G-protein-coupled receptor, a signaling mechanism is initiated that is mediated by a G protein. Each G protein consists of three subunits: α, β, and γ. Before a G protein is activated, it binds GDP.

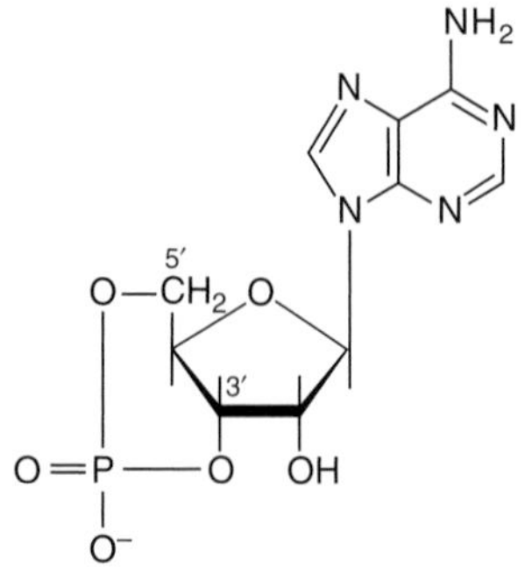

FIGURE 16.6

Structure of the Second Messenger Molecule Cyclic AMP (cAMP)

Note that cAMP, also referred to an adenosine 3′,5′-phosphate, is a diester that links the 3′ and 5′ carbons.

cAMP cAMP (Figure 16.6) is generated from ATP by adenylate cyclase in response to a hormone-receptor interaction. The G protein G_s stimulates cAMP synthesis when hormones such as glucagon, TSH, and epinephrine bind to their receptors. A G_s bound to an occupied receptor (signal initiation) undergoes GDP/GTP exchange, and the GTP-α_s subunit dissociates and activates adenylate cyclase (Figure 16.7). GTP hydrolysis terminates this association, and GDP-α_s recombines with GPCR-$\beta\gamma$ (primary signal termination). (In effect, α_s-GTP is a timing device. The rate at which the α subunit hydrolyzes GTP determines the signal's duration.) The activated adenylate cyclase synthesizes a number of cAMP molecules (signal amplification), which diffuse into the cytoplasm, where they bind to and activate cAMP-dependent protein kinase (PKA). Active PKA then phosphorylates and thereby alters the catalytic activity of key regulatory enzymes. The cAMP is quickly hydrolyzed by phosphodiesterase (secondary signal termination).

The target proteins affected by cAMP depend on the cell type. In addition, several hormones may activate the same G protein. Therefore different hormones may elicit the same effect. For example, glycogen degradation in liver cells is initiated by both epinephrine and glucagon.

Some hormones inhibit adenylate cyclase activity. Such molecules depress cellular protein phosphorylation reactions because their receptors interact with G_i protein. When G_i is activated, its α_i subunit dissociates from the $\beta\gamma$ dimer and prevents the activation of adenylate cyclase. For example, because its receptors in adipocytes are associated with G_i, PGE_1 (prostaglandin E_1) depresses lipolysis. (Recall that lipolysis is stimulated by glucagon and epinephrine.)

THE PHOSPHATIDYLINOSITOL CYCLE, DAG, AND CALCIUM The phosphatidylinositol cycle (Figure 16.8) mediates the actions of hormones and growth factors. Examples include acetylcholine (e.g., insulin secretion in pancreatic cells), vasopressin, TRH, GRH, and epinephrine (α_i receptors). Phosphatidylinositol-4,5-bisphosphate (PIP_2) is cleaved by phospholipase C to form the second messengers **DAG (diacylglycerol)** and **IP_3 (inositol-1,4, 5-trisphosphate)**. A hormone-receptor complex induces the activation of a G protein, which in turn activates phospholipase C. Several types of G protein may be involved in the phosphatidylinositol cycle. For example, G_Q (shown in Figure 16.8) mediates the actions of vasopressin.

The DAG product of the phospholipase C–catalyzed reaction activates protein kinase C. Several protein kinase C activities have been identified. Depending on the cell, activated protein kinase C phosphorylates specific regulatory enzymes, thereby activating or inactivating them.

Once generated, IP_3 diffuses to the calcisome (SER), where it binds to a receptor. The IP_3 receptor is a calcium channel. Cytoplasmic calcium levels then rise as calcium ions flow through the activated open channel. Calcium ions are involved in the regulation of a large number of cellular processes including contributing to the activation of plasma membrane–associated protein kinase C. Because calcium levels are still relatively low even when the calcium release mechanism has been activated (approximately 10^{-6} M), the calcium-binding sites on calcium-regulated proteins must have a high affinity for the ion. Several calcium-binding proteins modulate the activity of other proteins in the presence of calcium. Calmodulin, a type of calcium-binding protein, mediates many calcium-regulated reactions. In fact, calmodulin is a regulatory subunit for some enzymes (e.g., phosphorylase kinase, which converts phosphorylase b to phosphorylase a in glycogen metabolism).

cGMP cGMP (guanosine 3′-5′-phosphate) is synthesized from GTP by guanylate cyclase. Two types of guanylate cyclase are involved in signal transduction. In one type, which is membrane-bound, the extracellular domain of the enzyme is a hormone receptor. The other type is a cytoplasmic enzyme.

FIGURE 16.7

The Adenylate Cyclase Second Messenger System That Controls Glycogenolysis

When the receptor is unoccupied, the G_s protein α_s subunit has GDP bound and is complexed with the β, γ dimer. The binding of hormone (1) activates the receptor and leads to replacement of GDP with GTP (2). The activated subunit interacts with and activates adenylate cyclase. (3) The cAMP produced binds to cAMP-dependent protein kinase. Signal transduction ends when the ligand leaves the receptor, the bound GTP is hydrolyzed to GDP by the GTPase activity within the α_s subunit, and the α_s subunit dissociates from adenylate cyclase. Cyclic AMP is deactivated by hydrolysis to AMP, a reaction catalyzed by phosphodiesterase. (4) The α_s subunit then reassociates with the β,γ dimer.

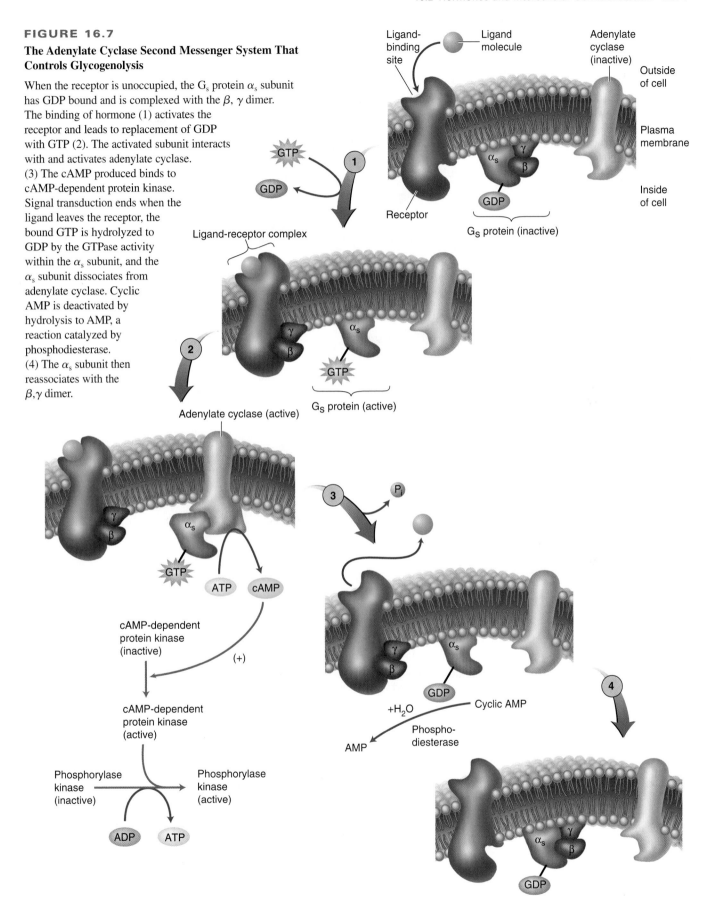

FIGURE 16.8

The Phosphatidylinositol Pathway

The binding of certain hormones to their receptor activates the α subunit of a G protein. The α subunit then activates phospholipase C, which cleaves IP₃ from PIP₂, leaving DAG in the membrane. DAG, acting with phosphatidylserine (PS) and Ca²⁺, activates protein kinase C, which subsequently phosphorylates key regulators. IP₃ binds to receptors on the SER, opening Ca²⁺ channels. Then Ca²⁺ moves into the cytoplasm and activates additional targets.

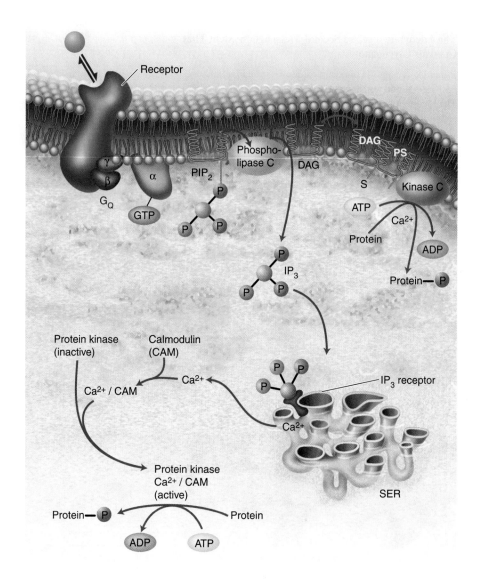

Two types of molecule are now known to activate membrane-bound guanylate cyclase: atrial natriuretic peptide and bacterial enterotoxin. *Atrial natriuretic factor* (ANF) is a peptide that is released from heart atrial cells in response to increased blood volume. The biological effects of ANF, that is, lowering of blood pressure via vasodilation and diuresis (increased urine excretion), are mediated by cGMP. ANF activates guanylate cyclase in several cell types. In one type, those in the kidney's collecting tubules, ANF-stimulated cGMP synthesis increases renal excretion of Na⁺ and water.

The binding of *enterotoxin* (produced by several bacterial species) to another type of guanylate cyclase found in the plasma membrane of intestinal cells causes diarrhea. For example, one form of traveler's diarrhea is caused by a strain of *E. coli* that produces *heat-stable enterotoxin*. The binding of this toxin to an enterocyte plasma membrane receptor linked to guanylate cyclase triggers excessive secretion of electrolytes and water into the lumen of the small intestine.

Cytoplasmic guanylate cyclase possesses a heme prosthetic group. The enzyme is activated by Ca²⁺, so any rise in cytoplasmic Ca²⁺ causes cGMP synthesis. This guanylate cyclase activity is activated by NO. In several cell types cGMP stimulates the functioning of ion channels.

Receptor tyrosine kinases (RTKs) are a family of transmembrane receptors that bind ligands such as insulin, epidermal growth factor (EGF), platelet-derived growth factor (PDGF), and insulin-like growth factor I (IGF-I). Although there are several structural differences among members of this group, they do possess the following features in common: an external domain that binds specific extracellular ligands, a transmembrane segment, and a cytoplasmic catalytic domain with tyrosine kinase activity. When a ligand binds to the external domain, a conformational change in the receptor protein activates the tyrosine kinase domain. The tyrosine kinase activity initiates a phosphorylation cascade that begins with an autophosphorylation of the tyrosine kinase domain. Because most research efforts have been devoted to the insulin receptor, its structure and proposed functions are described next.

The insulin receptor (Figure 16.9) is a transmembrane glycoprotein composed of two types of subunit connected by disulfide bridges. Two large α subunits (130 kD) extend extracellularly, where they form the insulin-binding site. Each of the two β subunits (90 kD) contains a transmembrane segment and a tyrosine kinase domain.

Insulin binding activates receptor tyrosine kinase activity and causes a phosphorylation cascade that modulates various intracellular proteins. Insulin receptor substrate 1 (IRS1) is one of several proteins that are phosphorylated by the activated insulin receptor. Phosphorylated IRS1 then binds to and activates several proteins, including phosphatidylinositol-3-kinase, as illustrated in Figure 16.10.) Phosphatidylinositol-3-kinase then phosphorylates PIP_2 (phosphatidylinositol 4,5-bisphosphate), a minor constituent of cell membrane, to form PIP_3 (phosphatidylinositol 3,4,5-trisphosphate). The subsequent activation of PIP_3-dependent protein kinase initiates a phosphorylation cascade that leads to the activation of several kinases (e.g., PKB and PKC), which in turn trigger pathways that lead to changes in the expression of several genes. One of the functions of PKB is to facilitate the translocation of the GLUT4 transporter to the plasma membrane of adipocytes and muscle cells.

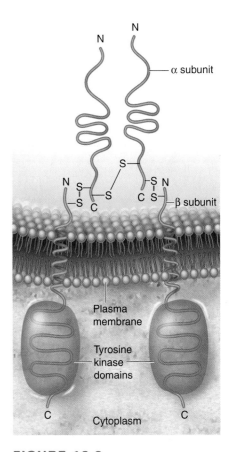

FIGURE 16.9

The Insulin Receptor

The insulin receptor is a dimer composed of two pairs of α and β subunits. The subunits are connected to each other by disulfide bridges.

KEY CONCEPTS

- There are two major types of cell-surface receptor that bind to hormone molecules: G-protein-linked receptors and receptor tyrosine kinases.
- The G proteins activated by GPCRs utilize one or more second messenger molecules to transduce the original signal into a signal cascade.
- Receptor tyrosine kinases activate phosphorylation cascades when they undergo autophosphorylation triggered by the binding of signal molecules.

QUESTION 16.5

In a hypothetical cAMP-mediated signal transduction cascade, the GTP-α_s/adenylate cyclase interaction following a single hormone-receptor binding event lasts for 2.3 seconds. The catalytic rate (turnover number) for the adenylate cyclase in question is 350 cAMP molecules produced per second. How many cAMP molecules would be produced if five hormone-receptor binding events were to occur before the hormone molecule dissipates in the bloodstream? What is the amplification effect of this step in the signaling pathway?

QUESTION 16.6

Explain the sequence of events that occurs when epinephrine triggers the synthesis of cAMP. Once formed, cAMP breaks down rapidly. Why is this an important feature for a second messenger in a signal transduction process?

QUESTION 16.7

The A subunit of cholera toxin causes the cAMP-mediated opening of chloride channels. A massive diarrhea results because the hydrolysis of GTP is prevented. Describe why this inhibition leads to the diarrhea.

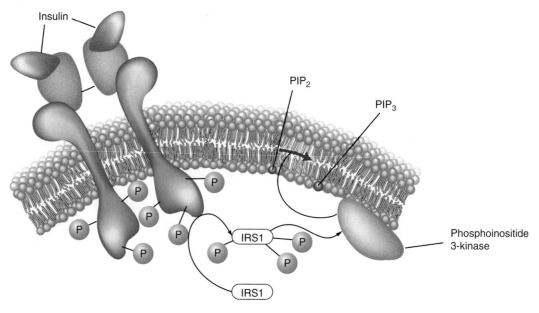

FIGURE 16.10

Simplified Model of Insulin Signaling

The binding of insulin to its tyrosine kinase receptor causes autophosphorylation reactions. The activated insulin dimer subsequently phosphorylates an initial set of substrate molecules. Only one is illustrated in this figure: insulin receptor substrate 1 (IRS1). The newly phosphorylated IRS1 then binds to and activates phosphoinositide 3-kinase, which then phosphorylates PIP_2 to form PIP_3. PIP_3 subsequently binds to and activates PIP_3-dependent protein kinase, which in turn activates, via phosphorylation, various kinases (e.g., PKB and PKC). These latter molecules continue the signal cascade, ending with alterations in gene expression.

$$CH_2 — O — NO_2$$
$$CH — O — NO_2$$
$$CH_2 — O — NO_2$$
Nitroglycerin

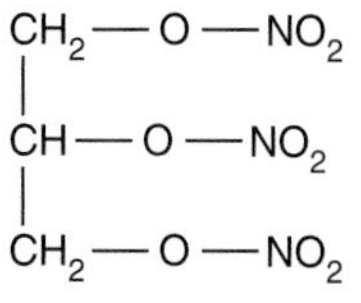

FIGURE 16.11

Nitroglycerin

QUESTION 16.8

Angina pectoris is a very painful symptom of coronary artery disease. In this condition, which is caused by the narrowing of the coronary arteries, insufficient oxygen flow during exertion causes chest pain that radiates to the neck and left arm. Traditionally, angina pectoris has been treated with nitroglycerin. It is now believed that nitroglycerin (Figure 16.11) relieves the pain because it acts to promote vasodilation (increased blood vessel diameter) throughout the body. Vasodilation, which results from smooth muscle relaxation, reduces blood pressure, which in turn reduces the heart's work load. Suggest how nitroglycerin relieves angina pectoris. [*Hint*: Cytoplasmic calcium ions stimulate muscle contraction.]

QUESTION 16.9

Cancer often results from a multistage process involving an initiating event (mediated by a viral infection or a carcinogenic chemical) followed by exposure to tumor promoters. Tumor promoters, a group of molecules that stimulate cell proliferation, cannot induce tumor formation by themselves. The phorbol esters, found in croton oil (obtained from the seeds of the croton plant, *Croton tiglium*), are potent tumor promoters. (Other examples of tumor promoters include asbestos and several components of tobacco smoke.) In one of the tumor-promoting actions of the phorbol esters, these molecules mimic the actions of DAG. In contrast to DAG, the phorbol esters are not easily disposed of. Explain the possible biochemical consequences of phorbol esters in an "initiated" cell. What enzyme is activated by both DAG and phorbol esters?

QUESTION 16.9 (CONT.)

Phorbol myrsitate acetate (PMA)
A phorbol ester

QUESTION 16.10

The term *diabetes*, derived from the Greek word *diabeinein* ("to go to excess"), was first used by Aretaeus (A.D. 81–138) to identify a group of symptoms that included intolerable thirst and "a liquefaction of the flesh and limbs into urine." After reviewing the accompanying Biochemistry in Perspective box (pp. 606–608), explain the physiological and biochemical basis for Aretaeus's findings.

Growth Factors

The survival of multicellular organisms requires that cell growth and cell division (mitosis) be rigorously controlled. The conditions that regulate these processes have not yet been completely resolved. However, a variety of hormonelike polypeptides and proteins, the **growth factors**, and some cytokines regulate the growth, differentiation, and proliferation of various cells. Often, the actions of several growth factors are required to promote cellular responses. Growth factors differ from hormones in that they are synthesized by a variety of cell types rather than by specialized glandular cells. Examples of mammalian growth factors include epidermal growth factor (EGF), platelet-derived growth factor (PDGF), and the somatomedins. The term **cytokine**, although sometimes used interchangeably with "growth factor," has traditionally described proteins produced by blood-forming cells and immune system cells. Cytokines may stimulate or inhibit cell growth or proliferation. Interleukins and interferons are examples of cytokines.

Epidermal growth factor (EGF) (6.4 kD), one of the first cellular growth factors identified, is a **mitogen** (a stimulator of cell division) for a large number of epithelial cells, such as epidermal and gastrointestinal lining cells. EGF triggers cell division when it binds to plasma membrane EGF receptors, which are transmembrane tyrosine kinases structurally similar to insulin receptors.

Platelet-derived growth factor (PDGF) (31 kD) is secreted by blood platelets during the clotting reaction. Acting with EGF, PDGF stimulates mitosis in fibroblasts and other nearby cells during wound healing. PDGF also promotes collagen synthesis in fibroblasts.

The **somatomedins** are a group of polypeptides that mediate the growth-promoting actions of GH. Produced in the liver and in a variety of other tissue

(Continued on p. 608)

BIOCHEMISTRY IN PERSPECTIVE

Diabetes Mellitus

Why does diabetes mellitus, a metabolic disease, damage the entire body? Diabetes mellitus is a group of devastating metabolic diseases caused by insufficient insulin synthesis, increased insulin destruction, or ineffective insulin action. All its metabolic effects result when the body's cells fail to acquire glucose from the blood. The metabolic imbalances that occur have serious, if not life-threatening, consequences (Figure 16A).

In insulin-dependent diabetes mellitus (IDDM), also called type I diabetes, inadequate amounts of insulin are secreted because the β cells of the pancreas have been

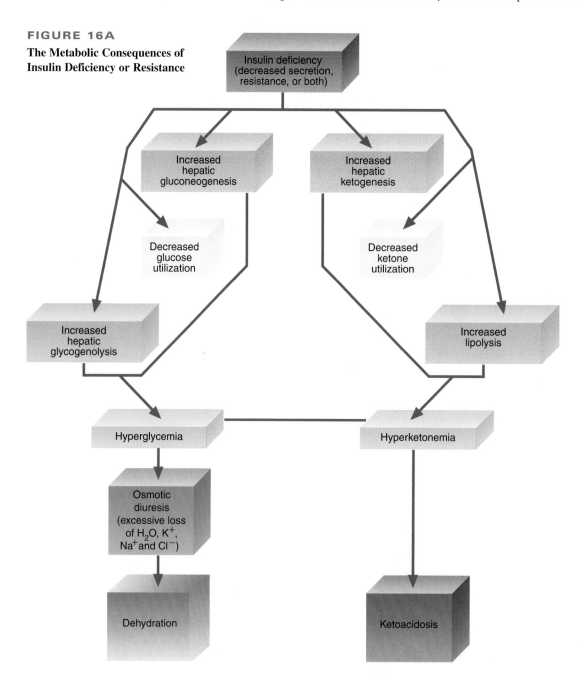

FIGURE 16A

The Metabolic Consequences of Insulin Deficiency or Resistance

BIOCHEMISTRY IN PERSPECTIVE cont

destroyed. Because IDDM usually occurs before the age of 20, it was until recently referred to as juvenile-onset diabetes. Non-insulin-dependent diabetes mellitus (NIDDM), also called type II or adult-onset diabetes, is caused by the insensitivity of target tissues to insulin. Although these forms of diabetes share some features, they differ significantly in others. Once quite rare, diabetes is now the third leading cause of death in the United States, where it afflicts at least 5% of the population.

The most obvious symptom of diabetes is **hyperglycemia** (high blood glucose levels), caused by inadequate cellular uptake of glucose. Among diabetics the severity of hyperglycemia may vary considerably. Because the kidney's capacity to reabsorb glucose from the urinary filtrate is limited, glucose appears in the urine (**glucosuria**). (The kidneys filter blood and then reabsorb substances such as glucose, amino acids, and ions from the filtrate.) Glucosuria results in **osmotic diuresis**, a process in which an excessive loss of water and electrolytes (Na^+, K^+, and Cl^-) is caused by the presence of solute in the filtrate. In severe cases, despite the consumption of large volumes of fluid, patients may become dehydrated.

Without insulin to regulate fuel metabolism, the three principal target tissues of this hormone (liver, adipose tissue, and muscle) fail to absorb nutrients appropriately. Instead, these tissues function as if the body were undergoing starvation. In the liver, gluconeogenesis accelerates because large amounts of amino acids are mobilized from muscle and synthesis of gluconeogenic enzymes (Section 8.2) increases. Glycogenolysis (ordinarily suppressed by insulin) produces additional glucose. The liver delivers these glucose molecules to an already hyperglycemic bloodstream. Increased lipolysis (Section 12.1) in adipose tissue (caused by the unopposed action of glucagon) releases large quantities of fatty acids into blood. In the liver, because these molecules are degraded by β-oxidation in combination with low concentrations of oxaloacetate (caused by excessive gluconeogenesis), large amounts of acetyl-CoA, the substrate for forming ketone bodies, are produced. Fatty acids not used to generate energy or ketone bodies are used in VLDL synthesis. This process causes hyperlipoproteinemia (high blood concentrations of lipoproteins) because lipoprotein lipase synthesis is depressed when insulin is lacking. In effect, many of the cells of diabetics "starve in the midst of plenty." In severe IDDM cases, despite increased appetite and food consumption, the failure to use glucose effectively eventually causes a decrease in body weight.

Insulin-Dependent Diabetes

In most cases of insulin-dependent diabetes, the insulin-producing β cells have been destroyed by the immune system. Although the symptoms of IDDM often manifest themselves abruptly, it now appears that β-cell destruction is caused by an inflammatory process over several years. The symptoms are not obvious until virtually all insulin-producing capacity has been destroyed. As in other inflammatory and autoimmune processes, β-cell destruction is initiated when an antibody binds to a cell-surface antigen. One of the most common autoantibodies found in type I diabetes is now believed to bind specifically to an antigen with glutamate decarboxylase activity. (Recall that glutamate decarboxylase catalyzes the synthesis of GABA from glutamate.) Autoantibodies to insulin and the tyrosine phosphatase IA-2 have also been detected; the significance of this phenomenon is unknown. Tyrosine phosphatase IA-2, one of several enzymes that remove phosphate groups from specific phosphotyrosine residues in key regulatory proteins, is found only in brain and in insulin-producing cells in the pancreas.

The most serious acute symptom of type I diabetes is **ketoacidosis**. Elevated concentrations of ketones in the blood (**ketosis**) and low blood pH along with hyperglycemia cause excessive water losses. (The odor of acetone on a patient's breath is characteristic of ketoacidosis.) Ketoacidosis and dehydration, if left untreated, can lead to coma and death. IDDM patients are treated with injections of insulin obtained from animals or from recombinant DNA technology. Before Frederick Banting and Charles Best discovered insulin in 1922, most type I diabetics died within a year after being diagnosed. Although exogenous insulin prolongs life, it is not a cure. Most diabetics have a shortened life span because of the long-term complications of their disease.

Many researchers now believe that IDDM is caused by both genetic and environmental factors. Although the precise cause is still unknown, individuals who have inherited certain genetic markers are at high risk for developing the disease. Certain HLA antigens, that is, HLA-DR3 and HLA-DR4, are found in a large majority of type I diabetics. The *HLA* or *histocompatibility antigens*, present on the surface of most of the body's cells, play an important role in determining how the immune system will react to foreign substances or cells. If a type I patient with both HLA-DR3 and HLA-DR4 antigens has an identical twin, the twin's risk for developing the disease is approximately 50%. That the risk is not 100% suggests that some environmental exposure is required for developing type I diabetes.

Non-Insulin-Dependent Diabetes

Non-insulin-dependent diabetes is a milder disease than the insulin-dependent form. Its onset is slow, often occurring after the age of 40. In contrast to type I patients, most individuals with type II diabetes have normal or often elevated blood levels of insulin. For a variety of reasons, type II patients are resistant to insulin. The most common cause of insulin resistance is the downregulation of insulin receptors. (Defective insulin receptors or improper insulin receptor processing also causes type II diabetes in some patients.) Approximately 85% of type II diabetics are obese. Because obesity itself promotes tissue insensitivity to insulin, individuals who are prone to this form of diabetes are at risk for the disease when they gain weight.

Treatment of NIDDM usually consists of diet control and exercise. Often, obese patients become more sensitive to insulin

BIOCHEMISTRY IN PERSPECTIVE cont

(i.e., there is an upregulation of insulin receptors) when they lose weight. Because sustained muscular activity increases the uptake of glucose without requiring insulin, exercise also decreases hyperglycemia. In some cases, oral hypoglycemic agents (OHAs), such as the sulfonylureas, are used. These drugs stimulate insulin release and, therefore, reduce gluconeogenesis and glycogenolysis in the liver and increase transport of glucose into insulin-sensitive body cells.

When the failure of type II diabetic patients to control hyperglycemia is accompanied by other serious medical conditions (e.g., renal insufficiency, myocardial infarction, or infections), a serious metabolic state referred to as **hyperosmolar hyperglycemic non-ketosis** (HHNK) can result. (Ketoacidosis is rare in type II diabetes.) Because of the additional metabolic stress, insulin resistance is exacerbated, and blood glucose levels rise. The patient may then become severely dehydrated. The resulting lower blood volume depresses renal function, which causes further increases in blood glucose concentrations. Eventually, the patient becomes comatose. Because the onset is slow, it may not be recognized until the dehydration is severe. (This is especially true for elderly diabetics, who often have a depressed thirst mechanism.) For this reason, HHNK is often more life-threatening than ketoacidosis.

Long-Term Complications of Diabetes

Despite the efforts of physicians and patients to control the symptoms of diabetes, few diabetics avoid the long-term consequences of their disease. Diabetics are especially prone to develop kidney

failure, myocardial infarction, stroke, blindness, and neuropathy. In *diabetic neuropathy*, nerve damage causes the loss of sensory and motor functions. In addition, circulatory problems often cause gangrene, which leads to tens of thousands of amputations annually.

Most diabetic complications stem from damage to the vascular system. For example, damaged capillaries in the eye and kidney lead to blindness and kidney damage, respectively. Similarly, the accelerated form of atherosclerosis found in diabetics leads to serious cases of myocardial infarction and stroke. Most of this damage results from glycation (p. 238), the nonenzymatic covalent modification of biomolecules that occurs when blood glucose levels are high.

Diabetics are also damaged by another consequence of hyperglycemia: increased metabolism by the sorbitol pathway. In some cells that do not require insulin for glucose uptake (e.g., Schwann cells in peripheral nerve and ocular tissues such as lens epithelial cells), the hexose enters by facilitated transport down its concentration gradient. Glucose is then converted to sorbitol (p. 235) by the NADPH-requiring enzyme aldose dehydrogenase. The oxidation of some sorbitol molecules to form fructose by sorbitol dehydrogenase is coupled to the reduction of NAD^+. Increased concentrations of NADH promote the reduction of pyruvate to form lactate. The accumulation of sorbitol and the redox changes caused by sorbitol oxidation are associated with several pathological changes in diabetics, for example, nerve damage and cataract formation. The physiological significance of the sorbitol pathway in normal cells is unknown.

SUMMARY: Diabetes is an example of how a single defect (the inability to synthesize or respond to insulin) in a complex biological system can cause devastating damage.

cells (e.g., muscle, fibroblasts, bone, and kidney) when GH binds to its cell surface receptor, the somatomedins are the major stimulators of growth in animals. The somatomedins are secreted by the liver into the bloodstream, in contrast to the paracrine action exhibited by other somatomedin-producing cells. In addition to stimulating cell division, the somatomedins promote (but to a lesser degree) the same metabolic processes as does the hormone insulin (e.g., glucose transport and fat synthesis). For this reason the two somatomedins found in humans have been renamed **insulin-like growth factors I and II** (IGF-I and IGF-II). Like other polypeptide growth factors, the somatomedins trigger intracellular processes by binding to cell surface receptors. Not surprisingly, somatomedin receptors are also tyrosine kinases.

Interleukin 2 (IL-2) (13 kD) is a member of a group of cytokines that regulate the immune system in addition to promoting cell growth and differentiation. IL-2 is secreted as a result of binding by activated T cells to a specific antigen-presenting cell. These cells are also stimulated to produce IL-2 receptors. The binding of IL-2 to these receptors stimulates cell division so that numerous identical T cells are produced. This process, as well as other aspects of the immune response, continues until the antigen has been eliminated from the body.

Several cytokines are growth inhibitors. The **interferons** are a group of polypeptides produced by a variety of cells in response to several stimuli, such as antigens, mitogens, viral infections, and certain tumors. The type I interferons protect cells from viral infection by stimulating the phosphorylation and inactivation of a protein factor required to initiate protein synthesis. Type II interferons, produced by T lymphocytes, inhibit the growth of cancerous cells in addition to having several immunoregulatory effects. As the name implies, the **tumor necrosis factors** (TNF) are toxic to tumor cells. Both TNF-α (produced by antigen-activated phagocytic white blood cells) and TNF-β (produced by activated T cells) suppress cell division. They may also have a role in regulating several developmental processes.

Steroid and Thyroid Hormone Mechanisms

The signal transduction mechanisms of hydrophobic hormone molecules such as the steroid and thyroid hormones result in changes in gene expression. In other words, the action of each type of hormone molecule causes the switching on or off of a specific set of genes, which in turn causes changes in the pattern of proteins that an affected cell produces. Steroid and thyroid hormones are transported in the blood to their target cells bound to several types of protein. Examples of steroid transport proteins include transcortin (also called corticosteroid-binding globulin), sex hormone–binding protein, and albumin. In addition to albumin, the thyroid hormones are transported by thyroid-binding globulin and thyroid-binding prealbumin.

Once they have reached their target cells, hydrophobic hormone molecules dissociate from their transport proteins, diffuse through the plasma membrane, and bind to their intracellular receptors (Figure 16.12). These receptors are high-affinity ligand-binding molecules that belong to a large family of structurally similar DNA-binding proteins. Depending on the type of hormone involved, initial binding to receptors may occur within the cytoplasm (e.g., glucocorticoid) or the nucleus (e.g., estrogen, androgens, and thyroid hormone). In the absence of hormone, several types of receptor have been observed to form complexes with other proteins. For example, unoccupied glucocorticoid receptors are found in the cytoplasm bound to chaperone proteins, such as heat shock protein 90 (hsp90). (Recall that heat shock proteins are so named because of their increased synthesis in response to cellular stressors such as elevated temperature.) The chaperone proteins block the receptor's DNA-binding site when the hormonal ligand is not present. When the hormone binds to its receptor, the chaperones dissociate and the receptor-ligand complex migrates to the nucleus as a homodimer.

Within the nucleus, each hormone-receptor complex binds to specific DNA segments called **hormone response elements** (HRE). The binding of the hormone-receptor complex to the base sequence of an HRE via zinc finger domains (see p. 149) in the receptor either enhances or diminishes the transcription of a specific gene. The same hormone-receptor complex can bind to and influence the transcription of as many as 50 to 100 different genes, therefore inducing global changes in cellular function.

KEY CONCEPT

Growth factors and cytokines are a group of hormonelike polypeptides and proteins that influence cell growth, proliferation, and differentiation.

KEY CONCEPT

Hydrophobic hormones such as the steroids and thyroid hormones diffuse across cellular membranes and bind to intracellular receptors.

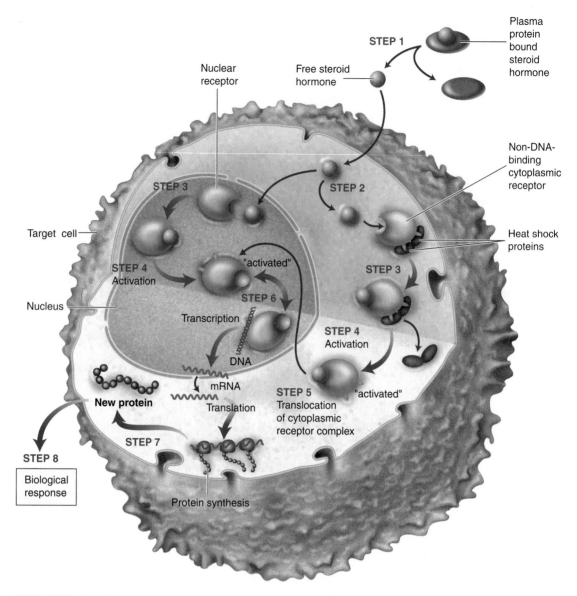

FIGURE 16.12

Model of Steroid Hormone Action Within a Target Cell

Steroid hormones are transported in blood associated with plasma proteins. When they reach a cell and are released (1), the hormone molecules diffuse through the plasma membrane, where they bind to receptor molecules in cytoplasm (2) or nucleus (3). After activation (4), a cytoplasmic hormone-receptor complex, a dimer (not shown), migrates to the nucleus (5). The binding of an activated hormone-receptor complex to HRE sequences within DNA (6) results in a change in the rate of transcription of specific genes, and therefore, in the pattern of proteins (7) that the cell produces. The net effect of the steroid hormone (8) is a change in the metabolic functioning of the cell.

16.3 METABOLISM IN THE MAMMALIAN BODY: DIVISION OF LABOR

Each of the major metabolic pathways that sustain life in multicellular organisms has been covered in this textbook. Any true understanding of metabolism, however, requires a more integrated approach. The feeding-fasting cycle, the self-regulating mechanism by which the mammalian body extracts energy and nutrients from food, is a well-understood process. A brief review of its operation provides an opportunity to observe biochemical reactions as they actually occur. Although the feeding-fasting cycle is a complex process, it is made more accessible with a brief overview of the roles of each body organ. The feeding-fasting cycle itself is covered in Section 16.4.

Each organ in the mammalian body contributes to the individual's function in several ways. For example, some organs are consumers of energy so that they may perform certain energy-driven tasks (e.g., muscle contraction). Other organs, such as those in the digestive tract, are responsible for efficiently supplying energy-rich nutrient molecules for use elsewhere. Information in the form of signal molecules (e.g., hormones and neurotransmitters) is utilized to regulate the balance between energy generation and energy expenditure. The roles of several systems and organs in relation to their metabolic contributions are discussed next.

Gastrointestinal Tract

The most obvious role of the organs of the gastrointestinal (GI) tract (Figure 1.2) is the digestion of nutrients such as carbohydrates, lipids, and proteins into molecules that are small enough to be absorbed (sugars, fatty acids, glycerol, and amino acids) by the enterocytes of the small intestine. As described earlier (p. 415), enterocytes then transport these molecules (and water, minerals, vitamins, and other substances) into the blood and lymph, which carry them throughout the body.

The enterocytes require enormous amounts of energy to support active transport and lipoprotein synthesis. Although some glucose is partially oxidized (leaving the enterocytes as lactate or alanine), most of the energy is supplied by glutamine. During the digestive process, enterocytes obtain glutamine from degraded dietary protein. Under fasting conditions, glutamine is acquired from arterial blood. Enterocytes also use some glutamine to form Δ^1-pyrroline-5-carboxylate, which is ultimately converted to proline. Other products of glutamine metabolism include lactate, citrate, ornithine, and citrulline. The liver receives blood containing dietary nutrients plus these products of glutamine metabolism. It uses lactate and alanine to synthesize glucose for export and glycogen for storage. The glucose in blood is preferentially delivered to glucose-dependent tissues (e.g., brain, red blood cells, and adrenal medulla).

The GI tract also produces a wide variety of protein hormones that provide the body with information concerning nutrient levels. In addition to insulin, prominent examples include ghrelin (Ghr), peptide YY (PYY), cholecystokinin (CCK), and glucagon-like peptide 1 (GLP-1). Ghrelin, produced by cells in the stomach and small intestine, stimulates appetite (food intake), whereas insulin, PYY, CCK, and GLP-1 promote satiety (i.e., inhibit food intake).

Liver

The liver performs a stunning variety of metabolic activities. In addition to its key roles in carbohydrate, lipid, and amino acid metabolism, the liver monitors and regulates the chemical composition of blood and synthesizes several plasma proteins. The liver distributes several types of nutrient to other parts of the body. Because of its metabolic flexibility, the liver reduces the fluctuations in nutrient availability caused by drastic dietary changes and intermittent feeding and fasting. For example, a sudden shift from a high-carbohydrate diet to one that is rich in proteins increases (within hours) the synthesis of the enzymes required for amino acid metabolism. Finally, the liver plays a critically important protective role in processing foreign molecules.

Muscle

Skeletal muscle, which is specialized to perform intermittent mechanical work, typically constitutes about one-half the body's mass. Therefore skeletal muscle consumes a large fraction of generated energy. The energy sources that provide

ATP for muscle contraction are glucose from its own store of glycogen and from the bloodstream, fatty acids from adipose tissue, and ketone bodies from the liver. During fasting and prolonged starvation, some skeletal muscle protein is degraded to provide amino acids (e.g., alanine) to the liver for gluconeogenesis.

In contrast to skeletal muscle, cardiac muscle must continuously contract to sustain blood flow throughout the body. To maintain its continuous operation, cardiac muscle relies on glucose in the fed state and fatty acids in the fasting state. It is not surprising, therefore, that cardiac muscle is densely packed with mitochondria. It can also use other energy sources, such as ketone bodies, pyruvate, and lactate. Lactate is produced only in small quantities in cardiac muscle because the isozyme of lactate dehydrogenase found in this tissue is inhibited by large concentrations of its substrate, pyruvate. The limited production of lactate means that glycolysis alone cannot be sustained in cardiac muscle.

Adipose Tissue

The role of adipose tissue is primarily the storage of energy in the form of triacylglycerols. A typical human stores enough energy in adipose tissue to sustain life for several weeks to several months. Depending on current physiological conditions, adipocytes store fat derived from the diet and liver metabolism or degrade stored fat to supply fatty acids and glycerol to the circulation. Recall that these metabolic activities are regulated by several hormones (i.e., insulin, glucagon, and epinephrine).

Adipocytes and macrophages within adipose tissue secrete a variety of peptide hormones, referred to as the adipokines. The adipokines include leptin and adiponectin. Leptin is a satiety-inducing protein, secreted into the bloodstream primarily by adipose tissue in proportion to adipose tissue mass. Adiponectin enhances glucose-stimulated insulin secretion and cellular responses to insulin (e.g., promoting fatty acid oxidation in skeletal muscle and suppressing glucose production by liver) primarily by activating AMP-dependent protein kinase (AMPK). Adiponectin blood levels are inversely proportional to body weight.

Brain

The brain ultimately directs most metabolic processes in the body. Sensory information from numerous sources is integrated in several areas in the brain. These areas then direct the activities of the motor neurons that innervate muscles and glands. Much of the body's hormonal activity is controlled either directly or indirectly by the hypothalamus and the pituitary gland. The hypothalamus plays a pivotal role in energy balance as it integrates hormonal and neural signals and nutrient levels to promote or suppress feeding behavior.

Like the heart, the brain does not provide energy to other organs or tissues. Under normal conditions, the brain uses glucose as its sole fuel. It is noteworthy that the adult brain, about 2% of body mass, typically consumes about 20% of the body's energy resources. The brain is highly dependent on a continuous supply of glucose in the blood because it stores little glycogen. During prolonged starvation, the brain can adapt to using ketone bodies as an energy source.

Kidney

The kidney has several important functions that contribute significantly to maintaining a stable internal environment. These include the following:

1. filtration of blood plasma, which results in the excretion of water-soluble waste products (e.g., urea and certain foreign compounds)

2. reabsorption of electrolytes, sugars, and amino acids from the filtrate

3. regulation of blood pH

4. regulation of the body's water content

Considering the functions of the kidney, it is not surprising that much of the energy generated in this organ is consumed by transport processes. Energy is provided largely by fatty acids and glucose. Under normal conditions, the small amounts of glucose formed by gluconeogenesis are used only within certain kidney cells. The rate of gluconeogenesis increases during starvation and acidosis. The kidney uses glutamine and glutamate (via glutaminase and glutamate dehydrogenase, respectively) to generate ammonia, which is used in pH regulation. (Recall that NH_3 reversibly combines with H^+ to form NH_4^+.) The carbon skeleton of glutamine and glutamate can then be used by the kidney as a source of energy.

KEY CONCEPT

Each organ in the mammalian system contributes to the body's overall function.

QUESTION 16.11

Describe two functions concerned with nutrient metabolism for each of the following organs:

a. intestine

b. liver

c. muscle

d. adipose tissue

e. kidney

f. brain

16.4 THE FEEDING-FASTING CYCLE

Despite their consistent requirements for energy and biosynthetic precursor molecules, mammals consume food only intermittently. This is possible because of elaborate mechanisms for storing and mobilizing energy-rich molecules derived from food (Figure 16.13). The changes in the status of various biochemical pathways during transitions between feeding and fasting illustrate metabolic integration and the profound regulatory influence of hormones. Substrate concentrations are also an important factor in metabolism. The terms *postprandial* and *postabsorption* are often used in discussions of the feeding-fasting cycle. In the **postprandial** state, which occurs directly after a meal has been digested or absorbed, blood nutrient levels are elevated above those in the fasting phase. During the **postabsorptive** state, for example, after an overnight fast, nutrient levels in blood are low.

The Feeding Phase

As the feeding phase begins, food is propelled along the gastrointestinal tract by muscle contractions initiated and controlled by the enteric (intestinal) nervous system. As it moves through the organs, food is broken into smaller particles and exposed to enzymes. Ultimately, the products of digestion (consisting largely of sugars, fatty acids, glycerol, and amino acids) are absorbed by the small intestine and transported into the blood and lymph. This phase is regulated by interactions between enzyme-producing cells of the digestive organs, the nervous system, and several hormones. The enteric nervous system, which is influenced by the parasympathetic and sympathetic nerves, is responsible for the waves of smooth muscle contraction that propel food along the tract, as well as for regulating the secretions of several digestive structures (e.g., from salivary and

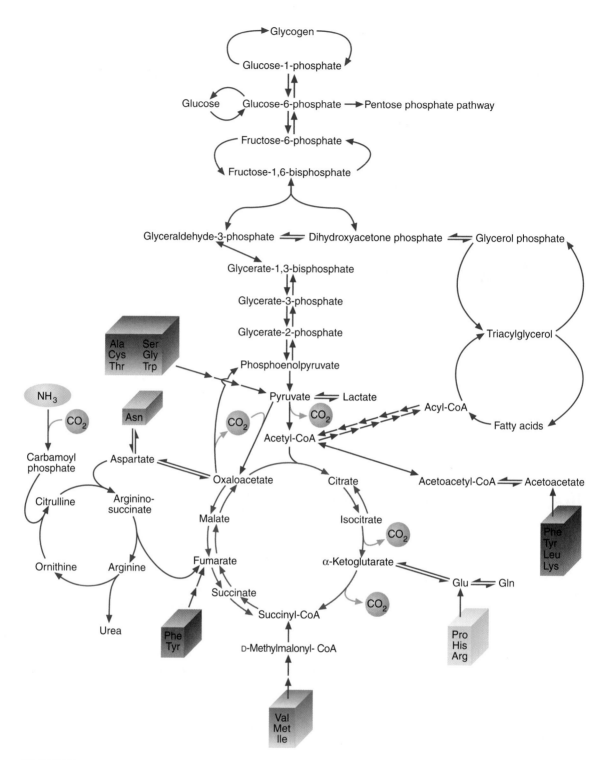

FIGURE 16.13

Nutrient Metabolism in Mammals

Despite the variability of the mammalian diet, these organisms usually provide their cells with adequate nutrients. Control mechanisms that regulate biochemical pathways are responsible for this phenomenon.

gastric glands). Hormones such as gastrin, secretin, and cholecystokinin also contribute to the digestive process (see Table 16.1). They do so by stimulating the secretion of enzymes or digestive aids such as bicarbonate and bile.

The early postprandial state is illustrated in Figure 16.14. As described, sugars and amino acids are absorbed and transported by the portal blood to the liver. The portal blood also contains a high level of lactate, a product of enterocyte metabolism. Most lipid molecules are transported from the small intestine in lymph as chylomicrons. Chylomicrons pass into the bloodstream, which carries them to tissues such as muscle and adipose tissue. After most triacylglycerol molecules have been removed from chylomicrons, these structures, now referred to as *chylomicron remnants*, are then taken up by the liver. The phospholipid, protein, cholesterol, and few remaining triacylglycerol molecules are then degraded or reused. For example, cholesterol is used to synthesize bile acids, and fatty acids are used in new phospholipid synthesis. Phospholipids, as well as other newly synthesized lipid and protein molecules, are then incorporated into lipoproteins for export to other tissues.

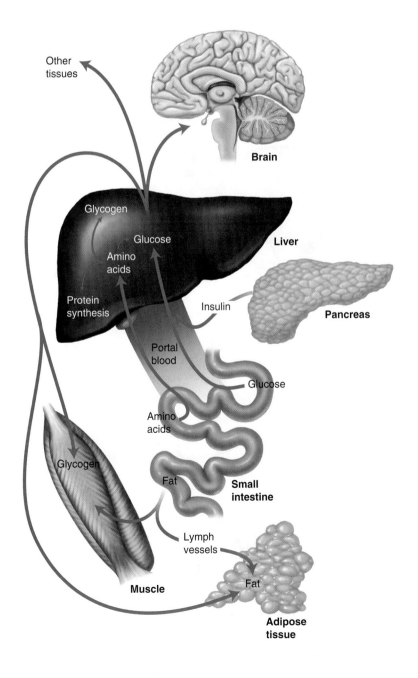

FIGURE 16.14

The Early Postprandial State

The primary substrates for glycogen synthesis in liver are amino acids and lactate (not shown) derived from portal blood. Note that the primary use for glucose in fat cells is as the precursor of glycerol. Fat cells do not carry out significant *de novo* fatty acid synthesis.

As glucose moves through the blood from the small intestine to the liver, the β cells within the pancreas are stimulated to release insulin. (High blood glucose and insulin levels depress glucagon secretion by the pancreatic α cells. The opposing effects of insulin and glucagon on glucose and fat metabolism are illustrated in Figure 16.15). Insulin release triggers several processes that ensure the storage of nutrients. These include glucose uptake by muscle and adipose tissue, glycogenesis in liver and muscle, fat synthesis in liver, fat storage in adipocytes, and gluconeogenesis (using excess amino acids and lactate). Recall that in the liver, most glycogen and fatty acids are synthesized from three-carbon molecules such as lactate, not directly from blood glucose. In addition, insulin influences amino acid metabolism. For example, insulin promotes the transport of amino acids into the cells (especially liver and muscle cells). In general, insulin stimulates protein synthesis in most tissues.

Although the effects of insulin on postprandial metabolism are profound, other factors (e.g., substrate supply and allosteric effectors) also affect the rate and degree to which these processes occur. For example, elevated levels of fatty acids in blood promote lipogenesis in adipose tissue. Regulation by several allosteric effectors further ensures that competing pathways do not occur simultaneously; in many cell types, for example, fatty acid synthesis is promoted by citrate (an activator of acetyl-CoA carboxylase), whereas fatty acid oxidation is depressed by malonyl-CoA (an inhibitor of carnitine acyltransferase I activity).

FIGURE 16.15

The Opposing Effects of Insulin and Glucagon on Blood Glucose Levels

In general, insulin promotes anabolic processes (e.g., fat synthesis, glycogenesis, and protein synthesis). Glucagon raises blood glucose levels by promoting glycogenolysis in liver and protein degradation in muscle. It also promotes lipolysis.

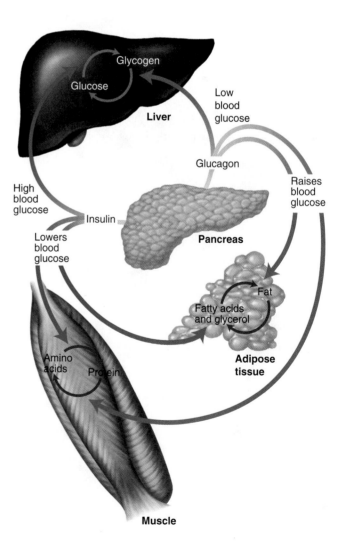

The Fasting Phase

The early postabsorptive state (Figure 16.16) of the feeding-fasting cycle begins as the nutrient flow from the intestine diminishes. As blood glucose and insulin levels fall back to normal, glucagon is released. Glucagon acts to prevent hypoglycemia by promoting glycogenolysis and gluconeogenesis in liver. Decreased insulin reduces energy storage in several tissues and leads to increased lipolysis and the release of amino acids such as alanine and glutamine from muscle. (Recall that several tissues use fatty acids in preference to glucose. Glycerol and alanine are substrates for gluconeogenesis, and glutamine is an energy source for enterocytes.)

If a fast becomes prolonged (e.g., overnight), several metabolic strategies maintain blood glucose levels. Increased mobilization of fatty acids from adipose tissue during the postabsorptive state is stimulated by norepinephrine. These fatty acids provide an alternative to glucose for muscle. (Reduced skeletal muscle consumption of glucose spares its use for brain. Recall that glucose is normally the

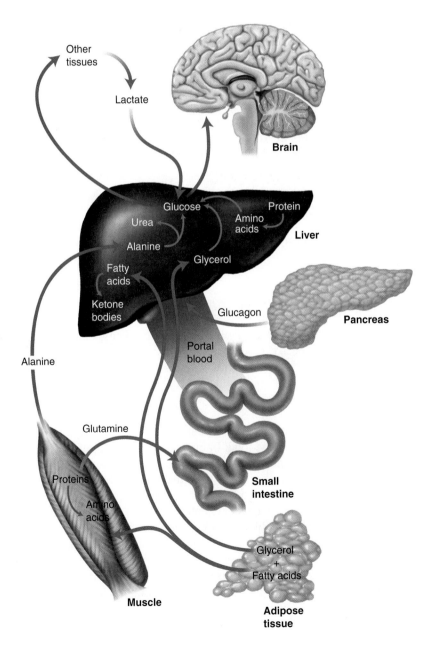

FIGURE 16.16
The Early Postabsorptive State

FIGURE 16.17

Plasma Levels of Fatty Acids, Glucose, and Ketone Bodies During the Early Days of Starvation

The concentrations of fatty acids and ketones increase. In contrast, glucose levels decrease.

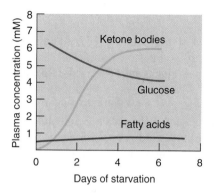

KEY CONCEPTS

- During the feeding phase, food is consumed, digested, and absorbed.
- Absorbed nutrients are then transported to the organs, where they are either used or stored.
- During fasting, several metabolic strategies maintain blood glucose levels.

only fuel source in brain.) In addition, the action of glucagon increases gluconeogenesis, using amino acids derived from muscle. (During fasting, insulin levels decline significantly.)

Under conditions of extraordinarily prolonged fasting (starvation), the body makes metabolic changes to ensure that adequate amounts of blood glucose are available to sustain energy production in the brain and other glucose-requiring cells. Additionally, fatty acids from adipose tissue and ketone bodies from liver are mobilized to sustain the other tissues. Because glycogen is depleted after several hours of fasting, gluconeogenesis plays a critical role in providing sufficient glucose. During early starvation, large amounts of amino acids from muscle are used for this purpose. However, after several weeks, the breakdown of muscle protein declines significantly because the brain is using ketone bodies as a fuel source. Figure 16.17 illustrates the plasma levels of glucose, fatty acids, and ketone bodies as starvation proceeds for several days.

QUESTION 16.12

Explain the metabolic changes that occur during starvation. What appears to be the principal purpose for the preferential degradation of muscle tissue during starvation?

QUESTION 16.13

Explain the changes in liver metabolism that occur when blood glucose levels drop after a meal has been digested.

Feeding Behavior

Feeding behavior is the complex mechanism by which animals, including humans, seek out and consume food. In mammals, regulation of feeding behavior involves hormonal and neural signals from peripheral organs (e.g., the GI tract and adipose tissue) and sensory input from the external environment (e.g., the sight, smell, and taste of palatable food) that together are integrated in the brain to regulate appetite and the body's metabolic processes (Figure 16.18).

For mammals, which have high energy requirements, the consumption of sufficient food to ensure the energy needed to sustain life is of critical importance. To this end, mammals have evolved a robust food-seeking system involving several neuronal pathways and numerous signaling molecules. In addition to providing a mechanism for balancing energy consumption and utilization, the mammalian brain links appetite systems to taste, olfaction, and reward systems to create a powerful drive that ensures survival.

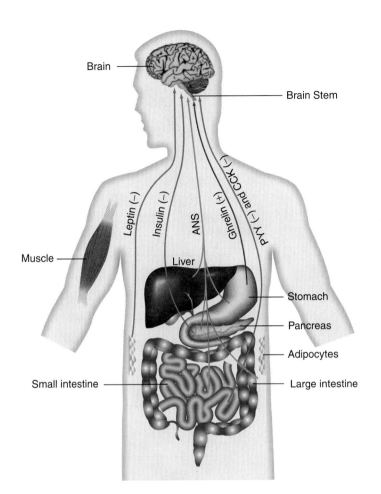

FIGURE 16.18
Feeding Behavior in Humans

Appetite and satiety in humans are regulated by hormonal and neural signals from peripheral organs. Peptide hormones such as PYY and CCK (produced by cells in the GI tract), insulin (produced by pancreatic β cells) and leptin (produced by adipose tissue), inhibit appetite (−); that is, they promote satiety. Ghrelin (produced by cells in the stomach and small intestine) stimulates appetite (+). Neuronal pathways of the autonomic nervous system (ANS) such as the vagus nerve continuously supply the brain with information related to the status of the body's internal organs.

Although appetite regulation is still not completely understood, it is clear that the neural circuits that control appetite are in the hypothalamus, located in the ventral (underside) portion of the vertebrate brain, and in the brain stem. Despite its small size, the hypothalamus, one of the most evolutionarily conserved regions of the brain, has a wide array of functions, including control of body temperature, electrolyte balance, monitoring of nutrient levels (e.g., blood glucose), and several aspects of emotional behavior. The primary neurons that control feeding behavior are in the *arcuate nucleus* (ARC) of the hypothalamus (Figure 16.19). Activation of ARC neurons that produce NPY (neuropeptide Y) and AgRP (agouti-related peptide) stimulate appetite, whereas stimulation of POMC (pro-opiomelanocortin) cells that produce α-MSH (α-melanocyte-stimulating hormone) suppress appetite. Under normal conditions, when leptin levels rise, indicating that the body's energy resources are sufficient, NPY/AgRP neurons are inhibited and POMC neurons are activated. In this circumstance appetite for food is depressed.

In contrast, increased food consumption, triggered by the activation of NPY/AgRP neurons and the inhibition of POMC neurons, results from falling leptin levels (caused by weight loss). All these neurons send signals to other neurons in the hypothalamus (referred to collectively as second-order neurons) that in turn relay them to other parts of the brain. Among these targets is the *nucleus tractus solitarius* (NTS) within the brain stem, which integrates this information with appetite-regulating signals from the GI tract via the *vagus nerve*. The result, depending on circumstances, range from increased appetite to a sense of satiety.

Insulin also reduces food intake via the NPY/AgRP and POMC neurons, although to a lesser extent than leptin. Other important appetite-regulating hormones include ghrelin (Ghr) and PYY. Ghr, an orexigenic (appetite stimulating)

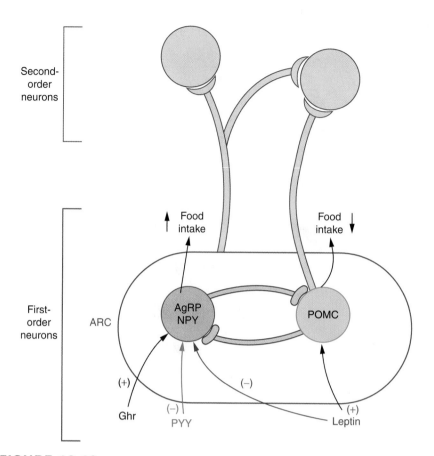

FIGURE 16.19

Appetite-Regulating Neurons in the Arcuate Nucleus (ARC)

Within the ARC of the hypothalamus, there are two sets of appetite-regulating neurons with opposing effects. Activation of AgRP/NPY neurons by Ghr increases appetite and energy-requiring metabolic processes. When POMC neurons are activated by leptin (and insulin to a lesser extent), appetite is depressed. Appetite is also negatively affected by the actions of leptin and PYY on AgRP/NPY neurons. Signals from AgRP/NPY and POMC neurons are communicated via second-order neurons to other parts of the hypothalamus and then other brain centers. Signals from these centers are subsequently sent to the NTS in the brain stem, where they are integrated with neural signals from the GI tract and other organs.

molecule released by cells within the stomach, as well as the intestine, activates NPY/AgRP neurons. PYY, produced by cells in the small intestine and colon, inhibits NPY/AgRP neurons.

The integration of disparate appetite-regulating hormonal and nutrient signals received within the hypothalamus appears to be mediated by AMPK. The signal transduction events that cause neurons to fire and release NPY and AgRP neurotransmitter molecules are triggered by activation of AMPK by appetite-promoting molecules such as ghrelin. These molecules bind to cell-surface receptors in neurons in ARC and other hypothalamic regions, in combination with low blood glucose levels in the process that leads to appetite stimulation. When appetite-inhibiting hormones such as leptin and insulin bind to their cell-surface receptors, AMPK activity is inhibited, with the result that NPY/ArGP neurons are inhibited.

BIOCHEMISTRY IN PERSPECTIVE

Obesity and the Metabolic Syndrome

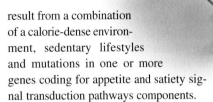

Why are so many humans predisposed to obesity in the modern world? Obesity, excess body weight that results from an imbalance between energy intake and energy expenditure, is a worldwide epidemic. If the human brain is so adept at balancing appetite, satiety, and physical activity, why has obesity become such a threat to health in the past 30 years? The answer lies in the long human struggle for survival. For most of its existence, *Homo sapiens* has had physically challenging hunter-gatherer and agrarian lifestyles and has been threatened by famines. Robust appetites provided sufficient calories required to sustain life and, when combined with the physiological capacity to store large quantities of fat, had significant survival value. During a famine, individuals with large fat stores can go without food for many weeks and have a better chance to survive than lean people.

In contrast, an increasing proportion of the modern world's population live under vastly different circumstances (availability of inexpensive, calorie-dense food combined with sedentary lifestyles). As a result, there has been significant weight gain among individuals genetically predisposed to what have now become problems with body weight regulation. For many people, a calorie-restricted environment combined with vigorous physical work would mask such vulnerability. Several rare mutations, linked to monogenic disorders of body weight regulation (Figure 16B), provide insight into human feeding behavior. For example, obese patients with leptin or leptin receptor defects have insatiable appetites. In Prader-Willi syndrome, another very rare disorder, exceptionally high blood levels of ghrelin drive an insatiable hunger. Although, by one estimate, at least 40% of the factors causing obesity can be attributed to genetics, the precise causes of most cases remains unknown. It has been suggested that obesity may result from a combination of a calorie-dense environment, sedentary lifestyles and mutations in one or more genes coding for appetite and satiety signal transduction pathways components.

In addition to the discomfort and severe social stigma associated with obesity, excessive body weight is a serious risk factor for a variety of illnesses such as hypertension, heart disease, several forms of cancer, osteoarthritis, and diabetes mellitus. Obesity is now recognized as a contributing factor in the metabolic syndrome.

Metabolic Syndrome

Metabolic syndrome is the term used to describe a cluster of clinical disorders that include, in addition to obesity, hypertension, dyslipidemia (high blood levels of total cholesterol and triacylglycerol, and low HDL levels), and insulin resistance. Insulin resistance, one of the earliest manifestations of the metabolic syndrome, originates from excess levels of free fatty acids (FFA) in blood. Expanding adipocytes, especially those of visceral (abdominal) adipose tissue, release fatty acids into the bloodstream. As the FFA levels in blood rise, these molecules begin accumulating in other cells. In insulin-sensitive tissue (muscle, liver, and pancreatic β cells) FFA disrupt signal transduction pathways. Among the consequences of this process, called *lipotoxicity*, are dyslipidemia and hyperinsulinemia (high blood levels of insulin) caused by FFA-stimulated insulin secretion. Other effects include excess glucose production in liver and inhibited insulin-mediated glucose uptake by muscle. When adipocytes also become insulin-resistant, there is an increase in lipolysis that results in the release of more FFA into the bloodstream.

As obesity progresses, adipose tissues develop a chronic state of low-level inflammation. Apparently, overstimulated adipocytes undergo physiological changes that cause ER stress and oxidative stress. Eventually, adipose tissue, now infiltrated with macrophages, releases inflammatory cytokines. (e.g., TNF-α, IL-6 and C-reactive protein). These and other cytokines also contribute to vascular disease by increasing insulin resistance. In genetically predisposed individuals, the metabolic syndrome may develop into type 2 diabetes (see earlier Biochemistry in Perspective box in this chapter, pp. 606–608).

FIGURE 16B
Obesity and Genetics
These mice have identical genomes except that the leptin gene was deleted from the mouse on the left.

SUMMARY: Natural selection in response to the rigors of chronic food scarcity has left many humans with the propensity to gain weight when food is plentiful. The body's inability to cope with lipotoxicity, caused by excessive body weight, can result in metabolic syndrome.

Chapter**Summary**

1. Multicellular organisms require sophisticated regulatory mechanisms to ensure that all their cells, tissues, and organs cooperate.

2. Hormones are molecules organisms use to convey information between cells. When target cells are distant from the hormone-producing cell, such molecules are called endocrine hormones. To ensure proper control of metabolism, the synthesis and secretion of many mammalian hormones are regulated by a complex cascade mechanism ultimately controlled by the central nervous system. In addition, a negative feedback mechanism precisely controls various hormone syntheses. A variety of diseases are caused by either overproduction or underproduction of a specific hormone or by the insensitivity of target cells.

3. Growth factors and some cytokines are polypeptides that regulate the growth, differentiation, or proliferation of various cells. They differ from hormones in that they are often produced by a variety of cell types rather than by specialized glandular cells.

4. Signal molecules initiate their effects in the target cell by binding to a specific receptor. Polar molecules, such as amines and peptides, bind to cell-surface receptors. They alter the activities of several enzymes and/or transport mechanisms in the cell. G-protein-coupled receptors use one or more of the second messengers (cAMP, cGMP, IP$_3$, DAG, and Ca^{2+}) to mediate the primary signal's effect on the target cell, and this has a significant amplifying effect on the signal transduction pathway. Tyrosine kinase receptors do not directly involve the generation of a second messenger. The nonpolar steroid and thyroid hormones diffuse through the lipid bilayer and bind to intracellular receptors. The hormone-receptor complex subsequently binds to a DNA sequence referred to as a hormone response element (HRE). The binding of a hormone-receptor complex to an HRE enhances or diminishes the expression of specific genes.

5. The feeding-fasting cycle illustrates how a variety of organs contribute via hormones and neurotransmitters to the acquisition of food molecules and their use. Feeding behavior is a mechanism by which animals seek out and consume food. The goal is to maintain balance between energy acquisition and energy expenditure. The hypothalamus contains the critical neural circuits that control appetite and satiety.

 Take your learning further by visiting the **companion website** for Biochemistry at **www.oup.com/us/mckee** where you can complete a multiple-choice quiz on integration of metabolism to help you prepare for exams.

Suggested**Readings**

Badman, M. K., and Flier, J. S., The Gut and Energy Balance: Visceral Allies in the Obesity Wars, *Science* 307:1909–1914, 2005.

Bajzer, M., and Seeley, R. J., Obesity and Gut Flora, *Nature* 444:1009–1010, 2006.

Basciano, H., Federico, L., and Adeli, K., Fructose, Insulin Resistance, and Metabolic Dyslipidemia, *Nutr. Metabol.* 2:5–18, 2005.

Cock, T.-A., Houten, S. M., and Auwerx, J., Peroxisome Proliferator-Activated Receptor-γ: Too Much of a Good Thing Causes Harm, *EMBO Rep.* 5(2):142–147, 2004.

Duncan, D. E., The Covert Plague, *Discover* 26(12):60–66, 2005.

Eldor, R., and Raz, I., Lipotoxicity versus Adipotoxicity—The Deleterious Effects of Adipose Tissue on Beta Cells in the Pathogenesis of Type 2 Diabetes *Diabetes Res. Clin. Prac.* 75S:S3–S8, 2006.

Feinman, R. D., and Fine, E. J., "A Calorie Is a Calorie" Violates the Second Law of Thermodynamics, *Nutr. J.* 3:9–13, 2004.

Friedman, J. M., Modern Science versus the Stigma of Obesity, *Nat. Med.* 10(6):563–569, 2004.

Kola, B., Boscaro, M., Rutter, G. A., Grossman, A. B., and Korbonits, M, Expanding Role of AMPK in Endocrinology, *Trends Endocrinol. Metab.* 17(5):205–215, 2006.

Murphy, K. G., and Bloom, S. R., Gut Hormones and the Regulation of Energy Homeostasis, *Nature* 444:854–859, 2006.

Muoio, D. M., and Newgard, C. B., Insulin Resistance Takes a Trip Through the ER, *Science* 306:425–426, 2004.

Stanley, S., Wynne, K., McGowan, B., and Bloom, S., Hormonal Regulation of Food Intake, *Physiol. Rev.* 85:1131–1158, 2005.

Wiederkehr, A., and Wollheim, C. B., Minireview: Implication of Mitochondria in Insulin Secretion and Action, *Endocrinplogy* 147(6):2643–2649, 2006.

Xue, B., and Kahn, B. B., AMPK Integrates Nutrient and Hormonal Signals to Regulate Food Intake and Energy Balance Through Effects in the Hypothalamus and Peripheral Tissues, *J. Physiol.* 574(1):73–83, 2006.

Key**Words**

adiponectin, *612*

cytokine, *605*

DAG, *600*

desensitization, *598*

downregulation, *598*

endocrine, *594*

epidermal growth factor, *605*

G protein, *599*

ghrelin, *619*

glucosuria, *607*

G-protein-coupled receptor, *599*

growth factor, *605*

guanine nucleotide exchange factor, *599*

heat shock protein, *609*

hormone response element, *609*

hyperglycemia, *607*

hyperosmolar hyperglycemic

nonketosis, *608*

insulin resistance, *598*

insulin-like growth factor, *608*

interferon, *609*

interleukin-2, *609*

IP$_3$, *600*

Review**Questions**

These questions are designed to test your knowledge of the key concepts discussed in this chapter, before moving on to the next chapter. You may like to compare your answers to the solutions provided in the back of the book and in the accompanying Study Guide.

1. Define the following terms:
   a. ketoacidosis
   b. hyperlipoproteinemia
   c. target cell
   d. growth factor
   e. chylomicron remnant

2. Which organ carries out each of the following activities?
   a. pH regulation
   b. gluconeogenesis
   c. absorption
   d. neural and endocrine integration
   e. lipogenesis
   f. urea synthesis

3. Define the following terms:
   a. down regulation
   b. G protein
   c. GPCR
   d. DAG
   e. RTK

4. State the action of each of the following hormones:
   a. corticotropin
   b. insulin
   c. glucagon
   d. oxytocin
   e. LH

5. NADH is an important reducing agent in cellular catabolism, whereas NADPH is an important reducing agent in anabolism. Review previous chapters and show how the synthesis and degradation of these two molecules are interconnected.

6. Define the following terms:
   a. tumor promoter
   b. osmotic diuresis
   c. histocompatibility antigen
   d. ketosis
   e. cytokine

7. Briefly discuss the major classes of second messenger that are now recognized.

8. How do phorbol esters promote tumor growth?

9. After about 6 weeks of fasting, the production of urea is decreased. Explain.

10. Extreme thirst is a characteristic symptom of diabetes. Explain.

11. State the action(s) of each of the following signal molecules:
    a. vasopressin
    b. FSH
    c. leptin
    d. ghrelin
    e. adiponectin

12. During periods of prolonged exercise, muscles burn fat released from adipocytes in addition to glucose. Explain how the need for additional fatty acids by muscle is communicated to the adipocytes.

13. Define the following terms:
    a. mitogen
    b. phorbol ester
    c. heat shock protein
    d. postprandial
    e. enteric

14. Bodybuilders often take anabolic steroids to increase their muscle mass. How do these steroids achieve this effect? (Common side effects of anabolic steroid abuse include heart failure, violent behavior, and liver cancer.)

15. State the action(s) of each of the following signal molecules:
    a. insulin
    b. glucagon
    c. epidermal growth factor
    d. interferon
    e. tumor necrosis factor

16. What are four metabolic functions of the kidney? How are these affected by diabetes mellitus?

17. During periods of fasting, some muscle protein is depleted. How is this process initiated, and what happens to the amino acids in these proteins?

18. State the action(s) of each of the following signal molecules:
    a. cholecystokinin
    b. glucagon-like peptide 1
    c. triiodothyronine
    d. somatostatin
    e. CRH

19. The kidney has an unusually large demand for glutamine and glutamate. How does the metabolism of these compounds help maintain pH balance?

20. Define the following terms:
    a. hypothalamus
    b. orexigenic
    c. metabolic syndrome
    d. dyslipidemia
    e. interleukin

21. Hemoglobin molecules exposed to high levels of glucose are converted to glycosylated products. The most common, referred to as hemoglobin A_{1C} (HbA_{1C}) contains a β-chain glycosylated adduct. Because red blood cells last about 3 months, HbA_{1C} concentration is a useful measure of a patient's blood sugar control. In general terms, describe why and how HbA_{1C} forms.

22. State the action of each of the following signal molecules:
    a. progesterone
    b. TRH
    c. gastrin
    d. TSH
    e. interleukin

23. What are the most common sites on proteins that are phosphorylated during signal transduction cascades?

24. Ketoacidosis is a common feature of insulin-dependent diabetes mellitus, but not of insulin-independent diabetes mellitus. Explain.

25. Type 2 diabetics are often obese. Explain how obesity contributes to the onset of diabetes.

26. Because the prime early symptom of diabetes is a high level of blood glucose, insulin is often associated primarily with carbohydrate metabolism. List several other processes that are insulin-dependent.

27. During the first week of a prolonged diet there is a relatively rapid weight loss. In addition to water loss, what other factor contributes to this phenomenon?

ThoughtQuestions

These questions are designed to reinforce your understanding of all of the key concepts discussed in the book so far, including this chapter and all of the chapters before it. They may not have one right answer! The authors have provided possible solutions to these questions in the back of the book and in the accompanying Study Guide, for your reference.

28. The hypothalamus and the pituitary are two endocrine glands located near each other in the brain. The secretions of the hypothalamus exert a powerful effect on the pituitary, yet if the pituitary is surgically transported to a remote location such as the kidney, the secretions of the hypothalamus have no effect. Suggest a reason for this observation.

29. Explain how a second messenger works. Why use a second messenger rather than simply relying on the original hormone to produce the desired effect?

30. Dieters frequently fast in an attempt to reduce their weight. During these fasts, they often lose considerable muscle mass rather than fat. Why is this so?

31. You are being stalked by a large tiger. Explain how your metabolism responds to help you escape.

32. During fasting in humans, virtually all the glucose reserves are consumed in the first day. The brain requires glucose to function and adjusts only slowly to other energy sources. Explain how the body supplies the glucose required by the brain.

33. In uncontrolled diabetes, levels of hydrogen ions are elevated. Explain how these ions are generated.

34. Why is it important for hormones to act at low concentrations and be metabolized quickly?

35. Hormones are often synthesized and stored in an inactive form within secretory vesicles. Secretion usually occurs only when the hormone-producing cell is stimulated. Explain the advantages that this process has over making the hormone molecules as they are needed.

36. Steroid hormones are often present in cells in low concentrations. This makes them difficult to isolate and identify. It is sometimes easier to isolate the proteins to which they bind by using affinity chromatography. (Refer to the "Biochemistry in the Lab" box entitled Protein Technology in Chapter 5, pp. 172–178). Explain how you would use this technique to isolate a protein suspected of steroid hormone binding.

37. Explain why skeletal muscle cells lack receptors for glucagon.

38. Skeletal muscle cannot synthesize fatty acids, yet it produces the enzyme acetyl-CoA carboxylase. Explain the role of this enzyme.

39. Long-term weight loss, as the result of dieting (calorie restriction), has a failure rate of about 95%. Review the basic principles of systems biology and explain how the body resists conscious efforts to lose weight.

40. Animals can convert glucose to fat, but not the reverse. Explain.

41. Describe the effects of starvation on urea production.

42. Describe the effect of leptin on feeding behavior and weight control.

43. The binding of small amounts of hormones to target cell receptors triggers an intricate signal cascade. Why is the signal cascade necessary? Why not just have a simple molecular mechanism between the hormone and changes in gene expression?

44. Several hormone may activate the same G protein. Therefore, different hormones may have the same effect. For example, glycogen degradation is initiated by both epinephrine and glucagon. Why is overlap of function an advantage?

45. As a result of social and economic changes after World War II, the Pima Indians of Arizona began to adopt a "westernized" lifestyle that included high-calorie diets and sedentary occupations. Soon afterward, obesity and type II diabetes became common. (A genetically related group, Pima Indians in Mexico, who live in remote mountain villages and maintain the diet and activity levels of their ancestors, have a low rate of diabetes and obesity.) Researchers have discovered that IRS1 levels are reduced in obese individuals from the Arizona Pima population, as compared with lean individuals. It was also shown that specific mutations in the IRS1 gene in combination with obesity led to a 50% reduction in insulin sensitivity. Explain this phenomenon.

46. Pima Indians are encouraged to exercise regularly to delay the onset of diabetes and/or improve diabetic symptoms. What impact does vigorous physical exercise have on their health?

47. The following are the energy reserves of the body: blood glucose, liver glycogen, muscle glycogen, adipose fat, and muscle protein. Comment on their importance in the body's energy economy, their calorie content, and their relative levels in an average body.

48. The fat store of a normal 150-pound (about 68 kg) man is about 1.5 kg or 141,000 cal. Assuming that during a prolonged fast such a person "burns" 2000 cal/day, determine how long this reserve will last.

49. During starvation conditions, death usually occurs before the body's energy reserves are totally exhausted. Suggest a reason for this phenomenon.

50. Hyperinsulinism, the result of the injection of an excessive insulin dose or an insulin-secreting tumor, can result in brain damage. Explain.

Nucleic Acids

The Genome All of the genetic information required to sustain living processes is encoded in the molecular structure of DNA. In eukaryotes most DNA is densely packed in chromosomes inside the nucleus.

625

Overview

THE DETERMINATION OF THE STRUCTURE OF DEOXYRIBONUCLEIC ACID BY JAMES WATSON AND FRANCIS CRICK IN THE EARLY 1950s WAS THE culmination of research that had begun almost a century before. The consequences of this seminal work are still unfolding as the genetic information of various organisms, including humans, is gradually being revealed. The astonishingly complex structure and functioning of DNA and its companion molecule, RNA, continue to fascinate researchers and students alike.

For countless centuries, humans have observed inheritance patterns without understanding the mechanisms that transmit physical traits and developmental processes from parent to offspring. Many human cultures have used such observations to improve their economic status, as in the breeding of domesticated animals or seed crops. It was not until the nineteenth century that the scientific investigation of inheritance, now referred to as **genetics**, began. By the beginning of the twentieth century, scientists generally recognized that physical traits are inherited as discrete units (later called "genes") and that chromosomes within the nucleus are the repositories of genetic information. Eventually, the chemical composition of chromosomes was elucidated and, after many decades of investigation, deoxyribonucleic acid (DNA) was identified as the genetic information. In the decades that followed the 1953 discovery of DNA structure by James Watson and Francis Crick (Figure 17.1), a new science emerged. **Molecular biology** is devoted to the investigation of gene structure and genetic information processing. Using the technologies developed by molecular biologists and biochemists, life scientists have studied the

FIGURE 17.1

The First Complete Structural Model of DNA

When James Watson (left) and Francis Crick discovered the structure of DNA in 1953 using Rosalind Franklin's data, they were research students at the Henry Cavendish Laboratory of Cambridge University.

processes by which living organisms organize and process genetic information. This work has revealed the following principles.

1. DNA directs the functioning of living cells and is transmitted to offspring. DNA is composed of two polynucleotide strands that form a double helix (Figure 17.2). The information in DNA is encoded in the form of the sequence of purine and pyrimidine bases (Refer to Figure 14.23). A **gene** is a DNA sequence that contains the base sequence information necessary to code for a gene product (a polypeptide or one of several types of RNA molecules) and regulatory sequences that control the production of the gene product. The complete DNA base sequence in an organism is referred to as the **genome**. DNA synthesis, referred to as **replication**, involves the complementary pairing of purine and pyrimidine bases in the two polynucleotide strands that comprise the DNA helix. The physiological and genetic function of DNA requires the synthesis of relatively error-free copies. Consequently, most organisms employ several DNA repair mechanisms.

2. The mechanism by which genetic information is decoded and used to direct cellular processes begins with the synthesis of another type of nucleic acid, ribonucleic acid (RNA). Its synthesis, referred to as **transcription** (Figure 17.3a), involves the complementary pairing of ribonucleotide bases with the bases in a DNA molecule. Each newly synthesized RNA molecule is called a **transcript**. The term **transcriptome** designates the complete set of RNA molecules that are transcribed from a cell's genome.

3. Several types of RNA participate directly in the synthesis of the enzymes, structural proteins, and other polypeptides required for the manufacture of all other biomolecules needed in organismal function. The base sequence of each messenger RNA (mRNA) specifies the primary sequence of a specific polypeptide. Ribosomal RNA (rRNA) molecules are components of the ribosomes. Each transfer RNA (tRNA) molecule is covalently bound to a specific amino acid and delivers it to the ribosome for

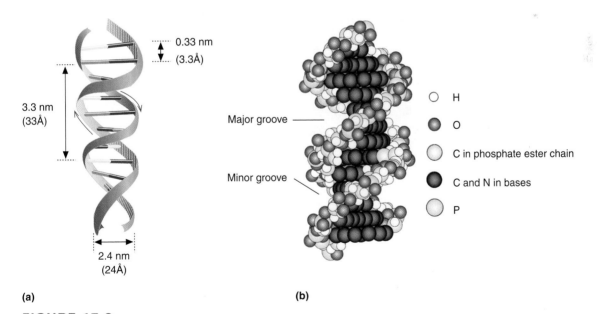

(a)   (b)

FIGURE 17.2

Two Models of DNA Structure

(a) The DNA double helix is represented as a spiral ladder; this is the conformation originally proposed by Watson and Crick and now known as B-DNA. (For the structural properties of three forms of DNA, see Table 17.1, later.) The sides of the spiral ladder represent the sugar-phosphate backbones. The rungs represent the base pairs. (b) In a space-filling model, the sugar-phosphate backbones are represented by colored spheres. The base pairs consist of horizontal arrangements of dark blue spheres. Wide and narrow grooves are created by twisting the two strands around each other.

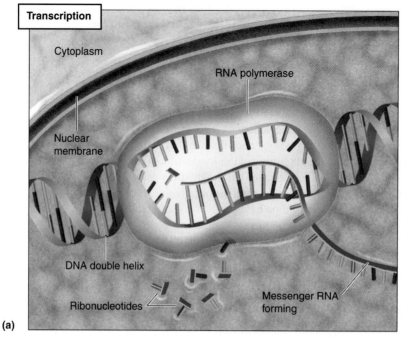

(a)

(b)

FIGURE 17.3

An Overview of Genetic Information Flow

The genetic information in DNA is converted into the linear sequence of amino acids in polypeptides in a two-phase process. During *transcription* (a), RNA molecules are synthesized from a DNA strand through complementary base pairing between the bases in DNA and the bases in free ribonucleoside triphosphate molecules. During the second phase, called *translation* (b), mRNA molecules bind to ribosomes that are composed of rRNA and ribosomal proteins. Transfer RNA–aminoacyl complexes position their amino acid cargo in the catalytic site within the ribosome in a process that involves complementary base pairing between the mRNA base triplets called codons and tRNA base triplets called anticodons. When the amino acids are correctly positioned within the catalytic site, a peptide bond is formed. After the mRNA molecule moves relative to the ribosome, a new codon enters the ribosome's catalytic site and base-pairs with the appropriate anticodon on another aminoacyl-tRNA complex. As an appropriate a stop codon in the mRNA enters the catalytic site, the newly formed polypeptide is released from the ribosome.

incorporation into a polypeptide chain. Protein synthesis, called **translation** (Figure 17.3b), occurs within ribosomes, the ribonucleoprotein molecular machines that translate the base sequences of mRNAs into the amino acid sequences of polypeptides. The entire set of proteins synthesized by a cell is referred to as the **proteome**.

4. **Gene expression** is the set of mechanisms whereby cells control the timing of gene product synthesis in response to environmental or developmental cues. A vast array consisting of proteins, called transcription factors, and RNA molecules, called *noncoding RNA molecules* (ncRNAs), regulate gene expression when they bind to specific DNA sequences. The term **metabolome** refers to the sum total of all the low-molecular-weight metabolite molecules produced by a cell as the result of its gene expression pattern.

The flow of genetic information can be summarized by the following sequence called the *central dogma*:

This diagram illustrates that the genetic information encoded in DNA base sequences flows from DNA, a molecule that is self-replicated during cell division (indicated by the arrow encircling DNA), to RNA, which specifies the primary structure of proteins. As originally conceived, the central dogma asserted that genetic information flows in one direction only: from DNA to RNA to protein. Several years ago, however, an important exception to the central dogma was revealed. Some of the viruses that possess RNA genomes also possess an enzyme activity referred to as *reverse transcriptase*. Once such a virus has infected a host cell, the reverse transcriptase acts to copy viral RNA to form a DNA copy. The viral DNA is then inserted into a host chromosome. One such virus is HIV, discussed in the Biochemistry in Perspective box entitled Viral "Lifestyles" (pp. 665–670).

Chapter 17 focuses on the structure of the nucleic acids. It begins with a description of DNA structure and the investigations that led to its discovery. This is followed by a discussion of current knowledge of genome and chromosome structure, as well as the structure and roles of the several forms of RNA. Chapter 17 ends with the description of viruses, macromolecular complexes composed of nucleic acid and proteins that are cellular parasites. In the following chapter (Chapter 18) several aspects of nucleic acid synthesis and function (i.e., DNA replication and transcription) are discussed. Protein synthesis (translation) is described in Chapter 19. The strategies and techniques that are routinely used to isolate, purify, characterize, and manipulate nucleic acids are described in Biochemistry in the Lab boxes in Chapters 17 and 18 (Nucleic Acid Methods, pp. 653–657, and Genomics, pp. 700–707).

17.1 DNA

DNA consists of two polynucleotide strands that wind around each other to form a right-handed double helix (Figure 17.2). The structure of DNA is so distinctive that this molecule is often referred to as *the double helix*. As described earlier (Sections 1.3 and 14.3), each nucleotide monomer in DNA is composed of a nitrogenous base (either a purine or a pyrimidine), a deoxyribose sugar, and phosphate. The mononucleotides are linked to each other by $3',5'$-phosphodiester bonds. These bonds join the $5'$-hydroxyl group of the deoxyribose of one nucleotide to the $3'$-hydroxyl group of the sugar unit of another nucleotide through a phosphate group (Figure 17.4). The antiparallel orientation of the two polynucleotide strands allows hydrogen bonds to form between the nitrogenous bases

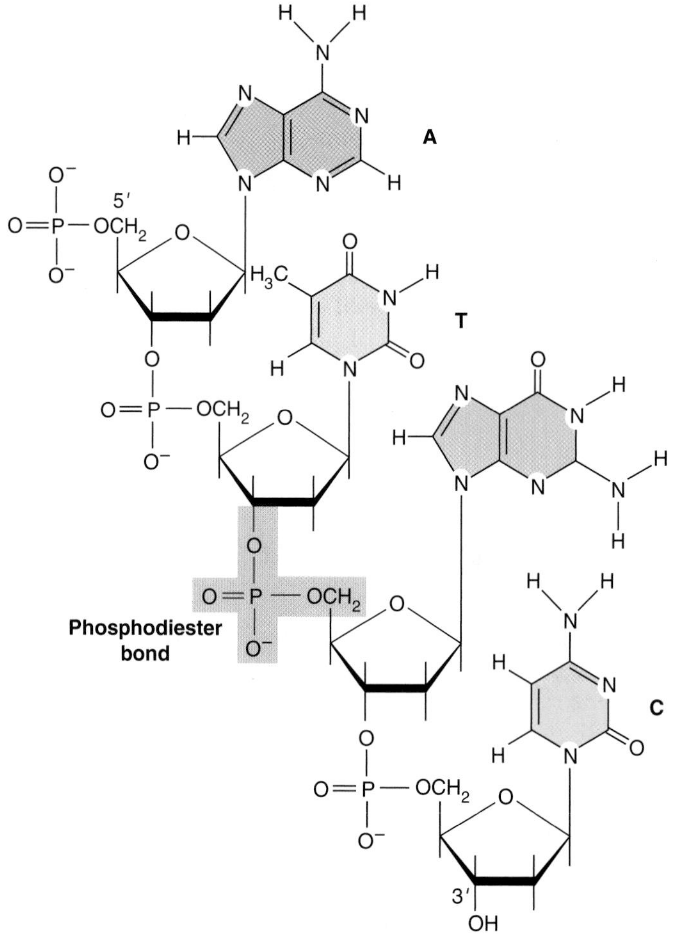

FIGURE 17.4

DNA Strand Structure

In each DNA strand the deoxyribonucleotide residues are connected to each other by 3′, 5′-phosphodiester linkage. The sequence of the strand section illustrated in this figure is 5′-ATGC-3′.

FIGURE 17.5

DNA Structure

In this short segment of DNA, the bases are shown in orange and the sugars are blue. Each base pair is held together by either two or three hydrogen bonds. The two polynucleotide strands are antiparallel. Because of base pairing, the order of bases in one strand determines the order of bases along the other.

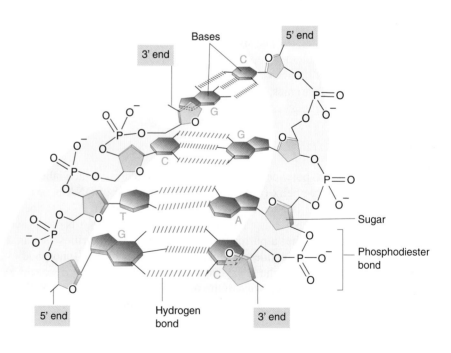

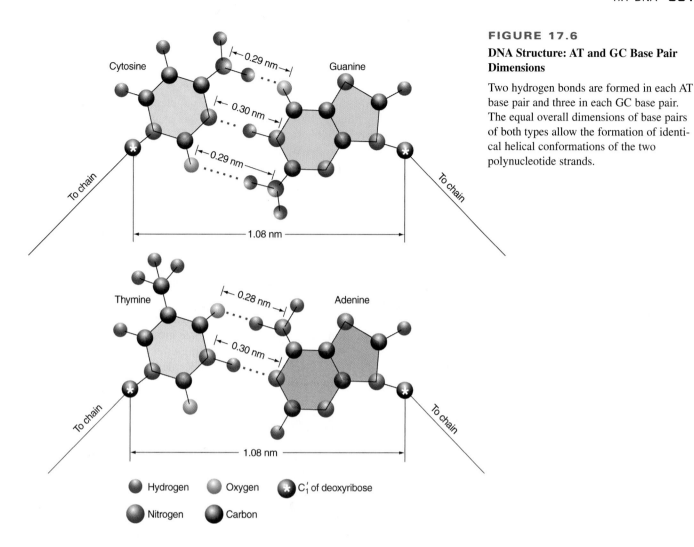

FIGURE 17.6

DNA Structure: AT and GC Base Pair Dimensions

Two hydrogen bonds are formed in each AT base pair and three in each GC base pair. The equal overall dimensions of base pairs of both types allow the formation of identical helical conformations of the two polynucleotide strands.

that are oriented toward the helix interior (Figure 17.5). There are two types of base pair (bp) in DNA: (1) adenine (a purine) pairs with thymine (a pyrimidine) (AT pair), and (2) the purine guanine pairs with the pyrimidine cytosine (GC pair). The overall structure of DNA resembles a twisted staircase, because each base pair is oriented at an angle to the long axis of the helix. The dimensions of crystalline DNA have been precisely measured.

1. One turn of the double helix spans 3.3 nm and consists of approximately 10.3 base pairs. (Changes in pH and salt concentrations affect these values slightly.)

2. The diameter of the double helix is 2.4 nm. Note that the interior space of the double helix is only suitable for base pairing a purine and a pyrimidine. Pairing two pyrimidines would create a gap, and two purines would not fit in the interior space of the double helix.

3. The distance between adjacent base pairs is 0.33 nm. The relative dimensions of both types of base pair are illustrated in Figure 17.6.

As befits its genetic information storage role in living processes, DNA is a relatively stable molecule. Several types of noncovalent interactions contribute to the stability of its helical structure.

1. **Hydrophobic interactions**. The base ring π cloud of electrons between stacked purine and pyrimidine bases is relatively nonpolar. The clustering of the base components of nucleotides within the double helix is a stabilizing factor in the three-dimensional macromolecule because it minimizes their interactions with water, thereby increasing overall entropy.

2. **Hydrogen bonds**. The base pairs, on close approach, form a preferred set of hydrogen bonds, three between GC pairs and two between AT pairs. The cumulative "zippering" effect of these hydrogen bonds keeps the strands in correct complementary orientation.

3. **Base stacking**. Once the antiparallel polynucleotide strands have been brought together by base pairing, the parallel stacking of the nearly planar heterocyclic bases stabilizes the molecule because of the cumulative effect of weak van der Waals forces.

4. **Hydration**. As with proteins, water stabilizes the three-dimensional structure of nucleic acids. DNA molecules bind a significant number of water molecules. The water content of B-DNA, the conformation illustrated in Figure 17.2, is about 30% by weight. Water molecules bind to phosphate groups, ribose 3′- and 5′-oxygen atoms, and electronegative atoms in the nucleotide bases. When measured under laboratory conditions, each nucleotide in B-DNA binds about 18 to 19 water molecules. Each phosphate can bind a maximum of 6 water molecules.

5. **Electrostatic interactions**. DNA's external surface, referred to as the *sugar-phosphate backbone*, possesses negatively charged phosphate groups. Repulsion between nearby phosphate groups, a potentially destabilizing force, is minimized by the shielding effects of water on divalent cations such as Mg^{2+}, and polycationic molecules such as the polyamines and histones (see p. 645).

KEY CONCEPT

DNA is a relatively stable molecule composed of two antiparallel polynucleotide strands wound around each other to form a right-handed double helix.

QUESTION 17.1

Describe how each type of noncovalent interaction contributes to the stability of DNA's helical structure.

QUESTION 17.2

When DNA is heated, it denatures; that is, the strands separate because hydrogen bonds are broken. The higher the temperature, the larger the number of hydrogen bonds that are broken. After reviewing DNA base pair structure, determine which of the following molecules will denature first as the temperature is raised. Explain your reasoning.

    a. 5′-GCATTTCGGCGCGTTA-3′
       3′-CGTAAAGCCGCGCAAT-5′
    b. 5′-ATTGCGCTTATATGCT-3′
       3′-TAACGCGAATATACGA-5′

DNA Structure: The Nature of Mutation

DNA is eminently suited for information storage. However, despite its several stabilizing structural features, the molecule is vulnerable to certain types of disruptive force. Solvent collisions, thermal fluctuations, and other spontaneous disruptive processes can result in mutations, or permanent changes in the base sequence of DNA molecules. In addition, a wide variety of xenobiotics, both natural and human-made, are known to alter DNA structure. Several examples of well-researched mutagenic factors are discussed.

As described (Figure 14.24), tautomeric shifts are spontaneous changes in nucleotide base structure that interconvert amino and imino groups and keto and enol groups. Usually tautomeric shifts have little effect on DNA structure. However, if tautomers form during DNA replication, base mispairings may result. For example, the imino form of adenine will not base-pair with thymine. Instead, it forms a base pair with cytosine (Figure 17.7). If this pairing is not corrected

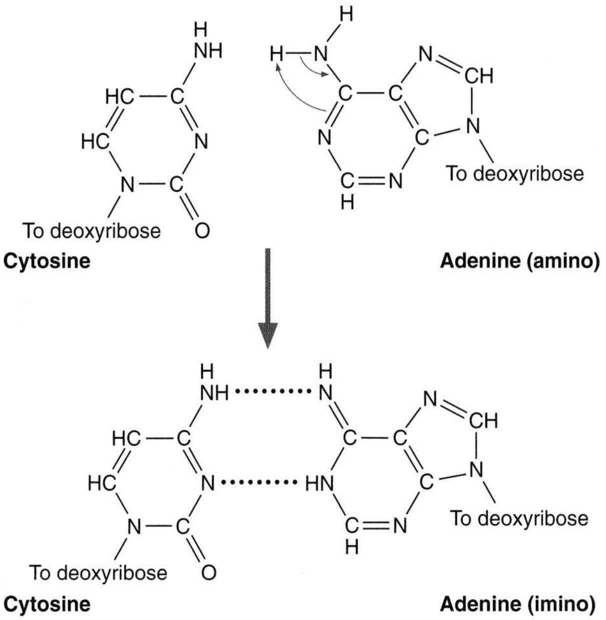

FIGURE 17.7

A Tautomeric Shift Causes a Transition Mutation

As adenine undergoes a tautomeric shift, its imino form can base-pair with cytosine. The transition shows up in the second generation of DNA replication when cytosine base-pairs with guanine. In this manner an A-T base pair is replaced by a C-G base pair.

immediately, a transition mutation results because cytosine has been incorporated during the replication process in a position that should carry thymine. In a **transition mutation** a pyrimidine base is substituted for another pyrimidine, or a purine is substituted for another purine. Transition mutations are **point mutations**, DNA base sequence changes that involve a single base pair. In the example described, an AT base pair is replaced by a GC base pair in the second generation of DNA replication.

Several spontaneous hydrolytic reactions also cause DNA damage. For example, it has been estimated that several thousand purine bases are lost daily from the DNA in each human cell. In depurination reactions the N-glycosyl linkage between a purine base and deoxyribose is cleaved. The protonation of N-3 and N-7 of guanine promotes hydrolysis. If repair mechanisms do not replace the purine nucleotide, a point mutation will result in the next round of DNA replication. Similarly, bases can be spontaneously deaminated. For example, the deaminated product of cytosine converts to uracil via a tautomeric shift. Eventually, what should be a CG base pair is converted to an AT base pair. (Uracil is similar in structure to thymine.)

Some types of ionizing radiation (e.g., UV, X-rays, and γ-*rays)* can alter DNA structure. Low radiation levels may cause mutation; high levels can be lethal. Radiation-induced damage caused by a free radical mechanism (either abstraction of hydrogen atoms or the creation of •OH and other ROS) can promote radiation-induced damage, including strand breaks, DNA-protein cross-linking (e.g., via thymine-tyrosine linkages), ring openings, and base modifications. The hydroxyl radical, formed by the radiolysis of water, as well as oxidative stress is known to cause some strand breakage and numerous base modifications (e.g., thymine glycol, 5-hydroxymethyluracil, and 8-hydroxyguanine).

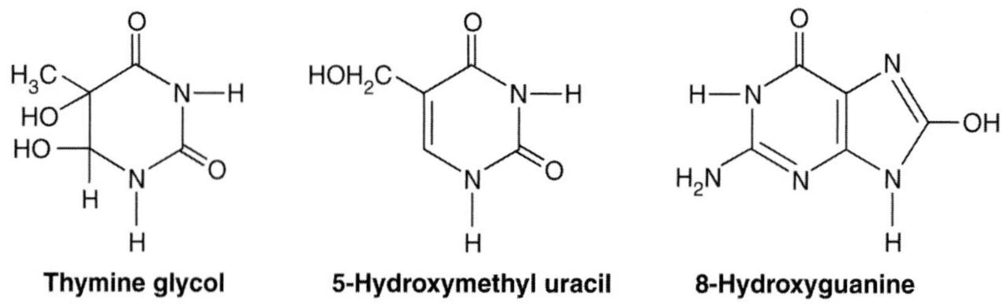

Thymine glycol　　　　**5-Hydroxymethyl uracil**　　　　**8-Hydroxyguanine**

8-Hydroxydeoxyguanosine (8-OHdG) levels in urine are used to measure the body's production of ROS. Smoking, for example, results in increases of 8-OHdG excretion by as much as 50%.

The most common UV-induced products, created by energy absorption by double bonds, are pyrimidine (thymine) dimers (Figure 17.8). The helix distortion that results from dimer formation stalls DNA synthesis.

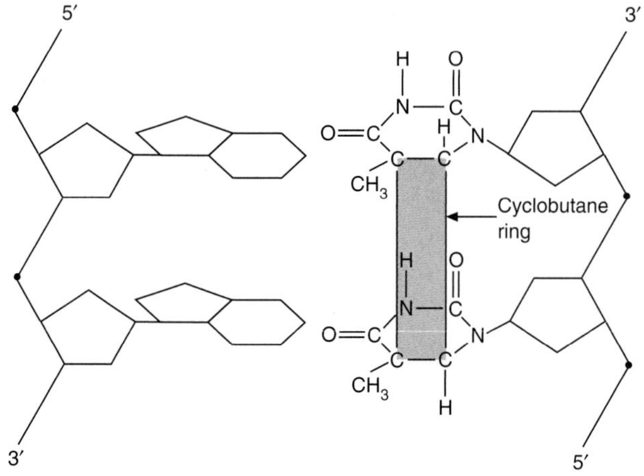

FIGURE 17.8

Thymine Dimer Structure

Adjacent thymines form dimers with high efficiency after absorbing UV light.

A large number of xenobiotics can damage DNA. The most important of these molecules belong to the following classes:

1. **Base analogues**. Because their structures are similar to the normal nucleotide bases, **base analogues** can be incorporated into DNA. For example, caffeine is a base analogue of thymine. Because it can base-pair with guanine, caffeine incorporation can cause a transition mutation.

2. **Alkylating agents**. **Alkylation** is the process in which electrophilic ("electron-loving") substances attack molecules that possess an unshared pair of electrons. When electrophiles react with such molecules, they usually add carbon-containing alkyl groups. Adenine and guanine are especially susceptible to alkylation, although thymine and cytosine can also be affected. Alkylated bases often pair incorrectly (e.g., methylguanine with thymine instead of with cytosine), leading to possible transition mutations on subsequent rounds of replication. In the case of methylguanine, a GC pair becomes an AT pair. Transversion mutations may also occur when the alkylating group is bulky. (In a **transversion mutation**, another type of point mutation, a pyrimidine is substituted for a purine or vice versa.) Alkylations can also promote tautomer formation, which may result in transition mutations. Examples of alkylating agents include dimethylsulfate and dimethylnitrosamine. Mitomycin C is a bifunctional alkylating agent. It can prevent DNA synthesis by cross-linking guanine bases. DNA repair mechanisms (Section 18.1) usually remove the abnormal base ring prior to replication, thus reducing the mutation frequency.

3. **Nonalkylating agents**. A variety of chemicals other than the alkylating agents can modify DNA structure. Nitrous acid (HNO_2), derived from the nitrosamines and from sodium nitrite ($NaNO_2$), deaminates bases. (Both nitrosamines and $NaNO_2$ are found in processed meats and in any foodstuffs preserved with nitrite pickling salt.) HNO_2 converts adenine, guanine, and cytosine to hypoxanthine, xanthine, and uracil, respectively. Another category of mutagens consists of the aromatic polycyclic hydrocarbons (e.g., benzo[a]pyrene found in cigarette smoke, a source of nitrosamines as well). Once consumed, these molecules can be converted to highly reactive derivatives by biotransformation reactions such as those catalyzed by cytochrome P_{450}. The reactive derivatives can then form adducts of most bases. Damage occurs primarily because this chemical modification prevents base pairing.

4. **Intercalating agents**. Certain planar molecules can distort DNA because they insert themselves (intercalate) between the stacked base pairs of the double helix (Figure 17.9). Either adjacent base pairs are deleted or new base pairs are inserted. If not corrected, deletions or insertions cause so-called frame-shift mutations (described in Chapter 19). In addition, chromosomes may break. The acridine dyes are examples of intercalating agents. Quinacrine is an acridine dye used to treat malaria and intestinal tapeworms.

KEY CONCEPT

DNA is vulnerable to certain types of disruptive force that can result in mutations, permanent changes in its base sequence.

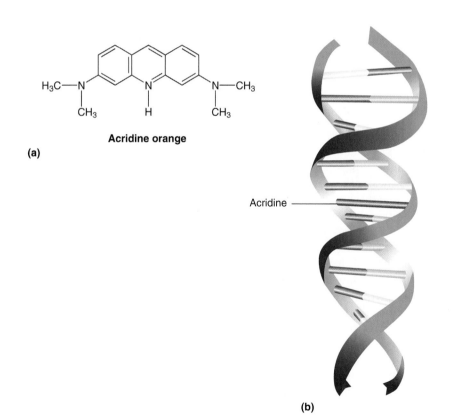

Acridine orange

(a)

Acridine

(b)

FIGURE 17.9

Intercalating Agent

An intercalating agent is a planar molecule that can insert between base pairs in DNA. The intercalating agent acridine orange (a), an acridine dye, inserts between two adjacent base pairs (b), and interferes with DNA replication. When acridine orange is bound to DNA illumination with UV light causes a green fluorescence. (Refer to Section 13.2 for a description of fluorescence.)

QUESTION 17.3

How will each of the following substances or conditions affect DNA structure?
a. ethanol b. heat c. dimethylsulfate d. nitrous acid e. quinacrine

QUESTION 17.4

Consider each of the following compounds. To what class of DNA-damaging xenobiotics does each belong?

Caffeine   **Benzo[a]pyrene**   **Ethyl chloride**

QUESTION 17.5

The accumulation of oxidative DNA damage now appears to be a major cause of aging in mammals. Animals that have high metabolic rates (i.e., use large amounts of oxygen) or excrete large amounts of modified bases in the urine typically have shorter life spans. The excretion of relatively large amounts of oxidized bases indicates a reduced capacity to prevent oxidative damage. Despite substantial evidence that oxygen radicals damage DNA, the actual radicals that cause the damage are still not clear. Suggest possible culprits in addition to the hydroxyl radical. Some tissues sustain more oxidative damage than others. For example, the human brain is believed to sustain more oxidative damage than most other tissues during an average life span. Suggest two reasons for this phenomenon.

DNA Structure: From Mendel's Garden to Watson and Crick

To modern eyes, the structure of DNA is both elegant and obvious. The DNA molecule is now a cultural icon, its high-tech uses synonymous with the concept of information storage and retrieval. The correct structure of DNA was proposed in 1953 by James Watson and Francis Crick. The investigation that led to this remarkable discovery is instructive for several reasons. First, as often happens in scientific research, the road to the elucidation of DNA structure was long, frustrating, and tortuous. Living organisms are so complex that discerning any aspect of their function is extraordinarily difficult. Moreover, scientists (and other humans) have a propensity to reject or ignore new information that does not fit comfortably with currently popular ideologies. This latter problem is probably unavoidable, because the scientific method requires a certain degree of skepticism. (How does one differentiate, for example, between unproven breakthrough concepts and erroneous ideas?) However, skepticism can often be confused with an unimaginative adherence to the status quo. Albert Szent-Gyorgyi (Nobel Prize in Physiology or Medicine, 1937), who identified ascorbic acid as vitamin C and made significant contributions to the elucidation of muscle contraction and the citric acid cycle, once remarked, "Discovery consists in seeing what everyone else has seen and thinking what nobody else has thought."

A second, more concrete reason for the length of the discovery process is that the development of new concepts often requires the integration of information from several scientific disciplines. For example, the DNA model was based on discoveries in descriptive and experimental biology, genetics, organic chemistry, and physics. Significant scientific advancements are usually made by imaginative and industrious individuals who have the good fortune to work when sufficient information and technology are available for solving the scientific problems that interest them. The most talented of these investigators often help create new technologies.

The scientific revolution that eventually led to the DNA model began quietly in the abbey garden of an obscure Austrian monk named Gregor Mendel. Mendel discovered the basic rules of inheritance by cultivating pea plants, keeping detailed records, and making inferences from his data. In 1865 Mendel published the results of his breeding experiments in the *Journal of the Brunn Natural History Society*. Although he sent copies of this publication to eminent biologists throughout Europe, his work was ignored until 1900. In that year, several botanists independently rediscovered Mendel's paper and recognized its significance. This long delay was due largely to the descriptive nature of nineteenth-century biology; few biologists were familiar with the mathematics that Mendel had used to analyze his data. By 1900, many biologists not only were trained in mathematics but also had a frame of reference for Mendel's principles, since many of the details of meiosis, mitosis, and fertilization had become common knowledge.

Amazingly, the substance that constitutes the inheritable units that Mendel referred to in his work was being investigated almost simultaneously. The discovery of "nuclein," later renamed nucleic acid, was reported in 1869 by Friedrich Miescher, a Swiss pathologist. Working with the nuclei of pus cells, Miescher extracted nuclein and discovered that it was acidic and contained a large amount of phosphate. (Although Joseph Lister had published his findings on antiseptic surgery in 1867, hospitals continued to be a rich source of pus for many years to come.) Interestingly, Miescher came to believe (erroneously) that nuclein was a phosphate storage compound.

The chemical composition of DNA was determined largely by Albrecht Kossel between 1882 and 1897 and P. A. Levene in the 1920s as suitable analytical techniques were developed. Levene, however, mistakenly believed that DNA was a small and relatively simple molecule. His concept, referred to as the *tetranucleotide hypothesis*, significantly retarded further investigations of DNA. Instead, proteins (the other major component of nuclei) were viewed as the probable carrier of genetic information. (By the end of the nineteenth century it was commonly accepted that the nucleus contains the genetic information.)

While geneticists focused on investigating the mechanisms of heredity and chemists elucidated the structures of nucleic acid components, microbiologists developed methods of studying bacterial cultures. In 1928, while investigating a deadly epidemic of pneumonia in Britain, Fred Griffith performed a remarkable series of experiments with two pneumococcal strains (Figure 17.10). One bacterial strain, called the smooth form (or type S) because it is covered with a polysaccharide capsule, is pathogenic. The rough form (or type R) lacks the capsule and is nonpathogenic. Griffith observed that mice inoculated with a mixture of live R and heat-killed S bacteria died. He was amazed when live S bacteria were isolated from the dead mice. Although the transformation of R bacteria into S bacteria was confirmed in other laboratories, Griffith's discovery was greeted with considerable skepticism. (Indeed, the concept of transmission of genetic information between bacterial cells was not accepted until the 1950s.)

In 1944 Oswald Avery and his colleagues Colin MacLeod and Maclyn McCarty reported their painstaking isolation and identification of the transforming agent in Griffith's experiment as DNA. Not everyone accepted this conclusion because the DNA sample described had a trace of protein impurities. Avery and McCarty later demonstrated that the digestion of DNA by deoxyribonuclease (DNase) inactivates the transforming agent. Eventually, it was determined that the DNA that transforms R pneumococci into the S form codes for an enzyme required to synthesize the gelatinous polysaccharide capsule. Such capsules protect the bacteria from the animal's immune system and increase adherence and colonization of host tissues.

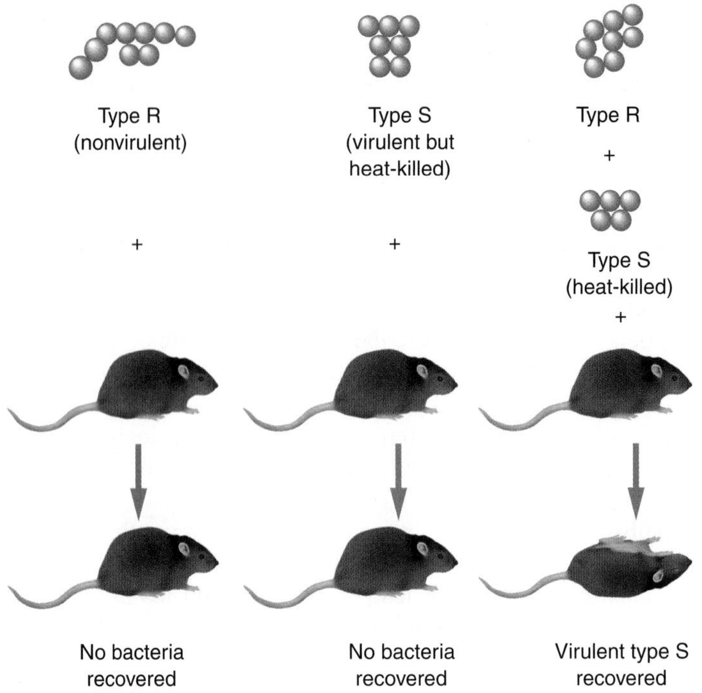

FIGURE 17.10

Fred Griffith's Experiment

In Griffith's experiment, a mixture of nonvirulent R pneumococi and virulent heat-killed S pneumococci killed mice. Virulent S pneumococci were recovered from the dead mice.

Another experiment that confirmed DNA as the genetic material was reported by Alfred Hershey and Martha Chase in 1952 (Figure 17.11). Using T2 bacteriophage, Hershey and Chase demonstrated the separate functions of viral nucleic acid and protein. (*Bacteriophage*, sometimes called *phage*, is a type of virus that infects bacteria.) When T2 phage infects an *Escherichia coli* cell, the bacterium is directed to synthesize several hundred new viruses. Within 30 minutes after infection, the cell dies as it bursts open, thus releasing the viral progeny. In the first phase of their experiment, Hershey and Chase incubated bacteria infected with T2 phage in a culture medium containing ^{35}S (to label protein) and ^{32}P (to label DNA). In the second phase the radioactively labeled virus was harvested and allowed to infect nonlabeled bacteria. Immediately afterward, the infected bacterial culture was subjected to shearing stress in a Waring blender. This treatment removed the phage from its attachment sites on the external surface of the bacterial cell wall. After separation from empty viral particles by centrifugation, the bacteria were analyzed for radioactivity. The cells were found to contain ^{32}P (thus confirming the role of DNA in "transforming" the bacteria into virus producers), whereas most of the ^{35}S remained in the supernatant. In addition, samples of the labeled infected bacteria produced some ^{32}P-labeled viral progeny.

By the early 1950s it had become clear that DNA was the genetic material. Because researchers also recognized that genetic information was critical for all living processes, determining the structure of DNA became an obvious priority. Linus Pauling (California Institute of Technology), Maurice Wilkins and Rosalind Franklin (King's College, London), and Watson and Crick (Cambridge University) were all working toward this goal. The structure proposed by Watson and Crick in the April 25,1953, issue of *Nature* was based on their scale model.

Considering how the scientific community responded to other concepts involving DNA as the genetic material, acceptance of the Watson-Crick structure was unusually rapid: Watson, Crick, and Wilkins were awarded the Nobel Prize in Chemistry in 1962.

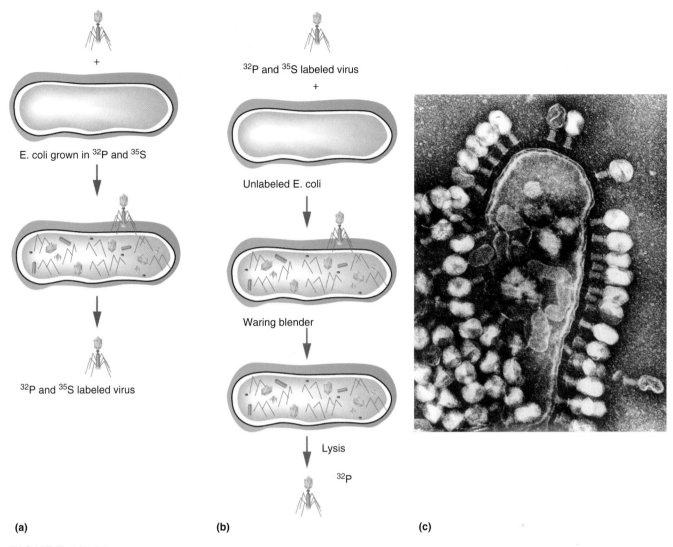

FIGURE 17.11

The Hershey-Chase Experiment

(a) Bacteriophage with radio-labeled phage were prepared by inoculating the virus into bacterial cultures containing the radioisotopes.
(b) When nonlabeled bacteria were infected with labeled phage, only the labeled phage DNA is found within the bacteria. Some of the ^{32}P label is found within viral progeny.(c) View of T4 bacteriophage lysing an *E. coli* cell.

The information used to construct this DNA model included the following:

1. The chemical structures and molecular dimensions of deoxyribose, the nitrogenous bases, and phosphate.

2. The 1:1 ratios of adenine to thymine and guanine to cytosine in the DNA isolated from a wide variety of species investigated by Erwin Chargaff between 1948 and 1952. (These 1:1 relationships are sometimes referred to as **Chargaff's rules**.)

3. Superb X-ray diffraction studies performed by Rosalind Franklin (Figure 17.12) indicating that DNA is a symmetrical molecule and probably a helix.

4. The diameter and pitch of the helix estimated by Wilkins and his colleague Alex Stokes from other X-ray diffraction studies.

5. The more recent demonstration by Linus Pauling that protein, another complex class of molecule, could exist in a helical conformation.

FIGURE 17.12

X-Ray Diffraction study of DNA by Rosalind Franklin and R. Gosling

Note the symmetry of the X-Ray diffraction pattern.

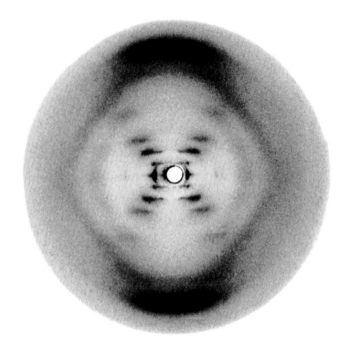

KEY CONCEPT

The model of DNA structure proposed by James Watson and Francis Crick in 1953 was based on information derived from the efforts of many individuals.

DNA Structure: Variations on a Theme

The structure discovered by Watson and Crick, referred to as **B-DNA**, represents the sodium salt of DNA under highly humid conditions. DNA can assume different conformations because deoxyribose is flexible and the C^1-N-glycosidic linkage rotates. (Recall that furanose rings have a puckered conformation.)

When DNA becomes partially dehydrated—that is, when the number of water molecules bound to each of the nucleotides drops to about 13 to 14—the molecule assumes the A form (Figure 17.13 and Table 17.1). In **A-DNA**, the base pairs are no longer at right angles to the helical axis. Instead, they tilt 20° away from the horizontal. In addition, the distance between adjacent base pairs is slightly reduced, with 11 bp per helical turn instead of the 10.4 bp found in the B form. Each turn of the double helix occurs in 2.5 nm, instead of 3.3 nm, and the molecule's diameter swells to approximately 2.6 nm from the 2.4 nm observed in B-DNA. The A form of DNA is observed when it is extracted with solvents such as ethanol. The significance of A-DNA under cellular conditions is that the structure of RNA duplexes and RNA/DNA duplexes formed during transcription resembles the A-DNA structure.

The Z form of DNA (named for its "zigzag" conformation) radically departs from the B form. **Z-DNA** ($D = 1.8$ nm), which is considerably slimmer than B-DNA ($D = 2.4$ nm), is twisted into a left-handed spiral with 12 bp per turn. Each turn of Z-DNA occurs in 4.5 nm, compared with 3.3 nm for B-DNA. DNA segments with alternating purine and pyrimidine bases (especially CGCGCG) are most likely to adopt a Z configuration. In Z-DNA, the bases stack in a left-handed staggered dimeric pattern, which gives the DNA a zigzag appearance and its flattened, nongrooved surface. Regions of DNA rich in GC repeats are often regulatory, binding specific proteins that initiate or block transcription. Although the physiological significance of Z-DNA is unclear, it is known that certain physiologically relevant processes such as methylation and negative supercoiling (discussed on p. 642) stabilize the Z form. In addition, short segments have been observed to form as the result of torsional strain during transcription.

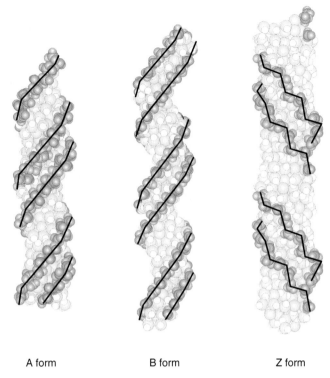

A form   B form   Z form

FIGURE 17.13

A-DNA, B-DNA, and Z-DNA

Because DNA is a flexible molecule, it can assume different conformational forms depending on its base sequence and/or isolation conditions. Each molecular form in the figure possesses the same number of base pairs. Refer to Table 17.1 for the dimensions of these three DNA structures.

TABLE 17.1 Selected Structural Properties of B-, A-, and Z-DNA

	B-DNA (Watson-Crick Structure)	A-DNA	Z-DNA
Helix diameter	2.4 nm	2.6 nm	1.8 nm
Base pairs per helical turn	10	11	12
Helix rise per base pair	3.3 Å	2.3 Å	3.8 Å
Helix rotation	Right-handed	Right-handed	Left-handed

Certain segments of DNA have been observed to have higher-order structures. Cruciforms are an important example. As their name implies, *cruciforms* are cross-like structures. They are likely to form when a DNA sequence contains a palindrome, a sequence that provides the same information whether it is read forward or backward (e.g., "MADAM, I'M ADAM."). In contrast to language palindromes, the "letters" are read in one direction on one of the complementary strands of DNA and in the opposite direction on the other strand. One-half of the palindrome on each strand is complementary to the other half. The DNA sequences that form palindromes, which may consist of several bases or thousands of bases, are called *inverted repeats*. In one proposed mechanism, cruciform formation begins with a small bubble, or *protocruciform*, and progresses as intrastrand base pairing occurs.

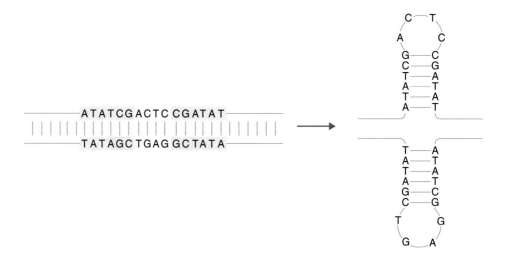

Packaging large DNA molecules to fit into cells requires DNA supercoiling. To undergo supercoiling, a single strand of double-helical DNA must be broken, or "nicked," and then either overwound or underwound before resealing. (A "nick" is a break in a single strand of double-helical DNA.) Small changes in DNA shape depend on sequence. For example, four sequential AT pairs produce a bend in the molecule. Significant bending or wrapping around associated proteins, however, requires supercoiling.

DNA Supercoiling

DNA supercoiling, once considered to be an artifact of DNA extraction techniques, is now known to facilitate several biological processes. Examples include packaging DNA into a compact form, as well as replicating and transcribing DNA (Chapter 18). Because DNA supercoiling is a dynamic three-dimensional process, the information that two-dimensional illustrations can convey is limited. To understand supercoiling, therefore, consider the following thought experiment. A long, linear DNA molecule is laid on a flat surface. Then the ends are brought together and sealed to form an unpuckered circle (Figure 17.14a). Because this molecule is sealed without under- or overwinding, the helix is said to be relaxed, and it remains flat on a surface. If the relaxed circular DNA molecule is held and twisted a few times, it takes the shape shown in Figure 17.14b. When this twisted molecule is returned to the flat surface and made to lie on the plane, it rotates to eliminate the twist.

When DNA is underwound, it twists to the right to relieve strain, with *negative supercoiling* as the result. Most naturally occurring DNA molecules are negatively supercoüed. Each winds around itself to form an interwound supercoil (Figure 17.15). While a DNA molecule is in this conformation, it is storing potential energy in the form of torque (force that tends to cause rotation). The stored energy, in turn, facilitates strand separation during processes such as DNA replication and transcription (Figure 17.16). Overwound DNA twists to the left to relieve stress and is positively supercoiled. Supercoils that form during strand separation in DNA replication interfere with the replication machinery. They are removed by enzymes called topoisomerases (e.g., DNA gyrase in *E. coli*). which make reversible cuts that allow the supercoiled DNA segments to unwind.

Chromosomes and Chromatin

DNA, which contains the genes (the units of heredity), is packaged into structures called chromosomes. The term **chromosome** originally referred only to the dense, dark-staining structures visible within eukaryotic cells during meiosis or mitosis. However, this term is now also used to describe the DNA molecules that occur in prokaryotic cells. The physical structure and genetic organization of prokaryotic and eukaryotic chromosomes are significantly different.

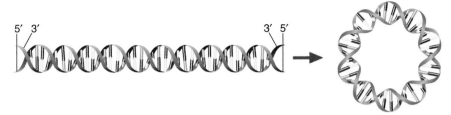

Linear double-stranded DNA molecule

Circular DNA molecule

(a)

FIGURE 17.14

Linear and Circular DNA and DNA Winding

(a) The formation of a relaxed circular DNA molecule. (b) When a relaxed molecule is twisted, it reverts to its flat structure upon release.

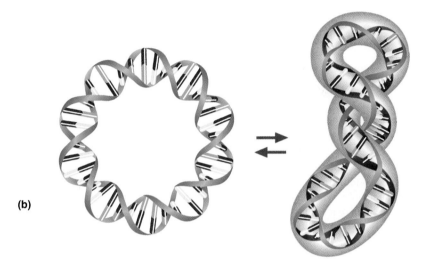

(b)

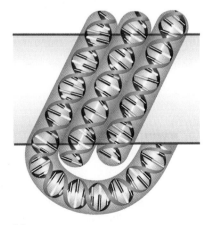

(a)   **(b)**

FIGURE 17.15

Supercoils

Supercoils occur in two major forms: (a) toroidal and (b) interwound.

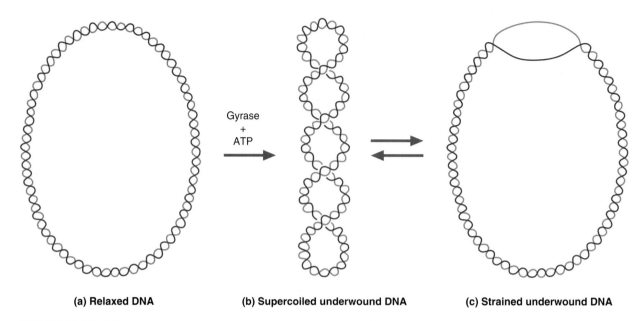

(a) Relaxed DNA   **(b) Supercoiled underwound DNA**   **(c) Strained underwound DNA**

FIGURE 17.16

Effect of Strain on a Circular DNA Molecule

When a negatively supercoiled DNA molecule is forced to lie in a plane, the strain relieved by the formation of the negative supercoiling is reintroduced. Breakage and re-formation of a phosphodiester linkage allows the conversion of a relaxed circular form (a) to the negatively super-coiled form (b). The strain relieved by the supercoiling process will be reintroduced when the underwound molecule is forced to lie in a plane (c).

PROKARYOTES In prokaryotes such as *E. coli*, a chromosome is a circular DNA molecule that is extensively looped and coiled so that it can be compressed into a relatively small space (1 μm × 2 μm). Yet the information in this highly condensed molecule must be readily accessible. The *E. coli* chromosome (circumference 1.6 μm) consists of a supercoiled DNA that is complexed with a protein core (Figure 17.17).

In this structure, called the nucleoid, the chromosome is attached to the protein core in at least 40 places. This structural feature produces a series of loops that limit the unraveling of supercoiled DNA if a strand break is introduced. Compression is further enhanced by packaging with architectural proteins that bind bacterial DNA and facilitate bending and supercoiling. In archaeal prokaryotes, DNA is packaged with histones that have structures similar to eukaryotic histones.

In addition, the polyamines (polycationic molecules such as spermidine and spermine) assist in attaining the chromosome's highly compressed structure.

$$\overset{+}{H_3N}-CH_2-CH_2-CH_2-CH_2-\overset{+}{NH_2}-CH_2-CH_2-CH_2-\overset{+}{NH_3}$$
Spermidine

$$\overset{+}{H_3N}-CH_2-CH_2-CH_2-\overset{+}{NH_2}-CH_2-CH_2-CH_2-CH_2-\overset{+}{NH_2}-CH_2-CH_2-CH_2-\overset{+}{NH_3}$$
Spermine

The positively charged polyamines bind to the negatively charged DNA backbone, thus overcoming the charge repulsion between adjacent DNA coils.

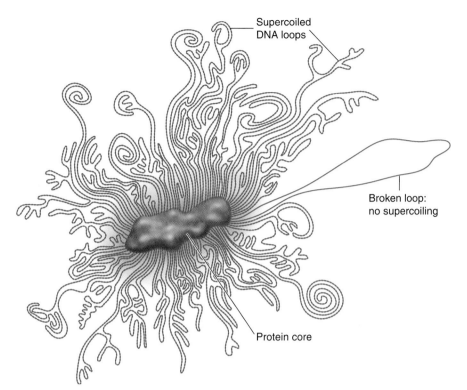

Supercoiled
DNA loops

Broken loop:
no supercoiling

Protein core

FIGURE 17.17

The E. coli Chromosome

The circular *E. Coli* chromosome is complexed with a protein core. Because the chromosome (3×10^6 bp) is highly supercoiled, the entire chromosome complex measures only 2 μm across. The attachment of each of the DNA loops to the protein core may prevent the unraveling of the entire supercoiled chromosome when a strand breaks.

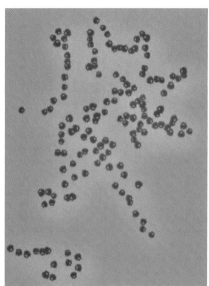

FIGURE 17.18

Electron Micrograph of Chromatin

A chromatin specimen spread out for electron microscopy. Note the "bead on a necklace" structure of this DNA-containing material.

EUKARYOTES In comparison to prokaryotes, the eukaryotes possess genomes that are extraordinarily large. Depending on species, the chromosomes of eukaryotes vary in both length and number. For example, humans possess 23 pairs of chromosomes and have a haploid genome of approximately 3 billion bp. The fruit fly *Drosophila melanogaster* has four chromosome pairs with 180 million bp, and corn (*Zea mays*) has 10 chromosome pairs with a total of 6.6 billion bp.

Each eukaryotic chromosome consists of a single linear DNA molecule complexed with histone proteins to form **nucleohistone**. The term **chromatin** is used to describe the complex of DNA and histones. Several types of nonhistone protein are also associated with chromatin. Examples include DNA replication and repair enzymes and a vast number of transcription factors of different types.

The histones are a group of small basic proteins found in all eukaryotes. The binding of histones to DNA results in the formation of **nucleosomes**, which are the structural units of eukaryotic chromosomes. Consisting of five major classes (H1, H2A, H2B, H3, and H4), the histones are similar in their primary structure among eukaryotic species. In electron micrographs, chromatin has a beaded appearance (Figure 17.18). Each of the "beads" is a nucleosome, which is composed of a positively supercoiled segment of DNA forming a toroidal coil around an octameric histone core (two copies each of H2A, H2B, H3, and H4).

Each of the highly conserved core histones (Figure 17.9a) contains a common structural feature called the *histone fold*: three α-helices separated by two short unstructured segments. The N-terminal tails of core histones consist of between 25 and 40 amino acid residues that protrude from nucleosomes. Covalent modifications of tail residues of several types (e.g., acetylation and methylation) alter the structural and functional properties of the histones, which in turn modify the accessibility of DNA to transcription factors. Such modifications are referred to as *epigenetic* modifications. The histone core forms when two sets of H2A and H2B form two head-to-tail heterodimers and H3 and H4 histones form two sets

COMPANION

GW

WEBSITE Visit the companion website at www.oup.com/us/mckee to read the **Biochemistry in Perspective** box on **epigenetics and the epigenome: genetic inheritance beyond DNA base sequences.**

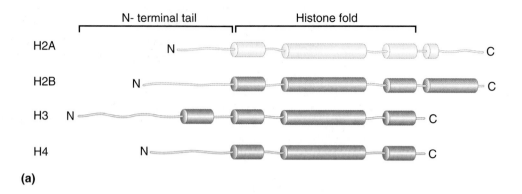

(a)

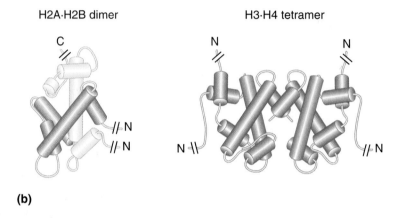

(b)

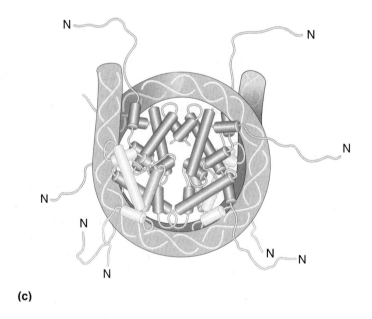

(c)

FIGURE 17.19

The Core Histones

Each nucleosome contains eight histone molecules: two each of H2A, H2B, H3, and H4. Each of these molecules contains a globular domain referred to as the histone fold and a long unstructured N-terminal domain. (a) These domains are illustarated as linear molecules with cylinders representing α-helices. (b) The formation of the histone core begins with the association of H2A and H2B to form a dimer and two molecules each of H3 and H4 combine to form a tetramer. The H3$_2$•H4$_2$ tetramer then binds to DNA. The nucleosome structure is complete (c) when 2 H2A•H2B dimers bind to the tetramer.

of head-to-tail heterodimers. The H3•H4 heterodimers then associate to form a
$H3_2•H4_2$ tetramer (Figure 17.19b). The assembly of the nucleosome begins when
the $H3_2•H4_2$ tetramer binds to DNA. When the two H2A•H2B dimers associate
with the tetramer, nucleosome assembly is complete (Figure 17.19c). One molecule of histone H1 binds to the nucleosome where the DNA enters and exits, and
acts as a clamp that prevents nucleosome unraveling (Figure 17.20). Approximately
145 bp are in contact with each histone octamer. Connection between adjacent
nucleosomes is by means of an additional 60 bp of linker DNA.

FIGURE 17.20

Histon H1

The binding of H1 to two different
sites on the nucleosome DNA
stablizes the wrapping of DNA
around the histone octomer.

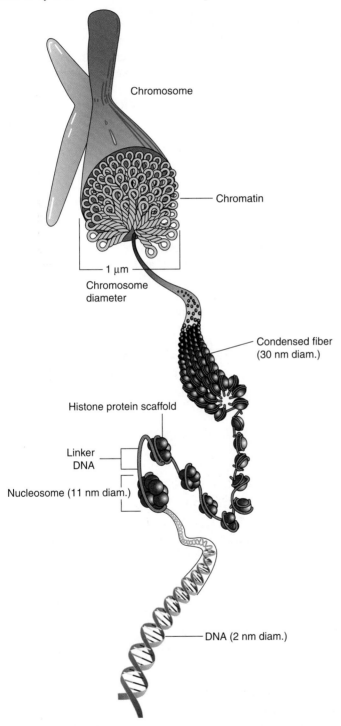

FIGURE 17.21

Chromatin

Nuclear chromatin contains many levels of coiled structure.

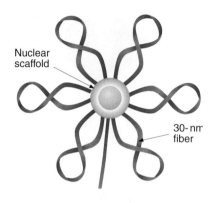

FIGURE 17.22

Chromatin

In one proposal for the structure of 200-nm filaments, the 30 mm fiber is looped and attached to a nuclear scaffold composed of protein.

In anticipation of cell division, chromatin is compacted (about 10,000-fold) to form chromosomes. The nucleosomes are coiled into a higher-order of structure referred to as the *30 nm fiber* (Figure 17.21). The 30 nm fiber is further coiled to form *200 nm filaments*. The three-dimensional structure of the 200 nm filaments contains numerous supercoiled loops attached to a central protein complex referred to as a nuclear scaffold (Figure 17.22). During interphase of the cell cycle, chromatin is observed in two forms. **Heterochromatin** is so highly condensed that it is transcriptionally inactive. A small portion of each cell's heterochromatin occurs in all of an individual organism's cells. Other portions of heterochromatin differ in a tissue-specific pattern. **Euchromatin**, a less condensed form of chromatin, has varying levels of transcriptional activity. Transcriptionally active euchromatin is the least condensed. Inactive euchromatin is somewhat more condensed, but not as much as heterochromatin. The mechanism by which chromatin reversibly condenses is unresolved, but histone covalent modifications are believed to play a significant role.

ORGANELLE DNA Mitochondria and chloroplasts are semiautonomous organelles; that is, they possess DNA and their own version of protein-synthesizing machinery. These organelles, both of which reproduce by binary fission, also require a substantial contribution of proteins and other molecules that are coded for by the nuclear genome. For example, mitochondrial DNA (mtDNA) codes for 2 rRNAs, 22 tRNAs, and several proteins, most of which are used for electron transport. The remainder of mitochondrial proteins are synthesized in the cytoplasm and transported into the mitochondria. Similarly, the chloroplast genome codes for several types of RNA and certain proteins, many of which are directly associated with photosynthesis. The activities of nuclear and organelle genomes are highly coordinated. Consequently, their individual contributions to organelle function are often difficult to discern. Because mitochondria and chloroplasts are now believed to be the descendants of free-living bacteria, it is not surprising that they are susceptible to the actions of antibiotics (e.g., chloramphenicol and erythromycin) if their concentrations are sufficiently high. Many of these substances are (or have been) used in clinical practice because they inhibit some aspect of bacterial genome function.

KEY CONCEPTS

- Each prokaryotic chromosome consists of a supercoiled circular DNA molecule complexed to a protein core.
- Each eukaryotic chromosome consists of a single linear DNA molecule that is complexed with histones to form nucleohistone.

QUESTION 17.6

Compare the structural features that distinguish B-DNA from A-DNA and Z-DNA. What is known about the functional properties of these variants of B-DNA, the Watson-Crick structure?

QUESTION 17.7

What are the major protein components of prokaryotic and eukaryotic chromosomes? What are their functions?

QUESTION 17.8

Explain the hierarchical relationships among the following: genomes, genes, nucleosomes, chromosomes, and chromatin.

QUESTION 17.9

Describe the evidence that James Watson and Francis Crick used to construct their model of DNA structure.

Genome Structure

The genome of each living organism is the full set of inherited instructions required to sustain all living processes, that is, the organism's operating system. Within each genome are the genes, the units of inheritance that determine the primary structure of gene products (polypeptides and RNA molecules). Genomes differ in size, shape, and sequence complexity. Genome size—the number of base-paired nucleotides, which is loosely related to organismal complexity—varies over an enormous range from less than 10^6 bp in some species of *Mycoplasma* (the smallest known bacteria) to greater than 10^{10} bp in certain plants. Most prokaryotic genomes are smaller than those of eukaryotes. In contrast to prokaryotic genomes, which typically consist of single circular DNA molecules, eukaryotic genomes are divided into two or more linear DNA molecules. The most significant difference between prokaryotic and eukaryotic genomes, however, is the vastly larger information-coding capacity of eukaryotic DNA. Amazingly, the majority of eukaryotic sequences do not code for gene products. For this reason, each type of genome will be considered separately. Short segments of the genomes of several eukaryotes are compared with that of *E. coli* in Figure 17.23.

PROKARYOTIC GENOMES Investigations of prokaryotic chromosomes, especially those of several strains of *E. coli*, have revealed the following features.

1. **Genome size.** As described, most prokaryotic genomes are relatively small, having considerably fewer genes than those of eukaryotes. The *E. coli* chromosome contains about 4.6 megabases, which code for about 4300 genes [1 megabase (Mb) = 1×10^6 bases].

2. **Coding capacity.** The genes of prokaryotes are compact and continuous; that is, they contain little, if any, noncoding DNA either between or within gene sequences. This is in sharp contrast to eukaryotic DNA, in which a significant percentage of DNA can be in noncoding form.

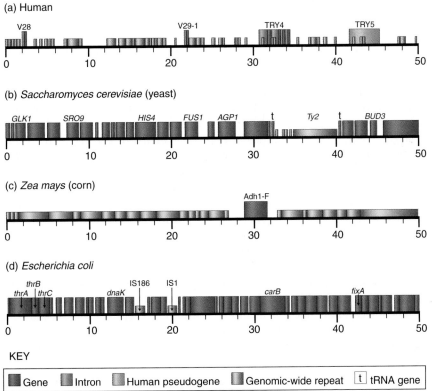

FIGURE 17.23

Comparison of 50 kb Segments of the Genomes of Selected Eukaryotes with the Prokaryote *E. coli* Genome.

As indicated, the genomes of organisms such as (a) humans, (b) Saccharomyces cerevisiae, (c) corn, and (d) *E coli* can vary considerably in their complexity and gene density. Genes are indicated by letters and/or numbers. Humans and other complex eukaryotes have genes that are interrupted with sequences such as introns and nonfunctional sequences called pseudogenes that resemble true genes. Bacteria have few if any genome-wide repeats (repetitive, noncoding segments).

3. **Gene expression**. The regulation of many functionally related genes is enhanced by organizing them into operons. An **operon** is a set of linked genes the transcription of which is regulated as a unit. About one-fourth of the genes of *E. coli* are organized into operons.

Recall that prokaryotes also often possess additional small pieces of DNA (see p. 42). called plasmids, which are usually, but not always, circular. Plasmids typically have genes that are not present on the main chromosome. Although these genes are seldom essential for bacterial growth and survival, they may code for biomolecules that provide the cell with a growth or survival advantage: antibiotic resistance, unique metabolic capacities (e.g., nitrogen fixation; degradation of unique energy sources such as aromatic compounds) or virulence (e.g., toxins or other factors that undermine host defense mechanisms).

EUKARYOTIC GENOMES The organization of genetic information in eukaryotic chromosomes has proven to be substantially more complex than that observed in prokaryotes. Eukaryotic nuclear genomes possess the following unique features:

1. **Genome size**. Eukaryotic genomes tend to be substantially larger than those of prokaryotes. However, in the higher eukaryotes genome size is not necessarily a measure of the complexity of the organism. For example, the haploid genome of humans is 3200 Mb. The genomes of peas and the salamander are 5000 and 90,000 Mb, respectively.

2. **Coding capacity**. Although there is enormous coding capacity, the majority of DNA sequences in eukaryotes do not appear to have coding functions; that is, they do not possess intact regulatory regions that initiate transcription (the production of RNA transcripts). The functions of these noncoding sequences are unknown; some of them probably have regulatory or structural roles. It has been estimated that no more than 1.5% of the human genome codes for proteins.

3. **Coding continuity**. Most eukaryotic genes investigated so far are discontinuous. Noncoding sequences (called **introns** or intervening sequences) are interspersed between sequences called **exons** (expressed sequences), which code for a gene product (any of various RNA molecules, some of which dictate the translation of proteins). Intron sequences, which may together be significantly longer than the exons in a protein coding gene, are removed from primary RNA transcripts by a splicing mechanism (Section 18.2) to produce a functional RNA molecule.

The existence of exons and introns enables eukaryotes to produce more than one polypeptide from each gene. By utilizing a process called alternative splicing (p. 722), various combinations of exons can be joined together to form a series of mRNAs. For example, the random rearrangements of gene sequences that encode the antigen receptors of the immunoglobulins have a major role in generating the millions of lymphocytes and antibodies produced by the mammalian immune system. One of the more surprising results of genome research and the genome projects (multinational scientific endeavors that determine the complete genome sequences of selected organisms) is the realization that genes make up such a small portion of eukaryotic genomes. Most eukaryotic DNA sequences are noncoding; that is, they do not specify the primary structures of polypeptides or RNAs. **Intergenic sequences**, those that do not code for gene products, have often in the past been referred to as "junk DNA" because they were perceived to serve no useful functions. Although the functions of most intergenic sequences are still unresolved, the sequences themselves have been analyzed and categorized. The discussion that follows focuses on DNA sequences of the types that occur in the human genome.

Of the approximately 3200 Mb of the human genome, approximately 38% comprise genes and related sequences (Figure 17.24) The portion of these

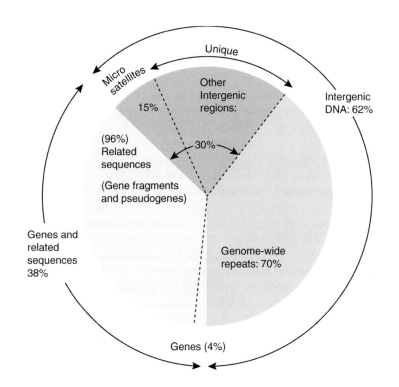

FIGURE 17.24

The Human Genome

The majority of DNA sequences in the human genome do not code for gene products. Approximately 62% of these sequences are intergenic DNA, which in turn can be divided into genome-wide repeats, microsatellites, and other intergenic regions. Thirty-eight percent of human sequences are genes and other related sequences of which only 4% code for gene products. Gene-related sequences account for the rest of these sequences.

sequences that code for gene products (polypeptides and functional RNAs) is approximately 4%. By varying estimates there are between 25,000 and 40,000 human genes. (Note that such a wide range of estimates is the result of the use of different algorithms in the analysis of genomic sequences. An algorithm is a set of rules followed by computer software that is designed to identify certain DNA sequence patterns.) Of the total number of genes identified, fewer than 2500 code for RNA molecules: the remainder code for proteins. The functions of almost one-quarter of known protein-coding genes (Figure 17.25) are related to DNA synthesis and repair, and gene expression. Signal transduction proteins are coded for by about 21% of genes and about 17% code for general biochemical functions of cells (i.e., metabolic enzymes). The remaining genes, about 38%, code for an array of proteins involved in transport processes (e.g., ion channels), protein folding (molecular chaperones and proteosomal subunits), structural proteins (e.g., actin, myosin, tubulin, and their accessory proteins), and immunological proteins (e.g., antibodies). Related gene sequences include *pseudogenes* (inactivated, non-functional gene copies) and gene fragments.

A little over 60% of the human genome consists of intergenic sequences. Although their significance is not understood, several types of repetitive sequence have been identified and investigated. There are two general classes: tandem repeats and interspersed genome-wide repeats. Each is briefly described.

Tandem repeats are DNA sequences in which multiple copies are arranged next to each other. These sequences were originally referred to as **satellite DNA** because they form a separate or "satellite" band when genomic DNA is broken into pieces and centrifuged to separate the fragments by density gradient centrifugation (see the box entitled Biochemistry in the Lab: Nucleic Acid Methods, pp. 653–657). The lengths of the repeated sequences vary from under 10 bp to over 2000 bp. Total lengths of the tandem repeats often vary between 10^5 and 10^7 bp. Certain types of tandem repeat apparently play structural roles in **centromeres** (the structures that contain kinetochores, which attach chromosomes to the mitotic spindle during mitosis and meiosis) and **telomeres** (structures at the ends of chromosomes that buffer the loss of critical coding sequences after a round of DNA replication). Two relatively small repetitive sequence types are

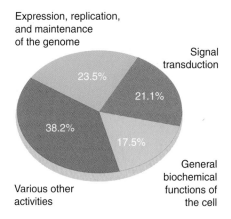

Expression, replication, and maintenance of the genome

Signal transduction

General biochemical functions of the cell

Various other activities

FIGURE 17.25

Human Proteins Coding Genes

Classifications of known protein-coding genes. It has been estimated that the functions of 13,000 of a possible 30,000 genes are as yet unknown.

referred to as minisatellites and microsatellites. **Minisatellites** have tandemly repeated sequences of 10 to 100 bp with total lengths between 10^2 and 10^5 bp. Telomeres contain minisatellite clusters. In **microsatellites**, also referred to as single-sequence repeats (SSR), there is a core sequence of 1 to 4 bp that is tandemly repeated from 10 to 100 times. The functions of these repetitive sequences are for the most part unknown. Because of their large number in genomes and because they are pleomorphic (i.e., vary with each individual organism), minisatellites and microsatellites are used as markers in genetic disease diagnosis, in kinship and population studies and in forensic investigations (see the box entitled Biochemistry in Perspective-Forensic Investigations, pp. 658–659).

As their name implies, **interspersed genome-wide repeats** are repetitive sequences that are scattered around the genome. Most of these sequences are the result of **transposition** (Section 18.1), a mechanism whereby certain DNA sequences, referred to as **mobile genetic elements**, can be duplicated and enabled to move within the genome. **Transposable DNA elements**, referred to as **transposons**, excise themselves and then insert at another site. More commonly, however, transposition mechanisms involve an RNA transcript intermediate. These latter DNA elements are called **RNA transposons** or **retrotransposons**.

Retrotransposons with lengths greater than 5 kb are called **LINEs** (long interspersed nuclear elements). LINE sequences contain a strong *promoter* (a base sequence upstream of a gene that is required for transcription initiation), an integration sequence (a base sequence required for insertion into another DNA molecule), and the coding sequences for transposition enzymes. LINEs have undergone duplication and mutation over time, and only a small percentage of them are at all functional. One in every 1200 mutations in humans is estimated to be the result of a LINE insertion. One example is hemophilia A (a blood-clotting disorder), which results when a LINE sequence inserts into the gene for clotting factor VIII.

SINEs (short interspersed nuclear elements) are less than 500 bp; although they contain insertional sequences, they cannot undergo transposition without the aid of a functional LINE sequence. SINEs have greatly expanded over time to comprise 11% of the human genome with over a million copies. The only SINE insertion linked to human disease involves the *Alu element* that mediates chromosome rearrangements, insertions, deletions, and recombinations. Mutations mediated by the *Alu* subfamily of SINEs have accounted for more than 20 different human genetic diseases. *Alu*-mediated insertions have been observed to cause hemophilia B (defective clotting factor IX), and unequal recombinations due to Alu mediated insertions have been linked to Lesch-Nyhan disease (p. 547) and Tay-Sachs disease (p. 385). *Recombination* (Section 18.1) is the shuffling of DNA sequences via crossing over of homologous chromosomes in the cells that give rise to egg or sperm cells. Unequal recombination is an aberrant process that causes insertion or deletion mutations.

KEY CONCEPTS

- In each organism's genome, the information required to direct living processes is organized for efficient storage and use.
- Genomes from different types of organism differ in their sizes and levels of complexity.

QUESTION 17.10

Define the following terms:
   a. tandem repeats
   b. centromere
   c. satellite DNA
   d. introns
   e. exons
   f. microsatellites
   g. transposition

QUESTION 17.11

Compare the sizes and coding capacity of prokaryotic genomes with those of eukaryotes. What other features distinguish them?

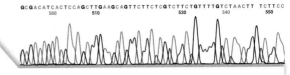

BIOCHEMISTRY IN THE LAB
Nucleic Acid Methods

The techniques used in the isolation, purification, and characterization of biomolecules take advantage of their physical and chemical properties. Most of the techniques used in nucleic acid research are based on differences in molecular weight or shape, base sequences, or complementary base pairing. Techniques such as chromatography, electrophoresis, and ultracentrifugation, which have been used successfully in protein research, have also been adapted to use with nucleic acids. In addition, other techniques have been developed that exploit the unique properties of nucleic acids. For example, under certain conditions DNA duplexes reversibly melt (separate) and reanneal (base-pair to form a duplex again). One of several techniques that exploit this phenomenon, called *Southern blotting*, is often used to locate specific (and often rare) nucleic acid sequences. After brief descriptions of several techniques used to purify and characterize nucleic acids, the common method for determining DNA sequences is outlined. More complex techniques are described in Chapter 18 in the Biochemistry in the Lab box entitled Genomics (pp. 700–707).

Once bacterial cells have been ruptured or eukaryotic nuclei have been isolated, their nucleic acids are extracted and deproteinized. This can be accomplished by several methods. Bacterial nucleic acid is often precipitated by treating cell preparations with alkali and lysozyme (an enzyme that degrades bacterial cell walls by breaking glycosidic bonds). Partially degraded protein is extracted by using certain solvent combinations (e.g., phenol and chloroform). Similarly, eukaryotic nuclei can be treated with detergents or solvents to release their nucleic acids. Depending on which type of nucleic acid is being isolated, specific enzymes are used to remove the other type. For example, RNase removes RNA from nucleic acid preparations, leaving DNA intact. DNA is further purified by centrifugation.

All nucleic acid samples must be handled carefully. First, nucleic acids are susceptible to the actions of a group of enzymes called nucleases. In addition to a variety of such enzymes that are released during cell extraction, nucleases can be introduced from the environment: for example, the experimenter's hands. Second, high-molecular-weight nucleic acids, primarily DNA, are sensitive to shearing stress. Purification procedures, therefore, must be gentle, applying as little mechanical stress as possible.

Techniques Adapted from Use with Other Biomolecules

Many of the techniques used in protein purification procedures have also been adapted for use with nucleic acids. For example, several types of chromatography (e.g., ion-exchange, gel filtration, and affinity) have been used in several stages of nucleic acid purification and in the isolation of individual nucleic acid sequences. Because of its speed, HPLC has replaced many slower chromatographic separation techniques applied to small samples.

A type of column chromatography that uses a calcium phosphate gel called **hydroxyapatite** has been especially useful in nucleic acid research. Because hydroxyapatite binds to double-stranded nucleic acid molecules more tenaciously than to single-stranded molecules, double-stranded DNA (dsDNA) can be effectively separated from single-stranded DNA (ssDNA), RNA, and protein contaminants by eluting the column with increasing concentrations of phosphate buffer. The use of hydroxyapatite columns has been largely replaced by a form of affinity chromatography in which the column matrix molecules have been covalently bonded to avidin, a small protein that binds specifically to biotin. The matrix can be prepared with a specific biotinylated ligand such as an ssDNA probe (known sequence). If a mixture containing ssDNA is applied to the column, specific sequences complementary to the probe will bind, and the rest of the sample will pass through. The bound target molecules can then be eluted from the column in nearly pure form.

The movement of nucleic acid molecules in an electric field depends on both their molecular weight and their three-dimensional structure. However, because DNA molecules often have relatively high molecular weights, their capacity to penetrate some gel preparations (e.g., polyacrylamide) is limited. Although DNA sequences with less than 500 bp can be separated by polyacrylamide gels with especially large pore sizes, more porous gels must be used with larger DNA molecules. Agarose gels, which are composed of a cross-linked polysaccharide, are used to separate DNA molecules with lengths between 500 bp and approximately 150 kilobases (kb). Larger sequences are now isolated with a variation of agarose gel electrophoresis in which two electric fields (perpendicular to each other) are alternately turned on and off. DNA molecules reorient themselves each time the electric field alternates, resulting in a very efficient and precise separation of heterogeneous groups of DNA molecules. The reorientation times decrease with the size of the DNA.

Density gradient centrifugation with cesium chloride (CsCl) has been widely used in nucleic acid research (see the Biochemistry in the Lab box entitled Cell Technology, pp. 67–68). At high speeds, a linear gradient of CsCl is established. Mixtures of DNA, RNA, and protein migrating through this gradient separate into discrete bands at positions where their densities are equal to the density of the CsCl. DNA molecules with high guanine and cytosine content are more dense than those with a higher proportion of adenine and thymine. This difference helps separate heterogeneous mixtures of DNA fragments.

BIOCHEMISTRY **IN THE LAB** cont

Techniques That Exploit the Unique Structural Features of the Nucleic Acids

Several unique properties of the nucleic acids (e.g., absorption of UV light at specific wavelengths and their tendency to reversibly form double-stranded complexes) are exploited in nucleic acid research. Several applications of these properties are briefly discussed.

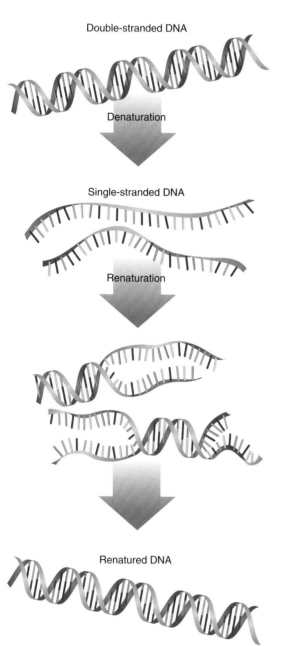

Double-stranded DNA

Denaturation

Single-stranded DNA

Renaturation

Renatured DNA

FIGURE 17A

Denaturation and Renaturation of DNA

Under appropriate conditions, DNA that has been denatured can renature; that is, strands with complementary sequences will re-form into a double helix.

Because of their aromatic structures, the purine and pyrimidine bases absorb UV light. At pH 7 this absorption is especially strong at 260 nm. However, when the nitrogenous bases are incorporated into polynucleotide sequences, various noncovalent forces promote close interactions between them. This decreases their absorption of UV light. This **hypochromic effect** is an invaluable aid in studies involving nucleic acid. For example, absorption changes are routinely used to detect the disruption of the double-stranded structure of DNA or the hydrolytic cleavage of polynucleotide strands by enzymes.

The binding forces that hold the complementary strands of DNA together can be disrupted. This process, referred to as **denaturation** (Figure 17A), is promoted by heat, low salt concentrations, and extremes in pH. (Because it is easily controlled, heating is the most common denaturing method in nucleic acid investigations.) When a DNA solution is slowly heated, absorption at 260 nm remains constant until a threshold temperature is reached. Then the sample's absorbance increases (Figure 17B). The absorbance change is caused by the unstacking of bases and the disruption of base pairing. The temperature at which one-half of a DNA sample is denatured, referred to as the melting temperature (T_m), varies among DNA molecules according to their base compositions. [Recall that DNA stability is affected by the number of hydrogen bonds between GC and AT pairs and base stacking interactions (see p. 632). More energy is required,

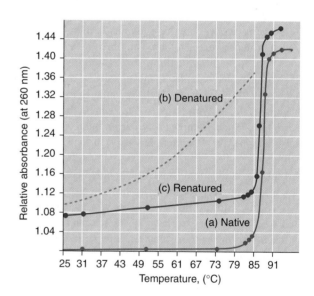

FIGURE 17B

DNA Denaturation

(a) When native DNA is heated, its absorbance does not change until a specific temperature is reached. The "melting temperature" T_m of a DNA molecule varies with its base composition. (b) When the denatured DNA is chilled, its absorbance falls, but along a different curve. It does not return to its original absorbance value. (c) Reannealed (renatured) DNA can be prepared by maintaining the temperature at 25°C below the denaturing temperature for an extended period.

▶▶

BIOCHEMISTRY IN THE LAB cont

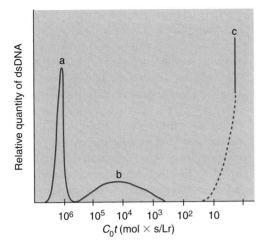

FIGURE 17C

DNA Sequence Pattern of the Mouse Genome

The degree of repetitiveness in the segments of total mouse DNA is determined by measuring C_0t values for several fractions of the genome. The dashed portion of the curve indicates estimated values. The reassociation kinetics of eukaryotic genomes typically reveal three primary classes of sequences: repetitive sequences that reanneal quickly (a), sequences of intermediate complexity (b), and unique sequences that reanneal slowly (c).

FIGURE 17D

Southern Blotting

(1) DNA analysis begins with its digestion by a restriction enzyme. (2) DNA fragments are separated by agarose gel electrophoresis. (3) The DNA fragments are transferred to nitrocellulose filter paper under denaturing conditions. (4) The ssDNA on the nitrocellulose filter paper is hybridized with a radioactively labeled ssDNA probe. (5) Any hybridized DNA can be visualized by autoradiography.

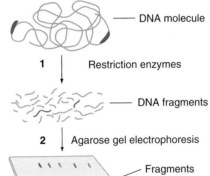

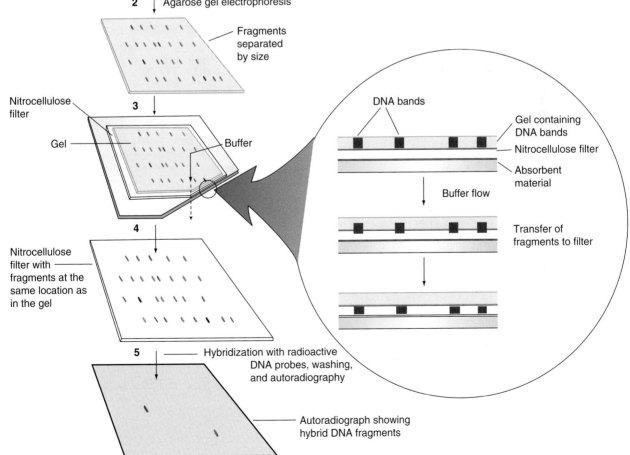

BIOCHEMISTRY IN THE LAB cont

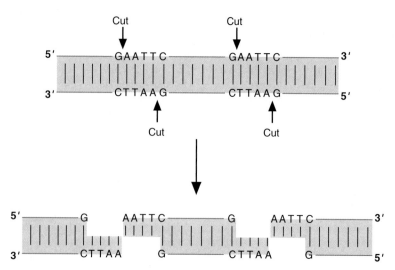

FIGURE 17E

Restriction Enzymes

Restriction endonucleases are enzymes isolated from bacteria that cut DNA at specific sequences. In this example the enzyme EcoRI (obtained from *E. coli*) makes staggered cuts that result in the formation of "sticky ends." A "sticky end" is a single-stranded terminus on a double-stranded DNA fragment. Sticky ends facilitate the formation of recombinant DNA because a single-stranded segment on one DNA fragment can anneal with a complementary sticky end on another DNA fragment. Some restriction enzymes make what are called "blunt cuts." For a discussion of recombinant DNA, see box Chapter 18 (Biochemistry in the Lab, Genomics, pp. 700–707).

therefore, to "melt" DNA molecules with high G and C content.] If the separated DNA strands are held at a temperature approximately 25°C below the T_m for an extended time, renaturation is possible. Renaturation, or reannealing, does not occur instantaneously because the strands explore various configurations until they achieve the most stable one (i.e., the one having paired complementary regions).

DNA melting is extraordinarily useful in nucleic acid **hybridization**. Single-stranded DNA from different sources associates (or "hybridizes") if there is a significant sequence homology (i.e., structural similarity). If a DNA sample is sheared into small uniform pieces, the rate of reannealing depends on the concentration of DNA strands and on the structural similarities between them. Reannealing rates have revealed valuable

FIGURE 17F

The Sanger Chain Termination Method

A specific primer is chosen so that DNA synthesis will begin at the point of interest. DNA synthesis continues until a radioactive dideoxynucleotide is incorporated and the chain terminates. Afterward, the products of the reactions are separated by gel electrophoresis and analyzed by autoradiography. The fragments migrate according to size. The sequence is determined by "reading" the gel.

information about genome structure. For example, organisms vary in the number and types of unique sequences their genomes contain. (A unique DNA sequence occurs only once per haploid genome.) The relative number of unique and repeated sequences can be determined by constructing a C_0t curve, which provides a measure of renaturation (where C_0 is the concentration of ssDNA in moles per liter and t is elapsed time in seconds). From C_0t curves, scientists have demonstrated that the velocity of reannealing declines as genomes become larger and more complex. The frequency of unique and repeated DNA sequences determined by measuring reannealing rates for the mouse genome is shown in Figure 17C.

Hybridization can also be used to locate and/or identify specific genes or other DNA sequences. For example, ssDNA from two different sources (e.g., tumor cells and normal cells) can be screened for sequence differences. If one set of ssDNA is biotinylated, then the double-stranded hybrids bind to an avidin column. If any unhybridized sequence is present, it passes through the column. Then it can be isolated and identified. In **Southern blotting** (Figure

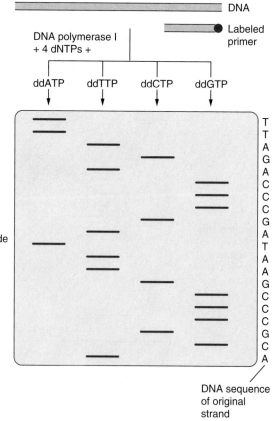

BIOCHEMISTRY IN THE LAB cont

17D) radioactively labeled DNA or RNA probes (sequences with known identities) locate a complementary sequence in the midst of a DNA digest, which typically contains a large number of heterogeneous DNA fragments. A DNA digest is obtained by treating a DNA sample with restriction enzymes that cut at specific nucleotide sequences (Figure 17E). (Produced by bacterial cells, restriction enzymes protect bacteria against viral infection by cleaving viral DNA at specific sequences.) Once the DNA sample has been digested, the fragments are separated by agarose gel electrophoresis according to their sizes. After the gel has been soaked in 0.5 M NaOH, a process that converts dsDNA to ssDNA, the DNA fragments are transferred to nitrocellulose filter paper by placing them on a wet sponge in a tray with a high salt buffer. (Nitrocellulose has the unique property of binding strongly to ssDNA.) Absorbent dry filter paper is placed in direct contact with the nitrocelluose filter/agarose gel sandwich. As buffer is drawn through the gel and filter paper by capillary action, the DNA is transferred and becomes permanently bound to the nitrocellulose filter. (The transfer of DNA to the filter is the "blotting" referred to in the name of this technique.) Subsequently, the nitrocellulose filter is exposed to the radioactively labeled probe, which binds to any ssDNA with a complementary sequence. For example, an mRNA that codes for β-globin binds specifically to the β-globin gene, even though β-globin mRNA lacks the introns present in the gene. Apparently, there is sufficient base pairing between the two single-stranded molecules to locate the gene.

DNA Sequencing

The determination of DNA nucleotide sequences has provided valuable insights in biochemistry, medical science, and evolutionary biology. The analysis of long DNA sequences begins with the formation of smaller fragments by means of one type of restriction enzyme. Each fragment is then sequenced independently by the chain-terminating method. As with protein primary structure determinations, these steps are repeated with a different set of polynucleotide fragments (generated by another type of restriction enzyme) that overlap the first set. Sequence information from both sets of experiments then orders the fragments into a complete sequence.

In DNA sequencing by the **chain-terminating method** (Figure 17F), developed by Frederick Sanger, restriction enzymes cleave large DNA segments into smaller fragments. Each fragment is separated into two strands, one of which is used as a template to produce a complementary copy. The sample is further divided into four test tubes. To each tube is added the substances required for DNA synthesis (e.g., the enzyme DNA polymerase and the four deoxyribonucleotide triphosphates and a ^{32}P-labeled primer (a short segment of a complementary DNA strand). The investigator determines the site at which sequencing is to start by selecting the appropriate primer.

In addition to a template, DNA synthesis materials, and a primer, each of the four tubes contains a different 2′, 3′-dideoxynucleotide derivative. (The dideoxy derivatives are synthetic nucleotide analogues in which the hydroxy groups on the 2′- and 3′-carbons have been replaced with hydrogens.) Dideoxynucleotides can be incorporated into a growing polynucleotide chain, but they cannot form a phosphodiester linkage with another nucleotide. Consequently, when dideoxynucleotides are incorporated, they terminate the chain. Because small amounts of the dideoxynucleotides are used, they are randomly incorporated into growing polynucleotide strands. Each tube, therefore, contains a mixture of DNA fragments containing strands of different lengths. Each newly synthesized strand ends in a dideoxynucleotide residue. The reaction products in each tube are separated by gel electrophoresis and analyzed together by autoradiography. Each band in the autoradiogram corresponds to a polynucleotide that differs in length by one nucleotide from the one that precedes it in any of the four lanes of the autoradiogram. Note that the smallest polynucleotide appears on the bottom of the gel, because it moves more quickly than larger molecules.

An automated version of the Sanger method allows rapid and efficient sequencing of DNA samples. Instead of radiolabeled primers, it uses fluorescent tagged dideoxynucleotides. Because each dideoxy analogue fluoresces a different color, the entire procedure is carried out in a single test tube. Afterward the reaction products are loaded and run on a single electrophoresis gel. A detector then scans the gel, and a computer determines the sequence of the colored bands (Figure 17G).

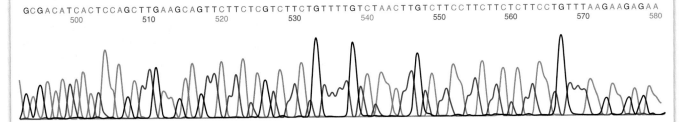

GCGACATCACTCCAGCTTGAAGCAGTTCTTCTCGTCTTCTGTTTTGTCTAACTTGTCTTCCTTCTTCTCTTCCTGTTTAAGAAGAGAA

500　　　510　　　520　　　530　　　540　　　550　　　560　　　570　　　580

FIGURE 17G

Automated DNA Sequencing.

With the use of fluorescent tags on the dideoxynucleotides, a detector can scan a gel quickly and determine the sequence from the order of the colors in the bands.

BIOCHEMISTRY IN PERSPECTIVE

Forensic Investigations

How is DNA analysis used in the investigation of violent crime? DNA persists for many years in dried biological specimens (e.g., blood, saliva, hair, and semen) and in bone. Consequently, DNA can be used as evidence in any type of forensic investigation in which such specimens are available. DNA analysis techniques that are typically used to ascertain the identity of victims and/or perpetrators of violent crimes are referred to as **DNA typing**, or profiling. DNA typing involves the analysis of several highly variable sequences called markers. By using sets of variable sequences, investigators can provide unique, identifying genetic profiles for each individual human.

In numerous court cases since the 1990s, DNA typing has provided decisive information concerning defendants' presence at a crime scene, or their absence. The techniques now available differ in their capacity to differentiate between individuals and in the speed with which results can be obtained. **DNA fingerprinting**, introduced in 1985 by the British geneticist Alec Jeffreys, is a variation of Southern blotting. In this technique, the banding characteristics of DNA minisatellites (see p. 652) from different individuals are compared—for example, crime scene specimen DNA with samples from suspects (Figure 17H). When the quantity of DNA extracted from a crime scene sample is too minute to analyze, it is amplified by means of the *polymerase chain reaction* (PCR), a technique that is used to amplify the number of copies of DNA in a tiny sample. Up to 10^9 copies can be obtained. (Refer to p. 703.) Consequently, DNA from a single cell is now sufficient for DNA fingerprint analysis. The entire genome in each sample is isolated and treated with a restriction enzyme. (Restriction enzymes are a group of enzymes produced by bacteria that cut DNA molecules at specific base sequence sites. In bacterial cells, they prevent, or "restrict," the replication of foreign, usually viral, DNA.)

Because of genetic variations, the DNA in minisatellite sequences from each individual fragments differently. (Such genetic differences are called **restriction fragment length polymorphisms**, or **RFLPs.**) After the restriction fragments have been separated according to size by agarose gel electrophoresis and transferred to nitrocellulose filter paper, they are exposed to radiolabeled probes. Because the sizes of the fragments that bind to these probes differ from one individual to the next, banding patterns have been used successfully to either convict or acquit in criminal cases.

Although RFLP testing is an accurate method, it does have limitations. Among these are the substantial amounts of time (6–8 weeks), labor, and expertise required to obtain DNA profiles. A newer methodology that analyzes **short tandem repeats (STR)**, (DNA sequences with between 2 and 4 bp repeats, called microsatellites, see p. 652), has significantly greater discriminating power than RFLP and is relatively rapid (several hours).

Moreover, STR sequences are sufficiently robust that STR analysis can often be successfully used to analyze degraded specimens. After DNA has been extracted from a specimen, several target STR sequences are amplified by PCR and linked to fluorescent dye molecules.

In the United States, 13 core polymorphic (highly variable) DNA markers, called *loci*, are used to generate genetic profiles and to distinguish between individuals. A **DNA profile**, which results when the PCR products are separated in an electrophoretic gel, consists of the pattern and the number of repeats of each target sequence visible on the gel. Fluorescence detection increases the sensitivity of the technique. Unlike RFLP, STR-based DNA typing is easily automated. If the DNA profiles from individual samples are compared and determined to be identical, the samples are said to be a match. If compared

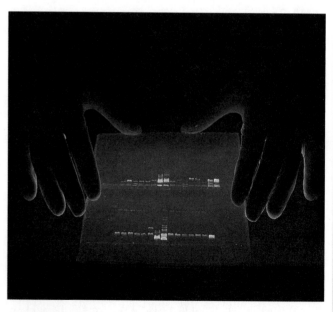

FIGURE 17H

Forensic Use of DNA Fingerprinting

In many RFLP (restriction fragment length polymorphism) analysis of biological evidence collected at a crime scene provides conclusive proof that an accused or convicted person did or did not leave unique DNA traces at that crime scene.

profiles are not identical, they are said to have come from different sources. The results are reported in terms of the probabilities of a random match (the chance that a randomly selected person from the population will have an identical DNA profile to that of the specimen of interest such as that left at a crime scene). The use of multiple markers and the sensitivity of the methodology reduce the random match probability to at least 1 in several billion.

SUMMARY: Forensic scientists use a technique called PCR to amplify crime scene DNA in order to generate the unique genetic profile that distinguishes one individual from all others.

17.2 RNA

For many years, the ribonucleic acids have been regarded as a class of polynucleotides involved in some aspect of protein synthesis. Rapidly accumulating evidence, however, has revealed that RNAs are amazingly versatile molecules that perform functions previously thought to be exclusively performed by proteins. Examples include gene expression regulation, cell differentiation, genome imprinting, and catalysis. The functional properties of the RNAs are due to their structure. The primary structure of polyribonucleotides is similar to that of their DNA counterparts, but there are several differences.

1. The sugar moiety of RNA is ribose instead of deoxyribose in DNA. The presence of the $2'$-OH group of ribose makes RNA more reactive than DNA.

2. The nitrogenous bases in RNA differ somewhat from those observed in DNA. Instead of thymine, RNA molecules use uracil. In addition, the bases in some RNA molecules are modified by a variety of enzymes (e.g., methylases, thiolases, and deaminases).

3. In contrast to the double helix of DNA, RNA exists as a single strand. For this reason, RNA can coil back on itself and form unique and often quite complex three-dimensional structures (Figure 17.26). The shape of these structures is determined by complementary base pairing by specific RNA sequences, as well as by base stacking and interactions between double-stranded regions (formed from single-stranded regions of the same molecule or between single-stranded regions of neighboring molecules) and free loops of RNA. Base-pairing rules apply in double-stranded regions where A-U, G-U, and G-C pairing occurs. Loop or single-stranded RNA (ssRNA) regions contain a number of modified bases in mature non-mRNA molecules. The base composition of RNA does not follow Chargaff's rules because RNA molecules are single-stranded.

4. RNA molecules have catalytic properties because of the complex three-dimensional structures with binding pockets that they can form. The majority of catalytic RNA molecules, which are referred to as **ribozymes**, catalyze self-cleavages or the cleavage of other RNAs. The most notable example of ribozyme activity, however, is peptide bond formation within ribosomes. A magnesium ion cofactor is usually required in RNA catalysis because Mg^{2+} stabilizes transition states.

The most prominent types of RNA are transfer RNA, ribosomal RNA and messenger RNA. The structure and function of each of these molecules is discussed

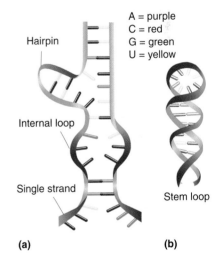

Hairpin

Internal loop

Single strand

A = purple
C = red
G = green
U = yellow

Stem loop

(a)   (b)

FIGURE 17.26

Secondary Structure of RNA

(a) Three of the many different types of secondary structure that occur in RNA molecules. (b) Stem loop. Note that RNA hairpins and stem loops form because of the inverted repeat sequences of DNA palindromes. (see p. 641).

next. Examples of several other types of RNA, referred to as noncoding RNAs (ncRNAs), are also provided.

Transfer RNA

Transfer RNA (tRNA) molecules transport amino acids to ribosomes for assembly into proteins and comprise about 15% of cellular RNA. The average length of a tRNA molecule is 75 nucleotides. Each tRNA molecule becomes bound to a specific amino acid. Consequently, cells possess at least one type of tRNA for each of the 20 standard amino acids. The three-dimensional structure of tRNA molecules, which resembles a warped cloverleaf (Figure 17.27), results primarily from extensive intrachain base pairing. tRNA molecules contain a variety of modified bases. Examples include pseudouridine, 4-thiouridine, 1-methylguanosine, and dihydrouridine:

| Ribose **Pseudouridine** | Ribose **4-Thiouridine** | Ribose **1-Methylguanosine** | Ribose **Dihydrouridine** |

The structure of tRNA allows it to perform two critical functions involving the 3'-terminus and the anticodon loop. The 3'-terminus forms a covalent bond to a

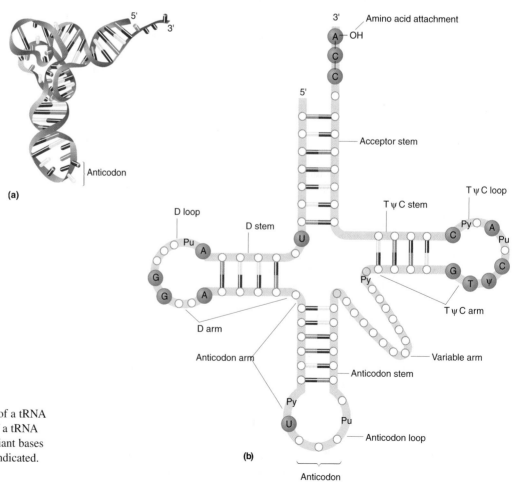

FIGURE 17.27

Transfer RNA

(a) Three-dimensional structure of a tRNA molecule. (b) Schematic view of a tRNA molecule. The positions of invariant bases and bases that seldom vary are indicated.

specific amino acid. (This specificity is achieved because the set of enzymes called the *aminoacyl-tRNA synthetases* link each amino acid to its tRNA.) The *anticodon loop* contains a three-base-pair sequence that is complementary to the DNA triplet code for the specific amino acid. The conformational relationship between the 3'-terminus and the anticodon loop allows the tRNA to align its attached amino acid properly during protein synthesis. (This process is discussed in Chapter 19.) tRNAs also possess three other prominent structural features, referred to as the D loop, the TψC loop, and the variable loop. (The Greek letter psi, ψ, stands for the modified base pseudouridine.) These structures facilitate the specific binding to the appropriate aminoacyl-tRNA synthetase and the appropriate alignment of the aminoacyl-tRNA within the nucleoprotein scaffold of the ribosome. The *D loop* is so named because it contains dihydrouridine. Similarly, the TψC loop contains the base sequence thymine, pseudouridine, and cytosine. tRNAs can be classified on the basis of the length of their *variable loop*. The majority (approximately 80%) of tRNAs have variable loops with four to five nucleotides, whereas the others have variable loops with as many as 20 nucleotides.

Ribosomal RNA

Ribosomal RNA (rRNA) is the most abundant form of RNA in living cells. (In most cells, rRNA constitutes approximately 80% of the total RNA.) The secondary structure of rRNA is extraordinarily complex (Figure 17.28). Although there are species differences in the primary nucleotide sequences of rRNA, the overall three-dimensional structure of this class of molecules is conserved. As its name suggests, rRNA is a component of ribosomes.

As described, ribosomes are cytoplasmic ribonucleoprotein complexes that synthesize proteins. The ribosomes of prokaryotes and eukaryotes are similar in shape and function, although they differ in size and in their chemical composition. Both types of ribosome consist of two subunits of unequal size, which are usually referred to in terms of their S values. [The Svedberg (or sedimentation) unit, S, is a measure of sedimentation velocity in a centrifuge. Because sedimentation velocity depends on the molecular weight and the shape of a particle, S values are not necessarily additive.] Prokaryotic ribosomes (70 S) are composed of a 50 S subunit and a 30 S subunit, whereas ribosomes of eukaryotes (80 S) contain a 60 S subunit and a 40 S subunit.

Several different kinds of rRNA and protein are found in each type of ribosomal subunit. The large ribosomal subunit of *E. coli*, for example, contains 5 S and 23 S rRNAs and 34 polypeptides. The small ribosomal subunit of *E. coli* contains a 16 S rRNA and 21 polypeptides. A typical large eukaryotic ribosomal subunit contains three rRNAs (5 S, 5.8 S, and 28 S) and 49 polypeptides; the small subunit contains an 18 S rRNA and approximately 30 polypeptides. The rRNA serves as a scaffold for the self-assembly of proteins to form the native ribosomal subunit and appears to have enzymatic functions, particularly in prokaryotes. (Naked rRNA can slowly carry out normal ribosomal function.)

Messenger RNA

As its name suggests, **messenger RNA** (mRNA) is the carrier of genetic information from DNA for the synthesis of protein. mRNA molecules, which typically constitute approximately 5% of cellular RNA, vary considerably in size. For example, the number of bases in mRNA from *E. coli* varies from 500 to 6000.

Prokaryotic mRNA and eukaryotic mRNA differ in several respects. First, many prokaryotic mRNAs are *polycistronic*; that is, they contain coding information for several polypeptide chains. In contrast, eukaryotic mRNA typically codes for a single polypeptide and is therefore referred to as *monocistronic*. (A **cistron** is a DNA sequence that contains the coding information for a polypeptide and several signals that are required for ribosome function.) Second, prokaryotic and eukaryotic mRNAs are processed differently. In contrast to prokaryotic mRNAs, which are translated into protein by ribosomes during or immediately after they

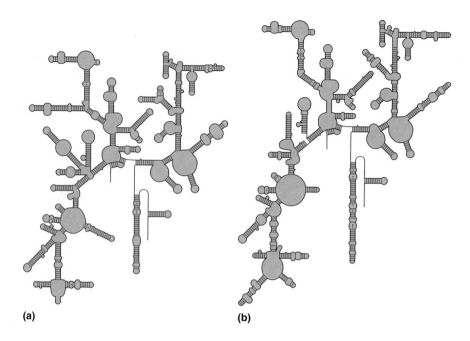

FIGURE 17.28

rRNA Structure

Although their sequences differ, the three-dimensional structure of these rRNAs from (a) *E. coli* and (b) *Saccharomyces cerevisiae* (yeast) appear remarkably similar.

(a)   **(b)**

are synthesized, eukaryotic mRNAs are modified extensively. These modifications include capping (linkage of 7-methylguanosine to the 5'-terminal residue), splicing (removal of introns), and the attachment of an adenylate polymer referred to as a poly (A) tail. (Each of these processes is described in Chapter 18.)

Noncoding RNA

As previously mentioned, the total set of RNA transcripts produced by a cell is called the transcriptome. Analyses of transcriptomes have revealed a surprisingly large number of functional noncoding (nc) RNAs. Current knowledge, although incomplete, suggests that ncRNAs form an extensive and sophisticated genome regulatory network. Among the most important of the ncRNAs are micro RNA, small interfering RNA, small nucleolar RNA, and small nuclear RNA. The **micro RNAs (miRNAs)** and the **small interfering RNAs (siRNAs)** are among the shortest of the ncRNAs, and both are involved in an RNA-degrading process called **RNA interference (RNAi)**. RNAi, once referred to as RNA silencing, probably arose as a defense mechanism. It is used to inactivate or degrade RNA molecules for two purposes: (1) the regulation of gene expression in developmental processes by means of inactivating selected genes and (2) protection of cells from viral RNA genomes.

Small nucleolar RNAs (snoRNAs) are single-stranded RNAs containing 70 to 1000 nucleotides; they facilitate chemical modifications of rRNA within the nucleolus. Encoded within the introns of rRNA genes, snoRNAs are a component of the small nucleolar ribonucleoprotein (snoRNP). The function of snoRNAs (over 100 in humans) is to guide the snoRNP via base pairing to the specific sequence site on a target rRNA. Modifications that occur during rRNA processing include methylation of the 2'OH of ribose and the isomerization of uridine to form pseudouridine.

Uridine   **Pseudouridine**

There are five **small nuclear RNAs** (snRNAs): Ul, U2, U4, U5, and U6. Composed of 100 to 300 nucleotides, the snRNAs combine with several proteins to form small nuclear ribonucleoproteins (snRNPs), often called "snurps." Together with several other proteins, the snRNPs form a molecular machine called the spliceosome because of its function: splicing is a key step in the processing of eukaryotic mRNAs. **Spliceosomes** excise introns from pre-mRNA and then join exons together.

QUESTION 17.12

When a gene is transcribed, only one DNA strand acts as a template for the synthesis of the RNA molecule. This strand is referred to as an **antisense** (or noncoding) strand; the nontranscribed DNA strand is called the **sense** (or coding) strand. The base sequence of the sense strand is the DNA version of the mRNA used to synthesize the polypeptide product of the gene. The antisense RNA (the transcript of the antisense DNA strand) plays a role in transcriptional and translational regulation. An antisense RNA can anneal specifically to a corresponding mRNA and prevent translation.

Because mRNA-antisense RNA binding is so specific, antisense molecules are considered to be promising research tools. Numerous investigators are using antisense RNA molecules to study eukaryotic function by selectively turning on and off the activities of specific genes. This so-called reverse genetics is also useful in medical research. Although serious problems have been encountered in antisense research (e.g., the inefficiency of inserting oligonucleotides into living cells and high manufacturing costs), antisense technology has already provided valuable insight into the mechanisms of cancer and viral infections.

Consider the following sense DNA sequence:

5′-GCATTCGAATTGCAGACTCCTGCAATTCGGCAAT-3′

Determine the sequence of its complementary strand. Then determine the mRNA and antisense RNA sequences. (Recall that in RNA structure, U is substituted for T. So A in a DNA strand is paired with a U as RNA is synthesized.)

17.3 VIRUSES

Viruses lack most of the properties that distinguish life from nonlife. For example, viruses cannot carry on metabolic activities on their own. Yet under the appropriate conditions they can wreak havoc on living organisms. Often described as obligate intracellular parasites, viruses can also be viewed as mobile genetic elements because of their structure, since each consists of a piece of nucleic acid enclosed in a protective coat. Once a virus has infected a host cell, its nucleic acid can hijack the cell's nucleic acid and protein-synthesizing machinery. As viral components accumulate, complete new viral particles are produced and then released from the host cell. In many circumstances, so many new viruses are produced that the host cell lyses (ruptures). Alternatively, the viral nucleic acid may insert itself into a host chromosome, resulting in transformation of the cell, as discussed in the Biochemistry in Perspective box entitled Viral "Lifestyles" (pp. 665–670).

Viruses have fascinated biochemists ever since their existence was suspected near the end of the nineteenth century. Driven in large part by the knowledge that viruses have a role in numerous diseases, viral research has benefited biochemistry enormously. Because a virus subverts normal cell function to produce new virus, a viral infection can provide unique insight into cellular metabolism. For example, the infection of animal cells has provided invaluable and relatively

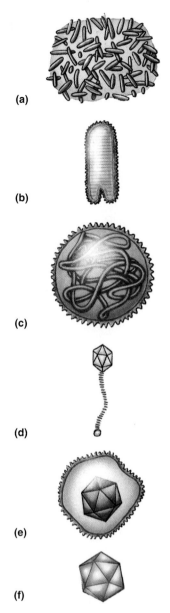

(a)

(b)

(c)

(d)

(e)

(f)

FIGURE 17.29

Representative Viruses

(a) Pox virus. (b) Rhabdovirus. (c) Mumps virus. (d) Flexible-tailed bacteriophage. (e) Herpes virus. (f) Papilloma (wart) virus.

KEY CONCEPTS

• Viruses are composed of nucleic acid enclosed in a protective coat. The nucleic acid may be a single- or double-stranded DNA or RNA.

• In simple viruses the protective coat, called a capsid, is composed of protein.

• In more complex viruses the nucleocapsid, composed of nucleic acid and protein, is surrounded by a membranous envelope derived from host cell membrane.

unambiguous information about the mechanisms that glycosylate newly synthesized proteins. In addition, several eukaryotic genetic mechanisms have been elucidated with the aid of viruses and/or viral enzymes. Viral research has also provided substantial information concerning prokaryotic and eukaryotic genome structure and carcinogenesis (the mechanisms by which normal cells are transformed into cancerous cells). For example, recent research has revealed that infection with one or more types of human papillomavirus is a prerequisite to the development of cervical cancer. Finally, viruses have been invaluable in the development of recombinant DNA technology.

The Structure of Viruses

An enormous number of viruses have been identified since 1892, when the Russian researcher Dmitri Ivanovski first isolated the tobacco mosaic virus. Because the origins and evolutionary history of viruses are unclear, their scientific classification has been difficult. Often, viruses have been assigned to groups according to such properties as their microscopic appearance (e.g., rhabdoviruses have a bullet shape), the anatomic structures from which they were first isolated (e.g., adenoviruses were discovered in the adenoids, a type of lymphoid tissue) or the symptoms they produce in a host organism (e.g., the herpes viruses cause rashes that spread and papillomavirus causes the formation of papillomas, a type of wart). In recent years, scientists have attempted to develop a systematic classification system based primarily on viral structure, although several other factors are also important (e.g., host and disease caused).

Viruses occur in a bewildering array of sizes and shapes. Virions (complete viral particles) range from 10 nm to approximately 400 nm in diameter. Although most viruses are too small to be seen with the light microscope, a few (e.g., the pox viruses) can be visualized because they are as large as the smallest bacteria.

Simple virions are composed of a *capsid* (a protein coat made of interlocking protein molecules called capsomeres), which encloses nucleic acid. (The term *nucleocapsid* is often used to describe the complex formed by the capsid and the nucleic acid.) Most capsids are either helical or icosahedral (20-sided structures composed of triangular capsomeres). The nucleic acid component of virions is either DNA or RNA. Although most viruses possess double-stranded DNA (dsDNA) or single-stranded RNA (ssRNA), examples with ssDNA and dsRNA genomes have also been observed. There are two types of ssRNA genome. A *positive-sense* RNA genome [(+)-ssRNA] acts as a giant mRNA; that is, it directs the synthesis of a long polypeptide that is cleaved and processed into smaller molecules. A *negative-sense* RNA genome [(−)-ssRNA] is complementary in base sequence to the mRNA that directs the synthesis of viral proteins. Viruses that employ (−)-ssRNA genomes must provide an enzyme, referred to as a reverse transcriptase, that synthesizes the mRNA.

In more complex viruses, the nucleocapsid is surrounded by a membrane envelope, which usually arises from the host cell nuclear or plasma membranes. Envelope proteins, coded for by the viral genome, are inserted into the envelope membrane during virion assembly. Proteins that protrude from the surface of the envelope, called spikes, are believed to mediate the attachment of the virus to the host cell. Representative viruses are illustrated in Figure 17.29.

BIOCHEMISTRY IN PERSPECTIVE

Viral "Lifestyles"

How do viruses infect and then disrupt host cells? Despite the diversity in the structures of viruses and the types of host cell that are infected, there are several basic steps in the life cycle of all viruses: infection (penetration of the virion or its nucleic acid into the host cell), replication (expression of the viral genome), maturation (assembly of viral components into virions), and release (the emission of new virions from the host cell). Because viruses usually possess only enough genetic information to specify the synthesis of their own components, each type must exploit some of the normal metabolic reactions of its host cell to complete the life cycle. For this reason there are numerous variations on these basic steps. This point can be illustrated by comparing the life cycles of two well-researched viruses: the T4 bacteriophage and the human immunodeficiency virus (HIV).

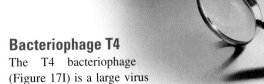

Bacteriophage T4

The T4 bacteriophage (Figure 17I) is a large virus with an icosahedral head and a long, complex tail similar in structure to T2 (p. 638). The head contains dsDNA, and the tail attaches to the host cell and injects the viral DNA into the host cell.

The life cycle of T4 (Figure 17J) begins with adsorbing the virion to the surface of an *E. coli* cell. The entire virion cannot penetrate into the cell's interior because the bacterial cell wall is rigid. Instead, the DNA is injected by flexing and constricting the tail apparatus. Once the viral DNA has entered the cell, the infective process is complete, and the next phase (replication) begins.

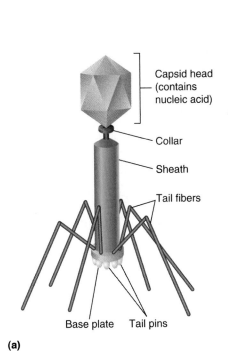

(a)

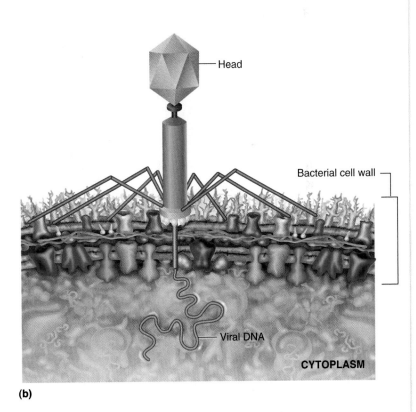

(b)

FIGURE 17I

The T4 Baceteriophage

(a) The structure of an intact T4 bacteriophage. (b) Penetration of the cell wall and injection of viral DNA (vDNA) into the host cell by the bacteriophage. The vDNA directs the host cell to synthesize about 30 proteins that facilitate new virion synthesis.

BIOCHEMISTRY **IN PERSPECTIVE** cont

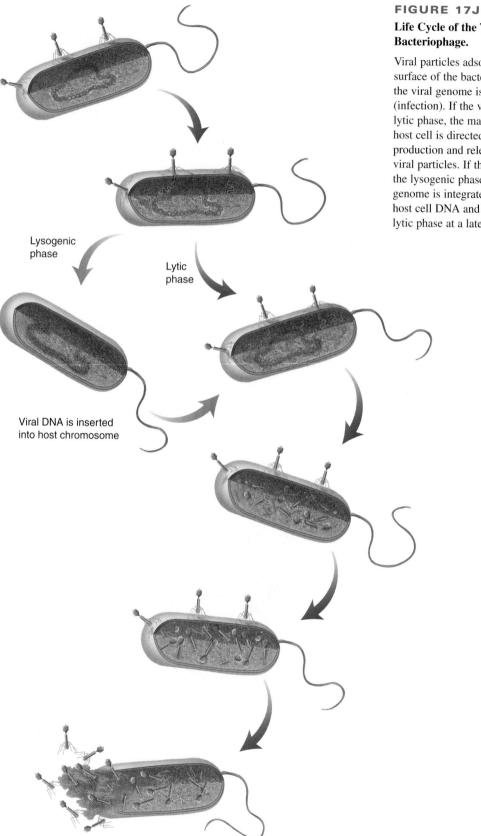

Lysogenic
phase

Lytic
phase

Viral DNA is inserted
into host chromosome

FIGURE 17J

**Life Cycle of the T4
Bacteriophage.**

Viral particles adsorb onto the
surface of the bacterial cell and
the viral genome is injected
(infection). If the virus enters the
lytic phase, the machinery of the
host cell is directed toward the
production and release of new
viral particles. If the virus enters
the lysogenic phase, the viral
genome is integrated into the
host cell DNA and may enter the
lytic phase at a later time.

▶▶

BIOCHEMISTRY IN PERSPECTIVE cont

Within 2 minutes after the injection of T4 phage DNA into an *E. coli* cell, synthesis of host DNA, RNA, and protein stops and phage mRNA synthesis begins. Phage mRNA codes for the synthesis of capsid proteins and some of the enzymes required for the replication of the viral genome and the assembly of virion components. In addition, other enzymes are synthesized that weaken the cell wall of the host, allowing release of the new phage for new rounds of infection. Approximately 22 minutes after the injection of viral DNA (vDNA), the host cell, now filled with several hundred new virions, lyses. Upon release, the virions attach to nearby bacteria, thus initiating new infections.

Bacteriophage that initiate this so-called **lytic cycle** are referred to as *virulent* because they destroy their host cells. Many phage, however, do not initially kill their hosts. So-called *temperate* or *lysogenic* phage integrate their genome into that of the host cell. (The term **lysogeny** describes a condition in which the phage genome is integrated into a host chromosome.) The integrated viral genome, called the **prophage**, is copied along with host DNA during cell division for an indefinite time. Occasionally, lysogenic phage enter a *lytic* phase. Certain external conditions, such as UV or ionizing radiation, activate the prophage, which directs the synthesis of new virions. Sometimes, a lysing bacterial cell releases a few virions that contain some bacterial DNA along with the phage DNA. When such a virion infects a new host cell, this DNA is introduced into the host genome. This process is referred to as **transduction**.

HIV

The human immunodeficiency virus (HIV) is the causative agent of acquired immune deficiency syndrome (AIDS). Left untreated, AIDS is a lethal condition because HIV destroys the body's immune system, rendering it defenseless against disease-causing organisms (e.g., bacteria, protozoa, and fungi, as well as other viruses) in addition to some forms of cancer.

HIV (Figure 17K) belongs to a unique group of RNA viruses called the retroviruses. **Retroviruses** are so named because they contain an enzymatic activity reverse transcriptase, which catalyzes the synthesis of a DNA copy of an ssRNA genome. A typical retrovirus consists of an RNA genome enclosed in a protein capsid. Wrapped around the capsid is a membranous envelope that is formed from a host cell lipid bilayer.

In the reproductive cycle of the retrovirus HIV (Figure 17L), the infective process begins when the virus binds to a host cell. Binding, which occurs between viral surface glycoproteins and specific plasma membrane receptors, initiates a fusion process between host cell membrane and viral membrane. Subsequently, the viral capsid is released into the cytoplasm and the viral reverse transcriptase catalyzes the synthesis of a DNA strand that is complementary to the viral RNA. This enzymatic activity also catalyzes the conversion of the single-stranded DNA into a double-stranded molecule. The double-stranded DNA version of the viral genome is then translocated into the nucleus, where

it integrates into a host chromosome. The integrated proviral genome, acting like a prophage, is replicated each time the cell undergoes DNA synthesis. The mRNA transcripts produced when the viral genome is transcribed direct the synthesis of numerous copies of viral proteins. New virus, created as copies of the viral RNA genome are packaged with viral proteins, is released from the host cell by a "budding" process.

HIV contains a cylindrical core within its capsid. In addition, to two copies of its (+)-ssRNA genome, the core contains several enzymes: reverse transcriptase, ribonuclease, integrase, and protease. The RNA molecules are coated with multiple copies of two low-molecular-weight proteins, p7 and p9. (The numbers in these and other protein names indicate their kilodalton mass; for example, p7 is a protein with a mass of 7 kD.) The bullet-shaped core itself is composed of hundreds of copies of p6, and copies of p17 form an inner lining of the viral envelope. The envelope of HIV contains two major viral proteins, gpl20 and gp41, in addition to host proteins.

HIV infection occurs because of direct exposure of an individual's bloodstream to the body fluids of an infected person. Most HIV is transmitted through sexual contact, blood transfusions, and perinatal transmission from mother to child. Once HIV has entered the body, it is believed to infect cells that bear the CD4 antigen on their plasma membranes. The principal group of cells that are attacked by HIV are the T-4 helper lymphocytes of the immune system. T-4 helper cells play a critical role in regulating the activities of other immune system cells. It is now known that T-cell infection requires the interaction of the gpl20-CD4 complex with a chemokine receptor, which acts as a co-receptor. (The immune system chemotactic agents called chemokines stimulate T cells by binding to receptors on the T-cell plasma membrane.) In the early stages of infection, the co-receptor CCR5 (or less often CXCR4) helps HIV enter T cells; other cells that are known to be infected by HIV include some intestinal and nervous system cells. Recent evidence suggests that humans with two copies of a gene for a variant of the CCR5 receptor, CCR5-Δ32, are resistant to HIV infection. CCR5-Δ32 apparently provides protection against plague and smallpox as well, but for all three diseases, the portion of the population that benefits is relatively small.

Once the gp120 envelope protein of HIV has bound to the CD4 antigen and chemokine receptor on the T cell, the viral envelope fuses with the host cell's plasma membrane. The two RNA strands are released into the cytoplasm. Reverse transcriptase, a heterodimer with several enzymatic activities, then catalyzes the synthesis of an ssDNA using the vRNA as a template. The heterodimer's RNase activity then degrades the vRNA. The same protein produces a double-stranded vDNA by forming a complementary strand of the ssDNA. Viral integrase integrates the vDNA into a host cell chromosome. The proviral DNA remains latent until the specific infected T cell is activated by an immune response. The proviral DNA can then direct the cell to synthesize viral components. Newly synthesized viruses bud from the infected cell.

▶▶

BIOCHEMISTRY IN PERSPECTIVE   cont

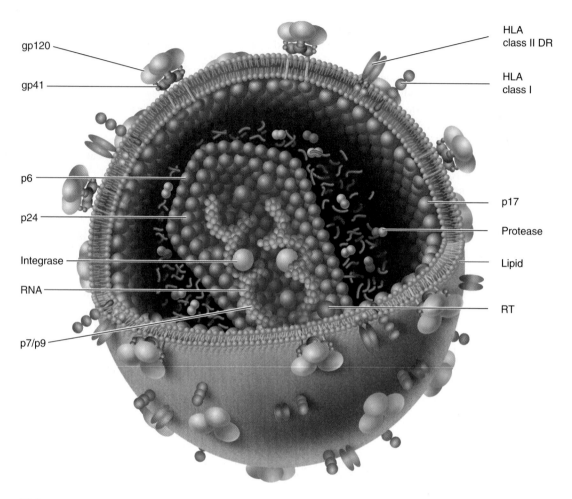

FIGURE 17K

HIV Structure

The surface of the virus is a lipid bilayer in which are embedded the viral glycoproteins gp 120 and gp41, as well as HLA (human leukocyte antigens) membrane proteins taken from host cells. (HLA proteins are signals that protect the viral particle from the immune system, which ordinarily searches out and destroys foreign invaders.) Lining the inside of the envelope are hundreds of copies of the structural protein p17. Two copies of the RNA genome are contained in a bullet-shape capsid composed of p6 and p24. Bound to the RNA are p7 and p9. Enzymes associated with viral genome are reverse transcriptase (RT), integrase, and protease.

Recent research efforts using microarray DNA chips (see the Biochemistry in the Lab box entitled Genomics in Chapter 18, pp. 700–707) have provided some of the details of the molecular mechanisms by which HIV disrupts the functioning of T-4 lymphocytes. By monitoring the mRNA expression of over 6000 genes simultaneously, researchers tracked the consequences of HIV infection. They observed that within 30 minutes of infection, the expression of 500 cellular genes had been suppressed and 200 had been activated. Within hours, host cell mRNA had largely been replaced by viral mRNA. The virus had crippled the cell's capacity to generate energy and repair virally inflicted DNA damage.

Cell death is triggered by several mechanisms that include the following:

1. Viral activation of the genes that induce apoptosis (programmed cell death), a normal cell mechanism by which cells respond to external signals such as those that occur in developmental processes.

2. The simultaneous budding of numerous viral particles from the cell membrane may tear the membrane and cause massive leakages that cannot be repaired.

BIOCHEMISTRY IN PERSPECTIVE cont

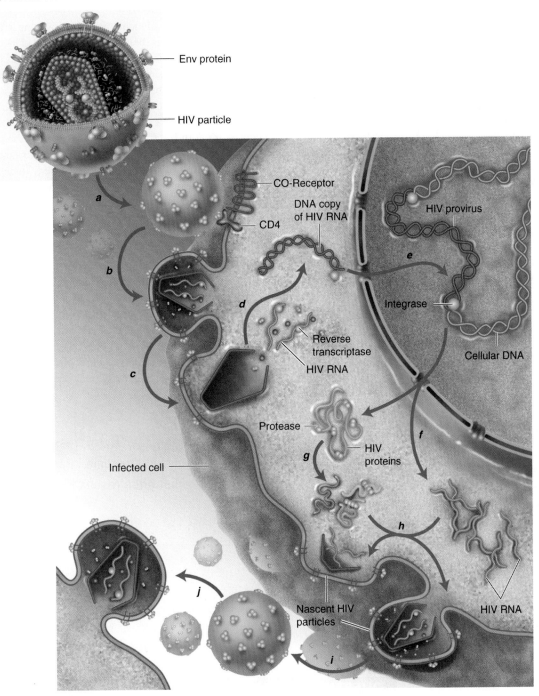

FIGURE 17L

Reproductive Cycle of HIV, a Retrovirus

After the viral particle binds to surface receptors on the host cell (a), its envelope fuses with the cell's plasma membrane (b), thus releasing the capsid and its contents (vRNA and several viral enzymes) into the cytoplasm (c). The viral enzyme reverse transcriptase catalyzes the synthesis of a DNA strand complementary to the vRNA (d), and then proceeds to form a second DNA strand that is complementary to the first. Subsequently the double-stranded viral DNA (vDNA) transfers to the nucleus, where it integrates itself into a host chromosome with the aid of a viral integrase (e). The provirus (the integrated viral genome) is replicated every time the cell synthesizes new DNA. Transcription of viral DNA results in the formation of two types of RNA transcript: RNA molecules that function as the viral genome (f) and molecules that code for the synthesis of viral protein (e.g., reverse transcriptase, capsid proteins, envelope proteins, and viral integrase) (g). The protein molecules are combined with the vRNA genome during the creation of new virus (h) that buds from the surface of the host cell (i) and then proceeds to infect other cells (j).

▶▶

BIOCHEMISTRY IN PERSPECTIVE cont

3. Massive release of new virus from a cell, directed by the provirus, may so deplete the cell that it disintegrates.

4. The binding of cell surface gpl20 molecules to CD4 receptors on nearby healthy cells leads to the formation of large, nonfunctional multinucleated cell masses called *syncytia*.

HIV infection progresses through several stages, which may vary considerably in length among individuals. Initial symptoms, which usually occur soon after the initial exposure to the virus and last for several weeks, include fever, lethargy, headache and other neurological complaints, diarrhea, and lymph node enlargement. (Antibodies to HIV are detectable during this period.) Exaggerated versions of these symptoms, referred to as the AIDS-related complex (ARC), may often recur. Eventually, the immune system becomes so compromised that the individual becomes susceptible to serious opportunistic diseases and is said to have developed AIDS. The time required for the development of AIDS may vary from 2 years to 8 or 10 years. For reasons that are not understood, a few patients do not develop AIDS even after 15 years of HIV infection. (It has recently been suggested that some of these individuals are infected with attenuated HIV variants.) Some of the most common AIDS-related diseases include *Pneumocystis carinii* pneumonia, cryptococcal

meningitis (inflammation of membranes that cover the brain and spinal cord), toxoplasmosis (brain lesions, heart and kidney damage, and fetal abnormalities), cytomegalovirus infections (pneumonia, kidney and liver damage, and blindness), and tuberculosis. HIV infection is also associated with several types of cancer, the most common of which is a rare skin cancer called Kaposi's sarcoma.

There is no cure for AIDS. Treatment seeks to suppress symptoms (e.g., antibiotics for the infections) and to slow viral reproduction. Mortality rates have decreased since 1995 because of the introduction of a treatment protocol called highly active antiretroviral therapy (HAART) that consists of combinations of drugs from the following categories: (1) nucleoside reverse transcriptase inhibitors (NRTIs: e.g., azidothymidine, also called zidovudine or AZT), (2) non-nucleoside reverse transcriptase inhibitors (NNRTIs: e.g., efavirenz) and protease inhibitors (e.g., indinavir). Both NRTIs and NNRTIs inhibit vDNA synthesis catalyzed by reverse transcriptase. Protease inhibitors are a class of drugs that prevent the processing of viral protein that is required for the assembly of new virions.

The development of an AIDS vaccine is problematic because the viral genome mutates frequently (i.e., its surface antigens become altered).

SUMMARY: Viral infection disrupts cell function. By supressing some cellular genes and activating others, the viral genome directs the host cell to produce new virus in a process that often results in cell death.

QUESTION 17.13

Recall that according to the central dogma, the flow of genetic information is from DNA to RNA and then to protein. Retroviruses are an exception to this rule. The alterations of the central dogma that are observed in retroviruses and other RNA viruses can be illustrated as follows:

Compare this illustration with that of the original central dogma (p. 629). Describe in your own words the implications of each component of these figures.

Chapter**Summary**

1. The information required for directing all living processes is stored in the nucleotide sequences of DNA. DNA is composed of two antiparallel polynucleotide strands that wind around each other to form a right-handed double helix. The deoxyribose-phosphodiester bonds form the backbones of the double helix, and the nucleotide bases project to its interior. The nucleotide base pairs form because of hydrogen bonding between the following bases: adenine with thymine, and cytosine with guanine.

2. Several types of noncovalent interaction contribute to the stability of DNA's structure: hydrophobic and van der Waal's interactions between stacked heterocyclic bases, hydrogen bonds between GC and AT base pairs, and hydration with water molecules. Decoding of the genetic information in DNA requires molecular machinery largely composed of proteins.

3. Mutations are changes in DNA structure, which may be caused by collisions with solvent molecules, thermal fluctuations, ROS, radiation, or xenobiotics. In a transition mutation, one pyrimidine base is substituted for another pyrimidine base or a purine base is substituted for another purine. In a transversion mutation, a pyrimidine base is substituted for a purine base, or vice versa.

4. DNA can have several conformations depending on the nucleotide sequence and the degree of hydration of the double helix. In addition to the classical structure determined by Watson and Crick (B-DNA), A-DNA and Z-DNA have also been observed. DNA supercoiling is now known to be a critical feature of several biological processes, such as DNA packaging, replication, and transcription.

5. Each eukaryotic chromosome is composed of nucleohistone, a complex formed by winding a DNA molecule around a histone octamer to form a nucleosome. Several types of histone covalent modification (e.g., acetylation and methylation) alter the structure and function of the histone components of nucleosomes. The DNA of mitochondria and chloroplasts is similar to the chromosomes found in prokaryotes.

6. A genome is the full set of the inherited instructions required to sustain an organism's living processes. Although there are some similarities, the genomes of prokaryotes and eukaryotes are substantially different in size, coding capacity, gene expression mechanisms, and coding continuity.

7. The majority of the DNA sequences in humans do not code for proteins or functional RNAs. There are two general classes of intergenic sequences: tandem repeats and interspersed genome-wide repeats. Mobile genetic elements can be duplicated and then move within the genome. Retrotransposons can cause disease by inserting into genes or regulatory sequences. Covalent modifications of DNA and histones can result in changes in gene expression.

8. RNA differs from DNA in that it contains ribose (instead of deoxyribose), has a somewhat different base composition, and is usually single-stranded. The forms of RNA involved in protein synthesis are transfer, ribosomal, and messenger RNAs. Transfer RNA molecules have specific amino acids attached to them by specific enzymes and transport these to the ribosome for incorporation into newly synthesized protein, where they are properly aligned during protein synthesis. The ribosomal RNAs are components of ribosomes, where they are the sites of catalytic activity. Messenger RNA contains within its nucleotide sequence the coding instructions for synthesizing a specific polypeptide. There are several classes of noncoding RNAs that have diverse roles in genome regulation and protection. Examples include miRNAs, siRNAs, snoRNAs, and snRNAs.

9. Viruses are obligate intracellular parasites. Although they are acellular and cannot carry out metabolic activities on their own, viruses can wreak havoc on living organisms. Each type of virus infects a specific type of host (or small set of hosts). This is possible because a virus can either inject its genome into the host cell or gain entrance for the entire viral particle. Each virus has the capacity to use the host cell's metabolic processes to manufacture new copies of itself, called virions. Viruses possess dsDNA, ssDNA, dsRNA, or ssRNA genomes.

10. The retrovirus HIV causes AIDS. Retroviruses are a class of RNA viruses that possess a reverse transcriptase activity that converts their RNA genome to a DNA molecule. This vDNA is then inserted into the host cell genome, causing a permanent infection. Eventually, HIV infection destroys the immune system of infected individuals.

 Take your learning further by visiting the **companion website** for Biochemistry at **www.oup.com/us/mckee** where you can complete a multiple-choice quiz on nucleic acids to help you prepare for exams.

Suggested**Readings**

Beck, S., and Olek, A. (eds.), *The Epigenome: Molecular Hide and Seek*, Wiley-VCH, Weinheim, Germany, 2003.

Biel, M., Wascholowski, V., and Giannis, A., Epigenetics—An Epicenter of Gene Regulation: Histones and Histone-Modifying Enzymes, *Angew. Chem. Int. Ed.* 44:3186–3216, 2005.

Brown, T. A., *Genomes 3*, Garland, New York, 2007.

Carmo-Fonseca, M., Platani, M., and Swedlow, J. R., Macromolecular Mobility Inside the Cell Nucleus, *Trends Cell Biol.* 12(11):491–495, 2002.

Cooke, M. S., and Evans, D. D., Reactive Oxygen Species: From DNA Damage to Disease, *Sci. Med.* 10(2):98–111, 2005.

Costa, F. F., Noncoding RNAs: Lost in Translation? *Gene* 382:1–10, 2007.

Costa, F. F., Noncoding RNAs: New Players in Eukaryotic Biology, *Gene* 357:83–94, 2005.

Cozzarelli, N. R., Cost, G. J., Nollman, M., Viard, T., and Stray, J. E., Giant Proteins That Move DNA: Bullies of the Genomic Playground, *Nat. Rev. Mol. Cell Biol.* 7(8):580–588, 2006.

Drucker, R., and Whitelaw, E., Retrotransposon-Derived Elements in the Mammalian Genome: A Potential Source of Disease, *J. Inherit. Metab. Dis.* 27:319–330, 2004.

Jablonka, E., and Lamb, M. J., *Evolution in Four Dimensions: Genetic, Epigenetic, Behavioral and Symbolic Variation in the History of Life*, MIT Press, Cambridge, Massachusetts, 2005.

Jirtle, R. L., and Weidman, J. R., Imprinted and More Equal, *Am. Sci.* 95(2):143–149, 2007.

Julian, M. M., Women in Crystallography, in Kass-Simon, G., and Fames, P. (eds.), *Women of Science: Righting the Record*, pp. 359–364, Indiana University Press, Bloomington and Indianapolis, 1990.

Kolata, G., *Flu: The Story of the Great Influenza Pandemic of 1918 and the Search for the Virus That Caused It*, Farrar, Straus, & Giroux, New York, 1999.

Lane, N., *Oxygen: The Molecule That Made the World*. Oxford University Press, Oxford, 2002.

Lucifero, D., Chaillet, J. R., and Trasler, J. M., Potential Significance of Genomic Imprinting Defects for Reproduction and Assisted Reproductive Technology, *Hum. Reprod. Update* l0(l):3–18, 2004.

Pauler, F. M., and Barlow, D. P., Imprinting Mechanisms—It Only Takes Two, *Genes Dev.* 20(10):1203–1206, 2006.

Sandman, K., and Reeve, J. N., Structure and Functional Relationships of Archaeal and Eukaryal Histones and Nucleosomes, *Arch. Microbiol.* 173:165–169, 2000.

Strachan, T., and Read, A. P., *Human Molecular Genetics 3*, Garland, New York, 2004.

Watson, J. D., Baker, T. A., Bell, S. P., Gann, A., Levine, M., and Losick, R., *Molecular Biology of the Gene*, Pearson Benjamin Cummings, San Francisco, 2004.

Key**Words**

A-DNA, *640*

alkylation, *634*

antisense strand, *663*

apoptosis, *668*

B-DNA, *640*

base analogues, *634*

centromere, *651*

chain-terminating method, *657*

Chargaff's rules, *639*

chromatin, *645*

chromosome, *642*

cistron, *661*

denaturation, *654*

DNA fingerprinting, *658*

DNA profile, *658*

DNA typing, *658*

epigenetic, *645*

exon, *650*

genes, *627*

genetics, *626*

genome, *627*

hybridization, *656*

hydroxyapatite, *653*

hypochromic effect, *654*

intercalating agent, *635*

intergenic sequence, *650*

interspersed genome-wide repeat, *652*

intron, *650*

LINE, *652*

lysogeny, *667*

lytic cycle, *667*

messenger RNA, *661*

metabolome, *629*

micro RNA, *662*

microribonucleoprotein, *000*

microsatellite, *652*

minisatellite, *652*

mobile genetic element, *652*

molecular biology, *626*

nonalkylating agent, *635*

noncoding RNA, *662*

nucleohistone, *645*

nucleosome, *645*

operon, *650*

point mutation, *633*

prophage, *667*

proteome, *629*

replication, *627*

restriction fragment length polymorphism, *658*

retrotransposons, *652*

retrovirus, *667*

ribosomal RNA, *661*

ribozyme, *659*

RNA-induced silencing complex, *662*

RNA transposons, *652*

satellite DNA, *651*

sense strand, *663*

short interfering RNA, *662*

short tandem repeats, *658*

SINE, *652*

small nuclear ribonucleoprotein particle, *663*

small nuclear RNA, *663*

small nucleolar RNA, *662*

Southern blotting, *656*

spliceosome, *663*

tandem repeats, *651*

telomere, *651*

transcript, *627*

transcription, *627*

transcription factor, *629*

transcriptome, *627*

transduction, *667*

transfer RNA, *660*

transition mutation, *633*

translation, *629*

transposable DNA elements, *652*

transposition, *652*

transposon, *652*

transversion mutation, *634*

Z-DNA, *640*

Review**Questions**

These questions are designed to test your knowledge of the key concepts discussed in this chapter, before moving on to the next chapter. You may like to compare your answers to the solutions provided in the back of the book and in the accompanying Study Guide.

1. Define the following terms:
   a. genetics
   b. replication
   c. transcription
   d. sugar-phosphate backbone
   e. bacteriophage

2. There are several structural forms of DNA. List and describe each form.

3. Describe the higher-order structure of DNA referred to as supercoiling.

4. Define the following terms:
   a. Chargaff's rules
   b. hypochromic effect
   c. transposon
   d. *Alu* family
   e. transcriptome

5. List three biological properties facilitated by supercoiling.

6. List three differences between eukaryotic and prokaryotic DNA.

7. Describe the structure of a nucleosome.

8. Define the following terms:
   a. satellite DNA
   b. splicing
   c. chain-terminating method
   d. DNA denaturation
   e. retrotransposon

9. Describe the structural differences between RNA and DNA.

10. What are the three most prominent forms of RNA? What roles do they play in cell function?

11. Z-DNA derives its name from the zig-zag conformation of phosphate groups. What features of the DNA molecule allow this distinctive structure to form?

12. There is one base pair for every 0.33 nm of DNA and the total contour length of all the DNA in a single human cell is 2 m. Calculate the number of base pairs in a single cell. Assuming that there are 10^{14} cells in the human body, calculate the total length of DNA. How does this estimate compare to the distance from the earth to the sun (1.5×10^8 km)?

13. Define the following terms:
   a. telomere
   b. minisatellite
   c. DNA profile
   d. intron
   e. exon

14. DNA sample contains 21% adenine. What is its complete percentage base composition?

15. The melting temperature of a DNA molecule increases as the GC content increases. Explain.

16. Define the following terms:
   a. transversion
   b. epigenetic
   c. intercalating agent
   d. LINE
   e. SINE

17. Organize the following into a hierarchy:
   a. chromosome
   b. gene
   c. nucleosome
   d. nucleotide base pair

18. Provide the complementary strand and the RNA transcription product for the following DNA segment:
   5'-AGGGGCCGTTATCGTT-3'

19. Define the following terms:
   a. transcriptome
   b. noncoding RNA
   c. small nucleolar RNA
   d microsatellite
   e. small interfering RNA

20. What are the polyamines, and what role do they play in DNA structure?

21. Define the following terms:
   a. sense strand
   b. hypochromic effect
   c. transposition
   d. transcript
   e. mobile genetic element

22. Describe the structural features that stabilize RNA molecules.

23. Describe how DNA methylation patterns are retained from one cell generation to the next. Refer to the online Biochemistry in Perspective box on epigenetics and the epigenome.

24. Which error would cause more damage to a cell: DNA replication or a DNA transcription error?

25. When an aromatic hydrocarbon intercalates between two stacked base pairs, what effect may there be on DNA structure?

26. What effects can intercalated hydrocarbon molecules have on DNA replication?

27. AZT is used in the treatment of AIDS. Examine its structure and suggest its means of action.

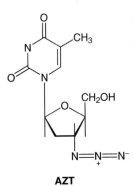

AZT

28. Chargaff's rules apply to DNA, but not RNA. Explain.

29. How and why are tRNA bases modified?

30. Describe how water affects DNA structure.

Thought Questions

These questions are designed to reinforce your understanding of all of the key concepts discussed in the book so far, including this chapter and all of the chapters before it. They may not have one right answer! The authors have provided possible solutions to these questions in the back of the book and in the accompanying Study Guide, for your reference.

31. Under physiological conditions, DNA ordinarily forms B-DNA. However, RNA hairpins and DNA-RNA hybrids adopt the structure of A-DNA. Considering the structural differences between DNA and RNA, explain this phenomenon.

32. What structural features of DNA cause the major groove and the minor groove to form?

33. Explain in general terms how polyamines aid in achieving the highly compressed structure of DNA.

34. In contrast to the double helix of DNA, RNA exists as a single strand. What effects does this have on the structure of RNA?

35. Jerome Vinograd found that when circular DNA from a polyoma virus is centrifuged, it separates into two distinct bands, one consisting of supercoiled DNA and the other of relaxed DNA. Explain how you would identify each band.

36. 5-Bromouracil is an analogue of thymine that usually pairs with adenine. However, 5-bromouracil frequently pairs with guanine. Explain.

37. The flow of genetic information is from DNA to RNA to protein. In certain viruses, the flow of information is from RNA to DNA. Does it appear possible for that information flow to begin with proteins? Explain.

38. HIV is a retrovirus. Why has the development of a vaccine to prevent HIV infection proven to be such a difficult undertaking?

39. You wish to isolate mitochondrial DNA without contamination with nuclear DNA. Describe how you would accomplish this task.

40. Unlike linker DNA and deproteinized DNA, DNA segments wrapped around histone cores are relatively resistant to the hydrolytic actions of nucleases. Explain.

41. The set of mRNAs present within a cell changes over time. Explain.

42. DNA and RNA are information-rich molecules. Explain the significance and implications of this statement.

43. A 10-year-old murder case in a small town has finally been solved, thanks to the efforts of a detective who took advantage of a new statewide DNA database containing samples of DNA from convicted felons. Explain how this case might have been solved, starting with the arrival of a crime scene unit at the murder scene. What technological advances made this case solvable?

44. Some living organisms are under considerable pressure to streamline their genomes for the sake of more efficient operation. As a result, the mitochondria of eukaryotic species have lost, to one degree or another, the overwhelming majority of their genes. During this process, several hundred mitochondrial genes were transferred to the nuclear genome. Yet mitochondria still retain a genome with the capacity to produce several electron transport proteins. Review mitochondrial electron transport and suggest a reason why these energy-generating organelles retained the genes to produce this set of molecules.

45. During the infective process, viruses may incorporate segments of the host genome into their structures. Explain how this phenomenon can contribute to the formation of new diseases or exacerbate existing pathologies.

46. It has been estimated that each phosphate group in B-DNA can form hydrogen bonds with six water molecules. Draw a diagram of a DNA phosphodiester linkage with its associated water molecules.

47. Alkylating agents are compounds that react with hydroxyl and amino groups to form alkylated molecules. For example:

Cytosine   ***N*-Methylaminocytosine**

When an individual is accidentally exposed to these substances, a variety of physiological systems are affected via genome alterations. Cancer (uncontrolled cell proliferation caused by genome damage) is not an uncommon result, in part, of a series of such exposures. Using your knowledge of the genome, suggest in general terms how cancer might occur.

48. Identical twin brothers begin life with identical genomes and epigenomes. How will this circumstance change with age? Suggest how these changes could be used as a forensic tool. [Hint: Refer to this chapters online Biochemistry in Perspective box on epigenetics and the epigenome at www....]

49. The suggestion has been made that DNA extracted from ancient fossils could be extracted, cloned, and used to resurrect an extinct species. Comment on the practicality of this idea.

50. Fluorouracil is a structural analogue of thymine. The fluorine promotes enolization. How is this effect used in the treatment of cancer?

CHAPTER 18

Genetic Information

OUTLINE

18.1 GENETIC INFORMATION: REPLICATION, REPAIR, AND RECOMBINATION
DNA Replication
DNA Repair
DNA Recombination

18.2 TRANSCRIPTION
Transcription in Prokaryotes
Transcription in Eukaryotes

18.3 GENE EXPRESSION
Gene Expression in Prokaryotes
Gene Expression in Eukaryotes

BIOCHEMISTRY IN THE LAB
Genomics

BIOCHEMISTRY IN PERSPECTIVE
Carcinogenesis

The Human Genome Project
The efforts of biochemists to uncover the mechanisms of heredity, begun over a century ago, have resulted in the sequencing of the human genome. Technology developed by biochemists and molecular biologists, such as electrophoresis, automated and computerized DNA sequencing machines, and bioinformatics, have made this extraordinary achievement possible. The Human Genome Project, performed by thousands of scientists in public and private laboratories around the world, took over 15 years to accomplish. Researchers are just beginning to interpret and utilize the resulting tidal wave of biological information to solve the medical and biological problems of humans.

675

Overview

WITHIN RECENT YEARS, THE INVESTIGATION OF GENETIC INHERITANCE THAT BEGAN WITH MENDEL HAS REACHED AN EXPLOSIVE STAGE. Knowledge concerning how living cells store, use, and inherit genetic information constantly increases. Biochemists and geneticists are now painstakingly revealing the principles of nucleic acid function. The practical consequences of work done in the 1980s, the 1990s, and the early years of the twenty-first century have been considerable. In addition to enabling numerous medical applications (e.g., diagnostic tests and therapies), recombinant DNA technology has allowed investigators access to the inner workings of living organisms in ways never before possible. Not only are long-standing questions about living processes more accessible to investigation, but amazingly rapid progress is being made in understanding the molecular basis of a wide variety of diseases such as cancer and heart disease.

Any successful information-based system involves conservation and transfer. The instructions required to produce a certain type of organization (e.g., the directions for building a house or for reproducing a living organism) must be stably stored to safeguard their accuracy and availability for use by the system. To be useful, however, information must also be transferred, that is, converted into a form that can be utilized. Living organisms have partitioned these functions as follows. DNA is a relatively stable molecule with structural features that maximize information storage and facilitate duplication. RNA molecules, more reactive than DNA, have numerous roles in protein synthesis and gene expression regulation. Finally, the proteins have diverse and flexible three-dimensional structures that perform most of the tasks that sustain the living state.

Although DNA contains the genetic information that drives living processes, it does not directly control cellular processes. The decoding of DNA base sequences requires molecular machinery, largely composed of proteins and powered by cellular energy resources. These machines bend, twist, unwind, and unzip DNA during replication and transcription. At first glance, the seemingly repetitious and regular structure of DNA makes it an unlikely partner for the productive binding of specific base sequences with appropriate proteins. However, considerable research efforts have revealed that the edges of base pairs within the major groove (and to a lesser extent the minor groove) of the double helix can participate in sequence-specific binding to proteins. Numerous contacts (often about 20 or so) involving hydrophobic interactions, hydrogen bonds, and ionic bonds between amino acid residues and nucleotide bases result in highly specific DNA-protein binding. Several examples of interactions between nucleotide bases and amino acid side chains are illustrated in Figure 18.1. The three-dimensional structures of most DNA-binding proteins analyzed thus far have surprisingly similar features. In addition to usually possessing a twofold axis of symmetry, many of these molecules can be separated into families (Figure 18.2) on the basis of the following structures: (1) helix-turn-helix, (2) helix-loop-helix, (3) leucine zipper, and (4) zinc finger. DNA-binding proteins, many of which are transcription factors, often form dimers. For example, a variety of transcription factors with leucine zipper motifs form dimers as their leucine-containing α-helices associate via van der Waals interactions.

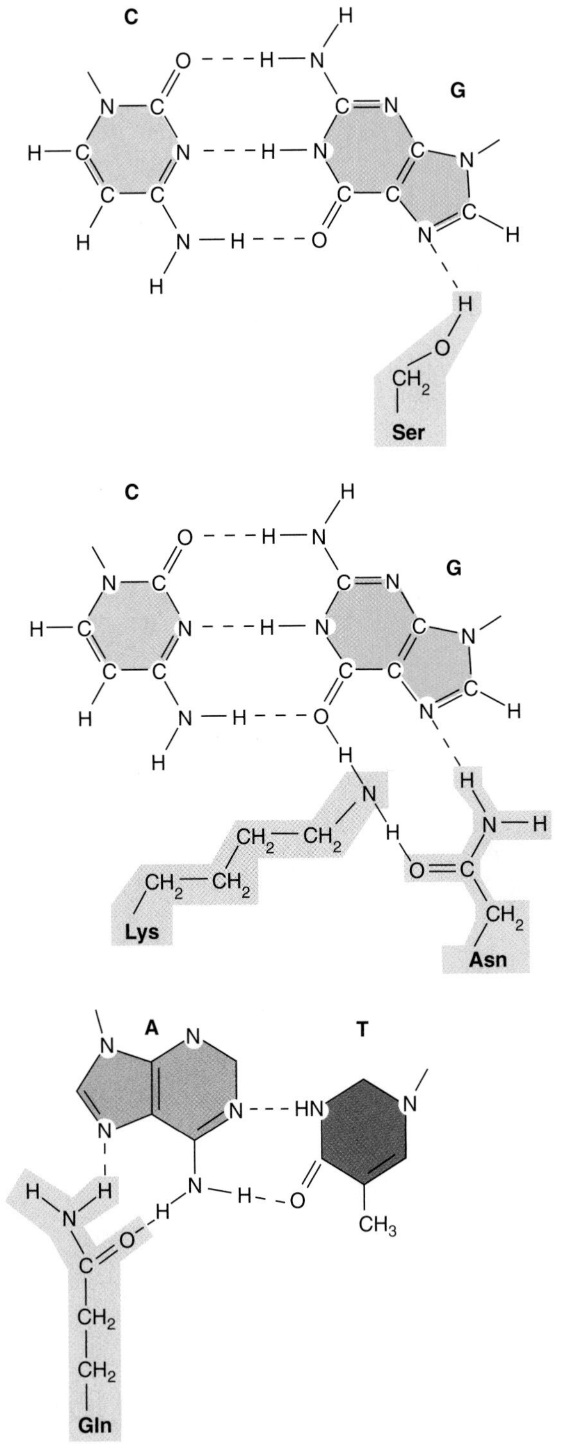

FIGURE 18.1
**Examples of Specific Amino
Acid–Nucleotide Base Interactions
During Protein-DNA Binding**

Chapter 18 provides an overview of the mechanisms that living organisms use to synthesize the nucleic acids DNA and RNA that direct cellular processes. The chapter begins with a discussion of several aspects of DNA replication, repair, and **recombination** (the reassortment of DNA sequences). This is followed by descriptions of the synthesis and processing of RNA. Also included is an overview of several of the tools of biotechnology that biochemists use to investigate living processes. Chapter 18 ends with a section devoted to gene expression, the mechanisms cells use to produce gene products in an orderly and timely manner.

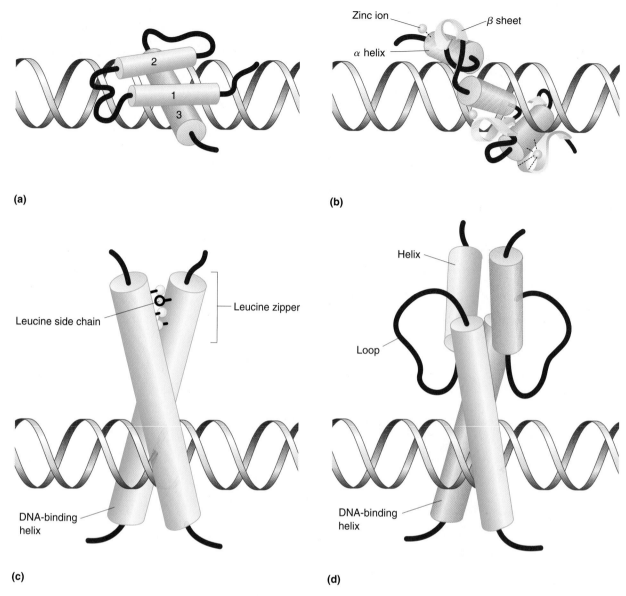

(a)

(b)

(c)

(d)

FIGURE 18.2

DNA-Protein Interactions

DNA-binding proteins contain specific structural motifs for interacting with DNA: (a) helix-turn-helix, (b) zinc fingers, (c) leucine zipper, and (d) helix-loop-helix.

18.1 GENETIC INFORMATION: REPLICATION, REPAIR, AND RECOMBINATION

Because of the strategic importance of DNA, all living organisms must possess the following features: rapid and accurate DNA synthesis and genetic stability maintained by effective DNA repair mechanisms. Paradoxically, the long-term survival of species also depends on genetic variations that allow adaptation to changing environments. In most species these variations arise predominantly from genetic recombination, although mutation also plays a role. In the following sections, the mechanisms that prokaryotes and eukaryotes use to achieve these goals are discussed. Prokaryotic genetic information processes are more completely understood than those of eukaryotes. As a result of their minimal growth requirements, short generation times, and relatively simple genetic composition, prokaryotes (especially *Escherichia coli*) are excellent subjects for investigations of genetic mechanisms. In contrast, multicellular eukaryotes possess several properties that hinder genetic investigations. The most formidable are long generation times (often months or years) and extraordinary difficulties in identifying gene products (e.g., enzymes or structural components). A common tactic in genetic research is to induce mutations, then observe changes in or the absence of a specific gene product. Because of the considerable complexity of higher organisms, this method has identified very few gene products in humans and other large mammals. Recombinant DNA techniques are being used to circumvent this obstacle.

DNA Replication

DNA replication must occur before every cell division. The mechanism by which DNA copies are produced is similar in all living organisms. After the two strands have separated, each serves as a template for the synthesis of a complementary strand (Figure 18.3). (In other words, each of the two new DNA molecules contains one old strand and one new strand.) This process, referred to as **semiconservative replication**, was first demonstrated in an elegant experiment (Figure 18.4)

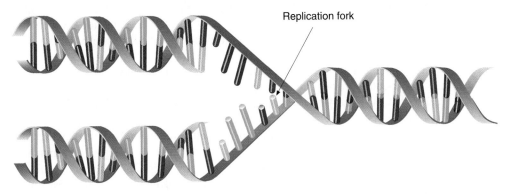

FIGURE 18.3

Semiconservative DNA Replication

As the double helix unwinds at the replication fork, each old strand serves as a template for the synthesis of a new strand.

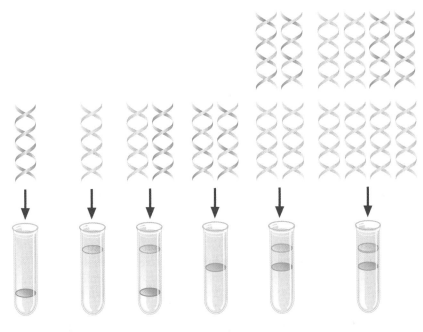

1. Heavy DNA   2. Light DNA   3. Mixture of 1 and 2   4. DNA one generation   5. DNA two generations   6. DNA three generations

FIGURE 18.4

The Meselson-Stahl Experiment

Centrifugation of *E. coli* DNA in CsCl can distinguish heavy DNA grown in ^{15}N media (1) from light DNA grown in ^{14}N media (2). A mixture is shown in (3). When *E. coli* cells enriched in ^{15}N are grown in ^{14}N for one generation, all genomic DNA is of intermediate density (4). After two generations, half the DNA is light and half is of intermediate density (5). After three generations, 75% of the DNA is light and 25% of the DNA is of intermediate density (6).

reported in 1958 by Matthew Meselson and Franklin Stahl. In this classic work, Meselson and Stahl took advantage of the increase in density of DNA labeled with the heavy nitrogen isotope ^{15}N (the most abundant nitrogen isotope is ^{14}N). After *E. coli* cells were grown for 14 generations in growth media whose nitrogen source consisted only of ^{15}NH$_4$Cl, the ^{15}N-containing cells were transferred to growth media containing the ^{14}N isotope. After one and two cell divisions, samples were removed. The DNA in each of these samples was isolated and analyzed by CsCl density gradient centrifugation. (Refer to the Biochemistry in the Lab box entitled Cell Technology, pp. 67–68, in Chapter 2, for a description of density gradient centrifugation.) Because this analytical method separates molecules based on size and shape or "density" pure ^{15}N-DNA and ^{14}N-DNA produce characteristic bands in centrifuged CsCl tubes (^{15}N-DNA is heavier than ^{14}N-DNA).

Upon centrifugation of the DNA isolated from ^{15}N-containing cells grown in ^{14}N medium for precisely one generation, only one band was observed. This band occurred halfway between where ^{15}N-DNA and ^{14}N-DNA bands would normally appear. It therefore seemed reasonable to assume that the new DNA was a hybrid molecule; that is, it contained one ^{15}N strand and one ^{14}N strand. (Any other means of replication would have created more than one band.) After two cell divisions, extracted DNA was resolved into two discrete bands of equal intensity, one made up of ^{14}N,^{14}N-DNA (light DNA) and one made up of hybrid molecules (^{14}N,^{15}N-DNA), a result that also supported the semiconservative model of DNA synthesis.

In the years since the Meselson and Stahl experiment, many of the details of DNA replication have been discovered. Until recently it was assumed that DNA replication machinery moved along a DNA "track" that is for the most part stationary. Recent research efforts have revealed that DNA replication occurs in specific nuclear or nucleoid compartments called **replication factories**, which are relatively stationary during the process of replication. The replication machinery within these factories performs as an energy-driven DNA pump. In prokaryotes, replication factories are attached to the cell membrane. In eukaryotes, replication factories assemble during the synthesis phase of the cell cycle (S. phase: see later, Figure 18.14) in association with the nuclear matrix.

DNA SYNTHESIS IN PROKARYOTES DNA replication in *E. coli* has proven to be a complex process that consists of several basic steps, each of which requires certain enzymatic activities: DNA unwinding, primer synthesis, and DNA polynucleotide synthesis. DNA fragment ligation and supercoiling control are also described.

DNA unwinding This function requires the presence of **helicases**, which as the name implies, are ATP-requiring enzymes that catalyze the unwinding of duplex DNA. The principal helicase in *E. coli* is DnaB, a hexamer that couples ATP hydrolysis to base pair separation.

Primer synthesis The formation of short RNA segments called **primers**, required for the initiation of DNA replication, is catalyzed by **primase**, an RNA polymerase. Primase is a 60 kD polypeptide product of the dnaG gene. A multienzyme complex containing primase and several auxiliary proteins is called the **primosome.**

DNA synthesis The synthesis of a complementary DNA strand in the $5' \rightarrow 3'$ direction by forming phosphodiester linkages between nucleotides base-paired to a template strand is catalyzed by large multienzyme complexes referred to as the DNA polymerases (Figure 18.5). In the current model for the catalytic mechanism of DNA polymerases, illustrated in Figure 18.6, the $3'$-hydroxyl oxygen is a nucleophile that attacks the α-phosphate of the incoming nucleotide to form a new P—O bond. DNA polymerase III (pol III) is the major

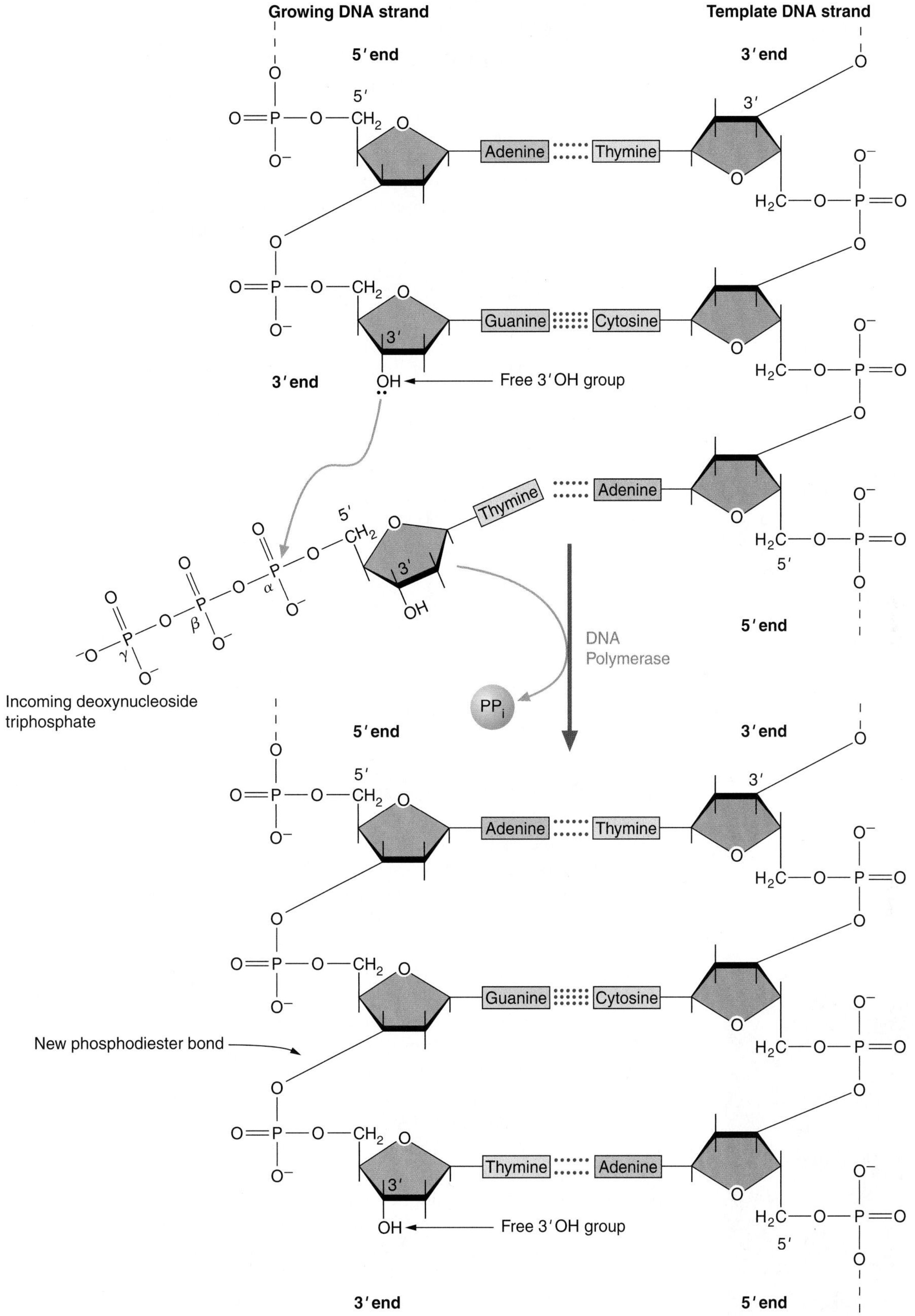

FIGURE 18.5

The DNA Polymerase Reaction

The essential feature of DNA synthesis is the formation of a phosphodiester linkage between a growing $5' \rightarrow 3'$ DNA strand and an incoming dNTP (deoxyribonucleoside triphosphate). The bond is created by a nucleophilic attack of the 3'-hydroxyl group of the terminal residue on the α-phosphate of the dNTP.

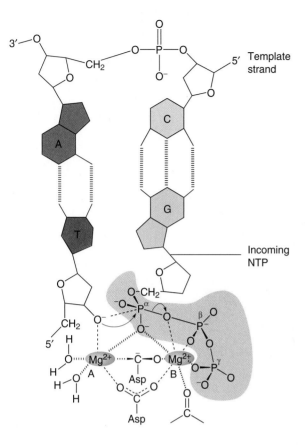

Template strand

Incoming NTP

FIGURE 18.6

Mechanism of DNA Polymerases

All DNA (and RNA) polymerases apparently use the same mechanism for template-driven nucleotide polymerization: an in-line nucleotidyl transfer. In this example from *E. coli* pol I, two Mg^{2+} ions (labeled A and B) are coordinated with the α-phosphate group of the incoming nucleotide (dNTP) and are themselves bridged by two aspartate side chain carboxylate groups. One ion, Mg^{2+} A, lowers the affinity of the 3'-hydroxyl oxygen for its hydrogen atom. The 3'-oxygen is a nucleophile that attacks the α-phosphate to form a new P—O bond. Both metal ions stabilize the negative charge of the transition state. By stabilizing the negatively charged pyrophosphate, Mg^{2+} B facilitates its departure.

TABLE 18.1   Subunits of DNA Polymerase III

		Subunit	Function
Holoenzyme	Core	α	$5' \rightarrow 3'$ polymerase
		ε	$3' \rightarrow 5'$ exonuclease
		θ	Unknown
		τ	Dimerization of core, ATPase
		β	Sliding clamp
	γ-Complex	γ	
		δ	Clamp loading unit that
		δ'	hydrolyzes ATP to drive
		χ	binding of β-protein to NA
		ψ	Also promotes processivity

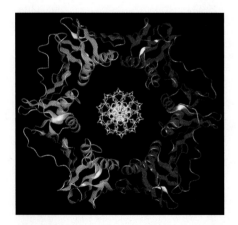

FIGURE 18.7

Cross Section of the β_2-Clamp of DNA Polymerase III

The β-protein is a dimer (shown in red and orange) that encircles the DNA and acts like a clamp.

DNA-synthesizing enzyme in prokaryotes. The pol III holoenzyme is composed of at least 10 subunits (Table 18.1). The core polymerase is formed from three subunits: α, ε, and θ. The β-protein (also called the sliding clamp protein, or the β_2-clamp) is composed of two subunits (Figure 18.7). It forms a donut-shaped ring around the template DNA strand. The γ-complex is composed of γ, δ, δ', χ, and ψ. Of these, γ, δ, and δ' contain an ATPase domain. In an ATP-dependent process, the γ-complex recognizes single DNA strands with primer and, acting as a **clamp loader**, transfers the β_2-clamp dimer to the core polymerase, where it forms a closed ring around the DNA strand. The inside diameter of the β_2-clamp is about 3.5 Å larger than dsDNA, large enough for hydrated DNA strands to slide through easily. The β_2-clamp promotes **processivity**; that is, it prevents frequent dissociation of the polymerase from the DNA template. The γ-complex is ejected in a process driven by ATP hydrolysis, and the pol III holoenzyme can proceed to replicate DNA. Note that the subunit τ allows two core enzyme complexes to form a dimer, which also improves processivity. The DNA replicating machine, called the **replisome**, is composed of two copies of the pol III holoenzyme, the primosome, and DNA unwinding proteins.

E. coli also possesses two other DNA polymerases. The first enzyme to be discovered was DNA polymerase I (pol I), also called the Kornberg enzyme after its discoverer Arthur Kornberg (Nobel Prize in Physiology or Medicine, 1959), is a DNA repair enzyme. Pol I also plays a role in the timely removal of RNA primer during replication after which it replaces the ribonucleotides with deoxyribonucleotides. DNA polymerase II (pol II) has an important role in DNA repair because it can reinitiate DNA synthesis beyond gaps caused by damaged segments. Both pol I and II synthesize DNA at rates too slow to be used in normal DNA synthesis, in which about 1000 nt are incorporated per second. In addition to a $5' \rightarrow 3'$ polymerizing activity, all three enzymes possess a $3' \rightarrow 5'$ exonuclease activity (e.g., the ε subunit in pol III), which has a proofreading function. (An **exonuclease** is an enzyme that removes nucleotides from an end of a polynucleotide strand.) Pol I also possesses a $5' \rightarrow 3'$ exonuclease activity.

Joining DNA fragments Frequently, during DNA replication (as well as DNA repair and recombination processes) DNA strand segments must be joined together. An enzyme called **DNA ligase** catalyzes the formation of a covalent phosphodiester bond between the $3'$-OH end of one segment with the $5'$-phosphate end of another segment.

Supercoiling control DNA *topoisomerases* prevent tangling of DNA strands. They function ahead of the replication machinery to relieve *torque* (rotary force) in the DNA that can slow down the replication process. The generation of torque is a very real possibility because the double helix unwinds rapidly (as many as 50 revolutions per second during bacterial DNA replication). Topoisomerases are enzymes that change the supercoiling of the DNA (see p. 642) by breaking one or both strands, which is followed by passing the DNA through the break and rejoining the strands. The terms *topoisomerase* and *topoisomers* (circular DNA molecules that differ only in their degree of supercoiling) are derived from *topology*, a branch of mathematics that examines any change in shape or position that can be achieved without cutting. A circular DNA molecule can be made to lie in a plane or can be supercoiled to lift it off the plane much like an overwound telephone cord. While the initial and final states look very different, they are topologically equivalent because they can be created without breaking the double helix or changing its duplex structure. When appropriately controlled, supercoiling can facilitate the unzipping of DNA molecules. Type I topoisomerases produce transient single-strand breaks in DNA; type II topoisomerases produce transient double-strand breaks. In prokaryotes, a type II topoisomerase called DNA gyrase helps to separate (decatanate) the replication products (i.e., linked circular chromosomes) and to create the ($-$) supercoils required for genome packaging. (In contrast, eukaryotic type II topoisomerases catalyze only the removal of superhelical tension.)

The replication of the circular *E. coli* chromosome (Figure 18.8) begins when there is a high ATP/ADP ratio and sufficient copies of DnaA, a DNA-binding protein. The initiation site on the *E. coli* chromosome is referred to as oriC. Once it has been initiated, replication proceeds in two directions. Helicases unwind the DNA duplex, two replisomes assemble, and replication proceeds outward in both directions. As the two sites of active DNA synthesis (referred to as **replication forks**) move farther away from each other, a "replication eye" forms. Because an *E. coli* chromosome contains one initiation site, it is considered a single replication unit. A replication unit, or **replicon**, is a DNA molecule (or DNA segment) that contains an initiation site and appropriate regulatory sequences.

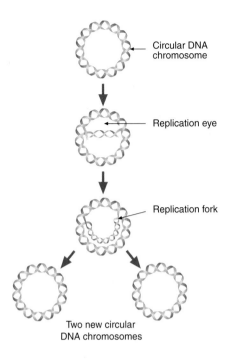

Circular DNA chromosome

Replication eye

Replication fork

Two new circular DNA chromosomes

FIGURE 18.8

Replication of Prokaryotic DNA

As DNA replication of a circular chromosome proceeds, two replication forks can be observed by using autoradiography. The structure that forms is called a replication eye.

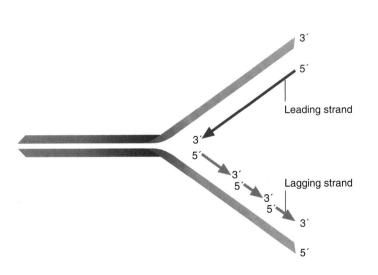

$3'$

$5'$

Leading strand

$3'$

$5'$

$3'$

$5'$

Lagging strand

$3'$

$5'$

$3'$

$5'$

FIGURE 18.9

DNA Replication at a Replication Fork

The $5' \rightarrow 3'$ synthesis of the leading strand is continuous. The lagging strand is also synthesized in the $5' \rightarrow 3'$ direction but in small segments (Okazaki fragments).

When DNA replication was first observed experimentally (using electron microscopy and autoradiography), investigators were confronted with a paradox. The bidirectional synthesis of DNA as it appeared in their research seemed to indicate that continuous synthesis occurs in the $5' \rightarrow 3'$ direction on one strand and in the $3' \rightarrow 5'$ direction on the other strand. (Recall that the DNA double helix has an antiparallel configuration.) However, all the enzymes that catalyze DNA synthesis do so in the $5' \rightarrow 3'$ direction only. It is now known that only one strand, referred to as the *leading strand*, is continuously synthesized in the $5' \rightarrow 3'$ direction. The other strand, the *lagging strand*, is also synthesized in the $5' \rightarrow 3'$ direction but in small pieces (Figure 18.9). Reiji Okazaki and his colleagues provided the experimental evidence for discontinuous DNA synthesis. Subsequently, these small pieces in the lagging strand, now called **Okazaki fragments**, are covalently linked together by DNA ligase. In prokaryotes such as *E. coli*, Okazaki fragments possess from 1000 to 2000 nucleotides.

The initiation of replication is a complex process involving several enzymes, as well as other proteins. Replication begins when *DnaA proteins* (52 kD) (Figure 18.10) bind to five to eight 9-bp sites, referred to as *DnaA boxes*, within the oriC sequence. Prokaryotic organisms vary in the number of DnaA boxes. *E. coli* has five DnaA boxes. The oligomerization of DnaA, which results in a nucleosome-like structure, requires ATP and the histonelike protein HU. As the DnaA-DNA complex forms, localized "melting" of the DNA duplex in a nearby region containing three 13 bp repeats causes a small segment of the double helix to open up (Figure 18.11). *DnaB* (a 300 kD helicase composed of six subunits),

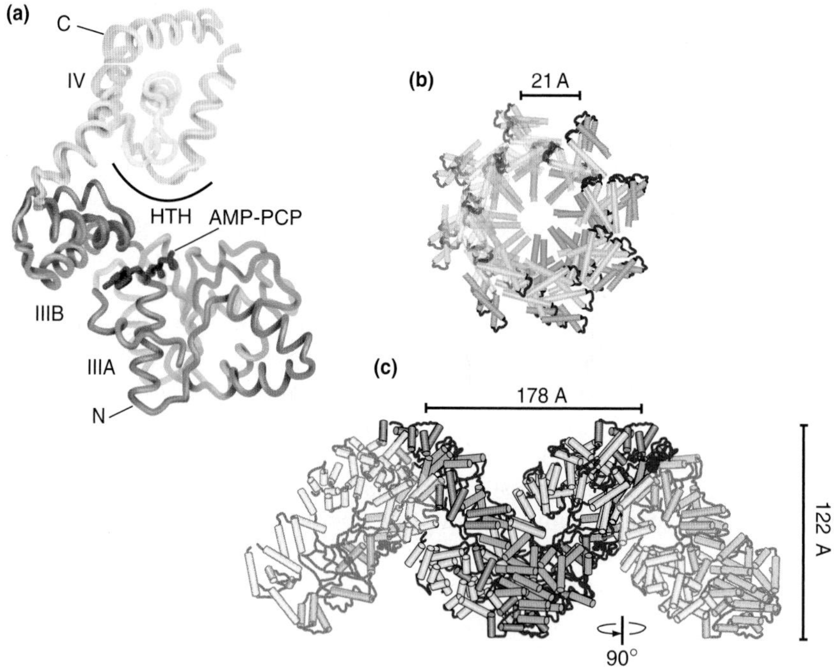

FIGURE 18.10

DnaA Structure

(a) DnaA consists of four domains: III (red and green) and IV (yellow) are illustrated. In this illustration AMP-PCP, an analogue of ATP (designated in dark blue) is bound in the ATP binding site of DnaA. When ATP binds to the ATP-binding site within domain IIIA of DnaA monomers, they undergo a conformational change that facilitates the formation of oligomers that bind, via domain IV, to highly conserved 9 bp sequences called Dna boxes (HTH = helix-turn-helix motif) (b) and (c) Top and side views, respectively, of DnaA oligomers. The architecture of the DnaA oligomer (an open right-handed helix) causes DNA that contains DnaA boxes to wrap around its exterior.

FIGURE 18.11

Replication Fork Formation

After DnaA and DnaB binding, the DnaB helicase separates the duplex DNA strands at two replication forks. Each DNA-binding domain (yellow) of DnaA binds to a 9 bp DnaA box. The binding of SSB to newly formed ssDNA prevents reassociation of the single strands. The helicase loader DnaC (not shown) facilitates the binding of DnaB helicase to DNA.

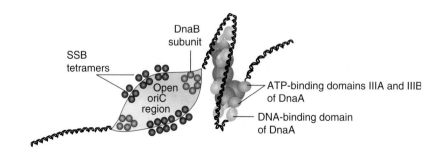

complexed with *DnaC* (29 kD), then enters the open oriC region. When DnaB is loaded onto the DNA, the DnaC is released. The replication fork moves forward as DnaB unwinds the helix. Topoisomerases relieve torque ahead of the replication machinery. As DNA unwinding proceeds, DnaA is displaced. The hydrolysis of ATP molecules bound within domain IIIA of DnaA causes the protein to revert to an inactive conformation that is incapable of binding DNA. The single strands are kept apart by the binding of numerous copies of single-stranded DNA-binding protein (SSB). The SSB, a tetramer, may also protect vulnerable ssDNA segments from attack by nucleases.

A model of DNA synthesis at a replication fork is illustrated in Figure 18.12. For pol III to initiate DNA synthesis, an RNA primer must be synthesized. On the leading strand, where DNA synthesis is continuous, primer formation occurs only once per replication fork. In contrast, the discontinuous synthesis on the lagging strand requires primer synthesis for each of the Okazaki fragments. The primosome travels along the lagging strand and stops and reverses direction at intervals to synthesize a short RNA primer. Subsequently, pol III synthesizes DNA beginning at the 3′ end of the primer. As lagging strand synthesis continues, the RNA primers are removed by pol I, which also synthesizes a complementary DNA segment. DNA ligase then joins the Okazaki fragments.

As illustrated in Figure 18.12, the synthesis of both the leading and lagging strands is coupled. The tandem operation of two pol III complexes requires that one strand (the lagging strand) be looped around the replisome. When the lagging strand pol III complex completes an Okazaki fragment, it releases the duplex DNA and the sliding clamp. As a result of its tether to the leading strand pol III (via the γ-complex), the lagging strand polymerase binds the newly synthesized RNA primer that is directly adjacent to the replication fork. This allows the primosome to move in and synthesize another RNA primer.

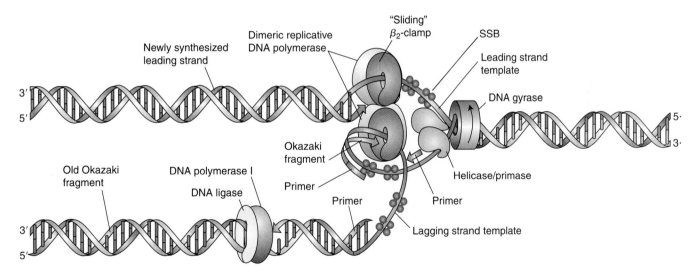

FIGURE 18.12

E. coli **DNA Replication Model**

The DNA duplex at the replication fork is unwound by DnaB and DNA gyrase (a topoisomerase). DNA gyrase uses ATP hydrolysis to introduce negative supercoils into DNA just ahead of the replication fork, which relieves torque, thereby returning the DNA molecule to a more relaxed state. As the helix unwinds, an RNA primer is synthesized on the lagging strand. New strand synthesis is catalyzed by two pol III holoenzymes tethered to each other by the τ-complex (not shown). Each pol III complex is linked to a β_2-clamp that tethers the polymerase to the template strand. Both leading and lagging strand synthesis move in the same direction because the lagging strand is looped out before it enters the pol III holoenzyme. When the lagging strand pol III complex completes an Okazaki fragment, it releases the strand. (Because of the alternate lengthening and shortening of the lagging strand, this replication model is referred to as the trombone model.) The primosome then synthesizes a new primer. Working together, pol I and DNA ligase remove the primer and fill and seal the gaps between the Okazaki fragments. Pol III rebinds the lagging strand at a new primer and begins the synthesis of a new Okazaki fragment.

Despite the complexity of DNA replication in *E. coli*, as well as its rate (as high as 1000 base pairs per second per replication fork), this process is amazingly accurate: approximately one error per 10^9 to 10^{10} base pairs per generation. This low error rate is largely a consequence of the precise nature of the copying process itself (i.e., complementary base pairing). Within the active site of the DNA polymerases there is a pocket that is precisely shaped to fit a nucleotide base pair in which a purine and a pyrimidine are properly aligned by hydrogen bonds and van der Waals interactions. If the nucleotide bases are mismatched, they do not fit into the pocket, and the incoming nucleotide usually leaves the site before the reaction occurs. Both pol III and pol I also proofread newly synthesized DNA. Most mispaired nucleotides are removed (by the $3' \rightarrow 5'$ exonuclease activities of pol III and pol I) and then replaced. Several postreplication repair mechanisms also contribute to the low error rate in DNA replication.

Replication ends when the replication forks meet on the other side of the circular chromosome at the termination site, the *ter* (τ) region. The ter region in *E. coli* is composed of six 20 bp terminator sites that, when bound to the DNA-binding protein *tus*, cause replication arrest. The asymmetric tus-ter complex prevents a replication fork from traveling in one direction but not the other via direction-dependent inhibition of DnaB helicase (Figure 18.13). After replication ends, the replisomes disassemble and the two daughter molecules are separated by a type II topoisomerase.

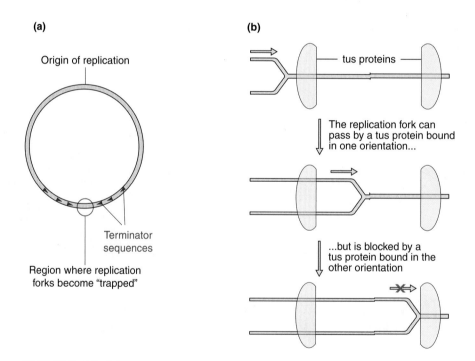

FIGURE 18.13

Role of tus in DNA Replication Termination in *E. coli*

(a) Within the ter region of the *E. coli* chromosome, there are six termination sequences. The arrowheads that designate these sequences indicate the direction that each sequence can be passed by a replication fork. (b) A ter sequence binds a pair of tus proteins, oriented in reverse directions. Depending on its binding orientation on a dsDNA ter segment, tus can prevent unwinding. In this diagram a replication fork is passing by the left-handed tus. The tus bound at this site is disrupted by the DnaB helicase, so the replication fork passes through unobstructed until it encounters the second tus protein. The second tus protein (on the right), oriented on the ter sequence in the opposite direction from the left-handed tus, blocks the replication fork by inhibiting DnaB.

DNA SYNTHESIS IN EUKARYOTES Although the principles of DNA replication in prokaryotes and eukaryotes have a great deal in common (e.g., semiconservative replication, the DNA polymerase mechanism, and bidirectional replicons), they also have significant differences. Not surprisingly, these differences appear to be related to the size and complexity of eukaryotic genomes.

Timing of replication In contrast to rapidly growing bacterial cells, in which replication occurs throughout most of the cell division cycle, eukaryotic replication is limited to a specific period referred to as the *S phase* (Figure 18.14). It is now known that eukaryotic cells produce certain proteins (Section 18.3) that regulate phase transitions within the cell cycle.

Replication rate DNA replication is significantly slower in eukaryotes than in prokaryotes. The eukaryotic rate is approximately 50 nucleotides per second per replication fork. (Recall that the rate in prokaryotes is about 20 times higher.) This discrepancy is presumably due, in part, to the complex structure of chromatin.

Replicons Despite the relative slowness of eukaryotic DNA synthesis, the replication process is relatively brief, considering the large sizes of eukaryotic genomes. For example, on the basis of the replication rate, the replication of an average eukaryotic chromosome (approximately 150 million bp) should take over a month to complete. Instead, this process usually is completed in several hours. Eukaryotes use multiple replicons to compress the replication of their large genomes into short periods (Figure 18.15). About every 40 kb along eukaryotic chromosomes, there is a site where replication machinery assembles. Humans have about 30,000 origins of replication.

Okazaki fragments From 100 to 200 nucleotides long, the Okazaki fragments of eukaryotes are significantly shorter than those in prokaryotes.

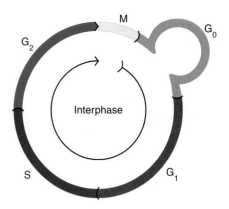

FIGURE 18.14

The Eukaryotic Cell Cycle

Interphase (the period between mitotic divisions) is divided into several phases. DNA replication occurs during the synthesis or S phase. The G_1 (first gap) phase is the time between mitosis and the beginning of the S phase. During the G_2 phase, protein synthesis increases as the cell readies itself for mitosis (M phase). After mitosis, many cells enter a resting phase (G_0).

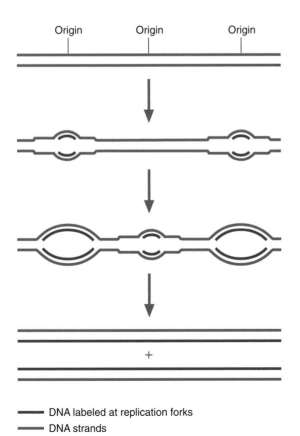

FIGURE 18.15

Multiple-Replicon Model of Eukaryotic Chromosomal DNA Replication

Short segment of a eukaryotic chromosome during replication.

DNA labeled at replication forks
DNA strands

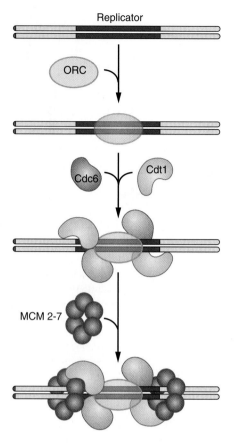

Replicator

ORC

Cdc6   Cdt1

MCM 2-7

FIGURE 18.16

Formation of a Preinitiation Replication Complex

Assembly of the preRC begins (during G_1 phase of the cell cycle) when the ORC binds to the replication origin sequence, sometimes referred to as the *replicator*. ORC proceeds to recruit Cdc6 and Cdt1. When the MCM complex subsequently binds, preRC formation is complete.

KEY CONCEPTS

- DNA is replicated by a semiconservative mechanism involving several enzymes.

- The leading strand is synthesized continuously in the $5' \rightarrow 3'$ direction.

- The lagging strand is synthesized in pieces in the $5' \rightarrow 3'$ direction; the pieces are then covalently linked.

In higher eukaryotes, replication begins with the sequential assembly of the **preinitiation replication complex (preRC)** (Figure 18.16). The formation of this complex involves a process that starts in early G_1 phase of the cell cycle when levels of cyclin-dependent kinase (cdk) and cell division cycle (cdc) proteins are low. The process, called licensing, limits DNA replication to once per cell cycle.

The components of preRC are assembled beginning when the **origin of replication complex (ORC)**, the subunits of which are analogues of DnaA, binds to the DNA initiation region or origin. Cdc6/Cdc18 and Cdt1 bind to ORC and recruit the **MCM complex** to the site. MCM is thought to be the major DNA helicase in eukaryotes. The recruitment of additional proteins of the cell cycle control machinery (e.g., cdk2-cyclin E and cdc45) allows for the DNA replication proteins to be loaded onto the replication fork and DNA synthesis to begin (Figure 18.17). The conversion of licensed preRC to an active initiation complex, requires the addition of pol α/primase, pol ε, and a number of accessory proteins, which occurs only at the onset of S phase. The cell cycle regulating kinases then phosphorylate and activate the components of the preRC. The proteins that bind to the ORC and complete the structure of the preRC are referred to as **replication licensing factors (RLFs)**.

When the initiation complex is active, newly phosphorylated MCM separates the DNA strands, each of which is then stabilized by **replication protein A (RPA)** (Figure 18.18). Replication is begun by primase, which synthesizes the RNA primers of the leading strand and each Okazaki fragment on the lagging strand. Polymerase α extends each primer by a short segment of DNA about 20 nucleotides long. DNA polymerase α is then displaced, and DNA polymerase δ continues the replication process. The attachment of DNA polymerase δ to each strand is controlled by **replication factor C (RFC)**, which is a clamp loader protein. After binding ATP, RFC then binds to PCNA, a clamp that is an accessory protein of DNA polymerase δ. The RFC/PCNA complex converts DNA polymerase δ into a processive enzyme. Finally, PCNA is loaded onto DNA and closes around it, triggering ATP hydrolysis

DNA replication of each chromosome continues until the replicons meet and fuse. When the replication machinery reaches the $3'$ end of the lagging strand, there is insufficient space to synthesize a new RNA primer. Incomplete lagging strand synthesis leaves the template strand without its complementary base pairs at the end of the chromosome. Chromosomes with $3'$-ssDNA overhangs are very susceptible to nuclease digestion and tend to fuse with each other, leading to chromosomal breakage during mitosis. Eukaryotic cells can compensate for this problem with **telomerase**, a ribonucleoprotein with reverse transcriptase activity, and an RNA molecule with a base sequence that is complementary to the TG-rich sequence of telomeres. Recall that telomeres (p. 651) are minisatellite sequences that occur at the ends of linear chromosomes. Telomerase uses the RNA base sequence to synthesize a single-stranded DNA to extend the $3'$ strand of the telomere (Figure 18.19). Afterward, the normal replication machinery synthesizes a primer and a new Okazaki fragment. The chromosome ends are then sequestered and stabilized by **telomere end-binding proteins (TEBPs)** that bind to GT-rich telomere sequences and **telomere repeat-binding factors (TRFs)** that secure the $3'$ overhang (now further away from critical coding sequences into a knotlike T-loop.

In most multicellular eukaryotes, telomerase is active only in germ cells (cells that give rise to eggs and sperm). In the human body during normal aging, the telomeres of somatic cells (differentiated cells not including eggs and sperm) shorten over time. Once telomeres are reduced to a critical length, chromosomes can no longer replicate. As a result, somatic cells eventually die. It is noteworthy that the fibroblasts (connective tissue cells) of patients with Hutchinson-Guilford progeria syndrome have abnormally short telomeres. These patients age rapidly, with death occurring in the preteen years. It is also known that telomerase is overexpressed in approximately 90% of all cancers.

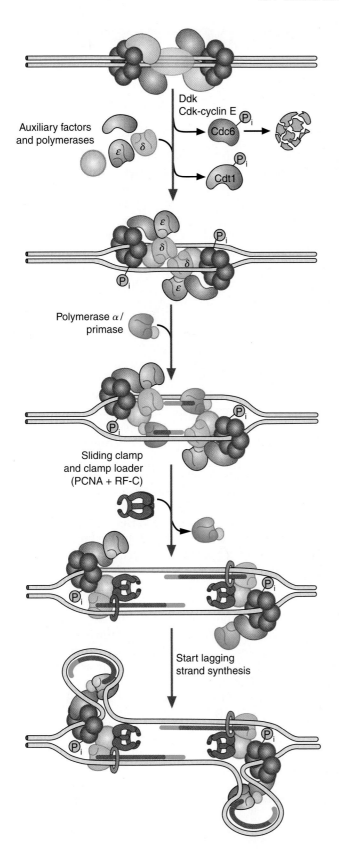

Auxiliary factors and polymerases

Ddk
Cdk-cyclin E

Cdc6

Cdt1

Polymerase α/primase

Sliding clamp and clamp loader (PCNA + RF-C)

Start lagging strand synthesis

FIGURE 18.17
Eukaryotic Replication Fork Formation

Ddk and Cdk-cyclin E trigger replication initiation by phosphorylating several proteins. Among the results is the release of Cdc6 and Cdtl from ORC. Upon recruitment of DNA polymerase δ and ε the initiation complex is complete. Polymerase α/primase is then recruited. Once the RNA primers have been synthesized on the leading strand and the primer sequence briefly extended by polymerase α, the clamp loader (RFC) binds to the sliding clamp (PCNA). The sliding clamp/clamp loader complex then transforms polymerase δ into a processive enzyme. After DNA has unwound sufficiently, lagging strand synthesis is initiated. In this illustration DNA polymerase ε is bound to the lagging strand. Although pol ε is believed to play an important role in DNA replication, its role is still uncertain.

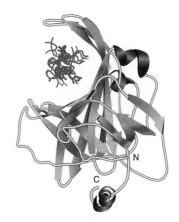

FIGURE 18.18
Replication Protein A Structure

Eukaryotes use RPA, a single-stranded DNA-binding protein, to prevent DNA strands from reannealing or being degraded by nucleases. The β-sheet in RPA forms a channel in which DNA (dark orange) binds.

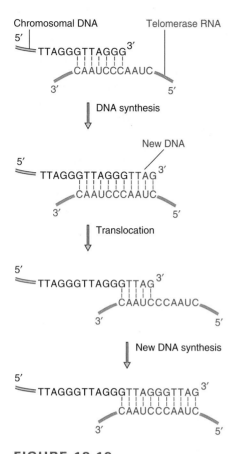

FIGURE 18.19

Telomerase-Catalyzed Extension of a Chromosome

The GT-rich, 3'-ssDNA end of a chromosome (the telomere): a portion of the RNA component of telomerase base pairs with this region, and the reverse transcriptase activity extends the template strand. The telomerase translocates to the end of the new segment, and the process is repeated until the ssDNA is long enough to accommodate the replication machinery and a primer. A new Okazaki fragment is then synthesized.

QUESTION 18.1

Compare the replication processes of prokaryotes and eukaryotes.

DNA Repair

The natural (or background) rate of spontaneous mutation is remarkably constant among species at about 0.1 to 1 mutation per gene per million gametes in every generation. This estimate might appear low, but in fact most mutations can be expected to reduce the viability of the organism. For example, humans lose approximately 50% of conceptions because of genetic factors.

Mutations take many forms, from point mutations (single base changes) to gross chromosomal abnormalities (Section 17.1). As described, they are caused by factors that include the properties of the bases themselves and various chemical processes (e.g., depurinations and oxidative stress), as well as the effects of xenobiotics. Not surprisingly, cells possess a wide range of DNA repair mechanisms.

The importance of maintaining the structural integrity of DNA is reflected in the variety of repair mechanisms employed by living cells. A few types of DNA damage can be repaired without the removal of nucleotides. For example, breaks in the phosphodiester linkages can be repaired by DNA ligase. In **photoreactivation repair**, or **light-induced repair**, pyrimidine dimers are restored to their original monomeric structures (Figure 18.20). In the presence of visible light, DNA photolyase cleaves the dimer, leaving the phosphodiester bonds intact. Light energy captured by the enzyme's flavin and pterin chromophores breaks the cyclobutane ring.

Most DNA repair mechanisms involve the removal of nucleotides. **Base excision repair** is a mechanism that removes and then replaces individual nucleotides whose bases have undergone damage (e.g., alkylation, deamination, or oxidation). One of several enzymes called **DNA glycosylases** cleaves the N-glycosidic linkage between the damaged base and the deoxyribose component of the nucleotide (Figure 18.21). The resulting **apurinic** or **apyrimidinic** (AP) sites are resolved through the action of nucleases that remove the deoxyribose residue. The gap is repaired by a DNA polymerase (pol I in bacteria and DNA polymerase β in mammals) and DNA ligase. In **nucleotide excision repair**, bulky (2–30 nt) lesions are removed, and the resulting gap is filled. The excision enzymes appear to recognize the physical distortion rather than specific base sequences. In *E. coli* (Figure 18.22), the excision nuclease (exinuclease), composed of Uvr A, B, and C, cuts the damaged DNA and removes a 12 to 13 nt ssDNA sequence containing the lesion. Uvr A identifies the damaged site (e.g., a thymine dimer) and associates with Uvr B to form a complex (A_2B). Next A_2B bends the DNA, and Uvr A dissociates from Uvr B–DNA. Uvr C then binds to Uvr B, cutting the

FIGURE 18.20

Photoreactivation Repair of Thymine Dimers

Light provides the energy for converting the dimer to two thymine monomers. No nucleotides are removed in this repair mechanism. Many species, including humans, do not possess DNA photolyase enzyme activity.

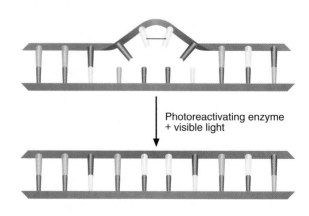

Photoreactivating enzyme + visible light

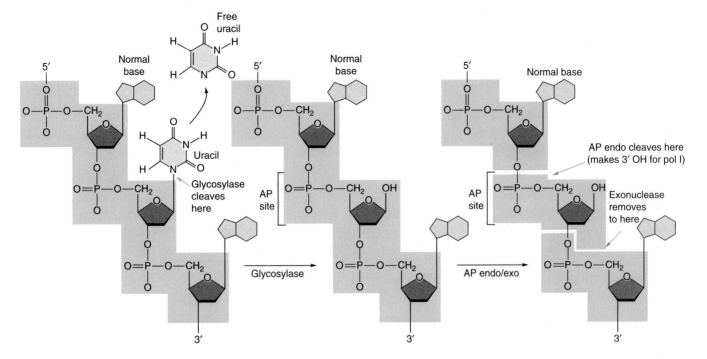

FIGURE 18.21

Base Excision Repair

DNA glycosylase hydrolyzes the N-glycosidic linkage to release the base (in this case, uracil). AP endonuclease cleaves the DNA backbone at the 5′ position of the AP (apyrimidic) site. An exonuclease cleaves the 3′ position of the AP site. A DNA polymerase then fills in the gap.

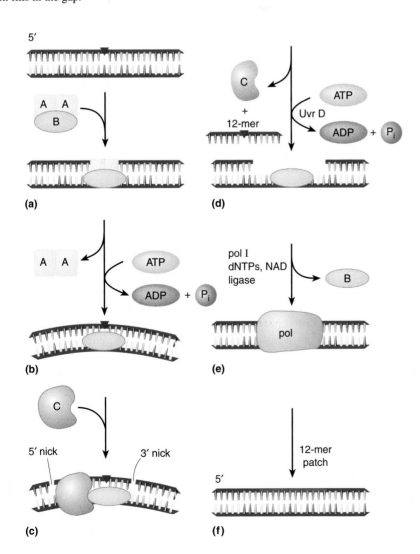

FIGURE 18.22

Excision Repair of a Thymine Dimer in *E. coli*

Uvr A, a damage recognition protein, detects helical distortion caused by DNA adducts such as thymine dimers (a). It then associates with Uvr B to form the A₂B complex. After binding to the damaged segments, A₂B forces DNA to bend. Uvr A then dissociates (b). The binding of the nuclease Uvr C to Uvr B (c) and the action of the helicase Uvr D (d) results in the excision of a 12-nucleotide DNA strand (12-mer). Then Uvr B is released (e), and the excision gap is repaired by pol I (f).

damaged DNA strand 4 or 5 nucleotides to the 3′ side of the thymine dimer. Then Uvr C cuts the strand 8 nucleotides to the 5′ side. Uvr D, a helicase, releases Uvr C and the thymine dimer–containing oligonucleotide. The excision gap is repaired by pol I and DNA ligase. Excision repair in humans is poorly understood and, probably is more complicated than the prokaryotic process.

Recombinational repair can eliminate certain types of damaged DNA sequence that are not eliminated before replication. For some structural damage, replication is interrupted because the replication complex detaches, moves beyond the damage, and reinitiates synthesis. Such gaps can be repaired by exchanging the corresponding segment of the homologous DNA molecule. When recombination is complete, the newly opened gap in the homologous molecule can easily be filled by DNA polymerase and DNA ligase, since an intact template is present. The remaining damaged segment can then be eliminated by other repair mechanisms. Recombinational repair closely resembles genetic recombination, which is discussed next.

KEY CONCEPTS

- DNA is constantly exposed to chemical and physical processes that alter its structure.
- Each organism's survival depends on its capacity to repair this structural damage.
- Examples include light-induced repair, base excision repair, nucleotide excision repair, and recombinational repair.

QUESTION 18.2

List four types of DNA repair. Explain the basic features of each.

DNA Recombination

Recombination, often referred to as genetic recombination, can be defined as the rearrangement of DNA sequences by exchanging segments from different molecules. The process of recombination, which produces new combinations of genes and gene fragments, is primarily responsible for diversity among living organisms. More important, the large number of variations made possible by recombination gives species opportunities to adapt to changing environments. In other words, genetic recombination is a principal source of the variations that make evolution possible.

There are two forms of recombination: general and site-specific. **General recombination**, which occurs between homologous DNA molecules, is most commonly observed during meiosis. (Recall that meiosis is the form of eukaryotic cell division in which haploid gametes are produced.) A similar process has been observed in some bacteria. In **site-specific recombination**, the exchange of sequences from different molecules requires only short regions of DNA homology. These regions are flanked by extensive nonhomologous sequences. Site-specific recombinations, which depend more on protein-DNA interactions than on sequence homology, occur throughout nature. For example, this mechanism is used by bacteriophage to integrate its genome into the *E. coli* chromosome. In eukaryotes, site-specific recombination is responsible for a wide variety of developmentally controlled gene rearrangements. Gene rearrangements are partially responsible for cell differentiation in complex multicellular organisms. One of the most interesting examples of gene rearrangement is the generation of antibody diversity in mammals. In the variation of site-specific recombination referred to as **transposition**, DNA sequences called **transposable elements** move from one chromosome or chromosomal region to another.

GENERAL RECOMBINATION General recombination requires the precise pairing of homologous DNA molecules. Figure 18.23 illustrates the currently accepted model for general recombination. This model, based on Robin Holliday's genetic investigations in fungi in 1964, involves the following essential steps:

1. Two homologous DNA molecules become paired.
2. Two of the DNA strands, one in each molecule, are cleaved.

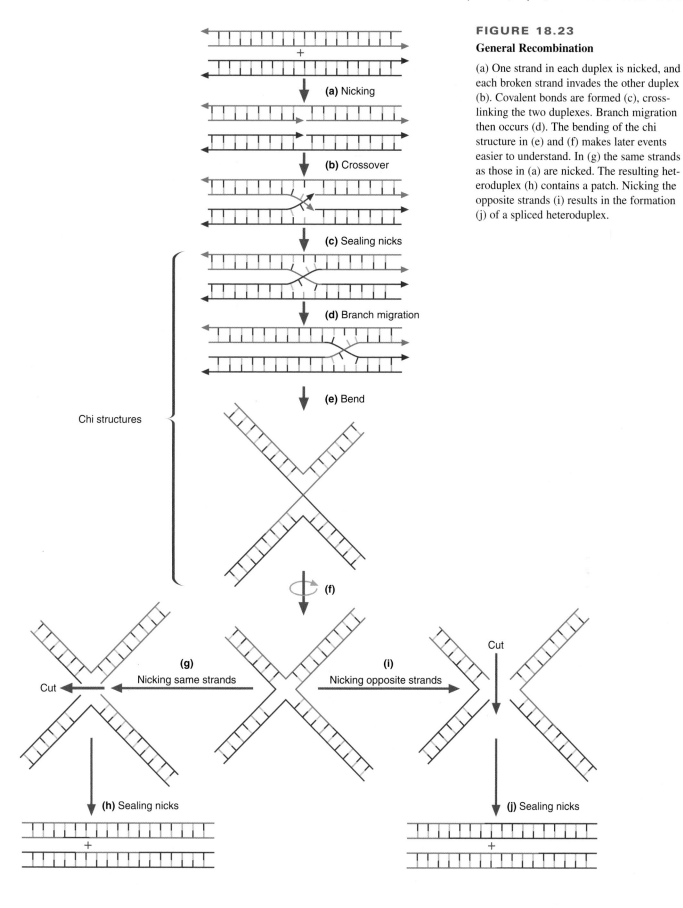

FIGURE 18.23

General Recombination

(a) One strand in each duplex is nicked, and each broken strand invades the other duplex (b). Covalent bonds are formed (c), cross-linking the two duplexes. Branch migration then occurs (d). The bending of the chi structure in (e) and (f) makes later events easier to understand. In (g) the same strands as those in (a) are nicked. The resulting heteroduplex (h) contains a patch. Nicking the opposite strands (i) results in the formation (j) of a spliced heteroduplex.

3. The two nicked strand segments cross over, thus forming a Holliday intermediate.

4. DNA ligase seals the cut ends.

5. Branch migration caused by base-pairing exchange leads to the transfer of a segment of DNA from one homologue to the other.

6. A second series of DNA strand cuts occurs.

7. DNA polymerase fills any gaps, and DNA ligase seals the cut strands.

Most investigations of general recombination have been done with bacteria. Certain mutations in *E. coli* have proved to be useful in determining the molecular details of this process. Recombination in *E. coli* is initiated by RecBCD, an enzyme complex that possesses both exonuclease and helicase activities. After binding to a DNA molecule, RecBCD cleaves one of the strands and proceeds to unwind the double helix until it reaches 5′-GCTGGTGG-3′ (the chi site), a sequence that occurs frequently in *E. coli* DNA. Strand exchange is effected by monomers of RecA, an ATPase that coats one of the strands. Powered by ATP, the RecA-coated strand segment then interacts with a nearby double-helical DNA segment with a homologous sequence, thus forming a triple helix. Branch migration (Figure 18. 24) is initiated as RuvA recognizes and binds to the Holliday junction. Two copies of RuvB, a hexamer with ATPase and helicase activities, then form a ring on either side of the junction. Branch migration is catalyzed by the RuvAB complex. This molecular machine separates, rotates, and pulls the strands in the two sets of helices, even after RecA dissociates. The migration ends when a specific sequence [5′-(A or T)TT (G or C)-3′] is reached. As RuvAB detaches, two RuvC proteins bind to the junction. Recombination ends as RuvC cleaves the crossover strands and the Holliday structure resolves itself to form two separate double-helical DNA molecules.

In bacteria, general recombination appears to be involved in several forms of intermicrobial DNA transfer.

1. **Transformation.** In **transformation**, naked DNA fragments enter a bacterial cell through a small opening in the cell wall and are introduced into the bacterial genome. (Recall Fred Griffith's experiment, see p. 637.)

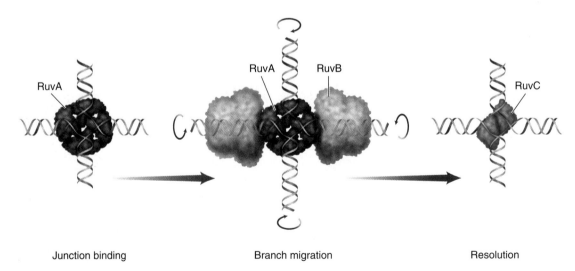

Junction binding   Branch migration   Resolution

FIGURE 18.24

Model of the Association of Ruv Proteins with a Holliday Junction

RuvA, a tetramer, first binds to the Holliday junction point. Two hexameric RuvB rings then form on both sides of the DNA/RuvA complex, with the DNA passing through the rings. Branch migration occurs as ATP hydrolysis drives the two RuvB rings to rotate the DNA helices in opposite directions. After branch migration, RuvA and RuvB detach and two RuvC proteins bind to the junction. RuvC, a nuclease, proceeds to cut the crossover strands, thus resolving the Holliday structure.

2. **Transduction**. **Transduction** occurs when bacteriophage inadvertently carry bacterial DNA to a recipient cell. After a suitable recombination, the cell uses the transduced DNA.

3. **Conjugation**. Certain bacterial species are known to engage in **conjugation**, an unconventional sexual mating that involves a donor cell and a recipient cell. The donor cell possesses a specialized plasmid (p. 42) that allows it to synthesize a sex pilus, a filamentous appendage that functions in a DNA exchange process. After the pilus has attached to the surface of the recipient cell, a fragment of the donor's genetic material is transferred. The transferred DNA segment can be integrated into the recipient's chromosome by recombination, or it may exist outside it in plasmid form.

General recombination in eukaryotes is believed to be similar to the process in prokaryotes. Several eukaryotic proteins have been discovered that closely resemble in both structure and function those observed in *E. coli*. For example, RAD51 and Dmcl are both RecA homologues and are required for repair of double-stranded breaks during homologous recombination in meiosis. The absence or loss of function of either of these genes leads to incorrect *synapsis*, or pairing of homologues during meiotic recombination.

Bacterial conjugation has medical consequences. For example, certain plasmids contain genes that code for toxins. The causative agent of a deadly form of food poisoning is *E. coli* 0157, a strain of *E. coli*, synthesizes a toxin that causes massive bloody diarrhea and kidney failure. This toxin is now believed to have originated in *Shigella*, another bacterium that causes dysentery. Similarly, the growing problem of antibiotic resistance is partly attributable to the spread of antibiotic-resistant genes among bacterial populations. Antibiotic resistance develops because antibiotics are overused in medical practice and in livestock feeds. Suggest a mechanism by which this extensive use promotes antibiotic resistance. [*Hint*: The high-level use of antibiotics acts as a selection pressure.]

SITE-SPECIFIC RECOMBINATION AND TRANSPOSITION Site-specific recombination relies on short segments of homologous DNA called **attachment (att) sites** or **insertional (IS) elements**. Recombination at these sites can lead to insertions, inversions, deletions, and translocations that may benefit or harm the cell. Examples of site-specific recombination include insertion of viral DNA into a host cell genome, acquisition of antibiotic resistance, phenotypic variation in plants, and antibody maturation in mammals. A simple case of insertion is illustrated by the integration of bacteriophage λ DNA into the *E. coli* chromosome (Figure 18.25). The process requires homologous att sites in the phage and bacterial genomes, a viral recombinase called λ integrase, and a bacterial gene product, the integration host factor (IHF). The mechanism follows the general model, with Holliday junction resolution resulting in the insertion of the λ genome into the bacterial chromosome.

Barbara McClintock, a geneticist working with Indian corn (maize), reported in the 1940s that mobile genetic elements were responsible for the phenotypic variation in corn kernel color. Acceptance of this idea was slow in coming because it required a paradigm shift away from the view of the genome as a static component of the cell. In 1967 transposable elements (**transposons** or "jumping genes") were confirmed in *E. coli*, and the concept of genome plasticity was finally accepted. Dr. McClintock received the Nobel Prize in Physiology or Medicine in 1983.

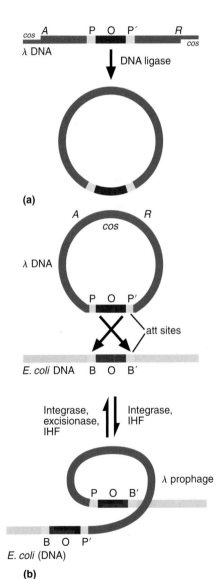

FIGURE 18.25

Insertion of the Bacteriophage λ Genome into the *E. coli* Chromosome

(a) The λ DNA circularizes as the single-stranded *cos* sequences anneal. (b) Insertion occurs through site-specific recombination between short homologous phage and bacterial sequences called attachment (att) sites.

The IS elements of simple prokaryotic transposons consist of the gene for the transposition enzyme **transposase**, flanked by short inverted terminal repeats that define the boundary of the transposon (Figure 18.26). (Recall that an inverted repeat is a sequence that is a reversed complement of another downstream sequence and that inverted repeats without intervening sequences constitute a palindrome, a DNA sequence whose complement reads the same in reverse.) More complicated bacterial transposons, called **composite transposons**, are made up of specific gene(s) (e.g., for antibiotic resistance) flanked by simple IS elements. Two transposition mechanisms have been observed, replicative and nonreplicative.

Replicative transposition In replicative transposition only one strand of the donor DNA is transferred to the target position, and replication followed by site-specific recombination results in the duplication of the transposon rather than insertion at the new site (Figure 18.27). An intermediate (the cointegrate) then forms that consists of the donor segment covalently linked to the target DNA. An additional enzyme, called resolvase, catalyzes a site-specific recombination that allows the resolution of the cointegrate into two separate molecules.

Nonreplicative transposition The transposase makes a double-stranded cut in the donor DNA and splices it into the staggered ssDNA cut ends of the target site (a "cut-and-paste" mechanism). The cell's DNA repair system fills the gaps in the target DNA, resulting in a short direct repeat on either side of the transposon insertion (Figure 18.28). Gap repair by homologous recombination can effectively

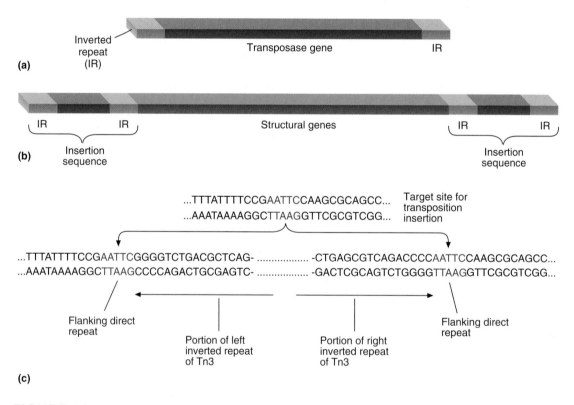

(a) Inverted repeat (IR) — Transposase gene — IR

(b) IR — IR — Insertion sequence — Structural genes — IR — IR — Insertion sequence

...TTTATTTTCCGAATTCCAAGCGCAGCC...
...AAATAAAAGGCTTAAGGTTCGCGTCGG...
Target site for transposition insertion

...TTTATTTTCCGAATTCGGGGTCTGACGCTCAG- .... -CTGAGCGTCAGACCCCAATTCCAAGCGCAGCC...
...AAATAAAAGGCTTAAGCCCCAGACTGCGAGTC- .... -GACTCGCAGTCTGGGGTTAAGGTTCGCGTCGG...

Flanking direct repeat

Portion of left inverted repeat of Tn3

Portion of right inverted repeat of Tn3

Flanking direct repeat

(c)

FIGURE 18.26

Bacterial Insertion Elements

(a) An insertion sequence. (b) A composite transposon. (c) Insertion of a transposon (Tn3) into bacterial DNA. The insertion process involves the duplication of the target site.

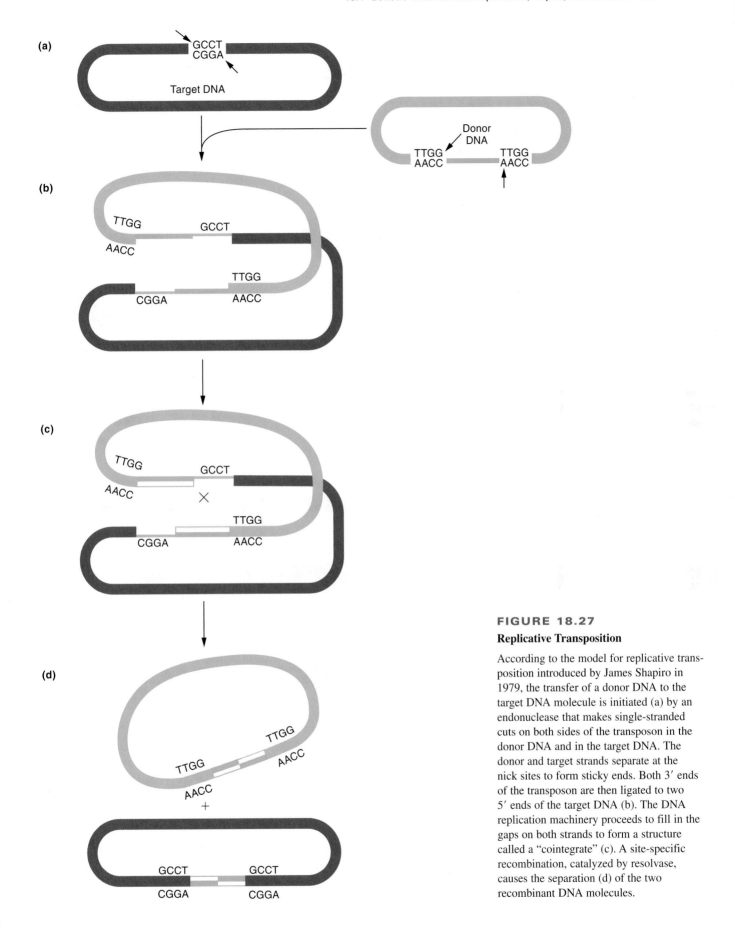

FIGURE 18.27

Replicative Transposition

According to the model for replicative transposition introduced by James Shapiro in 1979, the transfer of a donor DNA to the target DNA molecule is initiated (a) by an endonuclease that makes single-stranded cuts on both sides of the transposon in the donor DNA and in the target DNA. The donor and target strands separate at the nick sites to form sticky ends. Both 3′ ends of the transposon are then ligated to two 5′ ends of the target DNA (b). The DNA replication machinery proceeds to fill in the gaps on both strands to form a structure called a "cointegrate" (c). A site-specific recombination, catalyzed by resolvase, causes the separation (d) of the two recombinant DNA molecules.

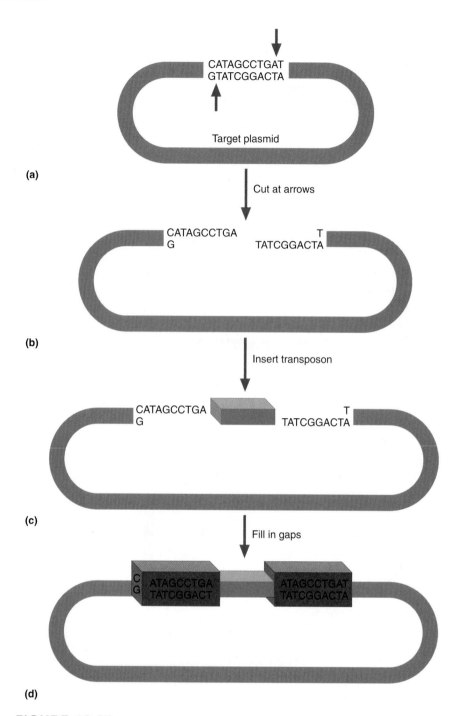

FIGURE 18.28

Nonreplicative Transposition

(a) Host DNA is cut (arrows) in a staggered fashion (b). (c) The transposon (blue) is covalently attached at both ends to a strand of host DNA. (d) After the gaps have been filled by a DNA polymerase activity, there are nine base pair repeats of host DNA (red) flanking the transposon.

replace the donor transposon, however, so it is often hard to confirm this type of transposition.

Some transposons found in eukaryotes resemble those found in bacteria. For example, the Ac element, the maize transposon that was first described by McClintock, is composed of a transposase gene flanked by short inverted repeats. (McClintock referred to the Ac transposon as a "controlling element" because it appeared to control the synthesis of the pigment anthocyanin in corn kernels.) Many other eukaryotic transposons, however, have somewhat different structures than those observed in bacteria. As described previously (p. 652), retrotransposons, also referred to as **retroposons** or **retroelements**, are a significant feature of eukaryotic genomes.

The transposition mechanism used by retrotransposons is virtually indistinguishable from that of the retrovirus life cycle. The transposition event for L1 elements, the only class of functional LINEs, is essentially a cut-and-paste process. It begins with transcription of the LI sequence by a cellular RNA polymerase. The RNA product is then translated in the cytoplasm by ribosomes. The retrotransposon proteins (transposase and reverse transcriptase) and the L1 RNA associate to form a complex and then reenter the nucleus. The RNA-protein complex binds to a target DNA sequence, usually a T-rich site. After the transposase has made a staggered cut in the DNA sequence, the reverse transcriptase synthesizes a DNA copy from the L1 RNA. The RNA is degraded, and a second DNA strand is synthesized. Insertion of the retrotransposon is complete when the staggered ends of the target DNA and the newly synthesized LI DNA element anneal and DNA ligase joins the ends. *Alu* elements, the most abundant as well as the only functional SINE, do not encode any functional molecules. *Alu* transposition depends instead on the assistance of an active L1 element. *Alu* elements, therefore, can be considered to be parasites of L1 retroelements.

Depending on the changes and their location, the effects of transposons can be viewed either as disruptive and damaging or as providing opportunities for genetic diversity. Some effects of transposition are observed as changes in gene expression, a topic that is discussed in Section 18.3.

KEY CONCEPTS

- In general recombination, the exchange of DNA sequences takes place between homologous sequences.
- In site-specific recombination, DNA-protein interactions are principally responsible for the exchange of nonhomologous sequences.
- The DNA sequences called transposons can move from one site in a genome to another.

QUESTION 18.4

One of the fascinating aspects of complex organisms such as mammals is the existence of gene families, groups of genes that code for the synthesis of a series of closely related proteins. For example, several different types of collagen are required for the proper structure and function of connective tissues. Similarly, there are several types of globin gene. It is currently believed that gene families originate from a rare event in which a DNA sequence is duplicated. Some gene duplications provide a selective advantage by providing larger quantities of important gene products. Alternatively, the two duplicate genes evolve independently. One copy continues to serve the same function, while the other eventually evolves to serve another function. Can you speculate about how gene duplications occur? Once a gene has been duplicated, what mechanisms introduce variations?

BIOCHEMISTRY IN THE LAB

Genomics

In the past few years, the life sciences have undergone an intellectual and experimental sea change created by the sequencing of the genomes of dozens of species, including that of humans. This newfound wealth of information provided by **genomics** (the large-scale analysis of entire genomes) holds promise not only for delivering unprecedented insights into the inner workings of living organisms, but also for previously unimaginable capacities to accelerate research into all areas of interest to humans (e.g., health and disease, biodiversity, and agriculture). However, determinations of the structure and identity of genes in living organisms tell us nothing about how they operate. The goal of the newly emerging science of **functional genomics** is to understand how biomolecules work together within functioning organisms. Currently, attempts at closing this knowledge gap involve a variety of new "whole-genome" technologies. For example, with DNA chips (glass or plastic wafers holding a microarray of different DNA sequence probes) researchers can simultaneously monitor the expression of thousands of genes in cultured cells. The following discussion describes the basic tools used in genomic technology.

The isolation, characterization, and manipulation of DNA sequences, now considered to be commonplace, is made possible by a series of techniques referred to as **recombinant DNA technology**. The essential feature of this technology is that DNA molecules obtained from various sources can be cut and spliced together. These techniques have revolutionized biology and biochemistry because they have made genomes more accessible to investigation than ever before. For example, the large number of DNA copies required in DNA sequencing methods have been obtainable through molecular cloning and (more recently) the polymerase chain reaction. Commercial applications of recombinant DNA techniques have revolutionized medical practice. For example, human gene products such as insulin and growth hormone, as well as certain vaccines and diagnostic tests, are now produced in large quantities by bacterial cells into which recombinant genes have been inserted. Currently, several research groups are investigating the use of recombinant techniques in human gene therapy, a process in which (it is hoped) defective genes can be replaced by their normal counterparts.

Figure 18A illustrates the basic features of recombinant DNA construction. The process begins by using a restriction enzyme (see the Biochemistry in the Lab box entitled Nucleic Acid Methods in Chapter 17, pp. 653–657) that generates sticky ends to cleave DNA from two different sources. The DNA fragments are then mixed under conditions that allow annealing (base pairing) between the sticky ends. Once base pairing has occurred, the fragments are covalently bonded together by DNA ligase. After recombinant DNA molecules have been isolated and purified, it is usually necessary to reproduce them so that sufficient quantities are available for further investigation. Molecular

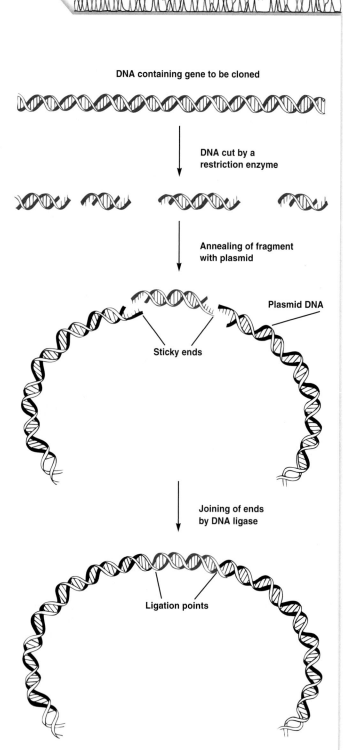

DNA containing gene to be cloned

DNA cut by a restriction enzyme

Annealing of fragment with plasmid

Plasmid DNA

Sticky ends

Joining of ends by DNA ligase

Ligation points

FIGURE 18A

Recombinant DNA Construction

Recombinant DNA molecules are created by treating DNA from two sources with restriction enzymes. Under hybridizing conditions, DNA fragments with sticky ends anneal together. Once base pairing has occurred, DNA ligase joins the fragments together.

▶▶

BIOCHEMISTRY IN THE LAB cont

cloning, a commonly used method for increasing the number of copies of DNA, is discussed next.

Molecular Cloning

In molecular cloning, a piece of DNA isolated from a donor cell (e.g., any animal or plant cell) is spliced into a vector. A **vector** is a DNA molecule capable of replication that is used to transport a foreign DNA sequence, often a gene, into a host cell.

The choice of vector depends on the size of donor DNA. For example, bacterial plasmids are often used to clone small pieces (15 kb) of DNA. Somewhat larger pieces (24 kb) are incorporated into bacteriophage λ vectors, whereas cosmid vectors are used for DNA fragments as large as 50 kb. Bacteriophage λ can be used as a vector because a substantial portion of its genome does not code for phage production and can therefore be removed. The removed viral DNA can then be replaced by foreign DNA. **Cosmids** are cloning vehicles that contain λ bacteriophage *cos* sites incorporated into plasmid DNA sequences with one or more selectable markers. The *cos* sites allow delivery to the host cell in a phage head, the plasmid DNA facilitates independent replication of the recombined unit, and the selectable markers permit detection of successful recombinants. Still larger pieces can be inserted into bacterial artificial chromosomes and yeast artificial chromosomes. **Bacterial artificial chromosomes**

(BAC), derived from a large *E. coli* plasmid called the F-factor, are used to clone DNA sequences as long as 300 kb. **Yeast artificial chromosomes** (YAC), which can accommodate up to 1000 kb, are constructed by using yeast DNA sequences that are autonomously replicating (i.e., contain a eukaryotic DNA replication origin).

As noted, forming recombinant DNA requires a restriction enzyme, which cuts the vector DNA (e.g., a plasmid) open (Figure 18B). After the sticky ends of the plasmid have annealed with those of the donor DNA, a DNA ligase activity joins the two molecules covalently. Then the recombinant vector is inserted into host cells.

In some circumstances the introduction of a cloning vector into a host cell is trivial. For example, phage vectors are designed to introduce recombinant DNA in an infective process called **transfection**, and some bacteria take up plasmids unaided. However, most host cells must be induced to take up foreign DNA. Several methods are used. In some prokaryotic and eukaryotic cells, the addition of Ca^{2+} to the medium promotes uptake. In others, a process called **electroporation**, in which cells are treated with an electric current, is used. One of the most effective methods for transforming animal and plant cells is the direct microinjection of genetic material. **Transgenic animals**, for example, are created by the microinjection of recombinant DNA into fertilized ova.

FIGURE 18B

DNA Cloning

In the cloning process, each clone is produced by introducing a recombinant molecule into a host cell, which then replicates the vector along with its own genome.

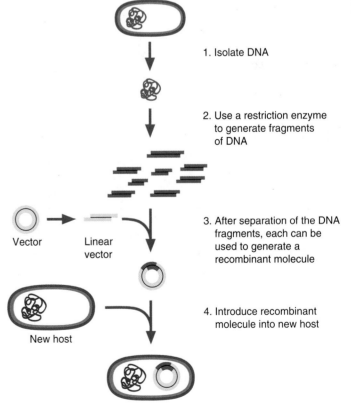

1. Isolate DNA

2. Use a restriction enzyme to generate fragments of DNA

Vector   Linear vector

3. After separation of the DNA fragments, each can be used to generate a recombinant molecule

New host

4. Introduce recombinant molecule into new host

BIOCHEMISTRY IN THE LAB cont

Once introduced, each type of cell replicates the recombinant DNA along with its own genome. Note that recombinant vectors must contain regulatory regions recognized by host cell enzymes.

As host cells that have been successfully transformed proliferate, they rapidly amplify the recombinant DNA. For example, under favorable conditions of nutrient availability and temperature, a single recombinant plasmid introduced into an *E. coli* cell can be replicated a billion times in about 11 hours. However, transformed and untransformed cells usually look exactly alike (see Figure 18C for an example of an exception to this rule). Consequently, researchers often design cloning protocols that use vectors with selectable **marker genes** (genes whose presence can be detected), to facilitate the identification of transformed cells. Antibiotic resistance genes, for example, are usually incorporated into the plasmid vectors introduced into bacteria. When the treated bacteria are plated out on a medium containing the antibiotic, only the transformed cells will grow. With eukaryotic organisms such as yeast, cells that lack an enzyme required to synthesize a nutrient may be used. For example, vectors containing the LEU2 gene are used to transform mutant yeast cells that lack a specific enzyme in the leucine biosynthetic pathway. Only cells that have successfully transformed are able to grow in a leucine-deficient medium.

In another approach, the **colony hybridization technique** (Figure 18D, bacteria are screened by using a radioactively labeled nucleic acid probe, an RNA molecule or a single-stranded

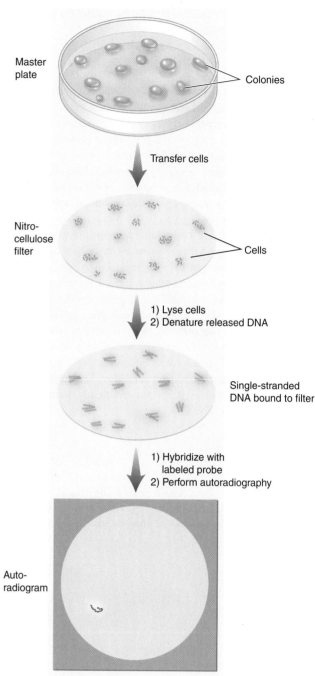

FIGURE 18D

Colony Hybridization

Bacterial cells are plated onto a solid medium that allows the growth of transformed cells only. When the colonies become visible, the plate is blotted with a nitrocellulose filter. The cells clinging to the filter are lysed, and the released DNA is denatured and deproteinized. Then a labeled probe is added and unhybridized probe molecules are washed away. Cells that possess DNA sequences that hybridize with the probe are identified by comparing the autoradiogram of the filter with the master plate.

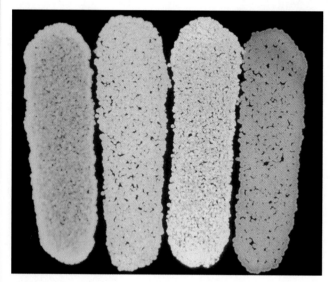

FIGURE 18C

Four Recombinant *E. coli* Cells, Each with a Different Variant of the Gene for Luciferase

Luciferase is an enzyme found in fireflies, mollusks, and several types of deep-sea fish. When in the presence of ATP and luciferin, luciferase catalyzes a light-emitting reaction. The luciferins are a group of bioluminescent compounds that emit light when they are oxidized by O_2 in a reaction catalyzed by a luciferase.

BIOCHEMISTRY IN THE LAB cont

DNA molecule with a sequence complementary to that of a specific sequence within the recombinant DNA. Bacterial cells are plated out onto solid media in petri dishes and allowed to grow into colonies. Each plate is then blotted with a nitrocellulose filter. (Some cells from each of the original colonies remain on the petri dishes.) The cells on the nitrocellulose filter are lysed, and the released DNA is treated so that hybridization with the probe can occur. Once nonhybridized probe molecules have been washed away, autoradiography (see the Biochemistry in the Lab box in Chapter 2 entitled Cell Technology, pp. 67–68) is used to identify the colonies on the master plate that possess the recombinant DNA of interest.

Polymerase Chain Reaction

Although cloning has been immensely useful in molecular biology, the **polymerase chain reaction** (PCR) is a more convenient method for obtaining large numbers of DNA copies. Using a heat-stable DNA polymerase from *Thermus aquaticus* (Taq polymerase), PCR can amplify any DNA sequence, provided the flanking sequences are known (Figure 18E). Flanking sequences must be known because PCR amplification requires primers. Priming sequences are produced by automated DNA-synthesizing machines.

PCR begins by adding *Taq* polymerase, the primers, and the ingredients for DNA replication to a heated sample of the target DNA. (Recall that heating DNA separates its strands.) As the mixture cools, the primers attach to their complementary sequences on either side of the target sequence. Each strand then serves as a template for DNA replication. At the end of this process, referred to as a *cycle*, the copies of the target sequence have been doubled. The process can be repeated indefinitely, synthesizing an extraordinary number of copies. By the end of 30 cycles, for example, a single DNA fragment has been amplified one billion times.

Genomic Libraries

Genomic libraries are collections of clones derived from fragments of entire chromosomes or genomes. They are used for a variety of purposes, the most important of which are the isolation of specific genes whose chromosomal location is unknown and in genome-wide sequencing efforts (gene mapping). Genomic libraries are produced in a process, referred to as **shotgun cloning**, in which a genome is randomly digested (Figure 18F). The range of fragment sizes, which is determined by the type of restriction enzyme and the experimental conditions chosen, must be compatible with the vector. To ensure that all sequences of interest are represented in the library, DNA samples are often only partially digested. The location of any gene can be identified if an appropriate probe is available.

In a variation of genomic libraries, collections of complementary DNA molecules called **cDNA libraries** are produced

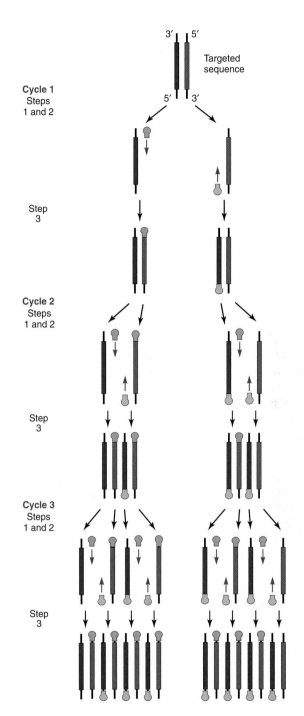

FIGURE 18E

Polymerase Chain Reaction

A single DNA molecule can be amplified millions of times by replicating a three-step cycle. In the first step the dsDNA sample is denatured by heating to 95°C. In step 2 the temperature is quickly lowered to 50°C and an oligonucleotide primer is added. The primer hybridizes to complementary sequences on the ends of the two strands. During step 3, DNA synthesis occurs as the temperature is raised to 70°C, the optimal temperature of *Taq* polymerase. The cycle is then repeated with both old and new strands serving as templates.

BIOCHEMISTRY IN THE LAB cont

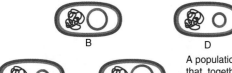

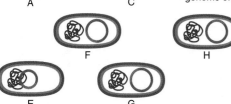

— Bacterium X

Isolate the total DNA from the organism

Generate genomic fragments

Join genomic fragments to vectors (—)

Introduce the recombinant molecules into *E. coli* cells

A population of *E. coli* cells that, together, contain all of the fragments of the genome of Bacterium X

FIGURE 18F

Creation of a DNA Library Using the Shotgun Method

After the organism's DNA is isolated from the organism and purified, it is cleaved into fragments with a restriction enzyme. The fragments are then randomly incorporated into vectors, and the recombinant molecules are introduced into cells such as *E. coli*. The collection of these cells is called a genome library.

from mRNA molecules by reverse transcription. This technique can be used to evaluate the transcriptome of certain cell types under specified circumstances. In other words, it is a method for determining which genes are expressed in a particular cell type. For example, with the use of DNA chip technology, gene expression in normal and diseased cells can be

investigated and compared. cDNA libraries are especially useful when eukaryotic DNA is cloned because mRNA molecules lack noncoding or intron sequences. Consequently, gene products can be more easily identified, and large amounts of gene product can be generated in bacteria, which cannot process introns.

Chromosome Walking

Overlapping DNA fragments of 20 to 40 kb from genomic libraries can be sequenced by means of chromosome walking (Figure 18G). First a radioactively labeled probe is used to identify and analyze clones containing the complementary sequence and any contiguous sequences. This process is repeated, using the end of the newly identified sequence as a probe. Chromosome walking continues in both directions until the entire molecule has been sequenced or a gene of interest finally located. A set of overlapping sequences is referred to as a **contig**. When eukaryotic genomes are analyzed, their large size often requires the use of large cloning vectors such as YACs and a technique called chromosomal jumping. In **chromosomal jumping** the overlapping clones contain DNA sequences of several hundred kb that are generated using restriction enzymes that cut infrequently.

DNA Microarrays

DNA microarrays, or DNA "chips," are used to analyze the expression of thousands of genes simultaneously (Figure 18H). Often no larger than a postage stamp, a DNA microarray consists of thousands, or hundreds of thousands, of oligonucleotides or ssDNA fragments attached to a glass or plastic support. At each position in the microarray the attached sequence, acting as a DNA probe, is designed to hybridize with a specific gene. In investigations of gene expression, an entire set of mRNA molecules from the cells of interest (i.e., the transcriptome) are reverse-transcribed into cDNA. The cDNA molecules, labeled with a fluorescent dye, are then incubated with a microarray under hybridizing conditions. After incubation, the microarray is washed to remove unhybridized molecules. Researchers then determine which genes are being expressed by identifying the positions on the microarray that are fluorescing. Scientists have used microscopes, photomultiplier tubes, and computer software to observe changes in gene expression under a variety of circumstances. Examples include comparisons of normal and cancerous cells, and cells exposed to different nutrients or signal molecules.

Genome Projects

The Human Genome Project was an intensive international effort to determine the nucleotide sequence of the entire human genome. As this goal was reached, the attention of researchers

BIOCHEMISTRY IN THE LAB cont

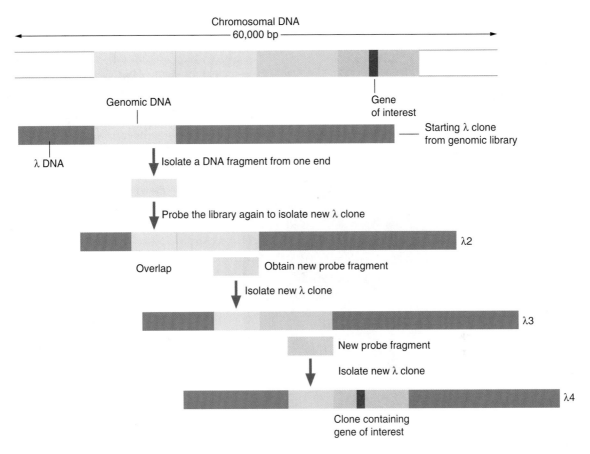

Chromosomal DNA
— 60,000 bp —

Genomic DNA

Gene of interest

λ DNA

Starting λ clone from genomic library

Isolate a DNA fragment from one end

Probe the library again to isolate new λ clone

λ2

Overlap

Obtain new probe fragment

Isolate new λ clone

λ3

New probe fragment

Isolate new λ clone

λ4

Clone containing gene of interest

FIGURE 18G

Chromosome Walking

In chromosome walking DNA clones, which contain overlapping sequences, are systematically identified. They may then be mapped and sequenced. Unknown genes may also be searched for. The process begins when DNA is cleaved into pieces and cloned. (In this example, bacteriophage λ vectors are used.) One end of the starting clone is labeled and used as a probe to identify the clone in the λ library that contains both that sequence and an adjacent sequence. Repeating this step results in the labeling of the end of the second clone, which is used as a probe to identify yet another overlapping clone. The process continues until a collection of clones that together contain all the sequences in the original DNA fragment has been obtained.

shifted to the **annotation** (i.e., functional identification) of the 30,000-odd human genes. Just as scientists have historically used structural and functional comparisons of other organisms in anatomy, biochemistry, physiology, and medicine to better understand human biology, the current effort to interpret human genome data is being aided immensely by comparisons with the information obtained in other genome projects. The genomes of well-researched organisms as diverse as bacteria (e.g., *E. coli*), yeast (e.g., *Saccharomyces cerevisiae*), the worm *Caenorhabditis elegans*, the fruit fly *Drosophila*, and various mammals (e.g., the mouse) have been used in genome structure analysis and in the assignment of recently discovered genes in other organisms.

When a genome is deciphered, no matter what type of organism it comes from, there are three basic steps in its analysis:

1. **Genetic linkage mapping**. Over most of the twentieth century geneticists constructed chromosome maps for selected organisms by analyzing recombination frequencies derived from genetic crosses. In humans the recombination frequency data used to determine the physical relationships and relative distances of genes on chromosomes were obtained by analyzing the pedigrees of large families. The low-resolution maps obtained from genetic mapping have been used as a framework for analyzing data obtained from more sophisticated methods.

BIOCHEMISTRY IN THE LAB cont

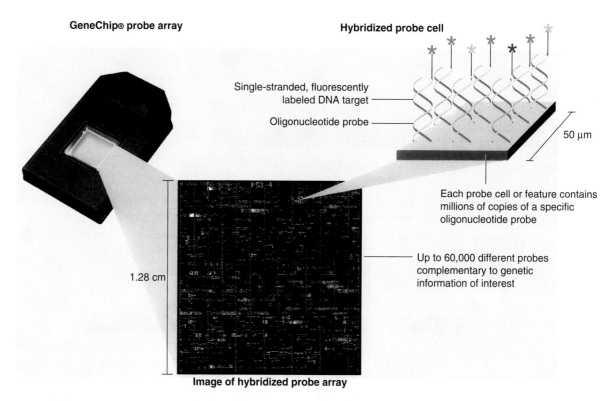

GeneChip® probe array

Hybridized probe cell

Single-stranded, fluorescently labeled DNA target

Oligonucleotide probe

50 µm

Each probe cell or feature contains millions of copies of a specific oligonucleotide probe

Up to 60,000 different probes complementary to genetic information of interest

1.28 cm

Image of hybridized probe array

FIGURE 18H

DNA Microarray Technology

DNA microarrays can be used to determine which genes are expressed in a specific cell type because each "chip" can accommodate from thousands to millions of DNA probes. (Oligonucleotide probes are synthesized on the chip surface by means of photolithographic techniques similar to those used in the manufacture of computer chips.) The microarray is incubated under hybridizing conditions with fluorescently labeled cDNA. The cDNA molecules are derived from mRNA extracted from the cells of interest.

2. **Physical mapping of genomes**. Physical maps are obtained by cleaving chromosomal DNA from the organism of interest with a restriction enzyme. Each DNA fragment is then cloned. The order of the fragments in the chromosome is determined by using probes made from known markers in the low-resolution genetic linkage map. The process is repeated with several other restriction enzymes so that ambiguities in the data can be resolved.

3. **DNA sequencing**. Individual cloned fragments from the entire chromosome or genome are sequenced. To make this feat feasible, DNA sequencing has been automated, as described in Chapter 17 (see the Biochemistry in the Lab box entitled Nucleic Acid Methods, pp. 653–657). The nucleotide sequence of an entire genome is determined by linking the results of sequencing experiments from overlapping clones.

Bioinformatics

The emergence of the *high-throughput* (i.e., rapid, high-volume, automated) technologies to analyze living organisms has created a vast amount of data on nucleic acid and polypeptide sequences. The information, which is collected from genome and proteome sequencing projects, and from microarray analysis of cell processes such as transcription, is placed in databases that are available to the scientific community. How do scientists analyze such enormous volumes of raw data? As a result of technological advances in computer science, applied mathematics, and statistics, the new science of **bioinformatics** has provided researchers with a powerful investigational tool. The use of computer algorithms has made a wide variety of previously intractable problems feasible, as the following examples illustrate.

1. Genes can be located by a process referred to as sequence inspection. Gene prediction programs utilize several clues to locate sequences that can potentially code for polypeptides called *open reading frames* (ORFs). ORFs are extended DNA sequences that could potentially code for a polypeptide. They begin with the three-base sequence UAG called a *start codon* and end with a *stop codon* of

BIOCHEMISTRY IN THE LAB cont

UAA, UAG, or UGA. Eukaryotic ORF scanning is complicated by the presence of introns, some of which are longer than the polypeptide domain–coding exons.

2. The capacity to search the genomes of hundreds of organisms for similarities among gene or regulatory sequences has provided invaluable insight into the relatedness of living organisms and the mechanisms used to sustain living processes.

3. Protein structure prediction has been facilitated by a method called homology modeling. Once a new protein-coding gene has been discovered, bioinformatic analysis is used to search among homologous or near-homologous molecules whose structure is already known.

4. In the field of evolutionary biology, bioinformatic programs have been used to trace the lineages of organisms based on rare events such as gene duplications and lateral gene transfer. (Lateral gene transfer is the transfer of genes between species.)

5. High-throughput gene expression analysis is now used to determine the genes involved in medical disorders (e.g., in comparisons of the transcription products of normal and cancerous cells).

6. Systems biologists' use of complex mathematical modeling combined with an ever-increasing source of biological data promises to significantly improve our understanding of life's operating systems.

■

18.2 TRANSCRIPTION

As with all aspects of nucleic acid function, the synthesis of RNA molecules is a very complex process involving a variety of enzymes and associated proteins. Recall that RNA molecules are transcribed from the cell's genes. As RNA synthesis proceeds, the incorporation of ribonucleotides is catalyzed by RNA polymerase, sometimes referred to as DNA-dependent RNA polymerase, or RNAP. The reaction catalyzed by all RNA polymerases is

$$NTP + (NMP)_n \longrightarrow (NMP)_{n+1} + PP_i$$

Because the nontemplate or plus (+) strand has the same base sequence as the RNA transcription product (except for the substitution of U for T), it is also called the **coding strand** (Figure 18.29). By convention, the direction of the gene, a segment of double-stranded DNA, is the same as the direction of the coding strand. Polymerization proceeds from the 5′ end to the 3′ end of the gene because the template DNA strand, also called the minus (−) strand, and the newly made RNA molecule are antiparallel. As noted, transcription generates several types of RNA, of which rRNA, tRNA, and mRNA are directly involved in protein synthesis (Chapter 19).

```
DNA                                           (+)
  5′— TTTGGACAACGTCCAGCGATC —3′   Nontemplate strand

  3′— AAACCTGTTGCAGGTCGCTAG —5′   Template strand
                                              (−)

RNA
  5′—UUUGGACAACGUCCAGCGAUC—3′
```

FIGURE 18.29

DNA Coding Strand

One of the two complementary DNA strands, referred to as the template (−) strand, is transcribed. The RNA transcript is identical in sequence to the nontemplate (+) or coding strand, except for the substitution of U for T.

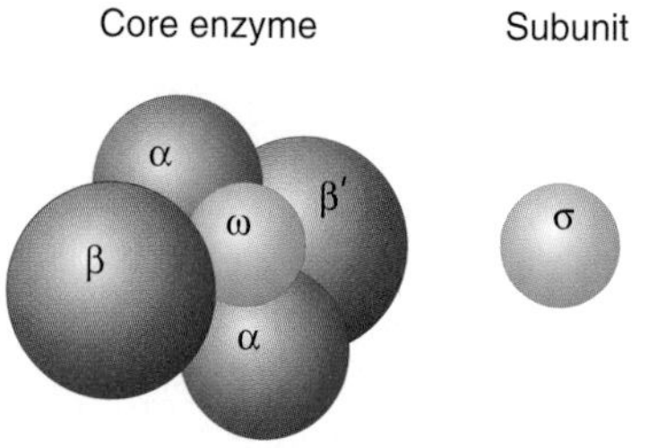

Core enzyme

α

β' ω

β

α

Subunit

σ

FIGURE 18.30

E. coli **RNA Polymerase**

The *E. coli* RNA polymerase consists of two α subunits and one each of β, and β' subunits. The transient binding of a σ subunit allows binding of the core enzyme to appropriate DNA sequences. Note that ω promotes the assembly of the core enzyme.

Transcription in Prokaryotes

The RNA polymerase (Figure 18.30) in *E. coli* catalyzes the synthesis of all RNA classes. The core enzyme (α_2, β, and β'), with a molecular weight of 370 kD, catalyzes RNA synthesis. Another protein, the ω subunit, promotes the assembly of the RNAP holoenzyme. The transient binding of the σ (sigma) factor to the core enzyme allows it to bind both the correct template strand and the proper site to initiate transcription. A variety of σ factors have been identified. For example, in *E. coli*, σ^{70} is involved in the transcription of most genes, whereas σ^{32} and σ^{28} promote the transcription of heat shock genes and the flagellin gene, respectively. (As its name suggests, flagellin is a protein component of bacterial flagella.) The superscript indicates the protein's molecular weight in kilodaltons.

The transcription of an *E. coli* gene is outlined in Figure 18.31. The process consists of three stages: initiation, elongation, and termination. Each is discussed briefly.

The initiation of transcription involves the binding of RNA polymerase to a **promoter**, a regulatory DNA sequence that is located upstream (i.e., toward the 5' end of a polynucleotide) of a gene. Although prokaryotic promoters are variable in size (from 20 to 200 bp), two short sequences at positions about 10 and 35 bp away from the transcription initiation site are remarkably similar among various bacterial species. There is a set of base sequences, called consensus sequences, at each of these sites. A **consensus sequence** represents an average

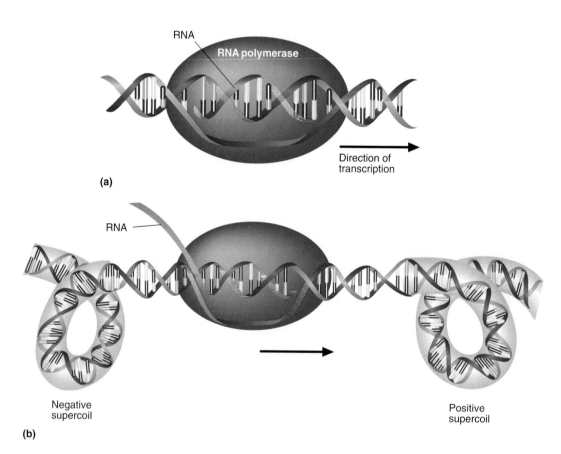

FIGURE 18.31

Transcription Initiation in *E. coli*

(a) A transcription bubble forms as a short DNA segment unwinds. An RNA-DNA hybrid forms as transcription progresses. The bubble moves to keep up with transcription as DNA unwinds ahead of it and rewinds behind it.
(b) Transcription induces coiling. Positive supercoils form ahead of the bubble, while negative supercoils form behind it.

of a number of closely related but nonidentical sequences. The sequences shown in Figure 18.32 are named in relation to the transcription starting point, the -35 region and the -10 region. (The -10 region is also called the *Pribnow box*, after its discoverer.) RNA polymerase slides along the DNA until it reaches a promoter sequence. Promoters vary widely in the efficiency with which they productively bind RNA polymerase. Transcription initiation rates between "strong" promoters (close to the consensus sequence) and "weak" promoters (far from the consensus sequence) may vary by as much as a thousandfold. Mutations within a promoter sequence usually weaken the promoter but can also convert a weak promoter in to a stronger one. Neither possibility is favorable. Once the RNAP holoenzyme (i.e., the core enzyme and its associated σ factor) has bound to the promoter region, a short DNA segment unwinds as the β' and σ subunits break the hydrogen bonds between 13 bp in the Pribnow box. Because the DNA strands are now separated, the enzyme-promoter complex is referred to as "open." Transcription begins with the binding of the first nucleoside triphosphate (usually ATP or GTP) to the RNA polymerase complex. A nucleophilic attack by the 3'-OH group of the first nucleoside triphosphate on the α-phosphate of a second nucleoside triphosphate (also positioned by base pairing in the adjacent site) causes the first phosphodiester bond to form. (Because the phosphate groups of the first molecule are not involved in this reaction, the 5' end of prokaryotic transcripts possesses a triphosphate group.) RNAP proceeds to catalyze additional phosphodiester bonds between ribonucleotides that are base-paired to the DNA template strand within the promoter. For the initiation phase to end successfully (i.e., for the RNAP holoenzyme to move away from the promoter), the growing RNA chain must reach a length of about 10 nt. Several attempts at "promoter clearance" usually fail, and the truncated transcripts are released and then degraded. When the transcribed sequence reaches a length of 10 nucleotides, the conformation of the RNA polymerase complex changes, the σ factor is released, and the initiation phase ends. As soon as an RNA polymerase has moved beyond the promoter site, another RNA polymerase can move in, bind to the site, and start another round of RNA synthesis.

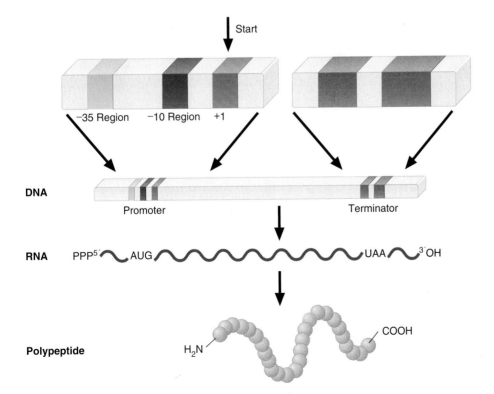

FIGURE 18.32

Typical *E. coli* Transcription Unit

If RNA polymerase can bind to the promoter, DNA transcription begins at $+1$, downstream from the promoter in bacteria. Translation of mRNA begins as soon as the ribosome-binding site on the transcribing mRNA is available.

Once the σ factor has detached and the affinity of the RNA polymerase complex for the promoter site has decreased, the elongation phase begins. The core RNA polymerase converts to an active transcription complex as it binds several accessory proteins. As RNA synthesis proceeds in the $5' \rightarrow 3'$ direction (Figure 18.31), the DNA unwinds ahead of the *transcription bubble* (the transiently unwound DNA segment, composed of 12 to 14 bp, in which an RNA-DNA hybrid has formed). At any one time there are the equivalent of about 30 bp of DNA within RNAP. The active site of the enzyme complex lies between the β and β' subunits. (Figure 18.33). As the dsDNA enters the enzyme and is separated into two strands, the template strand enters the active site through a channel. The nontemplate strand is looped away from the active site and travels in its own channel. When the template and nontemplate strands emerge from their separate channels, they re-form a double helix. Meanwhile, the growing RNA transcript exits through its own channel formed in the β and β' subunits. The unwinding action of RNA polymerase creates positive supercoils ahead of the transcription bubble and negative supercoils behind the bubble, which are resolved by topoisomerases. (As the bubble moves down the gene, it is said to move "downstream.") The incorporation of ribonucleotides continues until a termination signal is reached.

There are two types of transcription termination in bacteria: intrinsic termination and rho-dependent termination. In **intrinsic termination** (also referred to as **rho-independent termination**) RNA synthesis is terminated as the result of the transcription of the termination sequence that consists of an inverted repeat sequence followed by from 6 to 8 As (Figure 18.34). When the termination sequence is transcribed, the inverted repeat forms a stable hairpin that causes

FIGURE 18.33

Bacterial RNA Polymerase Model

In this model the components of RNAP are color-coded as follows: β (blue), β' (pink). The DNA template strand is red and the nontemplate strand is yellow. In (a) and (b) the double helix lies within a horizontal trough between the β and β' subunits. The black arrows indicate the direction of DNA movement within the enzyme. Rotation of the model by 21° reveals the growing RNA Transcript (gold) as it exits the RNAP.

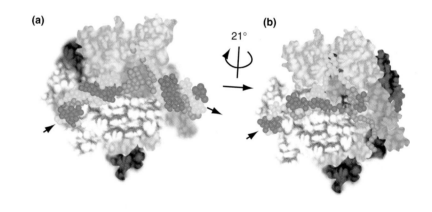

FIGURE 18.34

Intrinsic Termination

When the termination sequence (an inverted repeat followed by a series of As) in the template strand has been transcribed, the resulting RNA sequence forms into a stable hairpin (stem-loop) structure. After the disruption of DNA-RNA interactions by the hairpin, the RNA transcript is held on the template strand only by a short sequence of AU base pairs. The RNA molecule quickly dissociates because AU interactions are very weak.

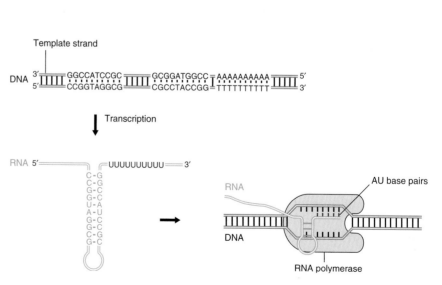

the RNA polymerase to slow or stop. The RNA transcript is then released because the RNA-DNA hybrid dissociates owing to weak base pair interactions between the As in a short polyadenylate [poly(A)] sequence that follows the inverted repeat and the complementary Us in the transcript. In **rho-dependent termination**, strong hairpins do not form and termination requires the aid of a protein known as the **rho factor**, an ATP-dependent helicase (Figure 18.35). Rho factor binds to a specific recognition sequence on the nascent mRNA strand upstream from the termination site. It then proceeds to unwind the RNA-DNA helix to release the transcript, and dislodge the polymerase.

In prokaryotes, mRNA is translated as soon as a ribosome-binding site is exposed (cotranscriptional translation). However, mature rRNA and tRNA molecules are produced from larger transcripts by posttranscriptional processing. The RNA processing reactions for *E. coli* rRNA are outlined in Figure 18.36. The *E. coli* genome contains several sets of the rRNA genes 16S, 23S, and 5S. (Each set of genes is called an **operon**.) In the primary processing step, the polycistronic 30S transcript is methylated and then cleaved by several RNases into a number of smaller segments. Further cleavage by different RNases produces mature rRNAs. A few tRNAs are also produced. The other tRNAs are produced from primary transcripts in a series of processing reactions in which they are trimmed down by several RNases. In the last step of tRNA processing, a large number of bases are altered by several modification reactions (e.g., deamination, methylation, and reduction).

Transcription in Eukaryotes

DNA transcription in prokaryotes and eukaryotes is similar is several respects. For example, the bacterial RNA polymerases and their eukaryotic counterparts are structurally similar and use a common transcription mechanism (e.g., promoter recognition, truncated transcripts, and promoter clearance). Also, although the initiation factors in bacteria and eukaryotes are only distantly related, they perform similar functions. These two types of organisms differ significantly, however, in the regulatory mechanisms that control gene expression. One of the most prominent examples of these differences is the limited access of the eukaryotic transcription machinery to DNA. Chromatin is usually at least partially condensed. Yet for transcription to occur, DNA must be sufficiently exposed and accessible for RNA polymerase activity. For DNA to be permissive to transcription, the histone tails must be acetylated by histone acetyltransferases (HATs), and the histone-DNA contacts must be weakened by **chromatin-remodeling complexes**. There are two classes of chromatin-remodeling complex. The SWI and SNF proteins (as a SWI/SNF complex) facilitate the release of the histone particle, while the NURF proteins release the contacts only enough to allow the histone particle to slide out of the way (Figure 18.37).

Principally because of the complexity of eukaryotic genomes, eukaryotic DNA transcription is not understood as completely as the prokaryotic process. However, the eukaryotic process is known to possess several unique features.

RNA POLYMERASE ACTIVITY Eukaryotes possess three nuclear RNA polymerases, each of which differs in the type of RNA synthesized, subunit structure, and relative amounts. RNA polymerase I, which is localized within the nucleolus, transcribes the larger rRNAs (28S, 18S, and 5.8S). The precursors of mRNA and most snRNAs are transcribed by RNA polymerase II, and RNA polymerase III is responsible for transcribing the precursors of the tRNAs, 5S rRNA, U6 snRNA, and the snoRNAs. Each polymerase possesses two large subunits and several (8–12) smaller subunits. For example, the two large subunits of RNA polymerase II, the enzyme that transcribes the majority of eukaryotic genes, have molecular weights of 215 and 139 kD. The number of smaller subunits varies among species; for example, plants possess eight, whereas

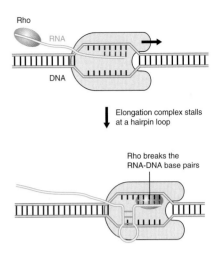

FIGURE 18.35

Rho-Dependent Termination

Rho factor is an ATP-dependent helicase that binds to a C-rich sequence in the RNA transcript. Once bound, rho moves along the transcript in the $5' \rightarrow 3'$ direction until it reaches the termination site now inside the transcription bubble. The RNA polymerase has stalled because, it is believed, of the formation of a weak hairpin. Rho disrupts DNA-RNA interactions by using ATP-derived energy to unwind the DNA-RNA hybrid, and the RNA transcript is released.

KEY CONCEPTS

- During transcription, an RNA molecule is synthesized from a DNA template.

- In prokaryotes this process involves a single RNA polymerase activity.

- Transcription is initiated when the RNA polymerase complex binds to a promoter sequence.

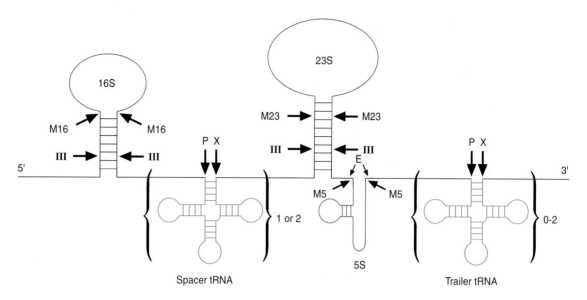

FIGURE 18.36

Ribosomal RNA Processing in *E. coli*

Each rRNA operon encodes a primary transcript that contains one copy each of 16S, 23S, and 5S rRNAs. Each transcript also encodes one or two spacer tRNAs and as many as two trailer tRNAs. Posttranscriptional processing involves numerous cleavage reactions catalyzed by various RNases and splicing reactions. (Individual RNases are identified by letters and/or numbers, e.g., M5, X, III.) RNase P is a ribozyme.

FIGURE 18.37

Chromatin Remodeling

The acetylation by HAT of the histone tails breaks their contacts with DNA. The core histones are then released from DNA contacts by sliding out of the way through the action of NURF proteins (a) to expose the promoter region on the DNA to the transcriptional machinery, or by a localized conformational change in chromatin (b) promoted by the SWI/SNF chromatin-remodeling complex.

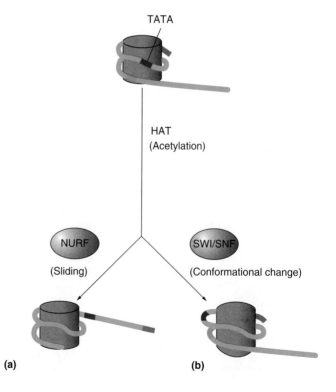

vertebrates have six. Some of the smaller subunits are also present in the other two RNA polymerases. In contrast to the prokaryotic RNA polymerase, the eukaryotic enzymes cannot initiate transcription themselves. Various transcription factors must be bound at the promoter before transcription can begin.

PROMOTERS The promoter sequences in eukaryotic DNA are larger, more complicated, and more variable than those of prokaryotes. Many promoters for RNA polymerase II contain consensus sequences, referred to as the TATA box, which occur about 25 to 30 bp upstream from the transcription initiation site. The binding of the transcription factor TFIID to the TATA box is the first step in the assembly of the RNA polymerase II transcription complex. The frequency of transcription initiation is often affected by binding certain transcription factors to upstream elements such as the *CAAT box* and the *GC box*. The activity of many promoters is affected by *enhancers*, regulatory sequences that may occur thousands of base pairs upstream or downstream of the gene they affect. (In yeast these sequences are called upstream activator sequences, or UAS.) The effects of enhancers can be complex. For example, a single gene may be controlled by the combined activities of several enhancers. Hormone response elements (Section 16.2) often act as enhancers.

RNA polymerase II (RNAP II) is the most investigated type of RNA polymerase in eukaryotes, largely because of its role in mRNA synthesis. The structural and functional properties of yeast RNAP II were determined by Roger Kornberg (Figure 18.38), who received the 2006 Nobel Prize in Chemistry for this work. In addition to the core enzyme (Figure 18.39), the RNAP II transcription

FIGURE 18.38

Roger Kornberg (right) won the 2006 Nobel Prize in Chemistry for his elucidation of the molecular mechanism of eukaryotic transcription. His father, Arthur Kornberg, who died in 2007, received the 1959 Nobel Prize in Physiology or Medicine for his work on prokaryotic DNA polymerases.

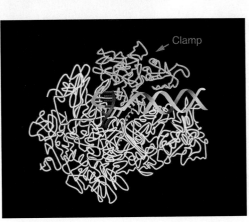

FIGURE 18.39

Yeast RNAP II Core Enzyme Structure

This view of the core RNAP II shows the polypeptide chains in white and orange (clamp protein), DNA (blue = template strand and green = nontemplate strand), and RNA (lavender).

machinery (3000 kD) consists of RNAP II, a set of five transcription factors referred to as *general transcription factors*, and a 20-subunit protein complex called Mediator. The general transcription factors TFIIB, TFIID, TFIIE, TFIIF, and TFIIH are the minimum number of additional proteins that are necessary for accurate transcription. They facilitate the recognition of the promoter, the construction of the preinitiation complex, and the ATP-dependent unwinding of DNA. Mediator is a protein complex that is required for the transcription of almost all RNAP II promoters. It has a structure with three domains: head, middle, and tail (Figure 18.40). Mediator is essentially a signal processor that acts as an adaptor between RNAP II and the transcription factors that are bound at positive and negative regulatory DNA sequences, which may be some distance away from the gene(s) they modulate. This sophisticated form of regulation is largely responsible for the intricate gene expression mechanisms in complex eukaryotic processes such as cell development and differentiation.

Once promoter clearance has been achieved, the RNAP II holoenzyme dissociates from Mediator and the elongation phase of transcription begins. The RNAP II complex continues the transcription process well past the functional end of the nascent transcript until a signal sequence (5'-AAUAAA-3') called a poly(A) sequence is reached. As soon as the poly(A) signal sequence is transcribed, several proteins now linked to the RNAP II complex bind to it and cause termination by cleaving the transcript about 10–30 nt downstream of the poly(A) sequence. A polyadenylate strand (100–250 adenylate residues), called a *poly(A) tail*, is synthesized by poly(A) polymerase and then covalently linked to the 3' terminus of mRNA transcripts.

RNA PROCESSING Posttranscriptional processing of RNA, especially that of mRNA, is different in prokaryotes and eukaryotes. In contrast to prokaryotic mRNA, which usually requires little or no processing, eukaryotic mRNAs are the products of extensive modification. Throughout processing, which begins shortly after transcription initiation, the nascent mRNA transcripts, called pre-mRNAs, become associated with about 20 different types of nuclear protein in ribonucleoprotein particles (hnRNP). Shortly after the transcription of the primary transcript begins, a modification of the 5' end called *capping* occurs.

FIGURE 18.40

The Yeast Mediator–RNA Polymerase II Holoenzyme Complex

The RNAP II preinitiation complex (RNAP II with associated transcription factors), shown in white, interacts with the crescent-shaped Mediator (shown in blue) via its head and middle domains. The tail domain of Mediator interacts with a regulatory protein that is bound to DNA.

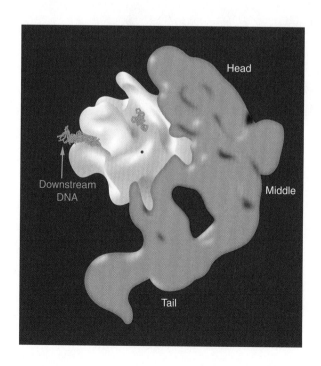

The cap structure (Figure 18.41), which consists of 7-methylguanosine linked to the mRNA through a triphosphate linkage, is synthesized when the pre-mRNA is about 30 nt long. It protects the 5′ end from exonucleases and promotes mRNA translation by ribosomes.

One of the more remarkable features of eukaryotic RNA processing is the removal of introns from RNA transcripts. In this process, called **RNA splicing**, introns are cut out and the exons are linked together to form a functional product. Most research efforts have been concerned with the splicing of pre-mRNAs, which is now described. The number of introns in the protein-coding eukaryotic genes varies widely, from one in the intron-containing genes in lower eukaryotes such as yeast to dozens or even hundreds in some mammalian genes. The human dystrophin gene, the largest in humans, is 2.4 million base pairs (Mb) long and contains 79 exons. The dystrophin-spliced mRNA (14 kb) codes for a structural protein with 3600 amino acid residues that connects the cytoskeleton with the plasma membrane and the extracellular matrix. Several mutations in the dystrophin gene cause the X-linked recessive disorder, Duchenne muscular dystrophy, which presents as progressive muscle degeneration.

RNA splicing takes place in a 4.8-megadalton (MD) RNA-protein complex called a **spliceosome**. Splicing occurs at certain conserved sequences, of which there are at least seven types. In eukaryotic nuclear pre-mRNA transcripts there are two intron types: GU-AG and AU-AC. In GU-AG introns, for example,

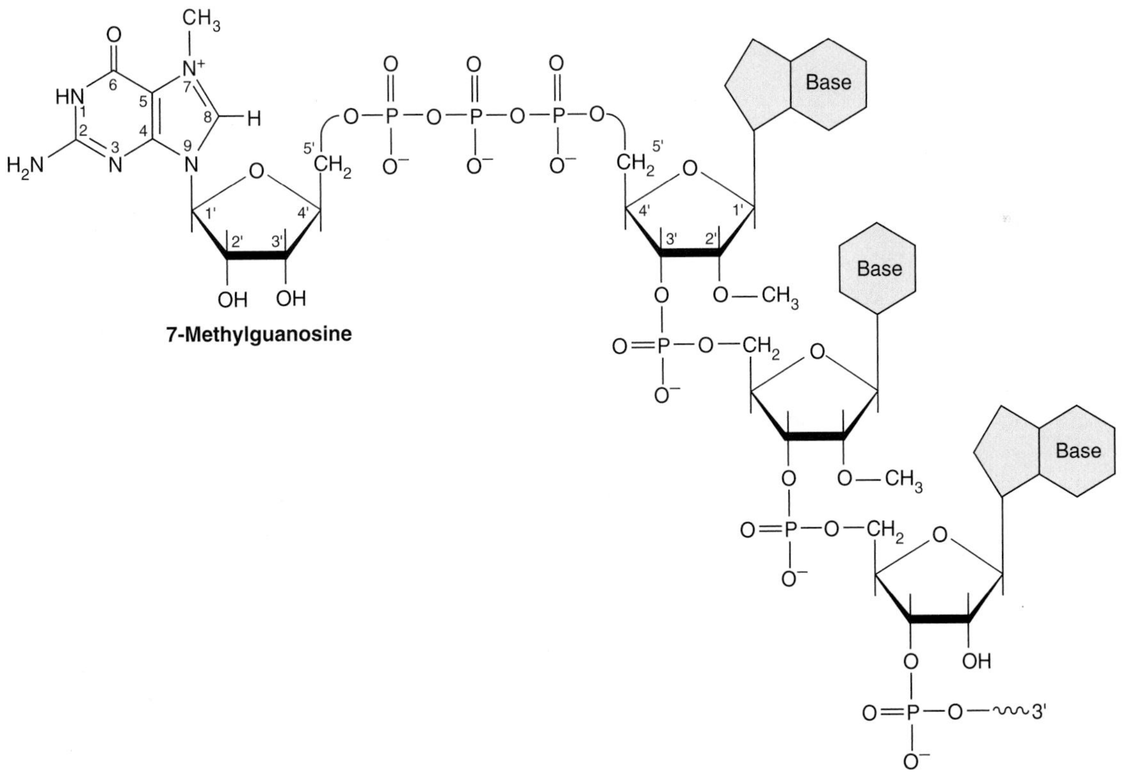

FIGURE 18.41

The Methylated Cap of Eukaryotic mRNA

The cap structure consists of a 7-methylguanosine attached to the 5′ end of an RNA molecule through a unique 5′ → 5′ linkage. The 2′-OH of the first two nucleotides of the transcript are methylated. The 5′ → 5′ bond formation is catalyzed by a guanyl transferase; N-7 methylation is catalyzed by guanine methyl transferase.

5'-GU-3' and 5'-AG-3' are the first and last dinucleotide sequences, respectively. The splicing event (Figure 18.42) occurs in two reactions:

1. A 2'-OH of an adenosine nucleotide within the intron, referred to as the "branch site," attacks a phosphate in the 5' splice site in a transesterification reaction. The intron forms a loop called a *lariat* because of the newly created 5'→2' phosphodiester bond.

2. The lariat is cleaved and the two exons are joined when the 3'-OH of the upstream exon attacks a phosphate that is adjacent to the lariat. Note that the 5' splice site is called the **donor site** and the 3' site is referred to as the **acceptor site**.

Recent evidence indicates that four active spliceosomes assemble with each pre-mRNA to form a **supraspliceosome** (21.1 MD). It has been suggested that the supraspliceosome increases the speed and efficiency of transcript splicing and provides opportunities for intron excision proofreading.

FIGURE 18.42

RNA Splicing

mRNA splicing begins with the nucleophilic attack of the 2'-OH of a specific adenosine on a phosphate in the 5'-splice site. A lariat is formed by a 2',5'-phosphodiester bond. In the next step, 3'-OH of exon 1 (acting as a nucleophile) attacks a phosphate adjacent to the lariat. This reaction releases the intron and ligates the two exons.

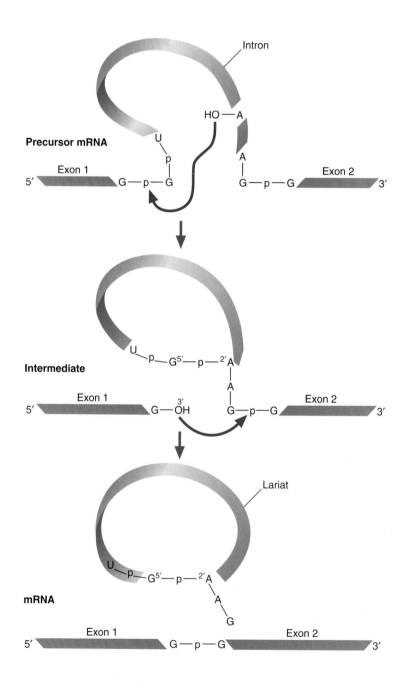

18.3 GENE EXPRESSION

Ultimately, the internal order most essential to living organisms requires the precise and timely regulation of gene expression. It is, after all, the capacity to switch genes on and off that enables cells to respond efficiently to a changing environment. In multicellular organisms, complex programmed patterns of gene expression are responsible for cell differentiation and intercellular cooperation.

The regulation of genes, as measured by their transcription rates, is the result of a complex hierarchy of control elements that coordinate the cell's metabolic activities. Some genes, referred to as **constitutive** or housekeeping **genes**, are routinely transcribed because they code for gene products (e.g., glucose-metabolizing enzymes, ribosomal proteins, and histones) required for cell function. In addition, in the differentiated cells of multicellular organisms, certain specialized proteins are produced that cannot be detected elsewhere (e.g., hemoglobin in red blood cells). Genes that are expressed only under certain circumstances are referred to as *inducible*. For example, the enzymes that are required for lactose metabolism in *E. coli* are synthesized only when lactose is actually present and glucose, the bacterium's preferred energy source, is absent.

Considering the obvious complexity of function observed in living organisms, it is not surprising that the regulation of gene expression has proven to be both remarkably complex and difficult to investigate. For many of the reasons stated, knowledge about prokaryotic gene expression is significantly more advanced than that of eukaryotes. Prokaryotic gene expression was originally investigated, in part, as a model for the study of the more complicated gene function of mammals. Although it is now recognized that the two genome types are vastly different in many respects, the prokaryotic work has provided many valuable insights into the mechanisms of gene expression. In general, prokaryotic gene expression involves the interaction of specific proteins (sometimes referred to as regulators) with DNA in the immediate vicinity of a transcription start site. Such interactions may have either a positive effect (i.e., transcription is initiated or increased) or a negative effect (i.e., transcription is blocked). In an interesting variation, the inhibition of a negative regulator (called a *repressor*) activates affected genes. (The inhibition of a repressor gene is referred to as derepression.) Eukaryotic gene expression uses these mechanisms and several others, including gene rearrangement and amplification and various complex transcriptional, RNA processing, and translational controls. In addition, the spatial separation of transcription and translation inherent in eukaryotic cells provides another opportunity for regulation: RNA transport control. Finally, eukaryotes (and prokaryotes) also regulate cell function through the modulation of proteins by covalent modifications.

This section describes, several examples of control of gene expression. The discussion of prokaryotic gene expression focuses on operons and riboswitches. An **operon** is a set of genes under the regulation of the same operator and promoter(s). The operator is a regulatory sequence that binds to specific repressor or activator proteins that modulate gene expression. The *lac* operon in *E. coli*, originally investigated by François Jacob and Jacques Monod in the 1950s, is the most thoroughly researched example. Bacteria also employ an RNA-based control mechanism called the **riboswitch**, made up of a specific untranslated sequence (UTS) within the mRNA along with the small metabolite to which it binds. The action of the riboswitch usually represses gene expression by terminating transcription, blocking translation, or self-destructing. Eukaryotic gene expression is less understood than that of prokaryotes such as *E. coli*. This circumstance is largely due to the more varied and complex regulatory mechanisms employed by eukaryotes, with their larger genomes and the additional regulatory opportunities afforded by cellular compartmentation. The section ends with a brief overview of growth factor–triggered gene expression.

Gene Expression in Prokaryotes

The highly regulated metabolism of prokaryotes such as *E. coli* allows these organisms to manage limited resources and to respond to a changing environment. The timely synthesis of enzymes and other gene products only when needed, combined with the rapid destruction of mRNAs by ribonucleases (RNases), prevents dissipation of energy and nutrients. At the genetic level, the control of inducible genes is often effected by the groups of linked structural and regulatory genes called operons. Investigations of operons, especially the *lac* operon, have provided substantial insight into how gene expression can be altered by environmental conditions. The *lac* operon and several types of riboswitch, a more recent discovery, are described.

THE *LAC* OPERON The *lac* operon (Figure 18.43) consists of a control element and structural genes that code for the enzymes of lactose metabolism. The control element contains the promoter site, which overlaps the operator site. The promoter site also contains the CAP site (described shortly). The structural genes Z, Y, and A specify the primary structure of β-galactosidase, lactose permease, and thiogalactoside transacetylase, respectively. β-Galactosidase catalyzes the hydrolysis of lactose, which yields the monosaccharides galactose and glucose, whereas lactose permease promotes lactose transport into the cell. The role of thiogalactose transacetylase is unclear, since lactose metabolism proceeds normally without it. A repressor gene *i*, directly adjacent to the *lac* operon, codes for the *lac* repressor protein, a tetramer that binds to the operator site with high affinity. (There are about 10 copies of *lac* repressor protein per cell.) The binding of the *lac* repressor to the operator prevents the functional binding of RNA polymerase to the promoter (Figure 18.44).

Without its inducer (allolactose, a β-1,6-isomer of lactose) the *lac* operon remains repressed because the *lac* repressor binds to the operator. When lactose becomes available, a few molecules are converted to allolactose by β-galactosidase. Allolactose then binds to the repressor, changing its conformation and promoting dissociation from the operator. Once the inactive repressor has diffused away from the operator, the transcription of the structural genes begins. The *lac* operon remains active until the lactose supply is consumed. Then the repressor reverts to its active form and rebinds to the operator.

If the *lac* operon is repressed in the absence of lactose, what is the source of allolactose? Transcription is never completely blocked, since the repressor protein occasionally detaches, with resulting synthesis of a low number of operon-coded proteins. So when the bacterial cell encounters lactose, there are a few molecules of lactose permease available to facilitate the transport of lactose into the cell, where it is converted to allolactose.

Glucose is the preferred carbon and energy source for *E. coli*. If both glucose and lactose are available, the glucose is metabolized first. Synthesis of the *lac* operon enzymes is induced only after the glucose has been consumed.

FIGURE 18.43

The *lac* Operon in *E. coli*

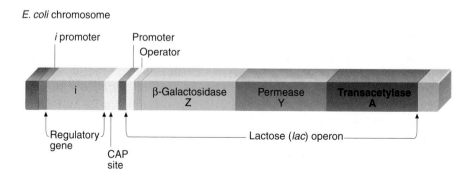

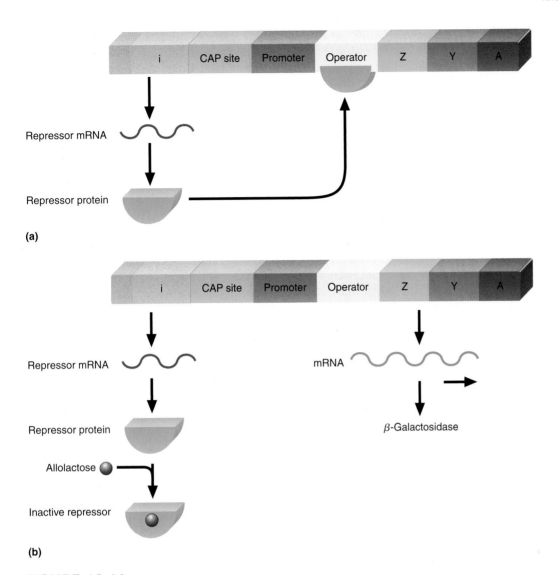

FIGURE 18.44

Function of the *lac* Operon

(a) The repressor gene *i* encodes a repressor that binds to the operator when lactose (the inducer) is not present. (b) When lactose is present, its isomer allolactose binds to repressor protein, thereby inactivating it. (Not shown: the effect of glucose on the *lac* operon. Refer to the text for a discussion of this topic.)

(This makes sense because glucose is more commonly available and has a central role in cellular metabolism. Why expend the energy to synthesize the enzymes required for the metabolism of other sugars if glucose is also available?) The delay in activating the *lac* operon is mediated by a catabolite gene activator protein (CAP). CAP is an allosteric homodimer that binds to the chromosome at a site directly upstream of the lac promoter when glucose is absent. CAP is an indicator of glucose concentration because it binds to cAMP. The cell's cAMP concentration increases when the cell has an energy deficit, that is, when the primary carbon source (glucose) is absent. The binding of cAMP to CAP, which occurs only when glucose is absent and cAMP levels are high, causes a conformational change that allows the protein to bind to the *lac* promoter. CAP binding promotes transcription by increasing the affinity of RNA polymerase for the *lac* promoter. In other words, CAP exerts a positive or activating control on lactose metabolism.

RIBOSWITCHES A **Riboswitch** is a metabolite-sensing domain in the 5′-untranslated region of mRNAs. Found mostly in bacteria, riboswitches monitor cellular concentrations of specific metabolites. Genes that contain riboswitches typically code for proteins that are involved either in the transport or synthesis of molecules that are expensive to produce, such as TPP (thiamine pyrophosphate) or FMN (flavin mononucleotide). Riboswitches are toggle switch–like devices that act as feedback inhibitors to prevent the wasteful acquisition of molecules that are already present in sufficient concentrations. They are composed of two structural elements: an *aptamer*, which directly binds the metabolite, and an *expression platform*, the gene expression regulator. When the aptamer binds the metabolite, it undergoes a structural change that in turn alters the structure of the expression platform. The most common results of this process are transcription and translation inhibition.

When TPP binds to its aptamer (Figure 18.45a) the riboswitch is converted from a structure that has an open translation initiation site to one in which the initiation site is sequestered in a hairpin loop, effectively blocking translation. The FMN riboswitch illustrated in Figure 18.45b terminates transcription when it reconfigures to form a terminator hairpin that prevents RNA polymerase binding. Examples of other riboswitches include those for cobalamin (coenzyme B$_{12}$), purines, and lysine. In rare instances, the riboswitch catalyzes a self-cleavage reaction (Figure 18.45c) that results in a decrease in mRNA copy number.

Gene Expression in Eukaryotes

Being vastly larger and more complicated than the genomes of prokaryotes, eukaryotic genomes are also substantially more intricately regulated, as the following example confirms. With the exception of mature red blood cells, human cells contain the same genome. Yet each of over 200 highly differentiated cell types expresses only a unique subset of genes that changes in response to intercellular signaling mechanisms and/or environmental cues. Like all multicellular

FIGURE 18.45

Riboswitches

(a) Translation prevention. When TPP binds to the aptamer, the riboswitch rearranges so that the expression platform forms a hairpin that blocks translation initiation. (b) Transcription termination. FMN binding causes the formation of a hairpin structure that halts transcription. (c) Self-cleavage. In rare instances, the binding of a metabolite, (in this case, the sugar GlcN6P) triggers a self-cleavage reaction.

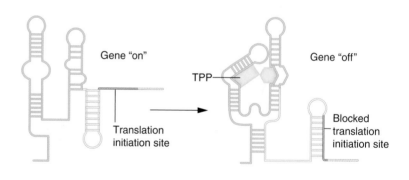

eukaryotes, human cells are the result of differential gene expression that occurs in an orderly and sequential process that began in the fertilized egg. In recent years progress in the investigation of eukaryotic gene expression has been made largely because of molecular cloning and other recombinant DNA techniques, as well as bioinformatics, described earlier in the Biochemistry in the Lab box entitled Genomics. Current evidence indicates that eukaryotic gene expression, as measured by changes in the amounts and activities of gene products produced, is regulated at the following levels: genomic control, transcriptional control (see pp. 711–716), RNA processing, RNA editing, RNA transport, and translational control. A brief description of these topics is followed by an overview of signal transduction-triggered gene expression.

GENOMIC CONTROL As described previously there appear to be two major influences on eukaryotic transcription initiation: chromatin structure and transcription factor–regulated RNA polymerase complex formation. Gene expression is affected by several types of change in the structural organization of the genome. Among the most commonly observed changes are DNA methylation, histone acetylation, and chromatin remodeling. A significant amount of gene regulation occurs through transcription initiation control. The particular set of proteins that assembles on a regulatory DNA sequence (Figures 18.46) is the result of the structure of the DNA sequence itself, as well as the gene regulatory proteins that are present in the cell (having been synthesized and/or activated by signal transduction processes) and their binding affinities for each other. Several proposed mechanisms for protein-mediated gene repression are illustrated in Figure 18.47.

Less common examples of genomic control include gene rearrangements and gene amplification. The differentiation of certain cells involves gene rearrangements, for example, the rearrangements of antibody genes in B lymphocytes (Figure 18.48). Transposition (see p. 695–699) is also believed to affect gene regulation. During certain stages in development, the requirement for specific gene products may be so great that the genes that code for their synthesis are selectively amplified. Amplification occurs via repeated rounds of replication within the amplified region. For example, the rRNA genes in various animals (most notably amphibians, insects, and fish) are amplified within immature egg cells (called oocytes).

COMPANION

W E B S I T E Visit the companion website at www.oup.com/us/mckee to read the Biochemistry in Perspective box on epigenetics for a description of the effects of DNA methylation and histone acetylation on genome structure.

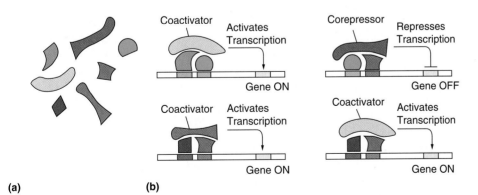

(a) **(b)**

FIGURE 18.46

Eukaryotic Gene Regulatory Proteins

(a) Seven regulatory proteins. (b) Various combinations of hypothetical regulatory proteins, capable of forming a diverse group of complexes because of their affinities for specific DNA sequences and for binding to each other. Note that one of the DNA-binding proteins is present in two complexes that activate transcription and one that represses transcription.

FIGURE 18.47

Proposed Mechanisms for Eukaryotic Gene Repression

(a) Competitive DNA binding. Transcription factor proteins compete for binding to the same regulatory sequence. (b) Masking the activation surface. Both activating and repressing proteins bind to DNA but at different sites. The repressor blocks transcription by binding to and masking activating sites on the activator. (c) Direct interaction with the general transcription factors. The repressor factor binds to a transcription factor bound to DNA, thus preventing a transcription complex from assembling.

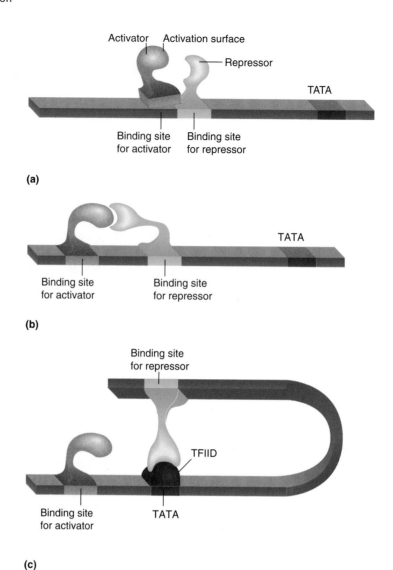

RNA PROCESSING Several types of eukaryotic RNA processing reaction have already been described (p. 714–716). Among the most important of these is *alternative splicing*, the joining of different combinations of exons to form cell-specific proteins (Figure 18.49). Tropomyosin is a protein found in a wide variety of cells (e.g., skeletal, smooth, and cardiac muscle, fibroblasts, and brain) (Figure 18.50). The vertebrate tropomyosin gene consists of 13 to 15 exons. Five of the exons are common to all isoforms of the protein, while the remaining exons are alternately used in different tropomyosin mRNAs. For example, rat striated muscle tropomyosin mRNA contains exons 3, 11, and 12, but not exons 2 or 13, while the smooth muscle isoform contains exons 2 and 3, but not exons 3, 11, and 12. This variation accounts in part for the differences in contractile fiber structure and function in these two muscle types. Rat brain tropomyosin mRNA lacks exons 2, 7, and 11 through 13 and functions in the actin-myosin cytoskeletal system in these noncontractile cells. The selection of alternative sites for polyadenylation also affects mRNA function. Such a change is involved in the switch, during the early phase of B-lymphocyte differentiation, from producing membrane-bound antibody to producing secreted antibody. In general, mRNAs with longer poly(A) tails are more stable, thereby increasing their opportunities for translation.

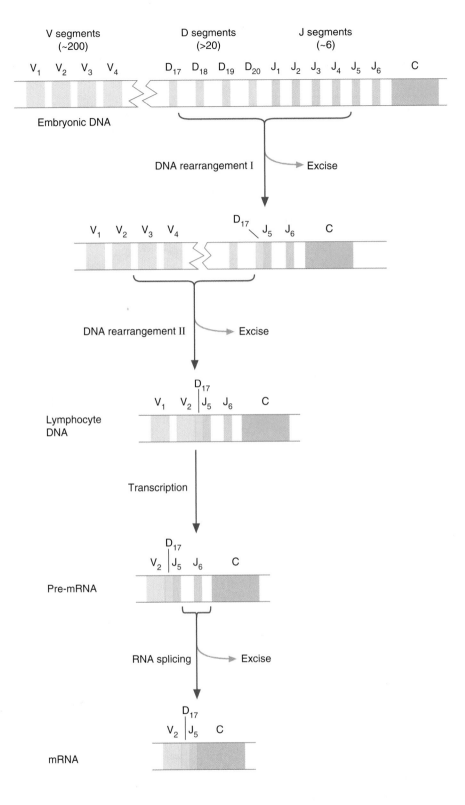

FIGURE 18.48
DNA Rearrangements

Each of the heavy chains of antibodies contains protein sequences derived from one variant each of V (variable), D (diversity), and J (joining) gene segments. The DNA that codes for each type of heavy (H) chain within lymphocytes is generated by the rearrangements of several types of the gene segment. After several rearrangements and excisions, a specific D segment is directly adjacent to specific V and J segments. After the newly created gene has been transcribed, several sequences separating the VDJ segment from the C (constant) segment are removed.

After transcription, base changes are effected by means of **RNA editing**. Alterations in mRNA base sequence can have several consequences. When they occur in the 5′ and 3′ UTRs, for example, translation initiation and RNA stability, respectively, may be affected. Other possibilities include the alteration of intron splice sites and changes in the amino acid sequence of the polypeptide product. Among the best-researched examples of RNA editing are C → U and A → I conversions, where I stands for inosine. The mRNA for the apolipoprotein

FIGURE 18.49

RNA Processing

The coding properties of an mRNA molecule depend on the types of processing event its precursor undergoes. Different polypeptides can be synthesized from splicing different combinations of exons from the same pre-mRNA transcript.

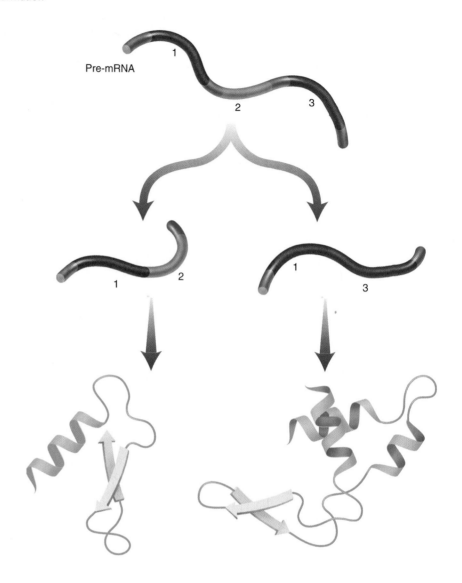

FIGURE 18.50

Alternate Splicing of the Tropomyosin Gene

Tropomyosin is a regulatory protein in muscle contraction. Alternate splicing of the primary transcript in other cells results in altered versions of tropomyosin that are utilized for different purposes.

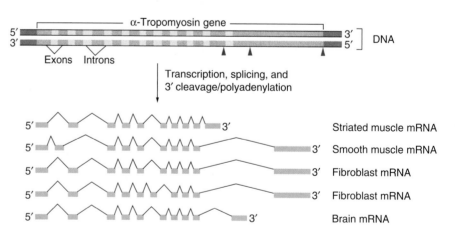

B-100 gene codes for a polypeptide with 4563 amino acid residues, which is a component of very-low-density lipoprotein (VLDL). Intestinal cells produce a shorter version of this molecule, namely, apolipoprotein B-48 (2153 amino acid residues), which becomes incorporated into the chylomicron particles produced by these cells. The cytosine in a CAA codon that specifies glutamine is converted by cytidine deaminase into a uracil. The new codon, UAA, is a stop signal in translation; hence a truncated polypeptide is produced during translation of the edited mRNA.

Among the best examples of an A→I transition occurs in some brain neurons, where a specific adenosine residue in the mRNA for a glutamate receptor subunit is deaminated by ADAR (adenosine deaminase acting on RNA). When an edited mRNA is read by a ribosome, the I base is read as a G. As a result, an arginine residue (codon CGA) is substituted for a glutamate (codon CAA), and the ion channel in the receptor becomes less permeable to Ca^{2+}. Transgenic mice in which the A → I transition was prevented were observed to develop a severe form of epilepsy.

RNA TRANSPORT Eukaryotes regulate molecular traffic into and out of the nucleus. Nuclear export signals, for example, capping (see p. 714) and association with specific proteins, are believed to control the transport of processed RNA molecules through nuclear pore complexes (Figure 2.21). Recent evidence indicates that export requires the binding of the 5′ end of the mRNA within the mRNP to cap-binding protein (CBP) and the TREX (transcription/export) complex, which is recruited in mammalian cells during the splicing process. As the mRNP complex is transported through the nuclear pore complex, certain proteins are removed and retained within the nucleus. Once an mRNP has reached the cytoplasm it undergoes remodeling (i.e., the remaining nuclear proteins are removed and exchanged for cytoplasmic RNP proteins). The latter process is driven to completion by GTP hydrolysis.

TRANSLATIONAL CONTROL Eukaryotic cells can respond to various stimuli (e.g., heat shock, viral infections, and cell cycle phase changes) by selectively altering protein synthesis. The covalent modification of several translation factors (nonribosomal proteins that assist in the translation process) has been observed to alter the overall protein synthesis rate and/or enhance the translation of specific mRNAs. For example, the phosphorylation of the protein eIF-2 prevents it from binding GTP, which is required for the initiation of protein synthesis.

SIGNAL TRANSDUCTION AND GENE EXPRESSION All cells respond to signals from their environment. They do so in part by altering gene expression patterns. Signal transduction–triggered changes in gene expression are initiated by the binding of a ligand to either a cell surface receptor or an intracellular receptor. The mechanisms by which signal molecules switch certain genes on or off are an intricate series of reactions and protein conformation changes that transmit information from the cell's environment to specific DNA sequences in the nucleus. Considering the enormous research efforts devoted to the investigation of cancer (See the "Biochemistry in Perspective" box entitled Carcinogenesis at the end of this chapter.), the best understood examples of such signal transduction pathways are those that affect cell division.

In contrast to single-cell organisms in which cell growth and cell division are governed largely by nutrient availability, the proliferation of cells in multicellular organisms is regulated by an elaborate intercellular network of signal molecules. Complicating features of intracellular signal transduction mechanisms that have been revealed by research efforts in cell proliferation include the following.

1. **Each type of signal may activate one or more pathways**. The mechanisms by which signal molecules alter gene expression often involve the simultaneous activation of several different pathways. This accounts for the changes in cell metabolism and appearance that accompany alterations of gene expression during development.

2. **Signal transduction pathways may converge or diverge**. Depending on circumstances, the activation of several types of receptor may result in the

same or overlapping responses. As mentioned earlier, signal molecules can also trigger several different pathways, any of which may also diverge.

In the eukaryotic cell cycle, cells repeatedly progress through each of the four phases (M, G_1, S, and G_2: refer to Figure 18.14). Investigations of mutant cell types reveal that checkpoints occur in G_1 (in yeast cells it is referred to as START), G_2, and M phases. The cell is prevented from entering the next phase until the conditions are optimal (e.g., sufficient cell growth in G_2 or alignment of chromosomes in M) and specific signals have been received. The fixed, rhythmic activities observed in cell division are regulated so that each phase is completed before the next one starts. Progression is accomplished by the alternating synthesis and degradation of a group of proteins called the cyclins. Cyclins, a group of regulatory proteins, bind to and activate the cyclin-dependent protein kinases (Cdks). The Cdks phosphorylate a variety of proteins that control the passage of cells though checkpoints in the cell cycle. Cell division regulation involves both positive and negative controls. Positive control is exerted largely by growth factors that bind to specialized cell receptors. The initiation of cell division typically requires binding a variety of such factors. Cell proliferation is inhibited by the protein products of *tumor suppressor genes*. Well-known examples of these genes include *Rb* (so named because of the role played by the loss of Rb gene function in the childhood eye cancer retinoblastoma) and the p53 gene (a cyclin chaperone that arrests cell cycle progression). The arrest of the cell cycle is prolonged when a certain amount of DNA damage has occurred, as in overexposure to radiation. If DNA repair mechanisms are incomplete, a complex mechanism involving p53 leads to programmed cell death, or **apoptosis**.

The positive effects exerted by growth factors are now believed to include gene expression that specifically overcomes the inhibitions at the cell cycle checkpoints, especially the G_1 checkpoint. The binding of growth factors to their cell surface receptors initiates a cascade of reactions that induces two classes of genes: early response genes and delayed response genes.

Early response genes These genes, which usually code for transcription factors, are rapidly activated, usually within 15 minutes. Among the best-characterized early response genes are the *jun, fos*, and *myc* protooncogenes. **Protooncogenes** are normal genes that, if mutated, can promote carcinogenesis. (Refer to the "Biochemistry in Perspective" box entitled Carcinogenesis). Each of the *jun* and *fos* protooncogene families codes for a series of transcription factors containing leucine zipper domains. Both Jun and Fos proteins form dimers that can bind DNA. Among the best-characterized of these is a Jun-Fos heterodimer, referred to as AP-1, which forms through a leucine zipper interaction. One member of a large class of transcription factors that possess the basic helix-loop-helix-leucine zipper (bHLHZip) DNA-protein binding motif is Myc. Members of the Myc protein family form homo- and heterodimers with themselves and with members of certain other transcription factor families. When Myc forms a heterodimer with a protein called Max, the expression of a large number of genes is affected. The products of some of Myc/Max target genes have a stimulatory effect on the cell cycle.

Delayed response genes These genes are induced by the activities of the transcription factors and other proteins produced or activated during the early response phase. Among the products of the delayed response genes are the Cdks, the cyclins, and other components required for cell division.

As mentioned earlier (see p. 600), many growth factors bind to tyrosine kinase receptors, and some of these are linked via G-protein-like mechanisms to DAG (diacylglycerol) and IP_3 (inositol trisphosphate) generation (Figure 16.6). Epidermal growth factor (EGF) is one growth factor of this type, and Figure 18.51 gives a brief overview of its role in the activation of the transcription factor AP-1.

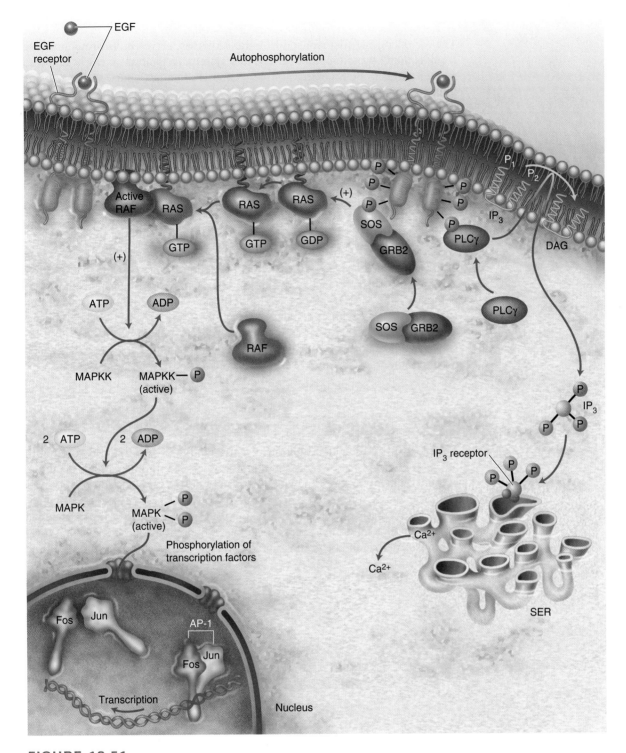

FIGURE 18.51

Eukaryotic Gene Expression Triggered by Growth Factor Binding

Signal transduction pathway illustrating selected events that are triggered when a growth factor (e.g., EGF) binds to its plasma membrane receptor. Subsequent changes in growth factor– or hormone-triggered gene expression that occur are typically mediated by several different mechanisms. This example illustrates only Ras and PLCγ activation. When EGF binds, it promotes dimerization of the receptor and autophosphorylation of the tyrosine residues on its cytoplasmic domain. When these residues are phosphorylated, the receptor binds a variety of cytoplasmic proteins. When GRB2 binds to the receptor, SOS (a GEF) is activated, and it in turn activates Ras by promoting the exchange of GDP for GTP. Activated Ras initiates a phosphorylation cascade by activating the protein kinase RAF, which in turn activates MAPKK. MAPKK activates MAPK, which in turn activates a number of transcription factors in the activation of a variety of transcription factors in the nucleus (e.g., fos and jun that form AP-1). When PLCγ binds to the EGF receptor, it catalyzes the cleavage of PIP_2 into IP_3 and DAG. IP_3 stimulates the release of Ca^{2+} into the cytoplasm. As illustrated in Figure 16.8, in the presence of Ca^{2+} and DAG, protein kinase C activates yet another series of protein kinases that in turn affect the function of several regulatory proteins.

The *ras* oncogene, first isolated from rat sarcomas, serves in its protooncogene form as the G-protein component of this system. Activation of ras is achieved when it binds to SOS/GRB2, a protein complex that has bound to the tyrosine kinase domain of the EGF receptor. SOS is a type of **guanine nucleotide–exchange factor (GEF)** that causes ras to release GDP and bind a GTP when it is in turn bound to the protein called GRB2. Ras becomes inactive when GTP is hydrolyzed, in a reaction catalyzed by **GTPase-activating proteins**, or GAPs.

Phospholipase Cγ(PLCγ) is also activated when it binds to the EGF receptor. Like its counterpart in G-protein–linked receptors (Figure 16.6), active PLCγ hydrolyzes PIP$_2$ to form DAG and IP$_3$. DAG activates protein kinase C (PKC), the enzyme that activates a variety of proteins involved in cell growth and proliferation.

Phosphorylation cascades are induced by both the activation of ras and increased levels of DAG and IP$_3$ in the cell following EGF binding. One of the key enzymes activated in the phosphorylation cascade is MAPKK (mitogen-activated protein kinase kinase). Activated MAPKK then phosphorylates both a tyrosine and a tryptophan of MAP kinase. (This unusual reaction appears to ensure that MAP kinase is activated only by MAPKK.) Active MAP kinase then phosphorylates a variety of cellular proteins. Among those are Jun, Fos, and Myc. Phosphorylated jun and fos proteins then combine to form AP-1, which promotes the transcription of several delayed response genes.

QUESTION 18.5

The mechanism by which light influences plant gene expression is referred to as *photomorphogenesis*. Because of serious technical problems with plant cell culture, relatively little is known about plant gene expression. However, certain DNA sequences, referred to as *light-responsive elements* (LRE), have been identified. On the basis of the gene expression patterns observed in animals, can you suggest (in general terms) a mechanism whereby light induces gene expression? [*Hint*: Recall that phytochrome is an important component of light-induced gene expression.]

BIOCHEMISTRY IN PERSPECTIVE

Carcinogenesis

What is cancer and what are the biochemical processes that facilitate the transformation of normal cells into those with cancerous properties? Cancer is a group of diseases in which genetically damaged cells proliferate autonomously. Such cells cannot respond to normal regulatory mechanisms that ensure the intercellular cooperation required in multicellular organisms. Consequently, they continue to proliferate, thereby robbing nearby normal cells of nutrients and eventually crowding surrounding healthy tissue. Depending on the damage they have sustained, abnormal cells may form either benign or malignant tumors. Benign tumors, which grow slowly and are limited to a specific location, are not considered cancerous and rarely cause death. In contrast, *malignant* tumors are often fatal because they can undergo *metastasis*, migration through blood or lymph vessels to distant locations throughout the body. Wherever new malignant tumors arise, they interfere with normal functions. When life-sustaining processes fail, patients die.

Cancers are classified by the tissues affected. The vast majority of cancerous tumors are carcinomas (tumors derived from epithelial tissue cells such as skin, various glands, breasts, and the lining of most internal organs). In the leukemias, which are cancers of the bone marrow, excessive leukocytes are produced. Similarly, the lymphocytes produced in the lymph nodes and spleen proliferate uncontrollably in the lymphomas. Tumors arising in connective tissue are called sarcomas. Despite the differences among the diseases in this diverse class, they also have several common characteristics, among which are the following.

1. **Cell culture properties**. When grown in culture, most tumor cells lack contact inhibition; that is, they grow to high density in highly disorganized masses. (Normal cells grow only in a single layer in culture and have defined borders.) In contrast to normal cells, growth and division in cancer cells tend to be growth factor–independent, and cancer cells often do not require attachment to a solid surface. The hallmark of cancer cells, however, is their immortality. Normal cells undergo cell division only a finite number of times, whereas cancer cells can proliferate indefinitely.

2. **Origin**. Each tumor originates from a single damaged cell. In other words, a tumor is a clone derived from a cell in which heritable changes have occurred. The genetic damage consists of mutations (e.g., point mutations, deletions, and inversions) and chromosomal rearrangements or losses. Such changes result in the loss or altered function

of molecules involved in cell growth or proliferation. Tumors typically develop over a long time and involve several independent types of genetic damage. (The risk of many types of cancer increases with age.)

The transformation process in which an apparently normal cell is converted or "transformed" into a malignant cell consists of three stages: initiation, promotion, and progression.

During the *initiation* phase of carcinogenesis, a permanent change in a cell's genome provides it with a growth advantage over its neighbors. Most initiating mutations affect protooncogenes or tumor suppressor genes. Protooncogenes code for a variety of growth factors, growth factor receptors, enzymes, or transcription factors that promote cell growth and/or cell division. Mutated versions of protooncogenes that promote abnormal cell proliferation are called **oncogenes** (Table 1). Because tumor suppressor genes suppress carcinogenesis, their loss also facilitates tumor development. Recall that *Rb* and *p53* are tumor suppressors. Other examples are *FCC* and *DCC*, which are both associated with susceptibility to colon cancer. The functions of most tumor suppressor genes are unknown. However, it now appears that *p53* codes for a 53kD protein that normally inhibits Cdk enzymes. Recent evidence indicates that other damaged or deleted tumor suppressor genes may code for enzymes involved in DNA repair mechanisms. The damage that alters function

TABLE 1   Selected Oncogenes*

Oncogene	Protooncogene Function
sis	Platelet-derived growth factor
erbB	Epidermal growth factor receptor
src	Tyrosine-specific protein kinase
raf	Serine/threonine-specific protein kinase
ras	GTP-binding protein
jun	Transcription factor
fos	Transcription factor
myc	Transcription factor

*Abnormal versions of protooncogenes that mediate cancerous transformations.

of protooncogenes and tumor suppressor genes is caused by the following.

1. **Carcinogenic chemicals**. Most cancer-causing chemicals are mutagenic; that is, they alter DNA structure. Some carcinogens (e.g., nitrogen mustard) are highly reactive electrophiles that attack electron-rich groups in DNA (as well as RNA and protein). Other carcinogens (e.g., benzo[*a*]pyrene) are actually procarcinogens, which are converted to active carcinogens by one or more enzyme-catalyzed reactions.

2. **Radiation**. Some radiation (UV, X-rays, and γ-rays) is carcinogenic. As noted, the damage inflicted on DNA includes single- and double-strand breaks, pyrimidine dimer formation, and the loss of both purine and pyrimidine bases. Radiation exposure also causes the formation of ROS, which may be responsible for most of radiation's carcinogenic effects.

3. **Viruses**. Viruses appear to contribute to the transformation process in several ways. Some introduce oncogenes into a host cell chromosome as they insert their genome. (Viral oncogenes are now recognized as sequences that are similar to normal cellular genes that have been picked up accidentally from a previous host cell. To distinguish viral oncogenes and their cellular counterparts, they are referred to as v-*onc* and c-*onc*, respectively.) Viruses can also affect the expression of cellular protooncogenes through insertional mutagenesis, a random process in which viral genome insertion inactivates a regulatory site or alters the protooncogene's coding sequence. Most virus-associated cancers have been detected in animals. Only a few human cancers have been proven to be associated with viral infection.

Tumor development can also be promoted by chemicals that do not alter DNA structure. So-called **tumor promoters** contribute to carcinogenesis by two principal methods. By activating components of intracellular signaling pathways, some molecules (e.g., the phorbol esters) provide the cell a growth advantage over its neighbors. (Recall that phorbol esters activate PKC because they mimic the actions of DAG.) The effects of many other tumor promoters are unknown but may involve transient effects such as increasing cellular Ca^{2+} levels or increasing synthesis of the enzymes that convert procarcinogens into carcinogens. Unlike initiating agents, the effects of tumor promoters are reversible. They produce permanent damage only with prolonged exposure after an affected cell has undergone an initiating mutation.

Following initiation and promotion, cells go through a process referred to as progression. During *progression*, genetically vulnerable precancerous cells, which already possess significant growth advantages over normal cells, are further damaged. Eventually, the continued exposure to carcinogens and promoters makes further random mutations inevitable. If these mutations affect cellular proliferative or differentiating capacity, then an affected cell may become sufficiently malignant to produce a tumor. A proposed sequence of the events in the development of colorectal cancer is outlined in Figure 18I.

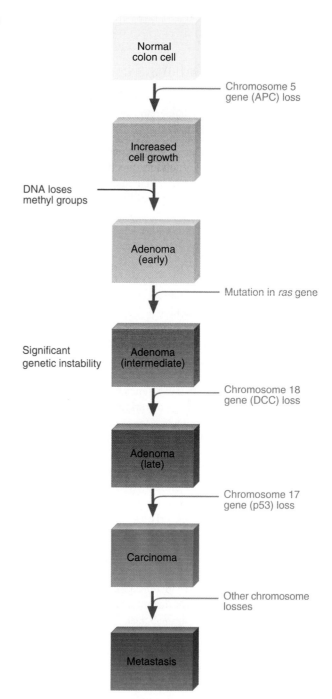

FIGURE 18I

The Development of Colorectal Cancer

Colorectal cancer develops over a long time. Because somatic cells are diploid, the loss of the tumor suppressor genes *APC, DCC,* and *p53* usually requires two mutations. Recent research suggests that mutations in genes involved in DNA repair processes may occur during the early stages of colon cancer. (An adenoma is a precancerous epithelial tumor.) Healthy diets that are high in folate, antioxidants, and certain polyunsaturated fatty acids found in fish provide significant protection against colorectal cancer.

▶▶

BIOCHEMISTRY IN PERSPECTIVE cont

Cancer is not only a genetic disease, it is presently appreciated as an epigenetic disease as well. Chromatin-associated gene silencing has been implicated in the progress of a number of cancers. Inappropriate silencing of tumor suppressor genes caused by abnormal methylation patterns in a regulatory sequence of these genes has been suggested as a contributing factor in carcinogenesis. Silencing of *MHL1*, a postreplication repair gene, by DNA methylation is seen in colorectal cancers. The CpG islands within these genes are normally unmethylated, and any level of methylation can inhibit transcription, or condense chromatin. (*CpG islands* are 1 to 2 kb sequences with a high density of 5′-CG-3′-dinucleotides. The cytosine residues in these sequences are often highly methylated.)

An Ounce of Prevention . . .

Because of the enormous cost and limited success of cancer therapy, it has become increasingly recognized that cancer prevention is cost-effective. Recent research indicates that the majority of cases of cancer are preventable. For example, over one-third of cancer mortality is directly caused by tobacco use, and another one-third of cancer deaths have been linked to inadequate diets. Tobacco smoke, which contains thousands of chemicals, many of which are either carcinogens or tumor promoters, is responsible

for most cases of lung cancer and contributes to cancers of the pancreas, bladder, and kidneys, among others. Diets that are high in fat and low in fiber content have been associated with increased incidence of cancers of the large bowel, breast, pancreas, and prostate. Other dietary risk factors include low consumption of fresh vegetables and fruit.

In addition to providing sufficient antioxidant vitamins, many vegetables (and fruits to a lesser extent) contain numerous nonnutritive components that actively inhibit carcinogenesis. Some carcinogenesis inhibitors (e.g., organosulfides), referred to as *blocking agents*, prevent carcinogens from reacting with DNA or inhibit the activity of tumor promoters. Other inhibitors, referred to as *suppressing agents* (e.g., inositol hexaphosphate), prevent the further development of neoplastic processes that are already in progress. Many nonnutritive food components (e.g., tannins and protease inhibitors) possess both blocking and suppressing effects. In general, these molecules very effectively protect against cancer because many of them inhibit the arachidonic acid cascade and oxidative damage. Apparently low-fat, high-fiber diets that are rich in raw or fresh vegetables that are leafy green (e.g., spinach), cruciferous (e.g., broccoli), and members of the allium family (e.g., onions), as well as fresh fruits, are a prudent choice for individuals seeking to reduce their risk of cancer.

 **SUMMARY:** Carcinogenesis is the process whereby cells with a growth advantage over their neighbors are transformed by mutations in the genes that control cell division into cells that no longer respond to regulatory signals.

Chapter**Summary**

1. DNA structure and function are so important that living organisms must possess efficient mechanisms for the rapid and accurate synthesis of this nucleic acid. DNA synthesis, referred to as replication, occurs by a semiconservative mechanism; that is, each of the two parental strands serves as a template to synthesize a new strand. Enzymatic activities required in DNA replication are used in DNA unwinding, primer synthesis, polynucleotide synthesis, supercoiling control, and ligation. Although the basic features of DNA replication are similar in prokaryotes and eukaryotes, there are significant differences (e.g., replication time and rate, replication origin numbers, Okazaki fragment size, and replication machinery structure).

2. There are several types of DNA repair mechanism. These include base excision and nucleotide excision repair, photoreactivation, and recombinational repair.

3. Genetic recombination, a process in which DNA sequences are exchanged between different DNA molecules, occurs in two forms. In general recombination, the exchange occurs between sequences in homologous chromosomes. In site-specific recombination, the exchange of sequences requires only short homologous sequences. DNA-protein interactions are principally responsible for the exchange of largely nonhomologous sequences.

4. Transposition, the movement of genetic elements (transposons) from one place to another within a genome, can cause genetic changes such as insertions, deletions, and translocations. The movement of retrotransposons, found in large numbers in eukaryotic genomes, can cause disease or can provide opportunities for genetic diversity.

5. The synthesis of RNA, referred to as DNA transcription, requires a variety of proteins. Transcription initiation

involves binding an RNA polymerase to a specific DNA sequence called a promoter. Regulation of transcription differs significantly between prokaryotes and eukaryotes. Examples of eukaryotic transcription processes not observed in prokaryotes include RNA processing features such as capping, poly(A) tail synthesis, and RNA splicing.

6. The control of transcription is still poorly understood. However, intensive research has disclosed many of the details in several examples of gene expression in both prokaryotes and eukaryotes. Prokaryotic gene expression regulation involves protein-based regulatory units called operons and RNA-based structures called riboswitches. Eukaryotes have a wide variety of mechanisms that control gene expression. Prominent examples include DNA methylation, histone covalent modification, and chromatin remodeling, RNA-processing reactions such as alternative splicing, and RNA editing, RNA transport, and translational controls.

 Take your learning further by visiting the **companion website** for Biochemistry at **www.oup.com/us/mckee** where you can complete a multiple-choice quiz on genetic information to help you prepare for exams.

SuggestedReadings

Azubel, M., Habid, N., Sperling, R., and Sperling, J., Native Spliceosomes Assemble with pre-mRNA to Form Supraspliceosomes, *J. Mol. Biol.* 356: 955–966, 2006.

Baralle, D., and Baralle, M., Splicing in Action: Assessing Disease-Causing Sequence Changes, *J. Med. Gene.* 42:737–748, 2005.

Brown, T. A., *Genomes 3*, Garland Science, New York, 2007.

Carthew, R. W., A New RNA Dimension to Genome Control, *Science* 313:305–306, 2006.

Cohen-Krausz, S., Sperling, R., and Sperling, J., Exploring the Architecture of the Intact Supraspliceosomes Using Electron Microscopy, *J. Mol. Biol.*, 328:319–327, 2007.

Collins, F. W., and Barker, A. D., Mapping the Cancer Genome, *Sci. Am.* 296(3):50–57, 2007.

Deininger, P. L., Moran, J. V., Batzer, M. S., and Kazazian, H. H., Mobile Elements and Mammalian Genome Evolution, *Curr. Opin. Gene. Dev.* 13:651–658, 2003.

Erzberger, J. P., Mott, M. L., and Berger, J. M., Structural Basis for ATP-Dependent DnaA Assembly and Replication—Origin Remodeling, *Nat. Struct. Mol, Biol.* 13(8):676–683, 2006.

Forsburg, S. L., Eukaryotic MCM Proteins: Beyond Replication Initiation, *Microbiol. Mol. Biol. Rev.* 68(1):109–131, 2004.

Kim, Y.-I., Nutritional Epigenetics: Impact of Folate Deficiency on DNA Methylation and Colon Cancer Susceptibility, *J. Nutr.* 135:2703–2709, 2005.

Mandel, M. and Breaker, R. R., Gene Regulation by Riboswitches, *Nat. Rev. Mol. Cell Biol.* 5:451–463, 2004.

O'Donnell, M., Replisome Architecture and Dynamics in *Escherichia coli, J. Biol. Sci.* 281(16):10653–10656, 2006.

Ray, D. A., Walker, J. A., Batzer, M. A., Mobile Element-Based Forensic Genomics, *Mutat. Res.* 616:24–33, 2007.

Steitz, T. A., A Mechanism for All Polymerases, *Nature* 391:231–232, 1998.

Watson, J. D., Baker, T. A., Bell, S. P., Gann, A., Levine, M., and Losick, R., *Molecular Biology of the Gene*, Pearson Benjamin Cummings, San Francisco, 2004.

Weinberg, R. A., *The Biology of Cancer*, Garland Science, New York, 2007.

KeyWords

acceptor site, *716*

annotation, *705*

apoptosis, *726*

bacterial artificial chromosome, *701*

base excision repair, *690*

β_2-clamp, *682*

bioinformatics, *706*

cDNA library, *703*

chromosomal jumping, *704*

clamp loader, *682*

chromatin remodeling, *711*

coding strand, *707*

colony hybridization technique, *702*

conjugation, *695*

consensus sequence, *708*

constitutive gene, *717*

contig, *704*

cosmid, *701*

DNA glycosylase, *690*

DNA ligase, *682*

DNA microarray, *704*

electroporation, *701*

exonuclease, *682*

functional genomics, *700*

general recombination, *692*

genomics, *700*

GTPase-activating protein, *728*

guanine nucleotide exchange facter, *728*

helicase, *680*

inverted repeat, *696*

light-induced repair, *690*

marker gene, *702*

MCM complex, *688*

mitogen, *728*

nucleotide excision repair, *690*

Okazaki fragment, *684*

oncogene, *729*

operon, *711*

origin of replication complex, *688*

palindrome, *696*

photoreactivation repair, *690*

polymerase chain reaction, *703*

preinitiation replication complex, *688*

primase, *680*

primer, *680*

primosome, *680*

processivity, *682*

Review**Questions**

These questions are designed to test your knowledge of the key concepts discussed in this chapter, before moving on to the next chapter. You may like to compare your answers to the solutions provided in the back of the book and in the accompanying Study Guide.

1. Define the following terms:
   a. recombination
   b. replisome
   c. replicon
   d. retroelement
   e. riboswitch

2. How does negative supercoiling promote the initiation of replication?

3. Explain how the Meselson-Stahl experiment supports the semiconservative model of DNA replication.

4. List and describe the steps in prokaryotic DNA replication. How does this process appear to differ from eukaryotic DNA replication?

5. Define the following terms:
   a. spliceosome
   b. telomerase
   c. transposition
   d. tumor promoter
   e. inverted repeat

6. Indicate the stage of DNA replication when each of the following enzymes is active:
   a. helicase
   b. primase
   c. DNA polymerases
   d. ligase
   e. topoisomerase
   f. DNA gyrase

7. DNA is polymerized in the $5' \rightarrow 3'$ direction. Demonstrate with the incorporation of three nucleotides into a single strand of DNA how the $5' \rightarrow 3'$ directionality is derived.

8. Define the following terms:
   a. replication licensing factor
   b. primosome
   c. processivity
   d. exonuclease
   e. consensus sequence

9. Mutations are caused by chemical and physical phenomena. Indicate the type of mutation that each of the following reactions or molecules might cause:
   a. ROS
   b. caffeine
   c. a small alkylating agent
   d. a large alkylating agent
   e. nitrous acid
   f. intercalating agents

10. How can viruses cause mutations?

11. Explain the significance of "jumping gens".

12. Define the following terms:
    a. DNA glycosylase
    b. clamp loader
    c. palindome
    d. primer
    e. protooncogene

13. Describe two forms of genetic recombination. What functions do they fulfill?

14. Although genetic variation is required for species to adapt to changes in their environment, most genetic changes are detrimental. Explain why genetic mutations are rarely beneficial.

15. General recombination occurs in bacteria, where it is involved in several types of intermicrobial DNA transfer. What are these types of transfer, and by what mechanisms do they occur?

16. Compare and contrast the mechanisms of replicative and non-replicative transposition.

17. Within cells, cytosine slowly converts to uracil. To what type of mutation would this lead in DNA molecules? Why is this not a problem in RNA?

18. A correlation has been found among species between life span and the efficiency of DNA repair systems. Suggest a reason why this is so.

19. Define the following terms:
    a. RNA editing
    b. electroporation
    c. annotation
    d. acceptor site
    e. polymerase chain reaction

20. What are the similarities and differences between cellular DNA replication and PCR?

21. Describe the purpose of marker genes in recombinant DNA technology.

22. Define and describe the roles of the following in replication:
    a. DNA ligase
    b. DNA polymerase III
    c. SSB proteins
    d. primase
    e. helicase
    f. RPA
    g. Okazaki fragments

23. Define and describe the roles of the following in transcription:
    a. transcription factors
    b. RNA polymerase
    c. promoter
    d. sigma factor
    e. enhancer
    f. TATA box

24. List the steps in the processing of a typical eukaryotic mRNA precursor that prepare for its functional role.

25. Describe the advantages and disadvantages for organisms that arrange genes in operons.

26. Determine the magnitude of amplification of a single DNA molecule that can be attained with PCR during five cycles.

27. Define the following terms:
    a. recombinational repair
    b. replication fork
    c. protooncogene
    d. apoptosis
    e. functional genomics

28. Many genes generate different products depending on the type of cell expressing the gene. How is this phenomenon accomplished?

29. Provide an example for the process described in Question 28.

30. During base excision repair, DNA glycosylase cleaves the N-glycosidic link between the altered base and the deoxyribose component of the nucleotide. Draw a typical nucleotide and indicate which bond is cleaved.

31. During certain stages of development, the requirement for certain gene products may require gene amplification. What purpose does gene amplification serve?

32. How are the genes referred to in Question 31 amplified?

33. Cells exposed to ultraviolet light develop thymine dimers and visible light reverses this damage. Explain how this form of DNA repair occurs.

34. RNA molecules are more reactive than DNA. Explain.

35. Define the following terms:
    a. DnaA box
    b. MCM complex
    c. contig
    d. shotgun method
    e. RFC

36. Describe how LINE 1 element transposition occurs.

37. Describe the function of telomere end-binding proteins.

38. Describe the function of telomere repeat-binding factors.

39. Describe how site-specific recombination occurs.

ThoughtQuestions

These questions are designed to reinforce your understanding of all of the key concepts discussed in the book so far, including this chapter and all of the chapters before it. They may not have one right answer! The authors have provided possible solutions to these questions in the back of the book and in the accompanying Study Guide, for your reference.

40. In the Meselson-Stahl experiment, why was a nitrogen isotope chosen rather than a carbon isotope?

41. In eukaryotes the DNA replication rate is 50 nucleotides per second. How long does the replication of a chromosome of 150 million base pairs take? If eukaryotic chromosomes were replicated like those of prokaryotes, the replication of a genome would take months. Actually, eukaryotic replication takes only several hours. How do eukaryotes achieve this high rate?

42. There appears to be insufficient genetic material to direct all the activities of several types of eukaryotic cell. Explain how genetic recombination, gene splicing and alternative RNA splicing help solve this problem.

43. Mustard gas is an extremely toxic substance that severely damages lung tissue when inhaled in large amounts. In small amounts, mustard gas is a mutagen and carcinogen. Considering that mustard gas is a bifunctional alkylating agent, explain how it mutates genes and impacts DNA replication.

44. Infection due to a rare, virulent strain of group A streptococcus has appeared relatively recently. In approximately 25 to 50% of these cases (reported in Great Britain and the United States), infection resulted in necrotizing fasciitis, a rapidly spreading destruction of flesh, often accompanied by hypotension (low blood pressure), organ failure, and toxic shock. If antibiotic treatment is not initiated within 3 days of exposure to the bacterium, gangrene and death may result. Similar cases were reported in the 1920s. However, these earlier cases had a significantly lower fatality rate, although antibiotics were not then available. (Physicians reported treating affected areas by washing with acidic solutions.) Group A streptococci are converted into the pathogenic form by becoming infected with a certain virus. This virus's genome contains a gene that codes for a tissue-destroying toxin. Can you describe in general terms how a viral infection might cause a permanent change in the pathogenicity of a group A streptococcus bacterium? Considering the apparent difference in virulence between the bacterium in the 1920s and the present, is there any method for determining whether the same strain of group A streptococcus is responsible for both sets of cases? Preserved specimens of infected tissue from these early cases are available. [*Hint*: Refer to the Biochemistry in the Lab box entitled Genomics.]

45. Adjacent pyrimidine bases in DNA form dimers with high efficiency after exposure to UV light. If these dimers are not

repaired, skin cancers can result. Melanin is a natural sunscreen produced by melanocytes, a type of skin cell, when the skin is exposed to sunlight. Individuals who spend long periods over many years developing a tan eventually acquire thick and highly wrinkled skin. Such individuals are also at high risk for skin cancer. Can you explain, in general terms, why these phenomena are related?

46. Phorbol esters have been observed to induce the transcription of AP-1–influenced genes. Explain how this process could occur. What are the consequences of AP-1 transcription? What role does intermittent exposure to phorbol esters have on an individual's health?

47. Because of overuse of antibiotics and/or weakened governmental surveillance of infectious disease, several diseases that had been thought to be no longer a threat to human health (e.g., pneumonia and tuberculosis) are rapidly becoming unmanageable. In several instances, so-called superbugs (microorganisms that are resistant to almost all known antibiotics) have been detected. How did this circumstance arise? What will happen if this process continues?

48. Retinoblastoma is a rare cancer in which tumors develop in the retina of the eye. The tumors arise because of the loss of the *Rb* gene, which codes for a tumor suppressor. Hereditary retinoblastoma usually appears in children who have inherited only one functional copy of *Rb*. Explain why the nonhereditary form of retinoblastoma usually occurs later in life.

49. Explain the difference between the potential effects on an individual organism of errors made during replication and those made during transcription.

50. Explain how a reverse transcriptase activity within a cell can result in gene amplification.

51. Use the tools described in Chapter 18 to describe in general terms how a researcher can map the genome of a newly discovered organism.

52. It was once thought that the DNA polymerase machinery moves along DNA in a manner analogous to a train on a track. Current evidence indicates that the polymerizing machinery is instead stationary, and DNA strands are pumped through the complex. What advantages does this stationary mechanism have?

53. Riboswitches are sensors that shut down transcription when excess metabolites are present. Describe the features of a riboswitch that is activated by excess metabolite concentrations.

54. DNA is the cellular repository of genetic information, and proteins are synthesized from RNA transcripts. Why don't cells simply skip the transcription step and use DNA directly in protein synthesis?

Protein Synthesis

OUTLINE

The Ribosome Ribosomes are ribonucleo-protein molecular machines that synthesize proteins in all living cells. In this illustration of the high-resolution structure of a complete bacterial 70S ribosome, proteins are illustrated in dark blue and magenta, and rRNA molecules are shown in cyan and gray. The tRNAs are orange and yellow. Note that the ribosome is primarily composed of rRNA, which perform most catalytic activities. Protein molecules largely serve supporting roles.

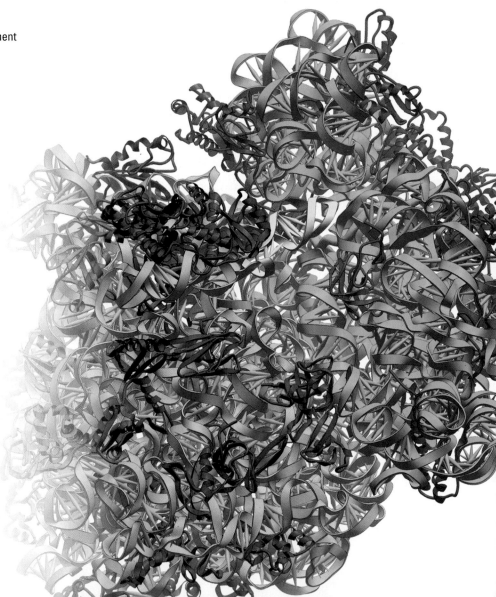

Overview

▼

PROTEINS ARE THE MOST DYNAMIC, NUMEROUS, AND VARIED CLASS OF
BIOMOLECULES. THE UNIQUENESS OF EACH CELL TYPE IS DUE ALMOST
entirely to the proteins it produces. It is not surprising, therefore, that a
relatively large amount of cellular energy is used in protein synthesis. Because
of their strategic importance in the cellular economy, the synthesis of proteins
is a regulated process. Although control is also of major importance at the
transcriptional level, control of the translation of genetic messages allows for
additional opportunities for regulation. This is especially true in multicellular
eukaryotes, whose complex lifestyles require amazingly diverse regulatory
mechanisms.

Protein synthesis is the process in which genetic information encoded in the
nucleic acids is translated into the 20 standard amino acid "alphabet'"
of polypeptides. In addition to translation (the mechanism by which a
nucleotide base sequence directs the polymerization of amino acids), protein
synthesis can be considered to include the processes of posttranslational
modification and targeting. *Posttranslational modification* consists of the
chemical alterations cells use to prepare polypeptides for their functional roles.
Several modifications assist in *targeting*, which directs newly synthesized
molecules to a specific intracellular or extracellular location.

In all, at least 100 different molecules are involved in protein synthesis. Among
the most important of these are the components of the ribosomes, large ribonucleoprotein machines that synthesize polypeptides. Each ribosome "reads" the
base sequence of an mRNA and, fueled by GTP, rapidly and precisely converts
this information into the amino acid sequence of a polypeptide. Speed is required
because organisms must respond expeditiously to ever-changing environmental
conditions. In prokaryotes such as *E. coli*, for example, a polypeptide of 100
residues is synthesized in less than 6 s. Eukaryotes are slower, at about two
residues per second. Precision in mRNA translation is critical because, as
described previously, the accurate folding, and therefore the proper functioning, of each polypeptide is determined by the molecule's primary sequence.

Discovered in the 1950s, ribosomes have been the subject of biochemical and
biophysical investigations ever since. An early revelation was that ribosome structure appears to be highly conserved. Although there are notable differences among
the ribosomes of various species, similarities in the three-dimensional structures
of rRNA and ribosomal proteins are remarkable. Despite differences in size, the
ribosomes of prokaryotes and eukaryotes are similar in their overall shapes and
core functions.

As the result of decades of intense work, an increasingly more detailed understanding of ribosomal structure and function is emerging. One of the unexpected
findings was that rRNA performs the critical functions of the ribosome. For
example, the catalytic activity that forms peptide bonds resides in an RNA
molecule. In addition, rRNA molecules also have roles in tRNA-mRNA docking,
ribosomal subunit association, proofreading, and the binding of translation factors.
Ribosomal proteins, for the most part, have supporting roles.

Chapter 19 provides an overview of protein synthesis. The chapter begins with
a discussion of the genetic code, the mechanism by which nucleic acid-base
sequences specify the amino acid sequences of polypeptides. This is followed
by discussions of protein synthesis as it occurs in both prokaryotes and eukaryotes

and a description of the mechanisms that convert polypeptides into their biologically active conformations. A critical feature of this process, protein folding, was described in Section 5.3. Chapter 19 ends with an introduction to **proteomics**, a relatively new technology being developed to characterize the protein products of the genome.

19.1 THE GENETIC CODE

It became apparent during early investigations of protein synthesis that translation is fundamentally different from the transcription process that precedes it. During transcription, the language of DNA sequences is converted to the closely related dialect of RNA sequences. During protein synthesis, however, a nucleic acid-base sequence is converted to a clearly different language (i.e., an amino acid sequence), hence the term *translation*. Because mRNA and amino acid molecules have little natural affinity for each other, researchers (e.g., Francis Crick) predicted that a series of adaptor molecules must mediate the translation process. This role was eventually assigned to tRNA molecules (Figure 17.27).

Before adaptor molecules could be identified, however, a more important problem had to be solved: deciphering the genetic code. The **genetic code** can be described as a coding dictionary that specifies a meaning for each base sequence. Once the importance of the genetic code was recognized, investigators speculated about its dimensions. Only four different bases (G, C, A, and U) occur in mRNA, and 20 amino acids must be specified. It therefore appeared reasonable that a combination of bases codes for each amino acid. A sequence of two bases would specify only a total of 16 amino acids (i.e., $4^2 = 16$). However, a three-base sequence provides more than sufficient base combinations for translation (i.e., $4^3 = 64$).

The first major breakthrough in assigning mRNA triplet base sequences (later referred to as **codons**) came in 1961, when Marshall Nirenberg and Heinrich Matthaei performed a series of experiments using an artificial test system containing an extract of *E. coli* fortified with nucleotides, amino acids, ATP, and GTP. They showed that poly(U) (a synthetic polynucleotide whose base components consist only of uracil) directed the synthesis of polyphenylalanine. Assuming that codons consist of a three-base sequence, Nirenberg and Matthaei surmised that UUU codes for the amino acid phenylalanine. Subsequently, they repeated their experiment using poly(A) and poly(C). Because polylysine and polyproline products resulted from these tests, the codons AAA and CCC were assigned to lysine and proline, respectively.

Most of the remaining codon assignments were determined by using synthetic polynucleotides with repeating sequences. Such molecules were constructed by enzymatically amplifying short chemically synthesized sequences. The resulting polypeptides, which contained repeating peptide segments, were then analyzed. The information obtained from this technique, devised by Har Gobind Khorana, was later supplemented with a strategy used by Nirenberg. This latter technique measured the capacity of specific trinucleotides to promote tRNA binding to ribosomes.

The codon assignments for the 64 possible trinucleotide sequences are presented in Table 19.1. Of these, 61 code for amino acids. Four codons serve as punctuation signals. UAA, UAG, and UGA are *stop* (polypeptide chain terminating) signals. AUG, the codon for methionine, also serves as a *start* signal (sometimes referred to as the *initiating codon*). The genetic code possesses the following properties.

1. **Specific**. Each codon is a signal for a specific amino acid.

2. **Degenerate**. Any coding system in which several signals have the same meaning is said to be degenerate. The genetic code is partially degenerate because most amino acids are coded for by several codons.

TABLE 19.1   The Genetic Code

Second Position

		U		C		A		G		
First position (5′ end)	U	UUU UUC	Phe	UCU UCC		UAU UAC	Tyr	UGU UGC	Cys	U C
		UUA UUG	Leu	UCA UCG	Ser	UAA UAG*	STOP	UGA* UGG	STOP Trp	A G
	C	CUU CUC	Leu	CCU CCC	Pro	CAU CAC	His	CGU CGC	Arg	U C
		CUA CUG		CCA CCG		CAA CAG	Gln	CGA CGG		A G
	A	AUU AUC	Ile	ACU ACC	Thr	AAU AAC	Asn	AGU AGC	Ser	U C
		AUA AUG	Met	ACA ACG		AAA AAG	Lys	AGA AGG	Arg	A G
	G	GUU GUC	Val	GCU GCC	Ala	GAU GAC	Asp	GGU GGC	Gly	U C
		GUA GUG		GCA GCG		GAA GAG	Glu	GGA GGG		A G

Third position (3′ end)

*The stop codons UGA and UAG are also used by some organisms and under specific conditions (described later in the Biochemistry in Perspective box entitled Context Dependent Coding Reassignment) to insert selenocysteine or pyrrolysine, respectively into a polypeptide sequence.

For example, leucine is coded for by six different codons (UUA, UUG, CUU, CUC, CUA, and CUG). In fact, methionine (AUG) and tryptophan (UGG) are the only amino acids that are coded for by a single codon.

3. **Nonoverlapping**. The mRNA coding sequence is "read" by a ribosome starting from the initiating codon (AUG) as a continuous sequence taken three bases at a time until a stop codon is reached. A set of contiguous triplet codons in an mRNA is called a **reading frame**.

4. **Almost universal**. The genetic code is used by the vast majority of organisms, but there are a few exceptions. Minor deviations have been observed in mitochondria. For example, instead of six arginine codons, there are only four. The remaining two codons (AGA and AGG) are instead used as stop codons. Since UGA, normally a stop codon, instead codes for tryptophan, mitochondria have four stop codons: UAG, UAA, AGA, and AGG. Similar minor changes have been observed in a few species of prokaryotes and eukaryotes such as protozoa and yeast.

It appears that the degenerate genetic code evolved to diminish the deleterious effects of point mutations; that is, base substitutions often result in the incorporation of the same or similar amino acids, as the following two examples involving leucine illustrate. Since all the codons with CU in the first two positions code for leucine, base substitutions at position 3 will have no effect. (Refer to the discussion of the wobble hypothesis, p. 740, for another aspect of this phenomenon.) An analysis of the genetic code also reveals that all of the codons with U in the second position code for hydrophobic amino acids. If the first base in the CUU codon (leucine) is replaced with A, another hydrophobic amino acid residue, isoleucine, will be substituted.

In certain circumstances, living organisms are not limited to the codon assignments of the genetic code. In two recently recognized examples, the nonstandard amino acids selenocysteine and pyrrolysine are coded for by stop codons. This phenomenon, called context-dependent codon reassignment, is described later in the Biochemistry in Perspective box on that topic (pp. 761–762).

KEY CONCEPT

The genetic code is a mechanism by which ribosomes translate nucleotide base sequences into the primary sequence of polypeptides.

QUESTION 19.1

As described earlier, DNA damage can cause deletion or insertion of base pairs. If a nucleotide base sequence of a coding region changes by any number of bases other than three base pairs, or multiples of 3, a frameshift mutation occurs. Depending on the location of the sequence change, such mutations can have serious effects. The following synthetic mRNA sequence codes for the beginning of a polypeptide:

5′-AUGUCUCCUACUGCUGACGAGGGAAGGAGGUGGCUUAUCAU-GUUU-3′

First, determine the amino acid sequence of the polypeptide. Then determine the types of mutation that have occurred in the following altered mRNA segments. What effect do these mutations have on the polypeptide products?

a. 5′-AUGUCUCCUACUUGCUGACGAGGGAAGGAGGUGGCUUAU-CAUGUUU-3′

b. 5′-AUGUCUCCUACUGCUGACGAGGGAGGAGGUGGCUUAUCAU-GUUU-3′

c. 5′-AUGUCUCCUACUGCUGACGAGGGAAGGAGGUGGCCCUUAU-CAUGUUU-3′

d. 5′-AUGUCUCCUACUGCUGACGGAAGGAGGUGGCUUAUCAU-GUUU-3′

Codon-Anticodon Interactions

Transfer RNA molecules are the "adaptors" that are required for the translation of the genetic message. Recall that each type of tRNA carries a specific amino acid (at the 3′ terminus) and possesses a three-base sequence called the **anti-codon**. The base pairing between the anticodon of the tRNA and an mRNA codon is responsible for the actual translation of the genetic information of structural genes. Although codon-anticodon pairings are antiparallel, both sequences are given in the 5′ → 3′ direction. For example, the codon UGC binds to the anticodon GCA (Figure 19.1).

Once the genetic code had been determined, researchers anticipated the identification of 61 types of tRNAs in living cells. Instead, they discovered that cells often operate with substantially fewer tRNAs than expected. Most cells possess about 50 tRNAs, although lower numbers have been observed. Further investigation of tRNAs revealed that the anticodon in some molecules contains uncommon nucleotides, such as inosinate (I), which typically occur at the third anticodon position. (In eukaryotes, A in the third anticodon position is deaminated to form I.) As tRNAs were investigated further, it became increasingly clear that some molecules recognize several codons. In 1966, after reviewing the evidence, Crick proposed a rational explanation, the **wobble hypothesis**.

The wobble hypothesis, which allows for multiple codon-anticodon interactions by individual tRNAs, is based principally on the following observations.

1. The first two base pairings in a codon-anticodon interaction confer most of the specificity required during translation. Recall that most redundant codons specifying a certain amino acid possess identical nucleotides in the first two positions. These interactions are standard (i.e., Watson-Crick) base pairings.

2. The interactions between the third codon and anticodon nucleotides are less stringent. In fact, nontraditional base pairs (i.e., non-Watson-Crick) often occur. For example, tRNAs containing G in the 5′ (or "wobble") position of the anticodon can pair with two different codons; that is, G can interact with either C or U. The same is true for U, which can interact with A or G (Figure 19.2a). When I is in the wobble position of an anticodon,

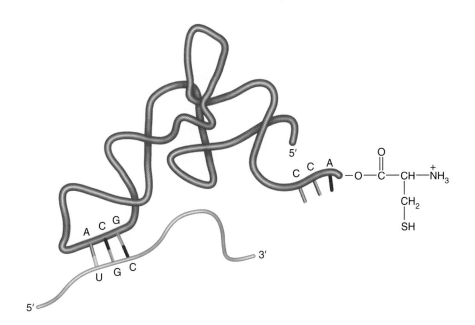

FIGURE 19.1

**Codon-Anticodon Base Pairing
of Cysteinyl-tRNA^{cys}**

The pairing of the codon UGC with the
anticodon GCA ensures that the amino acid
cysteine will be incorporated into a growing
polypeptide chain.

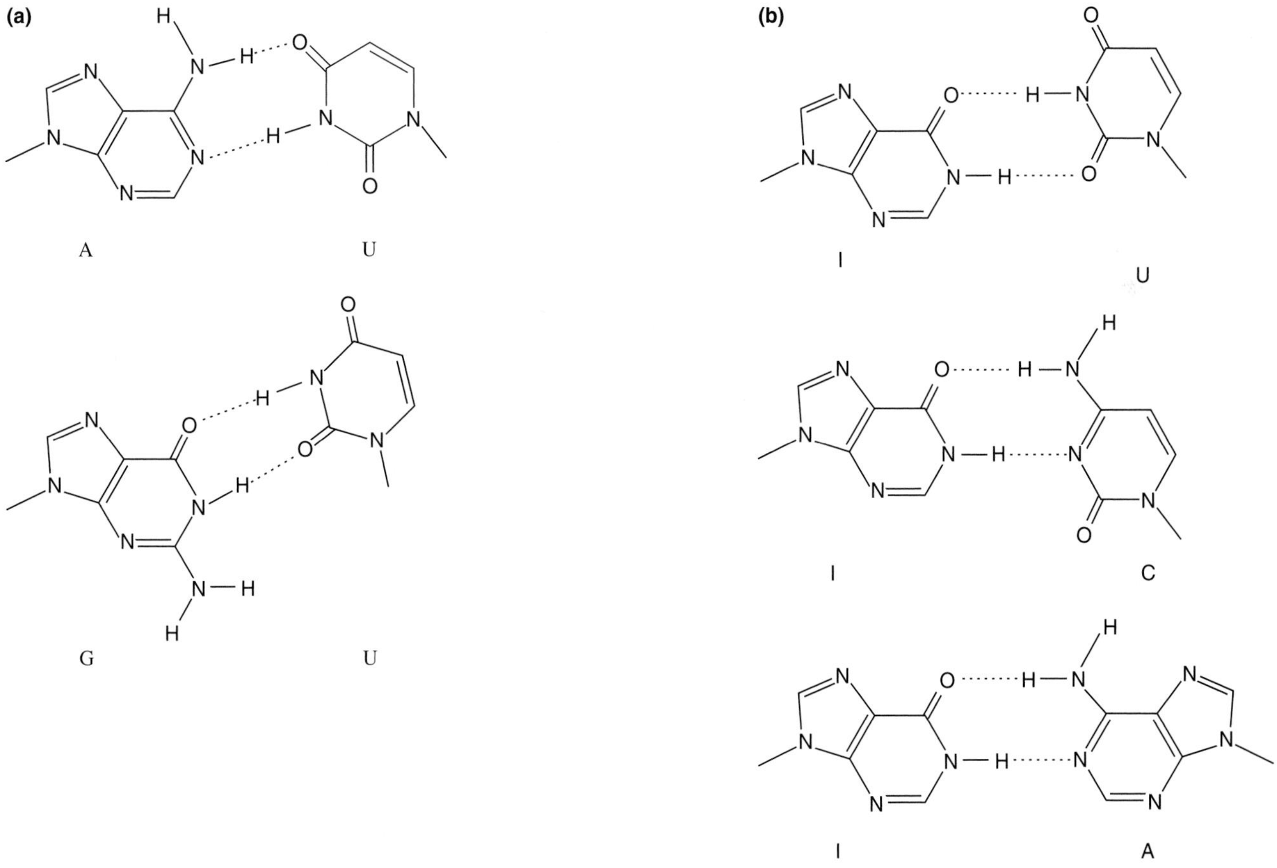

(a)

A   U

G   U

(b)

I   U

I   C

I   A

FIGURE 19.2

Wobble Base Pairs

Nonstandard base pairing is critical for the translation of the genetic code. Examples of
wobble base pairs include (a) AU and GU base pairs and (b) IU, IC, and IA base pairs.

KEY CONCEPTS

- The genetic code is translated through base-pairing interactions between mRNA codons and tRNA anticodons.
- The wobble hypothesis explains why cells usually have fewer tRNAs than expected.

a tRNA can base-pair with three different codons, because I can interact with U or A or C (Figure 19.2b).

A careful examination of the genetic code and the "wobble rules" indicates that all 61 codons can be translated with a minimum of 31 tRNAs. An additional tRNA for initiating protein synthesis brings the total to 32 tRNAs.

QUESTION 19.2

The sequence of a DNA segment is GGTTTA. What is the sequence of the tRNA anticodons?

QUESTION 19.3

The amino acid sequence for a short peptide is Tyr-Leu-Thr-Ala. What are the possible base sequences of the mRNA and the transcribed DNA strand that code for it? What are the anticodons?

The Aminoacyl-tRNA Synthetase Reaction

Although the accuracy of translation (approximately one error per 10^4 amino acids incorporated) is lower than those of DNA replication and transcription, it is remarkably higher than one would expect of such a complex process. The principal reasons for the accuracy with which amino acids are incorporated into polypeptides include codon-anticodon base pairing and the mechanism by which amino acids are attached to their cognate tRNAs. The attachment of amino acids to tRNAs, considered to be the first step in protein synthesis, is catalyzed by a group of enzymes called the aminoacyl-tRNA synthetases.

In most organisms there is at least one aminoacyl-tRNA synthetase for each of the 20 amino acids. (Each enzyme links its specific amino acid to any appropriate tRNA. This is important, because in most cells many amino acids have several cognate tRNAs each.) The process that links an amino acid to the 3′ terminus of the correct tRNA consists of two sequential reactions (Figure 19.3), both of which occur within the active site of the synthetase.

1. **Activation**. The synthetase first catalyzes the formation of aminoacyl-AMP. This reaction, which activates the amino acid by forming a high-energy mixed anhydride bond, is driven to completion through the hydrolysis of its other product, pyrophosphate. (An **anhydride** is a molecule containing two carbonyl groups linked through an oxygen atom. The term **mixed anhydride** describes an anhydride from two different acids, e.g., a carboxylic acid and phosphoric acid.)

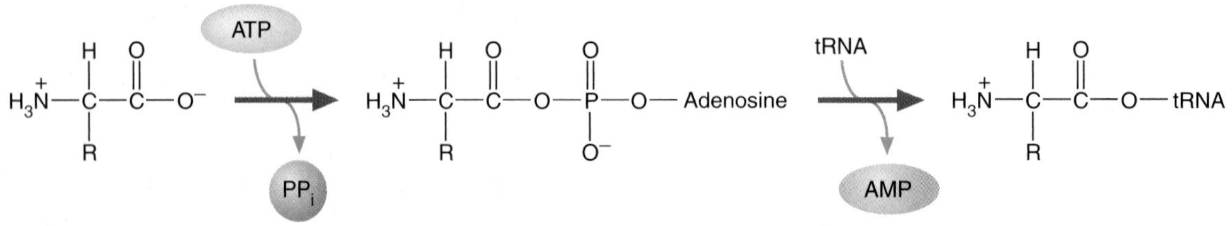

FIGURE 19.3

Formation of Aminoacyl-tRNA

Each aminoacyl-tRNA synthetase catalyzes two sequential reactions in which an amino acid is linked to the 3′-terminal ribose residue of the tRNA molecule.

2. **tRNA linkage**. A specific tRNA, also bound in the active site of the synthetase, becomes covalently bound to the aminoacyl group through an ester linkage. (Depending on the synthetase, the ester linkage may be through the 2'-OH or 3'-OH of the ribose moiety of the tRNA's 3'-terminal nucleotide. Subsequently, the aminoacyl group can migrate between the 2'-OH and 3'-OH groups. Only the 3'-aminoacyl esters are used during translation.) Although the aminoacyl ester linkage to the tRNA is lower in energy than the mixed anhydride of aminoacyl AMP, it still possesses sufficient energy to participate in acyl transfer reactions (peptide bond formation).

The sum of the reactions catalyzed by the aminoacyl-tRNA synthetases is as follows:

$$\text{Amino acid} + \text{ATP} + \text{tRNA} \rightarrow \text{aminoacyl-tRNA} + \text{AMP} + \text{PP}_i$$

The product PP_i is immediately hydrolyzed with a large loss of free energy. Consequently, tRNA charging is irreversible. Because AMP is a product of this reaction, the metabolic price for the linkage of each amino acid to its tRNA is the equivalent of the hydrolysis of two molecules of ATP to ADP and P_i.

The aminoacyl-tRNA synthetases are a diverse group of enzymes that vary in molecular weight, primary sequence, and number of subunits. Each enzyme efficiently produces a specific aminoacyl-tRNA product relatively accurately. As was mentioned, the specificity with which each of the synthetases binds the correct amino acid and its cognate tRNA is crucial for the fidelity of the translation process. Some amino acids can easily be differentiated by their size (e.g., tryptophan vs glycine) or the presence of positive or negative charges in their side chains (e.g., lysine and aspartate). Other amino acids, however, are more difficult to discriminate because their structures are similar. For example, isoleucine and valine differ only by a methylene group. Despite this difficulty, isoleucyl-tRNA^{ile} synthetase usually synthesizes the correct product. However, this enzyme occasionally also produces valyl-tRNA^{ile}. Isoleucyl-tRNA^{ile} synthetase, as well as several other synthetases, can correct such a mistake because it possesses a separate *proofreading* site. Because of its size, this site binds valyl-tRNA^{ile} and excludes the larger isoleucyl-tRNA^{ile}. After its binding in the proofreading site, the ester bond of valyl-tRNA^{ile} is hydrolyzed.

Aminoacyl-tRNA synthetases must also recognize and bind the correct (*cognate*) tRNA molecules. For some enzymes (e.g., glutaminyl-tRNA synthetase), anticodon structure is an important feature of the recognition process. However, several enzymes appear to recognize other tRNA structural elements, for example, the acceptor stem, in addition to or instead of the anticodon.

19.2 PROTEIN SYNTHESIS

An overview of protein synthesis is illustrated in Figure 19.4. Despite its complexity and the variations among species, the translation of a genetic message into the primary sequence of a polypeptide can be divided into three phases: initiation, elongation, and termination.

1. **Initiation**. Translation begins with **initiation**, when the small ribosomal subunit binds an mRNA. The anticodon of a specific tRNA, referred to as an *initiator tRNA*, then base-pairs with the initiation codon AUG on the mRNA. Initiation ends as the large ribosomal subunit combines with the small subunit. There are two sites on the complete ribosome for the codon-anticodon interactions involved in translation: the P (peptidyl) site (now occupied by the initiator tRNA) and the A (aminoacyl) site. In both prokaryotes and eukaryotes, mRNAs are read simultaneously by numerous ribosomes. An mRNA with several ribosomes bound to it is referred to

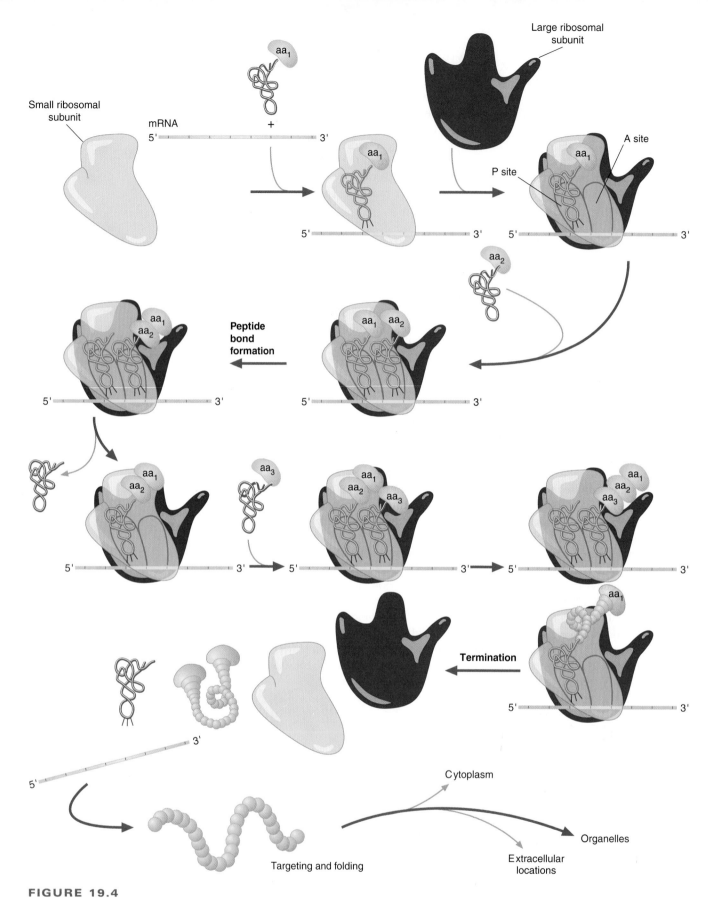

FIGURE 19.4

Protein Synthesis

No matter what the organism, translation consists of three phases: initiation, elongation, and termination. The elongation reactions, which include peptide bond formation and translocation, are repeated many times until a stop codon is reached. The numerous protein factors that facilitate each step in protein synthesis are different in prokaryotes and eukaryotes. Posttranslational reactions and targeting processes vary according to cell type.

as a **polysome**. In actively growing prokaryotes, for example, the ribosomes attached to an mRNA molecule may be separated from each other by as few as 80 nucleotides.

2. **Elongation**. During the **elongation** phase the polypeptide is synthesized according to the specifications of the genetic message. The mRNA base sequence is read in the $5' \rightarrow 3'$ direction, and polypeptide synthesis proceeds from the N-terminal to the C-terminal. Elongation begins as a second aminoacyl-tRNA becomes bound to the ribosome in the A site as the result of codon-anticodon base pairing. Peptide bond formation is then catalyzed by peptidyl transferase. During this reaction (referred to as *transpeptidation*) the α-amino group of the A-site amino acid (acting as a nucleophile) attacks the carbonyl group of the P-site amino acid (Figure 19.5). Because of peptide bond formation, both amino acids are now attached to the A-site tRNA. The now uncharged P-site tRNA is released from the ribosome. Within bacterial ribosomes, discharged tRNAs are released from another site, referred to as the E, or exit, site. Translocation is the mechanism whereby the ribosome moves along the mRNA and the next codon enters the A site. The A-site tRNA, now attached to the growing peptide chain moves into the P site. This series of steps, referred to as the *elongation cycle*, is repeated until a stop codon enters the A site.

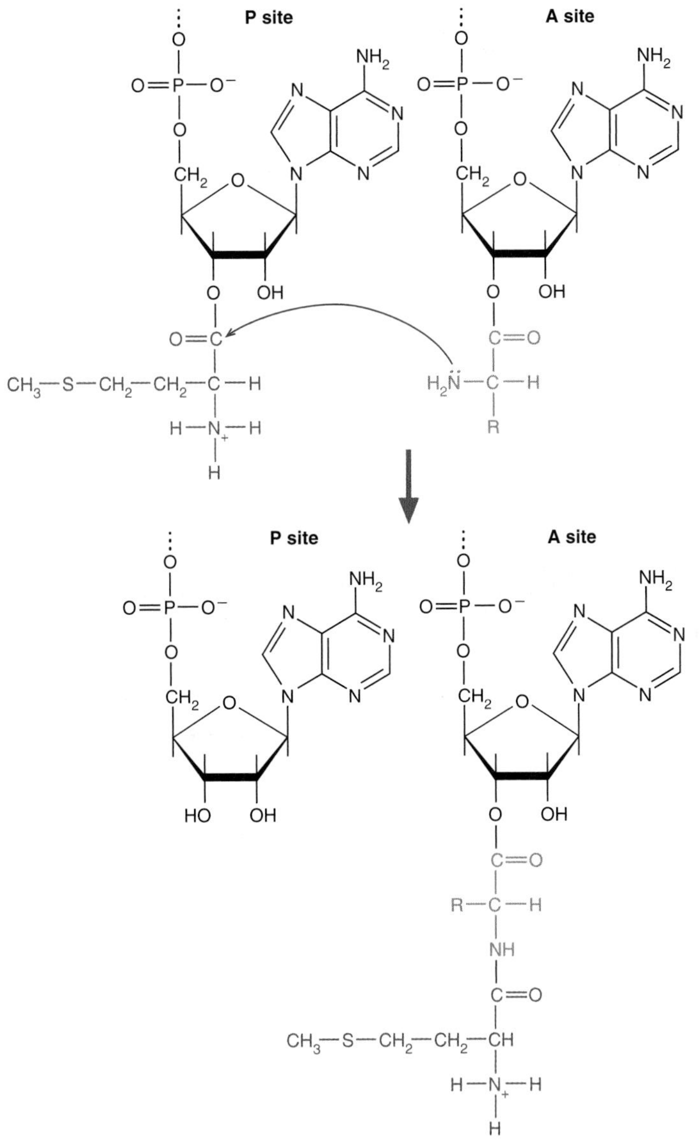

FIGURE 19.5

Peptide Bond Formation

During the first elongation cycle, peptide bonds form because of the nucleophilic attack of the A-site amino acid's amino group on the carbonyl carbon of the methionine residue in the P site. Because a peptide bond has formed, both amino acids are now attached to the A site tRNA.

3. **Termination**. During **termination** the polypeptide chain is released from the ribosome. Translation terminates because a stop codon cannot bind an aminoacyl-tRNA. Instead, a protein releasing factor binds to the A site. Subsequently, peptidyl transferase (acting as an esterase) hydrolyzes the bond connecting the now-completed polypeptide chain and the tRNA in the P site. Translation ends as the ribosome releases the mRNA and dissociates into the large and small subunits.

In addition to the ribosomal subunits, mRNA, and aminoacyl-tRNAs, translation requires an energy source (GTP) and a wide variety of protein factors. These factors perform several roles. Some have catalytic functions; others stabilize specific structures that form during translation. Translation factors are classified according to the phase of the translation process they affect, that is, initiation, elongation, or termination. The major differences between prokaryotic and eukaryotic translation appear to be due largely to the identity and functioning of these protein factors.

Regardless of the species, immediately after translation, some polypeptides fold into their final form without further modifications. Frequently, however, newly synthesized polypeptides are modified. These alterations, referred to as **posttranslational modifications**, can be considered to be the fourth phase of translation. They include removal of portions of the polypeptide by proteases, modification of the side chains of certain amino acid residues, and insertion of cofactors. Often, individual polypeptides then combine to form multisubunit proteins. Posttranslational modifications appear to serve two general purposes: (1) to prepare a polypeptide for its specific function and (2) to direct a polypeptide to a specific location, a process referred to as **targeting**. Targeting is an especially important process in eukaryotes because proteins must be precisely directed to a vast array of possible destinations. In addition to cytoplasm and the plasma membrane (the principal destinations in prokaryotes), eukaryotic proteins may be sent to a variety of organelles (e.g., mitochondria, chloroplasts, lysosomes, or peroxisomes).

Although there are many similarities between prokaryotic and eukaryotic protein synthesis, there are also notable differences. In fact, these differences are the basis for the therapeutic and research uses of several antibiotics (Table 19.2). Consequently, the details of prokaryotic and eukaryotic processes are discussed separately. Each discussion is followed by a brief description of mechanisms that control translation.

Prokaryotic Protein Synthesis

Translation in prokaryotes occurs on ribosomes composed of a 50S large subunit that consists of 23S and 5S rRNAs, 34 proteins, and a 30S small subunit having a 16S rRNA and 21 proteins. The large subunit contains the catalytic site for peptide bond formation, the peptidyl transferase center that is located within the 23S rRNA. The small subunit serves as a guide for the translation factors required to regulate the process. Figure 19.6 provides a three-dimensional reconstruction of a functioning *E. coli* ribosome.

KEY CONCEPTS

- Translation consists of three phases: initiation, elongation, and termination.
- After synthesis, many proteins are chemically modified and targeted to specific cellular or extracellular locations.

TABLE 19.2 Selected Antibiotic Inhibitors of Protein Synthesis

Antibiotic	Action
Chloramphenicol	Inhibits prokaryotic peptidyl transferase
Cycloheximide	Inhibits eukaryotic peptidyl transferase
Erythromycin	Inhibits prokaryotic peptide chain elongation
Streptomycin	Binding to 30S subunit causes mRNA misreading
Tetracycline	Binding to 30S subunit interferes with aminoacyl-tRNA binding

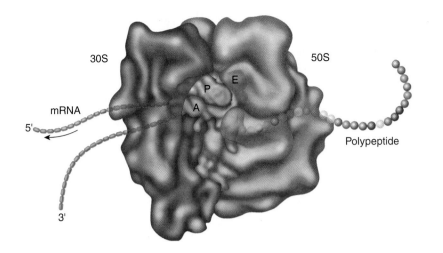

30S

50S

P

E

A

mRNA

5'

3'

Polypeptide

FIGURE 19.6

The Functional Ribosome

In this three-dimensional reconstruction of an *E. coli* ribosome during protein synthesis, the large and small subunits are shown in pink and orange, respectively. The relative positions of the mRNA, tRNAs, and the growing polypeptide chain are also illustrated. The tRNAs are identified as A and P to indicate their positions within the acyl and peptidyl sites where peptide bond formation occurs. The tRNA labeled E is in the exit position; that is, having discharged its amino acid during ongoing protein synthesis, it is in the process of leaving the ribosome. The movement of the mRNA through the ribosome is indicated by an arrow.

INITIATION As described, translation begins with the formation of an initiation complex (Figure 19.7). In prokaryotes this process requires three initiation factors (IFs): IF-1, IF-2, and IF-3. IF-3 binding to the 30S subunit prevents that subunit from binding prematurely to the 50S subunit. IF-1 then binds to the A site of the 30S subunit, thereby blocking it during initiation. As an mRNA binds to the 30S subunit, it is guided into a precise location (so that the initiation codon AUG is correctly positioned) by a purine-rich sequence referred to as the **Shine-Dalgarno sequence**. The Shine-Dalgarno sequence (named for its discoverers, John Shine and Lynn Dalgarno) occurs a short distance upstream from AUG. It binds to a complementary sequence contained in the 16S rRNA component of the 30S subunit. Base pairing between the Shine-Dalgarno sequence and the 30S subunit provides a mechanism for distinguishing a start codon from an internal methionine codon. Each gene on a polycistronic mRNA possesses its own Shine-Dalgarno sequence and an initiation codon. The translation of each gene appears to occur independently; that is, translation of the first gene in a polycistronic message may or may not be followed by the translation of subsequent genes.

In the next step in initiation, *IF-2* (a GTP-binding protein with a bound GTP) binds to the 30S subunit, where it promotes the binding of the initiating tRNA to the initiation codon of the mRNA. In its bound state, IF-2 interacts with the initiating tRNA and IF-1. The initiating tRNA in prokaryotes is *N*-formylmethionine-tRNA (fmet-tRNA^{fmet}), which is formed in the following process. After a special initiator tRNA has been charged with methionine, the amino acid residue is formylated in a reaction that requires N^{10}-THF. The enzyme that catalyzes this reaction binds met-tRNA^{fmet} but not met-tRNA^{met}. The initiation phase ends as the GTP molecule bound to IF-2 is hydrolyzed to GDP and P$_i$. GTP hydrolysis causes a conformational change that results in the binding of the 30S and 50S subunits. The three initiation factors, along with GDP and P$_i$, are released. The product of this process, the 70S ribosome, is now primed for elongation, the next phase in protein synthesis.

ELONGATION Elongation consists of three steps: (1) positioning an aminoacyl-tRNA in the A site, (2) peptide bond formation, and (3) translocation. As noted, these steps are referred to collectively as an elongation cycle.

The prokaryotic elongation process begins when an aminoacyl-tRNA, specified by the next codon, binds to the now-empty A site. Before it can be positioned in the A site, the aminoacyl-tRNA must bind EF-Tu-GTP. The elongation factor *EF-Tu* is a GTP-binding protein that positions aminoacyl-tRNA molecules in the A site. The EF-Tu also serves to protect the aminoacyl linkage from hydrolysis or premature or unregulated peptide bond formation.

FIGURE 19.7
Formation of the Prokaryotic Initiation Complex

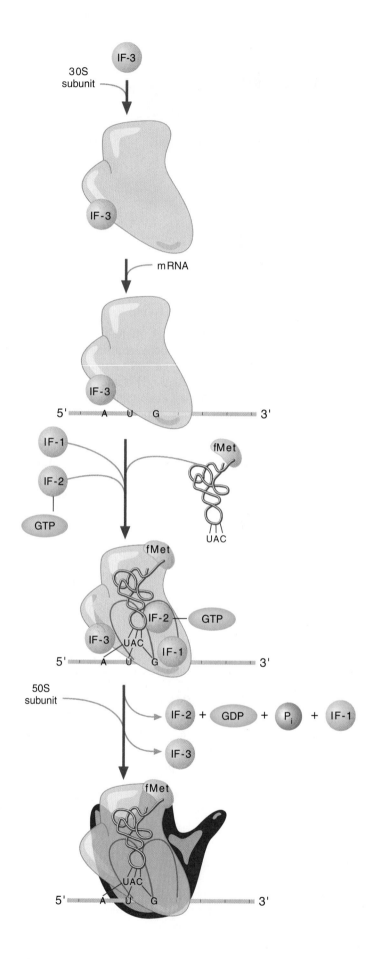

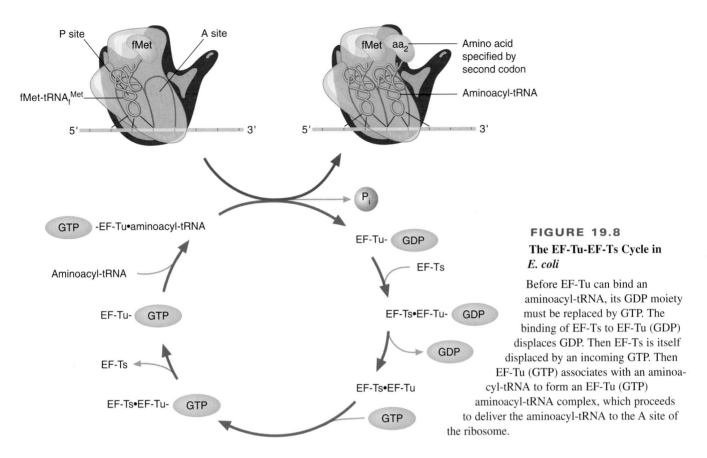

FIGURE 19.8

**The EF-Tu-EF-Ts Cycle in
*E. coli***

Before EF-Tu can bind an
aminoacyl-tRNA, its GDP moiety
must be replaced by GTP. The
binding of EF-Ts to EF-Tu (GDP)
displaces GDP. Then EF-Ts is itself
displaced by an incoming GTP. Then
EF-Tu (GTP) associates with an aminoacyl-tRNA to form an EF-Tu (GTP)
aminoacyl-tRNA complex, which proceeds
to deliver the aminoacyl-tRNA to the A site of
the ribosome.

After the anticodon of the aminoacyl-tRNA has been correctly paired to the
mRNA codon, the GTP bound to EF-Tu is hydrolyzed to GDP and P_i. GTP
hydrolysis releases EF-Tu-GDP from the ribosome. Then another elongation factor (called EF-Ts), acting as a **guanine nucleotide exchange factor (GEF)**,
promotes EF-Tu regeneration by displacing its GDP moiety. EF-Ts is then itself
displaced by an incoming GTP molecule (Figure 19.8). The newly formed EF-
Tu-GTP can then bind to a new aminoacyl-tRNA. (Insight into the structural
and functional properties of EF-Tu is provided shortly in the Biochemistry in
Perspective box entitled EF-Tu: A Motor Protein.)

After EF-Tu has delivered an aminoacyl-tRNA to the A site, the formation
of a peptide bond is catalyzed by peptidyl transferase, within the 23S rRNA component of the 50S subunit. Peptide bond formation occurs in a cleft in the 50S
ribosomal subunit on the side facing the 30S subunit. The mechanism whereby
the peptide bond is formed appears to occur via intrasubstrate proton shuttling
(Figure 19.9). The ribosome facilitates the reaction in several ways. These include
precise positioning of the substrates within the *peptidyl transferase center*, or
PTC (i.e., the acceptor ends of the A-site and P-site tRNAs become fixed in
place via interactions with residues of 23S rRNA), and an electrostatic environment that assists the proton shuttle process. The ribosome also reduces the free
energy requirements of the reaction by providing a relatively anhydrous environment in the active site that is essential for the formation of a highly polar
transition state. The energy required to drive this reaction is provided by the high-
energy ester bond linking the P-site amino acid to its tRNA. (During the first elongation cycle, this amino acid is formylmethionine.)

Immediately after peptide bond formation, the tRNA in the A site (referred
to as the peptidyl tRNA because it is linked to the growing peptide chain) is still
in the A site and the now-deacylated tRNA is in the P site. This phase of elongation is referred to as the *pretranslocation state*. Translocation, the movement
by the ribosome of the mRNA (so that a new codon-anticodon interaction can
occur in the A site) and the tRNAs from the A to P and P to E sites, respectively,

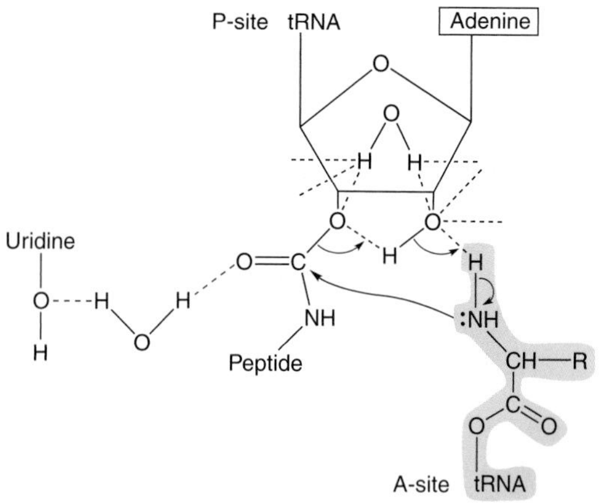

FIGURE 19.9

The Peptidyl Transferase Proton Shuttle Mechanism

The reaction begins with the nucleophilic attack of the α-amino nitrogen on the A-site aminoacyl group carbonyl carbon of the peptidyl chain (linked to P-site tRNA by an ester bond to 3′-OH group of the ribose of residue 76). During the first elongation cycle, a single *N*-formylmethionyl aminoacyl group is linked to the P-site tRNA. A six-membered transition state is thus created in which the 2′-OH group of the A-site ribose donates its proton to the adjacent 3′-oxygen atom as the latter receives an amino proton. The precise alignment of the substrates within the active site is fostered by hydrogen bonds between the ribose oxygens and a bridging water molecule (black dashed lines) and hydrogen bonds between the ribose 2′-hydroxyl oxygen and the bridging water molecule and certain cytosine and adenosine residues (not shown) of 23S rRNA (dashed lines). The carbonyl oxygen of the peptidyl group of the P-site tRNA is stabilized by a hydrogen bond, via a water molecule bridge, to the hydroxyl oxygen of a uridine residue of 23S rRNA.

requires the binding of another GTP-binding protein, referred to as EF-G. When EF-G-GTP binds near the A site, GTP hydrolysis is triggered. The resulting conformational change causes the two subunits to rotate in opposite directions in relation to each other. The rotational movement, which creates a space between the two subunits, accommodates the simultaneous transfer of the A-site peptidyl tRNA into the P site, and the associated shift of the mRNA moves the next codon into the A site. In the *posttranslocation state*, the deacylated tRNA is in the E site. The release of the deacylated tRNA from the E site occurs as the incoming EF-Tu aminoacyl-tRNA complex enters the A site and codon-anticodon interactions are initiated. After the release of EF-G-GDP, the ribosome is ready for the next elongation cycle. Elongation continues until a stop codon enters the A site.

TERMINATION The termination phase begins when a termination codon (UAA, UAG, or UGA) enters the A site. Three **release factors** (RF-1, RF-2, and RF-3) are involved in termination. Both RF-1 and RF-2 resemble tRNAs in shape and size. RF-1 recognizes the stop codons UAA and UAG, and RF-2 recognizes UAA and UGA. RF-3 is a GTPase that promotes the binding of RF-1 and RF-2 to the ribosome. The ribosome binds inactive RF-3-GDP, and acting as a GEF, converts it to the active form RF-3-GTP. The binding of the RFs alters ribosomal function. The peptidyl transferase, which is transiently transformed into an esterase, hydrolyzes the bond linking the completed polypeptide chain and the P-site tRNA. Hydrolysis of the GTP-bound to RF-3 triggers the release of RF-1 and RF-2. Following the polypeptide's release from the ribosome, the mRNA and tRNA also dissociate. The termination phase ends when the ribosome dissociates into its constituent subunits in a process that requires **ribosome recycling factor** (RRF), a tRNA-shaped protein that binds within the A site. The energy required to separate the two ribosomal subunits is supplied by EF-G-GTP. Then IF-3 binds to the small subunit to prevent premature rebinding by the large subunit.

POSTTRANSLATIONAL MODIFICATIONS As each **nascent** (newly synthesized) polypeptide emerges from the ribosome, it begins to fold into its final three-dimensional shape (see p. 159). As mentioned, most of these molecules also undergo a series of modifying reactions that prepare them for their functional role. Most of the information concerning posttranslational modifications has been obtained through research on eukaryotes. However, prokaryotic polypeptides are known to undergo several types of covalent alteration.

1. **Proteolytic processing**. Several cleavage reactions may occur. These include removing the formylmethionine residue and signal peptide sequences. **Signal peptides**, or leader peptides, are short peptide sequences, typically near the aminoterminal, that determine a polypeptide's destination. In bacteria, for example, a signal peptide is required to insert a polypeptide into the plasma membrane.

2. **Conjugation**. Most conjugated proteins in prokaryotes are lipoproteins. One recently discovered lipoprotein, found in the outer membrane of *E. coli*, is called Blc. It is a type of lipocalin produced under stressful conditions that assists in membrane biogenesis and repair. (The *lipocalins*, a group of proteins that bind hydrophobic ligands, were previously observed only in eukaryotes.) Although once considered rare in prokaryotes, several glycosylated proteins have also been identified in archaeans and in certain Gram-positive bacteria. The best-studied example is the cell surface glycoprotein of the halophilic (salt-loving) archaean *Halobacterium salinarium*.

3. **Methylation**. The group of enzymes referred to as the protein methyltransferases use *S*-adenosylmethionine to methylate certain proteins. For example, one type of methyltransferase found in *E. coli* and related bacteria methylates glutamate residues in membrane-bound chemoreceptors. The methyltransferase and a methylesterase are components in a methylation/demethylation process, which plays a role in a signal transduction mechanism involved in chemotaxis. (Recall that chemotaxis is the capacity of a living cell to respond to certain environmental cues by moving toward or away from specific molecules.)

4. **Phosphorylation**. In recent years, protein phosphorylation/dephosphorylation catalyzed by protein kinases and phosphatases has been revealed to be widespread among prokaryotes. Many of the purposes of these reactions remain unclear. However, roles for transient phosphorylation have been identified in chemotaxis and nitrogen metabolism regulation.

TRANSLATIONAL CONTROL MECHANISMS Protein synthesis is an exceptionally expensive process, costing four high-energy phosphate bonds per peptide bond (i.e., two bonds expended during tRNA charging and one each during A-site–tRNA binding and translocation). It is perhaps not surprising that enormous quantities of energy are involved. For example, approximately 90% of *E. coli* energy production used in the synthesis of macromolecules may be devoted to the manufacture of proteins. Although the speed and accuracy of translation require a high energy input, the cost would be even higher without metabolic control mechanisms. These mechanisms allow prokaryotic cells to compete with each other for limited nutritional resources.

In prokaryotes such as *E. coli*, most of the control of protein synthesis occurs at the level of transcription initiation. (Refer to Section 18.32 for a discussion

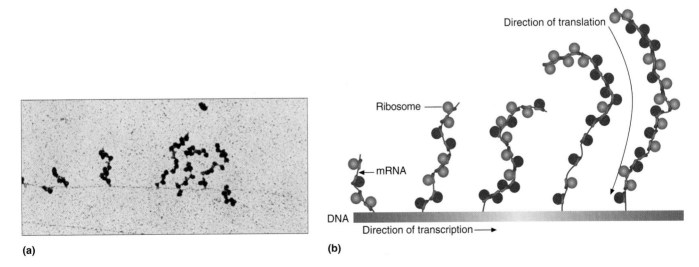

(a)

(b)

FIGURE 19.10

Transcription and Translation in *E. coli*

(a) An electron micrograph of *E. coli* transcription and translation. In *E. coli*, as in other prokaryotes, transcription and translation are directly coupled. (b) Diagram of (a). Note polyribosomes.

of the principles of prokaryotic transcriptional control.) This circumstance makes sense for several reasons. First, transcription and translation are spatially and temporally coupled; that is, translation is initiated shortly after transcription begins (Figure 19.10). Second, the lifetime of prokaryotic mRNA is usually relatively short. With half-lives of between 1 and 3 minutes, the types of mRNA produced in a cell can be quickly altered as environmental conditions change. Most mRNA molecules in *E. coli* are degraded by two exonucleases, referred to as RNase II and polynucleotide phosphorylase.

Despite the preeminence of transcriptional control mechanisms, the rates of prokaryotic mRNA translation also vary. A large portion of this variation is attributed to differences in Shine-Dalgarno sequences. Because Shine-Dalgarno sequences facilitate the selection of the initiation codon, sequence variations may affect the rate of translating genetic messages. For example, the gene products of the *lac* operon (β-galactosidase, galactose permease, and galactoside transacetylase) are not produced in equal quantities. Thiogalactoside transacetylase is produced at approximately one-fifth the rate of β-galactosidase (p. 718).

An interesting example of negative translational regulation in prokaryotes is provided by ribosomal protein synthesis. The approximately 55 proteins in prokaryotic ribosomes are coded for by genes located in 20 operons. Efficient bacterial growth requires that their synthesis be regulated in coordinate fashion among the operons and with rRNA synthesis. For example, in the P_{L11} operon, which contains the genes for the ribosomal proteins L1 and L11, excessive amounts of L1 (i.e., more L1 molecules than can bind available 23S rRNA) inhibit P_{L11} mRNA translation (Figure 19.11). Apparently, L1 can bind to either 23S rRNA or P_{L11} mRNA. In the absence of 23S rRNA, L1 inhibits the translation of its own operon by binding to the 5′ end of P_{L11} mRNA.

KEY CONCEPTS

- Prokaryotic protein synthesis is a rapid process involving several protein factors.
- Although most prokaryotic gene expression appears to be regulated by transcription initiation, several types of translational regulation have been detected.

FIGURE 19.11

A Negative Translational Control

The P_{L11} operon of *E. coli* is controlled by the level of one of its gene products. The ribosomal protein L1 can bind to 23S rRNA or its own mRNA. When excessive amounts of L1 accumulate, the L1 binds to the 5′ end of its own mRNA, thereby inhibiting the translation of the operon.

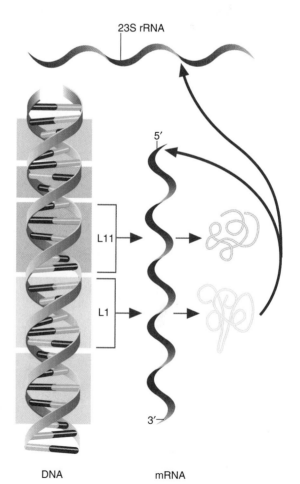

BIOCHEMISTRY IN PERSPECTIVE

EF-Tu: A Motor Protein

How do the structural features of the prokaryotic elongation factor EF-Tu facilitate its role as a motor protein?   EF-Tu (Figure 19A), the protein factor that positions aminoacyl-tRNA complexes in the A site of prokaryotic ribosomes, is a well-researched example of a GTP-binding motor protein. Recall that *motor proteins* (Section 2.1) use nucleotide hydrolysis to drive changes in their own conformations that promote ordered conformational changes in adjacent molecules or subunits. In other words, motor proteins, often called NTPases, function as mechanochemical transducers. These NTP hydrolysis–driven conformational changes, which principally occur in localized structural units called switches, alter the affinity of the NTPase for other molecules.

EF-Tu possesses three domains. Domain 1 contains a GTP binding site and two switch regions. Domain 2 is connected to domain 1 through a pliable peptide segment. In its active GTP-bound form (EF-Tu-GTP), the elongation factor possesses a binding site for an aminoacyl-tRNA. After aminoacyl-tRNA binding, the entire structure is referred to as the *ternary complex*. All three domains of EF-Tu are involved in tRNA binding. For example, the TψC stem of tRNA molecules (Figure 17.19) interacts with several amino acid residues in domain 3. The binding of aminoacyl-tRNA projects the anticodon away from the ternary complex so it is free to interact with mRNA codons.

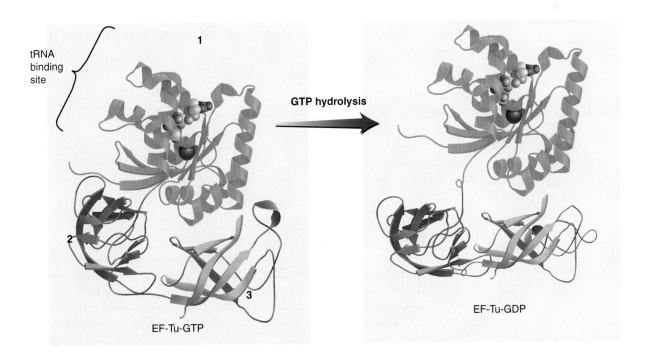

GTP hydrolysis

tRNA binding site

1

2

3

EF-Tu-GTP

EF-Tu-GDP

FIGURE 19A

EF-Tu

EF-Tu is a GTPase that positions aminoacyl-tRNA complexes within the A site of prokaryotic ribosomes during protein synthesis. The binding of a GTP molecule by EF-Tu causes domain conformation changes (not shown) that result in the creation of a binding cleft for an aminoacyl-tRNA complex. GTP hydrolysis causes domains 1 and 2 to move apart so that the aminoacyl-tRNA (not shown) is released. (The red sphere represents a magnesium ion.)

During protein synthesis, the interaction of EF-Tu-GDP (the inactive form) with EF-Ts releases GDP. The subsequent binding of GTP in the domain 1 nucleotide-binding site changes the conformation in the two switch regions. These changes bring domains 1 and 2 close together, forming a binding cleft. Once an aminoacyl-tRNA has been bound in the cleft, the ternary complex enters the ribosome where the aminoacyl-tRNA anticodon binds reversibly to an mRNA codon in the A site. When a ternary complex contains a cognate aminoacyl-tRNA, a conformation change in the ribosome triggers a conformation change in the EF-Tu nucleotide-binding site. The subsequent hydrolysis of GTP causes domains 1 and 2 to move apart, thus allowing the release of the aminoacyl-tRNA.

Switch region 1, which is believed to contain a β-hairpin in the inactive EF-Tu-GDP, is converted to a helix (the "switch helix") in EF-Tu-GTP. In switch region 2 (near the GTP-binding site) the γ-phosphate of GTP causes a conserved glycine residue to flip 180°, moving it 4.6 Å. The dramatic movement of this residue forces the switch helix to migrate four residues along the polypeptide chain, a process that changes the axis of rotation of the helix by 45° and results in a 46 Å movement of the most distal portion of domain 1. This feature of EF-Tu function has been described as a timing mechanism. The timer is activated when the ternary complex binds to the ribosome. The relatively slow rate of GTP hydrolysis provides sufficient time for the dissociation of incorrect codon-anticodon pairings. In contrast, the binding of a cognate aminoacyl-tRNA is so tight that there is sufficient time for GTP to undergo hydrolysis. Thus, the functioning of the ternary complex provides another mechanism for proofreading during translation in addition to that described for the aminoacyl-tRNA synthetases.

SUMMARY: EF-Tu is an NTPase that binds and hydrolyzes GTP. The binding of GTP to domain 1 of EF-Tu causes a change in conformation of the entire protein that facilitates the binding of an aa-tRNA. Once the aa-tRNA complex is positioned in the ribosome via codon-anticodon base pairing, GTP hydrolysis causes a conformational change that results in the release of the aa-tRNA.

Eukaryotic Protein Synthesis

Although the earliest investigations of protein synthesis (e.g., the discovery of aminoacyl-tRNA synthetases and the tRNAs) focused on mammalian cells, translation researchers directed their attention to bacteria in the 1960s. This change occurred for a variety of reasons, including the relative ease of culturing bacterial cells and the perception that bacterial gene expression is simpler and more accessible than that in the more complex eukaryotes. Only in the 1970s, when the principles of prokaryotic translation were understood, did the eukaryotic process again become a focus of attention. Not surprisingly, the large, complex genomes of eukaryotic cells (especially those in multicellular organisms) are now known to require sophisticated regulation of translation (Section 18.3). A significantly larger number of protein factors assist in translation and, the posttranslational modifications of eukaryotic polypeptides are more numerous than those observed in prokaryotes. Considering the structural complexity of eukaryotes, it is inevitable that polypeptide targeting mechanisms are also quite intricate.

In this section the features that distinguish the three phases of eukaryotic translation from its prokaryotic counterparts are described. This is followed by a discussion of several of the most prominent forms of eukaryotic posttranslational modification and targeting mechanisms. The section ends with a discussion of translational control mechanisms.

INITIATION Many the major differences between the prokaryotic and eukaryotic versions of protein synthesis can be observed during the initiation phase. Among the reasons for the additional complexity of eukaryotic initiation are the following.

1. **mRNA secondary structure**. Recall that eukaryotic mRNA is processed with the addition of a methylguanosine cap and a poly(A) tail and the removal of introns. In addition, eukaryotic mRNA does not associate with a ribosome until it has left the nucleus and, as a result, is free to interact with a number of cellular proteins. (An mRNA that is complexed with these proteins is sometimes referred to as a ribonucleoprotein particle.)

2. **mRNA scanning**. In contrast to prokaryotic mRNA, eukaryotic molecules lack Shine-Dalgarno sequences, which allow for the identification of the initiating AUG sequence. Instead, eukaryotic ribosomes "scan" each mRNA. This scanning is a complex (and poorly understood) process in which ribosomes bind to the capped 5′ end of the molecule and migrate in a 5′ → 3′ direction searching for a translation start site.

Eukaryotes use a more complex spectrum of initiation factors than prokaryotes. There are at least 12 eukaryotic initiating factors (eIFs), several of which possess numerous subunits. The functional roles of most of these factors are still under investigation.

Eukaryotic initiation (Figure 19.12) begins with the assembly of the preinitiation complex. This process begins when the small 40S ribosomal subunit binds to eIF-1 A, *eIF-2* (a GTP-binding protein), GTP, and an initiating species of methionyl-tRNA^{met} (met-tRNA$_i$). The binding of the initiating tRNA to the 40S subunit is mediated by eIF-2-GTP, which in turn is regenerated from inactive eIF-2-GDP by *eIF-2B*, a GEF. (Upon the release of GDP from eIF-2, GTP binding occurs.) The small (40S) subunit is prevented from binding to the large (60S) subunit during this phase of initiation because it is associated with *eIF-3*, a multisubunit protein. (Subunit assembly is also prevented by the association of eIF-6 with the 60S subunit.) The **43S preinitiation complex**, composed of the 40S subunit, eIF-1A, eIF-2-GTP, eIF-3, and methionyl-tRNA^{met}, is now able to bind to an mRNA. Most mRNAs cannot perform this step unless they have previously bound to a cap-binding complex. The **cap-binding complex (CBC)**, also referred to as eIF-4F, consists of eIF-4A (a helicase), eIF-4E (a translation initiation factor), and eIF-G (a scaffold protein). The binding of the CBC to the mRNA 5′-cap structure (Figure 19.12) is facilitated by structural features of the binding pocket in eIF-4E. Base stacking is an important stabilizing feature within this pocket: the aromatic ring of the 7-methylguanosine is sandwiched between two tryptophan residues. In addition, salt bridges form between the nitrogen atoms of the 7-methylated purine ring with the side chain carboxylate groups of glutamate or aspartate residues, and the oxygen atoms of the triphosphate group interact with the positively charged side chains of both lysine and arginine residues. After the CBC has bound to the mRNA cap structure, a helicase (eIF-4B) removes any secondary structures in the 5′-UTR that can interfere with translation initiation. The 3′-poly(A) tail of the mRNA is then brought into close proximity to the 5′-capped end by interactions between eIF-G and a **poly(A)-binding protein (PABP)** to form a circular mRNA molecule. The binding of the 43S preinitition complex to the 5′-UTR of the mRNA is mediated by the CBC. The completed initiation complex proceeds to scan the mRNA in search of the initiation codon 5′-AUG-3′ near the 5′ end.

In the final step of the initiation process, the initiation complex binds the 60S subunit (now dissociated from eIF-6). The formation of the complete ribosome (Figure 19.13), the 80S complex, involves the hydrolysis of the GTP bound to eIF-2. This process requires a subunit of eIF-5 that acts as a guanine nucleotide activating protein (GAP). The initiation factors eIF-2, eIF-3, eIF-4B, and eIF-1A are also released.

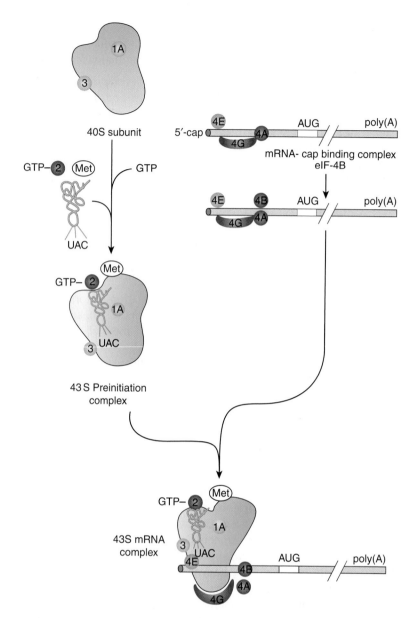

FIGURE 19.12

Assembly of the 43S mRNA Complex

In preparation for its role in polypeptide synthesis, the 40S subunit, previously bound to eIF-1A and eIF-3, now binds a complex of eIF-2-GTP (GTP-2) and met-tRNA^{met}. (The factor eIF-1A facilitates the recruitment of the GTP-2-met-tRNA^{met} complex.) Simultaneously, cap-binding complex (composed of eIF-4A (4A), eIF-4E (4E), and eIF-4G (4G) is binding to the 5′-cap structure of an mRNA. Once eIf-4B, a helicase, has removed secondary structure from the 5′ UTR, the 43S complex binds to the mRNA and begins to scan for the start codon, AUG. Although noted that eukaryotic initiation involves at least 30 different proteins, only the most important are mentioned in the text and illustrated in this figure.

ELONGATION Figure 19.14 illustrates the eukaryotic elongation cycle as it is currently understood. Several elongation factors (eEFs) are required during this phase of translation. For example, *eEF-1α* is a 50 kD polypeptide that mediates the binding of aminoacyl-tRNAs to the A site. After a complex has formed between eEF-1α, GTP, and the entering aminoacyl-tRNA, codon-anticodon interactions are initiated. If correct pairing occurs, eEF-1α hydrolyzes its bound

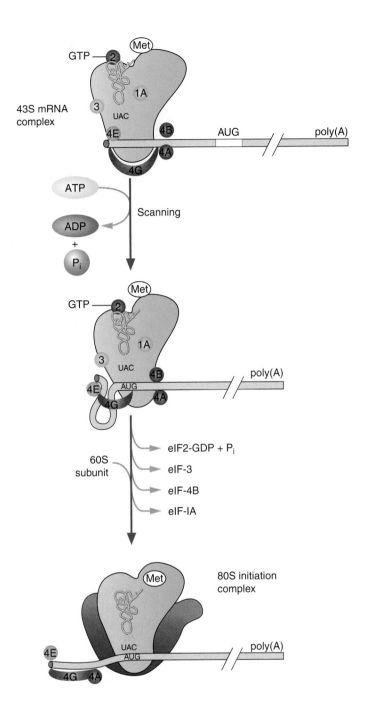

FIGURE 19.13

mRNA Scanning and 80S Initiation Complex Formation

The newly bound 43S complex moves in an ATP-requiring process facilitated by the CBC along the mRNA in the $5' \rightarrow 3'$ direction. Once a start codon has been recognized (i.e., once correct base pairing has occurred between the AUG codon and the anticodon of the initiation tRNA), the release of eIF-2, eIF-3, and eIF-4B is triggered. The subsequent binding of the 60S subunit results in the release of the remaining initiation factors (only eIF-1A is shown in this illustration) in a GTP-requiring process stimulated by eIF-5b (not shown).

GTP and exits the ribosome, leaving its aminoacyl-tRNA behind. If correct pairing does not occur, the complex leaves the A site, thereby preventing incorrect amino acid residues from being incorporated. This process has been referred to as **kinetic proofreading**.

During the next elongation step (i.e., peptide bond formation) the peptidyl transferase activity of the large ribosomal subunit catalyzes the nucleophilic attack of the A-site α-amino group on the carboxyl carbon of the P-site amino acid residue. Apparently eEF-1α dissociates from the ribosome immediately before transpeptidation. eEF-1β and eEF-1γ are GEFs that mediate the regeneration of eEF-1α by promoting an exchange of GDP for GTP.

Translocation in eukaryotes requires a 100 kD polypeptide referred to as *eEF-2*, which is also a GTP-binding protein. During translocation, eEF-2-GTP

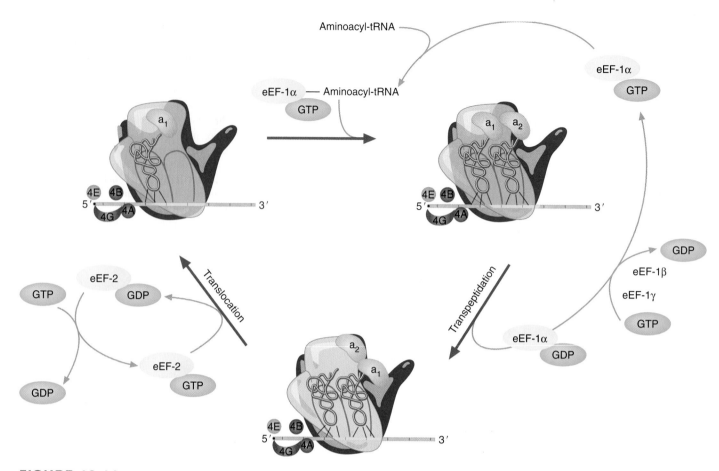

FIGURE 19.14

The Elongation Cycle in Eukaryotic Translation

Elongation comprises three phases: binding of an aminoacyl-tRNA to the A site, transpeptidation, and translocation.

binds to the ribosome. GTP is then hydrolyzed to GDP, and eEF-2-GDP is released. As noted, GTP hydrolysis provides the energy needed to physically move the ribosome along the mRNA. At the end of translocation, a new codon is exposed in the A site.

TERMINATION In eukaryotic cells, two release factors mediate the termination process: eRF-1 (a molecule that resembles the size and overall shape of a tRNA and recognizes and binds to stop codons) and eRF-3 (a GTPase). When a stop codon (UAG, UGA, or UAA) enters the active site, eRF-1 binds to it (Figure 19.15). As a result of this binding process, peptidyl transferase catalyzes the hydrolysis of the ester linkage between the polypeptide and the P-site tRNA. The hydrolysis of the GTP bound to eRF-3 is believed to then trigger the dissociation of eRF-1 from the ribosome. In eukaryotes, the function of ribosome recycling factor, the dissociation of prokaryotic ribosomal subunits, has been attributed to eRF-3.

The efficiency of eukaryotic translation (i.e., the number of polypeptides that can be synthesized per unit of time) is made possible to a large extent by the circular conformation of eukaryotic polysomes (Figure 19.16), As soon as ribosomal subunits and their associated protein factors are released, they are optimally positioned for recruitment into new ribosomes.

KEY CONCEPTS

- Eukaryotic protein synthesis, like its prokaryotic counterpart, has three phases: initiation, elongation, and termination.

- Features that are unique to eukaryotic translation include an abundance of protein factors that facilitate each step, cap-binding protein, and the formation of circular mRNA polysomes.

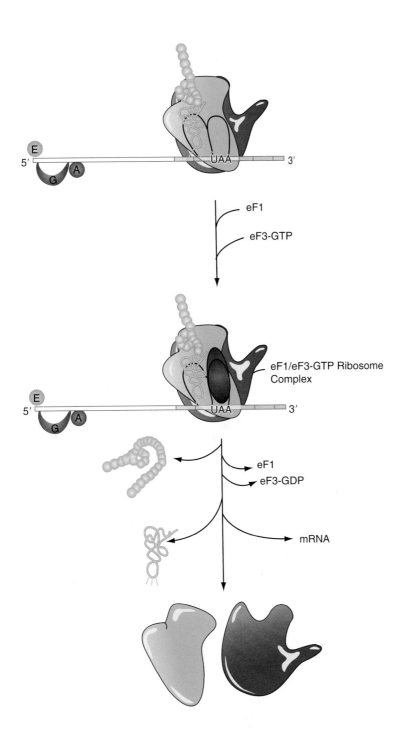

FIGURE 19.15
Eukaryotic Protein Synthesis Termination

As a stop codon (UAG, UGA, or UAA) enters and binds in the A site, release factor eRF-1 recognizes and binds to it. The eRF-1 binding promotes the conversion of the peptidyl transferase to a hydrolase that catalyzes the hydrolysis of the ester bond that links the now complete polypeptide chain to the P-site tRNA. The release of the polypeptide is followed by the dissociation of eRF-1 from the ribosome. The dissociation of the ribosomal subunits, which is mediated by eRF-3 and fueled by GTP hydrolysis, may be a function of eEF-3.

QUESTION 19.4

The eukaryotic elongation factor eEF-2 possesses a unique modification of a specific histidine residue called diphthamide. Mutant eukaryotic cells that cannot transform this histidine into diphthamide do not appear to be adversely affected. However, the ADP-ribosylation of the diphthamide residue of eEF-2 by toxins produced by *Corynebacterium diphtheriae* and *Pseudomonas aeruginosa* renders the factor inoperative.

QUESTION 19.4 (CONT.)

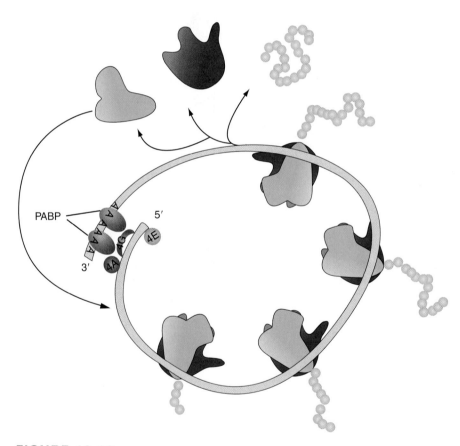

Cells die because they cannot synthesize proteins. The mechanism by which eEF-2 function is affected by ADP-ribosylation is unknown. Can you suggest any possibilities?

FIGURE 19.16

The Eukaryotic mRNA Polysome

Eukaryotic mRNA is circularized via an interaction between the 5′ UTR and the 3′-poly(A) tail that is mediated by PABP, the poly(A)-binding protein that links the poly(A) sequence to the CBC. As a result of this structural feature, when a polypeptide's synthesis is completed and the ribosomal subunits are released, the close proximity of these subunits to the 5′ cap facilitates immediate recruitment for another round of protein synthesis.

BIOCHEMISTRY IN PERSPECTIVE

Context-Dependent Coding Reassignment

How are selenocysteine and pyrrolysine, two nonstandard amino acids, incorporated into polypeptides during protein synthesis? There are two variations of the genetic code in which nonstandard amino acids are incorporated into polypeptides during protein synthesis. Selenocysteine and pyrrolysine (Figure 19B), now referred to as the twenty-first and twenty-second amino acids in proteins, are coded for by codons that are ordinarily stop signals. The indi-

rect mechanisms used, called *context-dependent codon reassignment*, for both amino acids are described.

Selenocysteine

The incorporation of selenocysteine into selenoproteins (e.g., glutathione peroxidase and thioredoxin reductase) is a widespread phenomenon that has been observed in Eubacteria, Archaea, and Eukarya. The following discussion is limited to mammals. Selenocysteine (sec) codon reassignment involves several molecules: a specific tRNA (tRNA[ser]sec), a seryl-tRNA synthetase with sec-generating ability, an mRNA-binding protein (SPB2), and a specialized elongation factor (EFsec). The seryl-tRNA synthetase binds tRNA[ser]sec, and serine is attached to its acceptor arm. The seryl moiety is converted to selenocysteine by a pyridoxal phosphate–containing enzyme called selenocysteine synthase to form sec tRNA[ser]sec. The coding of selenocysteine requires a SECIS (<u>se</u>leno<u>c</u>ysteine <u>i</u>nsertion <u>s</u>equence) element in the 3′ UTR of the mRNAs for all selenoproteins. The SECIS element rearranges into a loop-stem-bubble structure that recruits SBP2 to form a SECIS/SBP2 complex (Figure 19C). An AGU/AG sequence in the stem forms a nucleotide quartet that

FIGURE 19B
The Twenty-first and Twenty-second Amino Acids

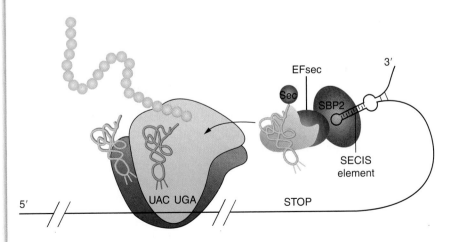

FIGURE 19C

Mechanism of Selenocysteine Incorporation into Eukaryotic Selenoproteins

The stop codon UGA is reassigned for selenocysteine incorporation by the interaction of a complex composed of a SECIS element bound to SBP2 (red), and EFsec (blue) that is charged with sec t-RNA[ser]sec (red and yellow) with the ribosome. The sec-tRNA[ser]sec complex is shown approaching the A site of the 80S ribosome. Once the sec t-RNA[ser]sec has been donated to the A site, a peptide bond forms between it and the nascent polypeptide.

BIOCHEMISTRY IN PERSPECTIVE cont

is the primary conserved sequence within the SECIS. One of the suggested functions of the sequence is stop codon suppression. Once formed the SECIS/SBP2 complex binds to EFsec, which has previously bound sec tRNA^{sec}. When the ribosomal A site becomes vacant (i.e., the UGA codon has moved into position), the SECIS element complex donates the sec tRNA^{sec}, after which a peptide bond forms. The selenocysteine residue is now incorporated into the polypeptide.

Pyrrolysine

Pyrrolysine is found in methyltransferases utilized by some methane-producing Archaea. A naturally occurring dipeptide composed of lysine linked via an ε-N-amide bond to 4-methylpyrroline-5-carboxylate, pyrrolysine is coded for by the stop codon UAG. The context-dependent coding of pyrrolysine involves a tRNA^{pyl} with a CUA anticodon and a tRNA synthetase that specifically binds pyrrolysine to its tRNA. In contrast to selenocysteine, pyrrolysine is synthesized before it is linked to a tRNA molecule. The precise mechanism whereby pyrrolysine is incorporated into protein is unknown. It has been noted that UAG codons are used significantly less often than other stop codons. A stem-loop PYLIS (pyrrolysine insertion sequence) element that is downstream of the UAG codon is believed to promote pyrrolysine insertion into methyltransferase polypeptides.

SUMMARY: In context-dependent codon reassignment, a specific tRNA, a tRNA synthetase, and other molecules are used to transform a stop codon into one that codes for the incorporation of a nonstandard amino acid.

POSTTRANSLATIONAL MODIFICATIONS IN EUKARYOTES Most nascent polypeptides undergo one or more types of covalent modification. These alterations, which may occur either during ongoing polypeptide synthesis or afterward, consist of reactions that modify the side chains of specific amino acid residues or break specific bonds. In general, posttranslational modifications prepare each molecule for its functional role and/or for folding into its native (i.e., biologically active) conformation. Over 200 different types of posttranslational processing reaction have been identified. Most of them occur in one of the following classes.

Proteolytic Cleavage The proteolytic processing of proteins is a common regulatory mechanism in eukaryotic cells. Typical examples of proteolytic cleavage (the hydrolysis of specific peptide bonds in certain proteins by proteases) include removal of the N-terminal methionine and signal peptides (see p. 750). Proteolytic cleavage is also used to convert inactive precursor proteins, called **proproteins**, to their active forms. Recall, for example, that certain enzymes, referred to as proenzymes or zymogens, are transformed into their active forms by cleavage of specific peptide bonds. The proteolytic processing of insulin (Figure 19.17) provides a well-researched example of the conversion of a polypeptide hormone into its active form. The inactive insulin precursor produced by removing the signal peptide is referred to as proinsulin. Inactive precursor proteins with removable signal peptides are called **preproteins**. The insulin precursor containing a signal peptide is referred to as preproinsulin.

Glycosylation A wide variety of eukaryotic proteins are glycosylated for structural and informational purposes (p. 254). In general, secreted proteins contain complex oligosaccharide species, whereas ER membrane proteins possess high-mannose species. The synthesis of the core N-linked oligosaccharide

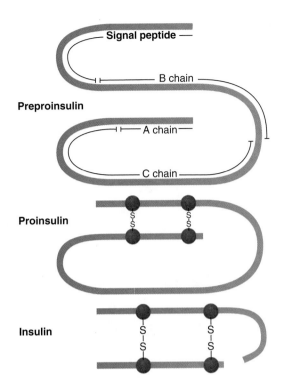

Preproinsulin — Signal peptide — B chain

Proinsulin — A chain — C chain

Insulin

FIGURE 19.17
Proteolytic Processing of Insulin

After the removal of the signal peptide, the peptide segment referred to as the C chain is removed by a specific proteolytic enzyme. Two disulfide bonds are also formed during insulin's posttranslational processing.

(common to all N-linked forms) is illustrated in Figure 19.18. The core oligosaccharide is assembled in association with phosphorylated dolichol. (Dolichol is a polyisoprenoid found within all cell membranes. Phosphorylated dolichol is found predominantly in ER membrane.)

Hydroxylation Hydroxylation of the amino acids proline and lysine is required for the structural integrity of the connective tissue proteins collagen (Section 5.3) and elastin. Additionally, 4-hydroxyproline is found in acetylcholinesterase (the enzyme that degrades the neurotransmitter acetylcholine) and complement (a complex series of serum proteins involved in the immune response). Three mixed-function oxygenases (prolyl-4-hydroxylase, prolyl-3-hydroxylase, and lysyl hydroxylase) located in the RER are responsible for hydroxylating certain proline and lysine residues. Substrate requirements are highly specific. For example, prolyl-4-hydroxylase hydroxylates only proline residues in the Y position of peptides containing Gly-X-Y sequences, whereas prolyl-3-hydroxylase requires Gly-Pro-4-Hyp sequences (Hyp stands for hydroxyproline; X and Y represent other amino acids). Hydroxylation of lysine occurs only when the sequence Gly-X-Lys is present. (Polypeptide hydroxylation by prolyl-3-hydroxylase and lysyl hydroxylase occurs only before helical structure forms.) The synthesis of 4-Hyp is illustrated in Figure 19.19. Ascorbic acid (vitamin C) is required to hydroxylate proline and lysine residues in collagen. Inadequate dietary intake of vitamin C can result in scurvy. The symptoms of scurvy (e.g., blood vessel fragility and poor wound healing) are effects of weak collagen fiber structure.

Phosphorylation Examples of the roles of protein phosphorylation in metabolic control and signal transduction have already been discussed. Protein phosphorylation may also play a critical (and interrelated) role in protein-protein interactions. For example, the autophosphorylation of tyrosine residues in PDGF receptors precedes the binding of cytoplasmic target proteins.

Lipophilic Modifications The covalent attachment of lipid moieties to proteins improves membrane-binding capacity and/or certain protein-protein interactions.

COMPANION

GW
WEBSITE Visit the companion website at www.oup.com/us/mckee to read the related Biochemistry in Perspective box on scurvy for Chapter 7.

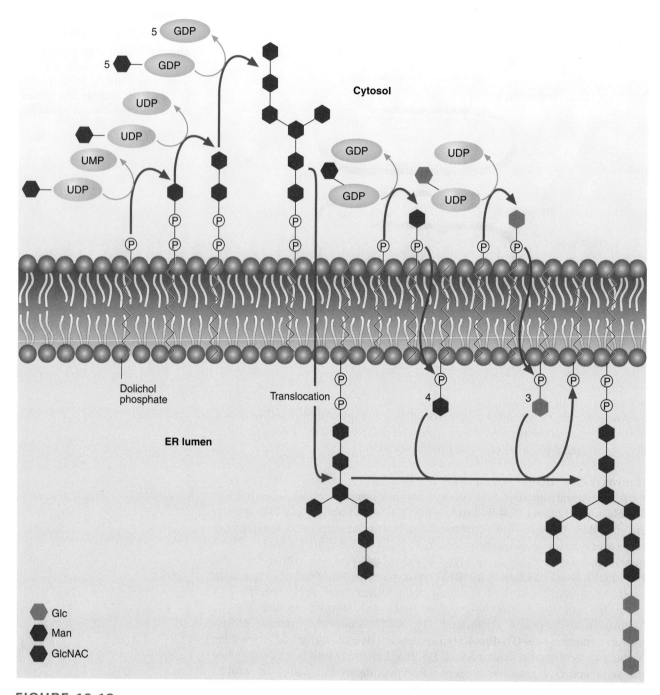

Cytosol

Dolichol
phosphate

Translocation

ER lumen

◇ Glc
◆ Man
⬡ GlcNAC

FIGURE 19.18

Synthesis of Dolichol-Linked Oligosaccharide

In the first step, GlcNAc-1-P is transferred from UDP-GlcNAc to dolichol phosphate (Dol-P). The next GlcNAc and the following five mannose residues are then transferred from nucleotide-activated forms. After the entire structure has flipped to the lumenal side of the membrane, each of the remaining sugars (four mannoses and three glucoses) is transferred first to Dol-P and then to the growing oligosaccharide. Then N-glycosylation of protein takes place in the endoplasmic reticulum (ER) in a one-step reaction catalyzed by a membrane-bound enzyme called glycosyl transferase.

Among the most common lipophilic modifications are acylation (the attachment of fatty acids) and prenylation (Section 11.1). Although the fatty acid myristate (14:0) is relatively rare in eukaryotic cells, myristoylation is one of the most common forms of acylation. N-Myristoylation (the covalent attachment of myristate by an amide bond to a polypeptide's amino terminal glycine residue) has been shown to increase the affinity of the α subunit of certain G proteins for membrane-bound β and γ subunits.

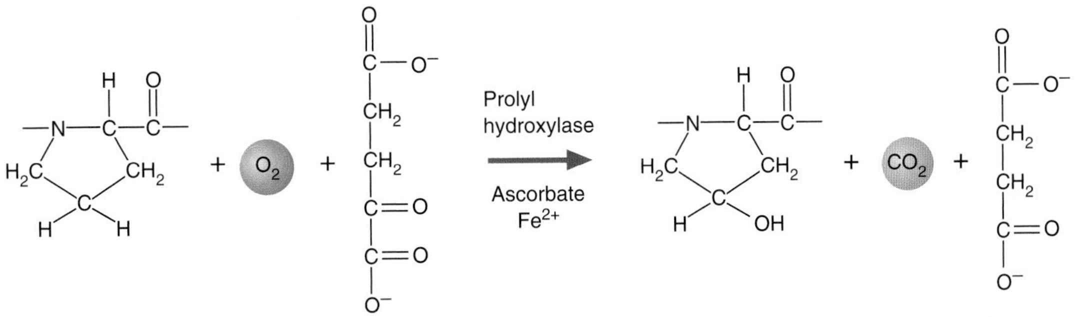

FIGURE 19.19

Hydroxylation of Proline

Ascorbic acid and ferrous iron are cofactors of prolyl-4-hydroxylase, the enzyme that catalyzes the hydroxylation of the C-4 position of certain prolyl residues in nascent polypeptides. Ascorbic acid, acting as a reducing agent, prevents the oxidation of the iron atom cofactor of the enzyme.

Methylation Protein methylation serves several purposes in eukaryotes. The methylation of altered aspartate residues by a specific type of methyltransferase promotes either the repair or the degradation of damaged proteins. Other methyltransferases catalyze reactions that alter the cellular roles of certain proteins. For example, methylated lysine residues have been found in such disparate proteins as ribulose-2,3-bisphosphate carboxylase, calmodulin, histones, certain ribosomal proteins, and cytochrome c. Other amino acid residues that may be methylated include histidine (e.g., histones, rhodopsin, and eEF-2) and arginine (e.g., heat shock proteins and ribosomal proteins).

Carboxylation Vitamin K-dependent carboxylation of glutamyl residues to form γ-carboxyglutamyl residues increases a protein's sensitivity to Ca^{2+}-dependent modulation. The carboxylation requires an NADPH-dependent reductase to convert phylloquinone, a vitamin K quinone (Figure 11.14a), to its hydroquinone form, a carboxylase that adds a carboxyl group to the γ-carbon of a glutamyl residue, as well as an epoxide reductase that converts the vitamin K-2,3-epoxide product of the carboxylation reaction to the original vitamin K quinone. The target protein must contain an appropriate signal sequence that binds to the carboxylase enzyme and a $(GluXXX)_n$ repeat ($n = 3–12$, X = other amino acids). Many of the known target proteins are involved in blood clotting (factors VII, IX, and X, and prothrombin). The anticoagulant coumadin acts by inhibiting the two reductases required in protein carboxylation.

Disulfide Bond Formation Disulfide bonds are generally found only in secretory proteins (e.g., insulin) and certain membrane proteins. (Recall that "disulfide bridges" are favored in the oxidizing environment outside the cell and confer considerable structural stability on the molecules that contain them.) As described earlier (Section 5.3), cytoplasmic proteins generally do not possess disulfide bonds because of the reducing conditions within cytoplasm. The ER has a nonreducing environment, so disulfide bonds form spontaneously in the RER as the nascent polypeptide emerges into the lumen. Although some proteins have disulfide bridges that form sequentially as the polypeptide enters the lumen (i.e., the first cysteine pairs with the second, the third residue pairs with the fourth, etc.), this is not true for many other molecules. Proper disulfide bond formation for these latter proteins is now presumed to be facilitated by **disulfide exchange**. During this process, disulfide bonds rapidly migrate from one position to another until the most stable structure is achieved. An ER enzymatic activity, referred to as protein disulfide isomerase (PDI), catalyzes this process. PDI also acts as a chaperone by rescuing wrongly folded polypeptides.

Protein Splicing Protein splicing (Figure 19.20) is a posttranslational mechanism in which an intervening peptide sequence is precisely excised from a nascent polypeptide. In this self-catalyzed, intramolecular reaction, a peptide bond is formed between the flanking amino-terminal and carboxy-terminal amino acid residues. The excised peptide segment is called an **intein**; the flanking segments that are spliced together to form the mature protein are called **exteins**. Protein splicing occurs without the aid of cofactors, metabolic energy sources, or auxiliary enzymes. First identified in 1990 during investigations of the catalytic subunit of an ATPase in the yeast *Saccharomyces cerevisiae*, protein splicing has been observed in certain unicellular organisms in all the domains of living organisms: Archaea, Bacteria, and Eukarya. The mechanism of the splicing reaction and the overall significance of protein splicing in cellular processes remain unresolved.

TARGETING Despite the vast complexities of eukaryotic cell structure and function, each newly synthesized polypeptide is normally directed to its proper destination. Considering that translation takes place in the cytoplasm (except for certain molecules that are produced within mitochondria and plastids) and that a wide variety of polypeptides must be directed to their proper locations, it is not surprising that the mechanisms by which cellular proteins are targeted are

FIGURE 19.20

Protein Splicing

Protein splicing is a posttranslational process in which an intein, a peptide sequence coded for by an intervening sequence in the gene, is precisely excised from the nascent polypeptide. In this auto-catalytic reaction, a peptide bond is formed that links the N-terminal and C-terminal amino acid residues of the flanking peptide sequences, called exteins, to form the mature protein.

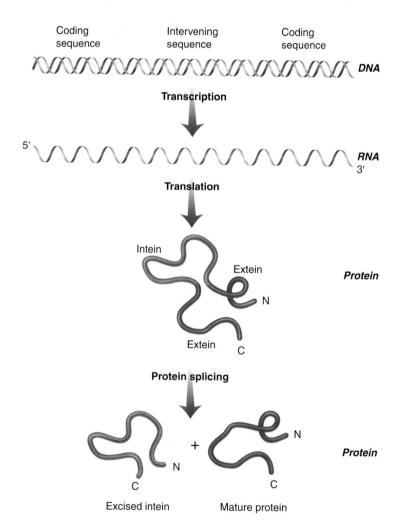

complex. Although this process is not yet completely understood, there appear to be two principal mechanisms by which polypeptides are directed to their correct locations: transcript localization and signal peptides. Each is briefly discussed.

It is generally recognized that cells often have asymmetrical protein distributions within the cytoplasm. For example, mature *Drosophila* eggs contain a gradient of bicoid, a protein that plays a critical role in gene regulation during development. A high concentration of bicoid in the anterior portion of the egg is required for the normal development of anterior body parts (i.e., head segments), whereas the low bicoid concentration in the posterior portion of the egg cytoplasm promotes the development of posterior body parts. If posterior cytoplasm is removed from one egg and substituted for anterior cytoplasm in a second egg, two sets of posterior body parts appear in the larva that develops from the recipient egg.

It is now believed that cytoplasmic protein gradients are created by **transcript localization**, that is, the binding of specific mRNA to receptors in certain cytoplasmic locations. mRNA cargo is moved to these locations along cytoskeletal filaments via attachment to the motor proteins kinesin, dynein, or myosin. It is known that bicoid mRNA is transported from nearby nurse cells into the developing oocyte (an immature egg cell) where it is carried along microtubules via a connection between the motor protein and its 3' UTR to the anterior ends of the cell. Once in the oocyte, bicoid mRNA binds via its 3' end to certain components of the anterior cytoskeleton. After the mature egg has been fertilized, translation of bicoid mRNA, coupled with protein diffusion, gives rise to the concentration gradient.

Polypeptides destined for secretion or for use in the plasma membrane or any of the membranous organelles must be specifically targeted to their proper location. Proteins of these types possess sorting signals referred to as signal peptides. Each signal peptide sequence facilitates the insertion of the polypeptide that contains it into an appropriate membrane. Signal peptides generally consist of a positively charged region followed by a central hydrophobic region and a more polar region. Although numerous signal peptides occur at the amino terminal, they may also occur elsewhere along the polypeptide.

The **signal hypothesis** was proposed by Gunter Blobel in 1975 to explain the translocation of polypeptides across RER membrane. Subsequent investigations revealed significant information concerning the insertion of polypeptides through the RER membrane. For this reason the discussion primarily focuses on this organelle. This is followed by a brief description of polypeptide uptake by other organelles.

As soon as about 70 amino acids have been incorporated into the polypeptide that emerges from a ribosome, a **signal recognition particle** (SRP) (a large complex consisting of six proteins and a 7S RNA molecule) binds to the ribosome (Figure 19.21). As a consequence of this binding, the EF-2 binding site is blocked and translation is temporarily arrested. The SRP then mediates binding of the ribosome to the RER via **docking protein**, a heterodimer also referred to as SRP receptor protein. Once binding to the RER has occurred, translation restarts, and the growing polypeptide inserts into the membrane. (The simultaneous translocation of a polypeptide during ongoing protein synthesis is sometimes referred to as **cotranslational transfer**.) As translation restarts, the SRP is released and an integral membrane protein complex, referred to as a **translocon**, is believed to mediate polypeptide translocation. The translocon consists of a hydrophilic transmembrane pore and several proteins that facilitate polypeptide translocation and processing. It is presumed that GTP hydrolysis provides the energy to push polypeptides across the RER membrane, because both the SRP and SRP receptor bind GTP. In **posttranslational translocation**, previously synthesized polypeptides are pulled across the RER membrane by an ATP-binding peripheral translocon-associated protein (hsp70).

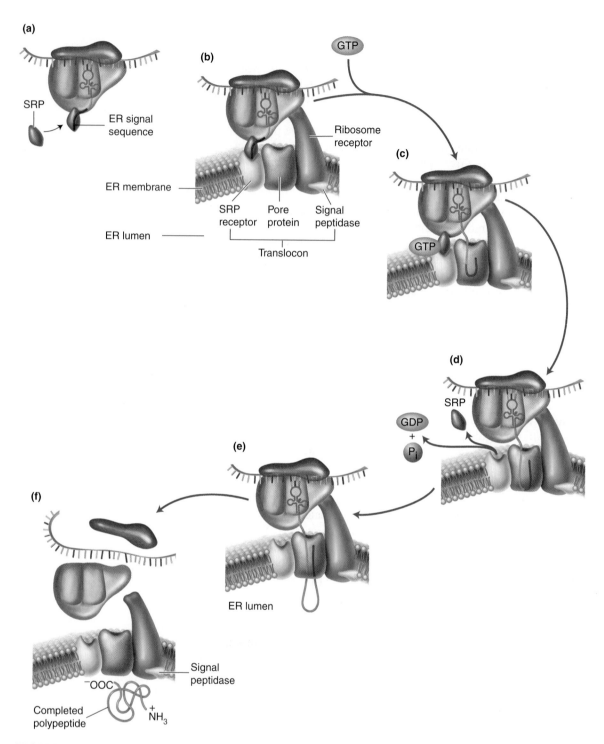

FIGURE 19.21

Cotranslational Transfer Across the RER Membrane

(a) When the nascent polypeptide is long enough to protrude from the ribosome, the SRP binds to the signal sequence, causing a transient cessation of translation. (b) The subsequent binding of SRP to the SRP receptor results in the binding of the ribosome to the translocon complex in the RER membrane. (c) Polypeptide synthesis begins again as GTP binds to the SRP–SRP receptor complex. GTP hydrolysis accompanies the binding of the signal sequence to the translocon and (d) the dissociation of SRP from its receptor. (e) The polypeptide continues to elongate until (f) translation is completed. The signal peptide is removed by signal peptidase in the RER lumen. The polypeptide is released into the lumen.

The fate of a targeted polypeptide depends on the location of the signal peptide and any other signal sequences. As illustrated in Figure 19.21 soluble secretory protein transmembrane transfer is usually followed by removal of an N-terminal signal peptide by signal peptidase, a process that releases the protein into the ER lumen. Such molecules usually undergo further posttranslational processing. The initial phase of the translocation of transmembrane proteins is similar to that of secretory proteins. For these molecules, the amino-terminal signal peptide serves as a *start signal* that remains bound in the membrane as the remaining polypeptide sequence is threaded through the membrane. So-called "single-pass" transmembrane proteins possess a *stop transfer signal* (or stop signal), which prevents further transfer across the membrane (Figure 19.22a). Membrane proteins with multiple membrane spanning segments (multipass) possess a series of alternating start and stop signals (Figure 19.22b).

Most proteins that are translocated into the RER are directed to other destinations. After undergoing initial posttranslational modifications, both soluble and membrane-bound proteins are transferred to the Golgi complex via transport vesicles that bud off from the ER and fuse with the *cis* face of the Golgi membrane (Figure 19.23). Proteins that ultimately reside in the ER possess retention signals. In most vertebrate cells this signal consists of the carboxy-terminal tetrapeptide Lys-Asp-Glu-Leu (KDEL sequence).

Within the Golgi complex, proteins undergo further modifications. For example, N-linked oligosaccharides are processed further, and O-linked glycosylation of certain serine and threonine residues occurs. Lysosomal proteins are targeted to the lysosomes by adding a mannose-6-phosphate residue. It is still unclear what signals direct secretory proteins to the cell surface (via exocytosis) or promote the delivery of plasma membrane proteins to their destination, although a "default mechanism" has been proposed. (In default mechanisms, the absence of a signal results in a specific sequence of events.) When protein modification is complete, transport vesicles exit from the trans face of the Golgi and move to their target locations.

As noted, although mitochondria and chloroplasts produce several of their own proteins, a variety of other proteins are produced on cytoplasmic ribosomes and subsequently imported. Again, specific signal sequences are required. The import of polypeptides into these organelles is complicated by the presence of several membranes. Consequently, polypeptide transfer in these organelles often involves several signal sequences. Figure 19.24 illustrates this import mechanism (the targeting of cytochrome c_1 to the inner membrane space of mitochondria).

TRANSLATION CONTROL MECHANISMS Eukaryotic translation control mechanisms are proving to be exceptionally complex, substantially more so than those observed in prokaryotes. In eukaryotes these mechanisms appear to occur on a continuum, from *global* controls (i.e., the translation of a wide variety of mRNAs is altered) to *specific* controls (i.e., the translation of a specific mRNA or small group of mRNAs is altered). Although most aspects of eukaryotic translational control are currently unresolved, the following features are believed to be important.

mRNA Export The spatial separation of transcription and translation afforded by the nuclear membrane appears to provide eukaryotes with significant opportunities for gene expression regulation. Only a small portion of RNA produced within the nucleus ever enters the cytoplasm. Most discarded RNA is probably nonfunctional introns. Export through the nuclear pore complex is known to be a carefully controlled, energy-driven process whose minimum requirements include the presence of a 5'-cap and a 3' poly(A) tail.

mRNA Stability In general, the translation rate of any mRNA species is related to its abundance, which is in turn dependent on its rates of both synthesis and

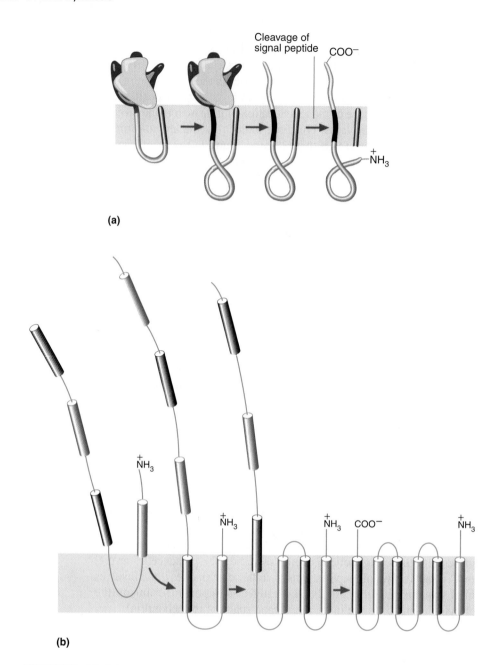

FIGURE 19.22

Cotranslational Transfer of Integral Membrane Proteins

(a) Transfer of a single-pass transmembrane protein. (b) Transfer of a multipass membrane protein. For the sake of clarity, the transfer apparatus has been omitted from the diagrams. In addition, the ribosome has been omitted from (b). The shaded segment is a signal peptide. The black segment is a stop transfer signal.

degradation. mRNA half-lives range from about 20 minutes to over 24 hours. Several features of mRNA structure are known to affect its stability, that is, its capacity to avoid degradation by various nucleases. The presence of certain sequences may confer resistance to nuclease action (e.g., palindromes that create hairpins), while other sequences may increase the likelihood of nuclease action, particularly if present in multiple copies. The binding of specific proteins to certain sequences can also affect mRNA stability. Finally, reversible adenylation

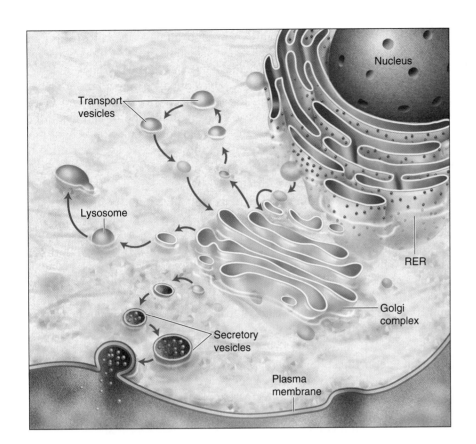

FIGURE 19.23

The ER, Golgi, and Plasma Membrane

Transport vesicles transfer new membrane components (protein and lipids) and secretory products from the ER to the Golgi complex, from one Golgi cisterna to another, and from the trans-Golgi network to other organelles (e.g., lysosomes) or to the plasma membrane.

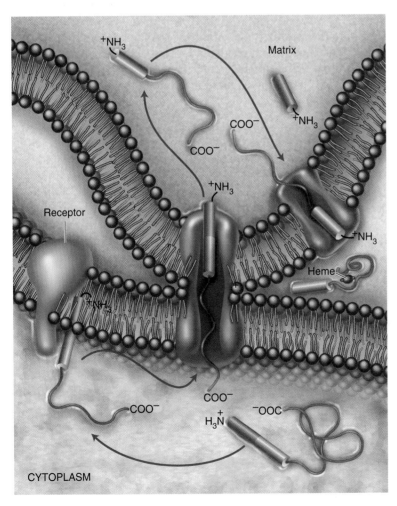

FIGURE 19.24

Posttranslational Transport of Cytochrome c_1 into a Mitochondrion

After its synthesis in cytoplasm, cytochrome c_1 must be translocated into the mitochondrial inner membrane space. (Recall that cytochrome c_1 is a component of complex III of the electron transport chain.) The targeting of cytochrome c_1 requires two sequences. The first targets the polypeptide to the matrix. This sequence is removed by a protease, and then the second sequence targets the molecule to the inner membrane space. The second targeting sequence is then also removed. After folding and binding a heme, the cytochrome c_1 molecule associates with complex III in the inner membrane.

and deadenylation of the 3' end of mRNA strongly influence both its stability and its translational activity. As described previously, most mRNAs transported into the cytoplasm have poly(A) tails containing between 100 and 200 nucleotides. As time passes, many poly(A) tails progressively shorten, bearing no fewer than 30 residues when the entire mRNA has been degraded. In certain circumstances the poly(A) tail of some mRNAs is selectively elongated or shortened. For example, mRNAs in mature oocytes are "masked" by removal of most of their poly(A) tail nucleotides. After fertilization, these mRNAs are reactivated by adding adenine nucleotides.

Negative Translational Control The translation of certain specific mRNAs is known to be blocked by binding repressor proteins to sequences near their 5' ends. A well-researched example is provided by ferritin synthesis control. Ferritin, an iron storage protein that is found predominantly in hepatocytes, is synthesized in response to high iron concentrations. Ferritin mRNA contains an iron response element (IRE) that binds an iron-binding repressor protein. When cellular iron concentrations are high, the iron atoms binding in large numbers to the repressor protein cause it to dissociate from the IRE. Then ferritin mRNA is translated.

Initiation Factor Phosphorylation The phosphorylation of eIF-2 in response to certain circumstances (e.g., heat shock, viral infections, and growth factor deprivation) has been observed to decrease protein synthesis generally. However, the translation of certain mRNAs has been observed to increase. For example, hsp (heat shock protein) synthesis increases in response to heat shock and other stressful conditions. The specific mechanisms are unknown.

Translational Frameshifting Certain mRNAs appear to contain structural information that, if activated, results in a +1 or −1 change in reading frame. This **translational frameshifting**, which has been most often observed in cells infected by retroviruses, allows more than one polypeptide to be synthesized from a single mRNA.

KEY CONCEPTS

- Eukaryotic protein synthesis is slower and more complex than that of its prokaryotic counterpart. In addition to requiring a larger number of translation factors and a more complex initiation mechanism, the eukaryotic process involves vastly more complicated posttranslational processing and targeting mechanisms.

- Eukaryotes use a wide spectrum of translational control mechanisms.

QUESTION 19.5

The mechanism involved in the posttranslational transport of proteins into chloroplasts has so far received only limited attention. However, the import of plastocyanin into the thylakoid lumen has been determined to require two import signals near the N-terminal of the newly synthesized protein. Assuming that chloroplast protein import resembles the import process for mitochondria, suggest a reasonable hypothesis to explain how plastocyanin (a lumen protein associated with the inner surface of the thylakoid membrane) is transported and processed. What enzymatic activities and transport structures do you expect are involved in this process?

GCGACATCACTCCAGCTTGAAGCAGTTCTTCTCGTCTTCTGTTTTGTCTAACTT TCTTCC

BIOCHEMISTRY IN THE LAB

Proteomics

The technology of proteomics is being developed to investigate the proteome, the functional output of the genome. The wealth of genetic data now available has provided primary sequence information for thousands of proteins. The identity and function of many of these molecules, however, are unknown. In addition, the structural alterations of many proteins cannot be predicted solely from nucleic acid sequences, either because of alternative mRNA splicing or because of the post-translational chemical modifications. As the sequencing efforts for many genome projects are coming to an end, the attention of biochemists and molecular biologists is shifting to the analysis of the protein products. The goals of proteomics are primarily twofold: to study the global changes in the expression of cellular proteins, and to determine the identity and functions of all the proteins in the proteomes of organisms. The potential applications for proteomic research are many and varied. In addition to providing opportunities to resolve basic biological problems (e.g., ascertaining the precise mechanisms by which cellular processes such as neuronal transport or mRNA splicing occur), proteomics-based technology has obvious uses in biomedical research. Examples of the latter include investigations of the causes and diagnosis of genetic and infectious diseases, and the development of drugs.

Proteomic Tools

The investigation of proteomes is currently focused primarily on developing accurate, but relatively fast methods for identifying and characterizing proteins. Among the oldest technologies used in proteomics are two-dimensional gel electrophoresis and mass spectrometry. The use of each is discussed briefly.

Two-Dimensional Gel Electrophoresis Protein expression is currently analyzed with two-dimensional (2-D) gels. Proteins are separated according to charge (within a pH gradient) in the first dimension of the gel, and on the basis of molecular mass in the second (Figure 19D). As many as 3000 individual proteins can be visualized on a 2-D gel. Protein analysis of multiple gels (e.g., determinations of presence, absence, or relative concentration of specific proteins in healthy and diseased cells) is accomplished by comparisons of 2-D gel images with proteome databases with the aid of specialized computer software. Although 2-D gel technology has been improved in both speed and capacity, it also has limitations. In addition to being labor intensive, 2-D gels are not useful in the evaluation of proteins of certain types, such as membrane proteins or proteins found in very low concentrations. Newer technologies are being developed to overcome these problems.

Mass Spectrometry As described previously (p. 177), **mass spectrometry** (MS) is a technique in which molecules are vaporized and then bombarded by a high-energy electron beam, which causes them to fragment as cations. As the ionized fragments enter the spectrometer, they pass through a strong magnetic field that separates them according to their mass-to-charge (*m/z*) ratio. Each type of molecule is identified by the pattern of fragments that is generated, each pattern or "fingerprint" being unique. Because proteins do not vaporize, they are instead digested and then dissolved in a volatile solvent and sprayed into the vacuum chamber of the mass spectrometer. The electron beam ionizes these peptide fragments, and the positively charged peptides are passed through the magnetic field. The peptide mass fingerprint that results is then compared to fragmentation information in protein databases. Although MS is highly accurate and automated, it is usually insufficient for identifying all the

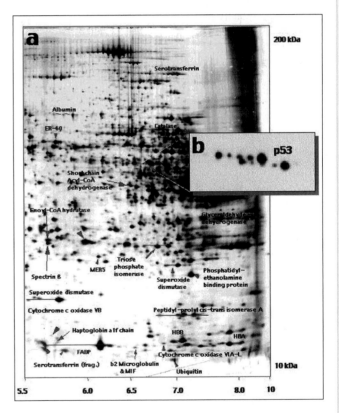

FIGURE 19D

Two-Dimensional Gel Pattern

A gel to which the extract of a liver sample had been added was run first in a pH gradient (isoelectric points) and then in SDS-PAGE to separate the proteins according to molecular mass. The insert shows an enlargement of multiple versions of p53, a protooncogene product, that was revealed by treatment with antibodies.

proteins in a sample. To improve protein identification, tandem MS (MS/MS), a method in which two mass spectrometers are linked in tandem, has been developed. In this technique the oligopeptide fragments produced in the first MS are transferred to the second MS, where they are further fragmented and analyzed. MS/MS is used to rapidly sequence proteins.

Despite the recent advances in proteomic research techniques, several problems are a serious barrier to accomplishing the enormous task of characterizing entire proteomes. Among the most important of these is the inefficiency of the methods available and the lack of a technique equivalent to PCR for amplification of proteins found in very small amounts.

■

Chapter**Summary**

1. Protein synthesis is a complex process in which information encoded in nucleic acids is translated into the primary sequence of proteins. During the translation phase of protein synthesis, the incorporation of each amino acid is specified by one or more triplet nucleotide base sequences, referred to as codons. The genetic code consists of 64 codons: 61 codons that specify the amino acids and three stop codons. Two of the stop codons can be used by some organisms to code for the nonstandard amino acids selenocysteine and pyrrolysine. Translation also involves the tRNAs, a set of molecules that act as carriers of the amino acids. The base-pairing interactions between codons and the anticodon base sequence of tRNAs result in the accurate translation of genetic messages. Translation consists of three phases: initiation, elongation, and termination. Each phase requires several types of protein factor. Although prokaryotic and eukaryotic translational mechanisms bear a striking resemblance to each other, they differ in several respects. One of the most notable differences is the identity, quantity, and function of the translation factors.

2. Protein synthesis involves a set of posttranslational modifications that prepare the molecule for its functional role,

   assist in folding, or target it to a specific destination. These covalent alterations include proteolytic processing, modification of certain amino acid side chains, and insertion of cofactors.

3. Prokaryotes and eukaryotes differ in their usage of translational control mechanisms. In addition to variations in Shine-Dalgarno sequences, prokaryotes use negative translational control, that is, the repression of the translation of a polycistronic mRNA by one of its products. In contrast, a wide variety of eukaryotic translational controls have been observed. These mechanisms range from global controls in which the translation rate of a large number of mRNAs is altered to specific controls in which the translation of a specific mRNA or small group of mRNAs is altered.

4. Proteomics is a technology that is used to investigate the proteome, the complete set of proteins produced from an organism's genome. The goals of proteomics are to study the global changes in the expression of cellular proteins over time and to determine the identity and the functions of all the proteins produced by organisms.

COMPANION GW WEBSITE Take your learning further by visiting the **companion website** for Biochemistry at **www.oup.com/us/mckee** where you can complete a multiple-choice quiz on protein synthesis to help you prepare for exams.

Suggested**Readings**

Algire, M. A., and Lorsch, J. R., Where to Begin? The Mechanism of Translation Initiation Codon Selection in Eukaryotes, *Curr. Opin. Chem. Biol.* 10:480–486, 2006.

Allen, G. S., and Frank, J., Structural Insights on the Translation Initiation Complex: Ghosts of a Universal Initiation Complex, *Mol. Microbiol.* 63(4):941–950, 2007.

Allmang, C., and Krol, A., Selenoprotein Synthesis: UGA Does Not End the Story, *Biochimie* 88:1561–1571, 2006.

Caban, R., and Copeland, P. R., Size Matters: A View of Selenocysteine Incorporation from the Ribosome, *Cell Mol. Life Sci.* 63:73–81, 2006.

DiGiulio, M., The Origin of the Genetic Code: Theories and Their Relationship, a Review, *Biosystems* 80:175–184, 2005.

Dong, Z., and Zhang, J.-T., Initiation Factor eIF3 and Regulation of mRNA Translation, Cell Growth, and Cancer, *Crit. Rev. Oncol. Hematol.* 59:169–180, 2006.

Fechter, P., and Brownlee, G. G., Recognition of mRNA Cap Structures by Viral and Cellular Proteins, *J. Gen. Virol.* 86:1239–1249, 2005.

Gregory, S. T., and Dahlberg, A. E., Peptide Bond Formation Is All About Proximity, *Nat. Struct. Mol. Biol.* 11(7):586–587, 2004.

Hatfield, D. L., and Gladyshev, V. N., How Selenium Has Altered Our Understanding of the Genetic Code, *Mol. Cell Biol.* 27(11):3565–3576. 2002.

Hauryliuk, V. V., GTPases of the Prokaryotic Translation Apparatus, *Mol. Biol.* 40(5):688–701, 2006.

Hirokawa, G., Demeshkina, N., Iwakura, N., Kaji, H., and Kaji, A., The Ribosome Recycling Step: Consensus on Controversy? *Trends Biochem. Sci.* 31(3):143–149, 2006.

Ibba, M., and Soll, D., Aminoacyl-tRNAs: Setting the Limits of the Genetic Code, *Genes Dev.* 18:731–739, 2004.

Jacobson, A., The End Justifies the Means, *Nat. Struct. Mol. Biol.* 12(6):474–475, 2005.

Nilsson, J., and Nissen, P., Elongation Factors on the Ribosome, *Curr. Opinion Struct. Biol.* 15:349–354, 2005.

Richter, J. D., and Sonenberg, N., Regulation of Cap-Dependent Translation by eIF4E Inhibitory Proteins, *Nature* 433:477–480, 2005.

Rodnina, M. V., Beringer, M., and Wintermeyer, W. How Ribosomes Make Peptide Bonds, *Trends Biochem. Sci.* 32(1):20–26, 2007.

Rospert, S., Ribosome Function: Governing the Fate of a Nascent Polypeptide, *Curr. Biol.* 14:R386–R388, 2004.

St. Johnston, D., Moving Messages: The Intracellular Localization of mRNAs, *Nat. Rev. Mol. Cell Biol.* 6(5):36–375, 2005.

Wilkinson, B., and Gilbert, H. F., Protein Disulfide Isomerase, *Biochim. Biophys. Acta* 1699:35–44, 2004.

Key**Words**

anhydride, *742*

anticodon, *740*

CAP-binding complex, *755*

codon, *738*

cotranslational transfer, *767*

disulfide exchange, *765*

docking protein, *767*

elongation, *745*

extein, *766*

genetic code, *738*

guanine nucleotide activation factor, *755*

guanine nucleotide exchange factor, *745*

initiation, *743*

initiation complex, *747*

intein, *766*

kinetic proofreading, *757*

mass spectrometry, *773*

mixed anhydride, *742*

molecular chaperone, *766*

nascent, *750*

open reading frame, *739*

poly(A) binding protein, *755*

polysome, *745*

posttranslational modification, *746*

posttranslational translocation, *767*

preinitiation complex, *755*

preproprotein, *762*

proprotein, *762*

protein splicing, *766*

proteomics, *738*

reading frame, *739*

releasing factor, *750*

ribosome recycling, *750*

SECIS element, *761*

Shine-Dalgarno sequence, *747*

signal hypothesis, *767*

signal peptide, *750*

signal recognition particle, *767*

targeting, *746*

termination, *746*

transcript localization, *767*

translational frameshifting, *772*

translocation, *745*

translocon, *767*

wobble hypothesis, *740*

Review**Questions**

These questions are designed to test your knowledge of the key concepts discussed in this chapter, before moving on to the next chapter. You may like to compare your answers to the solutions provided in the back of the book and in the accompanying Study Guide.

1. List and describe four properties of the genetic code.

2. What two observations prompted the wobble hypothesis?

3. Describe the two sequential reactions that occur in the active site of aminoacyl-tRNA synthetases.

4. Define the following terms:
   a. protein targeting
   b. scanning
   c. codon
   d. reading frame
   e. disulfide exchange

5. What are the major differences between eukaryotic and prokaryotic translation?

6. What are the major differences between eukaryotic and prokaryotic translation control mechanisms?

7. What are the three steps in the elongation cycle?

8. Briefly outline the major mechanisms used by eukaryotes to control translation.

9. Define the following terms:
   a. KDEL sequence
   b. signal peptide
   c. glycosylation
   d. genetic code
   e. context-dependent codon reassignment

10. Explain the differences among preproproteins, proproteins, and proteins.

11. Describe how kinetic proofreading takes place.

12. Describe the structure and function of the signal recognition particle.

13. Define the following terms:
    a. wobble
    b. anhydride
    c. aminoacyl-tRNA synthetase
    d. cognate tRNA
    e. polysome

14. Describe the function of the translocon in cotranslational transfer.
15. Describe how eukaryotic mRNA structure can affect translational control.
16. In general terms, describe the intracellular processing of a typical glycoprotein that is destined for secretion from a cell.
17. Describe the problems associated with determining a polypeptide's final three-dimensional shape using the primary structure of the molecule as a guide.
18. Define the following terms:
    a. translocation
    b. GAP
    c. GEF
    d. ribosome recycling factor
    e. nascent
19. Describe the process of protein splicing.
20. Why are tRNAs described as adaptor molecules?
21. What steps in the elongation cycle of protein synthesis require GTP hydrolysis? What role does it play in each step?
22. Name and explain the roles of the protein factors that participate in the initiation phase of prokaryotic protein synthesis.
23. Define the following terms:
    a. PABP
    b. CBC
    c. diphthamide
    d. SECIS element
    e. signal hypothesis
24. What types of chemical bond are involved in the following molecules?
    a. amino acids in polypeptides
    b. nucleotides in a polynucleotide strand
    c. codon-anticodon in translation
25. Explain the roles of the large and small subunits of ribosomes.
26. The term *translation* refers to which of the following?
    a. DNA → RNA
    b. RNA → DNA
    c. proteins → RNA
    d. RNA → proteins
27. List and describe the major classes of eukaryotic posttranslational modifications.
28. Explain the importance of the proper targeting of nascent polypeptides.
29. Explain the critically important roles of aminoacyl-tRNA synthetases in protein synthesis.
30. Define the following terms:
    a. intein
    b. extein
    c. translocon
    d. translational frameshifting
    e. mass spectrometry
31. Indicate the phase of protein synthesis during which each of the following processes occurs:
    a. A ribosomal subunit binds to a messenger RNA.
    b. The polypeptide is actually synthesized.
    c. The ribosome moves along the codon sequence.
    d. The ribosome dissociates into its subunits.
32. Determine the total amount of nucleotide bond energy that is required in the synthesis of the following tetrapeptide: Lys-Ala-Ser-Val.
33. Provide a DNA base sequence that could code for the following peptide: Ala-Ser-Phe-Tyr-Ser-Lys-Lys-Leu-Ala-Asp-Val-Ile.
34. What is the mRNA base sequence for the peptide in Question 33?
35. What would be the effect of a single base deletion in the codon for the second Ser residue in the DNA sequence that codes for the peptide in Question 33?
36. Eukaryotic protein synthesis is considerably slower than synthesis of prokaryotes. Explain.

ThoughtQuestions

These questions are designed to reinforce your understanding of all of the key concepts discussed in the book so far, including this chapter and all of the chapters before it. They may not have one right answer! The authors have provided possible solutions to these questions in the back of the book and in the accompanying Study Guide, for your reference.

37. The three-dimensional structures of ribosomal RNA and ribosomal protein are remarkably similar among species. Suggest reasons for these similarities.
38. Explain the significance of the following statement: The functioning of the aminoacyl-tRNA synthetases is referred to as the second genetic code.
39. Although aminoacyl-tRNA synthetases make few errors, occasionally an error does occur. How can these errors be detected and corrected?
40. What are the three phases of protein synthesis? Describe the principal events in each phase. What specific roles do translation factors play in both prokaryotic and eukaryotic translation processes?
41. Determine the codon sequence for the peptide sequence glycylserylcysteinylarginylalanine. How many possibilities are there?
42. Estimate the minimum number of ATP and GTP molecules required to polymerize 200 amino acids.
43. Discuss the role of GTP in the functioning of translation factors.
44. Posttranslational modifications serve several purposes. Discuss and give examples.
45. Describe how the base pairing between the Shine-Dalgarno sequence and the 30S subunit provides a mechanism for distinguishing a start codon from a methionine codon. What is the eukaryotic version of this mechanism?

46. Given an amino acid sequence for a polypeptide, can the base sequence for the mRNA that codes for it be predicted?

47. Because of the structural similarity between isoleucine and valine, the aminoacyl-tRNA synthetases that link them to their respective tRNAs possess proofreading sites. Examine the structures of the other α-amino acids and determine other sets of amino acids whose structural similarities might also require proofreading.

48. What advantages are there for synthesizing an inactive protein that must subsequently be activated by posttranslational modifications?

49. What factors ensure accuracy in protein synthesis? How does the level of accuracy usually attained in protein synthesis compare with that of replication or transcription?

50. Can you suggest a reason why ribosomes in all living organisms consist of two subunits and not one supramolecular complex?

51. A prokaryotic species is facing a new environmental stress that can be ameliorated by a catalytic activity that requires the side chain of a unique amino acid derivative called pyrovaline. How would such an organism develop a mechanism for the incorporation of this nonstandard amino acid into an enzyme molecule? What would be the properties of the molecules required to solve this problem?

Appendix

Solutions

CHAPTER 1

End-of-Chapter Questions

Review Questions

3

Eukaryotes are larger and considerably more complicated than prokaryotes. All multicellular organisms are eukaryotic.

6

a. Biochemistry: the study of the molecular basis of life.

b. Oxidation: loss of an electron.

c. Reduction: gain of an electron.

d. Active transport: movement of substances against a concentration gradient; energy is required.

e. Leaving group: a molecular group displaced during a reaction.

9

Cells use oxidation-reduction reactions to convert bond energy in biomolecules into higher energy ATP bonds. Energy is captured as electrons are transferred from reduced molecules to more oxidized ones.

12

Saturated hydrocarbons contain only carbon–carbon single bonds, whereas unsaturated carbons contain carbon–carbon double or triple bonds.

15

The molecules belong to the following classes:

a. Amino acid.

b. Sugar.

c. Fatty acid.

d. Nucleotide.

18

The primary functions of metabolism are acquisition and utilization of energy, synthesis of biomolecules, and removal of waste products.

21

Important ions found in living organisms are Na^+, K^+, Ca^{2+}, and Cl^-. Many polyatomic icons are also common, such as NH_4^+, PO_4^{3-}, and CO_3^{2-}.

24

The functions of polypeptides include transport, structural composition, defense, storage, regulation, movement, and catalysis (enzymes).

27

Nucleotides participate in energy-forming and energy-generating reactions. Much of the energy available to drive biochemical reactions is stored in ATP molecules.

30

In a hierarchical system, such as that found in living organisms, emergent properties (in any level in the structural hierarchy) have characteristics that cannot be predicted from the analysis of its component parts. For example, the structural and functional properties of the protein hemoglobin, which is composed of carbon, nitrogen, hydrogen, oxygen, and iron atoms, cannot be determined from analysis of those individual components.

32

Both human-designed complex systems (such as machines or factories) and living systems require raw materials (nutrients) and energy to manufacture components; systems of both types also produce waste products and heat. Machines can be designed to self-regulate upon receiving feedback from the environment (e.g., by monitoring temperature to determine heating or cooling needs). In contrast, living systems are self-sustaining; that is, they produce and repair all their own structural and functional components, and, via nucleic acids, they build the machines (enzymes) that make the components. Even the nucleic acids themselves are reproduced by living systems. This level of self-sustainability is not present in human-designed complex systems.

Thought Questions

36

The assumption that the biochemical processes in prokaryotes and eukaryotes are similar is safe only when we consider basic living processes (e.g., glycolysis and the general principles of genetic inheritance). Living organisms are so diverse in their adaptations that information acquired from research with prokaryotes must be judiciously interpreted in reference to eukaryotes.

39

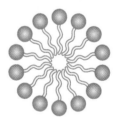
Micelle

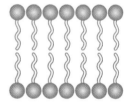
Bilayer segment

The structure of a micelle is illustrated in Figure 3.13 in your text. The phospholipid bilayer is a prominent feature of Figure 2.2 in your text.

42

In the normally functioning system the lipid molecule is not particularly toxic (i.e., its presence at normal levels does not interfere with brain function). In individuals that lack the key enzyme, this lipid begins to accumulate within neurons and other brain cells. Its physical presence interferes with the way nerve tracts are laid down. Neurological systems that require integration of brain functions (e.g., motor control and intelligence) begin to fail.

CHAPTER 2

In-Chapter Questions

2.1

The volume of a prokaryotic cell is calculated as follows:

$$\pi r^2 h = 3.14 \times (0.5 \ \mu m)^2 \times 2 \ \mu m = 1.57 \ \mu m^3$$

The volume of a eukaryotic cell is calculated as follows:

$$4/3 \ \pi r^3 = 4 \times (3.14 \times 10^3)/3 = 4200 \ \mu m^3$$

By dividing the volume of the hepatocyte by the volume of the prokaryotic cell ($4200 \ \mu m^3/1.57 \ \mu m^3$) the number of prokaryotic cells that would fit within the heptocyte is obtained: 2700

2.2

Without a means of disposal, the lipid molecules will accumulate in the cells. Cell function is eventually compromised and the cells die.

2.3

In response to neural signals originating in the retina of the chameleon's eyes, granules containing the appropriate pigments are released from melanosome-like organelles in the organism's skin. The change in skin color is the result of the ingestion of the pigment granules by keratinocytes.

2.4

Ten percent of 70 kg is 7 kg, which can be expressed as 7×10^6 mg. To calculate the weight of a single "average" mitochondrion, divide the estimated total weight of mitochondria by the estimated number of mitochondria (1×10^{16}):

$$\frac{7 \times 10^6 \ mg}{1 \times 10^{16}}$$

The answer (i.e., the weight of an average mitochondrion) is approximately

$$7 \times 10^{-10} \ mg, \ or \ 7 \times 10^{-7} \ mg. \ (1 \ mg = 1 \times 10^{-6} \ g)$$

2.5

Cell division involves the highly organized restructuring of the microtubules that form the mitotic spindle during the phases of mitosis. Microtubule function depends on dynamic instability, that is, the capacity to rapidly shorten and lengthen via polymerization/depolymermerization reactions. Cell division, which is unregulated in cancerous cells, is suppressed by taxol because this drug stabilizes microtubule structure.

End-of-Chapter Questions

Review Questions

3

Refer to Figure 2.6. The functions of the components of prokaryotic cells are as follows:

a. Nucleoid contains the bacterial chromosome.

b. Plasmid is the site of extrachromosomal DNA, often coding for special functions.

c. Cell wall provides protection and support.

d. Pili allow attachment to other cells.

e. Flagella allow locomotion.

6

a. Nucleus—eukaryotes

b. Plasma membrane—eukaryotes and prokaryotes

c. Endoplasmic reticulum—eukaryotes

d. Mitochondria—eukaryotes

e. Nucleolus—eukaryotes

f. Cytoskeleton—eukaryotes (*Note*: Prokaryotes do contain proteins that resemble eukaryotic cytoskeleton proteins in both structure and function, but they lack the network of microtubules, microfilaments, and intermediate fibers that comprise a cytoskeleton.)

9

The three phases of signal transduction in living organisms are reception, transduction, and response. Reception is the binding of an external signal molecule to a receptor on the cell surface. Transduction is the conversion of this extracellular message (i.e., the new signal molecule–receptor binding) to an intracellular message via a conformational change in the receptor. The response, the result of this internal message, is a cascade of events that involves covalent modifications of intracellular proteins and may include changes in enzyme activities, cell movement, and other cellular processes.

12

Lysosomes digest biomolecules of all types. In addition to the normal processing of cellular molecules, lysosomes destroy the components of foreign cells and other exogenous extracellular materials.

15

The highly developed framework of the cytoskeleton performs the following functions in eukaryotic cells: (1) maintenance of overall cell shape, (2) facilitation of coherent cellular movement, (3) provision of a supporting structure that guides the movement of organelles within the cell, and (4) service as a platform for enzyme and signal cascade complexes.

18

A cell wall provides an advantage of structural support to maintain cell shape and to protect against mechanical injury. Cells such as macrophages that require dramatic shape changes to function would be hindered (or disabled) by a rigid cell wall.

21

The principal function of the rough endoplasmic reticulum is the synthesis of membrane proteins and protein for export from the cell. Smooth endoplasmic reticulum, so named because it lacks attached ribosomes, is involved in lipid synthesis and biotransformation processes.

24

a. Nucleoplasm is the material within the nucleus that consists of proteins called lamins, which form a network of chromatin fibers.

b. The nuclear matrix, like the cytoskeleton, is a network of fibrous proteins that provide structural support and an organization framework for chromatin.

c. Autophagy is a cell's degradation of its own cellular components by acid hydrolases, digestive enzymes found in lysosomes.

d. Apoptosis is the genetically programmed series of events that lead to cell death.

e. Differential centrifugation is a cell fractionation technique in which homogenized cells are separated by centrifugal forces. Larger and denser components sediment at lower centrifugal forces, while smaller and less dense components sediment at higher centrifugal forces.

27

Motor proteins bind nucleotide molecules such as ATP or GTP. When these nucleotide molecules are hydrolyzed, they release energy that causes a specific change in the shape of the motor protein. This change causes specific shape changes in adjacent protein subunits, which transmit shape changes to other subunits as well. In this way the energy from the hydrolysis of ATP, for example, is converted into directed motion that can do work.

30

The intracellular and extracellular structures that protect the eukaryotic plasma membrane are the cell cortex and the extracellular matrix, respectively. The cell cortex, located on the inner surface of the plasma membrane, consists of microfilaments and other proteins in a three-dimensional meshwork that provides mechanical strength. The extracellular matrix is a gelatinous material of structural proteins and complex carbohydrates.

33

The nuclear envelope surrounds the nucleus and consists of two membranes strengthened by a network of nuclear lamins. The outer nuclear membrane is continuous with the RER (rough endoplasmic reticulum), the space between the membranes is continuous with the RER lumen, and the two membranes fuse at nuclear pores. By controlling transport into and out of the nucleus, the nuclear envelope separates DNA replication and transcription processes from the cytoplasm and allows for more sophisticated regulation of gene expression than would be possible otherwise.

Thought Questions

36

Because the thick mucoid coat prevents antibodies from binding to surface cellular structures used by the immune system for recognition, it interferes with the immune response.

39

This defect could lie anywhere in the synthesis scheme for the LDL receptor. The high cholesterol levels result because cholesterol is not being transferred into the cells and is building up in the blood, thus causing atherosclerosis. The most common defects are improper insertion of the receptor into the plasma membrane or a defective receptor.

42

Since the diameter of a spherical mycoplasma cell is 0.3 μm, the radius is 0.15 μm. The volume of a spherical mycoplasma cell: $V = 4\pi r^3/3 = (4)(3.14)(0.15\ \mu\text{m})^3/3 = 0.014\ \mu\text{m}^3$.

    Assuming that *E. coli* is cylindrical, with dimensions of 1 μm $\times$ 2 μm (i.e., a typical rod-shaped bacterium), its volume is $V = \pi r^2 h = (3.14)(0.5\ \mu\text{m})^2(2\ \mu\text{m}) = 1.6\ \mu\text{m}^3$.

    At 0.014 μm^3, the volume of a typical mycoplasma is significantly smaller than that of the cylindrical *E. coli* cell. To be more specific, the mycoplasma is 0.9 % of the size of an *E. coli*. Alternatively, the *E. coli* is about 114 times larger than the mycoplasma.

45

To trigger a response, all that is necessary is that the morphine molecule have a structure that resembles that of the endogenous molecule that ordinarily binds to the receptor, in this case an opiate peptide. Once morphine binds to the receptor, the cell response triggered is the same as the one that would occur if the endogenous signal molecule had bound to the receptor.

CHAPTER 3
In-Chapter Questions
3.1

The tetrahedral structures of the three molecules are as follows:

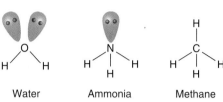

In the solid state of water, the oxygen atom has two electron pairs that form hydrogen bonds with neighboring water molecules. The nitrogen atom in ammonia has one unshared electron pair that can form a hydrogen bond with a neighboring ammonia molecule and methane has none. Note that the heats of fusion for these substances parallel the number of unshared electron pairs. Because of the ability of each ammonia molecule to form a hydrogen bond with a neighboring molecule, ammonia "ice" would be expected to be less dense than liquid ammonia.

3.2

From left to right in the illustration, the noncovalent interactions are ionic, hydrogen bonding, and van der Waals interactions.

3.3

Tendons and ligaments contain large amounts of collagen and other molecules that bind substantial amounts of structured water molecules. Water is an incompressible substance; that is, it cannot be forced to occupy a smaller space. As a result, structures containing large amounts of water can absorb relatively large amounts of force without damage.

3.4

The equilibrium shifts to the right to replace lost bicarbonate, and the acid concentration increases. The resulting condition is called acidosis.

End-of-Chapter Questions
Review Questions
3

The pK_a on the graph is approximately 6.4, so the effective buffer range is between 6.4 and 8.4.

6

Osmolarity = molarity x the extent of ionization (*i*, the number of ions produced per ionic compound). Na_3PO_4 dissociates into four ions. Assuming 85% ionization, the osmolarity of a 1.3 M solution of Na_3PO_4 would therefore be 1.3 $\times$ 4 x 0.85 = 4.4

9

$\pi = iMRT$ where $\pi = 0.01$ atm
$$i = 1$$
$$R = 0.0821\ \text{L atm/mol K}$$
$$T = 298\ \text{K}$$

Solving for *M*:
0.01 atm = (1) (0.0821 L atm/mol K) (298 K) (M)
M = 4.08 $\times$ 10^{-4} mol/L

Solving for the molecular weight of the protein:
0.056 g/0.030 L = 1.867 g/L
1.867 g = 4.08 $\times$ 10^{-4} mol
1 mol of the protein = 4575.98 g = 4600 g

12

Molecules b, c and d all would be expected to have a dipole moment.

15

Carbon dioxide is present in the blood in sufficient quantities to make it effective as a buffer. Phosphate concentration in blood is too low for this compound to be an effective buffer. Within cells, the phosphate concentration is much higher, and it can therefore act as an effective buffer.

19

a. Electrostatic interactions are noncovalent attractions between oppositely charged ions.

b. Equilibrium dialysis is a technique that determines ligand-binding properties of proteins by measuring how much ligand is taken up by a protein. The ligand and protein are initially separated by a semipermeable membrane.

c. A salt bridge is an electrostatic interaction in proteins between oppositely charged ionic groups, such as the ionized side groups of aspartic acid and lysine.

d. A micelle (Figure 3.13) is an aggregation of molecules that have a nonpolar and a polar component. The nonpolar domains avoid water, leaving the polar domains facing outward toward the surrounding water.

e. A weak acid is an acid that does not completely dissociate in water. Most carboxylic acids are weak acids.

21

The relationship between osmolarity and molarity is given by the equation $o = iM$, where o is osmolarity, i is the extent of ionization (number of ions produced per ionic compound), and M is the molarity.

24

$K_a = 6.3 \times 10^{-8}$, therefore, $pK_a = 7.2$
$pH = pK_a + \log[A^-]/[HA]$
      (where A^- = conjugate base of the weak acid HA)
$7.4 = 7.2 + \log[A^-]/[HA]$
$\log[A^-]/[HA] = 7.4 - 7.2 = 0.2$
   $[A^-]/[HA] = 1.58:1$ or $1.6:1$

27

For a substance to be used as a compatible solute, it must have solvent properties (polarity, ability to hydrogen-bond) that are similar to those of water; it must be nontoxic even in high concentrations; it must provide freezing point depression and osmoprotection; and it must interact with structured water in a way that stabilizes the three-dimensional structures of the cell membrane and of proteins.

30

When the concentrations of the weak acid and the conjugate base are equal, the Henderson-Hasselbalch equation simplifies to $pH = pK_a$ [since $[A^-]/[HA] = 1$, and $\log(1) = 0$]. The pK_a of acetic acid is 4.75, therefore the $pH = 4.75$.

$$pH = pK_a + \log\frac{[A^-]}{[HA]}$$

When $[A^-] = [HA]$,
$pH = pK_a + \log(1) = pK_a + 0$
$pH = pK_a = 4.75$ for acetic acid

33

The extreme electronegativity of the oxygen polarizes the O—H bond of water and makes the hydrogen electron deficient. Because the unshared pairs of electrons on the oxygen are available for bonding, an electrostatic interaction occurs.

Thought Questions

36

The salts dissolved in the seawater (a hypertonic solution) pull water out of plants and will cause them to die. This is the reverse of the normal flow of water from the environment into a plant.

39

In a liquid, the molecules must be free to move over one another. In the gelatin solution, each water molecule hydrogen-bonds with two segments of the protein, locking the protein chains and the water together. Because the water molecules are no longer able to move freely, the mixture becomes semirigid.

42

The conversion of glycogen to glucose creates an increase in osmotic pressure, and water would flow into the cell. To offset this rise in osmotic pressure, ions such as sodium and potassium are pumped out of the cell. These ions would be followed by water, thus restoring cell volume.

45

(a) In an anhydrous environment, the hydrophilic carboxyl and ammonium ions move to the interior and the hydrophobic alkyl groups move to the exterior of the polypeptide. (b) In a hydrated environment, the carboxyl and ammonium ions move to the exterior, whereas both alkyl groups move toward the interior of the polypeptide.

48

The micelle that forms would be U shaped, with the carboxyl groups projecting into the water and the hydrophobic "bottoms" of the "U" toward the center. It could also form a membrane-like structure with the chains being straight, with the carboxyl groups positioned on either side of the membrane.

51

The calculation is as follows: 335 J/g(1 g) + 4.25 J/g(1g)((100 − 0) + 2258 J/g(1 g) = 2597 J.

54

The molecule has polar and nonpolar components and would be expected to form a lipid bilayer. Alkyl chlorides are hydrophobic. As a result, one would expect a bilayer with the carboxyl groups projecting into the water and the chloride in the interior.

CHAPTER 4

In-Chapter Questions

4.1.

$\Delta G' = \Delta G^{\circ\prime} + RT \ln [ADP][P_i]/[ATP]$
where $R = 8.315 \times 10^{-3}$ kJ/mol•K
       $T = 310$ K
    $[ADP] = 0.00135$ M, $[ATP] = 0.004$ M,
       $[P_i] = 0.00465$ M
       $\Delta G^{\circ\prime} = -30.5$ kJ/mol

$\Delta G' = -30.5$ kJ/mol + (8.315 J/mol•K)(310)
       ln(0.00135 M)(0.00465 M)/(0.004 M)
$\Delta G' = -30.5 + 2.577$ (ln 0.00157)
     $= -30.5 - 16.64$
     $= -47.14$ kJ/mol = -47.1 kJ/mol

4.2

Amount of ATP required to walk a mile
= (100 kcal/mi)/7.3 kcal/mol
= 13.7 mol/mi $\times$ 507 g/mol = 6945.2 g/mi = 6950 g/mi

Amount of glucose required to produce 100 kcal through ATP
= (100 kcal)/(.04)(686 kcal/mol)
= 100 kcal/274.4 kcal/mol
= 0.36 mol
= 0.36 mol x 180 g/mol = 65.6 g of glucose

End-of-Chapter Questions
Review Questions

3

For a reaction to proceed to completion the total overall $\Delta G^{\circ\prime}$ must be negative, and there must be a common intermediate, in this case P_i. This is true of reaction b.

6

$$ATP + glutamate + NH_3 \longrightarrow ADP + P_i + glutamine$$
$$ATP + H_2O \longrightarrow ADP + P_i \quad \Delta G^{\circ\prime} = -30.5 \text{ kJ/mol}$$
$$Glutamine + H_2O \longrightarrow glutamate + NH_3$$
$$\Delta G^{\circ\prime} = -14.2 \text{ kJ/mol}$$

Reverse the second equation and add the $\Delta G^{\circ\prime}$ values.

$$ATP + H_2O \longrightarrow ADP + P_i \quad \Delta G^{\circ\prime} = -30.5 \text{ kJ/mol}$$
$$Glutamate + NH_3 \longrightarrow glutamine + H_2O$$
$$\Delta G^{\circ\prime} = +14.2 \text{ kJ/mol}$$
$$ATP + glutamate + NH_3 \longrightarrow ADP + P_i + glutamine$$
$$\Delta G^{\circ\prime} = -16.3 \text{ kJ/mol}$$

9

Under standard conditions the following statements are true: a and e.

12

The following statements are true: a, b, c, and f.

d. The sign and magnitude of ΔG give important information about the direction of a reaction, but no information about the rate.

e. At equilibrium, $\Delta G = 0$.

15

With an intermediate phosphoryl group transfer potential, ATP can accept a phosphate group from compounds that have a higher phosphoryl group transfer potential and transfer it to lower energy compounds. In other words, the number of compounds that can transfer a phosphate to or from ATP is maximized.

18

At equilibrium, $\Delta G^{\circ\prime} = -RT \ln K_{eq}$
$$-9700 \text{ J/mol} = -(8.315 \text{ J/mol} \cdot K)(298 \text{ K}) \ln K_{eq}$$
$$3.915 = \ln K_{eq}$$
$$K_{eq} = 50.1 = [glycerol][P_i]/[glycerol\text{-}3\text{-phosphate}]$$
$$50.1 = (1 \times 10^{-3} \text{ M})^2/[glycerol\text{-}3\text{-phosphate}]$$
$$[Glycerol\text{-}3\text{-phosphate}] = (1 \times 10^{-3} \text{ M})^2/50.1 = 2 \times 10^{-8} \text{ M}$$

21

At equilibrium: $\Delta G^{\circ\prime} = -RT \ln K_{eq}$

$$-13,800 \text{ J/mol} = -(8.315 \text{ J/mol} \cdot K)(298 \text{ K}) \ln K_{eq}$$
$$5.57 = \ln K_{eq}$$
$$K_{eq} = 262$$
$$K_{eq} = [glucose][P_i]/[glucose\text{-}6\text{-phosphate}]$$
$$[Glucose\text{-}6\text{-phosphate}] = 4 \text{ mM} = 4 \times 10^{-3} \text{ M}$$
$$262 = [glucose][P_i]/(4 \times 10^{-3} \text{ M})$$
$$1.05 = [glucose][P_i]$$
Assuming that $[glucose] = [P_i]$, $[P_i] = 1.02$ M.

24

$$\Delta G^{\circ\prime} = -13.8 \text{ kJ/mol}$$
$$K_{eq} = 9.2 \times 10^{-3} \text{ (from Review Question 23)}.$$
When $\Delta G = 0$, $\Delta G^{\circ\prime} = -RT \ln K_{eq}$
$$-13.8 \text{ kJ/mol} = (8.315 \times 10^{-3} \text{ kJ/mol} \cdot K)(T)(\ln 9.2 \times 10^{-3})$$
$$-13.8 \text{ kJ/mol} = (8.315 \times 10^{-3} \text{ kJ/mol} \cdot K)(-4.69)(T)$$
$$-13.8 \text{ kJ/mol} = -(0.039 \text{ kJ/mol} \cdot K)(T)$$
$$T = 354 \text{ K} = 81°C$$

Thought Questions

27

$$\Delta G' = -RT \ln K_{eq}$$
$$-16,700 \text{ J/mol} = -(8.315 \text{ J/mol} \cdot K)(298 \text{ K}) \ln K_{eq}$$
$$-16,700 = -2478 \ln K_{eq}$$
$$16,700/2478 = \ln K_{eq}$$
$$6.74 = \ln K_{eq}$$
$$K_{eq} = 851 \text{ [ATP][G-6-P]/[ATP] [glucose]}$$

30

ΔG is the most useful criterion of spontaneity because it reflects the change in entropy, which must increase for a reaction to be spontaneous.

33

$\Delta H^{\circ} = -88 \text{ kJ/mol}$ $\Delta S = 0.3 \text{ kJ/mol} \cdot K$ $T = 298 \text{ K}$
$$\Delta G^{\circ} = \Delta H^{\circ} - T\Delta S^{\circ}$$
$$\Delta G^{\circ} = (-88 \text{ kJ/mol}) - (298)(0.3 \text{ kJ/mol} \cdot K)$$
$$= -88 - 89.4$$
$$= -177.4 \text{ kJ/mol}$$
$$\Delta G^{\circ} = -RT \ln K_{eq}$$
$$(-177.4 \text{ kJ/mol})(1000 \text{ J/kJ}) = -(8.315 \text{ J/mol} \cdot K)(298 \text{ K}) \ln K_{eq}$$
$$71.6 = \ln K_{eq} \quad \text{and} \quad K_{eq} = 1.2 \times 10^{31}$$

The K_{eq} value indicates that the denaturation of the protein is virtually irreversible.

36

The balanced equation is

$$C_{17}H_{35}COOH + 26O_2 \longrightarrow 18CO_2 + 18H_2O$$
$$\Delta H = \Delta H_{products} - \Delta H_{reactants}$$
$$= 18 \text{ mol}(-94 \text{ kcal/mol}) + 18 \text{ mol}(-68.4 \text{ kcal/mol})$$
$$- (1 \text{ mol})(-211.4 \text{ kcal/mol})$$
$$= -1692 \text{ kcal} - 1231.2 \text{ kcal} + 211.4 \text{ kcal}$$
$$= -2711.8 \text{ kcal} = -2712 \text{ kcal}$$

39

The sulfur of the thioester is larger than oxygen of the alcohol. So sulfur can stabilize the unshared electron pairs more easily. Consequently, there is a larger difference in energy between the reactants and products, which translates into a larger ΔG.

CHAPTER 5

In-Chapter Questions

1

Amino acids a and b are neutral, nonpolar, c is basic, and d is an acidic amino acid.

2

Bacteria with surface polypeptides composed of D-amino acids are resistant to degradation because the proteases, the enzymes that immune system cells use to degrade protein in foreign cells, can only catalyze the hydrolysis of peptide bonds between L-amino acids. In other words, the active sites of proteases are stereospecific; that is; they can only effectively bind peptides composed of L-amino acids.

3

The number of possible tetrapeptides is $20^4 = 160,000$.

4

The structure of the penicillamine-cysteine disulfide is

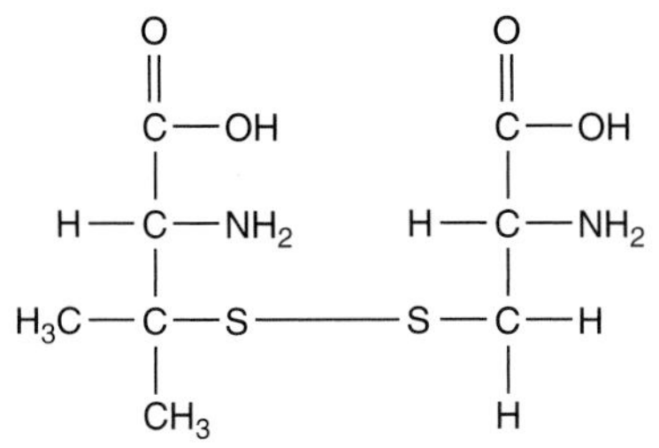

5

The complete structure of oxytocin is

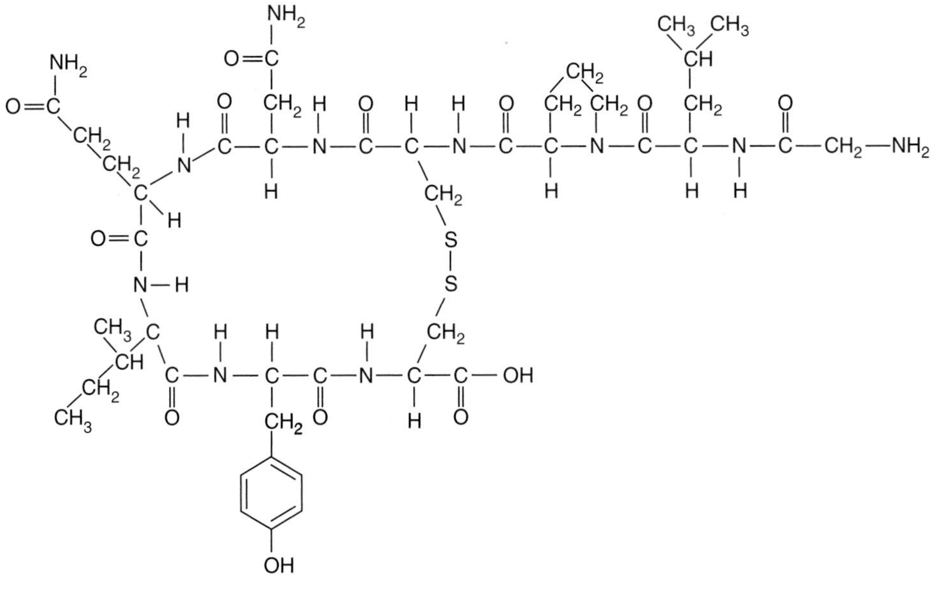

At pH 4 the terminal amino group of the glycine would be protonated to give the molecule a +1 charge. The isoelectric point of oxytocin is 5.6. Therefore at pH 9 the molecule will have a net negative charge.

6

The partial overlap in the biological properties of vasopressin and oxytocin can be explained in part by the Ile residue at position 3 of oxytocin. Presumably, given the overall similarities in size and the identical nature of six of the eight amino acid residues in the two molecules, the hydrophobic side chain of this Ile can reach and partially fit into the hydrophobic pocket of the vasopressin receptor. The degree of functional overlap between the two peptides is reduced by the leucine residue at position 8 of oxytocin because the side chain of this residue is not only neutral but smaller than the positively charged arginine residue of vasopressin. If the arginine residue of vasopressin is replaced by a lysine residue, it is expected that there will be a decrease in the molecule's binding properties because of the structural differences between the two side chains. This decrease would probably not be large, since the side chains are similar in length and are both positively charged.

7

The trait is recessive, and two copies of the aberrant gene are required for full expression of the disease. Primaquine induces the production of excess amounts of the strong oxidizing agent hydrogen peroxide. In the absence of sufficient amounts of the reducing agent NADPH, the peroxide molecules cause extensive damage to the cell. No, a higher than normal peroxide level in blood cells is damaging to the malarial parasite and is selected for in geographical regions where malaria occurs.

8

The solution to the first two illustrations is

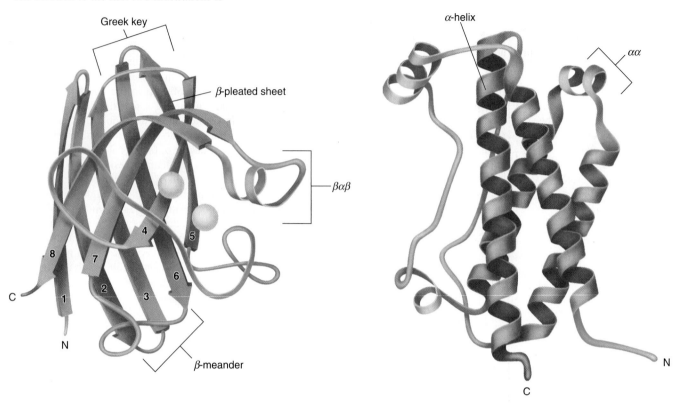

The solution to the third illustration is

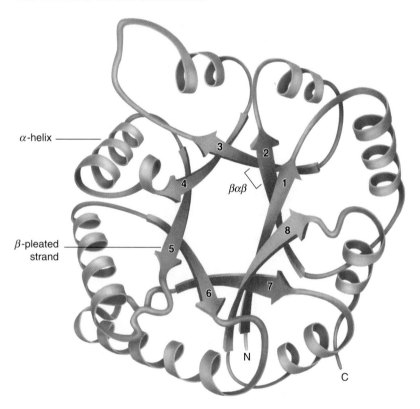

This protein is an example of an α/β-barrel structure.

9

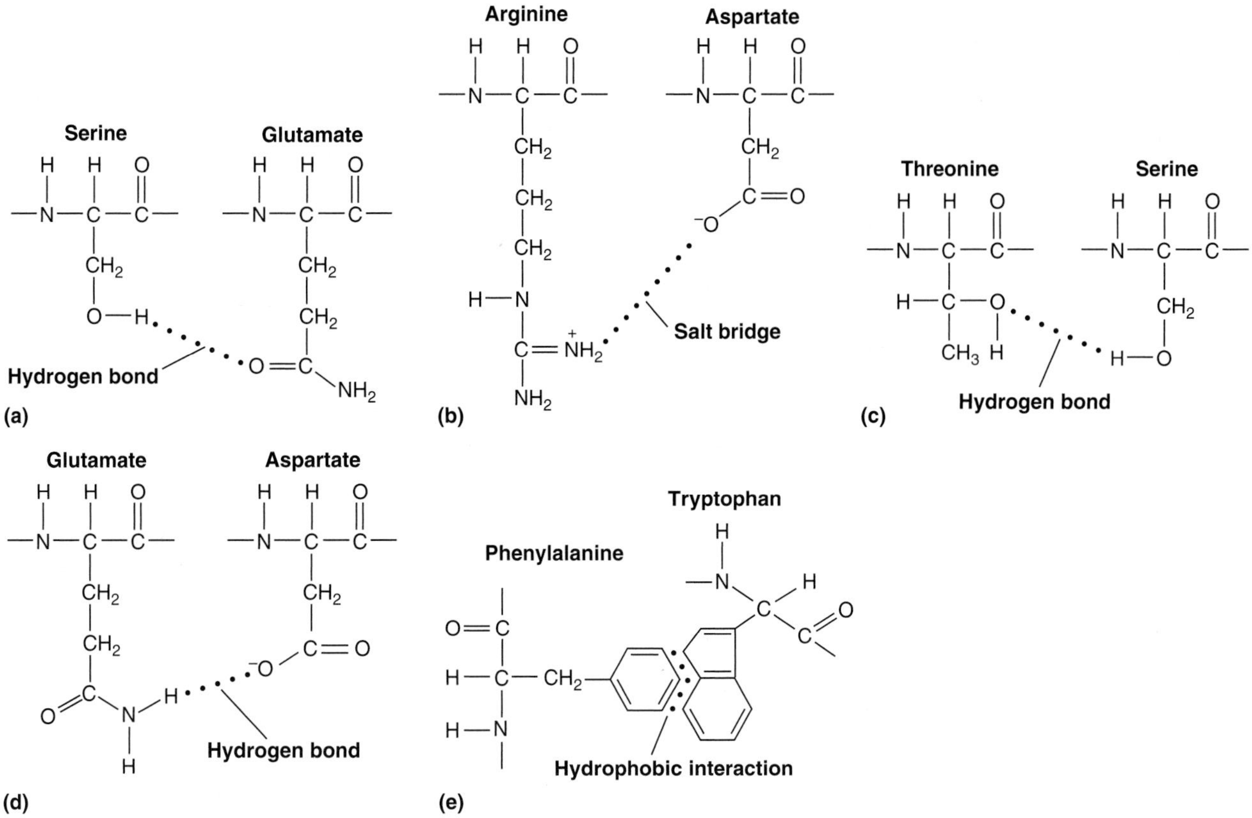

Serine

Glutamate

Hydrogen bond

(a)

Arginine

Aspartate

Salt bridge

(b)

Threonine

Serine

Hydrogen bond

(c)

Glutamate

Aspartate

Hydrogen bond

(d)

Phenylalanine

Tryptophan

Hydrophobic interaction

(e)

10

Collagen is a major structural protein found in connective tissues. Consequently, the failure of collagen molecules to form properly weakens these tissues causing diverse symptoms. Examples include cataracts, easily deformed bones, torn tendons and ligaments, and ruptured blood vessels.

11

BPG stabilizes deoxyhemoglobin. In the absence of BPG, oxyhemoglobin forms more easily. Fetal hemoglobin binds BPG poorly and, therefore, has a greater affinity for oxygen.

12

Myoglobin, composed of a single polypeptide, binds oxygen in a simple pattern—it binds the molecule tightly and releases it only when the cells' oxygen concentration is very low. The binding of oxygen by hemoglobin, a tetramer, has a more complicated sigmoidal pattern that is made possible by the noncovalent interactions among its four subunits.

End-of-Chapter Questions
Review Questions

3

Structure and net charge of arginine at various pH levels:

pH	Structure	Net Charge
1	$H_2N-C-N-CH_2-CH_2-CH_2-C-COOH$; $^+NH_2$; $^+NH_3$	+ 2
4	$H_2N-C-N-CH_2CH_2CH_2-CHCOO^-$; $^+NH_2$; $^+NH_3$	+ 1
7	$H_2N-C-NH-CH_2CH_2CH_2CH-COO^-$; $^+NH_2$; $^+NH_3$	+ 1
10	$H_2N-C-NH-CH_2CH_2CH_2-CHCOO^-$; $^+NH_2$; NH_2	0
12	$H_2N-C-NHCH_2CH_2CH_2-CH-COO^-$; $^+NH_2$; NH_2	0

6

The resonance forms of the peptide bond in glycylglycine are as follows:

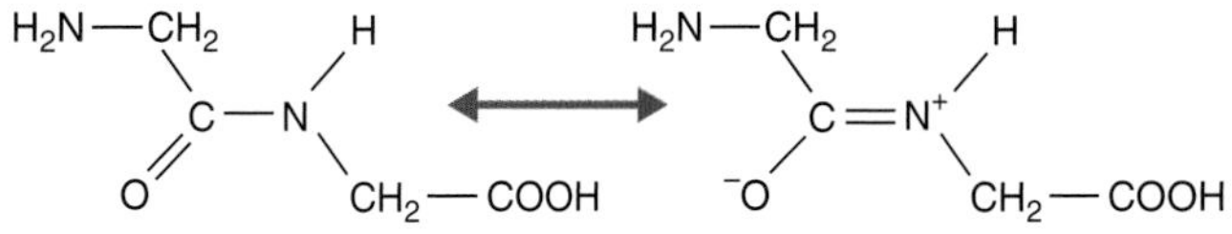

The partial double-bond character of the peptide bond makes it rigid and planar. Rotation around this bond is therefore hindered.

9

a. An asymmetric carbon has four different atoms (or groups) bonded to it.

b. Motor proteins are components of biological machines that bind nucleotides. Nucleotide hydrolysis drives precise changes in the protein's shape.

c. A prosthetic group is the nonprotein portion of a conjugated protein that is essential to the biological activity of the protein. A prosthetic group is often a complex organic molecule.

d. The primary structure of a polypeptide is its amino acid sequence.

e. A molten globule is the partially globular site of a folding polypeptide that resembles the molecule's native state.

12

The structural features of several amino acids do not foster α-helix formation. Because the R group of glycine is too small, the polypeptide chain becomes too flexible. Proline's rigid ring prevents the required rotation of the N—C bond. Sequences with larger numbers of amino acids with charged side chains (e.g., glutamate) and bulky side chains (e.g., tryptophan) are also incompatible with α-helix formation.

15

Refer to Biochemistry in the Lab (pp. 172–177) for protein purification techniques.

18

In sequencing a polypeptide, the next residue in a sequence is determined by the increase in height of the peak on the amino acid analyzer. If there is already a large amount of that amino acid present in the solution, it would be difficult, if not impossible, to detect any change in peak height. Small fragments have only a few amino acids, and this problem does not occur.

21

a. Electrophoresis is a class of techniques in which molecules are separated from each other because of differences in their net charge.

b. A molecular disease is a disease caused by a mutated gene.

c. The carbon next to the carboxyl group in an amino acid is the α-carbon.

d. The isoelectric point is the pH at which an amino acid or peptide is electrically neutral.

e. A peptide bond is an amide bond between two amino acids.

24

The primary driving force in protein folding is the requirement to achieve a low-energy state despite the decrease in entropy that occurs as the protein's three-dimensional structure becomes more ordered. Key considerations include the energy associated with different bond angles and bond rotation, the chemical properties of the amino acid side chains (e.g., whether or not the side chain will be charged at the cellular pH), and interactions between side chains. Of the noncovalent interactions that are possible, hydrophobic interactions are particularly important. (Recall that hydrophobic interactions are driven in part by the increase in entropy in the surrounding water molecules.)

27

a. A hydrophobic amino acid has a side group that is nonpolar.

b. A salt bridge is an electrostatic interaction in proteins between ionic groups of opposite charge.

c. A dabsyl amino acid is an N-terminal amino acid derivative that contains a dabsyl group, which is a fluorescent marker that selectively attaches to a protein's N-terminal amino acid. Since this marker can be analyzed by means of HPLC, this method allows identification of a protein's N-terminal amino acid.

d. Site-directed mutagenesis is a technique that introduces specific sequence changes into cloned genes.

e. Proteomics is the analysis of proteomes, which is the complete set of proteins produced within a cell under a given set of conditions.

Thought Questions

30

The large size of enzymes is required to stabilize the shape and functional properties of the active site and to shield it from extraneous molecules. In addition, structural features of the protein may function in recognition processes in signaling or binding to cellular structures.

33

The hydrophobic amino acid side chains are excluded from the water and tend to cluster together. This clustering holds portions of the polypeptides in a particular conformation.

36

The surface peptide bonds are not readily cleaved because chymotrypsin cleaves after hydrophobic residues and these residues are not abundant on the surface of the enzyme.

39

The amino acid analysis provides information concerning the kind and number of amino acids in the peptide. The carboxypeptidase and DNFB results show that the carboxyl and amino-terminal amino acids are both glycine. Finally, by overlapping the fragments, the sequence of amino acids can be determined. Remember to start with a fragment that ends with the N-terminal residue, in this case, glycine.

Gly–Leu, Gly–Leu–Glu, Glu–Gly–Pro, Gly–Pro–Met–Lys, Pro–Met–Lys–Lys, Lys–Glu, Lys–Glu–Thr–Phe–Leu, Thr–Phe–Leu–Leu–Gly, Leu–Leu–Gly

The overall structure then becomes

Gly–Leu–Glu–Gly–Pro–Met–Lys–Lys–Glu–Thr–Phe–Leu–Leu–Gly

42

The hydrophilic amino acids tyrosine, asparagine, serine, and histidine should all be on the surface.

45

A glycine residue with its small R group (an H atom) within a peptide molecule increases the molecule's rotational freedom. A polypeptide with a series of three chain-stiffening prolines in the positions ordinarily occupied by three glycine residues would become less flexible, a circumstance that would cause a change in conformation and probably loss of function.

48

Neurons are gently homogenized to fragment the membrane. This homogenate is ultracentrifuged and fractions taken. Each fraction is then assayed for endorphin-binding activity. After treatment with detergents to release the receptors from the membrane, the active fractions are passed through an affinity column to which endorphins are bound. The receptor is then eluted and characterized by analysis of its amino acid sequence by mass spectrometry. The three-dimensional structure of the receptor can be determined by X-ray crystallography.

51

The first step in the process is to develop an assay that allows the investigator to detect the peptides of interest during the purification protocol. Next, the peptides, as well as many other substances, are released from the pituitary gland by cell disruption. Low-molecular weight substances can then be removed by dialysis. The resulting material is then passed through an exclusion chromatography column in which the large-molecular weight materials elute first. The fractions containing the oxytocin and vasopressin are then passed though an affinity column, which will retain oxytocin or vasopressin, respectively. Electrophoresis can be used to assess the purity of each peptide.

CHAPTER 6

In-Chapter Questions

6.1

The amino acid residues forming the three-dimensional structure of the active site are chiral. As a result, the active site is chiral. It can bind only one isomeric form of a hexose sugar, in this case the D-isomer.

6.2

a. Isomerase

b. Transferase

c. Lyase

d. Oxidoreductase

e. Ligase

f. Hydrolase

6.3

The products of the degradation are the following compounds:

Methanol   **Phenylalanine**   **Aspartic acid**

Cleavage of the ester bond is catalyzed by an esterase; the amide bond is cleaved by a peptidase.

6.4

6.5

Dialysis removes the formaldehyde, formic acid, and methanol that build up in the bloodstream. The bicarbonate neutralizes the acid produced and helps offset the resultant acidosis. The ethanol competitively binds with the alcohol dehydrogenase. This slows the dchydrogenation of the methanol and allows time for the kidneys to excrete it.

6.6

The energy diagrams for the hypothetical reaction are as follows:

6.7

Menkes' syndrome—injections of copper salts into the blood would avoid intestinal malabsorption and provide the copper necessary to form adequate levels of ceruloplasmin and offset the symptoms of the disease.

6.8

Wilson's disease—zinc induces the synthesis of metallothioein, which has a high affinity for copper. Some organ damage can be averted because metallothionein sequesters copper and prevents this

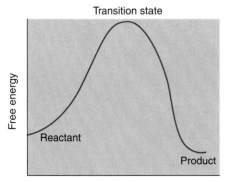

Transition state

Free energy

Reactant

Product

Progress of reaction
Hypothetical reaction without enzyme

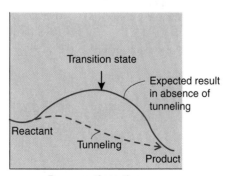

Transition state

Expected result in absence of tunneling

Reactant

Tunneling

Product

Progress of reaction
Hypothetical reaction
with enzyme and tunneling phenomenon

toxic metal from binding to and inactivating susceptible proteins and enzymes. Penicillamine forms a complex with copper in the blood. This complex is transported to the kidneys, where it is excreted.

6.9

a. Cofactor

b. Holoenzyme

c. Apoenzyme

d. Coenzyme

e. Coenzyme

6.10

The patient that failed to show improvement probably had a higher level of acetylating enzymes. The patient's dosage should be based on capacity to process the drug and not on body weight.

End-of-Chapter Questions

Review Questions

3

Cells regulate enzymatic reactions by using genetic control (certain key enzymes are synthesized in response to changing metabolic needs), covalent modification (certain enzymes are regulated by the reversible interconversion between their active and inactive forms, a process involving covalent changes in structure), allosteric regulation (binding effector molecules to pacemaker enzymes alters catalytic activity), and compartmentation (preventing wasteful "futile cycles" by physical separation of opposing biochemical processes within cells).

6.

Factors that contribute to enzyme catalysis include proximity and strain effects, electrostatic effects, acid-base catalysis, and covalent catalysis. Refer to page 204–206 for an explanation of each.

9

a. Oxidoreductase: an enzyme that catalyzes an oxidation-reduction reaction.

b. Lyase: an enzyme that catalyzes an addition or elimination reaction, adding a molecule across a double bond or removing a molecule to form a double bond.

c. Ligase: an enzyme that catalyzes the joining of two molecules, often using ATP for energy.

d. Transferase: an enzyme that catalyzes the transfer of a functional group from one molecule to another.

e. Hydrolase: an enzyme that catalyzes a nucleophilic substitution reaction where water is the nucleophile, resulting in a cleaved bond.

f. Isomerase: an enzyme that catalyzes the conversion of one isomer to another.

12

Refer to Table 6.3.

15

The activation energy for the reaction of glucose with molecular oxygen is quite high and, consequently, the reaction is relatively slow to occur.

18

At the start of a reaction, the concentrations of the reactants and products can be known precisely. Because equilibrium has not yet been established presumably only the forward reaction is taking place.

21

Enzymes lower the activation energy of a reaction by lowering the free energy of the transition state. The active site of the enzyme described most likely contains amino acid residues that stabilize the transition state with some or all of the following: electrostatic effects and noncovalent interactions, a shape that accepts the substrate yet eases strain in the transition state, and participation in the catalytic mechanism (e.g., by providing an acidic, basic, or nucleophilic residue to assist in acid-base or covalent catalysis). The active site should also place the substrate and reactants in close proximity to each other and in the proper orientation.

24

a. Metabolon: a complex of enzymes that share intermediates of a metabolic pathway so that the product of one enzyme is in close proximity to the active site of the next enzyme in the pathway.

b. *In vivo*: in a living cell or organism.

c. *In vitro*: in "glass" (i.e., a test tube); taking place under controlled laboratory conditions as opposed to within a living cell or organism.

d. In silico: resulting from computer simulation or computer modeling.

e. Metabolic flux: the rate of flow of metabolites, such as substrates, products, and intermediates, along biochemical pathways.

27

Compartmentation within eukaryotic cells is the physical separation of enzymes by a membrane (i.e., by containing certain enzymes within an organelle), or by attachment of enzymes to membranes or cytoskeletal filaments. Compartmentation (1) prevents competing reactions from occurring simultaneously and allows them to be regulated separately ("divide and control"), (2) reduces or removes diffusion barriers by locating enzymes and metabolites close to each other, (3) provides specialized reaction conditions (e.g., low pH) that would not be possible otherwise, and (4) protects other cellular components from potentially toxic reaction products ("damage control").

30

Amino acids that are capable of acting as acids or bases in enzyme catalysis are aspartic acid, glutamic acid, histidine, cysteine, tyrosine, lysine, and arginine. Of these amino acids, histidine is the most likely to be able to function as either an acid or a base, since its pK_R value, 6.0, is relatively close to physiological pH. At pH 7.6, histidine's R group would be completely ionized in aqueous solution. However, the pK_a may shift depending upon the environment of the active site. A relatively nonpolar active site would lower the pK_a to favor the neutral form of the R group (which could act as a base), while a polar active site would raise the pK_a to favor the charged form (which could act as an acid).

Thought Questions

33

Note the difficulty in estimating V_{max} for the uninhibited reaction. If 95 μmol/min is chosen for the maximum velocity, then K_m is 20 μM (the substrate concentration, [S], at half V_{max}, or 45 μmol/min). However, if 100 μmol/min is the maximum velocity, then $K_m = 25$ μM. The inhibited reaction appears to have a V_{max} of 16 μmol/min; $K_m = 15$ μM.

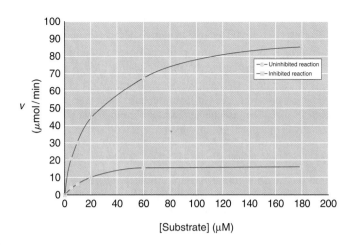

[Substrate] (μM)

36

Refer to Figure 6.18.

39.

$V_{max} = 10\ \mu m/s\ K_m = 5\ \mu m$

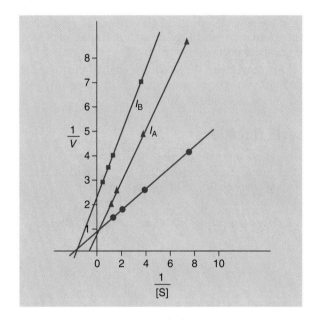

42

It is a hydrolase.

45

It is a transferase.

48

The aspartate carboxyl donates electrons to the histidine ring, polarizing it and raising the electron density on the nitrogen, making it a stronger base and raising the pK_a.

51

Alcohol dehydrogenase selectively metabolizes ethyl alcohol in preference to ethylene glycol. Treatment of the patient with ethyl alcohol will block the conversion of ethylene glycol to toxic metabolites. The ethylene glycol can then be excreted.

54

See Figure 6.18.

CHAPTER 7

In-Chapter Questions

7.1

a. Aldotetrose, b. Ketopentose, c. Ketohexose

7.3

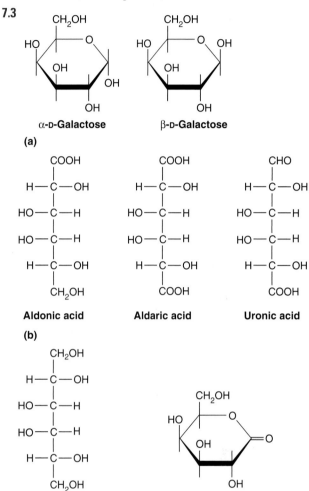

α-ᴅ-**Galactose**   β-ᴅ-**Galactose**

(a)

Aldonic acid   **Aldaric acid**   **Uronic acid**

(b)

Galactitol   δ-**Lactone of galactonic acid**

(c)   **(d)**

7.4

7.5

Carbohydrate   **Aglycone**

7.6

a. Glucose—reducing sugar
b. Fructose—reducing sugar
c. α-Methyl-D-glucoside—nonreducing
d. Sucrose—nonreducing

Sugars a and b are capable of mutarotation.

7.7

The larger, insoluble glycogen molecule makes a negligible contribution to the osmotic pressure of the cell. In contrast, each molecule of an equivalent number of glucose molecules contributes to osmotic pressure. If the glucose molecules were not linked to form glycogen, the cell would burst.

7.8

In an analog system, information is encoded as a continuous signal. For example, in old-fashioned clocks the hands move continuously and not in small steps. In analog systems, small fluctuations in information processing can be meaningful. The sugar code is an analog system because of microheterogeneity, the continuous spectrum of carbohydrate structures that cells can synthesize to encode biologically relevant signaling information. In a digital system, information is represented by discrete values ("digits") of a physical quantity. Digital clocks display time as a progression of discrete numbers with no intermediate values. The DNA code is a digital system composed of four digits (bases). During protein synthesis, each DNA base triplet that ultimately codes for an amino acid has a specific meaning.

End-of-Chapter Questions

Review Questions

3

In D-family sugars, the OH on the chiral carbon farthest from the carbonyl group is on the right side in a Fischer projection formula. So both (+)-glucose and (−)-fructose are D-sugars despite their rotation of plane-polarized light in opposite directions.

6

Heteroglycans are made up of more than one type of monosaccharide residue but homoglycans have only one. Examples of homoglycans and heteroglycans are starch and hyaluronic acid, respectively.

9

Starch and glycogen are both homoglycans containing glucose monomers linked by α-(1,4) glycosidic bonds with branch points connected by α-(1,6) glycosidic bonds. Glycogen, however, is much more highly branched than starch. Cellulose is a linear polymer of glucose linked by β-(1,4) glycosidic bonds.

12

a. Glycan: a polymer made from sugar molecules; a polysaccharide. Starch is a glycan made from α-D-glucose.
b. Heteroglycan: a glycan made from two or more types of monosaccharide.
c. Homoglycan: a glycan made from one type of monosaccharide. Cellulose, starch, glycogen, and chitin are all homoglycans.
d. GAG: glycosaminoglycan, a linear polymer with disaccharide repeating units; many of the sugar residues are amino derivatives and/or contain carboxyl and sulfate groups. Chondroitin sulfate is a GAG.
e. Syndecan: a class of proteoglycan that contains heparan-sulfate and chondroitin sulfate chains linked to a transmembrane protein.

15

a. Carboxylic acid and hydroxyl groups bind large amounts of water.
b. Hydrogen bonding is the primary type of bonding between water and glycosaminoglycans.

18

A reducing sugar reduces Cu(II) in Benedict's reagent. This reduction takes place because the hemiacetal portion of a sugar can form an aldehyde functional group, which can be oxidized to a carboxylic acid.

21

Numerous carbohydrate groups protect glycoproteins from denaturation because they are hydrophilic groups that surround the protein and protect against protease-catalyzed peptide bond cleavage.

24

a. Sugar code: information stored in the arrangement of various carbohydrate chains in glycoconjugates. For example, the glycosylation of cell membrane proteins and lipids places carbohydrate chains on the cell surface, where they can serve in cellular recognition.
b. Glycome: the total set of sugars and glycans that a cell or organism produces.
c. Glycoforms: forms of a glycoprotein that differ slightly in their glycan components.
d. Microheterogeneity: the production of glycoforms, possibly for cell- or tissue-specific purposes.
e. glycomics: the study of glycomes.

Thought Questions

27

The thick proteoglycan coat acts to protect bacteria by preventing the binding of antibodies to their surface antigens.

30

The structure of the sugars in Question 29 are as follows:

(a) α-D-**Mannopyranose**

(b) α-D-**Pyranoiduranic acid**

(c) α-D-**Fructopyranose**

(d) α-D-**Xylopyranose**

33

The water that is absorbed in large quantities by proteoglycans is incompressible. Therefore, tissues that contain proteoglycans in large amounts receive some protection against mechanical stress, i.e., the tissue resists deformation when pressure is applied.

36

The complete structures of the Type A and Type B antigens are as follows:

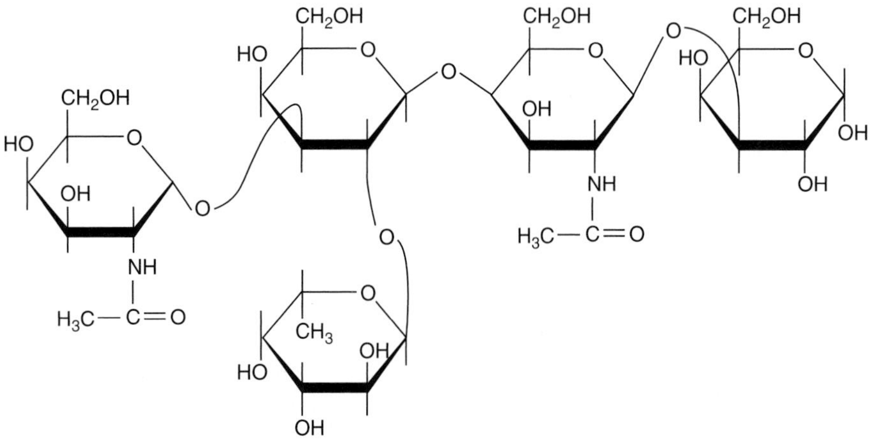

Type A

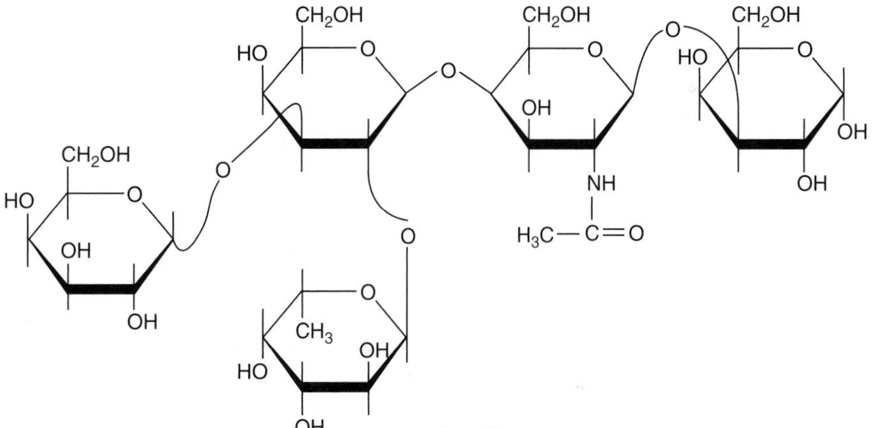

Type B

39

The two diasterisomeric products of the reduction of fructose are:

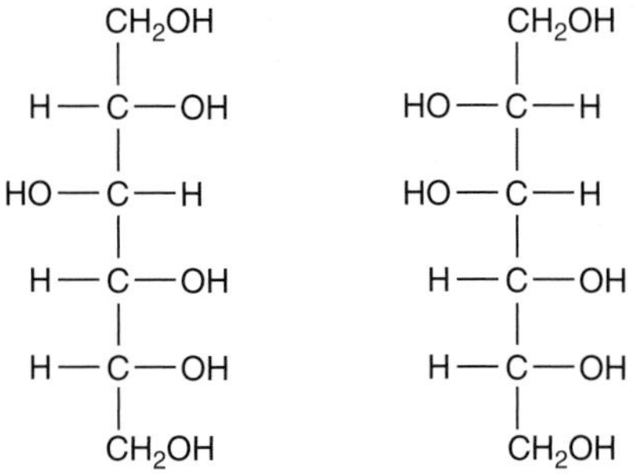

42

The structure of the oligosaccharide is as follows:

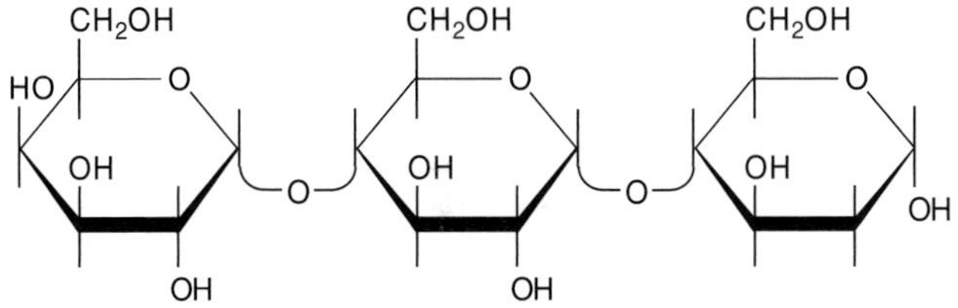

45

A free sugar has many OH groups that hydrogen bond and raise its boiling point above its decomposition temperature. Conversion of the sugar to a volatile derivative lowers its boiling point below its decomposition point and analysis by GC or GC/MS can be accomplished easily.

48

If glyceraldehyde forms a four-membered ring, the bond angles create significant strain. As a result the four membered ring does not form. The larger five-membered ring formed by the tetroses is relatively strain free and forms easily.

CHAPTER 8

In-Chapter Questions

8.1

The large excess of NADH that is produced by these reactions drives the conversion of pyruvate to lactate.

8.2

Chromium is acting as a cofactor.

8.3

In the absence of O_2, energy is produced only through glycolysis, an anaerobic process. Glycolysis produces less energy per glucose molecule than does aerobic respiration. Consequently, more glucose molecules must be metabolized to meet the energy needs of the cell. When O_2 is present, the flux of glucose through glycolysis is reduced.

8.4

At three strategic points, glycolytic and gluconeogenic reactions are catalyzed by different enzymes. For example, phosphofructokinase and fructose-l,6-diphosphatase catalyze opposing reactions. If both reactions occur simultaneously (i.e., in a futile cycle) to a significant extent, ATP hydrolysis in the reaction catalyzed by phospho-fructokinase releases large amounts of heat. If the heat is not quickly dissipated, an affected individual could die of hyperthermia.

8.5

In gluconeogenesis pyruvate is converted to oxaloacetate. NADH and H^+ are required to reduce glycerate 1,3-bisphosphate to glyceraldehyde-3-phosphate. NAD^+ is the oxidized form of NADH also produced in this reaction. ATP is needed to provide the energy to carboxylate pyruvate to oxaloacetate and phosphorylate glyceraldehyde-3-phosphate to glycerate-1,3-bisphosphate. Both of these reactions also produce ADP and P_i. GTP converts oxaloacetate to phosphoenolpyruvate. This reaction is also the source of GDP and P_i. Water is involved in the hydrolysis reactions of ATP to ADP and P_i, the conversion of phosphoenolpyruvate to 2-phosphoglycerate, and the hydrolysis of glucose-6-phosphate to glucose. Six protons are formed when four molecules of ATP and two molecules of GTP are hydrolyzed.

8.6

Without glucose-6-phosphatase activity, the individual cannot release glucose into the blood. Blood glucose levels must be maintained by frequent consumption of carbohydrate. Excess glucose-6-phosphate is converted to pyruvate, which is then reduced by NADH to form lactate.

8.7

The enzyme deficiencies prevent the breakdown of glycogen. Because the synthetic enzymes are active, some glycogen continues to be produced and causes liver enlargement. Because of the liver's strategic role in maintaining blood glucose, defective debranching enzyme causes hypoglycemia (low blood sugar).

End-of-Chapter Questions

Review Questions

3

Ribose-5-phosphate is an aldopentose with a phosphate group at the fifth carbon. Ribose-5-phosphate is a ketose with a carbonyl at the second carbon and a phosphate group at the fifth carbon.

6

Substrate-level phosphorylation is the synthesis of ATP (from ADP and P_i) that is coupled to the exergonic breakdown of a high-energy organic substrate. Examples of this process in glycolysis are the conversions of glycerate-1,3-bisphosphate to glycerate-3-phosphate (by phosphoglycerate kinase) and phosphoenolpyruvate to pyruvate (by pyruvate kinase).

9

Epinephrine promotes the conversion of glycogen to glucose by activating adenylate cyclase, an enzyme whose product, cAMP, initiates a reaction cascade that activates the glycogen degrading enzyme glycogen phosphorylase.

12

Gluconeogenesis occurs mainly in the liver. It is activated by processes that deplete blood glucose, such as fasting and exercise. Futile cycles are prevented by having the forward and reverse reactions catalyzed by different enzymes, both independently regulated.

15

a. Anaerobic: occurring in the absence of molecular oxygen.

b. Aerobic: requiring the presence of molecular oxygen.

c. Antioxidants: substances that prevent the oxidation of other molecules.

d. Glucose-alanine cycle: pyruvate is converted to alanine in muscle and transported to the liver, where the alanine is reconverted to pyruvate, thereby transferring an amino nitrogen from muscle amino acids to the liver, where it can be converted to urea for excretion.

e. Malate shuttle: a metabolic process in which oxaloacetate is transferred by reversible conversion to malate from a mitochondrion to the cytoplasm.

18

Refer to Figures 8.3 and 8.4, stages 1 and 2 reactions of glycolysis, where glucose is converted into pyruvate. Pyruvate may then be converted to ethanol by alcoholic fermentation:

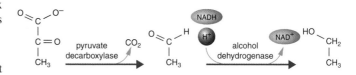

21

Severe hypoglycemia is so dangerous because brain cells (and red blood cells) rely solely on glucose for their energy needs. Severe hypoglycemia causes fainting, and requires *immediate* medical attention. When the condition is prolonged and/or recurrent, it may also result in long-term brain impairment or dysfunction.

24

In glycolysis, a dehydration occurs in the conversion of 2-phosphoglycerate to phosphoenolpyruvate, catalyzed by enolase.

27

The general effect of insulin is to lower blood glucose levels by initiating a process that leads to the inhibition of glycogenolysis and the activation of glycogenesis. Glucagon has the opposite effect: it raises blood glucose levels by initiating a process that leads to the activation of glycogenolysis and the inhibition of glycogenesis. Refer to Figure 8.20 for the effects of insulin and glucagon on the activation or inhibition of specific enzymes, including adenylate cyclase, which catalyzes the synthesis the secondary messenger cAMP.

30

In muscle, glycogen is synthesized to store glucose for energy, or metabolized when energy is needed. In the liver, glycogenesis and glycogenolysis are regulated to maintain blood glucose levels.

Thought Questions

33

In the synthesis of new glycogen molecules, a primer protein called glycogenin is used to initiate glycogen formation. Glucose is transferred from UDP-glucose to a specific tyrosine residue of the glycogenin. This glucose then serves as the starting point for a new growing glycogen molecule.

36

Two common oxidizing agents in anaerobic metabolism are NAD^+ and $NADP^+$.

39

When the cell membrane is compromised, the contents of the glycosome and the general cytoplasm mix and glycolysis proceeds unchecked. All the ATP in the cell is then used up and the cell dies.

42

When a phosphate group is transferred from the phosphoenolpyruvate to another group, an enol alcohol remains. This compound rapidly tautomerizes to a ketone thus driving the phosphoryl transfer reaction to completion.

45

If either NADPH or NADH were the sole hydrogen carrier it would be impossible to determine the need for carbon units in either the pentose phosphate pathway or glycolysis. A decrease in the concentration of either reducing agent activates enzymes that shunt carbon into its respective pathway.

48

The products of ethanol oxidation in the liver are acetaldehyde (a toxic molecule) and NADH. When excess NADH molecules are produced (above that required to generate energy) fat synthesis increases. Because the livers of children are not fully developed, the accumulation of fat easily damages them. Liver function can thereby be compromised.

CHAPTER 9

In-Chapter Questions

9.1

With a $\Delta E_0'$ value of $-0.345\ V$, the oxidation of NO_2^- is spontaneous as written. The oxidation of ethanol is not spontaneous as written because its $\Delta E_0'$ value is positive ($+0.275\ V$).

9.2

Reactions 3, 4, and 5 are redox reactions. In reaction 3, lactate is the reducing agent and NAD^+ is the oxidizing agent. In reaction 4, cyt b (Fe^{2+}) is the reducing agent and NO_2^- the oxidizing agent. In reaction 5, NADH is the reducing agent and CH_3CHO is the oxidizing agent.

9.3

The oxidation states of the functional group carbon (indicated in bold) are:

CH_3CH_2OH $0 - 1 - 1 + 1 = -1$
CH_3CHO $0 - 1 + 2 = +1$
CH_3COOH $0 + 1 + 2 = +3$

9.4

As it is incorporated into an organic molecule, the carbon atom in CO_2 is reduced.

9.5

Because of its symmetrical structure, a molecule of succinate derived from a ^{14}C-labeled acetyl-CoA is converted into two forms of oxaloacetate, one with a labeled methylene group and one with a labeled carbonyl group. ^{14}C-labeled CO_2 is not released until the third turn of the cycle when one-half of the original labeled

carbon is lost (the carbonyl group derived from acetyl-CoA). The labeled carbon is further scrambled when succinyl-CoA is converted to succinate during the third and fourth turns of the cycle.

9.6

Pyruvate carboxylase converts pyruvate to oxaloacetate. If the enzyme is inactive, concentrations of pyruvate in the system rise and pyruvate is converted by NADH to lactate. Excess lactate is then excreted in the urine.

9.7

Fluoroacetate is converted to fluoroacetyl-CoA. This substance then reacts with oxaloacetate to produce fluorocitrate. Fluorocitrate is toxic because it inhibits aconitase, the enzyme that normally converts citrate to isocitrate, hence the buildup of citrate. In the plant, fluoroacetate is stored in vacuoles away from the mitochondria.

End-of-Chapter Questions

Review Questions

3

a. Obligate anaerobes are organisms that not only do not use oxygen to generate energy but live in oxygen-deprived, reduced environments.

b. Aerotolerant anaerobes are organisms that generate energy via fermentation, but can exist in oxygen because they can protect themselves from oxygen's toxic effects.

c. Facultative anaerobes are organisms that, depending on the availability of oxygen, can generate energy via fermentation or aerobic respiration.

d. Obligate aerobes are organisms that require a continuous source of oxygen for energy generation.

6

Citrate, produced in the mitochondria by the citric acid cycle, is transported across the mitochondrial membrane into the cytoplasm. Once in the cytoplasm, citrate is cleaved by citrate lyase to acetyl-CoA and oxaloacetate. The acetyl-CoA is then used to synthesize fatty acids as well as other biomolecules. The oxaloacetate is reduced to malate and moved across the mitochondrial membrane, where it is reoxidized to oxaloacetate, the citric acid cycle intermediate.

9

a. Coenzyme A: an acyl carrier molecule that consists of a 3'-phosphate derivative of ADP linked to pantothenic acid via a phosphate ester bond; pantothenic acid is linked to β-mercaptoethylamine by an amide bond.

b. Amphibolic: referring to reactions or pathways that can function in both anabolic and catabolic processes.

c. Reactive oxygen species: a reactive derivative of molecular oxygen, including superoxide radical, hydrogen peroxide, the hydroxyl radical, and singlet oxygen.

d. Aerobes: organisms that utilize oxygen for energy generation; obligate aerobes require a continuous source of oxygen to survive.

e. NAD(P)$^+$ [nicotinamide adenine dinucleotide (phosphate)]: coenzymes that are important electron carriers for enzymes such as the dehydrogenases; $NADP^+$ has an additional phosphate attached to the 2'-OH group of adenosine.

12

Isocitrate dehydrogenase requires NAD^+ (or $NADP^+$) and the cofactor Mn^{2+}. The α-ketoglutarate dehydrogenase complex requires TPP (thiamine pyrophosphate), lipoic acid, CoASH (coenzyme A), and NAD^+. Succinate dehydrogenase requires FAD. Malate dehydrogenase requires NAD^+.

15

If a small amount of [1-^{14}C]glucose is added to an aerobic yeast culture, the radiolabel will appear in citrate molecules at the carboxyl carbon that is attached to citrate's third carbon atom. In glycolysis, the [1-^{14}C]glucose is split into two molecules of pyruvate, one of which contains the ^{14}C at its carboxyl carbon (C1). If the ^{14}C-labeled pyruvate is converted to acetyl-CoA by pyruvate dehydrogenase, the carboxyl carbon will be released as $^{14}CO_2$. However, if the pyruvate is carboxylated to form oxaloacetate, the ^{14}C will be C1 in oxaloacetate. Upon further reaction with acetyl-CoA, the ^{14}C becomes the carboxyl carbon attached to C3 of citrate.

18

The roles of the each enzyme, cofactor, and coenzyme of the pyruvate dehydrogenase complex are as follows. (1) Pyruvate decarboxylase (also called pyruvate dehydrogenase) decarboxylates pyruvate via the coenzyme TPP (thiamine pyrophosphate), to form HETPP (hydroxyethyl TPP) and CO_2. (2) Dihydrolipoyl transacetylase catalyzes the transfer of an acetyl group from HETPP to the coenzyme lipoic acid to regenerate TPP and form acetyl lipoic acid. A second transfer of this acetyl group from acetyl lipoic acid to the coenzyme CoASH forms acetyl CoA and dihydrolipoic acid. (3) Dihydrolipoyl dehydrogenase reoxidizes the dihydrolipoic acid to regenerate lipoic acid.

21

A high-carbohydrate diet results in the accumulation of acetyl-CoA and citric acid cycle intermediates that inhibit glycolysis and the citric acid cycle, in addition to activating the biosynthesis of fatty acids, which leads to the synthesis of triacylglycerols. Excess citrate, which inhibits the citric acid cycle, is transported out of the mitochondrion into the cytoplasm, and is cleaved to form acetyl Co-A and oxaloacetate. Acetyl-CoA can then be used in the synthesis of fatty acids and triacylglycerols. Citrate also activates the first reaction of fatty acid synthesis.

24

The steps in the citric acid cycle that are regulated are those catalyzed by citrate synthase, which converts acetyl-CoA and oxaloacetate into citrate; isocitrate dehydrogenase, which converts isocitrate to α-ketoglutarate; and α-ketoglutarate dehydrogenase, which forms succinyl-CoA. These steps represent important metabolic branch points. Acetyl-CoA may be used in biosynthetic pathways such as fatty acid synthesis. Also, acetyl-CoA activates pyruvate carboxylase (and inhibits pyruvate dehydrogenase), which activates the reaction of pyruvate with malate to form oxaloacetate (instead of more acetyl-CoA). Because citrate can penetrate the mitochondrial membrane and be cleaved to form acetyl-CoA and oxaloacetate, it therefore can be used to transport acetyl-CoA out of the mitochondrion for fatty acid synthesis. Citrate also inhibits PFK-1 in glycolysis and citrate synthase in the citric acid; cycle, it also activates the first step in fatty acid synthesis. α-Ketoglutarate plays an important role in amino acid metabolism and other metabolic processes. Succinyl-CoA inhibits citrate synthase and α-ketoglutarate dehydrogenase, and is required in the synthesis of porphyrins.

Thought Questions

27

After fructose has been absorbed from the intestine, it is transported to the liver, where it is converted to fructose-1-phosphate, which then enters the glycolytic cycle and is cleaved to form DHAP and glyceraldehyde. Its carbon atoms then are transformed into pyruvate. Pyruvate is converted to acetyl-CoA, which then can enter the citric acid cycle. Fructose that is taken up by adipose and muscle tissue enters the glycolytic pathway by being converted to fructose-6-phosphate by hexokinase.

30

a. $NADH + H^+ + \frac{1}{2} O_2 \longrightarrow NAD^+ + H_2O \quad \Delta E_0' = +1.14 \text{ V}$

$$\Delta G^{0'} = -nF\Delta E_0'$$
$$= (-2)(96{,}485 \text{ J/V} \cdot \text{mol})(+1.14 \text{ V})$$
$$= -219{,}985.8 \text{ J/mol}$$
$$= -220 \text{ kJ/mol}$$

b. $\text{Cyt c (Fe}^{2+}) + \frac{1}{2} O_2 \longrightarrow \text{Cyt c (Fe}^{3+}) + H_2O \quad \Delta E_0' = 0.58 \text{ V}$

$$\Delta G^{0'} = (-2)(96{,}485 \text{ J/V} \cdot \text{mol})(+0.58 \text{ V})$$
$$= -112 \text{ kJ/mol}$$

33

Acetate is converted to acetyl-CoA, which is then converted through the citric acid cycle to oxaloacetate. Via gluconeogenesis, oxaloacetate is converted to glucose. Excess glucose molecules are then stored as glycogen.

36

At low oxygen levels the cells switch over to glycolysis as a source of energy. As a result ATP levels fall. Without sufficient ATP, cells cannot maintain appropriate intracellular ion concentrations. Among the consequences of this phenomenon are rising calcium levels and the activation of calcium-dependent enzymes such as the phospholipases, which begin degrading membrane phospholipids. Osmotic pressure also rises causing cells to swell and leak their contents.

39

Pyruvate dehydrogenase catalyzes the conversion of pyruvate to acetyl-CoA, which is then used in the citric acid cycle. Lack of this enzyme would cause pyruvate to accumulate. Excess pyruvate molecules results in high lactate levels. Pyruvate is also converted by transamination to alanine, so the levels of this amino acid would also be expected to rise.

42

Common two-carbon fragments that could be present are acetate and acetyl-CoA.

CHAPTER 10

10.1

a. NADH

b. $FADH_2$

c. Cyt b (reduced)

d. NADH

e. NADH

10.2

DNP is a lipophilic molecule that binds reversibly with protons. It dissipates that proton gradient in mitochondria by transferring protons across the inner membrane. The uncoupling of electron transport from oxidative phosphorylation causes the energy from food to be dissipated as heat. DNP causes liver failure because of insufficient ATP synthesis in a metabolically demanding organ.

10.3
No, for ATP synthesis to occur, the proton concentration must be higher within the inside-out mitochondrial particles. ATP synthesis requires that protons move down a concentration gradient through the base of the ATP synthetase across the membrane.

10.4
Disregarding proton leakage and assuming that the glycerol phosphate shuttle is in operation, 38 ATP would be produced from the aerobic oxidation of a glucose molecule. If the malate shuttle is in operation, only 36 ATP would be produced.

10.5
Sucrose is a disaccharide composed of glucose and fructose. As described (p. 318) the oxidation of 1 mol of glucose yields a maximum of 31 mol of ATP. Fructose, which like glucose is also partially degraded by the glycolytic pathway, also yields a maximum of 31 mol ATP. The total maximum energy yield is 62 mol of ATP per mile of sucrose.

10.6
The larger selenium atom holds its electrons less tightly than sulfur. Selenium is more easily oxidized and therefore acts as a better scavenger for oxygen than does sulfur.

10.7
The SH groups reduce hydrogen peroxide or trap hydroxyl radicals to form water. An example of a nonsulfhydryl group–containing molecule that should be capable of this activity is vitamin C or any of a number of other antioxidants (carotenoids, flavonoids, tocopherols, etc.).

10.8
Low levels of G-6-PD in combination with a high level of oxidized GSH cause high oxidative stress. Without antioxidant protection, red cell membranes become fragile, a condition that eventually causes hemolytic anemia.

10.9
The phenolic groups of both molecules are responsible for their antioxidant activity because of the ease of formation of phenoxy radicals with subsequent neutralization of electron-deficient ROS.

End-of-Chapter Questions

Review Questions

3
Processes believed to be driven by mitochondrial electron transport are ATP synthesis, the pumping of calcium ions into the mitochondrial matrix, and the generation of heat by brown fat.

6
ATP synthesis requires the presence of a proton gradient across the mitochondrial membrane. Dinitrophenol, which has an ionizable hydrogen, can diffuse across the mitochondrial membrane. As it diffuses across the membrane, protons are transported from one side of the mitochondrial membrane to the other, thereby disrupting the proton gradient and interfering with ATP synthesis.

9
a. Thioredoxin (TRX): a thiol-containing protein that reduces protein disulfides (—S—S—) to sulfhydryls (2 —SH groups), restores function to proteins in which the disulfide form is inactive, and works in conjunction with peroxiredoxins to detoxify peroxides.

b. Catalase: an enzyme that catalyzes the conversion of peroxide (H_2O_2) to water and dioxygen.

c. β-Carotene: a plant pigment molecule that acts as an absorber of light energy and as an antioxidant.

d. α-tocopherol (vitamin E): a lipid-soluble molecule that acts as a radical scavenger in membranes; a phenolic antioxidant.

e. Q cycle: the transfer of two electrons from reduced coenzyme Q (UQH_2) to cytochrome c in complex III.

12.
a. RNS: reactive nitrogen species; examples include the peroxynitrite anion ($ONOO^-$) and the radicals nitric oxide and nitrogen dioxide.

b. Glutathione: a key cellular reducing agent; the reduced form is GSH, in which the —SH is from a cysteine residue; the oxidized form is GSSG, the result of the formation of a disulfide bond between two GSH molecules.

c. Submitochondrial particles: small membranous vesicles formed when mitochondria are subjected to sonication; used in ATP synthase research.

d. Dinitrophenol: an uncoupler that collapses the proton gradient by equalizing the proton concentration on both sides of membranes, resulting in the dissipation as heat of the energy derived from food molecules.

e. Torque: twisting force; term used to describe the force generated by the c subunit rotation in ATP synthase. This torque causes the rotation of the ε and γ subunits, which interact with the catalytic sites on the β subunits to interconvert between the conformations that result in the binding of ADP and P_i, followed by the synthesis and release of ATP.

15
The major enzyme defenses against oxidative stress are provided by superoxide dismutase, catalase, and glutathione peroxidase.

18
The advantage of dioxygen over the charged oxidizing agents, such as NO_3^-, Fe^{3+}, and SO_4^{2-}, is that O_2 readily diffuses across cell membranes. Unlike CO_2, which also diffuses across cell membranes, dioxygen is highly reactive and readily accepts electrons. Finally, O_2 is found almost everywhere on Earth's surface, and as such, is much more readily accessible than oxidizing agents such as sulfur.

21
The passage of K^+ across a membrane through an ionophore such as valinomycin lowers the electrical potential gradient. As a result, some of the energy that would have been used to synthesize ATP is dissipated as heat. This explains the rise in body temperature and sweating upon treatment with valinomycin.

24
When added to actively respiring mitochondria, azide inhibits cytochrome oxidase in complex IV, resulting in the accumulation of oxygen and H^+ on the matrix side, and reduced cytochrome c in the intermembrane space.

27

NADH dehydrogenase (Complex I) oxidizes NADH to NAD^+. When Complex I is inhibited, NADH accumulates (resulting in an increased $NADH:NAD^+$ ratio) and $FADH_2$, which is oxidized in Complex II, is unaffected.

Thought Questions

30

Glutamine is readily converted to glutamate which then enters the citric acid cycle as α-ketoglutarate. The first turn of the cycle liberates one CO_2 and produces 2 NADH, 1 $FADH_2$, and 1 ATP. The four-carbon product is consumed in two further turns of the cycle, each of which generates 3 NADH, 1 $FADH_2$, and 1 ATP. The total yield is 8 NADH, 3 $FADH_2$ and 3 ATP. Assuming that each NADH yields 2.5 ATP and each $FADH_2$ yields 1.5, a grand total of 27.5 mol ATP are generated.

33

Cyanide binds to cytochrome oxidase irreversibly. All the cytochromes in the ETC become reduced, and ATP synthesis ceases. In the case of dinitrophenol, the acidic phenol disrupts the proton gradient by shuttling protons across the membrane. None of the parts of the electron transport system are irreversibly altered by dinitrophenol. Removal of the phenol allows the system to restore itself.

36

Only the final cytochrome in the ETC has an open binding site that can bind the cyanide. The other iron atoms are all fully and strongly coordinated.

39

If oxygen is used as the electron acceptor for the ETC, the difference in voltage between NADH and water is 1.14 V. The complete reduction of nitrate by NADH would be $+1.18$ V. This is almost exactly the same as for oxygen. It would be reasonable to assume that the same amount of ATP would be produced.

42

Aconitase is an important enzyme in the citric acid cycle. It catalyzes the conversion of citrate to isocitrate. If this enzyme is deactivated, the citric acid cycle slows up or shuts down. As a result, electron transport is depressed and the concentration of ROS drops.

45

Oxygen acts as a radical and removes hydrogen from the alpha position of the amino acid. This produces a radical at the alpha position of the amino acid. This radical then reacts with another oxygen molecule to form an alkylperoxyl radical, $RCOO^{\bullet}$.

CHAPTER 11
In-Chapter Questions

11.1

The product of complete hydrogenation would be hard and therefore not useful as a margarine.

11.2

When soap and grease are mixed, the hydrophobic hydrocarbon tails of the soap insert (or dissolve) into the oil droplet. The oil droplet becomes coated with soap molecules. The hydrophilic portion of the soap molecules allows the soap-oil complex to be dispersed in water.

11.3

The phospholipid of the surfactant, whichh possesses a polar head group and two hydrophobic acyl groups, disrupts some of the intermolecular hydrogen bonds of the water, thereby decreasing the surface tension.

11.4

Carvone and camphor are monoterpenes; abscisic acid is a sesquiterpene.

11.5

Bile salts are structurally similar to soap in that they contain a polar head group (e.g., the charged amino acid residue glycine) and a hydrophobic tail (the steroid ring system).

11.6

a. Simple diffusion

b. Secondary active transport or facilitated diffusion

c. Primary active transport or exchanger protein

d. Primary active transport or gated channel

e. Fat molecules (triacylglycerals) are not directly transported across cell membranes; they must be hydrolyzed first.

f. Simple diffusion.

11.7

The main stabilizing feature of biological membranes is hydrophobic interactions among the molecules in the lipid bilayer. The phospholipids in the liqid bilayer. The phospholiqids in the lipid bilayer orient themselves so that their polar head groups interact with water. Proteins in the lipid bilayer interact favorably in their hydrophobic milieu because they typically have hydrophobic amino acid residues on their outer surfaces.

11.8

The transport mechanisms discussed in the chapter fit into the following categories:
sodium channel: uniporter
glucose permease: passive uniporter
Na^+-K^+-ATPase: antiporter

End-of-Chapter Questions
Review Questions

3

a. Peripheral protein: a protein that is on the surface of a cell membrane and does not penetrate the lipid bilayer.

b. Isoprenoid: one of a class of biomolecules that contain repeating five-carbon structural units known as isoprene units; examples include terpenes and steroids.

c. Omega-3 fatty acid: a fatty acid that contains an alkene between the third and fourth carbons from the last (omega) carbon on the hydrocarbon chain.

d. Endocannabinoid: a substance produced in the body that binds to the same receptor as tetrahydrocannabinol, a psychoactive drug. Anandamine is an endocannabinoid.

e. Neutral fats: a general term for triacylglycerols, which, as a class, are uncharged.

6

a. Phospholipids perform major roles as structural components of membranes, emulsifiers, and surface active agents.

b. Plant and animal membranes contain large amounts of sphingolipids.

c. Oils serve as an important energy reserve of fruits and seeds.

d. Waxes serve as protective coatings on the surface of leaves and stems, on the fur of animals, and on the shells of insects.

e. Steroids play an important structural role in animal membranes. Certain steroids act as hormones.

f. By acting as light-trapping pigments, carotenoids play an important role in photosynthesis.

9

The protein component of plasma lipoproteins serves to solubilize the lipoproteins in the blood. It also acts as a receptor that permits binding and uptake of lipoproteins by body cells.

12

For a phospholipid to move from one side of the bilayer to the other, the polar head must move through the hydrophobic portion of the phospholipid membrane. This process requires a significant amount of energy and is therefore relatively slow.

15

Prostaglandins have major recognized roles in all the following processes: reproduction, respiration, and inflammation. (*Note*: Thromboxanes promote blood clotting, while prostaglandins inhibit platelet aggregation, thus inhibiting blood clotting. Some prostaglandins also have vasodilating or vasoconstricting activity, which may affect blood pressure.)

18

The polar head region containing the ester and amide linkages are hydrophilic; the hydrocarbon tail region is hydrophobic.

21

a. Selective permeability is provided by the combination of the phospholipid hydrophobic chains, which create a barrier to ions and polar molecules, and transport proteins, which allow only specific polar and charged substances to cross the membrane.

b. The fluidity of the cell membrane and the properties of the phospholipid components that allows rapid lateral movement within the bilayer gives membranes self-sealing capability for small disruptions. When larger lesions occur, Ca^{2+} ions flow inward, triggering a repair process that involves vesicles, motor proteins, and membrane fusion proteins.

c. Membrane fluidity is largely determined by the degree of unsaturation in the long hydrocarbon chains of the phospholipid component. Fluidity is increased with an increase in unsaturation in the hydrocarbon chains, whereas an increase in saturated hydrocarbon chains decreases fluidity.

d. "Asymmetry of biological membranes" describes the differences between the lipid and protein composition of the inner surface of a bilayer as opposed to its outer surface.

e. The active transport of ions requires transmembrane ATP-hydrolyzing enzymes to provide the energy to fuel the movement of ions across the membrane and against a concentration gradient.

24

a. In primary active transport, ATP provides energy to transport a substance *against* a concentration gradient. The Na^+-K^+ ATPase pump is a primary transporter, since it transports substances across a membrane from lower to higher concentration. Secondary active transport uses the concentration gradients that were generated by primary active transport to drive the transport of a different substance against its concentration gradient. An example of secondary active transport is the use of the gradient created by the Na^+-K^+ ATPase pump to transport glucose.

b. Passive transport is the general term for the transport of substances across a membrane by diffusion. Since passive transport occurs *down* (or *with*) a concentration gradient (i.e., from an area of high concentration to low concentration), no energy is required. Facilitated diffusion is the passive transport of substances such as ions or polar molecules that are unable to cross the membrane alone and require the presence of a protein channel or carrier. For example, nonpolar organic molecules (such as steroid hormones) and carbon dioxide cross cell membranes by passive diffusion. The red blood cell glucose transporter is an example of a carrier, and Na^+ may diffuse through a membrane only through specific Na^+ channel proteins.

c. Both carrier-mediated and channel-mediated transport are examples of facilitated diffusion. In channel-mediated transport, an integral protein forms a channel through which a specific substance may pass. In carrier-mediated transport, the substance to be transported binds to the carrier protein and causes a conformational change. This change results in the substance crossing the membrane, where it is then released by the carrier protein. See (b) for examples.

27

Like saturated fatty acids, *trans* fatty acids, can have fully extended chains that pack together well, resulting in an increased melting point relative to unsaturated fatty acids that contain *cis*-alkenes.

30

In a nonpolar solvent, a phospholipid bilayer would be disrupted. There may be formation of inverted micelles, with the polar head groups in the center and the nonpolar tails facing outward, or an inverted bilayer, with the polar head groups in the center and the nonpolar tails facing outward.

Thought Questions

33

The fluidity of the membrane allows for flexible movement. Any breaks that do occur expose the hydrophobic core of the membrane to an aqueous environment. Hydrophobic interactions spontaneously move the broken ends together and, in combination with certain other components of cell membrane resealing mechanisms (e.g., cytoskeleton and calcium ions), the membrane reseals.

36

The carbohydrate portion of the glycolipid can form hydrogen bonds with the water. This carbohydrate is the polar group, and it is analogous to the charged portion of the phospholipid.

39

The ordered water molecules surrounding each phospholipid molecule are released from the polar heads. Order is lost and entropy increases.

42

High temperatures destabilize the membranes of prokaryotes adapted to "normal" temperatures. The membranes of thermophilic prokaryotes are resistant to high temperatures because they contain longer, saturated fatty acyl groups that pack closely together and lipids such as sterols or similar molecules that act as stiffening agents. In addition, they also contain glycero-ether lipids, which are difficult to hydrolyze.

45

The myelin sheath is primarily composed of hydrophobic molecules that are not good conductors of electricity. They serve to insulate electrically active nerve cells from the highly conductive aqueous environment.

CHAPTER 12

In-Chapter Questions

12.1

The triacylglycerols are emulsified in the small intestine by bile salts. They are then digested by lipases, the most important of which is pancreatic lipase. The products, fatty acids and monoacylglycerol, are transported into enterocytes and reconverted to triacylglycerol. Triacylglycerol is subsequently incorporated into chylomicrons, which are then transported into lymph via exocytosis and finally into the bloodstream for transport to the fat cells.

12.2

If there is a connection between female sex hormone levels and VLDL secretion, injection of estrogen into a male rat should elicit a timely and measurable increase in VLDL secretion. This process requires a concomitant increase in the synthesis of the components of VLDL, that is, apoproteins, triacylglycerols, phospholipid, and cholesterol. FABP synthesis should increase in response to the increased intracellular fatty acid concentrations.

12.3

a. Phospholipid

b. Acyl-CoA

c. Carnitine

12.4

Unlike the oxidation of glucose to form pyruvate, fatty acid oxidation, which involves the citric acid cycle and the electron transport system, cannot operate in the absence of O_2.

12.5

The yield from the oxidation of stearyl-CoA is calculated as follows:

8 FADH$_2$ × 1.5 ATP/FADH$_2$ =	12 ATP
8 NADH × 2.5 ATP/NADH =	20 ATP
9 Acetyl-CoA × 10 ATP/Acetyl-CoA =	90 ATP
	122 ATP

Two ATP are required to form stearyl-CoA from stearate to give a total of 120 ATP.

12.6

Propionyl-CoA can be reversibly converted to succinyl-CoA, an intermediate in the citric acid cycle. Oxaloacetate, a downstream intermediate of this cycle can be converted to PEP. PEP is then converted to glucose via gluconeogenesis. Adipic acid undergoes one round of β-oxidation to yield acetyl-CoA and succinyl-CoA. As just described, succinyl-CoA is sequentially converted to oxaloacetate, PEP, and then to glucose.

12.7

Because steroids inhibit the release of arachidonic acid, their use shuts down the synthesis of most if not all eicosanoid molecules, hence their reputation as potent anti-inflammatory agents. Aspirin inactivates cyclooxygenase and prevents the conversion of arachidonic acid to PGG$_2$ the precursor of prostaglandins and thromboxanes. Aspirin is not as effective an anti-inflammatory agent as the steroids because it shuts down only a portion of eicosanoid synthetic pathways.

12.8

Following the hydrolysis of sucrose, both monosaccharide products enter the bloodstream and travel to the liver, where fructose is converted to fructose-1-phosphate. Recall that the conversion of fructose-1-phosphate to glyceraldehyde-3-phosphate bypasses two regulatory steps. Consequently, more glycerol- phosphate and acetyl-CoA (the substrates for triacylglycerol synthesis) are produced. High blood glucose concentrations that result from this consumption of excessive amounts of sucrose trigger the release of larger than normal amounts of insulin. One of the functions of insulin is to promote fat synthesis.

12.9

a. β-Hydroxybutyrate is a product of ketone body metabolism

b. Malonyl-CoA is the product of the reaction of acetyl-CoA and carboxybiotin that occurs during fatty acid systhesis.

c. Biotin is a carrier of CO_2 in fatty acid synthesis and several other reactions.

d. Acetyl ACP delivers acetate to the synthetic machinery of fatty acid synthesis.

12.10

The higher activity of HMG-CoA reductase in obese patients in combination with a high-calorie diet increase the synthesis of cholesterol.

12.11

The labeled carbon atoms in cholesterol are as follows:

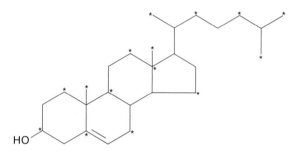

Cholesterol

12.12

The structure of cortisol differs from that of cortisone in that the C-11 hydroxyl group is replaced by a carbonyl group. Glycyrrhizic acid inhibits 11-β-hydroxysteroid dehydrogenase, which prevents the deactivation of cortisol. Cortisone is administered to Addison's disease patients because it is converted to cortisol by 11- β-hydroxysteroid dehydrogenase a reversible enzyme.

End-of-Chapter Questions

Review Questions

3

Peroxisomes have β-oxidation enzymes that are specific for long-chain fatty acids, whereas mitochondria possess enzymes that are specific for short and moderate chain length fatty acids. In addition, the first reaction in the peroxisomal pathway is catalyzed by a different enzyme than the mitochondrial pathway. The FADH$_2$ produced in the first peroxisomal reaction donates its electrons to O_2 directly (forming H_2O_2 instead of UQ, as in mitochondria). The processes are similar in that acetyl-CoA is derived from the oxidation of fatty acids.

6

In the short term, hormones alter the activity of preexisting regulatory enzyme molecules. For example, the binding of glucagon inhibits acetyl-CoA carboxylase. Long-term effects of hormones usually involve changes in the pattern of enzyme synthesis in target cells. For example, insulin promotes the synthesis of the enzymes involved in lipogenesis (e.g., acetyl-CoA carboxylase and fatty acid synthase).

9

Because of the presence of a methyl substituent on the β-carbon, the fatty acid first undergoes one cycle of α-oxidation. The resulting molecule, now shorter by one carbon atom, then undergoes one cycle of β-oxidation. The products of this latter process are two molecules of propionyl-CoA.

12

Enoyl-CoA isomerase converts the naturally occurring *cis* double bond at Δ^3 to a *trans* double bond at Δ^2, the correct position for the next round of β-oxidation.

15

a. Autoantibodies bind to surface antigens on the patient's own cells as if they were foreign. This binding results in an autoimmune disease.

b. Under low-glucose conditions, such as starvation or uncontrolled diabetes, ketone bodies (acetone, acetoacetate, or β-hydroxybutyrate, produced in the liver from excess acetyl-CoA) allow cardiac and skeletal muscle to convert excess acetyl-CoA from the β-oxidation of fatty acids into energy. During prolonged starvation, the brain may also use ketone bodies as an energy source.

c. Biotransformation converts toxic substances into less toxic metabolites. For example, toxic molecules that are hydrophobic may be converted into water-soluble derivatives for excretion.

d. Phase I reactions of biotransformation involve oxidoreductases and hydrolases to convert hydrophobic substances into more polar molecules.

e. During fatty acid synthesis, ACP (acyl carrier protein) binds acyl group intermediates through a thioester linkage with a phosphopantetheine group. (This is analogous to the function of CoASH in the β-oxidation of fatty acids.)

18

In Phase I of biotransformation, oxidoreductases and hydrolases catalyze reactions to increase the polarity of hydrophobic molecules. In Phase II, the water solubility of toxic molecules is dramatically improved by means of the conjugation of their functional groups with substances such as glucuronate, glutamate, sulfate, or glutathione. Phase III is the excretion of these biotransformed molecules.

21

All the lipid molecules are originally synthesized from the isoprene units in isopentenyl pyrophosphate molecules. Steroid and terpene molecules are assembled by head-to-tail condensation of these groups.

24

Conjugation reactions in the biotransformation process serve to make a toxic hydrophobic molecule much more water soluble so that it can be excreted.

27

During periods of prolonged starvation, when there is an excess of acetyl-CoA (from the β-oxidation of fatty acids) and very low

reserves of glucose, ketone bodies are formed to be metabolized for energy. When the concentration of acetoacetate is high, it decarboxylates to form acetone, which may be detected on the breath.

30

Insulin promotes triacylglycerol synthesis and the storage and uptake of fatty acids. Specifically, insulin inactivates hormone-sensitive lipase to prevent the hydrolysis of fats to glycerol and fatty acids, stimulates the release of VLDL from the liver, and activates lipoprotein lipase synthesis and transport to the endothelial cells serving fat and muscle tissue.

33

Since the $^{14}CO_2$ that is added to acetyl-CoA is removed in the reaction of malonyl-ACP with acetyl synthase to form acetoacetyl-ACP, no ^{14}C label appears in the eventual fatty acid products.

36

The β-oxidation of oleic acid results in approximately 118.5 ATP equivalents:
(7 FADH$_2$)(1.5 ATP/FADH$_2$) = 10.5 ATP
(8 NADH)(2.5 ATP/NADH) = 20 ATP
(9 Acetyl-CoA)(10 ATP/acetyl-CoA) = 90 ATP
Formation of oleoyl-CoA from oleic acid = – 2 ATP

39

Fatty acids are activated for β-oxidation and transported into the mitochondrial matrix by conversion to their CoASH derivatives. β-Oxidation takes place in the mitochondrial matrix. To cross both the outer and the inner mitochondrial membranes, the CoASH derivatives are needed. First, acyl-CoA synthase catalyzes the formation of the fatty acyl-CoA, which is released into the intermembrane space. The acyl group is then transferred to carnitine for transport across the inner membrane into the matrix.

Thought Questions

42

The potential consequences of faulty regulation include the creation of some level of futile cycling in which energy is wasted and the requirements of the cell for fatty acid synthesis and energy generation are compromised.

45

Severe dieting stimulates massive lipolysis. The large amounts of acetyl-CoA generated in this process trigger a vastly increased synthesis of the ketone bodies. When present in such large amounts, the ketone bodies overwhelm the buffering capacity of the blood and its pH falls.

48

In periods of fasting, blood glucose levels fall and glucagon and epinephrine are released. These hormones then bind to their respective adipocyte plasma membrane receptors. This binding initiates a cascade (cAMP activates protein kinase, which in turn activates hormone-sensitive lipase) that results in the release of fatty acids and glycerol into the blood.

51

The molecule is phytanic acid. The first phase of its metabolism is an α-oxidation. The product is pristanic acid, which then undergoes β-oxidation. Refer to Figure 12.11.

54

The ^{14}C label will appear in the mevalonate molecule as indicated by the asterisk:

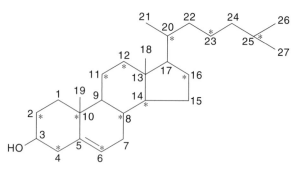

CHAPTER 13
In-Chapter Questions

13.1

a. LHCII is light-harvesting complex II; consists of a transmembrane protein that binds numerous chlorophyll a and chlorophyll b molecules and carotenoids; a major component of thylakoid membrane.

b. Lutein is a type of light-harvesting pigment found within the thylakoid membrane.

c. PSII is photosystem II; a complex composed of proteins and pigment molecules located in the stacked regions of thylakoid membrane that oxidizes water molecules and donates energized electrons to electron carriers that eventually reduce photosystem I.

d. MSP is manganese-stabilizing protein; the oxygen-evolving component of PSII; located in the stacked regions of thylakoid membrane.

e. CF_0CF_1 is the ATP synthase component of thylakoid membrane that penetrates into stroma.

f. P700 is a special pair of chlorophyll molecules within photosystem I that absorb light at 700 nm.

g. P680 is a special pair of chlorophyll molecules within photosystem II that absorb light at 680 nm.

h. O_2 generation occurs in the oxygen-evolving complex, which contains MSP and a critical tyrosine residue located in photosystem II. The electrons generated in the reduction of water are used to replace those transferred away from PSII when light energy is absorbed.

13.2

The energy of a photon is proportional to its frequency. Blue light has a higher frequency than green light and therefore has higher energy.

13.3

The presence of antenna pigments allows the light-harvesting systems of chloroplasts to collect energy from a wider range of frequencies than those absorbed by the chlorophylls. Becausee their absorption spectra overlap, the energy absorbed by the antenna pigments is quickly transferred to the critical chlorophylls of PSI and PSII.

13.4

Excessive light promotes the formation of ROS, whichh damage proteins such as D_1. β-Carotene is an antioxidant that prevents some of this damage.

13.5

a. Plastocyanin is a component of the cytochrome b_6f complex; a copper-containing protein that accepts electrons form plastoquinone.

b. β-Carotene is a carotenoid pigment that protects chlorophyll molecules from ROS.

c. Ferredoxin is a mobile, water-soluble protein that donates electrons to a flavoprotein called ferredoxin-NADP oxidoreductase.

d. Plastoquinone is a component of photosystem II that accepts electrons from pheophytin a to become plastoquinol.

e. Pheophytin a is a molecule similar in structure to chlorophyll that is a component of the electron transport pathway between PSII and PSI.

f. Lutein is a carotenoid that is a component of light-harvesting complexes.

13.6

The process of phosphorylation changes the conformation and the binding characteristics of the complex and thereby alters its capacity to bind to PSII.

13.7

Uncoupler molecules stop ATP synthesis in both mitochondria and chloroplasts because they destroy proton gradients.

13.8

Soaking the chloroplasts in an acidic solution lowers their internal pH. Treatment with base establishes an artificial pH gradient across the thylakoid membrane. As protons flow down this gradient ATP is synthesized.

13.9

Of the herbicides discussed, paraquat and DCMU are most hazardous to humans. Paraquat generates free radicals that can attack cell components. DCMU poisons the electron transport complex.

13.10

Reactions of the Calvin cycle strongly resemble the pentose phosphate pathway.

13.11

The hydrolysis of glycerate-1,3-bisphosphate generates 1 mol of ATP. Recall that aerobic respiration is stimulated by relatively high ADP concentrations and inhibited by relatively high ATP concentrations. Any measurable increase in ATP concentration has the effect of depressing aerobic respiration. Also recall that ATP is an inhibitor of PFK-1 and pyruvate kinase, enzymes required to channel carbon skeletons into the citric acid cycle.

13.12

Photorespiration, a wasteful process that evolves CO_2, is favored by low concentrations of CO_2 and high concentrations of O_2 and high temperatures. In C4 plants, CO_2 is incorporated into oxaloacetate at night within specialized mesophyll cells, when the risk of water loss is lower. In the morning the CO_2 released within bundle sheath cells is incorporated into sugar molecules. Because the CO_2 concentration is high within these cells compared to O_2 concentration, photorespiration is avoided. In CAM plants, CO_2 is also incorporated into oxaloacetate at night within mesophyll cells. Oxaloacetate is then reduced to form malale. In the morning, light stimulates the conversion of malate into pyruvate and CO_2, As in C4 plants, a high cellular concentration of CO_2 inhibits photorespiration.

End-of-Chapter Questions
Review Questions

3

The three primary photosynthetic pigments are (1) the chlorophylls, which absorb blue-violet and red wavelengths of light; (2) the carotenoids, which serve as antenna pigments and protect from ROS; and (3) the xanthophylls, which also serve as antenna pigments.

6

Excited molecules can return to the ground state by several means including (1) fluorescence (light is absorbed at one wavelength and emitted at a longer wavelength), (2) resonance energy transfer (energy is transferred to neighboring chromophores with overlapping absorption spectra), (3) oxidation-reduction (an excited electron returns to its ground state by reducing another molecule), (4) radiationless decay (an excited molecule returns to the ground state and loses its excess energy as heat). Of these processes, oxidation-reduction and resonance energy transfer are important in photosynthesis. Oxidation-reduction is important in the transfer of electrons through the electron carriers such as the quinones and iron-sulfur complexes. Resonance energy transfer passes light energy to chlorophyll from accessory pigments.

9

The net production of the dark, or light-independent, reactions of photosynthesis is one molecule of glyceraldehyde-3-phosphate. See page 486 for the reactions of the Calvin cycle.

12

The oxygen-evolving system is referred to as a clock because it involves five oxidation-reduction states that must be completed in order.

15

The optimal wavelengths for photosynthesis appear to be 400–500 nm and 600–700 nm.

18

a. Carbon fixation: the mechanism by which inorganic CO_2 is incorporated into organic molecules.

b. C3 metabolism: a photosynthetic pathway in which the first stable product is the three-carbon molecule glycerate-3-phosphate.

c. C4 metabolism: a photosynthetic pathway in which PEP is carboxylated to form oxaloacetate, a four-carbon molecule; C4 metabolism minimizes photorespiration.

d. CAM (crassulacean acid metabolism): a pathway that allows CO_2 uptake through stomata only at night and avoids water loss by closing stomata during daylight hours.

e. Phytochrome: a photoreceptor that regulates a number of plant growth and development processes, it is sensitive to red- and far-red light.

21

The chloroplast contains the thylakoid membranes in both appressed (stacked) and unappressed format. The ATPase is oriented in the membrane so that ATP synthesis is always exposed to the stromal compartment. LHCII and PSII are richly concentrated in the appressed regions to maximize light collection and electron transfer. PSI, which should not receive its excitation energy directly from PSII, is physically separated from it in the unappressed regions. Electron replacement of PSI and PSII is mediated by mobile carriers, so physical separation is not a problem.

24

The radio labeled ^{14}C in $^{14}CO_2$ becomes the first carbon of one of two glyceraldehyde-3-phosphate molecules produced after step 3 of the Calvin cycle. From there, the ^{14}C may be incorporated into ribulose-1,5-bisphosphate or used in the biosynthesis of starch, sucrose, or other metabolites.

27

The oxygen atoms in glucose molecules originate in CO_2.

30

When H_2S is the source of hydrogen atoms, the end products of photosynthesis are glucose and elemental sulfur. (The carbon and oxygen atoms in glucose are from CO_2 molecules.)

Thought Questions

33

Conjugation is a system of alternate double and single bonds. When light with sufficient energy strikes the π electrons of a conjugated system (or any double bond), an electron is promoted from the ground state to a higher energy state, referred to as the excited state. Conjugation lowers the energy difference between the ground state and the excited state; hence photons of lower energy are capable of achieving this transition.

36

High oxygen concentrations promote photorespiration.

39

Photorespiration is carried out by rubisco, an enzyme that has both oxygenase and carbon dioxide fixation activities. Under conditions of high carbon dioxide concentration, the oxygenase functions are repressed and photorespiration slows down.

42

The herbicides kill marine photosynthesizing organisms, thereby depressing worldwide O_2 production.

45

The spectrum of light that reaches the surface of the earth is richer in the blue region than in the ultraviolet. Also, pigments that absorb in the ultraviolet would be damaged or destroyed more easily by the more powerful ultraviolet.

48

Under conditions of high carbon dioxide concentration, the oxygenase activity of rubisco is depressed and photorespiration is decreased, and the amount of carbon fixed by photosynthesis will correspondingly increase. Since C4 plants do not have as much photorespiration, this effect would be less. Therefore, C3 plants would derive the most benefit.

CHAPTER 14
In-Chapter Questions

14.1

a. CH_3NH_2

b. NH_3

c. CH_3CH_3

14.2

Refer to Figure 14.1.

14.3

As their names suggest, hemoglobin and leghemoglobin are proteins in the globin superfamily. Recall that hemoglobin is an oxygen transport protein that contains a heme group, which binds reversibly with O_2. The heme in leghemoglobin also binds to O_2.

The function of leghemoglobin, the sequestration of oxygen molecules, can be deduced from the irreversible inactivation of the nitrogenase complex in root nodules by O_2.

14.4

The transamination products are as follows:

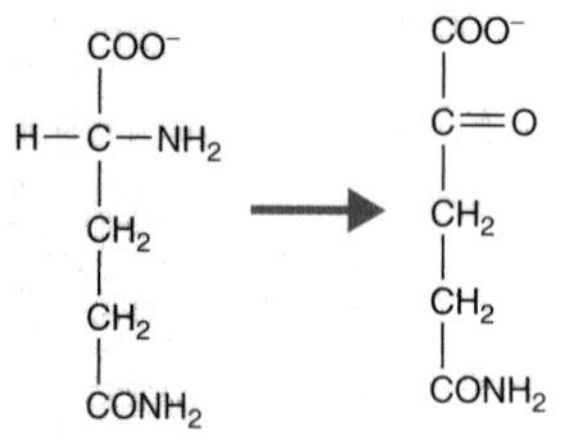

Glutamine
(a)

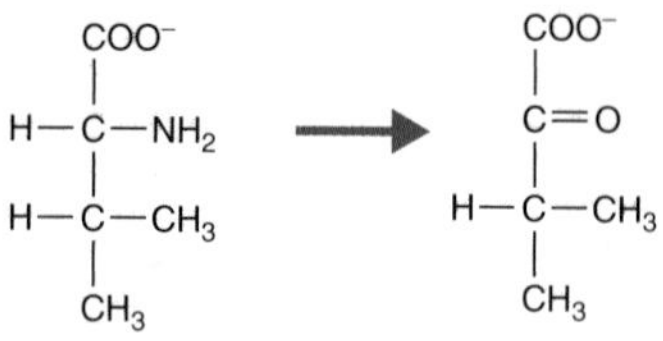

Isoleucine
(b)

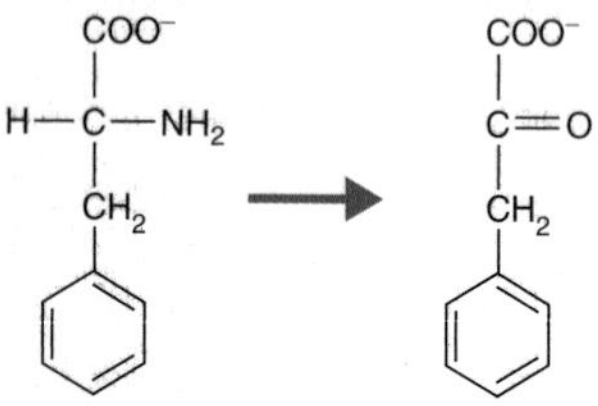

Phenylalanine
(c)

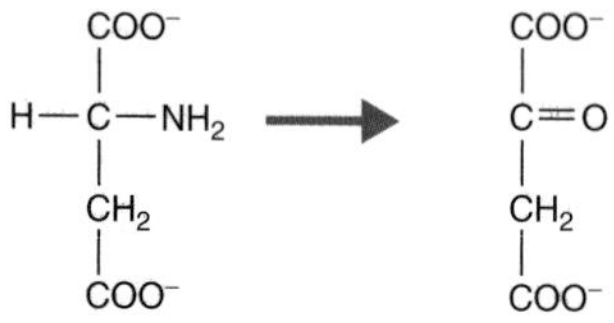

Aspartate
(d)

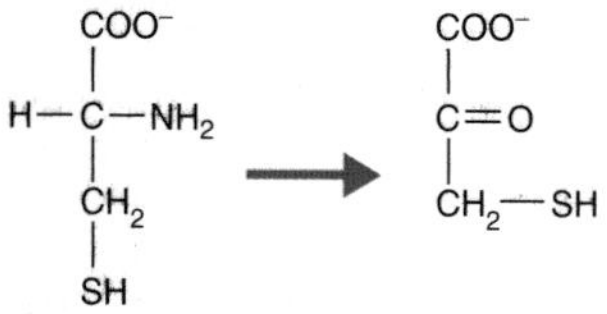

Aspartate
(e)

14.5

a. Glutamic acid
b. Tryptophan
c. Histidine
d. Tyrosine
e. Proline

14.6

Because of its close structural similarity to folic acid, methotrexate is a competitive inhibitor of the enzyme dihydrofolate reductase. (Recall that this enzyme converts folic acid to its biologically active form, THF.) Rapidly dividing cells require large amounts of folic acid. Methotrexate prevents the synthesis of THF, the one-carbon carrier required in nucleotide and amino acid synthesis. It is therefore toxic to rapidly dividing cells, especially those of certain tumors and normal cells that divide frequently such as hair and GI tract cells.

14.7

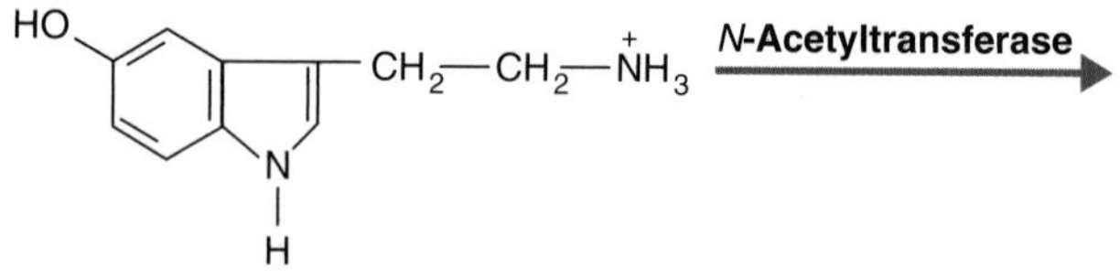
N-Acetyltransferase

Serotonin

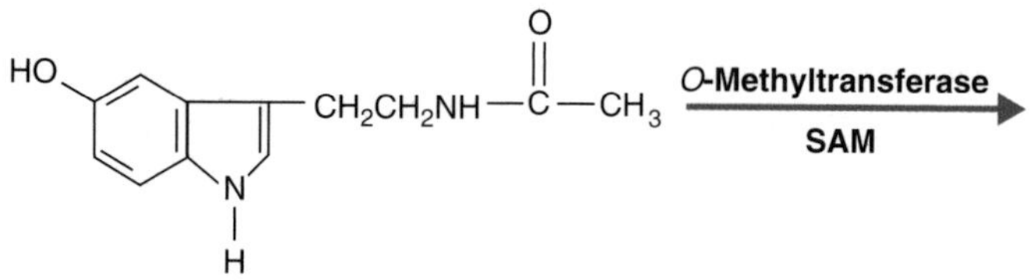
O-Methyltransferase
SAM

5-Hydroxy-N-acetyltryptamine

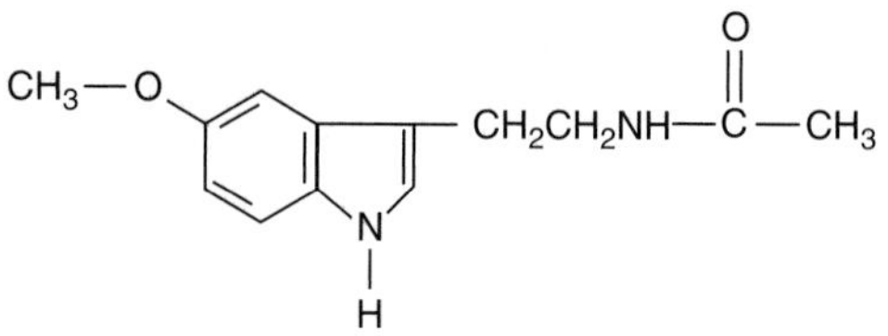

Melatonin

14.8

The reaction sequence is as follows:

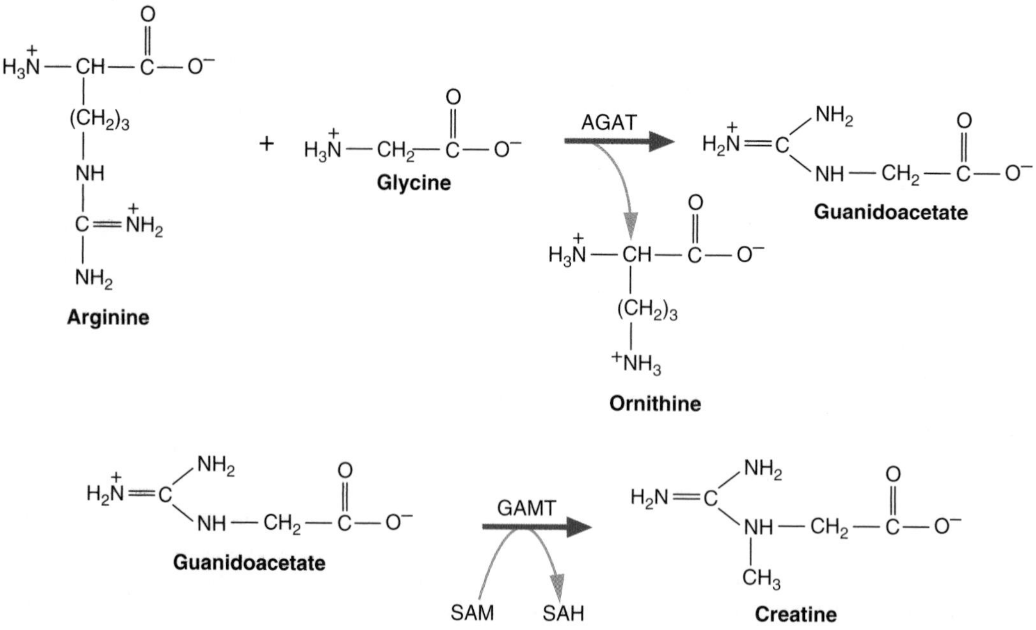
Arginine
Glycine
AGAT
Guanidoacetate
Ornithine
Guanidoacetate
GAMT
SAM
SAH
Creatine

Note that the secondary amino nitrogen alkylates more easily than the primary amino nitrogen in the guanidoacetate molecule.

14.9

a. Succinyl-CoA is condensed with glycine to produce δ-aminolevulinate (ALA), an intermediate in the heme biosynthetic pathway.

b. This molecule is the initial product of the condensation of succinyl-CoA and glycine to form ALA. ALA is formed when the carboxyl group is released.

c. Carbamoyl phosphate reacts with aspartate to form carbamoyl aspartate, a precursor of orotate, and subsequently UMP.

d. Phosphoribosylpyrophosphate reacts with orotate to give orotidine-5′-phosphate, a precursor of UMP.

End-of-Chapter Questions

Review Questions

3

While nitrogen incorporation into organic molecules is thermodynamically favored, the conditions in the biosphere (T, P, and pH) are such that transitions have low kinetic probability. Only a few organisms possess the machinery to surpass the energy barrier lo nitrogen compound synthesis.

6

Glutamate is synthesized from α-ketoglutarate by two means: (1) transamination, catalyzed by the aminotransferases (pyridoxal phosphate is a required coenzyme) and (2) direct amination, catalyzed by glutamate dehydrogenase. NADPH provides the reducing power for this reaction.

9

a. Blood-brain barrier: the specialized capillary endothelial cells that regulate the passage of substances from the blood into the central nervous system.

b. Catecholamine: one of a class of neurotransmitters derived from tyrosine; includes dopamine, norepinephrine, and epinephrine.

c. Tetrahydrobiopterin (BH_4): a cofactor required in hydroxylation reactions of aromatic amino acids; similar to folic acid.

d. L-DOPA (3,4-dihydroxyphenyl alanine): a neurotransmitter and catecholamine precursor produced by the hydroxylation of tyrosine.

e. Seasonal affective disorder: clinical depression triggered by decreased daylight in the autumn and winter.

12

a. Alanine belongs to the pyruvate family.

b. Phenylalanine belongs to the aromatic family.

c. Methionine belongs to the aspartate family.

d. Tryptophan belongs to the aromatic family.

e. Histidine belongs to the histidine family.

f. Serine belongs to the serine family.

15

Glutathione is involved in the synthesis of DNA, RNA, and the eicosanoids. It is also utilized as a reducing agent that protects cells from radiation and oxygen, and as a conjugating agent for environmental toxins. Glutathione is also believed to play a role in amino acid transport.

18

a. Heme oxygenase (HO): an ER enzyme that catalyzes the production of CO in the catabolism of heme.

b. Orotic aciduria: a rare genetic disease in which defective UMP synthase results in excessive urinary excretion of orotic acid.

c. Thioredoxin: a small redox protein with two —SH groups that mediates the transfer of electrons from NADPH to ribonucleotide reductase.

d. ALAS [γ-aminolevulinate (ALA) synthase]: an enzyme that catalyzes the first step of heme synthesis, condensing an enzyme in the biosynthesis of heme.

e. Creatine: a nitrogen-containing organic acid found primarily in muscle and brain; as phosphocreatine, it provides short-term storage of high-energy phosphate by transferring phosphate to ADP when energy demands are high.

21

The reactions involved in the synthesis of the purine are outlined in Figure 14.27.

24

Glutamate plays a central role in amino acid metabolism because it and α-ketoglutarate constitute one of the most common α-amino acid/α-keto acid pairs used in transamination reactions. Glutamate also serves as a precursor of several amino acids and as a component of polypeptides. Glutamine serves as the amino group donor in numerous biosynthetic reactions (e.g., purine, pyrimidine, and amino sugar synthesis), as a safe storage and transport form of ammonia, and as a component of polypeptides.

27

Reaction of pyridoxyl phosphate with alanine:

(a)

Pyridoxyl phosphate

Pyridoxamine phosphate

(b) **Reaction of Pyridoxamine phosphate with α-Ketoglutarate**

Pyridoxamine phosphate   **α-Ketoglutarate**

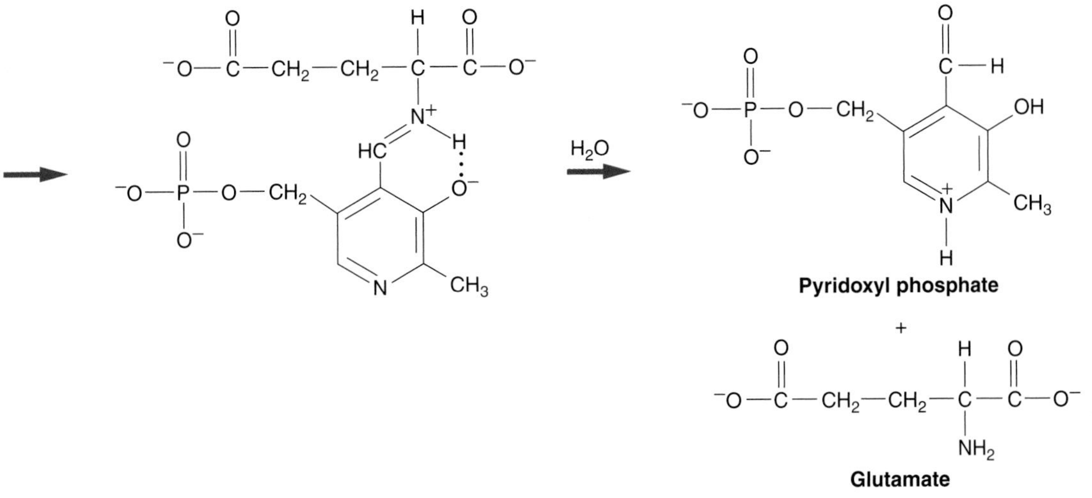

Pyridoxyl phosphate

Glutamate

30

Atmospheric nitrogen contains a nonpolar triple bond that must be broken to reduce N_2 to NH_3. Also, diatomic oxygen is a diradical. The inherent instability of radicals results in a molecule that is much more reactive (e.g., ready to accept electrons in a redox reaction), than the stable $N \equiv N$ triple bond, which contains no radicals.

33

The energy provided by ATP hydrolysis in the conversion of glutamate to glutamine makes the reaction energetically favorable. Note that this reaction also provides a means by which brain cells can decrease ammonia concentrations.

Thought Questions

36

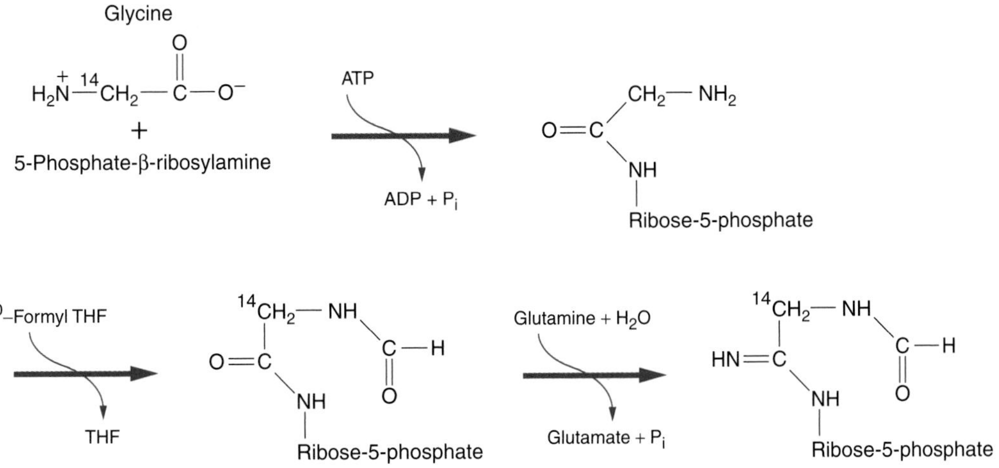

39

Glutamate is an excitatory neurotransmitter with stimulating effects on neurons that regulate bodily functions such as blood pressure and body temperature. Individuals who display symptoms after consuming monosodium glutamate apparently possess efficient mechanisms for transporting glutamate across the blood-brain barrier.

42

Radiolabel both the carbon and nitrogen of aspartate and glutamate. If the amino acid is used in the ring assembly, then both carbon and nitrogen should bear the label. If nitrogen exchange takes place, only the carbon atom will be labeled. In addition, isolating each intermediate allows the origin of each atom to be traced.

45

During the reaction catalyzed by serine hydroxymethyeltransferase the radiolabeled carbon atom of serine enters the THF pool as

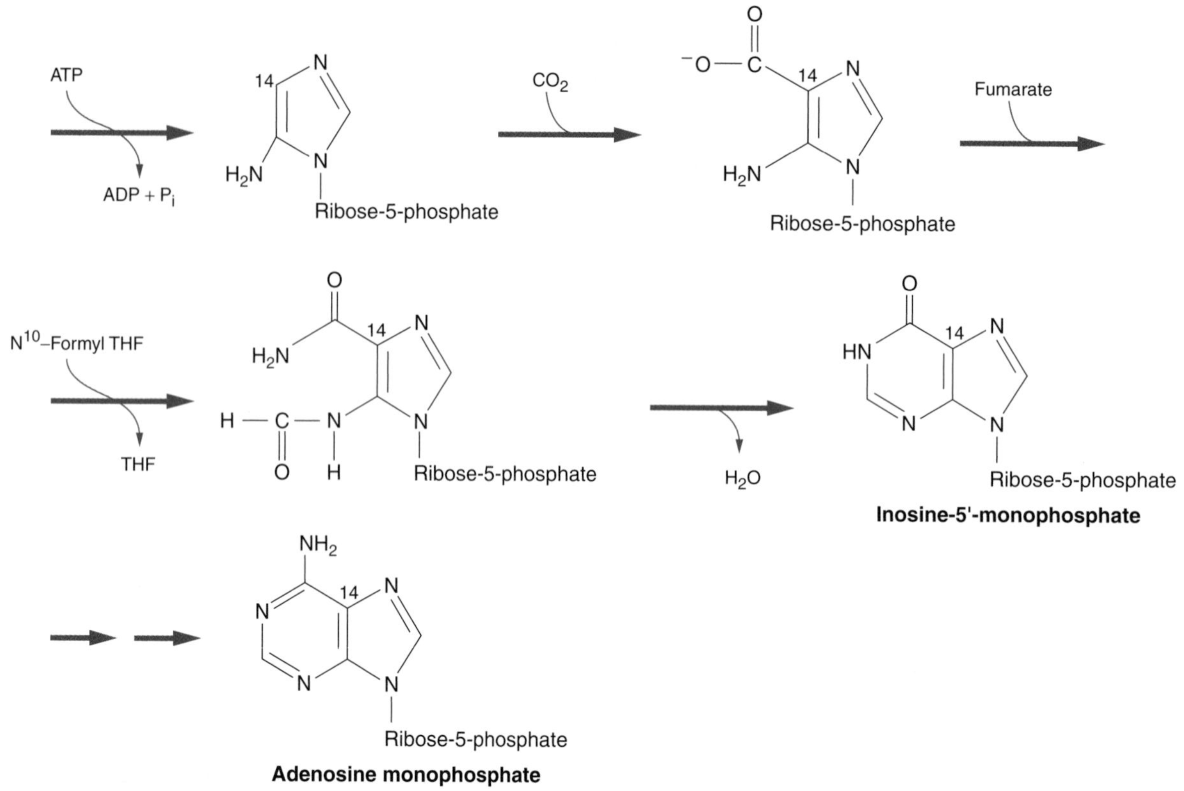

Inosine-5'-monophosphate

Adenosine monophosphate

N^5,N^{10}-methylene THF. Because N^5,N^{10}-methylene THF is reversibly converted to N^5,N^{10}-methynyl THF and N^{10}-formyl THF (the coenzyme used in purine synthesis) some radiolabeled carbon atoms will enter the purine synthetic pathway. Because N^{10}-formyl THF is a required coenzyme in the reactions in which 5'-phosphoribosyl-N-formyl-glycinamide and 5-phosphoribosy-1–4-carboxamide-5- foramidoimidazole are synthesized ^{14}C will appear as C-2 and C-8 of the purine ring (shown below).

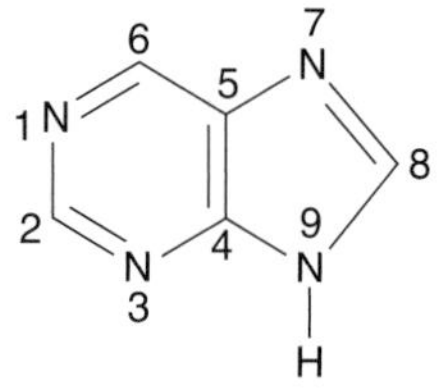

48

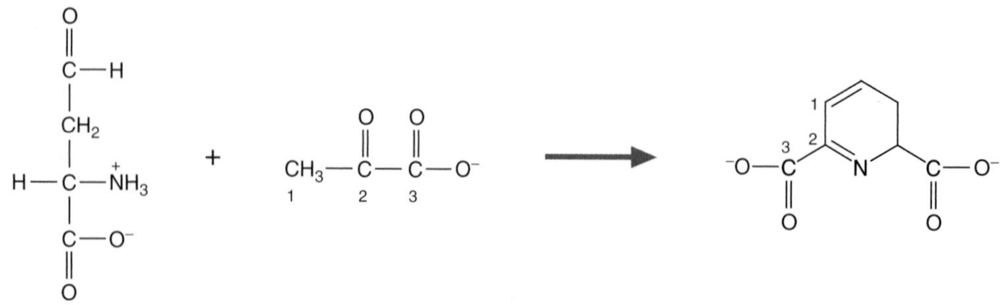

Aspartate β-semialdehyde   **Pyruvate**   **Dihydropicolinate**

CHAPTER 15

In-Chapter Questions

15.1
In newborn animals arginine will be an essential amino acid if the urea cycle is not yet fully functional.

15.2
Certain intestinal bacteria can release ammonia from urea molecules that diffuse across the membrane into the intestinal lumen. Treatment with antibiotics kills these organisms, thereby reducing blood ammonia concentration.

15.3

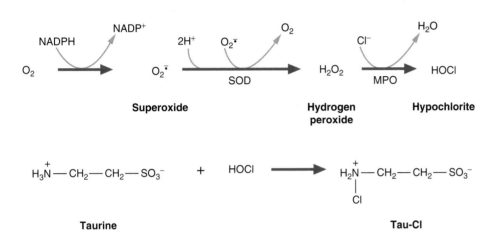

Crysteine Sulfate

$$^-O_2S\text{—}CH_2\text{—}CH_2\text{—}\overset{+}{N}H_3 + CO_2$$

Oxidation

$$^-O_2S\text{—}CH_2\text{—}CH_2\text{—}\overset{+}{N}H_3$$

$$^-O_3S\text{—}CH_2\text{—}CH_2\text{—}\overset{+}{N}H_3$$

Taurine

15.4

The reactions are as follows:

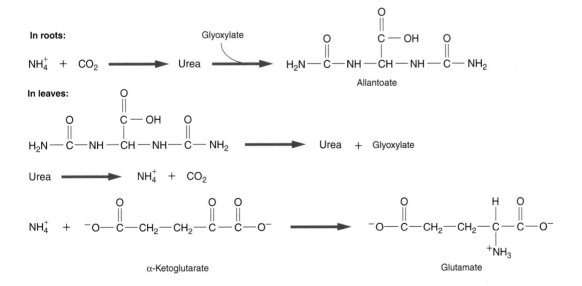

Superoxide　　　　　**Hydrogen peroxide**　　**Hypochlorite**

Taurine　　　　　　　　　　　　　　**Tau-Cl**

15.5

a. Dopamine is inactivated in an oxidation reaction catalyzed by MAO to form 3,4-dihydroxyphenylacetaldehyde.

b. Serotonin is inactivated in a two-step pathway: after serotonin is oxidized by MAO to form 5-hydroxyindole-3-acetaldehyde, the product is further oxidized by MAO to form 5-hydroxyindole-3-acetate.

c. Epinephrine is inactivated by MAO to form 3,4-dihydroxyphenylglycolaldehyde.

d. Norepinephrine is oxidized by MAO to form 3,4-dihydroxyphenylglycolaldehyde.

15.6

Acetylcholine is normally degraded rapidly by cholinesterase. Drugs that block the action of cholinesterase prevent this hydrolysis. Consequently, acetylcholine molecules remain in the synaptic cleft for an extended time. There they can rapidly and reversibly bind and rebind to a reduced number of functional acetylcholine receptors. This process promotes the depolarization of the muscle cells.

15.7

Gout is caused by high levels of uric acid. Animals that do not suffer from gout possess the enzyme urate oxidase, which converts uric acid to allantoin. Unlike uric acid, which is relatively insoluble in blood, allantoin readily dissolves and is easily excreted.

15.8

15.9

a. Urea is formed from ammonia, CO_2, and aspartate in the urea cycle.

b. Uric acid is the oxidation product of purines.

c. b-Alanine is produced in the degradative pathway of pyrimidines

15.10

Suggested catabolic reactions of β-alanine and β-aminoisobutyrate:

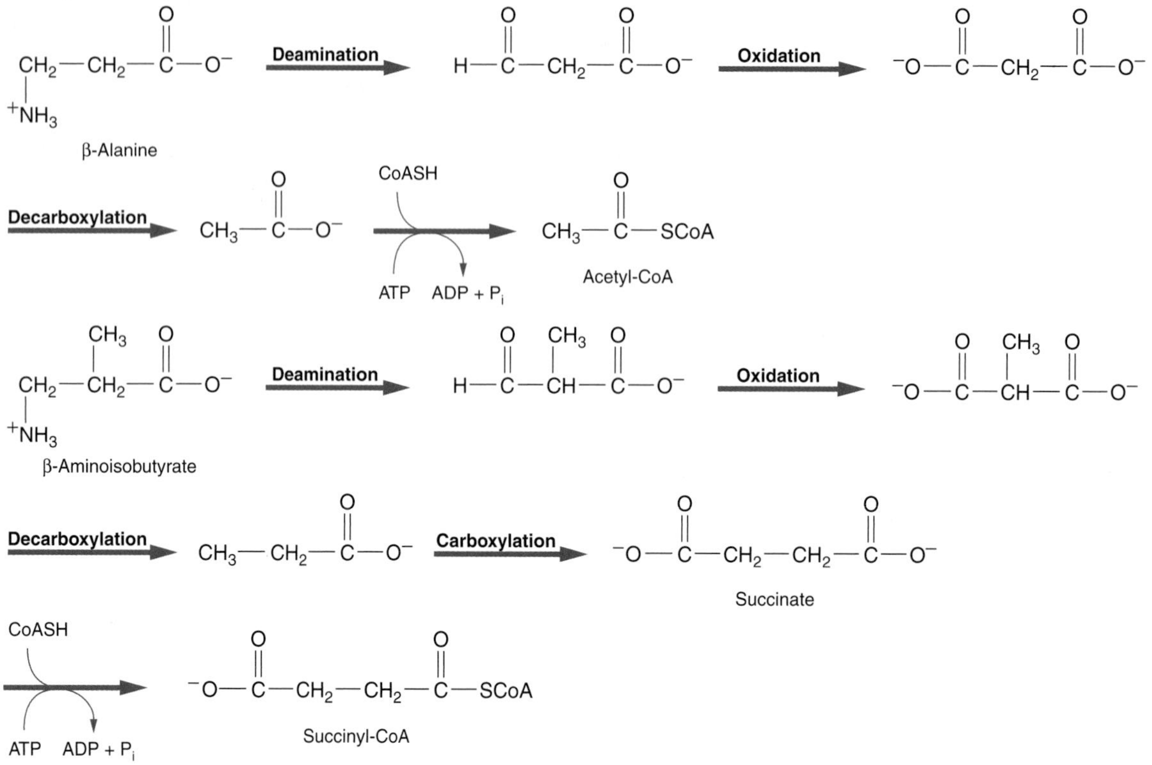

End-of-Chapter Questions

Review Questions

3

The process of protein turnover promotes metabolic flexibility, protects a cell from the accumulation of abnormal proteins, and is a key feature of organismal developmental processes.

6

The metabolic products of amino acid degradation are acetyl-CoA, acetoacetyl-CoA, pyruvate, α-ketoglutarate, succinyl-CoA, fumarate, and oxaloacetate. Ammonia is also a product.

9

a. Oligonucleotide: a short nucleic acid segment that contains fewer than 50 nucleotides.

b. Nuclease: an enzyme that hydrolyzes nucleic acid molecules to form oligonucleotides.

c. Urea cycle: a cyclic pathway in which waste ammonia molecules, CO_2, and aspartate molecules are converted to urea.

d. Ketogenic: describing a molecule whose carbon skeleton is a substrate for synthesizing fatty acids and ketone bodies.

e. Ubiquitin proteasomal system: a mechanism that degrades cellular proteins into short peptide fragments.

12

The first two reactions in the biochemical pathway that converts NH_4^+ to urea (i.e., the formation of carbamoyl phosphate and citrulline) occur in the mitochondrial matrix. Subsequent reactions that convert citrulline to ornithine and urea occur in the cytosol.

Both citrulline and ornithine are transported across the inner membrane by specific carriers.

15

Individuals with PKU lack phenylalanine hydroxylase (phenylalanine-4-monooxygenase) activity, so they cannot synthesize tyrosine from phenylalanine. Tyrosine is therefore an essential amino acid for these patients.

18

The term *Krebs bicycle* refers to two interlocking cyclic reaction pathways. The aspartate-arginosuccinate shunt of the citric acid cycle is responsible for regenerating the aspartate needed for the urea cycle from fumarate. The molecule the two cycles have in common is arginosuccinate.

21

For the degradation of lysine, see Figure 15.7.

24

The following compounds yield uric acid when degraded: DNA, FAD, and NAD^+, all of which contain purine.

27

A vegetarian diet excludes taurine. Since domestic cats cannot synthesize taurine, they must obtain it by consuming meat. Without taurine, domestic cats become listless and die prematurely.

30

Ketogenic amino acids that are degraded to form acetyl-CoA or acetoacetyl-CoA cannot be converted into glucose because (1) acetyl-CoA cannot be converted to pyruvate because the pyruvate dehydrogenase-catalyzed reaction is irreversible, and (2) when acetyl-CoA enters the citric acid cycle, two carbon atoms are lost as CO_2; thus no carbon atoms, which could have contributed to the formation of oxaloacetate (and eventually glucose), are added.

33

The nitrogen atom in position 3 on the uracil ring eventually ends up in urea molecules. (This nitrogen is between the two carbonyl carbons in uracil.)

36

The molecules that cause the urine odor that is characteristic of the maple syrup urine disease are the α-keto acids derived from leucine, isoleucine, and valine.

39

Humans cannot excrete waste nitrogen atoms as ammonia because of its toxicity. Urea is also toxic, but much less so than ammonia. The conversion of ammonia to urea not only allows the nitrogen to be transported and excreted in a much less toxic form but also prevents the large water loss that would be required by the excretion of ammonia.

Thought Questions

42

As the concentration of glutamate (as well as its deamination product ammonia) rises, the enzyme N-acetylglutamate synthase catalyzes the synthesis of N-acetylglutamate, an activator of carbamoyl phosphate synthetase I. The latter enzyme catalyzes the first committed step in urea synthesis.

45

These amino acids are intermediates in the urea cycle. Therefore, their addition stimulates the formation of urea.

48

Urea and uric acid, both of which are less toxic than ammonia, require significantly less water for their excretion than does ammonia. As a result, the use of these molecules facilitates water conservation in land animals. Osmotic pressure is an important factor. Since urea and uric acid contain two and four nitrogen atoms, respectively, less water is required (per nitrogen atom) for their excretion than that for NH_4^+.

51

The pathway whereby uracil is converted to β-alanine is as follows

Uracil → Dihydrouracil → β-Ureido-propionate → β-Alanine

The atoms in each molecule are numbered as indicated.

54

In the absence of insulin and/or insulin receptors, target tissues generate energy by means other than metabolizing glucose. Muscle proteins are degraded to amino acids, which are then used to generate energy for muscle contraction via the citric acid cycle.

CHAPTER 16
In-Chapter Questions

16.1

Several series of signal transduction components are illustrated in Figure 8.20. A prominent example series consist of glucagon (primary signal), glucagon receptor (receptor), adenylate cyclase (transducer), and activated protein kinase (response).

16.2

Cortisone, a steroid, is not degraded by the digestive system. In contrast, insulin, a protein, is inactivated because it is degraded to its amino acid components.

16.3

The types of noncovalent interactions that are involved in reversible binding of a signal molecule and its receptor include hydrogen bonding, hydrophobic interactions, and various types of electrostatic interactions, for example, salt bridges.

16.4

High levels of stress combined with lack of sleep or distorted wake/sleep cycles (e.g., shift work) lead to elevated levels of circulating cortisol. Cortisol depresses the synthesis and release of thymic hormones, resulting in compromised development of T lymphocytes and a depressed immune response. Light inhibits the production of melatonin, so its level falls when individuals have inadequate sleep. This reduces the melatonin-dependent inhibition of CRH release which amplifies the increase in circulating cortisol. The stress-induced immunosuppression is therefore enhanced and prolonged.

16.5

cAMP molecules would be produced by the binding of a single hormone molecule before it diffused away from the receptor site. The amplification factor for the hormone molecule is 350; that is, 350 cAMP molecules are produced for every hormone-receptor binding event.

16.6

Approximately 100,000 (or 10^5) molecules of target molecule (E_R) can be activated by a single molecule of hormone. cAMP is generated from ATP by adenylate cyclase when a hormone molecule binds to its receptor. The interaction between the receptor and adenylate cyclase is mediated by a G protein, G_s. As a consequence of hormone binding and the resulting conformational change, the receptor interacts with a nearby G_s protein. As G_s binds to the receptor, GDP dissociates. Then the binding of GTP to G_s allows one of its subunits to interact and stimulate adenylate cyclase, thus initiating cAMP synthesis. cAMP must break down quickly so the signaling mechanism can be precisely controlled.

16.7

The inhibition of GTP hydrolysis causes the subunit of G_s protein to continue activating adenylate cyclase. In intestinal cells, this enzyme activity opens chloride channels, causing loss of large amounts of chloride ions and water. The massive diarrhea caused by this process quickly leads to serious dehydration and electrolyte loss.

16.8

Nitroglycerin molecules are hydrolyzed in blood to yield NO. In turn, NO relaxes smooth muscle cells in the walls of blood vessels. It is believed that NO activates guanylate cyclase, which subsequently promotes the intracellular sequestration of calcium ions, thus allowing muscle cells to relax.

16.9

Both DAG and phorbol esters promote the activity of protein kinase C, which promotes cell growth and division. Phorbol esters provide initiated cells with a sustained growth advantage over normal cells. This condition is an early stage in carcinogenesis.

16.10

The high blood glucose levels in untreated diabetics result in the loss of increasingly large amounts of glucose along with water in the urine, a condition that causes dehydration. In the absence of usable glucose, the body rapidly degrades fats and proteins to generate energy, Hence Aretaeus's observation that in this disease excessive weight loss and excessive urination arc related.

16.11

a. The intestine engages in the digestion of food and transport of nutrients into blood.

b. Liver has a key role in carbohydrate, lipid, and amino acid metabolism and monitors and regulates blood composition.

c. During fasting or starvation, skeletal muscle provides amino acids to other organs. Cardiac muscle is the principal structural feature of the heart and is responsible for pumping blood carrying nutrients throughout the body.

d. Adipose tissue functions as energy storage and in the generation of fatty acids for energy generation and glycerol, a substrate for gluconeogenesis in the liver.

e. Kidney performs filtration of blood plasma and regulation of blood pH.

f. Brain directs metabolic processes of the body and integrates sensory information required to obtain food.

16.12

Long-term fasting or low-calorie diets arc interpreted by the brain as starvation. The brain responds by lowering the body's BMR. The majority of the energy is derived from fatty acid oxidation. The glucose needed for glucose-dependent tissues is generated via gluconeogenesis at the expense of muscle protein.

16.13

As blood glucose and insulin levels in blood drop back to normal, glucagon is released from the pancreas. Glucagon acts on the liver to prevent hypoglycemia by promoting glycogenolysis and gluconeogenesis. Glucagon stimulates glycogenolysis by triggering the synthesis of cAMP, which in turn initiates a cascade of reactions that lead to the activation of glycogen phosphorylase. Increased lipolysis, hydrolysis of fat molecules, provides glycerol molecules that are substrates for gluconeogenesis.

End-of-Chapter Questions

Review Questions

3

a. Downregulation: reduction of the number of cell surface receptors by endocytosis; receptor molecules that have been taken into the cell are eventually recycled to the cell surface or degraded; downregulation occurs in response to specific hormones to protect the cell against overstimulation by hormones.

b. G-proteins: multisubunit proteins that bind GTP and mediate transmembrane signaling.

c. GPCR (G-protein-coupled receptor): member of a large family of protein receptors transduce a wide variety of stimuli into intracellular signals; a GPCR that consist of an extracellular ligand-binding site, seven membrane-spanning helices, and intracellular segments that interact with G-proteins.

d. DAG (diacylglycerol): a second messenger that activates protein kinase C; formed by the cleavage of PIP_2 by phospholipase C.

e. RTK (receptor tyrosine kinase): a member of a family of transmembrane receptors with cytoplasmic tyrosine kinase activity, a transmembrane segment, and an external domain that binds specific extracellular ligands, such as insulin, epidermal growth factor, platelet-derived growth factor, and insulin-like growth factor.

6

a. Tumor promoter: a molecule that provides tumor cells a growth advantage over nearby cells.

b. Osmotic diuresis: a process in which the presence of solute in urinary filtrate causes an excessive loss of water and electrolytes.

c. Histocompatibility antigen (HLA): found on the surface of most of the body's cells; plays an important role in determining how the immune systems reacts to foreign substances or cells.

d. Ketosis: elevated concentrations of ketones in the blood.

e. Cytokines: proteins that may stimulate or inhibit cell growth or proliferation; produced by blood-forming cells and immune system cells.

9

For several weeks after the onset of fasting, blood glucose levels are maintained via gluconeogenesis. During most of this period, amino acids derived from the breakdown of muscle proteins are the major substrates for the process. Eventually, as muscle becomes depleted, the brain switches to ketone bodies as an energy source. Consequently, the production of urea (the molecule used to dispose of the amino groups of the amino acids) declines.

12

One consequence of physical activity is the activation of the sympathetic nervous system, which in turn stimulates the adrenal gland to secrete epinephrine and norepinephrine. These hormones then activate the adipocyte enzyme hormone-sensitive lipase, which catalyzes the hydrolysis of triacylglycerol molecules to form glycerol and the fatty acids used to drive muscle contraction.

15

a. Insulin: general anabolic effects, including glucose uptake and lipogenesis.

b. Glucagon: stimulates glycogenolysis and lipolysis.

c. Epidermal growth factor (EGF): stimulates cell division of a large number of epithelial cells, including epidermal and gastrointestinal lining cells.

d. Interferon: a growth inhibitor; type I protects cells from viral infection; type II inhibits the growth of cancer cells and also has several immunoregulatory effects.

e. Tumor necrosis factors (TNF): growth inhibitors that suppress cell division, are toxic to tumor cells, and may have a role in regulating several developmental processes.

18

a. Cholecystokinin: stimulates secretion of digestive enzymes and bile.

b. Glucagon-like peptide 1: promotes satiety.

c. Triiodothyronine (T_3): general stimulation of many cellular reactions.

d. Somatostatin: in the hypothalamus, inhibits secretion of GH and TSH; inhibits secretion of glucagon by the pancreas.

e. CRH (corticotropin-releasing hormone): stimulates secretion of corticotropin, which stimulates steroid synthesis in the adrenal cortex.

21

HbA_{1c} formation is a consequence of nonenzymatic glycosylation of hemoglobin that occurs in the presence of high blood glucose levels. In the Maillard reaction, the aldehyde group of glucose condenses with a free amino group in a protein to form a Schiff base. The Schiff base rearranges to form a stable ketoamine referred to as the Amadori product. The Amadori product subsequently destabilizes to form a reactive carbonyl-containing product that reacts with hemoglobin molecules to form an adduct such as HbA_{1c}.

24

In type I diabetes (insulin-dependent diabetes mellitus), no insulin is produced and the action of glucagon is unopposed. Glucagon causes an increase in lipolysis in adipocytes, leading to an excess of acetyl-CoA molecules, which are converted to ketone bodies. Ketoacidosis (an excess of ketone bodies in the blood with low blood pH) is rare in type II diabetes (insulin-independent diabetes mellitus), in which blood levels of insulin are normal or elevated but cells are resistant to insulin. Since the action of glucagon is not completely unopposed, and there is some (although reduced) glucose uptake by cells, lipolysis is not activated as it is in type I diabetes, and the excess production of ketone bodies does not occur.

27

In addition to water loss, contributing factors to the relatively rapid weight loss during the first week of a prolonged diet are depletion of glycogen stores, loss of muscle protein, and lipolysis of triacylgylcerols in adipocytes. Large amounts of amino acids from muscle protein are needed to provide glucose (via gluconeogenesis in the liver), the preferred energy source for the brain. Lipolysis releases fatty acids to provide an alternative energy source to glucose. Regarding the water loss, note that water bound by glycogen will be lost when glycogen is depleted. Also, a molecule of water is required to hydrolyze each glycosidic, ester, and peptide bond.

Thought Questions

30

Long-term fasting or low-calorie diets are interpreted by the body as starvation. The brain responds by lowering the body's BMR. One effect of this mechanism is that skeletal muscle mass is reduced because it is such a metabolically demanding tissue.

33

In uncontrolled diabetes mellitus, massive breakdown of fat reserve results in the production of large quantities of ketone bodies. Two of the ketone bodies (acetoacetic acid and β-hydroxybutyric acid) are weak acids. The release of hydrogen ions from large numbers of these molecules overwhelms the body's buffering capacity.

36

The steroid molecule is covalently bound to the matrix in a chromatographic column. The extract suspected of containing the steroid-binding protein is then passed through the column. Any proteins remaining on the column are eluted by changing the salt concentration of the eluting buffer. After isolation and purification, such proteins can be examined specifically for binding activity to the steroid.

39

Appetite regulation in humans is a set of complex and robust mechanisms that involve several areas of the brain, such as the hypothalamus. In response to calorie restriction, which is interpreted as starvation, the appetite centers of the brain respond by stimulating appetite, lowering the body's BMR (to conserve energy), and lowering energy expenditures (resulting in lethargy). As a result of these and other hormone- and peptide-triggered responses, continued achievement of weight loss and maintenance of a reduced weight over time become very difficult, if not impossible.

42

Leptin is a satiety-promoting adipokine that is synthesized in adipocytes in proportion to adipose tissue mass. Leptin's actions within the appetite center in the arcuate nucleus of the hypothalamus include stimulation of appetite-depressing POMC neurons and inhibition of the appetite-enhancing AgRP/NPY neurons. Weight control is promoted when leptin is produced in appropriate amounts and the leptin signaling pathway in the brain is working properly. If leptin is not produced in adequate amounts in relation to fat stores, or there is leptin resistance, weight control is dysfunctional.

45

IRS1 is one of numerous proteins in the signal transduction mechanisms triggered by insulin in target tissues. If IRS1 is defective or in low concentration, the expression of several genes will be altered. A defective IRS1 mechanism, combined with obesity, itself a risk factor for insulin resistance, results in dysfunctional insulin signaling pathway and abnormal insulin receptor processing. Type II diabetes is the foreseen consequence of these anomalies.

48

Under the stated conditions, the fat stores will last 70.5 days.

CHAPTER 17
In-Chapter Questions
17.1

The types of noncovalent interactions that stabilize DNA stmcture are hydrophobic interactions, hydrogen bonds, base stacking, and electrostation interactions. The cumulative zip-pering effect of the hydrogen bonds between base pairs keeps the strands in the correct complementary orientation. The parallel stacking of the nearly planar bases is a stabilizing factor because of the cumulative effect of weak van der Waals forces.

In electrostatic interactions the potentially destabilizing force of the phosphate group charges (repulsion between nearby negative charges) is offset by the binding of magnesium ions and polycationic molecules such as the polyamines and histones.

17.2

The cytosine-guanine base pair with its three hydrogen bonds is more stable than the adenine-thymine base pair. The more CG bp there are, the more stable the DNA molecule. Structure b, with the fewest CG bp, will therefore denature first.

17.3

a. Ethanol will disrupt the hydrogen bonding in the base pairs and denature the DNA.

b. Heat, which easily disrupts hydrogen bonds, will cause DNA chains to separate and denature.

c. Dimethylsulfate is an alkylating agent that can cause transversion and transition mutations.

d. Nitrous acid deaminates bases.

e. Quinaerine is an intercalating agent that can cause frame shift mutations.

17.4

a. Caffeine is a base analogue.

b. Benzo[*a*]pyrene is a nonalkylating agent that is easily converted into *a* highly reactive form that forms adducts with bases.

c. Ethyl chloride is an alkylating agent.

17.5

The brain is especially sensitive to oxidative stress because it uses a greater proportion of oxygen than other tissues. Consequently, the chance of oxidative damage is also high. In addition, when most types of brain cells are irreversibly damaged by ROS, they cannot be replaced. In addition to hydroxyl radicals, other ROS that can contribute to oxidative stress in the brain include superoxide, hydrogen peroxide, and singlet oxygen.

17.6

In A-DNA, the dehydrated form of DNA, the base pairs are no longer at right angles to the helical axis. Instead, they tilt 20 degrees away from the horizontal as compared to B-DNA. The distance between adjacent base pairs is slightly reduced, with 11 bp per helical turn instead of the 10.4 bp that occurs in the B-form. Each turn of the double helix of A-DNA occurs in 2.5 nm instead of the 3.4 nm of B-DNA. The diameters of A-DNA and B-DNA are 2.6 and 2.4 nm, respectively. The significance of A-DNA is unclear. It has been observed that its overall appearance resembles that of RNA duplexes and the RNA-DNA hybrids that form during transcription.

With a diameter of 1.8 nm, Z-DNA is considerably slimmer than B-DNA. It is twisted into a left-handed spiral with 12 bp per turn, each of which occurs in 4.5 nm instead of the 3.4 observed in B-DNA. Segments with alternating purine and pyrimidine bases are most likely to adopt the Z-DNA configuration. In Z-DNA the bases stack in a left-handed, staggered pattern that gives this form its flattened, non-grooved surface and zigzag appearance. The significance of Z-DNA is unresolved.

H-DNA (triple helix) segments can form when a poly-purine sequence is hydrogen-bonded to a polypyrimidine sequence. H-DNA, which has been observed to form under low pH conditions, is made possible by nonconventional, Hoogsteen base pairing. H-DNA may play a role in recombination.

17.7

In the prokaryotic chromosome there is a protein core to which the circular DNA molecule is attached. In addition, HU protein binds to the DNA and facilitates its bending and supercoiling. In eukaryotic chromosomes, DNA forms complexes with the histones to form nucleosomes. The polyamines are polycationic molecules that bind to negatively charged DNA so the latter molecule can overcome charge repulsions between adjacent coils during the compression process.

17.8

The genome is the total set of DNA-encoded genetic information in an organism. A chromosome is a DNA molecule, usually complexed with certain proteins. Chromatin is the partially decondensed form of eukaryotic chromosomes. Nucleosomes are the repeating structural units of eukaryotic chromosomes formed by the interaction of DNA with the histones. A gene is a DNA sequence that codes for a polypeptide or an RNA molecule.

17.9

The information used to construct this model included the following:

1. The chemical structures and molecular dimensions of deoxyribose, the nitrogenous bases, and phosphate.

2. The 1 : 1 ratios of adenine : thymine and guanine : cytosine in the DNA isolated from a wide variety of species investigated by Erwin Chargaff (**Chargaff's rules**).

3. X-ray diffraction studies performed by Rosalind Franklin indicating that DNA is a symmetrical molecule and probably a helix.

4. The diameter and pitch of the helix estimated by Wilkins and his colleague Alex Stokes from other X-ray diffraction studies.

5. The later demonstration by Linus Pauling that protein, another complex class of molecule, could exist in a helical conformation.

17.10

a. Tandem repeats are DNA sequences in which multiple copies are arranged next to each other; repeated sequence lengths vary from 10 bp to over 2000 bp.

b. Centromeres are the structures that attach eukaryotic chromosomes to the mitotic spindle during mitosis and meiosis.

c. Satellite DNA consists of DNA sequences arranged next to each other that form a distinct band when genomic DNA is digested and centrifuged; the original term for tandem repeats.

d. Introns are noncoding intervening DNA sequences in a split or interrupted gene that are excised during mRNA formation.

e. Exons are the coding regions in a split or interrupted eukaryotic gene.

f. Microsatellites are core DNA sequences of 2 to 4 bp that are tandemly repeated 10 to 20 times.

g. Transposition is the movement of a DNA segement from one site in the genome to another.

17.11

The genomes of prokaryotes are substantially smaller than those of eukaryotes. For example, the genome sizes of E. *coli* and humans are 4.6 and 30 Mb, respectively. Prokaryotic genomes are compact and continuous; that is, there are few, if any, noncoding DNA sequences. In contrast, eukaryotic DNA contains enormous amounts of noncoding sequences. Other distinguishing features of prokaryotic and eukaryotic DNA are the linkages of genes into operons in prokaryotes and intervening sequences in eukaryotic genes.

17.12

The antisense DNA sequence is 3′-CGTAAGCTTAACGTCT-GAGGACGTTAAGCCGTTA-5′; the mRNA sequence is 3′-CGUAAGCUUAACGUCUGAGGACGUUAAGCCGUUA-5′, The antisense RNA sequence is 3′-GCAUUCGAAUUGCA-GACUCCUGCAAUUCGGCAAU-5′.

17.13

In the original central dogma, the flow of genetic information is in one direction only, that is, from DNA to the RNA molecules, which then direct protein synthesis. The altered diagram indicates that the RNA genome of some viruses can replicate their RNA genomes (using a viral enzyme activity referred to as RNA-directed RNA polymerase) or undergo reverse transcription (i.e., synthesize DNA from an RNA sequence).

End-of-Chapter Questions

Review Questions

3

Supercoiling is a process in which DNA bends and twists to relieve torque, allowing DNA strands to be packaged in compact chromosomes.

6
Eukaryotic genomes are larger than those of prokaryotes. In contrast to prokaryotic genomes, which consist entirely of genes, the majority of eukaryotic DNA sequences do not appear to have coding functions. Unlike prokaryotic genes, most eukaryotic genes are not continuous (i.e., they usually contain introns).

9
RNA molecules differ from DNA in the following ways: (1) RNA contains ribose instead of deoxyribose, (2) the nitrogenous bases in RNA differ from those of DNA (e.g., uracil replaces thymine and several RNA bases are chemically modified), and (3) in contrast to the double helix of DNA, RNA is single-stranded.

12
There are approximately 6 million base pairs in a single human cell. Assuming that there are 10^{14} body cells, the total length of the DNA in the human body is approximately 2×10^{11} km. This estimated length is about 1000 times greater than the distance from the earth to the sun. (Note that 1 nm is 10^{-9} m.)

15
Guanine-cytosine pairs contain three hydrogen bonds, while adenine-thymine contain only two. The greater the number of hydrogen bonds holding the DNA strands together, the higher the melting point will be.

18
The complementary DNA strand (written in the standard 5′ to 3′ direction is 5′-AACGATAACGGCCCCT-3′. The RNA strand is 5′-AACGAUAACGGCCCCU-3′.

21
  a. Sense strand: the DNA strand that is not transcribed by RNA polymerase into an RNA molecule. The sense strand has the same sequence as an RNA molecule except that the base U is substituted for T.
  b. Hypochromic effect: the decrease in the absorption of UV light (260 nm) that occurs when purine and pyrimidine bases are incorporated into base pairs in polynucleotide sequences.
  c. Transposition: the movement of a piece of DNA from one site in a genome to another.
  d. Transcript: the newly synthesized RNA molecule that results from transcription.
  e. Mobile genetic element: a DNA sequence that undergoes transposition; that is, it can be duplicated and moved within the genome.

24
A DNA replication error will continue to be propagated as the replicated strands themselves undergo replication, and so on, for all subsequent "generations" of DNA molecules. A transcription error will result in only a small number of defective gene products, as opposed to all of the products of a mutated DNA sequence. Considering that a replication error, as opposed to a transcription error, is potentially permanent, a replication error would be expected to cause more cellular damage.

27
AZT (azidothimidine) substitutes an azide group for the 5′-hydroxyl group, which is needed to create the phosphate-sugar linkages of the DNA backbone. When AZT is incorporated into a growing DNA chain, the 5′-azide group prevents the addition of the next nucleotide and thus cuts short the reverse transcription process.

30
Water stabilizes DNA structure by binding to phosphate groups, deoxyribose 3′- and 5′-oxygen atoms, and electronegative atoms in the nucleotide bases. Also, the increased entropy of surrounding water molecules drives the hydrophobic interactions between the nucleotide bases within the helix.

Thought Questions

33
The polyamines are positively charged at pH 7, which promotes binding to the negative charges of the DNA backbone. Polyamine binding overcomes the mutual repulsion of the adjacent DNA chains, packing the chains more closely.

36
The electron-withdrawing effect of the bromine increases the likelihood of enol formation of uracil. This enol mimics the hydrogen-bonding pattern of cytosine. Therefore, this base can be paired with guanine.

39
Nuclei and mitochondria are separately obtained from source tissue by means of cell homogenization followed by density gradient centrifugation. The nucleic acids in each organellar fraction are then extracted with the aid of detergents, solvents, and proteases (to remove proieins), RNA is removed by treating each sample with RNase. The DNA from both types of organelle is then further purified by centrifugation.

42
Each base in the nucleotides of DNA and RNA has a specific shape that has unique information content. The enormous number of possible sequences of the nucleotides make possible a very large coding capacity that living organisms use to specify the molecular structure of all of their biomolecules.

45
Viruses frequently infect more than one host. Some of the nucleic acid sequences obtained from one or more hosts provide a virus particle with a selective advantage (e.g., making it easier to attack a host cell). For example, human influenza viruses frequently become more virulent after they have infected chickens and pigs.

48
During each brother's life, his epigenome will change in response to interactions with the environment. Examples include exposure to agents that change methylation of DNA bases. The older the brothers, the greater the differences will be. The two individuals could be distinguished by examining the epigenomic differences (e.g., by staining chromosomes to reveal DNA methylation patterns). This is not true of DNA base sequences, which would remain the same in both individuals.

CHAPTER 18
In-Chapter Questions

18.1
Briefly, prokaryotic DNA replication consists of DNA unwinding, RNA primer formation, DNA synthesis catalyzed by DNA polymerase and the joining of Okazaki fragments by DNA ligase. Prokaryotic DNA replication differs from the eukaryotic process in that prokaryotic replication is faster, the Okazaki fragments are longer and there is usually only one origin of replication per chromosome (eukaryotes have many per chromosome).

18.2
In excision repair short damaged sequences (e.g., thymine dimers) are excised and replaced with correct sequences. After an endonuclease

deletes the damaged single-stranded sequence, a DNA polymerase activity synthesizes a replacement sequence using the undamaged strand as a template. In photoreactivation repair a photoreactivating enzyme uses light energy to repair pyrimidine dimers. In recombinational repair damaged sequences are deleted. Repair involves an exchange of an appropriate segment of the homologous DNA molecule.

18.3

When antibiotics are used in large quantities, the bacterial cells that possess resistance genes (acquired through spontaneous mutations or through intermicrobial DNA transfer mechanisms such as conjugation, transduction, and transformation) survive and even flourish. Because of antibiotic use, which acts as a selection pressure, resistant organisms (once only a minor constituent of a microbial population) become the dominant cells in their ecological niche.

18.4

Most gene duplications are apparently a consequence of accidents during genetic recombination. Examples of possible causes of gene duplication are unequal crossing over during synapses and transposition. After a gene has been duplicated, random mutations and genetic recombination may introduce variations.

18.5

Because phytochrome has been demonstrated to mediate numerous light-induced plant processes, it appears reasonable to assume that it does so in part by interacting with light-response elements (LRE) in plant cell genomes. Presumably, phytochrome influences gene expression by binding, either alone or as part of a complex, to various LREs when its chromophore is activated by light.

End-of-Chapter Questions

Review Questions

3

The Meselson-Stahl experiment resulted in a single band of DNA on CsCl gradient centrifugation after one round of cell division. Its density was intermediate between heavy and light DNA. This could not have occurred unless the newly synthesized dsDNA contained an old "light" strand and a new "heavy" stand.

6

a. Helicase is an enzymatic activity that relieves torque generated by supercoiling ahead of the replication machinery.
b. Primase is an enzymatic activity that catalyzes the synthesis of RNA primers.
c. DNA polymerase is an enzymatic activity that catalyzes several reactions during DNA replication.
d. DNA ligase forms phosphodiester linkages between newly synthesized DNA fragments.
e. Topoisomerase is an enzymatic activity that prevents the tangling of DNA strands during DNA replication.
f. DNA gyrase facilitates the separation of DNA strands during prokaryotic replication.

9

a. ROS may cause single and double-strand breaks, pyrimidine dimers, and the loss of purine and pyrimidine bases.
b. Because caffeine is a base analogue of thymine, it can cause transition mutations.
c. Small alkylating agents attach to the nitrogen atoms of the purines and pyrimidines, destabilizing glycosidic linkages (leading to depurination), interfering with hydrogen bonding, and promoting both transversion and transition mutations.

d. Large alkylating agents have the same effects as small alkylating agents, but in addition they behave similarly to intercalating agents, leading to frameshift mutations and breakage of the DNA chain.
c. Nitrous acid deaminates bases. For example, cytosine is converted to uracil.
f. Intercalating agents cause deletion or insertion mutations.

12

a. DNA glycosylase: an enzyme that cleaves the N-glycosidic link between the damaged base and the deoxyribose component of a nucleotide during base excision repair.
b. Clamp loader: a protein that recognizes single DNA strands with primer, then transfers the β-clamp dimer to the core polymerase, where it forms a closed ring (or "clamp") around the DNA strand; in eukaryotes, replication factor C is a clamp loader protein that attaches DNA polymerase δ to each strand.
c. Palindrome: a sequence that provides the same information whether it is read forward or backward; DNA palindromes contain inverted repeat sequences. The top strand of DNA read 5'-3' is the same as the bottom strand read 5'-3'.
d. Primer: a short RNA segment required to initiate DNA synthesis.
e. Protooncogene: a normal gene that codes for molecules involved in cell cycle control, and that promotes carcinogenesis if mutated.

15

Intermicrobial DNA transfer mechanisms include transformation, transduction, and conjugation. In transformation, DNA fragments that enter the bacterial cell through cell wall openings and through a recombination event are inserted into the chromosome or a plasmid. In transduction, fragments of bacterial DNA are transferred to a recipient cell by a virus. Transduced DNA may insert into recipient cell DNA through recombination. In conjugation, a donor cell produces a sex pilus that allows DNA to transfer to a recipient cell.

18

Because DNA is constantly exposed to disruptive processes, its structural integrity is highly dependent on efficient repair mechanisms. The life span of an organism is dependent on the health of its constituent cells, which is in turn dependent on the timely and accurate expression of genetic information. Consequently, the capacity of the organisms in a species to maintain the integrity of DNA molecules is an important factor in determining life span.

21

Marker genes are useful in recombinant DNA technology because their function is known and their presence, which indicates that a successful recombinant event has occurred, is easily detected. For example, an antibiotic resistance gene, which codes for the synthesis of a substance that provides protection for a bacterium from the effects of an antibiotic, allows the growth of recombinant cells in a medium containing that antibiotic. Cells that do not contain the marker gene, that is, those in which the recombinant DNA is not present, do not survive.

24

The processing steps that prepare a typical eukoryotic mRNA for its functional role include capping (the linkage of a 7-methyl-guanosine to the 5' end), cleavage of mRNA and addition of a poly(A) tail to the 3' end, and splicing (the removal of introns).

27

a. Recombinational repair: a repair mechanism that can eliminate damaged DNA sequences of certain types that were not eliminated before replication; the undamaged parental strands recombine into the gap left after the removal of the damaged sequence.

b. Replication fork: the Y-shaped region of a DNA molecule that is undergoing replication; results from separation of two DNA strands.

c. Protooncogene: a normal gene that codes for molecules involved in cell cycle control and promotes carcinogenesis if mutated.

d. Apoptosis: programmed cell death.

e. Functional genomics: the scientific discipline devoted to elucidating how biomolecules work together within functioning organisms.

30

Refer to Figure 18.21.

33

In species that possess DNA photolyase, light energy captured by this enzyme's flavin and pterin chromophores is used to break the cyclobutane ring in a thymine dimer, thus converting the dimer back to two thymine monomers. The phosphodiester bonds are not affected. (Humans do not possess this enzyme.)

36

LINE 1 element (L1) transposition occurs by a "cut-and-paste" process in which the L1 is first transcribed to form L1 RNA, which then exits the nucleus to be translated. Retrotransposon proteins (transposase and reverse transcriptase) and the L1 RNA form a complex, reenter the nucleus, and bind to a target DNA sequence. Transposase makes a staggered cut in the DNA sequence, reverse transcriptase synthesizes a DNA copy from the L1 RNA, and a second DNA strand is synthesized. DNA ligase completes the insertion by joining the ends of the target DNA and the newly synthesized L1 DNA element.

39

Site-specific recombination is the exchange of DNA sequences that takes place at a specific site. It requires short, homologous DNA segments (attachment sites or insertional elements) that, once recognized, become base-paired to each other. Two of the strands are cleaved, cross over, and form a Holliday junction with Ruv proteins (refer to Figure 18.24). Branch migration occurs, the crossover strands are cut, and the Holliday junction is resolved, thus completing a site-specific recombination. The functions of enzymes involved include recognizing attachment sites and cutting and sealing the DNA strands.

Thought Questions

42

DNA sequence changing processes such as genetic recombination, gene splicing, and alternate RNA splicing can allow cells to alter gene expression and expand their repertoire of proteins. The best-known example is antibody production in lymphocytes. The rearrangement of several possible choices for each of a number of antibody gene segments via site-specific recombination results in the generation of an extremely large number of different antibody molecules.

45

Any exposure to UV light causes genetic damage in skin cells. The production of melanin, the "tanning" substance that absorbs the energy of UV light, is a response to damage that has already occurred. This damage, which accumulates over years of exposure to UV light, accelerates the aging process, hence the wrinkled and thickened skin. In some genetically predisposed individuals, the accumulating damage results in skin cancer.

48

Because the *Rb* gene codes for a tumor suppressor, retinoblastoma occurs only when both copies have been damaged or deleted. Usually a long period of time is required for random mutations to cause this event. In hereditary retinoblastoma, in which an affected individual possesses only one functional *Rb* gene, the time necessary for a random mutation to inactivate the second *Rb* gene is significantly less than that required for the inactivation of both genes that cause the nonhereditary version of the disease.

51

Refer to Biochemistry in the Lab: Fenomics, pp. 700–707.

54

The RNA "copies" of the DNA can be altered to generate protein diversity without changing the DNA template sequence. In addition, RNA molecules can be used repeatedly and then disposed of without damaging the original DNA. If the DNA were used directly, the master molecule could be damaged or destroyed. In addition, RNA molecules can easily leave the nucleoid or nucleus to travel to other parts of the cell.

CHAPTER 19
In-Chapter Questions

19.1

The amino acid sequence of the beginning of the polypeptide is Met–Ser–Pro–Thr–Ala–Asp–Glu–Gly–Arg–Arg–Trp–Leu–Ile–Met–Phe. The mutation types in the altered mRNA sequences are (a) insertion of one base, (b) deletion of one base, (c) insertion of two bases, (d) deletion of three bases. The consequences of these mutations are altered amino acid sequences of the polypeptides produced from mRNA. In (a), (b), and (c) a frame shift occurs. Therefore the amino acid sequences past the mutation are different. In (d) no frame shift occurs because three bases are deleted. In this case, the only difference between the normal polypeptide and the mutated version is the deletion of a single amino acid.

19.2

Assuming that the DNA sequence given is the coding strand, the mRNA sequence is 5′-GGUUUA-3′ and the anticodons are 5′-UAA-3′. If the DNA sequence is the template strand, the mRNA sequence is 5′-UAAACC-3′ and the anticodons are 5′-GGU-3′ and 5′-UUA-3′.

19.3

The possible choices for mRNA codon base sequences for the peptide are:

Tyr—Leu—Thr—Ala—

5′-UAU-3′	CUU	ACU	GCU
UAC	CUC	ACC	GCC
	CUA	ACA	GCA
	CUG	ACG	GCG
	UUA		
	UUG		

The possible choices for the DNA sequences that code for the peptide are:

Tyr—Leu—Thr—Ala—

3'-ATA-5'	GAA	TGA	CGA
ATG	GAG	TGG	CGG
	GAT	TGT	CGT
	GAC	TGC	CGC
	AAT		
	AAC		

The possible choices for the tRNA anticodons that code for the peptide are:

Tyr—Leu—Thr—Ala—

3'-AUA-5'	GAA	UGA	CGA
AUG	GAG	UGG	CGG
	GAU	UGU	CGU
	GAC	UGC	CGC
	AAU		
	AAG		

19.4

The formation of an ADP-ribosylated derivative of eEF-2 affects the three-dimensional structure of this protein factor. Presumably protein synthesis is arrested because the ability of eEF-2 to interact with or bind to one or more ribosomal components is altered.

19.5

After the synthesis of the plastocyanin precursor in cytoplasm, the first import signal mediates the transport of the protein into the chloroplast stroma. After this signal has been removed by a protease, a second import signal mediates the transfer of the protein into the thylakoid lumen. Plastocyanin then binds a copper atom, folds into its final three-dimensional structure, and associates with the thylakoid membrane.

End-of-Chapter Questions

Review Questions

3

The sequential reactions that occur within the active site of aminoacyl-tRNA synthesis are (1) the formation of aminoacyl-AMP, which contains a high-energy mixed anhydride bond and (2) linkage of the aminoacyl group to its specific tRNA.

6

The major differences between prokaryotic and eukaryotic translation are speed (the prokaryotic process is significantly faster), location (the eukaryotic process is not directly coupled to transcription as prokaryotic translation is), complexity (because of their complex lifestyles, eukaryotes possess complex mechanisms for regulatory protein synthesis: e.g., eukaryotic translation involves a significantly larger number of protein factors than prokaryotic translation), and posttranslational modifications (eukaryotic reactions appear to be considerably more complex and varied than those observed in prokaryotes).

9

a. KDEL sequence: this carboxy-terminal sequence of four amino acids (K-D-E-L = Lys-Asp-Glu-Leu) serves as a retention signal and occurs on nascent proteins destined for the ER.

b. Signal peptides: short peptide sequences that determine a polypeptide's destination, (e.g., by directing its insertion into a membrane).

c. Glycosylation: a posttranslational mechanism whereby carbohydrate groups are covalently attached to polypeptides.

d. Genetic code: the set of nucleotide base triplets (codons) that code for the amino acids in protein as well as for start and stop signals.

e. Context-dependent codon reassignment: variation in the genetic code in which a specific codon codes for a amino acid (or a stop signal) different from that which typically occurs. An example is the stop codon UAG that, in some methane-producing Archaea, codes for the nonstandard amino acid pyrolysine.

12

A signal recognition particle (SRP) is a large complex composed of protein and RNA that binds to a ribosome that has begun translating a polypeptide possessing a signal peptide component. Once the SRP has bound to the ribosome, translation is temporarily arrested. The SRP then mediates ribosomal binding to docking proteins on the surface of a membrane (e.g., RER membrane). Translation subsequently recommences, and the growing polypeptide inserts into the membrane.

15

The major differences between prokaryotic and eukaryotic translation control mechanisms are related to the complexity of eukaryotic gene expression. Features that distinguish eukaryotic translation include mRNA export (spatial separation of transcription and translation), mRNA stability (the half-lives of mRNA can be modulated), negative translational control (the translation of certain mRNAs can be blocked by the binding of specific repressor proteins), initiation factor phosphorylation (mRNA translation rates are altered by certain circumstances when eIF-2 is phosphorylated), and translational frameshifting (certain mRNAs can be frameshifted so that a different polypeptide is synthesized).

18

a. Translocation: movement of the ribosome along the mRNA during translation.

b. GAP (guanine nucleotide activating protein): a protein involved in the hydrolysis of the GTP bound to eIF-2 during the formation of a complete 80S ribosome.

c. GEF (guanine nucleotide exchange factor): promotes EF-Tu regeneration by displacing its GDP during prokaryotic elongation.

d. Ribosome recycling factor (RRF): a tRNA-shaped protein that binds within the A site and is needed in order to dissociate the ribosome into its constituent subunits, thus ending the termination phase in prokaryotes.

e. Nascent: newly synthesized.

21

GTP hydrolysis provides the energy that drives the movement of the peptidyl-tRNA from the A site to the P site in the ribosome. During elongation, GTP hydrolysis is also required for the incoming aminoacyl-tRNA to bind in the A site.

24

a. Amide linkage

b. Phosphodiester bond

c. Hydrogen bonds

27

The major classes of eukaryotic posttranslational modifications are proteolytic cleavage (the hydrolysis of specific peptide bonds), glycosylation (attachment of sugar residues to specific amino acid residues in the protein), hydroxylation (the adding of OH groups to proline and lysine residues), phosphorylation (the addition of phosphate groups to specific amino acid residues on a protein), lipophilic modification (covalent attachment of lipid groups to a protein), methylation (attachment of methyl groups), disulfide bond formation (formation of —S—S— bonds between cysteine residues), and protein splicing (a specific segment of a polypeptide is removed and the remaining ends are joined covalently by an amide linkage).

30

a. Intein: an excised peptide segment generated during protein splicing.

b. Extein: peptide segments that are spliced together to form a mature protein during protein splicing.

c. Translocon: an integral membrane protein complex that mediates polypeptide translocation.

d. Translational frameshifting: a +1 or –1 change in reading frame allowing more than one polypeptide to be synthesized from a single mRNA.

e. Mass spectrometry: a screening technique in which molecules are vaporized and then bombarded by a high-energy electron beam, causing them to fragment as cations.

33

The three-letter sequences listed below each amino acid in the first table are the possible mRNA sequences that code for that specific amino acid. For example, any of the four mRNA sequences listed below Ala will code for Ala. Thus, there are many correct answers to this question. Choose one three-letter sequence from each column to build an mRNA sequence that will code for this peptide. For example, one possible answer is:

mRNA: 5′-GCU UCU UUU UAU UCU AAA AAA UUA GCU GAU GUU AUU-3′

cDNA: 3′-CGA AGA AAA ATA AGA TTT TTT AAT CGA CTA CAA TAA-5′

Note that the sequence order must be reversed to write the sequence as "5′ ⟶ 3′":

5′-ATT AAC CTA AGC TAA TTT TTT AGA ATA AAA AGA AGC-3′

5′ ⟶ 3′ Possible Choices for mRNA Codon Base Sequences for This Peptide:

5′-Ala	Ser	Phe	Tyr	Ser	Lys	Lys	Leu	Ala	Asp	Val	Ile-3′
GCU	UCU	UUU	UAU	UCU	AAA	AAA	UUA	GCU	GAU	GUU	AUU
GCC	UCC	UUC	UAC	UCC	AAG	AAG	UUG	GCC	GAC	GUC	AUC
GCA	UCA			UCA			CUU	GCA		GUA	AUA
GCG	UCG			UCG			CUC	GCG		GUG	
	AGU			AGU			CUA				
	AGC			AGC			CUG				

3′ ⟶ 5′ Possible Choices for the DNA Sequences for This Polypeptide:

Ala	Ser	Phe	Tyr	Ser	Lys	Lys	Leu	Ala	Asp	Val	Ile
3′-CGA	AGA	AAA	ATA	AGA	TTT	TTT	AAT	CGA	CTA	CAA	TAA
3′-CGG	AGG	AAG	ATG	AGG	TTC	TTC	AAC	CGG	CTG	CAG	TAG
3′-CGT	AGT			AGT			GAA	CGT		CAT	TAT
3′-CGC	AGC			AGC			GAG	CGC		CAC	
	TCA			TCA			GAT				
	TCG			TCG			GAC				

36

Features of eukaryotic protein synthesis that help to account for the increased time required (as opposed to prokaryotic translation) include the greater quantity, variety, and functioning of eukaryotic translation factors (e.g., at least 12 IFs vs 3 for prokaryotes); additional processing of mRNA (addition of a cap and a poly(A) tail; removal of introns); and the increased quantity and variety of eukaryotic posttranslational modifications, such as hydroxylation, protein splicing, and disulfide bond formation. Also, because eukaryotic mRNA lacks Shine-Dalgarno sequences, eukaryotic ribosomes must search for a translation start site by binding to the capped 5′ end and moving toward the 3′ end.

Thought Questions

39

When errors in amino acid–tRNA binding do occur, they are usually the result of similarities in amino acid structure. Several aminoacyl-tRNA synthetases possess a separate proofreading site that binds incorrect aminoacyl-tRNA products and hydrolyzes them.

42

Four high-energy phosphate bonds are required to incorporate each amino acid into a polypeptide (i.e., 2 GTP and 2 ATP). The polymerization of 200 amino acids requires 400 GTP and 400 ATP.

45

Each Shine-Dalgarno sequence in a prokaryotic mRNA occurs near a start codon (AUG). The Shine-Dalgarno sequence provides a mechanism for promoting the correct alignment of the start codon on the ribosome (as opposed to a methionine codon) because it binds to a nearby complementary sequence in the 16S rRNA component of the 30S ribosome. Eukaryotic ribosomes identify the initiating AUG codon by binding to the capped 5′ end of the mRNA and scanning the molecules for a translation start site.

48

Preproproteins contain signal sequences that direct them to the ER for translocation and Golgi for modification. The cleavage of an inactive proprotein and other posttranslational modification processes ensures that the protein is active only when it has been targeted to its site of function.

51

The process would be similar to the pyrolysine insertion outlined in the text. Assuming the pyrovaline is available from a metabolic pathway, the following circumstances must obtain:

1. A codon is assigned to pyrovaline.
2. A tRNA with the requisite anticodon sequence is available.
3. An aminoacyl tRNA synthetase binds pyrovaline to its cognate tRNA.
4. A stem-loop or similar structure upstream of the newly assigned codon in the mRNA promotes the codon reassignment.
5. The tRNA bound to pyrovaline would then enters the ribosome, where its amino acid is incorporated into the protein.

Glossary

acceptor site In RNA, splicing the upstream 3'-OH splice site.

acetal The family of organic compounds with the general formula $RCH(OR')_2$; formed from the reaction of a hemiacetal with an alcohol.

acid A molecule that can donate hydrogen ions.

acidosis A condition in which the pH of the blood is below 7.35 for a prolonged time.

activation energy The threshold energy required to produce a chemical reaction.

active site The cleft in the surface of an enzyme where a substrate binds.

active transport The energy-requiring movement of molecules across a membrane against a concentration gradient.

acyl carrier protein A component of fatty acid synthase. Intermediates of fatty acid synthesis are linked to this molecule through a thioester linkage.

acyl group Any molecular group derived from a carboxylic acid by the removal of a hydroxyl group.

addition reaction A chemical reaction in which two molecules react to form a third and there are more groups attached to carbon atoms in the product.

adduct The product of an addition reaction.

adiponectin A peptide hormone that enhances glucose-stimulated insulin secretion and cellular responses to insulin.

A-DNA A short, compact DNA structure in which the base pairs are not at right angles to the helical axis; occurs when DNA becomes partially dehydrated.

aerobic metabolism The mechanism by which the chemical bond energy of food molecules is captured and used to drive the oxygen-dependent synthesis of adenosine triphosphate (ATP).

aerobic respiration The metabolic process in which oxygen is used to generate energy from food molecules.

aerotolerant anaerobe An organism that depends on fermentation for its energy needs and possesses protection from toxic oxygen metabolites; present in the form of detoxifying enzymes and antioxidant molecules.

affinity chromatography A technique in which proteins are isolated based on their capacity to bind to a specific ligand.

aldaric acid The product formed when the aldehyde and CH_2OH groups of a monosaccharide are oxidized to carboxylic acids.

alditol A sugar alcohol; the product of the reduction of the aldehyde or ketone group of a monosaccharide.

aldol cleavage A reverse of the aldol condensation.

aldol condensation An aldol addition reaction; the nucleophilic addition of a ketone enolate ion to an aldehyde to form a β-hydroxyketone, followed by the elimination of a water molecule.

aldonic acid The product of the oxidation of the aldehyde group of a monosaccharide.

aldose A monosaccharide with an aldehyde functional group.

alkaloid A member of a class of naturally occurring molecules that have one or more nitrogen-containing rings; many alkaloids have medicinal and other physiological effects.

alkalosis A condition in which the blood pH is above 7.45 for a prolonged period of time.

alkylation The introduction of an alkyl group into a molecule.

α-tocopherol A radical scavenger belonging to a class of compounds called phenolic antioxidants.

aliphatic hydrocarbon A nonaromatic hydrocarbon such as methane or cyclohexane.

allosteric enzyme An enzyme whose activity is affected by the binding of effector molecules.

allosteric transition The ligand-induced conformational change in a protein.

allostery The control of protein function through ligand-binding events.

Alzheimer's disease A progressive, fatal disease that is characterized by seriously impaired intellectual functions caused by neuronal death.

amethopterin A structural analogue of folate used to treat several types of cancer; also referred to as methotrexate.

amino acid pool The amino acid molecules that are immediately available in an organism for use in metabolic processes.

amino acid residue An amino acid that has been incorporated into a peptide molecule.

amphibolic pathway A metabolic pathway that functions in both anabolism and catabolism.

amphipathic molecule A molecule containing both polar and nonpolar domains.

amphoteric molecule A molecule that can act as both an acid and a base.

AMPK AMP-activated protein kinase; an important regulatory enzyme in energy metabolism.

amyloid deposits Insoluble extracellular proteinaceous debris found in the brains of patients with certain neurological diseases.

amylopectin A type of plant starch; a branched polymer containing $\alpha(1,4)$- and $\alpha(1,6)$-glycosidic linkages.

amylose A type of plant starch; an unbranched polymer of D-glucose residues linked with $\alpha(1,4)$-glycosidic linkages.

anabolic pathways A series of biochemical reactions in which large complex molecules are synthesized from smaller precursor molecules.

anaerobic organisms Organisms that do not use oxygen to generate energy.

anaerobic respiration The metabolic process in which species other than oxygen are the terminal electron acceptors in energy generation.

analogue A substance similar in structure to a naturally occurring molecule.

anaplerotic reaction A reaction that replenishes a substrate needed for a biochemical pathway.

anchor protein A molecule that facilitates the recruitment and assembly of specific sets of signal cascade proteins into complexes bound to the cytoskeleton.

anhydride The product of the condensation reaction between two carboxyl groups or two phosphate groups in which a molecule of water is eliminated.

annotation The functional identification of the genes of a genome.

antenna pigment A molecule that absorbs light energy and transfers it to a reaction center during photosynthesis.

anticodon A sequence of three ribonucleotides on a tRNA molecule that is complementary to a codon on the mRNA molecule; codon-anticodon binding results in the delivery of the correct amino acid to the site of protein synthesis.

antigen Any substance able to stimulate the immune system; generally a protein or large carbohydrate.

antioxidant A substance that prevents the oxidation of other molecules.

antiparallel Aligned in opposition.

antisense strand A noncoding DNA strand that is complementary to the base sequence of an mRNA molecule transcribed from the coding DNA strand.

anomer An isomer of a cyclic sugar that differs from another in its configuration about the hemiacetal or acetal carbon.

apoenzyme The protein portion of an enzyme that requires a cofactor to function in catalysis.

apoprotein A holoprotein without its prosthetic group.

apoptosis The genetically programmed series of events that lead to cell death.

apurinic site A nucleotide residue in a DNA molecule from which a purine base has been lost or removed.

apyrimidinic site A nucleotide residue in a DNA molecule from which a pyrimidine base has been lost or removed.

aquaporin A water channel protein.

archaea One of the three domains of living organisms: prokaryotic organisms that have the appearance of bacteria and many molecular properties that are similar to those of the eukaryotes.

aromatic hydrocarbon A molecule that contains a benzene ring or has properties similar to those exhibited by benzene.

asymmetric carbon A carbon bound to four different groups.

attachment (att) site A short DNA sequence that facilitates site-specific recombination; also refers to an IS element.

autocrine A hormonelike molecule that is active within the cell in which it is produced.

autoimmune disease A condition in which an immune response is directed against an individual patient's own tissues.

autopoiesis A system that is autonomous, self-organizing, and self-maintaining.

autotroph An organism that transforms light energy or chemical energy of various chemicals into the chemical bond energy of biomolecules.

β-carotene A plant pigment molecule that acts as an absorber of light energy and as an antioxidant.

β-oxidation The catabolic pathway in which most fatty acids are degraded; acetyl-CoA is formed as the bond between the α and β carbon atoms is broken.

β_2 clamp The protein complex that promotes processivity; that is, it prevents frequent dissociation of DNA polymerase from the DNA template.

B cell A B lymphocyte; a white blood cell that produces and secretes antibodies, the proteins that bind to foreign substances thereby initiating their destruction in the humoral immune response.

B-DNA The commonly found form of DNA, as the sodium salt under highly humid conditions.

bacteria One of the three domains of life: single-celled prokaryotes with diverse capacities to exploit their environments.

bacterial artificial chromosome A derivative of a large *E. coli* plasmid used to clone DNA sequences as long as 300 kb.

base A molecule that can accept hydrogen ions.

base analogue A molecule that resemble a normal DNA nucleotide base and can substitute for it during DNA replication, leading to mutation.

base excision repair A mechanism that removes and then replaces individual nucleotides in DNA whose bases have undergone various types of damage (e.g. alkylation, deamination or oxidation).

bile salts Amphipathic molecules with detergent properties that are important components of bile, a yellowish green liquid that aids in the digestion of fat; a conjugated derivative of the bile acids cholic acid and deoxycholic acid.

bioenergetics The study of energy transformations in living organisms.

biogenic amine An amino acid derivative that acts as a neurotransmitter (e.g., GABA and the catecholamines).

bioinformatics The computer-based field that facilitates the analysis of biological sequence data.

biomolecules Molecules that make up a living organism.

bioremediation The use of biological processes to decontaminate toxic waste sites.

biotranformation A series of enzyme-catalyzed processes in which toxic and/or hydrophobic molecules are converted into (usually) less toxic and more soluble metabolites.

branched-chain amino acid One of a group of essential amino acids (leucine, isoleucine, and valine) with branched carbon skeletons.

buffer A substance that resists large pH changes when small amounts of acids or bases are added; usually a solution that contains a weak acid and its conjugate base.

C3 plants Plants that produce glycerate-3-phosphate, a three-carbon molecule, as the first stable product of photosynthesis.

C4 metabolism A photosynthetic pathway in plants such as corn and sugarcane that produces a four-carbon molecule and avoids photorespiration.

C4 plants Plants that possess mechanisms that suppress photorespiration by separating rubisco (ribulose-1,5-bisphosphate carboxylase) from atmospheric O_2.

calvin cycle The major metabolic pathway by which carbon dioxide is incorporated into organic molecules

CAP binding complex (CBC) A protein complex that binds to capped mRNA molecules and facilitates their translation; consists of eIF-4A (a helicase), eIF-4E (a translation initiation factor), and eIF-G (a scaffold protein); also referred to as eIF-4F.

carbon fixation The biochemical process by which inorganic carbon dioxide is incorporated into organic molecules.

carbanion A carbon with a negative charge.

carbocation A carbon with a positive charge.

carotenoid An isoprenoid molecule that functions as a light-harvesting pigment and/or protects against reactive oxygen species (ROS).

carrier protein A membrane transport protein.

catabolic pathway A series of biochemical reactions in which a large complex molecule is degraded into smaller, simpler products; in some catabolic pathways, energy is captured.

catalyst A substance that enhances the rate of a chemical reaction but is not permanently altered by the reaction.

catecholamine One of a class of neurotransmitters derived from tyrosine; includes dopamine, norepinephrine, and epinephrine.

cDNA library A clone library of cDNA (complementary DNA) molecules produced from mRNA molecules by reverse transcription.

cell cortex The three-dimensional meshwork of proteins that reinforces the plasma membrane.

cell fractionation A technique involving homogenization and centrifugation that allows the study of cell organelles.

cellobiose A degradation product of cellulose; a disaccharide that contains two molecules of glucose linked by a $\beta(1,4)$-glycosidic bond.

cellular immunity Immune system processes mediated by T cells, a type of lymphocyte.

cellulose A polymer produced by plants that is composed of D-glucopyranose residues linked by $\beta(1,4)$-glycosidic bonds.

centromere A special region of repetitive DNA that plays a critical role in cell division; it holds the two sister chromatids together during prophase and metaphase of cell division.

chain-terminating method A technique for determining DNA base sequences that uses 2'-3'-dideoxy base analogues as chain-terminating inhibitors of DNA polymerase; also referred to as the Sanger method.

channel protein A membrane protein that contains a pore through which ions are transported.

chaperonins One of a family of molecules that control the folding and targeting of cell proteins.

Chargaff's rules A set of rules describing the base composition of DNA; posits the equality of the concentration of adenine and thymine and of cytosine and guanine.

chemiosmotic coupling theory ATP synthesis is coupled to electron transport by an electrochemical proton gradient across a membrane.

chemoautotroph An organism that transforms the energy of various chemicals into chemical bond energy.

chemoheterotroph An organism that uses preformed organic food molecules as its sole source of energy.

chemolithotroph An organism that transforms the energy in specific inorganic substances into chemical bond energy.

chemosynthesis The biochemical mechanism whereby chemical energy is extracted from certain minerals.

chiral carbon An asymmetric carbon in an molecule that has a mirror-image form.

chlorophyll A magnesium-containing green pigment molecule found in plants and photosynthetic bacteria that resembles heme; absorbs light energy in photosynthesis.

chloroplast A chlorophyll-containing plastid found in the cells of algae and higher plants.

chromatin The complex of DNA and histones found in the nucleus of eukaryotic cells.

chromatin remodeling complex A multisubunit complex that facilitates the release of the histones from nucleosomal DNA during transcription.

chromophore A molecular component that absorbs light of a specific frequency.

chromoplast A type of plastid in plants that accumulates the pigments that are responsible for the colors of leaves, flower petals, and fruits.

chromosomal jumping A technique used to isolate clones that contain discontinuous sequences from the same chromosome.

chromosome A very long DNA molecule associated with proteins that contains the genes of an organism.

chylomicron A large lipoprotein of extremely low density: transports dietary triacylglycerols and cholesteryl esters from the intestine to muscle and adipose tissue.

chylomicron remnants Chylomicrons after about 90% of the triacylglycerols have been removed by lipoprotein lipase.

circular dichroism A type of spectroscopy in which the relationship between molecular motion and structure is probed with electromagnetic radiation.

cis **isomer** An isomer in which two identical substituents are on the same side of the double bond.

cistron A DNA sequence that contains the coding information for a polypeptide and the signals required for ribosome function.

citric acid cycle A biochemical pathway that degrades the acetyl group of acetyl-CoA to CO_2 and H_2O as three molecules of NAD^+ and one molecule of FAD are reduced.

clamp loader The γ complex that recognizes single DNA strands with primer and transfers β_2-clamp dimer to the core polymerase.

coding strand The DNA strand that has the same base sequence as the RNA transcript (with thymine instead of uracil).

codon A sequence of three nucleotides in mRNA that directs the incorporation of an amino acid during protein synthesis or acts as a start or stop signal.

coenzyme A small organic molecule required in the catalytic mechanisms of certain enzymes.

cofactor The nonprotein component of an enzyme (either an inorganic ion or a coenzyme) required for catalysis.

colony hybridization technique A method used to identify bacterial colonies that possess a specific recombinant DNA sequence.

competitive inhibition A reversible type of enzyme inhibition in which they inhibitor molecule competes with the substrate for occupation of the active site.

composite transposon A bacterial transposon composed of a gene and flanking IS elements.

conjugate base The anion (or molecule) that results when a weak acid loses a proton.

conjugate redox pair An electron donor and its electron acceptor form: for example, NADH and NAD^+.

conjugated protein A protein that functions only when it carries other chemical groups attached by covalent linkages or by weak interactions.

conjugation Unconventional sexual mating between bacterial cells; a donor cell transfers a DNA segment into a recipient cell through a specialized pilus.

conjugation reaction A biochemical reaction that may improve the water solubility of a molecule by converting it to a derivative that contains a water-soluble group.

consensus sequence The average of several similar sequences: for example, the consensus sequence of the –10 box of *E. coli* promoter is TATAAT.

constitutive gene A routinely transcribed gene that codes for gene products required for basic cell functions.

contig One of a set of overlapping DNA sequences used to identify the base sequence of a region of DNA.

cooperative binding A mechanism in which binding of one ligand to a target molecule promotes the binding of other ligands.

Cori cycle A metabolic process in which lactate, produced in tissues such as muscle, is transferred to liver where it becomes a substrate in gluconeogenesis.

cosmid Cloning vehicles that contain the γ bacteriophage cos sites incorporated into plasmid DNA sequences with one or more selectable markers.

cotranslational transfer The insertion of a polypeptide across a membrane during ongoing protein synthesis.

covalent bond The sharing of electrons between atoms.

CpG CpG dinucleotides; methylated cytosines occur predominantly in 5'-CG-3' sequences.

CpG island Regions of the genome where CpGs constitute more than 50% of the bases.

Crassulacean acid metabolism A photosynthetic pathway that produces a four-carbon molecule (malate) in plants that live in hot, dry regions such as deserts.

cystic fibrosis An ultimately fatal autosomal recessive disease that is caused by the missing or defective chloride channel protein CFTR.

cystic fibrosis transmembrane conductance regulator (CFTR) The plasma membrane glycoprotein that functions as a chloride channel in epithelial cells.

cytochrome P_{450} system An electron transport system that consists of two enzymes (NADPH-cytochrome P_{450} reductase and cytochrome P_{450}); involved in the oxidative metabolism of many endogenous and exogenous substances.

cytokine A group of hormonelike polypeptides and proteins; also referred to as growth factors.

cytoskeleton A set of protein filaments (microtubules, microfilaments, and intermediate fibers) that maintains the cell's internal structure and allows organelles to move.

DAG Diacylglycerol.

de novo methyltransferase Methyltransferases that catalyze the methylation of unmodified CpGs.

decarboxylation The removal of a carboxylic group from a carboxylic acid as carbon dioxide.

degeneracy The capacity of structurally different system parts to perform the same or similar functions.

denaturation A disruption of protein or nucleic acid structure caused by exposure to heat or chemicals leading to loss of biological function.

density gradient centrifugation A technique in which cell fractions are further purified by centrifugation in a density gradient.

desensitization A process in which target cells adjust to changes in stimulation by decreasing the number of cell surface receptor or by inactivating those receptors.

detoxication The process by which a toxic molecule is converted to a more soluble (and usually less toxic) product.

detoxification Correction of a state of toxicity; the chemical reactions that produce sobriety in an enebriated person.

dialysis A laboratory technique in which a semipermeable membrane is used to separate small molecules from larger ones.

diastereomers A stereoisomer that is not an enantiomer (mirror-image isomer).

dicer A nuclease that cuts pre-microRNA into mature miRNAs or initiates the gene silencing process.

dictyosomes Term often used for the Golgi complex in plants.

differential centrifugation A cell fractionation technique in which homogenized cells are separated by centrifugal forces.

dipole A difference in charge between atoms in a molecule resulting from the unsymmetrical orientation of polar bonds.

disaccharide A glycoside composed of two monosaccharide residues.

dissipative system A system that facilitates the reduction of an energy gradient.

disulfide bridge A covalent bond formed between the sulfhydryl groups of two cysteine residues.

disulfide exchange An enzyme-catalyzed posttranslational process in which there is an interchange of disulfide bonds in a protein until the correct biologically relevant disulfide bonds are formed.

DNA fingerprinting A laboratory technique used to compare DNA banding patterns from different individuals.

DNA glycosylase A DNA repair enzyme that cleaves the N-glycosidic linkage between the damaged base and the deoxyribose component of the nucleotide.

DNA ligase An enzyme that catalyzes the formation of a covalent phosphodiester bond between the 3′-OH end of one segment with the 5′-phosphate end of another segment during DNA replication.

DNA microarray A DNA chip used to analyze the expression of thousands of genes simultaneously.

DNA profile A unique DNA pattern of repeats of target sequences that is separated in an electrophoresis gel; used to identify individuals.

DNA typing A DNA analysis technique used to identify individuals; involves the analysis of several highly variable sequences called markers.

docking protein An transmembrane protein of the rough endoplasmic reticulum that binds a signal recognition protein that is bound to a ribosome, thus triggering the resumption of protein synthesis; also referred to as signal recognition particle receptor protein.

donor site The 5′-splice site in the RNA splicing process.

downregulation The reduction in cell surface receptors in response to stimulation by specific hormone molecules.

dynein A motor protein associated with microtubules

effector A molecule whose binding to a protein alters the protein's activity.

eicosanoid A hormonelike molecule that contains 20 carbons; most are derived from arachidonic acid; examples include prostaglandins, thromboxanes, and leukotrienes.

electron acceptor Species that accepts electrons from an electron donor during a reaction.

electron donor Species that donates electrons to an electron acceptor during a reaction.

electron transport system A series of electron carrier proteins that bind reversibly to electrons at different energy levels.

electrophoresis A class of techniques in which molecules are separated from each other because of differences in their net charge.

electrophile An electron-deficient species that is preferentially attracted to a region of high electron density in another species during a chemical reaction.

electroporation A method of introducing a cloning vector into a host cell that involves treatment with an electrical current.

electrostatic interaction Noncovalent attraction between oppositely charged atoms or groups.

elimination reaction A chemical reaction in which a double bond is formed when atoms in a molecule are removed.

elongation The polypeptide chain growth phase during translation of an mRNA in a ribosome.

emergence New and unanticipated properties in each level of organization of a system that result from interactions among the components.

emergent property A new property conferred by the complexity and dynamics of the system.

enantiomer Mirror-image stereoisomers.

endergonic process A reaction that does not spontaneously go to completion; the standard free energy change is positive and the equilibrium constant is less than 1.

endocrine hormone A hormone secreted into the bloodstream that acts on distant target cells.

endocytosis A process in which a cell takes up solutes or particles by enclosing them in vesicles pinched off from its plasma membrane.

endomembrane system An extensive set of interconnecting internal membranes that divide the cell into functional compartments.

endoplasmic reticulum (ER) A series of membranous channels and sacs that provides a compartment separate from the cytoplasm for numerous chemical reactions.

endothermic reaction A reaction that requires energy.

enediol The intermediate formed during the isomerization reactions of monosaccharides. It contains a double bond with a hydroxyl group on each carbon of the double bond.

energy The capacity to do work.

energy transfer pathway A pathway that captures energy and transforms it into a form that organisms can use to drive biomolecular processes.

enthalpy The heat content of a system; in a biological system it is essentially equivalent to the total energy of the system.

entropy A measure of the randomness or disorder of a system; a measure of that part of the total energy in a system that is unavailable for useful work.

enzyme A biomolecule that catalyzes a biochemical reaction.

enzyme induction A process in which a signal molecule stimulates increased synthesis of a specific enzyme.

enzyme kinetics The study of the rates of enzyme-catalyzed reactions.

epidermal growth factor A protein that stimulates epithelial cells to undergo cell division.

epigenetic A covalent modification of DNA bases or histones that causes a change in gene expression.

epigenome The current epigenetic modifications within a cell.

epimer A molecule that differs from the configuration of another by one asymmetric carbon.

epimerization The reversible interconversion of epimers.

epimutation An alteration in the normal epigenetic pattern.

epoxide An ether in which the oxygen is incorporated into a three-membered ring.

essential amino acid An amino acid that cannot be synthesized by the body and must be supplied by the diet.

essential fatty acid A fatty acid that must be supplied in the diet because it cannot be synthesized by the body; linoleic and linolenic acids in humans.

euchromatin A less condensed form of chromatin that has varying levels of transcriptional activity.

eukarya One of the three domains of life: nucleus-containing single-celled and multicellular organisms.

eukaryotic cell A living cell that possesses a true nucleus.

exergonic process A reaction that spontaneously goes to completion as written; the standard free energy change is negative, and the equilibrium constant is greater than 1.

exon The region in a split or interrupted gene that codes for RNA and ends up in the final product (e.g., mRNA).

exonuclease An enzyme that removes nucleotides from the end of the polynucleotide strand.

exothermic reaction A reaction that releases heat.

extein Polypeptide segments that are spliced together to form a mature protein during protein splicing.

extracellular matrix (ECM) A gelatinous material, containing proteins and carbohydrates, that binds cells and tissues together.

extremophile An organism that lives under extreme conditions of temperature, pH, pressure, or ionic concentration that would easily kill most organisms.

extremozyme An enzyme that functions under extreme conditions of temperature, pressure, pH, and/or ionic concentration.

facilitated diffusion Diffusion of a substance across a membrane that is aided by a carrier.

facultative anaerobe An organism that possesses the capacity for detoxifying oxygen metabolites; energy is generated using oxygen, when available, as an electron acceptor.

fatty acid binding protein An intracellular water-soluble protein whose function is to bind and transport hydrophobic fatty acids.

feedback The mechanism in a self-regulating system in which the product of a process acts to modify the process.

feedback control The control of a self-regulating system (e.g., a metabolic process or pathway) in which product influence the output of the process.

fermentation An energy-yielding process in which organic molecules serve as both donors and acceptors of electron; the anaerobic degradation of sugars.

fibrous protein A protein composed of polypeptides arranged in long sheets or fibers.

flavin adenine dinucleotide (FAD) A tightly bound prosthetic group consisting of riboflavin, D-ribitol, and adenine that functions in the class of enzymes called flavoproteins.

flavin mononucleotide (FMN) A tightly bound prosthetic group consisting of a molecule of riboflavin and D-ribitol phosphate that functions in the class of enzymes called flavoproteins.

flavoprotein A conjugated protein in which the prosthetic group is either FMN or FAD.

fluid mosaic model The currently accepted model of cell membranes in which the membrane is a lipid bilayer with integral proteins buried in the lipid and peripheral proteins loosely attached to the membrane surface.

fluorescence A form of luminescence in which certain molecules can absorb light of one wavelength and emit light of another wavelength.

fold A core three-dimensional structure of a protein domain.

43S preinitiation complex A multisubunit complexcomposed of the 40S subunits

eIF-A, eIF-2-GTP, eIF-3, and methionyl-tRNA^{met} that binds to mRNA.

free energy The energy in a system available to do useful work.

free radical An atom or molecule that has an unpaired electron.

functional genomics The investigation of gene expression patterns.

functional group A group of atoms that undergoes characteristic reactions when attached to a carbon atom in an organic molecule or biomolecule

gasotransmitter An endogenous gaseous molecule that acts as a signal molecule.

gel filtration chromatography A technique used to separate molecules according to their size and shape that employs a column packed with a gelatinous polymer.

gene A DNA sequence that codes for a polypeptide, rRNA, or tRNA.

gene expression The mechanism by which living organisms regulate the flow of genetic information; the control of when and if genes are transcribed.

general recombination Recombination involving exchange of a pair of homologous DNA sequences; it can occur at any location on a chromosome.

genetic code The set of nucleotide base triplets (codons) that code for the amino acids in proteins as well as start and stop signals.

genetics The scientific investigation of inheritance.

genome The total genetic information possessed by an organism.

genomic imprinting The phenomenon in which expression of a gene is determined by the parent who contributed it; imprinting (DNA methylation) occurs during gametogenesis when the haploid genome is reprogrammed to match the sex of the parent.

genomics The investigation of entire genomes; the sequencing and characterization of genomes.

ghrelin A protein that stimulates appetite; produced by the cells of the stomach and small intestine.

globular protein A protein that adopts a globular shape.

glucagon A peptide hormone released from pancreatic α-cells; among its effects are increasing the level of glucose in blood via the breakdown of liver glycogen.

glucocorticoid A steroid hormone produced in the adrenal cortex that affects carbohydrate, protein, and lipid metabolism.

glucogenic Describing amino acids that are degraded to pyruvate or a citric acid intermediate; these amino acids are used as substrates in the synthesis of glucose in gluconeogenesis.

gluconeogenesis The synthesis of glucose from noncarbohydrate molecules.

glucose-alanine cycle A method of recycling α-keto acids between muscle and liver and for transporting ammonia to the liver.

glucosuria The presence of glucose in the urine.

glycan A polymer of monosaccharides; a polysaccharide.

glycerol phosphate shuttle A metabolic process that uses glycerol-3-phosphate to transfer electrons from NADH in the cytosol to mitochondrial FAD.

glycocalyx A layer on the external surface of many eukaryotic cells that contains substantial amounts of carbohydrate-containing molecules.

glycoconjugate A molecule that possesses covalently bound carbohydrate components (e.g., glycoproteins and glycolipids).

glycogen A glucose storage molecule in vertebrates; a branched polymer containing $\alpha(1,4)$- and $\alpha(1,6)$-glycosidic linkages.

glycogenesis A biochemical pathway that adds glucose to growing glycogen polymers when blood glucose levels are high.

glycogenolysis A biochemical pathway that removes glucose molecules from glycogen polymers when blood glucose levels are low.

glycolipid A glycosphingolipid; a molecule in which a monosacchraride, disaccharide, or oligosaccharide is attached to a ceramide through an O-glycosidic linkage.

glycolysis The enzymatic pathway that converts a glucose molecule into two molecules of pyruvate: the anaerobic process generates energy in the form of two ATP molecules and two NADH molecules.

glycome The total set of sugars and glycans that a cell or organism produces.

glycomics The investigation of the structural and functional properties of all the carbohydrate molecules produced by organisms.

glycoprotein A conjugated protein in which carbohydrate molecules are covalently bound.

glycosaminoglycan A long unbranched heteropolysaccharide chain composed of disaccharide repeating units.

glycoside The acetal of a sugar.

glycosidic link An acetal linikage formed between two monosaccharides.

glyoxylate cycle A modification of the citric acid cycle that occurs in plants, bacteria, and other eukaryotes: allows growth in these organims from two-carbon substrates such as ethanol, acetate, and acetyl-CoA.

Golgi apparatus (complex) A series of curved membranous sacs involved in packaging and distributing cell products to internal and external compartments.

G protein A protein that binds GTP, which activates the protein to perform a function; the hydrolysis of GTP to form GDP inactivates the G protein.

GPI (glycosylphosphatidylinositol) anchor A glycolipid used to link certain proteins to membrane, preferentially in lipid rafts.

G protein-coupled receptor A cell surface receptor that transduces the binding of a hormone or other signal molecule into an intracellular response via the activation of a G protein.

grana (pl) Stacks of thylakoid membrane.

granum (sing) The folded portion of the thylakoid membrane.

growth factor An extracellular polypeptide that stimulates cells to grow and/or undergo cell division.

GTPase-activating protein (GAP) A protein molecule that hydrolyzes GTP bound to a GTP-binding protein.

guanine nucleotide exchange factor (GEF) A protein that mediates a conformational change in the transmembrane region of a G-protein-coupled receptor and leads to GDP/GTP exchange during G-protein activation.

heat shock protein (hsp) A protein synthesized in response to stress (e.g., high temperature).

helicase An ATP-requiring enzyme that catalyzes the unwinding of duplex DNA.

hemiacetal One of the family of organic molecules with the general formula RCH(OR)OH that is formed by the reaction of one molecule of alcohol with an aldehyde.

hemiketal One of the family of organic molecules with the general formula RRC(OR)OH that is formed by the reaction of a molecule of alcohol with a ketone.

hemoprotein A conjugated protein in which heme, an iron-containing organic group, is the prosthetic group.

Henderson-Hasselbalch equation Kinetic rate expression that defines the relationship between pH, pK_a and the concentration of the weak acid and conjugate base components of a buffer solution.

heteroglycan A high-molecular-weight carbohydrate polymer that contains more than one kind of monosaccharide.

heterokaryon A structure formed from the fusion of the membranes of two different cells; used to demonstrate membrane fluidity.

heterotroph An organism that obtains energy by degrading preformed food molecules usually obtained by consuming other organisms.

high-density lipoprotein A type of lipoprotein with a high protein content that is believed to scavenge excess cholesterol from cell membranes and transport it to the liver.

holoenzyme A complete enzyme consisting of an apoenzyme plus a cofactor.

holoprotein An apoprotein combined with its prosthetic group.

homeostasis The capacity of living organisms to regulate metabolic processes despite variability in their internal and external environments.

homoglycan High-molecular-weight carbohydrate polymers that contain only one type of monosaccharide.

homologous polypeptide Protein molecules whose amino acid sequences are similar; implies a common evolutionary origin.

hormone A molecule produced by a specific cell that influences the function of distant target cells.

hormone response element A specific DNA sequence that binds hormone-receptor complexes; the binding of a hormone-receptor complex either enhances or diminishes the transcription of a specific gene.

hsp-60 One of a family of molecular chaperones that mediate protein folding by forming a large structure composed of two stacked seven-membered rings that facilitate the ATP-dependent folding of polypeptides; also called chaperonins or Cpn 60s.

hsp-70 One of a family of molecular chaperones that bind to and stabilize proteins during the early stages of the folding process.

humoral immune response The immunity that results from the presence of antibodies in blood and tissue fluid; also referred to as an antibody-mediated immunity.

Huntington's disease An inherited, fatal neurological disease caused by an excessively long polyglutamine sequence in the protein called huntingtin.

hybridization A laboratory technique in which fragments of single-stranded DNA from different sources anneal; the rate at which a DNA hybrid forms is a measure of the similarity of the two strands.

hydration A type of addition reaction in which water is added to a carbon—carbon double bond.

hydrocarbons Compounds that contain only carbon and hydrogen.

hydrogen bond The force of attraction between a hydrogen atom and a small highly electronegative atom (e.g., O or N) on another molecule or the same molecule.

hydrolase An enzyme that catalyzes reactions in which adding water cleaves bonds.

hydrolysis A chemical reaction in which molecules are cleaved by water.

hydrophobic interaction The association of nonpolar molecules when they are placed in water.

hydrophilic Describing molecules or portions thereof that dissolve easily in water; hydrophilic molecules possess positive or negative charges or contain relatively large numbers of electronegative oxygen or nitrogen atoms.

hydrophobic Describing molecules that do not dissolve in water and possess few if any electronegative atoms.

hydroxyapatite A calcium phosphate in gel form used in nucleic acid research; binds to double-stranded DNA more tenaciously than to single-stranded DNA.

hyperammonemia A potentially fatal elevation of the concentration of ammonium ions in the blood.

hyperglycemia Blood glucose levels that are higher than normal.

hyperosmolar hyperglycemic nonketosis Severe dehydration in non-insulin-dependent diabetics; caused by persistently high blood glucose levels.

hypertonic solution A concentrated solution with a high osmotic pressure.

hyperuricemia An abnormally high level of uric acid in the blood.

hypochromic effect The decrease in the absorption of UV light (260 nm) that occurs when purine and pyrimidine bases are incorporated into base pairs in polynucleotide sequences.

hypotonic solution A dilute solution with a low osmotic pressure.

hypoglycemia Blood glucose levels that are lower than normal.

imprinting control region (ICR) A common regulatory element that codes for a noncoding RNA and is rich in CpG islands.

inhibitor A molecule that reduces an enzyme's activity.

initiation The beginning phase of translation.

initiation complex The protein complex required to initiate the first step in the translation of ribosome-mediated mRNA.

inner membrane The innermost membrane of the nuclear envelope; contains integral proteins that are unique to the nucleus.

insertional element A short DNA sequence involved in site specific recombination; also called an IS element or att site.

insulin A peptide hormone released from pancreatic β-cells; among its many effects are the promotion of glucose uptake into the cells of certain target organs (liver, muscle and adipose tissue).

insulin-like growth factor (IGF) A protein in humans that mediates the growth-promoting actions of growth hormone; has insulin-like properties (e.g., promotes glucose transport and fat synthesis).

insulin resistance The insensitivity of tissues to insulin; a common cause is the down-regulation of insulin receptors.

integral protein A protein that is embedded within a membrane.

intein An excised peptide segment generated during protein splicing.

intercalating agents Planar molecules that insert themselves between base pairs; this action distorts the DNA chain.

interferon One of a group of glycoproteins that have nonspecific antiviral activity (e.g., stimulation of cells to produce antiviral proteins) that inhibits the synthesis of viral RNA and proteins and regulates the growth and differentiation of immune system cells.

intergenic sequence DNA sequences that do not code for gene products; often referred to as junk DNA.

interleukin 2 (IL-2) A member of a group of cytokines that regulate the immune system in addition to promoting cell growth and differentiation.

intermediate A species produced in the course of a reaction that exists for a finite period of time.

intermediate-density lipoprotein (IDL) A lipoprotein formed when a very low density lipoprotein shrinks in size and becomes more dense as a result of depletion of triacylglycerol, apolipoprotein, and phospholipid molecules.

intermediate filament A cytoskeletal component that provides cells with significant mechanical support; a flexible, strong, and relatively stable polymer (8–12 nm).

interspersed genome-wide repeats Repetitive DNA sequences that are scattered around the genome.

intrinsic termination Transcription termination that involves an RNA termination sequence that contains an inverted repeat sequence; also referred to as rho-independent termination.

intrinsically unstructured proteins (IUPs) Proteins that are partially or completely lacking in a stable three-dimensional stucture.

intron A noncoding intervening sequence in a split or interrupted gene; missing in the final RNA product.

inverted repeat A sequence that is a reversed complement of another downstream sequence; defines the boundary of a transposon.

ion-exchange chromatography A technique that separates molecules on the basis of their charge.

ionophore A substance that transports cations across membranes.

irreversible inhibition A form of enzyme inhibition in which an inhibitor molecule permanently impairs an enzyme, usually through binding via a covalent bond.

IP$_3$ Inositol-1,4,5-triphosphate; the IP$_3$ receptor is a calcium channel.

isoelectric point The pH at which a protein has no net charge.

isomers molecules with the same number and types of atoms.

isomerase An enzyme that catalyzes the conversion of one isomer to another.

isomerization reaction A reaction that involves the intermolecular shift of atoms or groups.

isoprenoid One of a class of biomolecules that contain repeating five-carbon structural units known as isoprene units: examples include terpenes and steroids.

isothermic reaction Reactions in which heat is not exchanged with the surroundings; $\Delta H = 0$.

isotonic solution A solution exactly the same particle concentration as that inside cells; there is no net movement of water in or out of the cells.

isozyme One of two or more forms of the same enzyme activity with similar amino acid sequences.

ketal The family of organic compounds with the general formula RRC(OR)$_2$; formed from the reaction of a hemiketal with an alcohol.

ketoacidosis Acidosis caused by the excessive accumulation of ketone bodies.

ketogenesis The condition in which excess acetyl-CoA molecules are converted to acetoacetate, β-hydroxybutyrate, and acetone (referred to collectively as the ketone bodies).

ketogenic Describing amino acids degraded to form acetyl-CoA or acetoacetyl-CoA.

ketone body One of three molecules (acetone, acetoacetate, or β-hydroxybutryate) that are produced in the liver from acetyl-CoA.

ketosis Accumulation of ketone bodies in blood and tissues.

kinetic proofreading A mechanism suggested to account for the precision of codon-anticodon pairing during translation: correct base pairing allows sufficient time for hydrolysis of GTP bound to an elongation factor.

kinesin A motor protein associated with microtubules.

Krebs bicycle A biochemical pathway in which the aspartate required in the urea cycle is generated from oxaloacetate, an intermediate in the citric acid cycle.

Krebs urea cycle The cyclic pathway that converts waste ammonia molecules along with carbon dioxide and aspartate into urea; named for its discoverer, Hans Krebs.

lactone A cyclic ester.

lactose A disaccharide found in milk; composed of one molecule of galactose linked in a $\beta(1,4)$-glycosidic bond to a molecule of glucose.

Le Chatelier's principle Law that states that when a system in equilibrium is disturbed, the equilibrium shifts to oppose the disturbance.

leaving group The group displaced during a nucleophilic substitution reaction.

lectin A carbohydrate-binding protein.

leptin A 16 kD satiety-inducing protein secreted into the bloodstream primarily by adipose tissue.

leukotriene A biologically active molecule derived from arachidonic acid; its synthesis is initiated by a peroxidation reaction.

ligand A molecule that binds to a specific site on a larger molecule.

ligase An enzyme that catalyzes the joining of two molecules.

light-independent reactions A photosynthetic pathway in which CO_2 is incorporated into carbohydrate that can occur in the absence of light; also referred to as the Calvin cycle.

light-induced repair DNA repair in which light energy is used to restore pyrimidine dimers to their original monomeric form; also referred to as photoreactivation repair.

light reactions The mechanism whereby electrons are energized and subsequently used in ATP and NADPH synthesis.

limit of resolution In microscopy, the minimum distance between two separate points that allows for their discrimination.

LINE (long interspersed nuclear elements) Retrotransposons with lengths greater than 5 kb that contain a strong promoter, an integration sequence, and the coding sequences for transposition enzymes.

lipid Any of a group of biomolecules that are soluble in nonpolar solvents and insoluble in water.

lipid bilayer A biomolecular lipid layer that constitutes the structural framework of cell membranes.

lipogenesis The biosynthesis of body fat (triacylglycerol).

lipoic acid A biomolecule that contains a carboxylate group and two sulfhydryl groups that are easily oxidized or reduced; functions as an acyl group carrier in pyruvate dehydrogenase complex and α-ketoglutarate dehydrogenase complex.

lipolysis The enzyme-catalyzed hydrolysis of triacylglycerol molecules.

lipoprotein A conjugated protein in which lipid molecules are the prosthetic groups; a protein-lipid complex that transports water-insoluble lipids in the blood.

lithotroph An organism that uses specific inorganic reactions to generate energy; also known as a chemolithotroph.

London dispersion force A temporary dipole-dipole interaction.

low-density lipoprotein A type of lipoprotein that contains cholesterol, triacylglycerols, and phospholipids; transports cholesterol to peripheral tissues.

lyase An enzyme that catalyzes the cleavage of C—O, C—C, or C—N bonds, thereby producing a product that contains a double bond.

lysogeny The integration of a viral genome into a host genome.

lysosome A saclike organelle capable of degrading most biomolecules.

lytic cycle A viral life cycle in which a virus destroys its host cell.

macromolecular crowding The dense packing of an enormous variety of macromolecules and other molecules within the interior of cells.

maintenance methyltransferase An enzyme that methylates newly synthesized DNA strands at sites opposite the methyl cytosine on the parental strand.

malate-aspartate shuttle A metabolic process in which the electrons from NADH in the cytosol are transferred to mitochondrial NAD$^+$.

malate shuttle A metabolic process in which oxaloacetate is transferred by reversible conversion to malate from a mitochondrion to the cytoplasm.

maltose A degradation product of starch hydrolysis; a disaccharide composed of two glucose molecules linked by an $\alpha(1,4)$-glycosidic bond.

marker gene A gene whose presence can be detected, facilitating the identification of transformed cells.

marker enzyme A enzyme known to be a reliable indicator of the presence of a specific organelle.

mass spectroscopy A technique in which molecules are vaporized and then bombarded by a high-energy electron beam, causing them to fragment as cations.

MCM (minichromosome maintenance complex) The major DNA helicase in eukaryotes.

membrane potential The potential difference across the membrane of living cells; usually measured in millivolts.

messenger RNA (mRNA) An RNA species produced by transcription that specifies the amino acid sequence of a polypeptide.

metabolic syndrome A cluster of clinical disorders that include obesity, hypertension, dyslipidemia, and insulin resistance.

metabolism The total of all chemical reactions in an organism.

metabolome The complete set of organic metabolites that are produced within a cell under the direction of the genome.

metalloprotein Conjugated proteins containing metal ions.

methotrexate A structural analogue of folate that is used in the treatment of several types of cancer; also called amethopterin.

methyl-CpG-binding protein (MeCP) Mediates chromatin-associated gene silencing by binding preferentially to the 5-MeCpG dinucleotides and recruiting histone deacetylose to the site along with histone methylases.

micelle An aggregation of molecules having a nonpolar and a polar component, leaving the polar domains facing the surrounding water.

microfilament A type of cytoskeletal fiber (5–7 nm) composed of polymers of globular actin (G-actin).

microribonucleoprotein (miRNP) A protein that suppresses the expression of specific genes by binding to a complementary site on an appropriate miRNA to silence translation.

microRNA (miRNA) One of various 22 nucleotide RNAs that regulate gene expression.

microsatellite DNA sequences of 2 to 4 bp that are tandemly repeated 10 to 20 times.

microsome A membranous vesicle derived from fragments of endoplasmic reticulum obtained by differential centrifugation.

microtubule A component of the cytoskeleton; composed of the protein tubulin.

mineralocorticoid A steroid hormone that regulates sodium and potassium metabolism.

minisatellite Tandemly repeated sequence of about 25 bp with total lengths between 10^2 and 10^5 bp.

mitochondrion (pl mitochondria) An organelle possessing two membranes in which aerobic respiration occurs.

mitogen A substance that stimulates cell division.

mixed anhydride An acid anhydride with two different R groups.

mixed terpenoid A biomolecule that is composed of a nonterpene component attached to an isoprenoid group.

mobile genetic element One of numerous DNA sequences that can be duplicated and move within the genome.

mobile phase The moving phase in chromatographic methods.

modular protein A protein thatcontains numerous duplicate or imperfect copies of one or more domains that are linked in series; also known as a mosaic protein.

modulator A ligand whose binding to an allosteric site of an enzyme alters the enzyme's activity.

module A component of a subsystem that performs a specific function.

molecular biology The science devoted to elucidating the structure and function of genomes.

molecular chaperone A molecule that assists in protein folding; most are heat shock proteins.

molecular disease A disease caused by a mutated gene.

molten globule A partially globular state of a folding polypeptide that resembles the molecule's native state.

monosaccharide A polyhydroxy aldehyde or ketone containing at least three carbon atoms.

monounsaturated Describing a fatty acid with a single double bond.

motif A unique combination of α-helix and β-pleated-sheet secondary structures that occurs in globular proteins; also known as a supersecondary structure.

motor protein Components of biological machines that bind nucleotides; nucleotide hydrolysis drives precise changes in the protein's shape.

multifunction protein A functional protein with two or more diverse and often unrelated functions.

mutarotation A spontaneous process in which the α and β forms of monosaccharides are readily interconverted.

mutation Any change in the nucleotide sequence of a gene.

myocardial infarction The interruption of the heart's blood supply leading to the death of cardiac muscle cells; a heart attack.

nascent Newly synthesized.

natively unfolded protein A functional protein with a complete lack of ordered structure.

negative cooperativity The binding of one ligand to a target molecule, decreasing the likelihood of subsequent ligand binding.

negative feedback A mechanism in a self-regulating system in which an accumulating product slows its own production.

nephrogenic diabetes insipidis An autosomal recessive disease in which the kidneys cannot produce concentrated urine.

neurotransmitter A molecule released at a nerve terminal that binds to and influences the function of other nerve cells or muscle cells.

neutral fat Triacylglycerol molecules.

N-**glycan** An asparagine-linked oligosaccharide.

nicotinamide adenine dinucleotide (NAD) A coenzyme form of nicotinic acid containing an *N*-ribosyl derivative of nicotinamide and adenosine linked through a pyrophosphate group; occurs as the oxidized form, NAD^+, and the reduced form, NADH, and is involved in electron transfer in a class of enzymes called dehydrogenases.

nicotinamide adenine dinucleotide phosphate (NADP) A coenzyme form of nicotinic acid containing an *N*-ribosyl derivative of nicotinamide and adenosine linked through a pyrophosphate group with an additional phosphate group attached at the $2'$-OH group of the adenosine sugar; occurs as the oxidized form $NADP^+$ and the reduced form NADPH, and is involved in electron transfer in a class of enzymes called dehydrogenases.

nitrogen fixation Conversion of molecular nitrogen (N_2) into a reduced biologically useful form (NH_3) by nitrogen-fixing microorganisms.

nonalkylating agents A variety of chemicals other than the alkylating agents that can modify DNA structure.

noncoding RNAs (ncRNAs) Types of RNA other than the RNAs involved in protein synthesis (i.e., tRNAs, rRNAs, and mRNAs) that act as an extensive genome regulatory network.

noncompetitive inhibition Inhibition of an enzyme in which the inhibitor binds to both the free enzyme and the enzyme-substrate complex.

nonessential amino acid An amino acid that can be synthesized by the body.

nonessential fatty acid A fatty acid that can be synthesized by the body.

nuclear envelope The double membrane that separates the nucleus from the cytoplasm.

nuclear matrix The cytoskeleton-like scaffold within the nucleus in which loops of chromatin are organized.

nuclease An enzyme that hydrolyzes nucleic acid molecules to form oligonucleotides.

nucleohistone DNA complexed with histone proteins.

nucleolus A structure revealed in the nucleus when the nucleus is stained with certain dyes: it plays a major role in the synthesis of ribosomal RNA.

nucleophile An electron-rich atom or molecule.

nucleophilic substitution A reaction in which a nucleophile substitutes for an atom or molecular group.

nucleoplasm The gelatinous substance within the nucleus that contains the cytoskeleton-like nuclear matrix and a network of chromatin fibers.

nucleoid In prokaryotes, an irregularly shaped region that contains a long circular DNA molecule.

nucleoside A biomolecule composed of a pentose sugar (ribose or deoxyribose) and a nitrogenous base.

nucleosome A repeating structural element in eukaryotic chromosomes composed of a core of eight histone molecules around which about 140 base pairs of DNA are wrapped; an additional 60 base pairs connect adjacent nucleosomes.

nucleotide excision repair Bulky lesions of 2 to 30 nucleotides are removed and the resulting gap filled; the excision enzymes appear to recognize the physical distortion rather than a specific base sequence.

obligate aerobe An organism that is highly dependent on oxygen for energy production.

obligate anaerobe An organism that grows only in the absence of oxygen.

O-**glycan** A mucin-type polysaccharide.

Okazaki fragment Any of a series of deoxyribonucleotide segments that are formed during discontinuous replication of one DNA strand as the other strand is continuously replicated.

oligomer A multisubunit protein in which some or all subunits are identical.

oligonucleotide A short nucleic acid segment that contains fewer than 50 nucleotides.

oligosaccharide An intermediate-sized carbohydrate composed of 2 to 10 monosaccharides.

omega-3-fatty acid α-Linolenic acid and its derivatives, such as eicosapentaenoic acid and docosahexaenoic acid.

omega-6-fatty acid Linoleic acid and its derivatives.

oncogene A mutated version of a protooncogene that promotes abnormal cell proliferation.

open reading frame (ORF) A series of triplet base sequences in mRNA that does not contain a stop codon.

operon A set of linked genes that are regulated as a unit.

opioid peptide A molecule produced in nervous tissue cells that relieves pain and produces pleasant sensations.

optical isomer A stereoisomer that possesses one or more chiral centers.

organelle A membrane-enclosed structure within a eukaryotic cell.

origin-of-replication complex (ORC) A protein complex that binds to the DNA replication origin during the initiation phase of DNA synthesis; contains analogues of the protein DnaA.

osmolytes An osmotically active substance synthesized by cells to restore osmotic balance.

osmosis The diffusion of solvent through a semipermeable membrane,

osmotic diuresis A process in which solutes in the urinary filtrate causes excessive loss of water and electrolytes.

osmotic pressure The pressure forcing the solvent, water, to flow across a membrane.

outer membrane The porous external membrane of the mitochondrion.

oxidation The removal of electrons.

oxidative phosphorylation The synthesis of ATP coupled to electron transport.

oxidation-reduction (redox) reaction A reaction involving the transfer of one or more electrons from one reactant to another.

oxidative stress Excessive production of reactive oxygen species.

oxidizing agent A substance that oxidizes (removes electrons from) another substance; the oxidizing agent is itself reduced in the process.

oxidoreductase An enzyme that catalyzes an oxidation-reduction reaction.

oxyanion A negatively charged oxygen atom.

palindrome A sequence that provides the same information whether it is read forward or backward; DNA palindromes contain inverted repeat sequences.

passive transport Transport of a substance across membranes that requires no direct input of energy.

Pasteur effect The observation that glucose consumption is greater under anaerobic conditions than when oxygen is utilized.

pentose phosphate pathway A biochemical pathway that produces NADPH, ribose and several other sugars.

peptide An amino acid polymer with fewer than 50 amino acid residues.

peptide bond An amide linkage in an amino acid polymer.

perinuclear space The space between the two membranes of the nuclear envelope.

peripheral protein A protein that is not embedded in the membrane but attached either by a covalent bond to a lipid molecule or by noncovalent interactions with a membrane protein or lipid.

pernicious anemia An illness caused by a deficiency of vitamin B_{12}; symptoms include low red blood cell count, weakness, and neurological disturbances.

pH optimum The pH value at which an enzyme's activity is maximal.

pH scale A measure of hydrogen ion concentration; pH is the negative log of the hydrogen ion concentration in moles per liter.

phagocytosis The engulfment of foreign or damaged cells by certain white blood cells.

phase I reaction A biotransformation reaction involving oxidoreductases and hydrolases that converts hydrophobic substances into more polar molecules.

phase II reaction A biotransformation reaction in which metabolites containing appropriate functional groups are conjugated with substances such as glucuronate, glutamate, sulfate, and glutathione.

3′-phosphoadenosine-5′-phosphosulfate A high-energy sulfate donor molecule used in the biosynthesis of the sulfatides; a type of glycolipid.

phosphoglyceride A type of lipid molecule found predominantly in membrane; composed of glycerol linked to two fatty acids, phosphate, and a polar group.

phospholipid An amphipathic molecule that has a hydrophobic domain (hydrocarbon chains of fatty acid residues) and a hydrophilic (a polar head group) domain; an important structural component of membranes.

phosphoprotein A conjugated protein in which phosphate is the prosthetic group.

phosphoryl group transfer potential The tendency of a phosphorylated molecule to undergo hydrolysis.

photoautotrophs Organisms that transform light energy (usually from the sun) into chemical bond energy.

photochemistry The study of chemical reactions that are initiated by light absorption.

photoheterotrophs Organisms that use both light and biomolecules as energy sources.

photophosphorylation The synthesis of ATP coupled to electron transport driven by light energy.

photoreactivation repair A mechanism to repair thymine dimers using the energy of visible light.

photorespiration A light-dependent process occurring in plant cells actively engaged in photosynthesis that consumes oxygen and liberates carbon dioxide.

photosynthesis The trapping of light energy and its conversion to the chemical energy required to incorporate carbon dioxide into organic molecules.

photosystem A photosynthetic mechanism composed of light-absorbing pigments.

plasma membrane The membrane that surrounds a cell, separating it from its external environment.

plasmid A circular double-stranded DNA molecule that can exist and replicate independently of a bacterial chromosome; plasmids are stably inherited but are not required for the host cell's growth and reproduction.

plastid An organelle found in plants, algae, and some protists that contain pigments and/or storage materials such as carbohydrate.

platelet-derived growth factor A protein secreted by blood platelets during clotting; stimulates mitosis during wound healing.

point mutation A change in a single nucleotide base in a DNA sequence.

polar An unequal distribution of electrons in a bond.

polar head group A molecular group that contains phosphate or other charged or polar groups.

poly(A) binding protein (PABP) Forms a circular mRNA molecule, during the initiation phase of eukaryotic translation, by interacting with the 3′-poly(A) tail and 5′-capped end of the mRNA and eIF-G, a translation initiation factor.

polymerase chain reaction (PCR) A laboratory technique that uses a heat-stable DNA polymerase to synthesize large quantities of specific nucleotide sequences from small amounts of DNA.

polypeptide An amino acid polymer with more than 50 amino acid residues.

polysaccharide A linear or branched polymer of monosaccharides linked by glycosidic bonds.

polysome An mRNA with several ribosomes bound to it.

polyunsaturated Describing a fatty acid with two or more double bonds, usually separated by a methylene group.

positive cooperativity The mechanism in which the binding of one ligand to a target molecule increases the likelihood of subsequent ligand binding.

positive feedback The mechanism in a self-regulating system in which the product of a reaction increases its own production.

postabsorptive The phase in the feeding-fasting cycle in which nutrient levels are low.

postprandial The phase in the feeding-fasting cycle immediately after a meal: blood nutrient levels are relatively high.

posttranslational modification One of a set of reactions that alter the structure of newly synthesized polypeptides.

posttranslational translocation The transfer of previously synthesized polypeptides across the membrane of the rough endoplasmic reticulum.

PPAR Peroxisome proliferator-activated receptors.

preinitiation complex A multisubunit protein complex, formed during the initiation phase of eukaryotic protein synthesis, that is able to bind to an mRNA.

prenylation The covalent attachment of prenyl groups (e.g., farnesyl and geranylgeranyl groups) to protein molecules.

preproprotein An inactive precursor protein with a removable signal peptide.

primary structure The amino acid sequence of a polypeptide.

primase An RNA polymerase that synthesizes short RNA segments, called primers, that are required for DNA synthesis.

primer A short RNA segment required to initiate DNA synthesis.

primosome A multienzyme complex involved in the synthesis of RNA primers at various intervals along the DNA template strand during *E. coli* DNA replication.

prion Proteinaceous infectious particle: believed to be a causative agent of several acquired neurodegenerative diseases (e.g., "mad cow" disease and Creutzfeld-t-Jakob disease).

processivity The prevention of frequent dissociation of a polymerase from the DNA template.

proenzyme An inactive precursor of an enzyme.

prokaryotic cell A living cell that lacks a nucleus.

promoter The sequence of nucleotides immediately before a gene that is recognized by RNA polymerase and signals the start point and direction of transcription.

prophage A viral genome integrated into host cell DNA.

proprotein An inactive precursor protein.

prostaglandin An arachidonic acid derivative that contains a cyclopentane ring with hydroxyl groups at C-11 and C-15.

prosthetic group The nonprotein portion of a conjugated protein that is essential to the biological activity of the protein; often a complex organic group.

protein A macromolecule composed of one or more polypeptides.

protein family A group of protein molecules that are related by amino acid sequence similarity.

protein folding The process in which an unorganized polypeptide acquires a highly organized and relatively stable three-dimensional structure.

protein splicing A posttranslational mechanism in which an intervening peptide sequence is precisely excised from a nascent polypeptide.

protein superfamily A large group of distantly related proteins; for example the globin superfamily includes the hemoglobins and myoglobins, which bind oxygen in blood and muscle cells, respectively, and the cytoglobins, which bind oxygen in the brain.

protein turnover The continuous degradation and resynthesis of protein in an organism.

proteoglycan A large molecule containing large numbers of glycosaminoglycan chains linked to a core protein molecule.

proteome The complete set of proteins produced within the cell.

proteomics The investigation of protein synthesis patterns and protein-protein interactions.

proteosome A multienzyme complex that degrades proteins linked to ubiquitin.

protocol The set of rules that specify how modules in a system interact.

protomer A component of an oligomer; may consist of one or more subunits.

protonmotive force The force arising from a gradient of protons and a membrane potential.

protooncogene A normal gene that codes for a protein involved in cell cycle regulation; promotes carcinogenesis if mutated.

purine A nitrogenous base with a two-ring structure; a component of nucleotides.

pyrimidine A nitrogenous base with a single-ring structure; a component of nucleotides.

Q cycle The movement of electrons from reduced coenzyme Q, UQH_2, to cytochrome c during electron transport.

quaternary structure Association of two or more folded polypeptides to form a functional protein.

racemization The interconversion of enantiomers.

radical An atom or molecule with an unpaired electron.

reaction center The membrane-bound protein complex in a photosynthesizing cell that mediates the conversion of light energy into chemical energy.

reaction mechanism Step-by-step description of a chemical reaction process.

reactive nitrogen species (RNS) Nitrogen-containing radicals often classified as ROS; the most important are nitric oxide, nitrogen dioxide, and peroxynitrite.

reactive oxygen species (ROS) A reactive derivative of molecular oxygen, including superoxide radical, hydrogen peroxide, the hydroxyl radical, and singlet oxygen.

reading frame A set of contiguous triplet codons in an mRNA molecule.

receptor protein A protein with binding sites for extracellular ligands (signal molecules).

receptor tyrosine kinase (RTK) A transmembrane receptor that contains a cytoplasmic domain with tyrosine kinase activity that is activated when a ligand is bound to the external domain.

recombinant DNA technology A series of techniques whose essential feature is that DNA molecules obtained from various sources can be cut and spliced together.

recombination A process in which DNA molecules are cut and rejoined in new combinations.

recombinational repair A repair mechanism that can eliminate certain types of damaged DNA sequences that were not eliminated before replication; the undamaged parental strands recombine into the gap left after the removal of the damaged sequence.

redox potential A measure of the tendency of an electron donor in a redox pair to lose an electron.

reducing agent A substance that reduces the oxidation number of another reactant; the reducing agent is itself oxidized in the process.

reducing sugar A sugar that can be oxidized by weak oxidizing agents.

reduction A decrease in oxidation number of an atom or molecule.

reduction potential The tendency for a specific substance to lose or gain electrons.

redundancy The use of duplicate parts in a fail-safe mechanism in a robust system.

releasing factor A protein involved in the termination phase of translation.

replication The process in which an exact copy of parental DNA is synthesized using the polynucleotide strands of the parental DNA as templates.

replication factor C A clamp loader protein that controls the attachment of DNA polymerase δ to each DNA strand.

replication factories Specific nuclear compartments (or nucleoids) in which DNA replication occurs.

replication fork The Y-shaped region of a DNA molecule that undergoes replication; results from separation of two DNA strands.

replication licensing factor The proteins that bind to the origin-of-replication complex (ORC) and complete the structure of the preRC.

replication protein A (RPA) A protein that stabilizes the separated DNA strands during replication.

replicon A unit of the genome that contains an origin for initiating replication.

replisome The large complex of polypeptides, including the primosome, that replicates DNA in *E coli*.

resonance energy transfer The transfer of energy from an excited molecule to a nearby molecule, thereby exciting the second molecule.

resonance hybrid A molecule with two or more alternative structures that differ only in the position of electrons.

respiration A biochemical process whereby fuel molecules are oxidized and their electrons are used to generate ATP.

respiratory burst An oxygen-consuming process in scavenger cells such as macrophages in which reactive oxygen species are generated and used to kill foreign or damaged cells.

respiratory control The control of aerobic respiration by ADP concentration.

response element A DNA sequence within the promoter of genes that are coordinately regulated; transcription is triggered when a specific hormone receptor complex binds.

restriction fragment length polymorphism (RFLP) One of a vast number of DNA sequence variations that can be used to identify individuals.

retroelement See retrotransposon.

retroposon See retrotransposon.

retrotransposon A subclass of transposons that use an RNA intermediate.

retrovirus One of a group of viruses with RNA genomes that carry the enzyme reverse transcriptase and form a DNA copy of their genome during the reproductive cycle.

rho-dependent termination Transcription termination in prokaryotes that requires rho factor

rho factor An ATP-dependent helicase involved in transcription termination in bacteria.

rho-independent termination A form of transcription termination in prokaryotes that does not involve rho factor; also referred to as intrinsic termination.

riboflavin Vitamin B_{12}.

ribosome A protein-RNA complex that is the site of protein biosynthesis.

ribosome recycling factor In bacteria, a tRNA-shaped protein; it binds within the A site and causes the dissociation of the ribosomal subunits after polypeptide synthesis.

ribosomal RNA (rRNA) The RNA present in ribosomes. Ribosomes contain several types of single-stranded ribosomal RNA that contribute to ribosome structures and are also directly involved in protein synthesis.

riboswitch An RNA-based control mechanism made up of a specific untranslated sequence within an mRNA. Usually, the action of the riboswitch, triggered by a change in its tertiary structure induced by binding to a ligand, is to repress translation.

RNA Ribonucleic acid; a single-stranded, unbranched macromolecule formed from ribonucleotides; a fundamental component in protein biosynthesis.

RNA editing The alteration of the base sequence in a newly synthesized mRNA molecule; bases may be chemically modified, deleted, or edited.

RNA-induced silencing complex (RISC) The antisense strand of the short interfering RNA bound to the RISC complex targets and anneals with the viral micro RNA; nucleases within the RISC then degrade the viral miRNA sequence.

RNA interference A cellular mechanism in which RNA molecules are degraded; functions in gene expression regulation and in defense against viral RNA genomes.

RNA splicing The process in which introns are cut out and the exons are linked together to form a functional RNA product.

RNA transposon A transposable element that uses a mechanism that involves an RNA transcript; also referred to as retrotransposon.

robust Describing a system that remains stable despite perturbations of diverse types.

rough endoplasmic reticulum (RER) A type of endoplasmic reticulum that has ribosomes bound to its external surface; nascent polypeptides, containing a signal peptide, are translocated through the RER membrane.

salt bridge An electrostatic interaction in proteins between ionic groups of opposite charge.

salting out The decrease in protein solubility caused by an increase in the ionic strength of the solution.

satellite DNA DNA sequences that are highly repetitive; when genomic DNA is digested and centrifuged, a satellite band forms.

saturated Describes molecule that contains no carbon–carbon double or triple bonds.

SDS-polyacrylamide gel electrophoresis A method for separating proteins or determining their molecular weights that employs the negatively charged detergent sodium dodecyl sulfate (SDS).

SECIS element **S**elenocysteine **i**nsertion **s**equence **e**lement; a base sequence required at the 3′-UTR (3′ untranslated region) of the mRNAs for all selenoproteins for the insertion of selenocysteine during translation.

second messenger A molecule that mediates the action of some hormones.

secondary structure The arrangement of a polypeptide chain into locally organized structures of α-helix and β-pleated sheet: secondary structure is maintained by hydrogen bonds between the amide hydrogen and the carbonyl oxygen of peptide bonds.

semiconservative replication DNA synthesis in which each polynucleotide strand serves as a template for the synthesis of a new strand.

sense strand The nontranscribed DNA strand; the DNA version of the mRNA used to synthesize the polypeptide product of a gene.

shine-dalgarno sequence A purine-rich sequence that occurs on an mRNA close to AUG (the initiation codon) that binds to a complementary sequence on the 30S ribosomal subunit, thereby promoting the formation of the correct preinitiation complex.

short interfering RNA (siRNA) A 20–25 nt noncoding RNA involved in RNA interference; a dsRNA that can specifically silence gene expression.

short tendem repeats DNA sequences with between 2 and 4 bp repeats; can be used to generate DNA profiles that distinguish among individuals.

shotgun cloning A cloning technique in which genomic libraries are created by the random digestion of a genome.

signal hypothesis A mechanism that explains how secreted or membrane proteins are synthesized on ribosomes bound to the rough endoplasmic reticulum; a sequence of amino acid residues on the nascent polypeptide chain that mediates the insertion of the polypeptide into the RER membrane.

signal peptide A short sequence typically near the amino terminal of a polypeptide that determines its insertion into a membrane of an organelle.

signal recognition particle A large multisubunit ribonucleoprotein complex that mediates the binding of the ribosome and the emerging signal peptide to the rough endoplasmic reticulum during protein synthesis; facilitates the passage of the growing polypeptide through the RER membane.

signal transduction The mechanisms by which extracellular signals are received, amplified, and converted to a cellular response.

simple diffusion A process in which each type of solute, propelled by random molecular motion, moves down a concentration gradient.

SINE (short interspersed nuclear elements) A repeating DNA sequence less than 500 bp long interspersed in mammalian genomes; SINES cannot undergo transposition without the aid of a functional LINE sequence.

site-directed mutagenesis A technique that introduces specific sequence changes into a cloned gene.

site-specific recombination Recombination of nonhomologous DNA sequences; a recombination mechanism between sequences with limited homology.

small nuclear ribonucleoprotein particle A complex of proteins and small nuclear RNA molecules that promotes RNA processing.

small nuclear RNA A small RNA molecule involved in the removal of introns from mRNA, rRNA and tRNA.

small nucleolar RNA (snRNA) An RNA component of nucleolar ribonucleoprotein that facilitates chemical modifications of rRNA.

smooth endoplasmic reticulum (SER) A type of endoplasmic reticulum involved in lipid synthesis and biotransformation processes.

solvation sphere A shell of water molecules that clusters around positive and negative ions.

somatomedin A polypeptide that mediates the growth promoting action of growth hormone.

southern blotting A laboratory technique in which radioactively labeled DNA or RNA profiles are used to locate a complementary sequence in a DNA digest.

spliceosome A multicomponent complex containing protein and RNA; used in the splicing phase of mRNA processing.

supraspliceosome Formed by four active spliceosomes and a pre-mRNA complex; it increases the speed and efficiency of transcript splicing and provides opportunities for intron excision proofreading.

sphingolipid A membrane lipid molecule that contains a long-chain amino alcohol and ceramide (a fatty acid derivative of sphingosine); an important component of plant and animal membranes.

sphingomyelin A type of phospholipid that contains sphingosine; the 1-hydroxyl group of ceramide is esterified to the phosphate group of phosphorylcholine or phosphorylethanolamine, and the amino group of sphingosine is in an amide linkage with a fatty acid.

spontaneous changes Physical or chemical processes that are accompanied by a release of energy.

SREBP (sterol regulatory element binding protein) One of several transcription factors that are membrane proteins in the endoplasmic reticulum or the Golgi apparatus.

standard reduction potential The measurement of the capacity of a substance to gain or lose electrons in a galvanic cell fitted with a standard hydrogen electrode set at 0.00V.

stationary phase The solid matrix in chromtographic techniques.

steady state A phase in an organism's life when the rate of anabolic processes is approximately equal to that of catabolic processes.

stereoisomer A molecule that has the same structural formula and bonding pattern as another but has a different arrangement of atoms in space.

steroid A derivative of triterpenes; contains four fused rings.

sterol carrier protein A cytoplasmic protein carrier for certain intermediates during cholesterol biosynthesis.

stroma A dense, enzyme-filled substance that surrounds the thylakoid membrane within the chloroplast.

stromal lamella A thylakoid membrane segment that interconnects two grana.

substrate The reactant in a chemical reaction that binds to an enzyme active site and is converted to a product.

substrate-level phosphorylation The synthesis of ATP from ADP by phosphorylation coupled with the exergonic breakdown of a high-energy organic substrate molecule.

subunit A polypeptide component of an oligomeric protein

sucrose A disaccharide composed of α-glucose and β-fructose residues linked through a glycosidic bond between both anomeric carbons.

supersecondary structure One of a set of specific combinations of α-helix and β-pleated-sheet structures of protein molecules.

systems biology A field of study based on engineering principles in which the interactions between the components of living organisms are investigated; complex data sets used by systems biologists come from genomics, proteomics, and experimental sources such as protein-protein interactions and biochemical reaction fluxes.

T cell A T lymphocyte; a white blood cell that bears antibody-like molecules on its surface and binds to and destroys foreign cells in cellular immunity.

tandem repeats DNA sequences in which multiple copies are arranged next to one another; lengths of repeated sequences vary from 10 bp to over 2000 bp.

target cell A cell that responds to the binding of a hormone or growth factor to a receptor protein.

targeting The process that directs newly synthesized proteins to their correct intracellular destination.

tautomer An isomer that differs from another in the location of a hydrogen atom and a double bond (e.g., keto-enol tautomers).

tautomerization Chemical reaction by which two tautomers are interconverted by the movement of a hydrogen atom and a double bond.

telomere A structure found at both end of a chromosome that buffers the loss of critical coding sequences after a round of DNA replication.

telomerase A ribonucleoprotein with an RNA component (a TG-rich repeat sequence) complementary to the telomere sequence.

telomere end-binding protein (TBP) A protein that binds to and stabilizes GT-rich telomere sequences.

telomere repeat-binding factor (TRBP) A protein that binds to and secures the 3′ overhang sequence of a telomere.

termination The phase of translation in which newly synthesized polypeptides are released from the ribosome.

terpene A member of a class of isoprenoids classified according to the number of isoprene residues it contains.

tertiary structure The globular three-dimensional structure of a polypeptide that results from interactions between the side chains (R groups) of the amino acid residues.

thermodynamics The study of energy and its interconversion.

thermogenin See uncoupling protein.

thiamine pyrophosphate The coenzyme form of thiamine; also called vitamin B_1.

thiolytic cleavage Cleavage of a carbon-sulfur bond.

thromboxane A derivative of arachidonic acid that contains a cyclic ester.

thylakoid lumen The internal compartment created by the formation of grana.

thylakoid membrane An intricately folded internal membrane within the chloroplast.

trans **isomer** An isomer in which two substituents are on opposite sides of the double bond.

transamination A reaction in which an amino group is transferred from one molecule to the α-carbon of an α-keto acid; the amino acid that donates the amino group is converted to the corresponding α-keto acid.

transcript An RNA molecule that is produced by the transcription of a DNA sequence.

transcription The process in which single-stranded RNA with a base sequence comlementary to the template strand of DNA is synthesized.

transcription factor A protein that regulates or initiates the synthesis of specific mRNAs by binding to DNA sequences called response elements.

transcript localization The binding of mRNAs to certain cellular structures within cytoplasm that permits the creation of protein gradients within the cell.

transcriptome The complete set of RNA molecules that are produced within a cell.

transduction The transfer of DNA segments between bacteria by bacteriophages.

transfection A mechanism by which bacteriophage inadvertently transfers bacterial chromosome or plasmid sequences to a new host cell.

transfer RNA (tRNA) A small RNA molecule, that binds to an amino acid and delivers it to the ribosome for incorporation into a polypeptide chain during translation.

transferase An enzyme that catalyzes the transfer of a functional group from one molecule to another.

transformation A process in whichDNA fragments enter a bacterial cell and are introduced into the bacterial genome.

transgenic animal An animal that results when recombinant DNA sequences are microinjected into a fertilized ovum.

transition mutation A DNA mutation that involves the substitution of a purine base by a different purine, or the substitution of a pyrimidine by a different pyrimidine.

transition state In catalysis, the unstable intermediate formed by the enzyme that has altered the substrate so that it now shares properties of both the substrate and the product.

translation Protein synthesis; the process by which the genetic message carried by mRNAs directs the synthesis of polypeptides with the aid of ribosomes and other cell constituents.

translational frame-shifting A +1 or −1 change in reading frame allows more than one polypeptide to be synthesized from a single mRNA.

translocation Movement of the ribosome along the mRNA during translation.

translocon An integral membrane protein that mediates translocation of a polypeptide.

transposase A prokaryotic transposition enzyme, coded for by a gene within an IS element.

transposable DNA elements A DNA sequence that excises itself and then inserts at another site.

transposition The movement of a DNA sequence from one site in a genome to another.

transposon (transposable element) A DNA segment that carries the genes required for transposition and moves about the chromosome; sometimes the name is reserved for transposable elements that also contain genes unrelated to transposition.

transulfuration pathway A biochemical pathway that converts methionine to cysteine.

transversion mutation A type of point mutation in which a pyrimidine is substituted for a purine and vice versa.

tumor necrosis factor (TNF) A protein, toxic to tumor cells, that suppresses cell division.

tumor promoter A molecule that provides cells with a growth advantage over nearby cells.

turnover The rate at which all molecules in a cell are degraded and replaced with newly synthesized molecules.

turnover number The number of molecules of substrate converted to product in each second per mole of enzyme.

ubiquination The covalent attachment of ubiquitin to proteins that are to be degraded.

ubiquitin A protein that is covalently attached by enzymes to proteins destined to be degraded.

ubiquitin proteosomal system An elaborate mechanism for the rapid destruction of proteins.

uncoupler A molecule that uncouples ATP synthesis from electron transport; it collapses a proton gradient by transporting protons across the membrane.

uncoupling protein A molecule that dissipates the proton gradient in mitochondria by translocating protons; also called thermogenin.

unsaturated molecule A molecule that contains one or more carbon–carbon double or triple bonds.

urea cycle A cyclic pathway in which waste ammonia molecules, carbon dioxide, and the amino nitrogen of aspartate molecules are converted to urea.

uronic acid The product formed when the terminal CH_2OH of a monosaccharide is oxidized.

van der Waals force A class of relatively weak, transient electrostatic interactions between permanent and/or induced dipoles.

vector A cloning vehicle into which a segment of foreign DNA can be spliced, ready for introduction into host cells and expression in them.

velocity The rate of a biochemical reaction; the change in the concentration of a reactant or product per unit time.

very-low-density lipoprotein (VLDL) A type of lipoprotein with a very high relative concentration of lipids; transports lipids to tissues.

vesicles Membranous sacs that bud off from a donor membrane and subsequently fuse with the membrane of another organelle or with the plasma membrane.

vesicular organelles Small spheroidal membranous sacs that contain substances that originated in the endoplasmic reticulum and Golgi apparatus or were brought into the cell by endocytosis.

vitamin An organic molecule required by organisms in minute quantities; some vitamins are coenzymes required for the function of certain cellular enzymes.

vitamin B_{12} A complex cobalt-containing molecule that is required for the N^5-methyl THF-dependent conversion of homocysteine to methionine.

wax A complex mixture of nonpolar lipids including wax esters.

weak acid An organic acid that does not completely dissociate in water.

weak base An organic base that has a small but measurable capacity to combine with hydrogen ions.

wobble hypothesis The hypothesis that explains why cells often have fewer tRNAs than expected; freedom in the pairing of the third base of the codon to the first base of the anticodon allows some tRNAs to pair with several codons.

work A physical change caused by a change in energy.

X inactivation The process by which female mammals silence one of their two X chromosomes.

yeast artificial chromosome A cloning vector that can accommodate up to 100 kb of DNA; contains eukaryotic sequences that function as centromeres, telomeres, and a replication origin.

Z DNA A form of DNA that is twisted into a left-handed spiral; named for the zigzag conformation, which is slimmer than that of B-DNA.

Z scheme A mechanism whereby electrons flow between photosystems II and I during photosynthesis.

Zymogen The inactive form of a proteolytic enzyme.

Zwitterion A neutral molecule that bears an equal number of positive and negative charges simultaneously.

Credits

PHOTOS

Chapter 1

1.1: © Woods Hole Oceanographic Institute, Woods Hole, MA.; **1.3:** © Susan Detwiler; **1.21a-c:** © & Courtesy of David S. Goodsell, the Scripps Research Institute.

Chapter 2

2.11: © Charles C. Brinton, Jr. and Judith Carnahan; **2.16:** ©Audrey M. Glauert & G. M. W. Cook; **2.20** © P. Schulz/Biology Media/Photo Researchers, Inc.; **2.21:** © Don Fawcett/Photo Researchers. Inc.; **2.22b:** © M. M. Perry and A. B. Gilbert. *J. of Cell Science* 39:257–272, 1979. Company of Biologists Ltd.; **2.23:** © Gopal Murti/Phototake; **2.24** © Gopal Murti/Visuals Unlimited; **2.25b:** © Don Fawcett/ Photo Researchers, Inc.; **2.26 a & b:** © C. R. Hackenbrook. Ultrastructural bases for metabolically linked mechanical activity in mitochondria. *J. of Cell Biology* 37: no. 2, 364–365 (1968). By permission American Society for Microbiology: © Rockefeller University Press; **Figure 2A:** © Don Fawcett/Photo Researchers. Inc.; **2.29a:** © J. W. Shuler/Photo Researchers. Inc.; **2.29b:** © J. L. Carson/ Custom Medical Stock Photo; **2.29c:** © Dr. Peter Dawson/Science Photo Library/Photo Researchers, Inc.

Chapter 3

Opener: NASA.

Chapter 4

Opener: © Image Source Black/Alamy.

Chapter 5

Opener: © & Courtesy of David S. Goodsell, the Scripps Research Institute; **5.15a & b:** From *Molecules of Life*, Purdue University; **5.35:** K. A. Piez in D. B. Wetlander. ed., *The Protein Folding Problem*, AAAS Selected Symposium 89, American Association for the Advancement of Science, Washington, D.C., 1984, pp. 47–61. Courtesy of the Collagen Corporation. Reproduced by permission of American Association for the Advancement of Science.

Chapter 7

7.19a: © Leonard Lessin/Peter Arnold, Inc.; **7.21, 7.22:** © Leonard Lessin/FBPA.

Chapter 8

Opener: © Stockbyte/Punchstock.

Chapter 10

10.13: From D. F. Parsons, *Science* 1963. 140: page 985. © 2001 American Association for the Advancement of Science.

Chapter 11

11.3a: © Leonard Lessin/Peter Arnold. Inc.; **11.3b:** © Leonard Lessin/FBPA; **11.5:** © Leonard Lessin/Peter Arnold, Inc.; **11.11:** © Leonard Lessin/FBPA; **11.16a:** © Leonard Lessin/Peter Arnold, Inc.

Chapter 12

Opener: © AFP/Getty Images.

Chapter 13

Opener: © Minden Pictures/Getty Images. Photographers: Michael & Patricia Fogden.

Chapter 17

Opener: From Elizabeth Pennisi, *The Human Genome*, Science Feb. 16 2001. 291: page 1177. © 2001 American Association for the Advancement of Science. Illustration by Cameron Slayden; **17.8:** From J. D. Watson, *The Double Helix*, p. 215, New York: Atheneum. © 1968 by J. D. Watson. A. C. Barrington photographer. Courtesy of Cold Spring Harbor Laboratory Archives; **17.11:** © Lee D. Simon, Department of Genetics, Rutgers University; **17.12:** R. E. Franklin and R. Gosling. Molecular configuration in sodium thymonucleate. *Nature* 171: 740–741. © 1953 Macmillan Magazines Ltd. Reprinted with permission from *Nature*.

Chapter 18

18.4: Reprinted from CELL, Vol. 79, 1994. pp 1233–1243, Krishna et al, "Crystal structure of…" © 2002, with permission from Elsevier Science; **18C:** © & courtesy of Dr. Keith V. Wood; **18H:** Courtesy of AFFYMETRIX, Inc.

Chapter 19

Opener: From Marat M. Yusupov, Gulnara Zh. Yusupova, Albion Baucom, Kate Lieberman, Thomas N. Earnest, J. H. D. Cate, and Harry F. Noller, *Crystal Structure of the Ribosome at 5.5 Å Resolution*, Science May 4 2001. 292: page 885. © 2001 American Association for the Advancement of Science; **19B:** Banks R. E., Dunn M. J., Hochstrasser D. F., Sanchez J.-Ch., Blackstock W., Pappin D. J., Selby P. J. "Proteomics: new perspectives, new biomedical opportunities." *The Lancet*, Vol. 356, No 9243, p. 1749–1756, November 18th, 2000. Reprinted with permission from Elsevier Science.

LINE ART

Chapter 2

2.4: Copyright © 1998 From Essential Cell Biology by Bruce Alberts, et al. Reproduced by permission of Routledge, Inc., part of the Taylor & Francis Group. **2.5** R. J. Ellis, "Macromolecular Crowding: Obvious but Underappreciated," *Trends in Biochemical Science.* 2004; **2.6:** From David S. Goodsell, The Machinery of Life. 1998. Copyright © 1998 Springer-Verlag. Reprinted with permission of Springer-Verlag/Germany. **2.7:** From M. Hoppert and F. Mayer. "Prokaryotes," American Scientist, Vol. 87. November–December 1999. Copyright © David Goodsell. Reprinted with permission. **2.8:** From David S. Goodsell, The Machinery of Life, 1998. Copyright © 1998 Springer-Verlag. Reprinted with permission of Springer-Verlag/Germany. **2.9:** From Prescott, et al., Microbiology 4/e. Copyright © 1999 by The McGraw-Hill Companies. This material is reproduced with permission of The McGraw-Hill Companies. **2.10a & b:** From David S. Goodsell, The Machinery of Life, 1998. Copyright © 1998 Springer-Verlag. Reprinted with

permission of Springer-Verlag/Germany. **2.12:** From Microbiology: An Introduction, by Gerard J. Tortora, Berdell R. Funke and Christine L. Case. Copyright © 1998 by The Benjamin/Cummings Publishing Company, Inc. Reprinted by permission of Addison Wesley Longman Publishers, Inc. **2.16:** From The World of the Cell, 4th ed. by Wayne M. Becker, Lewis J. Kleinsmith, and Jeff Hardin. Copyright © 2000 by Addison Wesley Longman, Inc. Reprinted by permission of Pearson Education, Inc. **2.17:** From The World of the Cell. 4th ed. by Wayne M. Becker, Lewis J. Kleinsmith, and Jeff Hardin. Copyright © 2000 by Addison Wesley Longman, Inc. Reprinted by permission of Pearson Education, Inc. **2.19:** From The World of the Cell, 4th ed. by Wayne M. Becker, Lewis J. Kleinsmith, and Jeff Hardin. Copyright © 2000 by Addison Wesley Longman, Inc. Reprinted by permission of Pearson Education, Inc. **2.21:** From The World of the Cell, 4th ed. by Wayne M. Becker, Lewis J. Kleinsmith, and Jeff Hardin. Copyright © 2000 by Addison Wesley Longman, Inc. Reprinted by permission of Pearson Education, Inc. **2.22a:** From Geoffrey Cooper, The Cell: A Molecular Approach, 1997. Reproduced with permission of Sinauer Associates, Inc. **2.30a & b:** From R. Lewis, Life 3/e. Copyright © 1998 by The McGraw-Hill Companies. This material is reproduced with permission of The McGraw-Hill Companies. **2.31:** Donald F. Ingber, © 1993 *J. of Cell Science*. **2A:** From Geoffrey Cooper, The Cell: A Molecular Approach, 1997. Reproduced with permission of Sinauer Associates, Inc. **2B:** From Lehninger Principles of Biochemistry by David Nelson and Michael Cox. Copyright © 2000, 1993, 1982 by Worth Publishers. Used with permission.

Chapter 3

3.1: From Silverberg, Chemistry 2/e. Copyright © The McGraw-Hill Companies. This material is reproduced with permission of The McGraw-Hill Companies. **3.8:** From R. Chang, Chemistry 7/e. Copyright © The McGraw-Hill Companies. This material is reproduced with permission of The McGraw-Hill Companies. **3.11:** Figure 6.27b from BIOLOGY, 7th ed. by Neil J. Campbell and Jane B. Reece. Copyright © 2005 by Pearson Education, Inc. Reprinted by permission.

Chapter 5

TA5.6: Copyright © 1999 from Introduction to Protein Structure by Carl Brandon and John Tooze. Reproduced by permission of Routledge, Inc., part of the Taylor & Francis Group; **5.15a & b:** Molecules of Life, Purdue University; **5.17** Copyright © 1999 from Introduction to Protein Structure by Carl Brandon and John Tooze. Reproduced by permission of Routledge, Inc., part of the Taylor & Francis Group; **5.19b:** Adapted from Biochemistry, by Reginal H. Garrett and Charles M. Grisham, 1996. Brooks/Cole Publishing. **5.21a:** With permission from the Annual Review of Biochemistry, Volume 45 © 1976 by Annual Reviews *www.AnnualReviews.org.* ; **5.21c,d, & e:** Copyright © 1999 from Introduction to Protein Structure by Carl Brandon and John Tooze. Reproduced by permission of Routledge, Inc., part of the Taylor & Francis Group; **5.22** D. Campbell and A. K. Downing, "NMR of Modular Proteins," Nature Structural Biology NMR Supplement. **5.24:** C. Reid and R. P. Rand, "Probing Protein Hydration and Conformational States in Solution," Biophysical Journal, 1997; **5.25:** From "Principles of Biochemistry with a Human Focus" 1st edition by GARRETT. 2002. Reprinted with permission of Brooks Cole, a division of Thomson Learning: www.thomsonrights.com. Fax 800 730–2215; **5.27:** "Intrinsically Unstructured Proteins and Their Functions," Nature Reviews: Molecular Cell Biology; **5.29a:** From S.T. Ferreira and F. G. Felice, "Protein Dynamics, Folding and Misfolding," FEBS Letters 498:129–134, 2001. Reprinted with permission of Elsevier Science BV; **5.29b:** From S.E. Ranford. "Protein Folding," Trends in Biochemical Sciences, Vol. 25, pp. 611–618. Copyright © 2000 Elsevier Science. Reprinted with permission from Elsevier Science; **5.30:** From S. E. Ranford, "Protein Folding," Trends in Biochemical Sciences, Vol. 25, pp. 611– 618. Copyright © 2000 Elsevier Science. Reprinted with permission from Elsevier Science. **5.31:** From Z. Xu, L. Horwich, and P. B. Stigler, "The Crystal Structure of the Asymmetric GroEl-Gros-Es-(ADP)7 Chaperonen Complex." Nature, Vol. 388, August 21, 1997, pp. 741–750. Reprinted with permission of Nature; **5.33:** Adapted from Biochemistry, by Reginal H. Garrett and Charles M. Grisham, 1996. Brooks/Cole Publishing; **5.35:** Figure from Kilmartin and Rossi-Bernardi, Physiological Review, Vol. 53, p. 884, 1973. Reprinted with permission of American Physiological Association. **5A:** Fig. 16.9, p. 462 from THE WORLD OF THE CELL, 6th ed. by Wayne M. Becker, Lewis J. Kleinsmith, and Jeff Hardin. Copyright 2006 by Pearson Education, Inc. Reprinted by permission; **5B:** "Cell Biology," Saunders: 2002, figure 39.3, p. 607; **5C:** Journal of General Physiology: The Sliding Filament Model: 1972–2004. Rockefeller University Press, figure 3 p. 649; **5D:** From Biochemistry, 2nd ed. by Christopher K. Mathews and K. E. Van Holde. Copyright © 1996 by The Benjamin/Cummings Publishing Company, Inc. Reprinted by permission of Pearson Education, Inc.

Chapter 6

6.2 a & b: From Biochemistry, 2nd ed. By Christopher K. Mathews and K. E. Van Holde. Copyright © 1996 by The Benjamin/Cummings Publishing Company, Inc. Reprinted by permission of Pearson Education, Inc.

Chapter 7

Opener: © Peter H. Seeberger, "Exploring Life's Sweet Spot," Nature: 437: 1239: October 27, 2005; **7.37:** From Cell Ultrastructure. 1st edition, by S. L. Wolfe © 1985. Reprinted with permission of Brooks/Cole, an imprint of the Wadsworth Group, a division of Thomson Learning. Fax 800 730–2215; **7.40:** From Lehninger Principles of Biochemistry by David Nelson and Michael Cox. Copyright © 2000, 1993, 1982 by Worth Publishers. Used with permission.

Chapter 8

8.8: Used with permission from the Journal of Chemical Education, Vol. 52, No. 6, 1975, pp. 370–373; Copyright © 1975, Division of Chemical Education, Inc.

Chapter 10

Opener: M.L. Hutcheon. T. M. Duncan. H. Ngai, and R. L. Cross, Proceedings of the National Academy of Sciences of the United States, Vol. 98, 2001. Reprinted with permission of Richard L. Cross; **10.4:** From Biochemistry, 2nd edition, by Garrett and Grisham © 1999. Reprinted with permission of Brooks/Cole, an imprint of the Wadsworth Group, a division of Thomson Learning. Fax 800 730–2215; **10.6:** Copyright © 1999 From Molecular Biology of the Cell by Bruce Alberts, et al. Reproduced by permission of Routledge, Inc., part of The Taylor & Francis Group; **10.8:** Figure from D. G. Nicholls and S. J. Ferguson, Bioenergetics 2/e. Copyright © Academic Press LTD, London. Reproduced by permission of the publisher. **10.14:** Reprinted from Trends in Biochemical Sciences, Vol. 22, Jung, Hill, Engelbrecht, pp. 420–423. Reprinted with permission from Elsevier Science; **10.15:** From Biochemistry, Update (with ThomsonNOW(T), InfoTrac® Printed Access Card) 3rd edition by GARRETT/GRISHAM. 2007. Reprinted with permission of Brooks/Cole, a division of Thommson Learning: www.thomsonrights.com. Fax 800 730–2215; **10.18:** From C. R. Scriver, et al., The Metabolic and Molecular Bases of Inherited Diseases. Copyright © 2001 by The McGraw-Hill Companies. This material is reproduced with permission of The McGraw-Hill Companies.

Chapter 11

Opener: From Geoffrey Cooper, The Cell: A Molecular Approach, 1997. Reproduced with permission of Sinauer Associates, Inc. **11.20:** From Geoffrey Cooper, The Cell: A Molecular Approach, 1997. Reproduced with permission of Sinauer Associates, Inc. **11.27:** From Geoffrey Cooper, The Cell: A Molecular Approach, 1997. Reproduced with permission of Sinauer Associates, Inc. **11.29:** From Lehninger Principles of Biochemistry, 4th edition, by David L. Nelson and Michael M. Cox. © 2004, W.H. Freeman. Used with permission. **11A:** From Thomas Zeuthen, Trends in Biochemical Sciences, Vol. 26, No. 2, pp. 77–79. Copyright © 2001 Elsevier Science. Reprinted with permission from Elsevier Science. **11B:** From Thomas Zeuthen, Trends in Biochemical Sciences, Vol. 26, No. 2, pp. 77–79. Copyright © 2001 Elsevier Science. Reprinted with permission from Elsevier Science.

Chapter 12

12.17: From Biochemistry, Update (with ThomsonNOW(T), InfoTrac® Printed Access Card) 3rd edition by GARRETT/GRISHAM. 2007. Reprinted with permission of Brooks/Cole, a division of Thommson Learning: www.thomsonrights.com. Fax 800 730–2215. **12.19:** With permission from the Annual Review of Biochemistry, Volume 52 © 1983 by Annual Reviews *www.AnnualReviews.org*.

Chapter 13

13.2: From The World of the Cell, 4th ed. by Wayne M. Becker, Lewis J. Kleinsmith, and Jeff Hardin. Copyright © 2000 by Addison Wesley Longman, Inc. Reprinted by permission of Pearson Education, Inc. **13.3:** Reprinted from Trends in Biochemical Sciences, Vol. 7, Anderson & Anderson. Copyright © 1982, with permission from Elsevier Science. **13.4:** From *The Complex Architecture of Oxygenic Photosynthesis* by Nathan Nelson. Nature Reviews: Molecular Cell Biology 5: 971–82. **13.5:** From *The Complex Architecture of Oxygenic Photosynthesis*

by Nathan Nelson. Nature Reviews: Molecular Cell Biology 5: 971–82. **13.6:** Reprinted from Trends in Biochemical Sciences, Vol. 22, Jung, Hill, Engelbrecht, pp. 420–423. Reprinted with permission from Elsevier Science. **13.14:** From Biochemistry by Berg, Tymoczko, and Stryer. © 2007, W.H. Freeman. Reprinted by permission. **13.15:** From *The Complex Architecture of Oxygenic Photosynthesis* by Nathan Nelson. Nature Reviews: Molecular Cell Biology 5: 971–82. **13.16:** Adapted from World of the Cell, 4/e by Wayne M. Becker, et al., 2000. **13.17:** Adapted from World of the Cell, 4/e by Wayne M. Becker, et al., 2000. **13.20:** From The World of the Cell, 4th ed. by Wayne M. Becker, Lewis J. Kleinsmith, and Jeff Hardin. Copyright © 2000 by Addison Wesley Longman, Inc. Reprinted by permission of Pearson Education, Inc. **13A:** [CREDIT PENDING]

Chapter 14

14.1: Reprinted with permission from Chemical Review, P. C. Dossantos et al, "Formation and Insertion of the Nitrogenase Iron-Molybedenum Cofactor." Copyright © 2004 American Chemical Society.

Chapter 15

Opener: © David Goodsell, the Scripps Research Institute. **15.2:** From "Mobilizing the Proteolytic Machine: Cell Biological Roles of Proteasome Activators and Inhibitors," by Martin Rechsteiner and Christopher P. Hill, Trends in Cellular Biology: 15 © 2005 Reprinted with permission of Elsevier Science.

Chapter 16

16.1: Figure from J. Koolman and K. H. Rahn, Color Atlas of Biochemistry, 1999. Reprinted with permission of Thieme Medical Publishers. **16.5:** From The World of the Cell, 4th ed. by Wayne M. Becker, Lewis J. Kleinsmith, and Jeff Hardin. Copyright © 2000 by Addison Wesley Longman, Inc. Reprinted by permission of Pearson Education, Inc. **16.7:** From The World of the Cell, 4th ed. by Wayne M. Becker, Lewis J. Kleinsmith, and Jeff Hardin. Copyright © 2000 by Addison Wesley Longman, Inc. Reprinted by permission of Pearson Education, Inc. **16.9:** From Geoffrey Cooper, The Cell: A Molecular Approach, 1997. Reproduced with permission of Sinauer Associates, Inc. **16.12:** From Thomas M. Devlin, Textbook of Biochemistry with Clinical Correlations, 1999. Copyright © 1999 John Wiley & Sons, Inc. This material is used by permission of John Wiley & Sons, Inc. **16B:** © Science VU/Jackson/Visuals Unlimited.

Chapter 17

17.1: From A. H. Fairlamb. "Brand New World of Post Genomes," Trends in Parasitology, Vol. 17, p. 255. Copyright © 2000 Elsevier Science. Reprinted with permission from Elsevier Science. **17.2:** From Feughelman, et al., "Molecular Structure of DNA." Nature, Vol. 175, May 14, 1955,. Reprinted with permission from Nature. **17.4:** Copyright © 1999 From Molecular Biology of the Cell by Bruce Alberts, et al. Reproduced by permission of Routledge, Inc., part of the Taylor & Francis Group. **17.11:** © Lee D. Simon, Department of Genetics, Rutgers University. **17.13:** From Principles of Biochemistry by A. Lehninger, D. Nelson and M. Cox. Copyright © 2000, 1993, 1982 by Worth Publishers. Used with permission. **17.16:** From Terence A. Brown, Genomes, 1999. Reprinted with permission of BIOS Scientific Publishers, Oxford, UK.

Index

Page numbers followed by f and t refer to figures and tables, respectively.

ROS. *See* Reactive oxygen species
Rotenone, 346, 346f
Rough endoplasmic reticulum
    animal cell, 45f
    eukaryotic cells, 48–49
    plant cell, 46f
Rubber, 387, 387t
Rubisco, 489, 491, 495–496
Ruv proteins, Holliday junction, 694f

S

Saccharomyces cerevisiae, ribosomal
    RNA (rRNA), 662f
Salt bridges, 76, 148, 150
Salt concentration, protein denatura-
    tion, 156
Salting out, 156, 172
SAM (S-adenosylmethionine), 210,
    211t
    one-carbon metabolism, 526, 528f,
        529
    pathway, 529f
Sanger chain termination method, 656f,
    657
Saponification, 379
Sarcoma, 729
Satellite DNA, 651
Saturated fatty acids, 14, 373, 373t
Scanning electron microscope (SEM),
    68
Schiff base, 138, 238
SDS–polyacrylamide gel electrophore-
    sis (SDS–PAGE), 174
Seafloor exploration, *Alvin*, 2, 3f
Seasonal affective disorder (SAD), 535
SECIS (selenocysteine insertion
    sequence) element, 761
Secondary structure, proteins, 142,
    146–147
Second law of thermodynamics, 106,
    107–108, 110
Second messengers, 594, 595f, 600,
    600f, 601f
Second-order kinetics, 191
Secretin, 596t
Secretory processes, lysosomes, 53
Secretory vesicles, 50f, 51
Selectins, 253, 256
Selenocysteine, 761–762
Self-assembly, cell, 36
Self-sealing, membranes, 397
Semiconservative replication, 679, 679f
Sense strand, 663
Sequencing, DNA, 657
Sequential model, allosteric, 217f, 218
Sequential reactions, enzyme kinetics,
    196
Serine, 11f, 126f, 127t, 132f, 513f, 515,
    517f
    forming acetyl–CoA, 568, 569f, 570f
    nonessential amino acid, 507t
Serine dehydrase, 563
Serine family, synthesis, 513f, 515,
    517f, 518
Serine proteases, 206, 212
Serotonin, 535
    amino acid derivative, 129f
    degradation, 579, 579f
    neurotransmitter, 533t
Sesquiterpenes, 386, 387t
Sex hormones, 390f
Shikimate pathway, 522, 522f

Shine–Dalgarno sequence, 747, 752,
    755
Short tandem repeats (STR), 658
Shotgun method, DNA library creation,
    703, 704f
Shuttle mechanisms
    electron transfer, 354f, 355f
    plant cells, 488–489
Sialic acid, 241, 241f
Sickle-cell hemoglobin, 144f, 145
Sickle-cell trait, 145
Sigma factor, 708, 708f
Signal cascade, 38, 66
Signal hypothesis, 767
Signal peptides, 750
Signal recognition particle, 767
Signal transduction
    cells, 37–38
    cytoskeleton, 66
    and gene expression, 725–726, 728
    hormones, 38, 130
    mechanisms, 594, 595f
    metabolism, 22–23
    stress-triggered, 98
Silk fibroin, molecular model, 166f
Simple diffusion, 402
SINEs (short interspersed nuclear
    elements), 652
Singer–Nicholson model, 394
Single-stranded DNA binding protein
    (SSB), 684f, 685
Singlet oxygen, 358–359
Site-directed mutagenesis, 159
Site-specific recombination, 692, 695,
    695f
β-Sitosterol, 391f
Slime layers, bacterial cell, 40
Small interfering RNAs (SiRNAs), 17,
    662
Small intestine, triacylglycerol
    digestion, 416f
Small nuclear ribonucleoproteins
    (snRNPs), snurps, 663
Small nuclear RNAs (snRNAs), 17,
    663
Small nucleolar RNAs (snoRNAs), 17,
    662
Smooth endoplasmic reticulum (SER),
    444
    animal cell, 45f
    eukaryotic cells, 48–49
    plant cell, 46f
Soapmaking, 379–380
Sodium, channel, 402, 533
Sodium hydroxide, 92f, 95f
Sodium-potassium pumps, 391, 402,
    403f
Sol-gel transitions, water-based
    material, 81–82
Solid state biochemistry, cytoskeleton,
    66
Solvation spheres, water, 80, 80f
Solvent properties, water, 79–88
Somatomedins, 605, 608
Somatostatin, 596t
Sorbitol, 235f
Southern blotting, DNA, 653, 655,
    656f
Space-filling models, 233
    adelyate kinase, 143f
    cholesterol, 389f
    DNA, 627f

E. coli chaperonin, 162f
    β-D-fructofuranose, 240f
    α-D-galactopyranose, 240f
    α-D-glucopyranose, 240f
    lysozyme, 183
    saturated and unsaturated fatty acids,
        374f
    sphingomyelin, 384f
    triacylglycerol, 378f
    water, 75f
Spearmint oil, 389
Specific activity, 195
Specificity, enzyme, 187
Specificity constant, 194
Spectrin, 399
Spermidine, 644
Spermine, 644
Sphingolipidoses, 385
Sphingolipids, 383–385, 396f, 446, 448
    storage diseases, 385–386
Sphingomyelin, 383, 384f, 447f
Sphingosine, 384f
Spliceosomes, 663, 715
Squalene, 387, 387t
    biosynthesis, 448f
    cholesterol synthesis from, 453f
    synthesis, 452f
SREBPs (sterol regulatory element
    binding proteins), 408–409, 442,
    456–458
SRE (sterol regulatory elements), 409
Stahl, Franklin, 679f, 680
Standard amino acids, 127
Standard free energy, hydrolysis of
    phosphorylated biomolecules,
    117t
Standard reduction potential, 311, 311t
Standard state, free energy, 111
Starch, 245–246, 245f, 246f
Start codon, 706
Start transfer signal, membrane pro-
    teins, 769
Start signal, codon, 738
Starvation, plasma levels during early
    days of, 425, 563, 618f
State function
    enthalpy, 107
    entropy, 108, 110
    Gibbs free energy change, 110
Stationary phase, 172
Stator, 349
Steady state, 593
Steady state assumption,
    Michaelis–Menten model, 193
Stearic acid, 373t, 374f
Stearoyl-CoA, desaturation, 436, 437f
Stem loop, RNA, 659f
Stereoisomers, 131
Steroids, 389–391, 448f
    hormone mechanisms, 609, 610f
    hormones, 594, 596t
    hormone synthesis, 458, 459f, 460f
    synthesis of selected, 458, 460f
Sterol, 390
Sterol carrier protein, 453
Sterol-sensing domain (SSD), 457
Stigmasterol, 391f
Stomata, 489
Stop codon, 706
Stop signals, codon, 738
Stop transfer signal, membrane
    proteins, 769

Strain effects, catalytic mechanisms,
    204
Streptokinase, 220–221
Streptomycin, 746t
Stress, physiological, 599
Stress adaptations, 98
Stress response, proteins, 141
Stroke, 367
Stroma, 59, 469, 472f
Stromal lamellae, 59, 469, 471f
Structured water, 80–81
Submitochondrial particles (SMP), 349
Substrate, 185
Substrate-accelerated death, 282
Substrate cycle, 285
Substrate-level phosphorylation, 324,
    273
Subunit, proteins, 150
Succinate, 327f, 331f, 332f
    path of electrons, 343f
Succinate dehydrogenase complex,
    341, 346t. *See also* Complex II
    (succinate dehydrogenase
    complex)
Succinyl-CoA, 324, 327f, 331f
    amino acids forming, 572–576
    citric acid cycle, 317f, 324, 327f,
        328f
    enzyme regulation by, 329
    propionyl-CoA to, 428f
Sucrose, 12, 243, 243f
Sugar code, 228, 254–257
Sugar-phosphate backbone, DNA, 632
Sugars, 11t, 12, 13, 14
Sulfatide, 384
Supercoiling, DNA, 642, 643f, 644f
Supercoiling control, DNA in
    prokaryotes, 682–686
Superfamilies, proteins, 141–142
Supernatant, cell fractionation, 67
Superoxide dismutases (SOD), 362
Superoxide radical, 358, 359f, 361f
Supersecondary structures, proteins, 147
Suppressing agents, 731
Supraspliceosome, 716
Surface active agent, 380
Surfactant, lung, 382
Swinging neck-lever model, myosin,
    158
Switch helix, 754
Symmetry model, *See* concerted
    model,
Symporters, 353, 405
Synapsis, 695
Synaptic vesicles, 533
Syndecans, 250
Systematic name, enzymes, 188
Systems biology, living organisms,
    24–26
    as integrated systems,

T

T_3 (triiodothyronine), 594, 595f, 596t
T_4 (thyroxine), 594, 595f, 596t
Tails, phospholipids, 35
Talose, 230f
Tandem mass spectrometry, 177
Tag polymerase, 703
Tandem repeats, 651
Target cells, 594, 598
Targeting, 737, 746, 766–767, 769
TATA box, 713

Names and Abbreviations of the Standard Amino Acids

Amino Acid	Three-Letter Abbreviations	One-Letter Abbreviations
Alanine	Ala	A
Arginine	Arg	R
Asparagine	Asn	N
Aspartic acid	Asp	D
Cysteine	Cys	C
Glutamic acid	Glu	E
Glutamine	Gln	Q
Glycine	Gly	G
Histidine	His	H
Isoleucine	Ile	I
Leucine	Leu	L
Lysine	Lys	K
Methionine	Met	M
Phenylalanine	Phe	F
Proline	Pro	P
Serine	Ser	S
Threonine	Thr	T
Tryptophan	Trp	W
Tyrosine	Tyr	Y
Valine	Val	V